# Biological and Medical Physics, Biomedical Engineering

More information about this series at http://www.springer.com/series/3740

# BIOLOGICAL AND MEDICAL PHYSICS, BIOMEDICAL ENGINEERING

The fields of biological and medical physics and biomedical engineering are broad, multidisciplinary and dynamic. They lie at the crossroads of frontier research in physics, biology, chemistry, and medicine. The Biological and Medical Physics, Biomedical Engineering Series is intended to be comprehensive, covering a broad range of topics important to the study of the physical, chemical and biological sciences. Its goal is to provide scientists and engineers with textbooks, monographs, and reference works to address the growing need for information.

Books in the series emphasize established and emergent areas of science including molecular, membrane, and mathematical biophysics; photosynthetic energy harvesting and conversion; information processing; physical principles of genetics; sensory communications; automata networks, neural networks, and cellular automata. Equally important will be coverage of applied aspects of biological and medical physics and biomedical engineering such as molecular electronic components and devices, biosensors, medicine, imaging, physical principles of renewable energy production, advanced prostheses, and environmental control and engineering.

## Editor-in-Chief:

Bernard S. Gerstman, Department of Physics, Florida International University, Miami, Florida, USA

## Editorial Board:

Jean-Luc Popot

# Membrane Proteins in Aqueous Solutions

## From Detergents to Amphipols

 Springer

Jean-Luc Popot
Institut de Biologie Physico-Chimique
Paris, France

ISSN 1618-7210          ISSN 2197-5647   (eBook)
Biological and Medical Physics, Biomedical Engineering
ISBN 978-3-030-10324-8          ISBN 978-3-319-73148-3   (eBook)
https://doi.org/10.1007/978-3-319-73148-3

Printed on acid-free paper

This Springer imprint is published by the registered company Springer International Publishing AG part of Springer Nature.
The registered company address is: Gewerbestrasse 11, 6330 Cham, Switzerland

*This book is dedicated to the memory of Annemarie Weber (1923–2012), one-summer teacher and lifelong friend.*

# Foreword

## The Emergence of Membrane Protein Biochemistry

Biochemistry aims to reveal the workings of biological molecules using the concepts of chemistry. While chemical conceptualizations have successfully revealed the operations of many soluble proteins and machines as large and complex as ribosomes, advances in such understanding have only recently begun to emerge for many of the macromolecular components of membranes. The very slow progress toward biochemical insights into membrane proteins and complexes can be understood as a methodological challenge, which is explored by Jean-Luc Popot in this excellent book *Membrane Proteins in Aqueous Solutions: From Detergents to Amphipols*. The challenge, which has only recently been overcome for most proteins, is to enable biochemical studies of structure and function by creating relatively stable, homogeneous preparations of proteins and complexes in aqueous solution, where they can be studied using the tools of biochemistry.

In the early 1970s, membrane protein biochemistry was a mystery. The basic lipid bilayer structure of membranes had been agreed upon, and there was also agreement that proteins must mediate membrane functions. Physiologists and molecular biologists had found that channels, transporters, permeases, and other functional elements mediate the translocation of ions, the conduction of nerve impulses, the transduction of energy, and many other key processes of living systems, but the chemistry of these functions was dramatically limited by the paucity of structural information. To illustrate the state of affairs, I note the idea of membrane protein structure shown in Fig. 1.

At the same time, a rapidly growing number of protein structural models had been established in chemical detail, and the biochemistry of a number of enzyme functions was understood from a chemical perspective that combined structural and functional data. Why did such a disparity exist between rapid advances in protein chemistry up to 1990 and the glacial pace of understanding membrane proteins?

The approaches that had been used were largely limited by the detergent strategies used to remove proteins from the lipid bilayer environments of membranes, and only a subset of the membrane menagerie could be examined as stable entities. Further, only a small subset of the menagerie could be crystallized for structural study, and only a still smaller subset allowed the resolution needed for the creation of chemically detailed models.

Two significant paths are now expanding our view: new methods of creating stable structures outside of the membrane context and a new method for examining their structures. The new structural method is single-particle electron cryomicroscopy (cryo-EM), and the new methods of preparation include nanodiscs and amphipols. A remarkable example of how such approaches combine is in the recent structure of a ligand-gated ion channel, TRPV1, stabilized in solution by an amphipol (Liao et al. 2013). This structure alone leapfrogs decades of work, mostly based on the "divide and conquer"

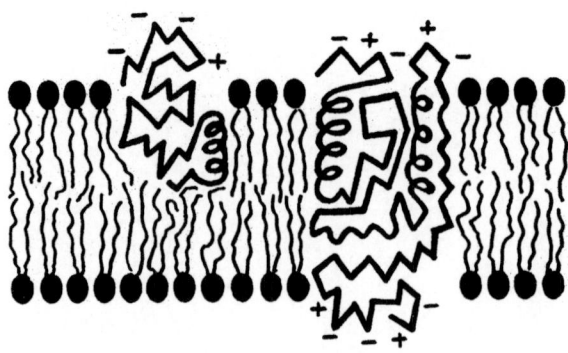

**Fig. 1** Structures of membrane proteins as imagined in 1972. The conceptualization lacks the insights on secondary structure that followed soon afterward when electron microscopy evidence for transmembrane $\alpha$-helices was obtained in 1975 (Henderson and Unwin 1975), and, later, a detailed X-ray crystallography structural model for the photosynthetic reaction center was elucidated in 1984 (Deisenhofer et al. 1984). Still later, crystallographic evidence for the $\beta$-barrel structural theme emerged in 1990 (Weiss et al. 1990) (Figure reproduced with permission from Singer and Nicolson 1972, © 1972 American Association for the Advancement of Science).

approach of examining separated parts of the structure – seeing the whole channel, including the transmembrane region, is an extraordinarily informative event!

To place the problems in perspective, Jean-Luc Popot explores basic principles of membrane architecture, from which point of view he develops the progressive insights of decades of advances in membrane protein solubilization, leading to an extensive treatment of the amphiphilic polymers as the most recent technology. In the text, a progressive set of issues are examined through the lens of the enabling technologies used: protein solubilization, purification, folding, and expression; followed by different methods that can be used: optical spectroscopy, NMR, crystallography, mass spectrometry, and electron microscopy; and then followed by practical uses for solubilized membrane proteins: surface immobilization and vaccines. Along the way, one is educated in the fundamental thinking that guides the applications, so that a reader will both appreciate the established possibilities and perhaps create new ones.

September 2017                                                                          Donald M. Engelman

## References

Deisenhofer, J., Epp, O., Miki, K., Huber, R., Michel, H. (1984) X-ray structure analysis of a membrane protein complex. Electron density map at 3 Å resolution and a model of the chromophores of the photosynthetic reaction center from *Rhodopseudomonas viridis*. *J. Mol. Biol.* **180**:385–398.

Henderson, R., Unwin, P.N.T. (1975) Three-dimensional model of purple membrane obtained by electron microscopy. *Nature* **257**:28–32.

Liao, M., Cao, E., Julius, D., Cheng, Y. (2013) Structure of the TRPV1 ion channel determined by electron cryo-microscopy. *Nature* **504**:107–112.

Singer, S.J., Nicolson, G.L. (1972) The fluid mosaic model of the structure of cell membranes. *Science* **175**:720–731.

Weiss, M.S., Wacker, T., Weckesser, J., Welte, W., Schulz, G.E. (1990) The three-dimensional structure of porin from *Rhodobacter capsulatus* at 3 Å resolution. *FEBS Lett.* **267**:268–272.

# Preface

Articles and reviews dealing with membrane proteins typically start with the following four statements:

(i) Membranes are an essential feature of living matter, playing key roles in compartmenting it and controlling exchanges of matter and information between and within cells.

(ii) Genes encoding transmembrane proteins (hereafter "membrane proteins," MPs) make up about one-third of genomes.

(iii) Our knowledge of the structure and function of MPs is essential both to understanding cell physiology and to controlling it; it is an absolute requirement in order to develop more efficient drugs, a majority of which have MPs as their target.

(iv) Yet this knowledge lags way behind that of soluble proteins, because MPs are much more difficult to produce, to purify, and to handle; as a result, MP structures, for instance, make up only a minute fraction of those hosted by the Protein Data Bank.

All of this is true, and it has spurred massive efforts to develop MP-specific methodologies. As will be seen in the course of this book, some of the results achieved are truly spectacular.

The central point we will tackle here is how to keep MPs in their native state after extracting them from their original environment, a problem that has haunted many a graduate student and prematurely aged many a principal investigator. Quite a few investigations can be carried out while keeping the target MP in the membrane medium it has evolved to function in, be it in the whole organism, in the cell, or in native membrane fragments. However, a detailed understanding of a MP's structure and function implies, in the quasi-totality of cases, to extract it from the membrane and to purify and handle it in aqueous solutions, a medium it is not adapted to. "Extracting" involves dispersing the molecules that make up a membrane, mainly lipids and proteins, so that they can be separated one from another and the one relevant MP or MP complex purified. MPs and lipids, however, are intimately associated in vivo. Dispersing them results, in nearly all cases, in the more or less rapid inactivation of the target protein.

As will be recalled in Chap. 2 of this book, dispersion is traditionally achieved using detergents. Detergents constitute a special class of surfactants – meaning, in our case, molecules with both hydrophilic and hydrophobic moieties – characterized by their ability to solubilize fats. The first detergents used in membrane biochemistry were industrial or natural detergents, and their use was highly empirical. Starting in the 1970s, membrane biochemists introduced themselves more deeply to the physical chemistry of surfactants, and in the 1980s organic chemists started synthesizing detergents specifically designed with membrane biology in sight. Comparing the effects of a series of well-defined detergents on the stability and functionality of MPs led to a better understanding of the structural parameters that contribute to make a detergent more or less destabilizing. As a result of this feedback, "milder" detergents were designed, synthesized, and validated, a process that continues to this day. In the course of these studies, it was realized that a major source of destabilization is that detergents tend to strip MPs off the lipids with which they are naturally associated. This initiated, starting essentially in the 1990s, attempts to replace the detergent environment with one that would be less aggressive.

MPs are naturally amphipathic, meaning that some of their surface is adapted to being solvated by water, some to being immersed in the highly hydrophobic membrane interior, where it is in intimate contact either with other MPs or with the hydrophobic moieties of lipids – most often with both. Making MPs water-soluble implies to cover these hydrophobic surfaces with a layer of surfactant, the hydrophobic moieties of which adsorb onto the protein, while the hydrophilic ones ensure the interface with water. Such is the role of detergents. Doing away with detergents imposes to replace them with other surfactants. One solution is to reinsert MPs into lipid vesicles or planar lipid films. This reconstitutes an environment that is similar to the native one, and is in general stabilizing. The large objects thus obtained are suitable for many functional studies, but neither for purification nor for most structural approaches. Non-detergent – "nonconventional" – media that complex MPs to form small (nanometric) particles are the main subject of this book. They include, in the order of their apparition on the biochemist's bench, detergent-stabilized lipid discs ("bicelles"), amphipathic peptides, amphipathic polymers ("amphipols"), fluorinated surfactants, protein-stabilized lipid discs ("nanodiscs"), etc., the properties, advantages, and limitations of all of which will be discussed in turn.

The book starts with a presentation of membrane proteins in their natural medium, with a discussion of their functions, their structure, the forces that stabilize them, and the interactions that they establish with their environment, in particular with lipids (Chap. 1). In Chap. 2, we consider the traditional approach to solubilizing MPs and handling them in aqueous solutions, namely the use of detergents, including a discussion on the mechanisms by which detergents destabilize (or are hypothesized to destabilize) MPs and a presentation of recently developed detergents. Chapter 3 introduces the various nonconventional approaches that have been developed, with an examination of which applications they are best suited to. Bicelles, nanodiscs, amphipathic peptides, and fluorinated surfactants are discussed in this chapter.

We then turn to amphipols. Chapter 4 describes their chemical structure, their synthesis, and their chemical-physical properties, with the stress put on their solution behavior. Chapter 5 examines the various ways to trap MPs with amphipols and the structure and properties of MP/amphipol complexes, the knowledge of which is essential to developing their applications in basic research, medicine, and the industry. Chapters 6, 7, 8, 9, 10, 11, 12, 13, 14, and 15 then discuss in turn each application that has been or is being validated: MP folding and in vitro synthesis, optical spectroscopy and solution studies, NMR, crystallography, electron microscopy, immobilization of MPs onto solid supports, mass spectrometry, and biomedical applications. Each chapter starts with a self-contained presentation of the principles of the technique to be discussed and the state of the art when applied to detergent-solubilized MPs, which should make it accessible even to readers with no familiarity with the approach considered. It comprises thematic tables that organize an exhaustive review of the literature so as to make it straightforward for the experimenter to identify those publications most relevant to her/his goals. Most technical chapters include one or several protocols, prepared and commented by experts in the use of amphipols for the application considered. Suggestions and speculations about future developments are offered at the end of each chapter and some more general comments at the end of the book.

It is my hope that this book, which I have struggled to make accessible to any member of a membrane biochemistry or biophysics laboratory, will help widen the use of original approaches that are easy to implement but whose enormous potential has only begun to be tapped.

Paris, France                                                                                      Jean-Luc Popot

# Acknowledgments

This book could not have been written without the help of many colleagues and friends.

First of all, I would like to thank the many colleagues who took on their time to provide me with protocols or contributions to the technical boxes and annexes, as well as for offering feedback on the first drafts of the chapters, among which are Antonio Calabrese, Marina Casiraghi, Laurent J. Catoire, Tassadite Dahmane, Christine Ebel, Fabrice Giusti, Christel Le Bon, Candice E. Paulsen, Martin Picard, Bernard Pucci, Giuseppe Zaccai, Francesca Zito, and Manuela Zoonens.

I am greatly indebted to Francis Haraux, who generously illustrated this book with his fanciful cartoons and, in the process, created the character of Cephalopol, whose adventures in physical chemistry, biochemistry, and biophysics have delighted me and many others.

I warmly thank Donald M. Engelman, who managed to squeeze in his busy schedule the writing of a foreword that places this book in its exact perspective.

Many colleagues sent me figures, unpublished papers, comments, and suggestions, or helped me dig up hard-to-find publications, etc., among which are Alison E. Ashcroft, Dominique Bagnard, Jean-Louis Banères, Edward A. Berry, Jaap Broos, Jean Cartaud, Pil Seok Chae, Yifan Cheng, Oliver B. Clarke, Didier Clenet, Melanie J. Cocco, Pierre-Jean Corringer, Gérard Crémel, Luis M. de la Maza, Eduardo Della Pia, Mark E. Dumont, Andreas Engel, Manuel Etzkorn, Yann Gohon, Valentin Gordeliy, Francesca Gubellini, Richard Henderson, David Julius, Werner Kühlbrandt, Jeremy H. Lakey, Tomas Laursen, Marc le Maire, Kevin Leonard, Konstantin S. Mineev, Milena Opačić, Krzysztof Palczewski, Eva Pebay-Peyroula, Jason D. Perlmutter, Daniel Picot, Vitaly Polovinkin, Gil G. Privé, Sheena E. Radford, Han Remaut, Charles R. Sanders, Aure Saulnier, Stephen G. Sligar, Frank Sobott, Christopher G. Tate, Christophe Tribet, Jeroen F. van Dyck, Dror E. Warschawski, Shinya Yoshikawa, and Thomas Zemb. My thanks also go to Bruno Miroux and Edith Godard, CNRS UMR 7099, Institut de Biologie Physico-Chimique, Paris, who did everything in their power to help me throughout this long period of writing.

I am grateful to the Springer team, Christopher T. Coughlin, Elias Greenbaum, and HoYing Fan, for their feedback, advice, and constant availability.

I do not forget the vital assistance of my faithful teapot, which brewed me, over the 3 years it took to put this book together, an estimated one metric ton of Darjeeling tea.

I feel nostalgic about my parents, who, a few years ago, would have been so happy to leaf together through a book they are distantly responsible for.

My final and warmest thanks go to my wife, Jacqueline Barra, for her endurance and support during all of this very long period, as well as for her help with creating many of the figures and hunting for permissions to reproduce the others. Without her patience and understanding, this book would never have made it to the printing press.

# Contents

# Symbols

| | |
|---|---|
| $A_\lambda$ | Absorbance at wavelength $\lambda$ |
| $Đ$ | Dispersity (formerly polydispersity index) of a polymer |
| $D_t$ | Translational diffusion coefficient |
| $M$ | Molar mass |
| $\langle M_n \rangle$ | Number-average molar mass of a polymer |
| $\langle M_w \rangle$ | Weight-average molar mass of a polymer |
| $N_A$ | Avogadro's number |
| pI | Isoelectric point |
| $P(r)$ or $P_r$ | Pair-distribution function |
| $R$ | Gas constant |
| $R_g$ | Radius of gyration |
| $R_g{}^\circ$ | Radius of gyration at infinite contrast |
| $R_H$ | Hydrodynamic radius |
| $R_S$ | Stokes radius |
| $s$ | Sedimentation coefficient |
| $s_{20,w}$ | Sedimentation coefficient at 20 °C in pure water |
| $T$ | Absolute temperature |
| $\bar{v}$ | Specific volume |
| $\langle X_n \rangle$ (formerly $DP$n) | Number-average degree of polymerization |
| $\Delta G^\circ$ | Standard free energy difference |
| $\Delta\varepsilon$ | Molar circular dichroism |
| $\varepsilon$ | Dielectric constant, extinction coefficient |
| $\eta$ | Viscosity |
| $[\theta]$ | Ellipticity ($[\theta] = 3298.2\ \Delta\varepsilon$) |
| $\lambda$ | Wavelength |
| $\mu^\circ$ | Standard chemical potential |
| $\rho$ | Scattering length density or hydrodynamic density of a particle |
| $\rho_0$ | Scattering length density or hydrodynamic density of a solvent |
| $\tau_c$ | Rotational correlation time |

# Membrane Proteins and Their Natural Environment

**Summary**

*This chapter offers a compact introduction to membrane proteins and their natural environment. An overview is presented of the cellular location and functions of membrane proteins, of lipid bilayers and the physical-chemical constraints they impose on membrane-spanning molecules, of the impact of these constraints on the structure of protein transmembrane regions, of lipid/protein interactions, and of membrane protein synthesis. Background information that is indispensable as a frame for the rest of the book is recalled, but the accent is put on notions that are essential to understanding how surfactants work and to optimizing their use. The nature and extent of conformational changes undergone by protein transmembrane regions during functional cycles are illustrated, taking as examples three membrane proteins, bacteriorhodopsin, the nicotinic acetylcholine receptor, and the sarcoplasmic reticulum calcium pump, whose stability and functionality in the presence of various surfactants have been studied in some detail and will be discussed in subsequent chapters.*

## 1.1 Introduction

Living organisms are made up of aqueous compartments that maintain and exploit concentration differences between them and with the outside world. Biological molecules and ions are kept from diffusing away, nutrients are imported and stored, by-products are expelled, and toxic molecules or ions are kept out. Differences in physical-chemical potentials between compartments are created and maintained. They are used to provide chemical and physical driving forces for an organism to sustain itself, act on its environment, and store and exchange energy and information. In its simplest form – from the point of view of compartmentalization – a living cell, such as a bacterium, is a little bag whose molecular content reproduces itself until enough material has accumulated to permit it to divide into two daughter cells, etc. In more complex organisms, the content of the bag can itself be compartmentalized, and cells can associate with one another and cooperate, exchange long-distance messages, etc.

Outside the cell is a liquid, water-based world. This may not seem so intuitive to us, humans, who breathe air and walk on soil, but the environment all cells live in is aqueous, be it the sea, freshwater, or the fluids stored inside our organisms. The inside of each cell, and each compartment within a cell, is also watery, and cells and multicellular organisms that must survive in dry environments, like spores, seeds, or tardigrades, go to extreme lengths to limit and survive partial desiccation. The fluid inside a cell, the cytoplasm, is a concentrated soup of molecules and ions, those imported by the cell, such as inorganic ions or nutrients, and those manufactured by it, be they small – sugars, for instance – or very large, such as the nucleic acids into the sequence of which is stored and thanks to which is exploited the genetic information, as well as the proteins that catalyze most biological reactions. The three-dimensional (3D) structure adopted by these *macromolecules* depends in large part on their interactions with water: for most water-soluble proteins, for instance, the main driving force that will initiate the positioning in space of their hundreds or thousands of amino-acid residues is the segregation of hydrophilic side chains, which oppose being deprived of water, from hydrophobic ones, which tend to segregate away from it. The force underlying this behavior is the strong interactions water molecules establish with one another. Unless those are replaced by even stronger ones, such as those that water molecules can establish with ions or strongly polar groups, foreign molecules or groups will be repelled and expelled from the water phase, an effect known as the *hydrophobic effect* (see e.g. Tanford 1980; Israelachvili 2011, and references therein). The hydrophobic effect is a powerful organizing phenomenon on which much of life as we know it is based.

The compartmentalization of living matter relies on the behavior of special classes of *amphipathic* molecules, that is in the case of living matter, molecules that comprise groups that interact favorably with water (called *hydrophilic*) and groups that are repelled by it (called *hydrophobic*). One can imagine, design, and put to work other types of amphipathic molecules, for instance, molecules that comprise groups that are soluble in hydrocarbons and groups that are soluble in fluorocarbons, all of them being highly insoluble in water (see Chap. 3, § 3.5). As used in this book, the term *amphipathic*, however, will always refer to molecules that comprise chemical groups with differential affinity for water. We have already noted that proteins are amphipathic, being comprised of both hydrophilic and hydrophobic groups, and that this combination is a major factor driving their 3D folding. As typical examples of smaller amphipathic biological molecules, Fig. 1.1 shows the structure of a selection of polar *lipids*. Each of these compounds comprises one or more polar group(s) and one or more apolar one(s), which generally are segregated at opposite ends of an elongated molecule. The polar end can carry charged moieties, e.g. phosphate, carboxylates or ammonium groups, or nonionic but hydrogen-bonding groups, such as hydroxyles. When exposed to water, these groups will tend to associate with it, because the polar interactions they establish with it overcome those of water molecules with themselves. The other end of most polar lipids is generally comprised of one, two (in most cases), or more hydrocarbon chains, either saturated or unsaturated, in which the partial charges carried by the hydrogen and carbon atoms are very small. Such groups are not polar enough to compete with the strong interactions that water molecules establish one with another, and they are repelled by it.

When a single molecule of lipid is let loose in a droplet of water, it will equilibrate between the bulk and, if available, the air/water interface. At the interface, the polar head group can stay in contact with water, whereas the hydrophobic acyl chains can be expulsed from it. Because less water molecules are perturbed, this state is at a lower standard chemical potential ($\mu°$) than is an isolated lipid molecule fully surrounded with water. An equilibrium establishes itself between entropy, which tends to dilute the molecules in the bulk, and the hydrophobic effect, which tends to push them to the surface. Water molecules are themselves at a higher standard chemical potential at the interface than they are in the bulk, because they cannot satisfy as efficiently the hydrogen bonds they form with their neighbors. Creating a certain area of air/water interface therefore costs energy, which is at the origin of the *surface tension* that tends to diminish the area of the interface: this is why a droplet of water deposited on a hydrophobic surface such as that of Teflon takes up a nearly spherical form.

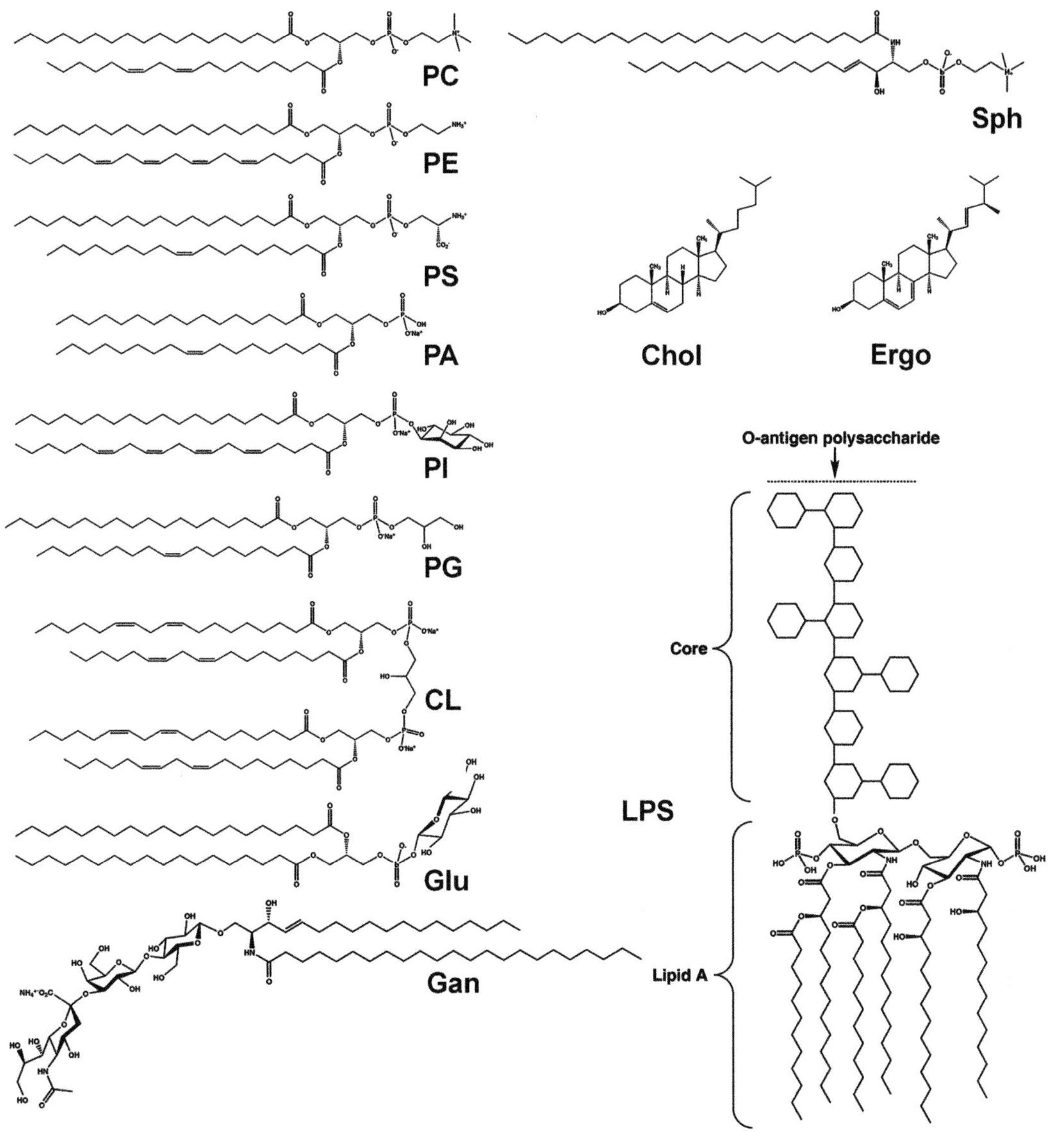

**Fig. 1.1** Chemical structures of some of the polar lipids commonly found in biological membranes. *Chol* cholesterol, *CL* cardiolipin, *Ergo* ergosterol, *Gan* ganglioside, *Glu* glucolipid, *LPS* lipopolysaccharide, *PA* phosphatidic acid, *PC* phosphatidylcholine, *PE* phosphatidylethanolamine, *PG* phosphatidylglycerol, *PI*, phosphatidylinositol, *PS* phosphatidylserine. The nature of the acyl chains is conveniently indicated by the notation $C_{m:n}$, where m stands for the number of carbons and n for that of double bonds.

By displacing water molecules from the interface, amphipathic molecules such as lipids lower the free energy cost of creating the air/water interface. The surface tension drops and, under the effect of gravity, the droplet sags. Lipids, and other amphipathic molecules, are therefore said to be surface-active compounds or *surfactants*. Nearly all biological molecules, including proteins, comprise hydrophobic and hydrophilic regions and, as a consequence, are surface-active. In the present book, however, the term "surfactant" will be restricted to relatively small molecules, such as lipids, detergents, or amphipols (APols), and not used for biological macromolecules.

## 1.2    Lipid Bilayers

Lipid molecules have a very low solubility in water as individual entities. When their concentration as monomers exceeds a very low limit called the *critical association concentration* (typically <1 nM; see Tanford 1980), they self-associate in such a way that their hydrophobic chains come together, leaving only the polar heads exposed to water. The 3D organizations that lipids can adopt in aqueous solutions depend, in particular, on their shape. In lipids such as phosphatidylcholine (PC; Fig. 1.1), the cross section of the molecules, taken normal to their long axis, is about the same at the level of the polar head and the acyl chains, making the molecule roughly cylindrical. When such molecules assemble side by side, they therefore tend to form a flat sheet, one face of which is hydrophilic, the other hydrophobic. Two such sheets whose hydrophobic faces come together form a *lipid bilayer*, in which the acyl chains are removed from the contact with water (Table 1.1, line ③; see e.g. Marčelja 1974). Because the edges of a patch of bilayer expose hydrophobic groups, they tend to come together, forming a closed vesicle. This simple mechanism is one of the essential features of life. Indeed, the mere exposure of such lipids to water will create vesicles and, therefore, define closed aqueous compartments. Whereas a lipid bilayer is very thin (typically 5–6 nm; see Fig. 1.3), its innermost region is extremely hydrophobic and will not let ions or polar molecules diffuse rapidly from one aqueous compartment to the other. It is therefore possible, given a minimal input of energy, to maintain concentration differences between the inside and outside of lipid vesicles, a process that is fundamental to biology, and on which is based a broad range of biological functions.

A lipid bilayer-based membrane, the *plasma membrane*, separates the aqueous solution inside the cells (the *cytosol*[1]) from the exterior medium. It is the only membrane in many bacteria, whether eubacteria or archaebacteria, as well as in enveloped viruses, where it has been ripped off the plasma membrane of the cell the virus budded from. In Gram-negative (Gram⁻) bacteria (so called because of their characteristic appearance following a staining procedure developed by bacteriologist H.G. Gram), a second membrane, the *outer membrane*, surrounds the cell, protecting the plasma membrane from direct exposure to the outside medium and defining an intermembrane space (the *periplasm*), in which resides the peptidoglycan, a mesh-like external cytoskeleton that protects the cell against osmotic shocks. The outer membrane contains proteins that select which molecules are imported into the periplasm and which are left out (see § 1.3). In eukaryotic cells, the inside of the cell (the *cytoplasm*) is divided into a multitude of compartments separated from the cytosol by lipid bilayer-based membranes (Fig. 1.2). Each of these compartments has its own membrane and luminal composition and set of functions. A constant traffic of vesicles, which bud from one compartment and fuse with another, ensures the transport, from one cell compartment to another or toward the plasma membrane and the extracellular medium, of water-soluble molecules contained within the compartments, as well as that of amphipathic molecules associated with their membrane.

Some of the compartments inside eukaryotic cells have a complex structure, being derived from symbiotic organisms that have become integrated and have evolved to fulfill special functions. Thus, *mitochondria*, in which respiration takes place, feature a double membrane, the outer one playing to some extent the role of an outer bacterial membrane. The inner one, in which the respiratory complexes that ensure electron transfer from reducing substrates to molecular oxygen reside, is convoluted into tubes or sacculi (the *cristae*) so as to increase the surface available (cf. Fig. 1.2). *Chloroplasts*, which are derived from photosynthetic bacteria, have a two-membrane exterior envelope, and their aqueous

---

[1] The cytosol is usefully defined as "that portion of the cell which is found in the supernatant fraction after centrifuging an homogenate at 105,000 × g for 1 hour" (Clegg 1983), that is, essentially, a solution devoid of cytoskeleton, membrane fragments, DNA, etc. but comprising most water-soluble proteins.

**Table 1.1**  Mean (dynamic) packing shapes of various surfactants and the structure they assemble into in aqueous solutions.

| | Critical packing parameter $v/a_0 \cdot l_c$ | Critical packing shape | Structures formed |
|---|---|---|---|
| ① | < 1/3 | Cone<br> | Spherical micelles<br> |
| ② | 1/3 – 1/2 | Truncated cone<br> | Cylindrical micelles<br> |
| ③ | ~1 | Cylinder<br> | Planar bilayers<br> |
| ④ | > 1 | Inverted truncated cone or wedge<br> | Inverted micelles<br> |

Adapted from Israelachvili (2011). © 2011 Elsevier Inc. All rights reserved. The critical parameter $v/a_0\, l_c$, where $v$, $a_0$, and $l_c$ are, respectively, the overall volume, polar head surface, and length of the hydrated molecule, taking into account thermally induced conformational excursions and electrostatic repulsion, defines the nature of the assemblies. Detergents, whose polar head is large as compared to their hydrophobic tail, form more or less spherical ①, lenticular, or cylindrical ② micelles; phosphatidylcholine and membrane-forming lipid mixtures, whose shape is, on average, close to cylindrical, assemble into bilayers ③; and lipids with a relatively small polar head, like phosphatidylethanolamine, when pure, form inverted micelles ④

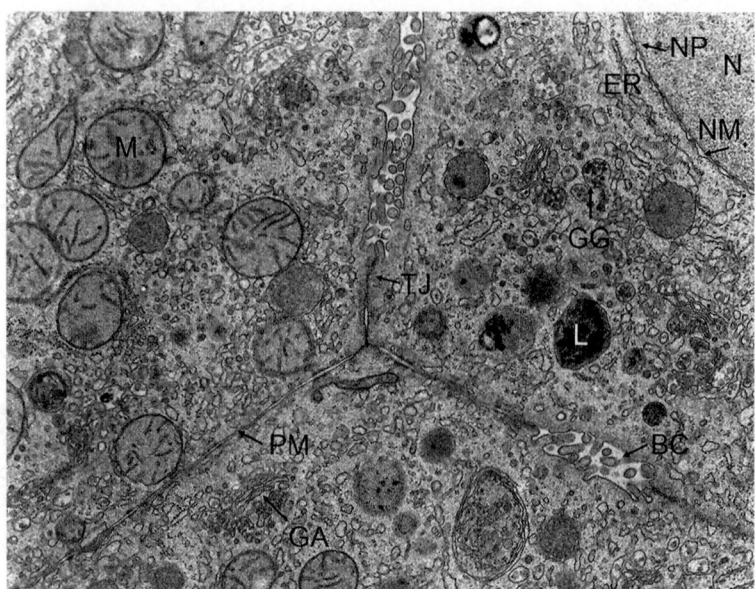

**Fig. 1.2** Electron micrograph of hepatic cells. Note the contacts established by the plasma membranes of neighboring cells and the many compartments inside each of them. *BC* bile canaliculi, *ER* endoplasmic reticulum, *GA* Golgi apparatus, *GG* glycogen granule, *L* lysosome, *M* mitochondrion, *N* nucleus, *NM* nuclear membrane, *NP* nuclear pore, *PM* plasma membrane, *TJ* tight junction (The micrograph is from the site http://medcell.med.yale.edu/histology/digestive_organs_lab/hepatocytes_em.php).

interior, the *matrix*, is packed with stacks of sacculae in which the transduction of light into chemical energy takes place, thanks to oxygenic photosynthesis.

Each of the membranes that compartmentalize a eukaryotic cell has its own lipid and protein composition, and its internal fluid also has a specific composition. For instance, the cytosol is a reducing medium, whereas the lumen of the endoplasmic reticulum is, as the exterior medium, oxidizing. The concentration of calcium ions is very low in the cytosol, very high in the lumen of the sarcoplasmic reticulum (SR); the electrochemical potential of $H_3O^+$ ions is higher in the lumen of chloroplasts and in the intermembrane space of mitochondria than in their matrix, etc.

Biological membranes do not comprise a single type of lipid but mixtures of them, in which both the polar head hydrophobic chains vary. The chains – most often two of them but sometimes more (cf. Fig. 1.1) – can be long or short (typically between 16- and 22-carbon long), saturated or unsaturated, sometimes branched (in archaebacteria). They can be associated to the polar head either by hydrolyzable ester functions (in eubacteria and eukaryotes) or non-hydrolyzable ether functions (in archaebacteria). The polar heads can be zwitterionic, as in PC or phosphatidylethanolamine (PE), and nonionic (formed of sugar residues) or carry a net negative charge, as in phosphatidylserine (PS) and phosphatidylglycerol (PG) (Fig. 1.1). In these *glycerophospholipids*, in bulk the most abundant lipids in animal cells, two acyl chains and a polar head are attached, respectively, to the first two and to the third one of the three hydroxyl functions of glycerol. In sphingomyelin (Sph), one acyl chain is linked by an amide bond to a ceramide group, which is comprised of an alkyl chain and a polar head (Fig. 1.1). Eukaryotic membranes also contain sterols, such as cholesterol, in animal cells, or ergosterol, in yeasts (Fig. 1.1). By themselves, sterols do not form bilayers, but they partition into them and modulate their thickness and fluidity. In a living cell, the composition of the lipids is adjusted so that they remain in a fluid phase. This is achieved by modulating the length and level of unsaturation of the chains so that they remain mobile with respect to one another and do not associate into a gel.

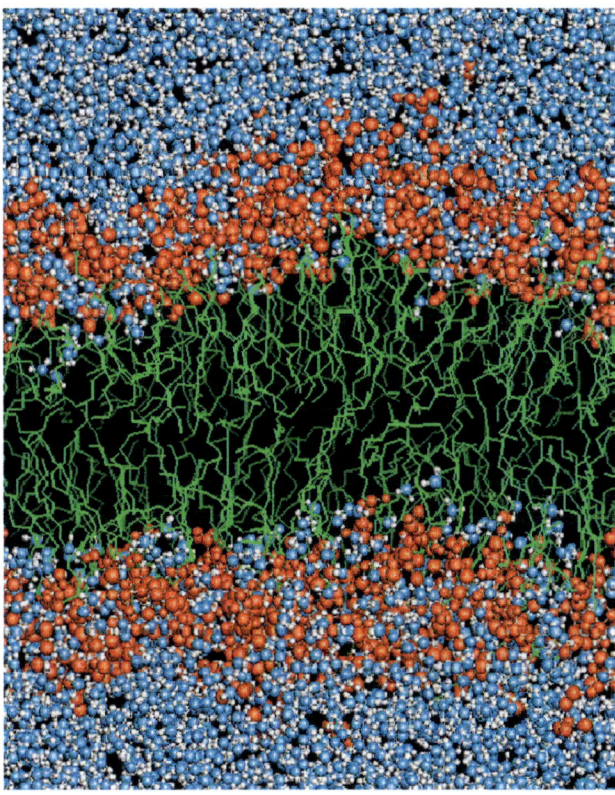

**Fig. 1.3** A simulation of a hydrated lipid bilayer. Snapshot from a molecular dynamics simulation of a fluid-phase, fully hydrated dimyristoylphosphatidylcholine (DMPC, diC$_{14:0}$PC) bilayer. Color coding: phospholipid head groups, *orange*; water hydrogens, *white*; water oxygens, *blue*; phospholipid hydrocarbon chains, *green*. The sizes of the phospholipid head-group atoms and the water molecules are reduced from their van der Waals size in order to permit seeing some distance into the structure, so that the interpenetration of the water and the phospholipid is visible. A partially disordered hydrocarbon layer ~30-Å thick is formed by the lipid hydrocarbon chains, where virtually no water molecules are present. Two head group layers, each ~15-Å thick, separate the hydrocarbon layer from the bulk water phase. The lowered concentration of water resulting from the high concentration of the head groups is evident (From Jakobsson 1997. © 1997 Published by Elsevier Ltd.).

Sterols contribute to smoothing out fluidity changes of membrane lipids over a broad range of temperatures. From one organism to another, and from one compartment to another within a single cell in a given organism, the lipid composition of membranes is extremely variable. Furthermore, the two monolayers (*leaflets*) that comprise a bilayer do not usually have the same composition: in eukaryotic plasma membranes, for instance, the outer monolayer is devoid of negatively charged lipids, which are confined to the inner one. This asymmetry is actively maintained by the cell.

A pure lipid bilayer, such as that shown in Fig. 1.3 as a molecular dynamics (MD) model, is kept together by the balance of multiple forces. At the level of the junction between the hydrophobic chains and the polar head groups, lipid molecules tend to press one against another as a result of the hydrophobic effect: indeed, the more accessible this region is to water, the more water molecules come into contact with the hydrophobic chains, which is energetically unfavorable. As a result of this effect, the head groups of the lipids, on the one hand, and their fatty acyl chains, on the other, are pressed one against another. The average surface occupied by each lipid molecule is the result of these forces balancing each other. A protein spanning the bilayer therefore experiences a different lateral pressure depending on the depth: it tends to be compressed in the head group and central acyl chain regions, and to expand at the junction between the two, which may modulate conformational equilibria (Marsh 1996). Molecules that dissolve into lipid bilayers may modify the profile of this pressure gradient and, thereby, affect protein function. The bilayer being free to expand or compress as a whole, the net overall pressure is just the atmospheric pressure.

The environments experienced by ions, small molecules, and macromolecules such as proteins in an aqueous phase vs. in the core of lipid membranes are so different that it has a dramatic effect on their solubility and their interactions with themselves and with one another. The main differences between the cytosol and membrane interior that are relevant to the organization and interactions of proteins are summarized in Table 1.2. A major factor is the difference in dielectric constant, $\varepsilon$: it is very high in water, a highly polarizable medium, and very low in the core of the bilayer. A consequence of this is that, as already mentioned, ions and polar molecules have a low solubility in the core of a lipid bilayer, another that ionic interactions are much stronger there than they are in water. In bulk water, a hydrogen bond between the $>$N–H group of a peptide bond and the O$=$C$<$ group of another, for instance, can be easily displaced by water molecules at almost no free energy cost, but it requires 4–6 kcal·mol$^{-1}$ to break it in the center of a bilayer (Table 1.2). In terms of an equilibrium constant, a difference of 5.4 kcal·mol$^{-1}$ entails, at 298 K, a factor of $\sim$10$^4$ in equilibrium constants in favor of the state where the hydrogen bond is satisfied. There is, therefore, an enormous thermodynamic pressure toward (i) excluding polar groups from the hydrophobic core of a lipid bilayer and (ii) bringing together complementary charges or partial charges that are being forced there, such as by forming salt bridges or hydrogen bonds (Table 1.2).

## 1.3    Membrane Protein Functions

Schematically, lipids provide a barrier against the free circulation of ions and polar molecules between the cytosol of a cell and the exterior fluid, as well as between the interior and exterior of cell compartments, whereas membrane-spanning proteins take care of actively transporting solutes or letting them diffuse passively in a controlled manner. These conceptually simple (but often mechanistically extremely sophisticated) operations are at the heart of a host of functions of membrane proteins (MPs), a few examples of which only can be listed here:

- Facilitating the passive diffusion of polar molecules or ions. Such is the role, for instance, of the porins of outer bacterial membranes, which act as a sieve, letting diffuse into the periplasm molecules that are smaller than a given size, with or without any specificity, such as that for sugars. Similarly, potassium channels in the plasma membrane of animal cells let K$^+$ ions diffuse from the cytosol, where they are more concentrated, toward the exterior of the cell. This sets up a transmembrane potential, negative inside, which increases until K$^+$ ions have reached their equilibrium electrochemical potential: when the transmembrane potential reaches, typically, 60–80 mV, the inward and outward passive fluxes of K$^+$ ions equilibrate.
- Regulated passive diffusion is a very frequent function. It is essential, for instance, in signaling. In the nerve or muscle plasma membrane exist voltage-sensitive Na$^+$-specific channels, which are normally closed but open in a transitory way when the membrane potential drops below a given threshold. The resulting massive entry of Na$^+$ ions entails an inversion of the membrane potential, which becomes positive inside. Thus, a partial and local depolarization, such as is created at a neuronal synapse or a neuromuscular junction by chemical signaling (see below), sets off a depolarization wave, the "action potential," which, by activating neighboring sodium channels, can propagate meters away the information that

**Table 1.2** Properties of the cytosolic and membrane environments that are relevant to protein folding and association.

| Property | Cytosol | Plasma membrane[a] |
|---|---|---|
| Solvent chemical homogeneity | Yes | No |
| Chemical groups available | HOH, ions, $-$SH | $-CH_3$, $-CH_2- = CH-$ |
| Isotropy | $\sim$Yes[b] | No |
| pH gradient | No | Yes[c] |
| Electric field $(V \cdot m^{-1})$ | $\sim$0[b] | $\sim 2 \times 10^c$ |
| Pressure gradient | No | Yes |
| Dielectric constant gradient | No | Yes |
| Redox potential gradient | No | Yes[c] |
| Volume or surface occupancy [protein/solvent (%)][d] | $\sim$17 | $\sim$35[e] |
| Separation between two proteins | | |
|     Distance (A) | $\sim$50 | $\sim$30–35 |
|     Intervening solvent molecules | $\sim$15–20 | $\sim$4 |
| Exchange time between solvent molecules(s)[f] | $\sim 10^{-11}$ | $\sim 10^{-7}$ |
| Viscosity at 20 °C $(\eta; N.s.m^{-2})$ | 0.001 | 0.1 |
| Dimensions | 3 | $\sim$2 |
| Translational diffusion[g] | | |
|     $D_{lat}$ $(m^2 \cdot s^{-1})$ | $\sim 10^{-10}$ | $\sim 10^{-11}$ |
|     Average range explored in 1 $\mu s$ $(\bar{x}; Å)$ | $\sim$250 | $\sim$50 |
| Dielectric constant $(\varepsilon)$ | 80 | $\sim$2[h] |
| $\Delta G°$ $(kcal \cdot mol^{-1})$ for: | | |
|     Breaking a main chain H-bond | $\sim$0 | +4–6[i] |
|     Deprotonating a Glu side chain (pH7) | $-$4 | >30[j] |
|     Opening a salt bridge | <1 | 60[k] |
|     Exposing one $Å^2$ of hydrophobic surface | +0.025[l] | $\sim$0 |
|     Exposing a Leu side chain to the solvent | +2.8[j] | $\sim$0 |
|     Associating two 50-kDa proteins (T$\Delta$S at 20 °C) | 8[m] | 5[m] |

The table and its annotations are from Popot and Engelman (2000)

[a]For properties that vary as a function of the depth in the membrane, the data correspond to those at the membrane center

[b]Note, however, that the cytosol is heavily encumbered (cf. Goodsell 1991, and Fig. 1.15, *left*)

[c]In most but not all membranes

[d]Estimated from data compiled by Goodsell (1991) for the cytosol and plasma membrane of an *E. coli* cell; calculations for the cytosol are based on Goodsell's estimates for the average size of a soluble protein and assume a 1:2.5 w/w ratio of RNA to protein (Alberts et al. 2015); calculations for the plasma membrane assume the average integral protein (often an oligomer) to comprise $\sim$12 transmembrane helices and to have about half of its volume buried into the membrane (cf. Popot and de Vitry 1990); estimates published in the literature vary from 17% to 50% (reviewed in Saxton 1989; see also Lavergne et al. 1992)

[e]Note that the percolation threshold for short-range diffusion of small molecules in the membrane plane is $\sim$50% (see Saxton 1989; Lavergne et al. 1992)

[f]In pure solvent

[g]For a middle-sized protein ($\sim$50 kDa) in either pure water or pure lipids; in the cytosol and in real membranes, diffusion coefficients vary with the distance range considered; see Peters (1986); Saxton (1989); Lavergne et al. (1992)

[h]From Dilger et al. (1982)

[i]From Allen (1975)

[j]Based on Engelman and Steitz (1981)

[k]From Honig and Hubbell (1984)

[l]From Reynolds et al. (1974)

[m]Based on Amzel (1997)

the cell has been stimulated. At the neuromuscular junction of vertebrates, motor nerve terminals, when they are invaded by an action potential, release a neurotransmitter, acetylcholine (ACh), which diffuses through the synaptic cleft to the muscle postsynaptic membrane, where it is recognized by a transmembrane (TM) receptor, the nicotinic ACh receptor (nAChR), which opens a relatively non-specific cation-selective channel (see § 1.6.2). The net entry into the cytosol of the muscle fiber of positive charges, principally carried by $Na^+$ ions, lowers the membrane potential of its plasma membrane to the level needed to activate the voltage-sensitive sodium channel and trigger an action potential, which propagates along its length. This in turn causes the release in the cytosol of calcium ions stored in the lumen of an internal compartment, the sarcoplasmic reticulum (SR), setting off muscle contraction. In tissues such as the liver or the eye lens, the cytosols of neighboring cells are put into communication one with another by TM channels, the connexons, which assemble laterally in the membrane plane into *gap junctions*. As a safety device, an abnormal increase of the cytosolic concentration of calcium in a cell, as happens if the integrity of its membrane is compromised, causes the connexons to close, disconnecting the affected cell from the rest of the tissue. Channels can also be activated mechanically. Such is the case of mechanosensitive channels, which, in bacteria, open in response to osmotic pressure changes that tend to expand the plasma membrane, thus releasing the pressure, or, in eukaryotes, play important roles in physiological processes such as touch, pain, or hearing.

• In the examples we have seen thus far, existing gradients are dissipated. They must be recreated, which is the role of pumps that tap various sources of energy. ATP is hydrolyzed to ADP and inorganic phosphate by pumps such as the plasma membrane $Na^+,K^+$-ATPase, which sets up the transmembrane $Na^+$ and $K^+$ gradients, or the $Ca^{2+}$-ATPase, which pumps back $Ca^{2+}$ ions into the SR lumen so as to end muscle contraction (see § 1.6.3). ATP is also hydrolyzed to operate the multidrug resistance pumps that extrude toxic compounds, such as antibiotics or anticancer drugs. ATP is regenerated from ADP and phosphate by the $F_1F_O$-ATP synthase, at the expense of dissipating a transmembrane proton electrochemical gradient. The free energy stored in the gradient is first converted, at the level of the TM complex $F_O$, into mechanical energy, which the $F_1$ domain, which lies in either the cytosol or the mitochondrial or chloroplastic matrix, converts into chemical energy by forcing the phosphorylation of ADP into ATP. In mitochondria, the proton gradient is created, in the course of respiration, by a series of membrane complexes that oxidize reducing molecules by transferring their electrons to $O_2$, releasing $H_2O$, and consuming and pumping protons in the process. In chloroplasts, the proton gradient is the result of oxygen-evolving photosynthesis, a process, also catalyzed by several TM complexes, in which the energy of photons is exploited to extract electrons and $H^+$ ions from water, releasing gaseous $O_2$ and reduced electron carriers. In some archaebacteria, a proton gradient is created, without involving any redox chemistry, by a single, small TM protein, bacteriorhodopsin (BR), which uses the light-induced isomerization of its cofactor, retinal, as a source of mechanical energy that sets off a series of transconformations whose end result is to extrude protons from the cytosol through the plasma membrane (see § 1.6.1).

*Membrane proteins in action*
© 2018 by Francis Haraux

- Various MPs convert ion gradients into mechanical energy and vice versa. For instance, protons flowing from the periplasm into the cytosol activate the rotary motors that move bacterial flagella, as well as molecular devices (among which the Tol and Ton systems) that transfer mechanical energy to MPs inserted into the outer membrane, allowing them to import nutrients, e.g. iron, against their electrochemical gradient.
- Gradients are also exploited by *secondary transporters*, which dissipate one to build another. $H^+$ or $Na^+$ gradients, for instance, can be used by symporters or antiporters to accumulate nutrients, such as sugars or amino acids, or expel drugs. The red blood cell anion exchanger ("Band 3") exchanges chloride ions against bicarbonate ones, which permits the removal of $CO_2$ from tissues and its release to the atmosphere in the lungs.
- Signaling is one of the major functions of MPs. Many of them indeed are located in the plasma membrane, at the border between the cytosol and the external medium, be it the world at large, with its resources and threats, or the intercellular medium, in metazoans, with the complex network of messages that organizes cell-to-cell collaboration. Signaling does not necessarily require the transfer through the membrane of molecules or ions. It can be achieved by such TM messages as oligomerization, reorganization of an oligomer, or transconformation of a monomer, which are caused by an external stimulus and detected in the cytosol. The first two types of rearrangements are typical, for instance, of receptors that detect nutrients or growth factors, the latter's dysfunction being at the origin of many cancers. Activation typically results in the phosphorylation or methylation of cytosolic domains, which is detected by cytosolic proteins. The third type is exemplified by the G protein-coupled receptors (GPCRs), a very large family of homologous MPs (close to 1000 members in humans) whose members, which share a common fold featuring seven TM helices, can detect light (rhodopsins), hormones, odorants, neurotransmitters, etc. GPCRs transmit information from the outside to the inside of the cell via ligand-induced conformational changes that result in the activation or inhibition of cytosolic proteins (for an overview, see Chap. 2, § 2.5.2).
- Many MPs are involved in one form or another of cell trafficking, such as cell division, or the budding or fusion of vesicles or cells. At the neuromuscular junction, for instance, synaptic vesicles in the nerve terminal are actively filled by membrane pumps with neurotransmitters, ATP, etc. When an action potential invades the terminal and depolarizes its plasma membrane, synaptic vesicles fuse with it, discharging their content in the synaptic cleft, thus

exciting the postsynaptic cell. Budding, migration, and fusion result in the transport of solutes and membrane components from one cell compartment to another. During their biosynthesis, MPs are first integrated into the endoplasmic reticulum (ER) membrane (see § 1.7.1). Those MPs that have been correctly modified posttranslationally and, if applicable, oligomerized accumulate into vesicles that bud from the ER and fuse with the *cis* cisternae of the Golgi apparatus, after which, by a succession of fission and fusion events, they move from one compartment of the Golgi to another and, finally, reach the plasma membrane. Some MPs from the plasma membrane are internalized when they have bound a hormone or a nutrient. Thus, the low-density lipoprotein (LDL) receptors, once loaded with the cargo they have picked from the blood, are internalized into *endocytic* vesicles that fuse with cytosolic compartments called *endosomes*, the internal pH of which is low (from pH 6.0–6.5 in early endosomes to pH 4.5–5.5 in late endosomes and lysosomes). This pH drop is the signal for LDL and its receptor to dissociate one from another. The receptor migrates back to the cell surface, whereas the LDL is transported to lysosomes, where it will be degraded and its components metabolized. This mechanism is exploited by many viruses, such as that of flu: hemagglutinin, a MP of the viral envelope, binds to cell-surface oligosaccharides containing sialic acids, upon which viral particles are transported to endosomes. At low pH, hemagglutinin undergoes a conformational change, exposing a hydrophobic segment that interacts with the endosomal membrane and brings about its fusion with the viral envelope, releasing the nucleic acid of the virus into the cytosol, where it will be transcribed and replicated. Newly synthesized viral envelope MPs are integrated into the ER and transported to the plasma membrane, where they bind the viral nucleic acid and its associated proteins. Viral particles bud and are released in the extracellular medium, launching a new cycle.

- MPs are involved in transferring soluble proteins across the plasma membrane of bacteria, across the two-membrane envelope of Gram$^-$ bacteria, across the membrane and into the lumen of the endoplasmic reticulum, into mitochondria and chloroplasts, etc., as well as in integrating into one or the other target membrane newly synthesized MPs, either in the course of their synthesis or posttranslationally (see § 1.7.1).

- Some membrane proteins are involved in structuring cells and tissues. "Band 3," for instance, the red blood cell anion exchanger, comprises, in addition to its TM domain, a cytosolic one that interacts with the cytoskeleton, playing a critical role in determining the mechanical properties of the cell. In the gap junctions, connexons integrated in the plasma membrane of two neighboring cells not only dock one onto the other to form an aqueous channel that spans the two plasma membranes, through which small molecules can diffuse from the cytosol of one cell to that of the other: they also mechanically associate the two cells one with another.

This quick and somewhat shallow overview of the tasks that MPs fulfill is quite patchy, but it gives an idea of what they must be able to achieve: form pores and regulate their specificity and their opening; recognize, bind, and import solutes, some of them small (ions, small molecules), some very large (proteins); transduce energy and information; associate mechanically membranes to each other or to inside or outside structures; and carry out the budding, transport and fusion of vesicles, etc. In the next section, we shall examine which structures have evolved to achieve such tasks. Forming a clear view of the kinds of structure adopted by MPs and what stabilizes them is indeed essential to understanding the problems encountered when handling them in aqueous solutions, and why certain types of surfactants are better than others at keeping those under control.

## 1.4  Membrane Protein Structure

### 1.4.1  Modes of Association with the Membrane

For reasons that will be briefly discussed below, and treated more at length in other chapters of the book, knowledge of the structure of MPs has long lagged behind that of soluble proteins, and it still does. The first electron-density map whose resolution was sufficient to show the secondary structure elements of a TM protein region, that of BR, was obtained in 1975 by Richard Henderson and Nigel Unwin using electron cryomicroscopy (cryo-EM) (Henderson and Unwin 1975). It showed the TM region to be comprised of a bundle of seven $\alpha$-helices oriented more or less normal to the plane of the membrane. It took 15 more years to bring the structure of BR, still by cryo-EM, to a resolution permitting an atomic model to be fitted into it (Henderson et al. 1990). The first X-ray structure of a MP, that of a bacterial photosynthetic reaction center comprising three MPs and many cofactors, was established in 1985 by Hartmut Michel and Johann Deisenhofer (Deisenhofer et al. 1985). Again, the dominant theme of the TM region was bundles of $\alpha$-helices. The first medium-resolution X-ray structure of a bacterial outer MP, a trimeric porin, was established in 1989. It showed that each monomer is comprised of a TM $\beta$-barrel (Weiss et al. 1989, 1991a, b). The first MP structures obtained by solution NMR, those of the $\alpha$-helix dimer of glycophorin A (MacKenzie et al. 1997) and of two small $\beta$-barrel outer membrane proteins (Arora et al. 2001; Fernández et al. 2002), date back to the turn of the century. Since then, the experimental determination of MP structures has picked up steam (Fig. 1.4), but the number of unique resolved structures remains only a small fraction of those deposited in the Protein Data Bank: ~500 vs. nearly 100,000, i.e. ~0.5%, which is very far from the ~30% of coding sequences TM proteins are expected to represent (Wallin and von Heine 1998).

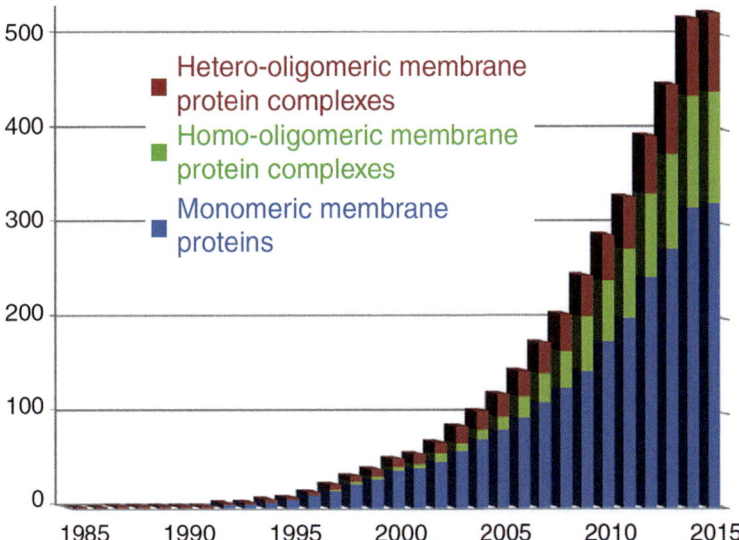

**Fig. 1.4**  Number of unique structures of membrane proteins deposited in the Protein Data Bank (PDB, www.rcsb.org) by the beginning of 2015. This number has been increasing exponentially since the first atomic model became available in 1985, but it still represents only a very small fraction of all the structures in the PDB (currently ~100,000). The majority of known MP structures are monomeric or from homo-oligomers. Multi-subunit membrane-protein complexes are particularly challenging to overexpress, and only 84 structures were available at the time of the study (From Zorman et al. 2015. © 2015 Elsevier Ltd. All rights reserved).

Nevertheless, the diversity of three-dimensional organizations that has now been revealed makes it possible to propose some general rules.

Before moving on to the TM organization of MPs and their interactions between themselves and with lipids, let us have a look at the various ways proteins can interact with membranes, as schematized in Fig. 1.5A:

- The polypeptide chain of some MPs does not come into contact with the lipids. Instead, the protein is merely anchored into the membrane via a covalently bound lipid, such as a glycolipid (①) or a fatty acyl chain (②).
- Some MPs do contact directly the membrane, but on one side of it only, interacting with it via α-helices that expose a very hydrophobic external surface (③). This arrangement appears to be rare, but, at variance with the next two ones, it is difficult to detect by sequence analysis and its prevalence may be underestimated.
- A very frequent type of MPs spans the membrane via a single TM α-helix (④, ⑥). The extramembrane regions of these proteins can be very small, limited to a few residues, or enormous, as in some growth factor receptors (cf., in Fig. 1.5B, the two very small subunits that comprise cytochrome $b_{559}$ and the large monomer of the homodimeric transferrin receptor).
- An equally large number of MPs span the membrane several times, featuring from 2 to more than 20 TM segments. In that case, the TM segments can be either α-helical, as shown in Fig. 1.5A, Scheme ⑦, and Fig. 1.5B, which is the case in all plasma membranes and in the

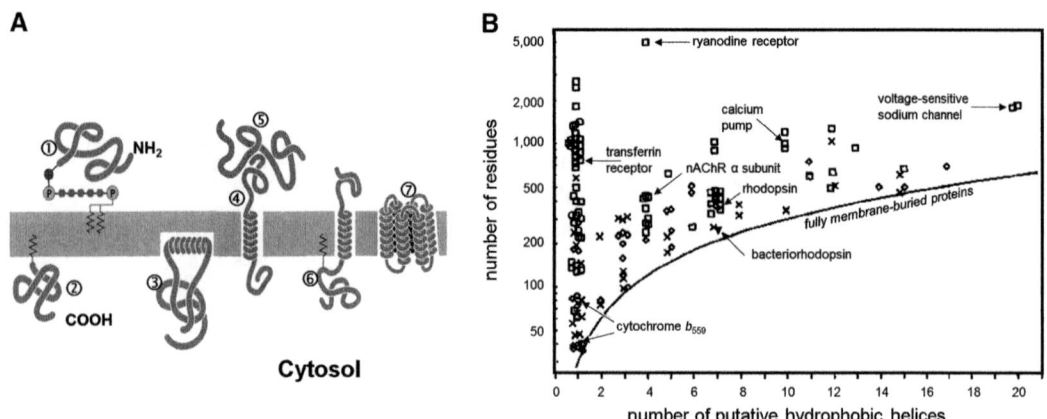

**Fig. 1.5** (**A**) The various ways proteins can be associated with membranes. See text (Modified from Alberts et al. 2015. © 2015, CCC Republication). (**B**) Length of MPs as a function of their number of putative hydrophobic transmembrane (TM) α-helices. Sequence analysis of putative eukaryotic MPs performed in 1990, before the structure of most MPs was known, yielded the number of extended hydrophobic segments, presumed to form TM helices, present in each polypeptide whose sequence had been established at the time. *Open squares*: proteins from membranes that are directly in contact with the cytosol (plasma membrane, endoplasmic and sarcoplasmic reticulum, retina sacculae, exocytotic vesicles). *Crosses*: proteins from the inner membrane of mitochondria. *Open diamonds*: proteins from the thylakoid membrane. The positions of the three MPs discussed in some detail in § 1.6, bacteriorhodopsin (BR; a prokaryotic protein; *solid triangle*), the α-subunit of the nicotinic acetylcholine receptor (nAChR), and the sarcoplasmic reticulum calcium pump (SERCA1a), are indicated, as well as that of a few other MPs that are mentioned in the text. The solid curve gives the approximate position of proteins that are essentially fully buried into the bilayer, assuming that ~30 residues are needed to span the full thickness of the bilayer (40–45 Å) and form one turn. The further a given MP lies above this line, the more extended its extramembrane domains (Adapted from Popot and de Vitry 1990).

membrane surrounding most compartments of eukaryotic cells, or folded into a $\beta$-barrel, as is the case for all MPs of bacterial outer membranes and a few MPs from the outer membranes of mitochondria and chloroplasts.

- Finally, many protein subunits are simply associated non-covalently to MPs themselves without being in direct contact with the lipids (Fig. 1.5A, Scheme ⑤). These proteins are called "extrinsic" MPs, because they can be removed by treatments, like extremes of pH or ionic strength, that do not disrupt the membrane itself.

Those MPs whose polypeptide chain is in direct contact with the hydrophobic core of the membrane (Fig. 1.5A, Schemes ③, ④, ⑥, ⑦) are called "integral" or "intrinsic" (Singer and Nicolson 1972). They cannot, with few exceptions, be extracted without resorting to strong surfactants, which will break the membrane apart (see Chap. 2). A nomenclature introduced by Günter Blobel (Blobel 1980) distinguishes those integral MPs that penetrate the lipid core but do not span it, called *monotopic* integral MPs (because they contact only one aqueous phase; Fig. 1.5A, Scheme ③), from those that span it once, and thus comprise one region in contact with each of the two aqueous phases separated by the membrane (called *bitopic* MPs; Schemes ④, ⑥; an admittedly somewhat confusing term given that these MPs sport a single TM segment, not two) and those that span it two or more times (called *polytopic* MPs; Scheme ⑦). Integral MPs can be further anchored to the membrane by covalently attached lipids (Scheme ⑥). Many MPs are organized into homo- or hetero-oligomers, which can comprise several subunits (sometimes tens of them), often associating bitopic and polytopic MPs to extrinsic ones (see e.g. Figs. 1.13 and 1.21). The complexes can themselves associate into supercomplexes, such as the mitochondrial "respirasome," which associates one copy of Complex I (the NADH:ubiquinone oxidoreductase), a dimer of Complex III (cytochrome $bc_1$), and one copy of Complex IV (cytochrome $c$ oxidase) (see Chap. 12, Fig. 12.17).

## 1.4.2 Structure of Transmembrane Protein Regions

In this book, we will be mostly concerned with bitopic and polytopic MPs. There are two major reasons for this focus:

(i) Very few data are currently available about applying APols and other nonconventional surfactants to studying monotopic integral MPs.

(ii) A major advantage of APols is their mildness toward lipid-embedded protein domains, which do not exist in anchored proteins; APols can certainly be useful, in specific cases, for handling anchored proteins in vitro (e.g. when the protein is vulnerable to detergents and the anchor cannot be deleted, or to attach these proteins onto solid supports via functionalized APols, to deliver them to membranes, etc.), but to date such experiments have been described only for TM proteins.

The following discussion focuses on TM protein regions, because it is those with which APols and other surfactants have been designed to interact. APols can also interact with extramembrane regions, a point that will be discussed in Chap. 5. The organization of these regions, however, raises few specific issues as compared to soluble proteins.

It is beyond the scope of the present book to present an exhaustive overview of MP structures, and only some general guidelines will be proposed, particularly where they are essential to understanding the mode of action of APols and other nonconventional surfactants, the reason(s) of their mildness toward MPs as compared to detergents, and as a help to solving problems that their use can

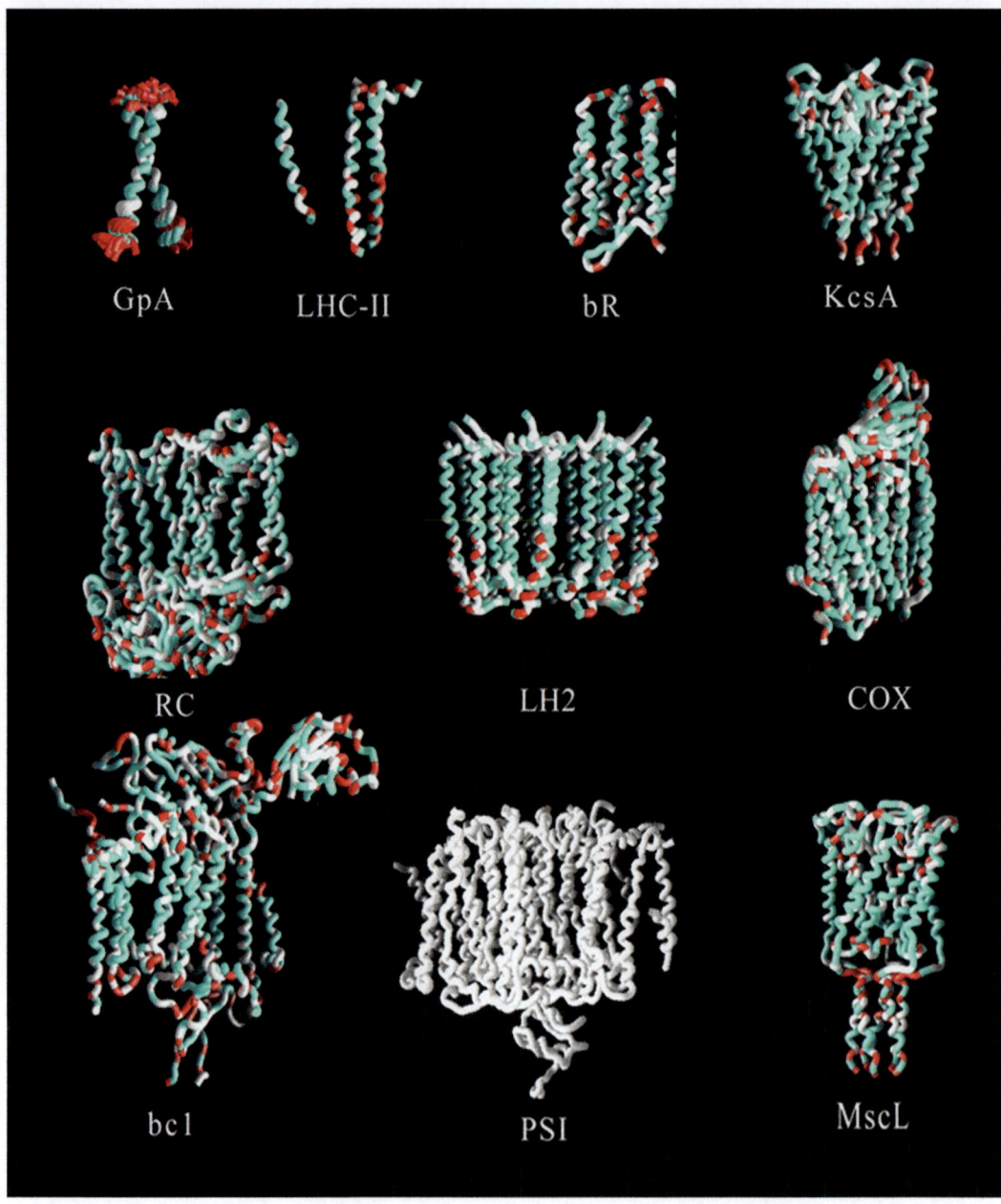

**Fig. 1.6** A gallery of the first established structures of membrane proteins whose TM region is organized as a bundle of α-helices, as they were available as protein database entries by June 1999. The main chain only is shown. Hydrophobic side-chain positions (A, V, L, I, F, M) are in *cyan*, strongly ionizable residue positions (D, E, R, K) in *red*, others in *white*. The marked hydrophobic character of TM helices is clear. Prosthetic groups are omitted, resulting in gaps. The proteins shown and their protein database files are as follows: *GpA* glycophorin A structure obtained by solution NMR (1AFO; MacKenzie et al. 1997), *LHC-II* eukaryotic light-harvesting Complex II (cryo-EM, 3.4-Å resolution; 1LHC; Kühlbrandt et al. 1994), *BR* bacteriorhodopsin from *Halobacterium salinarum* (cryo-EM, 3.0-Å resolution; 2AT9; Mitsuoka et al. 1999), *KcsA* potassium channel from *Streptomyces lividans* (X-ray diffraction, 3.2-Å resolution; 1BL8; Doyle et al. 1998), *RC* photosynthetic reaction center from *Rhodobacter sphaeroides* (X-ray diffraction, 2.2-Å resolution; 1AIJ; Stowell et al. 1997), *LH2* light-harvesting complex from *Rhodopseudomonas acidophila* (X-ray diffraction, 2.5-Å resolution; 1KZU; Prince et al. 1997), *COX* cytochrome *c* oxidase

encounter (for general reviews on MP structure, see e.g. White and Wimley 1999; Popot and Engelman 2000; Vinothkumar and Henderson 2010; Buchanan et al. 2012; White et al. 2018, and references therein). We will start with listing a few rules that are generally obeyed and then give some examples of more complex situations. In the next two sections, we will consider membrane protein/lipid interactions (§ 1.5) and the dynamics of TM protein regions (§ 1.6), both of which are essential to understanding the problems of instability most MPs face when extracted from their native membrane environment and how nonconventional surfactants can help mitigating them.

As more or less general rules, we can consider the following:

1. Protein surfaces exposed to the hydrophobic core of the membrane are highly hydrophobic. This is achieved by three mechanisms: (i) membrane-exposed segments span completely the hydrophobic core, leaving no loops exposed to it; (ii) they satisfy most of the hydrogen bonds that can be formed between the >N—H and O=C< groups of the main chain peptide bonds; and (iii) most exposed side chains are nonpolar (cf. Figs. 1.6 and 1.7). This is a natural consequence of the high cost of burying polar groups in the membrane core (Table 1.2). An exposed hydrogen-bonding side chain like that of serine residues will tend to either bond back with the TM surface of the protein or interact with complementary residues at the surface of other TM regions. This is why, for instance, an Ala → Glu mutation in the single TM $\alpha$-helix of the FGFR3 receptor of fibroblasts drives its dimerization, mimicking physiological activation and resulting in cancer (Li et al. 2006). Kinks and other irregularities in TM helices do exist, however, and often have important functional roles (see e.g. Popot and Engelman 2000 and examples shown below and in § 1.6).

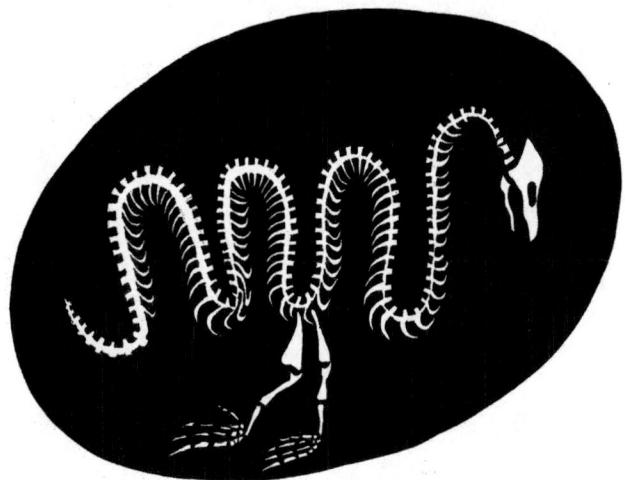

*Membrane protein structure revealed*
© 2018 by Francis Haraux

**Fig. 1.6** (continued) from *Paracoccus denitrificans* (X-ray diffraction, 2.7-Å resolution; 1AR1; Ostermeier et al. 1997), *bc1* cytochrome $bc_1$ complex from beef heart (TM subunits only; X-ray diffraction, 3.7-Å resolution; 3BCC; Xia et al. 1997), *PSI* Photosystem I reaction center from *Synechococcus elongatus* ($\alpha$ carbons only, side chains having not yet been identified at the time; X-ray diffraction, 4-Å resolution; 2PPS; Krauss et al. 1996), *MscL* mechanosensitive ion channel from *Mycobacterium tuberculosis* (X-ray diffraction, 3.5-Å resolution; 1MSL; Chang et al. 1998) (Figure from Popot and Engelman 2000).

2. A single α-helix, provided its central region is mostly hydrophobic, satisfies the above conditions, and many MPs feature only one, two, or three TM helices (Figs. 1.5B and 1.6). Such proteins can be monomeric, or assemble into homo- or hetero-oligomers. MPs whose TM segments are made of β-strands, on the contrary, must feature enough of them – that is, at least eight – to form a β-sheet which will close upon itself into a β-barrel, so that its TM edges can hydrogen bond together (cf. Fig. 1.7 and the TM structures of OmpA and OmpX shown in Chap. 10, Figs. 10.11 and 10.12). Note that the protein surfaces exposed to the membrane core are always made of only one or the other type of secondary structure that is either a single helix, or a bundle of helices, or a β-sheet closed upon itself into a β-barrel, or an assembly of β-barrels. No cases have been found of a partial barrel completed by a helix, for instance. Whereas it is frequent that α-helical regions that are linked into a single polypeptide in one MP are split into several subunits in another, homologous one, nevertheless yielding TM α-helix bundles with similar 3D structures (see examples cited in Popot and de Vitry 1990; Popot and Engelman 2000), no natural case is known (yet) of a MP whose TM β-barrel would be formed by β-strands contributed by several distinct subunits. Such proteins can however be created by genetic engineering, in vivo (Koebnik 1996) or in vitro (Debnath et al. 2010). The toxin α-hemolysin forms a 14-strand TM β-barrel by assembly of seven monomers, each of which contributes two β-strands (Fig. 1.7; Song et al. 1996).

3. Once a complete barrier against the membrane hydrophobic core has been formed, be it by a closed "hedge" of helices, a complete β-barrel, or an oligomer of barrels, other structures can

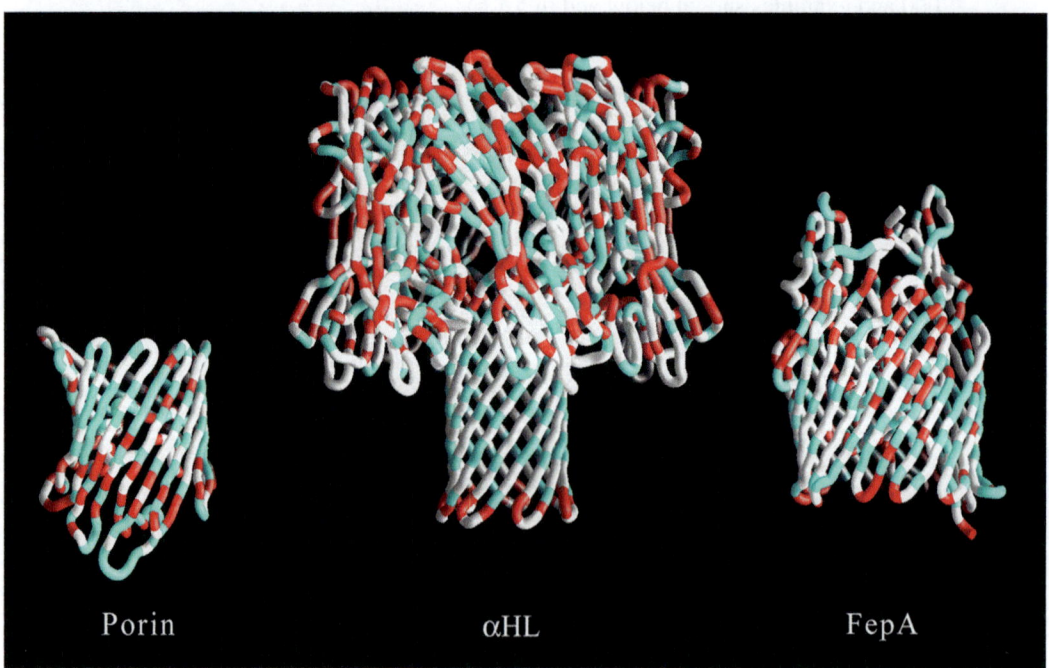

Porin                         αHL                         FepA

**Fig. 1.7** Examples of β-barrel structures. Three examples are shown for comparison with Fig. 1.6: *porin* from *Rhodobacter capsulatus* (X-ray diffraction, 1.8-Å resolution; 2POR; Weiss et al. 1991a, b), *αHL α*-hemolysin from *Staphylococcus aureus* (X-ray diffraction, 1.9-Å resolution; 7AHl; Song et al. 1996), and FepA ferric enterobactin receptor from *Escherichia coli* (X-ray diffraction, 2.4-Å resolution; 1FEP; Buchanan et al. 1999). As detailed in the legend to Fig. 1.6, hydrophobic groups are shown in *cyan* and strongly ionizable groups in *red* (From Popot and Engelman 2000).

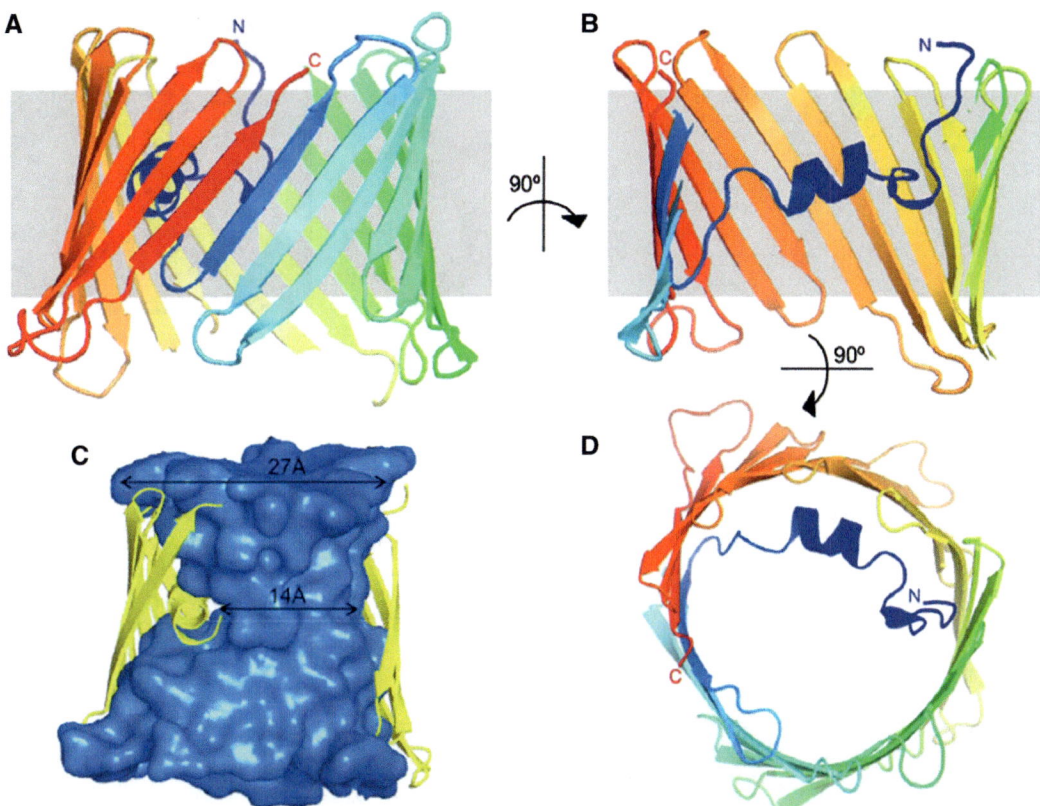

**Fig. 1.8** Crystal structure of mouse voltage-dependent anion channel 1 (VDAC1). (**A**) Ribbon representation of VDAC1 viewed along an axis parallel to the membrane plane. The VDAC1 protein structure is rainbow colored from the *N*-terminus in *blue* to the *C*-terminus in *red*. (**B**) Cross-section view of *A* rotated 90° clockwise around the $\beta$-barrel axis. $\beta$-strands 3–7 have been removed to illustrate the positioning of the *N*-terminal segment that forms an $\alpha$-helix inside the lumen of the barrel. (**C**) Cross-section view of *A* with $\beta$-strands 19 and 1–4 removed. The interior surface of the VDAC1 channel (*blue*) illustrates the contour of the pore. Dimensions at the entrance and along the narrowest point in the center of the pore are indicated. (**D**) Ribbon representation of VDAC1 viewed along an axis perpendicular to the membrane plane (same coloring as in *A*) (From Ujwal et al. 2008. © 2008 National Academy of Sciences).

form in the interior of the "corral" thus delimited, which is protected from contact with the lipids, whereas water can access to it from each side of the membrane. Thus, VDAC, the voltage-dependent mitochondrial anion channel, features a 19-strand $\beta$-barrel, inside which is located a partially disordered loop comprising a short helix (Hiller et al. 2008; Ujwal et al. 2008; Fig. 1.8).

The OmpF porin of *E. coli* is a homotrimer of 16-strand $\beta$-barrels (Cowan et al. 1992) (Fig. 1.9). It features a couple of structures that could not face the membrane core and are hidden within the protein: first, the $\beta$-strands closest to the central threefold axis, which are too short to span the thickness of the membrane hydrophobic core, face the interior of the trimer and are not exposed to the membrane (Fig. 1.9A); second, as in VDAC, the lumen of each of the three barrels contains a reentrant loop (Fig. 1.9B), whose size and composition determine the size of the pore contained in each protomer (Fig. 1.9C). The loop, which does not adopt a periodic secondary structure that would satisfy main chain hydrogen bonds, is hydrated by the water contained in the lumen of the channel.

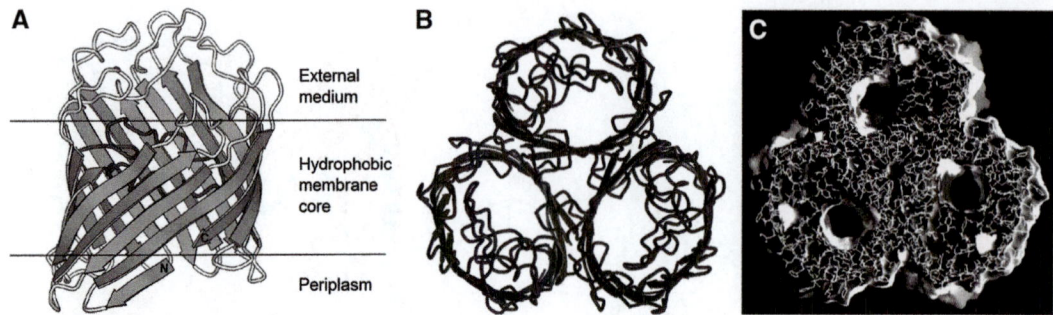

**Fig. 1.9** Transmembrane arrangement of a bacterial outer membrane porin, OmpF from *E. coli*. (**A**) General fold of a porin monomer. The large hollow β-barrel structure is formed by the antiparallel arrangement of 16 β-strands. The strands are connected by short loops or regular turns on the periplasmic rim (*bottom*), whereas long irregular loops face the cell exterior (*top*). The internal loop, which connects β-strands 5 and 6 and plunges inside the barrel, is highlighted in dark. The chain *N*- and *C*-termini interact one with another. The surface closest to the viewer, which faces the threefold symmetry axis of the trimer, is involved in subunit-subunit contacts. The approximate position of the membrane hydrophobic core is indicated (X-ray structure from Cowan et al. 1992). (**B**) Schematic representation of the OmpF trimer. The view is from the extracellular space along the molecular threefold symmetry axis. (**C**) Slice through the center of the OmpF trimer structure, represented by a stick model. The molecular surface is indicated so as to visualize more clearly the three pores (about 7 × 11 Å in size). Same view as in B. (Adapted from Schirmer 1998. © 1998 Academic Press. All rights reserved).

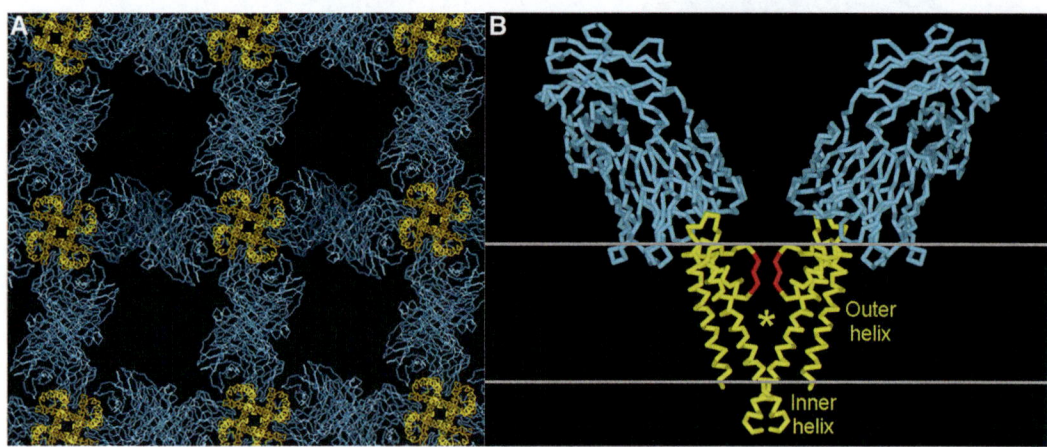

**Fig. 1.10** Two views of the tetrameric KcsA potassium channel from *Streptomyces lividans* (*yellow*) complexed by an F$_{ab}$ fragment (*cyan*). (**A**) Ribbon representation illustrating the packing of the complexes in the crystals and the contacts mediated by the F$_{ab}$ fragments. The homotetrameric channel is seen along the fourfold axis of symmetry, in projection onto what, in situ, would be the membrane plane. (**B**) View from the inside of the membrane, perpendicular to that in A. The extracellular side is on *top*. Only two of the four subunits are shown. Note the way the transmembrane helices – two per protomer – isolate from the membrane the pore-forming structural elements, to which each protomer contributes a short, reentrant helix and a loop, which forms the selectivity filter (*red*), and the water-filled central cavity (marked with an asterisk). The approximate position of the membrane core is indicated by the two gray lines (Adapted from Zhou et al. 2001. © 2001 Macmillan Publishers Limited, Nature. All rights reserved. See also Doyle et al. 1998).

Similarly, in ion channels such as the potassium channel KcsA from *Streptomyces lividans* (Doyle et al. 1998), the continuous fence of TM helices formed by the assembly of four two-TM helix monomers protects a central TM aqueous channel (Fig. 1.10). The channel contains a reentrant loop in which four extended strands provide the main chain oxygen atoms that can substitute for some of the

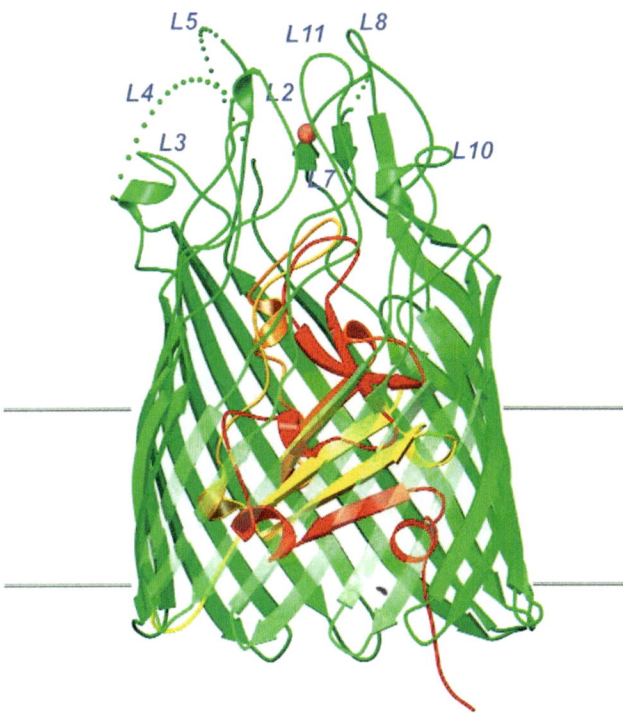

**Fig. 1.11**   Ribbon diagram of the bacterial iron importer FepA. The extracellular space is located at the *top* of the figure, the periplasmic space at the *bottom*. The position of the membrane bilayer is delineated by horizontal lines, as determined from the hydrophobic area found on the molecular surface. The putative position of the iron of ferric enterobactin is indicated by the *dark pink* sphere. Part of the barrel has been rendered transparent to reveal the *N*-terminal domain located in the lumen of the barrel (in *yellow, orange*, and *red*). All long extracellular loops are labeled; disordered parts of the loops are indicated by dotted lines (From Buchanan et al. 1999. © 1999 Macmillan Publishers Limited, Nature Structural Biology. All rights reserved).

water molecules that hydrate alkaline cations in the bulk and, by their geometry, define the specificity of the channel (Fig. 1.10C).

Large internal domains are found inside wide TM $\beta$-barrels, such as those of the bacterial iron importers FepA or FhuA. FepA is composed of two domains: (i) a 22-stranded antiparallel $\beta$-barrel, spanning the outer membrane (residues 154–724), and (ii) an *N*-terminal globular domain consisting of two long loops, several short $\beta$-strands, and single-turn helices (residues 1–153) that fold into the barrel, plugging the barrel pore (Fig. 1.11).

The Ste24p metalloprotease from the endoplasmic reticulum of *Saccharomyces cerevisiae* (ScSte24p), which is involved in the maturation of the mating pheromone **a**-factor, and the homologous human ZMPSTE24 protease, which processes the precursor form of the nuclear scaffold protein lamin A, contain a remarkably large intramembrane aqueous cavity (Fig. 1.12), whose functional role may be to serve as a water reservoir to facilitate substrate processing (Pryor et al. 2013; Clark et al. 2017).

Unexpected, surprising structures continue to be discovered, such as the very long (~110-Å) $\alpha$-helix, parallel to the membrane plane, that runs nearly all the way along the TM domain of mitochondrial Complex I and is presumed to be somehow involved in proton pumping (in *blue* in Fig. 1.13; Vinothkumar et al. 2014). Note the extreme complexity of Complex I, which, in its beef version, contains no less than 44 subunits for an overall mass of ~1 MDa. As mentioned above, and as

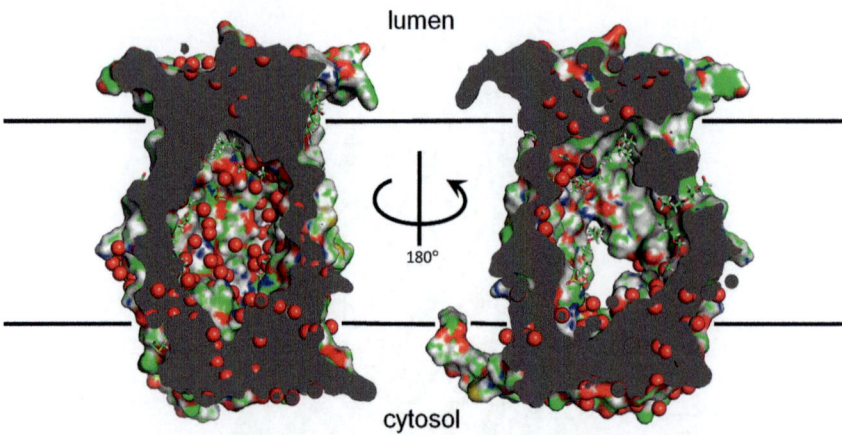

**Fig. 1.12** The water-filled cavity inside the transmembrane region of the human ZMPSTE24 protease. The protein has been cut open along a plane perpendicular to the probable plane of the membrane. Modeled lipids, detergents, and small molecules are shown as stick representations. Surfaces of modeled carbon atoms are shown in *green*, hydrogen in *gray*, nitrogen in *blue*, oxygen in *red*, and sulfur in *yellow*. The positions of modeled water molecules are shown as *red spheres*. The estimated position of the membrane's hydrophobic core is indicated (Adapted from Clark et al. 2017, courtesy of Mark E. Dumont. See also Pryor et al. 2013).

will be described in more detail in the chapter dedicated to electron microscopy (Chap. 12, § 12.3.3), Complex I is itself integrated into an even larger ensemble, the respirasome (mitochondrial supercomplex B), where it is associated to a dimer of Complex III (the cytochrome $bc_1$ complex) and a monomer of Complex IV (the cytochrome $c$ oxidase) (see Fig. 12.17). This can serve as a reminder that, when we handle "a membrane protein" in solution, as will be described in the next chapter, it may well be only a fraction of a larger ensemble that has not resisted solubilization (see e.g. Kühlbrandt 2015).

## 1.5    Membrane Protein/Lipid Interactions

### 1.5.1    The Fluid Mosaic Model

At the end of the 1940s, biological membranes were known to comprise both lipids and proteins, and the ability of lipids to spontaneously organize into bilayers had long been recognized. Nothing however was known of the sequence nor the 3D structure of MPs nor of their arrangement respective to lipids. A long controversy ensued (see e.g. Green et al. 1967), which was resolved, in the early 1970s, in favor of the fluid mosaic model proposed by Seymour J. Singer and Garth L. Nicolson (Singer and Nicolson 1972). According to this model, "the bulk of the phospholipid is organized as a discontinuous, fluid bilayer, although a small fraction of the lipid may interact specifically with the membrane proteins. The fluid mosaic structure is therefore formally analogous to a two-dimensional oriented solution of integral proteins (or lipoproteins) in the viscous phospholipid bilayer solvent" (Fig. 1.14A). Important in the elaboration of this model were the realization that a continuous (but for the inserted proteins) bilayer of lipids would provide the hydrophobic barrier needed to maintain compositional differences between the aqueous compartments separated by membranes and the fact that membrane proteins (or, at least, some of them) could redistribute rapidly in the membrane plane when, for instance, they were cross-linked by antibodies.

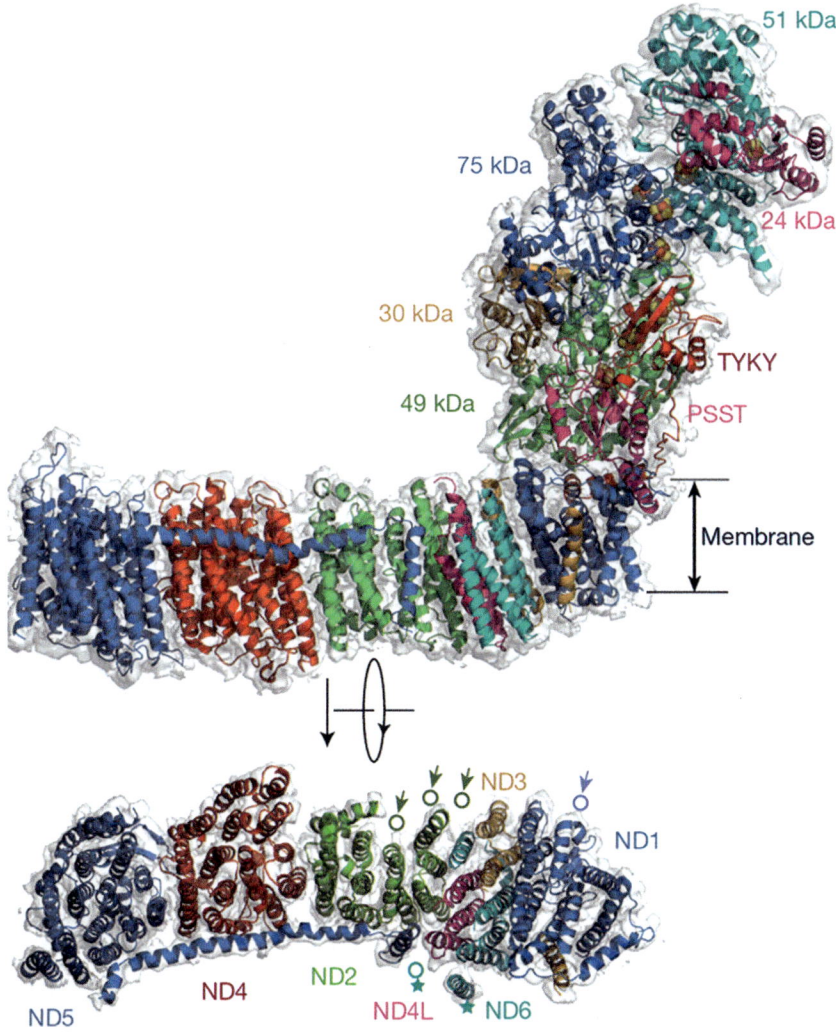

**Fig. 1.13** Structure of the core subunits of *Bos taurus* mitochondrial Complex I. Single-particle cryo-EM map at ~5-Å resolution. *Top.* Structural models of the 14 mammalian core subunits (*ribbon representation*) and their electron density (*transparent surface*); the subunits are colored individually and labeled with text in the same colors. The mitochondrial matrix is on *top. Bottom.* The seven membrane-bound mammalian core subunits, viewed from the matrix. *Arrows* indicate the positions of the four TM helices in *Thermus thermophilus* that are not present in *B. taurus*: three *N*-terminal TM helices in ND2 and one *C*-terminal TM helix in ND1. The position of TM helix 4 in ND6 is different in *B. taurus* and *T. thermophilus* (marked with *stars*). Note the very long amphipathic α-helix (in *blue*, ~110-Å long) that runs almost all the way along the TM region, parallel to the membrane plane, at the level of the lipid polar head region (From Vinothkumar et al. 2014. © 2014 Macmillan Publishers Limited, Nature. All rights reserved. See also Efremov et al. 2010; Baradaran et al. 2013).

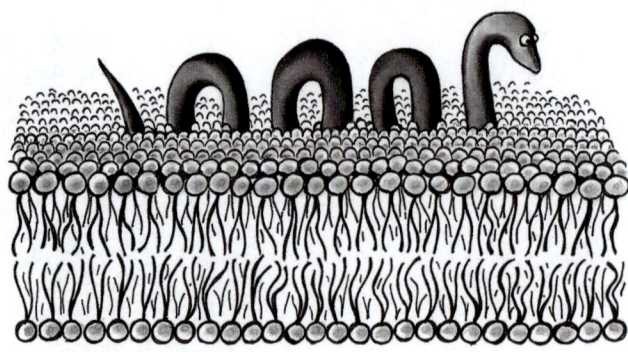

*The (erroneous) view of membrane proteins floating freely in a sea of lipids*
© 2018 by Francis Haraux

The fluid mosaic model has dominated the field of membrane biology for nearly four decades, and it is still the model most often represented in textbooks. Over the years, however, many of its limitations became clear (for discussions, see e.g. Engelman 2005; Nicolson 2014). For one thing, free diffusion of MPs in the plane of the membrane is more the exception than the rule, and it is often restricted to short distances. In eukaryotic cells, in particular, many MPs are anchored, or their diffusion is otherwise limited, due to interactions either with the cytoskeleton, within the cell, with the extracellular matrix, outside it, or, in the plane of the membrane, with lipids and other proteins. This has been taken into account in the modernized cartoon shown in Fig. 1.14B. A second important shortcoming of the original model is that it viewed MPs as free-floating in a two-dimensional "sea" of

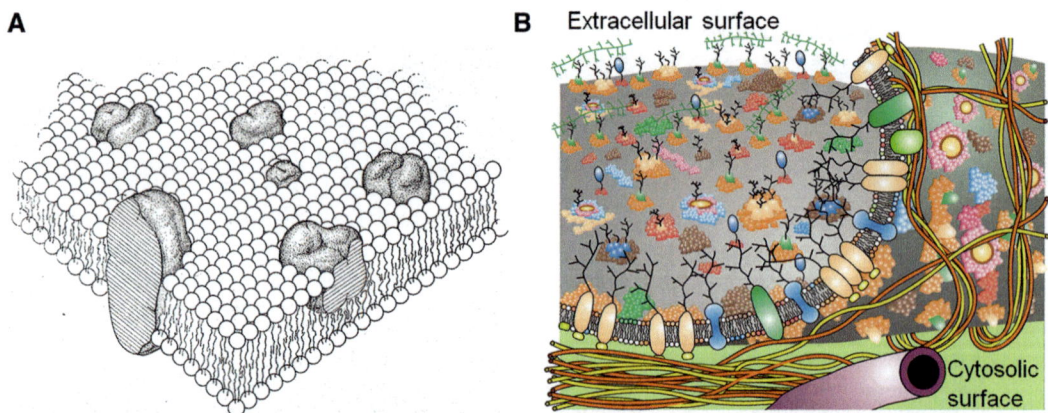

**Fig. 1.14** (**A**) The original cartoon schematizing the fluid mosaic model of biological membranes. From Singer and Nicolson (1972), reprinted with permission from the American Association for the Advancement of Science. (**B**) An updated version thereof, which takes into account information on the existence of membrane domains and the role of membrane-associated cytoskeletal and extracellular structures. Different integral MPs, lipids, and oligosaccharides are represented by different colors, and the membrane has been rolled over to show the inner membrane surface. Cytoskeletal fencing is apparent, which restricts the lateral diffusion of some but not all TM proteins. Other lateral diffusion restriction mechanisms are also represented, such as lipid domains, integral membrane protein complex formation (seen in the membrane cutaway), polysaccharide-glycoprotein associations (at the *far top left*), and direct or indirect attachment of inner surface membrane domains to cytoskeletal elements (at *lower left*). Although this figure suggests some possible integral membrane protein and lipid mobility restraint mechanisms, it is not meant to accurately represent the sizes or structures of integral membrane proteins, cytoskeletal structures, polysaccharides, lipids, submicro- or nano-sized domains or membrane-associated cytoskeletal structures, or their crowding in the membrane (From Nicolson 2014).

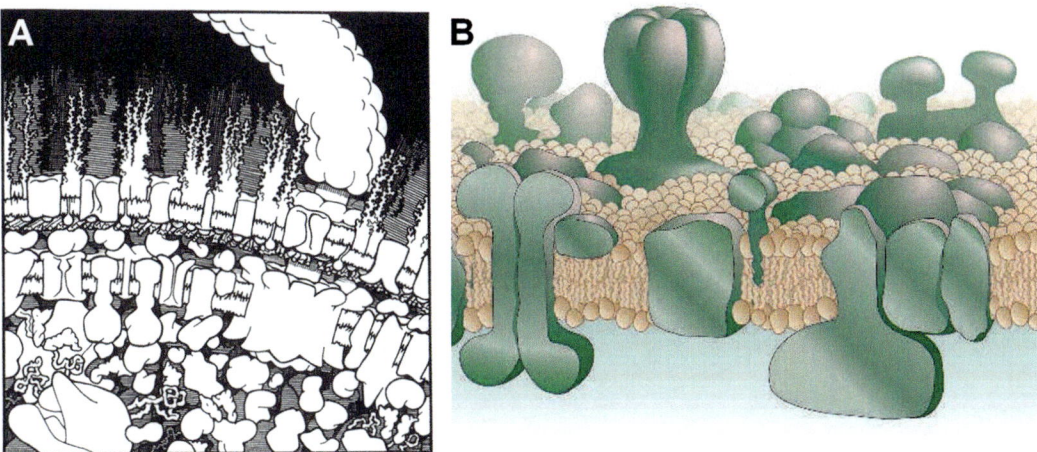

**Fig. 1.15** Two views of the crowding of membrane proteins in (**A**) the inner and outer membranes of an *E. coli* cell (from Goodsell 1991, © 2011 Elsevier Ltd. All rights reserved) and (**B**) a generic membrane (From Engelman 2005, © 2005 Macmillan Publishers Limited, Nature. All rights reserved).

lipids, with most MPs widely separated one from another (Fig. 1.14A). This is very far from the actual situation in real membranes, where MPs are crowded and, for many of them, interact one with another. Molecular crowding is partially taken into account into the cartoon of Fig. 1.14B, but even there it is underestimated. In fact, one can estimate, as more closely represented in the two cartoons of Fig. 1.15, that in most membranes, only about four layers of lipids typically separate one MP from its neighbors (Table 1.2), of which only two are "free," in the sense that they are not in direct contact with a protein. As we will see, lipids are found integrated in various ways into TM protein regions, and they may form the "glue" that keep together MP supercomplexes (below, and Chap. 12).

## 1.5.2   Bound Lipids

The "open sea" concept carried by the original fluid mosaic model has influenced, more or less consciously, the way biochemists and biophysicists have thought about lipids. Those have often been considered as a mere two-dimensional solvent, whose bulk physical properties, such as thickness, fluidity, deformability, charge distribution, internal variations of dielectric constant or local pressure, and so on, were felt to be important for the folding, assembly, stability, and function of MPs (for discussions, see e.g. Cevc and Marsh 1987; Lee 2004, 2011a, b; Andersen and Koeppe 2007; Marsh 2008; Phillips et al. 2009; Lundbaek et al. 2010; Anishkin et al. 2014, and references therein). These factors are undoubtedly important in the membrane-bound state of some MPs, as exemplified by their role in controlling, for instance, the opening and closing of mechanosensitive channels (see e.g. Battle et al. 2015; Teng et al. 2015, and references therein). However, their importance should not be overestimated. That the loss of membrane-induced physical constraints plays a role in the instability and/or dysfunction of detergent-solubilized MPs is more often invoked than documented (cf. Chap. 2). Whether this loss matters or not, and to which extent, is certainly protein dependent, given that many MPs are functional in their absence (Chaps. 2 and 5) and that a surprisingly high number of them can fold from a denatured to a functional state in non-bilayer environments (reviewed in Popot 2014; see Chap. 6).

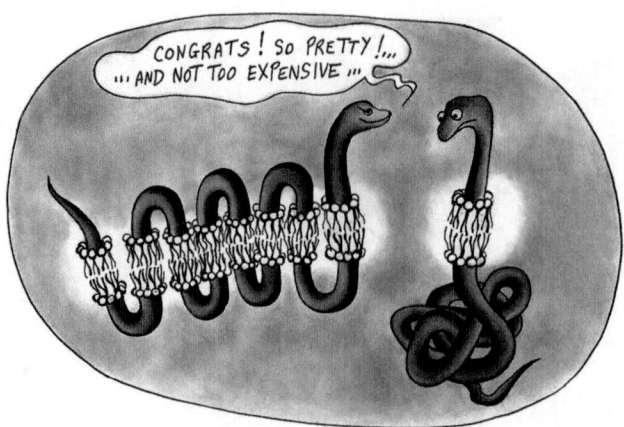

*Membrane protein/lipid interactions*
© 2018 by Francis Haraux

Detailed molecular interactions of individual lipids with specific binding sites at the surface of MPs were initially not given much attention. Biochemists knew of the frequent stabilizing effect of lipids on detergent-solubilized membrane proteins (see Chap. 2), but specific protein/lipid molecular interactions tended to be neglected as compared to general physical forces. Two further factors contributed to this relative lack of interest. First, crystal structures of MPs, when they became available, were most often obtained by crystallizing the proteins from detergent solutions, which are delipidating. Second, it takes a rather high resolution (typically at least ~3 Å, often much better) to identify protein-bound lipids with certainty and, in particular, to avoid confusing their acyl chains with the alkyl chains carried by most detergents (lipid polar heads are often disordered in crystals grown in detergent solutions and can be hard to identify).

With the advent of higher-resolution MP structures, and their multiplication, it has become increasingly evident that many if not most of them, when examined at a sufficient resolution, are found to carry lipids – often specific lipids – bound at specific positions (for general overviews, see e.g. Fyfe et al. 2001; Lee 2003, 2011a, b; Opekarová and Tanner 2003; Palsdottir and Hunte 2004; Hunte 2005; Qin et al. 2006; Hunte and Richers 2008; Marsh 2008; Smith 2012, and references therein). One can distinguish three types of cases (the borders between which are sometimes blurred; cf. Figs. 1.16, 1.17, and 1.20):

- Most frequently, lipids are found to be bound to the surface of the protein which, in situ, faces the core of the membrane. Their polar heads are generally located at the level where bilayer lipid polar heads would be expected to lie, and they are often in interaction with aromatic or cationic residues of the protein. Their hydrophobic chains – at least those that are not too disorganized to be observed – tend to lie in cranks or grooves at the surface of the protein, e.g. at the interface between two TM helices. Amino acid residues whose side chains line these grooves tend to be evolutionarily conserved. Some examples of surface-bound phospholipids are shown in Figs. 1.16, 1.17, and 1.18, bound cholesterol in Fig. 1.19. See also below (§ 1.6.3) the case of the PE molecule wedged between two TM helices in one of the conformations of the sarcoplasmic $Ca^{2+}$-ATPase. In crystal structures, cardiolipin (CL) is frequently found to be bound at well-defined positions at the surface of MPs from bacterial or mitochondrial membranes (see e.g. Fig. 1.22; reviewed in Planas-Iglesias et al. 2015). In the mitochondrial respirasome, CL is thought to mediate most of the interactions between complexes at the membrane level (Althoff et al. 2011) (see Chap. 12, § 12.3.3).

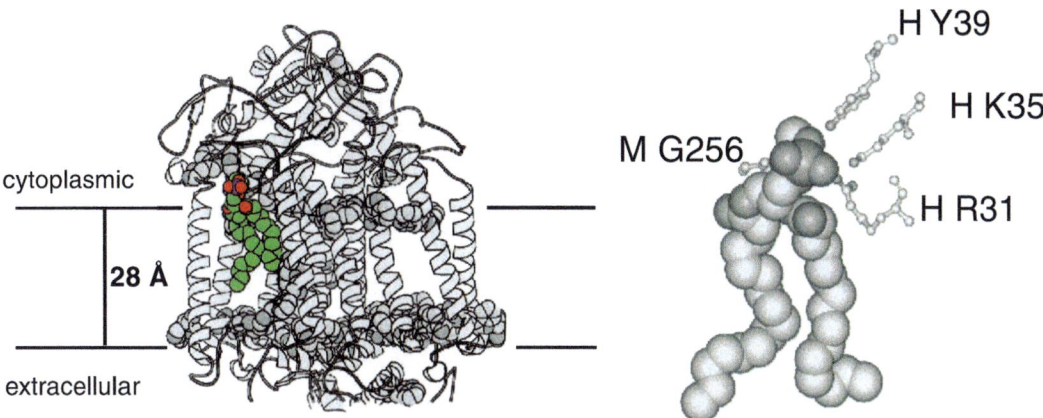

**Fig. 1.16** Crystal structure of the photosynthetic reaction center from *Thermochromatium tepidum* showing a bound phosphatidylethanolamine (PE) molecule. (**A**) The protein structure in ribbon representation, with the tryptophan residues and the bound PE (*green* and *red*) shown in space-filling representation. The location of the Trp residues and of the glycerol backbone of the bound PE molecule defines a hydrophobic thickness of ~28 Å for the bilayer. (**B**) The PE molecule (in space-filling representation) and the residues with which the lipid head group interacts. Note that these residues belong to two different subunits, H and M (Figure from Lee 2003, © 2003 Elsevier B.V. All rights reserved. Original data from Nogi et al. 2000, PDB file 1EYS).

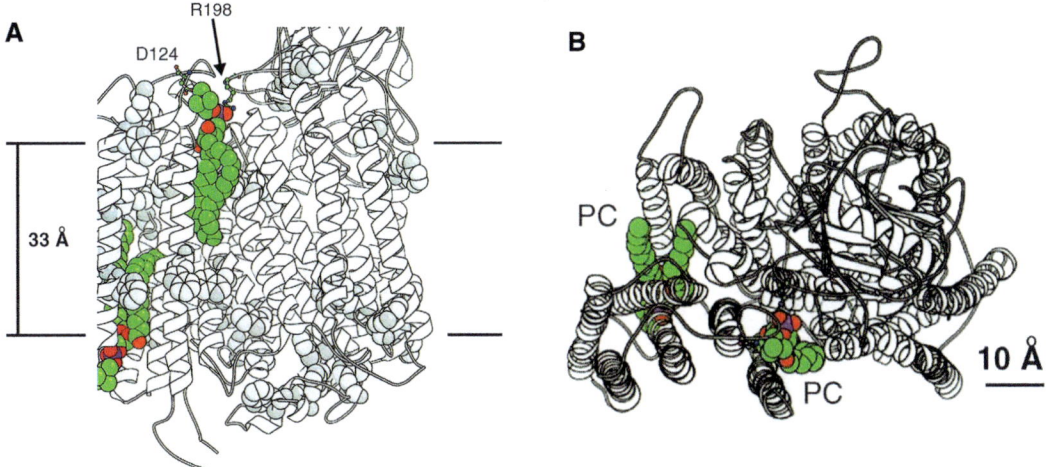

**Fig. 1.17** Binding sites for phosphatidylcholine (PC) on cytochrome *c* oxidase from *Paracoccus denitrificans*. (**A**) Part of the transmembrane region of the protein, showing the two bound PC molecules. On the periplasmic side (on *top*), salt bridges are formed between the phosphate and quaternary ammonium in the lipid head group and, respectively, Arg-198 in subunit II and Asp-124 in subunit III of the oxidase. (**B**) A view of the complex from the periplasm, showing how the PC molecules bind in deep grooves in the transmembrane surface of the protein (Figure from Lee 2003, © 2003 Elsevier B.V. All rights reserved. Original data from Iwata et al. 1995; Harrenga and Michel 1999, coordinates from PDB files 1QLE).

- Another frequently observed situation is that of lipids sandwiched between neighboring subunits. This is illustrated in Fig. 1.20 in the case of the tetrameric potassium channel KcsA. In this particular case, the four lipids found at subunit/subunit interfaces are phosphatidylglycerol molecules (PG), a negatively charged phospholipid (see Fig. 1.1) whose presence

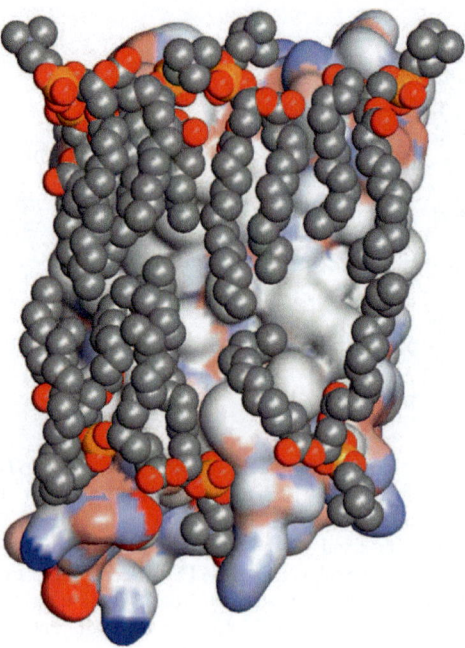

**Fig. 1.18**  Bound lipid molecules at the surface of aquaporin AQP0. In two-dimensional protein/lipid crystals, the surface of AQP0 is covered by a nearly continuous bilayer of well-defined lipid molecules, whose structure is distorted to match the rough surface of the protein. The lipids are shown in space-filling representation; the surface of the protein is colored according to atom charge, ranging from *red* for negative to *blue* for positive (Figure from Lee 2011a, b, © 2011 Elsevier Ltd. All rights reserved. Original data from Gonen et al. 2005, coordinates from PDB file 2B60. See also Hite et al. 2008).

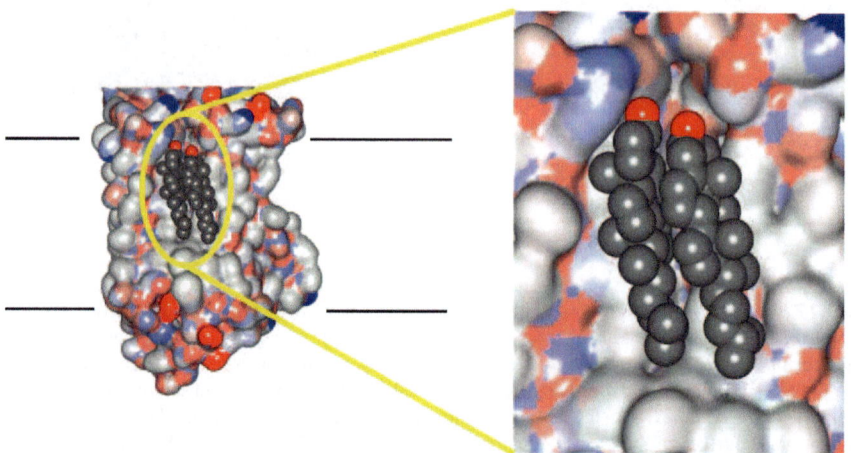

**Fig. 1.19**  Amphipathic molecules other than phospholipids can bind in clefts between transmembrane $\alpha$-helices. *Left*, two cholesterol molecules bound to the transmembrane surface of the $\beta_2$-adrenergic receptor are shown in space-filling representation, circled in *yellow*. *Right*, an expanded view of the binding site, in a cleft on the surface of the protein (Figure from Lee 2011a, b, © 2011 Elsevier Ltd. All rights reserved. Original data from Hanson et al. 2008, coordinates from PDB file 3D4S).

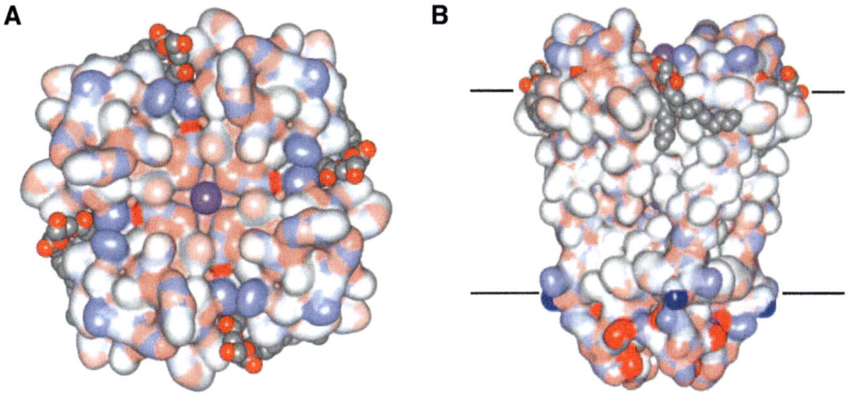

**Fig. 1.20** Lipid molecules buried at subunit-subunit interfaces. Phosphatidylglycerol (PG) molecules bound at protein/protein interfaces in the homotetrameric KcsA structure, shown in space-filling representation, in a view from the extracellular side of the membrane (**A**) and in a side view (**B**), with the approximate limits of the hydrophobic core of the bilayer indicated by the horizontal lines. A potassium ion (*purple*) moving through the central pore is shown in the center of the view in A. The presence of negatively charged lipids is known to be required for ion conduction through the KcsA potassium channel, suggesting that the PG molecules bound to KcsA are important for the function of the ion channel (Figure from Lee 2011a, b, © 2011 Elsevier Ltd. All rights reserved. Original data from Valiyaveetil et al. 2002, coordinates from PDB file 1K4C).

is known to be required for the channel to be functional. In both the cytochrome $bc_1$ (Lange et al. 2001) and cytochrome $b_6f$ (Kurisu et al. 2003; Stroebel et al. 2003) complexes, lipids stabilize the assembly of the single TM helix of the Rieske iron-sulfur protein with the rest of the complex. In some cases, e.g. subunit IV in the cytochrome $c$ oxidase from *Rhodobacter sphaeroides* (Svensson-Ek et al. 2002), interactions between a subunit and the rest of the complex it is part of appear to be entirely mediated by lipids (see Palsdottir and Hunte 2004, and references therein). As already mentioned, a continuous layer of lipids also seems to separate the TM regions of Complex I, cytochrome $bc_1$, and cytochrome $c$ oxidase in the mitochondrial respirasome (Althoff et al. 2011) (cf. Chap. 12, Fig. 12.17). Lipids similarly mediate the assembly of BR trimers into 2D crystals in the purple membrane patches from *Halobacterium salinarum* (§ 1.5.1).

• A less frequent but highly interesting case is that of lipids located inside a MP. The arrangement of lipids in the 2.5-Å resolution crystal structure of Photosystem I reaction center from *Synechococcus elongatus* offers three types of situation (Fig. 1.21). The reaction center as a whole is a supertrimer (Fig. 1.21A), each supermonomer comprising 12 protein subunits and no less than 127 cofactors: 96 chlorophylls, 2 phylloquinones, 3 $Fe_4S_4$ clusters, 22 carotenoids, and 4 lipids. The four lipids, shown in *turquoise* in Fig. 1.21B, are three PG molecules (numbered I, III, and IV) and a monogalactosyl diglyceride one (MGDG; numbered II). Lipid IV (a PG) occupies a rather classical location, being bound to the external, membrane-exposed surface of the trimer. The other three lipids are buried, lipids I (a PG) and II (the MGDG) inside each supermonomer and lipid III (a PG) at the interface between supermonomers (Fig. 1.21B). Lipid III is particularly remarkable: as shown in Fig. 1.21C, it does not simply fill a void between subunits; it acts as a bona fide cofactor, its phosphate group providing the fifth ligand to the magnesium atom of one of the chlorophylls. Six more lipids have been identified in the recent 2.8-Å structure of a PSI-LHCI supercomplex (Qin et al. 2015). Other examples of protein-buried lipids are discussed in Palsdottir and Hunte (2004).

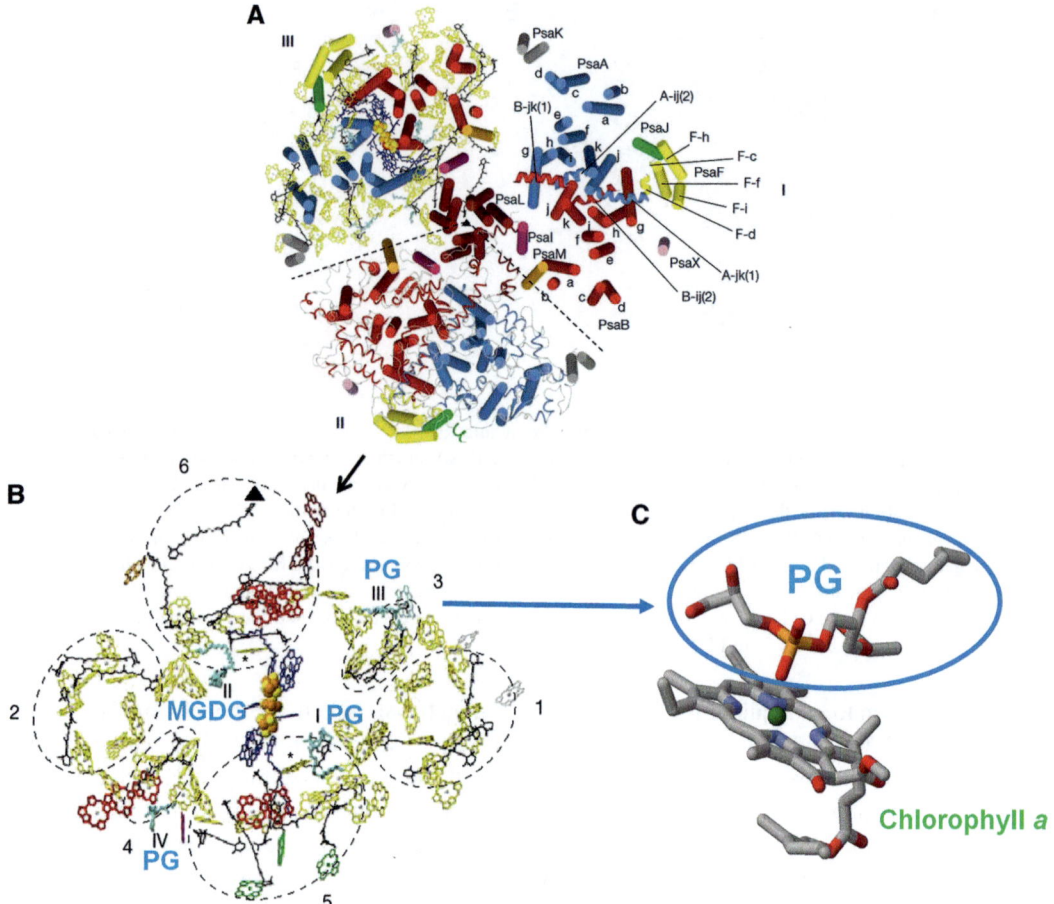

**Fig. 1.21** Lipids in the Photosystem I reaction center supertrimer from the thermophilic cyanobacterium *Synechococcus elongatus*. (**A**) Overall view of the supertrimer in projection on the plane of the membrane. *Dashed lines* mark the approximate limits of a supermonomer, a *black triangle* the threefold symmetry axis. (**B**) Overall view of a supermonomer, showing only the cofactors and lipids (the side chains of the antenna chlorophyll *a* molecules have been omitted). The four lipids are in *turquoise*: *PG* phosphatidyl-glycerol, *MGDG* monogalactosyl diglyceride. (**C**) Details of the liganding of the magnesium atom of a chlorophyll *a* molecule by the phosphate group of PG III (Adapted from Jordan et al. 2001. © 2001 Macmillan Publishers Limited, Nature. All rights reserved).

Comparison of the crystallographic structures of various cytochrome *c* oxidases indicates that lipid-binding sites can be conserved in the course of evolution and that, in detergent-solubilized oxidases, detergent alkyl chains may substitute to lipid acyl chains (Qin et al. 2006, 2007) (Fig. 1.22) (see also Tsukihara et al. 1996; Shinzawa-Itoh et al. 2007, for cytochrome *c* oxidase, and Palsdottir and Hunte 2004; Wenz et al. 2009; Vinothkumar 2011, for similar observations with the cytochrome $bc_1$ complex and bacterial GlpG). Examples of the same region on the protein surface being occupied by a hydrophobic chain in parallel MD simulations carried out either in a lipid or in a detergent environment have also been reported (see e.g. Rouse and Sansom 2015).

What these observations point to is that lipids should not be regarded as a featureless two-dimensional solvent, whose bulk physical properties only should be taken into consideration when considering their effects on integral MPs. Many of them bind at defined positions at the TM

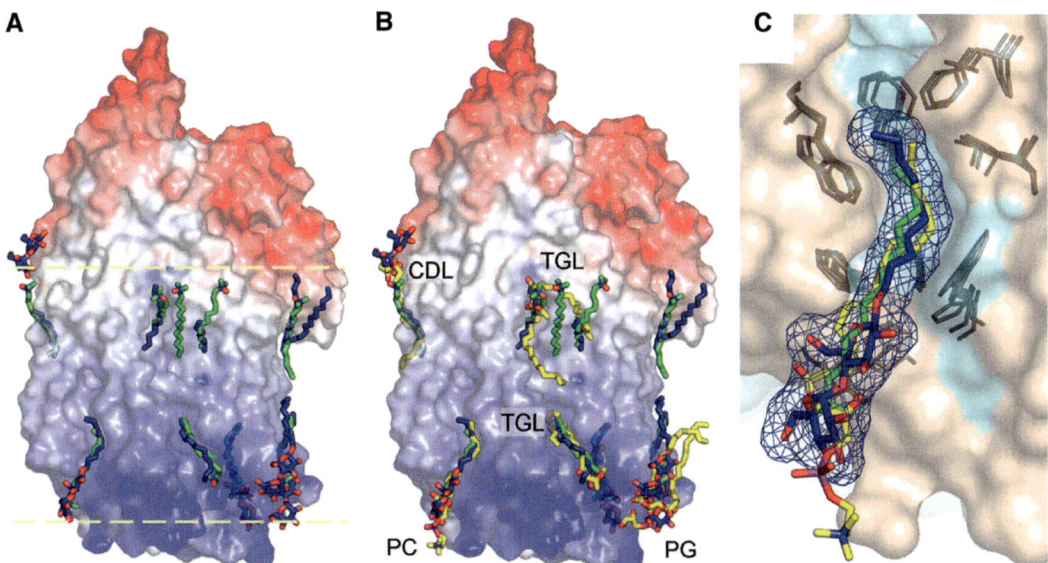

**Fig. 1.22** Surface representations of subunits I and II of various cytochrome *c* oxidases illustrating the position and superposition of detergent and lipid molecules in *Rhodobacter sphaeroides*, *Paracoccus denitrificans*, and bovine oxidase structures. (**A**) Molecular surface representations of the oxidase from *R. sphaeroides*, colored by relative electrostatic potential (*blue*, positive; *red*, negative), showing the detergent molecules, maltose head groups, and alkyl chains resolved in the structure (C atoms, *dark blue*; O atoms, *red*), superimposed on the LDAO detergent molecules resolved in the *P. denitrificans* structure (C, *green*; O, *red*; N, *blue*). The lipid bilayer region is indicated by two yellow dashed lines. (**B**) Lipid molecules in the bovine oxidase structure that occupy the same sites as alkyl chains resolved in *R. sphaeroides* and *P. denitrificans* structures (C, *yellow*). *PC* phosphatidylcholine, *PG* phosphatidyl-glycerol, *CDL* cardiolipin, *TGL* triacylglycerol. (**C**) Detailed view of one of the conserved lipid-binding sites, where an alkyl chain of PC resolved in the bovine oxidase occupies the same site as detergent alkyl chains resolved in the *R. sphaeroides* and *P. denitrificans* ones. The surface representation of subunits I and II of the *R. sphaeroides* oxidase is colored by subunits (subunit I, *cyan*; subunit II, *pinkish*). The decylmaltoside of the *R. sphaeroides* oxidase, the LDAO from the *P. denitrificans* one, and the partial lipid molecule identified in the bovine oxidase are colored by atom type as above. The mesh representation of the resolved decylmaltoside is shown in *blue*. Some well-conserved residues in all three structures surrounding the alkyl chain-binding site are colored *brownish* (From Qin et al. 2006. © 2006 National Academy of Sciences).

surface, if not in the interior of MPs. The molecular details of protein/lipid interactions define their binding sites, their specificity, and their modes of interaction. Many lipids, therefore, should be considered as cofactors rather than a mere solvent. As for any ligand, lipids will stabilize preferentially protein conformation(s) that provides them with the most affine binding sites. Thereby, they have the potential both to stabilize MPs and to influence their conformational equilibria and, in so doing, to modulate their function – a fact that has long been recognized by membrane biochemists (for a recent example of allosteric regulation of a GPCR by cholesterol, see Casiraghi et al. 2016). As will be discussed in Chap. 2, detergents compete with lipids for the hydrophobic TM surface of MPs. The resulting delipidation is a major cause of the destabilization that is generally observed following extraction of MPs from their native membrane environment using detergents.

## 1.6    Dynamics of Transmembrane Regions and the Function of Membrane Proteins

X-ray data provide us with static structures of MPs, but the presence of disordered regions often hints at local mobility, particularly in the water-exposed loops or domains. As a matter of fact, one of the tasks crystallographers frequently have to deal with, as is particularly vividly illustrated by studies of GPCRs, is to restrict this mobility in order for the protein to organize into well-ordered crystals. High-resolution cryo-EM indeed often reveals the existence of multiple accessible conformations of a given MP (see Chap. 12). Mobility is often essential to protein function, the movement of domains or loops relative one to another being a classical feature of catalytic or regulatory mechanisms. In the case of MPs, conformational changes in the TM regions are generally key elements of transport activity, or of the transmission of regulatory signals from one side of the membrane to the other. Large $\beta$-barrel MPs can signal the extracellular binding of a ligand by the change of conformation of a protein domain contained in the lumen of the barrel, which change is detected in the periplasm, initiating the import of the ligand (see e.g. Locher et al. 1998). In $\alpha$-helical MPs, relative displacements of helices or groups of helices are a frequent feature of functional cycles (see e.g. Martfeld et al. 2015).

Depending on the protein, transitions between conformational states in the course of the functional cycle may involve more or less important rearrangements, and they may or not affect the membrane-exposed surface of the protein. If they do, conformational changes may be affected or regulated by the membrane environment the protein is in contact with, e.g. by the lipid composition, or by interactions with small lipophilic molecules, TM peptides, or other MPs. Some MPs can function as more or less solid-state devices, e.g. the "antenna" complexes that collect photons and transfer the resulting excitons to reaction centers in photosynthesis. The functioning of other MPs requires, on the contrary, extensive conformational changes within the membrane, such as those leading to opening and closing of mechanosensitive channels. Another case of extensive rearrangement is that of receptors whose oligomerization state changes upon stimulation, e.g. growth factor receptors, which the binding of a ligand causes to dimerize. In this process, two monomers whose single TM helix was probably totally surrounded by lipids will see part of these interactions replaced with protein/protein ones. In between these two extremes lies a whole gamut of conformational changes that may be more or less restricted to the core of the protein or affect more or less extensively its lipid-exposed surface.

In the following sections, we will introduce three of the MPs that have served as models in the course of developing APols and their applications and that will recurrently appear in the upcoming chapters to illustrate and discuss, in particular, the effects of APols and other surfactants on MP stability, function, immobilization, or folding. Whereas the basic function of each of these proteins is simple, the details of their structure and their functional cycle can be quite intricate. In order to keep this section reasonably short, we will focus on a few selected issues, particularly inasmuch as they are important to understanding the observations described in the next chapters, and leave out many details, for which the reader is referred to specialized reviews.

### 1.6.1    Bacteriorhodopsin

BR is a small MP (~27 kDa), produced by the halophilic archaebacterium *H. salinarum*, which functions as a light-driven proton pump (for reviews, see e.g. Lanyi and Luecke 2001; Lanyi 2004; Andersson et al. 2009; Kandori 2015; Wickstrand et al. 2015; Brown and Ernst 2017). BR comprises seven TM $\alpha$-helices linked by short extramembrane loops (Fig. 1.24B). This helix bundle surrounds a cofactor, retinal (Fig. 1.23A), which is covalently but reversibly bound by a Schiff base (Fig. 1.23B) to

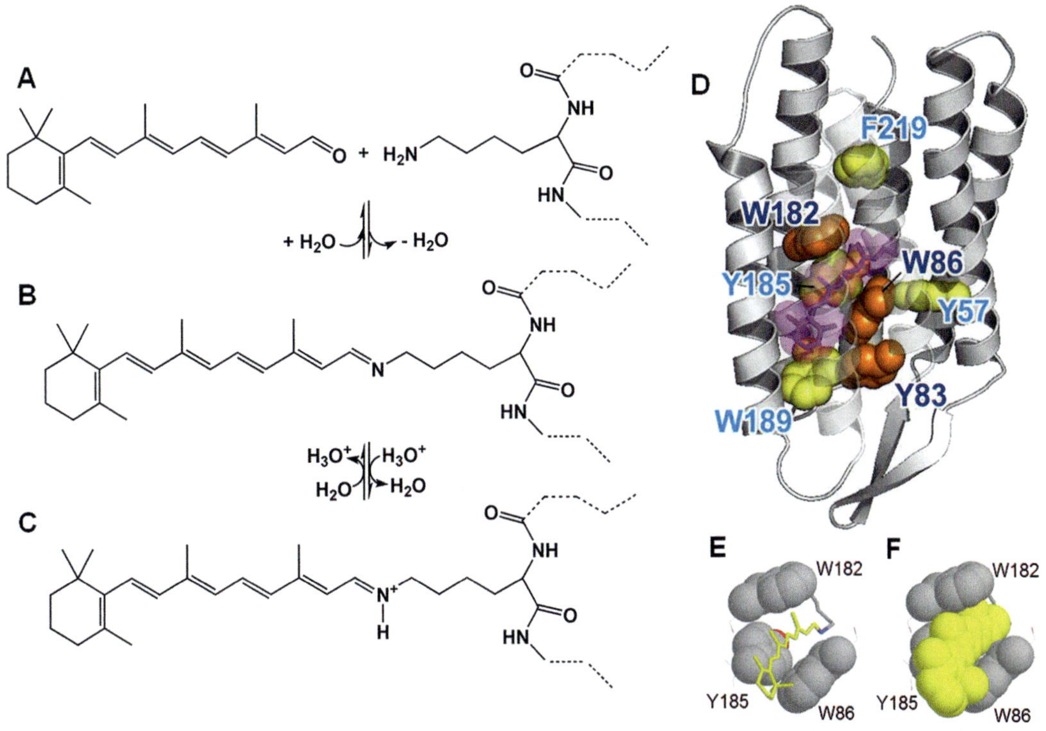

**Fig. 1.23** (**A**) Chemical structure of retinal and lysine 216. (**B**) Formation of an unprotonated Schiff base. (**C**) Protonation of the Schiff base. (**D**) Ribbon representation of the 3D structure of bacteriorhodopsin, viewed from within the membrane, with the front-most helices rendered transparent and the critical residues forming the retinal binding pocket in space-filling representation. The cytosol is on *top*. (**E**, **F**) Enlarged views of all-*trans* retinal (*yellow*), in either stick (**E**) or space-filling (**F**) representation, surrounded by aromatic residues W86, W182, and Y185 (PDB 1QM8) (Panels D–F are from Kandori 2015. © 2015 H. Kandori).

a lysine residue located in the seventh helix (helix G), about midway between the cytosolic and extracellular aqueous phases (Figs. 1.23D and 1.24B). In the resting state of "light-adapted" BR, retinal is in the all-*trans* conformation. Because the terminal ring at the free extremity of retinal is solidly held in a binding pocket comprising aromatic residues carried by helices C and F (Fig. 1.23E, F), any isomerization it undergoes will exercise on the protein a pressure tending to induce structural changes.

Free all-*trans* retinal has a strong peak of absorbance at ~380 nm. When bound to membrane-inserted BR by a Schiff base, which, in the resting state, is protonated (Fig. 1.23C), it absorbs maximally at ~570 nm, in the visible part of the spectrum, giving the protein its characteristic purple color. Light-induced isomerization of retinal from the all-*trans* to the 13-*cis* form initiates a conformational cycle during which the Schiff base transfers its proton to an aspartate residue (Asp85), and from there to the external medium, whereas another aspartate residue (Asp96) reprotonates it and picks up a proton from the cytosol (Fig. 1.24). The whole process requires 10–15 ms, and most of the many structural intermediates in the cycle can be distinguished spectroscopically (Fig. 1.24A). Of practical interest is the fact that, in the dark, a conformational equilibrium is reached in which part of the retinal is in the all-*trans* and part in the 13-*cis* conformation. This "dark-adapted" protein has a visible absorbance maximum at ~560 nm. Upon illumination, photocycling progressively converts all of the resting-state retinal to the all-*trans* form. The resulting ~10-nm red shift, which can be observed

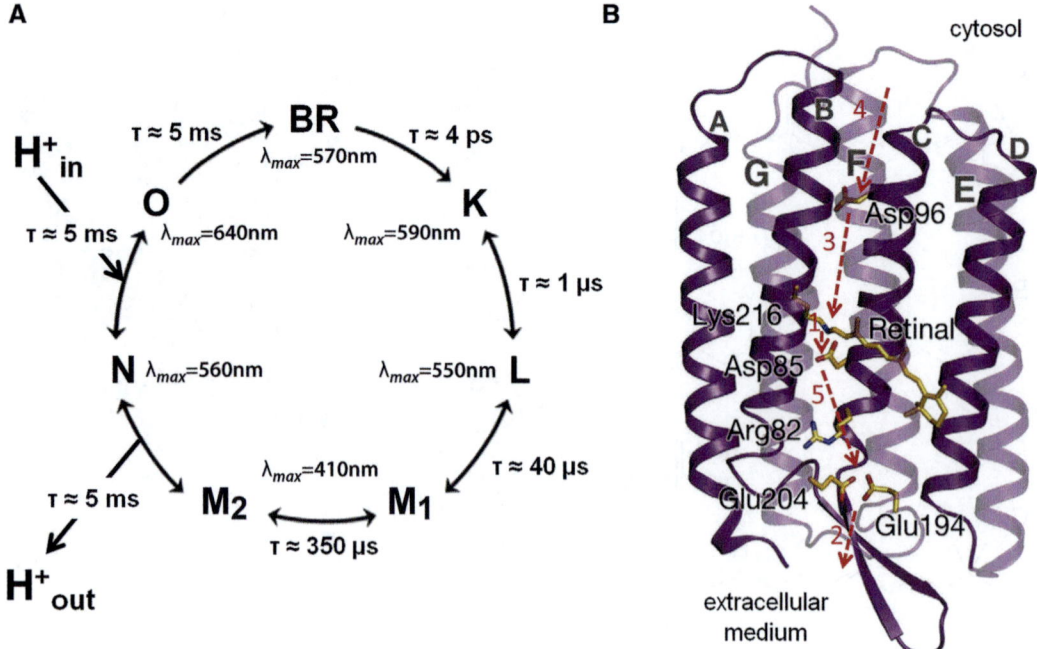

**Fig. 1.24** Overview of the photocycle of bacteriorhodopsin and the proton exchange steps. (**A**) Photocycle of BR with the spectral intermediates (K to O), their absorbance maximum (the two M states, in which the Schiff base is deprotonated, are indistinguishable by absorbance spectroscopy), and their approximate time scales. The absorbance of a photon by resting-state BR causes the all-*trans* retinal to isomerize to a 13-*cis* configuration, driving a sequence of structural changes within the protein that results in unidirectional proton transport. (**B**) 3D structure of BR, with the seven TM helices labeled from A to G, the location of key residues along the proton-translocation channel, and the accepted sequence of proton exchange events. *Step 1*: proton movement from the retinal Schiff base to Asp85. *Step 2*: proton release to the extracellular space from the proton release group consisting of Glu194, Glu204, and water molecules. *Step 3*: reprotonation of the Schiff base from Asp96. *Step 4*: proton uptake from the cytosol. *Step 5*: deprotonation of Asp85 via proton transfer to the proton release group after the 13-*cis* retinal has gone back to its all-*trans* resting-state configuration (Adapted from Wickstrand et al. 2015).

without having to measure proton pumping nor time-resolved absorbance changes, is a reliable indication that the protein is functional.

BR is functional as a monomer. However, when overexpressed in the plasma membrane of *H. salinarum*, it associates into two-dimensional crystals formed by the hexagonal arrangement of BR trimers and lipids, the lipids filling up the center of each trimer and forming a continuous layer between it and its six neighbors. These 2D crystals, called the "purple membrane," contain BR as the sole protein and can be easily purified in large amounts (hundreds of mg of pure BR can be obtained from a 10-L culture). As long as it remains integrated to the purple membrane, BR is remarkably stable. Denaturation, when it occurs, results in a color change from purple to yellow, due to the spontaneous hydrolysis of the Schiff base and the release of free retinal in the membrane or detergent environment (retinal is too hydrophobic to partition significantly in water). This ensemble of properties, combined with the small size of BR and the relative simplicity of its structure, has made it a favorite material for the development of biochemical and biophysical approaches to studying the structure and function of MPs in general. As will be described in the upcoming chapters, BR has proven a very precious test MP for the development of APols and other nonconventional surfactants.

Despite the lipid-mediated contacts between BR trimers, purple membrane crystals are extremely well ordered and diffract electrons to high resolution, which has permitted to establish the 3D structure of BR first at medium resolution (7 Å in the membrane plane), revealing the presence of seven TM helices (Henderson and Unwin 1975), and later to high resolution (3.5 Å in the membrane plane), permitting an atomic model to be built into the electron density map (Henderson et al. 1990; Grigorieff et al. 1996). BR however proved extremely reluctant to form well-ordered 3D crystals from detergent solutions, and it is only in lipid cubic phases (see Chap. 11, § 11.2.2.2) that such crystals could finally be obtained, from which the X-ray structure was solved to, initially, 2.5-Å resolution (Landau and Rosenbusch 1996; Pebay-Peyroula et al. 1997; Rummel et al. 1998). As of 2015, 84 X-ray diffraction, 6 electron diffraction, and 3 NMR structures of BR had been deposited in the Protein Data Bank, 21 X-ray structures reporting light-induced structural changes and changes induced by mutations, changes in pH, thermal annealing, or X-ray-induced photoreduction (reviewed in Wickstrand et al. 2015). The best of these structures reach <2-Å resolution, which permits, in particular, to visualize water molecules and protein-bound lipids.

An abundant, often contentious literature deals with the structural changes undergone by BR at each step of the photocycle, as well as with the structural effects of mutations created to either mimic one or the other of these states or to slow down the cycle so as to favor their observation, the ensemble of which is easily bewildering to the nonspecialized reader. A remarkable piece of meta-analysis has been carried out by Richard Neutze and colleagues to compare and classify all of the structures reported as of 2014, sort out reproducible observations from possible artifacts (such as those due to radiative damage, mixed states, crystal twining, crystallization conditions, etc.), and try to come up with a unified model (Fig. 1.25; Wickstrand et al. 2015; for earlier discussions, see e.g. Kühlbrandt 2000; Neutze et al. 2002; Hirai et al. 2009). According to this model, at the core of the photocycle is, as a consequence of the isomerization of retinal, a rearrangement of the Schiff base and its environment (including a key water molecule), the result of which is to (i) shift close to one another the pK of the Schiff base and that of Asp85, which are widely different in the resting state of BR, and (ii) move these two groups closer in space. These two changes allow the Schiff base proton to be transferred to Asp85 (*Step 1* in Fig. 1.24B), generating State $M_1$; after which, once the electrostatic attraction that existed between the positively charged Schiff base and the negatively charged carboxylate of Asp85 vanishes, the two groups separate (Fig. 1.25A, B). In the second part of the cycle (transition from $M_1$ to $M_2$), helix F moves outward (Fig. 1.25A), leading water molecules to transiently order in the cytoplasmic half of the channel, thereby facilitating the reprotonation of the Schiff base from Asp96 (*Step 3* in Fig. 1.24B), while other structural changes help control proton release from Asp85 and return to the ground state. Throughout the cycle, helix movements are limited to relatively modest displacements of helix F and part of helix C.

Also of relevance to some of the observations to be discussed later are two further pieces of data. First, Giuseppe Zaccai and his colleagues have shown, using neutron scattering measurements, that, in order to function, BR needs a "soft" environment. Stiffening either by lowering the temperature or as a result of dehydration will stop the photocycle (Ferrand et al. 1993; Lehnert et al. 1998; Zaccai 2000, 2004). Second, the photocycle of BR is sensitive to the chemical nature of its environment and, in particular, to the nature of the lipids surrounding it, among which squalene and phosphatidyl glycerophosphate (Dracheva et al. 1996; Joshi et al. 1998; Hendler and Dracheva 2001; Lee et al. 2015).

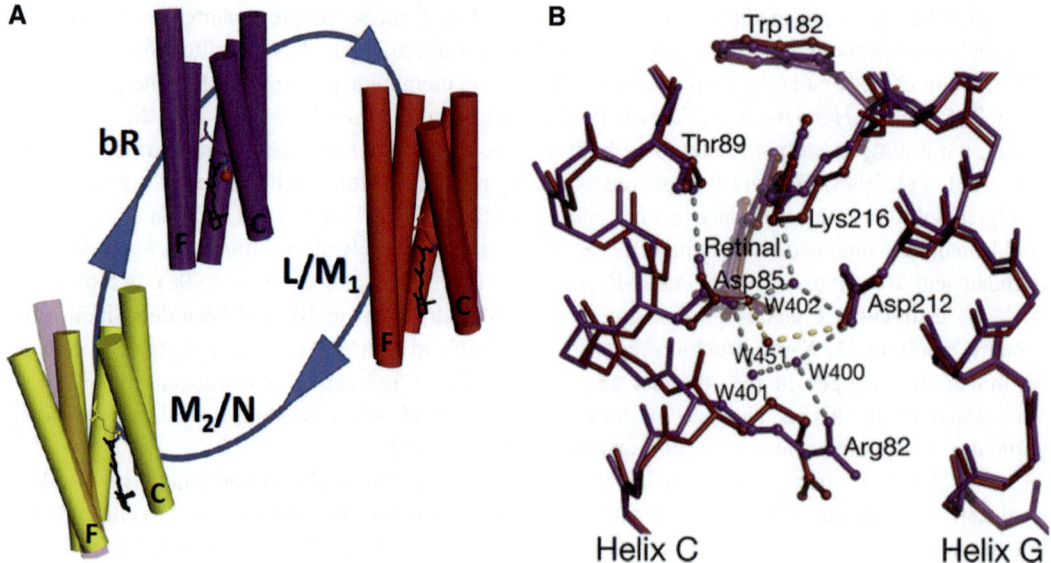

**Fig. 1.25** Overview of structural changes in the BR photocycle. According to Neutze and colleagues (Wickstrand et al. 2015), two major conformations are sufficient to understand the mechanism of proton pumping by BR: (i) the resting BR state in which water 402 forms hydrogen bonds to the Schiff base, Asp85, and Asp212 and (ii) the relaxed illuminated state, in which these bonds are disrupted and helix C moves toward helix G (**B**). These conformational changes set the stage for proton transport from the Schiff base to Asp85. A larger-scale displacement of helix F is however needed to facilitate reprotonation of the retinal from the cytoplasm (**A**). To illustrate helix movements, the resting-state structures of helices C and F are overlaid on the intermediate state conformations (states L/M$_1$, *red*; states M$_2$/N, *yellow*) as semitransparent *purple* helices. Helices A and B have been removed from the figure for clarity (Adapted from Wickstrand et al. 2015).

## 1.6.2    The Nicotinic Acetylcholine Receptor

The nicotinic acetylcholine receptor (nAChR) is present in the postsynaptic membrane at neuromuscular junctions and other synapses, including in the central nervous system (for recent reviews, see e.g. Changeux and Edelstein 2005; Changeux 2012; Corringer et al. 2012; Sine 2012; Cecchini and Changeux 2015). Upon binding the neurotransmitter acetylcholine (ACh) released in the synaptic cleft by the nerve terminal, it opens a TM cation-specific channel. As briefly mentioned above (§ 1.3), the net entry of positive charges that results, mainly due to Na$^+$ ions, depolarizes the membrane to the point that voltage-sensitive sodium channels are activated, setting off the propagation of an action potential. The concomitant entry of Ca$^{2+}$ ions provides an additional physiological signal (Pankratov and Lalo 2014).

Because of its physiological and historical importance (it is the first pharmacological receptor to ever have been isolated), the functional properties of the nAChR receptor and the mode of action of a large variety of its ligands have been studied in great details (for recent reviews, see e.g. Corringer et al. 2012; Sine 2012; Taly et al. 2014; Auerbach 2015; Changeux et al. 2015; Nemecz et al. 2016, and references therein). For the sake of the present discussion, it will be sufficient to mention the following (Scheme 1.1).

$$R \rightleftarrows A \rightleftarrows I \rightleftarrows D$$

**Scheme 1.1** The four (main) conformational states of the nicotinic acetylcholine receptor. The cation-specific channel is closed in the resting (**R**) and desensitized (**I, D**) states, open in the active (**A**) state. The affinity for acetylcholine increases in the order R < A < I < D (For detailed discussions, see e.g. Corringer et al. 2012; Sine 2012; Taly et al. 2014; Auerbach 2015; Changeux et al. 2015; Nemecz et al. 2016. See text).

In the absence of ligands, the nAChR is, in majority, in one (or a collection of) resting state(s) (**R**), in which the ion channel is closed. **R** is stabilized by antagonists like curare (hence their paralyzing effect). According to a model first developed in the laboratory of Jean-Pierre Changeux, binding of ACh induces a concerted, allosteric conformational change that results in populating the active state (**A**), thus opening the TM cation channel and inducing membrane depolarization. Upon extended application of ACh (hundreds of ms to minutes), two or more further changes of conformation take place, leading to fast or intermediate (**I**) and to slow (**D**) desensitized states, in which the channel is shut. The affinity of the binding sites for ACh increases in the order **R < A < I < D**, driving the successive transitions. A number of snake toxins, including $\alpha$-bungarotoxin ($\alpha$-Bgt) from the many-banded krait, *Bungarus multicinctus*, block access to the ACh-binding sites, paralyzing the snake's victims. The binding of $\alpha$-Bgt is non-covalent but nearly irreversible. The toxin can be labeled radioactively or fluorescently, making it a remarkable tool for innumerable investigations, examples of which will be shown in Chaps. 5 and 13. Numerous drugs, including local and general anesthetics, affect allosterically the receptor's transitions (see Corringer et al. 2012; Sine 2012; Taly et al. 2014; Auerbach 2015; Changeux et al. 2015; Nemecz et al. 2016), which are also dependent on the protein's membrane environment, in particular on the presence of anionic lipids and cholesterol (reviewed in Baenziger et al. 2015; Barrantes 2015; Hénault et al. 2015).

The neuromuscular form of the nAChR is a heteropentamer, comprising four types of homologous subunits in an $\alpha_2\beta\gamma\delta$ stoichiometry, with an overall MW of ~290 kDa (Fig. 1.26B). The *N*- and

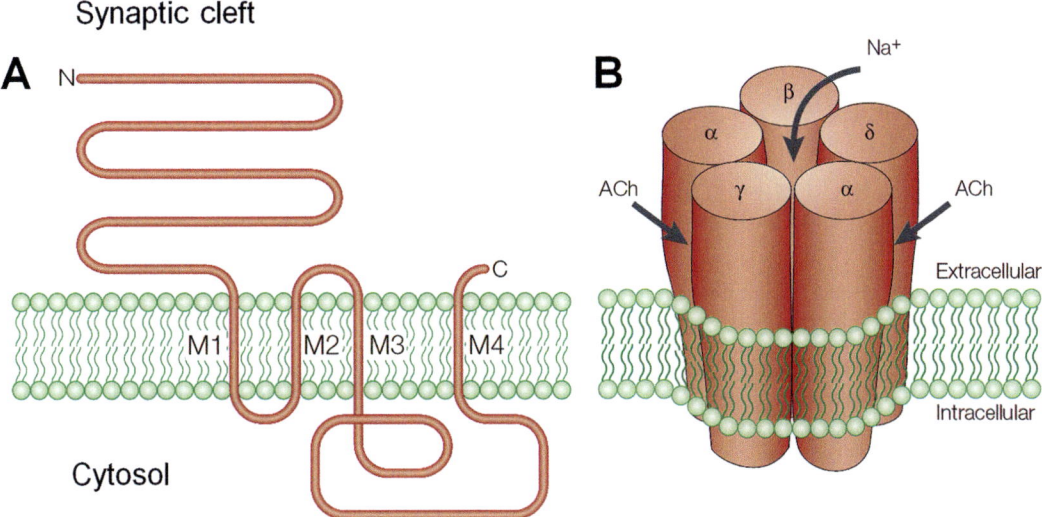

**Fig. 1.26** Schematic organization of pentameric ligand-gated ion channels (pLGICs). (**A**) Transmembrane topology of the subunits. (**B**) A schematic representation of the quaternary structure, common to all receptors of the family, showing the particular arrangement of the subunits in the heteropentameric muscle-type nicotinic acetylcholine (ACh) receptor, the location of the two ACh-binding sites (at the interfaces between an $\alpha$- and a $\gamma$-subunit and an $\alpha$- and a $\delta$-subunit), and the axial cation-conducting channel (From Karlin 2002. © 2002 Macmillan Publishers Limited, Nature. All rights reserved).

*C*-termini of each subunit lie in the extracellular space (Fig. 1.26A). Each subunit comprises a large extracellular region, a TM region comprising four TM helices, labeled M1 to M4, and an intracellular region formed by a large loop between M3 and M4 (Fig. 1.26B). The extracellular domain of each $\alpha$-subunit carries an ACh-binding site, which, in the $\alpha_2\beta\gamma\delta$ receptor, is located at the $\alpha/\gamma$ or $\alpha/\delta$ interface (Fig. 1.27, *bottom*). The short extracellular M2-M3 loop appears to play a crucial role in transmitting the conformational information from the extracellular to the TM region (see below). The cytosolic domain, which is absent in bacterial receptors from the same family, appears to be involved in further types of signaling and regulation and in assembly.

The muscle receptor is one among many different forms of nAChRs found in mammals, all of which result from the pentameric assembly of a variety of subunits, including several types of $\alpha$ and $\beta$ chains, which form various types of homo- or hetero-oligomers. The nAChR is the prototype of a large family of eukaryotic receptors, known as the Cys-loop receptors or, according to the most recent

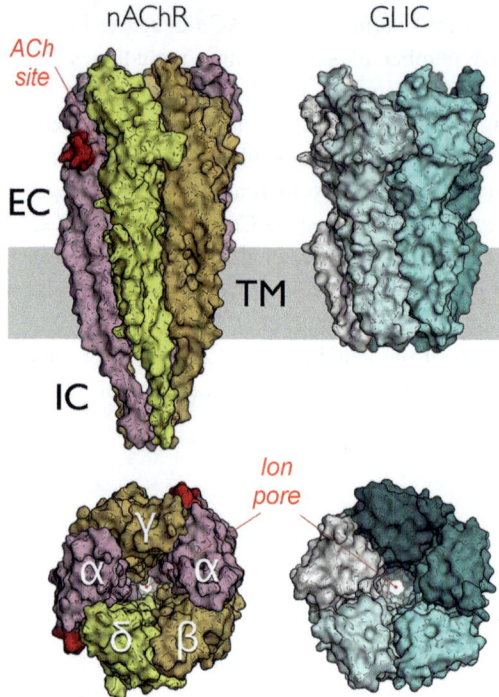

**Fig. 1.27** Overall organization of pentameric ligand-gated ion channels (pLGICs). To the *left*, the architecture of the hetero-oligomeric $\alpha_2\beta\gamma\delta$ muscle nAChR as visualized at ~4-Å resolution by cryo-EM analysis of 2D tubular crystals derived from the electric organ of *Torpedo marmorata* (Unwin 2005). To the *right*, the 2.9-Å resolution X-ray structure of the homopentameric prokaryotic channel GLIC from *Gloeobacter violaceus* established by Pierre-Jean Corringer and colleagues (Bocquet et al. 2009). On *top*, side views of the oligomers, with the position of the transmembrane region (TM) indicated in *gray*. Each subunit is color coded differently. *EC* extracellular region, *IC* intracellular region, *ACh site* one of the two binding sites of the neurotransmitter acetylcholine on the nAChR. These sites are located at the $\alpha/\gamma$ and $\alpha/\delta$ interfaces, with loop C (in *red*) acting as a binding-site lid. At the *bottom*, views of the TM domains from the extracellular side, along the axis of the channel, normal to the membrane plane. The ion pore lies on the symmetry or pseudosymmetry axis of the pentamers. The neurotransmitter binding sites and the ion pore are very distant (~60 Å) one from another, and the change of conformation of EC elicited by ACh binding is coupled to that of the TM region by an allosteric mechanism. Apart from the intracellular domain, which is absent in prokaryotes, the overall organizations of nAChR and the bacterial channel GLIC are very similar (From Cecchini and Changeux 2015. © 2015 Elsevier Ltd. All rights reserved).

nomenclature, the pentameric ligand-gated ion channels (pLGICs; a better term because prokaryotic homologues lack the characteristic cystine bridge of their eukaryotic counterparts). Other members of the family include the vertebrate $GABA_A$, glycine, and excitatory serotonin ($5\text{-}HT_3$) receptors, as well as an invertebrate glutamate-gated chloride channel (GluCl). The ACh and $5\text{-}HT_3$ receptors gate cationic channels, GluCl, and the $GABA_A$ and glycine receptors anionic ones. Orthologues have been discovered in prokaryotes, the bacterial channels GLIC and ELIC, which have played a critical role in crystallographic studies of the pLGIC superfamily (for reviews, see e.g. Corringer et al. 2012; Sine 2012; Nys et al. 2013; Unwin 2013; Taly et al. 2014; Cecchini and Changeux 2015; Changeux et al. 2015; Nemecz et al. 2016).

The overall organization common to all pLGICs was first established by cryo-EM studies, carried out by Nigel Unwin and his colleagues, of the nAChR from the electric ray, *Torpedo marmorata*. Torpedo electrocytes, which are evolutionarily derived from muscle fibers, produce massive amounts of receptors, which are densely packed in the postsynaptic membrane. The receptor present in purified postsynaptic membrane fragments can be prodded into organizing into tubular 2D crystals, in which it adopts a variety of helical arrangements suitable for crystallographic analysis. The study of cryo-EM images of the tubes has shown that the five subunits are arranged pseudosymmetrically around an axis normal to the membrane, along which runs the cation-selective channel (Brisson and Unwin 1985; see Unwin 2013, and references therein). The walls of the channel are formed by TM helices contributed by each subunit (Figs. 1.26, *right*, and 1.27, *left*).

The best electron density maps obtained by cryo-EM studies of Torpedo helical tubes are limited to ~4-Å resolution (Unwin 2005, 2013; Unwin and Fujiyoshi 2012). The nAChR receptor itself has resisted protracted attempts at 3D crystallization, but a host of high-resolution data have been obtained by (i) X-ray studies of a non-TM ACh-binding protein extracted from the central nervous system of the mollusc *Lymnaea stagnalis*, which is homologous to the extracellular region of the nAChR (Brejc et al. 2001; Smit et al. 2001); (ii) X-ray studies of the bacterial pLGICs ELIC, from *Erwinia chrysanthemi* (Hilf and Dutzler 2008; Spurny et al. 2012), and GLIC, from *Gloeobacter violaceus* (Bocquet et al. 2009; Hilf and Dutzler 2009; Nury et al. 2011; Prévost et al. 2012; Sauguet et al. 2013, 2014); (iii) X-ray studies of the GluCl channel from the invertebrate *Caenorhabditis elegans* (Hibbs and Gouaux 2011; Althoff et al. 2014), of the human $GABA_A$ receptor (Miller and Aricescu 2014) and of the mouse $5\text{-}HT_3$ receptor (Hassaine et al. 2014); (iv) single-particle cryo-EM studies of the glycine receptor from zebrafish (Du et al. 2015); and (v) the X-ray structure of a central nervous system ACh receptor, the $\alpha 4\beta 2$ nicotinic receptor (Morales-Perez et al. 2016). Taken together, these data provide a detailed view of how binding of the ligand controls the opening and closing of the channel. Whereas some details may differ from one protein to the other, the general picture that emerges from this ensemble of data gathered on disparate systems is generally very consistent.

In short, the binding of agonists induces a reorganization of the extramembrane region, which has been described as resulting from the composition of a radial movement (the pentamer "blooms" in the closed vs. the open state of the receptor; Fig. 1.28A) and a twisting movement (Fig. 1.28B). This entails a rearrangement of the interface between the extramembrane region and the TM helix bundle, pushing slightly outward the outermost part of the pore-lining M2 helices, which opens the pore.

In Fig. 1.29 is shown a model (Calimet et al. 2013; Taly et al. 2014) of the mechanism of signal transduction between the extramembrane and TM regions of pLGICs, based on MD simulations using the structures of the prokaryotic channels GLIC and ELIC and the eukaryotic channel GluCl. Residues belonging to the extramembrane region act on the M2-M3 loop of the TM helix bundle to control the position of the outermost part of helix M2, thereby opening and closing the pore. According to experimental data, the displacement of M2 is small and has a limited impact on the positions of the other helices, entailing only small movements at the interface with the membrane in GluCl and GlyR, and essentially none in GLIC.

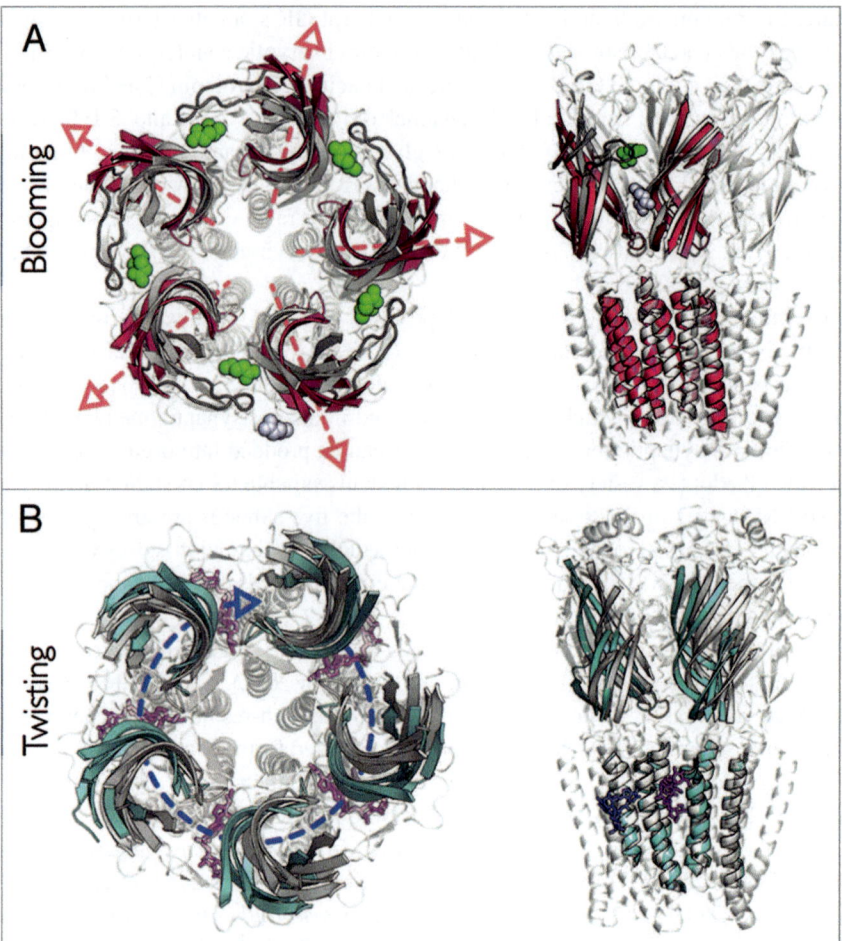

**Fig. 1.28** The blooming and twisting components of the isomerization underlying gating in pLGICs. The model is a composite of data obtained on GLIC and GluCl. (**A**) The blooming transition. The conformation of the A state as captured by the X-ray structure of GLIC at pH 4 (Sauguet et al. 2013) is shown in a cartoon representation in *light gray* with the C-loop closed on top of the orthosteric site in *darker gray*. For illustration, a hypothetical agonist bound to the extracellular domain is shown as green spheres; its coordinates correspond to those of L-glutamate in the active state of GluCl (Hibbs and Gouaux 2011) after optimal superposition of the TM regions of GLIC and GluCl. The position of the extracellular *β*-sandwiches in the resting state of pLGICs is shown in *pink*; coordinates were extracted from the crystal structure of GLIC pH 7 (Sauguet et al. 2014) and are shown after optimal superposition of the TM regions. The *pink dashed arrows* illustrate the direction of the blooming motion from the active to the resting state. The blooming transition results in a significant reshaping of the extracellular subunit/subunit interfaces, which open the orthosteric site and presumably reduce the affinity for the agonist. (**B**) The twisting transition. The conformation of the active state of pLGICs as captured by the X-ray structure of GluCl in complex with the allosteric agonist ivermectin (Hibbs and Gouaux 2011) is shown in *light gray*. Ivermectin bound at the subunit interfaces in the TM domain is shown as *magenta sticks*. The orientation of the extracellular *β*-sandwiches captured at the end of the twisting transition by molecular dynamics simulations of GluCl with ivermectin removed (Calimet et al. 2013) is shown in *cyan*; the coordinates of the channel taken after 100-ns relaxation without ivermectin are shown after optimal superposition of the TM regions. The *blue arrow* illustrates the direction of the twisting transition from the active (*untwisted*) to the resting (*twisted state*) (From Taly et al. 2014. See also Fig. 1.33).

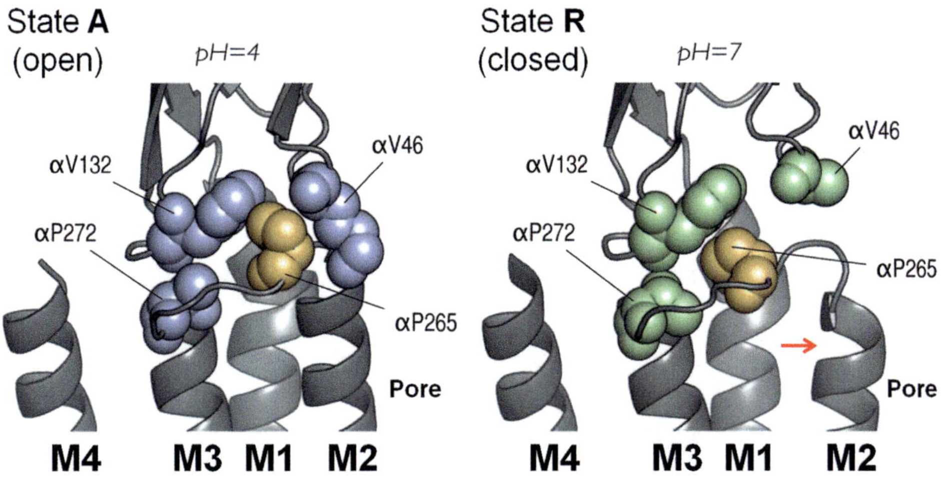

**Fig. 1.29** Coupling between the extracellular and TM regions of pLGICs. The interlocking of residues at the interface between the two regions is compared in the active **A** vs. the resting **R** state, based on the structures of GLIC at pH 4 (open, *left*) (Sauguet et al. 2013) and pH 7 (closed, *right*) (Sauguet et al. 2014). A few critical residues (Torpedo numbering) are shown as van der Waals spheres. V46 and V132 from the extracellular region (*blue* in the **A** state, and *green* in the **R** one) interact with an absolutely conserved proline residue in the M2-M3 loop of the TM region, P265 (*light orange*), to form a pin-in-socket assembly in the active state, which disassembles in the resting state. This governs the rearrangement of the TM helix bundle and, in particular, the movement of the outermost part of M2, opening and closing the TM pore (Adapted from Taly et al. 2014).

Electron density maps of the muscle-type nAChR have been obtained at medium resolution (~6-Å) shortly (~10 ms) after spraying with ACh Torpedo postsynaptic membrane fragments annealed into 2D tubular crystals, followed by quick-freezing and crystallographic analysis of cryo-EM images (Unwin and Fujiyoshi 2012). They suggest that the hetero-oligomeric nature of the pentamer induces a degree of asymmetry in the contribution of the various subunits to opening of the channel and that changes at the level of the protein/membrane interface remain very limited (Fig. 1.30).

But for the EM data on *T. marmorata* nAChR, all of the medium- to high-resolution data discussed thus far bear on homopentameric pLGICs whose natural ligands are not ACh. The X-ray structure of a heteropentameric ACh receptor has been recently obtained, that of the human central nervous system $\alpha4\beta2$ receptor (the numbers referring to the subunit subtype), whose subunit stoichiometry, in the structure solved, is $\alpha_2\beta_3$ (Morales-Perez et al. 2016). This receptor is the most abundant nAChR in the brain and is involved in nicotine addiction. Its structure was solved in complex with either nicotine or an iodinated nicotine analog and was suggested to represent a desensitized state, with the TM channel closed. In addition to providing the first view of a heteropentameric pLGIC (Fig. 1.31), and explaining why $\alpha/\beta$ interfaces bind ACh whereas $\beta/\alpha$ and $\beta/\beta$ ones do not, the structure offers, by comparison with previously established structures of other pLGICs in various conformations, further glimpses into the mechanism whereby transconformations of the extracellular domain induced by ACh binding are transduced into conformational transitions in the TM one (Fig. 1.32). As noted by the authors, the conclusions have to be taken with precaution, because all structures but those of *T. marmorata* nAChR have been obtained in a detergent environment, whereas the membrane environment is known to influence pLGIC function. Furthermore, whereas the overall structures of pLGICs are remarkably conserved throughout evolution, there are nevertheless differences between them, requiring caution when comparing closed, open, and partially or deeply desensitized conformations observed on different receptors (Nemecz et al. 2016).

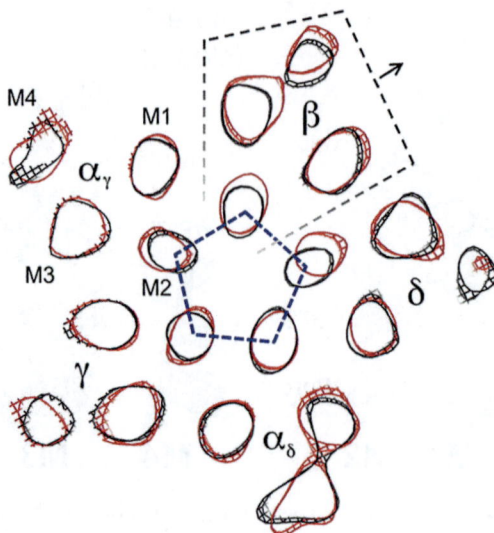

**Fig. 1.30** Comparison of closed- and open-channel densities from ~6-Å maps of Torpedo nAChR in a cross section through the extracellular leaflet of the lipid bilayer (where movements are largest); *black*, resting form; *red*, open form. Dashed lines highlight the pentagonally symmetric arrangement of M2 helices around the pore of the closed channel (*blue*) and the movement outward of all four helices of subunit $\beta$ when the channel opens (*black*). The mesh interval corresponds to 1 Å; all contours are at 1 $\sigma$ (From Unwin and Fujiyoshi 2012).

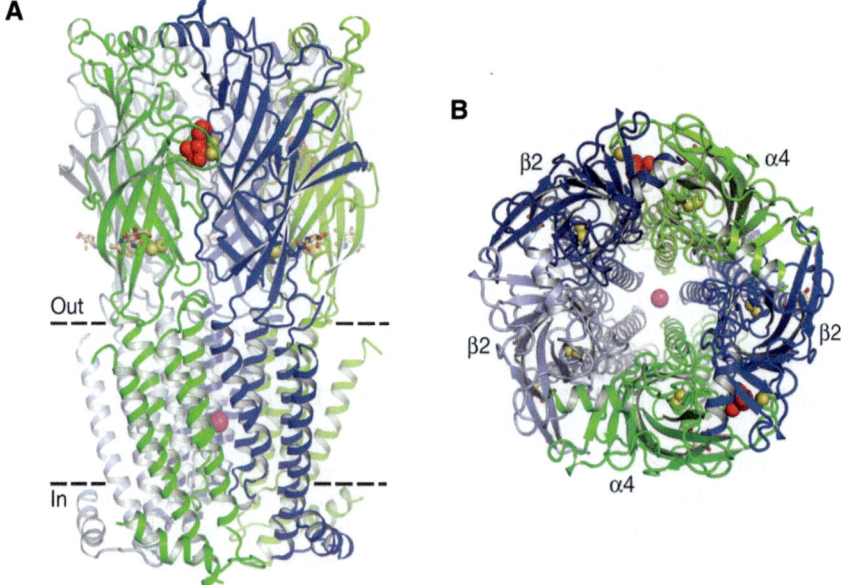

**Fig. 1.31** Architecture of the $\alpha 4 \beta 2$ nicotinic receptor. (**A**) View parallel to the plasma membrane. $\alpha 4$ subunits are in *green*, $\beta 2$ ones in *blue*. Nicotine (*red*) and sodium (*pink*) are represented as spheres. The Cys-loop and loop C disulfide bonds are shown as yellow spheres. *N*-linked glycans (*brown*) are shown as sticks. *Dashed lines* indicate the approximate position of the membrane. (**B**) View perpendicular to the plasma membrane, looking from the extracellular side (From Morales-Perez et al. 2016. © 2016 Macmillan Publishers Limited, Nature. All rights reserved).

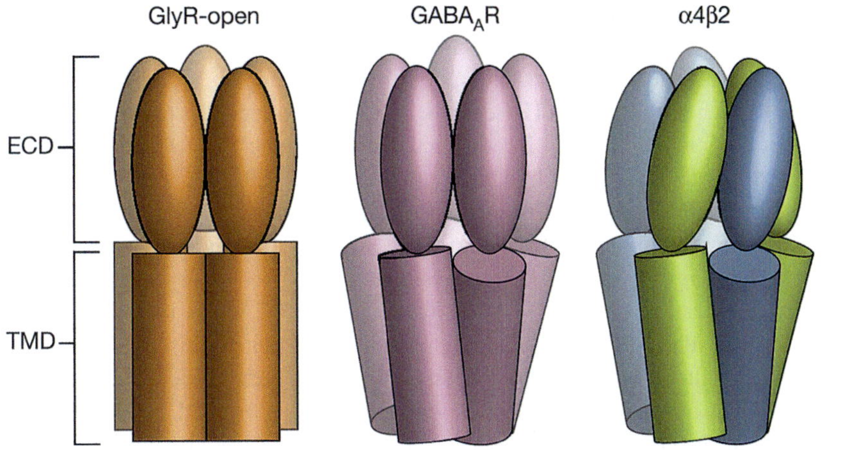

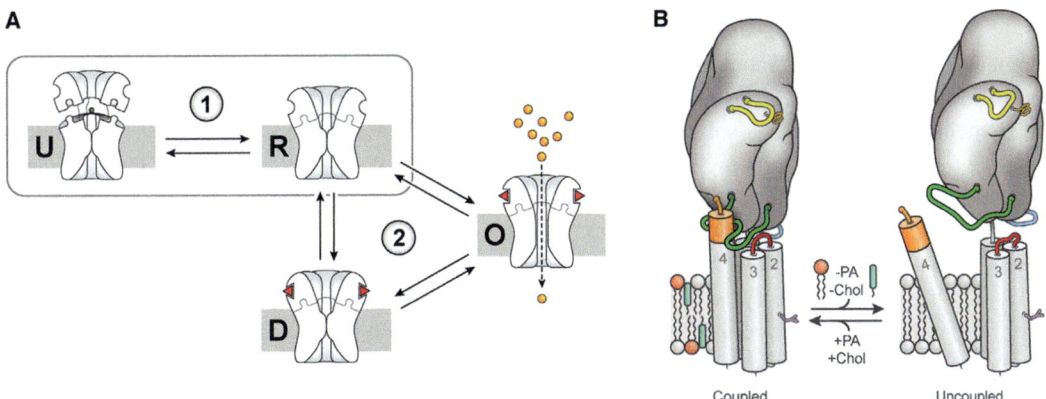

**Fig. 1.32** Conformational changes thought to underlie the desensitization of pLGICs. The cartoon illustrates the relative positions of the extracellular domains (ECD) and transmembrane domains (TMD) in the $\alpha 4\beta 2$ acetylcholine receptor (desensitized) compared to the open conformation of the glycine receptor (GlyR) and the partially desensitized conformation of the GABA$_A$ receptor (From Morales-Perez et al. 2016. © 2013 Macmillan Publishers Limited, part of Springer Nature. All rights reserved).

**Fig. 1.33** Decoupling of the nAChR in the absence of functionally critical lipids, cholesterol and phosphatidic acid (PA). (**A**) The nAChR from native Torpedo membranes undergoes agonist-induced conformational transitions from resting (**R**) to open (**O**) and then desensitized (**D**) conformations. The nAChR in reconstituted membranes also adopts an uncoupled conformation (**U**) that binds agonist but typically does not undergo the transition to the open or desensitized states. (**B**) Schematic diagram showing the current model of lipid-dependent uncoupling, illustrated using a single nAChR subunit. The lipid-exposed M4 TM helix likely plays a key role in sensing the lipid bilayer. In unfavorable membrane environments, the conformation of M4 may change, thus weakening interactions between the $\beta 1$-$\beta 2$ (*light blue*)/$\beta 6$-$\beta 7$ (*green*) loops of the extramembrane region and the M2–M3 linker (*red*) of the TM region (for details, see daCosta and Baenziger 2009) (From daCosta and Baenziger 2013. © 2013 Macmillan Publishers Limited, Nature. All rights reserved).

John E. Baenziger and colleagues have suggested that lipids regulate the function of the nAChR by affecting the interaction of helix M4, which lies at the protein/lipid interface (Fig. 1.30) with the rest of the TM region: in the absence of specific lipids (cholesterol and acidic lipids), M4 would move in such a way that conformational changes in the extracellular region would become uncoupled from those in the TM one (daCosta and Baenziger 2009, 2013; Baenziger et al. 2015; Hénault et al. 2015) (Fig. 1.33). This point will be further addressed when comparing the effects of detergents and APols on the allosteric transitions of the nAChR (Chap. 5, §§ 5.4 and 5.6).

### 1.6.3    The Sarcoplasmic Reticulum Calcium Pump

Primary ATPase transporters, in which transport is directly coupled to the hydrolysis of ATP, are categorized as P-, F-, and V-type ATPases and ABC transporters (Pedersen 2007). Among these, P-type ATPases comprise a family of cation transporters where the formation of an aspartylphosphorylated intermediate drives the active transport of cations. SERCA1a, the SR $Ca^{2+}$-ATPase, is the prototype of P-type ATPases and the first primary transporter whose 3D structure was solved by X-ray diffraction (Toyoshima et al. 2000; Toyoshima and Nomura 2002). Comparison of its stability and functionality in a detergent vs. an APol environment has played a key role in understanding the mechanisms of stabilization of MPs by APols (Chap. 5). SERCA1a is present in large amounts in skeletal muscle, particularly in the longitudinal tubules of fast-twitch muscle SR. It is responsible for terminating muscle contraction by pumping back $Ca^{2+}$ ions from the cytosol into the lumen of the SR, from which they had been released when the muscle fiber was excited. Various other isoforms of SERCA are present in other tissues and other cell compartments, including the plasma membrane and the Golgi apparatus. $Ca^{2+}$-ATPases are involved in maintaining the low cytosolic concentration of $Ca^{2+}$ that is essential to many signaling processes other than muscle contraction, such as neurotransmission, neurosecretion, egg fertilization, etc. (Carafoli 2002). All P-type ATPases are structurally alike, and their transport cycle is driven by the formation and hydrolysis of an "energy-rich" aspartylphosphorylated intermediate formed by reaction with ATP (hence their name; for a review, see Møller et al. 2010).

SERCA1a is a large, monomeric MP, with ten TM helices; three extramembrane (cytosolic) domains, called A (for "actuator"), N ("nucleotide binding"), and P ("phosphorylation"); and an overall MW of 110 kDa. The functional cycle – schematized in Fig. 1.34 – involves large relative movements between the cytosolic domains, coupled to extensive displacements of the TM helices relative to one another. Pumping is actuated by the phosphorylation of Asp351, in Domain P, at the expense of ATP, followed by hydrolysis. Each molecule of ATP consumed results in the thermodynamically uphill pumping of two $Ca^{2+}$ ions from the cytosol into the SR lumen and the downhill release of 2–3 luminal $H^+$ ions into the cytosol.

During its functional cycle, SERCA1a transits between State E1 (on the *right* in Fig. 1.34), with two $Ca^{2+}$ ions bound in the core of its TM region, and State E2 (on the *left*), in which they are replaced by protons. The following description of the main steps in the cycle is adapted from Møller et al. (2010). At the onset (*upper left* cartoon in Fig. 1.34), the $H_nE2$:ATP state carries an ATP molecule, bound to domain N (in *red* in Figs. 1.34 and 1.36). In the transition to state $Ca_2E1$-ATP (step ① in Fig. 1.34), two cytosolic $Ca^{2+}$ ions become strongly coordinated to TM sites formed by side-chain and main-chain groups carried by helices M4, M5, M6, and M8 (Fig. 1.35), releasing $n$ $H^+$ ions ($n = 2$–3) in the cytosol. The complexation of calcium ions forces rotational and translational changes in the position of these helices. In particular, helix M4, which carries Glu309, one of the residues forming $Ca^{2+}$-binding site II, is pushed by ~6 Å toward the cytosol (Toyoshima and Nomura 2002; Jensen et al. 2006; Sonntag et al. 2011). As a consequence, the cytosolic A-domain (in *yellow* in Figs. 1.34 and 1.36) is dislodged from its contacts with the N- and P-domains and moves to a new binding site on the N-domain. This permits the ATP molecule bound to N to closely approach Asp351, on the P-domain (in *blue* in Figs. 1.34 and 1.36), leading to the formation (step ②) of the $[Ca_2]E1{\sim}P$ state, where the bound $Ca^{2+}$ ions become occluded as a result of closure of the cytosolic entrance (Fig. 1.36). In the following step, where the P- and N-domain interactions are loosened because of the transfer of the $\gamma$-phosphate of ATP to Asp351, ADP is exchanged for ATP ③. Attracted by the phosphorylated P-domain, the A-domain pulls on its links with the M1, M2, and M3 helices, which opens a luminal channel. This leads to the release in the lumen of the two $Ca^{2+}$ ions, in exchange for $n$ luminal protons,

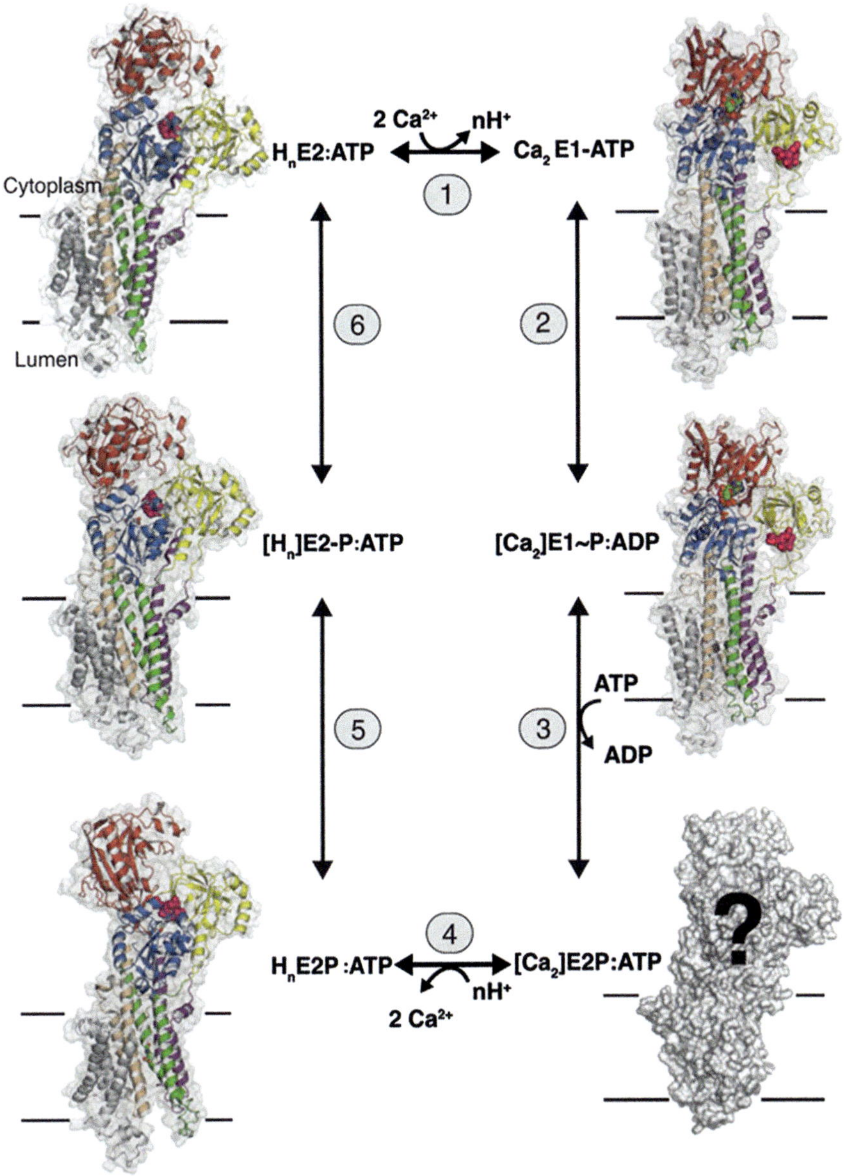

**Fig. 1.34** SERCA 1a structures representing key states of the transport cycle in terms of the following reactions: ① Exchange, on the cytosolic side, of *n* protons ($n = 2$–3) for two $Ca^{2+}$ ions. ② Phosphorylation by ATP, with the formation of the $[Ca_2]E1{\sim}P{:}ADP$ "energy-rich" intermediate with occluded (non-exchangeable) $Ca^{2+}$ (occlusion is noted by square brackets). ③ Conversion of $[Ca_2]E1{\sim}P{:}ADP$ to $[Ca_2]E2P{:}ATP$ after ADP/ATP exchange, with occluded $Ca^{2+}$ (structure still unknown). ④ Formation of the E2P ground state after luminal opening and the exchange of $Ca^{2+}$ with luminal protons. ⑤ Formation of the proton-occluded E2P transition state. ⑥ Dephosphorylation of E2P and return to the resting E2 state with bound protons and ATP. The structures are shown in gray transparent surface and in ribbon representations, with the A-domain in *yellow*, N-domain in *red*, P-domain in *blue*, TM helices M1–M2 in *purple*, M3–M4 in *green*, M5–M6 in *wheat*, and M7–M10 in *gray*. The TGES motif essential for hydrolysis of E2P is in *reddish* space-filling representation; $Ca^{2+}$-liganding residues 309, 771, and 796 in *sticks*; and $Ca^{2+}$ ions in *green* space-filling representation. Note the large relative displacements of TM helices that take place during the cycle, e.g. upon transition ①, during which two cytosolic $Ca^{2+}$ ions bind to TM sites and become occluded (see details in Fig. 1.36), and transition ④, which permits their release into the lumen (From Møller et al. 2010. Reproduced with permission of Cambridge University Press).

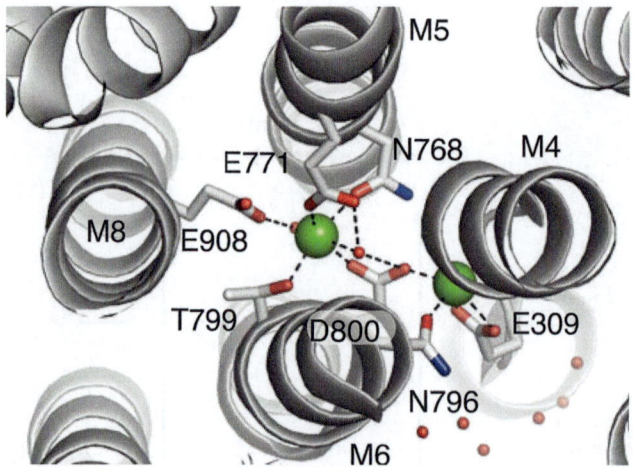

**Fig. 1.35** SERCA1a intramembrane $Ca^{2+}$-binding sites. Close-up view of the two $Ca^{2+}$-binding sites formed by side-chain and main-chain atoms of residues belonging to the M4, M5, M6, and M8 transmembrane helices, shown in ribbon representation with key coordinating side-chain residues in sticks. Water molecules are shown in *red*. The figure is based on the [$Ca_2$]E1:AMPPCP structure (pdb code 1T5S) (From Møller et al. 2010. Reproduced with permission of Cambridge University Press).

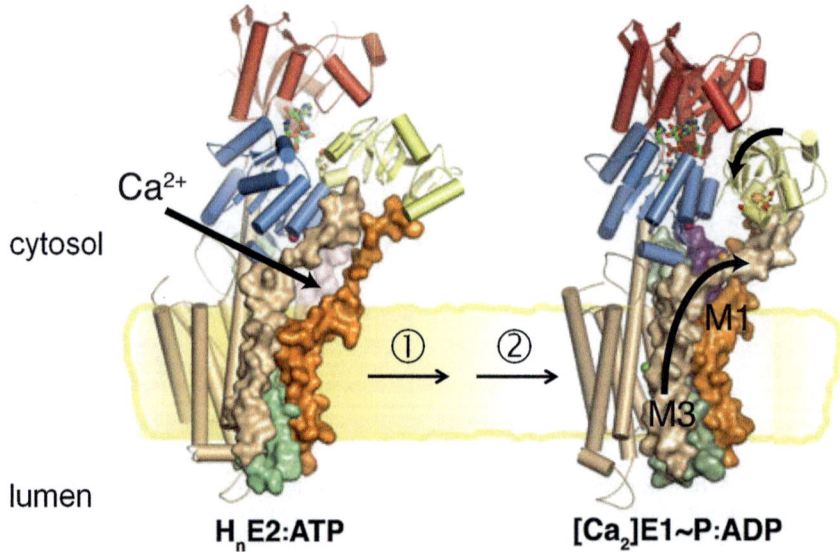

**Fig. 1.36** The $Ca^{2+}$- and ATP-induced E2 → E1P transition of SERCA1a (PDB structures 2C88 and 3BA6). Cartoon representations with helices M1 (*orange*), M2 (*magenta*), M3–M4 (*wheat*), and M5–M6 (*green*) shown in surface representation. In the E2 (calcium-free) state (*left*), the entrance to the $Ca^{2+}$-binding site(s) is widely open to the cytosol (*arrow*), due to a kink in helix M1. Binding of calcium in the presence of a nucleotide (*right*) results in the occlusion of the $Ca^{2+}$-binding sites, because the rotation of the A domain (*yellow*) entails a rearrangement of the cytosolic ends of helices M1 and M3, which block the access to the sites. The transitions ① and ② indicated refer to the scheme in Fig. 1.34 (Adapted from Møller et al. 2010. Reproduced with permission of Cambridge University Press).

which partially neutralize acidic residues at the binding sites. In this process, the pump switches back to the calcium-free E2 state ④. This is followed by closure of the luminal channel and occlusion of the bound protons ⑤. In the last step ⑥, Asp351 is dephosphorylated, leading to the reformation of $H_nE2$: ATP.

Even though neither the structure nor the function of the nAChR (§ 1.6.2) and SERCA has anything in common, it is worth pointing out some similarities and differences in the way each MP couples the conformations of its TM and extramembrane regions. In the case of the nAChR, the binding of acetylcholine (ACh) initiates a rotational rearrangement of the extracellular domains, which affects the position of TM helix M2, leading to the opening of the TM ion channel, more than 60 Å away. In the case of SERCA, the binding of cytosolic $Ca^{2+}$ ions to TM sites induces a rearrangement of the TM helix bundle, as a result of which extramembrane domains are forced to move with respect to each other, initiating the phosphorylation/dephosphorylation cycle that powers calcium pumping. Movements of the cytosolic domains in the course of this cycle, in turn, entail a further rearrangement of the TM helices, imposed by the links connecting cytosolic domain A to the M1–M3 helices, which results in the $Ca^{2+}$ ions being released into the SR lumen. Whereas TM rearrangements are much larger in SERCA than in pLGICs (compare Figs. 1.36 and 1.30), a key feature of the two protein families is the coupling of TM to extramembrane conformational changes. Such is not the case with BR, in which all key events occur within the TM region (§ 1.6.1). In visual rhodopsin, however, which shares the same cofactor, the helix movements initiated by the light-driven isomerization of retinal result not in proton pumping but in a rearrangement of the cytosolic region. This is detected by soluble proteins, which relay and amplify the signal. The same mechanism is exploited by class A (rhodopsin-like) GPCRs, whose cytosolic conformation changes as a result of the binding of extracellular ligands to a cleft formed by the TM helices, thus achieving TM signaling (see Chap. 2, § 2.5.2).

We have noted above that the two TM calcium-binding sites of SERCA are formed by residues carried by helices M4, M5, M6, and M8 (Toyoshima et al. 2000; Obara et al. 2005) (Fig. 1.35). As a result, SERCA1a is strongly stabilized by $Ca^{2+}$ against thermal- and detergent-induced denaturation (Møller et al. 1980; Merino et al. 1994), presumably because the bridging of TM helices by $Ca^{2+}$ ions opposes opening of the TM domain. This observation is of great interest when trying to understand the mechanism by which APols stabilize the calcium pump and, presumably, other MPs (Chap. 5, § 5.6)

A molecule of PE is bound in a cavity between TM helices M2 and M4 in the calcium-free conformation of the protein, acting as a wedge keeping these helices apart (Obara et al. 2005) (Fig. 1.37). In the calcium-bound form, the cavity is closed, so that the PE molecule must be displaced for $Ca^{2+}$ to bind (Obara et al. 2005). As noted by A.G. Lee (2011a, b), this could possibly provide an explanation for some of the effects of PE on $Ca^{2+}$-ATPase function (Starling et al. 1996). It may also perhaps be related to the high sensitivity to detergents exhibited by SERCA1a in its calcium-free form (see Chap. 5, § 5.6.1). A detailed analysis of this and other lipid-binding sites on SERCA1a is given by Drachmann et al. (2014).

In a recent development, contrast variation has permitted Chikashi Toyoshima and his colleagues to visualize the important rocking movements of SERCA1a with respect to the surrounding lipids and the rearrangement of molecular interactions, involving particularly tryptophan and basic residues that accompany the enzymatic cycle. These movements permit the hydrophobic surface of the TM domain to remain lipid-buried throughout (see Norimatsu et al. 2017; the online version of the article includes impressive videos of the rearrangements the protein and lipids undergo in the course of the cycle; see also comments in Sweadner 2017).

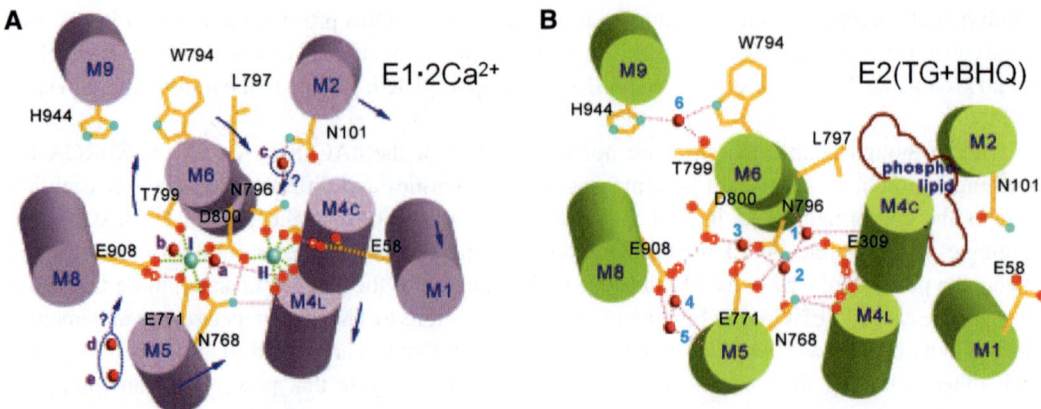

**Fig. 1.37** Schematic representation of the $Ca^{2+}$-binding sites in the calcium-bound state E1·2$Ca^{2+}$ (**A**) and calcium-free state E2 (**B**) of SERCA1a, the latter stabilized by the inhibitors thapsigargin (TG) and 2,5-di-*tert*-butyl-1,4-dihydroxybenzene (BHQ), which block the pump in a $H^+$-occluded state. *Cyan spheres* represent $Ca^{2+}$ ions; *red ones* represent water molecules. Bound protons appear as *red circles*. Arrows indicate the movements of transmembrane helices in the transition E1 → E2. *Dotted arrows* indicate potential movements of water molecules. *Dotted pink lines* indicate hydrogen bonds, and those in *light green* indicate $Ca^{2+}$ coordination. Note the presence of a PE molecule ("phospholipid") wedged between helices M2 and M4 in state E2. This molecule is absent in the calcium-bound state E1 (From Obara et al. 2005. © 2005 National Academy of Sciences).

## 1.7    Synthesis of Membrane Proteins

### 1.7.1    Natural Biosynthesis

In eukaryotic cells, where most MPs feature $\alpha$-helical TM regions, MPs are typically identified by the signal recognition particle (SRP) as their first hydrophobic segment emerges from the ribosome tunnel, which halts translation. The ribosome/SRP complex, along with mRNA and the *N*-terminal region of the MP, docks at the translocon, a TM complex that catalyzes insertion of the protein across the endoplasmic reticulum (ER) membrane (Fig. 1.38). SRP is released and synthesis resumes as a directional process in the course of which the nascent chain is inserted into a central channel in the translocon, or perhaps a lateral groove, in which it would be partially exposed to lipids (cf. Fig. 1.38; for reviews, see e.g. Park and Rapoport 2012; Collinson et al. 2015; Cymer et al. 2015).

Hydrophobic segments that will be TM helices in the final structure are oriented as a function of the distribution of charged residues around their extremities ("positive-inside" rule; see von Heijne 1986), which may involve interactions with the translocon, with lipids, and/or with already inserted helices, and released into the ER membrane as folded helices, probably singly or in hairpin pairs. The ensemble of helices that form a TM region pack together, which may be accompanied by the formation of reentrant loops and other non-TM features, helix reorientation, the binding of prosthetic groups or specific lipids, oligomerization, etc., to achieve the final 3D structure ("two-stage model"; see Popot and Engelman 1990; Engelman et al. 2003) (Fig. 1.39).

In bacteria, plasma membrane proteins, whose TM region is also $\alpha$-helical, are inserted and fold in a similar manner, except that translocation is sometimes posttranslational, in which case the protein is temporarily kept water soluble and its insertion is guided by specialized chaperones. In either case, translocation and insertion essentially proceed from the *N*- to the *C*-terminus of the MP, and the first

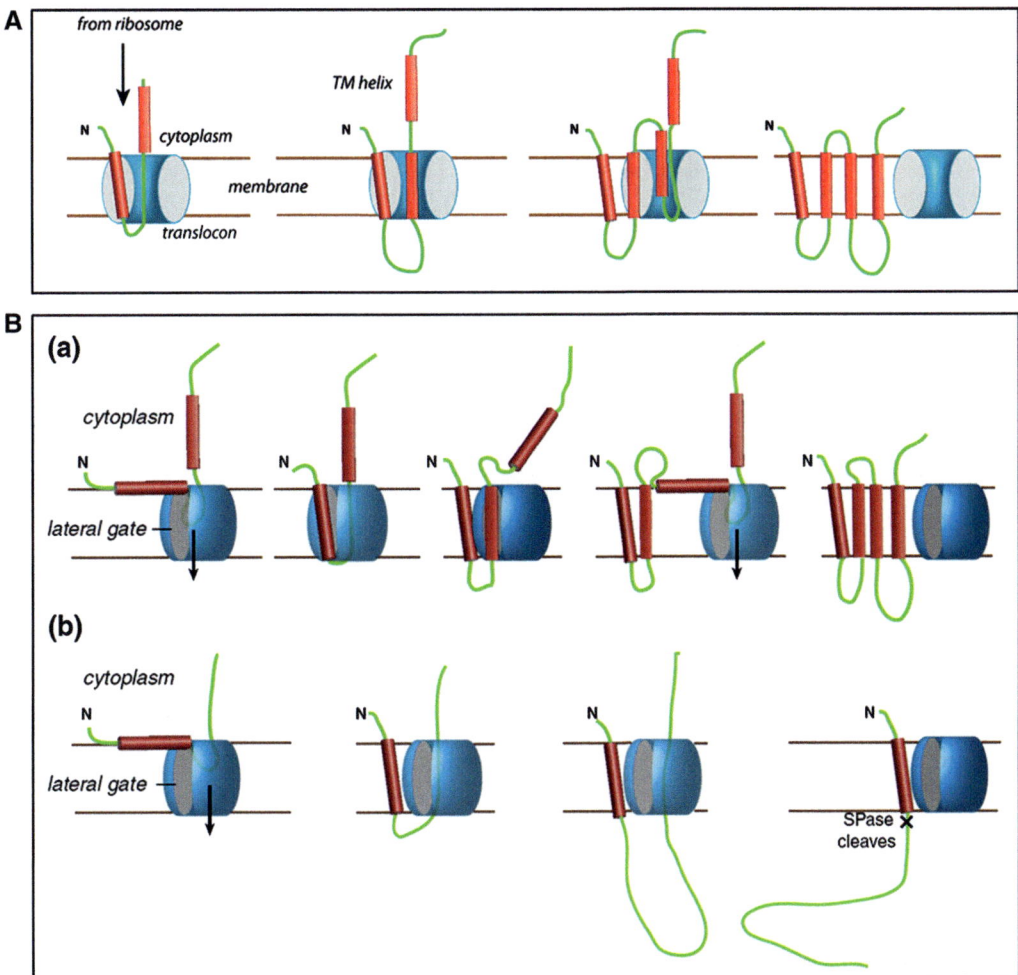

**Fig. 1.38** Schematic views of how the translocon may catalyze the insertion into the membrane of hydrophobic segments (*red*) and the translocation of hydrophilic ones (*green*). (**A**) A cartoon representing in broad terms current thinking about the insertion of multi-span proteins into membranes. Hydrophobic segments that are destined to span the membrane emerge from the ribosome and are threaded into the translocon's central channel, from which they partition into the membrane. (**B**) An alternative view of translocon-aided insertion of multi-span MPs and the secretion of soluble proteins propose that initial contact of future TM segments present in the nascent chain is with the membrane interface in the vicinity of the translocon and that the chain does not immediately thread into the translocon (a). Rather, the translocon provides a pathway for polar components of MPs to cross the membrane. The translocon does form a passageway through the membrane, as in model A, but it is suggested that it does this only for polar polypeptide segments in MPs and for secreted proteins, as shown in (b) (Adapted from Cymer et al. 2015. © 2015 Elsevier Ltd. All rights reserved).

helices are inserted in TM position, while the next ones either are still being synthesized, or they have been but have not yet interacted with the translocon (see Park and Rapoport 2012; Cymer et al. 2015).

As described in § 1.4, the TM regions of the proteins of the outer membrane of Gram⁻ bacteria fold into β-barrels. The protein is synthesized in the cytosol and then threaded through one or another translocon and secreted into the periplasm, where various chaperones and the Bam complex stabilize it, target it to the outer membrane, and catalyze its insertion (see Otzen and Andersen 2013; Pocanschi

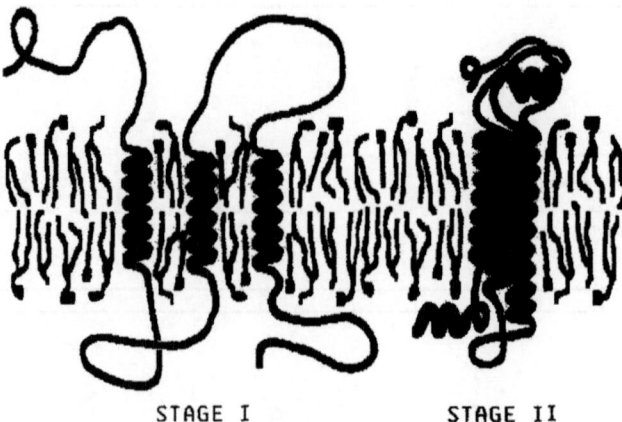

**Fig. 1.39** Two-stage model proposed for the folding of $\alpha$-helical integral membrane proteins. The first stage is the formation of independently stable trans-bilayer helices, principally in response to the hydrophobic effect and the formation of main-chain hydrogen bonds in the nonaqueous environment. The second stage is the interaction of the helices to form the tertiary fold of the polypeptide. Stage II can be accompanied by changes in the length and shape of helices, folding of the extramembrane domains, the formation of reentrant loops and channels, the binding of prosthetic groups and lipids, oligomerization, etc. (Reprinted with permission from Popot and Engelman 1990, © 1990 American Chemical Society).

et al. 2013; McMorran et al. 2014; Kleinschmidt 2015, and references therein). Reflecting their presumed evolutionary origins, some $\beta$-barrel proteins, such as VDAC (Fig. 1.8), are found in the outer membranes of chloroplasts and mitochondria.

A vast body of data, some of which are presented in Chaps. 6 and 7, suggests that neither translocons nor a membrane environment is needed for a MP to reach its functional 3D structure, which is primarily determined by (i) access to an amphipathic medium in which TM segments are protected from being exposed to water whereas extramembrane ones have access to it and (ii) interactions of the polypeptide chain with itself. This view is discussed in Popot (2014) and Popot and Engelman (2016) (for a summary, see Chap. 6, Box 6.3).

### 1.7.2  Overexpression

Most MPs are not expressed naturally in sufficient amounts for experimental purposes. There are four approaches to circumventing this problem:

(i) Overexpression in vivo and in situ, the protein being directed toward one of the membrane compartments of the cell. The main difficulty of this approach is that overexpression of MPs tends to be toxic and needs to be finely tuned in order not to kill the cells while achieving decent yields.

(ii) Overexpression in vivo in inclusion bodies (IBs). IBs are nontoxic particles that precipitate in the cytosol when overexpressed MPs are not targeted to a membrane. IBs permit to achieve very high yields. However, MPs in IBs are not properly folded. They must be solubilized under a denatured form in urea or sodium dodecyl sulfate and brought to their native state, which is rarely straightforward.

(iii) Cell-free expression (CFE) in vitro, in a lysate generally derived either from *E. coli* or from wheat germs. The protein may be either left to precipitate, and later solubilized with a detergent, or integrated into an accepting amphipathic medium, such as lipid vesicles, detergent micelles, APols, or nanodiscs.

(iv) For short MPs or MP fragments, one may also resort to chemical synthesis.

Routes (i) and (ii) are discussed in Chap. 6 along with the application of APols to folding denatured MPs, route (iii) in Chap. 7, along with their use for CFE. APols and other nonconventional surfactants have not been applied yet to chemical synthesis of MPs, which is therefore beyond the frame of this book.

# References

Alberts, B., Johnson, A., Lewis, J., Morgan, D., Raff, M., Roberts, K., Walter, P. (2015) *Molecular Biology of the Cell. Sixth Edition*, Garland Publishing, Inc., New York & London.

Allen, L.C. (1975) A model for the hydrogen bond. *Proc. Natl. Acad. Sci. USA* **72**:4701–4705.

Althoff, T., Hibbs, R.E., Banerjee, S., Gouaux, E. (2014) X-ray structures of GluCl in apo states reveal a gating mechanism of Cys-loop receptors. *Nature* **512**:333–337.

Althoff, T., Mills, D.J., Popot, J.-L., Kühlbrandt, W. (2011) Assembly of electron transport chain components in bovine mitochondrial supercomplex $I_1III_2IV_1$. *EMBO J.* **30**:4652–4664.

Amzel, L.M. (1997) Loss of translational entropy in binding, folding, and catalysis. *Proteins* **28**:144–149.

Andersen, O.S., Koeppe, R.E., 2nd. (2007) Bilayer thickness and membrane protein function: an energetic perspective. *Annu. Rev. Biophys. Biomol. Struct.* **36**:107–130.

Andersson, M., Malmerberg, E., Westenhoff, S., Katona, G., Cammarata, M., Wöhri, A.B., Johansson, L.C., Ewald, F., Eklund, M., Wulff, M., Davidsson, J., Neutze, R. (2009) Structural dynamics of light-driven proton pumps. *Structure* **17**:1265–1275.

Anishkin, A., Loukin, S.H., Teng, J., Kung, C. (2014) Feeling the hidden mechanical forces in lipid bilayer is an original sense. *Proc. Natl. Acad. Sci. USA* **111**:7898–7905.

Arora, A., Abildgaard, F., Bushweller, J.H., Tamm, L.K. (2001) Structure of outer membrane protein A transmembrane domain by NMR spectroscopy. *Nat. Struct. Biol.* **8**:334–338.

Auerbach, A. (2015) Agonist activation of a nicotinic acetylcholine receptor. *Neuropharmacology* **96**:150–156.

Baenziger, J.E., Hénault, C.M., Therien, J.P.D., Sun, J. (2015) Nicotinic acetylcholine receptor-lipid interactions: Mechanistic insight and biological function. *Biochim. Biophys. Acta* **1848**:1806–1817.

Baradaran, R., Berrisford, J.M., Minhas, G.S., Sazanov, L.A. (2013) Crystal structure of the entire respiratory complex I. *Nature* **494**:443–448.

Barrantes, F.J. (2015) Phylogenetic conservation of protein-lipid motifs in pentameric ligand-gated ion channels. *Biochim. Biophys. Acta* **1848**:1796–1805.

Battle, A.R., Ridone, P., Bavi, N., Nakayama, Y., Nikolaev, Y.A., Martinac, B. (2015) Lipid–protein interactions: Lessons learned from stress. *Biochim. Biophys. Acta* **1848**:1744–1756.

Blobel, G. (1980) Intracellular protein topogenesis. *Proc. Natl. Acad. Sci. USA* **77**:1496–1500.

Bocquet, N., Nury, H., Baaden, M., Le Poupon, C., Changeux, J.-P., Delarue, M., Corringer, P.-J. (2009) X-ray structure of a pentameric ligand-gated ion channel in an apparently open conformation. *Nature* **457**:111–114.

Brejc, K., van Dijk, W.J., Klaassen, R.V., Schuurmans, M., van Der Oost, J., Smit, A.B., Sixma, T.K. (2001) Crystal structure of an ACh-binding protein reveals the ligand-binding domain of nicotinic receptors. *Nature* **411**:269–276.

Brisson, A., Unwin, N. (1985) Quaternary structure of the acetylcholine receptor. *Nature* **315**:474–477.

Brown, L.S., Ernst, O.P. (2017) Recent advances in biophysical studies of rhodopsins – Oligomerization, folding, and structure. *Biochim. Biophys. Acta* **1865**:1512–1521.

Buchanan, S.K., Smith, B.S., Venkatramani, L., Xia, D., Esser, L., Palnitkar, M., Chakraborty, R., van der Helm, D., Deisenhofer, J. (1999) Crystal structure of the outer membrane active transporter FepA from *Escherichia coli*. *Nature Struct. Biol.* **6**:56–63.

Buchanan, S.K., Yamashita, S., Fleming, K.G. (2012) Structure and folding of outer membrane proteins, in: Tamm, L.K., ed., *Membranes*. Elsevier, Oxford:Academic Press, pp. 139–163.

Calimet, N., Simoes, M., Changeux, J.-P., Karplus, M., Talye, A., Cecchini, M. (2013) A gating mechanism of pentameric ligand-gated ion channels. *Proc. Natl. Acad. Sci. USA* **110**:E3987–3996.

Carafoli, E. (2002) Calcium signaling: a tale for all seasons. *Proc. Natl. Acad. Sci. USA* **99**:1115–1122.

Casiraghi, M., Damian, M., Lescop, E., Point, E., Moncoq, K., Morellet, N., Levy, D., Marie, J., Guittet, E., Banères, J.-L., Catoire, L.J. (2016) Functional modulation of a GPCR conformational landscape in a lipid bilayer. *J. Am. Chem. Soc.* **138**:11170–11175

Cecchini, M., Changeux, J.-P. (2015) The nicotinic acetylcholine receptor and its prokaryotic homologues: Structure, conformational transitions & allosteric modulation. *Neuropharmacology* **96**:137–149.

Cevc, G., Marsh, D. (1987) *Phospholipid Bilayers: Physical Principles and Models.* Wiley, New York, 442 p.

Chang, G., Spencer, R.H., Lee, A.T., Barclay, M.T., Rees, D.C. (1998) Structure of the MscL homolog from *Mycobacterium tuberculosis*: a gated mechanosensitive ion channel. *Science* **282**:2220–2226.

Changeux, J.-P. (2012) The nicotinic acetylcholine receptor: The founding father of the pentameric ligand-gated ion channel superfamily. *J. Biol. Chem.* **287**:40207–40215.

Changeux, J.-P., Corringer, P.-J., Maskos, U. (2015) The nicotinic acetylcholine receptor: From molecular biology to cognition. *Neuropharmacology* **96**:135–136.

Changeux, J.-P., Edelstein, S.J. (2005) *Nicotinic Acetylcholine Receptors: From Molecular Biology to Cognition.* Odile Jacob Publishing Corporation, New York, 284 p.

Clark, K.M., Jenkins, J.L., Fedoriw, N., Dumont, M.E. (2017) Human CaaX protease ZMPSTE24 expressed in yeast: Structure and inhibition by HIV protease inhibitors. *Protein Sci.* **26**:242–257.

Clegg, J.S. (1983) What is the cytosol? *Trends Biochem. Sci.* **8**:436–437.

Collinson, I., Corey, R.A., Allen, W.J. (2015) Channel crossing: how are proteins shipped across the bacterial plasma membrane? *Philos. Trans. R. Soc. B* **370**:20150025.

Corringer, P.-J., Poitevin, F., Prévost, M.S., Sauguet, L., Delarue, M., Changeux, J.-P. (2012) Structure and pharmacology of pentameric receptor channels: from bacteria to brain. *Structure* **20**:941–956.

Cowan, S.W., Schirmer, T., Rummel, G., Steiert, M., Ghosh, R., Pauptit, R.A., Jansonius, J.N., Rosenbusch, J.P. (1992) Crystal structures explain functional properties of two *E. coli* porins. *Nature* **358**:727–733.

Cymer, F., von Heijne, G., White, S.H. (2015) Mechanisms of integral membrane protein insertion and folding. *J. Mol. Biol.* **427**:999–1022.

daCosta, C.J.B., Baenziger, F.E. (2013) Gating of pentameric ligand-gated ion channels: Structural insights and ambiguities. *Structure* **21**:1271–1283.

daCosta, C.J.B., Baenziger, J.E. (2009) A lipid-dependent uncoupled conformation of the acetylcholine receptor. *J. Biol. Chem.* **284**:17819–17825.

Debnath, D., Nielsen, K.L., Otzen, D.E. (2010) *In vitro* association of fragments of a β-sheet membrane protein. *Biophys. Chem.* **148**:112–120.

Deisenhofer, J., Epp, O., Miki, K., Huber, R., Michel, H. (1985) Structure of the protein subunits in the photosynthetic reaction center of *Rhodopseudomonas viridis* at 3 Å resolution. *Nature* **318**:618–624.

Dilger, J.P., Fisher, L.R., Haydon, D.A. (1982) A critical comparison of electrical and optical methods for bilayer thickness determination. *Chem. Phys. Lipids* **30**:159–176.

Doyle, D.A., Cabral, J.M., Pfuetzner, R.A., Kuo, A., Gulbis, J.M., Cohen, S.L., Chait, B.T., MacKinnon, R. (1998) The structure of the potassium channel: Molecular basis of $K^+$ conduction and selectivity. *Science* **280**:69–77.

Dracheva, S., Bose, S., Hendler, R.W. (1996) Chemical and functional studies on the importance of purple membrane lipids in bacteriorhodopsin photocycle behavior. *FEBS Lett.* **382**:209–212.

Drachmann, N.D., Olesen, C., Møller, J.V., Guo, Z., Nissen, P., Bublitz, M. (2014) Comparing crystal structures of $Ca^{2+}$-ATPase in the presence of different lipids. *FEBS J.* **281**:4249–4262.

Du, J., Lü, W., Wu, S., Cheng, Y., Gouaux, E. (2015) Glycine receptor mechanism elucidated by electron cryo-microscopy. *Nature* **526**:224–229.

Efremov, R.G., Baradaran, R., Sazanov, L.A. (2010) The architecture of respiratory complex I. *Nature* **465**:441–445.

Engelman, D.M. (2005) Membranes are more mosaic than fluid. *Nature* **438**:578–580.

Engelman, D.M., Chen, Y., Chin, C.-N., Curran, R., Dixon, A.M., Dupuy, A., Lee, A., Lehnert, U., Mathews, E., Reshetnyak, Y., Senes, A., Popot, J.-L. (2003) Membrane protein folding: beyond the two-stage model. *FEBS Lett.* **555**:122–125.

Engelman, D.M., Steitz, T.A. (1981) The spontaneous insertion of proteins into and across membranes: the helical hairpin hypothesis. *Cell* **23**:411–422.

Fernández, C., Hilty, C., Wider, G., Wüthrich, K. (2002) Lipid-protein interactions in DHPC micelles containing the integral membrane protein OmpX investigated by NMR spectroscopy. *Proc. Natl. Acad. Sci. USA* **99**:13533–13537.

Ferrand, M., Dianoux, A.J., Petry, W., Zaccai, G. (1993) Thermal motions and function of bacteriorhodopsin in purple membranes: effects of temperature and hydration studied by neutron scattering. *Proc. Natl. Acad. Sci. USA* **90**:9668–9672.

Fyfe, P.K., McAuley, K.E., Roszak, A.W., Isaacs, N.W., Cogdell, R.J., Jones, M.R. (2001) Probing the interface between membrane proteins and membrane lipids by X-ray crystallography. *Trends Biochem. Sci.* **26**:106–112.

Gonen, T., Cheng, Y., Sliz, P., Hiroaki, Y., Fujiyoshi, Y., Harrison, S.C., Walz, T. (2005) Lipid-protein interactions in double-layered two-dimensional AQP0 crystals. *Nature* **438**:633–638.

Goodsell, D.S. (1991) Inside a living cell. *Trends Biochem. Sci.* **16**:203–206.

Green, D.E., Allmann, D.W., Bachmann, E., Baum, H., Kopaczyk, K., Korman, E.F., Lipton, S., MacLennan, D.H., McConnell, D.G., Perdue, J.F., Rieske, J.S., Tzagoloff, A. (1967) Formation of membranes by repeating units. *Arch. Biochem. Biophys.* **119**:312–335.

Grigorieff, N., Ceska, T.A., Downing, K.H., Baldwin, J.M., Henderson, R. (1996) Electron-crystallographic refinement of the structure of bacteriorhodopsin. *J. Mol. Biol.* **259**:393–421.

Hanson, M.A., Cherezov, V., Griffith, M.T., Roth, C.B., Jaakola, V.P., Chien, E.Y., Velasquez, J., Kuhn, P., Stevens, R.C. (2008) A specific cholesterol binding site is established by the 2.8 Å structure of the human $\beta_2$-adrenergic receptor. *Structure* **16**:897–905.

Harrenga, A., Michel, H. (1999) The cytochrome *c* oxidase from *Paracoccus denitrificans* does not change the metal center ligation upon reduction. *J. Biol. Chem.* **274**:33296–33299.

Hassaine, G., Deluz, C., Grasso, L., Wyss, R., Tol, M.B., Hovius, R., Graff, A., Stahlberg, H., Tomizaki, T., Desmyter, A., Moreau, C., Li, X.-D., Poitevin, F., Vogel, H., Nury, H. (2014) X-ray structure of the mouse serotonin 5-HT$_3$ receptor. *Nature* **512**:276–281.

Hénault, C.M., Sun, J., Therien, J.P.D., daCosta, C.J.B., Carswell, C.L., Labriola, J.M., Juranka, P.F., Baenziger, J.E. (2015) The role of the M4 lipid-sensor in the folding, trafficking, and allosteric modulation of nicotinic acetylcholine receptors. *Neuropharmacology* **96**:157–168.

Henderson, R., Baldwin, J.M., Ceska, T.A., Zemlin, F., Beckmann, E., Downing, K.H. (1990) Model for the structure of bacteriorhodopsin based on high resolution electron cryo-microscopy. *J. Mol. Biol.* **213**:899–929.

Henderson, R., Unwin, P.N.T. (1975) Three-dimensional model of purple membrane obtained by electron microscopy. *Nature* **257**:28–32.

Hendler, R.W., Dracheva, S. (2001) Importance of lipids for bacteriorhodopsin structure, photocycle, and function. *Biochemistry* **66**:1311–1314.

Hibbs, R.E., Gouaux, E. (2011) Principles of activation and permeation in an anion-selective Cys-loop receptor. *Nature* **474**:54–60.

Hilf, R.J., Dutzler, R. (2009) Structure of a potentially open state of a proton-activated pentameric ligand-gated ion channel. *Nature* **457**:115–118.

Hilf, R.J.C., Dutzler, R. (2008) X-ray structure of a prokaryotic pentameric ligand-gated ion channel. *Nature* **452**:375–379.

Hiller, S., Garces, R.G., Malia, T.J., Orekhov, V.Y., Colombini, M., Wagner, G. (2008) Solution structure of the integral human membrane protein VDAC-1 in detergent micelles. *Science* **321**:1206–1210.

Hirai, T., Subramaniam, S., Lanyi, J.K. (2009) Structural snapshots of conformational changes in a seven-helix membrane protein: lessons from bacteriorhodopsin. *Curr. Opin. Struct. Biol.* **19**:433–439.

Hite, R.K., Gonen, T., Harrison, S.C., Walz, T. (2008) Interactions of lipids with aquaporin-0 and other membrane proteins. *Pflügers Arch.* **456**:651–661.

Honig, B.H., Hubbell, W.L. (1984) Stability of "salt bridges" in membrane proteins. *Proc. Natl. Acad. Sci. USA* **81**:5412–5416.

Hunte, C. (2005) Specific protein-lipid interactions in membrane proteins. *Biochem. Soc. Trans.* **33**:938–942.

Hunte, C., Richers, S. (2008) Lipids and membrane protein structures. *Curr. Opin. Struct. Biol.* **18**:406–411.

Israelachvili, J.N. (2011) *Intermolecular and Surface Forces. 3rd ed.*, Elsevier/Academic Press, London, 706 p.

Iwata, S., Ostermeier, C., Ludwig, B., Michel, H. (1995) Structure at 2.8 Å resolution of cytochrome *c* oxidase from *Paracoccus denitrificans*. *Nature* **376**:660–669.

Jakobsson, E. (1997) Computer simulation studies of biological membranes: progress, promise and pitfalls. *Trends Biochem. Sci.* **22**:339–344.

Jensen, A.M.L., Sørensen, T.L.M., Olesen, C., Møller, J.V., Nissen, P. (2006) Modulatory and catalytic modes of ATP binding by the calcium pump. *EMBO J.* **25**:2305–2315.

Jordan, P., Fromme, P., Witt, H.T., Klukas, O., Saenger, W., Krauss, N. (2001) Three-dimensional structure of cyanobacterial photosystem I at 2.5 Å resolution. *Nature* **411**:909–917.

Joshi, M., Dracheva, S., Mukhopadhyay, A.K., Bose, S., Hendler, R.W. (1998) Importance of specific native lipids in controlling the photocycle of bacteriorhodopsin. *Biochemistry* **37**:14463–14470.

Kandori, H. (2015) Ion-pumping microbial rhodopsins. *Front. Mol. Biosci.* **2**:52.

Karlin, A. (2002) Emerging structure of the nicotinic acetylcholine receptors. *Nat. Rev. Neurosci.* **3**:102–114.

Kleinschmidt, J.H. (2015) Folding of β-barrel membrane proteins in lipid bilayers – Unassisted and assisted folding and insertion. *Biochim. Biophys. Acta* **1848**:1927–1943.

Koebnik, R. (1996) *In vivo* membrane assembly of split variants of the *E. coli* outer membrane protein OmpA. *EMBO J.* **15**:3529–3537.

Krauss, N., Schubert, W.-D., Klukas, O., Fromme, P., Witt, H.T., Saenger, W. (1996) Photosystem I at 4 Å resolution represents the first structural model of a joint photosynthetic reaction center and core antenna system. *Nature Struct. Biol.* **3**:965–973.

Kühlbrandt, W. (2000) Bacteriorhodopsin – the movie. *Nature* **406**:569–570.

Kühlbrandt, W. (2015) Structure and function of mitochondrial membrane protein complexes. *BMC Biol.* **13**:89.

Kühlbrandt, W., Wang, D.N., Fujiyoshi, Y. (1994) Atomic model of plant light-harvesting complex. *Nature* **367**:614–621.

Kurisu, G., Zhang, H., Smith, J.L., Cramer, W.A. (2003) Structure of the cytochrome $b_6 f$ complex of oxygenic photosynthesis: tuning the cavity. *Science* **302**:1009–1014.

Landau, E.M., Rosenbusch, J.P. (1996) Lipid cubic phases: A novel concept for the crystallization of membrane proteins. *Proc. Natl. Acad. Sci. USA* **93**:14532–14535.

Lange, C., Nett, J.H., Trumpower, B.L., Hunte, C. (2001) Specific roles of protein-phospholipid interactions in the yeast cytochrome $bc_1$ complex structure. *EMBO J.* **20**:6591–6600.

Lanyi, J.K. (2004) Bacteriorhodopsin. *Annu. Rev. Physiol.* **66**:665–688.

Lanyi, J.K., Luecke, H. (2001) Bacteriorhodopsin. *Curr. Opin. Struct. Biol.* **11**:415–419.

Lavergne, J., Bouchaud, J.-P., Joliot, P. (1992) Plastoquinone compartmentation in chloroplasts. II. Theoretical aspects. *Biochim. Biophys. Acta* **1101**:13–22.

Lee, A.G. (2003) Lipid-protein interactions in biological membranes: a structural perspective. *Biochim. Biophys. Acta* **1612**:1–40.

Lee, A.G. (2004) How lipids affect the activities of integral membrane proteins. *Biochim. Biophys. Acta* **1666**:62–87.

Lee, A.G. (2011a) Biological membranes: the importance of molecular detail. *Trends Biochem. Sci.* **36**:493–500.

Lee, A.G. (2011b) How to understand lipid-protein interactions in biological membranes, in: Yeagle, P., ed., *Structure of Biological Membranes*. 3rd edition, Taylor and Francis, Boca Raton, Florida, USA, pp. 273–313.

Lee, T.-Y., Yeh, V., Chuang, J., Chan, J.C.C., Chu, L.-K., Yu, T.-Y. (2015) Tuning the photocycle kinetics of bacteriorhodopsin in lipid nanodiscs. *Biophys. J.* **109**:1899–1906.

Lehnert, U., Réat, V., Weik, M., Zaccai, G., Pfister, C. (1998) Thermal motions in bacteriorhodopsin at different hydration levels studied by neutron scattering: correlation with kinetics and light-induced conformational changes. *Biophys. J.* **75**:1945–1952.

Li, E., You, M., Hristova, K. (2006) FGFR3 dimer stabilization due to a single amino acid pathogenic mutation. *J. Mol. Biol.* **356**:600–612.

Locher, K.P., Rees, B., Koebnik, R., Mitschler, A., Moulinier, L., Rosenbusch, J.P., Moras, D. (1998) Transmembrane signaling across the ligand-gated FhuA receptor: Crystal structures of free and ferrichrome-bound states reveal allosteric changes. *Cell* **95**:771–778.

Lundbaek, J.A., Koeppe, R.E., 2nd., Andersen, O.S. (2010) Amphiphile regulation of ion channel function by changes in the bilayer spring constant. *Proc. Natl. Acad. Sci. USA* **107**:1527–1530.

MacKenzie, K.R., Prestegard, J.H., Engelman, D.M. (1997) A transmembrane helix dimer: structure and implications. *Science* **276**:131–133.

Marčelja, S. (1974) Chain ordering in liquid crystals. II. Structure of bilayer membranes. *Biochim. Biophys. Acta* **367**:165–176.

Marsh, D. (1996) Lateral pressure in membranes. *Biochim. Biophys. Acta* **1286**:183–223.

Marsh, D. (2008) Protein modulation of lipids, and vice-versa, in membranes. *Biochim. Biophys. Acta* **1778**:1545–1575.

Martfeld, A.N., Rajagopalan, V., Greathouse, D.V., Koeppe, R.E., II (2015) Dynamic regulation of lipid-protein interactions. *Biochim. Biophys. Acta* **1848**:1849–1859.

McMorran, L.M., Brockwell, D.J., Radford, S.E. (2014) Mechanistic studies of the biogenesis and folding of outer membrane proteins *in vitro* and *in vivo*: What have we learned to date? *Arch. Biochem. Biophys.* **564**:265–280.

Merino, J.M., Møller, J.V., Gutiérrez-Merino, C. (1994) Thermal unfolding of monomeric $Ca^{2+},Mg^{2+}$-ATPase from sarcoplasmic reticulum of rabbit skeletal muscle. *FEBS Lett.* **343**:155–159.

Miller, P.S., Aricescu, A.R. (2014) Crystal structure of a human $GABA_A$ receptor. *Nature* **512**:270–275.

Mitsuoka, K., Hirai, T., Miyazawa, A., Kidera, A., Kimura, Y., Fujiyoshi, Y. (1999) The structure of bacteriorhodopsin at 3.0 Å resolution based on electron crystallography: implication of the charge distribution. *J. Mol. Biol.* **286**:861–882.

Møller, J.V., Lind, K.E., Andersen, J.P. (1980) Enzyme kinetics and substrate stabilization of detergent-solubilized and membraneous ($Ca^{2+} + Mg^{2+}$)-activated ATPase from sarcoplasmic reticulum. Effect of protein-protein interactions. *J. Biol. Chem.* **255**:1912–1920.

Møller, J.V., Olesen, C., Winther, A.M.L., Nissen, P. (2010) The sarcoplasmic $Ca^{2+}$-ATPase: design of a perfect chemi-osmotic pump. *Q. Rev. Biophys.* **43**:501–566.

Morales-Perez, C.L., Noviello, C.M., Hibbs, R.E. (2016) X-ray structure of the human α4β2 nicotinic receptor. *Nature* **538**:411–415.

Nemecz, Á., Prévost, M.S., Menny, A., Corringer, P.-J. (2016) Emerging molecular mechanisms of signal transduction in pentameric ligand-gated ion channels. *Neuron* **90**:452–470.

Neutze, R., Pebay-Peyroula, E., Edman, K., Royant, A., Navarro, J., Landau, E.M. (2002) Bacteriorhodopsin: a high-resolution structural view of vectorial proton transport. *Biochim. Biophys. Acta* **1565**:144–167.

Nicolson, G.L. (2014) The fluid-mosaic model of membrane structure: still relevant to understanding the structure, function and dynamics of biological membranes after more than 40 years. *Biochim. Biophys. Acta* **1838**:1451–1466.

Nogi, T., Fathir, I., Kobayashi, M., Nozawa, T., Miki, K. (2000) Crystal structures of photosynthetic reaction center and high-potential iron-sulfur protein from *Thermochromatium tepidum*: thermostability and electron transfer. *Proc. Natl. Acad. Sci. USA* **97**:13561–13566.

Norimatsu, Y., Hasegawa, K., Shimizu, N., Toyoshima, C. (2017) Protein-phospholipid interplay revealed with crystals of a calcium pump. *Nature* **545**:193–198.

Nury, H., Van Renterghem, C., Weng, Y., Tran, A., Baaden, M., Dufresne, V., Changeux, J.-P., Sonner, J.M., Delarue, M., Corringer, P.-J. (2011) X-ray structures of general anaesthetics bound to a pentameric ligand-gated ion channel. *Nature* **469**:428–431.

Nys, M., Kesters, D., Ulens, C. (2013) Structural insights into Cys-loop receptor function and ligand recognition. *Biochim. Biophys. Acta* **86**:1042–1053.

Obara, K., Miyashita, N., Xu, C., Toyoshima, I., Sugita, Y., Inesi, G., Toyoshima, C. (2005) Structural role of countertransport revealed in $Ca^{2+}$ pump crystal structure in the absence of $Ca^{2+}$. *Proc. Natl. Acad. Sci. USA* **102**:14489–14496.

Opekarová, M., Tanner, W. (2003) Specific lipid requirements of membrane proteins – a putative bottleneck in heterologous expression. *Biochim. Biophys. Acta* **1610**:11–22.

Ostermeier, C., Harrenga, A., Ermler, U., Michel, H. (1997) Structure at 2.7 Å resolution of the *Paracoccus denitrificans* two-subunit cytochrome *c* oxidase complexed with an antibody $F_v$ fragment. *Proc. Natl. Acad. Sci. USA* **94**:10547–10553.

Otzen, D.E., Andersen, K.K. (2013) Folding of outer membrane proteins. *Arch. Biochem. Biophys.* **531**:34–43.

Palsdottir, H., Hunte, C. (2004) Lipids in membrane protein structures. *Biochim. Biophys. Acta* **1666**:2–18.

Pankratov, Y., Lalo, U. (2014) Calcium permeability of ligand-gated $Ca^{2+}$ channels. *Eur. J. Pharmacol.* **739**:60–73.

Park, E., Rapoport, T.A. (2012) Mechanisms of Sec61/SecY-mediated protein translocation across membranes. *Annu. Rev. Biophys.* **41**:21–40.

Pebay-Peyroula, E., Rummel, G., Rosenbusch, J.P., Landau, E. (1997) X-ray structure of bacteriorhodopsin at 2.5 Å from microcrystals grown in lipidic cubic phases. *Science* **277**:1676–1881.

Pedersen, P.L. (2007) Transport ATPases into the year 2008: a brief overview related to types, structures, functions and roles in health and disease. *J. Bioenerg. Biomem.* **39**:349–355.

Peters, R. (1986) Fluorescence microphotolysis to measure nucleocytoplasmic transport and intracellular mobility. *Biochim. Biophys. Acta* **864**:305–359.

Phillips, R., Ursell, T., Wiggins, P., Sens, P. (2009) Emerging roles for lipids in shaping membrane protein function. *Nature* **459**:379–385.

Planas-Iglesias, J., Dwarakanath, H., Mohammadyani, D., Yanamala, N., Kagan, V.E., Klein-Seetharaman, J. (2015) Cardiolipin interactions with proteins. *Biophys. J.* **109**:1282–1294.

Pocanschi, C., Popot, J.-L., Kleinschmidt, J.H. (2013) Folding and stability of outer membrane protein A (OmpA) from *Escherichia coli* in an amphipathic polymer, amphipol A8-35. *Eur. Biophys. J.* **42**:103–118.

Popot, J.-L. (2014) Folding membrane proteins *in vitro*: A table and some comments. *Arch. Biochem. Biophys.* **564**:314–326.

Popot, J.-L., de Vitry, C. (1990) On the microassembly of integral membrane proteins. *Annu. Rev. Biophys. Biophys. Chem.* **19**:369–403.

Popot, J.-L., Engelman, D.M. (1990) Membrane protein folding and oligomerization: the two-stage model. *Biochemistry* **29**:4031–4037.

Popot, J.-L., Engelman, D.M. (2000) Helical membrane protein folding, stability and evolution. *Annu. Rev. Biochem.* **69**:881–923.

Popot, J.-L., Engelman, D.M. (2016) Membranes do not tell proteins how to fold. *Biochemistry* **55**:5–18.

Prévost, M.S., Sauguet, L., Nury, H., Van Renterghem, C., Huon, C., Poitevin, F., Baaden, M., Delarue, M., Corringer, P.-J. (2012) A locally closed conformation of a bacterial pentameric proton-gated ion channel. *Nat. Struct. Molec. Biol.* **19**:642–649.

Prince, S.M., Papiz, M.Z., Freer, A.A., McDermott, G., Hawthornwaite-Lawless, A.M., Cogdell, R.J., Isaacs, N.W. (1997) Apoprotein structure in the LH2 complex from *Rhodopseudomonas acidophila* strain 10050 : modular assembly and protein-pigment interactions. *J. Mol. Biol.* **268**:412–423.

Pryor, E.E., Jr., Horanyi, P.S., Clark, K.M., Fedoriw, N., Connelly, S.M., Koszelak-Rosenblum, M., Zhu, G., Malkowski, M.G., Wiener, M.C., Dumont, M.E. (2013) Structure of the integral membrane protein CAAX protease Ste24p. *Science* **339**:1600–1604.

Qin, L., Hiser, C., Mulichak, A., Garavito, R.M., Ferguson-Miller, S. (2006) Identification of conserved lipid/detergent-binding sites in a high-resolution structure of the membrane protein cytochrome *c* oxidase. *Proc. Natl. Acad. Sci. USA* **103**:16117–16122.

Qin, L., Sharpe, M.A., Garavito, R.M., Ferguson-Miller, S. (2007) Conserved lipid-binding sites in membrane proteins: a focus on cytochrome *c* oxidase. *Curr. Opin. Struct. Biol.* **17**:444–450.

Qin, X., Suga, M., Kuang, T., Shen, J.R. (2015) Photosynthesis. Structural basis for energy transfer pathways in the plant PSI-LHCI supercomplex. *Science* **348**:989–995.

Reynolds, J.A., Gilbert, D.B., Tanford, C. (1974) Empirical correlation between hydrophobic free energy and aqueous cavity surface area. *Proc. Natl. Acad. Sci. USA* **71**:2925–2927.

Rouse, S.L., Sansom, M.S.P. (2015) Interactions of lipids and detergents with a viral ion channel protein: Molecular dynamics simulation studies. *J. Phys. Chem. B* **119**:764–772.

Rummel, G., Hardmeyer, A., Widmer, C., Chiu, M.L., Nollert, P., Locher, K.P., Pedruzzi, I.I., Landau, E.M., Rosenbusch, J.P. (1998) Lipidic cubic phases: new matrices for the three-dimensional crystallization of membrane proteins. *J. Struct. Biol.* **121**:82–91.

Sauguet, L., Poitevin, F., Murail, S., Van Renterghem, C., Moraga-Cid, G., Malherbe, L., Thompson, A.W., Koehl, P., Corringer, P.-J., Baaden, M., Delarue, M. (2013) Structural basis for ion permeation mechanism in pentameric ligand-gated ion channels. *EMBO J.* **32**:728–741.

Sauguet, L., Shahsavar, A., Poitevin, F., Huon, C., Menny, A., Nemecz, À., Haouz, A., Changeux, J.-P., Corringer, P.-J., Delarue, M. (2014) Crystal structures of a pentameric ligand-gated ion channel provide a mechanism for activation. *Proc. Natl. Acad. Sci. USA* **111**:966–971.

Saxton, M.J. (1989) Lateral diffusion in an archipelago. Distance dependence of the diffusion coefficient. *Biophys. J.* **56**:615–622.

Schirmer, T. (1998) General and specific porins from bacterial outer membranes. *J. Struct. Biol.* **121**:101–109.

Shinzawa-Itoh, K., Aoyama, H., Muramoto, K., Terada, H., Kurauchi, T., Tadehara, Y., Yamasaki, A., Sugimura, T., Kurono, S., Tsujimoto, K., Mizushima, T., Yamashita, E., Tsukihara, T., Yoshikawa, S. (2007) Structures and physiological roles of 13 integral lipids of bovine heart cytochrome *c* oxidase. *EMBO J.* **26**:1713–1725.

Sine, S.M. (2012) End-plate acetylcholine receptor: Structure, mechanism, pharmacology, and disease. *Physiol. Rev.* **92**:1189–1234.

Singer, S.J., Nicolson, G.L. (1972) The fluid mosaic model of the structure of cell membranes. *Science* **175**:720–731.

Smit, A.B., Syed, N.I., Schaap, D., van Minnen, J., Klumpermank, J., Kits, K.S., Lodder, H., van der Schors, R.C., van Elk, R., Sorgedrager, B., Brejc, K., Sixma, T.K., Geraerts, W.P.M. (2001) A glia-derived acetylcholine-binding protein that modulates synaptic transmission. *Nature* **411**:261–268.

Smith, A.W. (2012) Lipid-protein interactions in biological membranes: A dynamic perspective. *Biochim. Biophys. Acta* **1818**:172–177.

Song, L., Hobaugh, M.R., Shustak, C., Cheley, S., Bayley, H., Gouaux, J.E. (1996) Structure of α-hemolysin, a heptameric transmembrane pore. *Science* **274**:1859–1866.

Sonntag, Y., Musgaard, M., Olesen, C., Schiøtt, B., Møller, J.V., Nissen, P., Thøgersen, L. (2011) Mutual adaptation of a membrane protein and its lipid bilayer during conformational changes. *Nat. Commun.* **2**:304.

Spurny, R., Ramerstorfer, J., Price, K., Brams, M., Ernst, M., Nury, H., Verheije, M., Legrand, P., Bertrand, D., Bertrand, S., Dougherty, D.A., de Esch, I.J.P., Corringer, P.-J., Sieghart, W., Lummis, S.C.R., Ulens, C. (2012) Pentameric ligand-gated ion channel ELIC is activated by GABA and modulated by benzodiazepines. *Proc. Natl. Acad. Sci. USA* **109**:E3028-E3034.

Starling, A.P., Dalton, K.A., East, J.M., Oliver, S., Lee, A.G. (1996) Effects of phosphatidylethanolamines on the activity of the $Ca^{2+}$-ATPase of sarcoplasmic reticulum. *Biochem. J.* **320**:309–314.

Stowell, M.H.B., McPhillips, T.M., Rees, D.C., Soltis, S.M., Abresch, E., Feher, G. (1997) Light-induced structural changes in photosynthetic reaction center: implication for mechanism of electron-proton transfer. *Science* **276**:812–816.

Stroebel, D., Choquet, Y., Popot, J.-L., Picot, D. (2003) An atypical haem in the cytochrome $b_6 f$ complex. *Nature* **426**:413–418.

Svensson-Ek, M., Abramson, J., Larsson, G., Tornroth, S., Brzezinski, P., Iwata, S. (2002) The X-ray crystal structures of wild-type and EQ(I-286) mutant cytochrome *c* oxidases from *Rhodobacter sphaeroides*. *J. Mol. Biol.* **321**:329–339.

Sweadner, K.J. (2017) An ion-transport enzyme that rocks. *Nature* **545**:162–164.

Taly, A., Hénin, J., Changeux, J.-P., Cecchini, M. (2014) Allosteric regulation of pentameric ligand-gated ion channels: an emerging mechanistic perspective. *Channels* **8**:350–360.

Tanford, C. (1980) *The Hydrophobic Effect: Formation of Micelles and Biological Membranes*. 2nd ed. John Wiley & Sons, New York, 233 p.

Teng, J., Loukin, S.H., Anishkin, A., Kung, C. (2015) The force-from-lipid (FFL) principle of mechanosensitivity, at large and in elements. *Pflügers Arch.* **467**:27–37.

Toyoshima, C., Nakasako, M., Nomura, H., Ogawa, H. (2000) Crystal structure of the calcium pump of sarcoplasmic reticulum at 2.6 Å resolution. *Nature* **405**:647–655.

Toyoshima, C., Nomura, H. (2002) Structural changes in the calcium pump accompanying the dissociation of calcium. *Nature* **418**:605–611.

Tsukihara, T., Aoyama, H., Yamashita, E., Tomizaki, T., Yamaguchi, H., Shinzawa-Itoh, K., Nakashima, R., Yaono, R., Yoshikawa, S. (1996) The whole structure of the 13-subunit oxidized cytochrome *c* oxidase at 2.8 Å. *Science* **272**:1136–1144.

Ujwal, R., Cascio, D., Colletier, J.-P., Faham, S., Zhang, J., Toro, L., Ping, P., Abramson, J. (2008) The crystal structure of mouse VDAC1 at 2.3 Å resolution reveals mechanistic insights into metabolite gating. *Proc. Natl. Acad. Sci. USA* **105**:17742–17747.

Unwin, N. (2005) Refined structure of the nicotinic acetylcholine receptor at 4 Å resolution. *J. Mol. Biol.* **346**:967–989.

Unwin, N. (2013) Nicotinic acetylcholine receptor and the structural basis of neuromuscular transmission: insights from *Torpedo* postsynaptic membranes. *Quart. Rev. Biophys.* **46**:283–322.

Unwin, N., Fujiyoshi, Y. (2012) Gating movement of acetylcholine receptor caught by plunge-freezing. *J. Mol. Biol.* **422**:616–634.

Valiyaveetil, F.I., Zhou, Y., MacKinnon, R. (2002) Lipids in the structure, folding, and function of the KcsA $K^+$ channel. *Biochemistry* **41**:10771–10777.

Vinothkumar, K.R. (2011) Structure of rhomboid protease in a lipid environment. *J. Mol. Biol.* **407**:232–247.

Vinothkumar, K.R., Henderson, R. (2010) Structures of membrane proteins. *Q. Rev. Biophys.* **43**:65–158.

Vinothkumar, K.R., Zhu, J., Hirst, J. (2014) Architecture of mammalian respiratory complex I. *Nature* **515**:80–84.

von Heijne, G. (1986) The distribution of positively charged residues in bacterial inner membrane proteins correlates with the trans-membrane topology. *EMBO J.* **5**:3021–3027.

Wallin, E., von Heijne, G. (1998) Genome-wide analysis of integral membrane proteins from eubacterial, archean, and eukaryotic organisms. *Protein Sci.* **7**:1029–1038.

Weiss, M.S., Abele, U., Weckesser, J., Welte, W., Schiltz, E., Schulz, G.E. (1991a) Molecular architecture and electrostatic properties of a bacterial porin. *Science* **254**:1627–1630.

Weiss, M.S., Kreusch, A., Schiltz, E., Nestel, U., Welte, W., Weckesser, J., Schulz, G.E. (1991b) The structure of porin from *Rhodobacter capsulatus* at 1.8 Å resolution. *FEBS Lett.* **280**:379–382.

Weiss, M.S., Wacker, T., Nestel, U., Woitzik, D., Weckesser, J., Kreutz, W., Welte, W., Schulz, G.E. (1989) The structure of porin from *Rhodobacter capsulatus* at 0.6 nm resolution. *FEBS Lett.* **256**:143–146.

Wenz, T., Hielscher, R., Hellwig, P., Schägger, H., Richers, S., Hunte, C. (2009) Role of phospholipids in respiratory cytochrome $bc_1$ complex catalysis and supercomplex formation. *Biochim. Biophys. Acta* **1787**:609–616.

White, S.H., von Heijne, G., Engelman, D.M. (2018) *Cell Boundaries: How Membranes and Their Proteins Work.* Garland, New York, *in preparation.*

White, S.H., Wimley, W.C. (1999) Membrane protein folding and stability : physical principles. *Annu. Rev. Biophys. Biomol. Struct.* **28**:319–365.

Wickstrand, C., Dods, R., Royant, A., Neutze, R. (2015) Bacteriorhodopsin: Would the real structural intermediates please stand up? *Biochim. Biophys. Acta* **1850**:536–553.

Xia, D., Yu, C.-A., Kim, H., Xia, J.-Z., Kachurin, A.M., Zhang, L., Yu, L., Deisenhofer, J. (1997) Crystal structure of the cytochrome $bc_1$ complex from bovine heart mitochondria. *Science* **277**:60–66.

Zaccai, G. (2000) Moist and soft, dry and stiff: a review of neutron experiments on hydration-dynamics-activity relations in the purple membrane of *Halobacterium salinarum*. *Biophys. Chem.* **86**:249–257.

Zaccai, G. (2004) The effect of water on protein dynamics. *Phil. Trans. R. Soc. Lond. B* **359**:1269–1275.

Zhou, Y., Morais-Cabral, J.H., Kaufman, A., MacKinnon, R. (2001) Chemistry of ion coordination and hydration revealed by a $K^+$ channel-$F_{ab}$ complex at 2.0 Å resolution. *Nature* **414**:43–48.

Zorman, S., Botte, M., Jiang, Q., Collinson, I., Schaffitzel, C. (2015) Advances and challenges of membrane-protein complex production. *Curr. Opin. Struct. Biol.* **32**:123–130.

# Extracting Membrane Proteins from Their Native Environment

**2**

**Summary**

*Whereas many functional studies and some structural ones can be carried out while keeping membrane proteins in their native environment, other investigations require them to be extracted from the membrane, purified, and handled in aqueous solutions. The usual route is to resort to detergents. Detergents are small amphiphiles that, in aqueous solutions, associate into particles, called micelles, into which lipids and membrane proteins can be solubilized. A major difficulty in the use of detergents is that they must be "strong" enough to disrupt membranes and disperse their components and "mild" enough not to inactivate the target proteins, a balance that is often quite difficult to achieve. In this chapter, we will review the mechanisms by which detergents extract membrane proteins, the nature of the complexes they form with them, the reasons why detergent-solubilized membrane proteins are often short-lived, and the measures that can be taken to confer them enough stability for experimental investigations to be carried out.*

## 2.1 Introduction

Many studies of membrane proteins (MPs) can be carried out in their cellular environment or on native membrane fragments. In the first case, the protein is exposed to the same or nearly the same aqueous and membrane environments as in the whole organism. In the second, a number of important interactions may have been lost, such as exposure to gradients of concentration of molecules and ions, of pH, or of redox potential, to the transmembrane electric field, to interactions with cytosolic or extracellular molecules, etc. These factors may affect the protein's function and, sometimes, its structure. Upon breaking cells open, MP cytosolic and extramembrane domains, for instance, become exposed to the same redox potential; in vivo, the former are exposed to a reducing medium and may comprise free sulfhydryls, whereas the latter, on the contrary, are exposed to an oxidizing environment and often feature disulfide bridges. Artifactual bridges between cytosolic cysteine residues may form upon breaking the cells open in an air-saturated buffer, whereas extracellular bridges may be reduced by the reductants freed from the cytosol. This can be prevented by homogenizing the tissue in the presence of iodoacetamide but at the cost of blocking free thiols that may be functionally important.

© Springer International Publishing AG, part of Springer Nature 2018
J. -L. Popot, *Membrane Proteins in Aqueous Solutions*, Biological and Medical Physics, Biomedical Engineering, https://doi.org/10.1007/978-3-319-73148-3_2

Intact cells and membrane fragments lend themselves to many functional and some structural approaches. Among the latter are, for instance, Förster resonance energy transfer (FRET) or electron paramagnetic resonance (EPR) studies, which can give information about conformational changes, oligomerization, interaction with soluble proteins, ligands, etc., or atomic force microscopy, single-molecule force spectroscopy, solid-state NMR, or some electron microscopy studies. In particularly favorable cases, the target MP assembles into dense patches in which it is the only or the principal protein. Some of these patches spontaneously form, or can be coaxed into forming, two-dimensional (2D) crystals in the membrane plane, making the proteins that form them accessible to crystallographic approaches. Such is the case, for instance, of bacteriorhodopsin (BR; see Chap. 1, § 1.6.1) and of the nicotinic acetylcholine receptor (nAChR; ibid., § 1.6.2).

Investigations of MPs integrated in their native membrane environment are extremely precious, because they provide benchmarks against which to gauge results obtained with more perturbed samples, e.g. in solution. It is exceedingly rare, however, that they suffice to provide an exhaustive understanding of the way a protein carries out its function. In particular, they seldom lead to high-resolution structures permitting atomic models of the protein's structure to be built. Furthermore, most spectroscopic approaches requiring pure samples cannot, usually, be carried out without first extracting the target MP from its native membrane and purifying it in aqueous solution – even if later experiments are carried out on reconstituted membrane-based samples, such as artificial vesicles, 2D crystals, or three-dimensional (3D) crystals formed of stacked 2D ones (see Chap. 11, § 11.2.2).

Solubilizing one's target MP and handling it in solution is most often an obligate prerequisite to purifying it, and this is usually where the troubles of the membrane biochemist begin.

## 2.2   Detergents

### 2.2.1   Chemical Structure

Detergents (from Latin *detergere*, to wipe away) are a special class of surfactants, whose distinctive characteristic is to be endowed with the ability to solubilize fats: whereas all detergents are surfactants, many surfactants – lipids, for instance – are non-solubilizing, and the distinction between the two classes of surfactants should be made[1]. As noted in the introduction to Chap. 1, nearly all biological molecules, including proteins, comprise hydrophobic and hydrophilic regions and, as a consequence, are surface active. In the present book, however, the term "surfactant" will be restricted to relatively small molecules, such as lipids, detergents, amphipathic peptides, or amphipols (APols), and not used for biological macromolecules, whereas "detergent" will be reserved to those surfactants that are able to solubilize fats. The dissociating character ("detergency") of some surfactants can be borderline, however, so that they are solubilizing with some membranes and/or under certain circumstances and not under others (see § 2.3.1).

Detergents are able to make fats water-soluble because of the special arrangements their molecules adopt when assembling in aqueous solutions. Most detergent molecules look like tadpoles: they are comprised of one polar "head" and one hydrophobic "tail," often an alkyl chain (Fig. 2.1)

---

[1] There is a cultural divide between physicists and biologists about the use of these two words. Physicists shun the word "detergent," which they associate to laundry, and term detergents "surfactants," without any special consideration of their solubilizing properties, which they usually have little use for. Biologists are not always familiar with the term "surfactant" and tend to call "detergents" even non-solubilizing surfactants. In this book, where we will constantly deal with solubilizing and non-solubilizing surfactants, the term "detergent" will be exclusively used to designate the former.

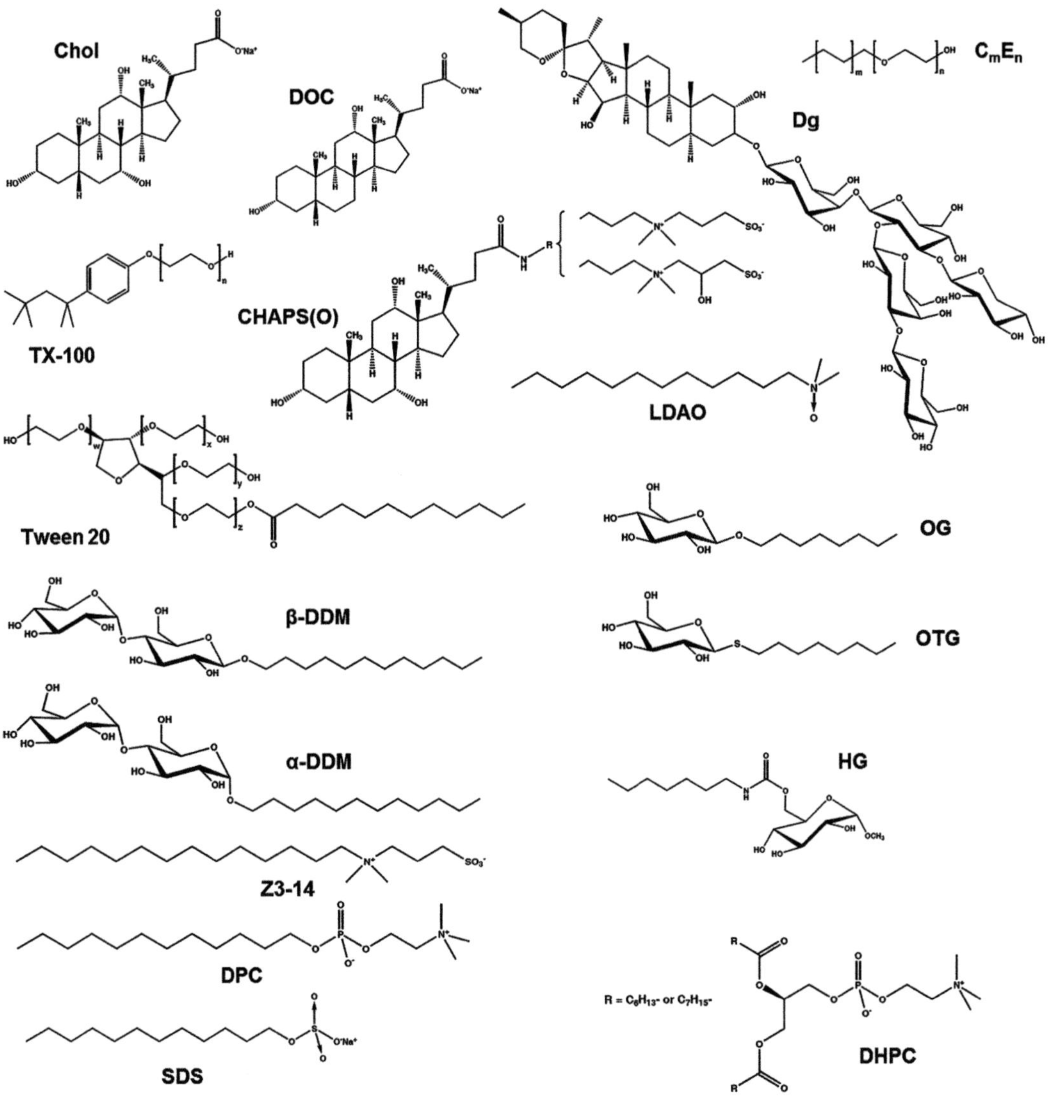

**Fig. 2.1** Chemical structure of some detergents commonly used in membrane biochemistry. CHAPS, 3-[(3-cholamidopropyl)dimethylammonio]-1-propanesulfonate (CMC = 6–10 mM); CHAPSO, 3-[(3-cholamidopropyl)-dimethylamino]-2-hydroxy-1-propane sulfonate (CMC ≈ 8 mM); $C_m E_n$, alkylpoly-oxyethylene, e.g. $C_8E_4$ (CMC ≈ 8.4 mM; Király et al. 1997), $C_{12}E_8$ (CMC = 0.07–0.1 mM), etc.; Chol, sodium cholate (CMC = 9–15 mM); $\alpha$- (or $\beta$-) DDM (= $C_{12}$-M), n-dodecyl-$\alpha$- (or $\beta$-) D-maltopyranoside (CMC ≈ 0.17 mM for $\beta$-DDM, ~0.15 mM for $\alpha$-DDM; Pagliano et al. 2012); Dg, digitonin (low CMC); DHPC, dihexanoyl- or diheptanoyl phosphatidylcholine (di$C_6$PC, CMC = 6–7 mM, and di$C_7$PC, CMC = 0.7–1.0 mM; Tausk et al. 1974); DOC, sodium deoxycholate (CMC = 2–6 mM); DPC, dodecylphosphocholine = fos-choline-12 (CMC ≈ 0.9 mM); HG, Hecameg (CMC ≈ 19.5 mM; Enzo Life Science web site); LDAO, lauryldimethylamine oxide (CMC = 1–2 mM); OG, n-octyl-$\beta$-D-glucopyranoside (CMC = 20–25 mM); OTG, n-octyl-$\beta$-D-thioglucopyranoside (CMC ≈ 9 mM); SDS, sodium dodecylsulfate (CMC = 7–10 mM); TX-100, Triton X-100 (CMC = 0.2–0.9 mM); Z3–Z14, Zwittergent 3–14 (CMC = 0.1–0.4 mM) (note that in lipid and detergent names, "lauryl" is often used for "dodecyl"). Values of CMC are for solutions in water or dilute salt solutions and are from Neugebauer (1988) and Bhairi and Mohan (2007), unless otherwise indicated.

(for reviews, see e.g. Neugebauer 1990; Sadaf et al. 2015). At variance with bilayer-forming lipids, the polar head of detergents, taking into account its dynamics, electrostatic repulsion, etc., is more bulky than their hydrophobic moiety. Upon assembling side-by-side, detergent molecules therefore generate a convex rather than a flat interface with the aqueous solution. As a result, instead of forming extended sheets like membrane lipids do, they aggregate into closed objects called *micelles* (see Fig. 2.3). Depending on the relative size of the hydrophobic and hydrophilic moieties and other molecular properties, micelles can be spherical, oblate (lenticular), prolate (cigar-like), or cylindrical (see Table 1.1, lines 1 and 2, in Chap. 1). The surface of micelles is hydrophilic, their core hydrophobic. They provide the medium into which hydrophobic or amphipathic molecules, such as lipids, can dissolve and become water-soluble.

As illustrated in Fig. 2.1, the polar head of detergents can be ionic and carry a net electric charge, zwitterionic and globally neutral, or uncharged. In the latter case, water solubility results from the presence of hydrogen-bonding groups, such as the ether groups of polyoxyethylene-based detergents (Tritons, Tweens, $C_mE_n$, etc.) or hydroxyl groups, which are often carried by sugar moieties: octylglucoside (OG) and dodecylmaltoside (DDM) are among the most widely used detergents in membrane biochemistry.

Some detergents have more complex structures. $DiC_6$- and $diC_7PC$ (both of them, confusingly, customarily abbreviated DHPC in the literature) have the structure of a phospholipid, with two acyl chains attached to a glycerophosphocholine head group. However, the chains are so short that DHPC molecules behave as detergents, often used in solution NMR investigations of MPs, as well as in bicelle-forming mixtures (for a discussion, see Hauser 2000; see § 2.3.1 and Chap. 3, § 3.2).

Some natural detergents (bile salts such as cholate and deoxycholate and their conjugated derivatives – taurocholate, glycocholate, etc. – digitonin), or partially synthetic ones (CHAPS, CHAPSO), are derived from a steroid, hydrophobic, polycyclic nucleus carrying (i) a polar head and (ii) hydroxyl groups. The latter groups are concentrated on one face of the nucleus, making it laterally amphipathic (Fig. 2.1). As will be described below (§ 2.2.2), this confers to the mixtures of these detergents with lipids in aqueous solutions a very peculiar behavior as compared with less complex detergents. Some modern semisynthetic detergents ("facial detergents") have further built upon this kind of structure (§ 2.5.1).

## 2.2.2  Physical-Chemical Properties

Because of their great theoretical interest as models of self-organizing systems and their extreme practical importance, the physical-chemical properties of aqueous solutions of detergents have been extensively studied (for reviews, see e.g. Tanford 1972, 1980; Israelachvili 2011). In the present section, we will focus on those aspects that are particularly relevant to understanding issues to be discussed in the next chapters.

The aggregation behavior exhibited by detergents in aqueous solutions is schematized in Fig. 2.2. It results from the balance between entropy, which tends to disperse the molecules, and the hydrophobic effect, which tends to bring their hydrophobic tails together (for a very clear introduction to the thermodynamics of micelle formation, see Chap. 7 in Tanford 1980). The distribution of detergent molecules between free monomers and molecules associated into micelles obeys a classic monomer/oligomer equilibrium. However, because micelles usually comprise several tens of molecules (typically 50–100, but for micelles of pure bile salts and their derivatives, which comprise only a few), their formation resembles a phase transition: below a given concentration called the critical micellar concentration (CMC), entropy dominates and micelles are essentially nonexistent. Above the CMC, the concentration of monomers increases only very slowly, almost all new molecules added to the solution associating into micelles. As a result, the concentration of micellar detergent in

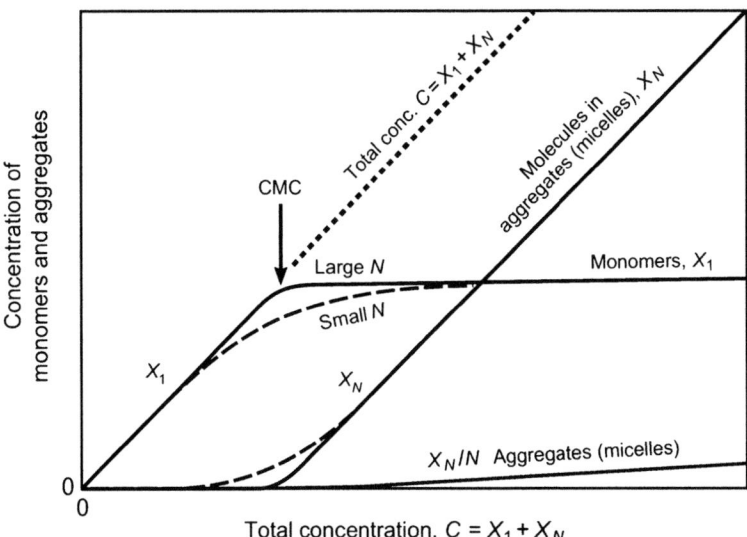

**Fig. 2.2** Evolution of the concentration of monomeric and micellar detergent in an aqueous solution as a function of the total concentration. The larger the aggregation number $N$, the sharper the transition at the CMC (From Israelachvili 2011. © 2011 Elsevier Inc. All rights reserved).

the solution above the CMC is very close to the total concentration minus the CMC (Fig. 2.2). As we will see below (§ 2.4.2), controlling the concentration of micellar detergent in one's samples is a critical factor in optimizing the stability of detergent-solubilized MPs.

Because detergents usually bear only one hydrophobic chain, and it is generally much shorter than lipid acyl chains – the most frequent lengths typically range between 8 and 12 carbons; see Fig. 2.1 – their CMC is much higher than the critical aggregation concentration of lipids: it lies most often in the 0.1–10 mM range (see legend to Fig. 2.1) vs. the nM range for the critical association concentration of lipids (Tanford 1980). The CMC is a critical factor when planning biochemistry experiments. It will determine, among others, what is the minimal concentration of detergent to use and by which techniques the detergent can be eliminated or exchanged: low-CMC detergents, for instance, cannot be easily eliminated by dialysis, because the concentration of monomers, the only species that crosses the membrane, is always low. This point will come up again when we will discuss the various ways a MP can be transferred from a detergent solution to APols (Chap. 5, § 5.2).

For nonionic detergents, the CMC is mostly determined by the length of the hydrophobic chain: the CMC of $n$-octyl-$\beta$-D-glucoside (OG), 20–25 mM in water, is similar to that of $C_8E_4$, ~8 mM, despite the wide dissimilarity of the polar heads (Fig. 2.1). A useful rule of thumb is that the CMC drops by a factor of ~10 for every pair of methylene groups added to the hydrophobic chain. Thus, $n$-dodecyl-$\beta$-D-maltoside (DDM or $C_{12}$-M) sports four more methylene groups than OG (Fig. 2.1), and its CMC is two orders of magnitude lower, ~0.16–0.19 mM in water. The polar head of DDM is twice bigger than that of OG. Because it is hydrated in both the monomeric and the micellar states, the size of the head does not affect the CMC very much (it does affect the shape of the micelles, though), but it determines the solubility of detergents in water: detergents with more hydrophobic tails require more hydrophilic heads.

As a rule, the solution properties of nonionic detergents are not very sensitive to experimental conditions, such as ionic strength or temperature. However, detergents with a polyoxyethylene (POE) polar head can be sensitive to temperature: raising the temperature favors the dehydration of POE, so that the head groups become less soluble and start interacting one with another. This brings about a phase separation between a micelle-poor and a micelle-rich phase. When the transition ("cloud point") occurs at a relatively low temperature, as for the Triton X-100 analog Triton X-114 (22 °C), this property can be used as a separation method, MPs accumulating in the detergent-rich phase. When working close to the CMC (see § 2.4.2, why one may want to do so), one should remain aware that changes in ionic strength and temperature can affect the CMC of even nonionic detergents (see e.g. Miyagishi et al. 2001; Molina-Bolívar et al. 2013, and references therein) and place one in the unpleasant situation of trying to work, unknowingly, at submicellar concentrations, with catastrophic results (the protein aggregates and, as a rule, precipitates).

When the polar heads carry charges, their electrostatic repulsion in micelles increases the CMC. This effect is already present for zwitterionic detergents, even though they carry no net charge: lauryldimethylamine oxide (LDAO), for instance, which carries the same alkyl chain as DDM but whose polar head bears partial charges (Fig. 2.1), has a CMC of 1–2 mM in water, 5–10× higher than that of DDM. The effect is much more pronounced for detergents carrying a net charge, like sodium dodecylsulfate (SDS). In pure water, the CMC of SDS is ~7–10 mM, almost two orders of magnitude above that of DDM, despite their bearing the same alkyl chain. As can be expected, the CMC of detergents with a charged polar head is very sensitive to the ionic strength: in 100–200 mM NaCl, the CMC of SDS drops to 1–2 mM (CMC data are from Neugebauer 1988, 1990; Király et al. 1997).

Micelles are not the nice well-ordered spherical assemblies shown in most cartoons:

- First, they are not always spherical. Depending on the shape of the molecule, they can be roughly spherical, ellipsoid – either prolate or oblate – or cylindrical, the relative size or apparent size (think e.g. of the modulation of the latter by the temperature or ionic strength) of the polar head and hydrophobic tail playing a major role in determining the average form (cf. Chap. 1, Table 1.1, rows 1 and 2). Apparently minor details of chemical structure, such as the head/tail orientation in the $\alpha$ vs. $\beta$ anomers of DDM (Fig. 2.1), can have important effects on the size and shape of the micelles (Abel et al. 2011).
- Second, micelles are not regular, and the hydrophobic moieties of the detergent molecules are not always totally shielded from water (cf. Fig. 2.3, *right*).

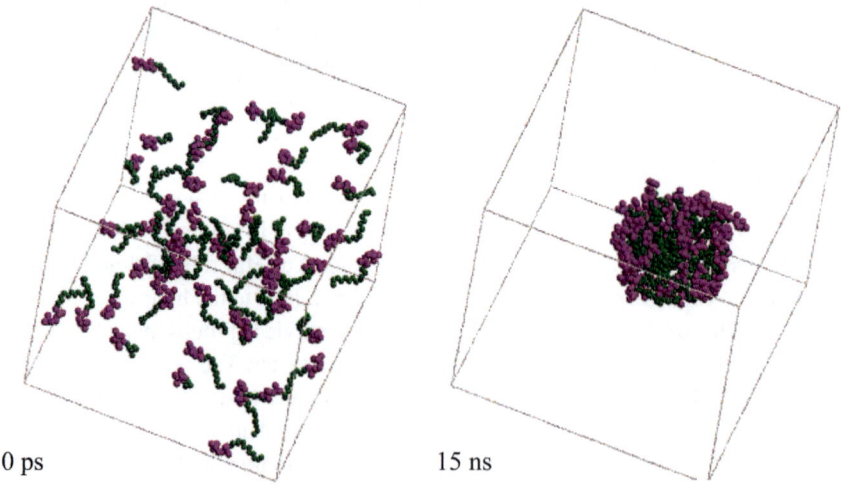

0 ps                                              15 ns

**Fig. 2.3** Spontaneous assembly of 54 molecules of dodecylphosphocholine (DPC; see Fig. 2.1) into a spherical micelle as seen in a molecular dynamics simulation. Snapshots are shown of the beginning and end of a 15 ns simulation. Head groups are drawn in *purple*, hydrophobic tails in *green*. Water is omitted for clarity. Note how, in the micelle, some hydrophobic groups are exposed to water (Reprinted with permission from Marrink et al. 2000, © 2000 American Chemical Society).

- Third, they are in a highly dynamic state: individual molecules deform on the 10–100 ps timescale, and they exchange their positions on the ns one (see e.g. Bogusz et al. 2001).
- Fourth, they are not of uniform size but distribute around an average one (see Tanford 1980).
- Fifth, they are in permanent exchange with the free monomers in the solution. In an OG micelle, for instance, the residency time of individual molecules is ~7 ns (Frindi et al. 1992). Given that the average micelle comprises ~92 molecules, this means that each ns more than 10 molecules leave the micelle, and, as the system is at equilibrium, as many enter it from the solution. The rate of exchange being essentially determined by the free energy cost of extracting the hydrophobic tail from the micelle core and exposing it to water, it drops by ca. one order of magnitude for each couple of methylene groups added to the tail. For DDM, one can therefore expect the residency time to be in the µs range (see Frindi et al. 1992). A DDM micelle, which comprises 110–140 molecules (le Maire et al. 2000), takes up and releases one every 10 ns or so.

*Taking membrane proteins out of their natural environment*
(© *2018 by Francis Haraux*)

Bile salts, like cholate and deoxycholate, and their semisynthetic derivatives, like CHAPS and CHAPSO (Fig. 2.1), have a more complex micellar behavior, because of their double amphipathic character: not only do they feature a longitudinal amphipathy, with a globally hydrophobic polycyclic ring and a polar head, but the ring itself is laterally amphipathic, due to the presence of hydroxyl moieties on one side only. The rings of several molecules therefore tend to interact side by side, forming small micelles that progressively grow in size (Carey and Small 1972). For the same reason, the mixed micelles these detergents form with lipids have a very peculiar arrangement (see § 2.3.1).

## 2.3   Solubilizing Membrane Proteins with Detergents

### 2.3.1   Solubilizing Biological Membranes with Detergents

The use of detergents to solubilize membrane proteins has long remained highly empirical. It is only since the beginning of the 1970s that a merge took place between the practical observations of biologists and the formalism and methods of investigation of physical chemists, to the great benefit of each party: biologists have introduced physical rationality in their use of detergents, and physical chemists and chemists have introduced biological objects in their experimental systems and mode of thinking. A nod is due to Charles Tanford, on the one hand, and to Kai Simons and Ari Helenius, on the other, for the key roles they played in this evolution (Tanford 1972, 1980; Helenius and Simons 1975; Tanford and Reynolds 1976; Helenius et al. 1979). Since then, a large number of reviews have been published on the use of detergents in membrane biology, only a few of which can be cited here (see e.g. Neugebauer 1990; Zulauf 1991; le Maire et al. 2000; Bowie 2001; Garavito and Ferguson-Miller 2001; Chevalier 2002; Gohon and Popot 2003; Seddon et al. 2004; Wiener 2004; Privé 2007; Arnold and Linke 2008; Duquesne and Sturgis 2010; Tate 2010; Zhang et al. 2011; Arachea et al. 2012; Lichtenberg et al. 2013; Otzen 2015; Sadaf et al. 2015; Champeil et al. 2016; Orwick-Rydmark et al. 2016). The following somewhat schematic presentation draws on this literature, as well as on the personal experience gathered over some 40 years of biochemical and biophysical work on a range of widely different membrane proteins, each of them raising its own specific problems.

Figure 2.4 schematizes the behavior of a system comprising, originally, a suspension in aqueous solution of biological membrane fragments, to which is added an increasing amount of detergent. Six steps are distinguished, depending on the detergent/membrane mass ratio reached.

In Step 1, the detergent is submicellar. It distributes between the membrane and the aqueous solution as a function of its partition coefficient, a critical parameter:

- Detergents bearing a highly hydrophobic chain have a partition coefficient that favors the membrane, for the same reason that they have a low CMC, namely the high free energy cost associated with exposing this chain to water. At a classical protein concentration of ~1 $g \cdot L^{-1}$, such detergents, e.g. DDM or Triton X-100, partition mainly in the membrane fragments. Their effect will therefore depend on the detergent/lipid mass ratio more than on the absolute

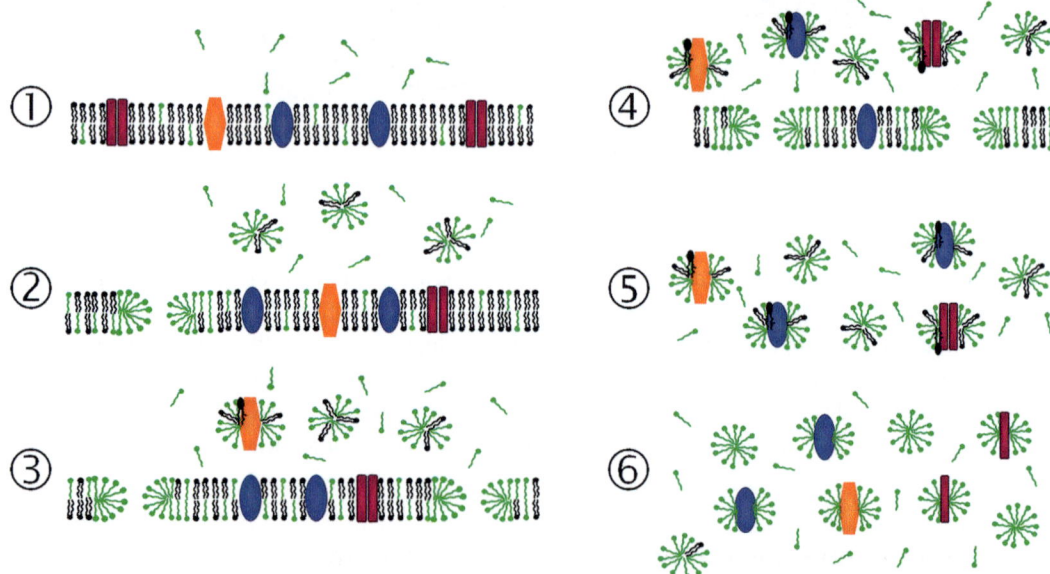

**Fig. 2.4** A schematic representation of the effects of adding increasing amounts of detergent to an aqueous suspension of membrane fragments. *Black*, membrane lipids; *green*, detergent molecules; *red*, *yellow*, and *blue*, three different transmembrane proteins. Steps 1–6, corresponding to increasing detergent/membrane mass ratios, are discussed in the text.

detergent concentration. It will, by the same token, be strongly dependent on the concentration of membranes in the sample, which must be kept constant from one experiment to the next if the solubilization conditions are to be reproducible.

- High-CMC detergents, e.g. OG or CHAPS, which have less hydrophobic tails, will tend, on the contrary, to partition in the aqueous phase more than in the membrane one. Their effect will therefore depend on the absolute concentration of detergent more than on that of membrane fragments, which bind little detergent.

Whatever the detergent, not much may be observed at that stage, but some extrinsic MPs may detach from the membrane, and the detergent that partitions into it may affect the function of some intrinsic proteins (for a careful analysis of why different detergents seem to have very different functional effects at that stage, whereas they actually act at similar molar ratios in the membrane, see the examination of the dependence of the inhibition of the $Na^+,K^+$-ATPase on detergent concentration in Brotherus et al. 1979).

At step 2, the concentration of detergent is high enough for mixed micelles to form. Because these micelles are not pure detergent but incorporate lipids extracted from the membrane, they are in equilibrium with a concentration of free detergent molecules slightly lower than the CMC. Most often, pores form in the membrane, which has two important consequences. First, the membrane becomes leaky, which may affect functional measurements (see e.g. in Chap. 5, § 5.6.1, the case of the sarcoplasmic $Ca^{2+}$-ATPase). Second, the detergent can use the rim of the pores to flip from the external leaflet of the membrane fragments – which usually form closed vesicles – to the inner leaflet. This is an important step, as membrane solubilization by detergents that do not undergo this process can be extremely slow (see e.g. Kragh-Hansen et al. 1998).

The ability of detergents to destabilize the target membrane and extract different types of lipids can vary widely from one detergent to another (see e.g. Banerjee et al. 1995). In some cases, the process may stop at stage 1 or 2. For instance, purple membrane fragments exposed to deoxycholate yield a contracted form of the native 2D BR crystals (see § 1.6.1), from which the lipid layer that ensures the contacts between BR trimers in the native membrane has been extracted (Henderson et al. 1982; Glaeser et al. 1985).

Mixed micelles are generally disordered. However, in the case of bile salts, the peculiar molecular structure of the detergent, whose hydrophobic moiety is itself laterally amphipathic (see § 2.2.1), causes the formation of disc-like micelles in which the detergent covers the rim of a small patch of bilayer, into which it partitions (see Carey and Small 1972; Marrink and Mark 2002) – a structure similar to that proposed for the bicelles formed by mixtures of lipids and DHPC, CHAPS, or CHAPSO (see e.g. Sanders and Landis 1995; Triba et al. 2005; Poget and Girvin 2007; Kim et al. 2009, and Chap. 3, § 3.2).

Step 3, sometimes very useful from a practical point of view, may see a selective extraction of some MPs. It may provide an opportunity to enrich one's sample with the target protein, whether in the supernatant or in the pellet. MPs solubilized at that stage find themselves associated with a lipid-rich mixture of detergent and lipids, which, as discussed below, is a stabilizing factor.

At step 4, a major fraction of the membrane is solubilized, at step 5, all or nearly all of it. Some detergents stop at step 4, whereas the "strongest" (most disruptive) ones reach step 5, the effect of course depending on the detergent's nature, its concentration and its mass ratio to the membrane fragments, the nature of the latter, but also on such factors as the temperature and buffer conditions: ionic strength, presence or not of calcium chelators, pH, redox potential, etc. If one's target protein is extracted quantitatively enough at step 4, there is no need to push the solubilization further, at the risk of inactivating it. Indeed, as more and more detergent is supplied, the mixture of detergent and lipids associated with the solubilized MPs becomes increasingly poorer in lipids, which is generally to be avoided (§ 2.4).

Step 6 sees the extracted proteins and lipids further diluted with detergent, which is most often detrimental: MPs become increasingly delipidated, oligomers may come apart, etc. (§ 2.4).

## 2.3.2  Membrane Protein/Detergent Complexes

The organization of MP/detergent complexes has long remained speculative. It was anticipated that a belt of detergent would adsorb onto the transmembrane (TM) region of MPs, ensuring the interface with the aqueous solution and making the proteins water-soluble. This view was supported by many indirect observations, such as measurements of the amount of bound detergent as a function of the size of the protein's TM region or radiation scattering solution studies of MP/detergent complexes (reviewed in le Maire et al. 2000). This model was directly validated by studying by neutron diffraction 3D crystals of MPs grown from detergent solutions. In X-ray diffraction experiments, detergents contrast poorly with the surrounding solution: the occasional well-organized molecules of detergent are hard to identify (see Chap. 1, § 1.5.2), whereas disordered bulk detergent is invisible. Neutron scattering and diffraction, however, can play on scattering length contrast (see Chap. 9, Box 9.2) to reveal specific components. At 40% $D_2O$, proteins are contrast matched and contribute very little to the diffraction pattern, whereas the hydrogenated hydrophobic tails of detergents contrast strongly with the solution. Specific deuteration, when it can be implemented (as regards detergents, see Hiruma-Shimizu et al. 2015), can be used to enhance or lower the contrast of chosen components or regions. This approach has been used to determine the distribution of the detergent in MP 3D crystals (Roth et al. 1989, 1991; Pebay-Peyroula et al. 1995; Penel et al. 1998; Prince et al. 2003; Snijder et al. 2003).

As an example, Fig. 2.5B shows an excerpt from a study in which either unlabeled OG or decyl-$N,N'$-dimethyl amine oxide with a deuterated alkyl chain was used to crystallize trimeric OmpF porin.

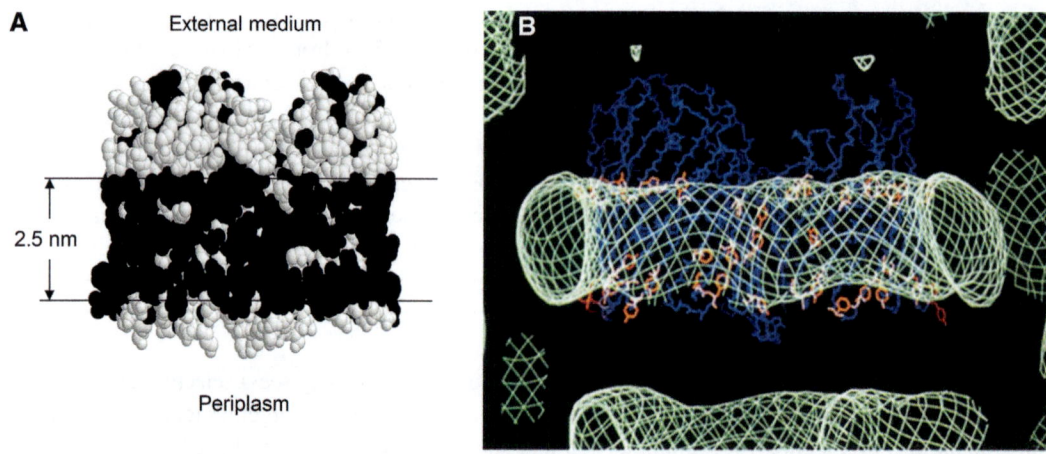

**Fig. 2.5** Detergent binding to the transmembrane region of trimeric OmpF in three-dimensional crystals. (**A**) Atomic structure of the outer membrane trimeric protein OmpF (PDB accession code 2OMF; based on data in Cowan et al. 1992). Strongly hydrophobic and aromatic amino acid side chains (Leu, Ile, Val, Ala, Met, Pro, Phe, Tyr, Trp), shown in *black*, form a ~2.5-nm high belt, which, in situ, faces the hydrophobic interior of the membrane (*horizontal lines*) (From Gohon and Popot 2003). (**B**) The distribution of octyl-$\beta$-D-glucoside hydrophobic tails around trimeric OmpF, as determined by neutron crystallography (*green cage*), fits exactly into the area delimited by the rings of aromatic side chains (*orange*) at the hydrophobic/polar boundaries of OmpF's surface, which probably correspond to the limits of the acyl chains of the lipid membrane in situ. The porin polypeptide backbone is shown in *dark blue* (From Pebay-Peyroula et al. 1995, © 1995 Elsevier Ltd. All rights reserved). The composite figure is from Gohon and Popot (2003) © 2013 Elsevier Ltd. All rights reserved.

The crystals were then studied by neutron diffraction at different contrasts, so as to determine the relative arrangements of the protein and detergent (Pebay-Peyroula et al. 1995). The contrast match point of detergent polar heads (i.e. the $D_2O/H_2O$ ratio at which their scattering length density is equal to that of the solvent and their contribution to neutron scattering or diffraction vanishes; see Chap. 9, Box 9.2) depends on their chemical composition and fraction of exchangeable hydrogens. In the case of the polar head of OG, its contrast match point lies at ~52% $D_2O$, so that at 40% $D_2O$, its contrast with the solution is weak, and it is mainly the octyl chains that contribute to building up the diffraction pattern (see Timmins et al. 1994). Figure 2.5B shows the belt of OG surrounding the TM region of OmpF in tetragonal crystals. This distribution is compared, in Fig. 2.5A, with that of hydrophobic and aromatic residues (in *black*) at the surface of the protein (based on data in Cowan et al. 1992). These residues form a belt ~2.5 nm high, which correspond to the expected thickness of the acyl-chain region in *Escherichia coli*'s outer membrane. Comparison of the two sets of data establishes that, as expected, the detergent substitutes for the lipids at the hydrophobic TM surface of the protein, thus forming with it a water-soluble complex. The same conclusion has been reached in a number of other neutron diffraction studies using other MPs, detergents, and/or crystal forms (see Roth et al. 1989, 1991; Penel et al. 1998; Prince et al. 2003; Snijder et al. 2003), as well as in X-ray diffraction studies in which the detergent belt was made visible by loading it under high pressure with xenon or krypton (Sauer et al. 2002) or by increasing the electron density of the solvent (Norimatsu et al. 2017).

MD simulation data complement our view of the organization of MP/detergent complexes. They indicate that, within the detergent belt, the arrangement of detergent molecules can be extremely different from one detergent to the next. Thus, in complexes between detergents and the BM2 protein from the influenza B virus, most DDM molecules tend to lie with their long axis parallel to the TM surface of the protein (Fig. 2.6C, C'), as do the acyl chains of dipalmitoyl PC in parallel simulations (Fig. 2.6A, A'), whereas DHPC (diC₆PC) molecules stand mostly on end, with their terminal methyl groups in contact with the protein (Fig. 2.6B, B') (Rouse and Sansom 2015). The arrangement of DHPC is reminiscent of that postulated to form the rim of bicelles (see Chap. 3, § 3.2).

MD simulation data are consistent with X-ray data (Chap. 1, Fig. 1.22C) in showing detergent alkyl chains taking the place of lipid acyl ones (Fig. 2.7).

The fact that detergents substitute to the membrane environment by adsorbing onto the TM surface of MPs is therefore well established. Unfortunately, it is equally well established that a MP solubilized in detergent solution is, as a rule, much less stable than it is in its natural environment.

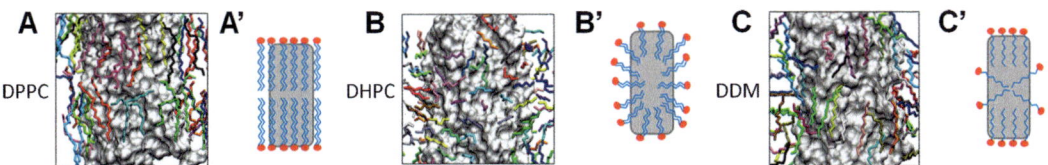

**Fig. 2.6** Interaction of lipid (dipalmitoylphosphatidylcholine; DPPC) or detergent – either DDM or dihexanoylphosphatidylcholine (DHPC) – with the transmembrane surface of the BM2 protein from influenza B virus in MD simulations. (**A–C**) Simulation results. The protein is displayed as a gray surface. Hydrophobic chains of the lipid and detergents are shown in stick representation (colored per molecule). The positions of the tails are the average positions adopted over the final 10 ns of each simulation performed at 323 K. (**A'–C'**) A (highly schematic) representation of the modes of interaction of each surfactant with the surface of the protein (Adapted from Rouse and Sansom 2015).

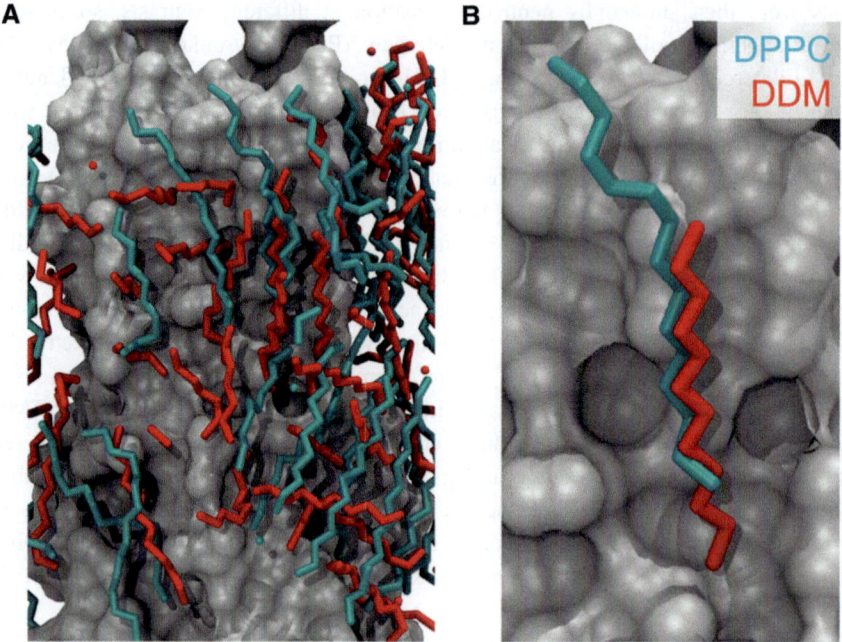

**Fig. 2.7** Interaction of the hydrophobic chains of DPPC and DDM with the surface of the BM2 protein as seen in MD simulations. (**A**) Superimposition of the most highly occupied positions of the hydrophobic chains of DPPC (*cyan*) and DDM (*red*) during 10 ns of simulation. (**B**) An example of the same region of the protein surface occupied by a hydrophobic tail in the two simulations. The protein is displayed as a gray surface in each case (From Rouse and Sansom 2015).

## 2.4    Why Are Membrane Proteins Unstable in Detergent Solutions?

### 2.4.1    Instability of Detergent-Solubilized Membrane Proteins Is a General Phenomenon

It is a sad fact of life, and the curse of the membrane biochemist, that most MPs start inactivating, more or less rapidly, as soon as they are solubilized.

This is illustrated in Fig. 2.8, taking BR as an example (for an introduction to BR, see Chap. 1, § 1.6.1). When it is part of the purple membrane, its native environment, BR is extraordinarily stable (see e.g. Brouillette et al. 1987): a suspension of purple membrane in water can be heated for 20 min at 70 °C without any loss of the absorbance at ~554 nm that is characteristic of the native holoprotein (Fig. 2.8A). On the contrary, once solubilized in *n*-octyl-$\beta$-D-thioglucoside (OTG; see Fig. 2.1), BR is stable only until ~30 °C, some signs of denaturation appearing already at 40 °C: loss of absorbance at 554 nm and apparition of an absorbance peak, due to free retinal, around 380 nm, and of some turbidity, due to the aggregation of the denatured protein (Fig. 2.8B). As expected, this increased sensitivity to thermal denaturation translates into a shorter shelf life at constant temperature: whereas purple membrane can be kept nearly indefinitely at room temperature, in the dark, as a suspension in water, OTG-solubilized BR loses almost 30% of its absorbance at 554 nm after 1 day and is nearly totally denatured after 6 days (Fig. 2.8C).

The case of BR is somewhat extreme given the extraordinary stability of this protein in its native membrane, but, because of it, it is particularly telling. This behavior is very general. Very few MPs can

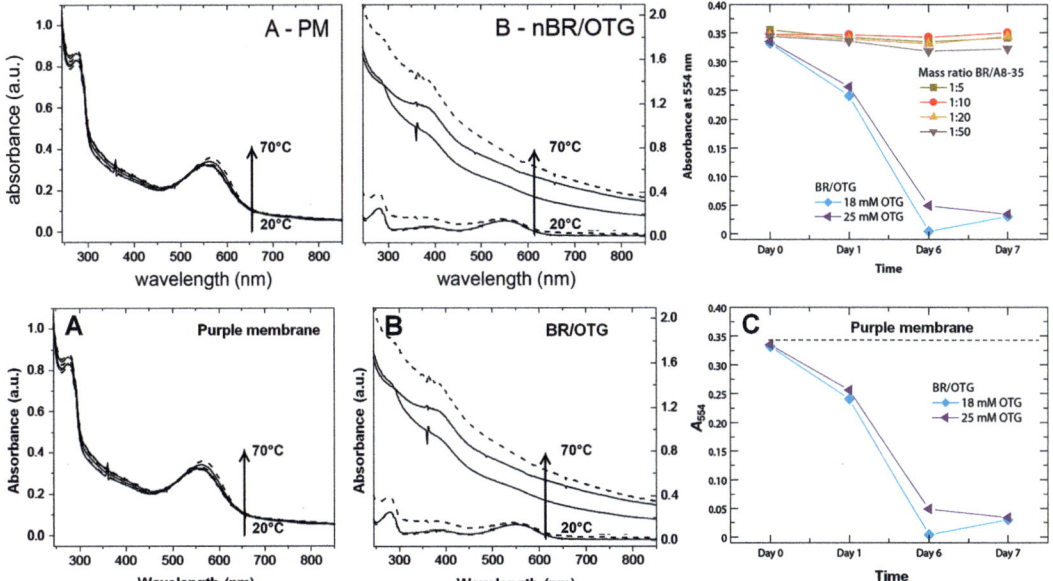

**Fig. 2.8** Destabilization of bacteriorhodopsin (BR) following purple membrane (PM) solubilization with octylthioglucoside (OTG). (**A, B**) Thermal denaturation of BR (**A**) in its native environment (in water) and (**B**) after solubilization in OTG (final OTG concentration, 18 mM, 100 mM NaCl, 25 mM sodium phosphate buffer, pH 7.0). UV-visible spectra were recorded after incubation for 20 min at the temperature indicated. They are represented by alternate solid and dashed lines at 10 °C intervals from 20 to 70 °C (From Dahmane et al. 2013). (**C**) Time stability of BR in OTG. PM was solubilized with OTG and stored in the dark at room temperature at the OTG concentrations indicated in the same buffer as above ([BR] = 0.22 g·L$^{-1}$). The absorbance at 554 nm, which is proportional to the concentration of the holoprotein, was followed as a function of time. The *dashed line* is a reminder that, when PM is stored under the same conditions, BR is perfectly stable (From Dahmane 2007; Popot et al. 2011).

be extracted with detergents without becoming unstable (bacterial outer membrane porins providing some exceptions). Membrane biochemists, as a rule, spend months if not years trying to identify the detergent and conditions that will endow their target protein with a modicum of stability, and, nevertheless, insufficient stability is a recurrent source of limitations as to which experiments can be carried out. To take two examples, crystallization attempts become very difficult if the target protein is not stable for at least a few days, preferably weeks, in concentrated solutions kept at 4 °C. Solution NMR measurements are generally carried out at room temperature or above it, to increase the tumbling rate of the MP/detergent complexes and improve the resolution of the spectra, and they also require concentrated solutions. They therefore consume huge amounts of costly isotopically labeled material if the samples must be discarded after only a few hours of measurements.

Many explanations have been proposed for the inactivating character of detergents. Some of them involve the loss of the physical constraints applied to the protein by the membrane environment (e.g. the gradient of lateral pressure; see Chap. 1, § 1.2). Whereas no experimental proof of this proposal has been provided, molecular dynamics (MD) simulations do indicate that the dynamics of MPs can increase slightly (typically by a factor $\leq 1.5\times$) following transfer from a lipid to a detergent environment (see e.g. Bond and Sansom 2003; Patargias et al. 2005; Perlmutter et al. 2014; Rouse and Sansom 2015; for a dissenting view, see Frey et al. 2017 and, about the lesser ordering of amino acid side chains exposed to a detergent rather than a lipid environment, Hite et al. 2008; see also Hong and Bowie 2011). There are good reasons to think that faster conformational dynamics might indeed favor denaturation and that one of the mechanisms by which APols stabilize MPs as compared to detergents is to damp this dynamics (see Chap. 5, § 5.6). It is sometimes argued, without, to my knowledge, any experimental demonstration having ever been brought, that fluctuations in the protein-adsorbed detergent belt, by leading some regions of the TM surface of MPs to become transiently exposed to water, would be destabilizing. It is somewhat hard to think of this as a cause of denaturation, because the time scale on which detergent molecules exchange positions is very short (ns; see § 2.2.2 and Bond et al. 2006), leaving no time for the protein to react to such events by more than transient side-chain reorientations.

My personal prejudice, based on years of experience fighting MP instability, is that, whereas some of these physical effects may well contribute to it, the major factor is probably the dissociating character of detergents, namely their ability to compete with stabilizing protein/protein and protein/lipid interactions (Gohon and Popot 2003; Popot 2010). I will illustrate this point with some experiments carried out using the cytochrome $b_6f$ complex as a model system. As a workhorse for developing stabilizing methods, the $b_6f$ complex is an excellent if somewhat frustrating MP to work with, because of its extreme sensitivity to detergents.

### 2.4.2　A Field Case: Inactivation of the Cytochrome $b_6f$ Complex by Detergents

Cytochrome $b_6f$ is a complex from the photosynthetic membrane of plants, green algae, and cyanobacteria. It transfers electrons from the two-electron, lipid-soluble donor plastoquinol ($PQH_2$), which has been reduced by photosystem II, to the one-electron, water-soluble acceptor plastocyanin, which will deliver them to photosystem I (for a recent review, see e.g. Saif Hasan et al. 2013). In so doing, cytochrome $b_6f$ contributes to building up the proton electrochemical potential that powers the synthesis of ATP by the $F_1F_O$-ATPase.

The $b_6f$ complex, as illustrated here by that from the fresh-water unicellular alga *Chlamydomonas reinhardtii*, is a superdimer, each monomer comprising eight different subunits (Pierre et al. 1995; Stroebel et al. 2003) (Figs. 2.9 and 2.13), three of which, cytochrome $b_6$, cytochrome $f$,

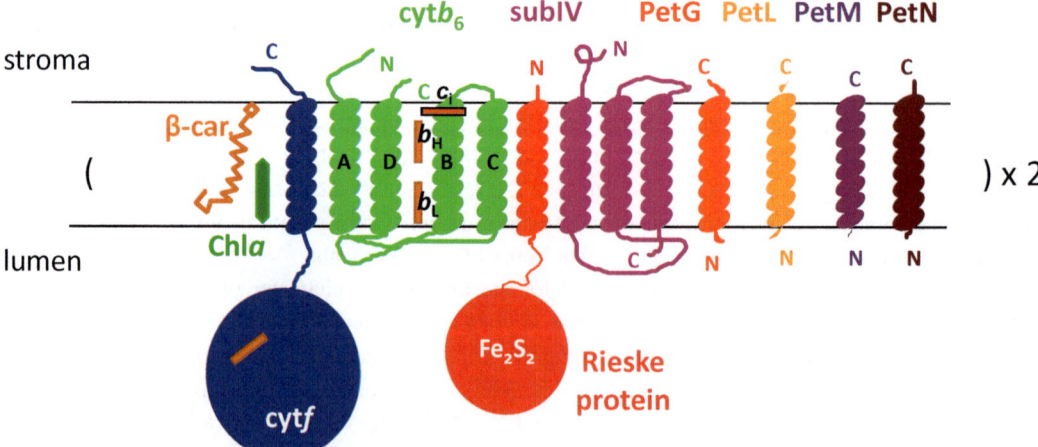

**Fig. 2.9** Subunit and cofactor composition and transmembrane organization of the cytochrome $b_6f$ complex from *Chlamydomonas reinhardtii*. The relative position given to the various components of the complex is arbitrary and does not reflect that in the 3D structure. $\beta$-car, $\beta$-carotene; Chl$a$, chlorophyll $a$ (Based on data in Pierre et al. 1995; Stroebel et al. 2003).

and the so-called Rieske iron-sulfur protein, are directly involved in electron transfer. Cytochrome $b_6$ is a polytopic MP, with four TM $\alpha$-helices, which carries two $b$-type and one $c$-type hemes; cytochrome $f$ and the Rieske protein are bitopic MPs, each of them featuring a single TM helix and a functional domain, located in the lumen of the chloroplast thylakoids, which carries, respectively, a $c$-type heme and an $Fe_2S_2$ iron-sulfur center. The functional cycle of the $b_6f$ complex is somewhat complicated and need not be described in detail here. For the sake of the present discussion, it is sufficient to say that one of its branches involves the transfer of electrons in the following sequence: $PQH_2 \rightarrow b_6 \rightarrow$ Rieske $\rightarrow f \rightarrow$ plastocyanin. Whereas all electron transfers involve tunneling, that from $b_6$ to $f$ via the $Fe_2S_2$ center of the Rieske protein is peculiar, because it requires a physical movement of the Rieske protein's extramembrane domain from a site on cytochrome $b_6$ to one on cytochrome $f$.

Early attempts at purifying the complex from *C. reinhardtii* revealed a great instability in detergent solutions (Pierre et al. 1995; Breyton et al. 1997). Indeed, exposure to detergent leads to the detachment of the Rieske protein from the complex – and, therefore, to complete inactivation – followed by the disaggregation of the superdimer into supermonomers that have lost both the Rieske protein, one of the small subunits, PetL, and a cofactor, chlorophyll *a*. This is illustrated in Fig. 2.10A, gradient to the right, and Fig. 2.10B, lower series of SDS-PAGE lanes. The various steps experimentally identified in the progressive disaggregation of the *C. reinhardtii* complex are schematized in Fig. 2.10C.

Two ways around this problem were identified:

- A first approach is to expose the complex to a minimal concentration of detergent. This is illustrated in Fig. 2.10A, gradient to the left, and Fig. 2.10B, upper row of lanes: decreasing the DDM concentration from 3 or 5 to 0.2 mM, which is very close to the CMC of ~0.17 mM, slowed down the disaggregation process sufficiently for the complex to stay mostly intact during the 4 h (Fig. 2.10A) or 3 h (Fig. 2.10B) of ultracentrifugation.
- A second approach, illustrated in Fig. 2.11, is to supplement the detergent solution with lipids. The detergent used in these experiments was Hecameg (HG), a short-chain analog of octylglucoside (Fig. 2.1) with a high CMC of ~19.5 mM under the conditions used. Under usual experimental conditions, the concentration of HG was kept at 20 mM, very close to the CMC. In the experiments of Fig. 2.11, it was deliberately raised to very high concentrations, either 50 or 100 mM, resulting in the rapid loss of electron transfer activity, which, after 30 min at 4 °C, fell to ~25% or ~16% of its initial value, respectively. The inactivation process could be considerably slowed down by supplementing the preparations with egg phosphatidylcholine (egg PC) at a 1:10 molar ratio of lipid to micellar detergent and even more so when raising this ratio to 1:5 (Fig. 2.11).

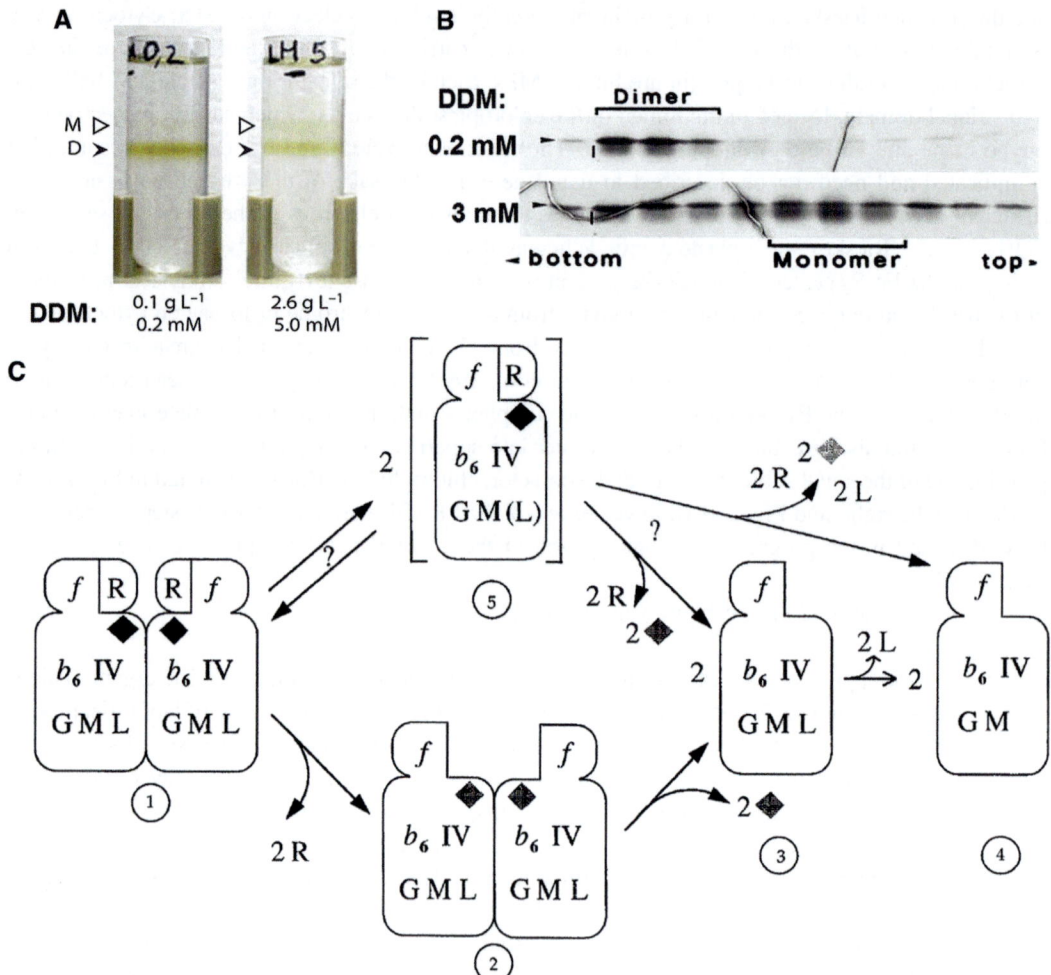

**Fig. 2.10** Behavior of the cytochrome $b_6f$ complex in 10–30% sucrose gradients containing $n$-dodecyl-$\beta$-D-maltoside (DDM) at concentrations either just above its CMC of ~0.17 mM or well above it. (**A**) In 0.2 mM DDM, the complex remains primarily a superdimer (D), whereas in the presence of 5.0 mM DDM, it starts to fragment into monomers (M) (From Breyton et al. 2009). (**B**) Similar experiments were carried out in the presence of either 0.2 mM or 3 mM DDM. Fractions were collected and analyzed by SDS-PAGE and the gels stained with silver. The areas shown correspond to the position of migration of the Rieske protein (sharp band marked with an *arrowhead*) and of cytochrome $b_6$ (fuzzy band marked with a *vertical bar*). In the presence of 0.2 mM DDM, the superdimer remains intact, and the Rieske protein and $b_6$ are found in the same fractions. In the presence of 3 mM DDM, the superdimer fragments into free Rieske protein, which trails throughout the gradient, and an essentially Rieske-less supermonomer. (**C**) Steps in the detergent-induced disaggregation of *C. reinhardtii* $b_6f$ complex. All forms represented have been observed experimentally but for the hypothetical intermediate ⑤. R, G, M, and L refer to the Rieske, PetG, PetM, and PetL subunits, respectively (neither the existence of PetN nor the transmembrane character of the Rieske protein had been recognized at the time of these experiments); a *dark-gray diamond* symbolizes the chlorophyll *a* molecule now known to be associated with cytochrome $b_6$ (Stroebel et al. 2003) (This figure was originally published in Breyton et al. 1997, © The American Society for Biochemistry and Molecular Biochemistry).

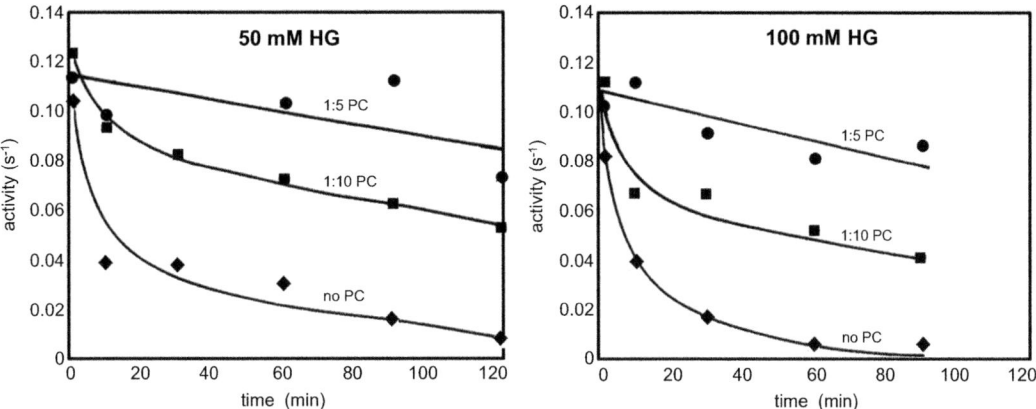

**Fig. 2.11** Kinetics of inactivation of the $b_6 f$ complex incubated with high concentrations of detergent supplemented or not with lipids. The purified complex was incubated with either 50 mM or 100 mM of Hecameg (HG; CMC $\approx$ 19.5 mM under the conditions used) containing either no added lipids (♦) or a molar ratio of egg phosphatidylcholine (PC) to micellar detergent of 1:5 (●) or 1:10 (■). The electron transfer activity from plastoquinol to plastocyanin was followed as a function of time (This figure was originally published in Breyton et al. 1997, © The American Society for Biochemistry and Molecular Biochemistry).

*Stabilizing solubilized membrane proteins by supplementing detergents with lipids*
(© *2018 by Francis Haraux*)

### 2.4.3 The Dissociating Character of Detergents as a Major Cause of Membrane Protein Inactivation

Such observations are by no means peculiar to the $b_6 f$ complex. They are, actually, more the rule than the exception. Thus, the functional reconstitution of the purified nicotinic acetylcholine receptor (nAChR; introduced in Chap. 1, § 1.6.2) into lipid vesicles remained unsuccessful for many years for want of supplementing with lipids the detergent solutions used for its purification (Huganir et al. 1979; Heidmann et al. 1980; Popot et al. 1981). Delipidation has been shown to be deleterious to the sarcoplasmic reticulum calcium pump SERCA1a (described in § 1.6.3) (Lund et al. 1989), to the serotonin $5\text{-HT}_{1A}$ G protein-coupled receptor (Banerjee et al. 1995), or to the red cell glucose transporter Glut1 (Haneskog et al. 1996). Achieving a decent stability is of course critical for crystallization attempts. The $b_6 f$ complex from *C. reinhardtii* could be crystallized only by working very close to the CMC of DDM and minimizing the number and duration of the purification steps, so that crystallization drops were set up on the day of purification, leaving as little time as possible for fragmentation to set in (Stroebel et al. 2003). Similar precautions against excessive delipidation were essential to crystallizing rhodopsin (Palczewski et al. 2000). Crystallizing the $b_6 f$ complex from the thermophilic cyanobacterium *Mastigocladus laminosus* required the addition of lipids (Kurisu et al. 2003), as was also the case for SERCA1a (Toyoshima et al. 2000) or for the lactose permease from *E. coli* (Guan et al. 2006). Inactivation of the shark rectal gland $\text{Na}^+,\text{K}^+\text{-ATPase}$ by $C_{12}E_8$ is critically dependent on the concentration of detergent (Esmann 1986). Such examples could be multiplied (see Hunte and Richers 2008 and references therein). As a general rule, the use of detergents with a low CMC makes it easier to limit the volume of free micelles, which act as a hydrophobic sink, pushing association/dissociation equilibria in the direction of dissociation.

These observations call for some comments:

- Supplementing a detergent solution with substantial amounts of lipids, as in the experiments of Fig. 2.11, can be expected to change somewhat the physical properties of the belt of surfactant the MP is associated with. The effect will be equivalent to increasing, on average, the critical packing parameter $v/a_0 \cdot l_c$ of the detergent, increasing the radius of curvature of the surfactant/water interface, and pushing the equilibrium shape of the micelles away from the spherical toward an elliptical shape, as would be the case if using a detergent with a smaller polar head (see Chap. 1, Table 1.1, rows 1 and 2).
- When the concentration of detergent is changed in the presence of no or very low amounts of lipids, as in the lipid-free experiments of Fig. 2.11, however, little effect can be expected, as a rule, on the properties of micelles. Indeed, most of the detergents used in biology have been selected so that their phase diagram comprises a broad $L_1$ phase (the free micelle phase), in which micellar properties depend little on the concentration. Such is the case of DDM, the detergent used in the experiments of Fig. 2.10, which remains in the $L_1$ phase in water up to ~45% w/w (Fig. 2.12) (Warr et al. 1986). As regards the aggregation number $N$ of DDM micelles, and therefore their size, they are nearly independent of concentration (from 2.5 to 10 $g \cdot L^{-1}$) and temperature (from 16 to 60 °C), and the micelles interact very little one with another (Aoudia and Zana 1998; see also Salvay et al. 2007). Under such conditions, it is hard to argue that it is a physical effect that destabilizes the protein when the concentration of DDM is raised.
- The same reasoning holds when the protein is exposed to a vast excess of mixed micelles with a given composition. In the two series of experiments of Fig. 2.11, whatever residual native lipids were present were diluted, if not bound to the protein, in a vast excess of PC/HG mixed

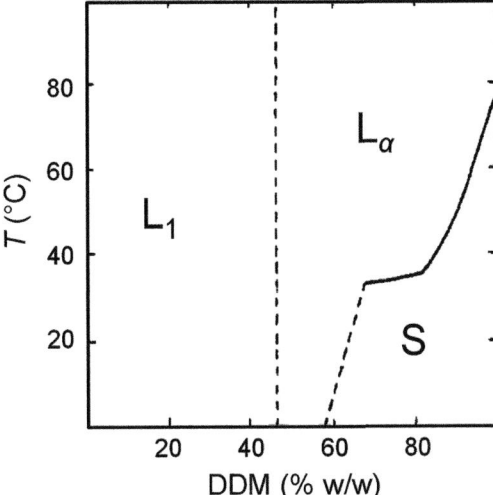

**Fig. 2.12** Phase diagram of aqueous solutions of *n*-dodecyl-*β*-D-maltoside. The various regions are labeled as follows: $L_1$ isotropic (micellar) solution, $L_\alpha$ lamellar (anisotropic) phase, and *S* solid. *Dashed lines* denote phase boundaries which were located with limited accuracy (Reprinted with permission from Warr et al. 1986, © 1986 American Chemical Society).

micelles. The CMC of pure HG is variously quoted to lie between ~16.5 mM (Ruiz et al. 1994), ~19.5 mM (Enzo Life Science web site), and 20–25 mM (Vegatec web site). Given that the concentration of free monomers drops somewhat in the presence of lipids, one can estimate that the concentration of micellar detergent was in the range of 30–35 mM in the presence of 50 mM HG (*left*) and of 80–85 mM in the presence of 100 mM HG (*right*). The rate of inactivation was faster in the second case, with, for instance, after 30 min and at a 1:10 PC/HG ratio, ~1/3 inactivation at 50 mM HG vs. ~1/2 at 100 mM. Under these two sets of conditions, the composition of the micelles is virtually identical, so the environment the complex experiences has the same physical properties. A change in those cannot be invoked to explain the faster inactivation at 100 mM HG.

- Under these three sets of circumstances, what does change with the concentration of surfactants is the ratio of detergent to protein and, if applicable, to lipids: as this ratio increases, any equilibrium between protein/protein, protein/lipid, protein/detergent, lipid/ lipid, and lipid/detergent interactions becomes displaced in favor of the detergent. This can affect, directly or indirectly, interactions between TM protein segments, whether within subunits or between subunits. Lipid molecules and protein subunits that interact to stabilize the complex will tend to be diluted in the micelles and replaced by detergent. Detergent molecules are, as a rule, smaller than lipid ones, and, with some exceptions like the LMNG series (see Fig. 2.15), they usually comprise a single hydrophobic chain. As a consequence, (i) they may not have the same "clamping" effect that a two-chain lipid may have when it straddles two TM segments or subunits, and (ii) they may more easily intrude into cranks in the protein structure, weakening protein/protein interactions and extracting bound lipids and cofactors (cf. Chap. 1, § 1.5.2). The protective effect of lipids, in turn, may be due (i) to stabilizing interactions they establish with the protein and (ii) to their blocking or slowing down the intrusion of detergent molecules into protein cranks or clefts.

It is to be noted that the protective effect of lipids is sometimes specific, such as that of cholesterol or cholesterol hemisuccinate on G protein-coupled receptors (GPCRs; reviewed in Lee 2011a, b), sometimes rather unspecific: the $b_6 f$ complex, for instance, is stabilized by egg PC (Fig. 2.11), even though *Chlamydomonas* thylakoid membranes do not comprise any PC (Vieler et al. 2007). It is not rare that, in X-ray structures, the fatty acyl chains of lipids can be discerned but the polar heads are fuzzy, suggesting that, in such cases, interactions with the protein at this level are weak (see e.g. Luecke et al. 1999; Qin et al. 2006). One may perhaps speculate that, under such circumstances, the binding of the two acyl chains of a foreign lipid into neighboring grooves at the surface of the protein suffices to ensure a degree of stabilization, without the nature of the polar head being of great importance. For further discussion of these aspects, the reader is referred, for example, to Garavito and Ferguson-Miller (2001) and Lee (2011a, b).

A closer look into what might happen when the cytochrome $b_6 f$ complex from *C. reinhardtii* is destabilized by detergents is offered by X-ray crystallography. Figure 2.13A presents a general view of the arrangement of the major subunits in the supercomplex. It shows, in particular, that the Rieske protein (in *green*), whose loss is the first detectable event in the inactivation of the complex by detergents (Fig. 2.10C), straddles the supercomplex, its extracellular luminal domain interacting with one of the two supermonomers and its TM $\alpha$-helical anchor with the other (Stroebel et al. 2003). This position makes it immediately obvious that the loss of this subunit may weaken the supercomplex, which, indeed, monomerizes very easily after the Rieske's removal, so that both events are generally concomitant (Fig. 2.10C; Breyton et al. 1997).

A close-up on the TM region (Fig. 2.13B) reveals two residual lipid molecules, one of them a sulfolipid, the other unidentified, and several detergent molecules. In this particular case, the detergent

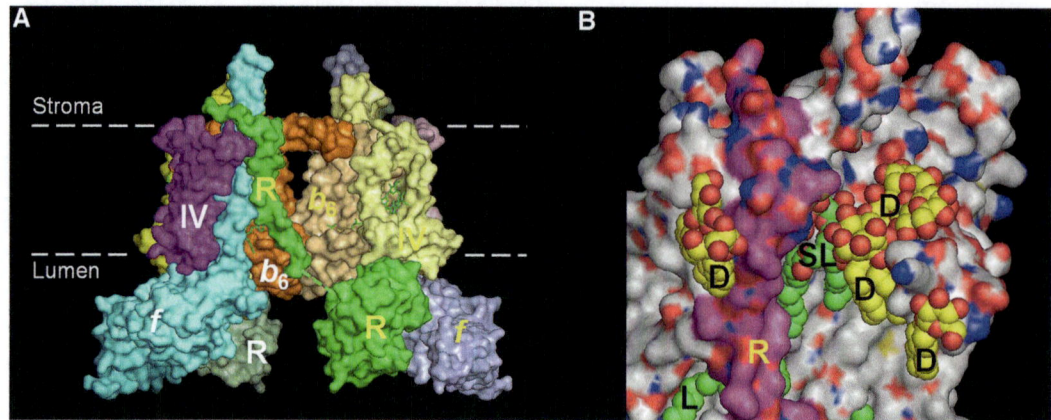

**Fig. 2.13**  Protein/lipid/detergent interactions in the vicinity of the cytochrome $b_6 f$ Rieske protein's transmembrane (TM) anchor as observed by X-ray crystallography. (**A**) General view of the cytochrome $b_6 f$ complex, showing the way the Rieske protein (*green*; R) straddles the superdimer, with its luminal extramembrane domain interacting with one of the supermonomers (whose subunits are labeled in *yellow*), its transmembrane (TM) anchor with the other (labeled in *white*). A disordered connecting loop between the luminal domain and the TM anchor, not visible in the X-ray structure, is indicated by *green dashes*. The approximate position of the thylakoid membrane is indicated by *dashed white lines*. Note how largely the anchor is exposed at the TM surface of the complex. PDB accession code 1Q90; Stroebel et al. (2003). (**B**) Close-up view of the stromal end of the Rieske protein's TM helix (*fuchsia*; R), surrounded by detergent molecules (*yellow* and *red*; D) and a couple of residual lipids (*green* and *red*; a sulfolipid, SL, and an unidentified lipid, L). The detergent is *trans*-4-(*trans*-4'-propylcyclohexyl)cyclohexyl-$\alpha$-D-maltoside (PCC-a-M; see Fig. 2.15) (Data from a 2.8-Å resolution X-ray map, courtesy of D. Picot; see Hovers et al. 2011).

is *trans*-4-(*trans*-4′-propylcyclohexyl)cyclohexyl-α-D-maltoside (PCC-a-M). The hydrophobic chain of PCC-a-M comprising two cyclohexyl rings (see Fig. 2.15) is much easier to distinguish from the lipid acyl chains than in the case of *n*-alkyl chains, even at the relatively limited resolution achieved here (2.8 Å; Hovers et al. 2011). The position of the two lipids, in contact both with the TM anchor of the Rieske protein and with the surface of the rest of the complex, strongly suggests that their loss is likely to weaken the association between them. Inactivation and monomerization of the $b_6f$ complex by detergent would therefore be initiated by the partitioning of the lipids into an excess of detergent micelles, thereby destabilizing the association of the Rieske protein with the complex. This loss would immediately inactivate the complex, the Rieske protein playing a key role in electron transfer, and would initiate its disaggregation. Such a mechanism readily explains why lowering the volume of the micellar phase and/or supplementing it with lipids both have a stabilizing effect on the complex.

We have seen in Chap. 1 (§ 1.5.2) that many structures of MPs and MP complexes show bound lipids that straddle TM segments, are in contact with two subunits, and/or are wedged in TM cranks or clefts, if not totally buried inside the protein (see e.g. Lee 2011 and other references cited in Chap. 1). It is my personal conviction that the mechanism I have described in some detail in the case of the $b_6f$ complex holds for many, if not most, MPs and that the major causes of MP inactivation by detergents are the insertion of detergent molecules into the TM structure (cf. Khelashvili et al. 2013; Lee et al. 2016; see Fig. 2.14) and the loss of subunits, of cofactors (for an example of the latter, see the case of photosystem II in de Vitry et al. 1991), and/or of lipids by dilution into detergent micelles (for earlier discussions, see e.g. Bowie 2001; Garavito and Ferguson-Miller 2001; Gohon and Popot 2003; Popot 2010; Tate 2010). It is this belief that prompted us to design APols, which we saw, initially, as a way to handle MPs in aqueous buffers in the total absence of surfactant beyond that bound to the protein. The story turned out to be more complicated, as will be described in Chap. 5, but this hypothesis did set us on the right track, and reduction of the "hydrophobic sink" did turn out to be one of the mechanisms, albeit not the only one, that account for the greater stability that most MPs exhibit when trapped in APols rather than being kept soluble by detergents.

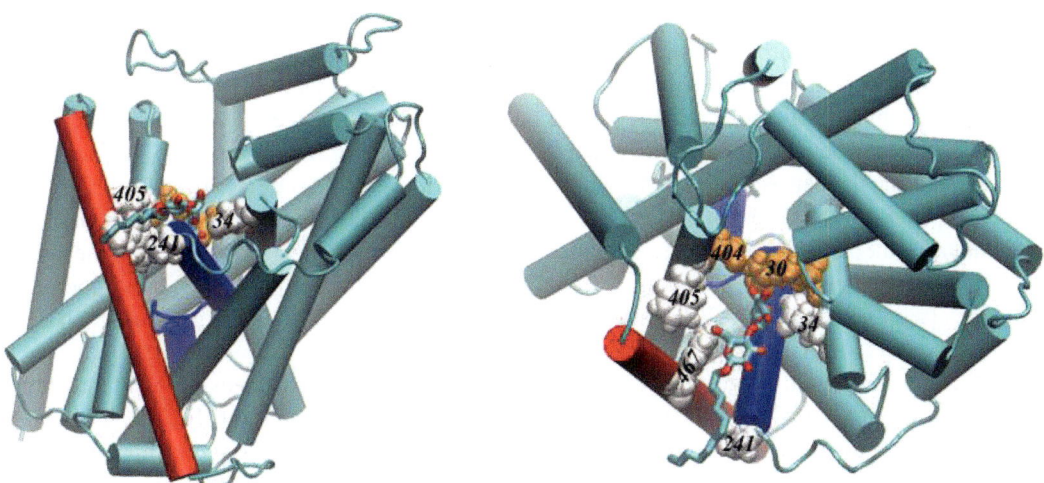

**Fig. 2.14** Molecular dynamics simulation showing the intrusion of a dodecylmaltoside molecule into the transmembrane region of the leucine transporter LeuT. Two views of a representative snapshot from an MD simulation showing DDM (in *licorice*) penetrating LeuT (in *cartoon*). TM helices 6 and 11 of the transporter are colored *blue* and *red*, respectively, and some key residues are shown in space-filling representation and labeled (From Khelashvili et al. 2013. http://pubs.acs.org/doi/abs/10.1021%. 2Fja405984v. For further reprinting, contact American Chemical Society).

## 2.4.4    Detergent-Induced Conformational Alterations

In the case of cytochrome $b_6 f$, exposure to detergents results in the progressive disaggregation of the complex. Alterations can be more subtle. With the multiplication of MP structures obtained by X-ray diffraction, NMR, and electron microscopy (EM) under a variety of conditions, such as in a bilayer environment vs. in complex with various detergents, evidence has accumulated that, whereas some MP structures are strictly identical in a lipid vs. a detergent environment (see e.g. Screpanti et al. 2006), environmental differences may affect not only the dynamics but also the average conformation of MPs: structures observed in detergent solution do not necessarily reflect faithfully the native structure(s) adopted in the membrane (for reviews and discussions, see e.g. Cross et al. 2011; Zhou and Cross 2013; Zoonens et al. 2013).

Zhou and Cross have carried out a systematic examination of the TM region of MP structures deposited in the Protein Data Bank. They note deviations from reasonable expectations, as well as divergences between structures obtained in different environments (natural membranes, artificial ones, detergent solutions, crystals) and using different techniques (solution-state and solid-state NMR, EM, X-ray crystallographic analysis of crystals obtained either in detergent solutions or in lipid cubic phases) (Zhou and Cross 2013). Some of these observations are relevant to the present discussion, inasmuch as they suggest ways in which the transfer from a natural membrane environment to a detergent one may affect MP structures, and plausible mechanisms underlying them. Whereas many structures are consistent from one system to another and appear reasonable from a structural point of view, suggesting that they do represent the native state (see e.g. Lanyi and Schobert 2007; Liao et al. 2012), others are more problematical. In some of them, TM $\alpha$-helices appear splayed out, leaving large interhelical gaps, and/or they are misoriented, with polar residues and residues expected to be involved in helix-helix interactions facing the exterior of the TM helix bundle, so that, in situ, they would be exposed to the membrane core. In other structures, some helices are too short to span a membrane or out of register with nearby helices. In many cases, it remains a matter of opinion whether the structure is actually aberrant or not: different structures for the same protein may well correspond to different functional states, particularly for transporters (see the example of the calcium pump SERCA1a in Chap. 1, § 1.6.3, Figs. 1.34 and 1.36). In some cases, however, it seems extremely likely that the native fold has indeed been disturbed. Possible causes invoked to explain these distortions include the loss of lateral pressure, attraction of TM hydrophilic side chains that normally face the interior of the TM helix bundle by the hydrophilic polar head region of detergent micelles, or the intercalation of detergent molecules between helices. It seems intuitive that, in the case or SERCA1a, for instance, the loss of the PE molecule that lies between helices M2 and M4 in the calcium-free E2 state (see Chap. 1, Fig. 1.37B) and its replacement by detergent molecules might well perturb the overall TM structure and, as a consequence, the arrangement of the extramembrane domains and the functionality of the whole protein. Detergent molecules intruding between TM segments are seen in many crystallographic structures (see Zhou and Cross 2013), as well as in some MD simulations (see e.g. Khelashvili et al. 2013; Lee et al. 2016; Fig. 2.14).

MD simulations do suggest that the transfer from a lipid to a detergent environment may bring about significant structural changes (see e.g. Cuthbertson et al. 2006; Choutko et al. 2011; Rodríguez-Ropero and Fioroni 2012). In a recent study combining MD simulations and functional measurements, it was concluded that solubilization with DPC (see Fig. 2.1) of the mitochondrial uncoupler UCP2 entails an artifactual widening of its TM region, turning it into a porin-like aqueous channel and abolishing its proton carrier activity (Zoonens et al. 2013).

It may seem paradoxical that, on the one hand, MPs are so often inactivated by detergents and can be so hard to handle in detergent solutions, and on the other hand, a large number of them (~100 as of 2014) have been folded or refolded in vitro using detergents and even more exotic media, which reproduce essentially none of the constraints exerted by biological membranes (reviewed in Popot 2014; cf. Chap. 6, § 6.2). The apparent paradox vanishes if one considers that folding or refolding is almost never carried out in pure detergent solutions but, typically, in lipid/detergent mixed micelles, in lipid vesicles, or in APols, the latter supplemented or not with lipids (see Table 1 in Popot 2014). This strongly suggests that it is the presence of detergent, more than the absence of a membrane, that compromises the achievement or the stability of the native fold. On the basis of these observations, it has been proposed that, for many if not most MPs, the determinants of folding come essentially from the amino acid sequence, not from physical constraints exerted by the environment. The latter can obviously modulate the structure and thereby the function of certain MPs (e.g. mechanosensitive channels), but in many, perhaps most, cases, it plays no critical role in determining the general 3D fold (discussed in Popot and Engelman 2016). Structural details, however, can be modulated by the environment and, in particular, by the binding of this or that lipid, which often plays important functional and/or regulatory roles.

## 2.5   Solutions to the Instability Problem

In § 2.4, we have described some of the problems associated with handling MPs in detergent solutions, discussed hypotheses about their origin, and presented some approaches to alleviating their impact, in particular limiting the volume of the hydrophobic sink and using lipid/detergent mixed micelles rather than pure detergent. There are, however, other ways, not necessarily exclusive one from the other, to stabilize MPs in aqueous environments. We can distribute them into three categories:

- A first approach is to optimize the structure of the detergent so as to preserve its ability to dissolve biological membranes and solubilize MPs while slowing down the rate at which it inactivates them (§ 2.5.1).
- A second approach is to either choose a more resistant homologue of the protein or to engineer it so as to adapt it to the detergent (§ 2.5.2).
- A third, somewhat radical approach is to handle the protein in the absence of detergent (§ 2.5.3).

### 2.5.1   Making Detergents Less Aggressive

Until the 1970s, biologists had to make do with detergents provided by nature (bile salts, digitonin) or synthetic detergents developed for industrial purposes (Triton, Tween, Lubrol, Brij, etc.). Even though some of them are still in use today in membrane biology laboratories (digitonin, because of its mildness; Triton X-100, because of its efficiency as a solubilizing agent; Tween as an adsorbent onto hydrophobic surfaces; etc.), each of these compounds presented its set of limitations, such as heterogeneity, the presence of impurities, the formation of large micelles, UV absorbance, harshness or, on the contrary, limited solubilizing power, etc. This led to the development of synthetic or semisynthetic detergents specifically designed for extracting and handling MPs. Among the most successful of the novel detergents that appeared in the mid-1970s to early 1980s (some of which had been synthesized years if not decades before but had not yet been discovered by membrane biochemists), one may cite octyl-$\beta$-D-glucoside (OG) (Stubbs et al. 1976; Keana and Roman 1978),

DDM (Knudsen and Hubbell 1978; Rosevear et al. 1980; VanAken et al. 1986), and CHAPS – a zwitterionic derivative of cholate (Hjelmeland et al. 1983) (see Fig. 2.1) – among tens of others.

***Developing less aggressive detergents***
*(© 2018 by Francis Haraux)*

With MP biochemistry maturing, the realization came that there is no silver bullet and that a detergent that is well tolerated by a given MP cannot be used with the next one, which encouraged diversifying existing molecules and experimenting with new structures. The development of MP crystallization and solution NMR brought its own demands. Crystallographers prefer their detergents to be nonionic, so as to avoid electrostatic repulsion between detergent belts, and to form small belts, so as to leave as much protein surface free as possible to facilitate the formation of crystalline contacts. They tend to use detergents with either sugar-based, polyoxyethylene-based, or aminoxyde-based polar heads, which are more favorable to crystallization. NMR spectroscopists also want to form as small MP/detergent complexes as possible, but they rather prefer them to repulse each other, so as to prevent aggregation, and, because they work at high concentrations of protein and detergent and relatively high temperature (to accelerate tumbling), they have a fine line to walk between keeping the protein monodisperse and denaturing it. They often resort to detergents with zwitterionic polar heads, such as dodecylphosphocholine (DPC) or dihexanoylphosphatidylcholine (DHPC, $diC_6PC$) (Fig. 2.1). As these activities developed, so did the panel of available detergents.

Developing a new detergent, validating its use for MP biochemistry and biophysics, setting up its industrial synthesis, and bringing it efficiently to the market are ripe with pitfalls. Many promising molecules have remained in drawers because their synthesis or purification was too complex; their phase diagram left insufficient leeway for varying the temperature, concentration, and/or buffer conditions; and their biochemical validation was insufficient or their marketing inefficient. Gaining a following on a cluttered market demands that the new molecule can boast of serious advantages. It also depends somewhat on chance, such as the new molecules being instrumental in a particularly successful and eye-catching project. Among recent, original proposals, one may perhaps, somewhat arbitrarily, mention tripod amphiphiles (Yu et al. 2000; Chae et al. 2008, 2013; Ehsan et al. 2017); facial amphiphiles (Zhang et al. 2007, 2011; Chae et al 2010; Lee et al. 2013) (see Fig. 2.15, di-*β*-D-maltoside cholane); DDM analogs with a melibiose polar head, yielding smaller micelles (Hutchison et al. 2017), or with hydrophobic tails that either carry a terminal cyclohexyl group (the Cymal series; see Ostermeier et al. 1997) or are branched (Hong et al. 2011; Zhang et al. 2011) or are comprised of

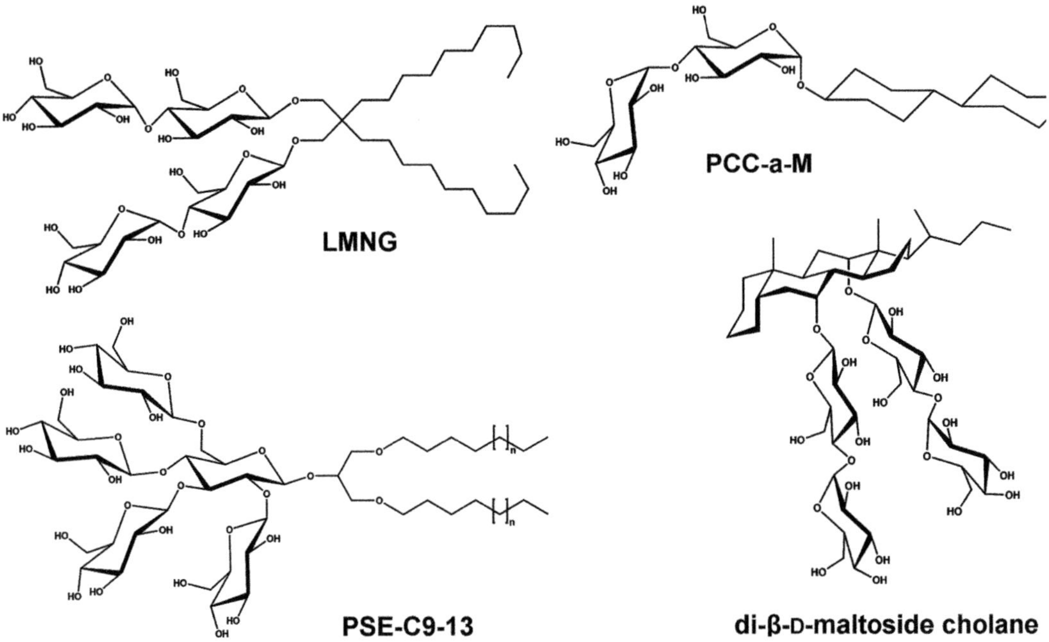

**Fig. 2.15** Four examples of recently developed detergents. LMNG, 2,2-didecylpropane-1,3-bis-$\beta$-D-maltopyranoside (CMC ≈ 10 μM; Chae et al. 2010b). PCC-a-M, *trans*-4-(*trans*-4'-propylcyclohexyl) cyclohexyl-$\alpha$-D-maltoside (CMC ≈ 36 μM; Hovers et al. 2011). PSE-C9-13, a series of pentasaccharide-bearing detergents; n = 3 → PSE-C9, n = 5 → PSE-C11, n = 7 → PSE-C13 (CMC ≈ 26 μM, ~4 μM, and ~1 μM, respectively; Ehsan et al. 2016). di-$\beta$-D-maltoside cholane, a facial amphiphile (CMC ≈ 0.1 mM; Zhang et al. 2007).

cycles, making it much more rigid than a linear alkyl chain (Hovers et al. 2011) (see Fig. 2.15, PCC-a-M); detergents derived from calixarene (Matar-Merheb et al. 2011) or from lithocholic acid or diosgenin (Chae et al. 2012); trisaccharide-based detergents (Pérez-Victoria et al. 2011; Sadaf et al. 2016); and others (see e.g. Das et al. 2017; Hussain et al. 2017). The potentialities of novel, underexploited natural surfactants are illustrated and discussed in Andersen and Otzen (2014); Otzen (2015). A marked tendency is to move toward larger heads and more rigid and/or encumbered hydrophobic chains. As a consequence, most novel detergents have low CMCs (cf. legend to Fig. 2.15). This, as we have seen, is a favorable factor in membrane biochemistry, because it makes it possible to limit the volume of the hydrophobic sink formed by the micellar phase.

A particularly successful family of novel detergents is that of maltose-neopentyl glycol (MNG) detergents, developed by Samuel H. Gellman and Pil-Seok Chae, which have been used, in particular, for handling and crystallizing GPCRs. These molecules, built around a central quaternary carbon atom originating from neopentyl glycol, carry two alkyl chains and two hydrophilic groups derived from maltose (see Fig. 2.15, LMNG). They have proven particularly mild to a variety of MPs (see Chae et al. 2010b, 2013, and references therein). The reason for their mildness is uncertain. One may perhaps suggest (i) that the way the two hydrophobic chains are grafted onto a highly encumbered carbon may make it more difficult for them to intrude deeply between MP TM segments and (ii) that the two hydrophobic chains may play the role of "clamps" tentatively attributed to two-chain lipids in § 2.4.3. Along the same line of thinking have been developed glucose-neopentyl glycol (GNG) detergents (Kellosalo et al. 2012; Chae et al. 2013), detergents derived from pentaerythritol (reviewed in Zhang et al. 2011), and detergents carrying two alkyl chains and either a highly branched pentasaccharidic polar head (Ehsan et al. 2016; see Fig. 2.15, PSE-C9-13) or three glucose moieties (Sadaf et al. 2016).

## 2.5.2    Making Membrane Proteins More Resistant

*Making membrane proteins more resistant*
(© *2018 by Francis Haraux*)

The surfactant is only one of the two partners in MP/detergent complexes. Stability can also be improved by selecting more resistant proteins. This is in principle beyond the frame of the present book. However, given that one of the primary interests of resorting to APols is to stabilize the target MP in aqueous solutions, it is important to know what the alternatives are, as well as to ponder whether they could be combined.

Working with more stable MPs than one's primary target, e.g. the human one, is classically achieved by looking for homologues of the target protein in prokaryotes, whose proteins are usually less fragile than their eukaryotic counterparts, and, among those, for MPs from thermophilic bacteria, which evolution has selected for remaining functional under extreme conditions. An alternative approach is to create and select in the laboratory detergent-resistant mutants of the target protein, a process that, for reasons to be described below, is known as "conformational thermostabilization" (Tate 2012). This approach was initiated in the 1990s and early 2000s with such MPs as the TM segment of the phage M13 coat protein (Deber et al. 1993; Wang and Deber 2000), $K^+$ channel (Cortes and Perozo 1997; Perozo et al. 1998), diacylglycerol kinase (DAGK; Lau et al. 1999; Zhou and Bowie 2000), and BR (for a review of these early efforts, see Bowie 2001). It has been applied with remarkable success to many GPCRs and has led, to date, to the crystallization and structure resolution of several of them (see for example, Warne et al. 2008; Lebon et al. 2011a, b). The approach looks very general and could certainly be extended to many other MPs. Before describing it, however, a short introduction to GPCRs is necessary in order to clarify the nature of the problem to be solved and the principle of the method. In addition, GPCRs will come up over and over again in subsequent chapters, and one objective of the present section is to provide a convenient memento to readers who are not overly familiar with them.

As mentioned in Chap. 1 (§ 1.3), GPCRs form a very large family of receptors (~800 members in man) that transmit information from the exterior of the cell to the cytosol by virtue of shifting their conformation upon ligand binding. They mediate most of our responses to hormones and neurotransmitters and are also responsible for our perception of the outside world through vision, olfaction, and taste. All GPCRs share a common evolutionary origin and a TM region folded into seven TM helices, with an extracellular *N*-terminus and intracellular *C*-terminus (Fig. 2.16; reviewed in Fredriksson et al. 2003). All GPCRs that have been thermostabilized yet belong to class A GPCRs, which are characterized by relatively short extramembrane loops and whose prototype is rhodopsin. Rhodopsin is an atypical GPCR in the sense that it is activated not by the binding of an extracellular ligand but by the light-induced isomerization of a covalently bound cofactor, retinal. In this, and in its

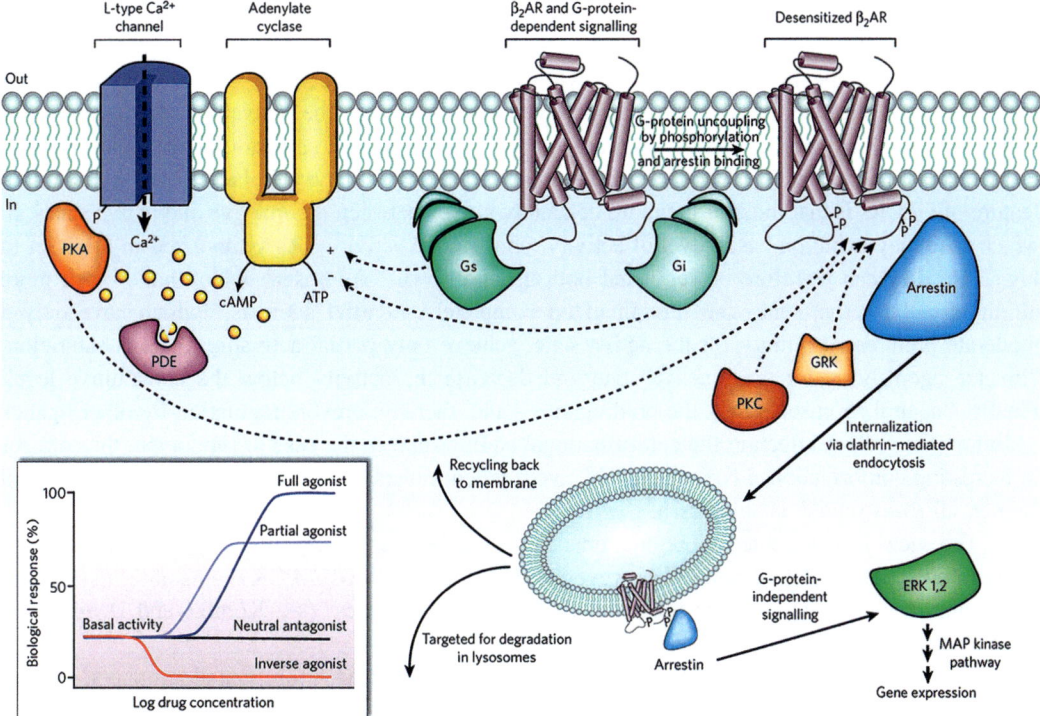

**Fig. 2.16** Signal transduction by G protein-coupled receptors. Diverse signaling pathways regulated by the type 2 $\beta$-adrenergic receptor ($\beta_2$AR). The $\beta_2$AR can activate two distinct G proteins, $G_{\alpha s}$ and $G_{\alpha i}$ (part of the $G_s$ and $G_i$ heterotrimers, respectively), which differentially regulate adenylate cyclase. Adenylate cyclase generates cyclic AMP (cAMP), which activates protein kinase A (PKA), a kinase that regulates the activity of several cellular proteins including the L-type $Ca^{2+}$ channel and the $\beta_2$AR. cAMP second messenger levels are downregulated by specific phosphodiesterase proteins (PDEs). Activation of the $\beta_2$AR also leads to phosphorylation by a G protein-coupled receptor kinase (GRK) and subsequent coupling to arrestin. Arrestin is a signaling and regulatory protein that promotes the activation of extracellular signal-regulated kinases (ERK), prevents the activation of G proteins, and promotes the internalization of the receptor through clathrin-coated pits. PKC, protein kinase C. The *inset* shows classification of ligand efficacy for GPCRs. Many GPCRs exhibit basal, agonist-independent activity. Inverse agonists inhibit this activity, and neutral antagonists have no effect beyond blocking access to the ligand-binding site. Agonists and partial agonists stimulate biological responses above the basal activity. Efficacy is not directly related to affinity; for example, a partial agonist can have a higher affinity for a GPCR than a full agonist (From Rosenbaum et al. 2009. © 2009 Macmillan Publishers Limited, Nature. All rights reserved).

general structure, it is similar to BR, but there is no trace of an evolutionary relationship between them, and their functions are completely different. Among GPCRs, rhodopsin is also peculiar in having a very low level of basal activity (see below), which endows our eyes with the possibility to detect even very weak visual signals. This feature turned out to be important in permitting its crystallization, because it reduces conformational variability. Other than that, rhodopsin is an archetypical GPCR. It was the first one whose 3D crystallographic structure was ever established, in 2000, by Krzysztof Palczewski and colleagues (Palczewski et al. 2000).

The signaling functions of GPCRs are illustrated in Fig. 2.16 in the case of the $\beta_2$ adrenergic receptor, whose natural ligands are adrenaline and noradrenaline. Similar in this respect to the pentameric ligand-gated ion channels (pLGICs), which we have discussed in Chap. 1 (§ 1.6.2), GPCRs are in equilibrium between two or more conformational states: binding of an extracellular ligand shifts the equilibrium toward the state(s) with the highest affinity for the ligand. However, at variance with pLGICs, conformational transitions do not control the opening and closing of a TM channel. Rather, the reorganization of the protein around the ligand-binding site is coupled, through the TM region, to the opening of a cleft in the cytoplasmic face of the receptor. Cytosolic proteins bind to it and propagate and amplify the signal. Two main routes are the activation of stimulatory or inhibitory heterotrimeric G proteins, on the one hand, and coupling to arrestin, on the other. Both of them are schematized in Fig. 2.16, along with some of the downstream cellular events. An important feature of GPCRs is that most of them are delicately poised between the inactive and active states, in which case they exhibit a basal level of activity (constitutive activity). As symbolized in the inset to Fig. 2.16, they can therefore be regulated both up and down. "Agonists," which have much more affinity for the active state, can turn them on completely. "Partial agonists," which have only a moderate preferential affinity for the active state, achieve only partial activation even at saturation. "Inverse agonists" favor the inactive state and decrease the activity below the constitutive level. Finally, "neutral agonists" block the binding site – and, thereby, prevent regulation by other ligands – without themselves affecting the conformational equilibrium. In the case of rhodopsin, the cofactor in its resting conformation, 11-*cis*-retinal, is a very potent inverse agonist, whereas the light-induced isomer, all-*trans*-retinal, is an agonist.

The view just presented of regulation being achieved simply by modulating the equilibrium between two conformational states is, however, widely oversimplified, GPCRs being able "to sample a continuum of conformations with relatively closely spaced energies" (see Kobilka and Deupi 2007, Rosenbaum et al. 2009, and references therein).

The conformational flexibility of GPCRs has been a nightmare for crystallographers, as (i) it considerably complicates obtaining crystals of receptors in a well-defined conformation and (ii) it probably contributes to their high sensitivity to detergents. Crystallizing GPCRs has required heroic efforts at stabilization, which typically have involved (i) crystallizing in the presence of a high-affinity ligand, in order to try to shift the equilibrium as completely as possible toward a single form, while stabilizing it against conformational excursions and (ii) various attempts at further stabilization, including binding of antibody $F_{ab}$ fragments, or camelid heavy chain antibodies (nanobodies), fusion of phage T4 lysozyme in the large cytosolic loop that links TM helices 5 and 6, or "conformational thermostabilization," which aims at accumulating mutations that improve the homogeneity and stability of the target receptor in detergent solution (reviewed in Rosenbaum et al. 2009; Tate 2012). Mutants can be selected either by directed evolution (Sarkar et al. 2008; Dodevski and Plückthun 2011) or by systematically scanning the sequence for stabilizing mutations and then combining enough of them to achieve the desired stability. We will take as an illustration the latter approach, whose principle and main results are reviewed with great clarity by one of its main promoters, Christopher G. Tate, in Tate (2012), on which the following summary is based (for an update, see Magnani et al. 2016).

To isolate thermostabilizing mutations by systematic scanning mutagenesis, point mutations are first introduced one by one at every position throughout the target receptor by substituting non-alanine residues with alanine ones and alanine residues with leucines. The mutated receptors are expressed in DH5$\alpha$ *E. coli* or in mammalian cells and membrane fragments collected, solubilized in DDM, and supplemented or not with a radioactive ligand. Thermostability is assessed by incubating the samples at the desired temperature for 30 min. Samples are then placed on ice and radioligands added if not already present and equilibrated. Receptor-bound and free radioligand are separated by gel filtration (for details on the procedure, see Magnani et al. 2008). Alanine/leucine-scanning mutagenesis was initially performed on three GPCRs, the $\beta_1$ adrenergic and $A_{2A}$ adenosine receptors (Serrano-Vega et al. 2008; Lebon et al. 2011a, b; Miller and Tate 2011) and a neurotensin receptor (NTS1) (Shibata et al. 2009). The thermostability of each mutant was determined using either radiolabeled inverse agonists ($\beta_1$ and $A_{2A}$ receptors) or radiolabeled agonists ($A_{2A}$ and NTS1 receptors). In each case, approximately 5–9% of the mutations tested were thermostabilizing, of which 60–90% were found in the TM region rather than in the loops. Each thermostabilizing point mutation typically improved the thermostability by only 1–3 °C, but within each scan there were generally one or two significantly better mutations. This first round of mutations permits to identify positions at which stabilizing mutations can be introduced. Further improvement can be obtained by substitution with residues other than alanines. Once individual thermostabilizing mutations have been identified, a selection of them are combined to generate an even more stable receptor. At the end of the day, stabilization, measured as the extension of lifetime in DDM as compared to wild type, was $125\times$ for the $\beta_1$ receptor (six point mutations) and $38\times$ (eight mutations) or $260\times$ (four mutations) for two different constructs of the $A_{2A}$ receptor. These two receptors were crystallized and yielded structures at resolutions varying between 2.3 and 3.4 Å, depending on the receptor and on the ligand used for co-crystallization (reviewed in Tate 2012).

Two points worth noting are (i) that the mutations lie outside the ligand-binding site and (ii) that, when the binding of various types of ligands to thermostabilized receptors is tested, it is found that there is little change of affinity for the ligand used during the selection, e.g. an inverse agonist, and a drop of affinity for ligands with the opposite pharmacology, e.g. an agonist. What the selection procedure achieves, therefore, is to stabilize one of the conformations the receptor can adopt by preference to the others, hence the term of "conformational thermostabilization." Another remarkable point is that, even though the screening process takes place in DDM, the resulting constructs are also more resistant to the short-chain detergents used for crystallization (see Lebon et al. 2011a, b).

The reasons why thermostabilized GPCRs are more stable than the wild type are complex and probably multifarious. In a recent review, analysis of MD data led to the conclusion that receptors are stabilized through a combination of factors including an increase in rigidity, a corresponding decrease in collective motions, reduced stress at specific residues, and the presence of ordered water molecules (Vaidehi et al. 2016; see also Lee et al. 2015). Given that the selection is carried out in the presence of lipids and DDM, it is probably difficult to exclude that reinforcement of their affinity for the surface of the native receptor may also play a role. An interesting open question is whether stabilization by mutations selected in the presence of detergent and stabilization by APols are or not additive.

### 2.5.3   What About Keeping Membrane Proteins Water-Soluble Without Using Detergents?

In the previous section, we have reviewed some attempts at increasing the stability of detergent-solubilized MPs by improving either the detergent or the protein. Can one, instead, just dispense with using any detergent? This is of course achieved when transferring the detergent-extracted protein to detergent-free lipid phases, such as lipid vesicles or cubic phases. Such experimental systems, however, do not lend themselves well to all experimental requirements, e.g. those of purification or solution NMR.

A particularly daring approach, which has often failed and sometimes succeeded, is to turn the target MP into a soluble one. This can be achieved chemically, by covalently linking to the protein hydrophilic groups such as polyethylene glycol (making it, however, impossible to crystallize) (see e.g. Sirokmán and Fasman 1993; Wei and Fasman 1995; Pomroy and Deber 1998). It can also be attempted genetically, by replacing the hydrophobic residues that normally face the membrane interior with hydrophilic ones and attempting folding in aqueous solutions. This is a difficult exercise which has many pitfalls, including unintended interactions that lead to aggregation. One successful case is that of the TM region of phospholamban, a homopentamer of single TM $\alpha$-helices. Making the membrane-exposed surface of the helix hydrophilic yielded water-soluble, helical pentamers, and further design refinement gave fully folded, soluble structures (Li et al. 2001; Slovic et al. 2005a, b). Similar efforts led to water-soluble versions of other self-associating single-TM complexes (see DeGrado et al. 2003), as well as of a tetrameric ion channel, KcsA, each monomer of which comprises two TM $\alpha$-helices (Slovic et al. 2004; Ma et al. 2008). Most dramatically, a water-soluble version of a GPCR, the $\mu$-opioid receptor, was shown to fold into a stable structure that binds naltrexone with an affinity close to that observed with the wild-type protein (Zhao et al. 2014). How general can such approaches be and how close to the native membrane-embedded structure are those of the MPs thus rendered water-soluble remain to be ascertained.

An alternative approach is to replace the detergent with less aggressive surfactants. Those need not necessarily be dissociating enough to extract the target MP from its native membrane – they may not be detergents – provided they keep MPs water-soluble after they have been extracted. This is how APols work. Before turning to APols, however, we will go through a broad survey of other nonconventional surfactant systems, including bicelles, nanodiscs, peptide-based systems, and fluorinated surfactants (Chap. 3). This will set the stage for a presentation and discussion of the original way APols work and what their advantages and drawbacks are as compared to other systems (Chaps. 4 and 5). Chapters 6, 7, 8, 9, 10, 11, 12, 13, 14, and 15 will then be devoted to describing and discussing the use of APols for various applications.

# References

Abel, S., Dupradeau, F.Y., Raman, E.P., MacKerell, A.D., Jr., Marchi, M. (2011) Molecular simulations of dodecyl-β-maltoside micelles in water: influence of the headgroup conformation and force field parameters. *J. Phys. Chem. B* **115**:487–499.

Andersen, K.K., Otzen, D.E. (2014) Folding of outer membrane protein A in the anionic biosurfactant rhamnolipid. *FEBS Lett.* **588**:1955–1960.

Aoudia, M., Zana, R. (1998) Aggregation behavior of sugar surfactants in aqueous solutions: Effects of temperature and the addition of nonionic polymers. *J. Colloid Interface Sci.* **206**:158–167.

Arachea, B.T., Sun, Z., Potente, N., Malik, R., Isailovic, D., Viola, R.E. (2012) Detergent selection for enhanced extraction of membrane proteins. *Prot. Expr. Purif.* **86**:12–20.

Arnold, T., Linke, D. (2008) The use of detergents to purify membrane proteins. *Curr. Protoc. Protein Sci.* 4:Unit 4.8.1–4.8.30.

Banerjee, P., Joo, J.B., Buse, J.T., Dawson, G. (1995) Differential solubilization of lipids along with membrane proteins by different classes of detergents. *Chem. Phys. Lipids* **77**:65–78.

Bhairi, S.M., Mohan, C. (2007) Detergents. A Guide to the Properties and Uses of Detergents in Biology and Biochemistry (Calbiochem booklet). EMD Biosciences, Darmstadt, 43 p.

Bogusz, S., Venable, R.M., Pastor, R.W. (2001) Molecular dynamics simulations of octylglucoside micelles: dynamic properties. *J. Phys. Chem. B* **105**:8312–8321.

Bond, P.J., Faraldo-Gómez, J.D., Deol, S.S., Sansom, M.S.P. (2006) Membrane protein dynamics and detergent interactions within a crystal: A simulation study of OmpA. *Proc. Natl. Acad. Sci. USA* **103**:9518–9523.

Bond, P.J., Sansom, M.S.P. (2003) Membrane protein dynamics versus environment: simulations of OmpA in a micelle and in a bilayer. *J. Mol. Biol.* **329**:1035–1053.

Bowie, J.U. (2001) Stabilizing membrane proteins. *Curr. Opin. Struct. Biol.* **11**:397–402.

Breyton, C., Gabel, F., Abla, M., Pierre, Y., Lebaupain, F., Durand, G., Popot, J.-L., Ebel, C., Pucci, B. (2009) Micellar and biochemical properties of (hemi)fluorinated surfactants are controlled by the size of the polar head. *Biophys. J.* **97**:1077–1086.

Breyton, C., Tribet, C., Olive, J., Dubacq, J.-P., Popot, J.-L. (1997) Dimer to monomer conversion of the cytochrome $b_6 f$ complex: causes and consequences. *J. Biol. Chem.* **272**:21892–21900.

Brotherus, J.R., Jost, P.C., Griffith, O.H., Hokin, L.E. (1979) Detergent inactivation of sodium- and potassium-activated adenosinetriphosphatase of the electric eel. *Biochemistry* **18**:5043–5050.

Brouillette, C.G., Muccio, D.D., Finney, T.K. (1987) pH dependence of bacteriorhodopsin thermal unfolding. *Biochemistry* **26**:7431–7438.

Carey, M.C., Small, D.M. (1972) Micelle formation by bile salts. Physical-chemical and thermodynamic considerations. *Arch. Intern. Med.* 130:506–527.

Chae, P.S., Gotfryd, K., Pacyna, J., Miercke, L.J.W., Rasmussen, S.G.F., Robbins, R.A., Rana, R.R., Loland, C.J., Kobilka, B.K., Stroud, R., Byrne, B., Gether, U., Gellman, S.H. (2010a) Tandem facial amphiphiles for membrane protein stabilization. *J. Am. Chem. Soc.* **132**:16750–16752.

Chae, P.S., Kruse, A.C., Gotfryd, K., Rana, R.R., Cho, K.H., Rasmussen, S.G., Bae, H.E., Chandra, R., Gether, U., Guan, L., Kobilka, B.K., Loland, C.J., Byrne, B., Gellman, S.H. (2013a) Novel tripod amphiphiles for membrane protein analysis. *Chemistry – Eur. J.* **19**:15645–15651.

Chae, P.S., Rana, R.R., Gotfryd, K., Rasmussen, S.G.F., Kruse, A.C., Cho, K.H., Capaldi, S., Carlsson, E., Kobilka, B.K., Loland, C.J., Gether, U., Banerjee, S., Byrne, B., Lee, J.K., Gellman, S.H. (2013b) Glucose-neopentyl glycol (GNG) amphiphiles for membrane protein study. *Chem. Commun.* **49**:2287–2289.

Chae, P.S., Rasmussen, S.G.F., Rana, R., Gotfryd, K., Chandra, R., Goren, M.A., Kruse, A.C., Nurva, S., Loland, C.J., Pierre, Y., Drew, D., Popot, J.-L., Picot, D., Fox, B.G., Guan, L., Gether, U., Byrne, B., Kobilka, B.K., Gellman, S.H. (2010b) Maltose-neopentyl glycol (MNG) amphiphiles for solubilization, stabilization and crystallization of membrane proteins. *Nat. Methods* **7**:1003–1008.

Chae, P.S., Rasmussen, S.G.F., Rana, R.R., Gotfryd, K., Kruse, A.C., Nurva, S., Loland, C.J., Guan, L., Gether, U., Byrne, B., Kobilka, B.K., Gellman, S.H. (2012) A new class of amphiphiles bearing rigid hydrophobic groups for solubilization and stabilization of membrane proteins. *Chemistry – Eur. J.* **18**:9485–9490.

Chae, P.S., Wander, M.J., Bowling, A.P., Laible, P.D., Gellman, S.H. (2008) Glycotripod amphiphiles for solubilization and stabilization of a membrane-protein superassembly: importance of branching in the hydrophilic portion. *ChemBioChem* **9**:1706–1709.

Champeil, P., Orlowski, S., Babin, S., Lund, S., le Maire, M., Møller, J., Lenoir, G., Montigny, C. (2016) A robust method to screen detergents for membrane protein stabilization, revisited. *Anal. Biochem.* **511**:31–35.

Chevalier, Y. (2002) New surfactants: new chemical functions and molecular architectures. *Curr. Opin. Colloid Interface Sci.* **7**:3–11.

Choutko, A., Glättli, A., Fernández, C., Hilty, C., Wüthrich, K., van Gunsteren, W.F. (2011) Membrane protein dynamics in different environments: simulation study of the outer membrane protein X in a lipid bilayer and in a micelle. *Eur. Biophys. J.* **40**:39–58.

Cortes, D.M., Perozo, E. (1997) Structural dynamics of the *Streptomyces lividans* $K^+$ channel (SKC1): oligomeric stoichiometry and stability. *Biochemistry* **36**:10343–10352.

Cowan, S.W., Schirmer, T., Rummel, G., Steiert, M., Ghosh, R., Pauptit, R.A., Jansonius, J.N., Rosenbusch, J.P. (1992) Crystal structures explain functional properties of two *E. coli* porins. *Nature* **358**:727–733.

Cross, T.A., Sharma, M., Yi, M., Zhou, H.-X. (2011) Influence of solubilizing environments on membrane protein structures. *Trends Biochem. Sci.* **36**:117–125.

Cuthbertson, J.M., Bond, P.J., Sansom, M.S.P. (2006) Transmembrane helix-helix interactions: Comparative simulations of the glycophorin A dimer. *Biochemistry* **45**:14298–14310.

Dahmane, T. (2007) Protéines membranaires et amphipols : stabilisation, fonction, renaturation, et développement d'amphipols sulfonatés pour la RMN des solutions. Thèse de Doctorat, Université Paris-7, Paris, 229 p.

Dahmane, T., Rappaport, F., Popot, J.-L. (2013) Amphipol-assisted folding of bacteriorhodopsin in the presence and absence of lipids. Functional consequences. *Eur. Biophys. J.* **42**:85–101.

Das, M., Du, Y., Ribeiro, O., Hariharan, P., Mortensen, J.S., Patra, D., Skiniotis, G., Loland, C.J., Guan, L., Kobilka, B.K., Byrne, B., Chae, P.S. (2017) Conformationally preorganized diastereomeric norbornane-based maltosides for membrane protein study: Implications of detergent kink for micellar properties. *J. Am. Chem. Soc.* **139**:3072–3081.

de Vitry, C., Diner, B.A., Popot, J.-L. (1991) Photosystem II particles from *Chlamydomonas reinhardtii*: purification, molecular weight, small subunit composition, protein phosphorylation. *J. Biol. Chem.* **266**:16614–16621.

Deber, C.M., Khan, A.R., Li, Z., Joensson, C., Glibowicka, M., Wang, J. (1993) Val → Ala mutations selectively alter helix-helix packing in the transmembrane segment of phage M13 coat protein. *Proc. Nat. Acad. Sci. USA* **90**:11648–11652.

DeGrado, W.F., Gratkowski, H., Lear, J.D. (2003) How do helix-helix interactions help determine the folds of membrane proteins? Perspectives from the study of homo-oligomeric helical bundles. *Protein Sci.* **12**:647–665.

Dodevski, I., Plückthun, A. (2011) Evolution of three human GPCRs for higher expression and stability. *J. Mol. Biol.* **408**:599–615.

Duquesne, K., Sturgis, J.N. (2010) Membrane protein solubilization. *Methods Mol. Biol.* **601**:205–217.

Ehsan, M., Du, Y., Scull, N.J., Tikhonova, E., Tarrasch, J., Mortensen, J.S., Loland, C.J., Skiniotis, G., Guan, L., Byrne, B., Kobilka, B.K., Chae, P.S. (2016) Highly branched pentasaccharide-bearing amphiphiles for membrane protein studies. *J. Am. Chem. Soc.* **138**:3789–3796.

Ehsan, M., Ghani, L., Du, Y., Hariharan, P., Mortensen, J.S., Ribeiro, O., Hu, H.L., Skiniotis, G., Loland, C.J., Guan, L., Kobilka, B.K., Byrne, B., Chae, P.S. (2017) New penta-saccharide-bearing tripod amphiphiles for membrane protein structure studies. *Analyst* **142**:3889–3898

Esmann, M. (1986) Solubilized (Na$^+$ + K$^+$)-ATPase from shark rectal gland and ox kidney – an inactivation study. *Biochim. Biophys. Acta* **857**:38–47.

Fredriksson, R., Lagerstrom, M.C., Lundin, L.G., Schioth, H.B. (2003) The G protein-coupled receptors in the human genome form five main families. Phylogenetic analysis, paralogon groups, and fingerprints. *Mol. Pharmacol.* **63**:1256–1272.

Frey, L., Lakomek, N.-A., Riek, R., Bibow, S. (2017) Micelles, bicelles, and nanodiscs: Comparing the impact of membrane mimetics on membrane protein backbone dynamics. *Angew. Chem. Int. Ed.* **56**:380–383.

Frindi, M., Michels, B., Zana, R. (1992) Ultrasonic absorption studies of surfactant exchange between micelles and bulk phase in aqueous micellar solutions of nonionic surfactants with a short alkyl chain. 3. Surfactants with a sugar head group. *J. Phys. Chem.* **96**:8137–8141.

Garavito, R.M., Ferguson-Miller, S. (2001) Detergents as tools in membrane biochemistry. *J. Biol. Chem.* **276**:32403–32406.

Glaeser, R.M., Jubb, J.S., Henderson, R. (1985) Structural comparison of native and deoxycholate-treated purple membrane. *Biophys. J.* **48**:775–780.

Gohon, Y., Popot, J.-L. (2003) Membrane protein-surfactant complexes. *Curr. Opin. Colloid Interface Sci.* **8**:15–22.

Guan, L., Smirnova, I.N., Verner, G., Nagamori, S., Kaback, H.R. (2006) Manipulating phospholipids for crystallization of a membrane transport protein. *Proc. Natl. Acad. Sci. USA* **103**:1723–1726.

Haneskog, L., Andersson, L., Brekkan, E., Englund, A.K., Kameyama, K., Liljas, L., Greijer, E., Fischbarg, J., Lundahl, P. (1996) Monomeric human red cell glucose transporter (Glut1) in non-ionic detergent solution and a semi-elliptical torus model for detergent binding to membrane proteins. *Biochim. Biophys. Acta.* **1282**:39–47.

Hauser, H. (2000) Short-chain phospholipids as detergents. *Biochim. Biophys. Acta.* **1508**:164–181.

Heidmann, T., Sobel, A., Popot, J.-L., Changeux, J.-P. (1980) Reconstitution of a functional acetylcholine receptor: conservation of the conformational and allosteric transitions and recovery of the permeability response; role of lipids. *Eur. J. Biochem.* **110**:35–55.

Helenius, A., McCaslin, D.R., Fries, E., Tanford, C. (1979) Properties of detergents. *Meth. Enzymol.* **56**:734–749.

Helenius, A., Simons, K. (1975) Solubilization of membranes by detergents. *Biochim. Biophys. Acta* **415**:29–79.

Henderson, R., Jubb, J.S., Rossmann, M.G. (1982) A contracted form of the trigonal purple membrane of *Halobacterium halobium*. *J. Mol. Biol.* **154**:501–514.

Hiruma-Shimizu, K., Shimizu, H., Thompson, G.S., Kalverda, A.P., Patching, S.G. (2015) Deuterated detergents for structural and functional studies of membrane proteins: Properties, chemical synthesis and applications. *Mol. Membr. Biol.* **32**:139–155.

Hite, R.K., Gonen, T., Harrison, S.C., Walz, T. (2008) Interactions of lipids with aquaporin-0 and other membrane proteins. *Pflügers Arch.* **456**:651–661.

Hjelmeland, L.M., Nebert, D.W., Osborne, J.C., Jr. (1983) Sulfobetaine derivatives of bile acids: nondenaturing surfactants for membrane biochemistry. *Anal. Biochem.* **130**:72–82.

Hong, H., Bowie, J.U. (2011) Dramatic destabilization of transmembrane helix interactions by features of natural membrane environments. *J. Am. Chem. Soc.* **133**:11389–11398.

Hong, W.-X., Baker, K.A., Ma, X., Stevens, R.C., Yeager, M., Zhang, Q. (2011) Design, synthesis and properties of branch-chained maltoside detergents for stabilization and crystallization of integral membrane proteins: Human connexin 26. *Langmuir* **26**:8690–8696.

Hovers, J., Potschies, M., Polidori, A., Pucci, B., Raynal, S., Bonneté, F., Serrano-Vega, M., Tate, C., Picot, D., Pierre, Y., Popot, J.-L., Nehmé, R., Bidet, M., Mus-Veteau, I., Bußkamp, H., Jung, K.-H., Marx, A., Timmins, P.A., Welte, W. (2011) A class of mild surfactants that keep integral membrane proteins water-soluble for functional studies and crystallization. *Mol. Memb. Biol.* **28**:171–181.

Huganir, R.L., Schell, M.A., Racker, E. (1979) Reconstitution of the purified acetylcholine receptor from *Torpedo californica*. *FEBS Lett.* **108**:155–160.

Hunte, C., Richers, S. (2008) Lipids and membrane protein structures. *Curr. Opin. Struct. Biol.* **18**:406–411.

Hussain, H., Du, Y., Tikhonova, E., Mortensen, J.S., Ribeiro, O., Santillan, C., Das, M., Ehsan, M., Loland, C.J., Guan, L., Kobilka, B.K., Byrne, B., Chae, P.S. (2017) Resorcinarene-based facial glycosides: Implication of detergent flexibility on membrane-protein stability. *Chemistry – Eur. J.* **23**:6724–6729.

Hutchison, J.M., Lu, Z., Li, G.C., Travis, B., Mittal, R., Deatherage, C.L., Sanders, C.R. (2017) Dodecyl-β-melibioside detergent micelles as a medium for membrane proteins. *Biochemistry* **56**:5481–5484.

Israelachvili, J.N. (2011) *Intermolecular and Surface Forces*, 3rd ed.. Elsevier/Academic Press, London, 706 p.

Keana, J.F., Roman, R.B. (1978) Improved synthesis of *n*-octyl-β-D-glucoside: a nonionic detergent of considerable potential in membrane biochemistry. *Membr. Biochem.* **1**:323–327.

Kellosalo, J., Kajander, T., Kogan, K., Pokharel, K., Goldman, A. (2012) The structure and catalytic cycle of a sodium-pumping pyrophosphatase. *Science* **337**:473–476.

Khelashvili, G., LeVine, M.V., Shi, L., Quick, M., Javitch, J.A., Weinstein, H. (2013) The membrane protein LeuT in micellar systems: Aggregation dynamics and detergent binding to the S2 site. *J. Am. Chem. Soc.* **135**:14266–14275.

Kim, H.M., Howell, S.C., Van Horn, W.D., Jeon, Y.H., Sanders, C.R. (2009) Recent advances in the application of solution NMR spectroscopy to multi-span integral membrane proteins. *Progr. Nucl. Magn. Reson. Spectrosc.* **55**:335–360.

Király, Z., Börner, R.H.K., Findenegg, G.H. (1997) Adsorption and aggregation of $C_8E_4$ and $C_8G_1$ nonionic surfactants on hydrophilic silica studied by calorimetry. *Langmuir* **13**:3308–3315.

Knudsen, P., Hubbell, W.L. (1978) Stability of rhodopsin in detergent solutions. *Membr. Biochem.* **1**:297–322.

Kobilka, B.K., Deupi, X. (2007) Conformational complexity of G-protein-coupled receptors. *Trends Pharmacol. Sci.* **28**:397–406.

Kragh-Hansen, U., le Maire, M., Møller, J.V. (1998) The mechanism of detergent solubilization of liposomes and protein-containing membranes. *Biophys. J.* **75**:2932–2946.

Kurisu, G., Zhang, H., Smith, J.L., Cramer, W.A. (2003) Structure of the cytochrome $b_6 f$ complex of oxygenic photosynthesis: tuning the cavity. *Science* **302**:1009–1014.

Lanyi, J.K., Schobert, B. (2007) Structural changes in the L photointermediate of bacteriorhodopsin. *J. Mol. Biol.* **365**:1379–1392.

Lau, F.W., Nauli, S., Zhou, Y., Bowie, J.U. (1999) Changing single side-chains can greatly enhance the resistance of a membrane protein to irreversible inactivation. *J. Mol. Biol.* **290**:559–564.

le Maire, M., Champeil, P., Møller, J.V. (2000) Interaction of membrane proteins and lipids with solubilizing detergents. *Biochim. Biophys. Acta* **1508**:86–111.

Lebon, G., Bennett, K., Jazayeri, A., Tate, C.G. (2011a) Thermostabilisation of an agonist-bound conformation of the human adenosine $A_{2A}$ receptor. *J. Mol. Biol.* **409**:298–310.

Lebon, G., Warne, T., Edwards, P.C., Bennett, K., Langmead, C.J., Leslie, A.G.W., Tate, C.G. (2011b) Agonist-bound adenosine $A_{2A}$ receptor structures reveal common features of GPCR activation. *Nature* **474**:521–525.

Lee, A.G. (2011a) Biological membranes: the importance of molecular detail. *Trends Biochem. Sci.* **36**:493–500.

Lee, A.G. (2011b) Lipid-protein interactions. *Biochem. Soc. Trans.* **39**:761–766.

Lee, S., Bhattacharya, S., Tate, C.G., Grisshammer, R., Vaidehi, N. (2015) Structural dynamics and thermostabilization of neurotensin receptor 1. *J. Phys. Chem. B* **119**:4917–4928.

Lee, S., Mao, A., Bhattacharya, S., Robertson, N., Grisshammer, R., Tate, C.G., Vaidehi, N. (2016) How do short chain nonionic detergents destabilize G protein-coupled receptors? *J. Am. Chem. Soc.* **138**:15425–15433.

Lee, S.C., Bennett, B.C., Hong, W.-X., Fua, Y., Baker, K.A., Marcoux, J., Robinson, C.V., Ward, A.B., Halpert, J.R., Stevens, R.C., Stout, C.D., Yeager, M.J., Zhang, Q. (2013) Steroid-based facial amphiphiles for stabilization and crystallization of membrane proteins. *Proc. Natl. Acad. Sci. USA* **110**:E1203–E1211.

Li, H., Cocco, M.J., Steitz, T.A., Engelman, D.M. (2001) Conversion of phospholamban into a soluble pentameric helix bundle. *Biochemistry* **40**:6636–6645.

Liao, J., Li, H., Zeng, W., Sauer, D.B., Belmares, R., Jiang, Y. (2012) Structural insight into the ion-exchange mechanism of the sodium/calcium exchanger. *Science* **335**:686–690.

Lichtenberg, D., Ahyayauch, H., Goñi, F.M. (2013) The mechanism of detergent solubilization of lipid bilayers. *Biophys. J.* **105**:289–299.

Luecke, H., Schobert, B., Richter, H.-T., Cartailler, J.-P., Lanyi, J.K. (1999) Structure of bacteriorhodopsin at 1.55 Å resolution. *J. Mol. Biol.* **291**:899–911.

Lund, S., Orlowski, S., de Foresta, B., Champeil, P., le Maire, M., Møller, J.V. (1989) Detergent structure and associated lipid as determinants in the stabilization of solubilized $Ca^{2+}$-ATPase from sarcoplasmic reticulum. *J. Biol. Chem.* **264**:4907–4915.

Ma, D., Tillman, T.S., Tang, P., Meirovitch, E., Eckenhoff, R., Carninie, A., Xu, Y. (2008) NMR studies of a channel protein without membranes: Structure and dynamics of water-solubilized KcsA. *Proc. Natl. Acad. Sci. USA* **105**:16537–16542.

Magnani, F., Serrano-Vega, M.J., Shibata, Y., Abdul-Hussein, S., Lebon, G., Miller-Gallacher, J., Singhal, A., Strege, A., Thomas, J.A., Tate, C.G. (2016) A mutagenesis and screening strategy to generate optimally thermostabilized membrane proteins for structural studies. *Nat. Protoc.* **11**:1554–1571.

Magnani, F., Shibata, Y., Serrano-Vega, M.J., Tate, C.G. (2008) Co-evolving stability and conformational homogeneity of the human adenosine $A_{2A}$ receptor. *Proc. Natl. Acad. Sci. USA* **105**:10744–10749.

Marrink, S.J., Mark, A.E. (2002) Molecular dynamics simulations of mixed micelles modeling human bile. *Biochemistry* **41**:5375–4382.

Marrink, S.J., Tieleman, D.P., Mark, A.E. (2000) Molecular dynamics simulation of the kinetics of spontaneous micelle formation. *J. Phys. Chem. B* **104**:12165–12173.

Matar-Merheb, R., Rhimi, M., Leydier, A., Huché, F., Galián, C., Desuzinges-Mandon, E., Ficheux, D., Flot, D., Aghajari, H., Kahn, R., Di Pietro, A., Jault, J.-M., Coleman, A.W., Falson, P. (2011) Structuring detergents for extracting and stabilizing functional membrane proteins. *PLoS One* **6**:e18036.

Miller, J.L., Tate, C.G. (2011) Engineering an ultra-thermostable $\beta_1$-adrenoceptor. *J. Mol. Biol.* **413**:628–638.

Miyagishi, S., Okada, K., Asakawa, T. (2001) Salt effect on critical micelle concentrations of nonionic surfactants, *N*-acyl-*N*-methylglucamides (MEGA-n). *J. Colloid Interface Sci.* **238**:91–95.

Molina-Bolívar, J.A., Hierrezuelo, J.M., Carnero Ruiz, C. (2013) Energetics of clouding and size effects in non-ionic surfactant mixtures: The influence of alkyl chain length and NaCl addition. *J. Chem. Thermodynamics* **57**:59–66.

Neugebauer, J.M. (1988) *A Guide to the Properties and Uses of Detergents in Biology and Biochemistry*. Calbiochem Co., La Jolla, 64 p.

Neugebauer, J.M. (1990) Detergents: An overview. *Meth. Enzymol.* **182**:239–253.

Norimatsu, Y., Hasegawa, K., Shimizu, N., Toyoshima, C. (2017) Protein-phospholipid interplay revealed with crystals of a calcium pump. *Nature* **545**:193–198.

Orwick-Rydmark, M., Arnold, T., Linke, D. (2016) The use of detergents to purify membrane proteins. *Curr. Protoc. Protein Sci.* **84**:4.8.1–4.8.34.

Ostermeier, C., Harrenga, A., Ermler, U., Michel, H. (1997) Structure at 2.7 Å resolution of the *Paracoccus denitrificans* two-subunit cytochrome *c* oxidase complexed with an antibody Fv fragment. *Proc. Natl. Acad. Sci. USA* **94**:10547–10553.

Otzen, D.E. (2015) Proteins in a brave new surfactant world. *Curr. Opin. Colloid Interface Sci.* **20**:161–169.

Pagliano, C., Barera, S., Chimirri, F., Saracco, G., Barber, J. (2012) Comparison of the α and β isomeric forms of the detergent *n*-dodecyl-D-maltoside for solubilizing photosynthetic complexes from pea thylakoid membranes. *Biochim. Biophys. Acta* **1817**:1506–1515.

Palczewski, K., Kumasaka, T., Hori, T., Behnke, C.A., Motoshima, H., Fox, B.A., Le Trong, I., Teller, D.C., Okada, T., Stenkamp, R.E., Yamamoto, M., Miyano, M. (2000) Crystal structure of rhodopsin: a G protein-coupled receptor. *Science* **289**:739–745.

Patargias, G., Bond, P.J., Deol, S.S., Sansom, M.S.P. (2005) Molecular dynamics simulations of GlpF in a micelle *vs.* in a bilayer: Conformational dynamics of a membrane protein as a function of environment. *J. Phys. Chem. B* **109**:575–582.

Pebay-Peyroula, E., Garavito, R.M., Rosenbusch, J.P., Zulauf, M., Timmins, P. (1995) Detergent structure in tetragonal crystals of OmpF porin. *Structure* **3**:1051–1059.

Penel, S., Pebay-Peyroula, E., Rosenbusch, J., Rummel, G., Schirmer, T., Timmins, P.A. (1998) Detergent binding in trigonal crystals of OmpF porin from *Escherichia coli. Biochimie* **80**:543–551.

Pérez-Victoria, I., Pérez-Victoria, F.J., Roldán-Vargas, S., García-Hernández, R., Carvalho, L., Castanys, S., Gamarro, F., Morales, J.C., Pérez-Victoria, J.M. (2011) Non-reducing trisaccharide fatty acid monoesters: novel detergents in membrane biochemistry. *Biochim. Biophys. Acta* **1808**:717–726.

Perlmutter, J.D., Popot, J.-L., Sachs, J.N. (2014) Molecular dynamics simulations of a membrane protein/amphipol complex. *J. Membr. Biol.* **247**:883–895.

Perozo, E., Cortes, D.M., Cuello, L.G. (1998) Three-dimensional architecture and gating mechanism of a $K^+$ channel studied by EPR spectroscopy. *Nat. Struct. Biol.* **5**:459–469.

Pierre, Y., Breyton, C., Kramer, D., Popot, J.-L. (1995) Purification and characterization of the cytochrome $b_6 f$ complex from *Chlamydomonas reinhardtii. J. Biol. Chem.* **270**:29342–29349.

Poget, S.F., Girvin, M.E. (2007) Solution NMR of membrane proteins in bilayer mimics: small is beautiful, but sometimes bigger is better. *Biochim. Biophys. Acta* **1768**:3098–3106.

Pomroy, N.C., Deber, C.M. (1998) Solubilization of hydrophobic peptides by reversible cysteine PEGylation. *Biochem. Biophys. Res. Commun.* **245**:618–621.

Popot, J.-L. (2010) Amphipols, nanodiscs, and fluorinated surfactants: Three non-conventional approaches to studying membrane proteins in aqueous solutions. *Annu. Rev. Biochem.* **79**:737–775.

Popot, J.-L. (2014) Folding membrane proteins *in vitro*: A table and some comments. *Arch. Biochem. Biophys.* **564**:314–326.

Popot, J.-L., Althoff, T., Bagnard, D., Banères, J.-L., Bazzacco, P., Billon-Denis, E., Catoire, L.J., Champeil, P., Charvolin, D., Cocco, M.J., Crémel, G., Dahmane, T., de la Maza, L.M., Ebel, C., Gabel, F., Giusti, F., Gohon, Y., Goormaghtigh, E., Guittet, E., Kleinschmidt, J.H., Kühlbrandt, W., Le Bon, C., Martinez, K.L., Picard, M., Pucci, B., Rappaport, F., Sachs, J.N., Tribet, C., van Heijenoort, C., Wien, F., Zito, F., Zoonens, M. (2011) Amphipols from A to Z. *Annu. Rev. Biophys.* **40**:379–408.

Popot, J.-L., Cartaud, J., Changeux, J.-P. (1981) Reconstitution of a functional acetylcholine receptor: incorporation into artificial lipid vesicles and pharmacology of the agonist-controlled permeability changes. *Eur. J. Biochem.* **118**:203–214.

Popot, J.-L., Engelman, D.M. (2016) Membranes do not tell proteins how to fold. *Biochemistry* **55**:5–18.

Prince, S.M., Howard, T.D., Myles, D.A., Wilkinson, C., Papiz, M.Z., Freer, A.A., Cogdell, R.J., Isaacs, N. (2003) Detergent structure in crystals of the integral membrane light-harvesting complex LH2 from *Rhodopseudomonas acidophila* strain 10050. *J. Mol. Biol.* **326**:307–315.

Privé, G.G. (2007) Detergents for the stabilization and crystallization of membrane proteins. *Methods* **41**:388–397.

Qin, L., Hiser, C., Mulichak, A., Garavito, R.M., Ferguson-Miller, S. (2006) Identification of conserved lipid/detergent-binding sites in a high-resolution structure of the membrane protein cytochrome *c* oxidase. *Proc. Natl. Acad. Sci. USA* **103**:16117–16122.

Rodríguez-Ropero, F., Fioroni, M. (2012) Structural and dynamical analysis of an engineered FhuA channel protein embedded into a lipid bilayer or a detergent belt. *J. Struct. Biol.* **177**:291–301.

Rosenbaum, D.M., Rasmussen, S.G.F., Kobilka, B.K. (2009) The structure and function of G protein-coupled receptors. *Nature* **459**:356–363.

Rosevear, P., VanAken, T., Baxter, J., Ferguson-Miller, S. (1980) Alkyl glycoside detergents: a simpler synthesis and their effects on kinetic and physical properties of cytochrome *c* oxidase. *Biochemistry* **19**:4108–4115.

Roth, M., Arnoux, B., Ducruix, A., Reiss-Husson, F. (1991) Structure of the detergent phase and protein-detergent interactions in crystals of the wild-type (strain Y) *Rhodobacter sphaeroides* photochemical reaction center. *Biochemistry* **30**:9403–9413.

Roth, M., Lewitt-Bentley, A., Michel, H., Deisenhofer, J., Huber, R., Oesterhelt, D. (1989) Detergent structure in crystals of a bacterial photosynthetic reaction center. *Nature* **340**:659–662.

Rouse, S.L., Sansom, M.S.P. (2015) Interactions of lipids and detergents with a viral ion channel protein: Molecular dynamics simulation studies. *J. Phys. Chem. B* **119**:764–772.

Ruiz, M.B., Prado, A., Goñi, F.M., Alonso, A. (1994) An assessment of the biochemical applications of the non-ionic surfactant Hecameg. *Biochim. Biophys. Acta* **1193**:301–306.

Sadaf, A., Cho, K.H., Byrne, B., Chae, P.S. (2015) Amphipathic agents for membrane protein study. *Meth. Enzymol.* **557**:57–94.

Sadaf, A., Mortensen, J.S., Capaldi, S., Tikhonova, E., Hariharan, P., Ribeiro, O., Loland, C.J., Guan, L., Byrne, B., Chae, P.S. (2016) A class of rigid linker-bearing glucosides for membrane protein structural study. *Chem. Sci.* **7**:1933–1939.

Saif Hasan, S., Yamashita, E., Cramer, W.A. (2013) Transmembrane signaling and assembly of the cytochrome $b_6f$-lipidic charge transfer complex. *Biochim. Biophys. Acta* **1827**:1295–1308.

Salvay, A.G., Santamaria, M., le Maire, M., Ebel, C. (2007) Analytical ultracentrifugation sedimentation velocity for the characterization of detergent-solubilized membrane proteins Ca$^{++}$-ATPase and ExbB. *J. Biol. Phys.* **33**:399–419.

Sanders II, C.R., Landis, G.C. (1995) Reconstitution of membrane proteins into lipid-rich bilayered mixed micelles for NMR studies. *Biochemistry* **34**:4030–4040.

Sarkar, C.A., Dodevski, I., Kenig, M., Dudli, S., Mohr, A., Hermans, E., Plückthun, A. (2008) Directed evolution of a G protein-coupled receptor for expression, stability, and binding selectivity. *Proc. Natl. Acad. Sci. USA* **105**:14808–14813.

Sauer, O., Roth, M., Schirmer, T., Rummel, G., Kratky, C. (2002) Low-resolution detergent tracing in protein crystals using xenon or krypton to enhance X-ray contrast. *Acta Crystallogr. D* **58**:60–69.

Screpanti, E., Padan, E., Rimon, A., Michel, H., Hunte, C. (2006) Crucial steps in the structure determination of the Na$^+$/H$^+$ antiporter NhaA in its native conformation. *J. Mol. Biol.* **362**:192–202.

Seddon, A.M., Curnow, P., Booth, P.J. (2004) Membrane proteins, lipids and detergents: not just a soap opera. *Biochim. Biophys. Acta* **1666**:105–117.

Serrano-Vega, M.J., Magnani, F., Shibata, Y., Tate, C.G. (2008) Conformational thermostabilization of the $\beta_1$-adrenergic receptor in a detergent-resistant form. *Proc. Natl. Acad. Sci. USA* **105**:877–882.

Shibata, Y., White, J.F., Serrano-Vega, M.J., Magnani, F., Aloia, A.L., Grisshammer, R., Tate, C.G. (2009) Thermostabilization of the neurotensin receptor NTS1. *J. Mol. Biol.* **390**:262–277.

Sirokmán, G., Fasman, G.D. (1993) Refolding and proton pumping activity of a polyethylene glycol-bacteriorhodopsin water-soluble conjugate. *Protein Sci.* **2**:1161–1170.

Slovic, A.M., Kono, H., Lear, J.D., Saven, J.G., DeGrado, W.F. (2004) Computational design of water-soluble analogues of the potassium channel KcsA. *Proc. Natl. Acad. Sci. USA* **101**:1828–1833.

Slovic, A.M., Lear, J.D., DeGrado, W.F. (2005a) *De novo* design of a pentameric coiled-coil: decoding the motif for tetramer versus pentamer formation in water-soluble phospholamban. *J. Pept. Res.* **65**:312–321.

Slovic, A.M., Stayrook, S.E., North, B., Degrado, W.F. (2005b) X-ray structure of a water-soluble analog of the membrane protein phospholamban: sequence determinants defining the topology of tetrameric and pentameric coiled coils. *J. Mol. Biol.* **348**:777–787.

Snijder, H.J., Timmins, P.A., Kalk, K.H., Dijkstra, B.W. (2003) Detergent organisation in crystals of monomeric outer membrane phospholipase A. *J. Struct. Biol.* **141**:122–131.

Stroebel, D., Choquet, Y., Popot, J.-L., Picot, D. (2003) An atypical haem in the cytochrome $b_6f$ complex. *Nature* **426**:413–418.

Stubbs, G.W., Smith, H.G., Jr., Litman, B.J. (1976) Alkyl glucosides as effective solubilizing agents for bovine rhodopsin. A comparison with several commonly used detergents. *Biochim. Biophys. Acta* **426**:46–56.

Tanford, C. (1972) Micelle Shape and Size. *J. Phys. Chem.* **76**:3020–3024.

Tanford, C. (1980) *The Hydrophobic Effect: Formation of Micelles and Biological Membranes*, 2nd ed.. Wiley, New York, 233 p.

Tanford, C., Reynolds, J.A. (1976) Characterization of membrane proteins in detergent solutions. *Biochim. Biophys. Acta* **457**:133–170.

Tate, C.G. (2010) Practical considerations of membrane protein instability for purification and crystallisation, in: Mus-Veteau, I., ed., *Membrane Protein Expression*. The Humana Press, Totowa, New Jersey, USA, pp. 187–203.

Tate, C.G. (2012) A crystal-clear solution for determining G-protein-coupled receptor structures. *Trends Biochem. Sci.* **37**:343–352.

Tausk, R.J.M., Karmiggelt, J., Oudshoorn, C., Overbeek, J.T.G. (1974) Physical chemical studies of short-chain lecithin homologues. I.: Influence of the chain length of the fatty acid ester and of electrolytes on the critical micelle concentration. *Biophys. Chem.* **1**:175–183.

Timmins, P., Pebay-Peyroula, E., Welte, W. (1994) Detergent organisation in solutions and in crystals of membrane proteins. *Biophys. Chem.* **53**:27–36.

Toyoshima, C., Nakasako, M., Nomura, H., Ogawa, H. (2000) Crystal structure of the calcium pump of sarcoplasmic reticulum at 2.6 Å resolution. *Nature* **405**:647–655.

Triba, M.N., Warschawski, D.E., Devaux, P.F. (2005) Reinvestigation by phosphorus NMR of lipid distribution in bicelles. *Biophys. J.* **88**:1887–1901.

Vaidehi, N., Grisshammer, R., Tate, C.G. (2016) How can mutations thermostabilize G protein-coupled receptors? *Trends Pharmacol. Sci.* **37**:37–46.

VanAken, T., Foxall-VanAken, S., Castleman, S., Ferguson-Miller, S. (1986) Alkyl glycoside detergents: synthesis and applications to the study of membrane proteins. *Methods Enzymol.* **125**:27–35.

Vieler, A., Wilhelm, C., Goss, R., Süss, R., Schiller, J. (2007) The lipid composition of the unicellular green alga *Chlamydomonas reinhardtii* and the diatom *Cyclotella meneghiniana* investigated by MALDI-TOF MS and TLC. *Chem. Phys. Lipids* **150**:143–155.

Wang, C., Deber, C.M. (2000) Peptide mimics of the M13 coat protein transmembrane segment. Retention of helix-helix interaction motifs. *J. Biol. Chem.* **275**:16155–16159.

Warne, T., Serrano-Vega, M.J., Baker, J.G., Moukhametzianov, R., Edwards, P.C., Henderson, R., Leslie, A.G.W., Tate, C.G., Schertler, G.F.X. (2008) Structure of a $\beta_1$-adrenergic G protein-coupled receptor. *Nature* **454**:486–491.

Warr, G.G., Drummond, C.J., Grieser, F., Ninham, B.W., Evans, D.F. (1986) Aqueous solution properties of nonionic *n*-dodecyl-β-D-maltoside micelles. *J. Phys. Chem.* **90**:4581–4586.

Wei, J., Fasman, G.D. (1995) A poly(ethylene glycol) water-soluble conjugate of porin: refolding to the native state. *Biochemistry* **34**:6408–6415.

Wiener, M.C. (2004) A pedestrian guide to membrane protein crystallization. *Methods* **34**:364–372.

Yu, S.M., McQuade, D.T., Quinn, M.A., Hackenberger, C.P., Krebs, M.P., Polans, A.S., Gellman, S.H. (2000) An improved tripod amphiphile for membrane protein solubilization. *Protein Sci.* **9**:2518–2527.

Zhang, Q., Ma, X., Ward, A., Hong, W.X., Jaakola, V.P., Stevens, R.C., Finn, M.G., Chang, G. (2007) Designing facial amphiphiles for the stabilization of integral membrane proteins. *Angew. Chem. Int. Ed. Engl.* **46**:7023–7025.

Zhang, Q., Tao, H., Hong, W.-X. (2011) New amphiphiles for membrane protein structural biology. *Methods* **55**:318–323.

Zhao, X., Perez-Aguilar, J.M., Matsunaga, F., Lerner, M., Xi, J., Selling, B., Johnson, A.T., Jr., Saven, J.G., Liu, R. (2014) Characterization of a computationally designed water-soluble human μ-opioid receptor variant using available structural information. *Anesthesiology* **121**:866–875.

Zhou, H.X., Cross, T.A. (2013) Influences of membrane mimetic environments on membrane protein structures. *Annu. Rev. Biophys.* **42**:361–392.

Zhou, Y., Bowie, J.U. (2000) Building a thermostable membrane protein. *J. Biol. Chem.* **275**:6975–6979.

Zoonens, M., Comer, J., Masscheleyn, S., Pebay-Peyroula, E., Chipot, C.J., Miroux, B., Dehez, F. (2013) Dangerous liaisons between detergents and membrane proteins. The case of mitochondrial uncoupling protein 2. *J. Am. Chem. Soc.* **135**:15174–15182.

Zulauf, M. (1991) Detergent phenomena in membrane protein crystallization, in: Michel, H., ed., *Crystallization of Membrane Proteins*. CRC Press, Boca Raton, pp. 53–72.

References

# Alternatives to Detergents for Handling Membrane Proteins in Aqueous Solutions

**3**

## Summary

*Attempts at substituting detergents with other surfactants for handling membrane proteins (MPs) in aqueous solutions have a long history. They are based on three main incentives: (i) trying to improve the stability of solubilized MPs; (ii) providing them with an environment that, in its physical characteristics and/or its chemical composition, is closer to the natural environment; and (iii) making them accessible to technologies that are difficult or impossible to implement in the presence of detergents. A first route is to reinsert the protein in a lipid bilayer, most often closed upon itself in the form of lipid vesicles, sometimes forming a planar "black lipid membrane." This approach is obligatory when functional assays require the protein to have access to two distinct aqueous compartments, but the objects formed are large, if not macroscopic, and do not lend themselves well to most biophysical investigations. A second route is to substitute totally or partially the detergent with other surfactants while forming water-soluble particles of nanometric dimensions. The use of specially developed amphipathic polymers called amphipols is one such approach, which will be described in detail in Chaps. 4 and 5, but it is far from being the only one. In order to provide a broader view of which systems are available to the experimenter, the present chapter reviews the four principal alternatives to detergents and amphipols: (i) bicelles, which are mixtures of lipids and detergents or short-chain lipids that, under appropriate conditions, form disc-shaped bilayer fragments into which MPs can integrate; (ii) nanodiscs, whose basic concept is similar to that of bicelles, but in which the rim of the bilayer disc is stabilized by specially engineered proteins; (iii) peptides or lipopeptides, which can either interact directly with the MP to be solubilized or stabilize MP/lipid complexes; and (iv) fluorinated surfactants, which resemble detergents in their chemical structure but whose hydrophobic chains contain fluorine atoms, which make them lyophobic (poorly miscible with hydrocarbons); this renders them less disruptive of the protein/protein and protein/lipid interactions that stabilize MPs.*

© Springer International Publishing AG, part of Springer Nature 2018
J. -L. Popot, *Membrane Proteins in Aqueous Solutions*, Biological and Medical Physics, Biomedical Engineering, https://doi.org/10.1007/978-3-319-73148-3_3

## 3.1　Introduction

In Chap. 2, we have discussed the mechanisms that underlie the frequent instability of membrane proteins (MPs) in detergent solutions and argued that a key factor is the loss of protein/protein and protein/lipid interactions that stabilize the protein in situ. We have mentioned four possible countermeasures: (i) to transfer the protein to lipid vesicles or other detergent-free lipid phases, (ii) to genetically modify it so as to render it detergent-resistant, (iii) to turn it into a water-soluble protein, and (iv) to develop particularly mild detergents. The systems described in the present chapter draw on approaches *i* and *iv*. As in approach *i*, they aim to preserve protein/lipid interactions and, for some of them, to reconstitute, around the protein, an environment chemically and physically as similar as possible to that it experiences in vivo while generally avoiding the formation of large structures. As in approach *iv*, they often rely of the development of original surfactants whose structure and properties are more and more remote from those of classical detergents.

The following four approaches will be considered:

(i) *Bicelles*, that is discoidal fragments of lipid bilayer whose rim is stabilized by molecules of detergent or short-chain lipid;

(ii) *Nanodiscs*, which also comprise a discoidal bilayer fragment but are stabilized by specially designed amphipathic proteins;

(iii) *Peptide-based systems*, which come in a variety of flavors, some of them attempting to mimic the proteins that stabilize nanodiscs, whereas others simply aim at replacing detergents;

(iv) *Fluorinated surfactants*, which are detergent-like molecules whose hydrophobic moiety or moieties incorporate(s) fluorine, which limits their ability to disrupt protein/protein and protein/lipid interactions.

For general reviews covering these and other systems, see e.g. Popot (2010), Warschawski et al. (2011), Zhang et al. (2011), Serebryany et al. (2012), Catoire et al. (2014), Otzen (2015), Tu et al. (2016), and Mineev and Nadezhdin (2017).

*Alternatives to detergents*
(© *2018 by Francis Haraux*)

## 3.2   Bicelles

In the field of MP studies, bicelles have been used mostly for NMR investigations and for MP crystallization, with a smattering of other applications. Because their use for NMR is intimately linked to their mode of formation and phase diagram, we will deal with these two aspects along with NMR applications.

### 3.2.1   The Formation and Phase Diagram of Bicelles and Their Use in Membrane Protein NMR Spectroscopy

It has long been known that certain mixtures of lipids and detergent do not distribute more or less randomly within mixed micelles but organize into patches of lipid bilayers surrounded and saturated with the detergent. Such is the case of phosphatidylcholine/bile salt mixtures. In the case of bile salts, this phenomenon is attributed to the laterally amphipathic nature of the sterol ring, one face of which is rendered hydrophilic by the presence of hydroxyl moieties. This behavior explains the ability of bile salts to dissolve high mass ratios of lipids (Small 1971; Carey and Small 1972). Before the advent, in the 1980s, of sugar-based detergents such as octylglucoside (OG) and dodecylmaltoside (DDM), bile salts were, along with Triton X-100 and Lubrol WX, among the "non-denaturing" detergents most frequently used by biochemists. However, the peculiar structure of lipid/bile salt mixed micelles was not exploited by membrane biophysicists until the 1990s, when it was realized, most notably under the impulsion of James H. Prestegard and Charles R. Sanders, (i) that the size of these assemblies could be modulated by playing on the lipid/detergent ratio; (ii) that they could be obtained with other combinations of surfactants, e.g. dimyristoylphosphatidylcholine (DMPC, $diC_{14:0}PC$)/$diC_6PC$ mixtures (Sanders and Schwonek 1992); (iii) that the discs thus obtained, dubbed "bicelles" – short for "bilayered micelles" (Sanders and Landis 1995) – could provide a bilayer-like environment to MPs; and (iv) that, if large enough, they would, as shown previously for non-biological systems (reviewed in Forrest and Reeves 1981), align in the magnetic field of NMR spectrometers (Ram and Prestegard 1988; Sanders and Prestegard 1990; Sanders and Schwonek 1992; Sanders et al. 1994; Sanders and Landis 1995; Prosser et al. 1996, 1998; for reviews, see e.g. Sanders and Prosser 1998; Opella and Marassi 2004; Sanders and Sönnichsen 2006; Wang 2008; Kim et al. 2009; Qureshi and Goto 2011; Warschawski et al. 2011; Dürr et al. 2013; Catoire et al. 2014; for a historical perspective, see Sanders 2008).

Bicelles are rather easily obtained by mixing in aqueous solution one or more lipids and the surfactant, generally CHAPSO or $diC_6PC$, that will form the rim (Fig. 3.1). Solubilization may be slow and may have to be accelerated by cycles of heating and cooling, freeze/thawing, and/or shearing (see e.g. Sanders et al. 1994; Mäler and Gräslund 2009; Ujwal and Bowie 2011). This is somewhat time-consuming, but, once formed, a preparation of bicelles can be stored frozen and used repeatedly. MPs are usually incorporated by diluting concentrated bicelles with a detergent solution of the protein (see e.g. Ujwal and Bowie 2011), but, depending on the protein, it is also possible to reconstitute it in lipid vesicles formed by the long-chain lipid and dissolve them with the detergent, to add the long-chain lipid to a solution of the protein in the detergent, to add buffer to a dried mixture of the protein and the two surfactants, etc. (see e.g. Triba et al. 2006; De Angelis and Opella 2007; Poget and Girvin 2007; Mäler and Gräslund 2009; Dürr et al. 2012; Morrison and Henzler-Wildman 2012, and references therein). In Barrett et al. (2012), the polyhistidine-tagged target MP (the transmembrane C-terminal domain of the amyloid precursor protein) was bound to an Ni:NTA column in detergent solution, the solution exchanged for DMPC/$diC_6PC$ bicelles, and the bicelle-integrated protein eluted with a bicelle-containing solution.

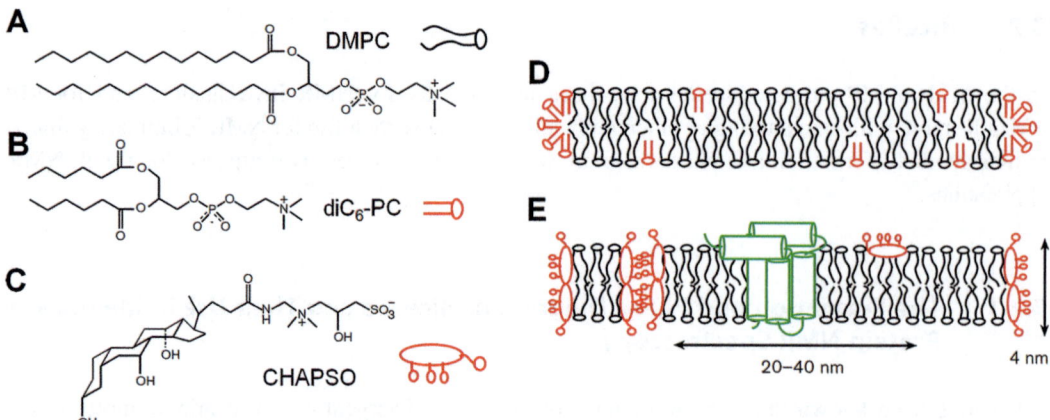

**Fig. 3.1** Components and cross-section models for small bicelles. (**A–C**) The chemical structures of dimyristoylphosphatidylcholine (diC$_{14:0}$PC, DMPC), diC$_6$PC, and CHAPSO. (**D**) A model DMPC/diC$_6$PC bicelle. (**E**) A model DMPC/CHAPSO bicelle containing an integral membrane protein (in *green*). Above the gel/liquid crystal transition for DMPC (~24 °C; Mabrey and Sturtevant 1976), the two components become partially miscible, and the small surfactant distributes between the rim and the plane of the bicelle. In the case of CHAPSO, there are at least two possible modes of bilayer interaction (Small 1971; Carey and Small 1972; Muller 1981): surface-associated and transmembrane (in the form of an oligomer in which the hydroxyl groups either face each other, as shown here, or line an aqueous pore). The bicelles are drawn approximately to scale on the basis of the known dimensions of $\alpha$-helices, liquid crystalline phosphatidylcholine molecules (Lewis and Engelman 1983), the thickness of DMPC bilayers (ibid.), and the diameter of bicelles (Chung and Prestegard 1993; Vold and Prosser 1996). The diameter illustrated is 25 nm (to scale) but can be varied experimentally by adjusting the lipid/small surfactant ratio (cf. Fig. 3.3) (From Sanders and Prosser 1998. © 1998 Elsevier Science Ltd. All rights reserved).

The functionality and stability of bicelle-integrated MPs are important points to establish before engaging into time-consuming structural investigations. Thus, diacylglycerol kinase (DAGK) has been shown to be functional in DMPC/CHAPSO and dipalmitoylphosphatidylcholine (diC$_{16:0}$PC, DPPC)/CHAPSO bicelles, by preference to diC$_{12:0}$PC/CHAPSO or diC$_6$PC-based bicelles (Czerski and Sanders 2000). Smr, the small multidrug-resistance protein from *Staphylococcus aureus*, binds its ligand tetraphenylphosphonium more efficiently in DMPC/diC$_6$PC bicelles than it does in most detergents (Poget et al. 2007). The achievement of a native-like three-dimensional conformation, long-term stability, and ability to bind specific drugs of Smr required careful tuning of bicelle composition (Poget and Girvin 2007). The DMPC/diC$_6$PC system is nowadays the most frequently used for NMR measurements and that which tends to yield the best data (Kim et al. 2009). Many variations exist, however, bearing on the nature of the lipid(s), of the detergent, on doping with paramagnetic ions (see below), and so forth (for recent examples, see e.g. Beaugrand et al. 2016, Mineev et al. 2017, Smrt et al. 2017, and references therein, and for an overview, Dürr et al. 2013).

An interesting property of bicelles, of particular value for NMR studies, is that large enough bicelles align in magnetic fields (Fig. 3.2). The degree of orientation depends on several factors, such as their overall concentration, the temperature, and their size, the latter depending on the ratio of large to small surfactant. Two cases of figures are encountered. Pure bicelles orient with their planes parallel to the magnetic field, as illustrated in Fig. 3.2, *center*. By doping them with paramagnetic ions, often with the help of a small fraction of lipids tagged so as to bind these ions, the orientation can flip to the orthogonal one, which is more convenient for NMR experiments, in which the bicelle plane is perpendicular to the magnetic field (Fig. 3.2, *right*; see e.g. Sanders et al. 1994; Prosser et al. 1996, 1998; Vold and Prosser 1996; Tiburu et al. 2001). Solid-state NMR analysis of such samples permits to

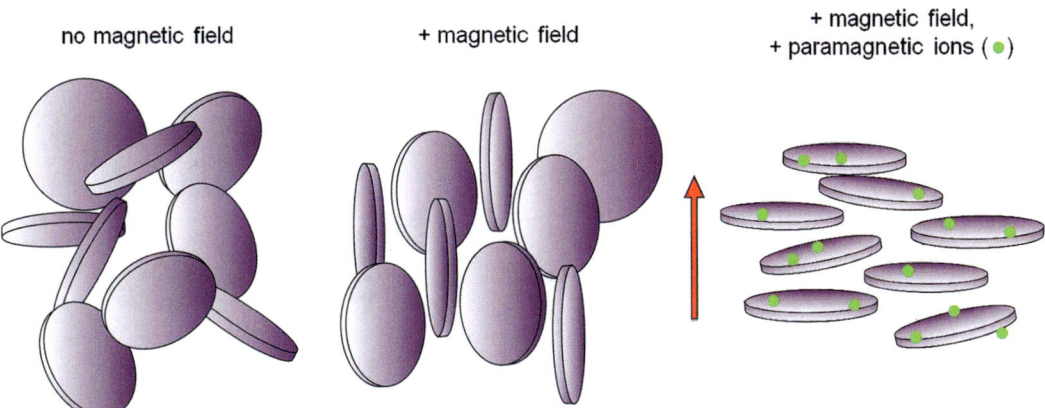

**Fig. 3.2** The alignment of large bicelles in a magnetic field. Pure bicelles organize with their plane parallel to the magnetic field (*red arrow*). By doping them with lanthanide ions (*green dots*), it is also possible to induce them to align with their planes normal to the field (Prosser et al. 1996, 1998). In this case, however, the exact structure adopted is uncertain (Adapted from Sanders and Prosser 1998. © 1998 Elsevier Science Ltd. All rights reserved).

determine the angle that the axis of an $\alpha$-helix or $\beta$-barrel makes with the plane of the bicelle and to place constraints on the angle that the axes of two helices make with one another. Magnetic alignment also provides an access to dipolar coupling and chemical shift anisotropy measurements, which can be directly employed for protein structural determination (see e.g. Ketchem et al. 1993, where alignment was achieved by the glass plate method).

The phase diagram of bicelle-forming surfactant mixtures is complex, because the structure and properties of the assemblies formed depend on the ratio between the small and large surfactants, on their nature, on their overall concentration, on the temperature, the presence of salts, and so forth (see e.g. Sternin et al. 2001; Harroun et al. 2005; Triba et al. 2005; van Dam et al. 2006; Ujwal and Bowie 2011; Warschawski et al. 2011; Dürr et al. 2012, 2013; Liebau et al. 2016). A phase diagram of DMPC/ diC$_6$PC mixtures as a function of temperature and the mass ratio $q$ of DMPC to diC$_6$PC, established by phosphorus NMR, is shown in Fig. 3.3, *left*, and some of the structures adopted schematically depicted in Fig. 3.3, *right*. Bicelles in which the two lipids are fully segregated (B) occupy only part of the diagram (below the melting temperature $T_m$ of DMPC, where its miscibility with diC$_6$PC is minimal). The other structures found include mixed micelles (M$_m$), at very low $q$; mixed bicelles (B$_m$), in the bilayer region of which the two lipids partially mix; perforated vesicles or sheets (V$_p$); and non-perforated vesicles (V$_m$) in which the two lipids are totally mixed. Raising the temperature of a given mixture (*red arrow* in Fig. 3.3, *left*) increases the miscibility of the two lipids, which takes the preparation through the various types of structures. Care should therefore be brought to choosing experimental conditions so as to remain in that region of the mixture's phase diagram where bicelles of the desired size will form, e.g. the small so-called "isotropic" bicelles – meaning "tumbling isotropically" – used for solution NMR (B in Fig. 3.3) or the large mixed bicelles (B$_m$ in Fig. 3.3) that align best in magnetic fields and are used for solid-state NMR applications. Excessive dilution should be avoided, for instance, or the bicelles will turn into vesicles. This is a serious experimental constraint, avoided by nanodiscs or amphipols. On the contrary, the transition from bicelles to perforated or non-perforated vesicles or sheets probably plays a useful role when crystallizing MPs from a bicellar preparation (see § 3.2.2). The way detergent solutions, isotropically tumbling and aligned bicelles, and lipid vesicles complement each other in the study of the structure and dynamics of MPs and their amenability to the various forms of NMR spectroscopy are schematically illustrated in Fig. 3.4.

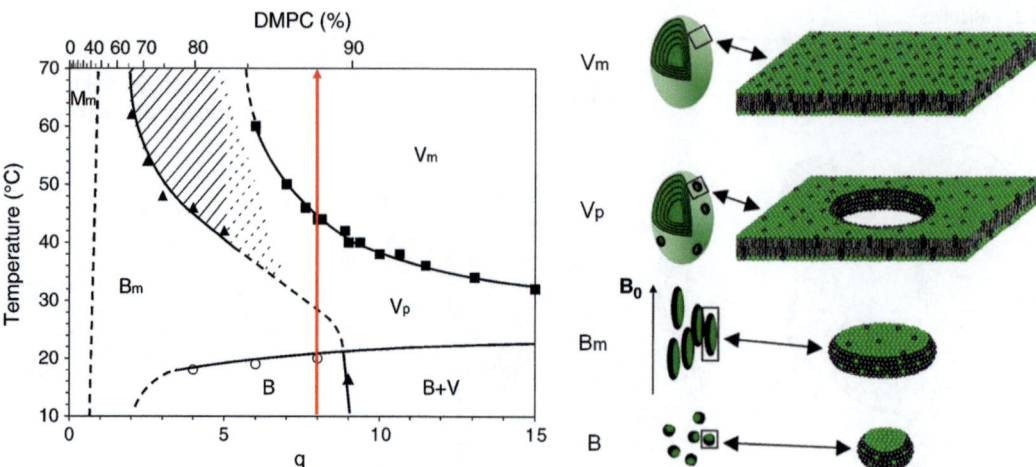

**Fig. 3.3** Left. Tentative temperature/composition diagram of DMPC/diC$_6$PC mixtures, at a concentration in water of 25% (w/w), based on $^{31}$P NMR. The lipid composition is given by $q$, the mass ratio of DMPC to diC$_6$PC. The diagram is incomplete, as some domain boundaries could not be precisely determined (*dashed lines*). The limits determined with *open circles*, *solid triangles*, and *solid squares* correspond, respectively, to $T_m$, the melting temperature of DMPC; $T_v$, above which vesicles start to form; and $T_h$, above which holes disappear. Below $T_m$ (the melting temperature of the acyl chains of DMPC), perfectly segregated bicelles (B) are found, coexisting with DMPC vesicles (V) at high $q$ values. Above $T_m$, diC$_6$PC molecules enter the bilayer and start to form mixed bicelles (B$_m$). Above $T_v$, multilamellar vesicles form, starting with perforated vesicles and/or sheets (V$_p$), coexisting with the remaining mixed bicelles (*shaded zone*). Finally, above $T_h$, bicelles and holes disappear because the DMPC bilayer is no longer saturated with diC$_6$PC, and only mixed vesicles (V$_m$) remain. For very low $q$ values, mixed bicelles probably turn into mixed micelles (M$_m$), without any lipid segregation. Raising from 10 to 70 °C, the temperature of a $q = 8$ mixture (*red arrow*) takes it through the various structures shown to the right. Right. Cartoon representation of some of the structures adopted by DMPC/diC$_6$PC mixtures as a function of temperature and the mass ratio of DMPC to diC$_6$PC, as determined by $^{31}$P NMR. *Light green* lipids represent DMPC, *dark green* lipids diC$_6$PC. The structures formed are represented at different scales (Adapted from Triba et al. 2005, © 2005 The Biophysical Society. Published by Elsevier Inc. All rights reserved. and Warschawski et al. 2011, © 2011 Elsevier B.V. All rights reserved).

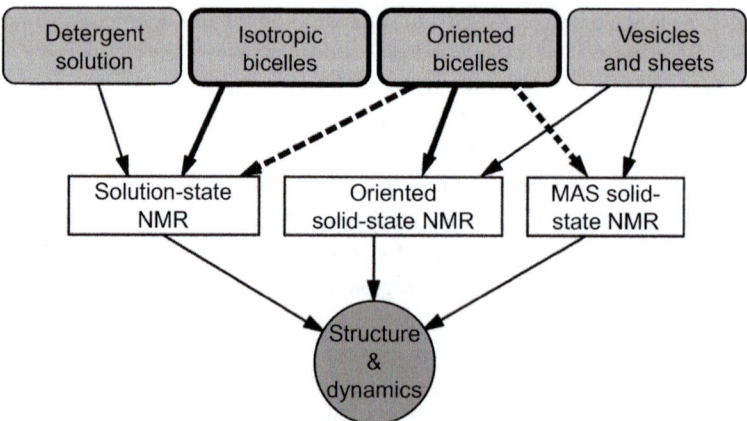

**Fig. 3.4** A schematic overview of the complementarity between detergent solutions, isotropically tumbling bicelles, aligned bicelles, and lipid vesicles, and the various forms of NMR spectroscopy used for studying the structure and dynamics of membrane proteins. Oriented solid-state samples can be obtained either with aligned bicelles or by spreading vesicles onto glass plates so as to form stacks of parallel sheets. MAS: magic-angle spinning (Adapted with permission from Dürr et al. 2012, © 2012 American Chemical Society).

Bicelles have been extensively used for NMR studies of molecules that interact with membranes, such as drugs, peptides that adsorb onto or insert into them, or very simple, typically dimeric bitopic MP fragments (see e.g. Bocharov et al. 2007, 2010), as well as that of soluble molecules that orient in the presence of oriented micelles (reviewed by Prosser et al. 2006 and Dürr et al. 2013). Of the ~140 NMR structures of TM proteins or peptides listed on www.drorlist.com/nmr/MPNMR.html, the site maintained by Dror E. Warschawski, the determination of only a dozen has resorted to bicelles, often in combination with other media (see Warschawski et al. 2011). The molecular mass of the largest of those (cytochrome $b_5$) does not exceed 13 kDa. To the outside observer, a rapid survey of the combined use of bicelles and NMR to study MP structure leaves the impression that establishing de novo a structure using only data collected with bicelles is quite difficult and not the strongest point of the approach. Indeed, despite 25 years of work on bicelles, all solution-NMR MP structures whose monomers comprise more than a couple of TM segments have been obtained in detergent solutions (see www.drorlist.com/nmr/MPNMR.html; reviewed by Marcotte and Auger 2005; Poget and Girvin 2007; Kim et al. 2009; Nietlispach and Gautier 2011; Warschawski et al. 2011). It is probably telling that the first NMR structure of a G protein-coupled receptor (GPCR) was derived from solid-state NMR data collected on liposome-reconstituted preparations rotated at the magic angle (Park et al. 2012), even though the group that carried out this study, that of Stanley J. Opella, has been a major contributor to the development of bicelle technology (see e.g. Park et al. 2006, 2011a, c).

Bicelle-based NMR seems of greater use in a host of situations where it permits to complement and extend data collected either by solution and/or solid-state NMR in other media, crystallography, or electron microscopy (EM). Because bicelles comprise a bilayer region, in which MPs preferentially reside (*cf.* Lee et al. 2008), they can be used to examine whether the protein structure is or not affected by a purely detergent environment, a procedure often resorted to (see e.g. Fanucci et al. 2003; Howell et al. 2005; Poget et al. 2007; Poget and Girvin 2007; Lau et al. 2008; Gautier et al. 2010), and to determine the tilt of $\alpha$-helices or $\beta$-barrels with respect to the bilayer plane (see e.g. Lindberg et al. 2003; De Angelis et al. 2004; Howell et al. 2005; Triba et al. 2006; Mahalakshmi and Marassi 2008), an information that is generally missing in X-ray and EM structures and always absent from solution-NMR ones. Bicelles are also useful for examining conformational transitions that may not be easily accessible to crystallography and may be perturbed by detergents (see e.g. Morrison et al. 2011; Gustavsson et al. 2012) or for studying the consequences of the binding of drugs to MPs in a more physiological environment than a detergent belt can provide (see e.g. Cui et al. 2010; Park et al. 2011b). They also provide an opportunity to study MP/lipid interactions and their effects on MP function, structure, and oligomerization under much better conditions than in detergent solutions. Furthermore, the fact that they permit to study the same MP in a continuous series of environments that range from pure detergent solution to isotropic bicelles, both of them accessible to solution NMR, and to aligned large bicelles and sheets or vesicles, which are a medium of choice for solid-state NMR (Fig. 3.3), offers a unique opportunity to check and complete the information gathered using one approach by that collected using another. This is an edge that bicelles have over nanodiscs, which, as will be discussed in § 3.3, present a very interesting alternative when it comes to studying MPs in a lipid bilayer environment by solution NMR but whose size can be adjusted only between rather narrow limits.

## 3.2.2 Bicelles and Membrane Protein Crystallography

It is somewhat ironic that, while it is their use for NMR studies that provided the primary impetus to the development of bicelles as hosts for MPs, the most important contribution bicelles have made to structural studies of polytopic MPs is in the field of crystallography. Indeed, in the early 1990s, Salem

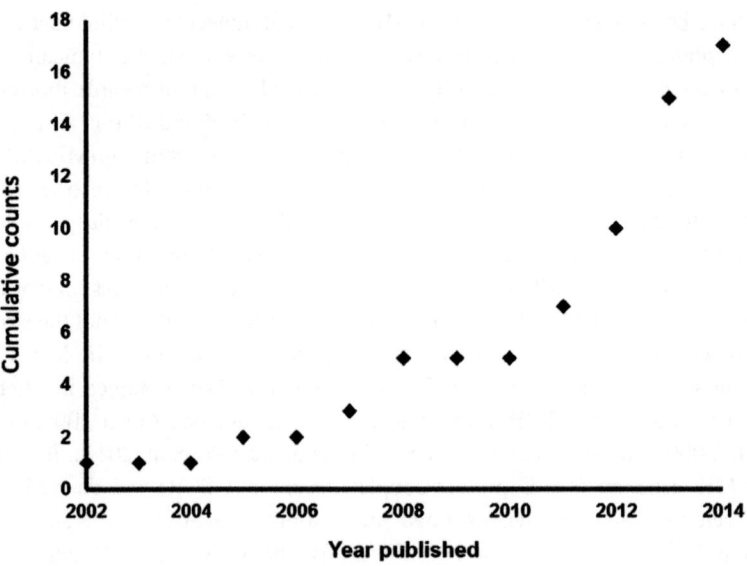

**Fig. 3.5** Cumulative number of membrane protein structures solved by X-ray crystallography using the bicelle method (From Poulos et al. 2015, © 2015 Elsevier Inc. All rights reserved).

Faham and James U. Bowie established, using as a model MP the usual victim of such forays into unknown territories, bacteriorhodopsin (BR), that bicelles provide a favorable starting point for MP crystallization (Faham and Bowie 2002; Faham et al. 2005). The approach was then extended to a GPCR, the $\beta_2$ adrenergic receptor (Rasmussen et al. 2007), to the voltage-dependent anion channel 1 (Ujwal et al. 2008), to xanthorhodopsin (Luecke et al. 2008), to the rhomboid protease (Vinothkumar 2011), to BamA, a bacterial outer membrane protein (Noinaj et al. 2013), etc. Since 2012, a dozen MPs covering all structural types have been crystallized from bicelles (Fig. 3.5; reviewed in Johansson et al. 2009; Ujwal and Bowie 2011; Agah and Faham 2012; Loll 2014; Poulos et al. 2015). Crystallizing BR from bicelles has been observed to yield a different crystal form (with different functional properties) than in lipid mesophases (Sanii and El-Sayed 2005; Sanii et al. 2005).

Crystallization from bicelles is facilitated by the fact that the fluidity of the preparations is higher at low than at high temperature, because bicelles are smaller (Fig. 3.3). On the one hand, this makes pipetting the preparations a lot easier than with the highly viscous cubic or sponge lipid mesophases that provide the other approach to crystallizing MPs in a lipid environment (see Chap. 11, § 11.2.2.2). Raising the temperature, on the other hand, will favor the formation of large bicelles and sheets, which is probably important to allow MP crystallization. As noted by Patrick J. Loll in a perceptive discussion (Loll 2014), MP crystallization from bicellar preparations appears to be underused given its advantages, a relative neglect that results probably more from a lack of familiarity with the approach than from technical causes.

### 3.2.3   Other Applications of Bicelles in Membrane Biology

NMR and crystallography are the two fields of MP biology to which bicelles have contributed most, but bicellar preparations have a broader potential and have also been used, for instance, in electron paramagnetic resonance (EPR) studies (see e.g. Fanucci et al. 2003; Lindberg et al. 2003; Ghimire et al. 2011; Gruene et al. 2011; Nusair et al. 2012) and in optical spectroscopy, particularly circular dichroism (CD) (Triba et al. 2006; McKibbin et al. 2007, 2009), as a vehicle to deliver a MP to oocyte membranes (Kang et al. 2010) and as the recipient surfactant in MP cell-free synthesis (Lyukmanova et al. 2011; Uhlemann et al. 2012) (*cf.* Chap. 7). Bicelles also have applications in galenics (see Dürr et al. 2012, 2013, and references therein).

## 3.3 Nanodiscs

Nanodiscs (NDs) are arguably the most important original contribution made to the handling of MPs in aqueous solutions over the past 20 years. Although their inception has a totally different origin from that of bicelles, they are conceptually related. In both cases, the underlying idea was to provide MPs with a bilayer environment while forming particles of nanometric dimensions. Bicelles, as recalled in the preceding section, originated from the study of the organization of the composite micelles formed by mixtures of phospholipids and bile salts, which play a key role in digestion. NDs are an outcome of basic studies on high-density lipoproteins (HDLs), which are natural vehicles for transporting lipids in the blood and delivering them to cells. In a sense, human physiology, which has to cope with the problem of transporting water-insoluble lipids in aqueous bodily fluids, has handed us two distinct solutions to this problem, each of which has been adapted by membrane biochemists to their own ends.

Since its inception at the end of the 1990s (Carlson et al. 1997; Bayburt et al. 1998, 2002; Denisov et al. 2004), the field of ND-based MP studies has become enormous, with, by the end of 2016, "over 550 publications that use the nanodisc technology to advance the understanding of MPs" (Denisov and Sligar 2017). The goal of this section is not to provide an exhaustive overview of this vast field, which has been covered in many reviews (see e.g. Nath et al. 2007; Borch and Hamann 2009; Ritchie et al. 2009; Bayburt and Sligar 2010; Popot 2010; Malhotra and Alder 2014; Denisov and Sligar 2016). A recent review by Ilia G. Denisov and Stephen G. Sligar is particularly to be commended for its exhaustiveness and the critical distance it strives to maintain (Denisov and Sligar 2017). The accent of the present section is put on the nature and properties of NDs as a MP environment rather than on the results obtained using them, of which only a few examples will be given. As is the case throughout this book as regards this and other surfactant systems, my objective is to give the reader elements for pondering the advantages and limitations of NDs as tools to investigate this or that particular biological problem.

### 3.3.1 High-Density Lipoproteins

The use of NDs for handling MPs in aqueous solutions emerged as a largely unforeseen outcome of research carried out on HDL particles by Ana Jonas, Stephen G. Sligar, and colleagues (see Jonas et al. 1991, Carlson et al. 1997, and references therein). The mature HDL is a complex comprising a protein, apolipoprotein A-I (apoA-I), present as two to seven copies per particle, often accompanied by other proteins, which stabilizes and keeps water-soluble a cargo of phospholipids, cholesterol, cholesteryl esters, and triglycerides. In its mature form, the complex has a globular shape ("spherical" HDL, sHDL). How exactly the lipids and proteins are assembled has been the object of protracted research and controversies and remains debated (for recent reviews, see Phillips 2013; Gogonea 2016). One of the difficulties is that HDLs are highly malleable, and the number of protein molecules and their arrangement in the particle adjust to the size of the lipid cargo so that different experiments can suggest different models without any of them being de facto erroneous.

Of particular relevance to the development of NDs is the fact that the formation of HDL particles starts with that of a so-called "nascent" form (nHDL), in which apolipoproteins (two to four apoA-I

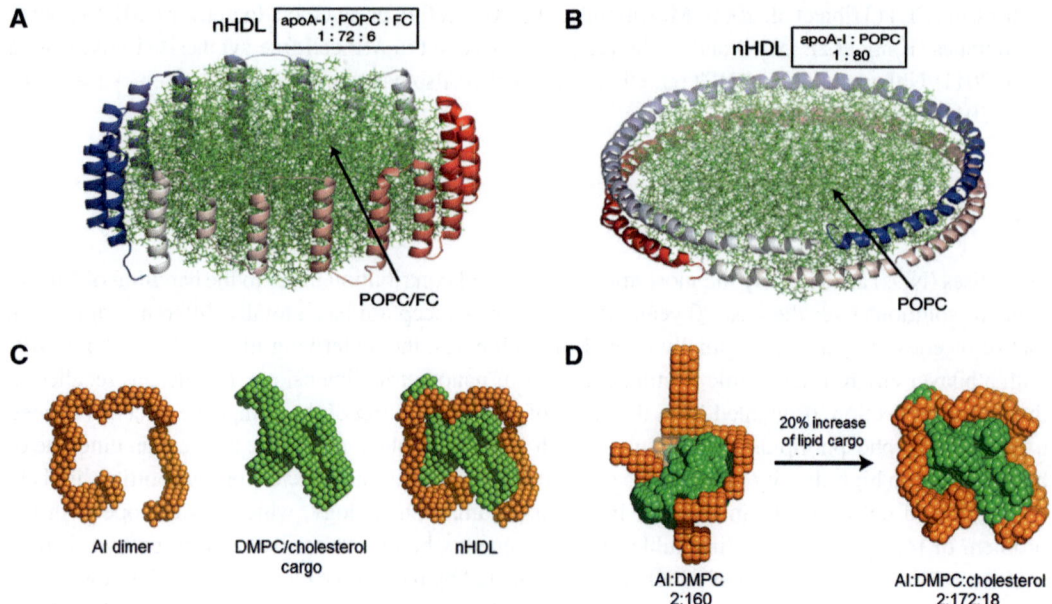

**Fig. 3.6** Models of nascent high-density lipoprotein particles (nHDL). (**A, B**) Discoidal models. The ApoA-I chains are shown in cartoon representation and colored with gradient *red/blue* (*N*-terminus in solid color, *C*-terminus in faded color). (**A**) The picket fence model (Jonas et al. 1989; Wald et al. 1990a, b; Nolte and Atkinson 1992; Phillips et al. 1997). (**B**) The apoA-I double-belt model proposed by Segrest et al. (1999). (**C**) Low-resolution structures obtained by fitting small-angle neutron scattering (SANS) curves calculated from bead models to the experimental scattering curves of the apoA-I dimer (*orange*), obtained by contrasting out the lipids (*left*), and to those of the lipid core (*green*), obtained by contrasting out the protein (*center*). Combining the two models yields a composite model of the nHDL particle (*right*). The particle contains two apoA-I molecules, 172 dimyristoylphosphatidylcholine (DMPC) molecules, and 18 cholesterol molecules (Gogonea et al. 2013). (**D**) A model for quantized nHDL particle expansion through recruitment of a previously lipid-free apoA-I loop in response to the increasing volume of the lipid cargo, based on SANS data for a low-lipid particle (two apoA-I molecules, 160 DMPC molecules) and a high-lipid one (same as in **C**; Gogonea et al. 2013). The composite figure is put together from extracts of various figures in Gogonea et al. (2016), © Gogonea (2016).

chains or a mixture of apoA-I and apoA-II ones) are combined with a bilayer-forming mixture of lipids, namely phospholipids and free cholesterol. In its simplest form, nHDL arranges into what is often described as a discoidal structure in which the lipids form a bilayer patch, whose rim is protected from the contact with water by a double circle of amphipathic α-helices formed by two copies of apoA-I (Fig. 3.6B). Alternative models in which the protein dimer wraps around the lipids without closing upon itself into a full circle have also been proposed, such as that in Fig. 3.6C, based on small-angle neutron scattering (SANS) data, using contrast variation to distinguish between the protein and lipid components (see Chap. 9, § 9.3.8). A model – also based on SANS data – of how apoA-I could rearrange as a function of the bulk of its lipid cargo by unfolding initially lipid-free extensions is shown in Fig. 3.6D. According to both SANS (Gogonea et al. 2013) and H/D exchange mass spectrometry (MS) data (Sevugan Chetty et al. 2012), either most or only part of the apoA-I dimer interacts with the lipids, depending on the hydrophobic surface to be screened from the water phase. Such observations have been important in guiding the design of engineered versions of apoA-I during the development of the ND system. H/D-MS experiments show the structure of apoA-I in nHDL to be highly dynamic, with α-helices unfolding and reforming on the second or subsecond time scale (Sevugan Chetty et al. 2012).

How nHDL grows into mature sHDL is beyond the frame of the present book but is worth mentioning cursorily. A cytosolic enzyme, lecithin cholesterol acetyltransferase, transforms free cholesterol into cholesteryl esters, which are highly hydrophobic and move to the core of the particle, joined by equally hydrophobic triglycerides delivered by other proteins. As the particle grows, phospholipid acyl chains tend to become exposed to its surface, which entails the recruitment of additional molecules of apoA-I and/or other proteins. The end result of this process is a heterogeneous collection of particles of various sizes, comprising variable proportions of protein and of bilayer-forming and non-bilayer-forming lipids, illustrating, once more, the flexibility and adaptability of apoA-I.

At the onset of their work (Carlson et al. 1997; Bayburt et al. 1998, 2002), S. G. Sligar and his colleagues intended to use nHDL to stabilize in a water-soluble form well-defined patches of lipid bilayer, whose surface would be used for in vitro experiments. It took some engineering of the apoA-I sequence before well-defined lipid patches suitable for incorporating and studying guest MPs were obtained (Denisov et al. 2004) and the efficiency, generality, and versatility of the approach became apparent (Denisov and Sligar 2017).

## 3.3.2 The Formation and Structure of Nanodiscs

NDs are generally obtained following a protocol derived from that used to form reconstituted nHDL (Jonas 1986; Jonas et al. 1990, 1991): an engineered version of apoA-I is mixed with lipids in detergent solution in the proportion appropriate to the size of the discs one aims to produce, and the detergent is removed. In the original procedure (Bayburt et al. 1998), the protein destined to stabilize the discs (hereafter called "membrane scaffold protein," MSP) was human apoA-I and the lipid DPPC ($diC_{16:0}PC$). The detergent was sodium cholate, which was removed by dialysis. A target MP, P450 reductase, was also included and ended up trapped within the NDs. Nowadays, NDs are systematically prepared using specially engineered MSPs.

As reviewed in Viegas et al. (2016) and Denisov and Sligar (2017), all of these parameters can be varied. Cholate can be replaced with almost any detergent, including OG, Triton X-100, decyl- or DDM, $diC_6PC$, CHAPS, and even sodium dodecylsulfate (SDS), an obvious caveat being that any MP that one aims to capture should not denature in it or must be able to refold when the detergent is eliminated. In practice, it is frequent that the mixture of MSPs and lipids solubilized in one detergent be mixed with the guest MP solubilized in another so that one deals with a mixture of detergents. Various protocols can be used for detergent removal. For detergents with a low CMC, like DDM, adsorption onto Bio-Beads has to be preferred to dialysis, which is too slow. A vast variety of lipids and lipid mixtures have been resorted to, including synthetic lipids such as DMPC, DPPC, and palmitoyloleyl-phosphatidylcholine (POPC, $C_{16:0}$, $C_{18:1}PC$), as well as mixtures comprising charged phospholipids and lipid mixtures from natural sources, such as *Escherichia coli* polar or total lipids, egg PC, or asolectin (soybean lipids). A key point is to carefully adjust the MSP/lipid ratio, so as to obtain homogeneous preparations of NDs with the desired size. Detailed protocols are available from the Sligar laboratory web site (http://sligarlab.life.uiuc.edu/nanodisc/protocols.html; see also Chap. 5, § 5.9.3, Protocol 5.3, about transferring a MP from amphipols to NDs).

ApoA-I itself does not produce monodisperse nHDL particles (Durbin and Jonas 1997; Li et al. 2004). Considerable efforts have therefore been invested to engineer it so as to obtain optimized MSPs yielding homogeneous NDs of a variety of sizes and carrying a variety of tags (Fig. 3.7). Five types of modifications have been experimented with. First, it was soon recognized that some of the *N*-terminal amino acid residues do not take part in forming the ND belt, which led to the development of a series of MSPs from which the first 11 or 22 residues of apoA-I have been deleted (e.g. MSP1D1 in Fig. 3.7)

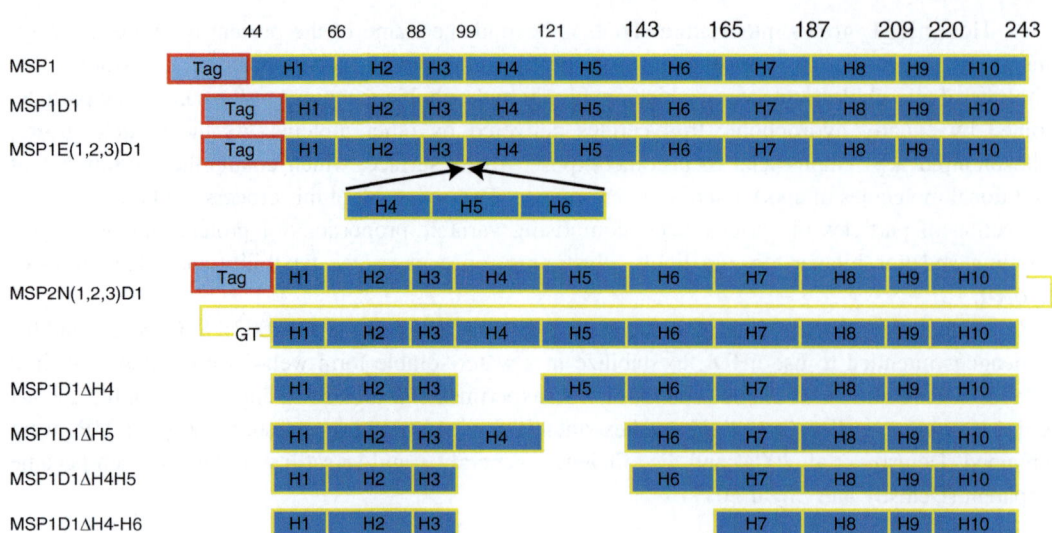

**Fig. 3.7** Schematic illustration of membrane scaffold proteins (MSPs) described in Grinkova et al. (2010) and Hagn et al. (2013). MSP1D1 and extended MSPs with several *N*-terminal affinity tags as well as *C*-terminal modifications with biotin and FLAG-tag are available. In MSP2, two copies of MSP1 are covalently linked (Reprinted with permission from Denisov and Sligar 2017, © 2017 American Chemical Society).

(Denisov et al. 2004). Second, by duplicating some of the sequence segments that fold into amphipathic α-helices when associating with lipids, MSPs can be made to stabilize larger discs, up to 17 nm in diameter as compared to the "standard" 10 nm ones, which permits to host larger MPs (MSP1E (1,2,3)D1 in Fig. 3.7) (Grinkova et al. 2010). Larger discs, however, tend to be less stable than standard ones, often collapsing to spherical aggregates (Denisov and Sligar 2017). Conversely, MSPs from which some helix-forming segments have been deleted, such as the MSP1D1Δ series in Fig. 3.7, generate smaller discs, with overall diameters in the 6 to 8 nm range, which increases the tumbling rate and improves the quality of solution NMR spectra (Hagn et al. 2013; Puthenveetil and Vinogradova 2013; Kucharska et al. 2015; Wang et al. 2015; see Fig. 3.12). These "mini-nanodiscs," however, tend to be less stable than classical NDs (Hagn et al. 2013), a problem that further MSP engineering might possibly solve, e.g. by inducing the formation of shorter helices (Puthenveetil et al. 2017). As a possible drawback, they offer guest MPs less of a bilayer-like environment and may possibly impose onto them constraints of their own (*cf.* the data shown in Bayburt et al. 2006, which suggest that a complete annulus of lipids is required for the BR trimer to be stable in NDs). Third, a duplicated version of MSP1, called MSP2, has been produced (Grinkova et al. 2010), which seems more difficult to handle but whose potentialities have not yet been fully explored. Fourth, a number of tags have been fused to the *N*- and/or the *C*-termini. Fifth, covalently circularized MSPs have been recently developed, which form more stable discs (Nasr et al. 2017).

The ND configuration may not correspond to the absolute free energy minimum of the MSP/lipid mixtures, given that the two components tend to separate irreversibly after 1 to 2 h incubation at 55 °C (Denisov et al. 2005). Nevertheless, NDs can be stored for months at 4 °C with minimal aggregation (Denisov and Sligar 2017). Virtually no exchange of MSPs between NDs is detected over periods of days to weeks (ibid.), and, according to Lai et al. (2015), lipids can be exchanged between NDs and bicelles but not between NDs. At the time scale of most experiments, NDs can therefore be considered as stable particles whose contents do not mix, which is one of their great assets (see § 3.3.5).

Lists of other proteins that can be used to form NDs, including other apolipoproteins, saposin A (see § 3.4.4), α-synuclein (Varkey et al. 2013; Eichmann et al. 2016, 2017), apolipophorins, and apomyoglobin, are given in Viegas et al. (2016) and Denisov and Sligar (2017). Projects are under way to develop amphipathic peptides designed to stabilize lipid bilayer patches (see § 3.4.5). In a recent development, disulfide bond formation protein B (DsbB), a bacterial plasma membrane protein featuring four TM α-helices, has been expressed in a soluble, functional form in the cytosol of *E. coli* after being fused to a truncated version of ApoA-I (Mizrachi et al. 2017). The stabilization of lipid discs and MPs by styrene-maleic acid copolymers (SMAs) will be discussed in Chaps. 4 and 5.

### 3.3.3  The Empty Nanodisc

The composition, organization, and dynamics of MP-free NDs have been extensively studied. Among the major questions investigated were the arrangement of the protein and the degree to which the behavior of the lipids faithfully reproduces that in protein-free lipid bilayers. The initial controversy between the "picket fence" model (Phillips et al. 1997), in which short helices have their axis normal to the plane of the disc (see Fig. 3.6A), and "belt" models (Fig. 3.6B), in which long helices run around its periphery (Wlodawer et al. 1979; Segrest et al. 1999), has been definitely resolved in favor of the latter by a host of indirect experimental data, MD simulations (Fig. 3.8), and, finally, direct experimental determination by NMR (Fig. 3.9) (see e.g. Koppaka et al. 1999; Li et al. 2006; Bibow et al. 2017; reviewed in Brouillette et al. 2001; Denisov and Sligar 2017). α-Helices are intrinsically rigid due to the network of hydrogen bonds that stabilizes them. Bends therefore occur preferentially at the weak

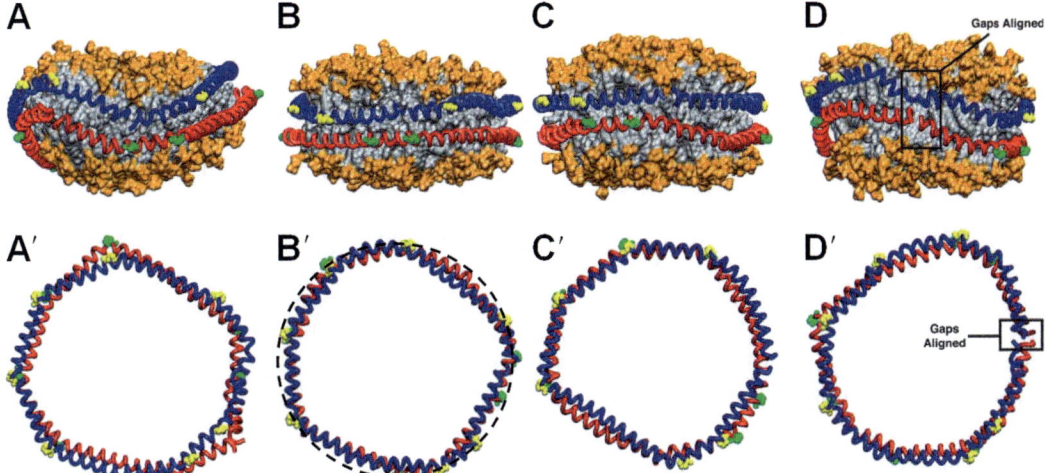

**Fig. 3.8** Membrane protein-free nanodiscs (NDs) as simulated by molecular dynamics (MD). (**A–D**) Side views of NDs after 4.2 ns of all-atom simulation. Each ND is comprised of 160 molecules of DPPC and two MSP molecules, respectively, MSP1 (**A**; 200 residues), MSP1 Δ(1–11) (MSP1D1; **B**; 189 residues), MSP1 Δ(1–22) (MSPID2; **C**; 178 residues), and MSP1 Δ(1–22)g (**D**; same sequence as in **C** but with the *N*-and *C*-termini of the two MSPs aligned, leaving a gap). MSPs are depicted in tube representation in *blue* and *red*. Prolines are highlighted in sphere representation in *yellow* and *green*. Lipid head groups are shown in *orange* and acyl chains in *gray*. (**A′–D′**). Top views. Lipids have been removed to reveal deviations from perfect circularity in the structure of MSPs, most noticeably the bends induced by the proline residues (From Shih et al. 2005, © 2005 The Biophysical Society. Published by Elsevier Inc. All rights reserved. A *dashed circle* has been fitted around MSPs in Panel B′ to facilitate comparison with Fig. 3.13B).

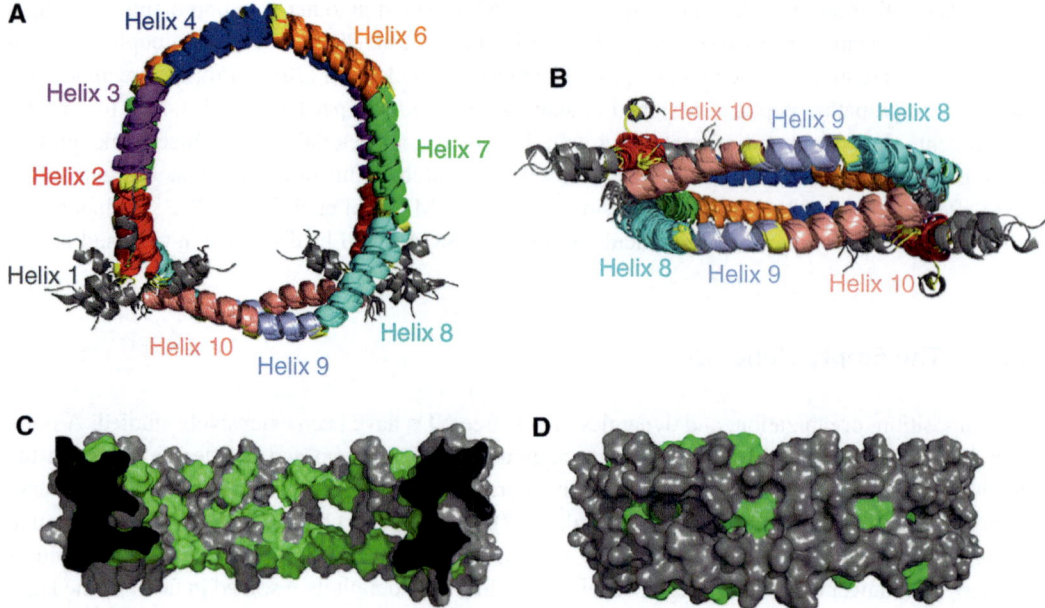

**Fig. 3.9** Experimental determination of the structure of the membrane scaffold proteins (MSPs) stabilizing a dimyristoylphosphatidylcholine (DMPC) nanodisc (ND). The structure was obtained by a combination of solution NMR, electron paramagnetic resonance (double electron-electron resonance (DEER)-derived distance restraints), and transmission electron microscopy data. (**A, B**) Top (**A**) and side (**B**) views of ten superimposed conformers of the MSPs. Two antiparallel MSPΔH5 molecules (a shortened version of apoA-I lacking the non-lipid-binding residues 1–54 as well as residues 121–142, which are proposed to form helix 5; *cf.* Fig. 3.7) *encircle* a DMPC bilayer patch (not shown) to form a structure with twofold symmetry. The nine helices of each monomer are individually color-coded and labeled. Pro or Gly residues (*yellow*) usually separate individual helices. (**C**) van der Waals surface representation of the MSPs cut in half to reveal the lipid-oriented interior, with hydrophobic residues colored in *green*. (**D**) A view of the solvent-exposed outside surface (From Bibow et al. 2017, © 2017 Macmillan Publishers Limited, Nature Structural Molecular Biology. All rights reserved).

points provided by proline and glycine residues. As a result, the equilibrium structure is polygonal rather than perfectly circular (Figs. 3.8 and 3.9).

A major asset of NDs being their ability to reproduce a bilayer environment, the state of the lipids has been the object of particular attention, using such techniques as NMR, molecular dynamics (MD), differential scanning calorimetry (DSC), laurdan fluorescence measurements, and SAXS (see e.g. Shaw et al. 2004; Denisov et al. 2005; Shih et al. 2005). The major conclusions can be summarized as follows. All experimental and MD data converge to describe the lipid component as a patch of bilayer. This is the major conclusion, the nuances that can be brought to this picture being second-order refinements. The bilayer nature of the lipid patch is supported experimentally, in particular, by the observation of phase transitions analogous to those observed with pure lipid bilayers, even though slightly higher transition temperatures and a lesser cooperativity indicate a degree of perturbation (Shaw et al. 2004; Denisov et al. 2005; Kijac et al. 2010). The latter is attributed (i) to the fact that a large number of lipids (see Fig. 3.12) are in contact with MSPs, and do not take part in the transition, and (ii) to the sensitivity of MSPs themselves to temperature changes (Denisov et al. 2005). As expected, phase transitions become sharper as the size of the NDs grows, more lipids taking part in them (Grinkova et al. 2010). There are other differences between those lipids that interact with MSPs and those at the center of the ND: the former tend to occupy a larger area per molecule and, therefore, to form a thinner film than those at the center (Siuda and Tieleman 2015). Overall, the number of lipid molecules that can be accommodated inside a ND is quite precisely predicted by the overall length of the amphipathic helices than stabilize them, assuming the MSP belt to be approximately circular and the average area occupied by lipids to be similar to that in liposomes (Denisov et al. 2004; Ritchie et al. 2009; Grinkova et al. 2010; Hagn et al. 2013), but lipids in NDs tend to be somewhat more compressed and have a lower conformational entropy (reviewed in Denisov and Sligar 2017). Indeed, both NMR data (Mörs et al. 2013) and MD simulations (Debnath and Schäfer 2015) show their dynamics to be restricted as compared to that in lipid vesicles (reviewed in Viegas et al. 2016).

### 3.3.4 Membrane Protein/Nanodisc Complexes

MPs are generally included into NDs at the time of their formation. A typical protocol is schematized in Fig. 3.10. In a first step, MSPs, lipids, and the guest MP are mixed in detergent solution (Fig. 3.10, *left*). If the objective is to obtain monomeric preparations of the target MP, care must be taken that there be an excess of NDs over the protein to be trapped. The amount of lipids must correspond to what the MSPs can accommodate, taking into account the volume that will be occupied by the TM region of the MPs (each TM $\alpha$-helix can be estimated to displace roughly five to seven phospholipids; Bayburt et al. 2006, 2007). The presence of too much or too little lipids will produce aggregates and/or non-planar and unstable discs (see e.g. Bayburt et al. 2002, 2006; Catte et al. 2006; Miyazaki et al. 2009; Shi et al. 2013; Goddard et al. 2015; Viegas et al. 2016). Small-scale trials followed by size-exclusion chromatography (SEC) analysis are needed in order to determine the ratios yielding the most homogeneous preparations of the right size. The guest protein, of course, has to be stable in the detergent used or to refold efficiently.

As noted above, the detergent is generally removed either by dialysis if its CMC makes it possible, as is the case for cholate, or by adsorption onto Bio-Beads. The preparation thus obtained is never perfectly homogeneous and is generally purified by size-exclusion chromatography, which, unless the MP is very bulky, yields a mixture of empty and MP-carrying NDs (Fig. 3.10, *center*). The two types of discs can then be separated, for example by immobilized metal affinity chromatography (IMC), if the guest protein carries a polyhistidine tag and the MSPs do not or it has been cleaved off (Fig. 3.10, *right*).

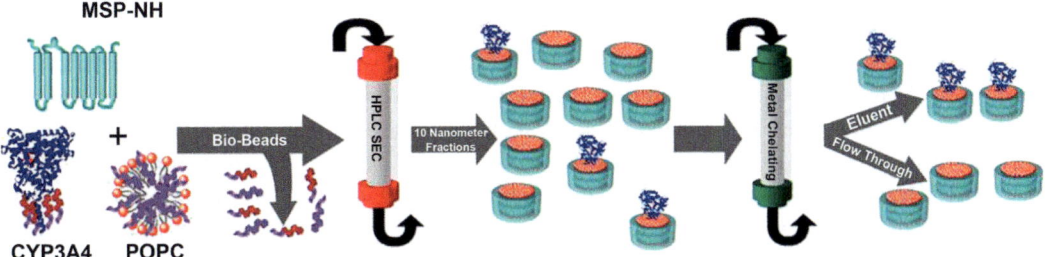

**Fig. 3.10** Schematic description of the preparation of membrane protein (MP)-containing nanodiscs (NDs). The guest MP (*dark blue*), lipids (*orange* and *gray*), and membrane scaffold proteins (*cyan*) are mixed in detergent solution in the appropriate ratio and the detergent(s) (wiggly molecules) removed either by dialysis or by adsorption onto polystyrene beads. Complexes of the correct size are purified by size-exclusion chromatography (SEC). Unless the bulk of the MP is sufficient for the discs that contain it to separate from empty discs during SEC, affinity chromatography may be used to eliminate empty discs (Reprinted with permission from Denisov and Sligar (2017), © 2017 American Chemical Society, after an original figure from Baas et al. (2004), © 2004 Elsevier Inc. All rights reserved).

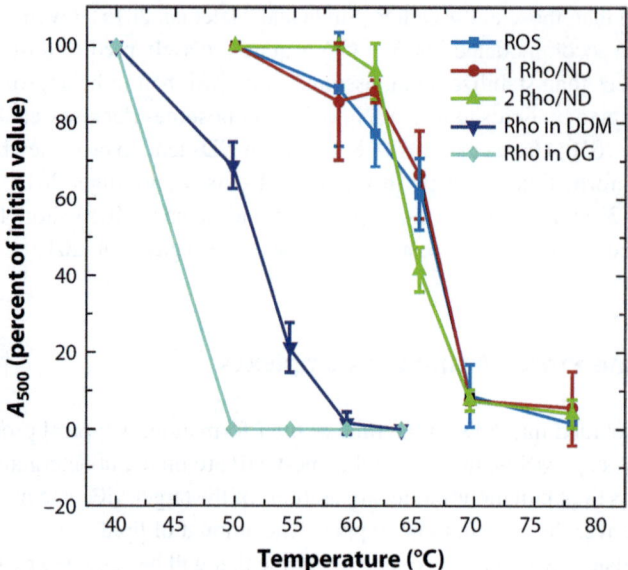

**Fig. 3.11** Thermostability of rhodopsin (Rho) upon integration into nanodiscs (NDs), as compared to that in rod outer segments (ROS; rhodopsin's native membrane environment), dodecylmaltoside (29 mM), or octylglucoside (OG; 51 mM). The NDs were comprised of hexahistidine-tagged zebrafish apoA-I and 1-palmitoyl-2-oleoyl-*sn*-glycero-3-phosphocholine (POPC) and contained either one or two copies of Rho per disc (Figure from Popot 2010, original data from Banerjee et al. 2008, ©2008 Elsevier Inc. All rights reserved).

Transferring a MP from a detergent to a ND environment generally results in stabilizing it (see e.g. Etzkorn et al. 2013; Rues et al. 2016), as illustrated in Fig. 3.11 in the case of rhodopsin (Banerjee et al. 2008).

It is desirable to keep in mind the relative sizes of the ND and the TM region of the MP it hosts. In Fig. 3.12, the dimensions of a standard ND (A; two copies of 189-residue MSP1D1, 60 lipids per monolayer) and of a "mini-ND" (B; two copies of 123-residue MSP1D1ΔH4H5H6 (Hagn et al. 2013), 27 lipids per monolayer (Denisov and Sligar 2017)) are compared to the TM dimensions of either a monomer or a trimer of BR. When a BR (or GPCR) monomer is hosted in a standard ND, it can remain in contact with lipids only (Fig. 3.12D). Most of these lipids interact with lipids that are themselves in contact with an MSP and whose conformation and dynamics are somewhat perturbed as compared to that in a lipid vesicle, but this situation, in which proteins are separated one from another by only two or three lipids molecules, is not too different from that which most MPs experience in a biological membrane, in which one can estimate that there is, on average, ~4 intervening lipids between each MP and its neighbors (Popot and Engelman 2000; see Chap. 1, Table 1.2). When a monomer of BR or GPCR is trapped in a small ND, however, or a trimer of BR in a standard ND, most lipids are in contact with both the host and the guest proteins, and protein-protein contacts are likely to occur (Figs. 3.12E, F). The resulting particle would resemble more a lipoprotein complex than a MP freely floating in a lipid bilayer. In practice, efficient incorporation of BR trimers occurs only in larger discs (Bayburt et al. 2006), and somewhat larger NDs than the small one schematized in Fig. 3.12B are used to trap monomeric BR (Hagn et al. 2013), leaving the proteins more leg room.

Even in the favorable case of a BR monomer inserted in a standard ND (Fig. 3.12D), however, MD simulations show that lipid-mediated interactions between BR and the two MSPs are sufficient to deform the latter (Fig. 3.13B), as compared to their equilibrium conformation in the absence of guest

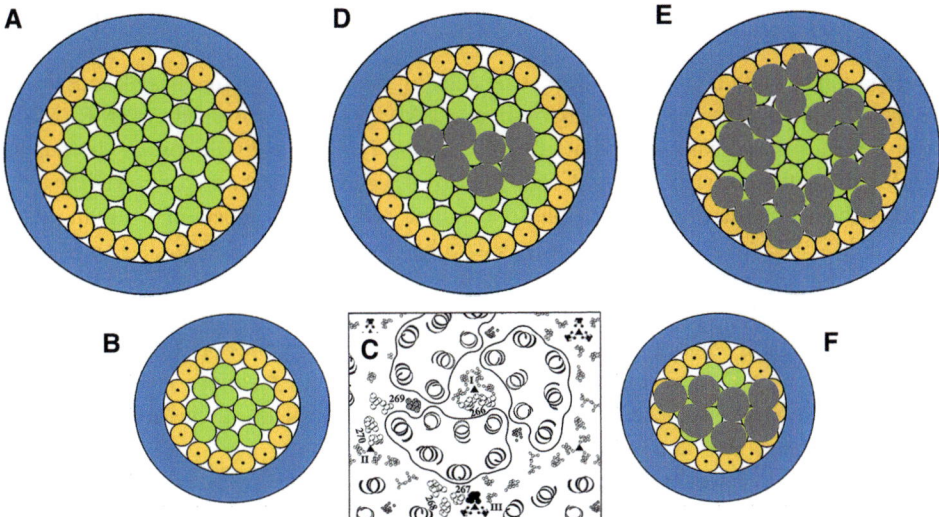

**Fig. 3.12** Large and small nanodiscs (NDs) and their size relative to that of membrane protein transmembrane (TM) regions. (**A, B**) Lipid packing in a "standard" ND (**A**; two copies of 189-residue MSP1D1, 60 lipids per monolayer; ~10 nm overall diameter) and in a small ND (**B**; two 123-residue MSP1D1ΔH4H5H6, from which helices 4–6 have been deleted, 27 lipids per monolayer; ~6.5 nm overall diameter; Hagn et al. 2013). Boundary lipids interacting with MSP are shown in *orange* and marked with *dots*. From Denisov and Sligar (2017). (**C**) A 1 nm *thick* slice through the cytoplasmic half of the TM region of trimeric bacteriorhodopsin (BR), shown to the same scale. The positions of the ten lipid molecules in the hexagonal unit cell of native purple membrane are indicated. Each BR monomer within the trimer is outlined by a *solid line* (From Grigorieff et al. 1996). (**D–F**) Superposition of a schematic view of the TM region of a BR trimer on a standard ND (**E**) and of that of a monomer on either a standard ND (**D**) or a small ND (**F**).

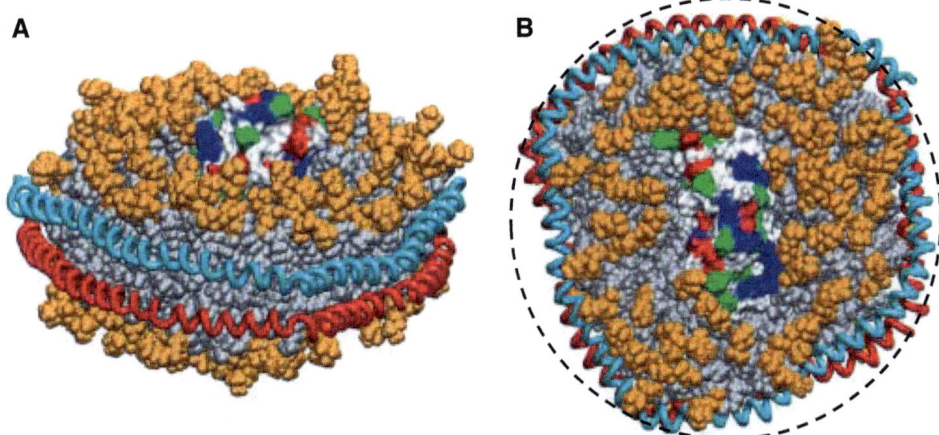

**Fig. 3.13** Side and top views of bacteriorhodopsin (BR) in a standard nanodisc (ND) comprising two copies of MSP1 Δ(1–11) (MSP1D1) after 4.5 ns of all-atom molecular dynamics simulation. Shown in *cyan* and *red* are the two MSPs surrounding the dipalmitoylphosphatidylcholine (DPPC) lipid bilayer in a belt-like fashion. DPPC molecules are shown in *orange* (polar heads) and *gray* (acyl chains). Embedded in the center of the lipid bilayer is a BR molecule shown using a surface representation colored according to residue properties (basic residues, *blue*; acid residues, *red*; other polar residues, *green*; nonpolar residues, *white*) (From Shih et al. 2005, © 2005 The Biophysical Society. Published by Elsevier Inc. All rights reserved. A *dashed circle* has been fitted around the ND in Panel **B** to facilitate comparison with the empty ND shown in Fig. 3.8B′).

protein (Fig. 3.8B′) (Shih et al. 2005). This is reflected in the restricted dynamics of both the protein and the lipids upon insertion of proteorhodopsin into DMPC NDs, as observed by NMR (Mörs et al. 2013). Stiffening may actually contribute to MP stabilization (for a discussion on the relationship between MP dynamics and stability, see Chap. 5, § 5.6). A detailed NMR comparison of the dynamics of OmpX either solubilized in dodecylphosphocholine (DPC) or trapped in medium-sized NDs (167-residue MSP1D1ΔH5, from which only helix 5 has been deleted; see Fig. 3.7) showed that, on the ps and ns time ranges, there are few differences at the level of the TM $\beta$-barrel and more in the loop regions (Hagn et al. 2013). A recent NMR study reports, surprisingly, that the TM $\beta$-strands of OmpX are *less* dynamic in DPC solution than they are in DMPC/diC$_6$PC bicelles or DMPC nanodiscs (Frey et al. 2017), whereas MD data suggest that the dynamics of rhodopsin/lipid interactions is slowed down in NDs as compared to bicelles (Vestergaard et al. 2015). Broader studies are clearly needed to sort out these various effects and assess their generality.

We have hitherto considered the case where a target MP has been purified in detergent solution before being transferred to NDs. It is also possible to trap a whole mixture of MPs, such as that present in a crude solubilization supernatant (Civjan et al. 2003; Duan et al. 2004) (Fig. 3.14). Target MPs or MP complexes can be purified from this mixture, e.g. by affinity chromatography, under conditions where their stability has every chance to be improved as compared to that in detergent solution and where they stand a better chance to retain at least part of their native bound lipids (see e.g. Mitra et al. 2013; Shirzad-Wasei et al. 2015; Gregersen et al. 2016). Alternatively, the whole mixture can be used as a soluble library of MPs (Marty et al. 2013; Roy et al. 2015; Wilcox et al. 2015), e.g. to identify MPs that bind to a given target, the one important constraint being that the TM region of the MPs or MP complexes of interest be small enough to fit inside a ND, or they will be lost at the trapping stage.

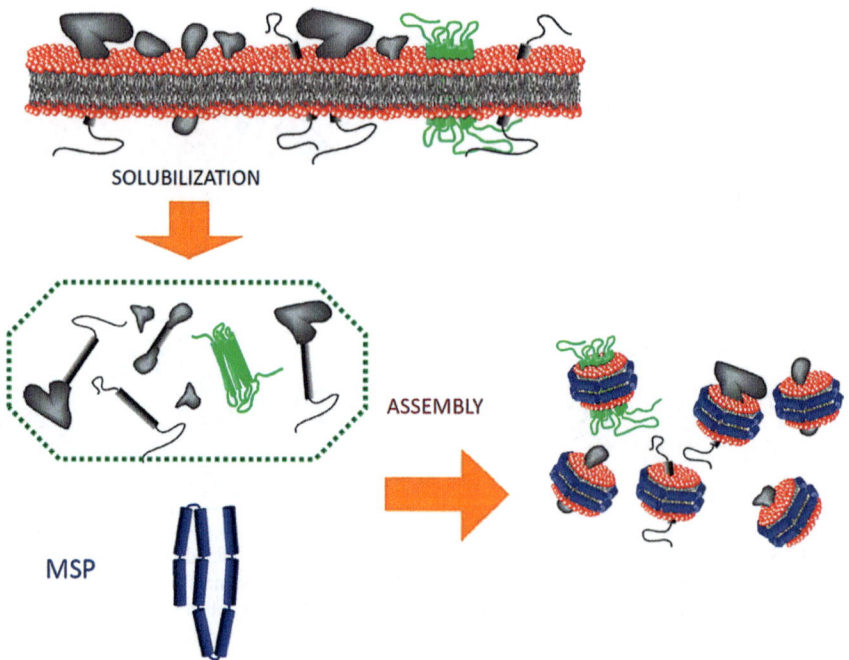

**Fig. 3.14** Libraries of water-soluble membrane protein can be generated by trapping in nanodiscs crude solubilization supernatants (Reprinted with permission from Denisov and Sligar 2017, © 2017 American Chemical Society).

Another way to produce MP/ND complexes is to use NDs as the host surfactant during MP in vitro cell-free synthesis (see e.g. Cappuccio et al. 2008; Katzen et al. 2008; Yang et al. 2011; Gao et al. 2012; Roos et al. 2012; Proverbio et al. 2013; Henrich et al. 2015, 2016, 2017; Rues et al. 2016).

Finally, a few cases have been described where the target MP was directly folded in NDs from a mixture of MP, MSPs, and lipids in SDS (Etzkorn et al. 2013; Shenkarev et al. 2013) (see below).

### 3.3.5 Nanodisc-Based Investigations of Membrane Proteins

ND-trapped MPs can be studied by most of the methods applicable to MP/detergent complexes, with the double advantage of increased stability and the presence of a lipid environment. Table 3.1 offers a selection of examples.

In discussing – briefly – the applications of NDs to MP structural and functional studies, one may perhaps consider three cases of figures: (i) situations where NDs have a clear edge over all or most of

**Table 3.1** A selection of works combining the use of nanodiscs with various biophysical or biochemical methodologies for producing and/or studying membrane proteins.

| Methodology | References |
|---|---|
| NMR | Kijac et al. (2007), Shenkarev et al. (2010), Park et al. (2011a), Qureshi and Goto (2011), Warschawski et al. (2011), Etzkorn et al. (2013), Hagn et al. 2013), Tzitzilonis et al. (2013), Malhotra and Alder (2014), Hagn and Wagner (2015), Kucharska et al. (2015), Mineev et al. (2015), Morgado et al. (2015), Casiraghi et al. (2016), Viegas et al. (2016), Mineev and Nadezhdin (2017), Nasr et al. (2017), and Puthenveetil et al. (2017) |
| EPR | Shin et al. (2014), Alvarez et al. (2015), Kang et al. (2015), and Georgieva (2017) |
| Absorbance, fluorescence, and vibrational optical spectroscopies | Denisov et al. (2007), Ranaghan et al. (2011), Tsukamoto et al. (2011), Taufik et al. (2013); Johnson et al. (2014), Shin et al. (2014), Mak et al. (2015), and Zoghbi and Altenberg (2017) |
| SANS and SAXS, neutron reflectivity | Bayburt et al. (2006), Wadsäter et al. (2012), Periasamy et al. (2013), and Skar-Gislinge et al. (2015) |
| Analytical ultracentrifugation | Inagaki and Ghirlando (2017) |
| Electron microscopy | Katayama et al. (2010), Ye et al. (2010), Frauenfeld et al. (2011), Gogol et al. (2013), Xu et al. (2013), Akkaladevi et al. (2015), Efremov et al. (2015), Grushin et al. (2015), Zhang et al. (2015), Daury et al. (2016), Gao et al. (2016), Gatsogiannis et al. (2016), Kedrov et al. (2016), Kumar et al. (2016), Lee et al. (2016), Matthies et al. (2016), Shen et al. (2016), and Nasr et al. (2017) |
| Mass spectrometry | Hopper et al. (2013), Landreh and Robinson (2015), Marty et al. (2016), and Henrich et al. (2017) |
| Immobilization onto solid supports, surface plasmon resonance and single-molecule studies, NMR-based drug screening | Bayburt and Sligar (2002), Shaw et al. (2007), Nath et al. (2008), Früh et al. (2011), Ritchie et al. (2011), Wadsäter et al. (2012), Laursen et al. (2014), Gillette et al. (2015), Lamichhane et al. (2015), and Hansen et al. (2016) |

(continued)

**Table 3.1** (continued)

| Methodology | References |
|---|---|
| Single-molecule force spectroscopy | Zocher et al. (2012) |
| Cell-free expression | Cappuccio et al. (2008), Katzen et al. (2008), Yang et al. (2011), Gao et al. (2012), Roos et al. (2012), Proverbio et al. (2013), Henrich et al. (2015, 2016, 2017), Rues et al. (2016) |
| (Re)folding MPs from a denatured state | Etzkorn et al. (2013) and Shenkarev et al. (2013) |
| X-ray crystallography | Nikolaev et al. (2017) and unpublished observations cited in Denisov and Sligar (2017) |

Based in part on Table 3.1 from Denisov and Sligar 2016, with additions and modifications

the other surfactants, (ii) situations where using one or another system is optional, and (iii) situations where the use of NDs presents particular challenges.

In the first category, one may cite two particular types of issues, namely (i) determining whether a given protein functions as a monomer or an oligomer and (ii) examining the role of the lipid environment. An example of the first circumstance is the long-standing controversy about whether GPCRs function as monomers or multimers. By preventing oligomerization of MPs trapped as monomers, NDs provide an unprecedented opportunity to investigate the functional roles of oligomerization. Complexes comprising either one or two rhodopsin molecules per standard ND were produced. Efficient transducin activation and isolation of a high-affinity transducin-metarhodopsin II complex was demonstrated for monodisperse monomeric rhodopsin. In the case of the two-rhodopsin population, only one of them was able to form a stable metarhodopsin II-G protein complex (Bayburt et al. 2007). These experiments thus provided clear evidence that monomeric rhodopsin is capable of fully coupling signal detection to transduction, a demonstration that is not easy to achieve unambiguously using other types of preparations. The demonstration was later extended to rhodopsin kinase (GRK1) phosphorylation and the binding of arrestin-1 (Bayburt et al. 2011). A similar demonstration has been achieved for the $\mu$-opioid GPCR (Kuszak et al. 2009). Two other nice examples of this type of investigations are a study of the role played by the formation of oligomers of dimers in TM signalization by chemoreceptors (Boldog et al. 2006) and the demonstration that talin activates unclustered integrins (Ye et al. 2010). Similarly, NDs have been used to demonstrate that SecYEG monomers are functional for pre-protein translocation (Taufik et al. 2013) and to compare the affinity of neutralizing antibodies for monomeric vs. trimeric HIV TM peptides (Reichart et al. 2016).

Another case of figure where the resort to NDs is, if not always compulsory, at least extremely useful is the influence of the lipid environment. For MPs that are small enough and integrated within large enough NDs, the latter offer an opportunity to study the interactions of the protein with its lipid environment under conditions of limited perturbation. To take a single example, BLT2, a GPCR involved in inflammatory processes, was initially folded from inclusion bodies using APols and the structure of its bound ligand, $LTB_4$, determined by solution NMR in a BLT2/APol complex (Catoire et al. 2010a; see Chap. 10, § 10.3.3.3). In order to study the allosteric modulation of BLT2 by cholesterol, however, the receptor was transferred from APols to NDs (Casiraghi et al. 2016; see Protocol 5.3 in Chap. 5, § 5.9.3). NDs have been used to examine the role of specific lipids in many other cases, for instance, in activating pre-protein transport by SecYEG/SecA complexes (Koch et al. 2016).

The interactions between the extramembrane domains of three proteins involved in membrane fusion, VAMP-2 (vesicle-associated membrane protein 2) and syntaxin-1a, on the one hand, and SNAREs (soluble N-ethylmaleimide-sensitive factor attachment protein receptors), on the other hand, have been studied by EPR and Förster resonance energy transfer (FRET) after trapping the first two

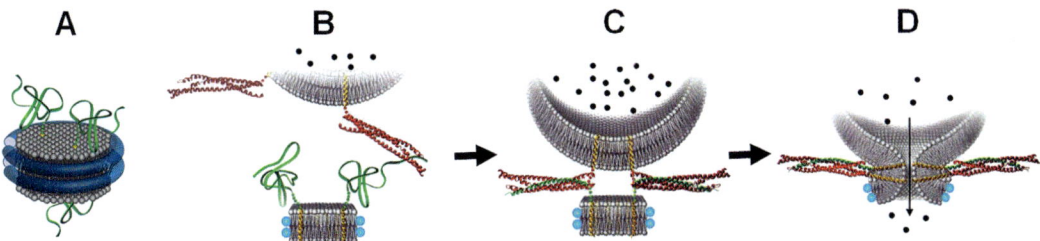

**Fig. 3.15** A liposome-nanodisc (ND) fusion assay. (**A**) VAMP-2 (*green*) is incorporated in the bilayer via its transmembrane domain (*yellow*) to form a v-disc that can be used for SNARE-mediated fusion assay. (**B–D**) The fusion process between a v-disc and a t-liposome. (**B**) VAMP-2 and t-SNARE (syntaxin and SNAP-25) form a trans-SNARE complex. (**C**) Zippering of proteins to form the SNARE complex brings the liposome and the ND together. This interaction enables the fusion between the two bilayers to occur. (**D**) Formation of a fusion pore between the liposome and the ND. The liposome content (*black dots*) is released in the outer medium through the fusion nanopore (Modified from Shi et al. 2013, © 2013 Macmillan Publishers Limited, part of Springer Nature. All rights reserved).

partners and the third one in distinct NDs (Shin et al. 2014). A related approach is to study the interactions between ND-trapped VAMP-2 and vesicle-bound SNAREs, resulting in the transient formation of a pore allowing transmitter release (Fig. 3.15) (Shi et al. 2012, 2013).

There are many cases where the choice of NDs over one or another stabilizing surfactant is more a matter of convenience, personal feeling, and case-by-case decision. As discussed in Chap. 2, interactions with lipids and the absence of detergent are most often stabilizing factors, but these conditions can be met in a variety of ways. To take the single example of solution NMR, good spectra can be obtained with some detergents, with APols or with NDs (see e.g. Raschle et al. 2009, 2010; Shenkarev et al. 2010; Park et al. 2011a; Qureshi and Goto 2011; Warschawski et al. 2011; Etzkorn et al. 2013; Hagn et al. 2013; Catoire et al. 2014; Elter et al. 2014; Kraft et al. 2015; Viegas et al. 2016, and data shown in Chap. 10). Detergents and APols have the benefit of simpler protocols for sample preparation, NDs and APols that of a higher MP stability, making it easier to collect data for extended periods at elevated temperatures, the latter accelerating tumbling and improving the resolution. Standard or large NDs provide the most bilayer-like environment, at the expense of resolution, whereas small NDs improve the latter while moving away from a true bilayer. The choice of one or another system will therefore depend on the characteristics of the target MP and the experimenter's priorities. The same discussion could be had regarding EM (*cf.* Chap. 12) and many other biophysical techniques. One cannot help thinking that sometimes experimenters inflict upon themselves unnecessary hardships mainly to be able to claim that they have handled their protein "in a lipid bilayer," when this is not necessarily essential to the demonstration sought nor totally true, and simpler systems could have made their life easier and their progress more rapid. An interesting approach is that of Alexander S. Arseniev and colleagues, who have used NDs as a benchmark in order to determine whether less cumbersome systems, such as detergent solutions, perturb or not the MP they study (Shenkarev et al. 2010). Cases where there are structural differences between a detergent-solubilized MP and the same protein trapped in a ND have been reported. Solution NMR indicates, for instance, that the TM $\beta$-strands of OmpX are up to two residues longer in small NDs than in certain detergent environments (Hagn et al. 2013). This is interpreted as a perturbation by the detergent of the native structure that exists in vivo. This interpretation is plausible, but one should not lose sight of the fact that even a ND is a perfect mimic neither of the bacterial outer membrane nor of an extended lipid bilayer and that it can exert its own constraints on the structure of the proteins it entraps.

A special case is that of X-ray crystallography. As discussed in Denisov and Sligar (2017), this is quite a challenging application for NDs, because residues belonging to the guest MP and MSP will generally not be positioned identically with respect to one another from one complex to the next, creating heterogeneity. Only poorly diffracting crystals of BR have been obtained to date (unpublished observations cited in Denisov and Sligar (2017)). When the MP sports huge extramembrane domains, crystallization may perhaps be achieved under conditions where all contacts form between those and the rotational disorder of the MSPs does not prevent the growth of well-ordered crystals. Another opportunity that does not seem to have been discussed yet would be to turn to advantage the inconvenience of trapping a MP in too narrow a ND: if protein/protein contacts between the guest MP and MSPs cannot be avoided, then perhaps the proteins may lock in the register permitting the most favorable contacts, generating homogeneous complexes. This might be an incentive to conduct crystallization trials using the smallest NDs that will accommodate the target MP. Appropriate engineering, like that of hydrogen bonds or salt bridges, could possibly help stabilizing a unique arrangement. Finally, a promising approach is to transfer ND-trapped MPs to lipidic mesophases. When BR/APol complexes are mixed with a mesophase, the protein and the APol part ways, diffuse separately, and highly ordered crystals of APol-free BR form (Polovinkin et al. 2014), providing a way to crystallize MPs that have been folded or stabilized using APols (*cf.* Chap. 11, § 11.2.2.2). The same process has recently been extended to ND-trapped ones (Nikolaev et al. 2017).

It is possible to fold denatured MPs to their native state while incorporating them into NDs. BR from *Halobacterium salinarum* has been refolded starting from a mixture of MSP1D1, DMPC, bacterio-opsin, and retinal dissolved in SDS and removing SDS with Bio-Beads (see the *Methods* section in Etzkorn et al. 2013). A similar procedure has been applied to BR from *Exiguobacterium sibiricum* and to the homotetrameric $K^+$ channel KcsA from *Streptomyces lividans* (Shenkarev et al. 2013). Extending this approach to other MPs, particularly complex ones, might not always be straightforward, given the many parasitic reactions that can take place and lead to aggregation and the large number of parameters to be optimized, but the procedure can obviously be made to work and is worth keeping in mind.

## 3.4    Amphipathic Peptides

Amphipathic peptides that act to destabilize membranes, such as antimicrobial peptides, bee venom melittin, or numerous bacterial toxins, are widespread in nature (for a review, see e.g. Peters et al. 2010). Their use as detergent substitutes to stabilize MPs in aqueous solutions was pioneered in the early 1990s (Dempsey and Sternberg 1991; Schafmeister et al. 1993) and has since branched in many different directions. Even though none of the approaches that have been investigated has yet become widely used, a rapid survey is in order. The structure of the peptides and the organization of the complexes they form with MPs and/or lipids are extremely variable (Table 3.2). One can categorize them as follows:

- Relatively long peptides (typically 20 to 25 residues long) that are designed to fold into laterally amphipathic $\alpha$-helices whose length is comparable to the thickness of the hydrophobic core of a lipid bilayer and whose hydrophobic face is expected to adsorb onto the hydrophobic transmembrane (TM) surface of MPs. They can be plain oligopeptides ("peptitergents"; Schafmeister et al. 1993; Table 3.2, line **1**; § 3.4.1) or endowed with bound fatty acyl chains meant to provide a "soft" interface between the peptide and the MP ("lipopeptide detergents"; McGregor et al. 2003; Table 3.2, line **2**; § 3.4.2).

**Table 3.2** A selection of publications describing amphipathic peptides designed to substitute for detergents.

| Line | Name | Structure | Section | Figure | References |
|---|---|---|---|---|---|
| 1 | Peptitergents (PDs) | Peptides designed to fold into amphipathic α-helices | § 3.4.1 | – | Schafmeister et al. (1993), Soomets et al. (1997), and Bavec et al. (1999) |
| 2 | Lipopeptide detergents (LPDs) | Peptides designed to fold into amphipathic α-helices, each of them carrying two fatty acyl chains | § 3.4.2 | 3.16 | McGregor et al. (2003), Kelly et al. (2005), Ho et al. (2008), and Privé (2009) |
| 3 | Designer peptide surfactants (DPSs; peptergents) | Short peptides (7–10 residues) with a hydrophobic "tail" and a hydrophilic "head" | § 3.4.3 | 3.17 | Santoso et al. (2002), Vauthey et al. (2002), Kiley et al. (2005), Yeh et al. (2005), Yang and Zhang (2006), Zhao et al. (2006), Matsumoto et al. (2009), Corin et al. (2011), Wang et al. (2011), and Koutsopoulos et al. (2012) |
| 4 | Saposin A-based complexes (picodiscs, Sal A discs, Salipro®) | A small protein (~9 kDa) whose amphipathic α-helices can form the rim of a small patch of lipid bilayer or adsorb onto the TM surface of MPs | § 3.4.4 | 3.19 | Popovic et al. (2012), Leney et al. (2015), Frauenfeld et al. (2016), Li et al. (2016a, b), and Flayhan et al. (2018) |
| 5 | Nanodisc-forming amphipathic peptides | Relatively long peptides (37 residues) that fold into amphipathic α-helices and associate with phospholipids to form ND-like complexes | § 3.4.5 | 3.23 | Park et al. (2011a), Zhao et al. (2013), Imura et al. (2014a, b), Midtgaard et al. (2014), Kariyazono et al. (2016), Kondo et al. (2016), Larsen et al. (2016), and Zhang et al. (2016) |

- Short peptides (6–8 residues) whose general organization is comparable to that of detergents: a short string of hydrophobic residues and a 1–2-residue polar "head" ("designer peptide surfactants"; Santoso et al. 2002, Vauthey et al. 2002; Table 3.2, line **3**; § 3.4.3).
- Interesting octyl-carrying, β-strand-forming octapeptides that have been shown to stabilize physically and biochemically the ABC exporter MsbA in the form of individual particles amenable to EM and to stabilize biochemically three other test MPs (Tao et al. 2013). They do not seem to have been further exploited.
- Saposin A, a small natural protein that folds into four amphipathic α-helices and can be used to trap lipids into "picodiscs," comprising only a few tens of lipids (Popovic et al. 2012), or to adsorb onto MP TM regions (Frauenfeld et al. 2016; Table 3.2, line **4**; § 3.4.4).
- Oligopeptides designed to assemble with lipids into nanodisc-like structures (Table 3.2, line **5**; § 3.4.5).
- Relatively long hydrophobized poly-γ-glutamic acid forming amphipathic polymers (Han et al. 2014). This approach will be discussed in Chap. 4 along with APols, to which it is closely related (see Chap. 4, Table 4.2, Study 4.22).

### 3.4.1    Peptitergents

The substitution of amphipathic peptides to apolipoprotein A to assemble lipoprotein particles was studied in the 1980s by Jere Segrest and collaborators (Anantharamaiah et al. 1985, 1990; Chung et al. 1985), but it seems that it is only in the 1990s that their application as substitutes to detergents for stabilizing MPs in aqueous solutions was first reported. Robert A. Stroud and his colleagues synthesized by solid-phase organic chemistry 24-residue peptides, which they called "peptitergents," designed to fold in water as laterally amphipathic helices (Table 3.2, Line **1**). Peptitergent $PD_1$ was shown by circular dichroism and X-ray crystallography to indeed fold into highly water-soluble $\alpha$-helices, which, upon crystallization, assembled into antiparallel helix bundles (Schafmeister et al. 1993). $PD_1$ was added to purified detergent solutions of two $\alpha$-helical MPs, BR (in nonyl-glucoside solution) and rhodopsin (in lauryldimethylamine oxide solution; LDAO), and a $\beta$-barrel one, PhoE porin (in OG), while diluting the detergent to <1/20 of its CMC. Under such conditions, PhoE precipitated completely, whereas ~85% of BR and ~60% of rhodopsin remained in solution under their native form (as ascertained from their UV-visible spectra) over a period of 2 days. Subsequent studies showed that $PD_1$ was efficient neither at directly extracting MPs from their native environment (Soomets et al. 1997; Bavec et al. 1999) nor at keeping rat cerebral cortical $Na^+/K^+$-exchanging ATPase water-soluble (Soomets et al. 1997). An original variation on the original strategy has been described, in which $PD_1$ was substituted, by genetic engineering, to the native TM hydrophobic helix of a P450 cytochrome, yielding a water-soluble enzyme suitable for $^1H$-NMR investigations (Schoch et al. 2003).

Peptitergents were initially developed with the view of applying them to MP crystallization (Schafmeister et al. 1993). This was a long shot, however, because the formation of well-ordered crystals of MP/peptitergent complexes would require that all complexes have strictly the same arrangement. No such crystals have been reported, and the study of peptitergents seems to have stopped.

### 3.4.2    Lipopeptide Detergents

Peptitergents provided a starting point for the development by Gilbert G. Privé and colleagues of a next generation of peptide-based surfactants called lipopeptide detergents (LPDs; Table 3.2, Line **2**). In LPDs, each 24-residue peptide is endowed with two fatty acyl chains, 12 to 20 carbon long, one bound to the *N*- and the other to the *C*-terminus (McGregor et al. 2003) (Fig. 3.16A). The rationale is that the acyl chains ought to preferentially interact with the hydrophobic face of the $\alpha$-helix formed by the peptide, providing a flexible, softer, more versatile surface for associating with the TM surface of target MPs. An additional potential bonus is that the structure of LPDs is expected to favor a MP/LPD arrangement in which the acyl chains align more or less with the protein's TM axis (Fig. 3.16C), much as lipid acyl chains do in membranes (Chap. 1). In aqueous solutions, LPDs do form $\alpha$-helices, which, according to both MD simulations (Kelly et al. 2005) and crystallographic data (Ho et al. 2008), assemble into bundles in which the alkyl chains run roughly parallel to helix axes and form the hydrophobic core of the bundle (Fig. 3.16B). LPDs can solubilize lipid vesicles (McGregor et al. 2003). They thus seem more detersive than PDs, but whether they can directly extract MPs from biological membranes has not been reported. Because they are expensive to produce, their recommended use is, rather, to trap and stabilize MPs that have been solubilized and purified in detergent solution. This has been demonstrated for a variety of MPs from both the $\alpha$-helical and

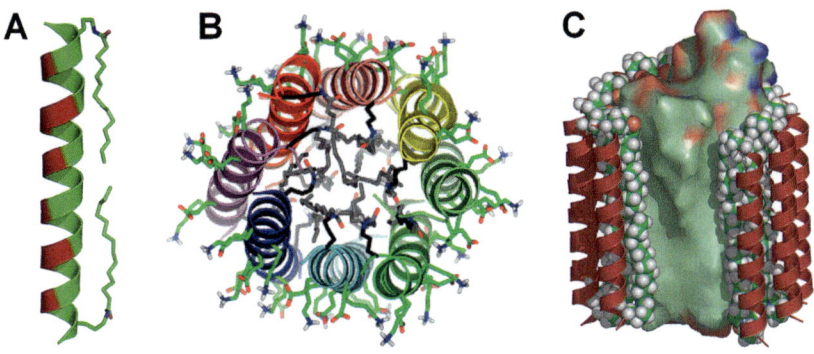

**Fig. 3.16** Lipopeptide detergents. (**A**) Schematic chemical structure. An LPD monomer consists of a 25-residue peptide designed to form an amphipathic $\alpha$-helix. The inner hydrophobic face (*green*) consists of alanines, the outer hydrophilic one (*red*) of polar residues. Ornithine residues at positions 2 and 24 are coupled to fatty acids that are designed to lie along the alanine face of the helix (From Privé 2009, © 2009 Elsevier Ltd. All rights reserved). (**B**) Molecular model, obtained by molecular dynamics, of an antiparallel bundle of eight LPD-12 molecules, each of them carrying two dodecyl chains. End-on view of the cylindrical assembly. Ornithine residues are drawn as sticks with *black* carbon atoms and are coupled via amide bonds to $C_{12}$ alkyl chains (*gray*). Lysine and glutamate residues face the exterior and are shown in stick representation with *green* carbons (From Kelly et al. 2005, © 2005 American Chemical Society). (**C**) Artist view of a proposed MP/LPD complex. The MP is represented by the solid surface. The peptide backbone of LPD-14 is represented by *red* ribbons and the ornithine residues and $C_{14}$ alkyl chains as space-filling spheres. The front-most LPD monomers are omitted for clarity (Reprinted with permission from McGregor et al. 2003, © 2003 Macmillan Publishers Limited, Nature Biotechnology. All rights reserved).

$\beta$-barrel classes (McGregor et al. 2003), including small monomeric proteins and large multisubunit enzyme complexes (unpublished data cited in Privé 2009).

Preliminary results with the $\beta$-barrel MP PagP demonstrated the potential of LPDs for solution NMR studies of MPs (McGregor et al. 2003). This approach has not been fully exploited yet, presumably hindered by high production costs. As regards X-ray crystallography, the same caveat can probably be lodged as with peptitergents, namely that in most cases the formation of well-ordered crystals of MP/LPD complexes would depend on the ability of LPDs to assemble (or reorganize) around MPs in exactly the same manner from one complex to the next, probably a harsh requirement. However, there seems to be no reason why LPDs could not be used as a shuttle to deliver MPs for crystallization in a lipid mesophase, as amphipol A8-35 (Polovinkin et al. 2014), styrene-maleic acid copolymers (Broecker et al. 2017), and NDs (Nikolaev et al. 2017) have been.

### 3.4.3 Designer Peptide Surfactants (Peptergents)

So-called designer peptide surfactants (DPSs) (also called "peptergents"; Yeh et al. 2005) have been developed by Shuguang Zhang and collaborators (Table 3.2, Line 3; reviewed in Koutsopoulos et al. 2012). At variance with peptitergents and lipopeptide detergents, those are short peptides, comprised of six to eight hydrophobic amino acid residues and one to two hydrophilic ones (Fig. 3.17). Rather than membrane-spanning peptides, they resemble detergents or phospholipids in their overall structure, and they are reasonably cheap to produce (typically < $ 30–50 a gram; Corin et al. 2011; Koutsopoulos

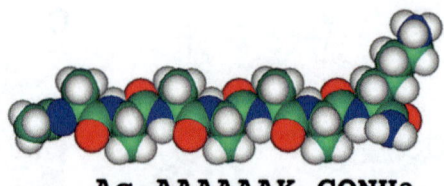

**Ac–AAAAAAK–CONH2**

**Fig. 3.17** Molecular model of a typical "designer peptide surfactant" (DPS). The hydrophobic moiety is generally comprised of alanine or valine residues, the hydrophilic one of lysine, arginine, or aspartate ones. The *N*- and *C*-termini are capped by acetylation and amidation, respectively. Each DPS is approximately 2–2.5 nm long, similar to phospholipids. Color code: *green*, carbon; *red*, oxygen; *blue*, nitrogen; *white*, hydrogen. The peptide shown was among the most efficient of those tested for stabilizing the Photosystem I reaction center in aqueous solutions (Matsumoto et al. 2009) (From Wang et al. 2011. © 2005 National Academy of Sciences, USA. See also Santoso et al. 2002).

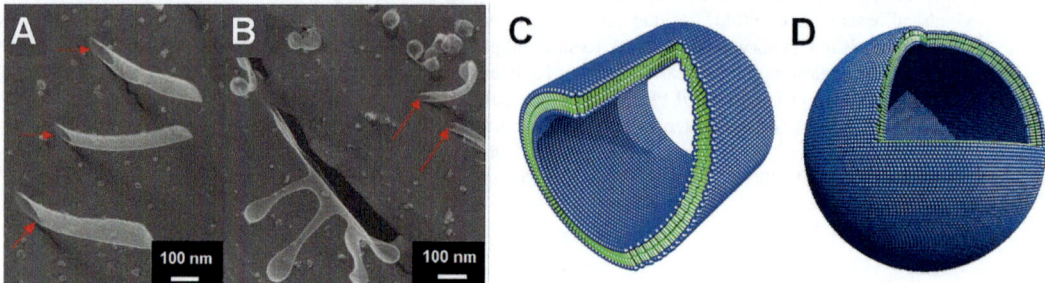

**Fig. 3.18** (**A**, **B**) Quick-freeze/deep-etch/platinum-coating transmission electron microscopy image of structures formed by designer peptide surfactants (DPSs) dissolved in water. The preparations contain both vesicles and hollow nanotubes, some of the latter branched and apparently budding. *Red arrows* point to tube openings (From Vauthey et al. 2002, © 2002 National Academy of Sciences, USA). (**C**, **D**) Proposed molecular models of nanotubes (**C**) and nanovesicles (**D**) formed by cationic DPSs at pH below their pI values. Color code: *blue*, positively charged amino acid heads; *green*, hydrophobic tail. The peptides pack so that the polar heads are exposed to water, sequestering the hydrophobic tails within the bilayer, much like in lipid bilayers. The diameters of the nanostructures are about 50–100 nm. Unlike lipid bilayers, in which the hydrophobic effects that expel the acyl chains from water are primarily responsible for the assembly, it is proposed that the packing of DPS bilayers involves backbone-to-backbone hydrogen bonds (Reprinted with permission from von Maltzahn et al. 2003, © 2003 American Chemical Society).

et al. 2012). They can directly extract glycerol-3-phosphate dehydrogenase (GlpD) from *Escherichia coli* membranes, although somewhat less effectively than the most efficient detergents (Yeh et al. 2005). GlpD was markedly more stable following DPS extraction than it is in OG solution. The size of the complexes was not reported. DPSs have been used as the recipient surfactant to keep G protein-coupled receptors (GPCRs) water-soluble during or after in vitro cell-free synthesis (Corin et al. 2011; Wang et al. 2011). Here, also, no size analysis was reported.

One caveat, when examining the solubilizing and stabilizing properties of DPSs, is that they are able to assemble into large structures, tubes, and vesicles, whose walls are thought to be comprised of peptide bilayers (Fig. 3.18). Whereas this basic organization evokes that of lipid bilayers, the peptide bilayer's internal chemistry is likely quite different, due to the formation of hydrogen bonds between peptide backbones and the higher rigidity of the peptides as compared to fatty acyl chains (von Maltzahn et al. 2003). When using DPSs to trap MPs, an important issue is therefore whether one is dealing with large aggregates or small complexes. Photosystem I reaction centers transferred from detergent solution to DPSs were found to be associated with large or small vesicles, depending on the peptide used (Matsumoto et al. 2009). The GPCR rhodopsin (Rho) was stabilized upon transfer from either DDM or OG to either detergent/DPS mixtures or pure DPSs, whether in the presence or absence of lipids (Zhao et al. 2006). Stabilization was markedly higher in the presence of lipids, strongly suggesting that Rho/lipid interactions persist in the presence of DPSs. The size of the objects formed was not reported.

DPSs can form micelles (Koutsopoulos et al. 2012), which would lead one to believe that small MP/DPS complexes can be obtained, but none seems to have been described yet, leaving it uncertain whether MP/DPS complexes can be used for NMR studies or crystallization attempts.

### 3.4.4    Stabilizing Membrane Proteins by Complexation with Saposin A ("Picodiscs," "Sap A discs," "Salipro® Nanoparticles")

Saposins are small (~80 residues), nonenzymatic proteins required for the breakdown of glycosphin-golipids within the lysosome (Kolter and Sandhoff 2005). Each saposin activates the breakdown of particular lipid substrates by facilitating the access of the lipid head groups to the active site of hydrolases. Saposins are thought to act by solubilizing the lipid substrates or simply by destabilizing the membrane structure. Ultracentrifugation and crystallographic studies of saposin A in aqueous solutions at neutral pH reveal a monomer folded as a bundle of four amphipathic $\alpha$-helices (Ahn et al. 2006) (Fig. 3.19A). At pH 4.8, as occurs in the lysosome lumen, saposin A interacts with liposomes, but does not remain bound to them, suggesting the formation of soluble protein/lipid complexes.

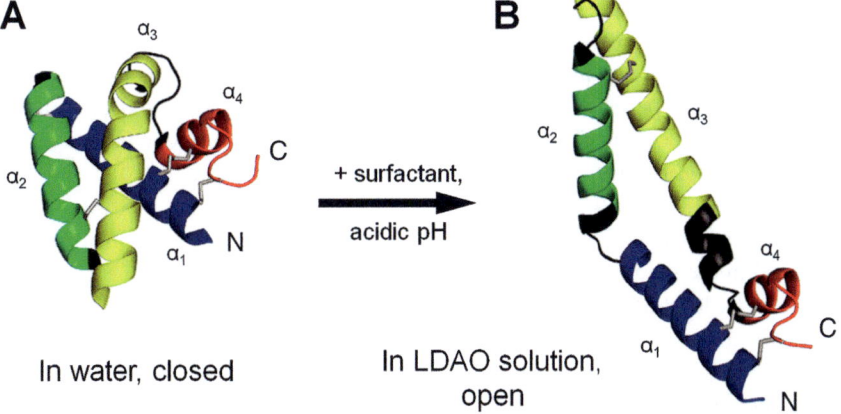

**Fig. 3.19**  Crystallographic structures of saposin A as a soluble monomer and as a detergent-bound dimer. (**A**) Ribbon representation of the soluble monomer of saposin A at pH 7. The four amphipathic helices ($\alpha_1$–$\alpha_4$) are colored *blue*, *green*, *yellow*, and *red*, respectively. The original structure is from Ahn et al. (2006). (**B**) Structure of a monomer in lauryldimethylamine oxide (LDAO) solution, pH 4.8. Helices $\alpha_1$ and $\alpha_4$ have been placed in the same orientation as in (**A**). The structure has opened like a jackknife, exposing the hydrophobic faces of the helices, which are buried in (**A**). Residues with significant conformational changes are shown in *black*. Three disulfide bridges associate $\alpha$-helices in pairs (*gray zigzags*). The arrangement of the two monomers in the saposin A dimer/LDAO complex is shown in Fig. 3.20 (From Popovic et al. 2012. © 2012 National Academy of Sciences, USA).

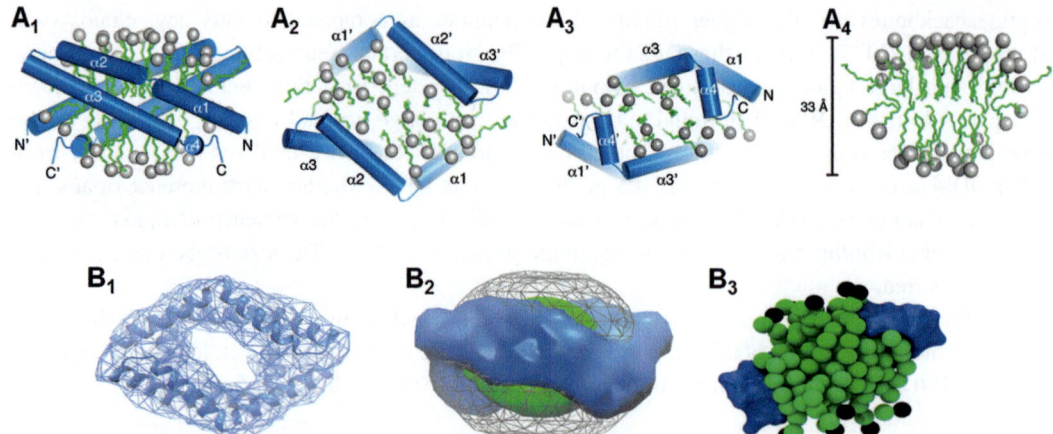

**Fig. 3.20** Crystallographic structure of saposin A/detergent complexes and proposed structure of saposin A/phospholipid picodiscs. (**A**) The saposin A/lauryldimethylamine oxide (LDAO) complex. The LDAO head groups are represented by *gray* spheres, the alkyl chains by *green* lines. (**A$_1$**) Side view of the complex. (**A$_2$**) View from the top of the dimer (relative to **A$_1$**). (**A$_3$**) View from the bottom of the dimer. (**A$_4$**) Assembly of the 40 LDAO molecules from the dimer. The view is similar to that in **A$_1$** but without the protein. (**B**) Proposed arrangement of the protein and lipids in a saposin A/palmitoyloleylphosphatidylcholine (POPC) picodisc, based on coarse-grained simulations. (**B$_1$**) The saposin A crystal structure, as observed in saposin A/LDAO complexes, is shown in ribbon representation superimposed over the spatial distribution of the protein from the simulations (*wire mesh*). Top view of the disc. (**B$_2$**) Side view of the saposin A/POPC disc from the simulations. The protein spatial distribution is shown in *solid blue* surface, the POPC acyl chains in *solid green* surface, and the POPC choline head group distribution in *gray wire frame*. (**B$_3$**) Snapshot of a representative side view from the coarse-grained simulation. The choline head groups of the lipids are represented by *black beads*; the acyl chains of the two POPC leaflets are colored *green*. The molecular surface of the saposin A chain located at the back of the complex is shown in *blue*. The front-most saposin A chain is omitted to reveal the lipidic core of the complex (From Popovic et al. 2012. © 2012 National Academy of Sciences, USA).

Indeed, following incubation of liposomes of various compositions with saposin A, SEC analysis reveals the presence of particles in the 35 to 45-kDa range, with hydrodynamic radii $R_H \approx 3.2$ nm (Popovic et al. 2012). The particles obtained by incubating saposin A with liposomes made from egg phosphatidylcholine (PC) contain $5 \pm 1$ lipids per saposin A chain, leading to the suggestion that they comprise two copies of saposin A and 8–12 lipids (Popovic et al. 2012).

The structure of saposin A/LDAO complexes was solved by X-ray diffraction and used as a starting point to construct models of saposin A/lipid particles by coarse-grained MD. The crystal structure obtained in the presence of detergent reveals an open fold (Fig. 3.19B) in which the protein exposes a concave hydrophobic surface, which is covered by LDAO acyl chains (Fig. 3.20A$_{1-3}$). Two V-shaped protein chains face each other and form an oval ring around a small patch of bilayer comprised of 40 well-ordered detergent molecules, 24 on one side, 16 on the other (Fig. 3.20A$_4$), with no direct protein/protein contacts between the two monomers (Fig. 3.20A$_{2-3}$) (Popovic et al. 2012).

It is worth noting that 40 LDAO molecules have approximately the same mass as the 8–12 lipids estimated to be present in a dimeric saposin A/PC complex, so that both a 2:40 saposin A/LDAO and a 2:10 saposin A/PC complex have masses of ~27 kDa (Popovic et al. 2012). Coarse-grained unconstrained MD simulations were used to construct 250 models of saposin A/PC 2:10 complexes. Figure 3.20B shows the result of one of the simulations with the closest configuration to that of the crystal structure, with the protein protomers arranged in a similar configuration to that in the saposin

A/LDAO complexes (Fig. 3.20B$_1$). As with LDAO, the lipid tails are held within the central cavity of the complex, where they interact with the hydrophobic faces of the saposin A protomers (Fig. 3.20B$_2$). A representative snapshot reveals that the lipids are arranged in two leaflets as a mini-bilayer (Fig. 3.20B$_3$), with more lipid molecules tending to lie in one leaflet than in the other. As noted by the authors of the study, the arrangement observed in the crystal structure may well represent only one of the configurations that saposin A/detergent or saposin A/lipid complexes may adopt in solution (Popovic et al. 2012). This hypothesis has been vindicated by more detailed experimental studies (Li et al. 2016b; see below). The so-called Sap A discs or picodiscs formed by saposin A-solubilized lipids have been exploited as a convenient formulation for presenting glycolipids to hydrolases or lipid-binding proteins (Leney et al. 2015; Li et al. 2016a).

Further studies, using catch-and-release electrospray ionization mass spectrometry (CaR-ESI-MS), size-exclusion chromatography/multi-angle laser light scattering (SEC-MALLS), and MD, do show that saposin A/PC complexes can adopt a variety of structures, some of them only transient (Fig. 3.21). More specifically, ESI-MS and SEC-MALLS data indicate that, at pH 4.8, saposin A/POPC complexes consist predominantly of saposin A dimers + 23–29 lipids, with $\langle MW \rangle = 38 \pm 3$ kDa and $\langle R_H \rangle \approx 3.1$ nm. In contrast, data acquired at pH 6.8 revealed that, in freshly prepared solutions, the complexes exist predominantly as saposin tetramers + 37–60 lipids ($\langle MW \rangle = 68.0 \pm 2.7$ kDa, $\langle R_H \rangle \approx 3.9$ nm). Over a period of hours, these complexes convert to trimers + 29–36 lipids ($\langle MW \rangle = 51.1 \pm 2.9$ kDa). MD suggests spheroidal structures for all complexes in solution (Fig. 3.21), which are essentially preserved during ESI-MS (Li et al. 2016b). These data suggest a certain flexibility of the arrangement of saposin A/lipid complexes and indicate that some complexes may accommodate more lipids than the crystal structure of the saposin A dimer/LDAO complex would predict.

Saposin A has been tested for its ability to trap and keep soluble three MPs, namely an archaeal mechanosensitive channel (T2, a putative homopentamer, 32.9 kDa, with four predicted TM helices per monomer), a bacterial peptide transporter (PepT$_{So2}$, a homotetramer, 56 kDa, with 14 TM helices per monomer), and the HIV-1 envelope glycoprotein (HIV-1 spike), which comprises an extrinsic

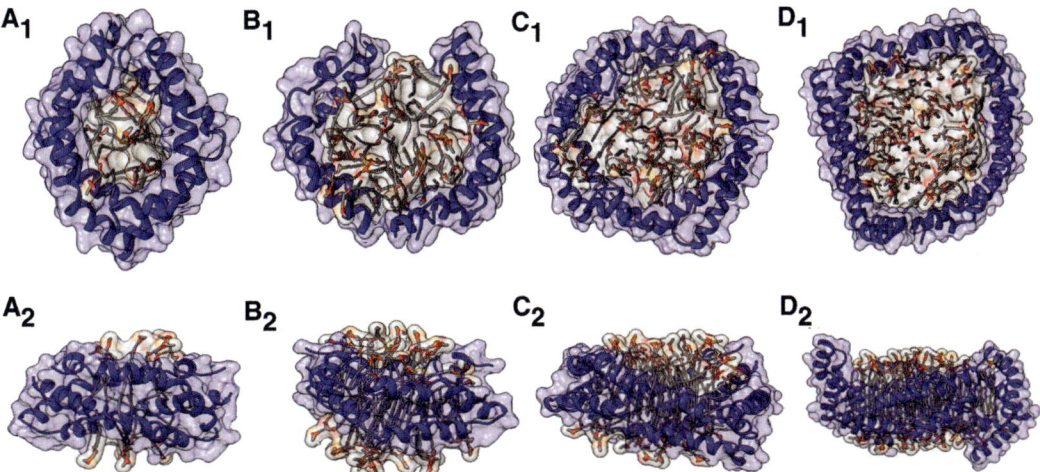

**Fig. 3.21** Averaged MD structures of saposin A/palmitoyloleylphosphatidylcholine (POPC) complexes in solution obtained after over 50 ns of full-atom simulations. Saposin A is shown as *purple* ribbons, POPC as *brownish* sticks. (**A**) Saposin A dimer +10 palmitoyloleylphosphatidylcholine (POPC) molecules (based on the model in Fig. 3.20B). (**B**) Saposin A dimer +26 POPC. (**C**) Saposin A trimer +33 POPC. (**D**) Saposin A tetramer +42 POPC. *Row 1*, top views; *row 2*, side views. Note that views are not all to the same scale (Reprinted with permission from Li et al. 2016b, © 2016 American Chemical Society).

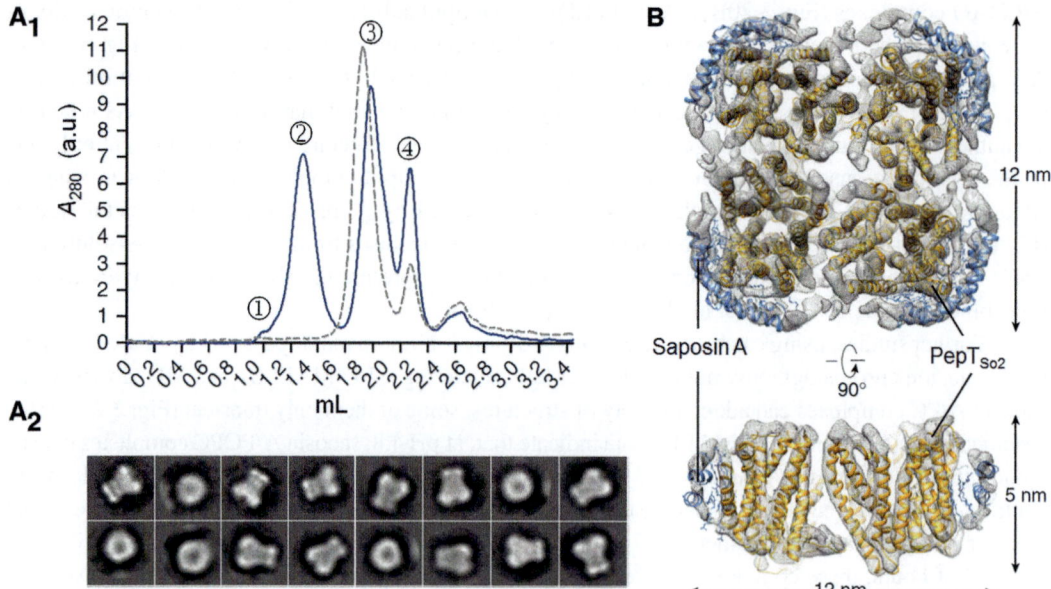

**Fig. 3.22** Analysis of saposin-trapped membrane proteins. (**A**) The archaebacterial T2 mechanosensitive channel. (**A₁**) Gel filtration analysis of the pentameric T2 channel incorporated into saposin/lipid nanoparticles (*solid blue line*) and of a protein-free control sample (*dashed gray line*). ① Void volume; ② T2/saposin A/lipid complexes; ③ saposin A/lipid complexes; ④ monomeric saposin A. (**A₂**) Selected two-dimensional class averages of negatively-stained T2/saposin A/lipid complexes. The side of each individual box measures 243 Å. (**B**) The bacterial PepT$_{So2}$ peptide transporter. Cryo-EM structure of PepT/saposin A/lipid nanoparticles. The three-dimensional density map of the bacterial transporter has been filtered to 6.5-Å resolution, showing the four PepT$_{So2}$ subunits and the tentative assignment of saposin A (in *yellow* and *blue*, respectively). *Top*, top view; *bottom*, side view cut perpendicularly to the plane of the membrane (Adapted from Frauenfeld et al. 2016, ©️ 2016 Macmillan Publishers Limited, Nature Methods. All rights reserved).

gp120 trimer and an integral gp41 one, each gp41 monomer featuring one TM $\alpha$-helix (Frauenfeld et al. 2016; in this work, saposin A/lipid complexes are called Salipro® – short for saposin-lipoprotein and a brand name – nanoparticles). Detergent-purified MPs were supplemented with saposin A in the presence (in the case of T2 and PepT$_{So2}$) or absence (HIV virus-like particles) of lipids, after which the detergent was removed by dilution and dialysis. SEC analysis showed T2 to be trapped into reasonably homogeneous complexes (Fig. 3.22A₁), which were analyzed by transmission EM after negative staining (Fig. 3.22A₂). The SEC profile of preparations of saposin-trapped PepT$_{So2}$ was similar. The transporter was significantly more stable after trapping (melting temperature $T_m = 72$ °C) than in nonyl-$\beta$-D-maltopyranoside solution ($T_m = 43$ °C). Single-particle cryo-EM was used to determine the structure of the complexes to 6.5-Å resolution. The cryo-EM density revealed a square-shaped particle with fourfold symmetry and overall dimensions compatible with each particle being comprised of a PepT$_{So2}$ tetramer and four saposin A molecules (Fig. 3.22B). The crystal structure of the peptide transporter could be directly docked into the cryo-EM density as a rigid body without modification. Whereas the TM helices of PepT$_{So2}$ were well resolved, saposin A and lipids were not. The dimensions of the density surrounding the transporter, however, are consistent with the crystal structure of saposin A. At the current low resolution for the saposin-lipid belt, it cannot be excluded that saposin A directly interacts with the TM helices of the transporter, but a loose contact mediated by lipids is also possible (Frauenfeld et al. 2016).

The HIV-1 spike is highly unstable in detergent solutions. In order to examine its stability in complex with saposin A, virus-like particles were incubated with detergent and saposin A, and the detergent was quickly removed using spin SEC columns. Whereas detergent-solubilized HIV-1 spikes dissociated completely within 30 min at 37 °C, saposin A-trapped spikes remained intact for up to 90 h at 37 °C, as judged by blue native-polyacrylamide gel electrophoresis (Frauenfeld et al. 2016).

Taken together, these data strongly suggest that saposin A provides a novel way to stabilize MPs in aqueous solutions in the absence of detergent. As compared to NDs, one may note two major differences: (i) on one hand, saposin A is highly versatile, being able to trap both very small (HIV-1 spikes, three TM helices) and very large (PepT$_{So2}$ tetramer, 56 TM helices) TM regions: the PepT$_{So2}$ tetramer would not fit into any of the existing NDs. (ii) On the other hand, the notion of trapping MPs within a bilayer patch disappears, it being not even certain that lipids are present at all in PepT$_{So2}$/saposin A complexes. Assuming they are, they do not occupy enough space to form a bilayer. Saposin A would therefore be akin to other surfactants that are better than detergents at preserving protein/protein and protein/lipid interactions, as is thought to be the case for APols (see Chap. 5), fluorinated surfactants (this Chapter, § 3.3.5), and, probably, many of the "mild" detergents (Chap. 2). What its advantages and drawbacks are over these other systems remain to be examined.

### 3.4.5 Peptide-Based Nanodiscs

A number of projects are under way to stabilize lipid bilayer patches not with full-length scaffold proteins as in classical NDs but with $\alpha$-helix-forming peptides derived or not from apolipoprotein A (see e.g. Park et al. 2011a; Zhao et al. 2013; Imura et al. 2014a, b; Midtgaard et al. 2014; Kariyazono et al. 2016; Kondo et al. 2016; Larsen et al. 2016; Zhang et al. 2016). In some cases, MP trapping has been demonstrated, e.g. that of BR (Larsen et al. 2016) or of a cytochrome P450/cytochrome $b_5$ complex (Zhang et al. 2016). Depending on the peptides used, it may or not be possible to adjust the peptide/lipid ratio so as to form more or less extended bilayer patches (see e.g. Park et al. 2011a; Kondo et al. 2016; Larsen et al. 2016). The development and validation of these interesting systems, an example of which is shown in Fig. 3.23, is only at its beginning.

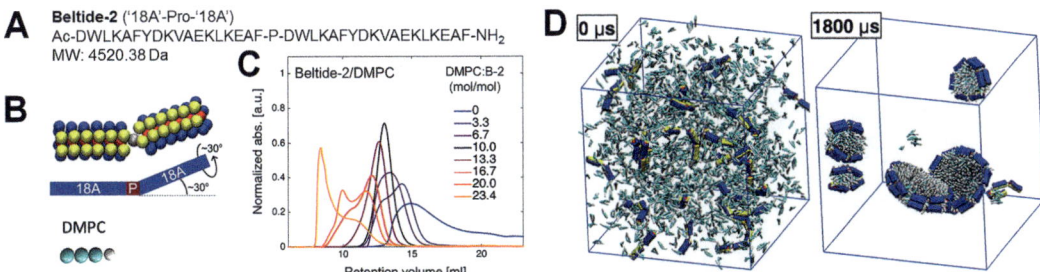

**Fig. 3.23** A two-part peptide that assembles spontaneously with lipids to form nanodiscs. (**A**) Sequence of "Beltide-2," a 37-residue peptide designed to fold into two 18-residue amphipathic $\alpha$-helices linked by a proline residue. The peptide was produced by solid-state synthesis. (**B**) Coarse-grained models of Beltide-2 and of DMPC. Hydrophilic peptide beads are in *blue* and hydrophobic beads in *yellow*. The central beads are *red*, and the linker bead is *gray*. The hydrophilic phospholipid head bead of DMPC is *white*, and the beads representing the hydrophobic tail group are *turquoise*. A sketch shows how Beltide-2 is expected to have a limited flexibility with an induced kink of ~30° and a fixed helix-unwinding twist. (**C**) Size-exclusion chromatography analysis of Beltide-2/DMPC particles at various lipid/peptide molar ratios. For a 10:1 ratio, a fairly homogeneous population of particles is observed (*black* chromatogram), whereas other ratios yield heterogeneous samples. The chromatograms were normalized by the area under the *curve*. (**D**) Coarse-grained molecular dynamics simulation snapshots for a Beltide-2/DMPC mixture (1:27 mol/mol) at time 0 and after 1.8 ms of simulation. The peptides and phospholipids self-assemble rapidly into small peptide/lipid particles that merge within ~2 ms to form peptide nanodiscs (Adapted from Larsen et al. 2016, published by the Royal Society of Chemistry).

## 3.5    Fluorinated Surfactants

Up till now, we have considered, in this chapter, surfactants or surfactant mixtures that provide solubilized MPs with an environment less aggressive than detergents and comprised of the same chemical moieties as those encountered in biological membranes: lipids, acyl chains, peptides, or small proteins. Fluorinated surfactants (FSs) exploit a different principle: their hydrophobic moieties are highly non-biological, and their mildness is due precisely to this peculiarity.

### 3.5.1    Background

The chemical structure of FSs resembles that of classical detergents (see Fig. 3.25), but their hydrophobic moieties contain fluorine atoms. The rationale behind the use of such compounds is the poor miscibility of hydrocarbons and fluorocarbons (see e.g. Kissa 1994; Mukerjee 1994; Nakano et al. 2002; Kirsch 2004; Riess 2005). This phenomenon is due to the fact that, because of the lesser polarizability of C–F *vs.* C–H bonds, van der Waals interactions between hydrocarbons and fluorocarbons are unfavorable, leading them to segregate one from another (Fig. 3.24). Surfactants with a perfluorinated hydrophobic moiety therefore do not partition well into lipid bilayers. As a result, most of them are not cytolytic and are unable to disperse biological membranes into their components: in a word, with rare exceptions (see below), they are not detergents (see e.g. Pavia et al. 1991; Chabaud et al. 1998; Barthélémy et al. 1999, 2002). The properties of FSs and their uses in MP studies have been reviewed in Breyton et al. (2010), Popot (2010), and Durand et al. (2014), to which the reader is referred for more detailed discussions, only a short account being presented here. The chemical formulae of a selection of fluorinated surfactants that have been developed over the past two decades in view of membrane biology applications are shown in Fig. 3.25.

Whereas they are widely used in other fields (see e.g. Kissa 2001), FSs, because most of them are not detergents and therefore unable to extract MPs from biological membranes, were long considered as presenting little interest in membrane biochemistry (with the exception of perfluoro-octanoate,

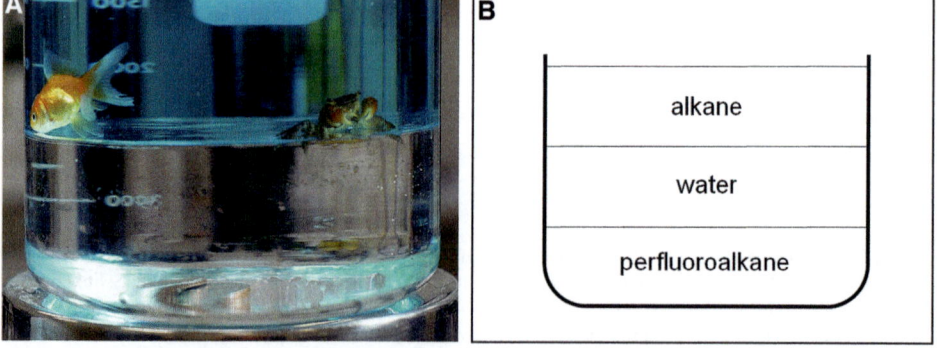

**Fig. 3.24** Perfluorocarbons are immiscible both with water and with hydrocarbons. In **A**, the top phase is water, the bottom phase perfluorodecalin. The photograph also illustrates the absence of toxicity of perfluorocarbons, which are used in many biomedical applications (perfluoro-octanoate, however, is slightly toxic; see Kennedy et al. 2004 and references therein) (From https://en.wikipedia.org/wiki/fluorocarbon). Panel **B** shows the behavior of a ternary mixture of alkane, fluoroalkane, and water: three phases form, superimposed as a function of their density, with each compound being only very sparingly soluble in any of the others. Similarly, micelles of fluorinated surfactants can coexist with bilayers formed of biological lipids.

which has been used for MP electrophoresis; Shepherd and Holzenburg 1995; Ramjeesingh et al. 1999). There were nevertheless two good reasons to expect that FSs, assuming that they could keep MPs soluble after they have been extracted with a classical detergent, might provide them with less destabilizing an environment: (i) lipids, subunits, and hydrophobic cofactors, having TM surfaces covered with methyl and methylene groups, should partition less favorably into FS micelle than into detergent ones, reducing the entropic drive toward dissociation, and (ii) because fluorinated alkyl chains are more bulky and more rigid than alkanes (see e.g. Kirsch 2004; Riess 2005), and they have little affinity for hydrogenated TM protein segments, they ought to be less efficient than detergents at disrupting stabilizing protein/protein and protein/lipid interactions. By the same token, however, it was to be feared that FSs might be ineffective as well at preventing MPs from aggregating, which early data seemed to bear out (Chabaud et al. 1998; Der Mardirossian et al. 1998). To try to improve interactions with the methyl group-covered TM surfaces of MPs while preserving the overall lyophobic ("lipid-fearing") character of FS micelles, a hydrogenated tip was therefore grafted onto the fluorinated tail, yielding "hemifluorinated surfactants" (HFSs) (cf. the structure of "HF-TAC" in Fig. 3.25A; Barthélémy et al. 2002; Breyton et al. 2004). Hereafter, FSs and HFSs will be collectively referred to as (H)FSs whenever a distinction between them does not need to be made.

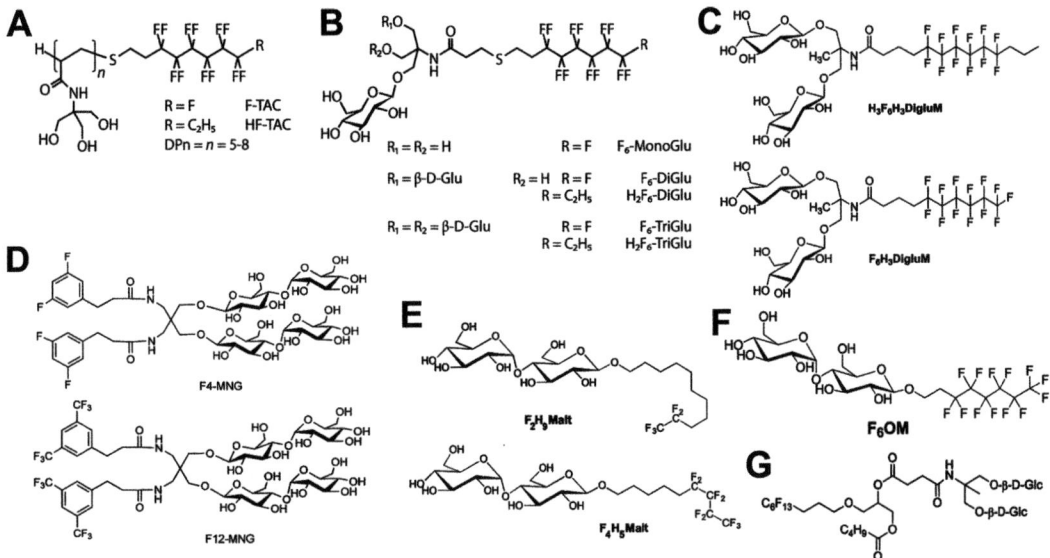

**Fig. 3.25** Chemical structure of some fluorinated surfactants developed for membrane biology. (**A**) F-TAC ($C_8F_{17}$-$C_2H_4$-S-poly-*tris*-(hydroxymethyl)aminomethane; Chabaud et al. 1998; Barthélémy et al. 1999; Breyton et al. 2004) and HF-TAC ($C_2H_5$-$C_6F_{12}$-$C_2H_4$-S-poly-*tris*-(hydroxymethyl)aminomethane; Barthélémy et al. 2002; Breyton et al. 2004). (**B**) (H)F-Mono-, Di-, and TriGlu; $F_6$- and $H_2F_6$-, as used in the text and in this figure, refer to hydrophobic moieties, where R = F and R = $C_2H_5$, respectively (Abla et al. 2008; Breyton et al. 2009). (**C**) $H_3F_6H_3$DigluM and $F_6H_3$DigluM (Glc: glucose residue). (From Abla et al. 2015). (**D**) Two hemifluorinated maltose-neopentyl glycol amphiphiles (F4-MNG and F12-MNG) (From Cho et al. 2013). (**E**) $F_2H_9\beta M$ and $F_4H_5\beta M$ (From Polidori et al. 2016). (**F**) Perfluoro-octylmaltoside (From Frotscher et al. 2015). (**G**) A hybrid double-chain surfactant possessing a diglucose (Glc) polar head group, a perfluorinated hexyl chain, and a hydrogenated butyl chain (From Legrand et al. 2016).

### 3.5.2    Novel Structures of Fluorinated Surfactants

The development of a wide range of novel (H)FSs specifically designed for MP biochemistry applications was undertaken, in the early 1990s, thanks to a long-term collaboration between our laboratory and that of Bernard Pucci, after which several other groups developed their own molecules.

The first molecules to be tested had as polar head formed by a short hydrophilic oligomer derived from tris-(hydroxymethyl)aminomethane (THAM) and a perfluorinated hydrophobic moiety ("F-TAC" in Fig. 3.25A; Chabaud et al. 1998; Barthélémy et al. 1999; Breyton et al. 2004). In a second stage, the tail was hemifluorinated ("HF-TAC" in Fig. 3.25A; Barthélémy et al. 1999; Chaudier et al. 2001, Breyton et al. 2004). These molecules yielded very promising results, judging from their ability to keep soluble and stabilize several test MPs, including BR and the cytochrome $b_6f$ complex (Chabaud et al. 1998; Breyton et al. 2004), and most of the early applications of (H)FSs were developed thanks to them.

The oligomeric polar head of (H)F-TAC – where "(H)F-" refers indifferently to the fluorinated or the hemifluorinated form – however, was a concern from the start because it is chemically poly-dispersed. From the biochemist's point of view, this means that (H)F-TAC batches consist of a mixture of molecules with slightly different properties and that this mixture will never be exactly the same from one batch to the next. Replacing the poly-THAM oligomer with a monodisperse head group turned out to be a highly frustrating endeavor. Indeed, grafting (hemi)fluorinated chains onto mono-disperse polar heads that, when associated to alkyl chains, yield efficient detergents – such as an aminoxyde (Chaudier et al. 2002), a monodisperse polyethylene glycol group (see Breyton et al. 2009), or saccharidic groups derived from galactose (Chabaud 1997; Chabaud et al. 1998), lactose (Lebaupain et al. 2006), or maltose (Polidori et al. 2006) – resulted in surfactants featuring unsatisfactory properties. A recurrent problem was that most of the molecules thus obtained tended to form huge polydisperse micelles by themselves and highly polydisperse MP/(H)FS complexes (for a discussion, see Breyton et al. 2009). This behavior suggested that the bulky hydrophobic moiety of (H)FSs requires a bulkier hydrophilic head than classical detergents do in order to create the overall molecular asymmetry that leads to the formation of small globular micelles (*cf.* Israelachvili et al. 1977, Tanford 1980, Israelachvili 2011; see Chap. 1, Table 1.1). A systematic investigation was therefore undertaken, in which polar heads carrying one, two, or three glucose moieties were grafted onto perfluorinated, hemifluorinated, or hydrogenated hydrophobic chains (Abla et al. 2008). This study led to the identification of two chemically defined (H)FSs, $F_6$-DiGlu and $H_2F_6$-DiGlu (Fig. 3.25B), which form with MPs small, well-defined complexes in which MPs are stabilized as compared to detergent solutions (Breyton et al. 2009).

Among the many new types of (H)FSs that have been described over the past decade, one may cite the following ones:

- Partially fluorinated surfactants, dubbed FASBs (fluorinated amidosulfobetaines), in which the tip of the hydrophobic chain is perfluorinated but a more or less extended hydrocarbon region is inserted between it and an amidosulfobetaine polar head (Starita-Geribaldi et al. 2007; Thebault et al. 2007).
- Hemifluorinated maltose-neopentyl glycol derivatives (HF-MNG), which carry fluorinated groups at the end of their two hydrocarbon arms (Fig. 3.25D; Cho et al. 2013).
- A hemifluorinated surfactant ($H_3F_6H_3DigluM$) featuring a branched diglucosylated polar head group and an apolar tail consisting of a perfluorohexane core terminated by a hydrogenated propyl tip (Fig. 3.25C; Abla et al. 2015).
- A partially fluorinated version of octylmaltoside called $F_6OM$ (Fig. 3.25F) and a variant thereof carrying a phosphocholine polar head ($F_6OPC$; Frotscher et al. 2015).

- Two maltose-based surfactants bearing either a perfluoroethyl ($F_2H_9$) or a perfluorobutyl ($F_4H_5$) tip at the end of their alkyl chain (Fig. 3.25E; Polidori et al. 2016).
- Two hybrid double-chain surfactants possessing a diglucose polar head group, a perfluorinated hexyl chain and a hydrogenated butyl chain (Fig. 3.25G; Legrand et al. 2016).

When developing novel (H)FS, care should be brought to optimize both the physical-chemical and the biochemical properties of the new molecules, which depend on the nature of both the apolar and the polar moieties. A typical example is, as mentioned above, that of the development of a series of (H)FSs sporting polar heads carrying either one, two, or three glucose moieties (Breyton et al. 2009; see Fig. 3.25B). Analytical ultracentrifugation and SANS data showed that molecules whose polar head bears a single glucosyl group form very large cylindrical micelles, whereas those with two or three glucose moieties form small, homogeneous, globular micelles. Surfactants that form cylindrical micelles are of limited use in membrane biology, because they do not give rise to the small MP/surfactant complexes that are required for such applications as purification, NMR, crystallization, or small-angle scattering studies. As regards the biochemical stability of trapped MPs, however, molecules bearing one or two glucose moieties were found to be stabilizing, whereas those with three moieties were destabilizing. Fluorinated and hemifluorinated surfactants with a two-glucose polar head thus appeared as the most promising molecules for biochemical applications and structural studies (Breyton et al. 2009). Similarly, a recent study showed that, of two surfactants carrying the same fluorinated chain, one was able and the other unable to destabilize lipid vesicles, depending on the nature of the polar head (Frotscher et al. 2015; Vargas et al. 2015).

### 3.5.3   Applications of Fluorinated Surfactants

After NDs (§ 3.3), amphipols (Chap. 5), and bicelles (§ 3.2), FSs are the nonconventional surfactants that have given rise to the most numerous applications, some of the best validated ones being listed in Table 3.3. Here is a rapid survey:

- MP stabilization, the rationales for which have been discussed above (Table 3.3, Line A).
- MP folding, taking advantage of the relative innocuousness of FSs, which interfere less than detergents do with those protein/protein interactions that determine the native fold (*ibid.*, Line B).

**Table 3.3** Some of the applications of micelle-forming surfactants carrying fluorinated or partially fluorinated hydrophobic chains.

| Line | Application | References |
|------|-------------|-----------|
| A | Stabilization | Chabaud et al. (1998), Breyton et al. (2004, 2009), Lebaupain et al. (2006), Polidori et al. (2006, 2016), Lebaupain (2007), Talbot et al. (2009), Joubert et al. (2010), Nehmé et al. (2010), Abla et al. (2012), and Cho et al. (2013) |
| B | Folding | Lebaupain (2007), Singh and Flowers II (2010), Kyrychenko et al. (2012b), Durand et al. (2014), and Frotscher et al. (2015) |
| C | Cell-free synthesis | Park et al. (2007, 2011), Breyton et al. (2009), and Blesneac et al. (2012) |
| D | Insertion into preformed membranes | Palchevskyy et al. (2006), Park et al. (2007), Posokhov et al. (2008), Rodnin et al. (2008), Raychaudhuri et al. (2011), and Kyrychenko et al. (2012a) |
| E | SANS | Abla et al. (2008, 2012), Breyton et al. (2009, 2013a, b); Durand et al. (2014) |

- MP cell-free in vitro synthesis, using FSs as a mild acceptor medium in which to let MPs fold as their hydrophobic segments emerge from the ribosome (*ibid.*, Line C).
- MP insertion into preformed lipid bilayers, the rationale there being that an FS solution above its CMC is able to keep MPs (or membrane-inserting toxins) water-soluble while respecting the integrity of the bilayer (*ibid.*, Line D).
- Small-angle neutron scattering. An interesting observation indeed is that the contrast-match point (see Chap. 9, Box 9.2) of $F_6$diGlu (Fig. 3.25B) and $F_6$diGluM (Abla et al. 2012) is very close to that of hydrogenated, unlabeled proteins. Appropriately adjusting the $D_2O$ content of the solution therefore permits to cancel the contribution of such a protein along with that of the surfactant, leaving the experimenter free to analyze the conformational changes of a deuterated protein interacting with the hydrogenated one (Table 3.3, Line E).

Because FSs are highly tensioactive and do not mix easily with hydrogenated surfactants, a monolayer of FSs at the air-water interface is essentially impenetrable to detergents. This has been exploited before to adsorb detergent-solubilized MPs under monolayers of appropriately tagged fluorinated lipids (Held et al. 1997; Vénien-Bryan et al. 1997; Lebeau et al. 2001; Fotinou et al. 2013) as well as to improve particle distribution in cryo-EM studies of ND-embedded or digitonin-solubilized MPs (Efremov et al. 2015; Gatsogiannis et al. 2016; Zhang and Chen 2016; Johnson and Chen 2017). FSs have proven useful as additives for MP crystallization (Efremov et al. 2010; Sobolev et al. 2013; Yoon et al. 2013; Shimada et al. 2017). Nitriloacetate-derivatized FSs have been synthesized for the same application (Petkova et al. 2007; Dauvergne et al. 2008; Kreutz et al. 2009). Newton black films (two air/water surfactant monolayers sandwiching a thin layer of aqueous solution, as in soap bubbles) of $C_6F_{13}$-SO-THAM and/or a nickel-bearing version thereof have been used to organize into two-dimensional layers a polyhistidine-tagged MP kept soluble by a hydrogenated detergent, whose arrangement was studied by X-ray reflectivity (Petkova et al. 2007).

(H)FS-trapped MPs could potentially be induced to form three-dimensional crystals, either in (H)FS solution or following transfer to lipidic mesophases, but this application remains to be validated (for discussions, see Breyton et al. 2009; Popot 2010; Durand et al. 2014). However, crystallization of a cytochrome *c* oxidase/cytochrome *c* complex has recently been achieved using a mixture of decylmaltoside and fluorinated octylmaltoside, where the use of hydrogenated detergents alone had been unsuccessful (Shimada et al. 2017).

As stated above (§ 3.5.1), with rare exceptions, such as perfluoro-octanoate (Shepherd and Holzenburg 1995; Ramjeesingh et al. 1999), FSs are unable to disperse biological membranes and extract MPs. A fluorinated derivative of octylmaltoside ($F_6$OM; Fig. 3.25F) has recently been described that is able to slowly solubilize POPC vesicles (Frotscher et al. 2015). The mixed micelles appear to be cylindrical, exhibiting large sizes when examined by dynamic light scattering. Extraction of MPs was not reported. Transfer to $F_6$OM was shown to facilitate refolding of the $\beta$-barrel MP outer membrane phospholipase A (OmpLA) from its urea-denatured form (Frotscher et al. 2015). As mentioned above, an analog of $F_6$OM with a phosphocholine polar head, $F_6$OPC, was found to be devoid of the solubilizing properties of $F_6$OM, underlining the critical role of the polar head in this process (Frotscher et al. 2015; Vargas et al. 2015).

The present section is by no means intended to present an exhaustive view of the properties and applications of (H)FSs. For more details about their physical-chemical properties and more in-depth discussions of their advantages and drawbacks for various applications, the reader is referred to earlier, more detailed reviews (Breyton et al. 2009; Popot 2010; Durand et al. 2014).

## 3.6   Nonconventional Surfactants Used for Handling Membrane Proteins in Aqueous Solutions: An Overview

At this point in the book, the reader could easily be forgiven for being somewhat bewildered: so many solutions offered for what seems essentially (but is not really) a single problem, handling MPs in aqueous solution under a stable form? We will come back to this issue in the final chapter on Conclusion and Perspectives. However, before embarking onto the discussion of what APols are and what they are good for, an orienting map and some guidelines could perhaps be useful.

A first point to keep in mind is that there is no panacea: not a single system will, at the current point in time (and, according to all probability, in the future), provide optimal conditions for any kind of experiment to be performed on any MP. To take a couple of obvious examples, stability is essential in some experiments, whereas the formation of small complexes is a prerequisite for others. Stability over days is a must for drug-screening experiments, for crystallization in solution, and, to a lesser extent, for NMR measurements and neutron scattering measurements, which typically require many hours of data collection. The formation of small complexes is an absolute requirement for solution NMR measurements, and the presence of as little bound surfactant as possible is essential for most solution crystallization experiments. Other experiments, such as most optical spectroscopy ones, or functional measurements, are generally not dependent on the protein being trapped in small complexes and can usually be completed in hours. The priority then shifts from long-term stability to the absence of interference with the measurements and/or as little perturbation as possible of the protein's environment.

Respecting as much as is possible the interactions that a MP establishes in situ with its neighbors is highly desirable. Some experiments, such as optical, EPR, and some NMR spectroscopy measurements, can be carried in vivo, which is, to an extent, ideal. However, for most structural explorations, and for many functional ones, the protein must be separated from most of the rest of the cell. Preserving an absolutely unchanged environment is something that cannot be done, even when working with "intact" membrane fragments: the lipid and protein composition and their asymmetry may be maintained, but not that of the medium on each side of the membrane – ion and small molecule composition, electrostatic and redox potential, interactions with intra- and extracellular solutes and matrices, and so forth. When extracting MPs from the membrane becomes unavoidable, as for purification, the problems worsen: some protein/protein interactions may have to be pried out, while others must be maintained. The bulk of the lipids has to go, whereas it is often essential that at least some of the protein/lipid interactions be preserved, or recreated.

Figure 3.26 presents a rough, somewhat subjective map of the way the various systems we have considered hitherto, as well as the APols we will discuss next, distribute as a function of two criteria: how aggressive the surfactant or surfactant mixture used is ("detergency") and how similar to the native environment is that experienced by the solubilized MP. Detergency is understood as the ability of the surfactant or surfactant mixture to separate membrane components one from another and, therefore, to extract MPs from biological membranes and to compete with natural protein/protein and protein/lipid interactions, a key factor in destabilizing MPs (see Chap. 2, § 2.4). Whether the environment resembles the natural one does not depend only on the presence of a bilayer but also, for less membrane-like systems, on the ability of the protein to retain or rebind lipids. Pure detergents used in the absence of lipids thus provide less native-like an environment than fluorinated surfactants or APols, whose structure is highly non-biological but which favor the retention of lipids.

Starting from the upper right corner of the map, liposomes – not a soluble system and, therefore, not discussed in this book – are arguably the most natural-looking system and one of the least aggressive. However, they cannot faithfully reproduce the asymmetry of lipid composition and

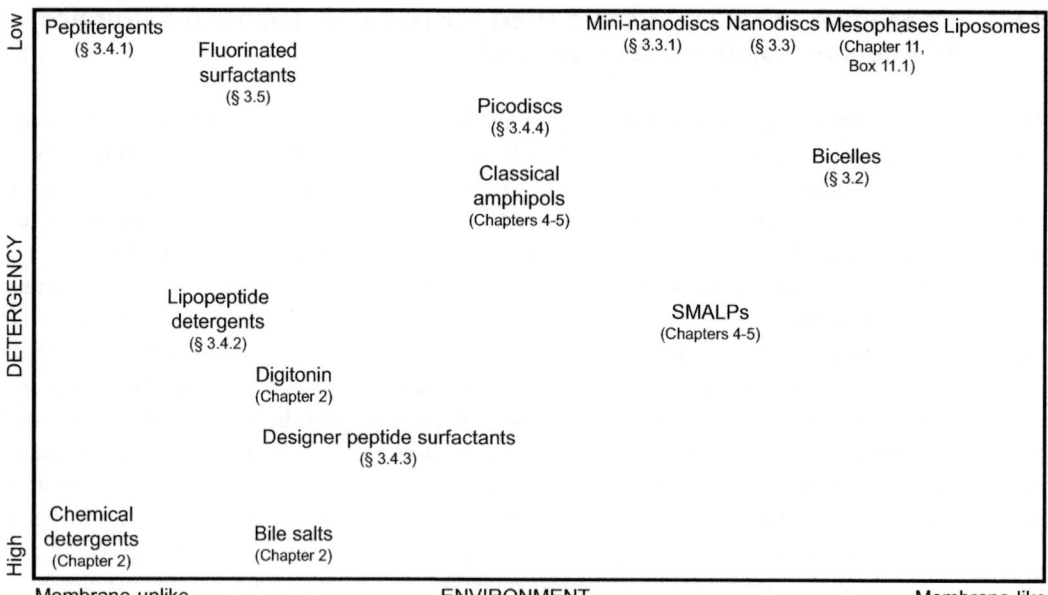

**Fig. 3.26** A rough cartography of the media used to handle membrane proteins (MPs) in aqueous environments. The chapter(s) or section where each class of surfactant is discussed is indicated. See text.

aqueous phase compositions that exist in vivo. Lipidic mesophases, also not a soluble system, add constraints on the lipid composition and the strong bend imposed on the bilayers they are comprised of. Among the systems discussed in this book, NDs provide the most native-like environment. However, lipid asymmetry is lost, and even though the lipids are arranged in a bilayer, the latter is perturbed, a perturbation that increases as one moves toward smaller discs ("mini-nanodiscs"). Bicelles present many of the advantages of NDs, but some of the detergent that forms their rim partitions into the lipid bilayer plane and has access to the protein. Saposin-based "picodiscs" contain few lipids and look more like lipoprotein complexes. As will be discussed in Chaps. 4 and 5, the case of SMALPs (styrene-maleic acid co-polymer lipoparticles) is complicated: on the one hand, they may look more native-like than NDs, because the protein is extracted along with native lipids; on the other, SMA is clearly more disruptive than the scaffold proteins of NDs, whether lipid asymmetry is retained is not known but looks improbable (see Chap. 5, § 5.3.1.2), the bilayer is more perturbed than in NDs, and it is likely that the protein can establish contacts with the polymer. Classical APols such as A8-35 can be looked at as milder forms of SMA (or SMAs as harsh APols). As NDs, they suffer from the disadvantage that, in most cases, MPs have to be extracted first with a detergent: they must either do without lipids, carry along the most essential of them, or rebind them upon trapping with APols. The same kind of remark can be done about fluorinated surfactants and peptitergents: whereas they are not detersive, they offer, by themselves, a most unnatural environment but can be supplemented with lipids. Lipopeptide detergents are an intermediate case, as they seem more detersive but provide a more lipid-like interaction surface. The case of "designer peptide surfactants" is currently open, as it does not seem to have been established that they can trap MPs in the form of small complexes.

Moving toward the lower left corner of the map, we enter the realm of more or less mild detergents. The mildest ones may have difficulties extracting MPs, but they tend – when it has been tested – to be less delipidating, digitonin being the archetype of a mild natural detergent. Bile salts and their derivatives tend to be efficient solubilizers (although not always), but the peculiar structure of their micelles makes them more apt to retain lipids in a more or less bilayer-like arrangement. Finally, a whole gamut of chemical detergents are available, remarkable progress having been achieved, in the past decades, to make them less and less aggressive (Chap. 2). When one's target MP can stand them, detergents remain those surfactants with which the smallest complexes can be formed, an important point in crystallogenesis and an essential one in solution NMR.

The map has to be taken with quite a few grains of salt. First, each class of surfactants is itself diverse. To take a couple of examples, "chemical detergents," as mentioned above, comprise a wide variety of "strong" detergents, like Triton X-100; "mild" detergents, like DDM; and "very mild" detergents, like maltose-neopentyl glycol (MNG) amphiphiles (see Chap. 2). "Classical amphipols" comprise both charged and uncharged species, whose mildness differs, and their respective ability to directly extract MPs has not been properly investigated (Chap. 5). Second, individual MPs exhibit different sensitivities to various surfactants, so that the map will vary depending on which target MP is considered. Third, there exists no comprehensive study encompassing a sufficient number of MPs, a sufficient number of criteria (solubility, stability, functionality, retention of lipids, and so forth), and the whole set of surfactants. In many cases, data are very sparse. Retention of MP/lipid interactions, for instance, is seldom directly measured and must be, in many cases, inferred from indirect information, such as the solubilizing properties of the surfactant and the stabilizing effect of lipids added to MP/surfactant complexes. The position assigned to a given class of surfactants is therefore based in part on indirect, piecemeal information. It is to some extent a matter of personal feeling and, therefore, open to discussion.

# References

Abla, M., Durand, G., Breyton, C., Raynal, S., Ebel, C., Pucci, B. (2012) A diglucosylated fluorinated surfactant to handle integral membrane proteins in aqueous solution. *J. Fluor Chem.* **134**:63–71.

Abla, M., Durand, G., Pucci, B. (2008) Glucose-based surfactants with hydrogenated, fluorinated, or hemifluorinated tails: synthesis and comparative physical-chemical characterization. *J. Org. Chem.* **73**:8142–8153.

Abla, M., Unger, S., Keller, S., Bonneté, F., Ebel, C., Pucci, B., Breyton, C., Durand, G. (2015) Micellar and biochemical properties of a propyl-ended fluorinated surfactant designed for membrane-protein study. *J. Colloid Interface Sci.* **445**:127–136.

Agah, S., Faham, S. (2012) Crystallization of membrane proteins in bicelles. *Methods Mol. Biol.* **914**:3–16.

Ahn, V.E., Leyko, P., Alattia, J.-R., Chen, L., Privé, G.G. (2006) Crystal structures of saposins A and C. *Protein Sci.* **15**:1849–1857.

Akkaladevi, N., Mukherjee, S., Katayama, H., Janowiak, B., Patel, D., Gogol, E.P., Pentelute, B.L., Collier, R.J., Fisher, M.T. (2015) Following Nature's lead: On the construction of membrane-inserted toxins in lipid bilayer nanodiscs. *J. Membr. Biol.* **248**:595–607.

Alvarez, F.J., Orelle, C., Huang, Y., Bajaj, R., Everly, R.M., Klug, C., Davidson, A.L. (2015) Full engagement of liganded maltose-binding protein stabilizes a semi-open ATP-binding cassette dimer in the maltose transporter. *Mol. Microbiol.* **98**:878–894.

Anantharamaiah, G.M., Brouillette, C.G., Engler, J.A., De Loof, H., Venkatachalapathi, Y.V., Boogaerts, J., Segrest, J.P. (1990) Role of amphipathic helices in HDL structure/function. *Adv. Exp. Med. Biol.* **285**:131–140.

Anantharamaiah, G.M., Jones, J.L., Brouillette, C.G., Schmidt, C.F., Chung, B.H., Hughes, T.A., Bhown, A.S., Segrest, J.P. (1985) Studies of synthetic peptide analogs of the amphipathic helix. Structure of complexes with dimyristoyl phosphatidylcholine. *J. Biol. Chem.* **260**:10248–10255.

Baas, B.J., Denisov, I.G., Sligar, S.G. (2004) Homotropic cooperativity of monomeric cytochrome P450 3A4 in a nanoscale native bilayer environment. *Arch. Biochem. Biophys.* **430**:218–228.

Banerjee, S., Huber, T., Sakmar, T.P. (2008) Rapid incorporation of functional rhodopsin into nanoscale apolipoprotein-bound bilayer (NABB) particles. *J. Mol. Biol.* **377**:1067–1081.

Barrett, P.J., Song, Y., Van Horn, W.D., Hustedt, E.J., Schafer, J.M., Hadziselimovic, A.H., Beel, A.J., Sanders, C.R. (2012) The amyloid precursor protein has a flexible transmembrane domain and binds cholesterol. *Science* **336**:1168–1171.

Barthélemy, P., Améduri, B., Chabaud, E., Popot, J.-L., Pucci, B. (1999) Synthesis and preliminary assessment of ethyl-terminated perfluoroalkyl slowdown surfactants derived from *tris*(hydroxymethyl)acrylamidomethane. *Org. Lett.* **1**:1689–1692.

Barthélemy, P., Tomao, V., Selb, J., Chaudier, Y., Pucci, B. (2002) Fluorocarbon-hydrocarbon non-ionic surfactant mixtures: a study of their miscibility. *Langmuir* **18**:2557–2563.

Bavec, A., Juréus, A., Cigić, B., Langel, U., Zorko, M. (1999) Peptitergent PD$_1$ affects the GTPase activity of rat brain cortical membranes. *Peptides* **20**:177–184.

Bayburt, T.H., Carlson, J.W., Sligar, S.G. (1998) Reconstitution and imaging of a membrane protein in a nanometer-size phospholipid bilayer. *J. Struct. Biol.* **123**:37–44.

Bayburt, T.H., Grinkova, Y.V., Sligar, S.G. (2002) Self-assembly of discoidal phospholipid bilayer nanoparticles with membrane scaffold proteins. *Nano Lett.* **2**:853–856.

Bayburt, T.H., Grinkova, Y.V., Sligar, S.G. (2006) Assembly of single bacteriorhodopsin trimers in bilayer nanodiscs. *Arch. Biochem. Biophys.* **450**:215–222.

Bayburt, T.H., Leitz, A.J., Xie, G., Oprian, D.D., Sligar, S.G. (2007) Transducin activation by nanoscale lipid bilayers containing one and two rhodopsins. *J. Biol. Chem.* **282**:14875–14881.

Bayburt, T.H., Sligar, S.G. (2002) Single-molecule height measurements on microsomal cytochrome P450 in nanometer-scale phospholipid bilayer disks. *Proc. Natl. Acad. Sci. USA* **99**:6725–6730.

Bayburt, T.H., Sligar, S.G. (2010) Membrane protein assembly into nanodiscs. *FEBS Lett.* **584**:1721–1727.

Bayburt, T.H., Vishnivetskiy, S.A., McLean, M.A., Morizumi, T., Huang, C.-C., Tesmer, J.J.G., Ernst, O.P., Sligar, S.G., Gurevich, V.V. (2011) Monomeric rhodopsin is sufficient for normal rhodopsin kinase (GRK1) phosphorylation and arrestin-1 binding. *J. Biol. Chem.* **286**:1420–1428.

Beaugrand, M., Arnold, A.A., Juneau, A., Balieiro Gambaro, A., Warschawski, D.E., Williamson, P.T.F., Marcotte, I. (2016) Magnetically oriented bicelles with monoalkylphosphocholines: versatile membrane mimetics for nuclear magnetic resonance applications. *Langmuir* **32**:13244–13251.

Bibow, S., Polyhach, Y., Eichmann, C., Chi, C.N., Kowal, J., Albiez, S., McLeod, R.A., Stahlberg, H., Jeschke, G., Güntert, P., Riek, R. (2017) Solution structure of discoidal high-density lipoprotein particles with a shortened apolipoprotein A-I. *Nat. Struct. Mol. Biol.* **24**:187–193.

Blesneac, I., Ravaud, S., Juillan-Binard, C., Barret, L.A., Zoonens, M., Polidori, A., Miroux, B., Pucci, B., Pebay-Peyroula, E. (2012) Production of UCP1, a membrane protein from the inner mitochondrial membrane, using the cell-free expression system in the presence of a fluorinated surfactant. *Biochim. Biophys. Acta* **1818**:798–805.

Bocharov, E.V., Pustovalova, Y.E., Pavlov, K.V., Volynsky, P.E., Goncharuk, M.V., Ermolyuk, Y.S., Karpunin, D.V., Schulga, A.A., Kirpichnikov, M.P., Efremov, R.G., Maslennikov, I.V., Arseniev, A.S. (2007) Unique dimeric structure of BNip3 transmembrane domain suggests membrane permeabilization as a cell death trigger. *J. Biol. Chem.* **282**:16256–16265.

Bocharov, E.V., Volynsky, P.E., Pavlov, K.V., Efremov, R.G., Arseniev, A.S. (2010) Structure elucidation of dimeric transmembrane domains of bitopic proteins. *Cell Adh. Migr.* **4**:284–298.

Boldog, T., Grimme, S., Li, M., Sligar, S.G., Hazelbauer, G.L. (2006) Nanodiscs separate chemoreceptor oligomeric states and reveal their signaling properties. *Proc. Natl. Acad. Sci. USA* **103**:11509–11514.

Borch, J., Hamann, T. (2009) The nanodisc: a novel tool for membrane protein studies. *Biol. Chem.* **390**:805–814.

Breyton, C., Chabaud, E., Chaudier, Y., Pucci, B., Popot, J.-L. (2004) Hemifluorinated surfactants: a non-dissociating environment for handling membrane proteins in aqueous solutions? *FEBS Lett.* **564**:312–318.

Breyton, C., Flayhan, A., Gabel, F., Lethier, M., Durand, G., Boulanger, P., Chamig, M., Ebel, C. (2013a) Assessing the conformation changes of pb5, the receptor binding protein of phage T5, upon binding to its *E. coli* receptor FhuA. *J. Biol. Chem.* **288**:30763–30772.

Breyton, C., Gabel, F., Abla, M., Pierre, Y., Lebaupain, F., Durand, G., Popot, J.-L., Ebel, C., Pucci, B. (2009) Micellar and biochemical properties of (hemi)fluorinated surfactants are controlled by the size of the polar head. *Biophys. J.* **97**:1077–1086.

Breyton, C., Gabel, F., Lethier, M., Flayhan, A., Durand, G., Jault, J.-M., Juillan-Binard, C., Imbert, Moulin, M., Ravaud S., Härtlein M., Ebel C. (2013b) Small angle neutron scattering for the study of solubilised membrane proteins. *Eur. Phys. J. E* **36**:71–86.

Breyton, C., Pucci, B., Popot, J.-L. (2010) Amphipols and fluorinated surfactants: two alternatives to detergents for studying membrane proteins *in vitro* in: Mus-Veteau, I., ed., *Heterologous expression of membrane proteins: Methods and protocols*. The Humana Press, Totowa, New Jersey, USA, pp. 219–245.

Broecker, J., Eger, B.T., Ernst, O.P. (2017) Crystallogenesis of membrane proteins mediated by polymer-bounded lipid nanodiscs. *Structure* **25**:384–392.

Brouillette, C.G., Anantharamaiah, G.M., Engler, J.A., Borhani, D.W. (2001) Structural models of human apolipoprotein A-I: a critical analysis and review. *Biochim. Biophys. Acta* **1531**:4–46.

Cappuccio, J.A., Blanchette, C.D., Sulchek, T.A., Arroyo, E.S., Kralj, J.M., Hinz, A.K., Kuhn, E.A., Chromy, B.A., Segelke, B.W., Rothschild, K.J., Fletcher, J.E., Katzen, F., Peterson, T.C., Kudlicki, W.A., Bench, G., Hoeprich, P. D., Coleman, M.A. (2008) Cell-free co-expression of functional membrane proteins and apolipoprotein, forming soluble nanolipoprotein particles. *Mol. Cell. Proteom.* **7**:2246–2253.

Carey, M.C., Small, D.M. (1972) Micelle formation by bile salts. Physical-chemical and thermodynamic considerations. *Arch. Intern. Med.* **130**:506–527.

Carlson, J.W., Jonas, A., Sligar, S.G. (1997) Imaging and manipulation of high-density lipoproteins. *Biophys. J.* **73**:1184–1189.

Casiraghi, M., Damian, M., Lescop, E., Point, E., Moncoq, K., Morellet, N., Levy, D., Marie, J., Guittet, E., Banères, J.-L., Catoire, L.J. (2016) Functional modulation of a GPCR conformational landscape in a lipid bilayer. *J. Am. Chem. Soc.* **138**:11170–11175

Catoire, L.J., Damian, M., Giusti, F., Martin, A., van Heijenoort, C., Popot, J.-L., Guittet, E., Banères, J.-L. (2010) Structure of a GPCR ligand in its receptor-bound state: leukotriene B$_4$ adopts a highly constrained conformation when associated to human BLT2. *J. Am. Chem. Soc.* **132**:9049–9057.

Catoire, L.J., Warnet, X.L., Warschawski, D.E. (2014) Micelles, bicelles, amphipols, nanodiscs, liposomes or intact cells: The hitch-hiker guide to the study of membrane proteins by NMR, in: Mus-Veteau, I., ed., *Membrane protein production for structural analysis.* Springer, pp. 315–346.

Catte, A., Patterson, J.C., Jones, M.K., Jerome, W.G., Bashtovyy, D., Su, Z., Gu, F., Chen, J., Aliste, M.P., Harvey, S.C., Li, L., Weinstein, G., Segrest, J.P. (2006) Novel changes in discoidal high density lipoprotein morphology: a molecular dynamics study. *Biophys. J.* **90**:4345–4360.

Chabaud, E. (1997) Application des tensioactifs fluorés à la manipulation *in vitro* des protéines membranaires. Rapport de D.E.A., Université Paris-VI.

Chabaud, E., Barthélémy, P., Mora, N., Popot, J.-L., Pucci, B. (1998) Stabilization of integral membrane proteins in aqueous solution using fluorinated surfactants. *Biochimie* **80**:515–530.

Chaudier, Y., Barthélémy, P., Pucci, B. (2001) Synthesis and preliminary assessment of hybrid hydrocarbon-fluorocarbon anionic and non-ionic surfactants. *Tetrahedron Lett.* **42**:3583–3585.

Chaudier, Y., Zito, F., Barthélémy, P., Stroebel, D., Améduri, B., Popot, J.-L., Pucci, B. (2002) Synthesis and preliminary biochemical assessment of ethyl-terminated perfluoroalkylamine oxide surfactants. *Bioorg. Med. Chem. Lett.* **12**:1587–1590.

Cho, K.H., Byrne, B., Chae, P.S. (2013) Hemifluorinated maltose-neopentyl glycol (HF-MNG) amphiphiles for membrane protein stabilisation. *ChemBioChem* **14**:452–455.

Chung, B.H., Anantharamaiah, G.M., Brouillette, C.G., Nishida, T., Segrest, J.P. (1985) Studies of synthetic peptide analogs of the amphipathic helix. Correlation of structure with function. *J. Biol. Chem.* **260**:10256–10262.

Chung, J., Prestegard, J.H. (1993) Characterization of field-ordered aqueous liquid crystals by NMR diffusion measurements. *J. Phys. Chem.* **97**:9837–9843.

Civjan, N.R., Bayburt, T.H., Schuler, M.A., Sligar, S.G. (2003) Direct solubilization of heterologously expressed membrane proteins by incorporation into nanoscale lipid bilayers. *BioTechniques* **35**:556–560, 562–563.

Corin, K., Baaske, P., Ravel, D.B., Song, J., Brown, E., Wang, X., Wienken, C.J., Jerabek-Willemsen, M., Duhr, S., Luo, Y., Braun, D., Zhang, S. (2011) Designer lipid-like peptides: a class of detergents for studying functional olfactory receptors using commercial cell-free systems. *PLoS ONE* **6**:e25067.

Cui, T.X., Canlas, C.G., Xu, Y., Tang, P. (2010) Anesthetic effects on the structure and dynamics of the second transmembrane domains of nAChR α4β2. *Biochim. Biophys. Acta* **1798**:161–166.

Czerski, L., Sanders, C.R. (2000) Functionality of a membrane protein in bicelles. *Anal. Biochem.* **284**:327–333.

Daury, L., Orange, F., Taveau, J.-C., Verchère, A., Monlezun, L., Gounou, C., Marreddy, R.K.R., Picard, M., Broutin, I., Pos, K.M., Lambert, O. (2016) Tripartite assembly of RND multidrug efflux pumps. *Nat. Commun.* **7**:10731.

Dauvergne, J., Polidori, A., Vénien-Bryan, C., Pucci, B. (2008) Synthesis of a hemifluorinated amphiphile designed for self-assembly and two-dimensional crystallization of membrane proteins. *Tet. Lett.* **49**:2247–2250.

De Angelis, A., Nevzorov, A., Park, S.H., Howell, S.C., Mrse, A.A., Opella, S.J. (2004) High-resolution NMR spectroscopy of membrane proteins in "aligned" bicelles. *J. Am. Chem. Soc.* **126**:15340–15341.

De Angelis, A.A., Opella, S.J. (2007) Bicelle samples for solid-state NMR of membrane proteins. *Nat. Protoc.* **2**:2332–2338.

Debnath, A., Schäfer, L.V. (2015) Structure and dynamics of phospholipid nanodiscs from all-atom and coarse-grained simulations. *J. Phys. Chem. B* **119**:6991–7002.

Dempsey, C.E., Sternberg, B. (1991) Reversible disc-micellization of dimyristoylphosphatidylcholine bilayers induced by melittin and [Ala-14]melittin. *Biochim. Biophys. Acta* **1061**:175–184.

Denisov, I.G., Baas, B.J., Grinkova, Y.V., Sligar, S.G. (2007) Cooperativity in cytochrome P450 3A4: linkages in substrate binding, spin state, uncoupling, and product formation. *J. Biol. Chem.* **282**:7066–7076.

Denisov, I.G., Grinkova, Y.V., Lazarides, A.A., Sligar, S.G. (2004) Directed self-assembly of monodisperse phospholipid bilayer nanodiscs with controlled size. *J. Am. Chem. Soc.* **126**:3477–3487.

Denisov, I.G., McLean, M.A., Shaw, A.W., Grinkova, Y.V., Sligar, S.G. (2005) Thermotropic phase transitions in soluble nanoscale lipid bilayers. *J. Phys. Chem. B.* **109**:15580–15588.

Denisov, I.G., Sligar, S.G. (2016) Nanodiscs for structural and functional studies of membrane proteins. *Nat. Struct. Mol. Biol.* **23**:481–486.

Denisov, I.G., Sligar, S.G. (2017) Nanodiscs in membrane biochemistry and biophysics. *Chem. Rev.* **117**:4669–4713.

Der Mardirossian, C., Krafft, M.-P., Gulik-Krzywicki, T., le Maire, M., Lederer, F. (1998) On the lack of protein-solubilizing properties of two perfluoroalkylated detergents, as tested with neutrophil plasma membranes. *Biochimie* **80**:531–541.

Duan, H., Civjan, N.R., Sligar, S.G., Schuler, M.A. (2004) Co-incorporation of heterologously expressed *Arabidopsis* cytochrome P450 and P450 reductase into soluble nanoscale lipid bilayers. *Arch. Biochem. Biophys.* **424**:141–153.

Durand, G., Abla, M., Ebel, C., Breyton, C. (2014) New amphiphiles to handle membrane proteins: *"Ménage à Trois"* between chemistry, physical chemistry, and biochemistry, in: Mus-Veteau, I., ed., *Membrane Proteins Production for Structural Analysis*. Springer, New York, Heidelberg, Dordrecht, London, pp. 205–251.

Durbin, D.M., Jonas, A. (1997) The effect of apolipoprotein A-II on the structure and function of apolipoprotein A-I in a homogeneous reconstituted high density lipoprotein particle. *J. Biol. Chem.* **272**:31333–31339.

Dürr, U.H., Gildenberg, M., Ramamoorthy, A. (2012) The magic of bicelles lights up membrane protein structure. *Chem. Rev.* **112**:6054–6074.

Dürr, U.H.N., Soong, R., Ramamoorthy, A. (2013) When detergent meets bilayer: Birth and coming of age of lipid bicelles. *Prog. Nucl. Magn. Reson. Spectrosc.* **69**:1–22.

Efremov, R.G., Baradaran, R., Sazanov, L.A. (2010) The architecture of respiratory complex I. *Nature* **465**:441–445.

Efremov, R.G., Leitner, A., Aebersold, R., Raunser, S. (2015) Architecture and conformational switch mechanism of the ryanodine receptor. *Nature* **517**:39–43.

Eichmann, C., Bibow, S., Riek, R. (2017) α-Synuclein lipoprotein nanoparticles. *Nanotech. Rev.* **6**:105–110.

Eichmann, C., Campioni, S., Kowal, J., Maslennikov, I., Gerez, J., Liu, X., Verasdonck, J., Nespovitaya, N., Choe, S., Meier, B.H., Picotti, P., Rizo, J., Stahlberg, H., Riek, R. (2016) Preparation and characterization of stable α-synuclein lipoprotein particles. *J. Biol. Chem.* **291**:8516–8527.

Elter, S., Raschle, T., Arens, S., Viegas, A., Gelev, V., Etzkorn, M., Wagner, G. (2014) The use of amphipols for NMR structural characterization of 7-TM proteins. *J. Membr. Biol.* **247**:957–964.

Etzkorn, M., Raschle, T., Hagn, F., Gelev, V., Rice, A.J., Walz, T., Wagner, G. (2013) Cell-free expressed bacteriorhodopsin in different soluble membrane mimetics: biophysical properties and NMR accessibility. *Structure* **21**:394–401.

Faham, S., Boulting, G.L., Massey, E.A., Yohannan, S., Yang, D., Bowie, J.U. (2005) Crystallization of bacteriorhodopsin from bicelle formulations at room temperature. *Protein Sci.* **14**:836–840.

Faham, S., Bowie, J.U. (2002) Bicelle crystallization: a new method for crystallizing membrane proteins yields a monomeric bacteriorhodopsin structure. *J. Mol. Biol.* **316**:1–6.

Fanucci, G.E., Lee, J.Y., Cafiso, D.S. (2003) Membrane mimetic environments alter the conformation of the outer membrane protein BtuB. *J. Am. Chem. Soc.* **125**:13932–13933.

Flayhan, A., Mertens, H.D.T., Ural-Blimke, Y., Molledo, M.M., Svergun, D.I., Loew, C. (2018) Saposin lipid nanoparticles: A highly versatile and modular tool for membrane protein research. *Structure* **26**:345–355.e345.

Forrest, B.J., Reeves, L.W. (1981) New lyotropic liquid crystals composed of finite nonspherical micelles. *Chem. Rev.* **81**:1–14.

Fotinou, C., Aittoniemi, J., de Wet, H., Polidori, A., Pucci, B., Sansom, M.S.P., Vénien-Bryan, C., Ashcroft, F.M. (2013) Tetrameric structure of SUR2B revealed by electron microscopy of oriented single particles. *FEBS J.* **280**:1051–1063.

Frauenfeld, J., Gumbart, J., van der Sluis, E.O., Funes, S., Gartmann, M., Beatrix, B., Mielke, T., Berninghausen, O., Becker, T., Schulten, K., Beckmann, R. (2011) Cryo-EM structure of the ribosome-SecYE complex in the membrane environment. *Nat. Struct. Mol. Biol.* **18**:614–621.

Frauenfeld, J., Löving, R., Armache, J.-P., Sonnen, A.F.-P., Guettou, F., Moberg, P., Zhu, L., Jegerschöld, C., Flayhan, A., Briggs, J.A.G., Garoff, H., Löw, C., Cheng, Y., Nordlund, P. (2016) A saposin-lipoprotein nanoparticle system for membrane proteins. *Nat. Meth.* **13**:345–351.

Frey, L., Lakomek, N.-A., Riek, R., Bibow, S. (2017) Micelles, bicelles, and nanodiscs: Comparing the impact of membrane mimetics on membrane protein backbone dynamics. *Angew. Chem. Int. Ed.* **56**:380–383.

Frotscher, E., Danielczak, B., Vargas, C., Meister, A., Durand, G., Keller, S. (2015) A fluorinated detergent for membrane-protein applications. *Angew. Chem. Int. Ed.* **17**:5069–5073.

Früh, V., IJzerman, A.P., Siegal, G. (2011) How to catch a membrane protein in action: a review of functional membrane protein immobilization strategies and their applications. *Chem. Rev.* **111**:640–656.

Gao, T., Petrlova, J., He, W., Huser, T., Kudlick, W., Voss, J., Coleman, M.A. (2012) Characterization of *de novo* synthesized GPCRs supported in nanolipoprotein discs. *PLoS ONE* **7**:e44911.

Gao, Y., Cao, E., Julius, D., Cheng, Y. (2016) TRPV1 structures in nanodiscs reveal mechanisms of ligand and lipid action. *Nature* **534**:347–351.

Gatsogiannis, C., Merino, F., Prumbaum, D., Roderer, D., Leidreiter, F., Meusch, D., Raunser, S. (2016) Membrane insertion of a Tc toxin in near-atomic detail. *Nat. Struct. Mol. Biol.* **23**:884–890.

Gautier, A., Mott, H.R., Bostock, M.J., Kirkpatrick, J.P., Nietlispach, D. (2010) Structure determination of the seven-helix transmembrane receptor sensory rhodopsin II by solution NMR spectroscopy. *Nat. Struct. Mol. Biol.* **17**:768–774.

Georgieva, E.R. (2017) Nanoscale lipid membrane mimetics in spin-labeling and electron paramagnetic resonance spectroscopy studies of protein structure and function. *Nanotech. Rev.* **6**:75–92.

Ghimire, H., Abu-Baker, S., Sahu, I.D., Zhou, A., Mayo, D.J., Lee, R.T., Lorigan, G.A. (2011) Probing the helical tilt and dynamic properties of membrane-bound phospholamban in magnetically aligned bicelles using electron paramagnetic resonance spectroscopy. *Biochim. Biophys. Acta* **1818**:645–650.

Gillette, W.K., Esposito, D., Abreu Blanco, M., Alexander, P., Bindu, L., Bittner, C., Chertov, O., Frank, P.H., Grose, C., Jones, J.E., Meng, Z., Perkins, S., Van, Q., Ghirlando, R., Fivash, M., Nissley, D.V., McCormick, F., Holderfield, M., Stephen, A.G. (2015) Farnesylated and methylated KRAS4b: high yield production of protein suitable for biophysical studies of prenylated protein-lipid interactions. *Sci. Rep.* **5**:15916.

Goddard, A.D., Dijkman, P.M., Adamson, R.J., Inácio dos Reis, R., Watts, A. (2015) Reconstitution of membrane proteins: A GPCR as an example. *Meth. Enzymol.* **556**:405–424.

Gogol, E.P., Akkaladevi, N., Szerszen, L., Mukherjee, S., Chollet-Hinton, L., Katayama, H., Pentelute, B.L., Collier, R. J., Fisher, M.T. (2013) Three dimensional structure of the anthrax toxin translocon-lethal factor complex by cryo-electron microscopy. *Prot. Sci.* **22**:586–594.

Gogonea, V. (2016) Structural insights into high-density lipoprotein: Old models and new facts. *Front. Pharmacol.* **6**:1–30.

Gogonea, V., Gerstenecker, G.S., Wu, Z., Lee, X., Topbas, C., Wagner, M.A., Tallant, T.C., Smith, J.D., Callow, P., Pipich, V., Malet, H., Schoehn, G., DiDonato, J.A., Hazen, S.L. (2013) The low-resolution structure of nHDL reconstituted with DMPC with and without cholesterol reveals a mechanism for particle expansion. *J. Lipid Res.* **54**:966–983.

Gregersen, J.L., Fedosova, N.U., Nissen, P., Boesen, T. (2016) Reconstitution of Na$^+$,K$^+$-ATPase in nanodiscs. *Methods Mol. Biol.* **1377**:403–409.

Grigorieff, N., Ceska, T.A., Downing, K.H., Baldwin, J.M., Henderson, R. (1996) Electron-crystallographic refinement of the structure of bacteriorhodopsin. *J. Mol. Biol.* **259**:393–421.

Grinkova, Y.V., Denisov, I.G., Sligar, S.G. (2010) Engineering extended membrane scaffold proteins for self-assembly of soluble nanoscale lipid bilayers. *Protein Eng. Des. Sel.* **23**:843–848.

Gruene, T., Cho, M.-K., Karyagina, I., Kim, H.-Y., Grosse, C., Giller, K., Zweckstetter, M., Becker, S. (2011) Integrated analysis of the conformation of a protein-linked spin label by crystallography, EPR and NMR spectroscopy. *J. Biomol. NMR* **49**:111–119.

Grushin, K., Miller, J., Dalm, D., Stoilova-McPhie, S. (2015) Factor VIII organisation on nanodiscs with different lipid composition. *Thromb. Haemost.* **113**:741–749.

Gustavsson, M., Traaseth, N.J., Veglia, G. (2012) Probing ground and excited states of phospholamban in model and native lipid membranes by magic angle spinning NMR spectroscopy. *Biochim. Biophys. Acta* **1818**:146–153.

Hagn, F., Etzkorn, M., Raschle, T., Wagner, G. (2013) Optimized phospholipid bilayer nanodiscs facilitate high-resolution structure determination of membrane proteins. *J. Am. Chem. Soc.* **135**:1919–1925.

Hagn, F., Wagner, G. (2015) Structure refinement and membrane positioning of selectively labeled OmpX in phospholipid nanodiscs. *J. Biomol. NMR* **61**:249–260.

Han, S.G., Na, J.H., Lee, W.K., Park, D., Oh, J., Yoon, S.H., Lee, C.K., Sung, M.H., Shin, Y.K., Yu, Y.G. (2014) An amphipathic polypeptide derived from poly-γ-glutamic acid for the stabilization of membrane proteins. *Prot. Sci.* **23**:1800–1807.

Hansen, R.W., Wang, X., Golab, A., Bornert, O., Oswald, C., Wagner, R., Martinez, K.L. (2016) Functional stability of the human κ-opioid receptor reconstituted in nanodiscs revealed by a time-resolved scintillation proximity assay. *PLoS ONE* **11**:e0150658.

Harroun, T.A., Koslowsky, M., Nieh, M.P., de Lannoy, C.F., Raghunathan, V.A., Katsaras, J. (2005) Comprehensive examination of mesophases formed by DMPC and DHPC mixtures. *Langmuir* **21**:5356–5361.

Held, P., Lach, F., Lebeau, L., Mioskowski, C. (1997) Synthesis and preliminary evaluation of a new class of fluorinated amphiphiles designed for in-plane immobilisation of biological macromolecules. *Tetrahedron Lett.* **38**:1937–1940.

Henrich, E., Dötsch, V., Bernhard, F. (2015) Screening for lipid requirements of membrane proteins by combining cell-free expression with nanodiscs. *Meth. Enzymol.* **556**:351–369.

Henrich, E., Ma, Y., Engels, I., Münch, D., Otten, C., Schneider, T., Henrichfreise, B., Sahl, H.G., Dötsch, V., Bernhard, F. (2016) Lipid requirements for the enzymatic activity of MraY translocases and *in vitro* reconstitution of Lipid II synthesis pathway. *J. Biol. Chem.* **291**:2535–2546.

Henrich, E., Peetz, O., Hein, C., LaGuerre, A., Hoffmann, B., Hoffmann, J., Dötsch, V., Bernhard, F., Morgner, N. (2017) Analyzing native membrane protein assembly in nanodiscs by combined non-covalent mass spectrometry and synthetic biology. *eLife* **6**:e20954.

Ho, D.N., Pomroy, N.C., Cuesta-Seijo, J.A., Privé, G.G. (2008) Crystal structure of a self-assembling lipopeptide detergent at 1.20 Å. *Proc. Natl. Acad. Sci. USA* **105**:12861–12866.

Hopper, J.T.S., Yu, Y.T.-C., Li, D., Raymond, A., Bostock, M., Liko, I., Mikhailov, V., Laganowsky, A., Benesch, J.L.P., Caffrey, M., Nietlispach, D., Robinson, C.V. (2013) Detergent-free mass spectrometry of membrane protein complexes. *Nat. Meth.* **10**:1206–1208.

Howell, S.C., Mesleh, M.F., Opella, S.J. (2005) NMR structure determination of a membrane protein with two transmembrane helices in micelles: MerF of the bacterial mercury detoxification system. *Biochemistry* **44**:5196–5206.

Imura, T., Tsukui, Y., Sakai, K., Sakai, H., Taira, T., Kitamoto, D. (2014a) Minimum amino acid residues of an α-helical peptide leading to lipid nanodisc formation. *J. Oleo Sci.* **63**:1203–1208.

Imura, T., Tsukui, Y., Taira, T., Aburai, K., Sakai, K., Sakai, H., Abe, M., Kitamoto, D. (2014b) Surfactant-like properties of an amphiphilic α-helical peptide leading to lipid nanodisc formation. *Langmuir* **30**:4752–4759.

Inagaki, S., Ghirlando, R. (2017) Nanodisc characterization by analytical ultracentrifugation. *Nanotech. Rev.* **6**:3–14.

Israelachvili, J.N. (2011) *Intermolecular and surface forces*, 3rd edition. Academic Press, London, 706 p.

Israelachvili, J.N., Mitchell, D.J., Ninham, B.W. (1977) Theory of self-assembly of lipid bilayers and vesicles. *Biochim. Biophys. Acta* **470**:185–201.

Johansson, L.C., Wöhri, A.B., Katona, G., Engström, S., Neutze, R. (2009) Membrane protein crystallization from lipidic phases. *Curr. Opin. Struct. Biol.* **19**:372–378.

Johnson, P.J., Halpin, A., Morizumi, T., Brown, L.S., Prokhorenko, V.I., Ernst, O.P., Dwayne Miller, R.J. (2014) The photocycle and ultrafast vibrational dynamics of bacteriorhodopsin in lipid nanodiscs. *Phys. Chem. Chem. Phys.* **16**:21310–21320

Johnson, Z.L., Chen, J. (2017) Structural basis of substrate recognition by the multidrug resistance protein MRP1. *Cell* **168**:1075–1085.

Jonas, A. (1986) Reconstitution of high-density lipoproteins. *Methods Enzymol.* **128**:553–582.

Jonas, A., Kezdy, K.E., Wald, J.H. (1989) Defined apolipoprotein A-I conformations in reconstituted high-density lipoprotein discs. *J. Biol. Chem.* **264**:4818–4824.

Jonas, A., von Eckardstein, A., Kézdy, K.E., Steinmetz, A., Assmann, G. (1991) Structural and functional properties of reconstituted high-density lipoprotein discs prepared with six apolipoprotein A-I variants. *J. Lipid Res.* **32**:97–106.

Jonas, A., Wald, J.H., Toohill, K.L., Krul, E.S., Kézdy, K.E. (1990) Apolipoprotein A-I structure and lipid properties in homogeneous, reconstituted spherical and discoidal high density lipoproteins. *J. Biol. Chem.* **265**:22123–22129.

Joubert, O., Nehmé, R., Bidet, M., Mus-Veteau, I. (2010) Heterologous expression of human membrane receptors in the yeast *Saccharomyces cerevisiae*, in: Mus-Veteau, I., ed., *Heterologous expression of membrane proteins*. Humana Press, New York, pp. 87–103.

Kang, C., Vanoye, C.G., Welch, R.C., Van Horn, W.D., Sanders, C.R. (2010) Functional delivery of a membrane protein into oocyte membranes using bicelles. *Biochemistry* **49**:653–655.

Kang, Y., Zhou, X.E., Gao, X., He, Y., Liu, W., Ishchenko, A., Barty, A., White, T.A., Yefanov, O., Han, G.W., Xu, Q., deWaal, P.W., Ke, J., Tan, M.H.E., Zhang, C., Moeller, A., West, G.M., Pascal, B.D., Van Eps, N., Caro, L.N., Vishnivetskiy, S.A., Lee, R.J., Suino-Powell, K.M., Gu, X., Pal, K., Ma, J., Zhi, X., Boutet, S., Williams, G.J., Messerschmidt, M., Gati, C., Zatsepin, N.A., Wang, D., James, D., Basu, S., Roy-Chowdhury, S., Conrad, C.E., Coe, J., Liu, H., Lisova, S., Kupitz, C., Grotjohann, I., Fromme, R., Jiang, Y., Tan, M., Yang, H., Li, J., Wang, M., Zheng, Z., Li, D., Howe, N., Zhao, Y., Standfuss, J., Diederichs, K., Dong, Y., Potter, C.S., Carragher, B., Caffrey, M., Jiang, H., Chapman, H.N., Spence, J.C.H., Fromme, P., Weierstall, U., Ernst, O.P., Katritch, V., Gurevich, V.V., Griffin, P.R., Hubbell, W.L., Stevens, R.C., Cherezov, V., Melcher, K., Xu, E. (2015) Crystal structure of rhodopsin bound to arrestin by femtosecond X-ray laser. *Nature* **523**:561–567.

Kariyazono, H., Nadai, R., Miyajima, R., Takechi-Haraya, Y., Baba, T., Shigenaga, A., Okuhira, K., Otaka, A., Saito, H. (2016) Formation of stable nanodiscs by bihelical apolipoprotein A-I mimetic peptide. *J. Pept. Sci.* **22**:116–122.

Katayama, H., Wang, J., Tama, F., Chollet, L., Gogol, E.P., Collier, R.J., Fisher, M.T. (2010) Three-dimensional structure of the anthrax toxin pore inserted into lipid nanodiscs and lipid vesicles. *Proc. Natl. Acad. Sci. USA* **107**:3453–3457.

Katzen, F., Fletcher, J.E., Yang, J.P., Kang, D., Peterson, T.C., Cappuccio, J.A., Blanchette, C.D., Sulchek, T., Chromy, B.A., Hoeprich, P.D., Coleman, M.A., Kudlicki, W. (2008) Insertion of membrane proteins into discoidal membranes using a cell-free protein expression approach. *J. Proteome Res.* **7**:3535–3542.

Kedrov, A., Wickles, S., Crevenna, A.H., van der Sluis, E.O., Buschauer, R., Berninghausen, O., Lamb, D.C., Beckmann, R. (2016) Structural dynamics of the YidC:ribosome complex during membrane protein biogenesis. *Cell Rep.* **17**:2934–2954.

Kelly, E., Privé, G.G., Tieleman, P.D. (2005) Molecular models of lipopeptide detergents: large coiled-coils with hydrocarbon interiors. *J. Am. Chem. Soc.* **127**:13446–13447.

Kennedy, G.L., Butenhoff, J.L., Olsen, G.W., O'Connor, J.C., Seacat, A.M., Perkins, R.G., Biegel, L.B., Murphy, S.R., Farrar, D.G. (2004) The toxicology of perfluorooctanoate. *Crit. Rev. Toxicol.* **34**:351–384.

Ketchem, R., Hu, W., Cross, T.A. (1993) High-resolution conformation of gramicidin A in a lipid bilayer by solid state NMR. *Science* **261**:1457–1460.

Kijac, A., Shih, A.Y., Nieuwkoop, A.J., Schulten, K., Sligar, S.G., Rienstra, C.M. (2010) Lipid-protein correlations in nanoscale phospholipid bilayers determined by solid-state nuclear magnetic resonance. *Biochemistry* **49**:9190–9198.

Kijac, A.Z., Li, Y., Sligar, S.G., Rienstra, C.M. (2007) Magic-angle spinning solid-state NMR spectroscopy of nanodisc-embedded human CYP3A4. *Biochemistry* **46**:13696–13703.

Kiley, P., Zhao, X., Vaughn, M., Baldo, M.A., Bruce, B.D., Zhang, S. (2005) Self-assembling peptide detergents stabilize isolated Photosystem I on a dry surface for an extended time. *PLoS Biol.* **3**:e230.

Kim, H.M., Howell, S.C., Van Horn, W.D., Jeon, Y.H., Sanders, C.R. (2009) Recent advances in the application of solution NMR spectroscopy to multi-span integral membrane proteins. *Progr. Nucl. Magn. Reson. Spectrosc.* **55**:335–360.

Kirsch, P. (2004) *Modern fluoroorganic chemistry: synthesis, reactivity, applications*. Wiley-VCH, Weinheim, 308 p.

Kissa, E. (1994) Structure of micelles and mesophases, in: Kissa, E., ed., *Fluorinated Surfactants: Synthesis, Properties, Applications*. Marcel Dekker, Inc., New York, pp. 264–282.

Kissa, E. (2001) *Fluorinated Surfactants and Repellents*, 2nd ed.. Marcel Dekker, New York, 615 p.

Koch, S., de Wit, J.G., Vos, I., Birkner, J.P., Gordiichuk, P., Herrmann, A., van Oijen, A.M., Driessen, A.J. (2016) Lipids activate SecA for high affinity binding to the SecYEG complex. *J. Biol. Chem.* **291**:22534–22543.

Kolter, T., Sandhoff, K. (2005) Principles of lysosomal membrane digestion: Stimulation of sphingolipid degradation by sphingolipid activator proteins and anionic lysosomal lipids. *Annu. Rev. Cell Dev. Biol.* **21**:81–103.

Kondo, H., Ikeda, K., Nakano, M. (2016) Formation of size-controlled, denaturation-resistant lipid nanodiscs by an amphiphilic self-polymerizing peptide. *Colloids Surf. B* **146**:423–430.

Koppaka, V., Silvestro, L., Engler, J.A., Brouillette, C.G., Axelsen, P.H. (1999) The structure of human lipoprotein A-I. Evidence for the "belt" model. *J. Biol. Chem.* **274**:14541–14544.

Koutsopoulos, S., Kaiser, L., Eriksson, H.M., Zhang, S. (2012) Designer peptide surfactants stabilize diverse functional membrane proteins. *Chem. Soc. Rev.* **41**:1721–1728.

Kraft, T.E., Hresko, R.C., Hruz, P.W. (2015) Expression, purification, and functional characterization of the insulin-responsive facilitative glucose transporter GLUT4. *Protein Sci.* **24**:2008–2019.

Kreutz, J.E., Li, L., Roach, L.S., Hatakeyama, T., Ismagilov, R.F., Rustem, F. (2009) Laterally mobile, functionalized self-assembled monolayers at the fluorous-aqueous interface in a plug-based microfluidic system: Characterization and testing with membrane protein crystallization. *J. Am. Chem. Soc.* **131**:6042–6043.

Kucharska, I., Edrington, T.C., Liang, B., Tamm, L.K. (2015) Optimizing nanodiscs and bicelles for solution NMR studies of two β-barrel membrane proteins. *J. Biomol. NMR* **61**:261–274.

Kumar, R.B., Zhu, L., Idborg, H., Radmark, O., Jakobsson, P., Rinaldo-Matthis, A., Hebert, H., Jegerschold, C. (2016) Structural and functional analysis of calcium ion mediated binding of 5-lipoxygenase to nanodiscs. *PLoS One* **11**: e0152116.

Kuszak, A.J., Pitchiaya, S., Anand, J.P., Mosberg, H.I., Walter, N.G., Sunahara, R.K. (2009) Purification and functional reconstitution of monomeric μ-opioid receptors: Allosteric modulation of agonist binding by $G_{i2}$. *J. Biol. Chem.* **284**:26732–26741.

Kyrychenko, A., Rodnin, M.V., Posokhov, Y.O., Holt, A., Pucci, B., Killian, J.A., Ladokhin, A.S. (2012a) Thermodynamic measurements of bilayer insertion of a single transmembrane helix chaperoned by fluorinated surfactants. *J. Mol. Biol.* **416**:328–334.

Kyrychenko, A., Rodnin, M.V., Vargas, M.U., Sharma, S.K., Durand, G., Pucci, B., Popot, J.-L., Ladokhin, A.S. (2012b) Folding of diphteria toxin T-domain in the presence of amphipols and fluorinated surfactants: Toward thermodynamic measurements of membrane protein folding. *Biochim. Biophys. Acta* **1818**:1006–1012.

Lai, G., Forti, K.M., Renthal, R. (2015) Kinetics of lipid mixing between bicelles and nanolipoprotein particles. *Biophys. J.* **197**:47–52.

Lamichhane, R., Liu, J.J., Pljevaljcic, G., White, K.L., van der Schans, E., Katritch, V., Stevens, R.C., Wüthrich, K., Millar, D.P. (2015) Single-molecule view of basal activity and activation mechanisms of the G protein-coupled receptor β2AR. *Proc. Natl. Acad. Sci. USA* **112**:14254–14259.

Landreh, M., Robinson, C.V. (2015) A new window into the molecular physiology of membrane proteins. *J. Physiol.* **593**:355–362.

Larsen, A.N., Sorensen, K.K., Johansen, N.T., Martel, A., Kirkensgaard, J.J., Jensen, K.J., Arleth, L., Midtgaard, S.R. (2016) Dimeric peptides with three different linkers self-assemble with phospholipids to form peptide nanodiscs that stabilize membrane proteins. *Soft Matter* **12**:5937–5949.

Lau, T.L., Partridge, A.W., Ginsberg, M.H., Ulmer, T.S. (2008) Structure of the integrin β3 transmembrane segment in phospholipid bicelles and detergent micelles. *Biochemistry* **47**:4008–4016.

Laursen, T., Singha, A., Rantzau, N., Tutkus, M., Borch, J., Hedegård, P., Stamou, D., Møller, B.L., Hatzakis, N.S. (2014) Single molecule activity measurements of cytochrome P450 oxidoreductase reveal the existence of two discrete functional states. *ACS Chem. Biol.* **9**:630–634.

Lebaupain, F. (2007) Développement de l'utilisation des tensioactifs fluorés pour la biochimie des protéines membranaires. Thèse de Doctorat, Université Paris-7, Paris, 254 p.

Lebaupain, F., Salvay, A.G., Olivier, B., Durand, G., Fabiano, A.-S., Michel, N., Popot, J.-L., Ebel, C., Breyton, C., Pucci, B. (2006) Lactobionamide surfactants with hydrogenated, hemifluorinated or perfluorinated tails: Physical-chemical and biochemical characterization. *Langmuir* **22**:8881–8890.

Lebeau, L., Lach, F., Venien-Bryan, C., Renault, A., Dietrich, J., Jahn, T., Palmgren, M.G., Kühlbrandt, W., Mioskowski, C. (2001) Two-dimensional crystallization of a membrane protein on a detergent-resistant lipid monolayer. *J. Mol. Biol.* **308**:639–647.

Lee, D., Walter, K.F., Brückner, A.K., Hilty, C., Becker, S., Griesinger, C. (2008) Bilayer in small bicelles revealed by lipid-protein interactions using NMR spectroscopy. *J. Am. Chem. Soc.* **130**:13822–13823.

Lee, H., Shingler, K.L., Organtini, L.J., Ashley, R.E., Makhov, A.M., Conway, J.F., Hafenstein, S. (2016) The novel asymmetric entry intermediate of a picornavirus captured with nanodiscs. *Sci. Adv.* **2**:e1501929.

Legrand, F., Breyton, C., Guillet, P., Ebel, C., Durand, G. (2016) Hybrid fluorinated and hydrogenated double-chain surfactants for handling membrane proteins. *J. Org. Chem.* **81**:681–688.

Leney, A.C., Rezaei Darestani, R., Li, J., Nikjah, S., Kitova, E.N., Zou, C., Cairo, C.W., Xiong, Z.J., Privé, G.G., Klassen, J.S. (2015) Picodiscs for facile protein-glycolipid interaction analysis. *Anal. Chem.* **87**:4402–4408.

Lewis, B.A., Engelman, D.M. (1983) Lipid bilayer thickness varies linearly with acyl chain length in fluid phosphatidylcholine vesicles. *J. Mol. Biol.* **166**:211–217.

Li, J., Fan, X., Kitova, E.N., Zou, C., Cairo, C.W., Eugenio, L., Ng, K.K.S., Xiong, Z.J., Privé, G.G., Klassen, J.S. (2016a) Screening glycolipids against proteins *in vitro* using picodiscs and catch-and-release electrospray ionization-mass spectrometry. *Anal. Chem.* **88**:4742–4750.

Li, J., Richards, M.R., Bagal, D., Campuzano, I.D.G., Kitova, E.N., Xiong, Z.J., Privé, G.G., Klassen, J.S. (2016b) Characterizing the size and composition of saposin A lipoprotein picodiscs. *Anal. Chem.* **88**:9524–9531.

Li, L., Chen, J., Mishra, V.K., Kurtz, J.A., Cao, D., Klon, A.E., Harvey, S.C., Anantharamaiah, G.M., Segrest, J.P. (2004) Double belt structure of discoidal high density lipoproteins: molecular basis for size heterogeneity. *J. Mol. Biol.* **343**:1293–1311.

Li, Y., Kijac, A.Z., Sligar, S.G., Rienstra, C.M. (2006) Structural analysis of nanoscale self-assembled discoidal lipid bilayers by solid-state NMR spectroscopy. *Biophys. J.* **91**:3819–3828.

Liebau, J., Ye, W., Mäler, L. (2016) Characterization of fast-tumbling isotropic bicelles by PFG diffusion NMR. *Magn. Reson. Chem.* **55**:395–404.

Lindberg, M., Biverståhl, H., Gräslund, A., Mäler, L. (2003) Structure and positioning comparison of two variants of penetratin in two different membrane mimicking systems by NMR. *Eur. J. Biochem.* **270**:3055–3063.

Loll, P.J. (2014) Membrane proteins, detergents and crystals: what is the state of the art? *Acta Crystallogr. F* **70**:1576–1583.

Luecke, H., Schobert, B., Stagno, J., Imasheva, E.S., Wang, J.M., Balashov, S.P., Lanyi, J.K. (2008) Crystallographic structure of xanthorhodopsin, the light-driven proton pump with a dual chromophore. *Proc. Natl. Acad. Sci. USA* **105**:16561–16565.

Lyukmanova, E.N., Shenkarev, Z.O., Khabibullina, N.F., Kopeina, G.S., Shulepko, M.A., Paramonov, A.S., Mineev, K.S., Tikhonov, R.V., Shingarova, L.N., Petrovskaya, L.E., Dolgikh, D.A., Arseniev, A.S., Kirpichnikov, M.P. (2011) Lipid-protein nanodisks for cell-free production of integral membrane proteins in a soluble and folded state: Comparison with detergent micelles, bicelles and liposomes. *Biochim. Biophys. Acta* **1818**:349–358.

Mabrey, S., Sturtevant, J.M. (1976) Investigation of phase transitions of lipids and lipid mixtures by high-sensitivity differential scanning calorimetry. *Proc. Natl. Acad. Sci. USA* **73**:3862–3866.

Mahalakshmi, R., Marassi, F.M. (2008) Orientation of the *Escherichia coli* outer membrane protein OmpX in phospholipid bilayer membranes determined by solid-State NMR. *Biochemistry* **47**:6531–6538.

Mak, P.J., Gregory, M.C., Denisov, I.G., Sligar, S.G., Kincaid, J.R. (2015) Unveiling the crucial intermediates in androgen production. *Proc. Natl. Acad. Sci. USA* **112**:15856–15861.

Mäler, L., Gräslund, A. (2009) Artificial membrane models for the study of macromolecular delivery. *Meth. Mol. Biol.* **480**:129–139.

Malhotra, K., Alder, N.N. (2014) Advances in the use of nanoscale bilayers to study membrane protein structure and function. *Biotechnol. Genet. Eng. Rev.* **30**:79–93.

Marcotte, I., Auger, M. (2005) Bicelles as model membranes for solid- and solution-state NMR studies of membrane peptides and proteins. *Concepts Magn. Reson.* **24A**:17–37.

Marty, M.T., Hoi, K.K., Robinson, C.V. (2016) Interfacing membrane mimetics with mass spectrometry. *Acc. Chem. Res.* **49**:2459–2467.

Marty, M.T., Wilcox, K.C., Klein, W.L., Sligar, S.G. (2013) Nanodisc-solubilized membrane protein library reflects the membrane proteome. *Anal. Bioanal. Chem.* **405**:4009–4016.

Matsumoto, K., Vaughn, M., Bruce, B.D., Koutsopoulos, S., Zhang, S. (2009) Designer peptide surfactants stabilize functional photosystem I membrane complex in aqueous solution for extended time. *J. Phys. Chem. B* **113**:75–83.

Matthies, D., Dalmas, O., Borgnia, M.J., Dominik, P.K., Merk, A., Rao, P., Reddy, B.G., Islam, S., Bartesaghi, A., Perozo, E., Subramaniam, S. (2016) Cryo-EM structures of the magnesium channel CorA reveal symmetry break upon gating. *Cell* **164**:747–756.

McGregor, C.-L., Chen, L., Pomroy, N.C., Hwang, P., Go, S., Chakrabartty, A., Privé, G.G. (2003) Lipopeptide detergents designed for the structural study of membrane proteins. *Nat. Biotechnol.* **21**:171–176.

McKibbin, C., Farmer, N.A., Edwards, P.C., Villa, C., Booth, P.J. (2009) Urea unfolding of opsin in phospholipid bicelles. *Photochem. Photobiol.* **85**:494–500.

McKibbin, C., Farmer, N.A., Jeans, C., Reeves, P.J., Khorana, H.G., Wallace, B.A., Edwards, P.C., Villa, C., Booth, P.J. (2007) Opsin stability and folding: modulation by phospholipid bicelles. *J. Mol. Biol.* **374**:1319–1332.

Midtgaard, S.R., Pedersen, M.C., Kirkensgaard, J.J.K., Sorensen, K.K., Mortensen, K., Jensen, K.J., Arleth, L. (2014) Self-assembling peptides form nanodiscs that stabilize membrane proteins. *Soft Matter* **10**:738–752.

Mineev, K.S., Goncharuk, S.A., Kuzmichev, P.K., Vilar, M., Arseniev, A.S. (2015) NMR dynamics of transmembrane and intracellular domains of p75NTR in lipid-protein nanodiscs. *Biophys. J.* **109**:772–782.

Mineev, K.S., Nadezhdin, K.D. (2017) Membrane mimetics for solution NMR studies of membrane proteins. *Nanotech. Rev.* **6**:15–32.

Mineev, K.S., Nadezhdin, K.D., Goncharuk, S.A., Arseniev, A.S. (2017) Facade detergents as bicelle rim-forming agents for solution NMR spectroscopy. *Nanotech. Rev.* **6**:93–103.

Mitra, N., Liu, Y., Liu, J., Serebryany, E., Mooney, V., DeVree, B.T., Sunahara, R.K., Yan, E.C.Y. (2013) Calcium-dependent ligand binding and G protein signaling of family B GPCR parathyroid hormone 1 receptor purified in nanodiscs. *ACS Chem. Biol.* **8**:617–625.

Miyazaki, M., Nakano, M., Fukuda, M., Handa, T. (2009) Smaller discoidal high-density lipoprotein particles form saddle surfaces, but not planar bilayers. *Biochemistry* **48**:7756–7763.

Mizrachi, D., Robinson, M.-P., Ren, G., Ke, N., Berkmen, M., DeLisa, M.P. (2017) A water-soluble DsbB variant that catalyzes disulfide-bond formation *in vivo. Nat. Chem. Biol.* **13**:1022–1028.

Morgado, L., Zeth, K., Burmann, B.M., Maier, T., Hiller, S. (2015) Characterization of the insertase BamA in three different membrane mimetics by solution NMR spectroscopy. *J. Biomol. NMR* **61**:333–345.

Morrison, E.A., DeKoster, G.T., Dutta, S., Vafabakhsh, R., Clarkson, M.W., Bahl, A., Kern, D., Ha, T., Henzler-Wildman, K.A. (2011) Antiparallel EmrE exports drugs by exchanging between asymmetric structures. *Nature* **481**:45–50.

Morrison, E.A., Henzler-Wildman, K.A. (2012) Reconstitution of integral membrane proteins into isotropic bicelles with improved sample stability and expanded lipid composition profile. *Biochim. Biophys. Acta* **1818**:814–820.

Mörs, K., Roos, C., Scholz, F., Wachtveitl, J., Dötsch, V., Bernhard, F., Glaubitz, C. (2013) Modified lipid and protein dynamics in nanodiscs. *Biochim. Biophys. Acta* **1828**:1222–1229.

Mukerjee, P. (1994) Fluorocarbon-hydrocarbon interactions in micelles and other lipid assemblies, at interfaces, and in solutions. *Colloids Surf. A* **84**:1–10.

Muller, K. (1981) Structural dimorphism in bile salt/lecithin mixed micelles. X-ray structural analysis. *Biochemistry* **20**:404–414.

Nakano, T.Y., Sugihara, G., Nakashima, T., Yu, S.C. (2002) Thermodynamic study of mixed hydrocarbon/fluorocarbon surfactant system by conductometric and fluorimetric techniques. *Langmuir* **18**:8777–8785.

Nasr, M.L., Baptista, D., Strauss, M., Sun, Z.J., Grigoriu, S., Huser, S., Plückthun, A., Hagn, F., Walz, T., Hogle, J.M., Wagner, G. (2017) Covalently circularized nanodiscs for studying membrane proteins and viral entry. *Nat. Meth.* **14**:49–52.

Nath, A., Atkins, W.M., Sligar, S.G. (2007) Applications of phospholipid bilayer nanodiscs in the study of membranes and membrane proteins. *Biochemistry* **46**:2059–2069.

Nath, A., Koo, P.K., Rhoades, E., Atkins, W.M. (2008) Allosteric effects on substrate dissociation from cytochrome P450 3A4 in nanodiscs observed by ensemble and single-molecule fluorescence spectroscopy. *J. Am. Chem. Soc.* **130**:15746–15747.

Nehmé, R., Joubert, O., Bidet, M., Lacombe, B., Polidori, A., Pucci, B., Mus-Veteau, I. (2010) Stability study of the human G protein-coupled receptor, Smoothened. *Biochim. Biophys. Acta* **1786**:1100–1110.

Nietlispach, D., Gautier, A. (2011) Solution NMR studies of polytopic alpha-helical membrane proteins. *Curr. Opin. Struct. Biol.* **21**:497–508.

Nikolaev, M., Round, E., Gushchin, I., Polovinkin, V., Balandin, T., Kuzmichev, P., Shevchenko, V., Borshchevskiy, V., Kuklin, A., Round, A., Bernhard, F., Willbold, D., Büldt, G., Gordeliy, V. (2017) Integral membrane proteins can be crystallized directly from nanodiscs. *Cryst. Growth Des.* **17**:945–948.

Noinaj, N., Kuszak, A.J., Gumbart, J.C., Lukacik, P., Chang, H., Easley, N.C., Lithgow, T., Buchanan, S.K. (2013) Structural insight into the biogenesis of β-barrel membrane proteins. *Nature* **501**:385–390.

Nolte, R.T., Atkinson, D. (1992) Conformational analysis of apolipoproteins A-I and E-3 based on primary sequence and circular dichroism. *Biophys. J.* **63**:1221–1239.

Nusair, N.A., Mayo, D.J., Dorozenski, T.D., Cardon, T.B., Inbaraj, J.J., Karp, E.S., Newstadt, J.P., Grosser, S.M., Lorigan, G.A. (2012) Time-resolved EPR immersion depth studies of a transmembrane peptide incorporated into bicelles. *Biochim. Biophys. Acta* **1818**:821–828.

Opella, S.J., Marassi, F.M. (2004) Structure determination of membrane proteins by NMR spectroscopy. *Chem. Rev.* **104**:3587–3606.

Otzen, D.E. (2015) Proteins in a brave new surfactant world. *Curr. Opin. Colloid Interface Sci.* **20**:161–169.

Palchevskyy, S.S., Posokhov, Y.O., Olivier, B., Popot, J.-L., Pucci, B., Ladokhin, A.S. (2006) Chaperoning of membrane protein insertion into lipid bilayers by hemifluorinated surfactants: application to diphtheria toxin. *Biochemistry* **45**:2629–2635.

Park, K.-H., Berrier, C., Lebaupain, F., Pucci, B., Popot, J.-L., Ghazi, A., Zito, F. (2007) Fluorinated and hemifluorinated surfactants as alternatives to detergents for membrane protein cell-free synthesis. *Biochem. J.* **403**:183–187.

Park, K.-H., Billon-Denis, E., Dahmane, T., Lebaupain, F., Pucci, B., Breyton, C., Zito, F. (2011) In the cauldron of cell-free synthesis of membrane proteins: Playing with new surfactants. *New Biotech.* **28**:255–261.

Park, S.H., Berkamp, S., Cook, G.A., Chan, M.K., Viadiu, H., Opella, S.J. (2011a) Nanodiscs *versus* macrodiscs for NMR of membrane proteins. *Biochemistry* **50**:8983–8985.

Park, S.H., Casagrande, F., Cho, L., Albrecht, L., Opella, S.J. (2011b) Interactions of interleukin-8 with the human chemokine receptor CXCR1 in phospholipid bilayers by NMR spectroscopy. *J. Mol. Biol.* **414**:194–203.

Park, S.H., Casagrande, F., Das, B.B., Albrecht, L., Chu, M., Opella, S.J. (2011c) Local and global dynamics of the G protein-coupled receptor CXCR1. *Biochemistry* **50**:2371–2380.

Park, S.H., Das, B.B., Casagrande, F., Tian, Y., Nothnagel, H.J., Chu, M., Kiefer, H., Maier, K., De Angelis, A.A., Marassi, F.M., Opella, S.J. (2012) Structure of the chemokine receptor CXCR1 in phospholipid bilayers. *Nature* **491**:770–783.

Park, S.H., Prytulla, S., De Angelis, A.A., Brown, J.M., Kiefer, H., Opella, S.J. (2006) High-resolution NMR spectroscopy of a GPCR in aligned bicelles. *J. Am. Chem. Soc.* **128**:7402–7403.

Pavia, A.A., Pucci, B., Riess, J.G., Zarif, L. (1991) New fluorinated biocompatible non-ionic telomeric amphiphiles bearing trishydroxymethyl groups. *Bioorg. Med. Chem. Letters* **1**:103–106.

Periasamy, A., Shadiac, N., Amalraj, A., Garajová, S., Nagarajan, Y., Waters, S., Mertens, H.D.T., Hrmova, M. (2013) Cell-free protein synthesis of membrane (1,3)-β-D-glucan (curdlan) synthase: co-translational insertion in liposomes and reconstitution in nanodiscs. *Biochim. Biophys. Acta* **1828**:743–757.

Peters, B.M., Shirtliff, M.E., Jabra-Rizk, M.A. (2010) Antimicrobial peptides: Primeval molecules or future drugs? *PLoS Pathog.* **6**:e1001067.

Petkova, V., Benattar, J.J., Zoonens, M., Zito, F., Popot, J.-L., Polidori, A., Jasseron, S., Pucci, B. (2007) Free-standing films of fluorinated surfactants as 2D matrices for organizing detergent-solubilized membrane proteins. *Langmuir* **23**:4303–4309.

Phillips, J.C., Wriggers, W., Li, Z., Jonas, A., Schulten, K. (1997) Predicting the structure of apolipoprotein A-I in reconstituted high-density lipoprotein disks. *Biophys. J.* **73**:2337–2346.

Phillips, M.C. (2013) New insights into the determination of HDL structure by apolipoproteins. *J. Lipid Res.* **54**:2034–2048.

Poget, S.F., Cahill, S.M., Girvin, M.E. (2007) Isotropic bicelles stabilize the functional form of a small multidrug-resistance pump for NMR structural studies. *J. Am. Chem. Soc.* **129**:2432–2433.

Poget, S.F., Girvin, M.E. (2007) Solution NMR of membrane proteins in bilayer mimics: small is beautiful, but sometimes bigger is better. *Biochim. Biophys. Acta* **1768**:3098–3106.

Polidori, A., Presset, M., Lebaupain, F., Améduri, B., Popot, J.-L., Breyton, C., Pucci, B. (2006) Fluorinated and hemifluorinated surfactants derived from maltose: Synthesis and application to handling membrane proteins in aqueous solution. *Bioorg. Med. Chem. Lett.* **16**:5827–5831.

Polidori, A., Raynal, S., Barret, L.-A., Dahani, M., Barrot-Ivolot, C., Jungas, C., Frotscher, E., Keller, S., Ebel, C., Breyton, C., Bonneté, F. (2016) Sparingly fluorinated maltoside-based surfactants for membrane-protein stabilization. *New J. Chem.* **40**:5364–5378.

Polovinkin, V., Gushchin, I., Balandin, T., Chervakov, P., Round, E., Shevchenko, V., Popov, A., Borshchevskiy, V., Popot, J.-L., Gordeliy, V. (2014) High-resolution structure of a membrane protein transferred from amphipol to a lipidic mesophase. *J. Membr. Biol.* **247**:997–1004.

Popot, J.-L. (2010) Amphipols, nanodiscs, and fluorinated surfactants: Three non-conventional approaches to studying membrane proteins in aqueous solutions. *Annu. Rev. Biochem.* **79**:737–775.

Popot, J.-L., Engelman, D.M. (2000) Helical membrane protein folding, stability and evolution. *Annu. Rev. Biochem.* **69**:881–923.

Popovic, K., Holyoake, J., Pomès, R., Privé, G.G. (2012) Structure of saposin A lipoprotein discs. *Proc. Natl. Acad. Sci. USA* **109**:2908–2912.

Posokhov, Y.O., Rodnin, M.V., Das, S.K., Pucci, B., Ladokhin, A.S. (2008) FCS study of the thermodynamics of membrane protein insertion into the lipid bilayer chaperoned by fluorinated surfactants. *Biophys. J.* **95**:L54-L56.

Poulos, S., Morgan, J.L., Zimmer, J., Faham, S. (2015) Bicelles coming of age: an empirical approach to bicelle crystallization. *Meth. Enzymol.* **557**:393–416.

Privé, G. (2009) Lipopeptide detergents for membrane protein studies. *Curr. Opin. Struct. Biol.* **19**:1–7.

Prosser, R.S., Evanics, F., Kitevski, J.L., Al-Abdul-Wahid, M.S. (2006) Current applications of bicelles in NMR studies of membrane-associated amphiphiles and proteins. *Biochemistry* **45**:8453–8465.

Prosser, R.S., Hunt, S.A., DiNatale, J.A., Vold, R.R. (1996) Magnetically aligned membrane model systems with positive order parameters: switching the sign of $S_{zz}$ with paramagnetic ions. *J. Am. Chem. Soc.* **118**:269–270.

Prosser, R.S., Hwang, J.S., Vold, R.R. (1998) Magnetically aligned phospholipid bilayers with positive ordering: a new model membrane system. *Biophys. J.* **74**:2405–2418.

Proverbio, D., Roos, C., Beyermann, M., Orbán, E., Dötsch, V., Bernhard, F. (2013) Functional properties of cell-free expressed human endothelin A and endothelin B receptors in artificial membrane environments. *Biochim. Biophys. Acta* **1828**:2182–2192.

Puthenveetil, R., Nguyen, K., Vinogradova, O. (2017) Nanodiscs and solution NMR: preparation, application and challenges. *Nanotech. Rev.* **6**:111–126.

Puthenveetil, R., Vinogradova, O. (2013) Optimization of the design and preparation of nanoscale phospholipid bilayers for its application to solution NMR. *Proteins: Struct. Funct. Bioinf.* **81**:1222–1231.

Qureshi, T., Goto, N.K. (2011) Contemporary methods in structure determination of membrane proteins by solution NMR. *Top. Curr. Chem.* **326**:123–185.

Ram, P., Prestegard, J.H. (1988) Magnetic field-induced ordering of bile salt/phospholipid micelles: new media for NMR structural investigations. *Biochim. Biophys. Acta* **940**:289–294.

Ramjeesingh, M., Huan, L.J., Garami, E., Bear, C.E. (1999) Novel method for evaluation of the oligomeric structure of membrane proteins. *Biochem. J.* **342**.

Ranaghan, M.J., Schwall, C.T., Alder, N.N., Birge, R.R. (2011) Green proteorhodopsin reconstituted into nanoscale phospholipid bilayers (nanodiscs) as photoactive monomers. *J. Am. Chem. Soc.* **133**:18318–18327.

Raschle, T., Hiller, S., Etzkorn, M., Wagner, G. (2010) Nonmicellar systems for solution NMR spectroscopy of membrane proteins. *Curr. Opin. Struct. Biol.* **20**:471–479.

Raschle, T., Hiller, S., Yu, T.Y., Rice, A.J., Walz, T., Wagner, G. (2009) Structural and functional characterization of the integral membrane protein VDAC-1 in lipid bilayer nanodiscs. *J. Am. Chem. Soc.* **131**:17777–17779.

Rasmussen, S.G., Choi, H.J., Rosenbaum, D.M., Kobilka, T.S., Thian, F.S., Edwards, P.C., Burghammer, M., Ratnala, V.R., Sanishvili, R., Fischetti, R.F., Schertler, G.F., Weis, W.I., Kobilka, B.K. (2007) Crystal structure of the human $\beta_2$ adrenergic G protein-coupled receptor. *Nature* **450**:383–387.

Raychaudhuri, P., Li, Q., Mason, A., Mikhailova, E., Heron, A.J., Bayley, H. (2011) Fluorinated amphiphiles control the insertion of α-hemolysin pores into lipid bilayers. *Biochemistry* **50**:1599–1606.

Reichart, T.M., Baksh, M.M., Rhee, J.-K., Fiedler, J.D., Sligar, S.G., Finn, M.G., Zwick, M.B., Dawson, P.E. (2016) Trimerization of the HIV transmembrane domain in lipid bilayers modulates broadly neutralizing antibody binding. *Angew. Chem. Int. Ed.* **55**:2688–2692.

Riess, J.G. (2005) Fluorous materials for biomedical uses, in: Gladysz, J.A., Curran, D.P., Horváth, I.T., eds., *Handbook of fluorous chemistry*. Wiley-VCH, Weinheim, pp. 521–573.

Ritchie, T.K., Grinkova, Y.V., Bayburt, T.H., Denisov, I.G., Zolnerciks, J.K., Atkins, W.M., Sligar, S.G. (2009) Reconstitution of membrane proteins in phospholipid bilayer nanodiscs. *Meth. Enzymol.* **464**:211–231.

Ritchie, T.K., Kwon, H., Atkins, W.M. (2011) Conformational analysis of human ATP-binding cassette transporter ABCB1 in lipid nanodiscs and inhibition by the antibodies MRK16 and UIC2. *J. Biol. Chem.* **286**:39489–39496.

Rodnin, M.V., Posokhov, Y.O., Contino-Pépin, C., Brettmann, J., Kyrychenko, A., Palchevskyy, S.S., Pucci, B., Ladokhin, A.S. (2008) Interactions of fluorinated surfactants with diphtheria toxin T-domain: testing new media for studies of membrane proteins. *Biophys. J.* **94**:4348–4357.

Roos, C., Zocher, M., Müller, D., Münch, D., Schneider, T., Sahl, H.G., Scholz, F., Wachtveitl, J., Ma, Y., Proverbio, D., Henrich, E., Dötsch, V., Bernhard, F. (2012) Characterization of co-translationally formed nanodisc complexes with small multidrug transporters, proteorhodopsin and with the *E. coli* MraY translocase. *Biochim. Biophys. Acta* **1818**:3898–3106.

Roy, J., Pondenis, H., Fan, T.M., Das, A. (2015) Direct capture of functional proteins from mammalian plasma membranes into nanodiscs. *Biochemistry* **54**:6299–6302.

Rues, R.-B., Dötsch, V., Bernhard, F. (2016) Co-translational formation and pharmacological characterization of *β*-adrenergic receptor/nanodisc complexes with different lipid environments. *Biochim. Biophys. Acta* **1858**:1306–1316.

Sanders, C.R. (2008) Development and application of bicelles for use in biological NMR and other biophysical studies, in: Webb, G.A., ed., *Modern Magnetic Resonance*. Springer, Dordrecht, pp. 233–239.

Sanders, C.R., Prosser, R.S. (1998) Bicelles: a model membrane system for all seasons? *Structure* **6**:1227–1234.

Sanders, C.R., Schwonek, J.P. (1992) Characterization of magnetically orientable bilayers in mixtures of dihexanoylphosphatidylcholine and dimyristoylphosphatidylcholine by solid-state NMR. *Biochemistry* **31**:8898–8905.

Sanders, C.R., Sönnichsen, F. (2006) Solution NMR of membrane proteins: practice and challenges. *Magn. Reson. Chem.* **44**:S24–S40.

Sanders II, C.R., Hare, B.J., Howard, K.P., Prestegard, J.H. (1994) Magnetically-oriented phospholipid micelles as a tool for the study of membrane-associated molecules. *Prog. NMR Spectrosc.* **26**:421–444.

Sanders II, C.R., Landis, G.C. (1995) Reconstitution of membrane proteins into lipid-rich bilayered mixed micelles for NMR studies. *Biochemistry* **34**:4030–4040.

Sanders II, C.R., Prestegard, J.H. (1990) Magnetically orientable phospholipid bilayers containing small amounts of a bile salt analogue, CHAPSO. *Biophys. J.* **58**:447–460.

Sanii, L.S., El-Sayed, M.A. (2005) Partial dehydration of the retinal binding pocket and proof for photochemical deprotonation of the retinal Schiff base in bicelle bacteriorhodopsin crystals. *Photochem. Photobiol.* **81**:1356–1360.

Sanii, L.S., Schill, A.W., Moran, C.E., El-Sayed, M.A. (2005) The protonation-deprotonation kinetics of the protonated Schiff base in bicelle bacteriorhodopsin crystals. *Biophys. J.* **89**:444–451.

Santoso, S., Hwang, W., Hartman, H., Zhang, S. (2002) Self-assembly of surfactant-like peptides with variable glycine tails to form nanotubes and nanovesicles. *Nano Lett.* **2**:687–691.

Schafmeister, C.E., Miercke, L.J.W., Stroud, R.A. (1993) Structure at 2.5 Å of a designed peptide that maintains solubility of membrane proteins. *Science* **262**:734–738.

Schoch, G.A., Attias, R., Belghazi, M., Dansette, P.M., Werck-Reichhart, D. (2003) Engineering of a water-soluble plant cytochrome P450, CYP73A1, and NMR-based orientation of natural and alternate substrates in the active site. *Plant Physiol.* **133**:1198–1208.

Segrest, J.P., Jones, M.K., Klon, A.E., Sheldahl, C.J., Hellinger, M., De Loof, H., Harvey, S.C. (1999) A detailed molecular belt model for apolipoprotein A-I in discoidal high-density lipoprotein. *J. Biol. Chem.* **274**:31755–31758.

Serebryany, E., Zhu, G.A., Yan, E.C.Y. (2012) Artificial membrane-like environments for *in vitro* studies of purified G-protein coupled receptors. *Biochim. Biophys. Acta* **1818**:225–233.

Sevugan Chetty, P., Mayne, L., Kan, Z.Y., Lund-Katz, S., Englander, S.W., Phillips, M.C. (2012) Apolipoprotein A-I helical structure and stability in discoidal high density lipoprotein (HDL) particles by hydrogen exchange and mass spectrometry. *Proc. Natl. Acad. Sci. USA* **109**:11687–11692.

Shaw, A.W., McLean, M.A., Sligar, S.G. (2004) Phospholipid phase transitions in homogeneous nanometer scale bilayers discs. *FEBS Lett.* **556**:260–264.

Shaw, A.W., Pureza, V.S., Sligar, S.G., Morrissey, J.H. (2007) The local phospholipid environment modulates the activation of blood clotting. *J. Biol. Chem.* **282**:6556–6563.

Shen, P.S., Yang, X., DeCaen, P.G., Liu, X., Bulkley, D., Clapham, D.E., Cao, E. (2016) The structure of the Polycystic Kidney Disease channel PKD2 in lipid nanodiscs. *Cell* **167**:763–773.

Shenkarev, Z.O., Lyukmanova, E.N., Butenko, I.O., Petrovskaya, L.E., Paramonov, A.S., Shulepko, M.A., Nekrasova, O.V., Kirpichnikov, M.P., Arseniev, A.S. (2013) Lipid-protein nanodiscs promote *in vitro* folding of transmembrane domains of multi-helical and multimeric membrane proteins. *Biochim. Biophys. Acta* **1828**:776–784.

Shenkarev, Z.O., Lyukmanova, E.N., Paramonov, A.S., Shingarova, L.N., Chupin, V.V., Kirpichnikov, M.P., Blommers, M.J., Arseniev, A.S. (2010) Lipid-protein nanodiscs as reference medium in detergent screening for high-resolution NMR studies of integral membrane proteins. *J. Am. Chem. Soc.* **132**:5628–5629.

Shepherd, F.H., Holzenburg, A. (1995) The potential of fluorinated surfactants in membrane biochemistry. *Anal. Biochem.* **224**:21–27.

Shi, L., Howan, K., Shen, Q.T., Wang, Y.J., Rothman, J.E., Pincet, F. (2013) Preparation and characterization of SNARE-containing nanodiscs and direct study of cargo release through fusion pores. *Nat. Protoc.* **8**:935–948.

Shi, L., Shen, Q.T., Kiel, A., Wang, J., Wang, H.W., Melia, T.J., Rothman, J.E., Pincet, F. (2012) SNARE proteins: one to fuse and three to keep the nascent fusion pore open. *Science* **335**:1355–1359.

Shih, A.Y., Denisov, I.G., Phillips, J.C., Sligar, S.G., Schulten, K. (2005) Molecular dynamics simulations of discoidal bilayers assembled from truncated human lipoproteins. *Biophys. J.* **88**:548–556.

Shimada, S., Shinzawa-Itoh, K., Baba, J., Aoe, S., Shimada, A., Yamashita, E., Kang, J., Tateno, M., Yoshikawa, S., Tsukihara, T. (2017) Complex structure of cytochrome $c$-cytochrome $c$ oxidase reveals a novel protein-protein interaction mode. *EMBO J.* **36**:291–300.

Shin, J., Lou, X., Kweon, D.-H., Shin, Y.-K. (2014) Multiple conformations of a single SNAREpin between two nanodisc membranes reveal diverse pre-fusion states. *Biochem. J.* **459**:95–102.

Shirzad-Wasei, N., Oostrum, J.V., Bovee-Geurts, P.H., Kusters, L.J., Bosman, G.J., DeGrip, W.J. (2015) Rapid transfer of overexpressed integral membrane protein from the host membrane into soluble lipid nanodiscs without previous purification. *Biol. Chem.* **396**:903–916.

Singh, R., Flowers, R.A., II (2010) Efficient protein renaturation using tunable hemifluorinated anionic surfactants as additives. *Chem. Commun.* **46**:276–278.

Siuda, I., Tieleman, D.P. (2015) Molecular models of nanodiscs. *J. Chem. Theory Comput.* **11**:4923–4932.

Skar-Gislinge, N., Kynde, S.A., Denisov, I.G., Ye, X., Lenov, I., Sligar, S.G., Arleth, L. (2015) Small-angle scattering determination of the shape and localization of human cytochrome P450 embedded in a phospholipid nanodisc environment. *Acta Crystallogr. D Biol. Crystallogr.* **71**:2412–2421.

Small, D.M. (1971) The physical chemistry of cholanic acids, in: P.P. Nair & D. Kritchevsky, eds., *The Bile Acids*, Plenum Press, pp. 249–356.

Smrt, S.T., Draney, A.W., Singaram, I., Lorieau, J.L. (2017) Structure and dynamics of membrane proteins and membrane associated proteins with native bicelles from eukaryotic tissues. *Biochemistry* **56**:5318–5327.

Sobolev, V., Edelman, M., Dym, O., Unger, T., Albeck, S., Kirma, M., Galili, G. (2013) Structure of ALD1, a plant-specific homologue of the universal diaminopimelate aminotransferase enzyme of lysine biosynthesis. *Acta Crystallogr. F* **69**:84–89.

Soomets, U., Kairane, C., Zilmer, M., Langel, U. (1997) Attempt to solubilize $Na^+/K^+$-exchanging ATPase with amphiphilic peptide $PD_1$. *Acta. Chem. Scand.* **51**:403–406.

Starita-Geribaldi, M., Thebault, P., Taffin de Givenchy, E., Guittard, F., Geribaldi, S. (2007) 2-DE using hemifluorinated surfactants. *Electrophoresis* **28**:2489–2497.

Sternin, E., Nizza, D., Gawrisch, K. (2001) Temperature dependence of DMPC/DHPC mixing in a bicellar solution and its structural implications. *Langmuir* **17**:2610–2616.

Talbot, J.-C., Dautant, A., Polidori, A., Pucci, B., Cohen-Bouhacina, T., Maali, A., Salin, B., Brèthes, D., Velours, J., Giraud, M.-F. (2009) Hydrogenated and fluorinated surfactants derived from *tris*(hydroxymethyl)-acrylamidomethane allow the purification of a highly active yeast $F_1F_O$ ATP synthase with an enhanced stability. *J. Bioenerg. Biomemb.* **41**:349–360.

Tanford, C. (1980) *The hydrophobic effect: formation of micelles and biological membranes*, 2nd ed.. John Wiley & Sons, New York, 233 p.

Tao, H., Lee, S.C., Moeller, A., Roy, R.S., Siu, F.Y., Zimmermann, J., Stevens, R.C., Potter, C.S., Carragher, B., Zhang, Q. (2013) Engineered nanostructured β-sheet peptides protect membrane proteins. *Nat. Methods* **10**:759–761.

Taufik, I., Kedrov, A., Exterkate, M., Driessen, A.J.M. (2013) Monitoring the activity of single translocons. *J. Mol. Biol.* **425**:4145–4153.

Thebault, P., Taffin de Givenchy, E., Starita-Geribaldi, M., Guittard, F., Geribaldi, S. (2007) Synthesis and surface properties of new semifluorinated sulfobetaines potentially usable for 2D-electrophoresis. *J. Fluorine Chem.* **128**:211–218.

Tiburu, E.K., Moton, D.M., Lorigan, G.A. (2001) Development of magnetically aligned phospholipid bilayers in mixtures of palmitoylstearoylphosphatidylcholine and dihexanoylphosphatidylcholine by solid-state NMR spectroscopy. *Biochim. Biophys. Acta* **1512**:206–214.

Triba, M.N., Warschawski, D.E., Devaux, P.F. (2005) Reinvestigation by phosphorus NMR of lipid distribution in bicelles. *Biophys. J.* **88**:1887–1901.

Triba, M.N., Zoonens, M., Popot, J.-L., Devaux, P.F., Warschawski, D.E. (2006) Reconstitution and alignment by a magnetic field of a β-barrel membrane protein in bicelles. *Eur. Biophys. J.* **35**:268–275.

Tsukamoto, H., Szundi, I., Lewis, J.W., Farrens, D.L., Kliger, D.S. (2011) Rhodopsin in nanodiscs has native membrane-like photointermediates. *Biochemistry* **50**:5086–5091.

Tu, Y., Peng, F., Adawy, A., Men, Y., Abdelmohsen, L.K.E.A., Wilson, D.A. (2016) Mimicking the cell: bio-Inspired functions of supramolecular assemblies. *Chem. Rev.* **116**:2023–2078.

Tzitzilonis, C., Eichmann, C., Maslennikov, I., Choe, S., Riek, R. (2013) Detergent/nanodisc screening for high-resolution NMR studies of an integral membrane protein containing a cytoplasmic domain. *PLoS One* **8**:e54378.

Uhlemann, E.M., Pierson, H.E., Fillingame, R.H., Dmitriev, O.Y. (2012) Cell-free synthesis of membrane subunits of ATP synthase in phospholipid bicelles: NMR shows subunit fold similar to the protein in the cell membrane. *Prot. Sci.* **21**:279–288.

Ujwal, R., Bowie, J.U. (2011) Crystallizing membrane proteins using lipidic bicelles. *Methods* **55**:337–341.

Ujwal, R., Cascio, D., Colletier, J.-P., Faham, S., Zhang, J., Toro, L., Ping, P., Abramson, J. (2008) The crystal structure of mouse VDAC1 at 2.3 Å resolution reveals mechanistic insights into metabolite gating. *Proc. Natl. Acad. Sci. USA* **105**:17742–17747.

van Dam, L., Karlsson, G., Edwards, K. (2006) Morphology of magnetically aligning DMPC/DHPC aggregates – perforated sheets, not disks. *Langmuir* **22**:3280–3285.

Vargas, C., Cuevas Arenas, R., Frotscher, E., Keller, S. (2015) Nanoparticle self-assembly in mixtures of phospholipids with styrene/maleic acid copolymers or fluorinated surfactants. *Nanoscale* **7**:20685–20696.

Varkey, J., Mizuno, N., Hegde, B.G., Cheng, N., Steven, A.C., Langen, R. (2013) α-Synuclein oligomers with broken helical conformation form lipoprotein nanoparticles. *J. Biol. Chem.* **288**:17620–17630.

Vauthey, S., Santoso, S., Gong, H., Watson, N., Zhang, S. (2002) Molecular self-assembly of surfactant-like peptides to form nanotubes and nanovesicles. *Proc. Natl. Acad. Sci. USA* **99**:5355–5360.

Vénien-Bryan, C., Balavoine, F., Toussaint, B., Mioskowski, C., Hewat, E., Helme, B., Vignais, P. (1997) Structural study of the response regulator HupR from *Rhodobacter capsulatus*. Electron microscopy of 2D crystals on a nickel-chelating lipid. *J. Mol. Biol.* **274**:687–692.

Vestergaard, M., Kraft, J.F., Vosegaard, T., Thøgersen, L., Schiøtt, B. (2015) Bicelles and other membrane mimics: Comparison of structure, properties, and dynamics from MD simulations. *J. Phys. Chem. B* **119**:15831–15843.

Viegas, A., Viennet, T., Etzkorn, M. (2016) The power, pitfalls and potential of the nanodisc system for NMR-based studies. *Biol. Chem.* **397**:1335–1354.

Vinothkumar, K.R. (2011) Structure of rhomboid protease in a lipid environment. *J. Mol. Biol.* **407**:232–247.

Vold, R.R., Prosser, R.S. (1996) Magnetically oriented phospholipid bilayered micelles for structural studies of polypeptides. Does the ideal bicelle exist? *J. Magn. Reson.* **B113**:267–271.

von Maltzahn, G., Vauthey, S., Santoso, S., Zhang, S. (2003) Positively charged surfactant-like peptides self-assemble into nanostructures. *Langmuir* **19**:4332–4337.

Wadsäter, M., Laursen, T., Singha, A., Hatzakis, N.S., Stamou, D., Barker, R., Mortensen, K., Feidenhans'l, R., Møller, B.L., Cárdenas, M. (2012) Monitoring shifts in the conformation equilibrium of the membrane protein cytochrome P450 reductase (POR) in nanodiscs. *J. Biol. Chem.* **287**:34596–34603.

Wald, J.H., Goormaghtigh, E., De Meutter, J., Ruysschaert, J.M., Jonas, A. (1990a) Investigation of the lipid domains and apolipoprotein orientation in reconstituted high-density lipoproteins by fluorescence and IR methods. *J. Biol. Chem.*:20044–20050.

Wald, J.H., Krul, E.S., Jonas, A. (1990b) Structure of apolipoprotein A-I in three homogeneous, reconstituted high-density lipoprotein particles. *J. Biol. Chem.* **265**:20037–20043.

Wang, G. (2008) NMR of membrane-associated peptides and proteins. *Curr. Protein Pept. Sci.* **9**:50–69.

Wang, X., Mu, Z., Li, Y., Bi, Y., Wang, Y. (2015) Smaller nanodiscs are suitable for studying protein lipid interactions by solution NMR. *Protein J.* **34**:205–211.

Wang, X.Q., Corin, K., Baaske, P., Wienken, C.J., Jerabek-Willemsen, M., Duhr, S., Braun, D., Zhang, S.G. (2011) Peptide surfactants for cell-free production of functional G protein-coupled receptors. *Proc. Natl. Acad. Sci. USA* **108**:9049–9054.

Warschawski, D.E., Arnold, A.A., Beaugrand, M., Gravel, A., Chartrand, E., Marcotte, I. (2011) Choosing membrane mimetics for NMR structural studies of transmembrane proteins. *Biochim. Biophys. Acta* **1808**:1957–1974.

Wilcox, K.C., Marunde, M.R., Das, A., Velasco, P.T., Kuhns, B.D., Marty, M.T., Jiang, H., Luan, C.H., Sligar, S.G., Klein, W.L. (2015) Nanoscale synaptic membrane mimetic allows unbiased high-throughput screen that targets binding sites for Alzheimer's-associated Ab oligomers. *PLoS One* **10**:e0125263.

Wlodawer, A., Segrest, J.P., Chung, B.H., Chiovetti, R., Jr., Weinstein, J.N. (1979) High-density lipoprotein recombinants: evidence for a bicycle tire micelle structure obtained by neutron scattering and electron microscopy. *FEBS Lett.* **104**:231–235.

Xu, X.P., Zhai, D., Kim, E., Swift, M., Reed, J.C., Volkmann, N., Hanein, D. (2013) Three-dimensional structure of Bax-mediated pores in membrane bilayers. *Cell Death Dis.* **4**:e683.

Yang, J.P., Cirico, T., Katzen, F., Peterson, T.C., Kudlicki, W. (2011) Cell-free synthesis of a functional G protein-coupled receptor complexed with nanometer scale bilayer discs. *BMC Biotechnol.* **11**:57.

Yang, S.J., Zhang, S.G. (2006) Self-assembling behavior of designer lipid-like peptides. *Supramol. Chem.* **18**:389–396.

Ye, F., Hu, G., Taylor, D., Ratnikov, B., Bobkov, A.A., McLean, M.A., Sligar, S.G., Taylor, K.A., Ginsberg, M.H. (2010) Recreation of the terminal events in physiological integrin activation. *J. Cell Biol.* **188**:157–173.

Yeh, J.I., Du, S., Tortajada, A., Paulo, J., Zhang, S. (2005) Peptergents: peptide detergents that improve stability and functionality of a membrane protein, glycerol-3-phosphate dehydrogenase. *Biochemistry* **44**:16912–16919.

Yoon, J.Y., Kim, J., An, D.R., Lee, S.J., Kim, H.S., Im, H.N., Yoon, H.J., Kim, J.Y., Kim, S.J., Han, B.W., Suh, S.W. (2013) Structural and functional characterization of HP0377, a thioredoxin-fold protein from *Helicobacter pylori*. *Acta Crystallogr. D* **69**:735–746.

Zhang, M., Huang, R., Ackermann, R., Im, S.-C., Waskell, L., Schwendeman, A., Ramamoorthy, A. (2016) Reconstitution of the Cytb5-CytP450 complex in nanodiscs for structural studies using NMR. *Angew. Chem. Int. Ed. Engl.* **55**:4497–4499.

Zhang, P., Ye, F., Bastidas, A.C., Kornev, A.P., Wu, J., Ginsberg, M.H., Taylor, S.S. (2015) An isoform-specific myristylation switch targets Type II PKA holoenzymes to membranes. *Structure* **23**:1563–1572.

Zhang, Q., Tao, H., Hong, W.-X. (2011) New amphiphiles for membrane protein structural biology. *Methods* **55**:318–323.

Zhang, Z., Chen, J. (2016) Atomic structure of the cystic fibrosis transmembrane conductance regulator. *Cell* **167**:1586–1597.

Zhao, X., Nagai, Y., Reeves, P.J., Kiley, P., Khorana, H.G., Zhang, S. (2006) Designer short peptide surfactants stabilize G protein-coupled receptor bovine rhodopsin. *Proc. Natl. Acad. Sci. USA* **103**:17707–17712.

Zhao, Y., Imura, T., Leman, L.J., Curtiss, L.K., Maryanoff, B.E., Ghadiri, M.R. (2013) Mimicry of high-density lipoprotein: functional peptide-lipid nanoparticles based on multivalent peptide constructs. *J. Am. Chem. Soc.* **133**:13414–13424.

Zocher, M., Roos, C., Wegmann, S., Bosshart, P.D., Dötsch, V., Bernhard, F., Müller, D.J. (2012) Single-molecule force spectroscopy from nanodiscs: An assay to quantify folding, stability, and interactions of native membrane proteins. *ACS Nano* **6**:961–971.

Zoghbi, M.E., Altenberg, G.A. (2017) Membrane protein reconstitution in nanodiscs for luminescence spectroscopy studies. *Nanotech. Rev.* **6**:33–46.

# Chemical Structure, Synthesis, and Physical-Chemical Properties of Amphipols

**4**

### Summary

*Many different chemical structures have been shown to yield amphipols (APols), that is amphipathic polymers that are able to keep individual membrane proteins (MPs) soluble in their native state under the form of small complexes. They have in common to be polydisperse (a consequence of their mode of synthesis, but, a priori, not a requirement), of moderate size (in general $\leq 20$ kDa), and to feature either interspersed or alternating hydrophilic and hydrophobic units or a succession of identical amphiphilic ones. The presence of a large number of hydrophilic moieties is essential to ensuring the solubility of the many hydrophobic groups that will interact with the transmembrane surface of MPs, and the properties of these moieties – such as pH- or calcium-dependent solubility – affect those of APols and MP/APol complexes. In aqueous solutions, most APols self-associate into small, micelle-like particles, of which hydrophobic moieties occupy the core and hydrophilic ones the surface. Whereas individual APol molecules are polydisperse, the particles they assemble into are nearly monodisperse, reflecting the existence of a thermodynamically optimal size. Because APols are relatively large molecules, grafting them with functional groups is generally possible without affecting their solution properties, which has permitted the development of a vast variety of labeled or tagged APols.*

## 4.1 Introduction

In the preceding chapter, we have examined a number of surfactants or surfactant mixtures that provide ways to handle membrane proteins (MPs) in detergent-free aqueous solutions (nanodiscs, amphipathic peptides, fluorinated surfactants, etc.) or while limiting their exposure to detergents (bicelles). Amphipols (APols) represent a further alternative. APols are usefully defined as "amphipathic polymers that are able to keep individual MPs soluble in their native form under the form of small complexes" (Popot 2010; Popot et al. 2011). Strictly speaking, such a definition could be applied to nanodiscs (NDs), peptitergents, or lipopeptides, given that proteins and peptides are polymers. However, the chemical structure of polypeptide chains endows them with the ability to adopt well-defined secondary structures, and for all three systems mentioned above, it is the resulting folded elements that, directly or indirectly, interact with the target MPs. Their self-organizing properties are therefore conceptually and practically vastly different from those of the synthetic polymers to which we will now turn, which is why they have been discussed separately in Chap. 3. An interesting, possibly intermediate case, which is currently difficult to classify, is that of amphipathic polymers obtained by grafting poly-$\gamma$-glutamic acid with octyl and glucosyl groups, which have been used to trap bacteriorhodopsin (BR) and a G protein-coupled receptor (GPCR) as water-soluble medium-sized complexes (Han et al. 2014, 2017). It has not been reported whether the peptide is structured or not. Under the assumption that it is not, these polymers have been included in the present chapter.

    The term "amphipol" was coined to distinguish these particular polymers from the very vast family of amphipathic polymers, most of which have structures that, e.g. because of their length, or of the distribution of hydrophilic and hydrophobic moieties, are not suited to stabilizing MPs under the form of small complexes. Amphipathic polymers are not only a fascinating object of study for physical chemists interested in molecular self-organization, they also have hosts of practical applications, such as for controlling the rheology of solutions, stabilizing emulsions, forming coats, delivering drugs, etc. APols represent only a minute subset of this vast family, as exemplified by the (sobering!) observation that, in a recent extensive review dedicated to the structure, synthesis, and applications of amphipathic polymers (Raffa et al. 2015), only one reference out of nearly 500 bears on APols.

    In the present chapter, we will review first (§ 4.2) the chemical structures of those polymers that have been validated as bona fide APols, leaving aside those that have not been shown to form small water-soluble complexes with MPs in their native state and therefore do not qualify as APols. For the same reason, "blocky copolymers," which feature long stretches of hydrophilic and hydrophobic units (see Fig. 4.9), are not included here because, in comparative studies using as model hydrophobic, non-membrane seed proteins called oleosins, they have been shown to form much larger complexes than APols do (Gohon et al. 2011). Similar observations were reported upon extraction of MPs with PreserveX™-QML (a proprietary mixture of blocky polymers), which yielded particles several hundreds of nm large (Trubetskoy et al. 2006). The issue is not settled yet in the case of random poly[N-(2-hydroxypropyl)-methacrylamide-co-dodecylmethacrylate] polymers, which have been shown to keep water-soluble dimers of a synthetic peptide mimicking the transmembrane (TM) anchor of glycophorin A (Stangl et al. 2014), forming complexes whose size and composition have not yet been reported. In § 4.3, we will examine those studies that provide information about the organization adopted by APols when they are dissolved in water. Finally, in § 4.4 we will review labeled and functionalized APols. The formation and properties of MP/APol complexes will be discussed in Chap. 5.

## 4.2    Amphipol Chemical Structure and Synthesis

In this section, we will examine what the basic structures of those polymers that have been proven to be effective as APols are. One point that should be raised and stressed from the onset is that there is a very, very long way to go from scribbling a potentially interesting chemical structure on the back of an envelope to turning it into a routine tool for biochemists and biophysicists to use. A first synthesis and test can be relatively easy and quick; debugging the synthesis, discovering and eliminating the pitfalls of contaminants and side products, developing a reliable and economically viable synthesis and purification route, assaying the new molecules in a large enough variety of conditions and on a broad enough panel of MPs, understanding the behavior of the MP/APol complexes thus formed, identifying their most promising applications, and convincing a good commercial partner to synthesize and market the molecules represent a long-term, labor-intensive, and hard-to-fund project. It is therefore advisable to shave one's many ideas with Occam's razor and retain only a limited number of basic structures, lest none of them be thoroughly explored. As a redeeming feature, once a basic structure has been carefully validated, it is (usually) a relatively simple matter to identify and fine-tune the chemistry required to develop an array of derivatives, each carrying one or another label or tag (§ 4.4). Most such modifications will not significantly affect the physical-chemical behavior of the polymer (see § 4.4.2). Therefore, the grinding work that has been invested upstream to validate the basic molecule and understand its advantages and drawbacks will usually remain exploitable.

In the following survey, we will successively consider (i) the original polyacrylate-based APols (§ 4.2.1), (ii) other ionic APols (§ 4.2.2), and (iii) non-ionic APols (§ 4.2.3). A selection of polymers that have been used to keep MPs soluble in aqueous solutions, whether they qualify as APols or not, is provided in Table 4.1, along with some key references.

### 4.2.1 Polyacrylate-Based Amphipols with Carboxylates as Their Hydrophilic Moieties

The concept of APols emerged from discussions between two polymerists, Roland Audebert and Christophe Tribet, and a membrane biologist, myself. R. Audebert's laboratory had been studying the properties of long poly(acrylic acid) (PAA) molecules sparsely modified with relatively long alkyl chains ($C_{12}$ or $C_{18}$). By themselves, these polymers disperse in water as individual molecules and form fluid solutions. However, in the presence of detergent, the alkyl chains partition into detergent micelles, creating nodes, and, thereby, a meshwork of polymers, turning the solution into a gel (see e.g. Sarrazin-Cartalas et al. 1994). R. Audebert was initially interested in forming nodes with MPs instead of micelles. I was only moderately enthused by the idea of using my hard-to-purify and fragile MPs to make gels, but much more so to devise ways to make them more stable in aqueous solutions. This led me to suggest making much shorter polymers, so that each protein would surround itself with several of them and cross-linking would be limited or avoided. The resulting complexes might be water-soluble, and, hopefully, eliminating the detergent would stabilize the protein (cf. Chap. 2). Because long hydrophilic loops extending into the solution would be a hindrance for many biophysical applications, such as crystallization or NMR, the alkyl chains had to be spaced much more closely than in Roland's original polymers, which meant that, in order to achieve a high enough aqueous solubility, they had to be short, which led us to settle for octyl chains. A highly flexible main chain, such as provided by PAA, was also a desirable feature, because it would facilitate a close coverage of the corrugated surface of the proteins and the formation of small particles. Because we were entering a totally unchartered territory, we decided to play both on the length of the polymers, which would determine how many of them would bind to a given protein and modulate the risks of cross-linking (cf. Yamamoto et al. 2000b), and on the charge density, which would affect the persistence length (rigidity) of the polymers and, perhaps, the stability of the protein. This led to the design of the four polymers, A8-35, A8-75, A34-35, and A34-75, described in the princeps publication (Tribet et al. 1996).

A8-35 (Fig. 4.1a) was obtained by grafting a commercial ~5-kDa PAA preparation first with octylamine, yielding A8-75, and then with isopropylamine (Fig. 4.2). The resulting amphipathic polymer comprises ~25 octyl chains per 100 monomers, ~40 isopropyl groups, and ~35 free carboxylates. Whereas carboxylates and octyl chains confer amphipathy to A8-35, the isopropyl groups limit its charge density, which, if it is too high, seems to negatively affect MP stability (see Chap. 5). Isopropyls do not contribute much to the global hydrophobicity at low or ambient temperature (Takei et al. 1993), but they do interact with the hydrophobic TM surface of MPs (see Chap. 5). The average mass per octyl chain of the sodium salt of A8-35 is close to 500 Da, similar to that of many detergents, but higher than that of those carrying octyl chains, such as octylglucoside or

**Table 4.1** A selection of polymers that have been used to handle membrane proteins.

| Polymer name [source, if commercially available] | Chemical structure | Figure | $\langle M_n \rangle$ | Comments | Selected references |
|---|---|---|---|---|---|
| A8-35 (noted 5-25C$_8$-40C$_3$ in some physical chemistry articles) [Anatrace; Jena Bioscience] | Short polyacrylate (~35 units) grafted with octylamine (~25%) and isopropylamine (~40%) | 4.1 | $\langle M_w \rangle \approx$ 8.6 kDa, $\langle M_n \rangle \approx$ 4.3 kDa[a], $Đ \approx 2$ | As of today the most extensively studied and most widely used APol. Its main limitations originate from its being charged and sensitive to low pH and multivalent cations. Particles and solution properties thoroughly characterized by SANS, SAXS, INS, AUC, DLS, SEC, FRET, surface tension measurements, and MD | Tribet et al. (1996), Ladavière et al. (2001, 2002), Gohon et al. (2004, 2006), Vial et al. (2007, 2009), Tribet and Vial (2008), Perlmutter et al. (2011), Giusti et al. (2012, 2014), Sverzhinsky et al. (2014), Tehei et al. (2014), and Watkinson et al. (2015) |
| A8-75 (also noted 5-25C$_8$) | Short polyacrylate (~35 units) grafted with octylamine (~25%) | | $\langle M_n \rangle \approx$ 4.1 kDa[a] | Properties appear similar to A8-35, but, due to its higher charge density, A8-75 may be less stabilizing to MPs | Tribet et al. (1996, 1997), Ladavière et al. (2001, 2002), Vial et al. (2005, 2007, 2009), Luccardini et al. (2006), Tribet and Vial (2008), Marie et al. (2014), and Watkinson et al. (2015) |
| A34-35 | Long polyacrylate (~140 units) grafted with octylamine (~25%) and isopropylamine (~40%) | | $\langle M_n \rangle \approx$ 17 kDa | Properties appear similar to those of A8-35. Not extensively used | Tribet et al. (1996) and Watkinson et al. (2015) |
| A34-75 | Long polyacrylate (~140 units) grafted with octylamine (~25%) | | $\langle M_n \rangle \approx$ 16 kDa | Properties appear similar to those of A8-75. Not extensively used | Tribet et al. (1996) and Watkinson et al. (2015) |
| THAM-based non-ionic APols | Telomers derived from *tris* (hydroxymethyl)-acrylamidomethane (THAM). Their solubility is provided by multiple hydroxyl groups | 4.7 | 3–28 kDa | The first attempt at creating non-ionic APols. The molecules provided a proof of concept, but were not soluble enough for routine use | Prata et al. (2001) |
| NAPols (glucose-based non-ionic APols) [Anatrace] | Telomers derived from THAM (10–90 units). Their solubility is provided by multiple glucose moieties. The first versions were heteropolymers, obtained either by co-telomerization or by grafting. Current NAPols are homotelomers with typically ~30 units, each carrying two glucose moieties and one undecyl chain | 4.7 | 8–60 kDa (typically ~13 kDa) | Entirely non-ionic polymers. Insensitive to pH and multivalent cations. NAPols are the only APols to date to have been validated for cell-free MP expression (along with NVoy) and isoelectrofocusing. Validated for NMR. Insensitive to pH and multivalent cations | Sharma et al. (2008, 2012), Bazzacco et al. (2009, 2012), and Watkinson et al. (2015) |

**Table 4.1** (continued)

| Polymer name [source, if commercially available] | Chemical structure | Figure | $\langle M_n \rangle$ | Comments | Selected references |
|---|---|---|---|---|---|
| PC-APols (phosphorylcholine-based APols C22-43 and C45-68) | Random copolymers of *n*-octyl-acrylamide, a phosphorylcholine-modified acrylamide carrying a secondary amine function, and, optionally, *N*-isopropylacrylamide | 4.1 | 22 or 45 kDa | Polycationic APols. Insensitive to pH and multivalent cations | Diab et al. (2007a, b), Tribet et al. (2009), and Basit et al. (2012) |
| SAPols (sulfonated APols) | Short polyacrylate (~35 units) grafted with *n*-octylamine (~25%) and taurine (~40%) | 4.1 | ~5.6 kDa | Polyanionic APols. Insensitive to pH and multivalent cations. Appear less stabilizing than A8-35. Validated for NMR and EM | Picard et al. (2006), Dahmane et al. (2011), Huynh et al. (2014), and Watkinson et al. (2015) |
| PMAL series [Anatrace] | Polymers carrying ammoniumamide, carboxylate, and dodecyl side chains in various proportions (~30 units) | 4.1 | ~12 kDa | Polymers carrying mixed charges. Shown to stabilize DAGK and deliver it to lipid vesicles. PMAL-C12 forms small water-soluble complexes with the sarcoplasmic calcium ATPase while protecting it from denaturation. PMAL-C8 has been used for cryo-EM single-particle studies. Also used to deliver to cell quantum dots and magnetic nanoparticles | Nagy et al. (2001), Gorzelle et al. (2002), Picard et al. (2006), Qi and Gao (2008), Qi et al. (2012), and Paulsen et al. (2015) |
| Poly(methacrylic acid) derivatives | Poly(methacrylic acid) grafted with octyl chains either randomly or in blocks | | 10–11 kDa (random form) | Used for studying interactions with lipid vesicles and for solubilizing hydrophobic non-transmembrane proteins (oleosins) | Liu et al. (2007), and Gohon et al. (2011) |
| p(HPMA)-co-p(LMA) polymers | Poly[*N*-(2-hydroxypropyl)-methacrylamide] polymers grafted with dodecylmethacrylate | | ~17 kDa | The most hydrophobic of these polymers have been shown to keep water-soluble dimers of an α-helical peptide mimicking the transmembrane anchor of glycophorin A | Stangl et al. (2014) |

(continued)

**Table 4.1** (continued)

| Polymer name [source, if commercially available] | Chemical structure | Figure | $\langle M_n \rangle$ | Comments | Selected references |
|---|---|---|---|---|---|
| SMALPs = Lipodisqs (styrene-maleic acid copolymer/lipid particles) [Malvern Cosmeceutics, Cray Valley; Polyscope Polymers (Xiran); Sigma-Aldrich] | Particles comprised of a styrene/maleic acid copolymer (SMA) and lipids | 4.1 | ~9.5 kDa (size of the 3:1 SMA polymer) | The copolymer can extract directly MPs from membranes, without the use of detergents, forming MP/lipid/polymer ternary complexes. Have been used for EPR distance measurements in a MP and for cryo-EM | Knowles et al. (2009), Jamshad et al. (2011, 2015a, b), Rajesh et al. (2011), Banerjee et al. (2012), Orwick-Rydmark et al. (2012), Orwick et al. (2012), Long et al. (2013), Sahu et al. (2013), Tanaka et al. (2015), Vargas et al. (2015), Zhang et al. (2015), Dörr et al. (2016), Logez et al. (2016), Wheatley et al. (2016), Cuevas Arenas et al. (2017), and Dominguez Pardo et al. (2017) |
| NVoy = NV10 [Expedeon] | Fructose-based polysaccharide carrying hydrophobic chains. The exact structure has not been released | 4.7 | ~5 kDa | No report yet on the size and dispersity of MP/NVoy complexes. Has been used for cell-free synthesis | Guild et al. (2011), and Klammt et al. (2011) |
| Amphibiopols | Hydrophobized derivatives of pullulan | | ~30 kDa | Do not qualify as APols, inasmuch as they cannot keep MPs soluble as small individual complexes. Have been used to stabilize suspensions of membrane fragments | Duval-Terrié et al. (2003), and Picard et al. (2004) |
| Hydrophobically grafted poly-γ-glutamic acid (APG) | Poly-γ-glutamic acid with ~35% of the carboxylates left free, ~41% grafted with octylamine, and ~24% grafted with glucosamine | | ~17 kDa | Whether APG is structured or not in MP/APG complexes is unknown. Forms ~80-kDa particles and traps BR and a GPCR under a functional form. Used to fold BR from SDS and to transfer a GPCR to preformed liposomes | Han et al. (2014) |
| Hydrophobically grafted poly-γ-glutamic acid (APG variant) | A variant of APG synthesized by coupling octylamine (~25%), glucosamine (~19%), and diethyl aminopropylamine (~25%) onto the carboxylic groups of poly-γ-glutamic acid, leaving ~31% of them free | | | A GPCR, human lysophosphatidic acid receptor 2, was stabilized by transfer from sarkosyl to the APG variant and its functionality assessed by measuring ligand-dependent dissociation from $G_{\alpha i3}$ | Han et al. (2017) |

**Table 4.1** (continued)

| Polymer name [source, if commercially available] | Chemical structure | Figure | $\langle M_n \rangle$ | Comments | Selected references |
|---|---|---|---|---|---|
| DIBMA = Sokalan CP9 | An alternating copolymer of diisobutylene and maleic acid | | $\langle M_w \rangle \approx$ 15.3 kDa, $\langle M_n \rangle \approx$ 8.4 kDa, $Đ \approx 1.8$ | DIBMA shows equal performance to SMA in solubilizing phospholipids, stabilizes outer membrane phospholipase A (OmpLA) under a functional form, and extracts proteins of various sizes directly from *E. coli* membranes. It has a milder effect on lipid acyl-chain order than SMA, does not interfere with optical spectroscopy in the far-UV range, and does not precipitate in the presence of millimolar concentrations of divalent cations | Oluwole et al. (2017) |

The table is updated from Zoonens and Popot (2014)

To prevent confusion in the literature, we have proposed to define amphipols as "amphipathic polymers that are able to keep individual MPs soluble (and native) under the form of small complexes" (Popot 2010; Popot et al. 2011). According to this definition, "amphibiopols" do not qualify as amphipols, and p(HPMA)-co-p(LMA) polymers and hydrophobically grafted poly-$\gamma$-glutamic acid remain to be validated as such. SMALPs can incorporate MPs into small (10–12 nm diameter) disc-like, lipid-containing particles, but also form much smaller complexes (see Chap. 5). MP/NVoy complexes also form small particles (see text)

*APol* amphipol, *AUC* analytical ultracentrifugation, *DAGK* diacylglycerol kinase, *DLS* dynamic light scattering, *EM* electron microscopy, *EPR* electron paramagnetic resonance, *FRET* Förster resonance energy transfer, *INS* inelastic neutron scattering, *MD* molecular dynamics, $\langle M_n \rangle$ number-average molecular mass, *MP* membrane protein, *NMR* nuclear magnetic resonance, *SANS* and *SAXS* small-angle neutron and X-ray scattering, respectively, *SEC* size exclusion chromatography

[a]The average mass of A8-35 (and, by extension, those of A8-75, A34-35, A34-75, and SAPols) has been recently revised; see Giusti et al. (2014)

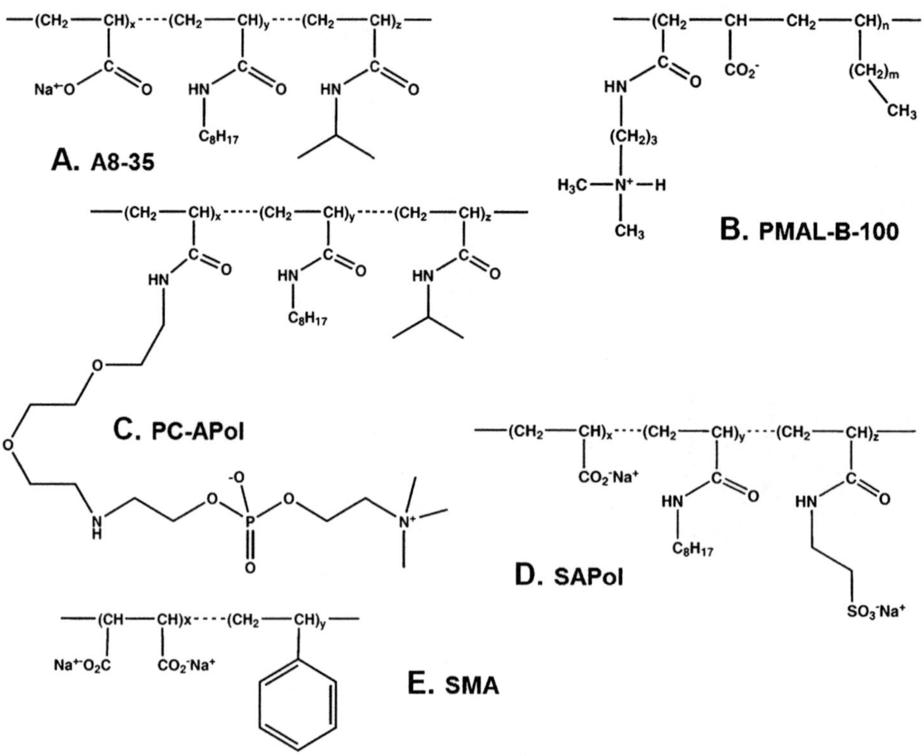

**Fig. 4.1** Chemical structure of some ionic amphipols. (**A**) *A8-35 and congeners:* a polyacrylate main chain randomly derivatized with octyl and, for some of them, isopropyl chains. APols from this series are denoted according to their approximate average molecular mass (for A8-35, $\langle M_w \rangle \approx 8.6$ kDa; see Giusti et al. 2014) and the percentage of carboxylates that have been left ungrafted (for A8-35, $x \approx 35\%$, $y \approx 25\%$, and $z \approx 40\%$; for its precursor, A8-75, $x \approx 75\%$, $y \approx 25\%$, and $z = 0\%$) (From Tribet et al. 1996). The octyl chains can be replaced with dodecyl ones (Y. Gohon and C. Tribet, unpublished observations). OAPA-20 (Nagy et al. 2001) has the same structure as A8-75, with $x \approx 80\%$, $y \approx 20\%$, and $z = 0\%$. Using the same nomenclature, it would be called A8-80. (**B**) *PMAL-B series:* polymers obtained by modification of an *alt*-copolymer (copolymer formed by a strict alternating sequence of two different monomers) of maleic anhydride and a long-chain olefin. Each repeating unit comprises a hydrophobic moiety and either a zwitterionic (structure shown) or an anionic (carboxylic) one. *m* can be 7, 11, or 15. The polymers can either combine the two types of units in a 1:1 ratio (PMAL-B-50) or incorporate only the zwitterionic one (*PMAL-B-100*) or the carboxylic one (PMAL-B-0). PMAL-C8, PMAL-C12, and PMAL-C16 denote PMAL-B-100 versions carrying $C_8$, $C_{12}$, or $C_{16}$ alkyl chains, respectively. The average mass of PMAL-B-50 is reported to be ~10 kDa. From Nagy et al. (2001), where "PMAL-B" denotes a version of PMAL-B-50 carrying $C_{12}$ chains. (**C**) *PC-APols:* phosphorylcholine-based zwitterionic APols. Their structures are similar to those of A8-35 and A8-75, except that the free carboxylates are replaced with a phosphorylcholine-bearing group. PC-APols with average masses of ~22 and ~45 kDa have been tested for MP stabilization (see Study 4.9 in Table 4.2) (From Diab et al. 2007a). (**D**) *SAPol:* same structure as A8-35, but the isopropylamine groups are replaced by taurine ones. Octyl chains can be replaced with dodecyl ones (From Dahmane et al. 2009). (**E**) *SMA:* a commercial styrene-maleic acid copolymer. The molar ratio of styrene to maleic acid units is either 2:1 or 3:1 and the average molecular mass 7.5–10 kDa (Orwick et al. 2012; Dörr et al. 2016).

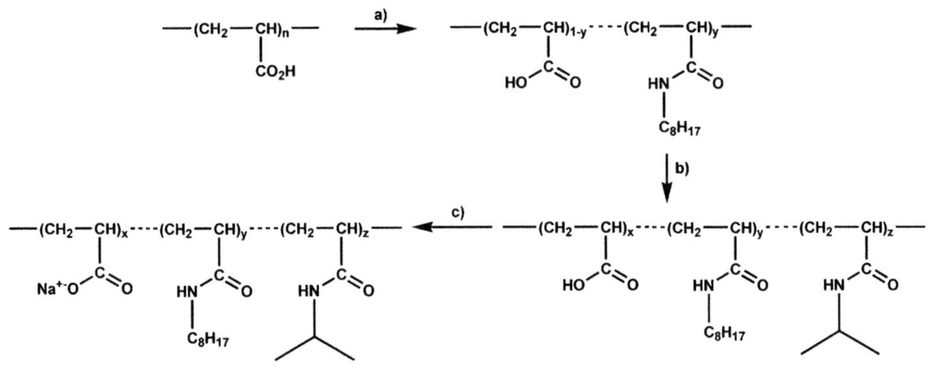

**Fig. 4.2** Synthesis of amphipol A8-35 by hydrophobic modification of a poly(acrylic acid) precursor (PAA). **a)** *n*-octylamine (0.25:1 molar ratio to PAA units), *N*-methyl-2-pyrrolidone (NMP)/dicyclohexylcarbodiimide (DCC), 60 °C for 1 h, then room temperature for 4 h; **b)** isopropylamine (0.40:1 molar ratio to PAA units), DCC/1-*N*-hydroxybenzotriazole (HOBt)/NMP, 50 °C for 1 h, then room temperature for 4 h; **c)** sodium methoxide (MeONa), followed by four cycles of precipitation in aqueous solution at pH < 2 and dissolution at pH > 8 (From Le Bon et al. 2014b. A detailed synthesis protocol prepared by Fabrice Giusti is given in § 4.5, along with some practical comments).

$C_8E_4$ (~300 Da) (see Chap. 2, Fig. 2.1). Above pH 7, the solubility of A8-35 in water and aqueous buffers is >30% w/w (>320 g·L$^{-1}$) (Gohon et al. 2006). Using NMR spectroscopy to document the vicinity between the various moieties, it has been shown that, under the conditions used, the grafts distribute randomly along the macromolecular chain (Magny et al. 1992). This polymer was named A8-35, where A stands for anionic, or acrylate, 8 for its approximate weight-average molecular mass, in kDa, and 35 for the percentage of carboxylates left free. In A8-75, there are ~25 octyl chains and ~75 free carboxylates per 100 monomers. A34-75 and A34-35 were obtained in the same way starting from a longer PAA. All of them were validated as APols in the original publication, inasmuch as they all efficiently trapped MPs and kept them soluble without denaturing them (Tribet et al. 1996). Only A8-35, however, has been heavily used for MP studies (for an overview, see Table 5.1 in Chap. 5). Estimates of its mass have varied. Using improved analytical methods (see below), its number-average length has been recently determined to be ~35 acrylate units, corresponding, for A8-35, to a number-average molecular mass $\langle M_n \rangle \approx 4.3$ kDa (Giusti et al. 2014). Given a dispersity (formerly called polydispersity index; see Gilbert et al. 2009 and § 4.6.1, Annex 4.1) $Đ \approx 2$, its weight-average mass $\langle M_w \rangle$ is ~8.6 kDa. Its length distribution is quite broad: upon hydrophobic size exclusion chromatography (SEC) analysis, the low-$R_S$ and high-$R_S$ half-height limits on each side of the maximum correspond to ~15 and ~200 units or ~1.2 and ~18 kDa, respectively (J. Rieger and F. Giusti, unpublished data) (about the definition and measurement of $\langle M_n \rangle$, $\langle M_w \rangle$, and $Đ$, see § 4.6.1, Annex 4.1). A8-75, with the same length distribution and octylamine density as A8-35, but no isopropylamine grafts, has been mostly used in early EM studies (see Chap. 12, Table 12.1), in some comparative mass spectrometry studies (Watkinson et al. 2015; see Chap. 14), and in studies of its interactions with lipid vesicles or cells (see Chap. 5, § 5.2.2.2).

*Amphipol synthesis and self-association*
(© 2018 by Francis Haraux)

OAPA-20, a commercial product used in later works (Nagy et al. 2001; Gorzelle et al. 2002), is essentially identical to A8-75 (it would be named A8-80 using the same nomenclature). A5-35, a shorter version of A8-35 whose weight-average mass was ~5.5 kDa, was tested on one MP, cytochrome $b_6 f$, and on a synthetic peptide mimicking the single TM $\alpha$-helix of glycophorin A and found to yield satisfying results (Gohon 1996, 2002; Popot et al. 2003), but was not studied in depth. Similarly, it was found (i) that a version of A8-35 in which octyl chains were replaced with dodecyl ones also yielded satisfying preparations (Y. Gohon and C. Tribet, unpublished observations) and (ii) that starting from a less polydisperse, noncommercial PAA preparation obtained by atom transfer radical polymerization (Davis and Matyjaszewski 2000) (see Annex 4.2, § 4.6.4.2) yielded A8-35 batches whose properties were undistinguishable from those of the more polydisperse preparations derived from commercial PAA (C. Tribet and F. Giusti, unpublished data cited in Le Bon et al. 2014b). For the reason evoked above – namely the desire to focus on a given type of structure and to exhaustively characterize its properties rather than to superficially examine too many different structures – these various avenues were not further pursued, despite their potential interest. The chemical characteristics of PMAL-B-0, a commercial anionic APol, will be discussed in § 4.2.2 along with those of other members of the PMAL series.

Determining and expressing the length distribution and average length of a mixture of polymers is a complicated matter (see § 4.6.1, Annex 4.1). The most recent determinations, using size exclusion chromatography (SEC) of methylated PAA in organic solvents and mass spectrometry (MS) analysis of the final product, indicate that the weight-average mass $\langle M_{\mathrm{w}} \rangle$ of A8-35 is ~8.6 kDa, which corresponds to ~70 monomers, carrying ~18 octyl chains and ~24 free carboxylates (Giusti et al. 2014). The exact value taken for the average mass affects the calculated number of chains that assemble to form a particle (§ 4.3) or that adsorb onto a given MP (Chap. 5), for which estimates of the number-average value $\langle M_{\mathrm{n}} \rangle \approx 4.3$ kDa should be used (see § 4.3.1.2.1 and Annex 4.1, § 4.6.1). These numbers, however, have little significance and in most cases are not relevant to designing and interpreting biochemical and biophysical experiments, where it is generally the mass that counts, whatever the number of chains involved. As will be further discussed in Chap. 5, it is highly advisable, in order to remain on firm ground, to express the composition of APol particles and MP/APol complexes in terms of mass or mass ratio rather than number of molecules or molar ratio, because the former can be more or less precisely determined, whereas the latter are intrinsically fuzzy, hard to determine, and will differ from one preparation of A8-35 to another if the starting PAA batches do not have exactly the same length distribution or if different amounts of shorter or longer polymers are lost during the synthesis or purification.

---

**Box 4.1   Introducing Biochemists to the Chemistry of Synthetic Polymers**

Biologists are familiar with polymers: proteins, nucleic acids, and polysaccharides. However, the mode of synthesis of biological polymers sets them apart from the synthetic polymers: biological polymers tend to have well-defined sequences and lengths, which is not the case of synthetic polymers, which are heterogeneous. Controlling, measuring, and expressing this heterogeneity calls for specific synthetic, analytical, and description techniques. Some of the terminology used in this chapter will be unfamiliar to most readers, such as weight- and number-average molar masses, as well as the dispersity or molar mass dispersion ($\langle M_w \rangle$, $\langle M_n \rangle$, and $Đ$, respectively). Indeed, synthetic polymers can only be described as populations of molecules that differ from each other by their length and, for most heteropolymers, by the sequence of the various units that comprise them. This is an ineluctable consequence of their mode of synthesis. Because of its complexity, this matter cannot be treated in a few lines. Two expert colleagues, Fabrice Giusti and Bernard Pucci, have kindly accepted to write up a non-exhaustive but helpful (we hope) overview of the most essential notions. In order not to break the continuity of the text, this presentation has been organized in a series of annexes collected at the end of the chapter (§ 4.6) and distributed as follows:

- *Annex 4.1* exposes in their broad lines some of the reasons from which originates the dispersity in size of synthetic polymers, explains the notions of average molar mass and dispersity, and describes the method that is most commonly used to measure each of these parameters and the way to express them.
- *Annex 4.2* A good understanding of how to limit dispersity requires a knowledge of the kinetic parameters that govern conventional radical polymerization (RP). This annex treats in more depth the topic of free radical polymerization and related kinetic aspects, providing a more detailed description of the origins of the dispersity, as well as setting the stage for Annex 4.3.
- *Annex 4.3* discusses the notion of degree of polymerization and its evolution in the course of a synthesis.
- *Annex 4.4* introduces and briefly discusses the two main processes that can be used to limit the breadth of polymer size distribution, namely telomerization and controlled radical polymerization (CRP).
- *Annex 4.5* discusses the benefits that, besides achieving a lower molecular dispersity, can be expected from the different CRP approaches.
- *Annex 4.6* describes ways in which functional groups can be grafted onto polymers.

---

A8-35 is commercially available (Table 4.1). Users, however, may want to prepare it themselves, e.g. to create new labeled forms or derivatives. The synthesis of A8-35 and its congeners (Fig. 4.2) is simple on paper. In practice, carrying it out properly requires both care and experience. Apparently trivial details may matter. Thus, a change in the source of poly(acrylic acid) (PAA) in the late 1990s resulted in a long spell of failures to produce A8-35 preparations with the nominal structure and behavior before the origin of the difficulties was identified and brought under control (for a discussion, see Gohon et al. 2006). Protocol 4.1, given in § 4.5 of this chapter, is based on published protocols (Tribet et al. 1996; Gohon et al. 2004, 2006; Giusti et al. 2014) and laboratory notes. It provides detailed guidelines for synthesizing A8-35, along with some caveats.

## 4.2.2 Other Ionic Amphipols

As will be described in § 4.3, one of the limitations of A8-35 and its congeners is that they owe their solubility to their carboxylate groups. They are therefore sensitive to factors that affect the solubility of the latter, such as a low pH or the presence of $Ca^{2+}$ ions. Protonation of some of the carboxylates lowers the solubility of A8-35 (Gohon et al. 2006), as does complexation by $Ca^{2+}$ and, to a much smaller extent, $Mg^{2+}$ ions (Picard et al. 2006; see § 4.3.2). Under such conditions, free A8-35 particles tend to aggregate and may precipitate, and so do MP/A8-35 complexes. The pH sensitivity is particularly annoying in solution NMR experiments, where the observation of water-exposed amide protons is easiest at a slightly acidic pH, which slows down their exchange with those of water. This led, quite rapidly, to attempts at diversifying APol structures and, in particular, at developing APols that would be pH- and $Ca^{2+}$-insensitive. Several routes have been successfully explored (Table 4.1). One of them is to replace the carboxylates with non-ionic moieties. The resulting non-ionic APols will be described in § 4.2.3. Another is to endow ionic APols with additional or alternative ionic groups that will keep them soluble even at low pH or in the presence of calcium ions, as described in the present section.

### 4.2.2.1 Sulfonated Amphipols

Sulfonated APols (SAPols) are variants of A8-35 in which the isopropylamine groups are replaced with taurine ones (Fig. 4.1d). In terms of charge density along the chain, they are therefore similar to A8-75. The development of SAPols was undertaken shortly after A8-35, and its congeners had been validated for use in NMR studies (Zoonens et al. 2005), and a first series of comparative biochemical experiments was published in 2006 (Picard et al. 2006). However, a full description of their synthesis and properties and the validation of their use for solution NMR studies had to wait until 2011 (Dahmane et al. 2011). The synthesis of SAPols will be described below in § 4.4.1, along with that of functionalized APols.

The purification of SAPols is more laborious than that of A8-35 (Dahmane et al. 2011). The purification of A8-35 and its congeners takes advantage of the fact that they are insoluble at acidic pH, so that several cycles of precipitation at pH < 2 followed by redissolution at pH > 8 make it straightforward to get rid of both hydrosoluble and poorly soluble contaminants, even on large batches (see § 4.5 Protocol 4.1). SAPols however cannot be precipitated in acidic solutions. They have to be separated from contaminants by large-scale SEC, which is much more time-consuming and has hampered their marketing.

At low pH, or in the presence of $Ca^{2+}$ ions, some or all of the charges carried by the carboxylate moieties of SAPols will be neutralized. Nevertheless, in keeping with the behavior of A8-35, which comprises only 35% of ungrafted carboxylates, the 40% of monomers grafted with taurine are sufficient to keep SAPols water-soluble down to pH −1 (Dahmane et al. 2011). Similarly, SAPol particles and MP/SAPol complexes are insensitive to the presence of divalent cations (Picard et al. 2006).

### 4.2.2.2 Phosphorylcholine-Based Amphipols

The structures of phosphorylcholine-based zwitterionic APols (PC-APols; Fig. 4.1C) are similar to those of A8-35 and A8-75, except that the free carboxylates are replaced with a phosphorylcholine-bearing zwitterionic group. Their synthesis follows quite a different route, as it does not rely on grafting a pre-existing PAA, but on (i) synthesizing a terpolymer precursor by reversible addition-fragmentation chain transfer (RAFT; see Chiefari et al. 1998 and Annex 4.6, § 4.6.4.2) and (ii) grafting phosphorylcholine groups onto it by reductive amination of phosphorylcholine glyceraldehyde (Diab et al. 2007b). PC-APols are not currently commercial.

Given their chemical structure, PC-APols are expected to carry no net charge at neutral and basic pH. At acidic pH, however, the secondary amine they carry will protonate and they will become cationic, which may bear on their use in isoelectrofocusing experiments. PC-APols with average masses in the range of ~22–45 kDa have been validated for MP trapping (Diab et al. 2007a). BR/PC-APol complexes remained soluble in aqueous media at pH $\geq$ 5, as well as in the presence of 1 M NaCl or 12 mM $Ca^{2+}$ ions (ibid.).

#### 4.2.2.3 PMAL Series

The first alternative APols to be validated following the original publication belong to the PMAL series (Nagy et al. 2001). They are synthesized according to a proprietary procedure whose details have not been released (see legend in Fig. 4.1). Their structure is significantly different from that of the APols discussed above: rather than featuring a random distribution of interspersed hydrophilic and hydrophobic groups (Fig. 4.3, *left*), PMAL polymers (Fig. 4.1B) are formed from a succession of pairs of groups, one of which is hydrophobic and the other hydrophilic (Fig. 4.3, *center*). The hydrophobic groups bear $C_8$, $C_{12}$, or $C_{16}$ alkyl chains. The nature of the hydrophilic groups depends on the type of PMAL. In PMAL-B-100, each polar group is comprised of one carboxylate and one ammoniumamide, as shown in Fig. 4.1. In PMAL-B-0, it comprises two carboxylates, making the polymer chemically similar to a grafted polyacrylate but for the regular alternation of two carboxylates and one alkyl chain. PMAL-B-50 features a 1:1 mixture of these two types of polar groups, presumed to be randomly distributed (Nagy et al. 2001; Gorzelle et al. 2002). PMAL-C8, PMAL-C12, and PMAL-C16 are versions of PMAL-B-100 carrying $C_8$, $C_{12}$, and $C_{16}$ alkyl chains, respectively.

#### 4.2.2.4 Styrene-Maleic Acid Copolymer

Styrene-maleic acid copolymer (SMA; Fig. 4.1e) belongs to a family of industrial polymers, whose use for handling MPs was first introduced by Knowles et al. (2009). SMA was not claimed by its promoters to be an APol. However, as we will see in this and the following chapter, it definitely qualifies as one, even though its solubilizing properties appear to set it apart.

SMA is obtained by copolymerization of maleic anhydride and styrene (Fig. 4.4). Its dispersity Đ is ~2.5 (Dörr et al. 2016), vs. ~2 for A8-35. The average mass of the molecules used in biochemistry is 7.5–10 kDa (ibid.). The distribution of the two types of monomers in the final polymer depends on their ratio: it is close to a regular alternation if this ratio is 1:1, less regular if they comprise maleic anhydride and styrene in a 1:2 or 1:3 ratio, as is the case of the preparations used in biochemistry and biophysics (ibid.).

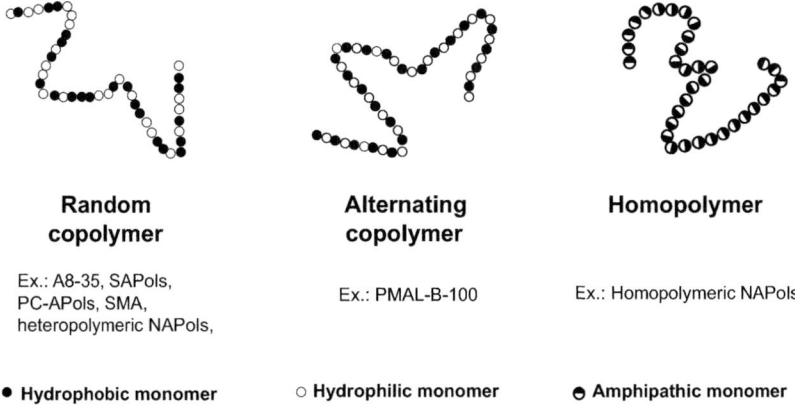

|  |  |  |
|---|---|---|
| **Random copolymer** | **Alternating copolymer** | **Homopolymer** |
| Ex.: A8-35, SAPols, PC-APols, SMA, heteropolymeric NAPols, | Ex.: PMAL-B-100 | Ex.: Homopolymeric NAPols |

● **Hydrophobic monomer**     ○ **Hydrophilic monomer**     ◑ **Amphipathic monomer**

**Fig. 4.3** A schematic representation of the distribution of hydrophilic and hydrophobic groups in various amphipols.

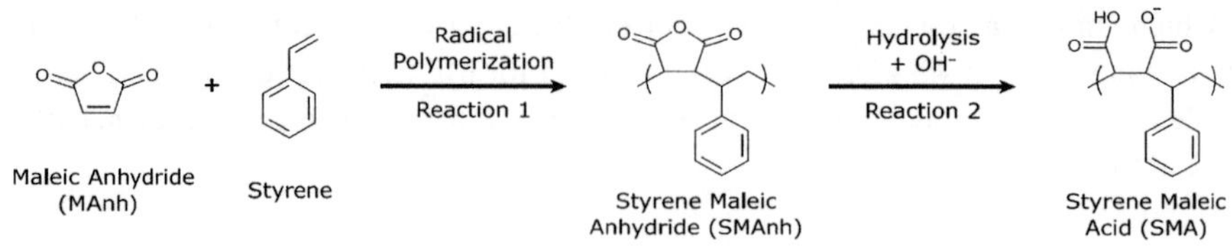

**Fig. 4.4** Schematic representation of the synthesis of styrene-maleic anhydride copolymer (Reaction 1) and the preparation of styrene-maleic acid copolymer (Reaction 2) . It is illustrated here for a 1:1 styrene-to-maleic anhydride/acid molar ratio, at which the two types of monomers tend to alternate. When styrene is present in excess, as is the case for preparations used in biochemistry, where the styrene/maleic acid molar ratio is ~2:1 or ~3:1, the monomer sequence distribution in the polymer becomes more complex. Reaction 1 is carried out by the manufacturer, Reaction 2 by the user (From Dörr et al. 2016).

SMA does not seem to have been used to trap lipid-free MPs. Rather, it has been exploited either to solubilize lipids or to directly extract from lipid vesicles or natural membrane MPs associated to lipids (Chap. 5). The resulting particles are referred to as SMALPs, styrene-maleic acid/lipid particles, or Lipodisqs®, a trade name. The major practical difference between SMA and classical APols like A8-35 is indeed that SMA is able to directly extract MPs, without recourse to detergents, whereas classical APols, with some possible exceptions, do not (see however Chap. 5, § 5.2.2.2). The mechanism of formation of SMALPs is discussed in Scheidelaar et al. (2015), Vargas et al. (2015), Zhang et al. (2015), Dörr et al. (2016), and Dominguez Pardo et al. (2017). Whether the complex SMA forms with MPs and lipids resemble nanodiscs (NDs) or bicelles more than MP/lipid/APol complexes (Dörr et al. 2016) is a debatable issue, which will be dealt with in Chap. 5, § 5.3.1.2.

DIBMA (= Sokalan CP9), an alternating copolymer of diisobutylene and maleic acid ($\langle M_w \rangle \approx$ 15.3 kDa, $\langle M_n \rangle \approx 8.4$ kDa, $Đ \approx 1.82$), has been recently advocated as a superior alternative to SMA (Oluwole et al. 2017). According to the data presented, DIBMA shows equal performance to SMA in solubilizing phospholipids, stabilizes outer membrane phospholipase A (OmpLA) under a functional form, and extracts proteins of various sizes directly from *E. coli* membranes (~70% of those extracted by DDM). Unlike SMA, DIBMA has only a mild effect on lipid acyl-chain order, does not interfere with optical spectroscopy in the far-UV range – it does not carry phenyl rings – and does not precipitate in the presence of millimolar concentrations of divalent cations. No report is provided about the effect of low pH. However, given the chemical structure of DIBMA, it can be expected to aggregate in acidic buffers.

### 4.2.2.5  Hydrophobically Grafted Poly-γ-Glutamic Acid (APG)

Poly-γ-glutamic acid (average molar mass ~10 kDa) was grafted to the level of ~41% of the carboxylates with octylamine and to ~24% with glucosamine, ~35% of the carboxylates being left free (Han et al. 2014). The resulting polymer (called APG, for amphipathic poly-γ-glutamic acid) assembles into rather broadly distributed ~80-kDa particles, comprising ~4–5 molecules and ~150 octyl chains. BR denatured in SDS was refolded in APG, and a GPCR, the type A endothelin receptor (ET$_A$), fused to a bacteriophage protein, was transferred to it from the detergent sarkosyl under a functional form, as judged by its binding of endothelin 1, forming rather polydisperse, ~300-kDa complexes. Upon mixing the complexes with liposomes, a fraction of the receptor associated with them (Han et al. 2014). A variant of APG has been recently described in which octylamine (~25%), glucosamine (~19%), and diethyl aminopropylamine (~25%) are coupled onto the carboxylic groups of poly-γ-glutamic acid, leaving ~31% of them free (Han et al. 2017). With only two papers published to date, it is somewhat early to weigh the pros and cons of this new system, but the first results appear promising, at least for APG (see Table 4.4).

### 4.2.3   Non-ionic Amphipols

There are several kinds of problems associated with the use of ionic polymers for handling MPs:

(i) Weak acids or bases, e.g. carboxylate groups, protonate or deprotonate depending on the pH of the solution, which changes the global charge of the polymer and can either severely diminish its solubility in water or turn a zwitterionic, globally neutral polymer into one with a net charge. To take two examples, the global charge of A8-35 or SMA diminishes as the pH drops below ~7 and some carboxylates protonate, making them more hydrophobic, whereas PC-APols, which are globally neutral at pH 7, become cationic at acidic pH.

(ii) A net charge will affect the behavior of the MP/polymer complex in such experiments as ion exchange chromatography or isoelectrofocusing.

(iii) Polymers with a higher charge density seem to stabilize MPs less efficiently (see Chap. 5), which is reminiscent of the case of charged detergents discussed in Chap. 2. This can easily be rationalized if one considers that disaggregation of a MP oligomer or expansion of a denaturing MP monomer will be favored by electrostatic repulsion between the layers of protein-bound surfactant.

(iv) No crystal structure of a MP has ever been solved using a globally charged detergent and only very few using zwitterionic, globally neutral detergents (see e.g. Moraes et al. 2014), making it dubious whether charged APols can be efficiently used for such a task (see Chap. 11).

These considerations have provided a strong impetus to the search for non-ionic APols, which was undertaken and carried out in collaboration with the organic chemist Bernard Pucci and his colleagues. This, however, proved to be quite a difficult and lengthy endeavor, because it is not straightforward to design, synthesize, and purify in a way that is reasonably economical and easy to scale up a polymer that features (i) a high density of hydrophobic moieties and (ii) enough non-ionic hydrophilic groups to endow it with the high solubility that is essential to its use in biochemistry and biophysics. Two types of non-ionic polar moieties are commonly found in the detergents used for biochemistry, the ether bonds of polyoxyethylene, as in Triton X-100 or $C_{12}E_8$, and the hydroxyl groups carried by sugar moieties, as in octylglucoside or dodecylmaltoside (see Chap. 2, Fig. 2.1). At this point, hydroxyl groups only have been tested with APols (Fig. 4.5).

#### 4.2.3.1   THAM-Based Non-ionic Amphipols

Most of the non-ionic APols validated to date are derived from *tris*(hydroxymethyl)acrylami-domethane (THAM; Fig. 4.6), an approach pioneered by Bernard Pucci and his collaborators. The three hydroxyl groups carried by THAM can be either left free, providing a moderately hydrophilic monomer, or one or more of them can be grafted with sugar moieties, increasing the hydrophilicity, and/or with an alkyl chain, creating an amphiphilic monomer. The synthesis of the first non-ionic APols ever to be tested (Fig. 4.5A) relied on the co-telomerization of (i) an amphipathic monomer carrying one *n*-heptyl or *n*-undecyl alkyl chain and (ii) a hydrophilic monomer carrying either three free hydroxyls or two free hydroxyls plus one grafted with D-galactose (Fig. 4.7). A range of telomers were synthesized, whose length varied from ~10 to ~130 monomers ($\langle M_n \rangle$ ranging from ~3 to ~28 kDa), using various combinations of the two types of amphipathic monomers (grafted with either a $C_7$ or a $C_{11}$ alkyl chain) and hydrophilic ones (ungrafted or carrying one D-galactosyl moiety). Their solubility was usually ~30–50 g·L$^{-1}$, which is significantly lower than that of A8-35 (>320 g·L$^{-1}$), but sufficient for carrying out test biochemical experiments. Those were run on two model MPs, BR and

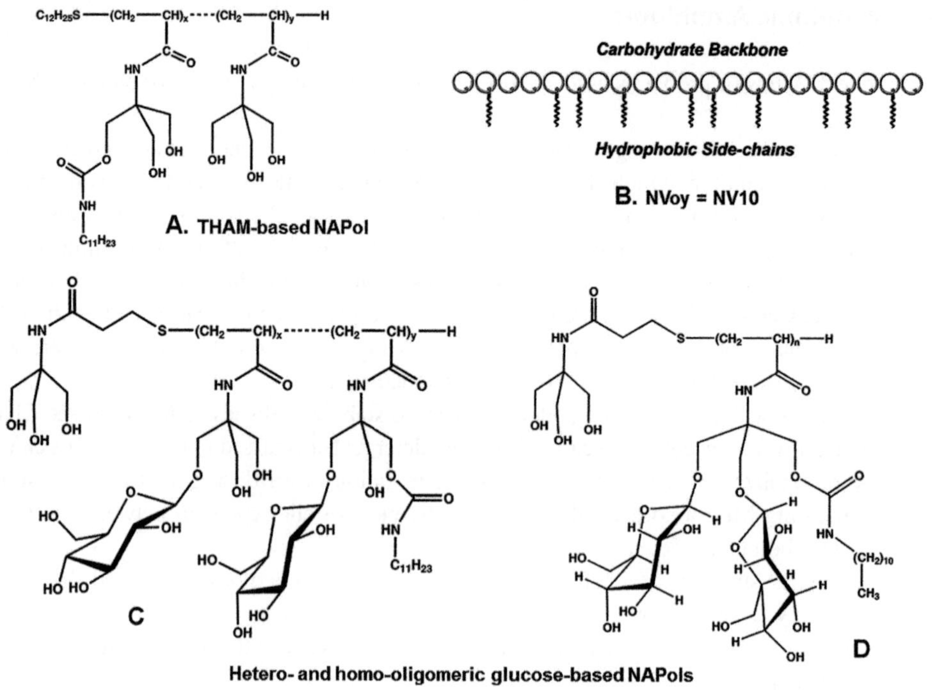

**Fig. 4.5** Chemical structure of some non-ionic amphipols. (**A**) *THAM-based non-ionic APols:* Non-ionic amphipols obtained by co-telomerization of hydrophilic and alkyl-grafted *tris*(hydroxymethyl)acrylamidomethane (THAM). Their solubility is ensured by the many hydroxyl groups they carry. In a variant structure, the hydrophilic monomers carried a galactosyl moiety (From Prata et al. 2001). (**B**) *NVoy = NV10:* a commercial polymer whose exact structure has not been released, described as "a linear, uncharged molecule composed of a polyfructose (25mer) backbone complemented with derivatized hydrophobic side chains with an overall molecular weight of 5 kDa" (Klammt et al. 2011). Cartoon released by the manufacturer. (**C, D**) *Glucose-based non-ionic APols (NAPols):* non-ionic APols obtained by grafting a glucosylated THAM-derived telomer with undecyl chains. As schematized in Fig. 4.8, heteropolymeric NAPols (**C**) can be obtained either by co-telomerization of hydrophilic and amphipathic monomers (Sharma et al. 2008) or by grafting alkyl chains onto a hydrophilic telomer (Bazzacco et al. 2009). Homopolymeric NAPols (**D**) are obtained by telomerization of an amphipathic monomer (Sharma et al. 2012).

cytochrome $b_6f$, the latter a particularly fragile, detergent-sensitive complex (Breyton et al. 1997) and therefore an excellent model to evaluate the mildness or harshness of novel surfactants (see e.g. Chae et al. 2010; Hovers et al. 2011). The best results, in terms of MP solubility, dispersity as estimated by sucrose gradient velocity sedimentation analysis, and stability over extended storage (2 weeks), were obtained with $C_{11}$-grafted polymers with a molecular mass of 3–5 kDa and a ratio of hydrophilic to amphipathic monomers of 3–3.5:1 (Prata et al. 2001).

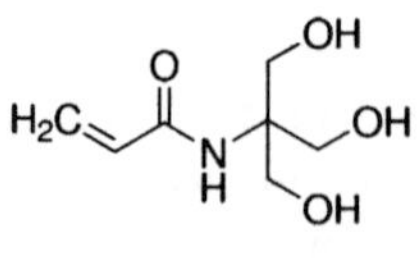

**Fig. 4.6** Chemical structure of *tris*(hydroxymethyl)acrylamidomethane (THAM).

These first studies established that non-ionic amphipathic polymers can be used to keep MPs soluble, stable, and functional. However, biochemical studies indicated that increasing the solubility of the polymers would be advantageous, which led to the development of a second family of THAM-

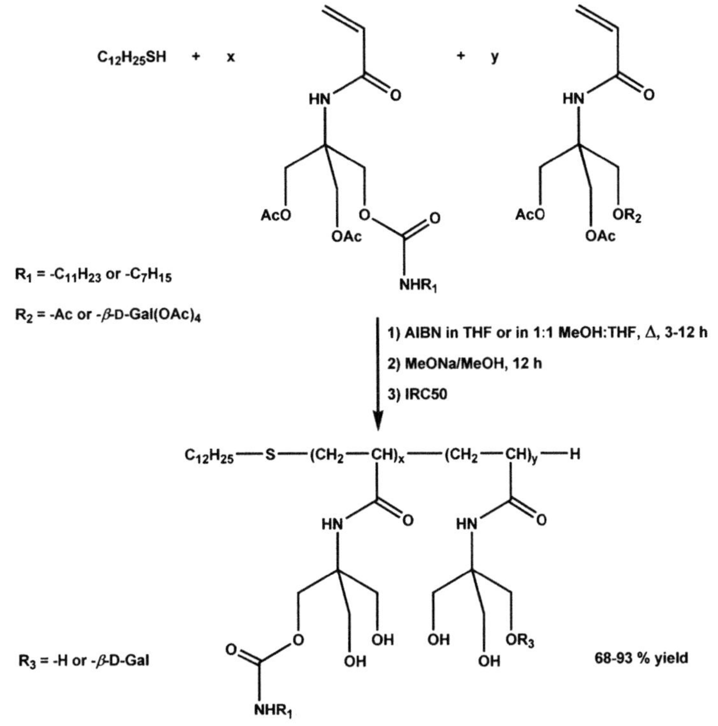

**Fig. 4.7** Synthetic scheme of the first THAM-derived amphipols. AIBN, azo-*bis*-isobutyronitrile; IRC 50, acidic resin; MeOH, methanol; MeONa, sodium methoxide; THF, tetrahydrofuran (From Prata et al. 2001, ©2001 John Wiley & Sons, Inc. All Rights Reserved).

derived APols, the solubility of which is enhanced by grafting both the hydrophilic and the amphipathic monomers with glucose moieties.

### 4.2.3.2 Glucose-Based Non-ionic Amphipols (NAPols)

Glucose-based non-ionic amphipols (hereafter simply designated as NAPols) were developed in three successive steps. The first series comprised copolymers obtained by co-telomerization of hydrophilic monomers carrying one $\beta$-D-glucose moiety and amphipathic monomers carrying one glucose moiety and one *n*-undecyl alkyl chain (Figs. 4.5C and 4.8, *left*, first series). Their molecular weight and hydrophilic/lipophilic balance were modulated by varying the molar ratio between the transfer reagent (a thiol) and the monomers and that between the hydrophilic and amphiphilic monomers, respectively, and were characterized by $^1$H NMR, UV spectroscopy, SEC in organic solvents, and Fourier-transform infrared (FTIR) spectroscopy. Their physical-chemical properties in aqueous solution were studied by dynamic light scattering (DLS), SEC, analytical ultracentrifugation (AUC), and surface tension measurements. All but one of the NAPols tested were found to be highly soluble in water ($>100$ g·L$^{-1}$). They assemble, within a large concentration range, into well-defined particles with a narrow size distribution. Varying the hydrophilic/amphiphilic monomer ratio in the range of 3.0–4.9, the degree of polymerization in the range of 51–78, and the resulting average molar mass in the range of 20–29 kDa had little incidence on their solution properties. NAPols were shown to efficiently keep soluble in aqueous solutions two test MPs, BR and the TM domain of *Escherichia coli*'s outer membrane protein A (tOmpA) (Sharma et al. 2008).

The synthesis of NAPols by co-telomerization (Sharma et al. 2008) is rather cumbersome, which prevents large-scale production and extensive biochemical studies. The molecules having been

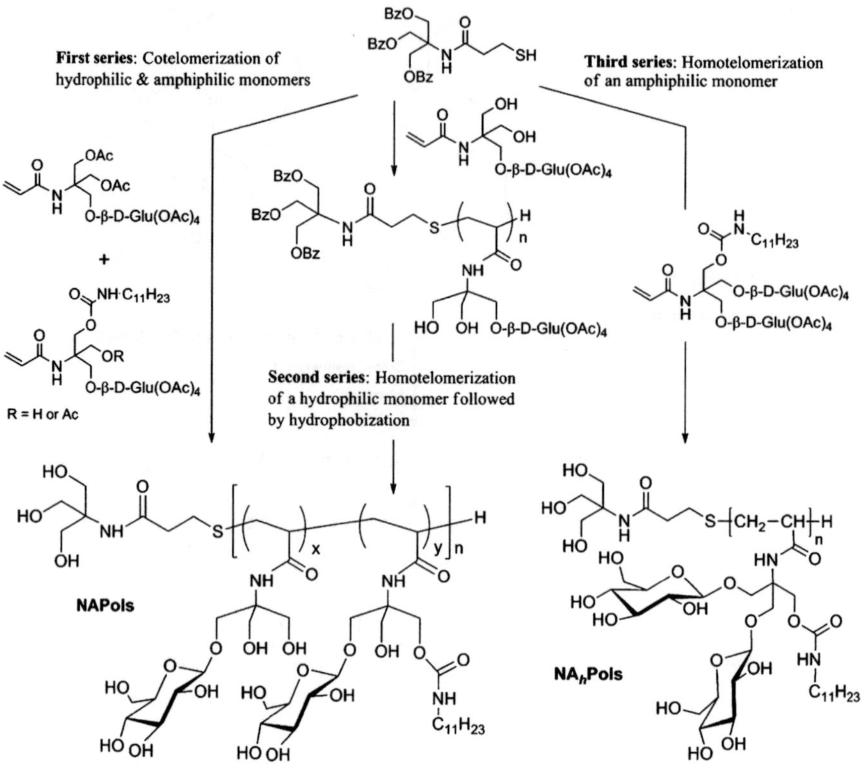

**Fig. 4.8** Three approaches to synthesizing glucose-based non-ionic amphipols. The first two routes yield copolymers comprising hydrophilic and amphipathic monomers. They are obtained either by co-telomerization (*left*, first series, Sharma et al. 2008) or by grafting alkyl chains onto a hydrophilic homopolymer (*center*, second series, Bazzacco et al. 2009). The third route, which involves free radical telomerization of a single, amphipathic monomer, yields homopolymers, each monomer comprising two β-D-glucose residues and a $C_{11}$ alkyl chain (*right*, third series, Bazzacco et al. 2012; Sharma et al. 2012) (The composite figure is reprinted with permission from Sharma et al. 2012, © 2012 American Chemical Society).

validated, a simpler synthesis route was developed, which relies on grafting alkyl chains onto a glucosylated homotelomer (Fig. 4.8, *center*, second series). The NAPols thus prepared are highly similar to those prepared by co-telomerization, but much easier to synthesize in bulk. As previously, they were tested on BR and tOmpA and found to strongly stabilize BR against denaturation (Bazzacco et al. 2009).

Finally, in a third series of experiments, the synthesis and structure were further simplified by resorting to free radical homotelomerization of a THAM-based monomer comprising a $C_{11}$ alkyl chain and two glucose moieties, using a thiol as transfer reagent (Figs. 4.5D and 4.8, *right*, third series). By controlling the thiol/monomer ratio, the number-average molecular mass of the polymers was varied from 8 to 63 kDa. Homopolymeric non-ionic APols were found to be highly soluble in water (up to 100 g·L$^{-1}$, above which solutions become viscous) and to self-organize, within a large concentration range, into small, compact particles with a narrow size distribution, regardless of the molecular mass of the polymer (see § 4.3.2). They proved able to trap and stabilize two test MPs, BR and the outer membrane protein X (OmpX) from *E. coli* (Sharma et al. 2012). Extensive biochemical and biophysical experiments have validated the use of homopolymeric NAPols for MP folding, cell-free synthesis, and solution NMR studies (Bazzacco et al. 2012; see Chaps. 6, 7 and 10, respectively).

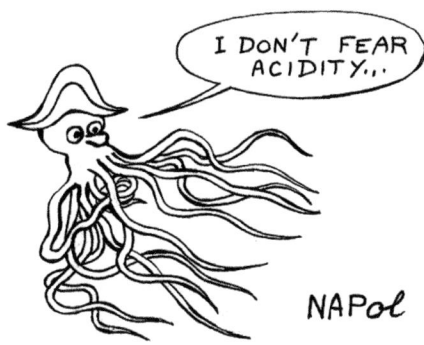

*At variance with A8-35, non-anionic amphipols (NAPols) can be used at low pH*
(© 2018 by Francis Haraux)

### 4.2.3.3   NVoy = NV10

A completely different non-ionic polymer, NVoy, also called NV10, has been used in two series of experiments as a recipient during cell-free synthesis of MPs (Guild et al. 2011; Klammt et al. 2011; see Chap. 7, § 7.3). NVoy is a commercial polymer, and neither its exact structure nor its synthesis appears to have been published. A scheme available from its manufacturer's web site is shown in Fig. 4.5B. According to Klammt et al. (2011), "NVoy is a linear, uncharged molecule composed of a polyfructose (25mer) backbone complemented with derivatized hydrophobic side chains with an overall molecular weight of 5 kDa." According to the same source, "static light scattering (SLS) coupled with refractive index (RI) and SEC [measurements] revealed that this polymer forms stable micelle-like multimers of approximately 112 kDa (22 polymer molecules) in aqueous solution." Two of the GPCRs expressed in the presence of NVoy were found to form small MP/NVoy complexes, as indicated by SEC and EM data, and to bind specific ligands (Klammt et al. 2011). There is therefore good evidence that NVoy qualifies as a non-ionic amphipol. Why it does, whereas "amphibiopols" (Duval-Terrié et al. 2003) do not (Picard et al. 2004) is perhaps related to the different nature of the glycosidic backbone: very short polyfructose, inulin-like chains (15 residues) in the case of NVoy (Klammt et al. 2011) and ~10× longer pullulan chains (~170 residues) in the case of "amphibiopols," with a different branching of the residues.

## 4.3   Self-Association Behavior of Amphipols in Aqueous Solutions

A prerequisite to the formation of small MP/polymer complexes is that the polymer does not form by itself too large particles when solubilized in water. Indeed, in our experience, all polymer preparations that self-organized into big particles were unable to form, once associated with MPs, the small objects that are desirable for most biophysical studies. A good knowledge of the size, organization, and dynamics of APol particles is important in many respects, be it to predict their behavior in separation experiments or as a model to understanding the properties of MP-associated APols.

Upon being solubilized in water, amphiphilic polymers tend to self-organize and, more often than not, to self-associate the hydrophobic effect pushing their hydrophobic moieties together, whereas hydrophilic groups tend to disperse in water. Which structures are thus formed depends on the distribution of the various groups along the polymer, two examples of which are shown in Fig. 4.9. Most APols have the behavior shown to the left of the figure. Depending on the length of the polymer

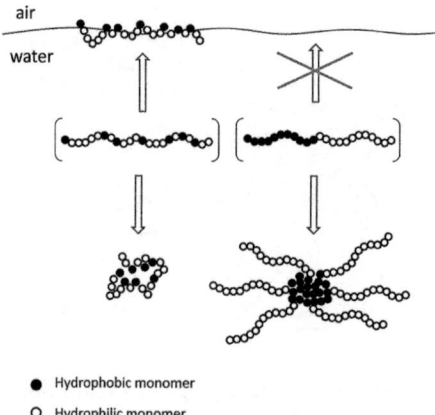

**Fig. 4.9** The different behavior in aqueous solution and at the air/water interface of amphipathic polymers with interspersed hydrophilic and hydrophobic groups, as is the case in amphipols *(left)* and of blocky copolymers *(right)* (Reprinted with permission from Raffa et al. 2015, © 2015 American Chemical Society).

and the length and density of the hydrophobic moieties, the particles formed can be unimolecular or multimolecular and have or not a tendency to associate one with another (see e.g. Noda and Morishima 1999; Yamamoto et al. 2000b; Di Cola et al. 2004; Sauvage et al. 2004). Those polymers that are currently used as APols are too small to form stable unimolecular particles, there being a deficit of hydrophobic groups to form a stable core and of hydrophilic groups to shield it from the aqueous solution. Several molecules therefore associate, until the number of hydrophobic and hydrophilic groups matches the constraints of forming a hydrophobic core and covering it with hydrophilic moieties (see § 4.3.1). Among the many APols described in § 4.2, only a handful have been studied by a large variety of physical techniques (Table 4.2). Among those, A8-35 is by far the most extensively characterized. We will therefore first review in some detail, in § 4.3.1, what is known of the formation, structure, and dynamics of A8-35 particles, taking it as the prototypical APol. In § 4.3.2, we will summarize more briefly what is known of the solution behavior of other APols.

### 4.3.1    Formation, Structure, and Dynamics of A8-35 Particles

#### 4.3.1.1    Critical Association Concentration

At extreme dilutions, entropy prevents the association of surfactant molecules, the entropic cost of bringing them together exceeding the free energy drop that results from shielding water-insoluble moieties from water. As the concentration is increased, a balance is reached when the two $\Delta G$ contributions cancel each other, and the first particles form. For detergents, the particles are called micelles, and the concentration at which micelles start forming is called the critical micellar concentration (CMC; see Chap. 2, § 2.2.2). The CMC of the detergents used in biochemistry lies typically between a few tens or hundreds of μM to a few tens of mM, corresponding roughly to 0.01–100 g·L$^{-1}$. For non-ionic detergents, the CMC depends essentially on the size of the hydrophobic moiety or moieties (ibid.). For self-associating polymers, various kinds of assemblies may form (Fig. 4.9), and the concentration at which the first of them appear is referred to as the critical association concentration (CAC). One can anticipate, intuitively, that, because each APol molecule carries many hydrophobic chains, the entropic cost of bringing enough of them together to form a hydrophobic core will be

**Table 4.2** An overview of studies describing the synthesis, structure, and properties of amphipols. Studies containing data on A8-35 are highlighted in *light blue*, those on SMALPs in *buff*, and those on glucosylated NAPols in *pink*.

| Study | Amphipol(s) | Description of synthesis | Characterization of amphipol particle solution properties by: | | | | | | | | | | Comments | References |
|---|---|---|---|---|---|---|---|---|---|---|---|---|---|---|
| | | | SANS | SG-AUC | SV-AUC | Eq-AUC | SEC | DLS | SLS | EM | NMR | Other methods | | |
| 4.1 | A8-35, A8-75, A34-35, A34-75 | ● | | | | | | | | | | | The first demonstration of the feasibility of trapping and stabilizing MPs using amphipathic polymers. The structure of A8-35 is given in Fig. 4.1a. | Tribet et al. (1996) |
| 4.2 | A8-75, [$^{14}$C]A8-75 | | | ● | | | | | | | | | A study of MP/A8-75 complexes including data on the migration of A8-75 particles in sucrose gradients. | Tribet et al. (1997) |
| 4.3 | *Tris*(hydroxymethyl) acrylamidomethane (THAM)-based non-ionic APol | ● | | | | | | | | | | | First description of trapping MPs using a non-ionic APol. The structure is given in Fig. 4.5a. | Prata et al. (2001) |
| 4.4 | OAPA-20, PMAL-B | ● | | | | | | | | | | | Description of the use of two APols to deliver a MP to lipid vesicles or to stabilize it. OAPA-20 is essentially identical to A8-75 (it would be named A8-80 using the nomenclature of Tribet et al. 1996). The structure of PMAL-B is given in Fig. 4.1b. | Nagy et al. (2001), Gorzelle et al. (2002) |
| 4.5 | A8-35, unlabeled and partially deuterated (DAPol) | | ● | | ● | ● | | | | | | Densimetry | Determination of the charge, specific volume and contrast-matching point of unlabeled and partially deuterated A8-35. | Gohon et al. (2004) |
| 4.6 | Unlabeled A8-35 and DAPol | ● | ● | | ● | ● | ● | ● | ● | | | | Determination of the mass, size, shape, aggregation number, $R_g$, $R_s$, and dispersity of A8-35 and DAPol particles. | Gohon et al. (2006) |
| 4.7 | A8-35, sulfonated APol (SAPol), PMAL-C12, PMAL*A*-C12 | | | | | ● | | | | | | | The effect of $Ca^{2+}$ and $Mg^{2+}$, the particles formed by various APols, is studied by SEC. | Picard et al. (2006) |
| 4.8 | PreserveX™-QML | | | | | | ● | | | | | | PreserveX™-QML, a commercial block copolymer, is used to disperse MP-containing membranes. The size of the various particles is studied by DLS. | Trubetskoy et al. (2006) |
| 4.9 | A8-35 and phosphorylcholine-based amphipols (PC-APols) | ● | | | | | | ● | ● | | | | A study of the thermodynamics of mixing APols with a neutral detergent either in the presence or absence of MPs, including light scattering data on the size of the pure and mixed particles. The properties of APols and MP/APol complexes are examined as a function of salt, pH, and the presence of divalent cations. | Diab et al. (2007a,b) |
| 4.10 | A8-35 | | | | | | | | | ● | | | In the course of a cryo-EM study of A8-35-trapped mitochondrial Complex I, a comparison was carried out of the thickness of the water films formed in the presence of either A8-35 or dodecylmaltoside (DDM). | Flötenmeyer et al. (2007) |

(continued)

**Table 4.2**  (continued)

| Study | Amphipol(s) | Description of synthesis | SANS | SG-AUC | SV-AUC | Eq-AUC | SEC | DLS | SLS | EM | NMR | Other methods | Comments | References |
|---|---|---|---|---|---|---|---|---|---|---|---|---|---|---|
| | | | | | | | | | | | | Characterization of amphipol particle solution properties by: | | |
| 4.11 | A8-35 and fluorescently labeled A8-35 (FAPol_NBD) | • | | | | | • | | | | | FRET | The formation, size, and stability of A8-35 particles and MP/A8-35 complexes were studied in the presence or absence of detergent or excess APols, providing insights into the rate of exchange of surfactants between particles. | Zoonens et al. (2007) |
| 4.12 | A8-35, DAPol, and [³H]A8-35 | • | | • | • | | | | | | | | In the course of a study of bacteriorhodopsin (BR)/A8-35 complexes, further data were collected on the behavior of A8-35 particles. The techniques listed are those relevant to these particles. | Gohon et al. (2008) |
| 4.13 | Heteropolymeric glucose-based non-ionic APols (NAPols) | • | | | • | | • | • | | | | Tensiometry | First-generation glucose-based NAPols were obtained by co-telomerization of hydrophilic and amphipathic monomers. | Sharma et al. (2008) |
| 4.14 | Glucosylated NA Pols similar to those in Study 4.13 | • | | | | | • | | | | | | Glucosylated NAPols similar to those in Study 4.13 were obtained by grafting hydrophobic chains onto a hydrophilic homopolymer. | Bazzacco et al. (2009) |
| 4.15 | Styrene-maleic acid copolymer (SMA) | • | | | | | | • | | • | • | EPR | The size of SMA/DMPC complexes (SMALPs) was studied by DLS and EM after negative staining and their organization by solution NMR and EPR. The structure of SMA is given in Fig. 4.1e and a scheme of the SMALP particles in Fig. 4.27a. | Knowles et al. (2009), Orwick et al. (2012) |
| 4.16 | Sulfonated APols (SAPols) | • | | | | | • | | | | | | Synthesis and properties of APols analogous to A8-35, in which the isopropyl groups are replaced by taurine. | Dahmane et al. (2011) |
| 4.17 | NVoy = NV10 | | | | | | • | | • | | | Coupled SEC, SLS, and RI measurements | NVoy, a commercial amphiphilic derivative of polyfructose (see scheme in Fig. 4.5b), was used to keep soluble MPs produced by cell-free synthesis (see Chap. 7). The particles were studied by SEC and SLS. | Guild et al. (2011), Klammt et al. (2011) |
| 4.18 | A8-35 | | | | | | | | | | | MD | The structure and dynamics of A8-35 particles were studied by all-atom and coarse-grained molecular dynamics simulations. | Perlmutter et al. (2011) |
| 4.19 | Homopolymeric glucose-based NAPols | • | • | | • | | • | • | | | | Densimetry, tensiometry | Homopolymeric glucose-based NAPols were obtained by telomerization of an amphipathic precursor. Their structure is shown in Fig. 4.5d. | Bazzacco et al. (2012), Sharma et al. (2012) |
| 4.20 | A8-35 and fluorescent versions thereof (FAPol_NBD and FAPol_rhod) | • | | | | | | | | | | Tensiometry, FRET | The critical aggregation concentration (CAC) of A8-35 was determined both by tensiometry and by FRET using two complementary fluorescent derivatives of A8-35. | Giusti et al. (2012) |

(continued)

**Table 4.2**  (continued)

| Study | Amphipol(s) | Description of synthesis | Characterization of amphipol particle solution properties by: | | | | | | | | | | Comments | References |
|---|---|---|---|---|---|---|---|---|---|---|---|---|---|---|
| | | | SANS | SG-AUC | SV-AUC | Eq-AUC | SEC | DLS | SLS | EM | NMR | Other methods | | |
| 4.21 | A8-35, A8-75, A34-35, A34-75, SAPols, NAPols | | | | | | | | | | | MS | The size and dispersity of APol monomers were studied by electrospray ionization-ion mobility-mass spectrometry (ESI-IM-MS). | Leney et al. (2012), Watkinson et al. (2015) |
| 4.22 | Poly-γ-glutamic acid grafted with glucosamine and octylamine (APG) | ● | | | | | ● | | | | | AFM, pyrene fluorescence | An amphipathic polypeptide was obtained by grafting poly-γ-glutamic acid with glucosamine and octylamine. The size of the particles was assessed by AFM and the CAC determined from pyrene partitioning. | Han et al. (2014) |
| 4.23 | A8-35, DAPol | | ● | | | | | | | | | SAXS | In the course of the study of a MP/A8-35 complex, SANS and SAXS data were collected on pure A8-35 and DAPol particles. | Sverzhinsky et al. (2014) |
| 4.24 | A8-35, DAPol | | | | | | | | | | | EINS, QENS | EINS, QENS, and MD were used to study the dynamics of A8-35 particles. Comparison between A8-35 and DAPol made it possible to separate experimentally the contributions of the backbone and side chains. | Tehei et al. (2014) |
| 4.25 | SMALPs | | ● | | | | | ● | | ● | ● | ATR-FTIR, DSC, ITC, FRET | An extensive series of multidisciplinary studies and reviews of the formation, size, shape, organization, and dynamics of DMPC/SMA particles. | Jamshad et al. (2015b), Scheidelaar et al. (2015), Tanaka et al. (2015), Vargas et al. (2015), Zhang et al. (2015), Dörr et al. (2016), Cuevas Arenas et al. (2017) |
| 4.26 | APG | ● | | | | | | | | ● | | AFM | APG is observed to assemble into rather polydisperse particles with an average mass of ~80 kDa, comprising on average 4–5 molecules and ~150 octyl chains. | Han et al. (2014) |

*AFM* atomic force microscopy, *APol* amphipol, *ATR-FTIR* attenuated total reflection Fourier-transform IR spectroscopy, *AUC* analytical ultracentrifugation, *DAPol* A8-35 with an unlabeled main chain and the octyl and isopropyl side chains perdeuterated, *DLS* dynamic light scattering, *DSC* differential scanning calorimetry, *EINS* elastic incoherent neutron scattering, *EM* electron microscopy, *EPR* electron paramagnetic resonance, *Eq-AUC* equilibrium AUC, *ESI-IM-MS* electrospray ionization-ion mobility-mass spectrometry, *FAPol$_{NBD}$ and FAPol$_{rhod}$* A8-35 fluorescently labeled with NBD (7-nitro-1,2,3-benzoxadiazole) or with rhodamine, respectively, *FRET* Förster resonance energy transfer, *INS* inelastic neutron scattering, *ITC* isothermal titration calorimetry, *MD* molecular dynamics, *MS* mass spectrometry, *NMR* nuclear magnetic resonance, *QENS* quasi-elastic neutron scattering, *RI* refractive index, *SANS and SAXS* small-angle neutron and X-ray scattering, respectively, *SAPol* a sulfonated amphipol, *SEC* size exclusion chromatography, *SG-AUC* sedimentation velocity AUC in sucrose gradients, *SLS* static light scattering, *SV-AUC* sedimentation velocity AUC. The table lists only studies relating to the properties of underivatized amphipols. Studies describing the synthesis, structure, and properties of functionalized amphipols are listed in Table 4.5. The chemical structures of selected ionic and non-ionic amphipols are shown in Figs. 4.1 and 4.5, respectively considerably reduced as compared to a detergent carrying a single copy of the same hydrophobic chain, leading to low CAC values. Inversely, for APols that carry net charges, like A8-35, bringing these charges together is electrostatically costly, which will raise the CAC, in the same way that sodium dodecyl sulfate (SDS) has a higher CMC than dodecylmaltoside despite carrying the same alkyl chain (Chap. 2).

Low CMCs or CACs can be rather tricky to measure. In the case of A8-35, two very distinct approaches were resorted to, which yielded reasonably consistent results (Giusti et al. 2012). The buffer used was Tris/HCl 20 mM, NaCl 100 mM, pH 8.0. At this pH, most of the carboxylates of A8-35 are known to be deprotonated (Gohon et al. 2004).

The first approach, illustrated in Fig. 4.10A, relies on measuring Förster resonance energy transfer (FRET) between two fluorescent versions of A8-35, one labeled with 7-nitro-1,2,3-benzoxadiazole (FAPol$_{NBD}$), the other with rhodamine (FAPol$_{rhod}$) (see Table 4.5). The Förster distance, below which energy transfer between the two fluorophores is highly efficient, lies between the radius and the diameter of A8-35 particles (Giusti et al. 2012). Below the CAC, FAPol$_{NBD}$ and FAPol$_{rhod}$ molecules diffuse separately, and they are too far away from one another for the energy captured by NBD to be transferred to rhodamine. Above the CAC, mixed particles form, and a FRET signal appears. Its intensity, $I_{FRET}$, increases linearly with $[C_{pol} - CAC]$, where $C_{pol}$ is the total concentration of polymer in the solution (Fig. 4.10B). Assuming CAC = 0.0016 g·L$^{-1}$, which is within or close to the error range in Fig. 4.10B, the ratio $I_{FRET}/[C_{pol} - CAC]$ is seen to remain constant

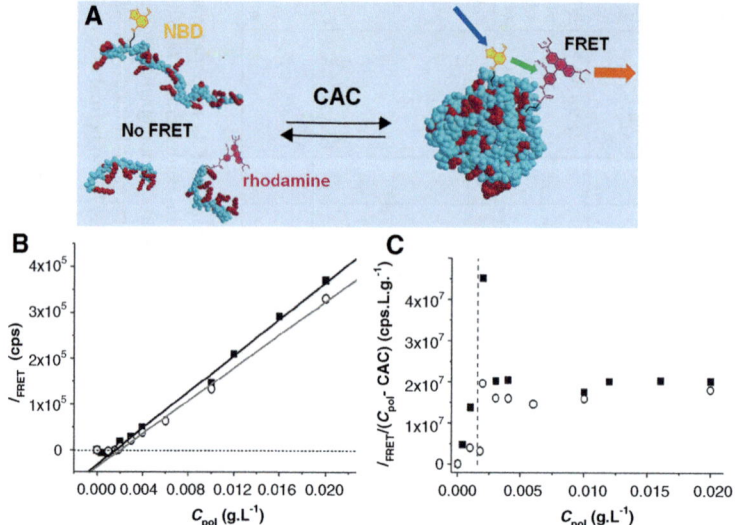

**Fig. 4.10** Determining the critical aggregation concentration (CAC) of A8-35 by Förster resonance energy transfer (FRET). (**A**) Principle of the experiment. A 1:1 mixture of a fluorescent version of A8-35 labeled with NBD (FAPol$_{NBD}$) and another labeled with rhodamine (FAPol$_{rhod}$) is excited at 476 nm ($\lambda_{exc}$ for NBD) and the intensity $I_{FRET}$ of fluorescence emission measured at 575 nm ($\lambda_{em}$ for rhodamine) (cf. Chap. 8, Fig. 8.9). The FRET signal is detected only above the CAC, when FAPol$_{NBD}$ and FAPol$_{rhod}$ molecules are close enough for energy transfer to occur. (**B**) Two independent experiments performed under the same conditions using samples prepared independently are noted by the symbols ■ and ○. $I_{FRET}$ is plotted as a function of $C_{pol}$, the total concentration of polymer. The straight lines extrapolate to $I_{FRET} = 0$ for $C_{pol} = 0.0018 \pm 0.0004$ and $0.002 \pm 0.0002$ g·L$^{-1}$, respectively. (**C**) $I_{FRET}/$ $[C_{pol} - CAC]$ plotted against $C_{pol}$, using a CAC value of 0.0016 g·L$^{-1}$ (within or close to the experimental error of the CAC values obtained by extrapolation of the data in **B**,) is roughly constant over a 10× range of concentration, consistent with the size of the particles remaining constant over that range (Reprinted with permission from Giusti et al. 2012, © 2012 American Chemical Society).

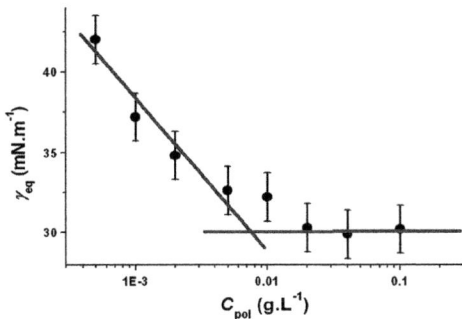

**Fig. 4.11** Surface tension of A8-35 solutions in Tris/HCl 20 mM, NaCl 100 mM, pH 8.0 buffer, measured by spinning drop tensiometry. Variation of the equilibrium surface tension $\gamma_{eq}$ (extrapolated to infinite equilibration time) with APol concentration. Lines are guides for the eye (Reprinted with permission from Giusti et al. 2012, © 2012 American Chemical Society).

over a tenfold range of concentrations (Fig. 4.10C). This is consistent with particles forming at the CAC and retaining the same size over the whole range of concentrations investigated.

The second approach to estimating the CAC of A8-35 is, classically, to determine the evolution of the surface tension of A8-35 solutions as a function of concentration. Above the CAC, the concentration of APol monomers increases much more slowly than under it, most of the molecules added to the solution associating into particles (for a discussion of the underlying thermodynamics, see Tanford 1980). As a result, the surface tension reaches a slowly descending plateau. The equilibrium surface tension of A8-35 solutions was determined by extrapolating to long times the variation of surface tension measured in a spinning drop tensiometer. One advantage of this method is the absence of contact between the air bubble and solid surfaces, which limits perturbations by contaminants, especially at the high dilutions required for APol solutions. At concentrations above 0.01 g·L$^{-1}$, the surface tension at equilibrium, $\gamma_{eq}$, reaches a plateau of $31 \pm 1$ mN·m$^{-1}$. At concentrations below $\sim$0.002 g·L$^{-1}$, $\gamma_{eq}$ decreases monotonously with increasing APol concentration (Fig. 4.11). In the case of monomeric surfactants, the crossover between these two regimes conventionally identifies, in a semilogarithmic plot, the threshold concentration at which self-assembly occurs. Given the experimental uncertainties involved in extrapolating $\gamma$ to an "infinite" equilibration time, as well as to the possible contribution of contaminants, a threshold concentration window of 0.004–0.008 g·L$^{-1}$ represents a reasonable estimate (Fig. 4.11). According to surface tension measurements, self-assembly therefore occurs slightly above the concentration of $\sim$0.002 g·L$^{-1}$ determined by FRET.

The low value of the CAC has important practical consequences. For instance, (i) in a solution of A8-35 at 1 g·L$^{-1}$, only 0.2–0.5% of the preparation is monomeric and able to cross a dialysis membrane; (ii) when running an A8-35 preparation over a SEC column, the peak of particles will be followed by a trail of monomers at the CAC; (iii) the rate of collision with MP/APol complexes in a solution containing 1 g·L$^{-1}$ free A8-35 will be $\sim$250$\times$ higher with particles than with individual molecules. Because each particle contains an average of $\sim$9 monomers (see next section), the probability of exchange between protein-bound and free A8-35 will be >2000$\times$ higher with free particles than with free monomers (see § 4.3.1.2.4, *Particle Interactions*). This has important implications regarding the way two MPs, or a MP and a highly hydrophobic ligand, may find each other and interact in APol solutions (see Chap. 5, § 5.4).

Except for APG, whose CAC was estimated to be $\sim$0.06 g·L$^{-1}$ (Han et al. 2014), the CAC of other APols has not yet been determined. For a similar number of alkyl chains per monomer, it can be expected to be lower than that for A8-35 if (i) the alkyl chains are longer and/or (ii) the net charge of the monomers is lower or the ionic strength higher. It will also drop, for a given buffer, charge density, and

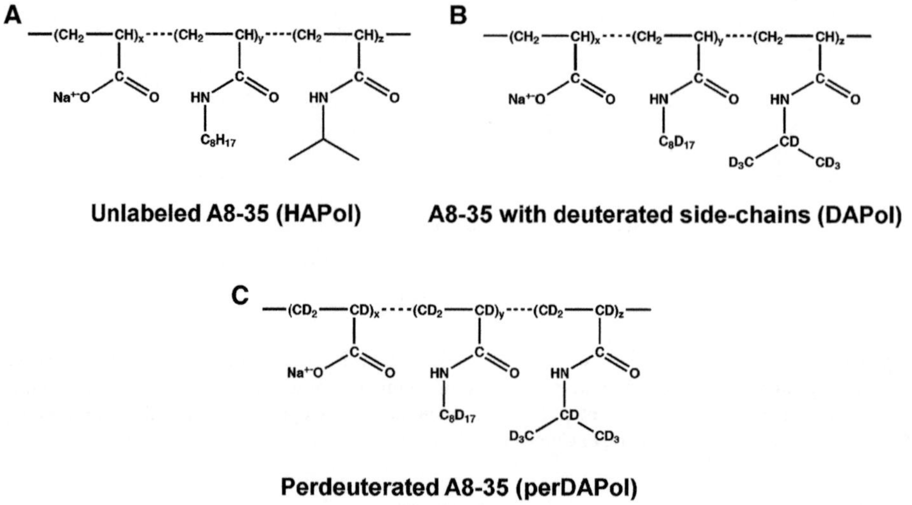

**Fig. 4.12** Chemical structures of (**A**) unlabeled A8-35 (HAPol), (**B**) A8-35 deuterated on the side chains only (DAPol; Gohon et al. 2004, 2006), and (**C**) perdeuterated A8.35 (perDAPol; Giusti et al. 2014).

type and density of alkyl chains, if the main chain is longer – e.g. A34-35 vs. A8-35 – because the entropic cost of bringing together enough alkyl chains to constitute the particle's core will be lower.

### 4.3.1.2 The Size, Shape, and Organization of A8-35 Particles

Table 4.2 provides an overview of those studies providing information about the structure of the various APols and the techniques that have been implemented to characterize the size, shape, organization, and dynamics of the particles they form in aqueous solutions. Studies that present data bearing on A8-35 particles are highlighted in light blue. They have resorted to the following forms of A8-35: (i) plain, unlabeled A8-35 (often noted HAPol in comparative studies with the next two forms), (ii) two deuterated forms noted DAPol (main chain hydrogenated, octyl and isopropyl side chains perdeuterated (Gohon et al. 2004, 2006) and perDAPol (perdeuterated A8-35; Giusti et al. 2014) (Fig. 4.12), (iii) a tritiated form ([³H]A8-35; Gohon et al. 2008), and (iv) several fluorescently labeled derivatives (FAPols; see § 4.4, Table 4.5).

In the following, we will summarize the main features of A8-35 particles, the reader being referred to the original articles for more details and discussion.

#### 4.3.1.2.1 Mass and Dispersity of Individual A8-35 Molecules

Determining the average mass and dispersity of individual A8-35 molecules is a relatively complex endeavor. It would be straightforward to derive an average mass if that of the starting PAA was known with certainty from the manufacturer, but such is unfortunately not the case. According to the most recent determination (Giusti et al. 2014), obtained by SEC analysis of permethylated PAA in a polar organic solvent (tetrahydrofuran), the number-average molecular mass $\langle M_n \rangle$ of the sodium salt of A8-35 is expected to be ~4.3 kDa, an estimate that is confirmed qualitatively by MS analysis of the final products (see below). For a dispersity $Đ \approx 2$, the weight-average mass $\langle M_w \rangle$ is therefore ~8.6 kDa, which, as noted above, corresponds to ~70 monomers, carrying, on average, ~18 octyl chains.[1]

---

[1] The early literature on this matter is somewhat confusing, because (*i*) older $\langle M_n \rangle$ determinations have given different results depending on the method used and (*ii*) some former calculations refer to $\langle M_n \rangle$, others to $\langle M_w \rangle$, which is ca. twice larger. Note that such estimates are useful in order to form an intuitive view of what A8-35 molecules look like, but reasoning in terms of APol mass or the number of octyl chains or labels per particle or MP/APol complex is safer.

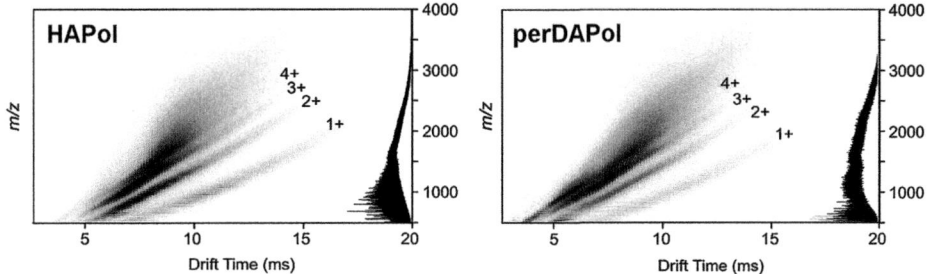

**Fig. 4.13** Electrospray ionization-ion mobility-mass spectrometry (ESI-IM-MS) spectra of unlabeled A8-35 (HAPol) and perdeuterated A8-35 (perDAPol) (1 g·L$^{-1}$ in 100 mM ammonium hydrogen carbonate, pH 8.0) (From Giusti et al. 2014).

The advent of electrospray ionization-ion mobility-mass spectrometry (ESI-IM-MS) has significantly improved the capability of MS to study complex polymeric mixtures (Weidner and Trimpin 2010; see Chap. 14, § 14.3.1). IM analysis separates ions based on their drift time through a neutral buffer gas under the influence of a weak electric field, the drift time depending on the ions' shape (collision cross section) and charge state (Kanu et al. 2008). Coupled with MS, this provides an extra dimension of separation that is particularly useful in the study of amphipols (see Leney et al. 2012; Giusti et al. 2014). Three-dimensional ESI-IM-MS spectra of unlabeled A8-35 (HAPol) and perdeuterated A8-35 (perDAPol) are shown in Fig. 4.13. Ions with +1, +2, +3, and +4 charge states that overlap in MS (spectrum at the right side of the panels) can be separated by IM, observed, and identified in the spectra, as indicated in the figure. The ESI-IM-MS data indicate that the two forms of A8-35 tested, HAPol, which was synthesized using commercial PAA, and perDAPol, which was prepared using a home-made perdeuterated PAA, have broad and very similar mass distributions. Note that, because species with higher mass-to-charge ratios (*m/z*) are less readily desolvated and detected, the distributions observed may be skewed toward low masses (Hernandez and Robinson 2007). The similar mass distributions observed in the two spectra indicate that A8-35 and perDAPol are highly comparable in terms of their dispersity and composition.

The weight-average $\langle M_w \rangle$ and number-average $\langle M_n \rangle$ molecular masses of each APol were calculated from the MS data after converting from *m/z* to mass. For A8-35, $\langle M_w \rangle = 4.8$ kDa and $\langle M_n \rangle = 3.8$ kDa, corresponding to a dispersity $Đ = \langle M_n \rangle / \langle M_w \rangle \approx 1.26$. For perDAPol, $\langle M_w \rangle = 5.2$ kDa, $\langle M_n \rangle = 4.2$ kDa, and $Đ \approx 1.24$. The most abundant ions in the spectra correspond to species with molecular masses of 3.4 kDa (A8-35) and 3.6 kDa (perDAPol). These values compare reasonably well with the mass estimates of A8-35 deduced from SEC analysis of the permethylated PAA, even though they may be skewed toward low masses, as noted above, due to high-mass polymer species potentially being ionized and detected less efficiently and to the inability to assign conclusively all the low-intensity high *m/z* species. The difference between the two values of $Đ$, 1.24 and 1.26, is much smaller than the experimental error, indicating a comparable molecular mass dispersity. These values are, however, much lower than that expected, $Đ \approx 2$, possibly due to the skewing effect noted above. The similarity of A8-35 and perDAPol in terms of chain length is confirmed by their mass ratios: 1.088 for $\langle M_w \rangle$, 1.101 for $\langle M_n \rangle$, and 1.059 for the most abundant ions, to be compared with the value of 1.081 expected from the extent of deuteration if the degrees of polymerization are identical.

As already noted, because A8-35 preparations are polydisperse and their average mass is difficult to determine with accuracy, and because both parameters will vary depending on the source of the PAA used in the synthesis, it is highly recommended, for the sake of accuracy and reproducibility, that APol concentrations and MP/APol ratios be expressed in g·L$^{-1}$ and in mass ratios, respectively, rather than in molarity and molar ratios.

### 4.3.1.2.2  Homogeneity, Mass, and Size of A8-35 Particles

Whereas individual A8-35 molecules are heterogeneous, they assemble into fairly homogeneous particles. This was first established by SEC analysis of aqueous solutions and further analyzed by DLS, small-angle neutron and X-ray scattering (SANS and SAXS), and AUC (Table 4.2 and Figs. 4.14 and 4.15). We will first discuss the main features that come out from these analyses and then examine some complications.

According to all of the five techniques mentioned above, A8-35 assembles into particles that are largely homogeneous. In SEC (Fig. 4.14A), the main peaks of particles formed by either unlabeled

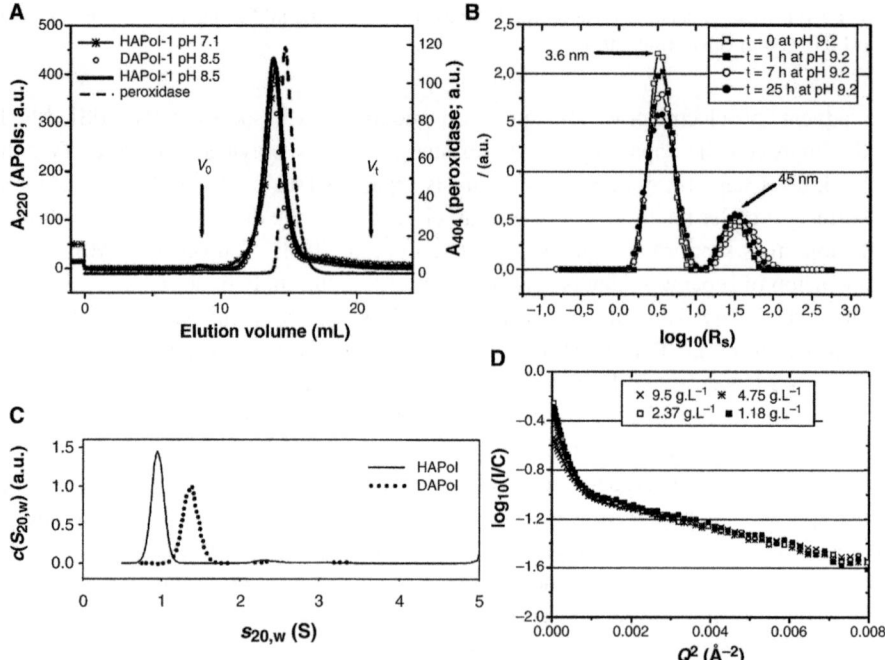

**Fig. 4.14** Solution behavior of A8-35 particles. (**A**) Size exclusion chromatography analysis of two batches of A8-35, one unlabeled (*HAPol-1*), the other with its side chains perdeuterated (*DAPol-1*). Arrows indicate the void ($V_0$) and total ($V_t$) volumes of the Superose 12 HR-10/30 column (determined using blue dextran and acetone, respectively). HAPol-1 and DAPol-1 were dissolved at 10 g·L$^{-1}$ in either 0.1 M NaCl, 0.02 M sodium phosphate, pH 7.1, or 0.1 M NaCl, 0.02 M Tris/HCl, pH 8.5. Elution profiles were scaled to the same maximum. The elution pattern of horseradish peroxidase ($M = 40$ kDa, $R_S = 3.0$ nm) is shown for comparison. The Stokes radius of both HAPol and DAPol particles, determined according to a calibration curve based on standard globular proteins, is 3.15 nm (Table 4.3). (**B**) Dynamic light scattering analysis of the size distribution of HAPol-1 particles and evolution of the distribution upon incubation for 0–25 h in 100 mM NaCl, 20 mM boric acid/NaOH, pH 9.2. buffer. All HAPol-1 concentrations were 10 g·L$^{-1}$. (**C**) Sedimentation velocity analytical ultracentrifugation analyses of HAPol-1 and DAPol-1. The two samples were analyzed at 5 °C and 60,000 rpm in 100 mM NaCl, 20 mM NaH$_2$PO$_4$/Na$_2$HPO$_4$, pH 7.1. Absorbance at 230 nm was recorded every 28 min for a total of 4.5 h and the profiles fitted using the program Sedfit (Alami et al. 2007), assuming a continuous distribution of particles. $c(s_{20,w})$ is the resulting $s_{20,w}$ distribution for HAPol-1 (*solid line*) and DAPol-1 (*dotted line*). (**D**) Small-angle neutron scattering by DAPol-1 solubilized at various concentrations in 100 mM NaCl, 20 mM NaH$_2$PO$_4$/Na$_2$HPO$_4$, pH 7.1. Intensities are normalized by the concentration and their logarithm plotted as a function of $Q^2$, the square of the momentum transfer $Q = (4\pi/\lambda) \sin(\theta/2)$, where $\theta$ is the scattering angle and $\lambda = 10$ Å is the neutron wavelength (Guinier plot; see Chap. 9, Box 9.2). The superimposition of the curves shows that there is no change in the size nor distribution of the particles between ~1 and ~10 g·L$^{-1}$ (All panels are reprinted with permission from Gohon et al. 2006, © 2006 American Chemical Society).

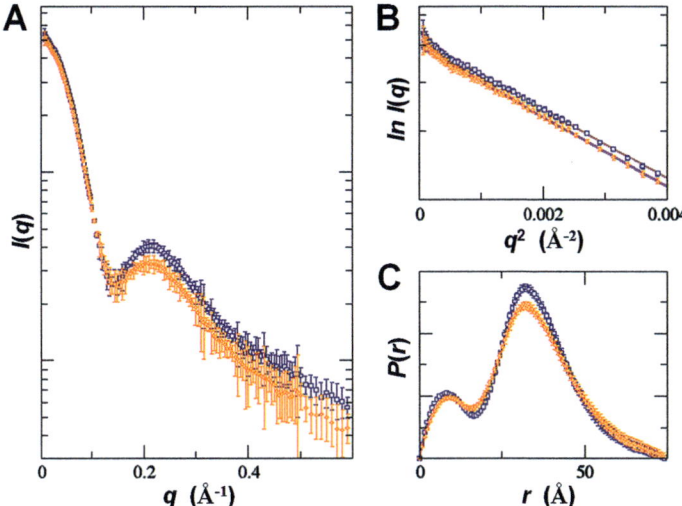

**Fig. 4.15** Comparison of small-angle X-ray scattering by hydrogenated and partially deuterated A8-35. (**A**) SAXS scattering data of HAPol (*purple squares*) and DAPol (*gold triangles*) from 9.1-g·L$^{-1}$solutions in 25 mM Tris/HCl buffer, 150 mM NaCl, pH 7.5. $I(q)$ is plotted vs. $q$, where $q = 4\pi \sin(\theta)/\lambda$, with $\lambda$ the wavelength of the X-rays (6 Å) and $2\theta$ the scattering angle from the incident beam. (**B**) Guinier plots of the data shown in **a**. The $R_g$ value deduced from the slope of the fits is ~2.4 nm, consistent with that determined by SANS (Table 4.3). (**C**) The $P(r)$ functions, derived from the data in **A**, describe the distribution of distances between scattering centers in the particle. Their maximal extension, ~7 nm, is consistent with the $R_S$ of the particles being distributed around an average value of 3.15 nm (Table 4.3) (From Sverzhinsky et al. 2014).

A8-35 (HAPol) or A8-35 deuterated on its side chains (DAPol) overlap perfectly. They are nearly symmetrical and almost, but not totally, as narrow as that formed by the globular proteins used as monodisperse standards: their half-height width is ~1.35 mL vs. 0.9–1.0 mL for albumin or horserad-ish peroxidase (Gohon et al. 2006). DLS reveals a main peak of small particles, with an average size similar to that estimated by SEC, along with a small proportion of larger contaminants (Fig. 4.14B). Although conspicuous in scattering experiments, the large objects represent a very small mass fraction of the samples (see below). In sedimentation velocity AUC (SV-AUC; Fig. 4.14C), the two preparations give a single peak, with sedimentation coefficients $s_{20,w}$ = 1.5 S for HAPol and 2.2 S for DAPol. The decrease of apparent molar mass with increasing angular velocity as well as inspection of the residuals of the fits confirm the conclusion from SEC data that the small particles are not strictly homogeneous: at the highest velocity, the largest of them end up in the pellet, which biases the average molar mass determination toward the smaller mass of those objects that remain in solution. Sample heterogeneity prevents the numerical estimate of systematic data noise or multi-run global analysis, leading to a mass estimate with a rather large uncertainty of 40 ± 5 kDa (Gohon et al. 2006). Finally, Guinier plots (Guinier and Fournet 1955; see Chap. 9, Box 9.2) of both SANS (Fig. 4.14D) and SAXS (Fig. 4.15) data are linear in the angular region (0.8 < $Q·R_g$ < 1.6) where their slope defines the radius of gyration $R_g$ of the particles, which is also consistent with a homogeneous population (polydisperse ones yield a curved plot; see Box 9.2 in Chap. 9). If the particles are spherical, their Stokes radius $R_S$ can be deduced from $R_g^\circ$, the radius of gyration at infinite contrast, as $R_S = R_g^\circ \times (5/3)^{1/2} = 3.1 \pm 0.25$ nm (Table 4.3), which perfectly matches the SEC determination. As will be discussed in the next section, the rapid drop of $I(Q)$ at high angles observed in SANS experiments is also consistent with the particles being roughly spherical, as are the results of molecular dynamics (MD) simulations (see § 4.3.1.2.3). The upward deviation of the Guinier plot at small angles confirms the presence of the

**Table 4.3**  Main characteristics of A8-35 particles.

| | | Samples | |
|---|---|---|---|
| | Estimated error | HAPol-1 | DAPol-1 |
| **Size exclusion chromatography** | | | |
| $R_{S(SEC)}$ (nm) | ±0.15 | 3.15 | 3.15 |
| **Static light scattering** | | | |
| $M$(kg·mol$^{-1}$) | ±5 | | |
| pH 6.8 | | 98 | |
| pH 7.5 | | 112 | |
| pH 8.0 | | 93 | |
| pH 9.2 | | 97 | |
| $A_2$(mL·g$^{-1}$)$^a$ | ±0.01 | | |
| pH 6.8 | | 15 | |
| pH 7.5 | | 13 | |
| pH 8.0 | | 30 | |
| pH 9.2 | | 50 | |
| **Dynamic light scattering** | | | |
| $R_{S(DLS)}$ (nm) | | | |
| pH 6.8 | | 3.6 | |
| pH 9.2 | | 3.6 | |
| **Small-angle neutron scattering** | | | |
| $R_g{}^\circ$ (nm)$^b$ | ±0.2 | 2.4 | 2.4 |
| $R_{s(SANS)}$ (nm)$^c$ | ±0.25 | 3.1 | 3.1 |
| $(\partial \rho_N / \partial C)_\mu$ (cm·g$^{-1}$)$^d$ | ±0.1 × 10$^{10}$ | 1.1 × 10$^{10}$ | 4.4 × 10$^{10}$ |
| $M_{SANS}$ (kg·mol$^{-1}$)$^e$ | ±1.5 | 46 | 34 |
| **Equilibrium ultracentrifugation** | | | |
| $M_{eq}$ (kg·mol$^{-1}$)$^f$ | ±7 | 28 | 42 |
| **Sedimentation velocity AUC** | | | |
| $S_{20,w}$ (S) (mass%)$^g$ | ±0.1 | | |
| 5 °C | | 1.5 (92%) | 2.2 (92%) |
| 20 °C | | 1.6 (80%) | 2.2 (70%) |
| $\phi'$ (mL·g$^{-1}$)$^h$ | ±0.004 | 0.866 | 0.820 |
| $S_{20,w}/(1 - \phi'\rho)$ (S) | ±0.1 | 12 | 12 |
| $M_{SV}$ (kg·mol$^{-1}$)$^i$ | ±5 | 42 | 44 |
| **Combined data** | | | |
| $\langle M \rangle$ (kg·mol$^{-1}$)$^j$ | ±5 | 39 | 40 |
| $\bar{v}_2$ (mL·g$^{-1}$)$^k$ | ±0.003 | 0.809 | 0.765 |
| $R_{min}$ (nm)$^l$ | ±0.1 | 2.3 | 2.3 |
| $R_{s(SEC)}/R_{min}$ | ±0.1 | 1.37 | 1.37 |
| $\delta$ (g H$_2$O/g APol)$^m$ | ±0.45 | 1.22 | 1.21 |

The table is from Gohon et al. (2006)

The table summarizes which information has been obtained by which of the following techniques: size exclusion chromatography (SEC), static light scattering (SLS), dynamic light scattering (DLS), small-angle neutron scattering (SANS), and equilibrium and sedimentation velocity analytical ultracentrifugation (Eq-AUC and SV-AUC, respectively) . The last section ("Combined Data") proposes a consensus picture of A8-35 particles, taking into account the relative accuracy and reliability of each technique. Two A8-35 preparations were used to gather the data, an unlabeled one ("HAPol-1") and a partially deuterated one ("DAPol-1")

$^a$Second virial coefficient. In order to avoid aggregation, SLS and DLS measurements carried out at pH 6.8 were performed immediately (minutes) after samples were diluted into the buffer

$^b$Radius of gyration at infinite contrast, $R_g{}^\circ$, deduced from Stuhrmann plots

$^c$Stokes radius, calculated from $R_g{}^\circ$ assuming the particles to be spherical

[d]Scattering length density in $H_2O$, from Gohon et al. (2004)
[e]Molecular mass calculated from the forward intensity at zero angle according to Jacrot and Zaccai (1981)
[f]Molecular mass calculated using data collected at 5 °C and 12,000 rpm and the whole sedimentation profile
[g]Sedimentation coefficient of the major population of small particles, representing the indicated mass fraction
[h]Operational parameter $\varphi'$ (apparent specific volume), from Gohon et al. (2004)
[i]Molecular mass calculated from $s_{20,w}$, $\varphi'$, and $R_S$ from SEC and SANS using the Svedberg equation
[j]Consensus number-average mass, $\langle M_n \rangle$, averaged from SANS and sedimentation velocity AUC estimates
[k]Partial specific volume, from Gohon et al. (2004)
[l]Radius of a sphere of dry APol with the consensus mass
[m]Maximal amount of associated water, estimated from $R_S/R_{min}$

large contaminants seen by DLS, and its amplitude is consistent with their small mass fraction (see below).

Stuhrmann plots, which examine the variations of $R_g^2$ vs. the inverse of the contrast between the particles and the solution – the latter being modulated by adjusting the $D_2O/H_2O$ ratio in the buffer – yield information about the organization of the various moieties inside the particles (Stuhrmann 1970). The plots for DAPol solutions are consistent with the hydrophobic alkyl chains occupying the core of the particles and the hydrophilic backbone the surface, as expected, whereas those for HAPol solutions are too noisy to be conclusive (Gohon 2001; Gohon et al. 2006).

Taken together, this ensemble of data is consistent with A8-35 particles being roughly spherical, having a mass of ~40 kDa, a radius of gyration $R_g \approx 2.4$ nm, and a Stokes radius $R_S \approx 3.15$ nm (Table 4.3). A particle of 40 kDa comprises 75–80 octyl chains, which is similar to the number of detergent alkyl chains in most micelles (Chap. 2), as well as to the number of alkyl chains that form hydrophobic clusters in sparingly modified polyacrylates (Petit-Agnely and Iliopoulos 1999; Petit-Agnely et al. 2000). The existence, within A8-35 particles, of a hydrophobic core whose size is similar to that in detergent micelles is expected on thermodynamic reasons, consistent with the Stuhrmann plots derived from SANS measurements on DAPol particles (Gohon et al. 2006), and strongly supported by MD calculations (Perlmutter et al. 2011; see next section).

Taking the number-average mass of A8-35 molecules, $\langle M_n \rangle$, to be ~4.3 kDa, the average particle is therefore comprised of ~9 mol. Given the broad mass distribution of the molecules, some particles will actually comprise only a few large ones and some 20 or more small ones. Comparison of the dry mass and volume of the particles suggests that they are highly hydrated, comprising ~1.2 g water per g of APol (Table 4.3). As will be further discussed below (§ 4.3.1.2.5), the size and shape of the particles do not appear to change as a function of concentration, in keeping with FRET data (Giusti et al. 2012), which also suggest that the size of the assemblies does not change much (Fig. 4.10). Table 4.3, reproduced from Gohon et al. 2006, sums up the main properties of A8-35 particles and the way each piece of information has been acquired.

It is interesting to note that neither the variable length of the chains nor the random distribution of octyl side groups in A8-35 hampers the formation of well-defined particles, even though those are not perfectly monodisperse. Particle polydispersity is more likely due to individual differences between molecules in the batch rather than to metastability or a broad equilibrium distribution of energetically nearly equivalent sizes around an optimal one, based on the following two arguments: (i) fractions selected by SEC from a batch that migrates as a broad peak will yield a narrow one upon being fractionated again (Gohon et al. 2006), suggestive of differences of composition between big and small particles, and (ii) exchange of molecules between small A8-35 particles is likely to be fast on the time scale of these experiments (see § 4.3.1.2.4), speaking against metastability. It does not seem that the variable length of individual molecules be a critical factor in generating particle polydispersity, since batches of A8-35 synthesized from precursor PAAs with either a broad ($Đ \approx 3.1$) or a very narrow ($Đ \approx 1.3$) dispersity yielded particles with similar size distributions (C. Tribet and F. Giusti, unpublished observations cited in Le Bon et al. 2014b). The hydrophobic/hydrophilic balance of individual molecules and, possibly, the degree of randomness of octyl chain distribution are more likely to determine the size of the resultant particles. This is an interesting observation, because it suggests that, by restraining the molecule-to-molecule variability of the density and/or distribution of octyl chains, narrower size distributions of APol particles and, as a consequence, of MP/APol complexes could probably be achieved, an important factor for crystallization attempts. This hypothesis is consistent with SEC analyses showing that the particles formed by homotelomeric non-ionic APols, which are comprised of molecules that vary in length but have a homogeneous composition, have a much narrower size distribution than those formed by their heterotelomeric analogs, whose composition and monomer distribution vary from molecule to molecule (see § 4.3.2).

Closer examination of the SEC data reveals some complications. First, there is evidence for some polydisperse material eluting well after the main peak, beyond the position at which elutes horseradish peroxidase, spreading almost to the total volume $V_t$ (Fig. 4.14A). This shallow, broad peak is observed in all A8-35 preparations. It has not been studied in detail, but is suspected to arise from APol molecules that are too small and/or not hydrophobic enough to assemble into particles. This region also comprises the contribution from molecules that dissociate from the particles as they progress through the column through APol-free buffer, but their concentration, equal to the CAC, is too low to account for all of the absorbance measured and should be constant throughout.

Second, there is a very small peak eluting in the void volume $V_0$ of the column, pointing to the presence of some very large objects. This population of particles is barely detectable in SEC, indicating that they represent a very small fraction of the mass of the sample. It becomes, however, prominent in radiation scattering experiments because, for a given molar concentration and composition, the scattered intensity increases as the square of the molecular mass of the scatterers. It is, as noted above, responsible for the peak at $R_S \approx 45$ nm in DLS experiments (Fig. 4.14B) and for the upshot in the SANS Guinier plots at very small angles (below 0.001 Å$^{-2}$; Fig. 4.14D). These objects are not visible in AUC experiments (Fig. 4.14C) because they sediment too rapidly as compared to the small particles. As shown by the SANS data, their proportion with respect to the small particles does not vary with the concentration of the preparations (Fig. 4.14D), indicating the absence of an equilibrium. The large objects, best characterized by DLS, can reach an average Stokes radius of up to ~60 nm and an average molecular mass in excess of 1 MDa. Although conspicuous in scattering experiments, they usually represent a very small mass fraction of the samples (~0.1% in the samples used in Gohon et al. 2006), and they do not associate with MPs (Gohon et al. 2008). They seldom interfere with experiments but can, if their presence is a hindrance, be removed by either SEC fractionation or ultracentrifugation.

As experience with A8-35 accumulated, a number of factors were found to compromise the quasi-monodispersity of A8-35 particles and, correlatively, that of MP/A8-35 complexes:

- Artifactual covalent binding to the polymer, during the synthesis, of a hydrophobic contaminant can be a major source of problems. Some early batches of A8-35 exhibited complex SEC elution patterns and curved Guinier plots (Gohon 2001; Gohon et al. 2006). This behavior was characteristic of samples with a high degree (typically >6%) of artifactual grafting of dicyclohexylurea (DCU), a by-product of the coupling reaction (see § 4.5, Protocol 4.1). The broad size distribution found in such batches suggests that some APol molecules, presumably because of a higher hydrophobicity and/or a nonrandom distribution of hydrophobic side

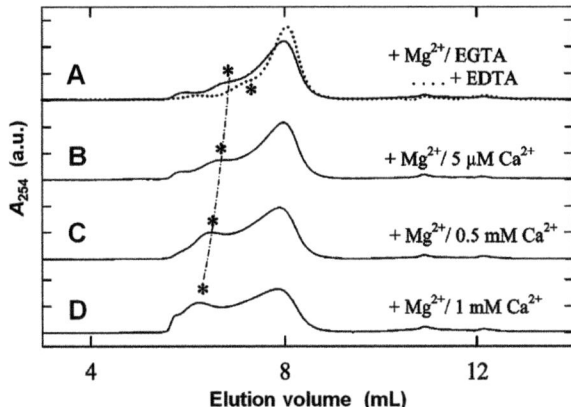

**Fig. 4.16** Effect of divalent cations on the dispersity of A8-35 particles. Size exclusion chromatography analysis of a (rather poor) batch of A8-35 in the presence or absence of various concentrations of $Ca^{2+}$ and of $Mg^{2+}$ (1 mM). 200-$\mu$L samples of A8-35 at 5 g·L$^{-1}$ were loaded onto a TSK 3000 SW column equilibrated at room temperature (23 °C). Experiments were carried out in 20 mM TES/NaOH buffer, 100 mM KCl, pH 7.0, with or without divalent cations and/or chelators, as indicated. Even in the presence of 0.5 mM EDTA (**A**, dotted line), the monodispersity of the A8-35 particles in these experiments was not excellent, either due to the artifactual grafting of some dicyclohexylurea (DCU; ~2%) or to the relatively low pH used. As chelators were omitted and/or divalent cations were supplemented, the size and abundance of aggregates (marked by *) increased dramatically (Reprinted with permission from Picard et al. 2006, © 2006 American Chemical Society).

chains, induce the formation of larger assemblies. Guidelines are provided in Gohon et al. (2006), about how to avoid this problem (cf. § 4.5, Protocol 4.1).

- Above pH 7, most of the free carboxylates carried by A8-35 are dissociated (Gohon et al. 2004). Acidification of the solutions below pH 7 causes some of them to protonate, increasing the hydrophobicity of A8-35 and promoting aggregation, which develops, even at a pH as high as 6.8 or 7.1, over a period of hours (Gohon et al. 2006). The process reverses in a few minutes if the pH is raised again. From a practical point of view, a procedure that yields stable solutions of well-dispersed small particles is to dissolve the sodium salt of A8-35 in distilled water. The pH of the resulting stock solutions is ~8.5. Stock solutions in water can be stored frozen to prevent bacterial growth. Dilute solutions should be buffered, preferably at pH $\geq$ 7.5, or kept under neutral gas so as to prevent acidification by atmospheric $CO_2$ (Gohon et al. 2006).

- Aggregation is induced by multivalent cations, particularly by $Ca^{2+}$ (Fig. 4.16). This polydispersity will also affect MP/A8-35 complexes. For this reason, it is advisable, whenever monodispersity is important, e.g. in radiation scattering experiments, to work in the presence of EDTA. EDTA has been shown to improve the resolution of 2D [$^{15}$N,$^{1}$H]-TROSY NMR spectra of A8-35-trapped OmpX, line widths being reduced by about one-third, on average (ca. $-13$ Hz), as compared to EDTA-free buffer (Catoire et al. 2010b; see Chap. 10).

#### 4.3.1.2.3    Shape and Internal Organization of A8-35 Particles

**SANS Analyses**

SV-AUC analyses are compatible with A8-35 particles being nearly spherical. A more direct indication is provided by SANS experiments, the variation of the intensity at high angles containing information about the shape of the scatterers. As shown in Fig. 4.17, the form factor, analyzed in terms of $Q^{-\alpha}$

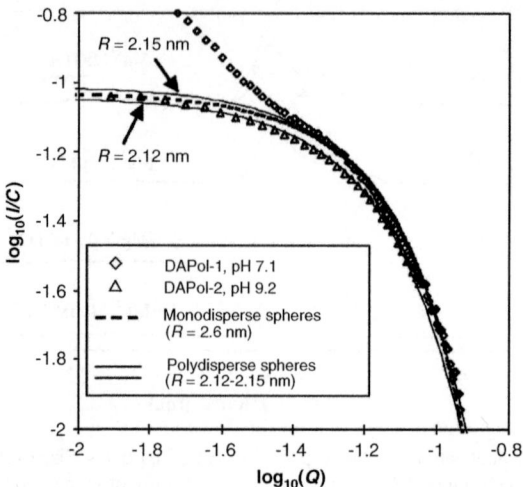

**Fig. 4.17** Analysis of small-angle neutron scattering by two batches of DAPol, DAPol-1, and DAPol-2, at intermediate values of $Q$. DAPol-1 ($\diamond$) was in 20 mM $NaH_2PO_4/Na_2HPO_4$, 100 mM NaCl, pH 7.1, DAPol-2 ($\triangle$) in 20 mM boric acid/NaOH, 100 mM NaCl, pH 9.2. Presumably due to the lower pH at which it stood for a week before the measurements, the sample of DAPol-1 was less monodisperse. The predicted $Q$-dependence of $I(Q)$ for random dispersions of homogeneous spheres with a radius of 2.6 nm ($R_g = 2.0$ nm) and for dissymmetric distributions of spheres (Griffith et al. 1987) with an average radius of 2.12 or 2.15 nm is plotted respectively in dashed and solid lines (Reprinted with permission from Gohon et al. 2006, © 2006 American Chemical Society).

variation, presents a rapid drop with $\alpha \approx 3.5$ in the $R_g \cdot Q$ range between 2 and 2.9. An exponent $\alpha \geq 3.5$ in the high-$Q$ region is reliably ascribed to the existence of a particle with a sharp interface (Lindner and Zemb 2002; Di Cola et al. 2004). The drop of intensity with rising $Q$ in the high-$Q$ range is close to that expected for monodisperse spheres with a radius of 2.6 nm (Fig. 4.17), in reasonable agreement with the estimates of Table 4.3. However, despite the fair match to SANS data observed at pH 9.2 (Fig. 4.17), the limited range of $Q$ and the uncertainty introduced by background subtraction preclude any strong conclusion to be drawn about the sphericity and dispersity of the particles (Gohon et al. 2006).

The conclusion that A8-35 particles present a well-defined surface, rather than a fluffy corona of loops and tails, presumably extends to the belts of MP-adsorbed APol, which is an important point for many biophysical applications (see Chaps. 5, 10, and 11).

### Molecular Dynamics Simulations

This conclusion is comforted by the result of MD simulations (Perlmutter et al. 2011). Models were constructed by letting four 10-kDa A8-35 chains with a random distribution of monomers spontaneously assemble into a 40-kDa particle. It was checked that using either twice shorter or twice longer chains does not affect the results (see below). The simulations were carried out in three steps. First, a series of all-atom MD (AAMD) simulations of the particle in solution was carried out, starting from an arbitrary initial configuration. Although AAMD simulations result in stable cohesive particles over a 45-ns simulation, the equilibration of the various moieties within the particle is limited. Coarse-grained MD (CGMD) was therefore resorted to, in a second step, using the AAMD simulations for parametrizing the CGMD model. For CGMD simulations, the various atoms are grouped as described below in Fig. 4.20. Because CGMD is computationally much less demanding than AAMD, it is possible to investigate processes on the microsecond time scale, including de novo particle assembly:

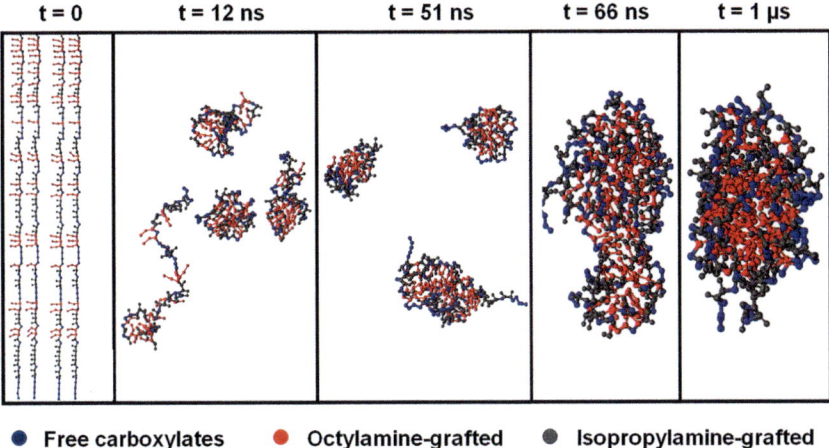

**Fig. 4.18** Snapshots from the CGMD simulation illustrating de novo particle assembly (*blue*, ungrafted monomers; *red*, octylamine-grafted ones; *gray*, isopropylamine-grafted ones; water and ions have been removed for clarity) (Reprinted with permission from Perlmutter et al. 2011, © 2011 American Chemical Society).

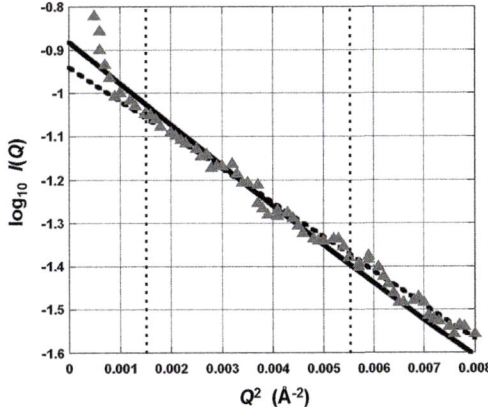

**Fig. 4.19** Comparison of experimental and predicted SANS data before and after coarse-grained simulation. Experimental data (from Gohon et al. 2006) are shown as *gray triangles*. Molecular dynamics data are shown after all-atom MD simulations without coarse-grained simulation (*solid line*) and after coarse-grained simulation (*dashed line*), the latter being necessary for the particle to reach its equilibrium structure. After coarse-grained simulation, the predicted Guinier plot matches exactly the experimental data, but for the low-angle upshot due to the presence of large particles in the experimental sample (see § 4.3.1.2.2), and yields the same radius of gyration, $R_g = 2.4$ nm. The vertical dotted lines delimit the Guinier region whose linear slope was used to calculate $R_g$ (Reprinted with permission from Perlmutter et al. 2011, © 2011 American Chemical Society).

as shown in Fig. 4.18, initially elongated molecules first collapse upon themselves, before coalescing into a globular particle.

A third set of simulations used reverse coarse-graining (rCG), yielding all-atom coordinates for the structures obtained by CGMD simulations. Excellent agreement was observed between rCG MD models and experimental SANS data, which is not the case for models obtained without resorting to CGMD (Fig. 4.19). It is notable that particles composed of chains either twice as long or half as long as

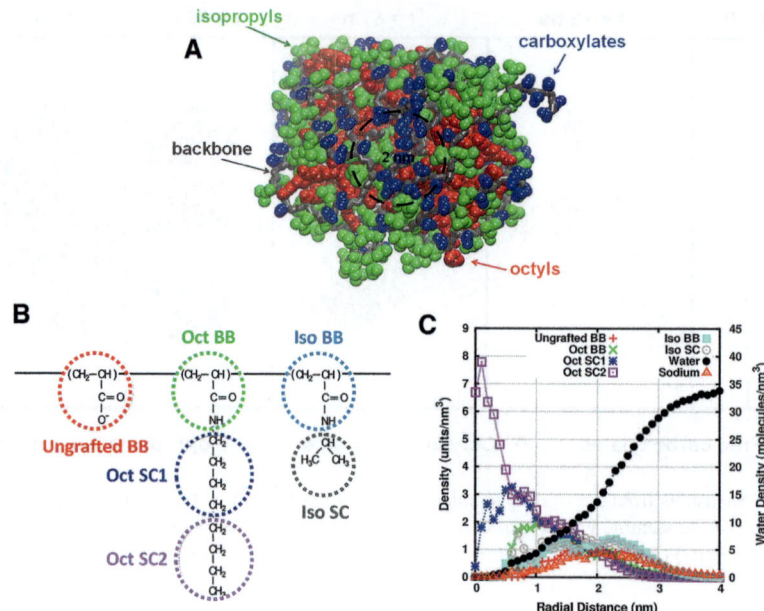

**Fig. 4.20** Internal organization of A8-35 particles as deduced from molecular dynamics (MD) simulations. (**A**) Snapshot image of an all-atom model of A8-35 particle formed by in silico self-assembly after 4 μs of coarse-grained MD calculation (CGMD) followed by reverse coarse-grained (rCG) simulation. Water, counterions, and APol hydrogens have been omitted for clarity. The backbone is in *dark gray*, free carboxylates in *blue*, isopropyl grafts in *green*, and octyl grafts in *red*. The *dashed line* is a circle with radius 1 nm (From Tehei et al. 2014). (**B**) Chemical structure of the ungrafted, octylamine-grafted, and isopropylamine-grafted units with the groups used in CGMD indicated: *ungrafted BB*, ungrafted backbone group; *Oct BB*, backbone group carrying an octyl side chain; and *Iso BB*, backbone group carrying an isopropyl side chain. Octyl side chains are divided into a proximal group, *Oct SC1*, and a distal group, *Oct SC2*. Isopropyl side chains form a single group, *Iso SC* (Adapted from Perlmutter et al. 2011). (**C**) Radial distribution of the various moieties of A8-35 in the particle and of water molecules and Na⁺ ions. Data extracted from all-atom MD trajectories after CGMD and rCG simulations (Perlmutter et al. 2011). The density of each type of coarse-grain group shown in panel **B** is plotted as a function of the distance from the particle center of gravity. The core of the particle is formed by the octyl chains (*pink* and *dark blue* symbols) and practically devoid of water. Free carboxylates and sodium ions (*red* and *orange* symbols), whose distributions overlap closely, cover the surface of the particle, whereas the backbone atoms of the octyl-grafted monomers (*green*) are pulled toward the core and occupy intermediate positions (The figure is adapted from Tehei et al. 2014).

those used in the main analysis, as well as particles made up of a single 40-kDa molecule, present similar values of $R_\mathrm{g}$ and ellipsoid axial ratios (Perlmutter et al. 2011). This is consistent with experimental data indicating that particles formed from chains that are, on average, either ~4× times longer (Tribet et al. 1996) or ~1.5× shorter (Gohon 1996) than those of A8-35 behave similarly upon SEC (C. Tribet, unpublished data) and appear equally efficient at keeping MPs water-soluble (Gohon 1996; Tribet et al. 1996).

Over the course of the rCG simulations, the structures relax on the atomistic scale, but their larger-scale molecular organization does not change. This organization is described by the radial component density distribution presented in Fig. 4.20. There is a very clear pattern of separation: the side chains of the octyl groups form a hydrophobic core surrounded by the ungrafted, charged carboxylates. For the CGMD simulations, each sequence was simulated three times, and the results were entirely consistent. Figure 4.20 shows the average of the three simulations. In the rCG simulations, the distributions of components taken from the CGMD coordinates are stable, and there is no change in the organization on the nanosecond time scale, even though there is a difference in sampling between the resolutions.

The character of the particle interior is of particular interest because it is likely to resemble that fraction of the APol belt that is directly in contact with the TM surface of MPs in MP/A8-35 complexes. As mentioned above, the available experimental data are limited: Stuhrmann plots of the data for DAPol do suggest the existence of a hydrophobic core, but the data collected on unlabeled A8-35 are inconclusive, possibly because of insufficient contrast (Gohon et al. 2006). The particles observed in the AAMD and rCG simulations have very different particle cores (Perlmutter et al. 2011). The AAMD simulations carried out in the first step of the MD work show a core that is to a surprising degree permeable to water. In contrast, in the rCG simulations there is practically no water in the particle center, which forms a hydrophobic, water-excluding domain (Fig. 4.20).

In summary, SANS and MD data are consistent in describing A8-35 particles as compact, roughly spherical, micelle-like particles, with a nearly water-free hydrophobic core and a well-defined hydrophilic surface providing a sharp interface with the solution. In the next section, we will examine what is known of the dynamics of these objects.

### 4.3.1.2.4 Dynamics of A8-35 Particles

The dynamics of A8-35 particles has been investigated by neutron scattering (Tehei et al. 2014), MD simulations (Perlmutter et al. 2011; Tehei et al. 2014), AUC (Gohon et al. 2006), and, indirectly, FRET (Zoonens et al. 2007). It comprises three fairly different aspects, which we will examine successively: (i) the internal dynamics of the particles, (ii) their diffusion in solution, and (iii) their interactions.

**Internal Dynamics of A8-35 Particles**

In the preceding two sections, we have dealt with coherent neutron scattering, which informs on structure. Neutrons that exchange energy with the target, and therefore experience a change of wavelength, provide precious experimental information on the dynamics of scattering nuclei on a picosecond to nanosecond time scale and an ångström length scale. Because the incoherent scattering cross section of $^1H$ is much larger than that of $^2H$ or any other nucleus present in A8-35, motions of different regions in the polymer can be discriminated using selective deuteration. The thermal dynamics of A8-35 particles has been examined by elastic incoherent neutron scattering (EINS) and quasi-elastic neutron scattering (QENS) using either unlabeled A8-35 (HAPol) or DAPol, in which the hydrogen atoms in the octyl and isopropyl side chains are replaced by deuterium (Fig. 4.12) (Tehei et al. 2014). The HAPol sample yielded information on all backbone and side-chain group motions, whereas the DAPol sample data were dominated by the backbone. Experiments were performed at 7 °C, a temperature that provides good signal-to-noise ratios in both EINS and QENS experiments, on the time scales of ~10 ps and ~18 ps (90 and 50 µeV resolution, respectively) and ~1 ns (0.9 µeV resolution). All experiments were performed in $D_2O$ in order to minimize the incoherent scattering contribution of the solvent.

The mean square displacements (MSDs) calculated from the EINS data include vibrational as well as conformational sampling motions in the 10 and 18 ps time scales (Fig. 4.21). At 7 °C, the MSD measured for DAPol (~0.5 Å$^2$) is half that for HAPol (~1 Å$^2$) (Fig. 4.21, *right*), indicating that the side chains contribute significantly larger conformational fluctuations than the backbone groups. MSDs have been measured on different time scales for proteins, lipids, and polysaccharides (Daniel et al. 1999; Zaccai 2000, 2011, 2013; Natali et al. 2004). HAPol values are significantly larger than protein MSDs and similar to those estimated for lipid MSDs under similar time scale, hydration, and temperature conditions.

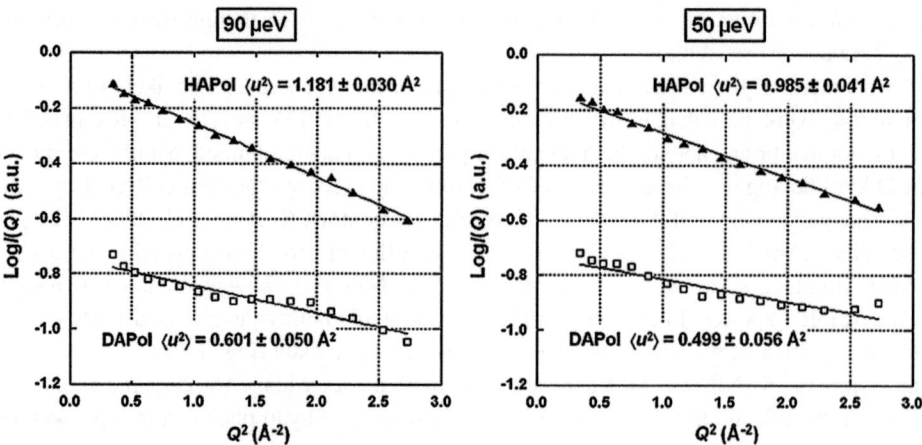

**Fig. 4.21** Elastic incoherent neutron scattering (EINS) by HAPol and DAPol. Logarithm of the normalized elastic intensity vs. scattering vector squared measured at 90 μeV (*left*) and 50 μeV (*right*), at 7 °C, on the IN6 time-of-flight spectrometer of the Institut Laue-Langevin, with corresponding linear fits. $\langle u^2 \rangle$ is the atomic mean square displacement (MSD), in $\text{Å}^2$ (From Tehei et al. 2014).

The MSD measured for HAPol can be extrapolated to longer time scales from data on other systems by assuming that similar motion classes are sampled. MSDs of lipids in bilayers, polysaccharides, and proteins increase by a factor of ~2 between the ps and ns time scales (Natali et al. 2005; Jasnin et al. 2010; Zaccai 2013). These data lead one to expect HAPol MSDs of ~2 $\text{Å}^2$ for longer sampling times, suggesting similar MSDs for lipids and the core of the A8-35 particle when adjusted to the ns time scale. This is roughly consistent with MD data, which however, as shown below, predict that the core of A8-35 particles is actually somewhat more viscous than that of a lipid bilayer (see Fig. 4.23).

The QENS analysis provides more information on the types of thermal motion that the CH, $CH_2$, and $CH_3$ groups are undergoing in A8-35 particles (Fig. 4.22), while confirming the higher mobility of the side chains when compared to the polymer backbone. Jump diffusion motions observed in the HAPol sample (diffusion between sites after a mean residence time at each site), with correlation times on the picosecond time scale, are similar to those interpreted from QENS studies as kink propagation in lipid chains (König and Sackmann 1996; Trapp et al. 2010). The *gauche-trans-gauche* kink is the simplest higher-order defect of lipid chains. The other motions in A8-35 particles detected by QENS most likely correspond to local reorientational dynamics of the backbone CH and $CH_2$ and side-chain $CH_2$ and $CH_3$ groups.

Similar conclusions were drawn from the analysis of MD trajectories (Perlmutter et al. 2011; Tehei et al. 2014). They show that, in keeping with INS results, side-chain $CH_2$ and $CH_3$ groups, including the methylene groups of the octyl chains, are much more mobile than the CH and $CH_2$ groups of the main chain (Fig. 4.23A). When a comparison is carried out between A8-35 particles, SDS micelles, and palmitoyloleoylphosphatidylcholine (POPC) bilayers, the mobility of the polar moieties is seen to decrease in the order SDS >> POPC > A8-35 (Fig. 4.23B) and that of the terminal methyl group of the hydrophobic chains in the order SDS > POPC > A8-35 (Fig. 4.23C), with the strongest differences in the polar regions. As will be discussed below, the higher viscosity of an A8-35 vs. a detergent environment probably plays a role both in the higher stability of APol-trapped vs. detergent-solubilized MPs and in the functional effects APols have on certain MPs (Chap. 5, § 5.6).

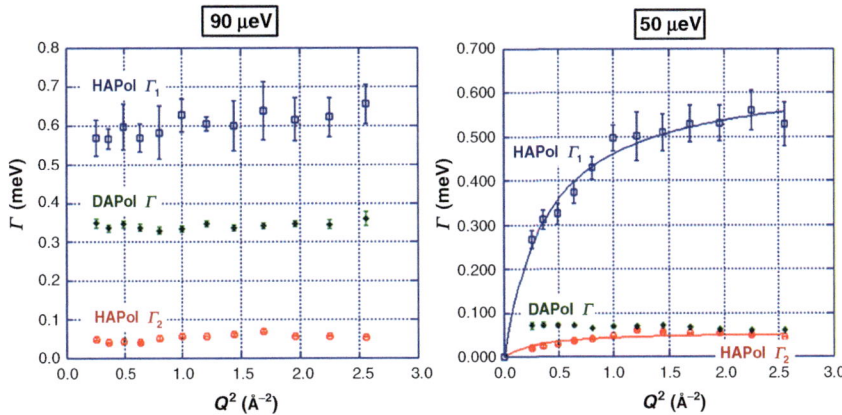

**Fig. 4.22** Analysis of quasi-elastic neutron scattering (QENS) spectra measured at 7 °C on the IN6 spectrometer of the Institut Laue-Langevin. *Left panel*, 90 μeV resolution; *right panel*, 50 μeV resolution. HAPol sample: half-widths at half-maximum, $\Gamma_1$ (*blue squares*) and $\Gamma_2$ (*red circles*), of the two Lorentzians fitted to the QENS spectrum, plotted as a function of $Q^2$. The "jump diffusion" model fit is shown for the 50 μeV resolution data (*right panel*). DAPol sample (*green diamonds*): $\Gamma$ value for the one-Lorentzian fit to the QENS spectrum for the two resolution conditions (From Tehei et al. 2014).

**Translational Diffusion Coefficient**

The translational diffusion coefficient of A8-35 particles can be deduced from QENS data on HAPol measured on a backscattering spectrometer (0.9 μeV resolution, ~1-ns time scale). The QENS spectrum is well fitted by a single Lorentzian, whose half-width at half-maximum, $\Gamma$, is plotted versus $Q^2$ in Fig. 4.24. The linear dependence of $\Gamma$ on $Q^2$ indicates that, on the time scale of ~1 ns, macromolecular translation of the A8-35 particle as a whole dominates the QENS spectrum. A translational diffusion coefficient $D_t = (4.45 \pm 0.11) \times 10^{-7}$ cm$^2 \cdot$ s$^{-1}$ can be calculated from the plot.

The SV-AUC data in Gohon et al. (2006), obtained at 1–10 g·L$^{-1}$, yield a translational diffusion coefficient at infinite dilution in water $D_0 = 6.8 \times 10^{-7}$ cm$^2 \cdot$s$^{-1}$. Correcting for the crowding and the viscosity of the solutions used in the QENS experiments, this corresponds to an expected diffusion coefficient at 240 g·L$^{-1}$ in D$_2$O of $4.16 \times 10^{-7}$ cm$^2 \cdot$s$^{-1}$, in fair agreement with the neutron scattering experimental value of $(4.45 \pm 0.11) \times 10^{-7}$ cm$^2 \cdot$s$^{-1}$ (Tehei et al. 2014) (Fig. 4.24). This is an important control, confirming that APol particles of similar properties to the ones characterized previously are still present at the high concentration required for the QENS measurements. It is quite remarkable that A8-35 particles remain essentially the same in the 1–10-g·L$^{-1}$ range (Gohon et al. 2006), at ~240 g·L$^{-1}$ (Tehei et al. 2014), and presumably, given the stability of the normalized FRET signal above the CAC (Fig. 4.10C), down to 0.002 g·L$^{-1}$ (Giusti et al. 2012), that is, over five orders of magnitude of concentration (see § 4.3.1.2.5).

**Particle Interactions**

Interactions between A8-35 particles have not been studied directly. However, the kinetics of exchange of A8-35 molecules between the solution and belts of MP-associated A8-35 have been examined by FRET (Zoonens et al. 2007). These experiments will be discussed below from the point of view of the dynamics of MP/APol complexes (Chap. 5, § 5.6). Extrapolating from these data to interactions in pure APol solutions, which seems reasonable, suggests that, at 0.2 g·L$^{-1}$ and in the presence of 100 mM NaCl, the content of the particles mixes within minutes, whereas it takes hours in the absence of salt, consistent with a mechanism involving collisions between charged entities (Zoonens et al. 2007). A simple reasoning suggests that these exchanges involve collisions between particles rather than

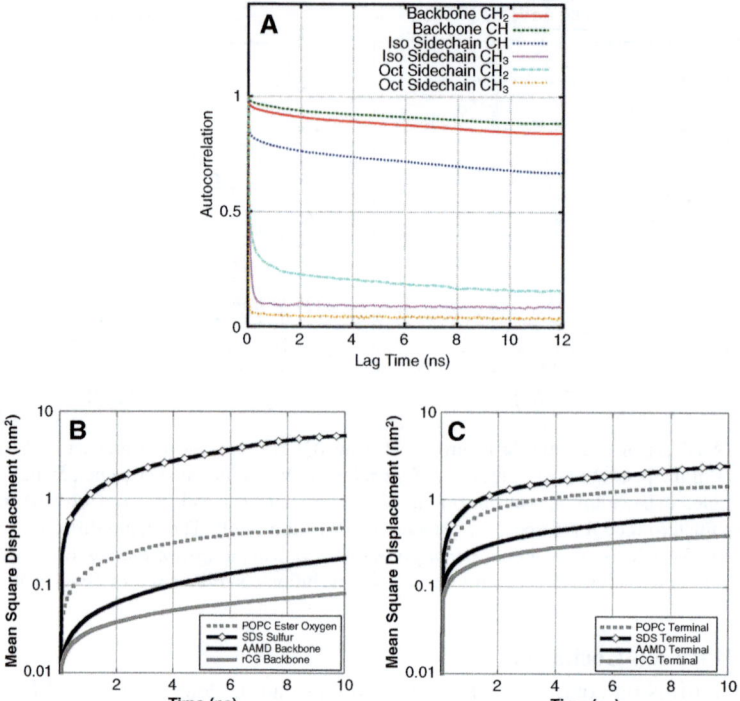

**Fig. 4.23** Dynamics of A8-35 particles as seen by molecular dynamics simulations. (**A**) Autocorrelation functions for different carbon-hydrogen bond vectors in the MD model of an A8-35 particle. Backbone CH and CH$_2$ groups (*red* and *green* curves, respectively) reorient much more slowly than side-chain CH$_2$ and CH$_3$ groups (*yellow*, *pink*, and *light blue* curves) (From Tehei et al. 2014). (**B, C**) Mean square displacement in A8-35 particles compared with palmitoyloleoylphosphatidylcholine (POPC) bilayers and sodium dodecyl sulfate (SDS) micelles (the latter data from Perlmutter and Sachs 2009a, b), for atoms from the polar moieties (**B**) and for the terminal methyl group of the hydrocarbon chain (**C**). AAMD, rCG: data from the initial all-atom and the final reversed coarse-grained simulations of A8-35 particles, respectively. SDS micelles provide the most fluid environment, followed by POPC bilayers and A8-35 particles, differences being particularly marked in the polar regions (Reprinted with permission from Perlmutter et al. 2011, © 2011 American Chemical Society).

between particles and individual A8-35 molecules. Indeed, at this concentration, which is ~100× above the CAC, ~99% of A8-35 is associated into particles and only ~1% present as free molecules. Taking the particles to comprise $n$ molecules, where $n \approx 9$, their molar concentration is therefore ~ $(100/n)\times$ that of the free molecules. MD simulations predict that free molecules collapse upon themselves (Fig. 4.18), so their overall shape can be expected not to be too different from that of the particles. Given that, for homothetic particles, $D_t$ is proportional to $M^{-1/3}$, free A8-35 molecules can be expected to diffuse only ~1.6× more rapidly than particles. Particles therefore must experience collisions with other particles ~$(100/1.6\,n) \times$ more often than collisions with free molecules. Because the former comprise $n$ molecules rather than one, particle-mediated exchange is likely to be ~100/ $1.6 = 60\times$ more efficient than that mediated by single molecules. This difference will become even more pronounced at the more usual concentrations of several g·L$^{-1}$.

The rate of adsorption of A8-35 at the air/water interface, examined by dynamic surface tension (DST) measurements using the maximum pressure bubble method (Fainerman et al. 1994; Miller et al. 1994), was found to be rapid (seconds) at concentrations >1 g·L$^{-1}$ (Giusti et al. 2012) (Fig. 4.25). Below this threshold, it decreases markedly with decreasing polymer concentration. As discussed in

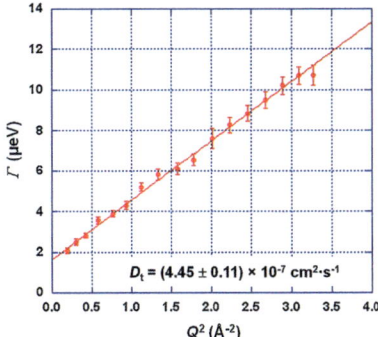

**Fig. 4.24** Half-width at half-maximum, $\Gamma$, of the Lorentzian fitted to the QENS spectrum of HAPol measured on the IN16 backscattering spectrometer of the Institut Laue-Langevin at 7 °C, plotted as a function of $Q^2$, and deduced value of the translational diffusion constant $D_t$ (From Tehei et al. 2014).

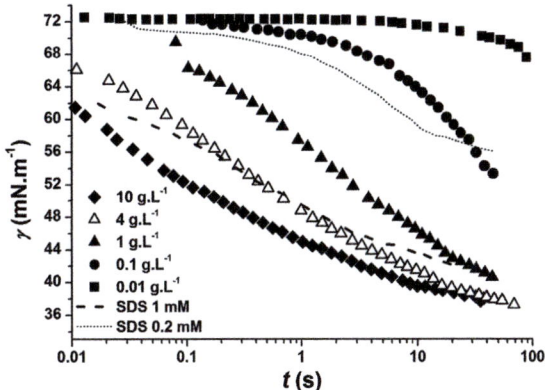

**Fig. 4.25** Evolution with time of the surface tension of A8-35 solutions measured by the maximum bubble pressure method. The buffer was Tris/HCl 20 mM, NaCl 100 mM, pH 8.0. A8-35 concentrations: 0.01 g·L$^{-1}$ (■), 0.1 g·L$^{-1}$ (●), 1 g·L$^{-1}$ (▲), 4 g·L$^{-1}$(△), 10 g·L$^{-1}$ (◆). Dotted and dashed lines are for solutions of sodium dodecyl sulfate in 100 mM NaCl, at 0.2 and 1 mM, respectively (1 g·L$^{-1}$ A8-35 corresponds to a concentration of octyl moieties of ~2 mM) (Reprinted with permission from Giusti et al. 2012, © 2012 American Chemical Society).

Chap. 12 (§ 12.3.1, Fig. 12.7), this phenomenon is probably of importance in single-particle electron microscopy experiments, because it affects the rate of thinning of water films in carbon film holes in the short time interval between blotting and quick-freezing, an important factor in controlling the spatial distribution of MP/APol complexes within the film (cf. Flötenmeyer et al. 2007).

#### 4.3.1.2.5   Data Pertaining to the Phase Diagram of A8-35

No extensive phase diagrams for A8-35 have been published yet, but there are numerous pieces of relevant information in the literature, some of which have been mentioned above. They are compiled here for convenience:

- A8-35 solubility in water and aqueous buffers is >30% w/w (>320 g·L$^{-1}$) (Gohon et al. 2006).
- In Tris/HCl 20 mM, NaCl 100 mM, pH 8.0 buffer, particles form above ~0.002 g·L$^{-1}$ (Giusti et al. 2012).

- According to FRET, DLS, AUC, and QENS, it seems likely, although not finally established, that the size of these particles does not vary between the CAC and 240 g·L$^{-1}$, that is, over five orders of magnitude (Gohon et al. 2006; Giusti et al. 2012; Tehei et al. 2014). Small particles thus seem to occupy a very large region of the phase diagram of A8-35, which is of great practical interest for experiments where MP/APol complexes must be highly concentrated, such as for EINS and QENS measurements (Tehei et al. 2014), in crystallization attempts (Charvolin et al. 2014) (Chap. 11), in solid-state NMR, or for certain types of light spectroscopy experiments (see e.g. Polovinkin et al. 2014; Chap. 8, § 8.2.2).
- At pH $\geq$ 7.0, most carboxylic moieties are deprotonated (Gohon et al. 2004).
- The mass of the particles does not change in the pH range 6.8–9.2 nor between 5 and 20 °C (Gohon et al. 2006).
- A moderate repulsion between particles is observed in the presence of 100 mM NaCl, the second virial coefficient, $A_2$, increasing from ~15 to ~50 mL·g$^{-1}$ between pH 6.8 and pH 9.2 (Gohon et al. 2006).
- Judging from the results of SAXS measurements carried out on MP/A8-35 complexes, the repulsion between particles probably vanishes between 300 and 500 mM NaCl, above which an attractive regime develops (Popot et al. 2003) (cf. Fig. 11.13 in Chap. 11).
- According to SANS data, an attractive regime also develops, in the presence of 100 mM NaCl, upon addition of 5–10% 4-kDa polyethylene glycol to a 20-g·L$^{-1}$ solution of MP/A8-35 complexes (Charvolin et al. 2014).
- Lowering the pH to $\leq$7 induces aggregation (Gohon et al. 2006).
- Multivalent ions induce aggregation (Picard et al. 2006).
- Given that MP/A8-35 complexes can be frozen without inactivating the protein nor causing it to aggregate (Gohon et al. 2008), it seems probable that the APol belt and, by extension, A8-35 particles are not strongly perturbed by freezing.

### 4.3.2 Solution Properties of Other Amphipols

A8-35 is the only APol whose solution properties have been extensively studied by such a broad variety of complementary methods. Data for other APols are most often limited to SEC or DLS estimates of the size of the particles they form in aqueous solutions, except for glucose-based NAPols and SMALPs (see Table 4.4). Table 4.4 also indicates which of the listed APols are (or can be expected to be) either sensitive or largely insensitive to aggregation in the presence of Ca$^{2+}$ ions or at acidic pH.

A8-75 and SAPols are two PAA-based APols with the same backbone length distribution and extent of grafting with octylamine as A8-35 (Fig. 4.1). However, they lack isopropylamine, the corresponding carboxylates being either left free (A8-75) or grafted with taurine (SAPols), resulting in a much higher charge density. The particles they form comigrate with A8-35 particles upon SEC (Dahmane et al. 2011) (Fig. 4.26A, B), suggesting, but not demonstrating, a similar size, mass, and size distribution (because the higher charge density may speed up elution, the particles may actually be somewhat smaller than those of A8-35). SAPol particles are insensitive to pH (Dahmane et al. 2011) (Fig. 4.26B) and largely insensitive to the presence of Ca$^{2+}$ or Mg$^{2+}$ ions (Picard et al. 2006) (Fig. 4.26D).

"Blocky" copolymers similar to A8-75 but derived from poly(methacrylic acid) and in which the octyl groups are distributed in a nonrandom manner have been observed to assemble into markedly

**Table 4.4** Particle size and known or expected sensitivity of various amphipols to aggregation at low pH and/or in the presence of $Ca^{2+}$ ions.

| Amphipol | Particle size $R_S$ (nm) | $\langle M \rangle$ (kDa) | Aggregation Low pH | $Ca^{2+}$ | Comments | References |
|---|---|---|---|---|---|---|
| A8-35 | ~3.15 | ~40 | + | + | $R_S$ from SEC. $R_S$ from DLS = 3.6 nm. $\langle M \rangle$ from combined techniques. See § 4.3.1.2, Table 4.3, and Fig. 4.19 | Gohon et al. (2006) |
| A8-75 | ~3.15 | | +[a] | +[a] | $R_S$ estimated from near comigration with A8-35 upon SEC | Dahmane et al. (2011) |
| SAPols | ~3.15 | | – | – | $R_S$ estimated from near comigration with A8-35 upon SEC | Dahmane et al. (2011) and Picard et al. (2006) |
| PC-APols (C22-43) | ~3.3 | | – | – | $R_S$ from DLS | Diab et al. (2007a, b) |
| PMAL-C12 | ~6 | | ? | + | $R_S$ estimated from SEC data in Fig. 2 in the reference cited | Picard et al. (2006) |
| SMALPs | ~5[b] | [c] | + | + | Pure SMA does not seem to form particles. The dimensions of SMALPs depend on the lipid/SMA ratio. A model mainly derived from SANS data is shown in Fig. 4.27 | Carazo et al. (2015), Grimaldo et al. (2015), Jamshad et al. (2015a, b), Lee et al. (2016), Scheidelaar et al. (2015), Zhang et al. (2015), Dörr et al. (2016), and Grethen et al. (2017) |
| Homotelomeric glucosylated NAPols (NA11 and NA29) | ~3 | ~50 | – | – | $R_S$ from DLS and SEC, $\langle M \rangle$ from AUC and SANS | Sharma et al. (2012) |
| NVoy | | ~112 | –[a] | –[a] | $\langle M \rangle$ from SLS coupled with refractive index measurements and SEC | Klammt et al. (2011) |
| APG | ~2.5–3 | ~80 | ?[d] | ?[d] | Based on estimates from SEC and AFM data | Han et al. (2014) |

Note that even though they remain or can be expected to remain soluble at low pH, the charge of APols such as SAPols, PC-APols, or PMAL-C12 will change as a function of pH: the net charge of SAPols diminishes at low pH, whereas PC-APols and PMAL-C12 become cationic
[a]Expected on the basis of the chemical structure
[b]For particles with a high SMA/lipid ratio
[c]Not stated; will depend on the SMA/lipid ratio
[d]Hard to predict, given that hydrophilic moieties comprise both carboxylates and glucosyl groups

larger particles, with $R_H$ typically in the range of 15–20 nm (Liu et al. 2007), and to trap hydrophobic, non-membrane proteins (oleosins) in the form of much larger complexes than A8-35 and A8-75 do (Gohon et al. 2011).

Phosphorylcholine-based zwitterionic APols (PC-APols), whose net charge is very low at neutral pH, also form particles of a similar size as those of A8-35 (Diab et al. 2007b), as estimated by DLS. Being denser, these particles migrate faster than A8-35 particles upon ultracentrifugation on

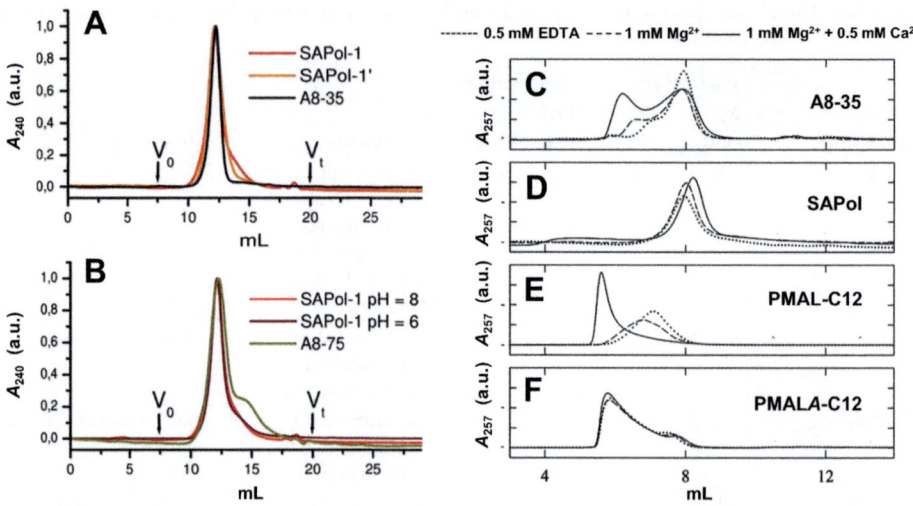

**Fig. 4.26** *Left.* Size exclusion chromatography of amphipols A8-35 and A8-75 and various batches of SAPols. Analyses were performed on a Superose 12 10/300GL column. Elution was carried out either with Tris/HCl buffer (pH = 8.0) or with phosphate buffer (pH = 6.0). (**A**) Two different batches of SAPol and one of A8-35. Elution at pH 8, detection at 240 nm. (**B**) SAPol-1 at pH 6 and 8 and A8-75 at pH 8. Detection at 240 nm. Chromatograms were normalized to the same maximal absorbance (a.u.: arbitrary units) (From Dahmane et al. 2011, © 2011 John Wiley & Sons, Inc., all rights reserved). *Right.* Effect of divalent cations on the aggregation of four different amphipols. Various APols were subjected to SEC in 20 mM *N*-[tris(hydroxymethyl)methyl]-2-aminoethanesulfonic acid (TES)/NaOH, 100 mM KCl, pH 7 buffer supplemented with either 0.5 mM EDTA (*dotted lines*), 1 mM $Mg^{2+}$ (*dashed lines*), or 1 mM $Mg^{2+}$ plus 0.5 mM $Ca^{2+}$ (*solid lines*) (Reprinted with permission from Picard et al. 2006, © 2006 American Chemical Society).

sucrose gradients (SG-AUC) (Diab et al. 2007b). MP/PC-APol complexes do not aggregate at pH 5 nor in the presence of either 1 M NaCl or 12 mM $Ca^{2+}$ (Diab et al. 2007a).

PMAL-C12 appears to form significantly bigger particles ($R_S \approx 6$ nm), sensitive to multivalent cations (Picard et al. 2006) (Fig. 4.26E). PMAL*A*-C12 (poly(maleic anhydride-*alt*-1-tetradecene) substituted with 3-(amidopropyl)dimethylamino-1-propane sulfonate, according to *Anatrace*'s catalogue) seems insensitive to multivalent cations, but forms very large aggregates (Picard et al. 2006) (Fig. 4.26F).

At variance with other APols, pure SMA, presumably because of its very short hydrophobic chains (phenyl rings) and high charge density, seems to form solutions of random-coiled individual molecules rather than collapsing or assembling into particles (Sauvage et al. 2004; Dörr et al. 2016). A large battery of approaches has recently been applied to studying the relatively small complexes that SMA forms with dimyristoylphosphatidylcholine (DMPC) (for a review, see Dörr et al. 2016). Note that there is good evidence that, as is the case for bicelles (see e.g. Sanders and Prosser 1998; Dürr et al. 2013), the size of the DMPC/SMA complexes depends on the ratio of the constituents (Carazo et al. 2015; Grimaldo et al. 2015; Zhang et al. 2015). In general, however, experiments are carried out in the presence of a large excess of SMA, so that the complexes studied have a minimal size (Dörr et al. 2016). DLS, EM, SANS, attenuated total reflection (ATR)-FTIR, differential scanning calorimetry (DSC), and NMR data have yielded a detailed picture of the size and shape of the complexes, as well as the arrangement of the various groups within them. The particles are comprised of a small patch of DMPC bilayer, whose acyl chains are covered with SMA (Fig. 4.27A), very much like the arrangement of long-chain vs. short-chain lipids in bicelles or of lipids vs. scaffold proteins in nanodiscs (Chap. 3).

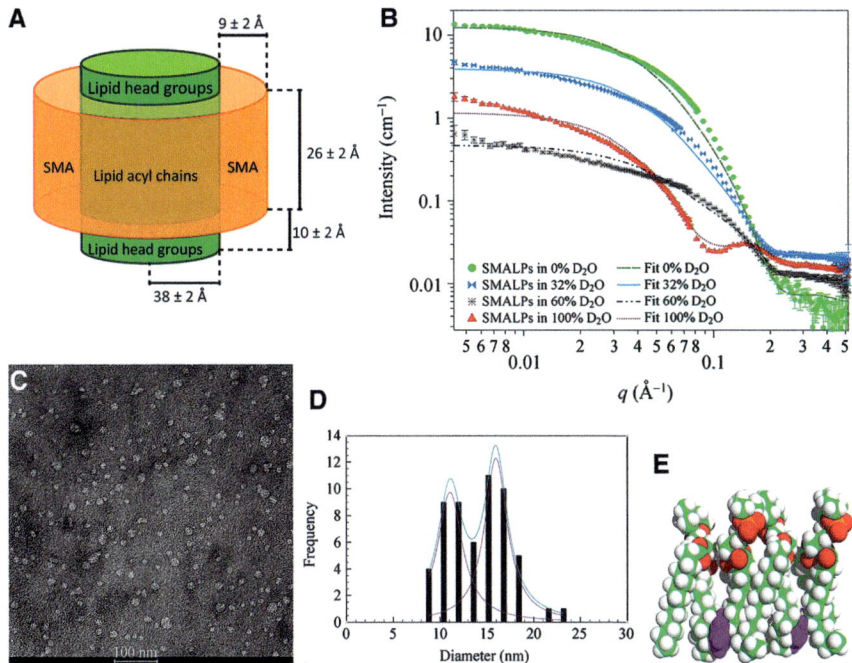

**Fig. 4.27** Arrangement of lipids and styrene-maleic acid copolymer (SMA) in SMALPs. (**A**) Dimensions of SMA/lipid particles (SMALPs) consisting of dimyristoylphosphatidylcholine (DMPC; *green*) and a SMA copolymer with a styrene-maleic acid ratio of 2:1 (*yellow*), as deduced from small-angle neutron scattering experiments (The figure, adapted from data in Jamshad et al. 2015b, is from Dörr et al. 2016). (**B**) Scattering patterns from SMALPs fitted to the model in **A**. Symbols represent the experimental results, lines the fitting. (**C**) Negative-stain transmission electron microscopy image of SMALPs. (**D**) Histogram showing the bimodal distribution of the diameters of SMALPs from the preparation shown in **C**, fitted by two Gaussians. (**E**) A molecular model showing the orientation and interactions of the phenyl moieties of SMA (*purple*) with the DMPC acyl chains (*green* and *white*) in SMALPs. All panels except **A** are from Jamshad et al. (2015b).

As expected, the carboxylates occupy the surface of the SMA belt. They seem to interact electrostatically with the head groups of lipids that reside in the outer layer of the nanodisc, whereas the phenyl moieties of SMA intercalate between the lipid acyl chains (Jamshad et al. 2015b; Dörr et al. 2016) (Fig. 4.27E). The model fits reasonably well experimental SANS data (Fig. 4.27B). Analysis of a population of SMALPs by EM after negative staining shows a bimodal distribution of diameters, with two maxima at $11.1 \pm 3.3$ nm and $16.0 \pm 3.0$ nm (Fig. 4.27C, D). These dimensions are larger than those observed in the SANS experiments, which suggest a maximum diameter of 9.8 nm (Fig. 4.27A). According to the authors, this disparity may reflect the different preparation methods for the two samples, the SANS data being averaged in buffered solution whereas the EM images are taken of individual particles from a dried sample stained with uranyl acetate (Jamshad et al. 2015b). Like A8-35, SMA aggregates at low pH and in the presence of multivalent cations (Gulati et al. 2014; Dörr et al. 2016).

Differences between SMA-bounded lipid patches and scaffold protein-bounded nanodiscs (NDs; Chap. 3, § 3.3) should not be overlooked. First, as mentioned above, whereas the size of NDs is determined by that of the scaffold proteins, that of SMA/lipid particles depends on the ratio of its components (Carazo et al. 2015; Grimaldo et al. 2015; Zhang et al. 2015; Grethen et al. 2017), much like that of bicelles depends on the lipid/detergent ratio, temperature, and concentration (Chap. 3,

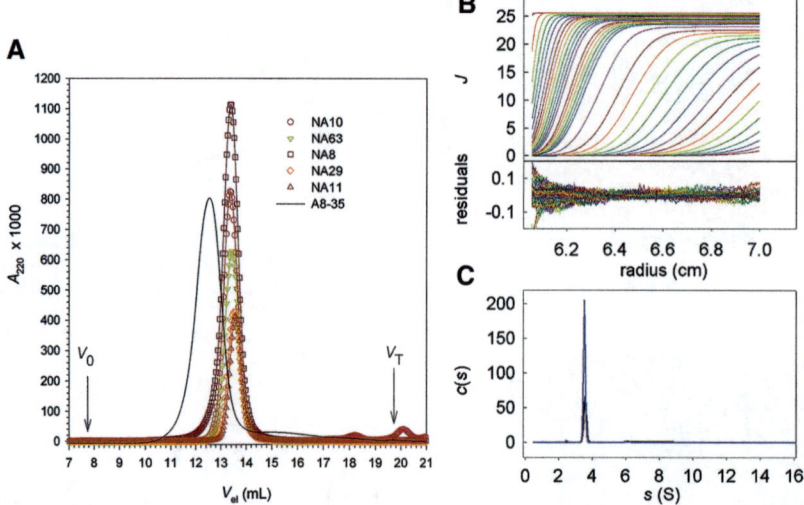

**Fig. 4.28** Size and mass analysis of the particles formed by glucose-based homotelomeric NAPols. (**A**) Size exclusion chromatography analysis of five NAPol batches with number-average molecular masses, $\langle M_n \rangle$, ranging from 8 to 63 kDa (noted NA8 to NA63), compared to A8-35. The polymers were dissolved at 10 g·L$^{-1}$ in 20 mM Tris/HCl buffer, 0.1 M NaCl, pH 8.0. Detection was done at 220 nm. $V_0$ (7.9 mL) and $V_T$ (20.1 mL) stand for the exclusion volume and the total volume of the column, respectively. (**B**) Sedimentation velocity analysis of NA29 in H$_2$O at 42,000 rpm and 20 °C. Detection with interference optics. *Top.* Superimposition of selected experimental profiles, corrected for systematic noises (*dots*) and of the corresponding models (derived from the $c(s)$ analysis; *lines*). *Bottom.* Residuals. (**C**) Superimposition of the distributions of NA29 particles obtained in H$_2$O at 9.80 (*blue*) and 4.93 (*black*) g·L$^{-1}$, showing a main species sedimenting at 3.55 and 3.60 S, respectively (Reprinted with permission from Sharma et al. 2012, © 2012 American Chemical Society).

§ 3.2). Second, phase transition and Raman spectroscopy studies (Orwick et al. 2012; Jamshad et al. 2015b; Tanaka et al. 2015; Oluwole et al. 2017) indicate that the lipid bilayer is more perturbed than it is in NDs, suggesting some degree of mixing between the polymer and the lipids (cf. Fig. 4.27E). Third, the lipid content of the various particles in a preparation is not segregated. It exchanges within seconds, with kinetics consistent with the occurrence of lateral fusion and fission events (Cuevas Arenas et al. 2017; Grethen et al. 2018). This observation suggests that SMALPs, at variance with NDs, do not prevent MPs nor lipids supposedly isolated in distinct patches from interacting one with another (which, depending on the experiments contemplated, can be either a useful feature or a nuisance). Preferential association of specific lipids with MPs extracted in SMALPs (see Dominguez Pardo et al. 2017 and references therein) thus must betray an equilibrium situation, rather than a kinetically frozen one (Cuevas Arenas et al. 2017). By the same token, it seems unlikely that SMALPs preserve the compositional asymmetry of the biological membranes the patches were extracted from.

Homotelomeric glucose-based non-ionic APols (NAPols) are highly soluble in water. Their solutions are transparent and remain fluid up to a concentration of 100 g·L$^{-1}$, above which their viscosity increases. NAPol particles have been studied by DLS, SEC, AUC, SANS, and densitometry (Sharma et al. 2012). Whatever the average molecular mass of the molecules, the particles they assemble into migrate upon SEC with the same $R_S$ and elute as though they were smaller than those formed by A8-35 (Fig. 4.28). This is probably due essentially to their carrying no charge, given that their $R_S$, when measured by DLS, is close to that of A8-35 particles, and their $R_g$, measured by SANS, is identical (Sharma et al. 2012). NAPols are significantly denser than A8-35: their specific volume is 0.771 mL·g$^{-1}$ (Sharma et al. 2012) vs. 0.866 mL·g$^{-1}$ for A8-35 (Gohon et al. 2004). As a result, the mass of NAPol particles is slightly higher as that of A8-35 ones (~50 kDa, as determined by AUC and SANS, vs. ~40 kDa). As for A8-35, each particle comprises, on average, ~75 alkyl chains (undecyl ones in the case of NAPols rather than octyl ones in the case of A8-35). Upon AUC, NAPol particles appear perfectly monodisperse (Fig. 4.28). Both SANS and AUC data are consistent with their being globular and compact.

It is worth noting that, upon SEC, the particles formed by homotelomeric NAPols elute as a very sharp peak, much narrower than that formed by A8-35 particles, suggesting a greater size homogeneity (Fig. 4.28A). This is confirmed by the narrowness of the $c(s)$ distribution in SV-AUC experiments (Fig. 4.28C). As noted above (§ 4.3.1.2.2), SEC analysis of fractions from a preparative SEC fractionation of A8-35 suggests that the variability of $R_S$ among particles may be due to individual molecules differing with respect to the extent or distribution of hydrophobic grafting. It is easy to imagine, for instance, that a large molecule that presents either a surfeit or a deficit of octyl chains will recruit respectively less or more smaller molecules in order for a core comprising 75–80 alkyl chains to form. At variance with A8-35, the molecules present in a batch of homotelomeric NAPol differ only in their length, not in the density nor distribution of alkyl chains. This could possibly lead to a narrower distribution of particle sizes. Consistent with this hypothesis, particles formed by heterotelomeric glucosylated NAPols, which are chemically heterogeneous, elute slightly later than A8-35 particles, as homotelomeric NAPol particles do, but they do not feature a narrower distribution (Sharma et al. 2008).

No detailed report appears to have been published on NVoy particles. According to unpublished data cited in the Supplementary Information to Klammt et al. (2011), their mass is ~112 kDa, corresponding to ~22 molecules.

We lack data to gather to which extent the other kinds of information that have been collected on A8-35 particles can be extended to other, less thoroughly studied APols. Certain tendencies can be surmised with some degree of confidence. For instance, it has been a constant observation, with a large spectrum of preparations of PAA-derived polymers, that samples that do not form by themselves small monodisperse particles do not yield small monodisperse MP/APol complexes (see Gohon et al. 2008). The behavior of the polymer alone, therefore, provides a first indication about its probable usefulness in biochemistry and biophysics. Symmetrically, the observation of small MP/polymer complexes can be taken as an indication that the polymer is likely to form by itself small particles. Some physical-chemical trends can probably be surmised. For instance, because particle assembly, as does micellization, forces polar moieties to come in close vicinity one to another, it is reasonable to expect that increasing the charge density along a given type of chain will raise the CAC (e.g. A8-75 or SAPols as compared to A8-35), whereas diminishing it or raising the ionic strength will have the reverse effect. There is a dearth of studies about the composition and organization of APol particles. However, it is striking that particles of A8-35 and glucose-based NAPols, whose chemical structures and physical properties are quite different, both comprise some 75–80 alkyl chains (Gohon et al. 2006; Sharma et al. 2012). This number is typical of detergent micelles (Chap. 2) and has also been encountered in studies of microdomain formation in other amphipathic polymers (Petit-Agnely and Iliopoulos 1999). This observation can perhaps provide a guideline when designing new APols or trying to guess at the self-organization properties of existing but poorly characterized ones. The sensitivity of APols to pH or multivalent cations, when it has not been experimentally established, can in general be surmised from the chemical structure (Table 4.4). Other features, such as solubility, viscosity, or the rate of exchange of molecules between particles, appear more difficult to anticipate.

## 4.4    Labeled and Functionalized Amphipols

One of the attractive characteristics of APols is that it is relatively easy to label them, isotopically or otherwise. Because, if they are not displaced by another surfactant, APols bind tightly to the TM surface of MPs (Chap. 5), they can therefore be used to label or tag non-covalently but essentially irreversibly any MP without having to modify it (Chap. 13). As previously observed for longer polymers (Ringsdorf et al. 1991; Morishima et al. 1995), a broad range of chemical modifications can be brought to the basic structure of APols without, as a rule, compromising the solubility of the molecules, their ability to self-associate and to adsorb onto MP TM surfaces, nor the solution properties of the resulting MP/APol complexes. Furthermore, because (i) each MP binds several APol molecules – BR, for instance, a small MP (27 kDa), binds ~54 kDa of A8-35 (Gohon et al. 2008) (see Chap. 5), i.e. ~12 molecules – and (ii) APols mix freely both in particles and at the surface of MPs (Zoonens et al. 2007), complexes carrying two or more functions can be formed by the simple device of trapping the protein with a mixture of APols (see e.g. Della Pia et al. 2014a, b; Le Bon et al. 2014a, and Chap. 13).

Labeled or tagged APols can be put to a multitude of uses, among which:

- Studying the solution properties of APols (this chapter) and MP/APol complexes (Chaps. 5 and 9); the miscibility of APols (Chap. 5, § 5.7); their biodistribution upon injection into living organisms (Chap. 15); their association with MPs and the composition, structure, and dynamics of MP/APol complexes (Chap. 5, §§ 5.3, 5.4, 5.5, and 5.6); the exchange of surfactants at the surface of MPs (Chap. 5, § 5.7); or the distribution of APols and MP/APol complexes in fractionation experiments (Chap. 5, § 5.2.1);
- Immobilizing MPs onto solid supports (Chap. 13);
- Modulating the contrast between APols and solvent or MPs in SANS or AUC experiments (Chap. 9) or improving NMR spectra (Chap. 10);
- Associating an adjuvant to a MP used as an immunogen (Chap. 15);

as well as many other applications still to be validated, some of which are tentatively indicated in Table 4.5.

Labeling or functionalizing APols can take various courses, each of which has its specific constraints and advantages regarding both the synthesis and the purification of the derivative. Once a basic APol structure has been validated, labeling it or endowing it with functional groups is, in general albeit not always, relatively straightforward (reviewed in Le Bon et al. 2014b). An overview of functionalized APols that have been validated for practical use is schematically presented in Fig. 4.29 and a list of labeled or functionalized variants of A8-35 and A8-75 in Table 4.5, along with some indications and references about the uses they have been or could be put to. As regards other types of APols, a biotinylated derivative of PC-APols has been described by Basit et al. (2012), a biotinylated NAPol by Ferrandez et al. (2014), and a thiolated SMA by Lindhoud et al. (2016).

### 4.4.1    Synthesis of Labeled or Tagged Derivatives of A8-35 and A8-75

The various routes toward labeled or functionalized derivatives of A8-35 and A8-75 have been recently reviewed and their respective fields of application, advantages, and drawbacks discussed by Le Bon et al. (2014b). Only a simplified overview is provided here. Further details on the chemistry of the syntheses are provided in Annex 4.6 (§ 4.6.6).

**Table 4.5** Tagged and labeled derivatives of A8-35 and A8-75 and their uses.

| Type of modification | Amphipol modified (short name of derivative) | Applications[a] | References[b] |
|---|---|---|---|
| **Isotopic labeling** | | | |
| $^{14}$C | A8-75 | Following A8-75 distribution and exchange | Tribet et al. (1997) |
| $^{3}$H | A8-35 | Following A8-35 distribution | Gohon et al. (2008) and Charvolin et al. (2014) |
| | | Evaluating MP/A8-35 mass ratio in complexes | |
| $^{2}$H (on side chains) | A8-35 (DAPol) | Eliminating the contribution of side-chain protons for NMR, EINS, or QENS measurements | Gohon et al. (2004, 2006, 2008), Zoonens et al. (2005), Catoire et al. (2009, 2010a), Etzkorn et al. (2014), Planchard et al. (2014), Sverzhinsky et al. (2014), and Tehei et al. (2014) |
| | | Contrast-matching in SANS, AUC | |
| $^{2}$H (perdeuteration) | A8-35 (perDAPol) | Eliminating protons for NMR measurements | Giusti et al. (2014) |
| | | Contrast-matching in SANS, AUC | |
| **Fluorescent labeling ($\lambda_{exc}$; $\lambda_{em}$)** | | | |
| Naphthalene (290 nm/310–370 nm) | A8-75 | Studying the interactions of A8-75 with lipid vesicles | Vial et al. (2005, 2009) |
| NBD (470 nm/530 nm) | A8-35 (FAPol$_{NBD}$) | Studying A8-35 distribution, binding, and exchange | Zoonens et al. (2007), Giusti et al. (2012), Le Bon et al. (2014a), and Sverzhinsky et al. (2014) |
| | | Determining the CAC of A8-35 by FRET with FAPol$_{rhod}$ | |
| Alexa Fluor 488 (490 nm/525 nm) | A8-35 (FAPol$_{AF488}$) | Unpublished data | Unpublished data |
| Fluorescein (495 nm/540 nm) | A8-35 (FAPol$_{fluo}$) | Mapping the transmembrane region of a MP by FRET with single-tryptophan mutants | Opačić et al. (2014b) |
| Rhodamine (555 nm/575 nm) | A8-35 (FAPol$_{rhod}$) | Examining the distribution of A8-35 upon delivery of a transmembrane peptide to cells in culture | Giusti et al. (2012) and Fernandez et al. (2014) |
| | | Determining the CAC of A8-35 by FRET with FAPol$_{NBD}$ | |
| | | Following the biodistribution of APols in mice | |
| Atto 647 (651 nm/667 nm) | A8-35 (FAPol$_{A647}$) | Following in vivo the distribution and elimination of A8-35 in mice | Unpublished results; see Chap. 15. |
| Alexa Fluor 647 (651 nm/668 nm) | A8-35 (FAPol$_{AF647}$) | Following in vivo distribution and elimination of A8-35 in mice | Della Pia et al. (2014a, b), Fernandez et al. (2014), and Le Bon et al. (2014a) |
| | | Tracer in APol-mediated immobilization of MPs | |

(continued)

**Table 4.5** (continued)

| Type of modification | Amphipol modified (short name of derivative) | Applications[a] | References[b] |
|---|---|---|---|
| **Tags, adjuvants** | | | |
| Biotin | A8-35 (BAPol) | Immobilizing MPs onto chips or beads for SPR or fluorescence measurements. See Chap. 13, §§ 13.2.1 and 13.2.2 | Charvolin et al. (2009), Della Pia et al. (2014a, b), and Ferrandez et al. (2014) |
| | | Selecting soluble protein binders against immobilized MPs. See Chap. 13, § 13.2.3 | Ferrandez et al. (2014) |
| | | Visualizing the APol belt in EM using monovalent avidin. See Chap. 13, § 13.3 | Perry et al. (2018) |
| Polyhistidine | A8-35 (HistAPol) | Reversibly immobilizing MPs onto $Ni^{2+}$- or $Co^{2+}$-bearing supports. See Chap. 13, § 13.2.4 | Giusti et al. (2015) |
| Randomly distributed imidazole moieties | A8-35 (ImidAPol) | Reversibly immobilizing MPs onto $Ni^{2+}$- or $Co^{2+}$-bearing supports. See Chap. 13, § 13.2.4 | Unpublished data cited in Le Bon et al. (2014a) |
| Oligodeoxynucleotide | A8-35 (OligAPol) | Immobilizing MPs onto DNA chips. See Chap. 13, § 13.2.4 | Le Bon et al. (2014a) |
| | | Vaccination, using the oligodeoxynucleotide as an adjuvant | |
| Thiomorpholine | A8-35 (SulfidAPol) | Immobilization of MPs onto gold surfaces, e.g. for surface plasmon resonance or surface-enhanced resonance spectroscopy. See Chap. 13, § 13.2.4 | Unpublished data |
| | | Attaching gold beads to APol belts for EM applications. See Chap. 13, § 13.3 | |
| Peptide EP67 | A8-35 (EP67-A8-35) | Vaccination, using the peptide as an adjuvant | Tifrea et al. (2018b) |
| Resiquimod | A8-35 (ResiqAPol) | Vaccination, using Resiquimod as an adjuvant | Tifrea et al. (2018a) |
| **Miscellaneous** | | | |
| Azobenzene | A8-75 | Photostimulated permeabilization of lipid vesicles and protein renaturation | Sebai et al. (2010, 2012) and Martin et al. (2015) |
| 4-Carboxytempo | A8-35 (SpinAPol) | Electron paramagnetic resonance, NMR, and fluorescence spectroscopy applications; protection against free radicals? | Unpublished data |
| N-Methylation | MethylAPol | A version of A8-35 carrying methylamide- rather than amide-linked side chains, for use in FTIR and other infrared spectroscopic studies | Unpublished data |

Updated from Le Bon et al. (2014b)

[a]Most of the applications listed have been validated in one or more of the studies whose references are given. A few of them are only contemplated or being explored

[b]References to synthesis and applications

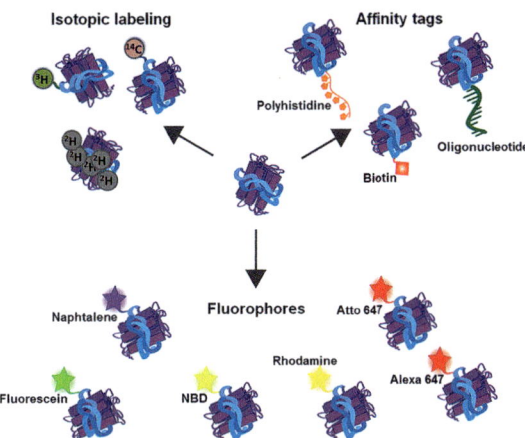

**Fig. 4.29**  A schematic overview of some of the labeled or functionalized APols whose use has been validated. See Table 4.5 for a more complete overview, details, and references (Adapted from Della Pia et al. 2014b).

Labeling or functionalizing A8-35 or A8-75 can be achieved in several ways, four of which are schematized in Fig. 4.30. A labeled or functionalized amine may be grafted onto the polymer backbone during (Fig. 4.30A) or after (Fig 4.30B–D) grafting hydrophobic side chains onto the PAA (HMPAS step, for "hydrophobically modified poly(acrylic acid) synthesis"). Grafting is performed after the synthesis if the moiety to be grafted is unstable under HMPAS conditions, or when its amine is not soluble in *N*-methylpyrrolidone (NMP). In such a case, labeling can be carried out in another organic solvent or in aqueous medium. In the latter case, grafting takes place at the surface of already formed A8-35 or A8-75 particles (Fig. 4.30C, D), and not on isolated macromolecular chains as is the case in NMP solution. This is expected to affect the distribution of the grafts (see below). Labeling can be performed either onto the polymer itself (Fig. 4.30A, C) or onto a pre-functionalized version of A8-35 bearing a reactive arm (Fig. 4.30B, D). Pre-functionalizing A8-35 presents a number of advantages: not only does it provide a route to the synthesis of labeled or tagged versions of A8-35, but it also ensures that whole sets of derivatives will feature exactly the same average length, length distribution, and octylamine and isopropylamine density and distribution – and, therefore, in principle, the same solution behavior. For this reason, this kind of general precursor has been dubbed "universal amphipol" (UAPol; somewhat of an overstatement!) (Zoonens et al. 2007). A very convenient type of UAPol is one carrying a free amine group (UAPol-NH$_2$), which can react with isocyanate, isothiocyanate, or activated ester derivatives of a probe. In principle, UAPol-NH$_2$ could also be used to attach a functional group by reductive amination (by reacting the amine with a carbonyl derivative), but this pathway offers limited prospects, most of the commercially available functionalized probes that are designed to be bound using reductive amination carrying an amino group (amine or hydrazine) rather than a carbonyl one.

On paper, a thiol-carrying APol (ThiAPol) could be a marvelous tool, because thiols react quantitatively with maleimide or alkyl halides to form a stable thioether bond and with activated thiols to form a covalent but reducible disulfide bond. A ThiAPol could be expected to provide higher yields of coupling than can be achieved with UAPol-NH$_2$. Moreover, the reversible character of the disulfide bond would offer the opportunity to endow APols with removable functions. Such APols could be used, for instance, to ensure the controlled release of drugs, probes, or other conjugates following exposure of functionalized ThiAPols to a reducing cellular environment such as the cytosol (Sauer et al. 2010) or the lysosomal lumen (Stefano et al. 2009). ThiAPols could also serve to immobilize

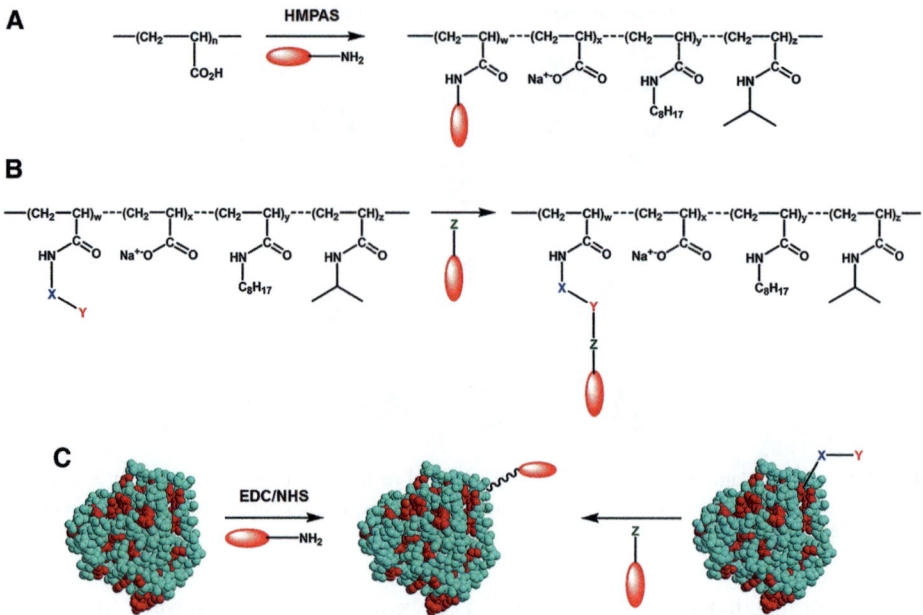

**Fig. 4.30** Four different approaches to labeling A8-35. (**A**) In *N*-methylpyrrolidone (NMP), by reacting an amino derivative of the probe with carboxylic groups of the poly(acrylic acid) at the stage of hydrophobic modification (HMPAS). (**B**) By reacting a pre-functionalized APol (UAPol) in organic medium with an activated probe. (**C, D**) By labeling A8-35 (**C**) or UAPol (**D**) particles in aqueous buffer. An amino derivative of the probe can be directly reacted with activated carboxyl groups of the particle, formed after treatment with ethyl-3-[3-dimethylaminopropyl]carbodiimide hydrochloride (EDC) in the presence of *N*-hydroxysuccinimide (NHS) (**C**). Alternatively, labeling may be achieved by reacting a UAPol particle with the appropriate probe derivative (**D**) (From Le Bon et al. 2014b).

MP/APol complexes onto gold surfaces, as is classically done with soluble proteins (Yoshimoto et al. 2008), integral MPs (Terrettaz et al. 2002), or whole cells (Roberts et al. 1998; Murphy et al. 2004). For all these reasons, a thiolated version of A8-35 appeared as a biochemist's dream. In practice, however, its synthesis turned out to be a chemist's nightmare, and a satisfactory route to producing it in good yield remains to be established (for a discussion, see Le Bon et al. 2014b). Labeling of SMA with cysteamine has been achieved, up to a level of ~0.6 thiol per polymer, and the resulting thiolated SMALPs labeled with thiol-reactive compounds (Lindhoud et al. 2016).

A number of derivatives of A8-35 and A8-75 have been obtained by incorporating the label or tag during PAA hydrophobization. They include several isotopically labeled versions (Tribet et al. 1997; Gohon et al. 2004, 2006, 2008; Giusti et al. 2014), a biotinylated derivative (BAPol) (Charvolin et al. 2009), and two derivatives carrying imidazole groups, in the form either of polyhistidine tags (HistAPol) (Giusti et al. 2015) or of distributed imidazole moieties (ImidAPol; unpublished data cited in Le Bon et al. 2014b) (see Table 4.5 and references therein).

When a high level of labeling is not required, as is the case, for instance, for fluorescent APols (FAPols), an efficient course is to derivatize an intermediate such as UAPol-NH$_2$. This is the method of choice when the fluorophore is too fragile for HMPAS, or insoluble in NMP, or when no aminated version of it is available. The synthesis of some FAPols, such as that of a naphthalene derivative of A8-75, can proceed by HMPAS. However, this route is harsh, particularly due to the final treatment with sodium methanolate, and few fluorophores can stand it. Rather, an activated version of the fluorophore, carrying either an isothiocyanate or an *N*-hydroxysuccinimidyl ester (NHS-ester), is usually reacted with the free amine functions of UAPol-NH$_2$. Provided a suitably functionalized version of the fluorophore to be grafted is commercially available, which is frequently the case, a single step is required for the synthesis, purification remaining, usually, the most time-consuming task. A large variety of FAPols has been synthesized and used for a wide variety of experiments, both in vitro and in vivo (Table 4.5 and Chap. 8, Fig. 8.9).

In a number of cases, labeling has to be carried out in aqueous solution, after A8-35 or A8-75 particles have already formed (routes C and D in Fig. 4.30). Such is the case for an oligonucleotide (ODN)-carrying version of A8-35 (OligAPol; Le Bon et al. 2014a). As compared to the syntheses mentioned above, coupling an ODN to an A8-35 particle presents special challenges, given that the reaction involves two large partners (~6.6 and ~40 kDa, respectively), both of them polyanions and, therefore, repulsing each other. Carboxylate functions of A8-35 were reacted with an amine-functionalized ODN in the presence of zero-length cross-linkers, ethyl-3-[3-dimethylaminopropyl] carbodiimide hydrochloride (EDC) and N-hydroxysuccinimide (NHS), leading to the formation of an amide function. The initial objective of 1 ODN per particle could not be reached, the grafting ratio being limited to ~0.5 ODN per A8-35 particle. This level of grafting was nevertheless sufficient to show that OligAPols assemble into well-behaved particles (see § 4.4.2) and to validate them as efficient tools to trap, handle, and immobilize MPs (Le Bon et al. 2014a; see Chap. 13).

The same approach was used to synthesize SAPols. Derived with 40% isopropylamine, using HMPAS, A8-75 yields A8-35 (Fig. 4.2). If isopropylamine is replaced with taurine, the end product is a sulfonated APol ("SAPol") (Fig. 4.1). However, SAPols cannot be efficiently synthesized by HMPAS because of the poor solubility of taurine in NMP. Taurine is therefore grafted in aqueous buffer, in the presence of EDC, onto the surface of already formed A8-75 particles (Dahmane et al. 2011) (route C). It can be assumed that taurine is grafted more or less randomly onto water-exposed carboxylates lying at their surface. Therefore, whereas the distribution of the octyl chains along the PAA backbone of SAPols is expected to be random (Magny et al. 1992), that of the taurine residues is likely biased, because carboxylate-rich regions of the chains stand a higher probability of reacting than octylamide-rich ones (cf. Fig. 4.18).

## 4.4.2   Solution Behavior of Labeled or Tagged Derivatives of A8-35 and A8-75

In general, labeling or tagging does not affect the solution properties of A8-35 or A8-75. Such is the case, for instance, for all FAPols (see e.g. Zoonens et al. 2007; Giusti et al. 2012; Opačić et al. 2014b) as well as for biotinylated A8-35 (unpublished data) or isotopically labeled APols, except, of course, as regards DAPol and perDAPol, whenever the properties measured depend on the mass or scattering length (cf. § 4.3.1.2.2).

Some tags however do affect the solution behavior of APol particles. Such is the case of the hexahistidine tag of HistAPol or the imidazole moieties of ImidAPol. The batch of ImidAPol most extensively investigated carried ~6 imidazole moieties per 100 PAA units, i.e. ~20 per 40-kDa particle. In basic aqueous solutions, its behavior resembles very much that of underivatized A8-35; in acidic solutions, however, ImidAPol precipitates quantitatively only when the pH is close to 3, due to the additional hydrophilicity provided, at acidic pH, by protonated imidazoles (F. Giusti, unpublished results). HistAPol was tagged to the level of 4–5 His$_6$ tags per particle, or 25–30 imidazole groups, i.e. a total level slightly higher than that of ImidAPol, and with quite a different distribution, since imidazoles, in HistAPol, come in groups of six. As is the case for ImidAPol, the solution properties of HistAPol under acidic conditions differ from those of plain A8-35, a large fraction of the polymer remaining in solution at pH < 5 (Giusti et al. 2015). At basic pH, HistAPol assembles into more heterogeneous and slightly larger particles than A8-35 or ImidAPol. From a practical point of view,

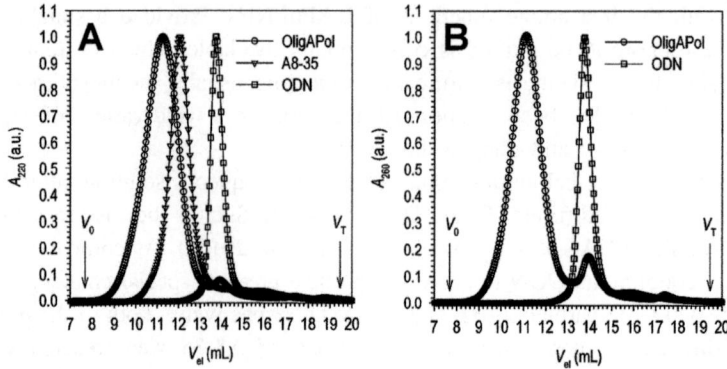

**Fig. 4.31** SEC analysis of purified OligAPol. Elution profiles of purified OligAPol and its precursors, unlabeled A8-35 and free oligonucleotide (*ODN*), recorded either at 220 nm (**A**), where both the unlabeled particles and the ODN are detected, or at 260 nm (**B**), where only the ODN is visible. The profiles have been normalized to the same maximum. a.u.: arbitrary units. $V_{el}$, elution volume; $V_0$, excluded volume; $V_T$, total volume (From Le Bon et al. 2014a, © 2014 Oxford University Press).

care has to be paid to carefully select the pH at which to precipitate ImidAPol and HistAPol during their purification. Above ~9 imidazole moieties per 100 PAA units (~30 per particle), the solubility of HistAPol at acidic pH increases to the point that purification by cycles of precipitation/resolubilization becomes impossible.

As mentioned above, OligAPol could be grafted only to the level of ~0.5 ODN per A8-35 particle. The mass contributed by the ODN, ~6.6 kDa, is large as compared to that of the average A8-35 molecule, but small as compared to that of A8-35 particles. Nevertheless, as shown in Fig. 4.31, it strongly affects the migration of the latter during SEC. When detection is carried out at 220 nm, where both the ODN and the polymer are visible, the OligAPol sample is seen to form a much broader peak than that formed by unlabeled A8-35, starting to elute well in advance of it and overlapping with it (Fig. 4.31A). At 260 nm, where only the free and grafted ODN absorbs, the OligAPol peak appears slightly narrower (Fig. 4.31B). The OligAPol peak detected at 220 nm (Fig. 4.31A) is comprised of two particle populations, (i) particles that do not carry any tag and (ii) tagged particles eluting well before them. Such a marked effect cannot be attributed to the relatively small mass excess contributed by the ODN, which by itself should barely affect the $R_S$ of the particles. It suggests that the ODN extends away from the surface of the APol particle, possibly due to electrostatic repulsion.[2] This behavior does not prevent the tagged molecules to assemble with untagged ones into well-behaved particles nor the use of OligAPol to trap and immobilize MPs (Le Bon et al. 2014a; see Chap. 13).

---

[2] Note that, because the exchange of molecules between particles is more rapid than their elution (§ 4.3.1.2.4), what is being followed in these SEC experiments is not really the migration of tagged particles, but that of particle-associated tagged molecules that "change horse" several times in the course of the experiment: they associate successively with several different particles, whose rate of migration increases or diminishes depending on whether they carry an ODN or not. Particles that are untagged at the time when they cross the detection beam have carried a tag about half of the time during their migration and therefore elute ahead of pure A8-35 particles. This explains why the OligAPol peak does not fully overlap that of untagged A8-35. Over time, some unlabeled particles that trail at the rear of the OligAPol peak lose their chance to pick an ODN again and are left behind, which explains why the OligAPol peak appears slightly broader at 220 than at 260 nm.

## 4.5     Protocol 4.1: Synthesis of A8-35

Whereas unlabeled A8-35 is commercially available, experimenters may want to use variants thereof that are either tagged or labeled isotopically or otherwise to suit their special needs. The synthesis of A8-35 is simple on paper, but it has its pitfalls. As has been described in § 4.3.1.2.2, polymers that do not match the nominal composition are at risk of presenting an abnormal solution behavior, which, in turn, will entail suboptimal properties of MP/A8-35 complexes, such as the formation of small aggregates (Gohon et al. 2008). This can be disastrous, for instance, for SANS or NMR experiments (see Chaps. 9 and 10, respectively). We present here a detailed protocol with comments and caveats (in italics and marked with a pointing hand ☞).

A8-35 is obtained by hydrophobization of a poly(acrylic acid) (PAA) precursor. This is achieved by successively grafting octylamine (yielding A8-75 as an intermediate) and isopropylamine onto the precursor dissolved in *N*-methyl-2-pyrrolidone (NMP), in the presence of *N,N'*-dicyclohexylcar-bodiimide (DCI) as activating agent (Tribet et al. 1996; Gohon et al. 2004, 2006). After purification, which is performed in aqueous media, the basic form of the polymer is obtained by neutralization in water with sodium hydroxide. The procedure can be decomposed in three steps: (i) PAA preparation and characterization, (ii) PAA modification, and (iii) purification and characterization of the final product.

### 4.5.1     Preparation and Characterization of the PAA Precursor

☞ *The starting material, PAA, is available from two different manufacturers, Aldrich and Acros, as 50% w/w aqueous solutions of a partial sodium salt form (Aldrich) or acidic form (Acros). The PAA from Aldrich presents a slightly different weight-average molar mass,* $\langle M_w \rangle$ *(see § 4.6.1, Annex 4.1, for definition), from that from Acros (~5 vs. ~5.5 kDa, respectively). Because of the broad dispersion of masses around the average one, this is without consequences on the properties of the final A8-35.*

**Disaggregation of the Precursor**
*Both commercial solutions are viscous and turbid, due to polymer aggregation. Extensive disaggregation is an essential preliminary step. Indeed, because hydrophobization must occur randomly, it is critical to achieve a thorough dispersion of the PAA chains in the reacting medium. This requires to break all weak intermacromolecular interactions, such as those resulting from H-bonding (Henke et al. 2011; Swift et al. 2016) and/or from chain entanglement (Harrington 2008). The disentanglement of the PAA chains can be achieved by heating at 90 °C a 10× diluted commercial solution, but better results are obtained by bringing the pH of a dilute solution to 10, followed by neutralization on an acidic resin column. PAA is recovered under its solid acidic form after removing water by freeze-drying. As expected (Yamamoto et al. 2000a, b), lyophilization does not lead back to polymer aggregation, and the resulting PAA readily yields limpid solutions in aqueous or organic polar solvents.*

**Characterization of the Disaggregated Precursor**
Dry mass. *Being hygroscopic, PAA is never totally water-free, even after freeze-drying. Because the chemical modification must be performed on a precisely known amount of polymer, the dry mass is commonly determined by total organic carbon analysis (TOC) or by simple acid/base titration of the carboxylic acid functions.*

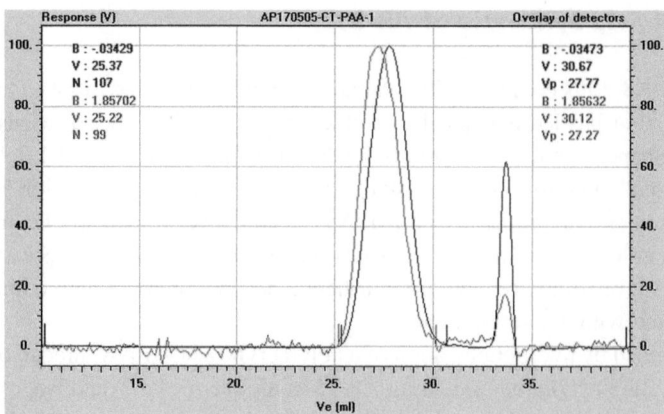

**Fig. 4.32** Gel permeation/size exclusion analysis of sodium polyacrylate (Aldrich) performed in 0.5 M LiNO$_3$ in water. Signals from refractive index (*blue* curve) and viscosity (*green* curve) detectors. Low molecular weight species (e.g. salts) elute between 33 and 34 mL (See Gohon et al. 2004, 2006 for detailed procedure. Figure courtesy of Christophe Tribet).

Size and homogeneity. *Because the size and size distribution of the PAA are not precisely known (the supplier provides only* $\langle M_w \rangle$*) and because the polymer has been handled in order to disaggregate it, its average size and size distribution must be evaluated by SEC (cf. Annex 4.1). SEC can be performed either in aqueous solution (Gohon et al. 2004, 2006) or in organic solvents (Giusti et al. 2014), the two approaches yielding significantly different results (for a discussion, see Giusti et al. 2014). Whichever approach is chosen, the polymer must migrate as a single, symmetric peak (Fig. 4.32).*

**Standard protocol for PAA disaggregation and characterization** (after Gohon et al. 2006; Giusti et al. 2014)

Disaggregation

Polyacrylic acid (PAA; MW 5000) was purchased from Aldrich (~50% w/w solution in water). 20 g of this turbid solution were diluted down to ~10% w/w in water purified on a Millipore Milli-Q Advantage A10 system (Milli-Q water) and supplemented with 5.6 g solid NaOH. The basic solution was poured into 500 mL of ethanol under vigorous stirring and the precipitated sodium polyacrylate recovered by filtration. After rinsing with 100 mL ethanol, the polymer was dissolved into 50 mL Milli-Q water and the solution filtered on Millipore membrane (0.22-μm cutoff) prior to being eluted in several runs on a home-packed column of Dowex 50WX2-200 or 50WX8-200 resin, hydrogen form. PAA was collected in the acidic fractions (2 ≤ pH ≤ 4, checked with pH paper). The sulfonic acid functions of the resin were regenerated after each run with 4 N HCl. The PAA fractions were pooled and the volume of the solution reduced to ~50 mL by evaporation under reduced pressure. The concentrated solution was freeze-dried, yielding ~9 g of PAA as a white powder. The extent of hydration, estimated by pH-metry (100 mg of PAA titrated with a 0.01 N NaOH solution), never exceeded 10%.

**Characterization**

Prior to SEC analysis, PAAs were modified by methylation of the carboxylic acid groups using trimethylsilyldiazomethane according to the method described by Couvreur et al. (2003). Briefly, 50 mg of PAA were dissolved in 10 mL of a 9:1 v/v tetrahydrofuran (THF)/methanol mixture and treated with 0.5 mL of trimethylsilyldiazomethane 2 M in diethyl ether (Presser and Hüfner 2004). The solvent was evaporated under reduced pressure and the poly(methyl acrylate) dried under vacuum for 12 h. Measurements were performed with a Viscotek TDAmax system from Malvern Instruments. The system comprises an integrated solvent and sample delivery module (GPCmax) and a Tetra Detector

Array (TDA) including a differential refractive index (RI) detector, a right-angle (90°) and a low-angle (7°) light scattering (LS) detector (RALS/LALS), a four-capillary differential viscometer, and a diode array UV detector. THF was used as the mobile phase at a flow rate of 1 mL·min$^{-1}$ and toluene as a flow rate marker. 100-µL samples of polymer solution (10 g·L$^{-1}$) were injected after filtration through a 0.45-µm pore-size membrane. The separation was carried out on three Polymer Laboratories columns (3 × PL gel; 5-µm Mixed C; 300 × 7.5 mm) and a guard column (PL gel 5 µm). Columns and detectors were kept at 40 °C. The OmniSEC 4.6.2 software was used for data acquisition and analysis. $\langle M_n \rangle$, $\langle M_w \rangle$, and the molar mass dispersity, $Đ = \langle M_w \rangle / \langle M_n \rangle$ (cf. § 4.6.1, Annex 4.1), were deduced from a calibration curve based on narrow polystyrene standards (from Polymer Standards Service-USA), using only the RI detector (simple detection). In addition, $\langle M_n \rangle$ was also calculated using the combined signals from RALS/LALS, RI, and the viscometer (triple detection).

## 4.5.2   Hydrophobic Modification of the PAA Precursor

☞ *Procedures for PAA hydrophobization, developed by* Wang et al. (1988), *were adapted by* Tribet et al. (1996), *to synthesize the first batches of A8-35. More detailed protocols are given in* Gohon et al. (2004, 2006) *and* Giusti et al. (2014). *The grafting of alkylamines onto the PAA dissolved in* N-*methyl-pyrrolidone (NMP) is mediated by activation of the carboxylic acid units by DCI. The mechanism of activation is given in Fig. 4.33. The molar ratio of each reagent must be strictly respected, because a slight drift of the chemical composition may have dramatic consequences on the water solubility and association behavior of the final APol. An additional reactant, HOBt, is introduced during the second*

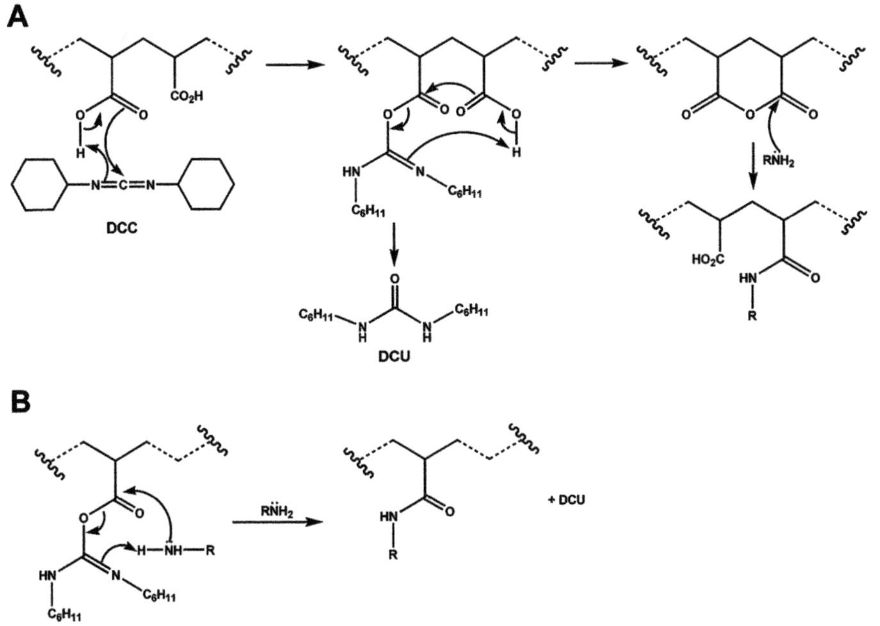

**Fig. 4.33** Simplified mechanism of DCI-mediated amidation. (**A**) When DCI is used in the presence of an excess of carboxylic acid functions, it dehydrates vicinal carboxylic acids to form an anhydride, releasing *N,N′*-dicyclohexylurea (DCU). The anhydride reacts with an amine to form an amide and a free carboxylate. In the early stage of the PAA modification, this mechanism is favored because of the high density of carboxylate units, as well as because a six-membered cyclic intermediate is generated. (**B**) Because the unstable *O*-acylurea intermediate (activated ester) is also reactive, it may directly react with the nucleophilic alkylamine when there is no vicinal carboxylate to form the anhydride. Most of the DCU formed in the reaction precipitates in the medium and can be removed by filtration.

*step (grafting of isopropylamine), because the reactivity of the remaining free carboxylates decreases drastically as the amount of grafted amines increases.*

**Standard Procedure for PAA Modification**

PAA (5 g, 69 mmol acid) was dissolved in 100 mL NMP over 4 h at 60 °C. Octylamine (2.22 g, 17.25 mmol in 15 mL NMP) was added, after which DCI (3.73 g, 18.11 mmol in 15 mL NMP) was supplemented dropwise. The mixture was allowed to react for 1 h at 60 °C, then brought to room temperature (RT), and kept under stirring for 3 h. After cooling at 0 °C, DCU was removed by filtration. 4 g (29.6 mmol) of 1-hydroxybenzotriazole (HOBt) were added to the filtrate and the mixture heated at 50 °C. After dropwise addition of isopropylamine (1.63 g, 27.6 mmol in 25 mL NMP), followed by DCI (6.26 g, 30.33 mmol in 30 mL NMP), the reaction medium was kept at 50 °C for 1 h and stirring carried on for 3 h at RT. The resulting acidic solution of A8-35 was cooled to 0 °C and filtered by suction.

At this stage, the A8-35 solution is contaminated by HOBt, residual DCU, and other undesirable by-products, which must be removed by purification.

### 4.5.3    Purification of A8-35

☞ *The purification procedure relies on the differential solubility in water of the acidic and basic forms of A8-35. When totally ionized (above pH 7), A8-35 is highly soluble in water ($>320$ g·L$^{-1}$; Gohon et al. 2006), whereas the protonated form is insoluble. Thus, if the purification is performed in a large excess of aqueous solution, the hydrophobic contaminants (DCU, essentially) can be removed by precipitation at high pH and the water-soluble impurities (HOBt and NMP) washed away from the A8-35 precipitate at pH < 7. Extensive purification is critical and must be carefully carried out. Because the solution behavior of polymers may depend on their history (see e.g. Selb and Gallot 1980; Yamamoto et al. 2000a, b), precipitation and dissolution are carried out quickly and at extreme pH values (0–1 and 10–12 for acidic and basic solutions, respectively). Reaching a satisfying level of purity requires four cycles of dissolution/precipitation. Salts are removed by dialyzing the final A8-35 solution against slightly basic water.*

**Standard Procedure for A8-35 Purification**

7.5 g (0.138 mol, two equivalents) sodium methanoate are added to the NMP solution which is diluted in large excess of water (1.5 L). After removal by filtration of residual DCU, the polymer is precipitated by adding dropwise the basic solution to 100 mL of 3 N HCl. After redissolution of the polymer in 500 mL of 0.3 N NaOH, precipitation is repeated according to the same procedure. The cycle is repeated two additional times. Finally, the basic solution (300 mL) is dialyzed against 2 L of $10^{-3}$ N NaOH for 2 days and then against Milli-Q water for 1 h and finally freeze-dried, yielding 7.74 g of product (90% yield with respect to the mass of A8-35 expected assuming 100% conversion of the purified PAA precursor; modification of 1 g of pure PAA should yield 1.72 g of A8-35, sodium salt).

### 4.5.4    Determination of the Chemical Composition of the Final Product

☞ *The chemical composition of the final product can be determined by $^1H$ and $^{13}C$ NMR spectroscopy.*

The $^1H$ NMR spectrum can be analyzed in three principal regions: 0.8–1.4 ppm (methyl and methylene protons of alkyl side groups), 2.6–1.4 ppm (backbone protons), and 2.9–4.1 ppm (methylene and methine protons of the side groups in $\alpha$ position to the amide function) (see Fig. 4.34). Octyl groups give rise to three peaks centered at 3.15 ppm, 1.3 ppm, and 0.85 ppm, isopropyl groups to peaks at 3.9–4.0 ppm and 1.1 ppm.

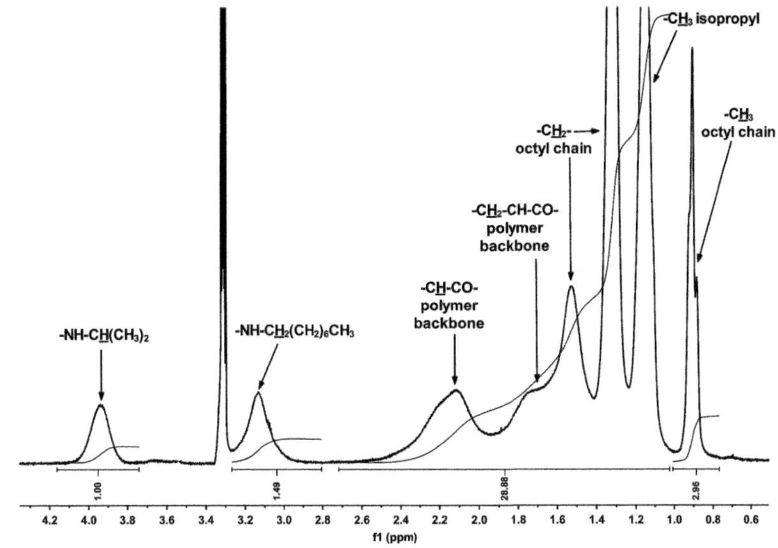

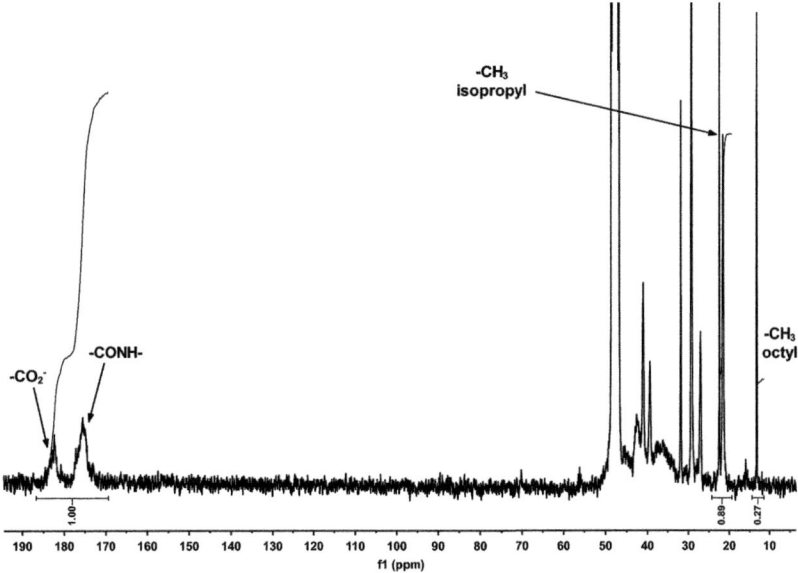

**Fig. 4.34** $^1$H NMR spectrum of A8-35 recorded in CD$_3$OD.

**Fig. 4.35** $^{13}$C NMR spectrum of A8-35 recorded in CD$_3$OD.

The fraction of isopropyl groups with respect to the initial acrylate ones is given by the ratio of the integral of the peaks of $\alpha$-methine signals at 4.0 ppm to the integral of the broad peak at 2.4–1.9 ppm, corresponding to the backbone methine protons. Because of the overlap between peaks at 1.3 ppm and 1.1 ppm, the integral of the signals of isopropyl group methyl protons is not accurate in this region and cannot be used to derive a degree of grafting.

The fraction of octyl groups is given by (i) half the ratio of the integral of the peaks of $\alpha$-methylene at 3.0 ppm to the backbone methine integral and (ii) the third of the ratio of the integral of methyl protons at 0.85 ppm to the same methine integral. The two estimates thus obtained do not differ by more than 10%.

$^{13}$C NMR provides further estimates of the grafting ratios (Fig. 4.35). The two carbonyl peaks in the 170–190 ppm region are due to the carboxylate and amide functions. The ratio of their integrals provides a reliable measurement of the ratio of the two functions. The ratios of the integral of the peak centered at 12 ppm (methyl group of the octyl side chain) and of half of that at 22 ppm (the two methyl groups of the isopropyl side chain) to the sum of the integrals of unmodified and modified carboxylates yield, respectively, the grafting ratios of octyl and isopropyl side chains, which are comparable to the estimates obtained by $^1$H NMR.

Further estimation of grafting ratios can be obtained by elemental analysis and by pH-metry (estimation of free carboxylates), the latter being performed in 80:20 v/v ethanol/water.

*Protocol prepared by Fabrice Giusti on the basis of* Tribet et al. (1996), Gohon et al. (2004, 2006), Giusti et al. (2014)*, and laboratory notes.*

## 4.6    Annexes

### 4.6.1    Annex 4.1. Determining and Expressing the Average Mass and Dispersity of Polymers

Determining the average mass of synthetic polymers is challenging, the main issue being that, whatever the synthesis pathway, a polymer preparation will always present a more or less broad molecular mass distribution. Synthetic polymers result from the successive incorporation of units called monomers, which react on themselves to form a macromolecular chain, much like, in vivo, amino acids associate one with another to yield the primary structure of a protein. However, at variance with biomacromolecules, whose number of units (sugars, amino acids, nucleotides, isopentenyl, etc.) and sequence are usually well-defined, the number of monomers that form a synthetic macromolecular chain can only be approximated by an average value and a rough estimation of deviations from this average. This is a consequence of the fact that chemical synthesis of long polymers cannot be controlled step by step, as most biological syntheses are, but comprises many sources of length variation.

#### 4.6.1.1    The Origin of Dispersity

Polymerization reactions fall into two main categories, chain-growth polymerization, where the reaction produces only macromolecular chains, and step-growth polymerization, where the reaction produces macromolecular chains and by-products called condensates (Flory 1953). Polycondensation (such as a diol or diamine reacting with a diacid, yielding respectively polyester and polyamide, with water as the condensate; Fig. 4.36A) is an example of step-growth process, whereas radical and ionic (cationic or anionic) polymerization are examples of chain-growth processes (Fig. 4.36B).

Both methods lead to heterogeneous mixtures. As an example, we will consider the case of free radical polymerization, a chain-growth process, with monomers endowed with a vinyl group (Fig. 4.37). Polymerization is initialized by an initiator (Ini$_2$), whose decomposition generates two free radicals (Ini$^\bullet$). The radicals react with a monomer (M) to form the first reactive adduct initiator monomer (Fig. 4.37, Step 1) (Mayo et al. 1951). Next, each adduct reacts with another monomer to form a di-adduct, which in turn reacts with yet another monomer, yielding a tri-adduct, etc., leading to the formation and growth of a macroradical. The growth step is called propagation (Fig. 4.37, Step 2). One would be tempted to postulate that the final size of the polymer can be simply defined by the ratio of the initial concentration of the monomer to that of the initiator. However, reality is more complicated, the growth of the macroradicals being limited by several random phenomena, namely the efficiency of the initiator, the randomness of growth from one polymer to the next, transfer reactions, and the vagaries of reaction termination.

The efficiency of the initiator is related to the fraction of Ini$^\bullet$ radicals that do react with a monomer to initiate the polymerization, which is not 100%. Further, the growth of the chains can be stopped by transfer reactions, which occur between a macroradical and a second partner, which can be either the solvent, a monomer, an initiator, or an impurity (Fig. 4.38). Those reactions that generate a macromolecule and a new free radical affect the size of the macromolecule in formation, without stopping the polymerization process, as the newly formed radical reacts with unreacted monomers to yield another macroradical. The process of polymerization is stopped by disappearance of the free radicals. This may occur according to various mechanisms, among which chain disproportionation and combination (Fig. 4.37, Step 3). Chain disproportionation involves the transfer of an atom (hydrogen, essentially) from a macroradical donor to a macroradical acceptor, which combination entails the covalent coupling of two macroradicals. Given the existence of these various factors, the final preparation is necessarily polydisperse.

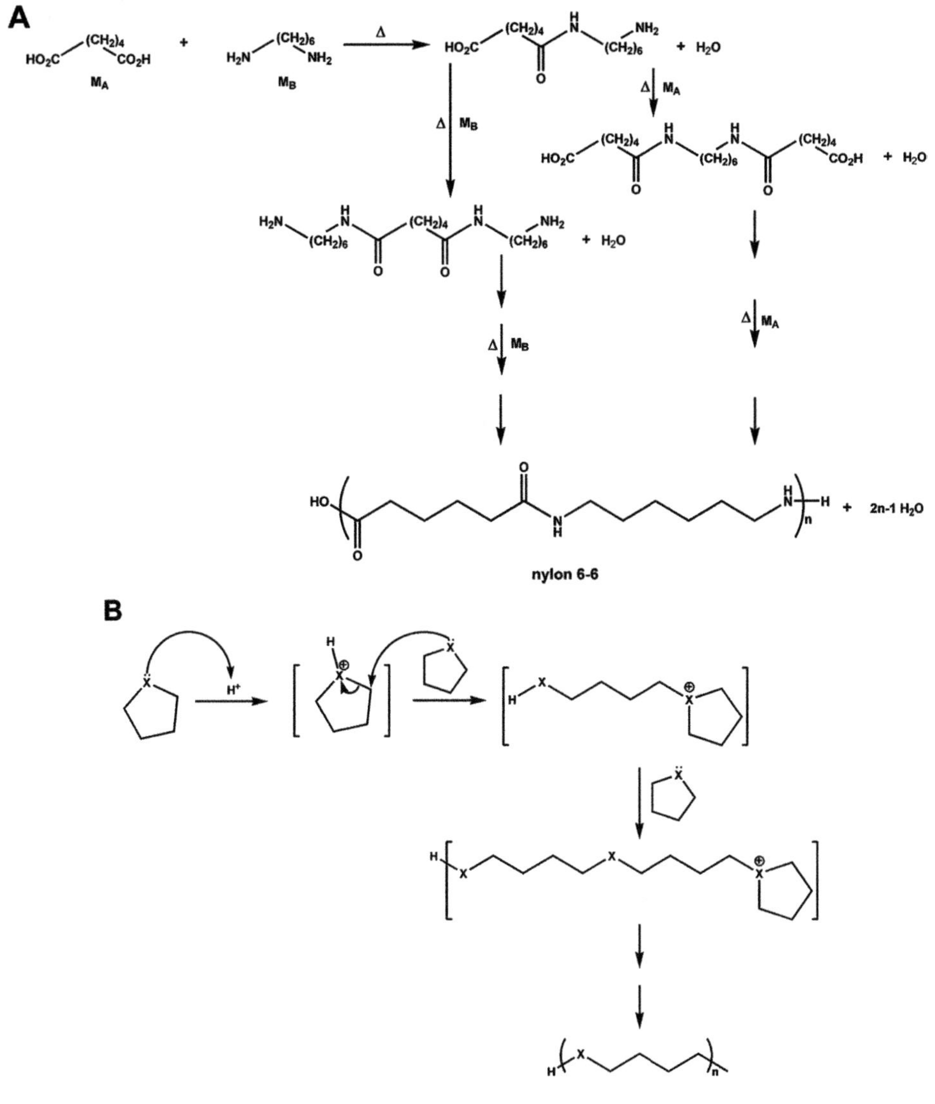

**Fig. 4.36** Schematic illustration of two different polymerization mechanisms. (**A**) A step-growth process, the synthesis of Nylon 6-6. Here, copolymerization by condensation of two symmetrically bi-functionalized monomers, namely adipic acid ($M_A$) and hexamethylenediamine ($M_B$), generates oligomers of increasing size and the condensate (water). (**B**) A cationic ring-opening reaction as an example of chain-growth process, generating only the macromolecular chain, without condensates.

### 4.6.1.2   How Can Size Dispersity Be Limited?

The chemist can resort to several strategies to try to limit the drift of polymer size. Various methods aiming at controlling, with more or less efficiency, the size of the polymers during their formation have been developed during the last three decades, both for chain-growth (reviewed in Starks 1974; Chen et al. 2009; Misha and Kumar 2012) and for step-growth polymerization (reviewed in Yokozawa and Yokoyama 2004). As regards the chain-growth process, most approaches consist in performing the polymerization in the presence of a reactant that controls the growth of the macroradicals (this point will be discussed in § 4.6.4, Annex 4.4). In the best cases, the dispersity will be considerably reduced, but it will never reach the ideal situation where all macromolecules in a preparation would have strictly the same size. Therefore, the molar mass of a given polymer will always be expressed as a statistic average, and these statistics are complicated.

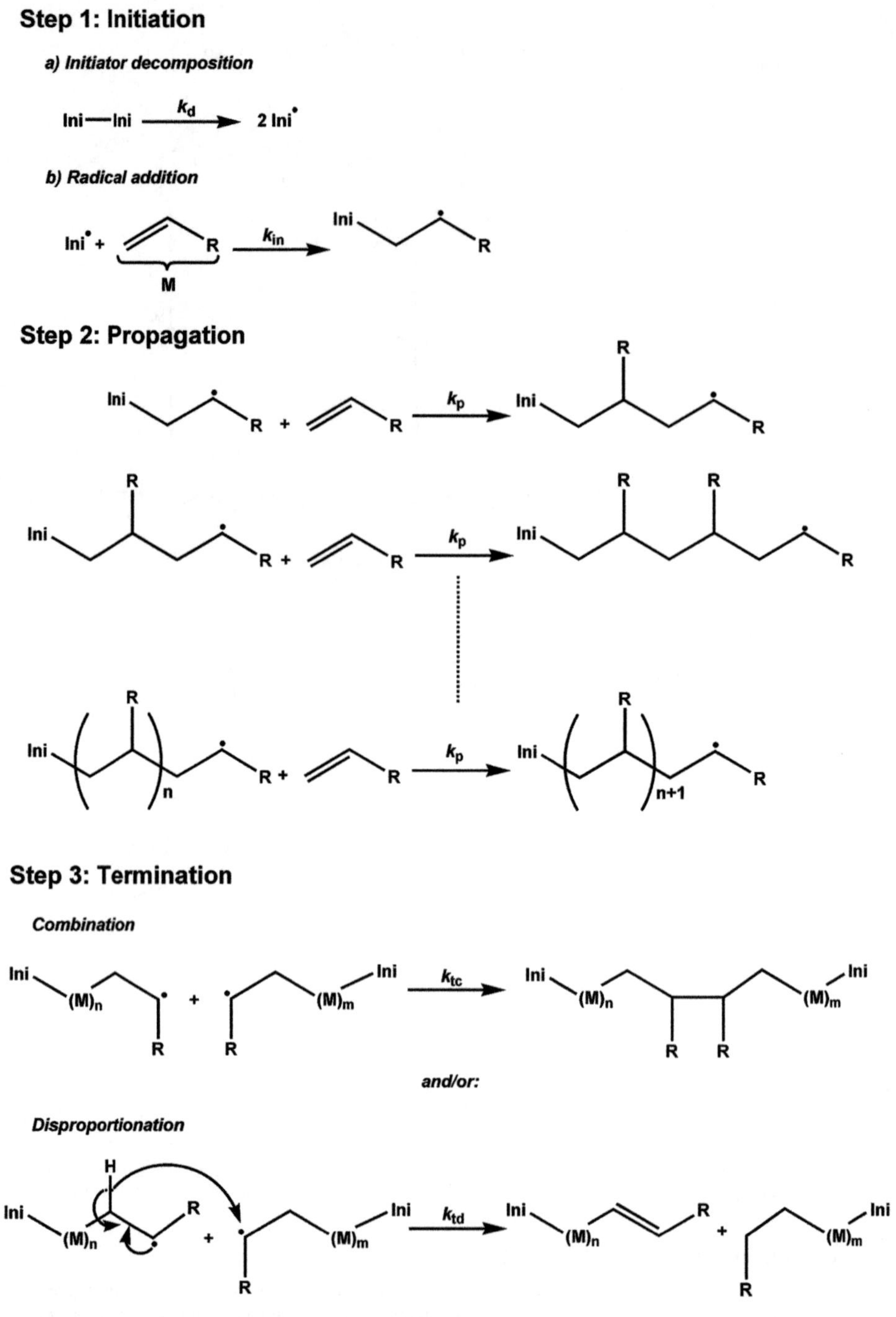

## Step 1: Initiation

### a) Initiator decomposition

$$\text{Ini—Ini} \xrightarrow{k_d} 2\,\text{Ini}^\bullet$$

### b) Radical addition

## Step 2: Propagation

## Step 3: Termination

### Combination

*and/or:*

### Disproportionation

**Fig. 4.37** Schematic representation of the conventional radical polymerization mechanism. The reaction is a three-step process that involves initiation, propagation, and termination. The kinetic of each event is characterized by a rate constant ($k_d$, $k_{in}$, $k_p$, $k_{tc}$, and $k_{td}$). Here, $\text{Ini}_2$ refers to any molecule susceptible to decompose under heat or UV irradiation to form free radical species ($\text{Ini}^\bullet$). Because free radical species are highly reactive ($k_{in} \ggg k_d$), the global rate of initiation depends strongly on the rate of decomposition of the initiator. According to Flory's hypothesis, except for the first adducts with $n < 5$, the kinetic constant of a given reaction is independent of the size of the molecule bearing the functional groups, so that $k_p$ remains the same for each propagation reaction. The global rate constant of the termination can be expressed as $k_{te} = a \cdot k_{tc} + (1 - a) \cdot k_{td}$ where $0 \leq a \leq 1$ is the fraction of termination taking place by combination.

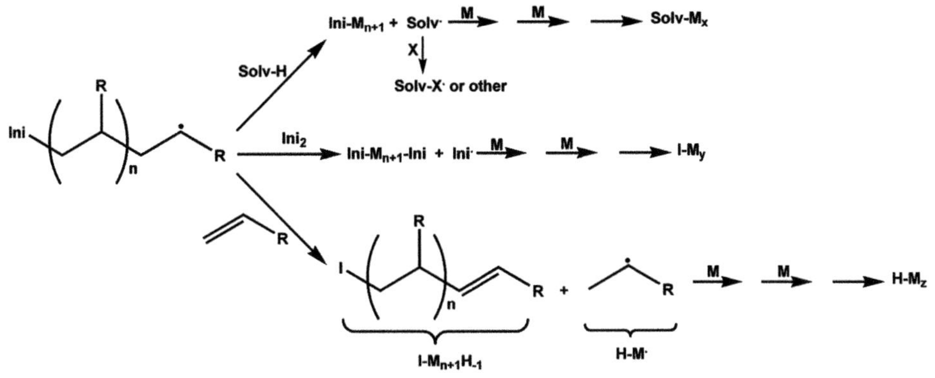

**Fig. 4.38** Schematic representation of the transfer side reactions taking place during radical polymerization. The growing macroradical may either capture a hydrogen atom (H$^{\bullet}$) from the solvent (Solv-H), exchange it with a molecule of monomer (M), or react with a molecule of initiator (Ini$_2$). The products of the side reactions are free radicals, which may react with different molecules present in the medium (solvent, initiator, monomer, polymer, etc.) or provide further radical polymerization reactions, yielding a mixture of heterogeneous-sized macromolecules (Ini-M$_{n+1}$ + Solv-M$_x$ + Ini-M$_y$ + H-M$_z$).

### 4.6.1.3 Expressing the Average Mass and Dispersity of Polymers

Whatever strategy has been adopted, the final product of any chemical polymerization is a mixture of macromolecules of variable length and mass. By performing appropriate analyses of a polymer sample (see below), different average values of the molar mass and statistical deviations around it can be obtained, the two most frequently used of which are the number-average and the weight-average masses, noted respectively $\langle M_n \rangle$ and $\langle M_w \rangle$, whose definitions will follow. These two values would be identical for a perfectly homogeneous sample, but they differ if the sample is heterogeneous, and their ratio gives a measure of the dispersity. The following summary is based on recent IUPAD recommendations about which nomenclature and notations appear preferable (Gilbert et al. 2009).

The study of colligative properties (those properties that vary as the number of molecules per unit volume, which can be measured, e.g. by cryoscopy, osmometry, ebullioscopy, etc.) yields the number-average molar mass of the polymer, $\langle M_n \rangle$. $\langle M_n \rangle$ represents the average molar mass of macromolecules weighed as a function of their abundance in the sample. If the sample comprises $N_i$ macromolecules of molar mass $M_i$, with $M_i$ stretching over a given range, $\langle M_n \rangle$ is defined as follows:

$$\langle M_n \rangle = \frac{\sum_i N_i M_i}{\sum_i N_i} \tag{4.1}$$

For a homopolymer, the number-average size or degree of polymerization $\langle X_n \rangle$ (formerly noted $\langle DP_n \rangle$) of the polymer is defined as:

$$\langle X_n \rangle = \frac{\langle M_n \rangle}{M_0} \tag{4.2}$$

where $M_0$ is the molar mass of the monomer.

Measuring the intrinsic properties of the sample (properties linked to the nature and the mass of the macromolecules, which can be studied, e.g. by light scattering or by viscometry) gives an estimate of the weight-average molar mass of the polymer, $\langle M_w \rangle$. The contribution to the mass of a given sample of those molecules whose molar mass is $M_i$ is given by $W_i = N_i \cdot M_i$. The weight-average mass is obtained by weighing the contribution of each polymer by its mass:

$$\langle M_w \rangle = \frac{\sum_i W_i M_i}{\sum_i W_i} = \frac{\sum_i N_i M_i^2}{\sum_i N_i M_i} \tag{4.3}$$

where $\Sigma_i N_i M_i$ is the total mass of the sample. The respective meanings of $\langle M_n \rangle$ and $\langle M_w \rangle$ are best perceived by considering that on each side of $\langle M_n \rangle$ lies an equal number of molecules, on each side of $\langle M_w \rangle$ an equal mass of them. It is therefore intuitive that for a mixture of small and big molecules, $\langle M_w \rangle$ will necessarily be larger than $\langle M_n \rangle$ (an example is given below).

For a homopolymer, the weight-average degree of polymerization ($\langle X_w \rangle$, formerly $\langle DP_w \rangle$) can be deduced from $\langle M_w \rangle$ as follows:

$$\langle X_w \rangle = \frac{\langle M_w \rangle}{M_0} \tag{4.4}$$

The dispersity in molar mass distribution, noted $Đ_M$, is related to $\langle M_n \rangle$ and $\langle M_w \rangle$ by:

$$Đ_M = \frac{\langle M_w \rangle}{\langle M_n \rangle} \tag{4.5}$$

$Đ_X$, the dispersity in size, is given by:

$$Đ_X = \frac{\langle X_w \rangle}{\langle X_n \rangle} \tag{4.6}$$

For a homopolymer, $Đ_M = Đ_X = Đ$. In the ideal case of perfect size homogeneity, $Đ$ would be equal to 1.

The meaning of $Đ$ is not that intuitive. As pointed out by Rane and Choi (2005), $Đ$ is related to the standard deviation around the average mass or length, but is *not* the standard deviation. (This is obvious given that the standard deviation for a perfectly homogeneous sample would be 0, not 1.) The standard deviation $S_n$ of $\langle M_n \rangle$ is related to the dispersity as follows:

$$\frac{S_n^2}{\langle M_n \rangle^2} = Đ - 1 \tag{4.7}$$

The meaning of $Đ$ can be illustrated by this example adapted from Dörr et al. (2016). Let us consider a distribution of three polymer molecules with molecular weights of 500, 1000, and 10,000 Da. In this example, $\langle M_w \rangle \approx 8800$ Da, and $\langle M_n \rangle \approx 3900$ Da, resulting in $Đ \approx 2.25$. For a typical polymer with $Đ \approx 2.5$, the polymer chains have a broad size distribution with the smallest and largest chains differing by more than one order of magnitude in molecular weight and thus chain length.

Two other mass averages that are occasionally used are the viscosity-average molar mass $\langle M_v \rangle$ and the sedimentation-average molar mass $\langle M_z \rangle$. Typically, the values obtained for the four averages relate to each other as $\langle M_z \rangle > \langle M_w \rangle > \langle M_v \rangle > \langle M_n \rangle$.

### 4.6.1.4　Determining the Average Mass and Dispersity of Polymers

$\langle M_n \rangle$ and $\langle M_w \rangle$ are most often determined by size exclusion chromatography (SEC). A good description of this approach is given in Meunier (1997) and Meunier et al. (2014). SEC experiments are performed on a HPLC instrument specifically equipped and dedicated to polymer mass analysis (Fig. 4.39). It comprises several (three to four) columns connected in series and several detection systems. The columns are calibrated with standard polymers of various average molar masses with a low dispersity ($Đ < 1.1$). The analysis requires knowledge of the intrinsic viscosity $[\eta]$ of the polymer (see below).

### 4.6.1.5　Calibration of Size Exclusion Chromatography Columns

A crucial part of the process is the choice of a suitable method of calibration, the most frequently used of which is called "universal calibration." Universal calibration relies on the fact that the hydrodynamic volume $V_H$, intrinsic viscosity $[\eta]$, and molecular mass $M$ of a polymer are linked by the following relation:

$$V_H \propto [\eta] \cdot M \tag{4.8}$$

Therefore, if two polymers, 1 (the standard) and 2 (the sample), have the same hydrodynamic volume and, therefore, elute simultaneously from a SEC column:

$$[\eta]_1 \cdot M_1 = [\eta]_2 \cdot M_2 \tag{4.9}$$

The intrinsic viscosity may be seen as the partial contribution of the solute to the whole viscosity $\eta$ of the solution. It is commonly expressed as follows:

$$[\eta] = \lim_{c \to 0} \frac{\eta - \eta_0}{\eta_0 \cdot c} \tag{4.10}$$

where $\eta_0$ is the viscosity of the solution in the absence of the solute and $c$ is the concentration of the solute in $g \cdot dL^{-1}$.

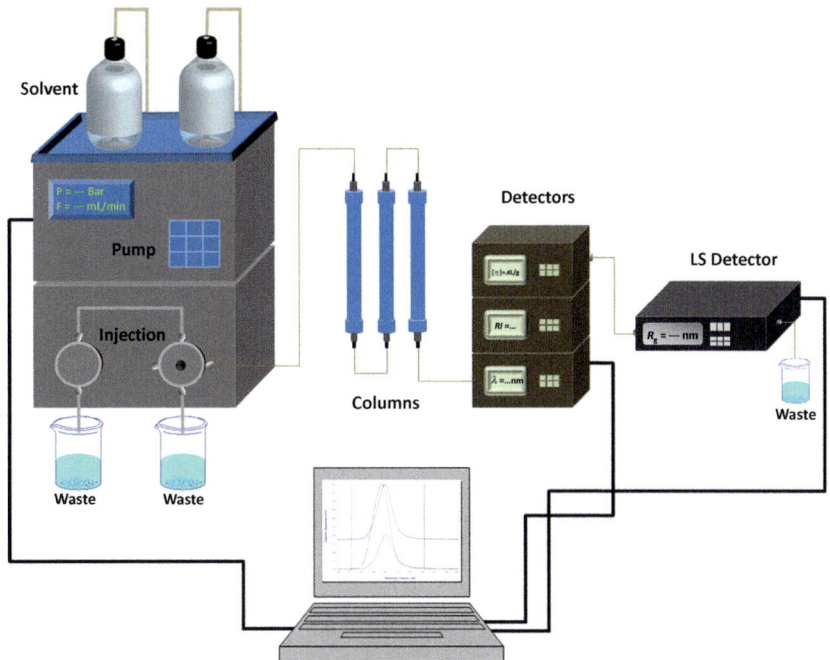

**Fig. 4.39** Schematic representation of a typical triple- or, optionally, quadruple-detection SEC HPLC instrument. The solvent, usually tetrahydrofuran (THF), is pumped and degassed online. The runs are performed isocratically (i.e. at constant solvent composition) at low flow rate ($<2$ mL·min$^{-1}$). The injected polymer is eluted through three thermostated coupled columns with different permeation cutoff in order to cover a broad range of molar masses. The elution is monitored by UV-visible absorbance measurements and/or by refractometry, both of which are sensitive to polymer concentration. A viscometer is placed after the concentration detector and gives the viscosity of the solution, from which the polymer's intrinsic viscosity $[\eta]$ is deduced. Formerly, a low-angle light scattering (LALS) detector was used as a third or fourth detector. It is now replaced by a multi-angle light scattering (MALS) detector. Light scattering yields the polymer's radius of gyration, which, for particles with a given geometry, is correlated to the molar mass. $[\eta]$ and the refractive index $RI$ are involved in the complex equations relating the molar mass to the intensity of light scattering. The coupling of three detectors measuring $RI$, $[\eta]$, and the intensity of light scattering therefore allows to determine the average molar mass of any unknown sample with good accuracy.

A calibration curve is obtained by plotting $\log([\eta]_1 \cdot M_1)$ vs. the elution volume of each standard. $M_2$ can then be deduced from Eq. 4.9 provided $[\eta]_2$ is known. $[\eta]_2$ can be either directly measured using an online viscometer or deduced from the Mark-Houwink-Sakurada equation (see e.g. Wagner 1987 and references therein):

$$[\eta] = kM^{\alpha} \tag{4.11}$$

where $k$ and $\alpha$ are constants that depend on polymer composition, temperature, and solvent. They are listed in handbooks for many polymers.

Combining Eqs. 4.9 and 4.11 gives:

$$M_2 = (k_1/k_2)^{1/\left(1+\alpha_2\right)} \cdot M_1{}^{R} \tag{4.12}$$

with $R = (1 + \alpha_1)(1 + \alpha_2)$.

Knowing $M_i$ for each position along the elution profile, $\langle M_n \rangle$ and $\langle M_w \rangle$ are given by:

$$\langle M_n \rangle = \frac{\sum_i h_i}{\sum_i h_i / M_i} \tag{4.13}$$

$$\langle M_w \rangle = \frac{\sum_i h_i M_i}{\sum_i M_i} \tag{4.14}$$

where $h_i$ is the height measured on the SEC curve (refractive index or UV absorbance) for each position $i$ along the elution profile.

An absolute value of the molar mass can also be directly obtained by passing the sample through a low- or multi-angle light scattering (LALS or MALS, respectively) detector.

The reader must keep in mind that all theories exposed above remain applicable only as long as the macromolecules composing the sample are not involved in any mechanism of self-association. Therefore, SEC experiments must be performed in dispersing media. For instance, the estimation of A8-35 average molar masses cannot be achieved in aqueous buffers.

## 4.6.2    Annex 4.2. Kinetics of Conventional Radical Polymerization and Origin of Size Dispersity

Radical polymerization (RP) is initiated slowly, because the rate ($R_{in}$) of the first-order reaction of the initiator (Ini$_2$) decomposition is low. This rate is given by the following relation:

$$R_{in} = f k_d [\text{Ini}_2] \tag{4.15}$$

where $R_{in}$ is the rate of initiation, $[\text{Ini}_2]$ is the initiator concentration, $f$ is the factor of efficiency (see Annex 4.1, Fig. 4.37a), and $k_d$ is the rate constant of decomposition, with $k_d \approx 10^{-3}$ to $10^{-6}$ s$^{-1}$ (Braunecker and Matyjasziewski 2007).

The consumption of Ini$_2$ as a function of time is given by:

$$[\text{Ini}_2] = [\text{Ini}_2]_0 e^{-k_d t} \tag{4.16}$$

where $[\text{In}_2]_0$ is the initial concentration of Ini$_2$.

The reactions of the resulting free radicals with the monomer (initiation followed by propagation) are very fast. $R_p$, the rate of propagation, is expressed as:

$$R_p = k_p [M][M^\bullet] \tag{4.17}$$

where $M$ is the monomer, $M^\bullet$ is a free macroradical, and $k_p$ is the rate constant of propagation, with $k_p \approx 10^2 - 10^4$ L$\cdot$mol$^{-1}\cdot$s$^{-1}$ (Braunecker and Matyjasziewski 2007). The size of macroradicals has little influence on their reactivity, so that, except for the first units ($\langle X_n \rangle < 5$), all macroradicals $M^\bullet$ exhibit the same $k_p$. Termination reactions (whose bimolecular rate constants are higher than propagation ones by several orders of magnitude, but which involve two radicals, present at very low concentrations, and, therefore, occur more rarely) may take place at each step of the process by combination and/or disproportionation (cf. Annex 4.1, Fig. 4.37c), which contributes to drastically broaden the distribution in the size of the resulting macromolecules. The ratio of the two kinds of termination reactions depends on such parameters as temperature, concentration of monomer, viscosity of the medium, physical-chemical properties of polymer, average degree of polymerization, etc. For example, an increase of viscosity favors disproportionation reaction. If initially the termination operated mainly by combination, this increases the dispersity.

$R_{te}$, the rate of termination, is given by:

$$R_{te} = k_{tc} [M_n^\bullet][M_m^\bullet] + k_{td} [M_n^\bullet][M_m^\bullet] = k_{te} [M_n^\bullet][M_m^\bullet] \tag{4.18}$$

where $M_n^\bullet$ and $M_m^\bullet$ are two different growing chains, $k_{tc}$ is the rate constant of termination by combination, $k_{td}$ is the rate constant of termination by disproportionation, and $k_{te} = a \cdot k_{tc} + b \cdot k_{td}$ is the global rate constant of termination with $a + b = 1$ and $k_{te} \approx 10^6 - 10^8$ L$\cdot$mol$^{-1}\cdot$s$^{-1}$ (Braunecker and Matyjasziewski 2007).

Eventual uncontrolled transfer reactions (see Fig. 4.38) may contribute to further increasing the size heterogeneity of the polymer. $R_{tr}$, the rate of transfer, is expressed as follows:

$$R_{tr} = \sum_i k_{tri} [M_n] [SH_i] \tag{4.19}$$

where $k_{tri}$ is the transfer rate constant of the transfer reagent $SH_i$.

Considering the overlap of the various phenomena (propagation, termination, and transfer) that occur more or less simultaneously, predicting the final average degree of polymerization and dispersity is far from an easy task.

### 4.6.3   Annex 4.3. Expression of the Instantaneous and Cumulative Average Degree of Polymerization in Conventional Radical Polymerization

The instantaneous number-average degree of polymerization ($\langle X_n \rangle$) and average kinetic chain length $\lambda$ are related by $\langle X_n \rangle = k \cdot \lambda$, with $k = 1$ for disproportionation reaction and $k = 2$ for combination reaction. $\lambda$ describes the average length of those chains that are produced at any given time during the progress of the reaction. It is given by Eq. 4.20 (Mayo 1943):

$$\lambda = \frac{R_p}{R_{in} + \sum_i R_{tri}} \tag{4.20}$$

By neglecting the transfer reactions, $\lambda$ can be expressed more simply as:

$$\lambda = \frac{R_p}{R_{in}} = \frac{k_p[M]}{2\sqrt{f k_d . k_t [I_2]}} = \lambda_0 \tag{4.21}$$

where $\lambda_0$ is the kinetic average chain length when transfer reactions are neglected.

The corresponding cumulative $\langle X_n \rangle$ is given by:

$$\langle X_n \rangle = \frac{\Delta[M]}{f\Delta[Ini_2]} = \frac{[M] - [M]_0}{f\Delta[Ini_2]} = \frac{\alpha \cdot [M]_0}{f\Delta[Ini_2]} = \langle X_0 \rangle \tag{4.22}$$

where $\langle X_0 \rangle$ is the cumulative average chain length when transfer reactions are neglected. $\Delta[Ini_2]$ can be easily deduced from Eq. 4.2, and $\Delta[M]$ can be obtained from the Tobolsky relation (Tobolsky 1958):

$$-\ln\frac{[M]}{[M]_0} = -\ln(1 - \alpha) = \frac{2k_p}{\sqrt{k_{te}}} \sqrt{\frac{f[Ini_2]_0}{k_d}} \left(1 - e^{-k_d t/2}\right) \tag{4.23}$$

where $\alpha = ([M]_0 - [M])/[M]_0$ is the extent of conversion of the monomer.

As an illustration, the evolution of $\langle X_0 \rangle$ and $\lambda_0$ when polymerizing styrene in bulk at two different temperatures has been plotted against time and $\alpha$ in Fig. 4.40, panels A–C.

Note that these equations do not give any information on the dispersity of the resulting polymer.

$\langle X_0 \rangle$ remains only a theoretical prediction. In practice, the drift from the theoretical model can be significant (see Fig. 4.40D). However, despite its somewhat virtual character, the concept of kinetic chain length allows to understand why molecular dispersity is an unavoidable issue in RP.

Besides transfer side reactions, some other phenomena related to the intrinsic properties of the medium reaction can further impact the size and size distribution of the chains. Such is the case, for instance, of the gel effect or Trommsdorff-Norrish effect (Marten and Hamielec 1982), which results from the increase of the viscosity during the reaction and the consecutive decrease of the rate of termination, due to the lowered rate of diffusion of the macroradicals.

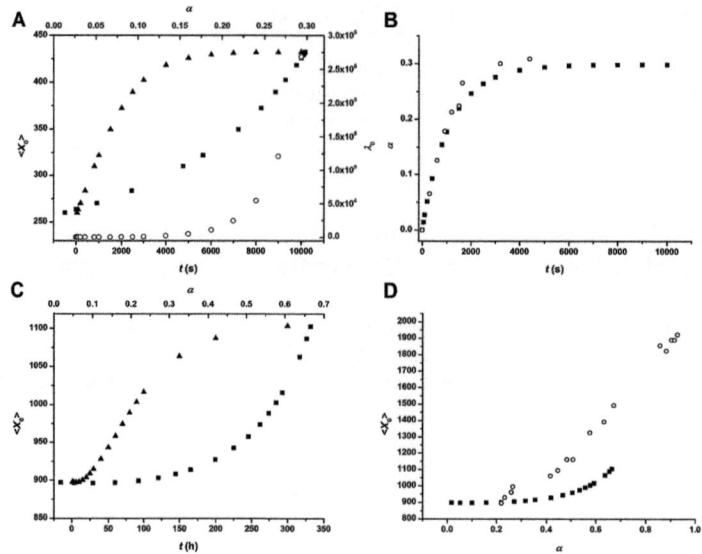

**Fig. 4.40** Polymerization kinetics in theory and in reality. (**A**) Plot of $\lambda_0$ vs. time ($\bigcirc$) and of $\langle X_0 \rangle$ vs. time (▲) and vs. $\alpha$ (■) for a simulated bulk polymerization of styrene ($[st]_0 = 8.7$ M) performed at 100 °C with $[In_2]_0 = 0.01$ M. Values of rate constants are as given in Tobolsky (1958). Because the initial ratio $[M]_0/[I_2]_0$ is high (870:1) and the concentration of monomer is still high after total consumption of the initiator, $\lambda_0$ and $\langle X_0 \rangle$ are expected to be very high. (**B**) A comparison of the plots of $\alpha$ vs. time according to Eqs. 4.8 and 4.9 (■) and to experimental results ($\bigcirc$) (Reprinted with permission from Tobolsky 1958, © 1958 American Chemical Society). Experimental results fit reasonably well the theoretical prediction. (**C**) Plots of $\langle X_0 \rangle$ vs. time (▲) and vs. $\alpha$ (■) for the same bulk polymerization of styrene simulated at 60 °C. Values of rate constants were given by Tefera et al. (1997). The rate of polymerization is slower than that at 100 °C, but the expected average molecular weight is higher. (**D**) Plots of $\langle X_0 \rangle$ vs. $\alpha$ for the simulated experience shown in panel **c** (■) and for the experimental results obtained by Marten and Hamielec (1982) ($\bigcirc$) (© 1982 John Wiley & Sons, Inc.). The marked difference ($\langle X_n \rangle \approx 2\langle X_0 \rangle$) could result from termination, which essentially proceeds by recombination with styrene, but also from the gel effect (see text).

### 4.6.4    Annex 4.4. How to Control the Size of Polymers Obtained by Radical Polymerization?

There are two main approaches to reducing size dispersity, telomerization (chain transfer) and controlled/living radical polymerization (CRP).

#### 4.6.4.1    Telomerization (Chain Transfer)

Telomerization has been widely used as a versatile tool to generate short polymers with $1.5 \leq Đ \leq 2$ (Starks 1974; Farina 1987). It has largely come out of use since CRP methods proved to be more reliable. The concept is based on controlling the transfer phenomena that occur concomitantly with radical propagation and impact polymer size, as described in Annex 4.1 (cf. Fig. 4.38). The impact of transfer reactions on the final size distribution can be best understood by referring to $\lambda$ (Annex 4.3, Eq. 4.20), which can also be expressed as follows (Mayo 1943):

$$\frac{1}{\lambda} = \frac{R_{in} + \sum_i R_{tr_i}}{R_p} = \frac{1}{\lambda_0} + \sum_i C_{T_i} \frac{(TA_i)}{(M)} \tag{4.24}$$

where $C_{tri} = k_{tri}/k_p$ is the transfer constant $C_{tr}$ of species $i$ to the macroradical in formation, whose concentration is $[TA_i]$. $k_{tri}$ and $k_p$ are the kinetic rate constants of transfer and propagation, respectively (see § 4.6.2, Annex 4.2).

Note that when $TA_i$ is a transfer agent, $C_{Ti}$ is high. $1/\lambda_0$ can be neglected and $1/\langle X_n \rangle$ can be expressed as follows:

$$\frac{1}{\langle X_n \rangle} = \frac{1}{k} \cdot \frac{1}{\lambda} \approx \frac{1}{k} \sum_i C_{T_i} \frac{(TA_i)}{(M)} \qquad (4.25)$$

with $k = 1$ when termination occurs by disproportionation and $k = 2$ when it operates by recombination (see § 4.6.3, Annex 4.3).

Starting from this simple observation, some chemists, in the early 1950s (reviewed in Starks 1974), developed a new method of radical polymerization called telomerization, where the reactions of polymerization were carried out in the presence of a reactant with a highly transfer efficiency called the telogen (where "telo" refers to the Greek τέλος, meaning "end," and "mer" to μηρός, "part"; the term "telomere" is also used to denote the short repetitive DNA sequences present at the extremities of the chromosomes). The method was found to be efficient only when the transfer constant ($C_T$) and the concentration of the telogen were high enough, so that the ratio of the telogen to monomer's concentration ($[TA]/[M]$) could be kept constant during the reaction. Ideally, $C_T \approx 1$ ensures that the value of $1/\langle X_n \rangle$ is only ruled by the ratio of the telogen to the monomer initial concentrations ($\langle X_n \rangle \approx [M]_0/[TA]_0$) and the molar mass distribution is consequently narrowed (Chung and Solomon 1992). According to Eq. 4.25, the length of the macromolecular chains generated by telomerization is expected to be short.

Transfer reactions may intervene during the initiation and the propagation steps without affecting the global kinetics of radical polymerization (see Annex 4.1, Fig. 4.38). The efficiency of the transfer is tightly linked to $C_T$, which depends on many parameters including the structure of the telogen and of the monomer, the solvent, and the temperature (Mayo 1943). The higher the value of $C_T$, the more efficient the transfer. For instance, alkanethiols were found to be efficient telogens for the radical telomerization of acrylic monomers (see e.g. Roy et al. 1972; Pucci et al. 1988; De La Fuente and Madruga 2000; Loubat and Boutevin 2000a, b).

A new telomerization method called catalytic chain transfer (CCT) emerged during the early 1980s (Farina 1987; Chung and Solomon 1992; Heuts et al. 2000). CCT involves the use of hematoporphyrin complexes as new transfer agents (Enikolopyan et al. 1981). These complexes are highly active ($C_T \approx 100$–$2500$) in catalytic amount (0.006 mole percent to the monomer), because they are not consumed during the polymerization reaction. Despite its high efficiency, the method has a restricted field of application (Enikolopyan et al. 1981; Pierik and van Herk 2003). Furthermore, the dispersity index remains close to 2.

During the 1990s, some chemists who were interested in synthesizing polymers of narrower molecular mass distribution developed and promoted the concept of controlled radical polymerization (CRP), aka living radical polymerization (Matyjaszewski and Spanswick 2005; Braunecker and Matyjasziewski 2007) (§ 4.6.4.2).

## 4.6.4.2    Controlled/Living Radical Polymerization (CRP)

CRP was developed on the basis of observations made during the synthesis of ionic (essentially anionic; cf. Fig. 4.41A) polymers. It was, till the early 1990s, the only reliable technique providing polymers with $Đ < 1.5$ (Greszta et al. 1994; Braunecker and Matyjasziewski 2007; Misha and Kumar 2012). CRP was pioneered during the 1980s, but did not become fully established before the 1990s (Greszta et al. 1994). Briefly, aiming at generating well-defined polymers by conventional radical polymerization (RP) stood as a challenge for a long time. The main limitations were inherent to the kinetics of RP: slow initiation, fast propagation, and very fast termination leading ineluctably to molecular heterogeneity (see above). Obviously, regarding the RP dilemma, the ideal radical polymerization should work as the anionic process (AP) does (see Fig. 4.41A), i.e. a fast initiation associated to a very low occurrence of transfer and termination reactions. For that reason, at variance with RP, which leads to growing chains that are exposed at each moment to irreversible termination (dead-end polymers), CRP was devised so as to yield polymers endowed with a "living character." Initially introduced to qualify the long-living growing chains observed in AP, the terminology "living" as employed in CRP refers to the reversible active-dormant state of macromolecules generated during the reaction and regulated by a thermodynamic equilibrium (Fig. 4.41B, C).

Under the name of CRP are regrouped a large variety of techniques, which can be distributed between the following three main categories:

- Nitroxide-mediated polymerization (NMP) (Bertin and Boutevin 1996);
- Atom transfer radical polymerization (ATRP) (Wang and Matyjaszewski 1995);
- Reversible addition-fragmentation chain transfer polymerization (RAFT) (Chiefari et al. 1998).

These three approaches are schematized in Fig. 4.41.

The mechanisms involved in CRP may differ significantly from conventional radical polymerization (RP), and there are also some noticeable differences between NMP, ATRP, and RAFT. We will not discuss in detail these points, which have been thoroughly reviewed in Braunecker and Matyjasziewski (2007). In the following paragraphs, we

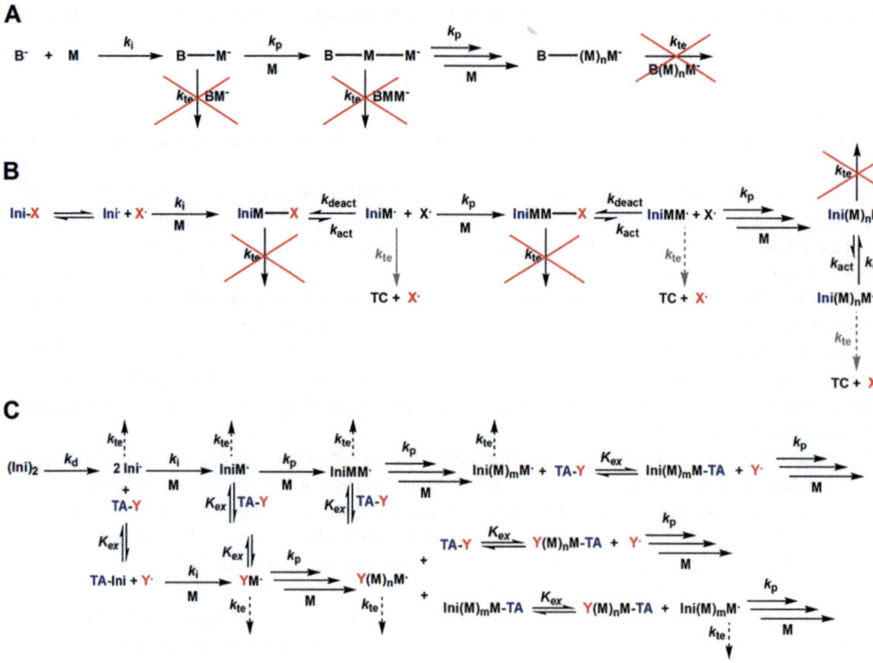

**Fig. 4.41** Schematic representation of three synthesis routes involving living radical polymerization. (**A**) Anionic polymerization. The anionic initiator ($B^\bullet$) is completely consumed at the first step of the reaction with the monomer ($M$) leading to the formation of the first adduct. Because of the coulombic repulsions, reactive centers cannot collide with each other to terminate, so that propagation is allowed to proceed till complete consumption of the monomer. Because each growing chain has been initiated at the early stage of the polymerization and has been allowed to grow for a comparable time, monomers are homogeneously incorporated into them, yielding polymers with a narrower molecular mass distribution than by conventional RP. (**B**) CRP performed according to a "persistent radical effect" (PRE) system. Fast exchange between active and dormant species with a lower concentration of active species and incorporation of a further monomer after reversible liberation of a transfer agent radical ($X^\bullet$). As mentioned for AP, fast initiation and low occurrence (*gray* to *dashed* arrows) of termination and transfer reactions yield polymers with low dispersity. At variance with AP, the low occurrence of termination does not result from electrostatic repulsions, but from the fact that the growing species reacts (ideally) with only a few monomer units (within a few milliseconds) before it is deactivated to the dormant state, IniMM-X (where it remains for several seconds). When they occur, termination reactions yield terminated chains (*TC*) and further $X^\bullet$. (**C**) CRP performed according to a "degenerative transfer" (DT). Fast interconversion of two different propagating species by reversible chain transfer of $Y$. Exchange may intervene at each step of the polymerization, lowering de facto the occurrence (*dashed* arrow) of termination reactions. Because it is thermodynamically neutral ($\Delta G^\circ \approx 0$), this exchange has been described as "degenerative" (Greszta et al. 1994). $k_p$, $k_{te}$, $k_{act}$, and $k_{deact}$ are the kinetic rate constants of, respectively, propagation, termination, activation, and deactivation. $K_{ex}$ is the equilibrium constant of exchange.

provide only a few essential elements, which can be easily understood by the lay reader provided he has been through Annex 4.1. For further details, we refer to publications listed in Table 4.6.

CRP has been developed on the basis of two distinct approaches designated as persistent radical effect (PRE) and degenerative transfer (DT).

In *PRE* (which concerns ATRP and NMP), polymerization is initiated quickly (in the presence of a specific initiator Ini-X), and all monomers react at an early stage of the reaction to form short chains that grow up to full conversion. Propagating macroradicals Ini$M_n^\bullet$ are quickly deactivated by species $X^\bullet$, leading to stable Ini$M_n$-X species. The dormant species (Ini*MM-X* in Fig. 4.41B) are activated in the presence of an appropriate catalyst or stimulus (see below) to reform the growing centers (Ini*MM*$^\bullet$). Macroradicals can propagate but also terminate. However, persistent radicals ($X^\bullet$) cannot terminate by reacting with each other, but only reversibly react with the growing species Ini$M_n^\bullet$. Thus, each termination reaction leads to the accumulation of $X^\bullet$, whose concentration increases with time. Consequently,

**Table 4.6** Some key references to the principles of controlled radical polymerization (CRP).

| CRP method | Reviews | Articles |
|---|---|---|
| NMP | Braunecker and Matyjasziewski (2007) and Grubbs (2011) | Georges et al. (1993), Veregin et al. (1993), and Bertin and Boutevin (1996) |
| ATRP | Braunecker and Matyjasziewski (2007). Pintauer and Matyjaszewski (2008), Matyjaszewski and Tsarevsky (2014), and Boyer et al. (2016) | Wang and Matyjaszewski (1995) and Buback et al. (2016) |
| RAFT | Barner-Kowollik et al. (2006), Braunecker and Matyjasziewski (2007), and Moad et al. (2012) | Chiefari et al. (1998), Feldermann et al. (2004), Meiser et al. (2011), Meiser and Buback (2012), and Buback et al. (2016) |

the concentration of macroradicals, as well as the probability of termination, decreases with time. The growing radicals therefore predominantly react with $X^{\bullet}$, which is present at much higher concentration, rather than with themselves (see Fig. 4.41B).

In such fast-initiated systems, $\langle X_n \rangle$ is expected to evolve as follows:

$$\langle X_n \rangle = \frac{[M]_0}{f[\text{Ini-}X]_0} \cdot \alpha \tag{4.26}$$

where $[\text{Ini-}X]_0$ is the initial concentration of the initiator, $f[\text{Ini-}X]_0$ represents the fraction of initiated chains at the early stage of the reaction, and $\alpha = ([M]_0 - [M])/[M]_0$ is the extent of conversion of the monomer (see § 4.6.3, Annex 4.3, Eq. 4.23). The molecular mass of the polymer is therefore expected to evolve linearly with the degree of conversion, which is the typical signature of a "living" synthesis system (Bertin and Boutevin 1996), whereas linearity can be observed only at low conversion when polymerization is performed by RP (see Annex 4.3, Fig. A4.3.1, and Braun 2009; Misha and Kumar 2012).

Synthesis systems based on *DT* (namely RAFT) follow typical RP kinetics with slow initiation and fast termination. The concentration of transfer agent is much larger than that of radical initiators, and its transfer constant is typically high (rate of transfer larger than rate of propagation, i.e. $C_T$ much higher than 1). Thus, the transfer agent plays the role of the dormant species. The monomer is consumed by a very small concentration of radicals which can terminate but also, with a higher probability, degeneratively exchange with the dormant species (the highly transfer-effective reagent TA-Y can be reversibly bound to the growing chains and indifferently exchanged between them). This reversible and rapid distribution of the transfer agent between active and dormant species allows to readjust the drift of the incorporated monomers' amount in each growing chain at the early stage of the polymerization (see Fig. 4.41C), resulting in the following linear relation:

$$\langle X_n \rangle = \frac{[M]_0}{[\text{TA}-Y]_0} \cdot \alpha \tag{4.27}$$

where the initial concentration of transfer agent is $[\text{TA}-Y]_0$.

Living polymerization is achieved by using specific reagents (some examples of structures of which are given in Fig. 4.42), which reversibly bind to the growing chain. Monomers are incorporated only either after homolytic rupture of this reversible link (meaning that, the covalent bond being symmetrically broken, one of the two electrons involved in the covalent link is captured by each of the two vicinal atoms, generating two radicals) or during a degenerative exchange process (reversible chain transfer). Whatever the mechanism, polymerization performed by CRP evolves more homogeneously than when performed by RP, leading to a significantly lower dispersity.

The choice of the method is dictated by the experimental conditions (temperature, solvent, type of monomer) and the structure to be obtained (see Annex 4.5). For instance, a limitation of NMP is that only a few successful examples have been reported when the method was applied to acrylate derivatives (Knoop and Studer 2003). Alternative approaches (mainly ATRP and RAFT) are often preferred for CRP of acrylic monomers (Braunecker and Matyjasziewski 2007; Grubbs 2011).

Because RAFT does not involve the use of toxic metal salts and it is compatible with a wide range of monomers, it was initially often preferred to ATRP (Boyer et al. 2016). However, constant adaptations and modernizations of the method have resulted in drastic limitation of the concentrations of metal complexes (Pintauer and Matyjaszewski 2008; Boyer et al. 2016). Emerging trends are even aiming at developing metal-free ATRP. One successful typical example has been recently reported, where ATRP was photoinduced (Treat et al. 2014). After 20 years of development and improvements, numerous ATRP-based methods have found applications for a wide range of monomers without the

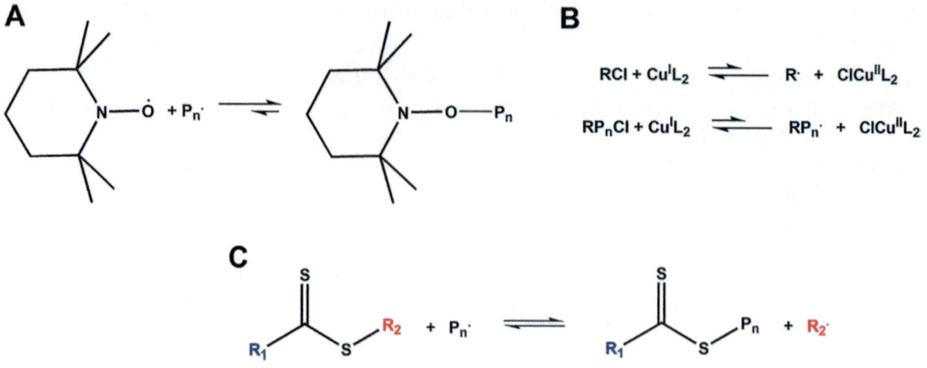

**Fig. 4.42** Examples of chemical structures of reactants used in controlled radical polymerization. (**A**) In nitroxide-mediated polymerization (NMP), the control of polymerization is mediated by the establishment of an equilibrium between a nitroxide, (2,2,6,6-tetramethylpiperidin-1-yl)oxyl (TEMPO) (*left*) and its alkoxyamine derivative (*right*). $P_n^{\bullet}$ is a growing chain. (**B**) In atom transfer radical polymerization (ATRP), control is ensured by a reversible reaction between an alkyl halide and a liganded (L being an organic ligand) copper salt. (**C**) In reversible addition-fragmentation chain transfer polymerization (RAFT), the size of the chains is controlled by exchange with a dithioester or a trithiocarbonate ($R_1 = R\text{-}S\text{-}$) (See references in Table 4.6 for detailed structures and mechanisms).

limitation in the nature of solvents and conventional specific conditions required by RP (Braunecker and Matyjasziewski 2007; Matyjaszewski and Tsarevsky 2014; Boyer et al. 2016). ATRP is nowadays one of the most widely used approaches to CRP.

Globally, CRP approaches represent efficient and reliable ways for synthesizing polymers by RP while limiting the recurrent problem of size heterogeneity, but they are also useful tools for achieving particular designs and the incorporation of functional moieties that had remained hitherto inaccessible. The main advances made in polymer design and chemistry since CRP became a standard polymerization method are shortly reviewed in the next Annex.

## 4.6.5 Annex 4.5. Polymer Topology and How to Control It

Prior to the development of CRP, the design and functionalization of polymers were severely restricted due to the limitations of RP and of the various techniques of polymer chemical modification. A brief description and comparison of the various strategies developed during the pre- and the post-CRP period is given in the following paragraphs.

Three main parameters, namely the topology, the composition, and the functionality, are to be considered when designing polymers. Whereas the synthesis of a linear homopolymer can be (in theory) easily achieved, difficulties tend to accumulate when aiming at building more complex structures comprising two or more types of comonomers. CRP has become the prevalent approach to try to circumvent them (Braunecker and Matyjasziewski 2007; Matyjaszewski and Tsarevsky 2014).

### 4.6.5.1 Polymer Topology

The topology (or architecture) of a polymer characterizes its three-dimensional architecture. This structure may be very simple (linear) or highly elaborate (branched, multiarm, etc.), particularly when the polymer is comprised of different macromolecular parts ("macromeres") (Fig. 4.43). CRP, thanks to its versatility, has become a favorite strategy for the synthesis of well-defined complex macrostructures, a task that would be very laborious if not downright out of reach if attempted by conventional RP. Indeed, the versatility of CRP lies here in the fact that macromolecules obtained by living polymerization are dormant species that can be easily activated to react with another target center. Because of the termination reactions, this is not possible with conventional RP, where the macromeres composing the polymer must be assembled with each other by chemical conjugation mediated by appropriate functional groups bound to each of them (see below).

The advantages and advances brought by CRP to the development of polymers with original topologies will not be further discussed in this section. For more information, see e.g. Matyjaszewski and Spanswick (2005), Boyer et al. (2009), and Matyjaszewski and Tsarevsky (2014).

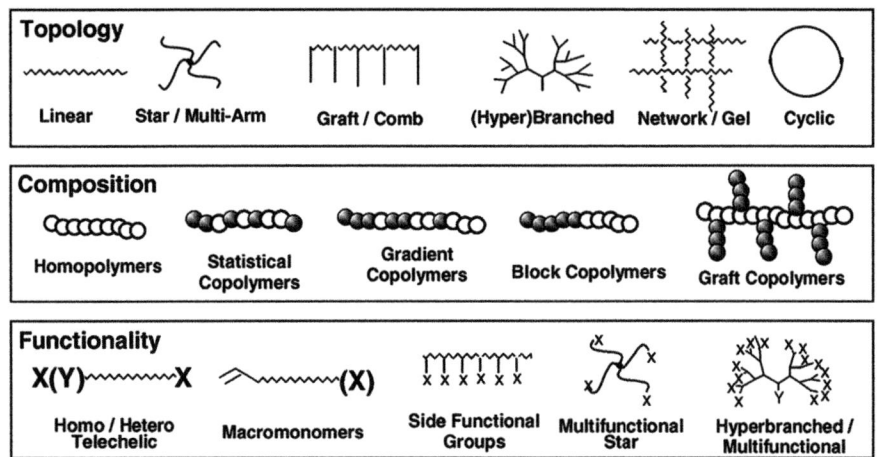

**Fig. 4.43** Various types of compositions, architectures, and functionalities accessible by CRP (From Braunecker and Matyjasziewski 2007, © 2007 Elsevier Ltd).

### 4.6.5.2  Polymer Composition (Microstructure)

Because the physical-chemical properties of a polymer are closely linked to its microstructure (or composition) (Liu et al. 2007), the distribution of the different comonomers along the polymer's backbone (see Fig. 4.44) must be known, and controlled as much as possible. Indeed, the physical properties of the final polymers may turn out to be very different from those expected if this parameter is not properly mastered. This problem has been identified quite early and characterized for RP during the first half of the twentieth century (Mayo and Lewis 1944; Lewis et al. 1948). For a given copolymer in formation, the evolution of the concentration of the two monomers $M_1$ and $M_2$ is given by:

$$\frac{d[M_1]}{d[M_2]} = \frac{[M_1]}{[M_2]} \cdot \frac{r_1[M_1] + [M_2]}{[M_1] + r_2[M_2]} \tag{4.28}$$

where $[M_1]$ and $[M_2]$ are the concentrations of unreacted monomers, $r_1$ is the ratio of the rate constants for the reaction of an $M_1$-type radical with $M_1$ and $M_2$, respectively, and $r_2$ is the ratio for reaction of an $M_2$-type radical with $M_2$ and $M_1$, respectively.

By introducing $f$ and $F$, the molar fractions of monomer in the feed and in the copolymer, respectively, Eq. 4.28 becomes:

$$F_1 = 1 - F_2 = \frac{d[M_1]}{d[M_1] + d[M_2]} = \frac{r_1 f_1^2 + f_1 f_2}{r_1 f_1^2 + 2f_1 f_2 + r_2 f_2^2} \tag{4.29}$$

$\langle F_1 \rangle$, the cumulative value of $F_1$ (i.e. the average value of the molar fraction of $M_1$ that has been incorporated into the various terminated and growing macromolecular chains formed since the initiation), may be expressed as:

$$\langle F_1 \rangle = \frac{f_{1,0} - f_{1,t}(1 - \alpha)}{\alpha} \tag{4.30}$$

where $f_{1,0}$ and $f_{1,t}$ are initial and instantaneous value of $f_1$, respectively, and $\alpha$ is as defined in Eq. 4.23.

Although Eqs. 4.28 and 4.29 were established in the context of RP, they are valid for all kinds of polymerization. Various rearrangements of Eq. 4.29 (not shown) have been proposed to deduce the value of $r_1$ and $r_2$ (Alfrey and Price 1947; Lewis et al. 1948; Fineman and Ross 1950; Kelen et al. 1980).

$F$ and $f$ are accessible by FTIR (Parambil et al. 2012), NMR (Bataille and Bourassa 1989; Cracowski et al. 2010; Parambil et al. 2012), or elemental analyses (Cracowski et al. 2010) of the feed and of the copolymer.

Five different cases are commonly considered:

- $r_1 \approx r_2 \approx 1$: generated radicals react randomly with both monomers, and the relative rates of monomer consumption are determined only by the relative monomer concentrations in the feed mixture. The composition of monomers in the copolymer is approximately identical to that in the feed, and the distribution of each monomer is expected to be random. Equation 4.28 becomes:

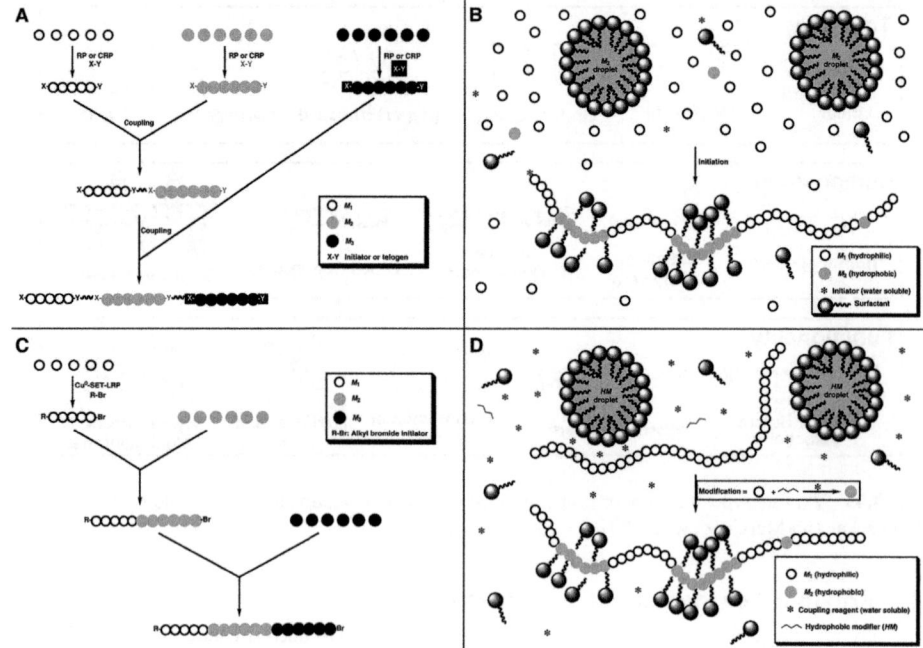

**Fig. 4.44** Schematic representation of the synthesis of blocky copolymers by different methods. (**A**) Iterative chain-end coupling of functionalized end-capped blocks obtained by CRP (see § 4.6.6.2 and Golas and Matyjaszewski 2010). (**B**) Copolymerization performed by RP or CRP in dispersed medium. For instance, copolymerization of a hydrophilic monomer with a hydrophobic monomer performed by RP in aqueous micellar solution (Candau et al. 1996; Volpert et al. 1996). (**C**) Iterative homopolymerization performed by single-electron transfer living radical polymerization (SET-LRP) mediated by copper metal (Soeriyadi et al. 2011; Alsubaie et al. 2014). (**D**) Hydrophobic modification of a hydrophilic homopolymer precursor performed in dispersed medium (Liu et al. 2007).

$$\frac{d[M_1]}{d[M_2]} = \frac{[M_1]}{[M_2]} \tag{4.31}$$

- $r_1 \approx r_2 \approx 0$ or $r_1 \cdot r_2 \approx 0$: each comonomer shows a strong preference for cross-propagation (reaction with the other comonomer), and in the extreme case, depending on the feed composition, the generated copolymer presents a nearly perfect alternated sequence (such as styrene and maleic anhydride, which are the two components of SMA, when present in a 1:1 ratio; see Di Cola et al. 2004). Equation 4.28 becomes:

$$\frac{d[M_1]}{d[M_2]} = 1 \tag{4.32}$$

- $r_1 < 1$ and $r_2 < 1$: the preference for cross-propagation is not absolute, and the copolymerization is ruled by the trend to a less pronounced alternation. For Eq. 4.31 to be verified, the composition (called azeotropic composition) in the feed must be as follows:

$$\frac{[M_1]}{[M_2]} = \frac{1 - r_2}{1 - r_1} \tag{4.33}$$

- $r_1 > 1$ and $r_2 < 1$: each monomer reacts preferentially with $M_1$, so the copolymer will be enriched in $M_1$. A special case is when $r_1 \cdot r_2 \approx 1$ and Eq. 4.28 becomes:

$$\frac{d[M_1]}{d[M_2]} = r_1 \frac{[M_1]}{[M_2]} \tag{4.34}$$

in which case there is a linear relationship between the composition of the feed and that of the polymer.

- $r_1 > 1$ and $r_2 > 1$: each monomer reacts preferentially with itself, and the copolymerization results in a mixture of two homopolymers.

Since the first paper of Mayo and Lewis (1944) till today (Kanellou et al. 2015), many works concerning the co- and ter-polymerization and the determination of the reactivity ratio of the monomers have been published, resulting in a helpful and wide range of data available to the scientific community.

Blocky, statistical, and alternating sequences are the three types of structures most commonly encountered. However, it is noticeable that, despite the numerous possibilities of structures that can be designed by varying the nature of comonomers, RP performed in conventional conditions generally leads to alternating or random copolymers, and optionally to a gradient composition, but most rarely to a blocky structure. Indeed, diblock or multiblock structures cannot be directly achieved via RP in homogeneous solutions: they require proper chain-end functionalization of two homopolymers to allow a specific coupling reaction between the two macromolecules (see Fig. 4.44A, § 4.6.6.2, and references in Bertin and Boutevin 1996).

When the comonomers are not miscible in the same medium and/or can be segregated in two different phases, blocky structures are preferentially obtained via RP or CRP performed in dispersed media, either as classical or inverse emulsions (Candau et al. 1996; Volpert et al. 1996), as micro-emulsions (Matyjaszewski and Tsarevsky 2014), or as solvent-free polymer mixtures (blends) (Ruzette and Leibler 2005) (Fig. 4.44B).

In principle, diblock structures could be obtained via living polymerization performed in bulk (Bertin and Boutevin 1996), because it is conceivable, under proper conditions, to react the dormant form of a homopolymer of a monomer $M_1$, obtained by CRP, with a second monomer $M_2$, so as to yield a $M_1$-co-$M_2$ diblock copolymer. Up till recently (and for reasons that will not be explained here), blocky structures of higher order (triblock or multiblock) remained achievable only by block conjugation following polymerization (see Fig. 4.44A and Golas and Matyjaszewski 2010). Thanks to the latest improvements of CRP, they are now accessible via iterative polymerization reactions (Soeriyadi et al. 2011; Alsubaie et al. 2014) (Fig. 4.44C), but multiblock copolymers of higher molecular weight than ~25 kDa remain inaccessible by this method (Alsubaie et al. 2014).

Alternatively, multiblock copolymers can be obtained by chemical modification of a homopolymer precursor performed in dispersed media (Liu et al. 2007; Gohon et al. 2011) (Fig. 4.44D). Although the effectiveness of this method has been demonstrated, its use is not widespread.

For further information regarding the development of original polymeric microstructures obtained by CRP, the reader is referred to reviews by Braunecker and Matyjasziewski (2007), Rizzardo and Solomon (2012), Matyjaszewski and Tsarevsky (2014), and Boyer et al. (2016).

## 4.6.6  Annex 4.6. Functionalizing Polymers

### 4.6.6.1  General Considerations

As the case of APols demonstrates, polymers and copolymers present interesting intrinsic properties, but further functionalities can be conferred onto them by endowing them with one or more appropriate moieties. This can be achieved according to several methods, which can be classified into the following four categories:

- *Polymerization starting from functional initiators or transfer agents.* The method consists in performing RP or CRP in the presence of a functionalized initiator or transfer agent, which will endow the polymers with the desired functional moiety (Fig. 4.45A).
- *Post-polymerization modification of end-group functionality.* This two-step procedure involves the polymerization by RP or CRP of a monomer in the presence of a functionalizable initiator or transfer agent and subsequent modification of the reactive extremities (Fig. 4.45B).
- *Introduction, during polymerization, of functional monomers* (Fig. 4.46A).
- *Post-polymerization modification of functionalizable monomers introduced into the polymer.* Following a similar approach, a functionalizable monomer is homo- or copolymerized via RP or CRP, and the functional polymer is obtained by modifying it (Fig. 4.46B).

Each method presents its advantages and drawbacks. Those must be carefully considered before opting for one or the other strategy, which must be adapted to the functionality to be introduced. For instance, introduction of a functional moiety that inhibits radical reactions can only be achieved after polymerization. Post-polymerization modifications can

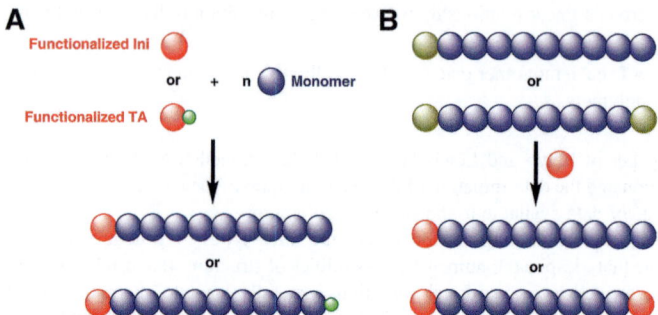

**Fig. 4.45** Functionalization of polymer extremities. (**A**) Use of a functionalized initiator (Ini) or transfer reagent (TA). If polymerization is performed by RP, the polymer can be functionalized at one or both chain extremities depending on whether termination operates via disproportionation or recombination. CRP performed in the presence of a functionalized transfer agent allows a dissymmetric functionalization of the two chain ends (see e.g. Pound et al. 2007; Boyer et al. 2009). (**B**) Functionalization of an end group after polymerization. The modification is achieved on one or both reactive extremities, which can be either the initiator, the transfer agent, or a terminal olefin if termination occurred via disproportionation (see e.g. Tolstyka et al. 2008; Heredia et al. 2009).

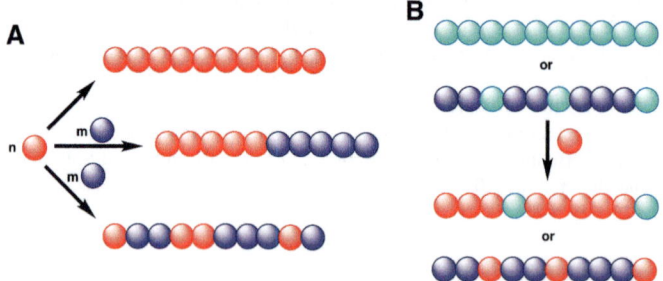

**Fig. 4.46** Incorporation of functional groups within the polymer chain. (**A**) Homopolymerization of a functional monomer (*red*) or either iterative or simultaneous copolymerization with a nonfunctional monomer (*blue*) generates, respectively, diblock or random copolymers (see e.g. Boyer et al. 2009; Grubbs 2011; Ceylan et al. 2017). (**B**) Chemical modification carried out on a functionalizable homo- or copolymer. The modification is selectively achieved on the reactive monomer (*teal*), yielding a functionalized monomer (*red*), whereas the monomers shown in *blue* are unreactive (see e.g. Le Bon et al. 2014b). As a rule, the extent to which the reaction will proceed is hard to predict, as it depends strongly on the reagents and the accessibility of the modifiable sites. The modification can be performed to completion (the case presented in this panel) for certain systems (Rossi et al. 2008; Zhao et al. 2013), whereas it cannot for some other (Giusti et al. 2012; Le Bon et al. 2014a).

target either the extremities or internal, lateral reactive sites, depending on the desired location and number of functional groups, etc.

Note that post-polymerization modification of polymer end chain would be easily conceivable for most polymers without involving specific and sophisticated methods. Indeed, because of termination mechanisms, a polymer obtained by RP is systematically prefunctionnalized at its extremities, either by the initiator (Ini-$M_{2n}$-Ini) or the initiator and an olefin (Ini-$M_n$-=) or an alkane, depending on whether termination operates via recombination or disproportionation. Because unreactive initiator moieties can be converted to more reactive group and because olefins are reactive toward

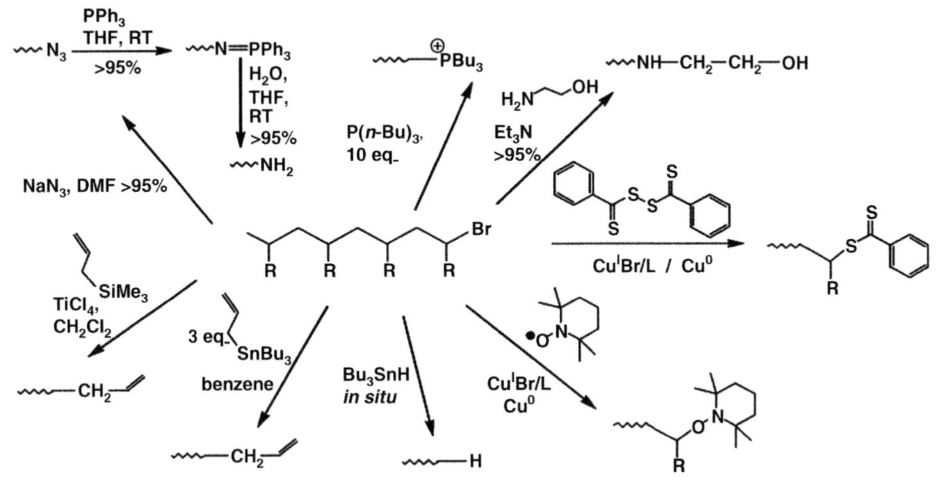

**Fig. 4.47** Various possibilities of functionalization of the brominated chain end of polymers obtained by ATRP (From Braunecker and Matyjasziewski 2007, © 2011 Elsevier Ltd. All rights reserved).

several functions (thiols, peroxides, other oxidants and reducers, etc.), this offers obvious opportunities for end-chain modification and subsequent end-group functionalization.

End-group modification mediated by CRP makes it possible to insert a more specific and homogeneous reactive group at each chain end (Fig. 4.47).

For more details, see Boyer et al. (2009, 2016) and Matyjaszewski and Tsarevsky (2014).

### 4.6.6.2 Possible Applications of Controlled Radical Polymerization to the Chemistry of Amphipols

A8-35, A8-75, their congeners A34-35 and A34-75, SAPols, etc. are obtained by hydrophobization of a poly(acrylic acid) (PAA) precursor (Tribet et al. 1996; Dahmane et al. 2009). Functionalized and labeled derivatives thereof are obtained either by incorporating functional moieties at the time of hydrophobization, or by grafting them on the APol either directly on the carboxylates or via a reactive group, typically an amine, incorporated into it (see main text, § 4.4.1). As a result of this procedure, all APols derived from a given PAA share the same average length and dispersity (with the reservation that some molecules, depending on their size and/or physical properties, may be preferentially lost during purification procedures).

Functional moieties incorporated at the time of hydrophobization are expected to be statistically distributed along the polymer chain (Magny et al. 1992), whereas incorporation by derivation, in aqueous solution, of assembled APol particles will likely be biased toward the most hydrophilic stretches of the sequence (§ 4.4.1). The dispersity of the original PAA, which is quite broad ($Đ ≈ 2$), will be that of the resulting APols (Gohon et al. 2004, 2006; Giusti et al. 2014).

If these characteristics are incompatible with one or another application, CRP methods could be resorted to in order to achieve either a narrower size distribution or a different distribution of functional groups, e.g. one limited to chain ends. In such a case, there seems to be little doubt that ATRP would be preferred to NMP or RAFT. For one thing, ATRP has already been used to synthesize PAAs and hydrophobized derivatives thereof with narrow molecular mass distributions (Liu et al. 2007; Gohon et al. 2011; C. Tribet and F. Giusti, unpublished data cited in Le Bon et al. 2014b). For another, ATRP is a highly versatile technique that makes it straightforward, in particular, to introduce various functional groups at the extremities of the macromolecular chains, some examples of which are shown in Fig. 4.45.

*Annexes prepared by Fabrice Giusti and Bernard Pucci.*

## References

Alami, M., Dalal, K., Lelj-Garolla, B., Sligar, S.G., Duong, F. (2007) Nanodiscs unravel the interaction between the SecYEG channel and its cytosolic partner SecA. *EMBO J.* **26**:1995–2004.

Alfrey, T., Price, C.C. (1947) Relative reactivities in vinyl copolymerization. *J. Polym. Sci.* **2**:101–106.

Alsubaie, F., Anastasaki, A., Wilson, P., Haddleton, D.M. (2014) Sequence-controlled multi-block copolymerization of acrylamides via aqueous SET-LRP at 0 °C. *Polym. Chem.* **6**:406–417.

Banerjee, S., Sen, K., Pal, T.K., Guha, S.K. (2012) Poly(styrene-co-maleic acid)-based pH-sensitive liposomes mediate cytosolic delivery of drugs for enhanced cancer chemotherapy. *Int. J. Pharm.* 436:786–797.

Barner-Kowollik, C., Buback, M., Charleux, B., Coote, M.L., Drache, M., Fukuda, T., Goto, A., Klumperman, B., Lowe, A.B., McLeary, J.B., Moad, G., Monteiro, M.J., Sanderson, R.D., Tonge, M.P., Vana, P. (2006) Mechanism and kinetics of dithiobenzoate-mediated RAFT polymerization. I. The current situation. *J. Polym. Sci.* **44**:5809–5831.

Basit, H., Sharma, S., Van der Heyden, A., Gondran, C., Breyton, C., Dumy, P., Winnik, F.M., Labbé, P. (2012) Amphipol-mediated surface immobilization of FhuA: a platform for label-free detection of the bacteriophage protein pb5. *Chem. Commun.* **48**:6037–6039.

Bataille, P., Bourassa, H. (1989) Determination of the reactivity parameters for the copolymerization of butyl acrylate with vinyl acetate. *J. Polym. Sci.* **27**:357–365.

Bazzacco, P., Billon-Denis, E., Sharma, K.S., Catoire, L.J., Mary, S., Le Bon, C., Point, E., Banères, J.-L., Durand, G., Zito, F., Pucci, B., Popot, J.-L. (2012) Non-ionic homopolymeric amphipols: Application to membrane protein folding, cell-free synthesis, and solution NMR. *Biochemistry* **51**:1416–1430.

Bazzacco, P., Sharma, K.S., Durand, G., Giusti, F., Ebel, C., Popot, J.-L., Pucci, B. (2009) Trapping and stabilization of integral membrane proteins by hydrophobically grafted glucose-based telomers. *Biomacromolecules* **10**:3317–3326.

Bertin, D., Boutevin, B. (1996) Controlled radical polymerization. *Polym. Bull.* **37**:337–344.

Boyer, C., Bulmus, V., Davis, T.P., Ladmiral, V., Liu, J., Perrier, S. (2009) Bioapplications of RAFT polymerization. *Chem. Rev.* **109**:5402–5436.

Boyer, C., Corrigan, N.A., Jung, K., Nguyen, D., Nguyen, T.-K., Adnan, N.N.M., Oliver, S., Shanmugam, S., Yeow, J. (2016) Copper-mediated living radical polymerization (atom transfer radical polymerization and copper (0) mediated polymerization): From fundamentals to bioapplications. *Chem. Rev.* **116**:1803–1949.

Braun, D. (2009) Origins and development of initiation of free radical polymerization processes. *Int. J. Polym. Sci.* **2009**: e893234.

Braunecker, W.A., Matyjasziewski, K. (2007) Controlled/living radical polymerization: Features, developments and perspectives. *Prog. Polym. Sci.* **32**:93–146.

Breyton, C., Tribet, C., Olive, J., Dubacq, J.-P., Popot, J.-L. (1997) Dimer to monomer conversion of the cytochrome $b_6 f$ complex: causes and consequences. *J. Biol. Chem.* **272**:21892–21900.

Buback, M., Schroeder, H., Kattner, H. (2016) Detailed kinetic and mechanistic insight into radical polymerization by spectroscopic techniques. *Macromolecules* **49**:3193–3213.

Candau, F., Volpert, E., Lacik, I., Selb, J. (1996) Free-radical polymerization in micellar media: Effect of microenvironment. *Macromol. Symp.* **111**:85–94.

Carazo, J.M., Sorzano, C.O.S., Otón, J., Marabini, R., Vargas, J. (2015) Three-dimensional reconstruction methods in single particle analysis from transmission electron microscopy data. *Arch. Biochem. Biophys.* **581**:39–48.

Catoire, L.J., Damian, M., Giusti, F., Martin, A., van Heijenoort, C., Popot, J.-L., Guittet, E., Banères, J.-L. (2010a) Structure of a GPCR ligand in its receptor-bound state: leukotriene B$_4$ adopts a highly constrained conformation when associated to human BLT2. *J. Am. Chem. Soc.* **132**:9049–9057.

Catoire, L.J., Zoonens, M., van Heijenoort, C., Giusti, F., Guittet, E., Popot, J.-L. (2010b) Solution NMR mapping of water-accessible residues in the transmembrane β-barrel of OmpX. *Eur. Biophys. J.* **39**:623–630.

Catoire, L.J., Zoonens, M., van Heijenoort, C., Giusti, F., Popot, J.-L., Guittet, E. (2009) Inter- and intramolecular contacts in a membrane protein/surfactant complex observed by heteronuclear dipole-to-dipole cross-relaxation. *J. Magn. Res.* **197**:91–95.

Ceylan, H., Yasa, I.C., Sitti, M. (2017) 3D chemical patterning of micromaterials for encoded functionality. *Adv. Mater.* **29**:1605072.

Chae, P.S., Rasmussen, S.G.F., Rana, R., Gotfryd, K., Chandra, R., Goren, M.A., Kruse, A.C., Nurva, S., Loland, C.J., Pierre, Y., Drew, D., Popot, J.-L., Picot, D., Fox, B.G., Guan, L., Gether, U., Byrne, B., Kobilka, B.K., Gellman, S.H. (2010) Maltose-neopentyl glycol (MNG) amphiphiles for solubilization, stabilization and crystallization of membrane proteins. *Nat. Methods* **7**:1003–1008.

Charvolin, D., Perez, J.-B., Rouvière, F., Giusti, F., Bazzacco, P., Abdine, A., Rappaport, F., Martinez, K.L., Popot, J.-L. (2009) The use of amphipols as universal molecular adapters to immobilize membrane proteins onto solid supports. *Proc. Natl. Acad. Sci. USA* **106**:405–410.

Charvolin, D., Picard, M., Huang, L.-S., Berry, E.A., Popot, J.-L. (2014) Solution behavior and crystallization of cytochrome $bc_1$ in the presence of amphipols. *J. Membr. Biol.* **247**:981–996.

Chen, X., Chen, L., Feng, J., Yao, Z., Qian, J. (2009) Toward polymer product design. I. Dynamic optimization of average molecular weights and polydispersity index in batch-free radical polymerization. *Ind. Eng. Chem. Res.* **48**:6739–6748.

Chiefari, J., Chong, Y.K., Ercole, F., Krstina, J., Jeffery, J., Le, T.P.T., Mayadunne, R.T.A., Meijs, G.F., Moad, C.L., Moad, G., Rizzardo, E., Thang, S.H. (1998) Living free-radical polymerization by reversible addition-fragmentation chain transfer: The RAFT process. *Macromolecules* **31**:5559–5562.

Chung, R.P.-T., Solomon, D.H. (1992) Recent developments in free-radical polymerization–a mini review. *Prog. Org. Coat.* **21**:227–254.

Couvreur, L., Lefay, C., Belleney, J., Charleux, B., Guerret, O., Magnet, S. (2003) First nitroxide-mediated controlled free-radical polymerization of acrylic acid. *Macromolecules* **36**:8260–8267.

Cracowski, J.-M., Montembault, V., Améduri, B. (2010) Free-radical copolymerization of 2,2,2-trifluoroethyl methacrylate and 2,2,2-trichloroethyl α-fluoroacrylate: Synthesis, kinetics of copolymerization, and characterization. *J. Polym. Sci.* **48**:2154–2161.

Cuevas Arenas, R., Danielczak, B., Martel, A., Porcar, L., Breyton, C., Ebel, C., Keller, S. (2017) Fast collisional lipid transfer among polymer-bounded nanodiscs. *Sci. Rep.* **7**:45875.

Dahmane, T., Giusti, F., Catoire, L.J., Popot, J.-L. (2011) Sulfonated amphipols: Synthesis, properties and applications. *Biopolymers* **95**:811–823.

Daniel, R.M., Finney, J.L., Réat, V., Dunn, R., Ferrand, M., Smith, J.C. (1999) Enzyme dynamics and activity: Time-scale dependence of dynamical transitions in glutamate dehydrogenase solution. *Biophys. J.* **77**:2184–2190.

Davis, K.A., Matyjaszewski, K. (2000) Atom transfer radical polymerization of *tert*-butyl acrylate and preparation of block copolymers. *Macromolecules* **33**:4039–4047.

De La Fuente, J.L., Madruga, E.L. (2000) Homopolymerization of methyl methacrylate and styrene: Determination of the chain-transfer constant from the Mayo equation and the number distribution for *n*-dodecanethiol. *J. Polym. Sci.* **38**:170–178.

Della Pia, E.A., Holm, J., Lloret, N., Le Bon, C., Popot, J.-L., Zoonens, M., Nygård, J., Martinez, K.L. (2014a) A step closer to membrane protein multiplexed nano-arrays using biotin-doped polypyrrole. *ACS Nano* **8**:1844–1853.

Della Pia, E.A., Westh Hansen, R., Zoonens, M., Martinez, K.L. (2014b) Functionalized amphipols: A versatile toolbox suitable for applications of membrane proteins in synthetic biology. *J. Membr. Biol.* **247**:815–826.

Di Cola, E., Plucktaveesak, N., Waigh, T.A., Colby, R.H., Tan, J.S., Pyckhout-Hintzen, W., Heenan, R.K. (2004) Structure and dynamics in aqueous solutions of amphiphilic sodium maleate-containing alternating copolymers. *Macromolecules* **37**:8457–8465.

Diab, C., Tribet, C., Gohon, Y., Popot, J.-L., Winnik, F.M. (2007a) Complexation of integral membrane proteins by phosphorylcholine-based amphipols. *Biochim. Biophys. Acta* **1768**:2737–2747.

Diab, C., Winnik, F.M., Tribet, C. (2007b) Enthalpy of interaction and binding isotherms of non-ionic surfactants onto micellar amphiphilic polymers (amphipols). *Langmuir* **23**:3025–3035.

Dominguez Pardo, J.J., Dörr, J.M., Iyer, A., Cox, R.C., Scheidelaar, S., Koorengevel, M.C., Subramaniam, V., Killian, J.A. (2017) Solubilization of lipids and lipid phases by the styrene-maleic acid copolymer. *Eur. Biophys. J.* **46**:91–101.

Dörr, J.M., Scheidelaar, S., Koorengevel, M.C., Dominguez, J.J., Schäfer, M., van Walree, C.A., Killian, J.A. (2016) The styrene-maleic acid copolymer: a versatile tool in membrane research. *Eur. Biophys. J.* **45**:3–21.

Dürr, U.H.N., Soong, R., Ramamoorthy, A. (2013) When detergent meets bilayer: Birth and coming of age of lipid bicelles. *Prog. Nucl. Magn. Reson. Spectrosc.* **69**:1–22.

Duval-Terrié, C., Cosette, P., Molle, G., Muller, G., Dé, E. (2003) Amphiphilic biopolymers (amphibiopols) as new surfactants for membrane protein solubilization. *Protein Sci.* **12**:681–689.

Enikolopyan, N.S., Smirnov, B.R., Ponomarev, G.V., Belgovskii, I.M. (1981) Catalyzed chain transfer to monomer in free radical polymerization. *J. Polym. Sci.* **19**:879–889.

Etzkorn, M., Zoonens, M., Catoire, L.J., Popot, J.-L., Hiller, S. (2014) How amphipols embed membrane proteins: Global solvent accessibility and interaction with a flexible protein terminus. *J. Membr. Biol.* **247**:965–970.

Fainerman, V.B., Miller, R., Joos, P. (1994) The measurement of dynamic surface-tension by the maximum bubble pressure method. *Colloid Polym. Sci.* **272**:731–739.

Farina, M. (1987) Chemistry and kinetics of the chain transfer reaction. *Makromol. Chem. Macromol. Symp.* **10-11**:255–272.

Feldermann, A., Coote, M.L., Stenzel, M.H., Davis, T.P., Barner-Kowollik, C. (2004) Consistent experimental and theoretical evidence for long-lived intermediate radicals in living free radical polymerization. *J. Am. Chem. Soc.* **126**:15915–15923.

Fernandez, A., Le Bon, C., Baumlin, N., Giusti, F., Crémel, G., Popot, J.-L., Bagnard, D. (2014) *In vivo* characterization of the biodistribution profile of amphipols. *J. Membr. Biol.* **247**:1043–1051.

Ferrandez, Y., Dezi, M., Bosco, M., Urvoas, A., Valério, M., Le Bon, C., Giusti, F., Broutin, I., Durand, G., Polidori, A., Popot, J.-L., Picard, M., Minard, P. (2014) Amphipol-mediated screening of molecular orthoses specific for membrane protein targets. *J. Membr. Biol.* **247**:925–940.

Fineman, M., Ross, S.D. (1950) Linear method for determining monomer reactivity ratios in copolymerization. *J. Polym. Sci.* **5**:259–262.

Flory, P.J. (1953) *Principles of polymer chemistry.* Cornell Univ. Press, Ithaca, NY, 688 p.

Flötenmeyer, M., Weiss, H., Tribet, C., Popot, J.-L., Leonard, K. (2007) The use of amphipathic polymers for cryo-electron microscopy of NADH:ubiquinone oxidoreductase (Complex I). *J. Microsc.* **227**:229–235.

Georges, M.K., Veregin, R.P.N., Kazmaier, P.M., Hamer, G.K. (1993) Narrow molecular weight resins by a free-radical polymerization process. *Macromolecules* **26**:2987–2988.

Gilbert, R.G., Hess, M., Jenkins, A.D., Jones, R.G., Kratochvíl, P., Stepto, R.F.T. (2009) Dispersity in polymer science (IUPAC Recommendations 2009). *Pure Appl. Chem.* **81**: 351–353.

Giusti, F., Kessler, P., Westh Hansen, R., Della Pia, E.A., Le Bon, C., Mourier, G., Popot, J.-L., Martinez, K.L., Zoonens, M. (2015) Synthesis of a polyhistidine-bearing amphipol and its use for immobilizing membrane proteins. *Biomacromolecules* **16**:3751–3761.

Giusti, F., Popot, J.-L., Tribet, C. (2012) Well-defined critical association concentration and rapid adsorption at the air/water interface of a short amphiphilic polymer, amphipol A8-35: A study by Förster resonance energy transfer and dynamic surface tension measurements. *Langmuir* **28**:10372–10380.

Giusti, F., Rieger, J., Catoire, L., Qian, S., Calabrese, A.N., Watkinson, T.G., Casiraghi, M., Radford, S.E., Ashcroft, A. E., Popot, J.-L. (2014) Synthesis, characterization and applications of a perdeuterated amphipol. *J. Membr. Biol.* **247**:909–924.

Gohon, Y. (1996) Etude des interactions entre un analogue du fragment transmembranaire de la glycophorine A et des polymères amphiphiles: les amphipols. Thèse de DEA, Université Paris VI, Paris, 28 p.

Gohon, Y. (2002) Etude structurale et fonctionnelle de deux protéines membranaires, la bactériorhodopsine et le récepteur nicotinique de l'acétylcholine, maintenues en solution aqueuse non détergente par des polymères amphiphiles. Thèse de Doctorat, Université Paris-VI, Paris, 467 p.

Gohon, Y., Dahmane, T., Ruigrok, R., Schuck, P., Charvolin, D., Rappaport, F., Timmins, P., Engelman, D.M., Tribet, C., Popot, J.-L., Ebel, C. (2008) Bacteriorhodopsin/amphipol complexes: structural and functional properties. *Biophys. J.* **94**:3523–3537.

Gohon, Y., Giusti, F., Prata, C., Charvolin, D., Timmins, P., Ebel, C., Tribet, C., Popot, J.-L. (2006) Well-defined nanoparticles formed by hydrophobic assembly of a short and polydisperse random terpolymer, amphipol A8-35. *Langmuir* **22**:1281–1290.

Gohon, Y., Pavlov, G., Timmins, P., Tribet, C., Popot, J.-L., Ebel, C. (2004) Partial specific volume and solvent interactions of amphipol A8-35. *Anal. Biochem.* **334**:318–334.

Gohon, Y., Vindigni, J.-D., Pallier, A., Wien, F., Celia, H., Giuliani, A., Tribet, C., Chardot, T., Briozzo, P. (2011) High water solubility and fold in amphipols of proteins with large hydrophobic regions: Oleosins and caleosin from seed lipid bodies. *Biochim. Biophys. Acta* **1808**:706–716.

Golas, P.L., Matyjaszewski, K. (2010) Marrying click chemistry with polymerization: expanding the scope of polymeric materials. *Chem. Soc. Rev.* **39**:1338–1354.

Gorzelle, B.M., Hoffman, A.K., Keyes, M.H., Gray, D.N., Ray, D.G., Sanders II, C.R. (2002) Amphipols can support the activity of a membrane enzyme. *J. Am. Chem. Soc.* **124**:11594–11595.

Greszta, D., Mardare, D., Matyjaszewski, K. (1994) "Living" radical polymerization. 1. Possibilities and limitations. *Macromolecules* **27**:638–644.

Grethen, A., Glueck, D., Keller, S. (2018) Role of coulombic repulsion in collisional lipid transfer among SMA(2:1)-bounded nanodiscs. *J. Membr. Biol.*, in the press.

Grethen, A., Oluwole, A.O., Danielczak, B., Vargas, C., Keller, S. (2017) Thermodynamics of nanodisc formation mediated by styrene/maleic acid (2:1) copolymer. *Sci. Rep.* **7**:11517.

Griffith, W.L., Triolo, R., Compere, A.L. (1987) Analytical scattering function of a polydisperse Percus-Yevick fluid with Schulz- ($\Gamma$-) distributed diameters. *Phys. Rev. A* **35**:2200–2206.

Grimaldo, M., Roosen-Runge, F., Hennig, M., Zanini, F., Zhang, F., Jalarvo, N., Zamponi, M., Schreiber, F., Seydel, T. (2015) Hierarchical molecular dynamics of bovine serum albumin in concentrated aqueous solution below and above thermal denaturation. *Phys. Chem. Chem. Phys.* **17**:4645–4655.

Grubbs, R.B. (2011) Nitroxide-mediated radical polymerization: Limitations and versatility. *Polym. Rev.* **51**:104–137.

Guild, K., Zhang, Y., Stacy, R., Mundt, E., Benbow, S., Green, A., Myler, P.J. (2011) Wheat germ cell-free expression system as a pathway to improve protein yield and solubility for the SSGCID pipeline. *Acta Crystallogr. Sect. F Struct. Biol. Cryst. Commun.* **67**:1027–1031.

Guinier, A., Fournet, G. (1955) *Small-angle scattering of X-rays.* John Wiley & sons, New York, 268 p.

Gulati, S., Jamshad, M., Knowles, T.J., Morrison, K.A., Downing, R., Cant, N., Collins, R., Koenderink, J.B., Ford, R. C., Overduin, M., Kerr, I.D., Dafforn, T.R., Rothnie, A.J. (2014) Detergent-free purification of ABC (ATP-binding-cassette) transporters. *Biochem. J.* **461**:269–278.

Han, S.G., Baek, S.I., Son, T.J., Lee, H., Kim, N.H., Yu, Y.G. (2017) Preparation of functional human lysophosphatidic acid receptor 2 using a P9* expression system and an amphipathic polymer and investigation of its *in vitro* binding preference to $G_\alpha$ proteins. *Biochem. Biophys. Res. Commun.* **487**:103–108.

Han, S.G., Na, J.H., Lee, W.K., Park, D., Oh, J., Yoon, S.H., Lee, C.K., Sung, M.H., Shin, Y.K., Yu, Y.G. (2014) An amphipathic polypeptide derived from poly-$\gamma$-glutamic acid for the stabilization of membrane proteins. *Prot. Sci.* **23**:1800–1807.

Harrington, J.C. (2008) Charge density effects on polyelectrolyte dynamic rheology. *J. Appl. Polym. Sci.* **107**:3310–3317.

Henke, A., Kadłubowski, S., Wolszczak, M., Ulański, P., Boyko, V., Schmidt, T., Arndt, K.-F., Rosiak, J.M. (2011) The structure and aggregation of hydrogen-bonded interpolymer complexes of poly(acrylic acid) with poly(N-vinylpyrrolidone) in dilute aqueous solution. *Macromol. Chem. Phys.* **212**:2529–2540.

Heredia, K.L., Grover, G.N., Tao, L., Maynard, H.D. (2009) Synthesis of heterotelechelic polymers for conjugation of two different proteins. *Macromolecules* **42**:2360–2367.

Hernandez, H., Robinson, C.V. (2007) Determining the stoichiometry and interactions of macromolecular assemblies from mass spectrometry. *Nat. Protocols* **2**:715–726.

Heuts, J.P.A., Muratore, L.M., Davis, T.P. (2000) Preparation and characterization of oligomeric terpolymers of styrene, methyl methacrylate and 2-hydroxyethyl methacrylate: A comparison of conventional and catalytic chain transfer. *Macromol. Chem. Phys.* **201**:2780–2788.

Hovers, J., Potschies, M., Polidori, A., Pucci, B., Raynal, S., Bonneté, F., Serrano-Vega, M., Tate, C., Picot, D., Pierre, Y., Popot, J.-L., Nehmé, R., Bidet, M., Mus-Veteau, I., Bußkamp, H., Jung, K.-H., Marx, A., Timmins, P.A., Welte, W. (2011) A class of mild surfactants that keep integral membrane proteins water-soluble for functional studies and crystallization. *Mol. Memb. Biol.* **28**:171–181.

Huynh, K.W., Cohen, M.R., Moiseenkova-Bell, V.Y. (2014) Application of amphipols for structure-functional analysis of TRP channels. *J. Membr. Biol.* **247**:843–851.

Jacrot, B., Zaccai, G. (1981) Determination of molecular weight by neutron scattering. *Biopolymers* **20**:2413–2426.

Jamshad, M., Charlton, J., Lin, Y.-P., Routledge, S.J., Bawa, Z., Knowles, T.J., Overduin, M., Dekker, N., Dafforn, T.R., Bill, R.M., Poyner, D.R., Wheatley, M. (2015a) G protein-coupled receptor solubilization and purification for biophysical analysis and functional studies, in the total absence of detergent. *Biosc. Rep.* **35**:e00188.

Jamshad, M., Grimard, V., Idini, I., Knowles, T.J., Dowle, M.R., Schofield, N., Sridhar, P., Lin, Y., Finka, R., Wheatley, M., Thomas, O.R.T., Palmer, R.E., Overduin, M., Govaerts, C., Ruysschaert, J.-M., Edler, K.J., Dafforn, T.R. (2015b) Structural analysis of a nanoparticle containing a lipid bilayer used for detergent-free extraction of membrane proteins. *Nano Res.* **8**:774–789.

Jamshad, M., Lin, Y.P., Knowles, T.J., Parslow, R.A., Harris, C., Wheatley, M., Poyner, D.R., Bill, R.M., Thomas, O.R., Overduin, M., Dafforn, T.R. (2011) Surfactant-free purification of membrane proteins with intact native membrane environment. *Biochem. Soc. Trans.* **39**:813–818.

Jasnin, M., van Eijck, L., Koza, M.M., J. Peters., Laguri, C., Lortat-Jacob, H., Zaccai, G. (2010) Dynamics of heparan sulfate explored by neutron scattering. *Phys. Chem. Chem. Phys.* **12**:3360–3362.

Kanellou, A., Spilioti, A., Theodosopoulos, G.V., Choinopoulos, I., Pitsikalis, M. (2015) Statistical copolymers of benzyl methacrylate and diethylaminoethyl methacrylate: monomer reactivity ratios and thermal properties. *J. Org. Inorg. Chem.* **1**:1–11.

Kanu, A.B., Dwivedi, P., Tam, M., Matz, L., Hill Jr., H.H. (2008) Ion mobility – mass spectrometry. *J. Mass Spectrom.* **43**:1–22.

Kelen, T., Tüdös, F., Turcsányi, B. (1980) Confidence intervals for copolymerization reactivity ratios determined by the Kelen-Tüdös method. *Polym. Bull.* **2**:71–76.

Klammt, C., Perrin, M.-H., Maslennikov, I., Renault, L., Krupa, M., Kwiatkowski, W., Stahlberg, H., Vale, W., Choe, S. (2011) Polymer-based cell-free expression of ligand-binding family B G-protein coupled receptors without detergents. *Prot. Sci.* **20**:1030–1041.

Knoop, C.A., Studer, A. (2003) Hydroxy- and silyloxy-substituted TEMPO derivatives for the living free-radical polymerization of styrene and n-butyl acrylate: Synthesis, kinetics, and mechanistic studies. *J. Am. Chem. Soc.* **125**:16327–16333.

Knowles, T.J., Finka, R., Smith, C., Lin, Y.-P., Dafforn, T., Overduin, M. (2009) Membrane proteins solubilized intact in lipid containing nanoparticles bounded by styrene-maleic acid copolymer. *J. Am. Chem. Soc.* **131**:7484–7485.

König, S., Sackmann, E. (1996) Molecular and collective dynamics of lipid bilayers. *Curr. Opin. Colloid Interface Sci.* **1**:78–82.

Ladavière, C., Toustou, M., Gulik-Krzywicki, T., Tribet, C. (2001) Slow reorganization of small phosphatidylcholine vesicles upon adsorption of amphiphilic polymers. *J. Colloid Interface Sci.* **241**:178–187.

Ladavière, C., Tribet, C., Cribier, S. (2002) Lateral organization of lipid membranes induced by amphiphilic polymer inclusions. *Langmuir* **18**:7320–7327.

Le Bon, C., Della Pia, E.A., Giusti, F., Lloret, N., Zoonens, M., Martinez, K.L., Popot, J.-L. (2014a) Synthesis of an oligonucleotide-derivatized amphipol and its use to trap and immobilize membrane proteins. *Nucleic Acids Res.* **42**: e83.

Le Bon, C., Popot, J.-L., Giusti, F. (2014b) Labeling and functionalizing amphipols for biological applications. *J. Membr. Biol.* **247**:797–814.

Lee, S.C., Khalid, S., Pollock, N.L., Knowles, T.J., Edler, K., Rothnie, A.J., Thomas, O.R.T., Dafforn, T.R. (2016) Encapsulated membrane proteins: A simplified system for molecular simulation. *Biochim. Biophys. Acta* **1858**:2549–2557.

Leney, A.C., McMorran, L.M., Radford, S.E., Ashcroft, A.E. (2012) Amphipathic polymers enable the study of functional membrane proteins in the gas phase. *Anal. Chem.* **84**:9841–9847.

Lewis, F.M., Walling, C., Cummings, W., Briggs, E.R., Mayo, F.R. (1948) Copolymerization. IV. Effects of temperature and solvents on monomer reactivity ratios. *J. Am. Chem. Soc.* **70**:1519–1523.

Lindhoud, S., Carvalho, V., Pronk, J.W., Aubin-Tam, M.E. (2016) SMA-SH: Modified styrene-maleic acid copolymer for functionalization of lipid nanodiscs. *Biomacromolecules* **17**:1516–1522.

Lindner, P., Zemb, T. (2002) *Neutrons, X-rays and light scattering methods applied to soft condensed matter.* Elsevier, Amsterdam, 552 p.

Liu, R.C.W., Pallier, A., Brestaz, M., Pantoustier, N., Tribet, C. (2007) Impact of polymer microstructure on the self-assembly of amphiphilic polymers in aqueous solutions. *Macromolecules* **40**:4276–4286.

Logez, C., Damian, M., Legros, C., Dupré, C., Guéry, M., Mary, S., Wagner, R., M'Kadmi, C., Nosjean, O., Fould, B., Marie, J., Fehrentz, J.A., Martinez, J., Ferry, G., Boutin, J.A., Banères, J.-L. (2016) Detergent-free isolation of functional G protein-coupled receptors into nanometric lipid particles. *Biochemistry* **55**:38–48.

Long, A.R., O'Brien, C.C., Malhotra, K., Schwall, C.T., Albert, A.D., Watts, A., Alder, N.N. (2013) A detergent-free strategy for the reconstitution of active enzyme complexes from native biological membranes into nanoscale discs. *BMC Biotechnol.* **13**:41.

Loubat, C., Boutevin, B. (2000a) Etude de la télomérisation du méthacrylate de méthyle par l'acide thioglycolique. Application à la détermination des probabilités de structures des télomères synthétisés. *Macromol. Chem. Phys.* **201**:2853–2860.

Loubat, C., Boutevin, B. (2000b) Telomerization of acrylic acid with thioglycolic acid. Effect of the solvent on the $C_T$ value. *Polym. Bull.* **44**:569–576.

Luccardini, C., Tribet, C., Vial, F., Marchi-Artzner, V., Dahan, M. (2006) Size, charge, and interactions with giant lipid vesicles of quantum dots coated with an amphiphilic macromolecule. *Langmuir* **22**:2304–2310.

Magny, B., Lafuma, F., Iliopoulos, I. (1992) Determination of microstructure of hydrophobically modified water-soluble polymers by $^{13}$C NMR. *Polymer* **33**:3151–3154.

Marie, E., Sagan, S., Cribier, S., Tribet, C. (2014) Amphiphilic macromolecules on cell membranes: from protective layers to controlled permeabilization. *J. Membr. Biol.* **247**:861–881.

Marten, F.L., Hamielec, A.E. (1982) High-conversion diffusion-controlled polymerization of styrene. *Int. J. Appl. Polym. Sci.* **27**:489–505.

Martin, N., Ruchmann, J., Tribet, C. (2015) Prevention of aggregation during refolding of carbonic anhydrase via Coulomb and hydrophobic complexation with octadecyl-modified or azobenzene-modified poly(acrylate) derivatives. *Langmuir* **31**:338–349.

Matyjaszewski, K., Spanswick, J. (2005) Controlled/living radical polymerization. *Mater. Today* **8**:26–33.

Matyjaszewski, K., Tsarevsky, N.V. (2014) Macromolecular engineering by atom transfer radical polymerization. *J. Am. Chem. Soc.* **136**:6513–6533.

Mayo, F.R. (1943) Chain transfer in the polymerization of styrene: The reaction of solvents with free radicals. *J. Am. Chem. Soc.* **65**:2324–2329.

Mayo, F.R., Gregg, R.A., Matheson, M.S. (1951) Chain transfer in the polymerization of styrene. VI. Chain transfer with styrene and benzoyl peroxide; the efficiency of initiation and the mechanism of chain termination. *J. Am. Chem. Soc.* **73**:1691–1700.

Mayo, F.R., Lewis, F.M. (1944) Copolymerization. I. A basis for comparing the behavior of monomers in copolymerization; the copolymerization of styrene and methyl methacrylate. *J. Am. Chem. Soc.* **66**:1594–1601.

Meiser, W., Barth, J., Buback, M., Kattner, H., Vana, P. (2011) EPR Measurement of fragmentation kinetics in dithiobenzoate-mediated RAFT polymerization. *Macromolecules* **44**:2474–2480.

Meiser, W., Buback, M. (2012) Assessing the RAFT equilibrium constant via model systems: An EPR study – Response to a comment. *Macromol. Rapid Commun.* **33**:1273–1279.

Meunier, D.M. (1997) Molecular weight determinations of polymers, in: Settle, F.A., ed., *The Handbook of Instrumental Techniques for Analytical Chemistry.* Prentice Hall, Upper Saddle River, New Jersey, USA, pp. 853–866.

Meunier, D.M., Lyons, J.W., Kiefer, J.J., Niu, Q.J., DeLong, L.M., Li, Y., Russo, P.S., Cueto, R., Edwin, N.J., Bouck, K. J., Silvis, H.C., Tucker, C.J., Kalantar, T.H. (2014) Determination of particle size distributions, molecular weight distributions, swelling, conformation, and morphology of dilute suspensions of cross-linked polymeric nanoparticles via size-exclusion chromatography/differential viscometry. *Macromolecules* **47**:6715–6729.

Miller, R., Joos, P., Fainerman, V.B. (1994) Dynamic surface and interfacial-tensions of surfactant and polymer-solutions. *Adv. Coll. Interf. Sci.* **49**:249–302.

Misha, V., Kumar, R. (2012) Living radical polymerization: a review. *J. Sci. Res. BHU* **56**:141–176.

Moad, G., Rizzardo, E., Thang, S.H. (2012) Living radical polymerization by the RAFT process – A third update. *Aust. J. Chem.* **65**:985–1076.

Moraes, I., Gwyndaf Evans, G., Sanchez-Weatherby, J., Newstead, S., Stewart, P.D.S. (2014) Membrane protein structure determination – The next generation. *Biochim. Biophys. Acta* **1838**:78–87.

Morishima, Y., Nomura, S., Ikeda, T., Seki, M., Kamachi, M. (1995) Characterization of unimolecular micelles of random copolymers of sodium 2-(acrylamido)-2-methylpropanesulfonate and methacrylamides bearing bulky hydrophobic substituents. *Macromolecules* **28**:2874–2881.

Murphy, W.L., Mercurius, K.O., Koide, S., Mrksich, M. (2004) Substrates for cell adhesion prepared via active site-directed immobilization of a protein domain. *Langmuir* **20**:1026–1030.

Nagy, J.K., Kuhn Hoffmann, A., Keyes, M.H., Gray, D.N., Oxenoid, K., Sanders, C.R. (2001) Use of amphipathic polymers to deliver a membrane protein to lipid bilayers. *FEBS Lett.* **501**:115–120.

Natali, F., Castellano, C., Pozzi, D., Congiu-Castellano, A. (2005) Dynamic properties of an oriented lipid/DNA complex studied by neutron scattering. *Biophys. J.* **88**:1081–1090.

Natali, F., Relini, A., Gliozzi, A., Rolandi, R., Cavatorta, P., Deriu, A., Fasano, A., Riccio, P. (2004) The influence of the lipid–protein interaction on the membrane dynamics. *Physica B: Condensed Matter* **350**:E623-E626.

Noda, T., Morishima, Y. (1999) Hydrophobic association of random copolymers of sodium 2-(acrylamido)-2-methylpropanesulfonate and dodecyl methacrylate in water as studied by fluorescence and dynamic light scattering. *Macromolecules* **32**:4631–4640.

Oluwole, A.O., Danielczak, B., Meister, A., Babalola, J.O., Vargas, C., Keller, S. (2017) Solubilization of membrane proteins into functional lipid-bilayer nanodiscs using a diisobutylene/maleic acid copolymer. *Angew. Chem. Int. Ed. Engl.* **56**:1919–1924.

Opačić, M., Durand, G., Bosco, M., Polidori, A., Popot, J.-L. (2014a) Amphipols and photosynthetic light-harvesting pigment-protein complexes. *J. Membr. Biol.* **247**:1031–1041.

Opačić, M., Giusti, F., Broos, J., Popot, J.-L. (2014b) Isolation of *Escherichia coli* mannitol permease, EII^mtl, trapped in amphipol A8-35 and fluorescein-labeled A8-35. *J. Membr. Biol.* **247**:1019–1030.

Orwick, M.C., Judge, P.J., Procek, J., Lindholm, L., Graziadei, A., Engel, A., Grobner, G., Watts, A. (2012) Detergent-free formation and physicochemical characterization of nanosized lipid-polymer complexes. *Angew. Chem. Int. Ed.* **51**:4653–4657.

Orwick-Rydmark, M., Lovett, J.E., Graziadei, A., Lindholm, L., Hicks, M.R., Watts, A. (2012) Detergent-free incorporation of a seven-transmembrane receptor protein into nanosized bilayer Lipodisq particles for functional and biophysical studies. *Nano Lett.* **12**:4687–4692.

Parambil, A.M., Puttaiahgowda, Y.M., Shankarappa, P. (2012) Copolymerization of *N*-vinyl pyrrolidone with methyl methacrylate by Ti(III)-DMG redox initiator. *Turk. J. Chem.* **36**:397–409.

Paulsen, C.E., Armache, J.-P., Gao, Y., Cheng, Y., Julius, D. (2015) Structure of the TRPA1 ion channel suggests regulatory mechanisms. *Nature* **520**:511–517.

Perlmutter, J.D., Drasler, W.J., Xie, W., Gao, J., Popot, J.-L., Sachs, J.N. (2011) All-atom and coarse-grained molecular dynamics simulations of a membrane protein stabilizing polymer. *Langmuir* **27**:10523–10537.

Perlmutter, J.D., Sachs, J.N. (2009a) Experimental verification of lipid bilayer structure through multi-scale modeling. *Biochim. Biophys. Acta* **1788**:2284–2290.

Perlmutter, J.D., Sachs, J.N. (2009b) Inhibiting lateral domain formation in lipid bilayers: simulations of alternative steroid headgroup chemistries. *J. Am. Chem. Soc.* **131**:1636–16363.

Perry, T., Souabni, H., Rapisarda, C., Fronzes, R., Giusti, F., Popot, J.-L., Zoonens, M., Gubellini, F. (2018) Visualizing transmembrane regions of protein complexes by electron microscopy using biotinylated amphipols. *Submitted for publication.*

Petit-Agnely, F., Iliopoulos, I. (1999) Aggregation mechanism of amphiphilic associating polymers studied by [19]F and [13]C nuclear magnetic resonance. *J. Phys. Chem. B* **103**:4803–4808.

Petit-Agnely, F., Iliopoulos, I., Zana, R. (2000) Hydrophobically modified sodium polyacrylates in aqueous solutions: Association mechanism and characterization of the aggregates by fluorescence probing. *Langmuir* **16**:9921–9927.

Picard, M., Dahmane, T., Garrigos, M., Gauron, C., Giusti, F., le Maire, M., Popot, J.-L., Champeil, P. (2006) Protective and inhibitory effects of various types of amphipols on the $Ca^{2+}$-ATPase from sarcoplasmic reticulum: a comparative study. *Biochemistry* **45**:1861–1869.

Picard, M., Duval-Terrié, C., Dé, E., Champeil, P. (2004) Stabilization of membranes upon interaction of amphipathic polymers with membrane proteins. *Protein Sci.* **13**:3056–3058.

Pierik, S.C.J., van Herk, A.M. (2003) Catalytic chain transfer copolymerization of methyl methacrylate and butyl acrylate. *Macromol. Chem. Phys.* **204**:1406–1418.

Pintauer, T., Matyjaszewski, K. (2008) Atom transfer radical addition and polymerization reactions catalyzed by ppm amounts of copper complexes. *Chem. Soc. Rev.* **37**:1087–1097.

Planchard, N., Point, E., Dahmane, T., Giusti, F., Renault, M., Le Bon, C., Durand, G., Milon, A., Guittet, E., Zoonens, M., Popot, J.-L., Catoire, L.J. (2014) The use of amphipols for solution NMR studies of membrane proteins: advantages and limitations as compared to other solubilizing media. *J. Membr. Biol.* **247**:827–842.

Polovinkin, V., Balandin, T., Volkov, O., Round, E., Borshchevskiy, V., Utrobin, P., von Stetten, D., Royant, A., Willbold, D., Arzumanyan, A., Popot, J.-L., Gordeliy, V. (2014) Nanoparticle surface-enhanced Raman scattering of bacteriorhodopsin stabilized by amphipol A8-35. *J. Membr. Biol.* **247**:971–980.

Popot, J.-L. (2010) Amphipols, nanodiscs, and fluorinated surfactants: Three non-conventional approaches to studying membrane proteins in aqueous solutions. *Annu. Rev. Biochem.* **79**:737–775.

Popot, J.-L., Althoff, T., Bagnard, D., Banères, J.-L., Bazzacco, P., Billon-Denis, E., Catoire, L.J., Champeil, P., Charvolin, D., Cocco, M.J., Crémel, G., Dahmane, T., de la Maza, L.M., Ebel, C., Gabel, F., Giusti, F., Gohon, Y., Goormaghtigh, E., Guittet, E., Kleinschmidt, J.H., Kühlbrandt, W., Le Bon, C., Martinez, K.L., Picard, M., Pucci, B., Rappaport, F., Sachs, J.N., Tribet, C., van Heijenoort, C., Wien, F., Zito, F., Zoonens, M. (2011) Amphipols from A to Z. *Annu. Rev. Biophys.* **40**:379–408.

Popot, J.-L., Berry, E.A., Charvolin, D., Creuzenet, C., Ebel, C., Engelman, D.M., Flötenmeyer, M., Giusti, F., Gohon, Y., Hervé, P., Hong, Q., Lakey, J.H., Leonard, K., Shuman, H.A., Timmins, P., Warschawski, D.E., Zito, F., Zoonens, M., Pucci, B., Tribet, C. (2003) Amphipols: polymeric surfactants for membrane biology research. *Cell. Mol. Life Sci.* **60**:1559–1574.

Pound, G., Aguesse, F., McLeary, J.B., Lange, R.F.M., Klumperman, B. (2007) Xanthate-mediated copolymerization of vinyl monomers for amphiphilic and double-hydrophilic block copolymers with poly(ethylene glycol). *Macromolecules* **40**:8861–8871.

Prata, C., Giusti, F., Gohon, Y., Pucci, B., Popot, J.-L., Tribet, C. (2001) Non-ionic amphiphilic polymers derived from *tris*(hydroxymethyl)-acrylamidomethane keep membrane proteins soluble and native in the absence of detergent. *Biopolymers* **56**:77–84.

Presser, A., Hüfner, A. (2004) Trimethylsilyldiazomethane. A mild and efficient reagent for the methylation of carboxylic acids and alcohols in natural products. *Monatshefte für Chem. Chem. Mon.* **135**:1015–1022.

Pucci, B., Ragonnet, B., Pavia, A.A. (1988) Télomères et cotélomères d'intérêt biologique et biomédical. II. Synthèse de télomères galactosylés fluorescents marqueurs potentiels des lectines membranaires. *Eur. Polym. J.* **24**:1087–1091.

Qi, L., Gao, X. (2008) Quantum dot-amphipol nanocomplex for intracellular delivery and real-time imaging of siRNA. *ACS Nano* **2**:1403–1410.

Qi, L., Wu, L., Zheng, S., Wang, Y., Fu, H., Cui, D. (2012) Cell-penetrating magnetic nanoparticles for highly efficient delivery and intracellular imaging of siRNA. *Biomacromolecules* **13**:2723–2730.

Raffa, P., Wever, D.A.Z., Picchioni, F., Broekhuis, A.A. (2015) Polymeric surfactants: Synthesis, properties, and links to applications. *Chem. Rev.* **115**:8504–8563.

Rajesh, S., Knowles, T.J., Overduin, M. (2011) Production of membrane proteins without cells or detergents. *New Biotechnol.* **28**:250–254.

Rane, S.S., Choi, P. (2005) Polydispersity index: How accurately does it measure the breadth of the molecular weight distribution? *Chem. Mater.* **17**:926–926.

Ringsdorf, H., Venzmer, J., Winnik, F.M. (1991) Fluorescence studies of hydrophobically modified poly(*N*-isopropylacrylamides) *Macromolecules* 24:1678–1686.

Rizzardo, E., Solomon, D.H. (2012) On the origins of nitroxide-mediated polymerization (NMP) and reversible addition–fragmentation chain transfer (RAFT). *Aust. J. Chem.* **65**:945–969.

Roberts, C., Chen, C., Mrksich, M., Martichonok, V., Ingber, D.E., Whitesides, G.M. (1998) Using mixed self-assembled monolayers presenting RGD and (EG)$_3$OH groups to characterize long-term attachment of bovine capillary endothelial cells to surfaces. *J. Am. Chem. Soc.* **120**:6548–6555.

Rossi, N.A.A., Zou, Y., Scott, M.D., Kizhakkedathu, J.N. (2008) RAFT synthesis of acrylic copolymers containing poly (ethylene glycol) and dioxolane functional groups: Toward well-defined aldehyde-containing copolymers for bioconjugation. *Macromolecules* **41**:5272–5282.

Roy, K.K., Pramanick, D., Palit, S.R. (1972) Application of dye techniques in the study of chain-transfer properties of thiols. *Makromol. Chem.* **153**:71–80.

Ruzette, A.-V., Leibler, L. (2005) Block copolymers in tomorrow's plastics. *Nat. Mater.* **4**:19–31.

Sahu, I.D., McCarrick, R.M., Troxel, K.R., Zhang, R., Smith, H.J., Dunagan, M.M., Swartz, M.S., Rajan, P.V., Kroncke, B.M., Sanders, C.R., Lorigan, G.A. (2013) DEER EPR measurements for membrane protein structures via bifunctional spin labels and lipodisq nanoparticles. *Biochemistry* **52**:6627–6632.

Sanders, C.R., Prosser, R.S. (1998) Bicelles: a model membrane system for all seasons? *Structure* **6**:1227–1234.

Sarrazin-Cartalas, A., Iliopoulos, I., Audebert, R., Olsson, U. (1994) Association and thermal gelation in mixtures of hydrophobically modified polyelectrolytes and nonionic surfactants. *Langmuir* **10**:1421–1426.

Sauer, A.M., Schlossbauer, A., Ruthardt, N., Cauda, V., Bein, T., Bräuchle, C. (2010) Role of endosomal escape for disulfide-based drug delivery from colloidal mesoporous silica evaluated by live-cell imaging. *Nano Lett.* **10**:3684–3691.

Sauvage, E., Plucktaveesak, N., Colby, R.H., Amos, D.A., Antalek, B., Schroeder, K.M., Tan, J.S. (2004) Amphiphilic maleic acid-containing alternating copolymers – 2. Dilute solution characterization by light scattering, intrinsic viscosity, and PGSE NMR spectroscopy. *J. Polym. Sci.* **42**:3584–3597.

Scheidelaar, S., Koorengevel, M.C., Pardo, J.D., Meeldijk, J.D., Breukink, E., Killian, J.A. (2015) Molecular model for the solubilization of membranes into nanodisks by styrene maleic acid copolymers. *Biophys. J.* **108**:279–290.

Sebai, S., Cribier, S., Karimi, A., Massotte, D., Tribet, T. (2010) Permeabilisation of lipid membranes and cells by a light-responsive copolymer. *Langmuir* **26**:14135–14141.

Sebai, S., Milioni, D., Walrant, A., Alves, I., Sagan, S., Huin, C., Auvray, L., Massotte, D., Cribier, S., Tribet, C. (2012) Photocontrol of the translocation of molecules, peptides, and quantum dots through cell and lipid membranes doped with azobenzene copolymers. *Angew. Chem. Int. Ed.* **51**:2132–2136.

Selb, J., Gallot, Y. (1980) Distinction entre les phénomènes d'agrégation et de micellisation présentés par des copolymères amphipathiques. Cas des copolymères polystyrène/polyvinylpyridinium en milieu aqueux. *Makromol. Chem.* **181**:809–822.

Sharma, K.S., Durand, G., Gabel, F., Bazzacco, P., Le Bon, C., Billon-Denis, E., Catoire, L.J., Popot, J.-L., Ebel, C., Pucci, B. (2012) Non-ionic amphiphilic homopolymers: Synthesis, solution properties, and biochemical validation. *Langmuir* **28**:4625–4639.

Sharma, K.S., Durand, G., Giusti, F., Olivier, B., Fabiano, A.-S., Bazzacco, P., Dahmane, T., Ebel, C., Popot, J.-L., Pucci, B. (2008) Glucose-based amphiphilic telomers designed to keep membrane proteins soluble in aqueous solutions: synthesis and physico-chemical characterization. *Langmuir* **24**:13581–13590.

Soeriyadi, A.H., Boyer, C., Nyström, F., Zetterlund, P.B., Whittaker, M.R. (2011) High-order multiblock copolymers via iterative Cu(0)-mediated radical polymerizations (SET-LRP): Toward biological precision. *J. Am. Chem. Soc.* **133**:11128–11131.

Stangl, M., Hemmelmann, M., Allmeroth, M., Zentel, R., Schneider, D. (2014) A minimal hydrophobicity is needed to employ amphiphilic p(HPMA)-co-p(LMA) random copolymers in membrane research. *Biochemistry* **53**:1410–1419.

Starks, C.M. (1974) *Free radical telomerization.* Academic Press/Elsevier, New York, 280 p.

Stefano, J.E., Hou, L.H., Honey, D., Kyazike, J., Park, A., Zhou, Q., Pan, C.Q., Edmunds, T. (2009) *In vitro* and *in vivo* evaluation of a non-carbohydrate targeting platform for lysosomal proteins. *J. Control. Release* **135**:113–118.

Stuhrmann, H.B. (1970) Interpretation of small-angle scattering functions of dilute solutions and gases. A representation of the structures related to a one-particle scattering function. *Acta Cryst. Sect. A* **26**:297–306.

Sverzhinsky, A., Qian, S., Yang, L., Allaire, M., Moraes, I., Ma, D., Chung, J.W., Zoonens, M., Popot, J.-L., Coulton, J.W. (2014) Amphipol-trapped ExbB−ExbD membrane protein complex from *Escherichia coli*: A biochemical and structural case study. *J. Membr. Biol.* **247**:1005–1018.

Swift, T., Swanson, L., Geoghegan, M., Rimmer, S. (2016) The pH-responsive behaviour of poly(acrylic acid) in aqueous solution is dependent on molar mass. *Soft Matter* **12**:2542–2549.

Takei, Y.G., Aoki, T., Sanui, K., Ogata, N., Okano, T., Sakurai, Y. (1993) Temperature-responsive bioconjugates. 1. Synthesis of temperature-responsive oligomers with reactive end groups and their coupling to biomolecules *Bioconjugate Chem.* **4**:42–46.

Tanaka, M., Hosotani, A., Tachibana, Y., Nakano, M., Iwasaki, K., Kawakami, T., Mukai, T. (2015) Preparation and characterization of reconstituted lipid-synthetic polymer discoidal particles. *Langmuir* **31**:12719–12626.

Tanford, C. (1980) *The Hydrophobic Effect: Formation of Micelles and Biological Membranes*, 2nd ed. John Wiley & Sons, New York, 233 p.

Tefera, N., Weickert, G., Westerterp, K.R. (1997) Modeling of free radical polymerization up to high conversion. II. Development of a mathematical model. *J. Appl. Polym. Sci.* **63**:1663–1680.

Tehei, M., Perlmutter, J.D., Giusti, F., Sachs, J.N., Zaccai, G., Popot, J.-L. (2014) Thermal fluctuations in amphipol A8-35 particles: A neutron scattering and molecular dynamics study. *J. Membr. Biol.* **247**:897–908.

Terrettaz, S., Ulrich, W.P., Vogel, H., Hong, Q., Dover, L.G., Lakey, J.H. (2002) Stable self-assembly of a protein engineering scaffold on gold surfaces. *Protein Sci.* **11**:1917–1925.

Tifrea, D.F., Pal, S., Le Bon, C., Giusti, F., Cocco, M.J., Zoonens, M., de la Maza, L.M. (2018a) Resiquimod conjugated with amphipols bound to the *Chlamydia muridarum* MOMP enhances protection against a mucosal challenge. *In preparation.*

Tifrea, D.F., Pal, S., Le Bon, C., Giusti, F., Zoonens, M., Cocco, M.J., Popot, J.-L., de la Maza, L.M. (2018b) Co-delivery of adjuvants and antigens using amphipol-linked adjuvants and adjuvant combinations enhance protection elicited by a membrane protein-based vaccine against a mucosal challenge with *Chlamydia. Submitted for publication.*

Tobolsky, A.V. (1958) Dead-end radical polymerization. *J. Am. Chem. Soc.* **80**:5927–5929.

Tolstyka, Z.P., Kopping, J.T., Maynard, H.D. (2008) Straightforward synthesis of cysteine-reactive telechelic polystyrene. *Macromolecules* **41**:599–606.

Trapp, M., Gutberlet, T., Juranyi, F., Unruh, T., Demé, B., Tehei, M., Peters, J. (2010) Hydration dependent studies of highly aligned multilayer lipid membranes by neutron scattering. *J. Chem. Phys.* **133**:164505.

Treat, N.J., Sprafke, H., Kramer, J.W., Clark, P.G., Barton, B.E., Read de Alaniz, J., Fors, B.P., Hawker, C.J. (2014) Metal-free atom transfer radical polymerization. *J. Am. Chem. Soc.* **136**:16096–16101.

Tribet, C., Audebert, R., Popot, J.-L. (1996) Amphipols: polymers that keep membrane proteins soluble in aqueous solutions. *Proc. Natl. Acad. Sci. USA* **93**:15047–15050.

Tribet, C., Audebert, R., Popot, J.-L. (1997) Stabilization of hydrophobic colloidal dispersions in water with amphiphilic polymers: Application to integral membrane proteins. *Langmuir* **13**:5570–5576.

Tribet, C., Diab, C., Dahmane, T., Zoonens, M., Popot, J.-L., Winnik, F.M. (2009) Thermodynamic characterization of the exchange of detergents and amphipols at the surfaces of integral membrane proteins. *Langmuir* **25**:12623–12634.

Tribet, C., Vial, F. (2008) Flexible macromolecules attached to lipid bilayers: impact on fluidity, curvature, permeability and stability of the membranes. *Soft Matter* **4**:68–81.

Trubetskoy, O.V., Finel, M., Burke, T.J., Trubetskoy, V.S. (2006) Evaluation of synthetic polymeric micelles as a stabilization medium for the handling of membrane proteins in pharmaceutical drug discovery. *J. Pharm. Pharmaceut. Sci.* **9**:271–280.

Vargas, C., Cuevas Arenas, R., Frotscher, E., Keller, S. (2015) Nanoparticle self-assembly in mixtures of phospholipids with styrene/maleic acid copolymers or fluorinated surfactants. *Nanoscale* **7**:20685–20696.

Veregin, R.P.N., Georges, M.K., Kazmaier, P.M., Hamer, G.K. (1993) Free radical polymerizations for narrow polydispersity resins: electron spin resonance studies of the kinetics and mechanism. *Macromolecules* **26**:5316–5320.

Vial, F., Cousin, F., Bouteiller, L., Tribet, C. (2009) Rate of permeabilization of giant vesicles by amphiphilic polyacrylates compared to the adsorption of these polymers onto large vesicles and tethered lipid bilayers. *Langmuir* **25**:7506–7513.

Vial, F., Oukhaled, A.G., Auvray, L., Tribet, C. (2007) Long-living channels of well-defined radius opened in lipid bilayers by polydisperse, hydrophobically-modified polyacrylic acids. *Soft Matter* **3**:75–78.

Vial, F., Rabhi, S., Tribet, C. (2005) Association of octyl-modified poly(acrylic acid) onto unilamellar vesicles of lipids and kinetics of vesicle disruption. *Langmuir* **21**:853–862.

Volpert, E., Selb, J., Candau, F. (1996) Influence of the hydrophobe structure on composition, microstructure, and rheology in associating polyacrylamides prepared by micellar copolymerization. *Macromolecules* **29**:1452–1463.

Wagner, H.L. (1987) The Mark–Houwink–Sakurada relation for poly(methyl methacrylate). *J. Phys. Chem. Ref. Data* **16**:165–173.

Wang, J.-S., Matyjaszewski, K. (1995) Controlled "living" radical polymerization. Atom transfer radical polymerization in the presence of transition-metal complexes. *J. Am. Chem. Soc.* **117**:5614–5615.

Wang, K.T., Iliopoulos, I., Audebert, R. (1988) Viscometric behavior of hydrophobically modified poly(sodium acrylate) *Polym. Bull.* **20**:577–582.

Watkinson, T.G., Calabrese, A.N., Giusti, F., Zoonens, M., Radford, S.E., Ashcroft, A.E. (2015) Systematic analysis of the use of amphipathic polymers for studies of outer membrane proteins using mass spectrometry. *Int. J. Mass Spectrom.* **391**:54–61.

Weidner, S.M., Trimpin, S. (2010) Mass spectrometry of synthetic polymers. *Anal. Chem.* **82**:4811–4829.

Wheatley, M., Charlton, J., Jamshad, M., Routledge, S.J., Bailey, S., La-Borde, P.J., Azam, M.T., Logan, R.T., Bill, R. M., Dafforn, T.R., Poyner, D.R. (2016) GPCR-styrene maleic acid lipid particles (GPCR-SMALPs): their nature and potential. *Biochem. Soc. Trans.* **44**:619–623.

Yamamoto, H., Hashidzume, A., Morishima, Y. (2000a) Micellization protocols for amphiphilic polyelectrolytes in water. How do polymers undergo intrapolymer associations? *Polym. J.* **32**:745–752.

Yamamoto, H., Tomatsu, I., Hashidzume, A., Morishima, Y. (2000b) Associative properties in water of copolymers of sodium 2-(acrylamido)-2-methylpropanesulfonate and methacrylamides substituted with alkyl groups of varying lengths. *Macromolecules* **33**:7852–7861.

Yokozawa, T., Yokoyama, A. (2004) Chain-growth polycondensation: living polymerization nature in polycondensation and approach to condensation polymer architecture. *Polym. J.* **36**:65–83.

Yoshimoto, K., Hirase, T., Nemoto, S., Hatta, T., Nagasaki, Y. (2008) Facile construction of sulfanyl-terminated poly (ethylene glycol)-brushed layer on a gold surface for protein immobilization by the combined use of sulfanyl-ended telechelic and semitelechelic poly(ethylene glycol)s. *Langmuir* **24**:9623–9629.

Zaccai, G. (2000) How soft is a protein? A protein dynamics force constant measured by neutron scattering. *Science* **288**:1604–1607.

Zaccai, G. (2011) Neutron scattering perspectives for protein dynamics. *J. Non-Cryst. Solids* **357**:615–621.

Zaccai, G. (2013) The ecology of protein dynamics. *Curr. Phys. Chem.* **3**:9–16.

Zhang, R., Sahu, I.D., Liu, L., Osatuke, A., Comer, R.G., Dabney-Smith, C., Lorigan, G.A. (2015) Characterizing the structure of lipodisq nanoparticles for membrane protein spectroscopic studies. *Biochim. Biophys. Acta* **1848**:329–333.

Zhao, J., Wang, H., Liu, J., Deng, L., Liu, J., Dong, A., Zhang, J. (2013) Comb-like amphiphilic copolymers bearing acetal-functionalized backbones with the ability of acid-triggered hydrophobic-to-hydrophilic transition as effective nanocarriers for intracellular release of curcumin. *Biomacromolecules* **14**:3973–3984.

Zoonens, M., Catoire, L.J., Giusti, F., Popot, J.-L. (2005) NMR study of a membrane protein in detergent-free aqueous solution. *Proc. Natl. Acad. Sci. USA* **102**:8893–8898.

Zoonens, M., Giusti, F., Zito, F., Popot, J.-L. (2007) Dynamics of membrane protein/amphipol association studied by Förster resonance energy transfer. Implications for *in vitro* studies of amphipol-stabilized membrane proteins. *Biochemistry* **46**:10392–10404.

Zoonens, M., Popot, J.-L. (2014) Amphipols for each season. *J. Membr. Biol.* **247**:759–796.

# Formation and Properties of Membrane Protein/Amphipol Complexes

**5**

**Summary**

*Complexes between membrane proteins (MPs) and amphipols (APols) can be obtained (i) by transferring to APols a detergent-solubilized MP, (ii) by directly extracting a membrane-bound one, (iii) by folding in APols a denatured MP, or (iv) by synthesizing it in vitro in the presence of APol. The complexes comprise a belt of APol that surrounds the transmembrane region of the protein and, if available, protein-bound lipids. Trapping with APols very generally maintains the activity of the protein, but a few cases of reversible inhibition have been observed. Most APol-trapped MPs are more stable, and generally much more so, than their detergent-solubilized counterparts. Three mechanisms appear to contribute to this stabilization: (i) reduction of the hydrophobic sink, (ii) the intrinsically less dissociating character of APols as compared to detergents, and (iii) damping by APols of MP dynamics, which raises the free energy barrier to unfolding. The latter phenomenon appears to be involved in the inhibition observed with some MPs, inasmuch as it can slow down conformational transitions. MPs trapped with APols can be transferred directly to detergent solutions, to other APols, to lipid vesicles, to lipidic mesophases, to black lipid films, or to cell membranes, as well as, indirectly, to nanodiscs or SMALPs.*

## 5.1 Introduction

This present chapter deals with the way to form membrane protein/amphipol (MP/APol) complexes and with a description of their general properties. Whereas it must be kept in mind that different MPs may be affected in different ways by being transferred from a membrane to an APol environment and that different APols have different properties, the detailed knowledge accumulated on the composition, structure, and dynamics of a few particularly well-studied complexes provides a good starting point to understanding the behavior of the others.

There are many ways to form membrane MP/APol complexes. In this chapter, we will be principally concerned with methods that start from the folded protein, whether or not it has been beforehand extracted from its native membrane. In Chaps. 6 and 7, we will examine two other approaches, in which the target protein is either folded from a denatured to a functional state using

© Springer International Publishing AG, part of Springer Nature 2018
J. -L. Popot, *Membrane Proteins in Aqueous Solutions*, Biological and Medical Physics,
Biomedical Engineering, https://doi.org/10.1007/978-3-319-73148-3_5

APols (Chap. 6) or synthesized in vitro with APols as the accepting medium (Chap. 7). The next two chapters will deal with the optical properties of MP/APol complexes (Chap. 8) and other solution studies (Chap. 9). Excerpts of these and other data to be presented in other chapters, particularly NMR and electron microscopy (EM) data (Chaps. 10 and 12, respectively), will be used in the present one to draw an integrated picture of MP/APol complexes.

In this chapter, we will also consider three further important aspects of working with APols:

(i) The effects of APols on the stability of MPs and the mechanisms underlying these effects;
(ii) Their influence on the dynamics and functional properties of MPs;
(iii) Transferring APol-trapped MPs from APols to other surfactants, such as other APols, detergents, nanodiscs, styrene maleic acid lipid particles (SMALPs), or lipid vesicles.

## 5.2    Forming Membrane Protein/Amphipol Complexes

At the time of this writing, close to 100 distinct integral MPs have been trapped with one or the other APol (most of which are compiled in Table 5.1), a number that has kept increasing steadily over the past 20 years (Fig. 5.1). With some 150 references, Table 5.1 looks quite formidable. However, it has been laid out so that the reader interested in finding out what has been published on which MP or type of MP using which APol can locate it at a glance. Such a search is extremely time-consuming, if not impossible, to do via databanks, as most articles do not include "amphipol" among their keywords, much less the specific APol that has been used. As regards applying specific techniques to studying APol-trapped MPs, some examples are given in Table 5.4, and exhaustive lists will be provided in the chapters dealing with each technique.

MPs that have been trapped in APols cover all types of functions and structures, monomeric or oligomeric, folded into $\alpha$-helix bundles or into $\beta$-barrels (summarized in Table 5.2), and their masses range from ~3 kDa (a single transmembrane (TM) $\alpha$-helix; see e.g. Stangl et al. 2014) to more than 1 MDa (e.g. the mitochondrial $I_1III_2IV_1$ supercomplex; Althoff et al. 2011) (Table 5.1, Column 5). This diversity provides a couple of interesting indications about the generality of the use of APols: first, they can trap and keep in their native state rugged MPs – most $\beta$-barrel ones – as well as fragile ones: most $\alpha$-helical MPs, including the particularly delicate G protein-coupled receptors (GPCRs); second, at variance with nanodiscs (NDs), there does not seem to be an upper limit as to the dimensions of the TM regions APols can encompass and keep water-soluble. One should note, however, that whereas all types of TM proteins have been trapped and kept water-soluble using APols (Table 5.2), no report has been published yet about using them with monotopic MPs. There does not seem to be any reason why this should be problematic. Indeed, bacterial sulfide-quinone reductase expressed in *E. coli* C41(DE3) and purified by immobilized-metal affinity chromatography (IMAC) in the presence of 0.03% *n*-dodecyl-$\beta$-$_\text{D}$-maltopyranoside (DDM) has been successfully transferred to A8-35, the detergent being removed using Bio-Beads. Sulfide-oxidizing activity was observed, albeit with a reduced $V_{max}$ (A.H. Abbas, F. Bouillaud and colleagues, unpublished observations).

Experiments in which APols have been used to stabilize other nanomaterials, such as quantum dots (Luccardini et al. 2006; Qi and Gao 2008; Sebai et al. 2012; Booth et al. 2013; Lim et al. 2015) or siRNA (Li et al. 2015b), are beyond the frame of this book and will not be reviewed here.

Given the many different APols that have been tested, the number of distinct MP/APol complexes that have been described reached >130 by the spring of 2017. Their distribution between the various structural types of MPs and of APols is summarized in Table 5.2.

**Table 5.1** Integral membrane proteins that have been shown to be kept soluble by amphipols. Proteins are listed in chronological order of first publication of their trapping by any APol. The bibliography covers the period 1996–2016. *Color coding of the cells*: (i) *initial state of the protein*, native or refolded to native-like state, *white*; unfolded, *light gray*; synthetic peptides, *light blue-gray*; (ii) *nature of the APol*, A8-35, *blue*; A8-75, *gray*; NAPols (glucosylated unless otherwise indicated), *pink*; PC-APols, *cyan*; SAPols, *green*; SMA, *tan*; other APols, *white*. (iii) Characterization of the native state of the protein: when functional tests (typically ligand binding or enzymatic activity) have been carried out on the APol-trapped protein (not after transfer to another environment), the cell is *salmon-colored*.

| Protein(s) (abbreviation) | Organism, membrane | Structural type[a] | Number of chains | Overall mass | State and environment before trapping | Amphipol[b] | Evidence for native state after trapping | References |
|---|---|---|---|---|---|---|---|---|
| Bacteriorhodopsin (BR) | Halobacterium salinarum, plasma membrane | α | 1 + retinal | 27 kDa | | A8-35 | Spectrum, photocycle, SEC, SV-AUC, SANS, NMR | Tribet et al. (1996, 2009), Gohon et al. (2008), Charvolin et al. (2009), Dahmane et al. (2009), Bechara et al. (2012), Sharma et al. (2012), Della Pia et al. (2014), Ferrandez et al. (2014), Giusti et al. (2014), Le Bon et al. (2014), and Polovinkin et al. (2014a, b) |
| | | | | | | A8-75 | Spectrum | Tribet et al. (1996) |
| | | | | | Native, OTG | A34-35, A34-75 | Spectrum | Tribet et al. (1996) |
| | | | | | | NAPols[c] | Spectrum, photocycle, SEC, SV-AUC, SANS | Prata et al. (2001), Sharma et al. (2008, 2012), Bazzacco et al. (2009, 2012), and Bechara et al. (2012) |
| | | | | | | PC-APols | Spectrum, SD-AUC | Diab et al. (2007) and Tribet et al. (2009) |
| | | | | | | SAPols | Spectrum, SEC | Dahmane et al. (2009) |
| | | | | | Native, OG | APG | Spectrum | Han et al. (2014) |
| | | | | | Native, ? | A8-35 | Raman and FTIR spectra | Kumar et al. (2016) |
| | | | | | Native, DMPC vesicles | SMA | Spectrum | Knowles et al. (2009), Orwick-Rydmark et al. (2012), Goddard et al. (2015), Lindhoud et al. (2016) see also Broecker et al. (2017) (BR from *Haloquadratum walsbyi*) |
| | | | | | | A8-35 | Spectrum, photocycle, SEC, NMR | Pocanschi et al. (2006), Dahmane et al. (2013), Etzkorn et al. (2013), and Elter et al. (2014) |
| | | | | | Denatured, SDS | NAPols | Spectrum | Bazzacco et al. (2012) |
| | | | | | | APG | Spectrum | Han et al. (2014) |
| | | | | | Denatured, TFE[c] | A8-35 | Spectrum | Dahmane et al. (2013) |
| | | | | | Nascent chain, cell-free synthesis | NAPols | Spectrum | Bazzacco et al. (2012) |
| Cytochrome $b_6 f$ | Chlamydomonas reinhardtii, thylakoids | α | 2 × 8 + prosthetic groups | 228 kDa | | A8-35 | Spectrum, SG-AUC, STEM | Tribet et al. (1996, 1997, 1998), Charvolin et al. (2009), and Bechara et al. (2012) |
| | | | | | | A8-75 | Spectrum | Tribet et al. (1996) |
| | | | | | Native, Hecameg or DDM | A34-35, A34-75 | | |
| | | | | | | A5-75 | | Gohon (1996) and Popot et al. (2003) |
| | | | | | | PC-APols | Spectrum, SG-AUC, electron transfer activity | Diab et al. (2007) |
| | | | | | | NAPols[c] | | Prata et al. (2001), Bazzacco et al. (2012), and Bechara et al. (2012) |
| | | | | | | SAPols | | Bazzacco et al. (2012) |
| Outer membrane protein F (OmpF) | Escherichia coli, outer membrane | β | 3 | 102 kDa | Native, C8-POE | A8-35 | SG-AUC, SEC, SDS-PAGE | Tribet et al. (1996) and Arunmanee et al. (2014) |
| | | | | | | A8-75 | SG-AUC | Tribet et al. (1996, 1997) |
| | | | | | | A34-35, A34-75 | | Tribet et al. (1996) |

| Protein(s) (abbreviation) | Organism, membrane | Structural type[a] | Number of chains | Overall mass | State and environment before trapping | Amphipol[b] | Evidence for native state after trapping | References |
|---|---|---|---|---|---|---|---|---|
| Photosynthetic reaction center | Rhodobacter sphaeroides, plasma membrane | α | 3 | 96 kDa | | A8-35 | SG-AUC | Tribet et al. (1996) |
| | | | | | Native, Hecameg | A8-75 | SG-AUC | Tribet et al. (1996, 1997) |
| | | | | | | A34-35, A34-75 | SG-AUC | Tribet et al. (1996) |
| | | | | | Native, plasma membrane | SMA | Spectrum, photochemistry | Swainsbury et al. (2014) |
| Glycophorin A transmembrane anchor (tGpA), monomer or dimer | Human, plasma membrane | α | 1 or 2 | ~4 or ~8 kDa | Synthetic peptide, urea | A5-75 | CD | Gohon (1996) and Popot et al. (2003) |
| | | | | | Synthetic peptides, TFE | A8-35 | CD, FRET, sequence-specific dimerization | Stangl et al. (2014) |
| F₁F₀ ATP synthase | E. coli, plasma membrane | α | 21–24 | ~560 kDa | Native, taurodeoxycholate | A8-75 | NS-EM | Wilkens and Capaldi (1998a,b), Wilkens (2000), and Wilkens et al. (2000) |
| Sarcoplasmic calcium pump (SERCA1a) | Oryctolagus cuniculus, sarcoplasmic reticulum | α | 1 | 110 kDa | Native, DDM | A8-35 | ATPase activity, SEC, SV-AUC | Champeil et al. (2000) and Picard et al. (2006) |
| | | | | | | SAPol | ATPase activity, SEC | Picard et al. (2006) |
| | | | | | | PMAL-C12[e], PMALA-C12 | ATPase activity, SEC | Picard et al. (2006) |
| Diacylglycerol kinase (DAGK) | E. coli, plasma membrane | α | 3 × 1 | 40 kDa | Native, Empigen or DM | OAPA-20[e], PMAL-B-50, PMAL-B-100[e] | SDS-PAGE after chemical cross-linking, ATPase activity, MS | Nagy et al. (2001), Gorzelle et al. (2002), and Hopper et al. (2013) |
| Photosystem I | Synechocystis PCC 6803, thylakoids | α | 3 × 12 + >120 cofactors | > 1 MDa | Native, DDM | A8-35 | Electrochemistry | Kievit and Brudvig (2001) |
| Nicotinic acetylcholine receptor (nAChR) | Torpedo marmorata, plasma membrane | α | 2 × 5 | 535 kDa | Native, CHAPS | A8-35 | Allosteric transitions, ligand and toxin binding | Martinez et al. (2002) and Charvolin et al. (2009) |
| Rhodopsin | Bos taurus, retina discs | α | 1 + retinal | 40 kDa | Native, CHAPS | A8-35 | Transition to meta-II state | C. Tribet, C. Creuzenet & JLP, unpublished observations cited in Popot et al. (2003) |
| | | | | | Native, DDM | A8-35 | NS-EM (rhodopsin/GRK5 complex) | He et al. (2017) |
| Cytochrome $bc_1$ | B. taurus, inner mitochondrial membrane | α | 2 × 11 | 490 kDa | Native, DDM | A8-35 | Catalysis of electron transfer, SEC, SG-AUC, SAXS | Popot et al. (2003), Charvolin et al. (2009), Charvolin et al. (2014), and Ferrandez et al. (2014) |
| | | | | | | NAPols | MS | Bechara et al. (2012) |
| Photosystem II reaction center | Pisum sativum, thylakoids | α | 2 × 17 | 550 kDa | Native, CHAPS or DDM | A8-35 | Photochemistry | A. Zehetner & H. Scheer, unpublished observation cited in Popot et al. (2003) |
| Photosystem II reaction center | Thermosynechococcus elongatus, thylakoids | α | 2 × 20 + cofactors | 450 kDa | Native, DDM | A8-35 | SEC, oxygen evolution | Nowaczyk et al. (2004) |
| Maltose transporter | E. coli, plasma membrane | α | 1 | 150 kDa | Native, OG, or direct extraction with A8-35 | A8-35 | ATPase activity | M. Zoonens & H.A. Shuman, unpublished observations cited in Popot et al. (2003) |

| Protein(s) (abbreviation) | Organism, membrane | Structural type[a] | Number of chains | Overall mass | State and environment before trapping | Amphipol[b] | Evidence for native state after trapping | References |
|---|---|---|---|---|---|---|---|---|
| Outer membrane protein A (OmpA) or its transmembrane domain (tOmpA) | E. coli or Klebsiella pneumonia; outer membrane | β | 1 | 35 kDa (OmpA) 19 kDa (tOmpA) | Refolded, C$_8$-POE or C$_8$E$_4$ | A8-35 | SDS-PAGE, NMR | Zoonens et al. (2005, 2007), Renault (2008), Sharma et al. (2008), Charvolin et al. (2009), Dahmane et al. (2009), Bechara et al. (2012), Della Pia et al. (2014), Le Bon et al. (2014), and Planchard et al. (2014) |
|  |  |  |  |  |  | NAPols | SDS-PAGE, NMR | Sharma et al. (2008) and Bazzacco et al. (2009) |
|  |  |  |  |  |  | SAPols | SDS-PAGE, NMR | Dahmane et al. (2009) |
|  |  |  |  |  | Denatured, urea | A8-35 | SDS-PAGE, tryptic digestion, fluorescence, pore formation in BLM | Pocanschi et al. (2006, 2013) |
|  |  |  |  |  |  | SAPols | SDS-PAGE | Dahmane et al. (2009) |
|  |  |  |  |  |  | A8-75 |  |  |
|  |  |  |  |  | Refolded, OG | A34-35, A34-75 NAPols SAPols | CD, ESI-IMS-MS | Watkinson et al. (2015) |
| FomA | Fusobacterium nucleatum, outer membrane | β | 1 | 40 kDa | Denatured, urea | A8-35 | SDS-PAGE, pore formation in BLM | Pocanschi et al. (2006) |
| NADH dehydrogenase (Complex I) | Neurospora crassa, inner mitochondrial membrane | α | ~35 + prosthetic groups | 1.1 MDa | Native, Triton X-100 | A8-35 | Cryo-EM | Flötenmeyer et al. (2007) |
| V-ATPase transmembrane peptides | Saccharomyces cerevisiae, plasma membrane | α | 1 | 3-4 kDa | Synthetic peptides, TFE | A8-35 | CD | Duarte et al. (2008)[f] |
| Outer membrane protein X (OmpX) | E. coli, outer membrane | β | 1 | 18.6 kDa | Refolded, C$_8$-POE or diC$_6$PC | A8-35 | SDS-PAGE, NMR, MD | Catoire et al. (2009, 2010b), Etzkorn et al. (2014), and Perlmutter et al. (2014) |
|  |  |  |  |  |  | NAPols | SDS-PAGE, NMR | Bazzacco et al. (2012) and Sharma et al. (2012) |
| Leukotriene BLT1 receptor | Human, plasma membrane | α | 1 | 38 kDa | Denatured, SDS | A8-35 | Ligand binding | Dahmane et al. (2009), Banères et al. (2011), and Mary et al. (2014) |
| Leukotriene BLT2 receptor | Mus musculus, plasma membrane | α | 1 | 41.5 kDa | Denatured, SDS | A8-35 | Ligand binding, ligand-induced conformational changes | Dahmane et al. (2009), Catoire et al. (2010a, 2011), Banères et al. (2011), Mary et al. (2014), and Casiraghi et al. (2016)[g] |
| Serotonin 5-HT$_{4a}$ receptor | M. musculus, plasma membrane | α | 1 | 44 kDa | Denatured, SDS | A8-35 | Ligand binding | Dahmane et al. (2009), Banères et al. (2011), and Mary et al. (2014) |
| Cannabinoid CB1 receptor | M. musculus, plasma membrane | α | 1 | 53 kDa | Denatured, SDS | A8-35 | Ligand binding | Dahmane et al. (2009), Banères et al. (2011), and Mary et al. (2014) |

| Protein(s) (abbreviation) | Organism, membrane | Structural type[a] | Number of chains | Overall mass | State and environment before trapping | Amphipol[b] | Evidence for native state after trapping | References |
|---|---|---|---|---|---|---|---|---|
| Lipid A palmitoyltransferase (PagP) | E. coli, outer membrane | β | 1 | 22 kDa | Native, DMPC vesicle | SMA | CD, phospholipase activity | Knowles et al. (2009) |
| | | | | | Denatured, urea | A8-35 | SDS-PAGE, SEC, CD, ESI-IMS-MS | Leney et al. (2012) |
| | | | | | Refolded, DDM or LDAO | A8-35 | CD, enzymatic activity, ESI-IMS-MS | Calabrese et al. (2015) and Watkinson et al. (2015) |
| | | | | | Refolded, LDAO | A8-75, A34-35, A34-75, NAPols, SAPols | | Watkinson et al. (2015) |
| Respirasome supercomplex $I_1III_2IV_1$ | B. taurus, mitochondrial inner membrane | α | Complex I + $bc_1$ dimer + cytochrome c oxidase | 1.7 MDa | Native, digitonin | A8-35 | SG-AUC, BN-PAGE, in-gel activity measurements, cryo-EM | Althoff et al. (2011) |
| | | | | | | | Enzymatic activity, absorbance and resonance Raman spectroscopies | Shinzawa-Itoh et al. (2016) |
| | | | | | Native, PCC-a-M | | Enzymatic activity, cryo-EM | Sousa et al. (2016) |
| Transient receptor potential ankyrin 1 ion channel (TRPA1) | M. musculus, plasma membrane | α | 4 | 512 kDa | Native, fos-choline 12 | A8-35 | CD, $Ca^{2+}$ flux after transfer to lipid vesicles, NS-EM | Cvetkov et al. (2011) and Huynh et al. (2014) |
| | Human, plasma membrane | α | 4 | 510 kDa | Native, MNG-3 | PMAL-C8 | Cryo-EM | Paulsen et al. (2015) |
| Major outer membrane protein (MOMP) | Chlamydia trachomatis, outer membrane | β | 3 | ~120 kDa | Native, Zwittergent 3-14 | A8-35 | SDS-PAGE, CD | Tifrea et al. (2011, 2014, 2018a,b), and Cocco et al. (2013) |
| Corticotrophin-releasing factor receptor 1 (CRFR1) | Human, plasma membrane | α | 1 | 47 kDa | Nascent chain, cell-free synthesis | NVoy | SEC, ligand binding | Klammt et al. (2011) |
| Corticotrophin-releasing factor receptor 2β (CRFR2β) | M. musculus, plasma membrane | α | 1 | 49 kDa | Nascent chain, cell-free synthesis | NVoy | SEC, ligand binding | |
| Ghrelin GHS-R1a receptor (GHS-R1a) | Human, plasma membrane | α | 1 | 41 kDa | Denatured, SDS | A8-35 | Ligand binding, G protein activation and arrestin recruitment | Damian et al. (2012)[g] |
| | | | | | | NAPols | | Bazzacco et al. (2012) |
| | | | | | Refolded, asolectin vesicles | SMA | | Logez et al. (2016) |
| Vasopressin type 2 receptor (V2R) | Human, plasma membrane | α | 1 | 42 kDa | Native, detergent | NAPols | Ligand binding, G protein activation and arrestin recruitment, conformational transitions | Rahmeh et al. (2012) |

| Protein(s) (abbreviation) | Organism, membrane | Structural type[a] | Number of chains | Overall mass | State and environment before trapping | Amphipol[b] | Evidence for native state after trapping | References |
|---|---|---|---|---|---|---|---|---|
| Ferrichrome-iron outer membrane receptor (FhuA) | E. coli, outer membrane | β | 1 | 82 kDa | Native, LDAO | PC-APol | Binding of viral protein | Basit et al. (2012) |
| T domain of diphtheria toxin | Corynebacterium diphtheriae, secreted in soluble form | (α) | 1 | ~19 kDa | Soluble form in detergent-free buffer | NAPols | A study of conformational transitions induced by low pH and/or exposure to NAPols | Kyrychenko et al. (2012) |
| Outer membrane protein T (OmpT) | E. coli, outer membrane | β | 1 | 35 kDa | Denatured, urea | A8-35 | SDS-PAGE, SEC, CD, ESI-IMS-MS | Leney et al. (2012) |
| | | | | | Refolded, DDM or LDAO | A8-35 | CD, protease activity, ESI-IMS-MS, fast photochemical oxidation of proteins (FPOP)-LC-MS/MS | Calabrese et al. (2015) and Watkinson et al. (2015, 2017) |
| | | | | | Refolded, LDAO | A8-75, A34-35, A34-75, NAPols, SAPols | | Watkinson et al. (2015) |
| Aquaporin SoPIP2;1 | Spinach, plasma membrane | α | 4 | 30 kDa | Native, OG | A8-35 | NS-EM | Vahedi-Faridi et al. (2013) |
| Capsaicin receptor (TRPV1) | Rattus norvegicus, plasma membrane | α | 4 | 380 kDa | Native, DDM | A8-35 | Cryo-EM | Cao et al. (2013), Liao et al. (2013, 2014) and Gao et al. (2016) |
| Dhurrin metabolon (CYP79A1, CYP71E1, cytochrome P450 oxidoreductase and subcomplexes) | Sorghum bicolor, endoplasmic reticulum | α | 2–3 | 140–200 kDa | Native, Triton X-100 | A8-35 | SEC, BN-PAGE, | Laursen et al. (2013) |
| | | | | | Native in Sorghum bicolor microsomes | SMA | enzymatic activity, transfer to liposomes | Laursen et al. (2016) |
| Cytochrome c oxidase (COX) | S. cerevisiae, inner mitochondrial membrane | α | 11 | ~197 kDa | Native, S. cerevisiae membranes | SMA | Enzymatic activity, lipid analysis, NS-EM | Long et al. (2013) and Smirnova et al. (2016) |
| Photoreceptor-specific ABC transporter (ABCA4) | B. taurus, retina discs | α | 1 | 257 kDa | Native, DDM | A8-35 | SEC, ATPase activity, ATP-induced conformational changes, NS-EM | Tsybovsky et al. (2013) and Zhang et al. (2015) |
| Peripherin-ROM1 complex | B. taurus, retina discs | α | 4 | 153 kDa | Native, DDM | A8-35 | NS-EM | Kevany et al. (2013) |
| KCNE1 (minK) | Human, plasma membrane | α | Uncertain | Uncertain | Refolded, POPC/POPG vesicles | SMA | EPR DEER measurements | Sahu et al. (2013) |

| Protein(s) (abbreviation) | Organism, membrane | Structural type[a] | Number of chains | Overall mass | State and environment before trapping | Amphipol[b] | Evidence for native state after trapping | References |
|---|---|---|---|---|---|---|---|---|
| Light-harvesting complex II (LHCII) | Arabidopsis thaliana, thylakoids | α | 3 | ~75 kDa | Native, α-DDM | A8-35 / NAPols | SEC, absorbance spectrum, CD, fluorescence decay | Liguori et al. (2013) and Opačić et al. (2014) |
| ExbB-ExbD complex | E. coli, plasma membrane | α | 6 | ~140 kDa | Native, DDM | A8-35 | BN-PAGE, SEC, SAXS, SV-AUC, NS-EM | Sverzhinsky et al. (2014) |
| Mannitol permease (EII^mtl) | E. coli, plasma membrane | α | 2 | 136 kDa | Native, DOC or C10-PEG | A8-35 | Fluorescence emission spectrum, mannitol phosphorylation activity | Opačić et al. (2014) |
| Potassium channel KcsA | Streptomyces lividans, plasma membrane | α | 4 | 41 kDa | Native, E. coli membranes | SMA | SEC, CD, SDS-PAGE, fluorescence spectroscopy | Dörr et al. (2014) |
| TRPV2 channel | Unspecified / Rabbit, plasma membrane | α | 4 | ~385 kDa | Native, detergent / Native, DDM/lipid mixture | A8-35 | Cryo-EM | Fan et al. (2014) / Zubcevic et al. (2016) |
| CsgG secretion channel | E. coli, outer membrane | β | 9 | ~260 kDa | Native, C8E4/LDAO/DMPC mixture | A8-35 | NS-EM, binding of CsgE | Goyal et al. (2014) |
| Transient receptor potential melastatin 1 channel (TRPM1) | Human, plasma membrane | α | 2 | ~360 kDa | Native, fos-choline-14 | A8-35 | BN-PAGE, SEC, cryo-EM | Agosto et al. (2014) |
| Melanocortin-2 receptor (MC2R) | Human, plasma membrane | α | 1 | ~34 kDa | Denatured, SDS | A8-35 | SEC, NMR (preliminary data) | Elter et al. (2014) |
| Melanocortin-4 receptor (MC4R) | Human, plasma membrane | α | 1 | ~37 kDa | Denatured, SDS | A8-35 | | |
| P-glycoprotein (ABCB1; PgP) | Human, plasma membrane | α | 1 | 170 kDa | Native, insect cell membranes | SMA | Ligand binding, AUC, SEC, cryo-EM | Gulati et al. (2014) |
| Multidrug-resistance protein 1 (ABCC1; MRP1) | Human, plasma membrane | α | 1 | 190 kDa | Native, H69AR or HEK cell membranes | SMA | Ligand binding | |
| Multidrug-resistance protein 4 (ABCC4; MRP4) | Human, plasma membrane | α | 1 | 150 kDa | Native, insect or HEK cell membranes | SMA | | |
| ABCG2 | Human, plasma membrane | α | 2 | ~140 kDa | Native, insect cell membranes | SMA | Ligand binding | |
| Cystic fibrosis transmembrane regulator (ABCC7; CFTR) | M. musculus, plasma membrane | α | 1 | ~170 kDa | Native, S. cerevisiae microsomes | SMA | | |

| Protein(s) (abbreviation) | Organism, membrane | Structural type[a] | Number of chains | Overall mass | State and environment before trapping | Amphipol[b] | Evidence for native state after trapping | References |
|---|---|---|---|---|---|---|---|---|
| Endothelin-1 receptor Type A (ET$_A$) | Human, plasma membrane | $\alpha$ | 1 | (fusion protein) | Native in sarkosyl | APG | Ligand binding | Han et al. (2014) |
| $\gamma$-Secretase complex | Human, endoplasmic reticulum | $\alpha$ | 4 | 200–240 kDa | Native, digitonin | A8-35 | Cryo-EM | Lu et al. (2014) and Bai et al. (2015) |
| FtsZ-PBP2-PBP2a divisome complexes | Staphylococcus aureus, plasma membrane | $\alpha$ | Unknown | Unknown | Native, S. aureus membranes | SMA | DLS, EM, flow cytometry | Paulin et al. (2014) |
| Ryanodine receptor 1 (RyR1) | O. cuniculus, sarcoplasmic reticulum | $\alpha$ | 4 | 2.3 MDa | Native, CHAPS | A8-35 | Cryo-EM | Baker et al. (2015) and Wei et al. (2016) |
| NCS1 benzyl-hydantoin transporter (Mhp1) | Microbacterium liquefaciens, plasma membrane | $\alpha$ | 1 | 54.6 kDa | Native, DDM | A8-35 | Ligand binding, ESI-IMS-MS | Calabrese et al. (2015) |
| Galactose permease (GalP) | Streptococcus thermophilus, plasma membrane | $\alpha$ | 1 | 51.7 kDa | Native, DDM | A8-35 | | |
| Mechanosensitive Piezo1 channel | M. musculus, plasma membrane | $\alpha$ | 3 | ~900 kDa | Native, $C_{12}E_{10}$ | A8-35 | Cryo-EM | Ge et al. (2015) |
| Adenosine A$_{2A}$ receptor (A$_2$AR) | Human, plasma membrane | $\alpha$ | 1 | 44.7 kDa | Native, Pichia pastoris or human HEK293T cell membranes | SMA | Ligand binding, CD, SEC, AUC | Jamshad et al. (2015) |
| Insulin-responsive facilitative glucose transporter (GLUT4) | R. norvegicus, plasma membrane | $\alpha$ | 1 | ~55 kDa | Native, LMNG | A8-35 | SEC, ligand binding | Kraft et al. (2015) |
| E. coli tyrosine kinase (ETK) | E. coli, plasma membrane | $\alpha$ | 1 | ~80 kDa | Native, E. coli membranes | SMA | NS-EM, ssNMR, auto-phosphorylation | Li et al. (2015) |
| Multidrug transporter AcrB | E. coli, plasma membrane | $\alpha$ | 3 | ~360 kDa | Native, E. coli membrane | SMA | Ligand binding, NS-EM | Postis et al. (2015) |
| Tetraspanins | Human, plasma membrane | $\alpha$ | variable | variable | Native, S. cerevisiae membranes | SMA | | Skaar et al. (2015) |
| Hepatitis C virus nonstructural protein 5A (NS5A) | Hepatitis C virus, human endoplasmic reticulum | $\alpha$ | 1 | 49 kDa | Refolded, fos-choline-12 | NAPol | NMR | Sólyom et al. (2015) |

| Protein(s) (abbreviation) | Organism, membrane | Structural type[a] | Number of chains | Overall mass | State and environment before trapping | Amphipol[b] | Evidence for native state after trapping | References |
|---|---|---|---|---|---|---|---|---|
| Bacterial translocon (SecYEG) | E. coli, plasma membrane | α | 3 | ~75 kDa | Native, E. coli membrane | SMA | Ligand binding | Prabudiansyah et al. (2015) |
| Voltage-gated potassium channel Kv1.3 | Human, plasma membrane | α | 4 × 1 | ~232 kDa | Native, fos-choline-12 | A8-35 | BN-PAGE, NS-EM | Spear et al. (2015) |
| Serotonin 5-HT$_{3a}$ receptor | Human, plasma membrane | α | 5 × 1 | ~260 kDa | Native, DDM or C$_{12}$E$_9$ | A8-35 | NS-EM | Wu et al. (2015) |
| Transient receptor potential melastatin 4 channel (TRPM4) | Human, plasma membrane | α | 4 × 1 | ~668 kDa | Native, DDM or LMNG | A8-35 | BN-PAGE, SEC, NS-EM | Constantine et al. (2016) |
| Melatonin receptor (MT1R) | Human, plasma membrane | α | 1 | 39 kDa | Native, Pichia pastoris membranes | SMA | Ligand binding, G protein activation and arrestin recruitment | Logez et al. (2016) |
| F$_1$F$_o$ ATP synthase dimer and sub-complexes | Polytomella sp., inner mitochondrial membrane | α | 2 × 17 | ~1.6 MDa | Native, detergent | A8-35 | Progressive dislocation of the superdimer by high concentrations of A8-35 examined by BN-PAGE | Vázquez-Acevedo et al. (2016) |
| STRA6 receptor for retinol uptake | Zebrafish, plasma membrane | α | 2 × 1 (+ calmodulin) | ~180 kDa | Native, LMNG | A8-35 | Cryo-EM | Chen et al. (2016) |
| Leucine zipper-EF-hand-containing transmembrane protein 1 (LETM1; residues 111-698) | Mouse, mitochondrial inner membrane | α | 6 × 1 | ~404 kDa | Native, fos-choline-12 | A8-35 | NS-EM | Shao et al. (2016) |
| Polycystic Kidney Disease Channel-2 (PKD2) | Human, plasma membrane | α | 4 × 1 | ~325 kDa | Native, DDM | A8-35 | Cryo-EM | Shen et al. (2016) |
| Holotranslocon (SecYEG, SecDF-YajC, YidC) | E. coli plasma membrane | α | 7 | ~250 kDa | Native, DDM | A8-35 | Cryo-EM, SANS | Botte et al. (2016) |
| Engineered AcrAB-TolC multidrug efflux pump | E. coli plasma and outer membranes | α + β | (12; partially fused) | ~760 kDa | Native, DDM | A8-35 | Cryo-EM | Jeong et al. (2016) |
| HasR heme transporter and HasR/HasA complex | Serratia marcescens outer membrane | β | 1 and 2 | ~95 and ~115 kDa | Native, Zwittergent 3-14 | A8-35 | NS-EM | Wojtowicz et al. (2016) |

| Protein(s) (abbreviation) | Organism, membrane | Structural type[a] | Number of chains | Overall mass | State and environment before trapping | Amphipol[b] | Evidence for native state after trapping | References |
|---|---|---|---|---|---|---|---|---|
| Cation diffusion facilitator CzcD | *Capriavidus metallidurans* CH34 | α | 2 × 1 | ~68 kDa | Native, *E. coli* plasma membrane | SMA | DLS, solid-state NMR | Bersch et al. (2017) |
| Lysophosphatidic acid receptor 2 (LPA2) | Human, plasma membrane | α | 1 | Fusion protein | Native, sarkosyl | Variant of APG | Ligand-dependent G protein binding | Han et al. (2017) |
| Cyclic-nucleotide-gated channel (TAX-4) | *Caenorhabditis elegans*, plasma membrane | α | 4 × 1 | ~335 kDa | Native, LMNG + lipids | A8-35 | Cryo-EM | Li et al. (2017) |
| Outer membrane phospholipase A (OmpLA) | *E. coli*, outer membrane | β | 1 | ~31 kDa | Native, proteoliposomes | DIBMA | Enzymatic activity, CD, NS-EM, DLS, SEC | Oluwole et al. (2017) |
| Equilibrative nucleoside transporter-1 (hENT1) | Human, plasma and mitochondrial membranes | α | 1 | ~50 kDa | Native, Sf9 membranes | SMA | Ligand binding, SEC | Rehan et al. (2017) |
| Pneumolysin | *Streptococcus pneumoniae*, secreted toxin | β | 42 | ~2.2 MDa | Native, Cymal-6 | A8-35 | Cryo-EM | van Pee et al. (2017) |
| AcrABZ-TolC multidrug efflux pump | *E. coli* plasma and outer membranes | α + β | 12 | ~770 kDa | Native, DDM | A8-35 | Cryo-EM | Wang et al. (2017) |
| Transient receptor potential channel polycystin-2 (PC2) | Human, | α | 4 | ~440 kDa | Native, LMNG/CHS | A8-35 | Cryo-EM | Wilkes et al. (2017) |

*Abbreviations*: *α-DDM* n-dodecyl-α-D-maltoside, *APG* amphipathic poly-γ-glutamic acid, *APol* amphipol, *BLM* black lipid membrane, *BN-PAGE* blue-native polyacrylamide gel electrophoresis, *C₈E₄* tetraethylene glycol monooctyl ether, *C₈-POE* octylpolyoxyethylene, *C₁₀-PEG* decylpoly(ethyleneglycol) 300, *C₁₂E₁₀* polyoxyethylene(10)dodecyl ether, *CD* circular dichroism, *CHAPS* 3-((3-cholamidopropyl) dimethylammonio)-1-propanesulfonate, *CHAPSO* 3-[(3-cholamidopropyl)dimethylammonio)-2-hydroxy-1-propanesulfonate, *CHS* cholesterol hemisuccinate, *cryo-EM* electron cryo-microscopy, *DDM* n-dodecyl-β-D-maltoside, *DIBMA* diisobutylene/maleic acid copolymer, *DLS* dynamic light scattering, *DM* n-decyl-β-D-maltoside, *diC₆PC* 1,2-dihexanoyl-sn-glycero-3-phosphocholine, *DMPC* 1,2-dimyristoyl-sn-glycero-3-phosphocholine, *DMSO* dimethyl sulfoxide, *Empigen* a trademark for n-dodecyl-N,N-dimethylglycine, *DOC* sodium deoxycholate, *EPR DEER* double electron-electron resonance electron paramagnetic resonance, *ESI-IMS-MS* electrospray ionization mass spectrometry coupled with ion mobility spectrometry, *fos-choline-12* dodecylphosphorylcholine, *FRET* Förster resonance energy transfer, *GPCR* G protein-coupled receptor, *H69AR* small-cell lung cancer cell line, *Hecameg* 6-O-(N-heptylcarbamoyl)-methyl-α-D-glycopyranoside, *HEK* human embryonic kidney cell line, *LDAO* N,N-dimethyldo-decylamine-N-oxide, *LMNG* lauryl maltose neopentyl glycol, 2,2-didecylpropane-1,3-bis-β-D-maltopyranoside, *MD* molecular dynamics simulations; MNG-3 = LMNG, *NAPols* non-ionic APols (glucosylated unless otherwise indicated), *NMR* nuclear magnetic resonance, *NS-EM* electron microscopy in negative stain, *OG* n-octyl-β-D-glucopyranoside, *OTG* n-octyl-β-D-thioglucopyranoside, *POPC* 1-palmitoyl-2-oleoyl-sn-glycero-3-phosphocholine, *POPG* 1-palmitoyl-2-oleoyl-sn-glycero-3-phospho-(1'-rac-glycerol), *SANS* small-angle neutron scattering, *SAPols* sulfonated APols, *SAXS* small-angle X-ray scattering, *SDS* sodium dodecyl sulfate, *SDS-PAGE* polyacryl-amide gel electrophoresis in the presence of SDS, *SEC* size exclusion chromatography, *SG-AUC* sucrose gradient sedimentation velocity analytical ultracentrifugation, *SMA* styrene-maleic acid copolymer, *SV-AUC* sedimentation velocity analytical ultracentrifugation, *STEM* scanning transmission electron microscopy, *ssNMR* solid-state NMR, *TFE* trifluoroethanol

[a]Structural type of the transmembrane region, either $\alpha$-helical or $\beta$-stranded

[b]Including tagged or labeled derivatives of the basic structure

[c]Includes THAM-based and glucose-based NAPols; "NAPols," otherwise, refers only to any of the various forms of glucose-based NAPols

[d]A reproducible protocol could not be established for refolding BR by direct transfer from TFE to A8-35, but an artifactual result is highly improbable (see text)

[e]OAPA-20 is almost identical to A8-75; PMAL-B-100 and PMAL-C12 are two different names for the same polymer (see Gorzelle et al. 2002 and Chap. 4)

[f]The peptides, which are $\alpha$-helical in lipid bilayers, SDS, TFE, DMSO, and, presumably, in the native V-ATPase, adopted a $\beta$-strand structure in A8-35, presumably due to a mid-segment arginine interacting with the carboxylate groups of the APol

[g]In Damian et al. (2012) and Casiraghi et al. (2016), the GPCR was folded in A8-35 and then transferred to nanodiscs for functional measurements; see § 5.7

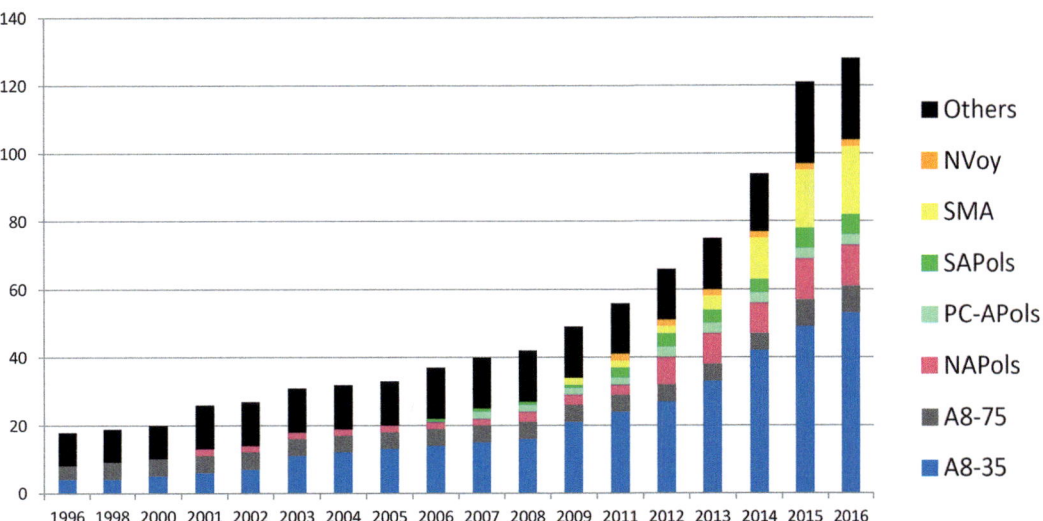

**Fig. 5.1** Cumulative number of membrane protein/amphipol complexes described over the years. Each MP/APol complex is counted only once, on the year it was first reported, but a given protein, e.g. BR, will appear as many times as its trapping by a different APol has been described. The cumulative numbers shown are therefore those of distinct MP/APol complexes, not of proteins, the total number of which is currently close to 100 (see Table 5.1; two relevant papers that appeared too late to be incorporated in the table and statistics are Chiu et al. 2017 and Jin et al. 2017). The level of description of the complexes varies widely. In the simplest case, it was only demonstrated that a given APol can keep a given MP soluble in the absence of detergent. Other complexes have been characterized by extensive biochemical and biophysical studies (see Table 5.4, where the various investigations performed are indicated). The structure of over a dozen APol-trapped MPs has been solved to near-atomic resolution by single-particle electron cryomicroscopy, permitting atomic models to be built and, in many cases, the APol layer to be visualized (see Chap. 12). "Others" refer to several little-used APols (cells left colorless in Column 7 of Table 5.1). Data from Table 5.1.

**Table 5.2** Distribution of membrane protein/amphipol complexes between the various structural types of MPs and of APols.

| Amphipol | Structural type of the transmembrane domain | | | | Total |
|---|---|---|---|---|---|
| | α (monomeric) | α (oligomeric) | β (monomeric) | β (oligomeric) | |
| A8-35 | 16 | 35 | 6 | 6 | **62** |
| A8-75 | 1 | 3 | 3 | 1 | **8** |
| NAPols | 5 | 3 | 4 | 0 | **12** |
| PC-APols | 1 | 1 | 1 | 0 | **3** |
| SAPols | 2 | 1 | 3 | 0 | **6** |
| SMA | 11 | 9 | 2 | 0 | **22** |
| NVoy | 2 | 0 | 0 | 0 | **2** |
| Others | 1 | 10 | 6 | 2 | **19** |
| **Total** | **39** | **62** | **25** | **9** | **135** |

The data cover the period 1996–2016. As in Fig. 5.1, a given MP appears in the table as many times as it has been trapped in a different APol. The background colors of the cells in Column 1 refer to the code used in Column 7 of Table 5.1 (Data from Table 5.1).

As mentioned above, there are many ways to form MP/APol complexes. They fall into the following categories:

(i) Transferring a native MP to APols from a detergent solution;
(ii) Folding a denatured MP or a peptide by transfer to APols;
(iii) Directly extracting native MPs from synthetic or natural membranes;
(iv) Expressing MPs in vitro and folding them in the presence of an APol and the absence of any membrane.

Table 5.3 presents an overview of the most frequently used approaches, sorted by the environment the protein was initially in (Column 1), the chemical structure of the APol it was transferred to (Column 2), and the method used for effecting the transfer (Column 3).

**Table 5.3** Methods of formation of membrane protein/amphipol complexes. Only initial environments that have been used more than once are included; see other cases in Table 5.1, Column 6. In the case of environments and transfer methods that have been used recurrently, only the first and some typical articles are cited. Other references can be found in Table 5.1.

| Initial environment | Amphipol[a] | Method of transfer | Comments | References |
|---|---|---|---|---|
| **Methods starting from a native protein** | | | | |
| $C_8$-POE, $C_8E_4$ (CMC ≈ 7–12 mM) | A8-35, NAPols | Dilution[b] | Under its CMC, the deter-gent distributes between monomers in the aqueous solution and monomers diluted into the APol belt surrounding the protein and protein-free APol particles (see Box 5.3). It can be removed by dialysis, SEC, adsorption onto Bio-Beads, etc. | Tribet et al. (1996), Sharma et al. (2008), and Bazzacco et al. (2009) |
| | | Bio-Beads[c] | APols do not detectably adsorb onto Bio-Beads (cf. Fig. 5.5). Can be combined with dialysis, dilution/concentration cycles, SEC, IMAC, etc. in order to remove any remaining detergent (but see, in § 5.2.1, the effect of removing the free APol particles, as occurs with SEC and IMAC, on the dispersity of the complexes) | Zoonens et al. (2005, 2007), Catoire et al. (2009, 2010b), and Charvolin et al. (2009) |
| CHAPS (CMC ≈ 6–10 mM) | A8-35 | Dilution | See comments given for $C_8$-POE/$C_8E_4$ | Martinez et al. (2002) |
| | | Bio-Beads | | Charvolin et al. (2009) |
| $DiC_6PC$ CMC ≈ 6–7 mM | NAPols | Bio-Beads | See comments given for $C_8$-POE/$C_8E_4$ | Bazzacco et al. (2012) and Sharma et al. (2012) |
| Digitonin (very low, poorly defined CMC) | A8-35 | Sequestration of digitonin by γ-cyclodextrin | Elimination by adsorption or complexation is required given the very low CMC of digitonin (<0.5 mM) | Althoff et al. (2011) |
| | | Bio-Beads | | Lu et al. (2014) |

(continued)

**Table 5.3**   (continued)

| Initial environment | Amphipol[a] | Method of transfer | Comments | References |
|---|---|---|---|---|
| Dodecylmaltoside (DDM) (CMC ≈ 0.17 mM) | A8-35, PMAL-C12, SAPols | Dilution | Seldom applicable because of the low CMC of DDM | Champeil et al. (2000) and Picard et al. (2006) |
| | A8-35, NAPols | Bio-Beads | More generally applicable than dilution | Charvolin et al. (2009), Bazzacco et al. (2012), Cao et al. (2013), and Liao et al. (2013) |
| | A8-35 | SEC | The protein in DDM solution was supplemented with A8-35, concentrated by ultrafiltration, and separated from DDM by SEC in the absence of surfactant | Kevany et al. (2013) and Tsybovsky et al. (2013) |
| Hecameg (CMC ≈ 19.5 mM) | A8-35, NAPols | Dilution | See comments given for $C_8$-POE/$C_8E_4$ | Tribet et al. (1996) and Prata et al. (2001) |
| Octylglucoside (OG) (CMC ≈ 20–25 mM) Octylthioglucoside (OTG) (CMC ≈ 9 mM) | A8-35, NAPols | Dilution | See comments given for $C_8$-POE/$C_8E_4$ | Tribet et al. (1996), Prata et al. (2001), Sharma et al. (2008), and Bazzacco et al. (2009, 2012) |
| | | Bio-Beads | | Gohon et al. (2008), Charvolin et al. (2009), and Sharma et al. (2012) |
| Sarkosyl (CMC ≈ 13.7 mM) | APG and a variant thereof | Dialysis | According to SEC analysis, transfer from sarkosyl to APG is accompanied by a large size increase | Han et al. (2014, 2017) |
| Biological membranes or proteoliposomes | SMA (A8-35) | Direct extraction | Essentially specific to SMA. Cases of direct extraction by A8-35 have been observed, but not studied in detail (see text, § 5.2.2.2, and Box 5.1) | Knowles et al. (2009), Jamshad et al. (2011), Orwick-Rydmark et al. (2012), Gulati et al. (2014), Postis et al. (2015), Dörr et al. (2016), Lee and Pollock (2016), and Logez et al. (2016) |
| **Methods starting from an unfolded protein or a synthetic peptide** | | | | |
| SDS | A8-35, SAPols, NAPols | Precipitation of potassium dodecyl sulfate (PDS) | The method of choice for refolding α-helical MPs from SDS. A membrane protein unfolded in organic solvent can be first transferred to SDS, where it acquires some α-helicity, and then folded by transfer to APols. See Chap. 6 | Pocanschi et al. (2006b), Dahmane et al. (2009, 2011, 2013), Catoire et al. (2010a), Bazzacco et al. (2012), Elter et al. (2014), and Popot (2014) |
| | A8-35 | PDS precipitation, dialysis, dilution, or Bio-Beads | These two articles present interesting comparisons of the yields and degrees of monodispersity achieved using various procedures for folding BR by transfer from SDS to A8-35. See Chap. 6 | Dahmane et al. (2013) and Elter et al. (2014) |

(continued)

**Table 5.3** (continued)

| Initial environment | Amphipol[a] | Method of transfer | Comments | References |
|---|---|---|---|---|
| TFE | A8-35 | Dialysis, dilution, or freeze-drying a mixture of peptide or BR with APol in TFE or TFE + ethanol, followed by resuspension into aqueous buffer | The first two studies where transfer from TFE was tested were unsatisfying, the first one because, for reasons unrelated to the transfer procedure, the expected α-helical structure was not achieved, and the second because of irreproducibility. However, both studies indicated that this method can be made to work. In Stangl et al. 2014, it has been successfully applied to trapping in A8-35 the transmembrane helix dimer of glycophorin A. See Chap. 6 | Duarte et al. (2008), Dahmane et al. (2013), and Stangl et al. (2014) |
| Urea | A8-35, SAPols | Dilution into urea-free buffer | Dilution can be rapid or slow. See Chap. 6 | Pocanschi et al. (2006a, b, 2013) and Dahmane et al. (2009) |
| Nascent polypeptide emerging from ribosome tunnel | NVoy, NAPols | Cell-free synthesis in the presence of APol | Polyanionic APols (A8-35, SAPols, SMA) inhibit membrane protein synthesis (Park et al. 2011; Periasamy et al. 2013). See Chap. 7 | Klammt et al. (2011) and Bazzacco et al. (2012) |

*Abbreviations: APG* amphipathic poly-γ-glutamic acid, *APol* amphipol, $C_8E_4$ octyltetraoxyethylene, $C_8$-*POE* octylpolyoxyethylene, *CMC* critical micellar concentration, *diC_6PC* 1,2-dihexanoyl-*sn*-glycerol-3-phosphocholine, *IMAC* immobilized-metal affinity chromatography, *NAPols* non-ionic, THAM-derived APols (any kind), *PDS* potassium dodecyl sulfate, *POE* polyoxyethylene, *SAPols* sulfonated APols, *SDS* sodium dodecyl sulfate, *SEC* size exclusion chromatography, *SMA* styrene-maleic acid copolymer, *TFE* trifluoroethanol, *THAM tris*(hydroxymethyl)acrylamidomethane
[a]Including tagged or labeled derivatives of the basic structure
[b]Dilution under the CMC of the detergent
[c]Adsorption of the detergent onto Bio-Beads

The most widely used route to date has been to start from a detergent solution of the target MP – either in a purified form or not – and replace the detergent belt that keeps it soluble (Chap. 2) with an APol one. A standard procedure is described in § 5.9, Protocol 5.1. As in the preceding chapter, we will start with reviewing what has been established using A8-35, by far the most extensively studied APol, and then point out what is known of the similarities and differences with other types of APols.

## 5.2.1    Transferring Native Membrane Proteins from Detergent Solution to Amphipols

Starting from a native or refolded MP in non-denaturing detergent solution, the most usual procedure for transfer to APols starts with forming ternary MP/detergent/APol complexes, which happens spontaneously upon supplementing the MP/detergent solution with APols (Fig. 5.2, step I → II). Mixing of the surfactants, both at the surface of the protein and in protein-free mixed particles, takes place very rapidly: within the time of injection in isothermal titration calorimetry (ITC) (Tribet et al. 2009; Fig. 5.3), in less than a second as measured in stopped-flow experiments (Zoonens et al. 2007; see § 5.6, Fig. 5.40). It is usual, however, to let the mixture stand for 15–30 min before proceeding to the removal of the detergent.

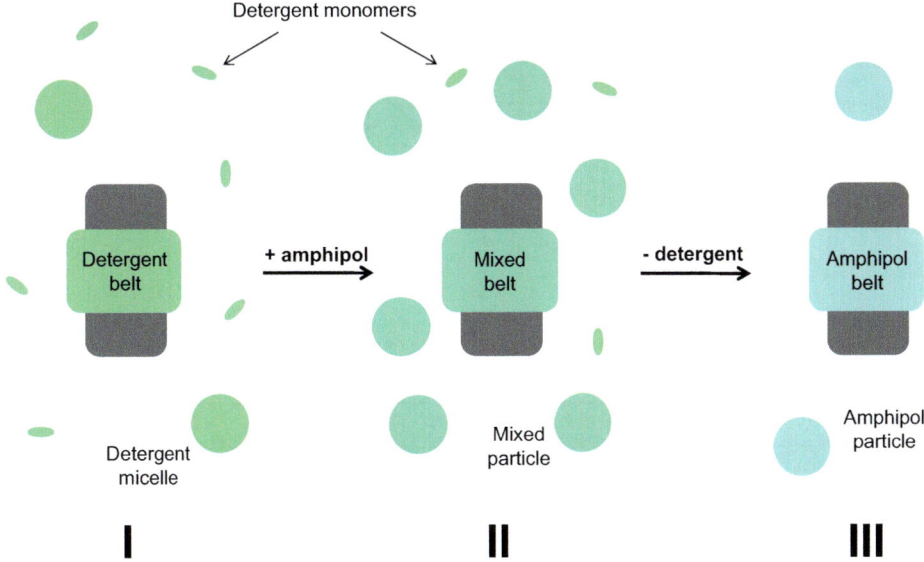

**Fig. 5.2** A schematic view of the transfer of a membrane protein (*gray*) from a detergent environment (*green*) to an amphipol one (*blue*). (**I**) In the initial state, the protein is kept soluble by a belt of detergent adsorbed onto its TM surface (cf. Chap. 2). The detergent distributes between free monomers, whose concentration is about equal to the CMC, free micelles, and the protein-adsorbed belt. (**II**) Following addition of APol, the two surfactants mix, forming mixed particles and a mixed protein-adsorbed belt (*blue-green*). If, as in this example, the volumes of APol and non-monomeric detergent are about the same, the concentration of monomeric detergent is expected to drop to ~½ of the CMC (see Box 5.3). (**III**) The detergent is then removed by any of the methods described in the first part of Table 5.2, e.g. adsorption onto polystyrene beads, leaving a mixture of MP/APol complexes and APol particles.

The duration of the incubation seems to have only limited effects on the outcome of the transfer procedure (Tribet et al. 1997). Two points, however, may be worth keeping in mind. First, in many cases, the mixture also contains lipids, whose rebinding to the protein upon diluting the detergent with APol (§ 5.3.1.2; see e.g. Martinez et al. 2002; Dahmane et al. 2013) may be slower than the mixing of detergent and APol. Second, it is essential that APol be added before the detergent is removed: direct dilution of a MP in detergent solution into an APol solution leads to the formation of aggregates (Fig. 5.4), indicating that, upon dilution of the detergent under its CMC in the presence of APol, protein/protein contacts form more rapidly than protein/APol ones and that the APol is inefficient (or slow) at dissociating the resulting aggregates (Tribet et al. 1997).

*Extracting a membrane protein with a detergent and transferring it to an amphipol*
(© *2018 by Francis Haraux*)

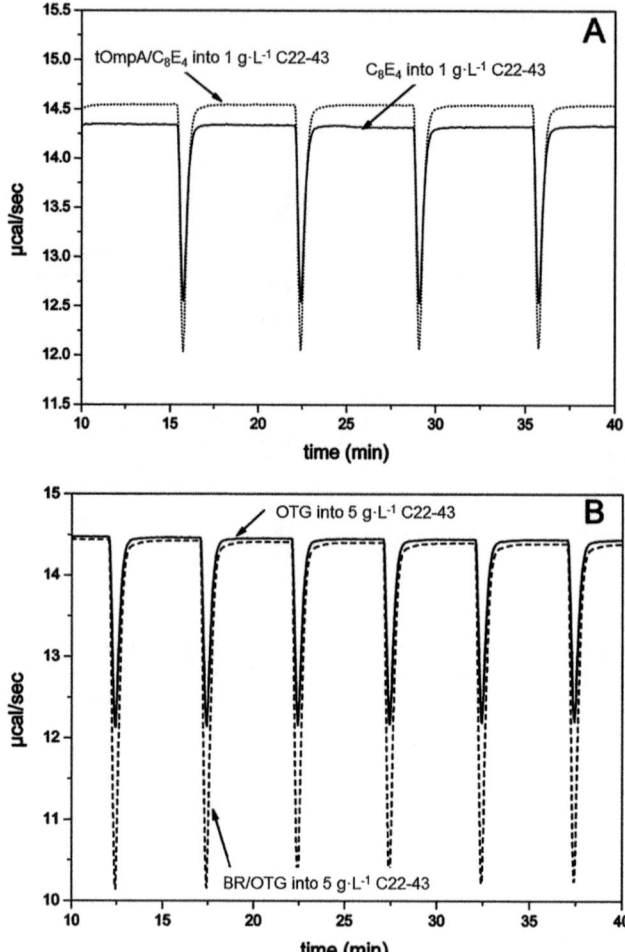

**Fig. 5.3** Thermograms of the calorimetric titration of membrane protein/detergent complexes into phosphorylcholine-based amphipols (PC-APols). (**A**) Titration of pure $C_8E_4$ and of tOmpA/$C_8E_4$ complexes into 1 $g \cdot L^{-1}$ PC-APol solutions. (**B**) Titration of pure OTG and of BR/OTG complexes into 5 $g \cdot L^{-1}$ PC-APol. Note that the kinetics of mixing are the same whether pure detergent or MP/detergent complexes are injected, indicating that exchange of detergent for APol at the surface of the protein is faster than the time resolution of the instrument (<1 min) (Reprinted with permission from Tribet et al. 2009, © 2009 American Chemical Society).

In the ternary mixture (stage II in Fig. 5.2), detergent and APol mix about ideally, both in the free particles and at the surface of the protein (Zoonens et al. 2007; Tribet et al. 2009). The critical aggregation concentration (CAC) of APols is so low (~0.002 $g \cdot L^{-1}$ for A8-35 (Giusti et al. 2012), possibly lower for non-ionic APols; see Chap. 4, § 4.3.1.1) that the concentration of free APol molecules is negligible. Because the monomeric detergent in the aqueous phase is in equilibrium with that in the mixed particles, where it is diluted by the APol, its chemical potential is lower than in a pure detergent solution, and its concentration is expected to drop. One can estimate, for instance, that it will be about half the CMC if the particles comprise about equal volumes of APol and detergent, as represented in Fig. 5.2 (see § 5.6, Box 5.3). This actually depends on detergent/detergent vs. detergent/APol interactions in the particles and may be more complex, for instance, for a charged detergent mixed with a non-ionic APol. As will be discussed in § 5.6, the function of the protein may be affected by being transferred from a pure detergent to a mixed detergent/APol environment, and its stability may improve as compared to that in pure detergent (see e.g. in § 5.6, Fig. 5.33, the case of the sarcoplasmic calcium ATPase examined in Champeil et al. 2000).

In the second step (II → III in Fig. 5.2), the detergent is removed from the mixture. It does not seem that the chemical structure nor physical-chemical properties of the detergent pose any obstacle to its removal: Table 5.1 lists 23 detergents and a few detergent mixtures from which MPs have been successfully transferred to APols. Some of these detergents have high CMCs (OG, CHAPS, etc.), some very low ones (digitonin, LMNG, etc.). Detergents that have been used more than once are listed in Table 5.3, along with the methods used for their removal. Whereas it seems that any detergent can be used, not any method of removal will work with any detergent. Detergents with a high CMC, for instance, can be dialyzed away, whereas this process is very slow with low-CMC ones, since only the monomers, which often represent a very small fraction of the detergent present in the preparation, can cross the dialysis membrane. In practice, dialysis is mostly used in a second step, after the concentration of the detergent has been lowered under its CMC, under which circumstances its monomers distribute between the aqueous phase and the APol belts and particles as a function of their partition coefficient and the volume of each phase (for an early analysis of this phenomenon, using membrane fragments, see Brotherus et al. 1979).

When starting from a MP in a solution of detergent sufficiently close to its CMC, a useful technique is to dilute the MP/APol/detergent mixture with surfactant-free buffer well under the CMC of the detergent. This does not physically remove the detergent from the sample, but it displaces it from the protein-bound surfactant belt: if the final concentration of free detergent is 1/20 the CMC, for instance, the concentration of detergent in the belt will be in a volume ratio of ~1:19 to that of the APol (cf. § 5.6, Box 5.3). This procedure is very useful to determine, typically after a quick centrifugation, how much APol is required to keep the protein soluble (see e.g. Fig. 5.4 and Tribet et al. 1996; Dahmane et al. 2009) or to compare the functional properties of the protein in a detergent environment vs. one comprised of nearly pure APol (see e.g. in § 5.4, Fig. 5.25, the case of the nicotinic acetylcholine receptor described in Martinez et al. 2002). Following dilution, the detergent can be

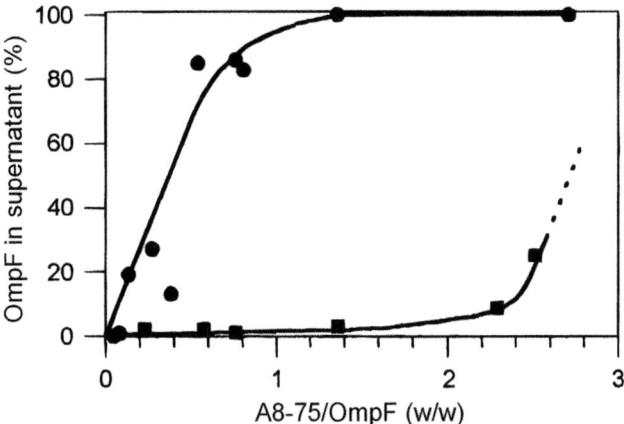

**Fig. 5.4** Solubility of membrane protein/amphipol complexes in aqueous solution as a function of the polymer/protein ratio and the order of APol addition and detergent dilution. A solution of OmpF porin at 3.5 μM in 40 mM octyl-POE solution was either supplemented with A8-75 at the indicated mass ratios and diluted under the CMC of the detergent in surfactant-free buffer (●) or diluted directly into a buffer containing identical amounts of APol (■). After a 10-min incubation at 4 °C, the samples were centrifuged for 30 min at 4 °C in the A-110 rotor of an Airfuge (Beckman) at 20 psi (~140 kPa, ~210,000 × g). The protein in the supernatant was titrated by spectrophotometry (Reprinted with permission from Tribet et al. 1997, © 1997 American Chemical Society).

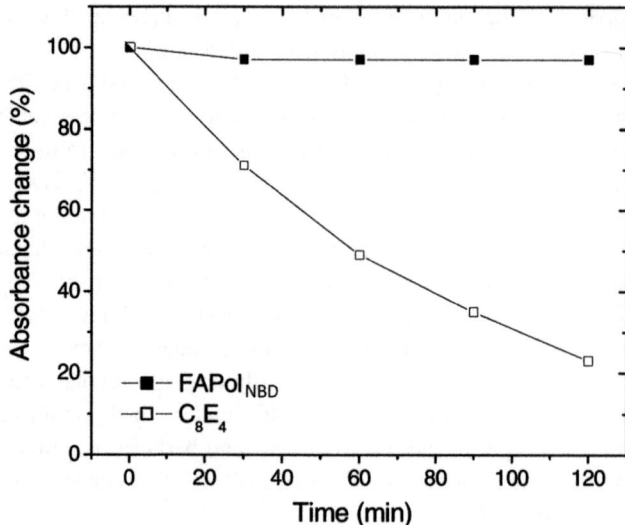

**Fig. 5.5** Adsorption of $C_8E_4$ and A8-35 onto polystyrene beads. The loss of surfactant by adsorption onto Bio-Beads SM2 was followed (in two parallel experiments) by UV-visible measurements at 205 nm for the $C_8E_4$ sample and at 475 nm for that containing NBD-labeled A8-35 (FAPol$_{NBD}$). The initial concentrations of detergent and FAPol$_{NBD}$ were 6 g·L$^{-1}$ and 1 g·L$^{-1}$, respectively. The surfactant/beads mass ratio was 1:10 in both cases. The buffer was 20 mM Tris/HCl, pH 8. Because of the high background at 205 nm, data for $C_8E_4$ are less accurate than those for FAPol$_{NBD}$ (Reprinted with permission from Zoonens et al. 2007, © 2007 American Chemical Society).

physically eliminated by dialysis, adsorption onto Bio-Beads, ultrafiltration, gel filtration, sucrose gradient centrifugation, etc. Note that the quick exchange afforded by dilution can be preferable to the slow one that takes place when using Bio-Beads to remove the detergent, without first diluting it. Such is the case when BR is transferred from OTG to A8-75: its integrity is preserved after dilution followed by Bio-Beads, whereas it denatures when using Bio-Beads without dilution (M. Zoonens, personal communication). This is likely due to a destabilizing effect of A8-75/OTG mixtures, to which BR is exposed for an extended period of time in the second case, but not in the first one.

For low-CMC detergents, however, the total concentration of detergent (bound + free) is generally too much above the CMC for dilution to be very practical, the resulting samples being too dilute. A common procedure is to incubate the ternary mixture with a material, typically polystyrene beads, such as Bio-Beads SM2, onto which the detergent will adsorb. Because most of the surface of the beads is located in narrow cranks, APol particles do not have access to it, and their adsorption is minimal (Zoonens et al. 2007) (Fig. 5.5). Digitonin can be specifically removed by adsorption onto $\gamma$-cyclodextrin (Althoff et al. 2011).

For reasons to be discussed below, APol is always added to the detergent solution in excess (typically 2–3×) over what the protein will actually bind. Removing the detergent by dialysis or adsorption does not separate the extra, free particles of APol from MP/APol complexes (step III in Fig. 5.2), because neither is adsorbed nor crosses the dialysis membrane. The situation is different when the detergent is removed by running the sample through a size exclusion chromatography (SEC) column or in a sucrose gradient (Fig. 5.6A) or by immobilizing the protein on an affinity column and washing it with surfactant-free buffer (Fig. 5.6C). Under such conditions, MP/APol complexes are generally separated from free APol particles. This does not, as a rule, compromise their solubility, but it can affect their dispersity. This phenomenon is illustrated in Fig. 5.6 in the cases of a very large oligomeric complex, cytochrome $bc_1$, which does not comigrate with free APol particles during

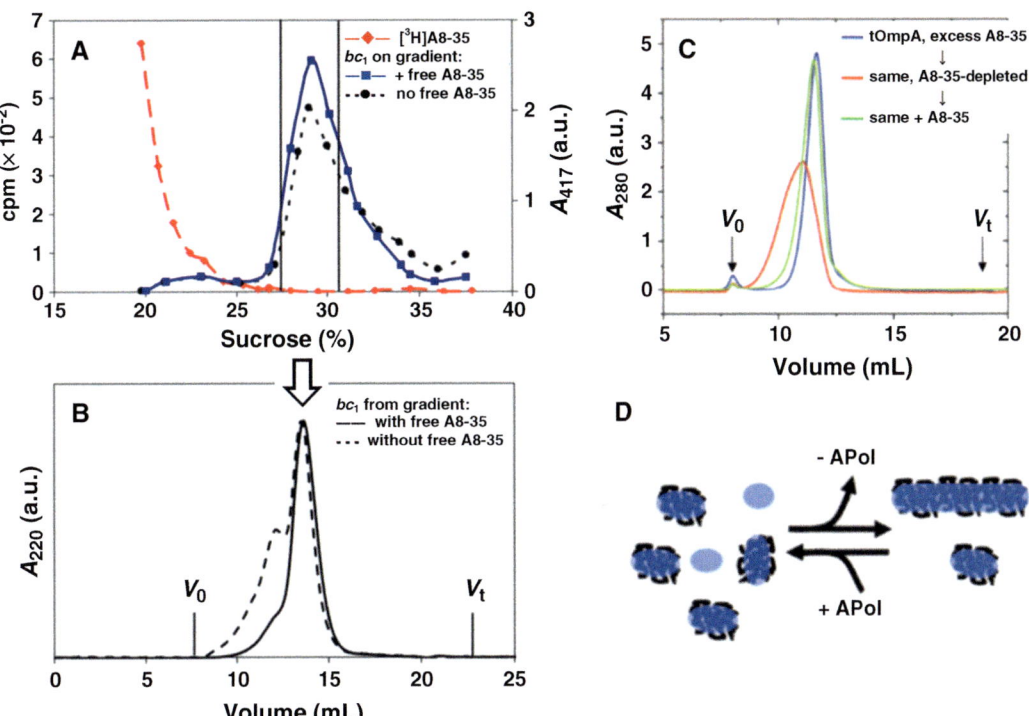

**Fig. 5.6** Effect of removing free amphipol from solutions of membrane protein/amphipol complexes. (**A**) Separating free A8-35 from cytochrome $bc_1$/A8-35 complexes by ultracentrifugation in 20–40% sucrose gradients. The complexes were layered onto either an APol-free gradient (*black circles*) or a gradient containing 0.1 g·L$^{-1}$ free A8-35 (*blue squares*) and centrifuged for 16 h at 150,000× g at 4 °C. The concentration of cytochrome $c_1$ in each fraction was determined from its absorbance at 417 nm. As a control, free [$^3$H]A8-35 was layered onto an APol-free gradient, centrifuged under the same conditions, and the radioactivity of the fractions counted (*red dashes*), showing that free APol deposited on the top of the gradient is well separated from $bc_1$/APol complexes. (**B**) The three most concentrated fractions from the gradients in Panel A (between the vertical bars), containing or not free A8-35, were pooled, washed free from sucrose by dialysis, and injected onto a size exclusion column in an APol-free buffer. Elution profiles were analyzed at 220 nm (From Charvolin et al. 2014). (**C**) SEC profiles of tOmpA/A8-35 complexes after trapping with an excess of APol (*blue curve*), after separation of the tOmpA/A8-35 complexes thus obtained from free A8-35 particles by immobilized-metal affinity chromatography (*red curve*), and after adding back free A8-35 to the latter sample (*green curve*) (Adapted from Zoonens et al. 2007). (**D**) Interpretation of the data in Panels B and C: the equilibrium between protein/protein and protein/APol interactions is shifted one way or the other depending on the volume of the APol "phase" (From Zoonens and Popot 2014).

sucrose gradient centrifugation (Fig. 5.6A, B; Charvolin et al. 2014) and a very small monomeric MP, the TM $\beta$-barrel of *Escherichia coli* OmpA (tOmpA), which can be separated from free APol particles by immobilized-metal affinity chromatography (IMAC) (Fig. 5.6C; Zoonens et al. 2007). It has also been observed for other MPs, such as bacteriorhodopsin (BR; Gohon et al. 2008) and the trimeric porin OmpF (Arunmanee et al. 2014). In the cases of the $bc_1$ complex and of tOmpA, SEC analysis of preparations depleted from free APol showed the formation of small oligomers (Fig. 5.6B, C). If free APol is added back to APol-depleted preparations, the oligomers dissociate (Fig. 5.6C). The aggregates formed by BR and OmpF upon APol depletion have been characterized by EM. They result from the side-by-side association of MP/APol complexes via the TM surfaces of the proteins and tend to form what is probably helicoidal filaments, in some cases two-dimensional (2D) crystals

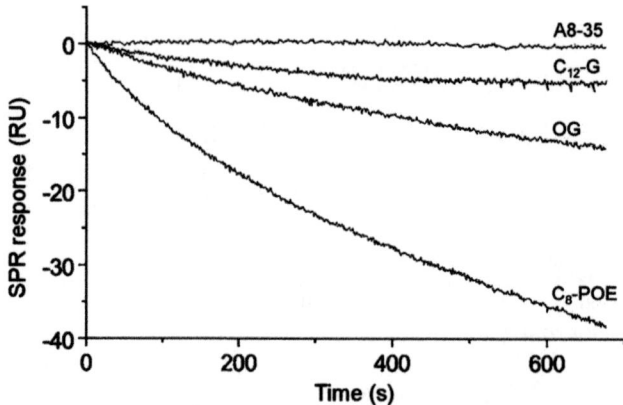

**Fig. 5.7** Dissociation rates of various surfactants from immobilized OmpF as examined by surface plasmon resonance (SPR). An OmpF mutant carrying a single cysteine in a periplasmic turn was labeled with biotin maleimide and immobilized, at the level of 150 resonance units (RU), on the surface of a chip bearing covalently attached streptavidin. The buffer contained 10 g·L$^{-1}$ octylpolyoxyethylene (C$_8$-POE), 10 mM HEPES, 15 mM NaCl, and 3.4 mM EDTA, pH 7.4. The chip was washed with a 10-g·L$^{-1}$ solution of $n$-dodecyl-$\beta$-D-glucopyranoside (C$_{12}$-G), $n$-octyl-$\beta$-D-glucopyranoside (OG), or $n$-octyl-polyoxy-ethylene (C$_8$-POE) or a 1-g·L$^{-1}$ solution of A8-35 and the exchange of surfactant followed by SPR at a flow rate of 5 mL·min$^{-1}$. Once a stable baseline had been achieved, i.e. after 30–60 min (zero RU value on this graph), the solution was replaced with surfactant-free buffer ($t = 0$) and the dissociation of the surfactant monitored by SPR. The signal from a blank surface bearing only streptavidin and treated in the same manner was subtracted from the raw data, so that the curves shown represent the evolution over time of the amount of surfactant that is actually bound to OmpF (Data from Q. Hong and J.H. Lakey, unpublished observations cited in Popot et al. 2003).

(see Chap. 12, § 12.3.1). Cross-linking data suggest the same behavior of diacylglycerol kinase (DAGK) complexed by either OAPA-20 or PMAL-B when the excess APol is removed (Nagy et al. 2001).

This behavior may seem paradoxical when one knows that, once adsorbed onto the TM surface of a MP, APols do not desorb significantly, even over extended periods of time, unless they are displaced by another surfactant. This can be observed either with FRET or ITC experiments (Zoonens et al. 2007; Tribet et al. 2009) or by immobilizing an APol-trapped MP onto a surface plasmon resonance (SPR) chip via a protein-bound biotin tag and washing it extensively with surfactant-free buffer (Fig. 5.7). At variance with the desorption of detergent that occurs during a similar process, no measurable desorption of the APol is observed (Fig. 5.7). One might therefore expect that depleting MP/APol preparations from free APols should not change significantly the composition of the complexes, and yet it induces aggregation (Fig. 5.6).

The aggregation observed when free APol is removed from a bulk solution of MP/APol complexes is, nevertheless, rather straightforward to rationalize. A plausible interpretation is schematized in Fig. 5.6D. When an immobilized MP/APol complex is exposed to a surfactant-free solution, APol desorption is slowed down by the high free energy cost of partially exposing to water the hydrophobic TM surface of the protein. In solution, however, MPs have the option of replacing MP/APol contacts with MP/MP ones, creating oligomers and leaving some APol molecules free to desorb. When free APol is fed back, the equilibrium is displaced again in the direction of isolated MP/APol complexes (Fig. 5.6C).

This observation is of great practical interest. On the one hand, it indicates that experiments in which the homogeneity of MP/APol complexes is essential, such as small-angle radiation scattering, must be conducted in the presence of an excess of APol (see Chap. 9, § 9.3.8). On the other hand, and more speculatively, it is conceivable that one might harness the tendency to self-organization of APol-depleted MP/APol complexes to generate structures amenable to image analysis by EM (see Chap. 12, § 12.3.2).

Transferring MPs from detergent solutions to other APols than A8-35 or A8-75 is carried out along similar ways (Table 5.3), but, except for OAPA-20 and PMAL-B (Nagy et al. 2001), the effects of removing free APol have not been analyzed. They may well vary from one APol to the next, because they are likely determined by the balance of forces between MP/APol, MP/MP, and APol/APol interactions. One may perhaps expect, for instance, that MPs complexed by APols more highly charged than A8-35, such as A8-75 or SAPols, will show less tendency to aggregate upon APol depletion, whereas complexes with APols carrying no net charge, such as NAPols, might be more prone to aggregation. This, however, is purely speculative and needs to be experimentally investigated.

Back transfer from APols to detergents or transfer to other surfactants such as other APols, nanodiscs, lipid vesicles, or biological membranes is also possible and will be discussed in § 5.7.

Transfer from detergent solution is not the only way to form MP/APol complexes. As noted above, alternative routes are (i) to directly extract MPs with APols; (ii) to transfer denatured MPs from a denaturing solution to APols, where they will fold; and (iii) to synthesize the protein in vitro in the presence of APol. The first approach has been validated mostly with styrene-maleic acid copolymers (SMAs); the second with classical APols such as A8-35, SAPols, or NAPols; and the third with uncharged APols, namely NAPols and NVoy (Table 5.3).

## 5.2.2   Direct Extraction of Proteins from Membranes

Direct extraction of MPs using APols has been mostly carried out with SMAs. As discussed in § 5.2.2.2, there are indications that certain MPs (or, perhaps more likely, proteins from certain membranes) can be extracted using A8-35, but this line of research remains to be developed.

*Direct extraction of membrane proteins by amphipols*
*(© 2018 by Francis Haraux)*

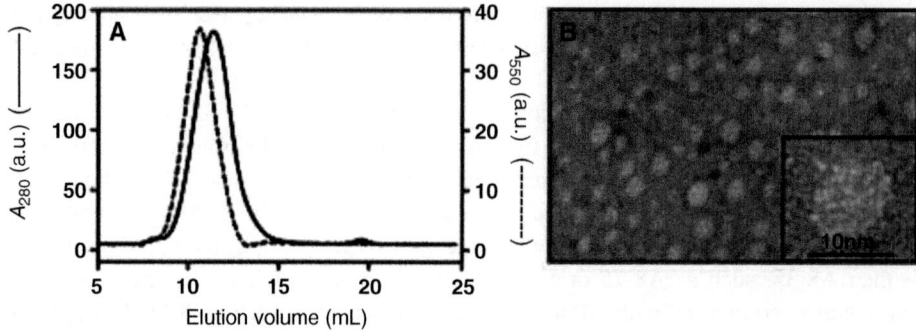

**Fig. 5.8** Characterization of membrane protein/SMA complexes. (**A**) Size exclusion chromatography of PagP (*solid line*) and BR (*dashed line*) incorporated into SMALPs. Absorbance was measured at 280 nm and 550 nm, respectively. (**B**) Transmission electron micrograph of uranyl acetate-stained SMALPs (×100,000), with the insert showing a single nanoparticle (Reprinted with permission from Knowles et al. 2009, © 2009 American Chemical Society).

### 5.2.2.1 Styrene-Maleic Acid Copolymers

The use of styrene-maleic acid copolymer (SMA) to disperse lipids into bicelle-like SMA/lipid particles (SMALPs, also called by the trade name Lipodisqs®) has been introduced in Chap. 4, § 4.3.2. When MP-containing native or artificial membranes are supplemented with SMA, MP/lipid/SMA particles are formed. In the princeps demonstration that SMA can be used to extract MPs from a lipid environment, by T.R. Dafforn, M. Overduin, and their collaborators (Knowles et al. 2009), detergent-solubilized BR and lipid A palmitoyltransferase (PagP) were reconstituted in dimyristoylphosphatidylcholine (DMPC) vesicles and the resulting proteoliposomes dissolved with SMA. The particles were characterized by SEC, DLS, and EM after negative staining (Fig. 5.8) and the native state of the proteins ascertained by spectroscopy (absorbance, CD, and Fourier-transform infrared spectroscopy) and, in the case of PagP, enzymatic activity. The approach has since been extended to some 20 different MPs (Tables 5.1 and 5.2), the largest TM region thus entrapped to date being that of the AcrB trimer, which features 36 TM α-helices (Postis et al. 2015). Some MPs were extracted from artificial vesicles, some directly from native biological membranes (Tables 5.1 and 5.3; reviewed in Jamshad et al. 2011, 2015a, b; Rajesh et al. 2011; Malhotra and Alder 2014; Dörr et al. 2016; Lee et al. 2016; Lee and Pollock 2016; Wheatley et al. 2016). Some MPs are relatively difficult to solubilize from native membranes, which has been attributed to a low lipid/protein ratio or tight lipid packing (Dörr et al. 2016). For such proteins, extraction can be facilitated by the addition of DMPC, as was done to extract BR from native purple membrane (Knowles et al. 2009; Orwick-Rydmark et al. 2012). At this point, direct solubilization of β-barrel proteins by SMA from the outer membranes of bacteria, chloroplasts, or mitochondria has not been reported.

A particularly interesting but, at this stage, poorly documented issue is to which extent SMA extraction can be used to study complexes of MPs that inactivate upon being extracted with detergents. Current data suggest that this will likely be a case-by-case situation. In *Staphylococcus aureus*, SMA was successfully used to extract the divisome PBP2/PBP2a penicillin-binding complex and to demonstrate that the drug resistance modifier (-)-epicatechin gallate alters the spatial relationship between the two proteins (Paulin et al. 2014). On the contrary, the association of *E. coli* SecYEG with YidC and with SecDFyajC that is known to exist in vivo could not be evidenced following extraction of his-tagged SecY by SMA, whereas that between YidC and SecDFyajC was (Prabudiansyah et al. 2015). Similarly, even though cytochrome *c* oxidase forms a supercomplex with cytochrome $bc_1$ in the mitochondrial inner membrane of *S. cerevisiae*, the $bc_1$ was not found in oxidase-containing SMALPs, whereas those did contain loosely bound respiratory supercomplex factors (Smirnova et al. 2016). A recent study reported that, upon SMA solubilization of sorghum microsomes followed by affinity purification, more than 130 proteins where found to be associated to NADPH-dependent cytochrome P450 oxidoreductase, including two cytochrome P450 enzymes specifically involved in the synthesis of dhurrin, a compound that releases cyanide as a defense against predators. Functional data support the existence of a metabolon (a complex in which metabolites are channeled from one enzyme to the next) involving these three P450s, all of them bitopic MPs, and a soluble partner (Laursen et al. 2016). It will be very interesting to examine more in depth which complexes can be extracted intact under which conditions, which not, and why.

It has been reported that there seems to be a limit to the size of the complexes that can be encapsulated in SMALPs (Lee et al. 2016). This is rather unexpected, given that, by varying the ratio of SMA to lipids, it is possible to form very large discs (~90–100 nm in diameter) (Li et al. 2015a; Zhang et al. 2015b).

SMA extracts MPs from native membranes along with a complement of lipids, another very interesting characteristic that will be discussed in § 5.3.1.2.

A diisobutylene/maleic acid copolymer (DIBMA) has recently been shown to exhibit similar performances to SMA in solubilizing phospholipids, stabilizing an integral membrane enzyme, and extracting MPs from biomembranes, with advantages linked to the absence of phenyl rings: no absorbance in near-UV and less perturbation of the lipids (Oluwole et al. 2017).

### 5.2.2.2  Can Membrane Proteins Be Directly Solubilized Using Polyacrylate-Based Amphipols?

In the initial stages of the development of conventional APols (A8-35 and other polyacrylate-based APols), attempts were made to use them to directly extract MPs from native membranes, namely the purple membrane from *Halobacterium salinarum* and thylakoid membranes from *Chlamydomonas reinhardtii*, two membranes whose protein content is very high. They showed very little efficiency, leading to the current approach of extracting the target protein with a detergent and then transferring it to APols. Over the years, however, a couple of exceptions were encountered. Using A8-35, it was observed that, according to immunoblots, close to half of the $FGK_2$ maltose transporter could be extracted directly from *E. coli* plasma membranes, that is, about as much as could be solubilized by DDM under the same conditions (M. Zoonens and H.A. Shuman, unpublished data quoted in Popot et al. 2003). Similarly, the human insulin receptor was totally extracted by A8-35 from the Chinese hamster ovary cells where it had been overexpressed (Gérard Crémel and colleagues, unpublished data quoted in Popot et al. 2003). A summary of these two sets of experiments is provided in Box 5.1. Preliminary studies with *C. reinhardtii* thylakoid membranes using mixtures of APols and detergent in various proportions showed that photosynthetic complexes can be extracted to various degrees (Bazzacco 2009), possibly a way to gently extract fragile supercomplexes. Whole proteomes, including MPs, have been extracted from human cells using a high concentration of A8-35 combined with sonication (Ning et al. 2013), but no functional studies were carried out (see Chap. 14, § 14.4).

---

**Box 5.1   Direct Extraction of Membrane Proteins by A8-35**

In Popot et al. (2003), it was briefly mentioned that two sets of preliminary experiments indicated that certain MPs (or MPs from certain membranes) can be directly extracted with A8-35. Some details are provided below.

**B5.1.1. Direct extraction of insulin receptors from Chinese hamster ovary (CHO) cells**

In order to test whether the insulin receptor expressed in CHO cells and solubilized with Triton X-100 (TX-100) can be transferred to A8-35 under a functional form, the TX-100-solubilized receptor was immobilized on a wheat-germ agglutinin (WGA) column (Leray et al. 1993), washed, and eluted by 0.3 M acetylglucosamine in the presence either of CHAPS or of A8-35. The receptor was observed to elute faster in the presence of A8-35 than of CHAPS. Determination of the protein content and radioactive insulin binding in the eluate indicated that the specific activity was twice

**Box 5.1  (continued)**

higher in A8-35 than in CHAPS. Insulin-dependent autophosphorylation was evaluated by SDS-polyacrylamide gel electrophoresis and blotting, followed by antiphosphotyrosine revelation. The basal activity was observed to be higher in A8-35 than in CHAPS. Upon ultracentrifugation on sucrose gradients, the A8-35-trapped receptor migrated faster and as an apparently narrower band than in the presence of Triton X-100 (Fig. 5.9). Altogether, these observations indicate that the insulin receptor can be kept in solution by A8-35 under a functional and relatively monodisperse form.

It was then tested whether the receptor could be directly extracted by A8-35. CHO cells expressing insulin receptors were exposed for 30 min to concentrations of either A8-35 or Triton X-100 ranging from 0% to 0.3%, at 4 °C, in 20 mL 50 mM Hepes, 150 mM NaCl, pH 7.6, supplemented with protease inhibitors. Insoluble material was pelleted by centrifugation at 100,000 × g. Protein content and insulin binding in the pellets were determined. As shown in Fig. 5.10, for a given concentration, the ability of the two surfactants to solubilize insulin receptors is very similar, resulting in complete solubilization of insulin binding sites at ~0.03% surfactant.

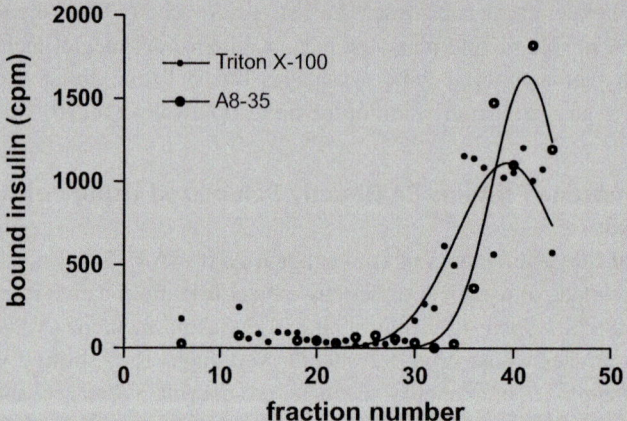

**Fig. 5.9** Comparison of the migration in sucrose gradients of the insulin receptor either solubilized in Triton X-100 or trapped in A8-35 (G. Crémel, T. Corbière, C. Dziukala, and P. Hubert, unpublished data).

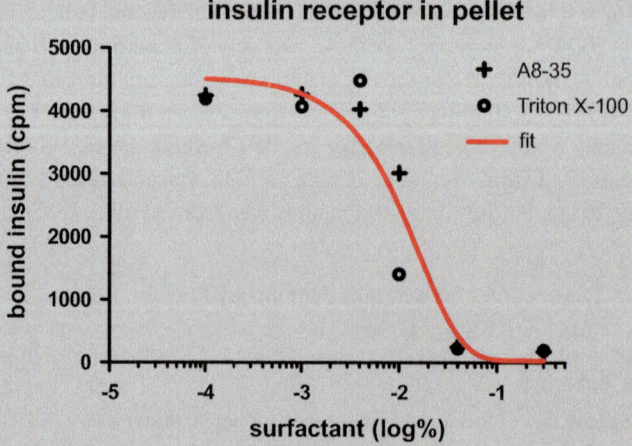

**Fig. 5.10** Extraction of the insulin receptor from CHO cells by either Triton X-100 or A8-35. The amount of receptor that can be pelleted at 100,000 × g is plotted as a function of the logarithm of the concentration of surfactant (G. Crémel, T. Corbière, C. Dziukala, and P. Hubert, unpublished data).

**Box 5.1    (continued)**

Whereas the protein content in the supernatants was identical (~11 mg·L$^{-1}$ in both cases at 0.3% surfactant), the specific binding activity was higher in A8-35 (5700 cpm per μg protein) compared to TX-100 (3900 cpm per μg protein). This effect is not due to a deleterious effect of Triton X-100 on the binding activity (Leray et al. 1992). The increase of specific binding activity in A8-35- vs. TX-100-extracted receptor is the same as observed when TX-100-solubilized receptor is transferred to A8-35, which suggests a structural rearrangement of the receptor, such as could result, for instance and hypothetically, from the preservation or recovery of some critical receptor/lipid interactions.

In summary, these preliminary data strongly suggest that A8-35 is able to efficiently solubilize the insulin receptor, improving insulin binding and preserving insulin-dependent autophosphorylation.

**B5.1.2. Direct extraction of the maltose transporter from *E. coli* plasma membrane**

The maltose transporter belongs to the ABC transporter superfamily. It comprises two different integral MPs, MalF and MalG, and two copies of a cytoplasmic protein, MalK, able to hydrolyze ATP. The three subunits MalK, MalF, and MalG, the latter fused to the glutathione S-transferase affinity tag (GST-MalG), were overexpressed in *E. coli* (NT169 strain). Direct extraction of the FGK$_2$ complex by A8-35 was evidenced after incubating bacterial membranes at various A8-35 concentrations. The presence of the three subunits of the FGK$_2$ complex in soluble and insoluble material was detected with immunoblots (Fig. 5.11).

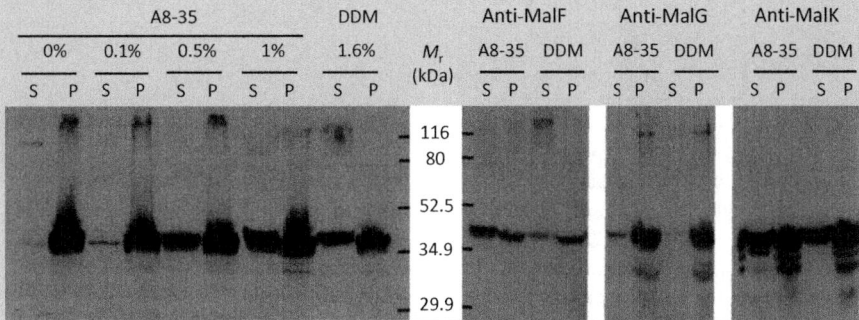

**Fig. 5.11**  Solubilization by A8-35 and by dodecylmaltoside (DDM) of the FGK$_2$ maltose transporter complex from *Escherichia coli* plasma membrane. Bacterial membranes at a protein concentration of 20 g·L$^{-1}$ were diluted five times in 20 mM Tris/HCl buffer, pH 7.0, 5 mM MgCl$_2$, 10 mg·L$^{-1}$ of phenylmethylsulfonyl fluoride (PMSF), 1 mM DTT, 20% glycerol, and A8-35 at final concentrations varying from 0% to 1% w/v. Detergent-solubilized membranes with 1.6% DDM were used as a control. After 30 min of incubation on ice, the samples were centrifuged at 100,000 × *g* for 30 min at 4 °C. The supernatants and pellets were analyzed using immunoblots, which were revealed with either a mix of the antibodies raised against the three subunits (*left panel*; 0–1% w/v A8-35) or each antibody individually (*right panel*; 1% w/v A8-35) (M. Zoonens and H.A. Shuman, unpublished data).

The solubilization of FGK$_2$ increased with the concentration of A8-35 and reached a plateau between 0.5% and 1% A8-35. The total amount of FGK$_2$ solubilized in A8-35 (roughly half of the material present in the membrane preparation) was comparable to that extracted by 1.6% DDM (Fig. 5.11, last two lanes in the *left* panel; see Davidson and Nikaido 1991; Reich-Slotky et al. 2000). The three subunits were present in the soluble fraction. However, a poorer solubilization of GST-MalG was observed in both A8-35 and DDM. This may be due to the presence of the GST-tag fused to MalG, which could lead to a high proportion of misfolded, non-assembled forms.

In order to check on the integrity of the FGK$_2$ complex in the sample solubilized with 1% A8-35, the supernatant was added to a glutathione agarose resin and incubated for 2 h at 4 °C under

**Box 5.1   (continued)**

gentle shaking. The resin was then transferred into a column and washed four times with buffer containing 50 mM Tris/HCl, pH 8, 5 mM $MgCl_2$, 10 mg·$L^{-1}$ PMSF, 1 mM DTT, and 10% glycerol, containing either 0.01% DDM (DDM sample) or no surfactant (A8-35 one). Elution was performed with the same buffers supplemented with 10 mM of glutathione. The elution fractions were analyzed by SDS-PAGE (Fig. 5.12).

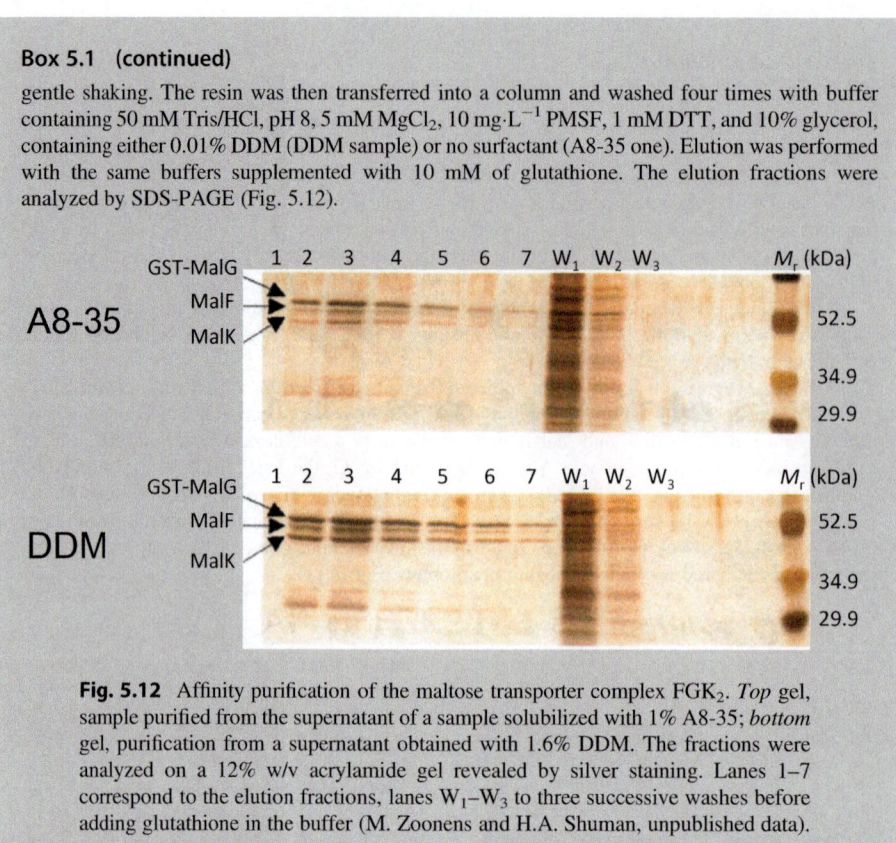

**Fig. 5.12**  Affinity purification of the maltose transporter complex $FGK_2$. *Top* gel, sample purified from the supernatant of a sample solubilized with 1% A8-35; *bottom* gel, purification from a supernatant obtained with 1.6% DDM. The fractions were analyzed on a 12% w/v acrylamide gel revealed by silver staining. Lanes 1–7 correspond to the elution fractions, lanes $W_1$–$W_3$ to three successive washes before adding glutathione in the buffer (M. Zoonens and H.A. Shuman, unpublished data).

The co-elution of MalF and MalK with the tagged GST-MalG indicates that the $FGK_2$ complex purified in A8-35 is fully associated.

Christophe Tribet and colleagues have conducted extensive studies of the association of APols and other polymers with lipid vesicles and resulting phenomena, such as permeabilization, lateral segregation of the lipids and polymer, and vesicle disruption, using either conventional polyacrylate-based APols (Ladavière et al. 2001, 2002; Tribet and Vial 2008; Vial et al. 2005, 2007, 2009) or versions thereof carrying photoexcitable side chains, whose hydrophobicity can be modulated by light (Sebai et al. 2010, 2012) (for a review, see Marie et al. 2014). Most relevant to the present discussion is the observation that A8-35 is able to disperse small unilamellar vesicles of either egg phosphatidyl-choline (egg PC)/dipalmitoyl phosphatidic acid (DPPA) or dipalmitoyl phosphatidylcholine (DPPC)/DPPA into small particles ($R_H = 8$–10 nm), as ascertained by both DLS and EM (Ladavière et al. 2001; Vial et al. 2005) – possibly discoidal mixed micelles resembling SMA/lipid particles. At 25 °C, however, the kinetics can be extremely slow (days). It seems that A8-35 and its congeners can indeed be used to directly extract MPs in some cases, but that only certain combinations of membranes, MPs, APols, and experimental conditions lead to a reasonably quantitative extraction. This question would deserve to be further investigated.

A molecular dynamics (MD) study has been published of the interactions of a complex of C8-PMAL-B-100 and siRNA with lipid bilayers (Li et al. 2015b).

### 5.2.3   Folding Membrane Proteins in Amphipols

As shown in Table 5.1, a dozen MPs have been transferred and folded in APols starting from a denatured state. The denaturing medium is typically sodium dodecyl sulfate (SDS) for $\alpha$-helical proteins (BR and seven G protein-coupled receptors (GPCRs)) and urea for $\beta$-barrel ones (four bacterial outer membrane proteins). The dodecyl sulfate is typically removed by precipitating it as its potassium salt, PDS, whereas the urea is diluted. The procedure has been validated for A8-35 (Pocanschi et al. 2006b; for a recent review, see Le Bon et al. 2018), SAPols (Dahmane et al. 2009), NAPols (Bazzacco et al. 2012), and amphipathic poly-$\gamma$-glutamic acid (APG) (Han et al. 2014). A few synthetic peptides have been transferred to A8-35 from either urea or trifluoroethanol (Gohon 1996; Duarte et al. 2008; Stangl et al. 2014). In comparative studies, it was shown that, provided lipids have not been removed, the functional properties of BR refolded in A8-35 are exactly the same as those of native BR trapped in the same APols starting from a solution in non-denaturing detergent (Dahmane et al. 2013). Similarly, GPCRs folded in APols from SDS-solubilized inclusion bodies exhibit the same pharmacological properties as native receptors (Dahmane et al. 2009; Banères et al. 2011; Bazzacco et al. 2012; Mary et al. 2014; Logez et al. 2016). Folding MPs to their native form is one of the most promising uses of APols and will be discussed in detail in Chap. 6.

*Amphipol-assisted folding of membrane proteins*
(© 2018 by Francis Haraux)

Alternatively, MPs can be synthesized in vitro using APols as the accepting medium. However, polyanionic APols such as A8-35, SAPols, and SMA have all been found to inhibit cell-free synthesis of MPs, possibly by interacting with cationic sites at the surface of the translation apparatus (Park et al. 2011; Periasamy et al. 2013). Only glucose-based NAPols (Bazzacco et al. 2012) and NVoy (Guild et al. 2011; Klammt et al. 2011), both of which are non-ionic, have been validated to date for use in MP cell-free expression. Using APols for this application will be discussed in Chap. 7.

*Amphipol-assisted cell-free synthesis of membrane proteins*
(© *2018 by Francis Haraux*)

## 5.3    Composition, Organization, Dynamics, and Solution Properties of Membrane Protein/Amphipol Complexes

MP/APol complexes have been studied by a vast number of biochemical and biophysical techniques, including compositional analysis, small-angle neutron and X-ray scattering (respectively SANS and SAXS), sucrose gradient analytical ultracentrifugation (SG-AUC), sedimentation velocity and equilibrium analytical ultracentrifugation (SV-AUC and Eq-AUC), SEC, DLS, CD, EM after negative staining (NS-EM) and electron cryomicroscopy (cryo-EM), solution NMR, mass spectrometry (MS), and many other approaches. Studies that present data on their composition, organization, solution properties, and dynamics are summarized in Table 5.4.

The most complete data have been collected on BR/A8-35 complexes, with which we will start (Study 5.16 in Table 5.4). Purple membrane (PM) was solubilized in octylthioglucoside (OTG) and BR transferred, by adsorbing the detergent onto Bio-Beads, either to plain A8-35 or to isotopically labeled versions thereof, namely either DAPol, in which the octyl and isopropyl side chains are perdeuterated (Chap. 4, Fig. 4.12), or [$^3$H]A8-35. The native state and functionality of the protein were established by examining its UV-visible spectrum and photocycle. The particles were characterized by compositional analysis, SG-, SV-, and Eq-AUC, SANS, and SEC (Gohon et al. 2008). This study was complicated by the fact that, when it was carried out, not all aspects of the synthesis of A8-35 had been brought under control. As a result, unknown to the authors, some batches were rendered more hydrophobic than they ought to have been by the artifactual binding of dicyclohexylurea (DCU), a side product of the grafting reaction, making some BR/A8-35 complexes, particularly the BR/DAPol ones, prone to aggregation (see Chap. 4, § 4.2.1, and Protocol 4.1). While seriously complicating the study, this had the useful consequence of calling attention to the fact that a good monodispersity of free APol particles is a reliable predictor of the capacity of an APol preparation to form monodisperse MP/APol complexes (Gohon et al. 2008).

**Table 5.4** A selection of studies providing data on the composition, structure, and solution properties of membrane protein/amphipol complexes. Studies are listed in chronological order of first publication. The table is not exhaustive, particularly as regards SEC studies. Studies providing data on MP/A8-35 complexes are highlighted in *blue*, those on MP/A8-75 complexes in *gray*, those on SMALPs in *buff*, those on PC-APol complexes in *cyan*, those on MP/glucosylated NAPols in *pink*, those on MP/SAPol complexes in *green*, and those involving other APols in *white*. For more detailed information about the proteins, the reader is referred to Table 5.1.

| Study | Amphipol(s) | Protein(s)[a] | Characterization of membrane protein/amphipol particle solution properties by: | | | | | | | | | | | Comments | References |
|---|---|---|---|---|---|---|---|---|---|---|---|---|---|---|---|
| | | | Compositional analysis | SANS | SG-AUC | SV-AUC | Eq-AUC | SEC | DLS | CD | EM | NMR | Other methods | | |
| 5.1 | A8-35, A8-75, A34-35, A34-75 | BR, OmpF, $b_6f$ bacterial photosynthetic reaction center | | | • | | | | | | | | | The first article showing that APols can be used to trap MPs under a functional form as largely mono-disperse particle, while respecting the oligomeric state observed in detergent solution. | Tribet et al. (1996) |
| 5.2 | A5-75 | $b_6f$, tGpA | | | • | | | | | • | | | | Data unpublished in article form, showing that A5-75, a short version of A8-75, can be used to trap cytochrome $b_6f$ and a synthetic peptide mimicking the transmembrane α-helix of glycophorin A. The $b_6f$ retains its native state and the peptide adopts an α-helical configuration. | Gohon (1996) and Popot et al. (2003) |
| 5.3 | A8-75 | OmpF, bacterial photosynthetic reaction center, $b_6f$ | • | | • | | | | | | | | | An analysis of the MP/APol mass ratio in the complexes, their behavior in the presence and absence of free APol, and the exchange of protein-bound APol for free APol or detergent. The physical stability of MP/APol complexes in the absence of free APol is demonstrated. | Tribet et al. (1997) |
| 5.4 | A8-35, A8-75 | $b_6f$ | • | | • | | | | | | | | Cryo-STEM | Cytochrome $b_6f$ was trapped in its monomeric and dimeric forms with either A8-35 or A8-75. The mass of each type of particle, as estimated by determination of the lipid, APol, and protein contents, was found to be consistent with that determined by STEM. The stability of the $b_6f$ dimer was examined in the presence or absence of excess lipid and/or APol. | Tribet et al. (1998) |
| 5.5 | A8-35 | SERCA1a | | | | • | | • | | | | | | The sarcoplasmic calcium ATPase was trapped in A8-35 and the size and solution behavior of the complexes studied by SEC and SV-AUC in the presence and absence of either DDM or Ca²⁺. The functional effects of A8-35 on the ATPase were also studied. | Champeil et al. (2000) |

(continued)

**Table 5.4**  (continued)

| Study | Amphipol(s) | Protein(s)[a] | Characterization of membrane protein/amphipol particle solution properties by: | | | | | | | | | | | Comments | References |
|---|---|---|---|---|---|---|---|---|---|---|---|---|---|---|---|---|
| | | | Compositional analysis | SANS | SG-AUC | SV-AUC | Eq-AUC | SEC | DLS | CD | EM | NMR | Other methods | | |
| 5.6 | OAPA-20 | DAGK | | | | | | | | | | | | Diacylglycerol kinase was trapped either with OAPA-20 (≈ A8-75) or with PMAL-B (= PMAL-C12) and its state of oligomerization and aggregation studied by cross-linking with glutaraldehyde followed by SDS-PAGE. In addition, the transfer of DAGK from APols to lipid vesicles was demonstrated. | Nagy et al. (2001) |
| | PMAL-B (= PMAL-C12) | | | | | | | | | | | | Cross-linking + SDS-PAGE | | |
| 5.7 | A8-35 | nAChR | | | • | | | | | | | | | The nicotinic acetylcholine receptor was trapped in A8-35 and the migration of its monomeric and dimeric forms in sucrose gradients compared to that in the presence of CHAPS. Allosteric transitions induced by the binding of an analog of acetylcholine were studied by fluorescence spectroscopy. | Martinez et al. (2002) |
| 5.8 | A8-35 | tOmpA | | | | | | | | | | • | | The complexes formed between A8-35 and the transmembrane β-barrel domain of OmpA have been studied by solution NMR. The rotational correlation time $\tau_c$ of the particles has been determined and the points of contact between the two partners have been mapped. | Zoonens et al. (2005) |
| 5.9 | A8-35 | OmpA, FomA, BR | | | | | | • | | • | | | | Three membrane proteins have been folded in A8-35 and their final state studied by SDS-PAGE, CD, and functional measurements. | Pocanschi et al. (2006) |
| 5.10 | A8-35 | BR, $b_6f$ | | | • | | | | | | | | ITC | BR and cytochrome b6 f were trapped in either A8-35 or various APols and the complexes studied by centrifugation on sucrose gradients and visible spectroscopy. The effects of pH, high salt, and Ca²⁺ ions have been examined, as well as the enthalpy change associated with substituting APol for detergent at the surface of BR. | Diab et al. (2007) |
| | PC-APols | | | | | | | | | | | | | | |
| 5.11 | A8-35 | tOmpA | | | | • | • | • | | | | | FRET | The stability of tOmpA/A8-35 complexes upon removal of the free APol has been studied, as well as the kinetics of exchange of a protein-bound fluorescent APol for an unlabeled APol or for detergent. | Zoonens et al. (2007) |
| 5.12 | A8-35 | BR | • | • | | • | • | • | | | • | | | An extensive examination of the composition, size, and internal organization of BR/lipid/A8-35 particles, of the functional state of BR, and of the effect of removing excess APol from the solutions. | Gohon et al. (2008) |

| Study | Amphipol(s) | Protein(s)[a] | Characterization of membrane protein/amphipol particle solution properties by: | | | | | | | | | | | Comments | References |
|---|---|---|---|---|---|---|---|---|---|---|---|---|---|---|---|
| | | | Compositional analysis | SANS | SG-AUC | SV-AUC | Eq-AUC | SEC | DLS | CD | EM | NMR | Other methods | | |
| 5.13 | Glucosylated NAPols | BR, tOmpA, GHS-R1a | • | • | | • | | | | | | • | | Glucosylated NAPols were tested on three MPs, BR, tOmpA, and the ghrelin receptor GHS-R1a. The native state of the proteins was demonstrated by functional (BR, GHS-R1a) or solution NMR (tOmpA) measurements. The mass and dimensions of the complexes were established by SEC, SV-AUC, and SANS. | Bazzacco et al. (2009, 2012) and Sharma et al. (2012) |
| 5.14 | A8-35 | OmpX | | | | | | | | | | • | MS | An NMR analysis of protein/APol contacts in OmpX/A8-35 complexes. | Catoire et al. (2009) |
| 5.15 | SMA | BR, PagP | | | | | | • | | • | • | | | The first demonstration that styrene-maleic acid copolymers can be used to directly extract MPs from membranes and keep them soluble in small SMA-bounded lipid discs. | Knowles et al. (2009) |
| 5.16 | A8-35 / PC-APols | BR, tOmpA | | | | | | | | | | | ITC | A thermodynamic study of the exchange of detergents and APols at the surface of MPs. The enthalpy of APol/detergent exchange on the hydrophobic surface of IMPs is negligibly small, an indication of the similarity of the molecular interactions of IMPs with the two types of amphiphiles. The enhanced stability against dilution of MP/APol complexes, compared to MP/detergent ones, originates from the difference in entropy gain achieved upon release in water the surfactant. | Tribet et al. (2009) |
| 5.17 | A8-35 | OmpX | | | | | | | | | | • | | An NMR investigation of the accessibility to water and dynamics of A8-35-trapped OmpX. EDTA is shown to accelerate the tumbling time of the particles and improve the resolution of the signals. | Catoire et al. (2010b) |
| 5.18 | A8-35 | Mitochondrial supercomplex $I_1III_2IV_1$ (respirasome) | | | | | | | | | • | | BN-PAGE | A cryo-EM study of A8-35-trapped supercomplex $I_1III_2IV_1$, the largest MP yet stabilized in APols, showing the first direct visualization of a MP-bound APol belt. | Althoff et al. (2011) |
| 5.19 | SAPols | BR, tOmpA | | | | | | • | | | | • | SDS-PAGE | Solution NMR spectra of tOmpA trapped with sulfonated APols have a resolution comparable to that of A8-3-trapped tOmpA, indicating a similar particle size. They can be recorded at low pH and high temperature without denaturing the protein. | Dahmane et al. (2009) |
| 5.20 | NVoy | CRFR1, CRFR2β | | | | | | • | | | • | • | | Two GPCRs expressed in vitro in the presence of NVoy were shown to be functional and form small particles amenable to solution NMR investigations. | Klammt et al. (2011) |

(continued)

**Table 5.4** (continued)

| Study | Amphipol(s) | Protein(s)[a] | Compositional analysis | SANS | SG-AUC | SV-AUC | Eq-AUC | SEC | DLS | CD | EM | NMR | Other methods | Comments | References |
|---|---|---|---|---|---|---|---|---|---|---|---|---|---|---|---|
| **5.21** | A8-35 / NAPols | BR, tOmpA, $b_6f$, $bc_1$ | | | | | | | | | | | MS | MALDI-TOF-MS analysis of eight MP/APol complexes confirms that APols do not interfere with determining the mass of MPs (cf. Study 5.14) and that protein-bound lipids retained in the complexes can be identified. | Bechara et al. (2012) |
| **5.22** | A8-35 | OmpT, PagP | | | | | | | | | | | MS | Electrospray ionization mass spectrometry coupled with ion mobility spectrometry (ESI-IMS-MS) analysis of MP/A8-35 complexes shows that the two proteins tested are properly folded and can be transferred to the gas phase without denaturation. | Leney et al. (2012) |
| **5.23** | SMA | BR | | | | | | • | • | • | | | EPR | BR extracted with SMA from DMPC vesicles and labeled with nitroxides shows absorbance, CD, and double electron-electron resonance (DEER) EPR spectra compatible with it being in its native state. | Orwick-Rydmark et al. (2012) |
| **5.24** | A8-35 | TRPV1 ion channel | | | | | | | | | • | | | A series of high-resolution cryo-EM studies showing the first detailed views of the APol belt surrounding the transmembrane region of a MP. | Cao et al. (2013) and Liao et al. (2013, 2014) |
| **5.25** | A8-35 | BR | | | | | | • | | | | | Photocycle measurements | Comparative refolding studies evidencing the functional effects of the reassociation of lipids with the renatured protein. | Dahmane et al. (2013) |
| **5.26** | A8-35 | BR | | | | | | • | | | | • | | A comparison of the size, stability, and NMR spectra of BR in DDM, A8-35, or nanodiscs. | Etzkorn et al. (2013) |
| **5.27** | A8-35 | Peripherin-ROM1 complex | | | | | | | | | • | | | 3D reconstruction of the complex from images of negatively stained particles permits a tentative identification of the A8-35 belt. | Kevany et al. (2013) |
| **5.28** | A8-35 | ABCA4 transporter | | | | | | | | | • | | | 3D reconstruction of the complex from images of negatively stained particles permit, a tentative identification of the A8-35 belt. | Tsybovsky et al. (2013) |
| **5.29** | A8-35 | SoPIP2:1 tetramer | | | | | | | | | • | | | A comparison of the structure of the SoPIP2:1 tetramer and its associated surfactant belt in OG, DDM, LMNG, and A8-35, from 3D reconstruction of images of negatively stained particles. | Vahedi-Faridi et al. (2013) |
| **5.30** | SMA | Cytochrome $c$ oxidase (Complex IV) | • | | | | | • | | | • | | BN-PAGE | Complex IV was extracted from *S. cerevisiae* mitochondrial inner membranes with SMA, forming disc-like particles with a diameter of ~12 nm. | Long et al. (2013) |
| **5.31** | A8-35 | OmpF | | | | | | • | • | | • | | | Outer membrane protein F stabilized with minimal amphipol forms linear arrays and lipopolysaccharide-dependent 2D crystals. | Arummanee et al. (2014) |

| Study | Amphipol(s) | Protein(s)[a] | Characterization of membrane protein/amphipol particle solution properties by: | | | | | | | | | | | Comments | References |
|---|---|---|---|---|---|---|---|---|---|---|---|---|---|---|---|---|
| | | | Compositional analysis | SANS | SG-AUC | SV-AUC | Eq-AUC | SEC | DLS | CD | EM | NMR | Other methods | | |
| 5.32 | A8-35 | Cytochrome $bc_1$ complex | • | • | • | | | • | | | | | SAXS | A study of the interactions in solution and crystallization of $bc_1$/A8-35 complexes and of the effect of adding salt or polyethylene glycol and of APol depletion. | Charvolin et al. (2014) see also Popot et al. (2003) |
| 5.33 | A8-35 | OmpX | | | | | | | | | | • | | A comparison of the structure, particle size, and accessibility from the solvent of OmpX trapped in A8-35 or solubilized in $diC_6PC$. | Etzkorn et al. (2014) |
| 5.34 | SMA | PgP | | | | | | • | | • | • | | | A low-resolution (~3.5 nm) cryo-EM study of PgP extracted from insect cell membranes with SMA. | Gulati et al. (2014) |
| 5.35 | A8-35 | OmpX | | | | | | | | | | | MD | Molecular dynamics simulations of an OmpX/A8-35 complex. | Perlmutter et al. (2014) |
| 5.36 | A8-35 | ExbB–ExbD complex | • | • | | • | | • | | | • | | SAXS | An extensive study of ExbB-ExbD/A8-35 complexes. | Sverzhinsky et al. (2014) |
| 5.37 | SMA | Photosynthetic reaction center | • | | | | | | | | • | | Absorbance spectroscopy | A study of the functionality and stability of bacterial photosynthetic reaction centers extracted with SMA and of the size and composition of the resulting particles. | Swainsbury et al. (2014) |
| 5.38 | PMAL-C8 | TRPA1 | | | | | | | | | • | | | A single-particle cryo-EM studies of TRPA1 showing the protein-bound APol belt at 6-Å resolution. | Paulsen et al. (2015) |
| 5.39 | SMA | AcrB | • | | | • | | | | | • | | | The AcrB trimer was extracted from E. coli membranes with SMA. SV-AUC and EM imaging after negative staining show that the particles can form dimers associated via the SMA/lipid belts in an ionic strength-dependent manner. Lipid analysis shows that the particles contain only about half the amount of lipids needed to fully protect the transmembrane region from contact with SMA. | Postis et al. (2015) |
| 5.40 | A8-35 | PagP, OmpT, Mhp1, GalP | | | | | | | | | | | ESI-IMS-MS | A comparative study of the conformational stability of MPs when transferred to the gas phase either from DDM or from A8-35. | Calabrese et al. (2015) see also Watkinson et al. (2015) |

(continued)

**Table 5.4** (continued)

| Study | Amphipol(s) | Protein(s)[a] | Characterization of membrane protein/amphipol particle solution properties by: | | | | | | | | | | | Comments | References |
|---|---|---|---|---|---|---|---|---|---|---|---|---|---|---|---|
| | | | Compositional analysis | SANS | SG-AUC | SV-AUC | Eq-AUC | SEC | DLS | CD | EM | NMR | Other methods | | |
| 5.41 | SMA | COX (Complex IV) | • | | | | | • | | | | • | | Complex IV (cytochrome *c* oxidase) was extracted from *S. cerevisiae* mitochondria by SMA and the particles characterized by SEC, NS-EM, and qualitative lipid analysis. | Smirnova et al. (2016) |
| 5.42 | A8-35 | STRA6 | | | | | | • | | | | • | | Reconstruction at 3.9-Å resolution visualizes the layer of protein-adsorbed A8-35. | Chen et al. (2016) |
| 5.43 | A8-35 | TRPM4 | | | | | | • | | • | • | | BN-PAGE, cross-linking | Demonstration of the tetrameric nature of TRPM4. | Constantine et al. (2016) |
| 5.44 | DIBMA | OmpLA | | | | | | • | • | • | • | • | Raman scattering | Demonstration that diisobutylene/maleic acid (DIBMA) copolymer can be used alternatively to SMA to solubilize phospholipids and a membrane enzyme, OmpLA. | Oluwole et al. (2017) |

*Abbreviations:* 3D three-dimensional, *AFM* atomic force microscopy, *APol* amphipol, *BN-PAGE* blue-native polyacrylamide gel electrophoreses, *CD* circular dichroism, *DAPol* A8-35 with an unlabeled main chain and the octyl and isopropyl side chains perdeuterated, *DDM* n-dodecyl-β-D-maltoside, *DIBMA* diisobutylene/maleic acid copolymer, *DM* n-decyl-β-D-maltoside, *DEER* double electron-electron resonance, *DLS* dynamic light scattering, *EM* electron microscopy, *EPR* electron paramagnetic resonance, *Eq-AUC* equilibrium analytical ultracentrifugation, *ESI-IMS-MS* electrospray ionization-ion mobility spectrometry-mass spectrometry, *FRET* Förster resonance energy transfer, *GPCR* G protein-coupled receptor, *ITC* isothermal titration calorimetry, *LMNG* lauryl maltose neopentyl glycol, 2,2-didecylpropane-1,3-*bis*-β-D-maltopyranoside, *MALDI* matrix-assisted laser desorption/ionization, *MD* molecular dynamics, *MS* mass spectrometry, *NMR* nuclear magnetic resonance, *OG* n-octyl-β-D-glucopyranoside, *SANS and SAXS* small-angle neutron and X-ray scattering, respectively, *SAPol* a sulfonated amphipol, *SEC* size exclusion chromatography, *SG-AUC* sedimentation velocity analytical ultracentrifugation in sucrose gradients, *SV-AUC* sedimentation velocity analytical ultracentrifugation, *TOF* time of flight

[a]Membrane protein short names are given in the first column of Table 5.1

### 5.3.1   Particle Composition

Purple membrane solubilized with OTG was ultracentrifuged. The supernatant was supplemented with either unlabeled, deuterated, or tritiated A8-35 and the detergent removed by adsorption onto Bio-Beads. The composition of the resulting BR/A8-35 particles was studied by (i) lipid analysis, (ii) measurement of the content in $[^3H]A8-35$, (iii) spectrophotometric determination of the content in native and total protein, and (iv) determination by SEC, SANS, and AUC of the contrast match point, mass, size, shape, and density of the particles. These properties depend on the mass of each constituent present in the particles and provide internal controls as to the consistency of the measurements (Gohon et al. 2008). The results are summarized in Table 5.5, along with those collected on a few other MP/APol complexes. These data are compared, whenever possible, with available data about the composition of MP/DDM complexes.

Both the APol and lipid contents of the particles deserve some discussion.

**Table 5.5** Composition of some membrane protein/amphipol (MP/APol) complexes and comparison of the number of APol $n$-alkyl chains bound with that in MP/dodecylmaltoside (DDM) complexes (Updated from Popot et al. 2003).

| Membrane protein/amphipol complexes | | | | | | | | MP/DDM complexes | | |
|---|---|---|---|---|---|---|---|---|---|---|
| Protein (TM structure) | APol | MP (kDa) | APol (kDa) | Lipids (kDa) | APol/MP (g/g) | $n$-alkyl chains per MP/APol complex[a] | References | DDM/MP (g/g) | $C_{12}$ chains per MP/DDM complex | References |
| BR (7 $\alpha$-helices) | A8-35 | 27 | ~54 | ~10 | ~2 | ~110 | Gohon et al. (2008) | ~4.1 | ~208 | Møller and le Maire (1993) |
| Cytochrome $b_6f$ complex (dimer) (2 × 12 $\alpha$-helices) | A8-75 | 228 | ~46 | + | ~0.22 | ~93 | Tribet et al. (1997, 1998) | ~0.65 | ~260 | Breyton et al. (1997) |
| Photosynthetic reaction center (*Rhodobacter sphaeroides*) (11 $\alpha$-helices) | A8-75 | 102 | ~41 | ? | ~0.41 | ~83 | Tribet et al. (1997) | ~0.90 | ~148 | Møller and le Maire (1993) |
| OmpF trimer (3 × 16 $\beta$-strands) | A8-75 | 96 | ~41 | ? | ~0.43 | ~83 | Tribet et al. (1997) | | | |
| Nicotinic acetylcholine receptor dimer (2 × 20 $\alpha$-helices) | A8-35 | 535 | ~150 | + | ~028 | ~303 | Martinez et al. (2002) | | | |
| Transmembrane domain of OmpA (tOmpA) (8 $\beta$-strands) | A8-35 | 19 | 31–45[b] | – | 1.6–2.3 | 63–89[b] | Zoonens et al. (2007) and Perlmutter et al. (2014) | | | |
| CRFR1 and CRFR2$\beta$ (7 $\alpha$-helices) | NVoy | 47–49 | ~100 | – | ~2.1 | ?[c] | Klammt et al. (2011) | | | |
| BR (7 $\alpha$-helices) | NAPol | 27 | 97 | ~10[d] | ~3.6 | ~136 | Sharma et al. (2012) | ~4.1 | ~208 | Møller and le Maire (1993) |
| Cytochrome $bc_1$ complex (dimer) (2 × 13 $\alpha$-helices) | A8-35 | 490 | ~54 | + | ~0.11 | ~110 | Charvolin et al. (2014) | | | |
| ExbB$_4$/ExbD$_2$ (14 $\alpha$-helices) | A8-35 | 139 | 167–208 | + | 1.2–1.5 | 337–420 | Sverzhinsky et al. (2014) | ~0.59 | ~160 | Sverzhinsky et al. (2014) |

[a]Octyl chains but for NAPols, which carry undecyl chains
[b]The lowest value is based on experimental measurements (Zoonens et al. 2007) and is likely to be an underestimate (see § 5.3.1.1); the highest one is based on molecular dynamics studies of OmpX/A8-35 models (Perlmutter et al. 2014) and may be an overestimate (see § 5.3.3)
[c]The mass of NVoy per alkyl chain has not been released
[d]Assumed to be identical to that in BR/A8-35 complexes

### 5.3.1.1   Amphipol vs. Detergent Binding

Table 5.5 summarizes existing data about the composition of MP/APol complexes (no data are currently available about the protein/polymer ratio in MP/SMA and MP/NVoy complexes). When a comparison of APol vs. detergent (DDM) binding is possible, one notes a marked tendency for MPs to bind less APols than DDM, whether surfactant binding is expressed in mass ratio to the protein or as the number of $n$-alkyl chains associated with it. BR/A8-35 complexes, for instance, comprise twice less surfactant than BR/DDM ones: ~2 g A8-35 per g protein, vs. ~4 g DDM, and ~110 A8-35 octyl chains per complex, vs. ~210 DDM dodecyl chains. This tendency is found with other complexes, with the single exception of the $ExbB_4/ExbD_2/A8-35$ ones, which appear to bind much more surfactant than $ExbB_4/ExbD_2/DDM$ complexes (Table 5.5). BR/NAPol complexes comprise, in mass, about as much surfactant as BR/DDM ones, but this reflects the higher relative mass of the glucosylated polar moieties of NAPol, the number of undecyl chains per BR/NAPol complex (~136) being similar to that of octyl chains in BR/A8-35 ones (~110).

It must be stressed that APol binding data have to be taken with some caution, because they have seldom been collected under ideal conditions: as has been discussed above, when MP/APol complexes are separated from free APol, they tend to aggregate, which must be accompanied by the release of some free APol. This probably explains why the experimentally measured amount of A8-35 that remains associated with the TM region of OmpA (tOmpA) following removal of free APol by IMAC (Zoonens et al. 2007) proves insufficient to totally surround the hydrophobic surface of a MP of very similar size, OmpX, in MD simulations (Perlmutter et al. 2014; see § 5.3.3). In Fig. 5.13, the number of

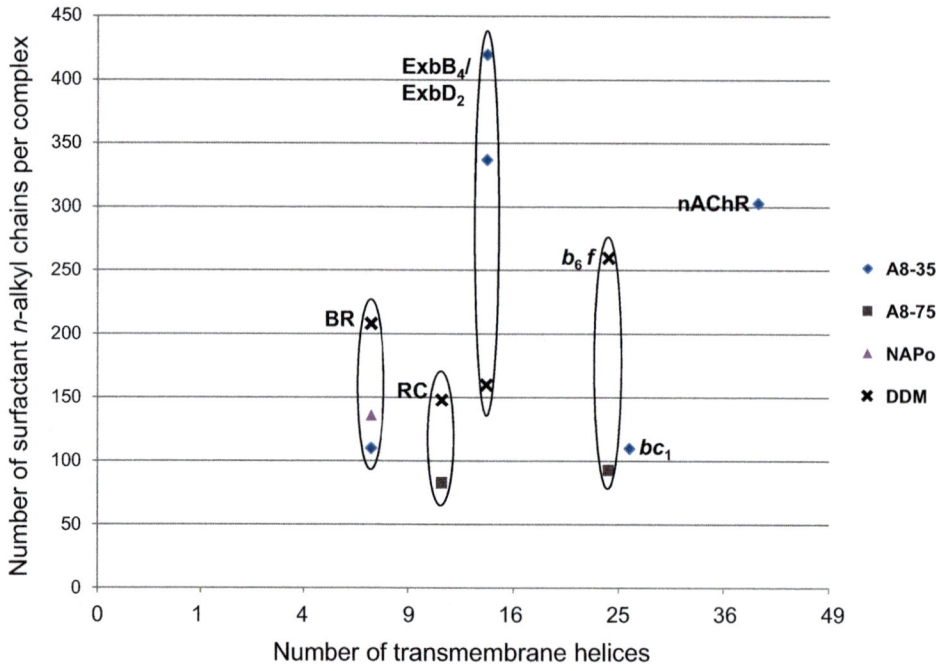

**Fig. 5.13** Amphipol and dodecylmaltoside (DDM) binding by membrane proteins with an $\alpha$-helical transmembrane (TM) region plotted as a function of the number of TM helices, the latter distributed on a square-root scale. The amount of surfactant bound is expressed in terms of $n$-alkyl chains, namely octyl chains for A8-35 and A8-75, undecyl chains for NAPols, and dodecyl chains for DDM. Membrane proteins: BR, bacteriorhodopsin from *H. salinarum*; RC, photosynthetic reaction center from *R. sphaeroides*; $ExbB_4/ExbD_2$ complex from *E. coli*; $b_6f$ complex from *C. reinhardtii*; $bc_1$ complex from *B. taurus*; nAChR, nicotinic acetylcholine receptor (dimeric form) from *T. marmorata* (Data from Table 5.5).

*n*-alkyl chains associated with a given α-helical MP is plotted as a function of the square root of the number of TM helices in the protein, which can be taken to be roughly proportional to their membrane-exposed surface. Whereas it is obvious that this is only a very crude measure of their hydrophobic TM surface, a more or less linear relationship should be expected. The considerable scatter of the data clearly suggests imprecise measurements of APol binding. More systematic measurements under better controlled conditions would be highly desirable. A standard procedure is proposed in § 5.13, Protocol 5.2.

Despite this caveat, it seems clear that there is a tendency for less APol *n*-alkyl chains than DDM ones to bind to a given MP (Fig. 5.13), even though these chains are shorter in APols than in DDM and could, a priori, cover less of the protein's hydrophobic surface. This suggests a different organization of the alkyl chains at the surface of the protein. Indeed, according to NMR data, the octyl chains of A8-35 appear to lie sidewise at the surface of the protein, thereby occupying a larger surface than if they were more or less perpendicular to it (Catoire et al. 2009; Planchard et al. 2014; Fig. 5.14). Furthermore, both NMR data (Planchard et al. 2014; Fig. 5.14) and MD data (Perlmutter et al. 2014; Fig. 5.15) indicate that, in the case of A8-35, part of the protein surface is in contact with isopropyl chains (Fig. 5.15), which may contribute to explain why less *n*-alkyl chains are required to shield this surface from water than in the case of DDM.

### 5.3.1.2  Lipid Binding

Quantitative data about the presence of lipids in MP/APol complexes are scarce. The most detailed analysis is that of BR/A8-35 complexes obtained by solubilizing purple membrane fragments in OTG, centrifuging, supplementing the supernatant with A8-35, and eliminating OTG by adsorption onto Bio-Beads. BR/A8-35 complexes were then separated from free APol particles by centrifugation onto sucrose gradients, so as to eliminate those lipids that would not be associated to the complexes. Thin-layer chromatography analyses revealed no difference between lipids extracted from PM or from purified BR/A8-35 complexes and were similar to literature profiles (Corcelli et al. 2000). Phosphate determination performed on two different preparations of BR/APol complexes indicated the presence of 5.2 and 4.5 moles of phospholipid per mole of BR, to be compared with 4.6 mol/mol in PM. These values are within experimental error of each other and of those previously reported for PM: 6.4, 5, and 4.3 mol/mol (see Renner et al. 2005, and references therein). This indicates that BR/APol complexes retain – or, more likely, rebind – all or almost all of the lipids originally present in the *purple* membrane, i.e. ~0.38 g lipid per g BR (Gohon et al. 2008).

It is interesting to view these data in the context of more recent experiments, in which BR was either trapped from OTG-solubilized PM, as in the above experiments, or refolded in A8-35 either from SDS-solubilized PM – that is in the presence of PM lipids – or from bacterio-opsin (BO; the apoprotein) that had been delipidated in organic solvents before being transferred to SDS (Dahmane et al. 2013). Photocycle measurements show a clear difference between, on the one hand, OTG-solubilized BR and BR refolded in A8-35 in the absence of lipids and, on the other hand, BR trapped in A8-35 from OTG-solubilized PM or BR refolded in A8-35 in the presence of PM lipids (Dahmane et al. 2013). The simplest interpretation of these data is that, in OTG, lipids are separated from BR, or some critical lipids do not interact with it as they do in PM, whereas, after transfer to A8-35, they rebind to BR. Note that, even though A8-35-trapped BR appears to rebind a full complement of lipids, those do not suffice to form a complete annulus. Indeed, in the purple membrane, BR molecules are assembled into trimers. The trimers are separated from each other by lipids, and their central cavity filled with lipids, but individual BR molecules in a trimer interact both via protein/lipid and protein/protein contacts (Grigorieff et al. 1996; cf. Chap. 3, Fig. 3.12C). The membrane therefore does not contain enough lipids to form a complete annulus around each BR monomer.

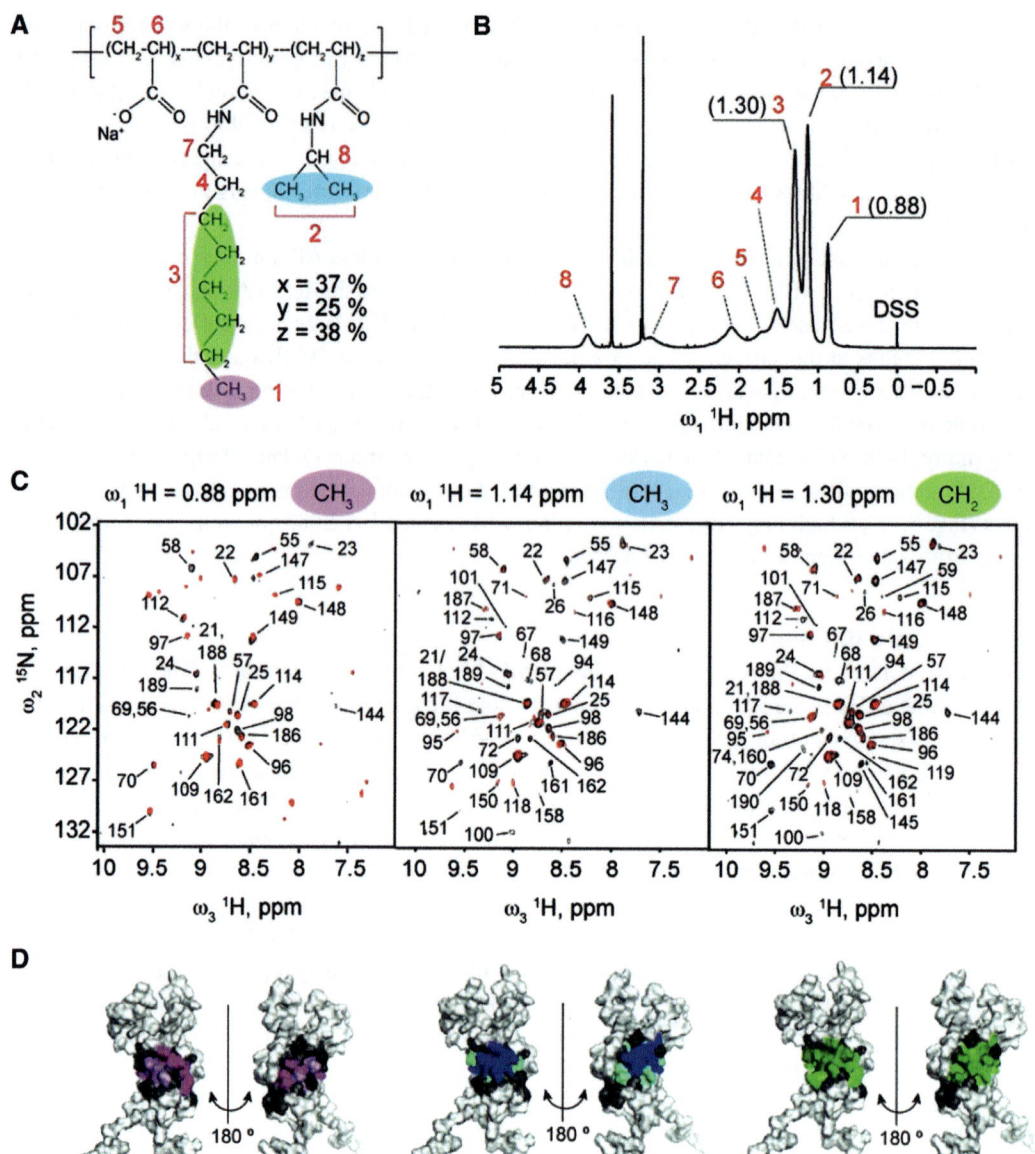

**Fig. 5.14** Analysis of intermolecular contacts between the transmembrane region of *Klebsiella pneumoniae* OmpA (Kp-tOmpA) and A8-35. (**A**) Chemical structure of A8-35, with the numbering and color code for the various types of protons indicated. (**B**) 1D $^1$H spectrum of [u-$^2$H,$^{13}$C,$^{15}$N]Kp-tOmpA/A8-35 complexes in 20 mM NaH$_2$PO$_4$ buffer (pH/D = 7.9) containing 100 mM NaCl, 10 mM EDTA, and 10% D$_2$O. NMR assignments of A8-35 resonances (labels *1–8*) are reported on the spectrum. $^1$H chemical shifts of side-chain resonances are indicated in ppm units by reference to 4,4-dimethyl-4-silapentane-1-sulfonic acid (DSS). (**C**) Selected 2D [$^{15}$N,$^1$H$^N$] planes from the 3D $^{15}$N-edited ($^1$H,$^1$H) HSQC-NOESY-TROSY spectra using 100 ms (*red*) and 200 ms (*black*) NOESY mixing times obtained on [u-$^2$H,$^{13}$C,$^{15}$N] Kp-tOmpA/A8-35 complexes in 20 mM NaH$_2$PO$_4$ buffer (pD = 7.9) containing 100 mM NaCl, 10 mM EDTA, and 100% D$_2$O. 2D [$^{15}$N,$^1$H$^N$] planes were extracted at the $^1$H frequencies indicated at the top of the spectra. (**D**) Distribution of intermolecular NOE signals throughout the TM region of Kp-tOmpA (PDB accession code: 2K0L). *Color code*: *light gray*, residue not detected; *dark gray*, residue for which intermolecular NOE interactions could not be detected; *light color*, residues for which intermolecular NOEs are only detected in 3D NOESY using long mixing times (e.g. 200 ms); *dark color*, residues for which intermolecular NOE signals are detected in 3D NOESY using short mixing times (e.g. 100 ms). The colors correspond to the A8-35 moieties shown in Panel A (From Planchard et al. 2014).

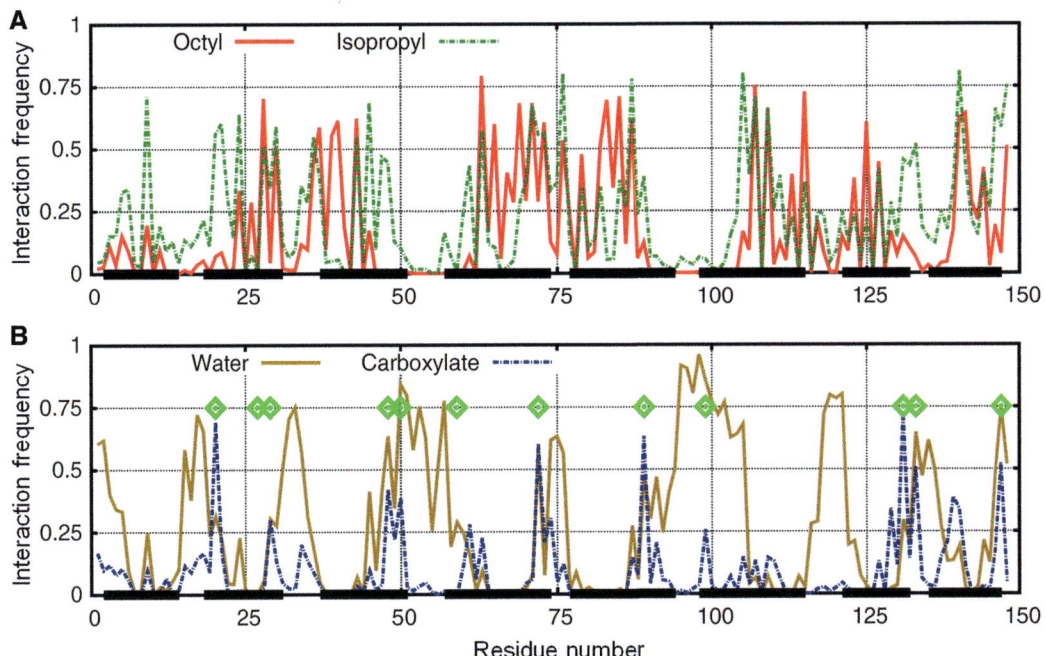

**Fig. 5.15** Frequency with which each amino acid interacts with various A8-35 moieties or with water in OmpX/A8-35 complexes, as deduced from MD simulations. For clarity, Panel (**A**) presents hydrophobic contacts and Panel (**B**) hydrophilic ones. Thick *black lines* indicate the location of the TM *β*-strands, *green diamonds* that of basic amino acids (Lys and Arg) (From Perlmutter et al. 2014).

In the case of the nicotinic acetylcholine receptor (nAChR), functional data can also be interpreted as indicating that transfer from detergent (CHAPS) to APol (A8-35) allows lipids to rebind, but the evidence is indirect and the interpretation is not unique (Martinez et al. 2002; see § 5.4, Fig. 5.25).

MPs are often claimed to be extracted by SMA along with a complete annulus of membrane lipids. The "cookie cutter" simile occasionally used to describe this process is amusing, but it is misleading, the reality being far more complex. In most studies, the amount of lipids is simply estimated from the size of the Lipodisqs, as deduced from either EM images of negatively stained preparations or from DLS measurements (see e.g. Long et al. 2013). In some studies, a qualitative analysis was performed. In most cases, it shows that the lipid composition of the particles is similar to that of the membranes the proteins were extracted from (Long et al. 2013; Dörr et al. 2014; Swainsbury et al. 2014; cf. Fig. 5.16A, B). In the case of the SecYEG complex, however, the particles were found to be enriched in negatively charged lipids (which are known to be important functionally) (Fig. 5.16C, D), whereas the lipids accompanying other MPs extracted from the same *E. coli* membrane had the same composition as the membrane (Prabudiansyah et al. 2015). Following expression in Sf9 insect cells of the human equilibrative nucleoside transporter-1 (hENT1), extraction with SMA, and lipid analysis, it was observed that polyunsaturated lipids were specifically excluded from the complexes (Rehan et al. 2017).

In the few cases where quantitative analyses were performed, the results are case-dependent. In the case of bacterial photosynthetic reaction centers, it was estimated that the lipids extracted along with the protein (~150 of them) are sufficient to form three layers around it (Swainsbury et al. 2014). The dimer of the cation diffusion facilitator CzcD was found to retain 32–35 phospholipids, which were estimated to be able to form about 1 layer around its TM region (Bersch et al. 2017). In the case of

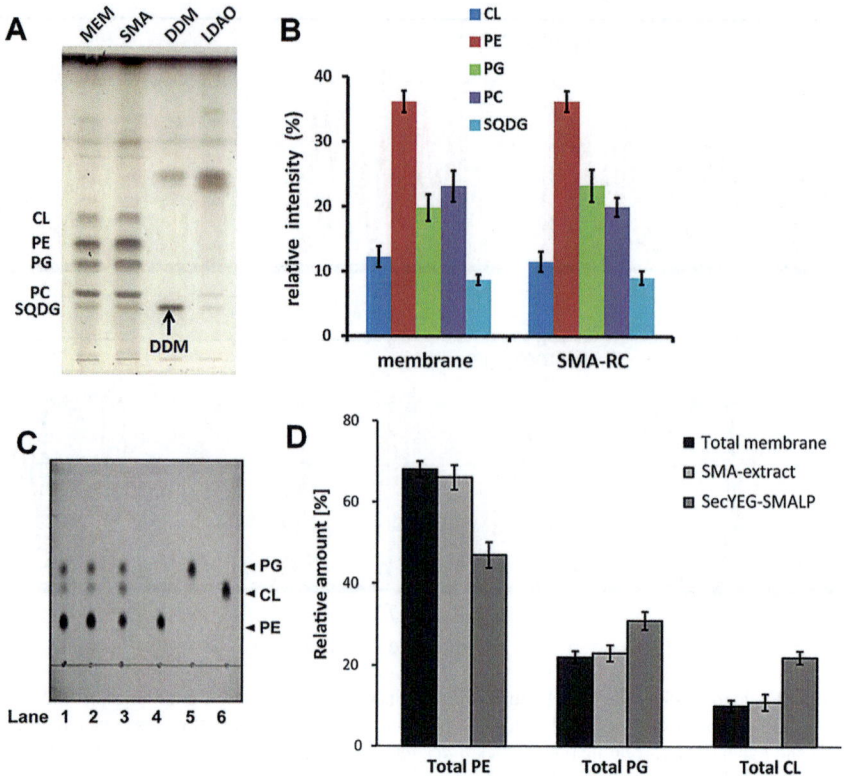

**Fig. 5.16** Lipid composition of some membrane protein/lipid/SMA particles as compared to that of the membrane the protein was extracted from. (**A, B**) Lipid content of membranes containing bacterial photosynthetic reaction centers (RC) and of RC-containing SMALPs. (**A**) Lipid profiles determined by thin-layer chromatography (TLC). Lipids were identified by running pure samples of each as a standard (not shown). Bands above the labeled lipids are attributed to RC pigments. Dodecylmaltoside (DDM) was visualized, but lauryldimethylaminoxide (LDAO) did not stain. Additional bands in the DDM and LDAO profiles are unidentified. *Abbreviations: CL* cardiolipin, *PE* phosphatidylethanolamine, *PG* phosphatidyl-glycerol, *PC* phosphatidylcholine, *SQDG* sulfoquinovosyl diacylglycerol. (**B**) Relative populations of lipids in intact membranes and SMA/lipid nanodiscs as deduced from six independent lipid extractions, quantified by densitometry (From Swainsbury et al. 2014). (**C, D**) Lipid analysis of SMA-trapped SecYEG. (**C**) Phospholipids extracted from the samples were separated by TLC and visualized with molybdenum blue reagent. The retention factor ($R_f$) of phospholipids in the total membrane (lane 1), SMA-extracted membrane protein fraction (lane 2), and purified SMA-trapped SecYEG (lane 3) were compared with the standards PE (lane 4), PG (lane 5), and CL (lane 6). (**D**) Liquid chromatography-mass spectrometry analysis of the total PE, PG, and CL species in total membrane (*black*), SMA-extracted membrane protein fraction (*light gray*), and purified SMA-trapped SecYEG (*dark gray*). Error bars are standard deviation from three experiments (From Prabudiansyah et al. 2015, © 2015 Elsevier).

AcrB, only ~40 lipids are retained, which was deemed insufficient to form a single layer around the protein, implying that the latter must come into contact with the styrene groups of the polymer (Postis et al. 2015). A similar situation prevails for SMA-trapped P-glycoprotein 1 (PgP), which is accompanied by only ~11 lipids (Knowles et al. 2009), in keeping with cryo-EM data, which do not show the presence of a lipid disc (Gulati et al. 2014). Similarly, hENT1/SMA complexes contain only 18 phospholipid molecules per protein (Rehan et al. 2017). The similarity between a BR/lipid/A8-35 complex, in which the protein has rebound all ten lipids it was in contact with in PM (Gohon et al. 2008), and a PgP/lipid/SMA one, in which it has retained ~11, is more striking than the difference. Attractive as the notion may seem, distinguishing MP/A8-35 and MP/SMA complexes as non-bilayer vs. bilayer systems (Dörr et al. 2016) is not necessarily warranted, at least certainly not in all cases. It seems likely that, as is the case for bicelles (see Chap. 3, § 3.2), the size of the discs, and, therefore, the protein/lipid ratio, is modulated by experimental conditions, in particular the SMA/lipid ratio in the final preparation, as indeed observed (Li et al. 2015a; Zhang et al. 2015b; Grethen et al. 2017).

To which extent the composition of the lipids and their organization in SMALPs, in particular their distribution between the two monolayers, faithfully mimics the situation experienced in vivo by the MPs they host remains to be ascertained. A key observation, discussed in Chap. 4 (§ 4.3.2), is the fact that the lipid content of SMALPs exchanges very rapidly (within seconds), probably according to a bulk mechanism that involves lateral fusion and fission events (Cuevas Arenas et al. 2017; Grethen et al. 2018). Given that SMA seems to have no preference for extracting specific lipids as long as they are in the fluid phase (Dominguez Pardo et al. 2017), these observations suggest that the preferential association of specific lipids with MPs extracted in SMALPs that has been observed in some experiments (Prabudiansyah et al. 2015; Rehan et al. 2017) reflects an equilibrium situation, not a kinetically frozen one. By the same token, one should expect that lipids that, in the original membrane, were located either in the outer or in the inner monolayer will rapidly mix and that whatever may remain of the original compositional asymmetry be due to specific binding to the protein.

### 5.3.2   Particle Size and Organization

Detailed structural studies have been carried out on BR/A8-35 complexes, using, in particular, SANS, Eq-AUC, and SV-AUC (Study 5.16 in Table 5.4). An introduction to these techniques and their application to MP/APol complexes is provided in Chap. 9. In both cases, advantage was taken of the possibility to modulate the contrast of the APol with the solvent by deuterating its side chains, which increases both its buoying density and its neutron scattering length density. The density of the solvent was modulated thanks to the use of $H_2O$, $D_2O$, and $D_2{}^{18}O$ (Gohon et al. 2008). These studies will be described in Chap. 9, § 9.3. They show that BR and its associated lipids occupy, as expected, the center of the particles, whose overall mass is ~90 kDa (Table 5.5), the APol forming a belt adsorbed onto the TM region of the protein/lipid complex. The thickness of the belt – its extension away from the surface – can be estimated to ~1.7 nm (see Chap. 9, Fig. 9.18). For the most part, these measurements, obtained by a wide range of approaches, provide consistent estimates of the composition, size, and mass of the complexes. Combining composition and $s$-values yields $R_S \approx 3.6$ nm, whereas Eq-AUC and SV-AUC analyses give $R_S \approx 3.8$ nm. SANS yields an $R_g{}^\circ$ value (radius of gyration at infinite contrast) of ~3.0 nm, which, for spherical particles, would correspond to $R_S \approx 3.8$ nm. By EM after negative staining (NS-EM), one observes particle half-widths and half-lengths of ~3.2 and ~4 nm. The outlier is SEC, which provides significantly larger estimates ($R_S \approx 5.0 \pm 0.15$ nm). Other observations suggest that, for reasons that are probably electrostatic in origin, since the same phenomenon is not observed with NAPols (Sharma et al. 2012; see below), SEC tends to overestimate the hydrodynamic radius of MP/A8-35 complexes as compared to MP/neutral detergent ones. Indeed, SEC (Zoonens et al. 2007) yields much larger apparent $R_S$ differences between tOmpA/A8-35 complexes ($R_S \approx 4.3$–3.7 nm) and tOmpA/dihexanoylphosphatidylcholine (diC$_6$PC) ones ($R_S \approx 2.6$ nm) than is indicated by an NMR study of their respective rotational correlation times, $\tau_c$ (Zoonens et al. 2005). The NMR data suggest that SEC overestimates the $R_S$ of tOmpA/A8-35 particles by as much as 30–50% (Zoonens et al. 2007), a difference close to that (~30%) observed for BR/A8-35 complexes between SEC data, on the one hand, and SANS and AUC ones, on the other (Gohon et al. 2008). Why such a phenomenon does not affect the determination by SEC of the Stokes radius of pure A8-35 particles (Chap. 4, § 4.3.1.2.2) appears paradoxical and remains to be understood. A model of the BR/lipid/A8-35 complex based on compositional, AUC and SANS data is shown in Chap. 9, Fig. 9.18.

A similar study (Study 5.17 in Table 5.4) has been carried out on complexes between BR and a glucosylated NAPol (Sharma et al. 2012). The conclusions were essentially the same, except that (i) the mass of APol bound is significantly higher (~97 kDa, putting the total mass of the complex at ~135 kDa), even though the content in APol $n$-alkyl chains is only slightly higher (~136 $C_{11}$ chains for BR/NAPol complexes vs. ~110 $C_8$ ones for BR/A8-35 ones; Table 5.5); (ii) due to the higher density of NAPols, the specific volume of the complexes is lower (0.791 mL·g$^{-1}$ vs. 0.856 mL·g$^{-1}$ for BR/A8-35 complexes); and (iii) the $R_S$ is slightly higher (4.1 nm) and is identical whether determined by AUC or by SEC. The NAPol belt that keeps BR soluble is therefore slightly thicker than that formed by A8-35. A model of BR/lipid/NAPol complexes based on this ensemble of data is shown in Chap. 9, Fig. 9.19.

An interesting comparison has been carried out in Study 5.56 (Table 5.4) between SAXS and SANS data and low-resolution EM images of A8-35-trapped ExbB$_4$/ExbD$_2$ complexes (Sverzhinsky et al. 2014; see Chap. 9, § 9.3.8.3, Fig. 9.21).

Single-particle electron cryomicroscopy studies have revealed the APol layer surrounding the TM region of several MPs, many of them at high, near-atomic resolution (see e.g. Althoff et al. 2011; Liao et al. 2013, 2014; Lu et al. 2014; Paulsen et al. 2015; Chen et al. 2016), a recent example of which is shown in Fig. 5.20. These reconstructions confirm the conclusions from earlier, less direct analyses. They will be discussed in Chap. 12, which is devoted to the use of APols for EM (Fig. 5.17).

The question of the relative size of MP/APol vs. MP/detergent complexes is rather muddled. As will be discussed in Chap. 10, which is devoted to the application of APols to NMR, MP complexes with the small detergents used in solution NMR, such as $C_8E_4$ or diC$_6$PC, tumble more rapidly than MP/A8-35 complexes, indicative of a smaller $R_S$. The question is less clear as regards MP/A8-35

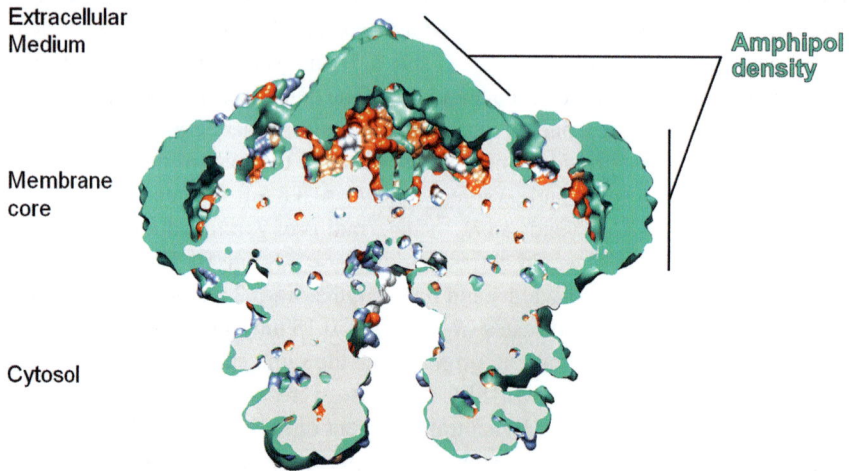

**Fig. 5.17** Space-filling representation of the zebrafish STRA6 retinol receptor/calmodulin complex, trapped with A8-35, determined to 3.9-Å resolution by single-particle electron cryomicroscopy. The image shows a slice through the middle of the outer cleft that protrudes in the extracellular space. The A8-35 layer is shown in *teal*. Note that it does not cover only the transmembrane region, but seems to extend over the extracellular cleft. The composition of this extracellular density is not actually known. The central cavity being on the whole extremely hydrophobic and having lateral passages connecting it to the bulk lipid, it could be filled with lipid-like molecules, to which A8-35 would likely adsorb (Figure courtesy of Oliver Clarke and Filippo Mancia. See Chen et al. 2016).

vs. MP/DDM complexes, but the most reliable data seem to point to MP/A8-35 complexes being smaller than MP/DDM ones. Some elements of discussion are provided in Box 5.2.

**Box 5.2  Are Membrane Protein/A8-35 Complexes Actually Bigger Than Membrane Protein/Dodecylmaltoside Ones?**

There is a seeming paradox in the fact that most comparisons of the composition of MP/A8-35 vs. MP/DDM complexes indicate that a given protein binds less APol than detergent (Table 5.5 and Fig. 5.13), whereas a couple of estimates of the size of the two types of complexes conclude that the former are bigger than the latter. Historically, the first comparison was carried out using as a model the sarcoplasmic $Ca^{2+}$-ATPase, SERCA1a (Champeil et al. 2000; Fig. 5.18). Experimental conditions were not ideal, because it had not been realized, at the time, that artifactual grafting of dicyclohexylurea increased the hydrophobicity of certain A8-35 batches, leading MP/APol complexes to form small oligomers (Gohon et al. 2004, 2006), nor that traces of $Ca^{2+}$ caused the complexes to aggregate (Picard et al. 2006; see Chap. 4, § 4.3.1.2.1). Nevertheless, the data did suggest that SERCA1a/A8-35 complexes are bigger than SERCA1a/DDM ones (Fig. 5.18).

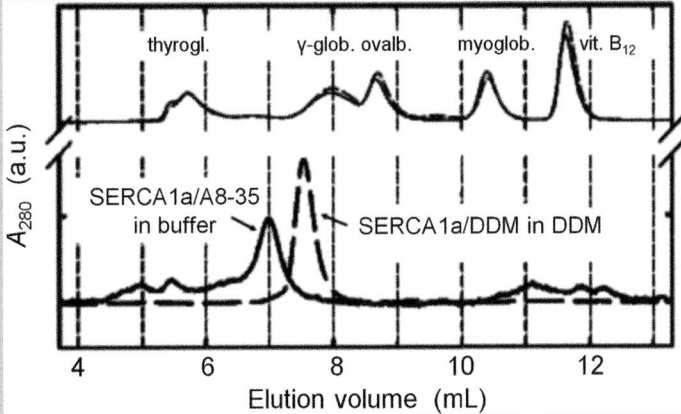

**Fig. 5.18** Size exclusion chromatography of SERCA1a either trapped with A8-35 or solubilized in DDM. *Solid thick line*, chromatography in the absence of detergent of delipidated SERCA1a/A8-35 complexes; *dashed thick line*, chromatography in the presence of 1 $g \cdot L^{-1}$ DDM of a delipidated SERCA1a/DDM complex without amphipol. The *thin lines* correspond to gel filtration standard proteins, eluted in the absence or presence of detergent (This research was originally published in Champeil et al. 2000, © The American Society for Biochemistry and Molecular Biology).

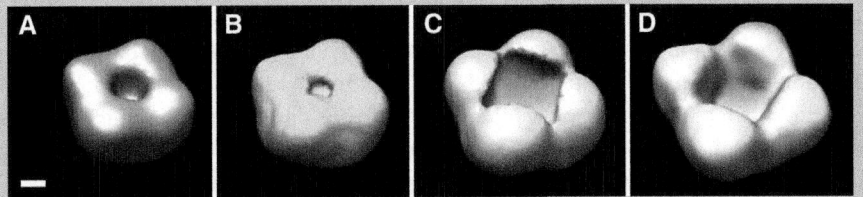

**Fig. 5.19** 3D maps of SoPIP2;1 tetramers complexed by (**A**) OG, (**B**) DDM, (**C**) LMNG, and (**D**) A8-35. Isocontours include overall masses of 190 kDa (A), 220 kDa (B), 280 kDa (C), and 340 kDa (D). The scale bar represents 20 Å (From Vahedi-Faridi et al. 2013).

This conclusion, however, was invalidated when the comparison of $R_S$ values obtained for BR/A8-35 complexes by SEC, SANS, and AUC showed that SEC grossly overestimates the $R_S$ of

**Box 5.2 (continued)**

the complexes (Gohon et al. 2008), presumably because of electrostatic interactions between the resin and the polyanionic APol, given that no such inconsistency is observed with non-ionic APols (Sharma et al. 2012) (see main text).

In a comparative single-particle study of negatively stained plant aquaporin SoPIP2;1 tetramers complexed by various surfactants, it was concluded that the surfactant belt is thicker in APol-trapped complexes than in detergent solution: $11 \pm 2$ Å in OG, $13 \pm 2$ Å in DDM, $15 \pm 2$ Å in LMNG, and $19 \pm 3$ Å in A8-35 (Vahedi-Faridi et al. 2013; Fig. 5.19). However, the definition of the envelope of the reconstructed image depended on an estimate of the amount of bound surfactant. In the case of A8-35, this was taken to be ~220 kDa (taking the measured binding to BR as a basis and assuming the mass of APol bound to be proportional to that of the protein), whereas, on the basis of the data collected in Table 5.5 and Fig. 5.13, ~90 kDa would seem more likely given the size of the TM region of aquaporins. The thickness of the belt deduced from this work is therefore almost certain to be overestimated.

Based on the density of the dry surfactants, it is difficult to understand why MP/A8-35 complexes should be larger than MP/DDM ones, given that they generally comprise much less surfactant and that the specific volumes of A8-35 and DDM are comparable ($0.809$ mL·g$^{-1}$ for A8-35 (Gohon et al. 2004), $0.81$–$0.837$ mL·g$^{-1}$ for DDM (le Maire et al. 2000)). However, hydration must be taken into account. Water binding by DDM has been estimated, by either equilibrium ultracentrifugation or SEC, to be in the range of $0.26$–$0.34$ g water per g detergent (de Vitry et al. 1991; Møller and le Maire 1993), that of A8-35, by a combination of biophysical measurements, to be $1.22 \pm 0.45$ g water per g polymer (Gohon et al. 2004). Combining the extremes of each range, one concludes that the volume of hydrated DDM associated to 1 g of BR is comprised between $3.9$ and $4.7$ cm$^3$, that of hydrated A8-35 between $2.4$ and $3.3$ cm$^3$. One would therefore expect the BR-bound belt of A8-35 to be roughly between $0.5\times$ and $0.8\times$ as thick as the DDM one.

Consistent with this view, a comparative NMR study of BR in DDM solution, DMPC nanodiscs, or A8-35 concluded that the NMR spectra were comparable and all three systems were adequate for solution NMR measurements (Etzkorn et al. 2013; cf. Chap. 10, Fig. 10.14). The work included estimates of $M$ and $\tau_c$ values, which, at variance with SEC data, were consistent with BR/A8-35 complexes being smaller than BR/DDM ones: $M \approx 93$ kDa in A8-35 vs. ~128 kDa in DDM, $\tau_c = 31.7$ vs. $41.4$ ns. Because the BR/A8-35 preparation was partially aggregated, the data are likely to actually overestimate the size of BR/A8-35 monomers. Indeed, on the basis of direct measurements using [$^3$H]A8-35, the mass of the monomeric complexes would be expected to be ~81 kDa (Gohon et al. 2008), assuming the lipid-free BR used by Etzkorn et al. (2013) to bind as much APol as the lipoprotein complexes studied by Gohon et al. (2008), which is probably itself an overestimate.

Everything considered, it seems therefore reasonable to conclude that the most reliable data seem to point to MP/A8-35 complexes being smaller than MP/DDM ones.

The most detailed structural study to date of a MP extracted by SMA is that of the AcrB trimer (Study 5.39 in Table 5.4). As mentioned above, this preparation contains relatively few lipids, in insufficient number to form a full annulus (Postis et al. 2015). In keeping with these analyses, fitting the crystallographic structure of AcrB into the low-resolution envelope obtained by single-particle analysis of EM images of the negatively stained complex reveals some extra material surrounding the TM region of the trimer, attributed to the polymer and lipids, but no lipid disc (Fig. 5.20).

Relatively little is known of the properties of MP/NVoy complexes. In Study 5.27 (Table 5.4), two NVoy-trapped GPCRs, CRFR1 and CRFR2$\beta$, of respective molecular masses, ~47 and ~49 kDa, were studied by SEC combined with ultraviolet (UV) absorbance, static light scattering, and refractive index measurements (cf. Fig. 9.9 in Chap. 9). It was concluded that NVoy-trapped CRFR1 and CRFR2$\beta$ migrate as monomers, with experimental molecular masses of ~40 and ~48 kDa, respectively. Each of them binds ~20 NVoy molecules, i.e. ~100 kDa of polymer (Klammt et al. 2011). This is about twice more than the mass of A8-35 (~54 kDa) bound by BR, a protein that also features seven TM helices, and comparable that of BR-bound NAPol, another glycosylated polymer (~97 kDa) (Table 5.5).

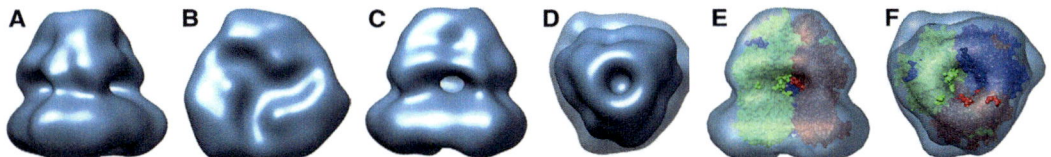

**Fig. 5.20** (**A–D**) Surface views of the AcrB/lipid/SMA complex reconstructed by single-particle analysis of electron microscopy images of negatively stained complexes. Threefold symmetry can be seen from the base (**B**, cytoplasmic face) and top (**D**, periplasmic face) of the structure. (**E, F**) Fitting of the AcrB crystal structure (PDB 1IWG; Murakami et al. 2002) into the EM reconstruction, as seen from the side (**E**) and top (**F**). Extra density can be seen surrounding the TM region of AcrB, which is attributed to the SMA/phospholipid belt (From Postis et al. 2015).

### 5.3.3   Protein/Polymer Interactions

Most of the information on MP/APol interactions is derived from either NMR measurements or MD simulations. NMR data will be discussed in detail in Chap. 10. They show, in brief, that contacts between the polymer and the protein, as deduced from magnetization transfer between the two partners, are essentially limited to the TM surface of the latter (Zoonens et al. 2005; Catoire et al. 2009; Planchard et al. 2014). The polymer restricts water accessibility to this surface, as observed by $^1H/^2H$ exchange (Catoire et al. 2010b) or sensitivity to a water-soluble paramagnetic agent (Etzkorn et al. 2014). The relaxivities of OmpX amide protons in A8-35 correlate remarkably well with those in $diC_6PC$, showing that both surfactants adsorb specifically onto the hydrophobic surface of OmpX and with a similar distribution (Etzkorn et al. 2014; see e.g. Fig. 10.13 in Chap. 10).

Fast photochemical oxidation coupled with liquid chromatography and MS/MS has been used to compare the accessibility of the surface of OmpT, either kept in solution by DDM or trapped with A8-35, to hydroxyl radicals generated by laser photolysis of $H_2O_2$ (Watkinson et al. 2017). The study shows that the distribution of the two surfactants is qualitatively similar, with the same residues being protected in each case. However, a quantitative analysis of the degree of modification indicates that A8-35 affords a better protection to the surface of the $\beta$-barrel that extends into the extracellular space beyond the TM region, suggesting either the existence of extramembrane contacts or an effect on the dynamics of the protein (see § 5.6), whereas the DDM belt appears to cover better some regions of the TM surface on the periplasmic side of the membrane (Watkinson et al. 2017; see Fig. 14.14 in Chap. 14). These experiments will be described in more detail in Chap. 14, § 14.3.3.

Experimental data have been complemented by MD simulations of an A8-35-trapped MP, OmpX (Perlmutter et al. 2014), using the parametrization and procedures developed previously for the simulation of A8-35 particles (Perlmutter et al. 2011; see Chap. 4, § 4.3.1.2.3). Because no experimental measurement of the mass of A8-35 bound by OmpX is yet available, a first step of the MD analysis was to determine the minimal amount of polymer needed to cover the whole TM surface, in conformity with NMR data (Catoire et al. 2009, 2010b; Etzkorn et al. 2014). One copy of OmpX was placed in a simulation box, supplemented with increasing amounts of A8-35, and left to equilibrate. For practical reasons, A8-35 increments from one simulation to the next were set to the relatively high mass of ~11 kDa (~90 polymer units). As shown in Fig. 5.21, three increments (~33 kDa) did adsorb onto the TM surface of OmpX, but did not suffice to cover it entirely; four increments (~44 kDa) achieved complete coverage, whereas five increments (~55 kDa) resulted only in increasing the thickness of the APol belt. The value of 44 kDa is significantly higher than the value of ~25 kDa determined for the binding of A8-35 to tOmpA, a protein with a similar structure and size as OmpX (Zoonens 2004; Zoonens et al. 2007). The experimental value, however, for reasons discussed in § 5.3.1.1, is likely to be an underestimate (Zoonens et al. 2007). On the contrary, the MD value of 44 kDa may be a slight excess as compared to what actually binds in vitro, given that using smaller increments might have led to a lower, more accurate figure. This may possibly explain some features of the model.

Figure 5.21 shows snapshots from the end of the assembly simulations. The APol moieties interacting with the hydrophobic surface of the protein appear to be largely comprised of octyl chains, whereas the outer surface of the APol belt is rich in ungrafted carboxylates. This distribution is shown more quantitatively in Fig. 5.22, where the density of each type of group is plotted as a function of the distance to the axis of OmpX's $\beta$-barrel. The peak of distribution of octyl groups lies at ~1.7 nm from the axis, that of isopropyl chains at ~2 nm, and that of carboxylates at ~2.2 nm. In keeping with experimental results, the thickness of the APol belt is ~1.5–2 nm.

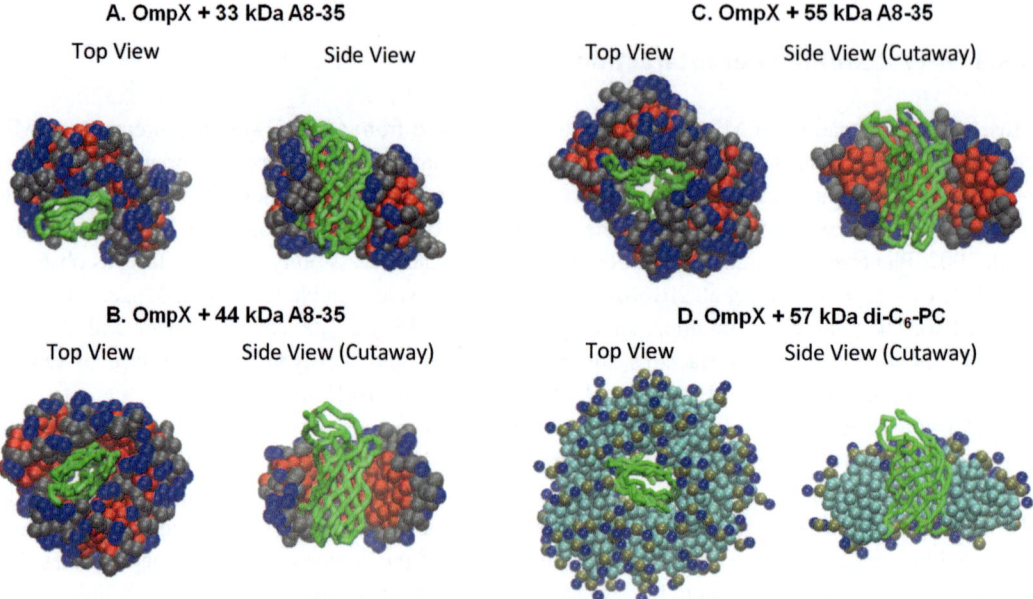

**Fig. 5.21** Snapshots illustrating the configuration of OmpX/A8-35 and OmpX/diC$_6$PC complexes after 10 μs of coarse-grained molecular dynamics. For each system, the *left* figure shows the top view (corresponding to the extracellular surface of the protein), the *right* figure side view with cutaway. *Color code*: *green*, OmpX; *red*, octyl chains of A8-35; *gray*, isopropyl chains of A8-35; *blue*, carboxylates of A8-35 and choline moieties of diC$_6$PC; *tan*, phosphate moieties of diC$_6$PC; *cyan*, methylene groups of diC$_6$PC (From Perlmutter et al. 2014).

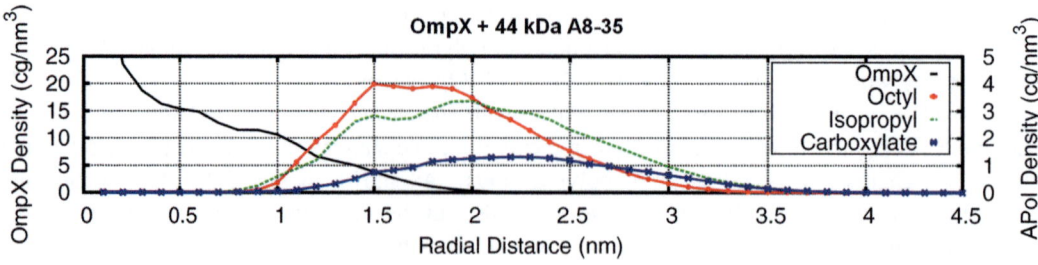

**Fig. 5.22** Radial density of various A8-35 moieties in OmpX/A8-35 complexes calculated by aligning the axis of the $\beta$-barrel with the $z$-axis and recording the radial density of components in a 2-nm-thick cylindrical slab surrounding the central region of the barrel. The density is expressed in terms of the number of coarse-grained (cg) segments per nm$^3$ (From Perlmutter et al. 2014).

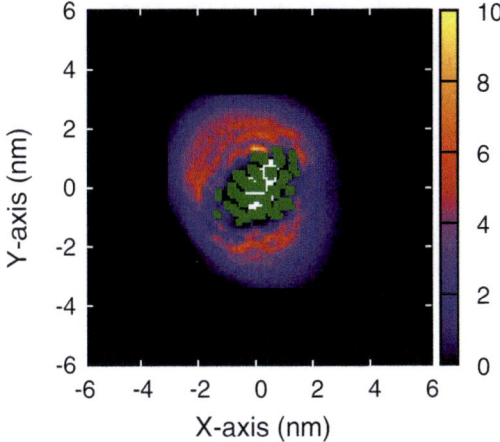

**Fig. 5.23** Time-averaged two-dimensional APol densities in a 2-nm-thick cylindrical slab surrounding the central region of OmpX's $\beta$-barrel, showing that the polymer density is not radially symmetric. The density is expressed in terms of the number of coarse-grained (cg) segments per $nm^2$. *Green* indicates regions of significant protein density (7.5 cg segments per $nm^2$). The surface of the APol belt is approximately given by the outermost *blue ring* (Adapted from Perlmutter et al. 2014).

The strong overlap of the distributions of the various APol moieties seen in Fig. 5.22 is in part an illusion due to the radial averaging of a layer whose thickness is not uniform around a protein whose surface is not a perfect cylinder (Fig. 5.23). Figure 5.21B provides a better impression of the way the various moieties distribute around the protein. Whether the nonuniform thickness of the APol belt is a real feature of the actual complexes or an artifact resulting from the presence of an excess APol in the model is not certain. It is, however, reminiscent of the bulges observed by cryo-EM in the belt of A8-35 surrounding the respirasome (Althoff et al. 2011; see Chap. 12, Fig. 12.17). This issue will be discussed in Chap. 12, § 12.3.3. That the model is a fair representation of the real complex is suggested by a comparison of the accessibility of amide groups estimated in silico by MD and that determined experimentally, based on the rate of $^1H/^2H$ exchange as followed by NMR (Catoire et al. 2010b): as shown in Fig. 5.24, *top*, the accessibility as calculated from the MD data follows remarkably well that measured by NMR. It is also quite similar to that observed when OmpX is embedded in a bilayer of dioleoylphosphatidylcholine (DOPC; Fig. 5.24, *bottom*), whereas the first and last two TM $\beta$-strands are less well protected in $diC_6PC$ (Fig. 5.24, *middle*).

As discussed in § 5.3.1.1, MD simulations indicate, in agreement with NMR data, that both octyl and isopropyl side chains interact with the hydrophobic TM surface of OmpX and exclusively with it (Fig. 5.15A). Such is not the case of the free carboxylates, which transiently interact with the basic residues located in the extramembrane loops and turns (Fig. 5.15B). This is a reminder that experimental data do indicate that APols can interact with water-exposed regions of MPs (cf. Watkinson et al. 2017; § 5.3.3), as well as with non-membrane proteins (see e.g. Ma et al. 2014; Martin et al. 2014, 2015, and references therein), particularly if they are basic (see Champeil et al. 2000). See also the distribution proposed for A8-35 at the surface of the zebrafish STRA6 retinol receptor/calmodulin complex, where the APol seems to cover a presumably lipid-filled water-exposed pocket (Chen et al. 2016; Fig. 5.17).

A largely unexplained discrepancy between MD and experimental data concerns the radius of gyration, $R_g$, of OmpX/A8-35 vs. OmpX/$diC_6PC$ complexes. Calculated values for OmpX/A8-35 complexes increase with the amount of APol bound, from slightly less than 2 nm with 22 kDa APol, ~2.3 nm at 44 kDa, up to slightly more than 2.5 nm at 66 kDa APol, which is certainly an excess (Perlmutter et al. 2014). All of these values are smaller than that, ~3.1 nm, found in MD simulations of the OmpX/$diC_6PC$ complex (cf. Fig. 5.21). Yet, experimentally, OmpX/$diC_6PC$ complexes appear to

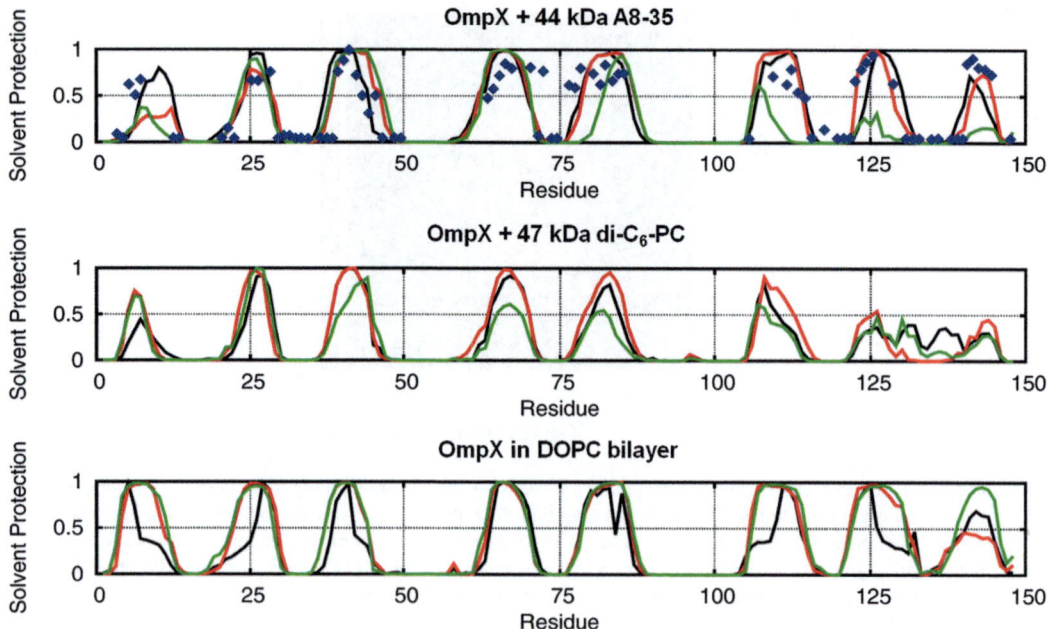

**Fig. 5.24** Relative solvent protection of the backbone of OmpX in A8-35, diC$_6$PC, or a bilayer of dioleoylphosphatidylcholine (DOPC), as deduced from MD simulations. Each *solid line* represents an independent simulation. The experimental accessibility of amide protons in OmpX/A8-35 complexes, based on the rate of $^1$H/$^2$H exchange (Catoire et al. 2010b), is shown as *blue diamonds* (From Perlmutter et al. 2014).

tumble more rapidly than OmpX/A8-35 ones (Fernández et al. 2001; Catoire et al. 2010b). Several causes may contribute to this apparent contradiction. It may be that some factors that contribute to slowing down the tumbling of MP/APol complexes have not yet been identified. Such was the case, in early experiments (Zoonens et al. 2005), of the presence of traces of Ca$^{2+}$, which, presumably, bridged some of the complexes – a problem later identified and eliminated by adding EDTA to NMR buffers (Catoire et al. 2010b). A higher complement of bound water, for instance, could contribute to slowing down the tumbling of MP/A8-35 complexes. Factors that affect the resolution of NMR data will be discussed in Chap. 10. Note that it cannot be excluded either that the MD model of OmpX/diC$_6$PC complexes may just have too much diC$_6$PC bound.

MD simulations have been exploited to examine the effects of transferring OmpX from a lipid bilayer to a detergent or APol environment on its dynamics (Perlmutter et al. 2014). This issue will be discussed in § 5.6, after we have considered the effects of APols on the stability and functionality of APol-trapped MPs.

## 5.4  Functionality of Amphipol-Trapped Membrane Proteins

Functional studies have been carried out with close to 40 different APol-trapped MPs. These data are compiled in Table 5.1, where the cells summarizing the evidence for the preservation of the native state (last-but-one column) are salmon-colored when functional evidence is available. This evidence is usually in the form of either enzymatic measurements or ligand-binding ones. Even though there are some exceptions, which will be discussed below, the overwhelming evidence is that most proteins remain functional after transfer to APols and that the binding of small ligands, toxins, and antibodies is unimpaired as compared to that observed in biological membranes. A few examples will suffice to illustrate the kind of evidence that has been gathered, as well as some caveats. For other specific cases, the reader is referred to Table 5.1 and the references given therein.

BR, an archaebacterial light-driven proton pump (see Chap. 1, § 1.6.1), accomplishes its entire photocycle after being trapped in A8-35 (Pocanschi et al. 2006a, b; Gohon et al. 2008; Charvolin et al. 2009; Dahmane et al. 2013) or in NAPols (Bazzacco et al. 2012). Compared to the kinetics observed in the purple membrane, the first steps of the photocycle are accelerated whether BR is solubilized in detergent or trapped in A8-35. For the detergent-solubilized state, this effect has been attributed to a conformational relaxation that brings Asp-85, the first proton acceptor, closer to the Schiff base donor (Milder et al. 1991). The same is probably true of the APol-trapped state (Dahmane et al. 2013). The last steps of the photocycle feature similar kinetics in APol and in PM, whereas in detergent solution the return to the fundamental state is slower. However, late kinetics similar to those in detergent solution are observed for BR refolded in A8-35 in the absence of lipids (Dahmane et al. 2013). In keeping with earlier observations on partially delipidated and relipidated purple membrane (Joshi et al. 1998), these data strongly suggest that the restoration of native-like protein/lipid interactions upon transferring BR from detergent to APol is responsible for the recovery of PM-like late kinetics (Dahmane et al. 2013).

The nicotinic acetylcholine receptor (nAChR) is a ligand-gated channel of the pLGIC family. Located in the postsynaptic membrane at neuronal synapses and the neuromuscular junction, it transduces a presynaptic chemical signal, the release of acetylcholine in the synaptic cleft, into an electrical one, due to the permeabilization to cations of the postsynaptic membrane (see Chap. 1, § 1.6.2). The allosteric equilibrium between the resting and desensitized states of the nAChR that exists in the absence of the neurotransmitter is strongly perturbed when receptor-rich postsynaptic membranes from the electric organ of *Torpedo marmorata* are solubilized in detergent (Changeux et al. 1980; compare Fig. 5.25A, B with Fig. 5.25C, D). Membrane-like equilibrium and transition kinetics are recovered when a CHAPS-solubilized preparation is supplemented with A8-35 and diluted below the CMC of CHAPS (Martinez et al. 2002; Fig. 5.25E, F). Lipids are known to modulate the function of the nAChR (Chap. 1, § 1.6.2; see daCosta and Baenziger 2009, Baenziger et al. 2015, Hénault et al. 2015, and references therein). As for BR, one possible explanation for the return to membrane-like allosteric properties upon transfer to APols could therefore be that this transfer allows lipids to rebind to critical allosteric sites at the TM surface of the receptor. Because detergents may also perturb allosteric equilibria by binding to the TM domain of the nAChR, much as anesthetics do (see Forman et al. 2015), it is equally possible, however, that the return to a membrane-like allosteric equilibrium be due to the dissociation of the detergent, whether it is replaced by lipids or by APol. Both factors can come into play, of course, if dilution of the detergent allows lipids to rebind to sites from which they had been displaced.

Eight G protein-coupled receptors (GPCRs) have been studied in more or less detail after trapping or folding in one or the other APol (Table 5.1; for reviews, see Banères et al. 2011; Mary et al. 2014). Functional characterization is often limited to measuring the specificity and affinity of ligand binding. Such is the case for the leukotriene BLT1, serotonin 5-HT$_{4(a)}$, and cannabinoid CB1 receptors, all of them expressed as inclusion bodies and folded by transfer from SDS to A8-35 (Dahmane et al. 2009; Banères et al. 2011; Mary et al. 2014). In the case of the leukotriene BLT2 receptor, expressed and folded in the same way, it was observed that G protein activation and arrestin recruitment were inhibited or slowed down by A8-35, but not by NAPols. Conformational equilibria of the A8-35-trapped BLT2 receptor were characterized by ligand binding and by NMR, and the receptor-bound conformation of the ligand, leukotriene B$_4$, established by NMR (Dahmane et al. 2009; Catoire et al. 2010a, 2011; Banères et al. 2011; Mary et al. 2014). The ghrelin GHS-R1a receptor was folded in NAPols and ligand binding, conformational transitions, G protein activation, and arrestin recruitment characterized (Bazzacco et al. 2012). Interestingly, the ghrelin receptor folded in APol presents the

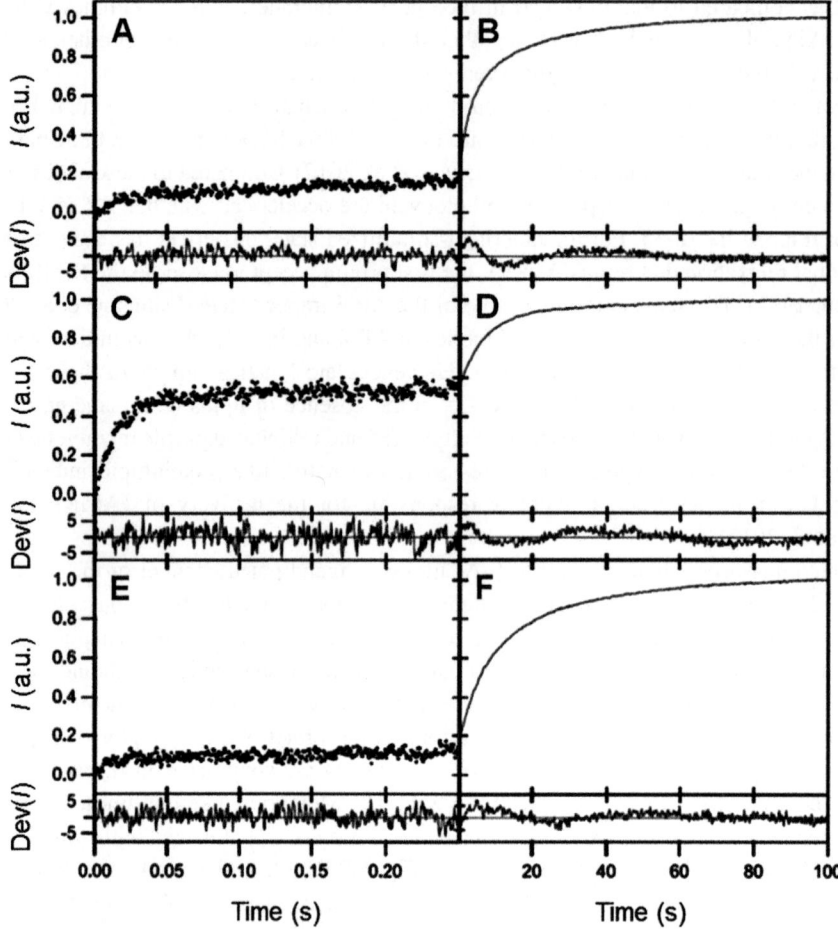

**Fig. 5.25** Allosteric transitions of the nicotinic acetylcholine receptor in three different environments. Kinetics of binding of a fluorescent ligand to nAChR in (**A**, **B**) native membrane fragments from *Torpedo marmorata* electric organ; (**C**, **D**) after solubilization in detergent solution (CHAPS); (**E**, **F**) after addition of A8-35 and dilution below the CMC of CHAPS. In its native membrane environment, the nAChR pre-exists to the addition of ligands in an equilibrium between a low-affinity resting state and high-affinity, inactive state(s), in a proportion of about 9:1. Upon addition of a low concentration of fluorescent agonist, only the high-affinity state(s) binds the ligand (panel **A**), relaxation of the resting state conformation to high-affinity ones occurring more slowly (panel **B**). After solubilization, the ratio between high- and low-affinity states in the absence of ligand becomes about 1:1, explaining the higher level of fast binding seen in Panel **C**. When most of CHAPS is replaced by A8-35 in the environment of the receptor, the allosteric equilibrium comes back to a situation similar to that in the membrane (**E**, **F**) (From Martinez et al. 2002, © 2015 Federation of European Biochemical Societies).

same high level of constitutive activity as that measured after heterologous expression in HEK cells (Bazzacco et al. 2012). The GHS-R1a and melatonin MT1R receptors were extracted by SMA either from asolectin vesicles, into which GHS-R1a had been integrated following expression in inclusion bodies and in vitro folding in A8-35, or from native membranes from *Pichia pastoris*, into which MT1R had been inserted in vivo. They were characterized by their ligand binding, conformational transitions, G protein activation, and arrestin recruitment (Logez et al. 2016). The vasopressin V2R receptor was expressed in insect cells under its native form, solubilized in a detergent mixture, transferred to NAPols, and characterized by its ligand binding, conformational transitions, G protein activation, and arrestin recruitment (Rahmeh et al. 2012). In short, it seems that GPCRs can be trapped by any APol under a functional form and bind their ligands with the same specificity and affinity as measured in a membrane environment. However, A8-35 is observed to inhibit or slow down G protein activation and arrestin recruitment, whereas NAPols and SMA do not. The difference between A8-35 and NAPols may probably be attributed to the polyanionic character of A8-35 vs. the non-ionic one of NAPols. It is however surprising, if electrostatic interactions are involved, that the equally polyanionic SMA be not inhibitory as well (Logez et al. 2016). A possible interpretation is that more abundant lipids in SMALPs keep the polymer further away from the receptor.

As a rule, no interference is observed with the binding of small water-soluble ligands to APol-trapped MPs, such as that of $Ca^{2+}$ and ATP to SERCA1a (Champeil et al. 2000), that of small acetylcholine analogs to the nAChR (Martinez et al. 2002; Charvolin et al. 2009), or that of various ligands to GPCRs (Dahmane et al. 2009; Catoire et al. 2010a, 2011; Banères et al. 2011; Bazzacco et al. 2012; Rahmeh et al. 2012). Two GPCRs expressed in vitro in the presence of NVoy, CRFR1, and CRFR2$\beta$ have been shown to bind their ligands (Klammt et al. 2011), as well as three SMA-extracted GPCRs, the adenosine $A_{2A}$ receptor ($A_{2A}R$), ghrelin GHS-R1a receptor, and melatonin MT1R receptor (Jamshad et al. 2015a, b; Logez et al. 2016). It is to be noted that A8-35 does not interfere with the binding of leukotriene LTB$_4$ to the BLT1 or BLT2 receptors (Dahmane et al. 2009; Catoire et al. 2010a), even though, given the hydrophobicity of LTB$_4$, its binding site must be itself quite hydrophobic and could in principle attract APol octyl chains. APols do not block either the binding of large water-soluble partners, such as that of $\alpha$-bungarotoxin ($\alpha$-Bgt) (8 kDa) to the nAChR (Charvolin et al. 2009), that of bacteriophage T5 protein pb5 (68 kDa) to FhuA (Basit et al. 2012), nor the recognition of several MP targets by synthetic proteins called $\alpha$Reps (15–20 kDa) (Ferrandez et al. 2014; see Chap. 13, § 13.2.3) or by antibodies (~150 kDa) (Charvolin et al. 2009; Tifrea et al. 2011; Le Bon et al. 2014), nor, as mentioned above, the interaction of GPCRs with G proteins and arrestin (Bazzacco et al. 2012; Rahmeh et al. 2012). In the latter case, however, it has been observed, as mentioned above, that interactions are less efficient with A8-35-trapped than with NAPol-trapped GPCRs. Also, specific binding of cationic ligands can be difficult to measure in the presence of A8-35, because of a high background of non-specific binding (Ferrandez et al. 2014; Mary et al. 2014, and unpublished data by various groups). Experiments in which retinal was added to bacterio-opsin (BO) refolded in the presence of A8-35 indicate that this very hydrophobic ligand can be delivered a posteriori to the apoprotein (Dahmane et al. 2013). Retinal presumably moves from free APol particles, where it must partition, to the protein-bound APol belt during collisions between the particles and BO/APol complexes and then inserts itself into the $\alpha$-helix bundle. The covalent Schiff base then forms spontaneously (Huang et al. 1981).

Many MPs exhibit enzymatic activity after trapping with APols (Table 5.1). Thus, *E. coli* diacylglycerol kinase (DAGK) retains full enzymatic activity upon transfer from decylmaltoside to PMAL-B-100 (Gorzelle et al. 2002). The bacterial outer membrane enzyme PagP retains phospholipase activity after being trapped in SMALPs (Knowles et al. 2009), as well as in A8-35, A8-75, A34-35, A34-75, SAPols, and NAPols (Leney et al. 2012; Calabrese et al. 2015; Watkinson et al. 2015). The transmembrane domain of the bacterial EII$^{mtl}$ mannitol permease performs the transphosphorylation from phosphoenolpyruvate to mannitol more rapidly after trapping in A8-35 than it does in detergent solution (Opačić et al. 2014b). Similarly, the basal ATPase activity of ABCA4, a photoreceptor-specific ABC transporter, is higher after trapping with A8-35 than in detergent solution (Tsybovsky et al. 2013). Cytochrome $bc_1$ transfers electrons from ubiquinol to oxidized cytochrome $c$ at comparable rates whether solubilized in DDM or trapped by A8-35 (Charvolin et al. 2014). Other enzymatic reactions that have been evidenced include the protease activity of OmpT in A8-35, A8-75, A34-35, A34-75, SAPols, and NAPols (Leney et al. 2012; Calabrese et al. 2015; Watkinson et al. 2015), and the autophosphorylation of SMA-trapped *E. coli* tyrosine kinase (Li et al. 2015a). Nucleotide binding by several SMA-trapped eukaryotic ABC transporters could be demonstrated, but not their ATPase activity, because SMALPs are destabilized by divalent metal ions (Gulati et al. 2014).

Redox reactions have been evidenced with several photosystems, respiratory complexes, and other redox enzymes trapped in A8-35 (Kievit and Brudvig 2001; Nowaczyk et al. 2004; Althoff et al. 2011; Laursen et al. 2013; Charvolin et al. 2014; Shinzawa-Itoh et al. 2016) or SMALPs (Laursen et al. 2016). The oxidation of reduced cytochrome $c$ by SMA-trapped cytochrome $c$ oxidase can be observed only after depleting the system of free SMA, which interferes with the binding of cytochrome $c$ to the oxidase (Smirnova et al. 2016).

A particularly interesting case is that of SERCA1a, the $Ca^{2+}$-ATPase, which has been studied in some details. SERCA1a, which is located in the sarcoplasmic reticulum of fast-twitch muscle, extracts calcium ions from the cytosol. This relaxes actomyosin and, thereby, brings muscle contraction to an end (see Chap. 1, § 1.6.3). Several APols (A8-35, PMAL-C12, PMAL$A$-C12, and SAPols) have been found to both protect SERCA1a against inactivation (see § 5.6) and to slow down its turnover (Champeil et al. 2000; Picard et al. 2006). Intermediate effects are observed in mixtures of APols and detergent. Trivial mechanisms, such as sequestration of $Ca^{2+}$ or interference with ATP binding, do not account for the inhibition (Champeil et al. 2000; Picard et al. 2006). As will be discussed in § 5.6, both the inhibition and stabilization of SERCA1a, and their intriguing correlation, can probably be traced to a common cause, namely the influence of APols on protein dynamics.

Inhibition by A8-35 of the ATPase activity of the $F_1F_O$ ATP synthase, reversible upon addition of an excess of detergent, has also been reported, but not studied in any details (Wilkens et al. 2000).

## 5.5 Biochemical Stability of Amphipol-Trapped Membrane Proteins

The fact that transferring a MP from detergent solution to APols stabilizes it is a very general even though not absolutely universal observation (see e.g. Tribet et al. 1996; Champeil et al. 2000; Picard et al. 2006; Gohon et al. 2008; Dahmane et al. 2009; Tifrea et al. 2011; Bazzacco et al. 2012; Cocco et al. 2013; Dahmane et al. 2013; Etzkorn et al. 2013; Pocanschi et al. 2013; Huynh et al. 2014; reviewed in Kleinschmidt and Popot 2014; Zoonens and Popot 2014; Le Bon et al. 2018). It is illustrated in Fig. 5.27 in the case of BR. Note that, as discussed in Chap. 2, § 2.4, one of the mechanisms by which surfactants destabilize MPs is by diluting subunits or bound lipids in the "phase" represented by the associated surfactant. In order to compare the intrinsic inactivating or stabilizing properties of surfactants, one must therefore, as in the case of Fig. 5.26, pay attention to the volume of non-monomeric, associated surfactant the protein is exposed to, so as to distinguish effects that merely result from differences in the volume of the "hydrophobic sink" (see Chap. 2, § 2.4) from those that reflect molecular interactions between the protein and the surfactant molecules that form the layer surrounding its TM region.

The effects of trapping a fixed quantity of BR with increasing amounts of APols have been examined in some details (Dahmane 2007). Native BR was extracted from PM with OTG along with PM lipids (Gohon et al. 2008). Upon trapping it with A8-35 at BR/APol mass ratios ranging from 1:5 to 1:50, ternary BR/lipid/APol complexes formed. Control samples were stored in either 18 mM or 25 mM OTG. Under the experimental conditions used, the volume of the hydrophobic sink (non-monomeric surfactant) was roughly comparable in 18 or 25 mM OTG and at BR/A8-35 ratios of 1:10 or 1:20, respectively. At room temperature, BR is stable for at least a week whatever the concentration of APol, whereas in OTG it denatures almost totally over the same period (Fig. 5.26, *left*). Even at 40 °C, BR trapped at low (1:5–1:10 w/w) APol ratios is highly stabilized by A8-35, denaturing by <10% over a week (Fig. 5.26, *right*). At this temperature, the protein in OTG is totally inactivated in less than a day. Whereas a large excess of APols (ratio 1:50) is well tolerated at 4 °C and

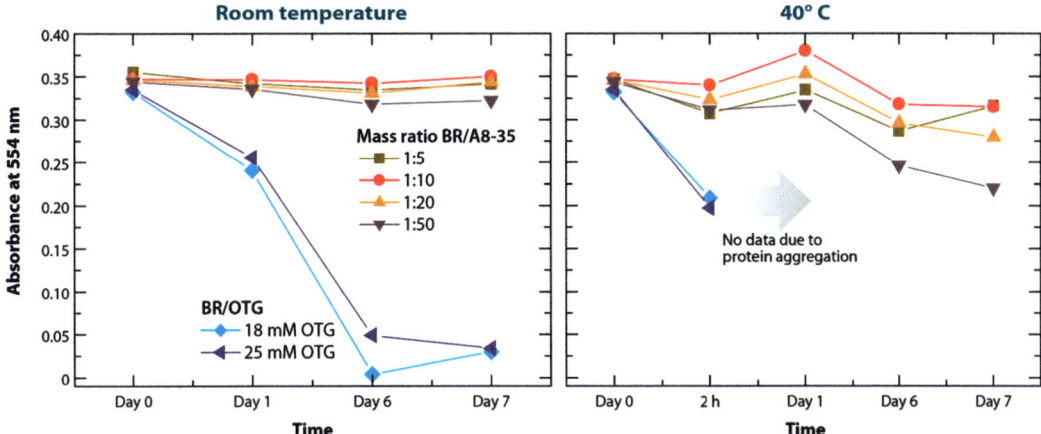

**Fig. 5.26** Time stability of bacteriorhodopsin (BR) in A8-35 vs. octylthioglucoside (OTG). BR was extracted with OTG from *Halobacterium salinarum* purple membrane (PM), along with PM lipids, trapped in A8-35 (Gohon et al. 2008) at various BR/A8-35 mass ratios, and stored in the dark either at room temperature or at 40 °C in a buffer comprised of 100 mM NaCl and 20 mM sodium phosphate, pH 7.0 ([BR] = 0.22 g·L$^{-1}$). Its absorbance at 554 nm, which is proportional to the concentration of the holoprotein, was followed as a function of time. Control samples were kept in 18- or 25-mM OTG solutions (total OTG concentration, including bound detergent). These two concentrations correspond to roughly the same mass concentration of non-monomeric surfactant as that of A8-35 in the samples trapped respectively at 1:10 and 1:20 BR/A8-35 mass ratios. The absence of data points in OTG at 40 °C past 2 h is due to the aggregation of the protein, accompanied by complete bleaching (From Popot et al. 2011, adapted from Dahmane 2007).

at room temperature, at 40 °C it does affect the stability of BR, ~1/3 of which is denatured after 6 days (Fig. 5.26, *right*). This effect is most likely a consequence of delipidation, which is favored by increasing the volume of the hydrophobic sink. Inactivation is nevertheless very slow compared to that in OTG.

The increased thermostability of APol-trapped BR is further illustrated in Fig. 5.27: A8-35-trapped BR stands well being exposed at 60 °C for 20 min (Fig. 5.27B), whereas BR in OTG denatures at 40 °C (Fig. 5.27A; Dahmane et al. 2013). Figure 5.27C, D illustrates that BR that has been denatured in SDS and refolded in A8-35 in the presence of the lipids present in the purple membrane is as stable as native BR trapped in the presence of lipids and slightly more stable than BR refolded in A8-35 in the absence of lipids (Dahmane et al. 2013). This increased stability translates into a very long shelf life: BR/A8-35/lipid preparations at a mass ratio of 1:5 (i.e. with ~3 g free APol per g BR) do not exhibit any denaturation even after 6 months of storage at 4 °C (Gohon et al. 2008).

Of practical interest is the fact that, at variance with detergent-solubilized BR, BR/A8-35 complexes can be frozen and thawed without denaturing the protein (Gohon et al. 2008) and even lyophilized (C. Le Bon and M. Zoonens, unpublished data). As will be discussed in Chap. 15, § 15.2, resistance to freezing and/or lyophilization is of great importance for field use of MP-based vaccines. It is also possible to prepare highly concentrated samples, in which BR is native, by precipitating the complexes via APol/APol interactions, e.g. by acidifying (Gohon 2002) or supplementing with Ca$^{2+}$ ions (M. Zoonens, unpublished data) preparations of MP/A8-35 complexes, or by bridging with tetravalent avidin complexes of BR with biotinylated A8-35 (Bazzacco 2009). Acid precipitation has been used to concentrate A8-35-based proteome extracts (Ning et al. 2014; see Chap. 14, § 14.4). BR/ZnO hybrid films have been engineered as a potential sensing element for low-temperature detection of ethanol vapor by depositing BR/A8-35 complexes onto a film of ZnO particles and drying the mixed film under vacuum at room temperature. Preservation of the native state of BR was assessed by Raman and FTIR measurements (Kumar et al. 2016).

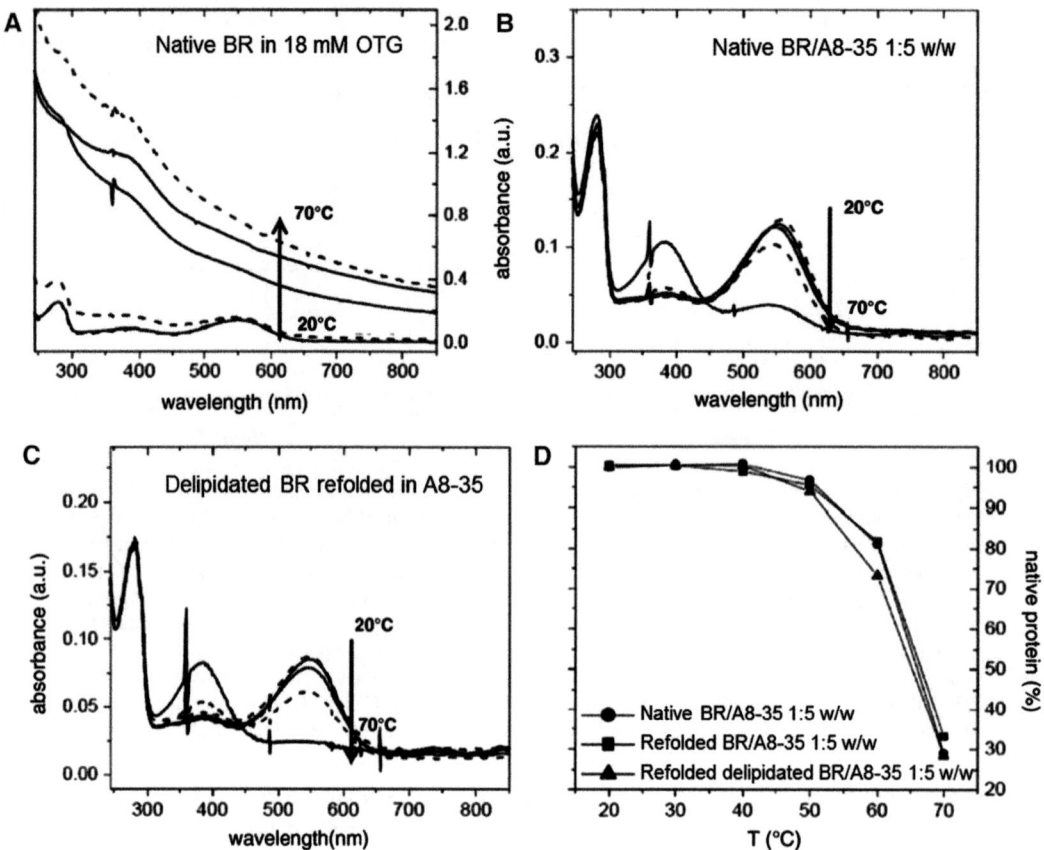

**Fig. 5.27** Thermal denaturation of bacteriorhodopsin (BR) in a detergent vs. an amphipol environment. (**A**) Native BR solubilized from purple membrane (PM) and stored in 18 mM octylthioglucoside (OTG). (**B**) The same preparation after being transferred from OTG to A8-35 at a 1:5 BR/A8-35 mass ratio. (**C**) A preparation obtained by solubilizing PM in organic solvent, separating the denatured apoprotein from retinal and lipids, transferring it to SDS, and refolding it in A8-35 in the presence of retinal and absence of lipids. In all three cases, UV-visible spectra were recorded after incubation for 20 min at the temperature indicated. They are represented by, alternately, solid and dashed lines at 10 °C intervals from 20 °C (first *dashed line*) to 70 °C (last *solid line*). The increase in turbidity in A is due to aggregation of the denatured protein, the peak at ~382 nm in all panels to the release of free retinal. (**D**) Comparison of the thermal denaturation of native, A8-35-trapped BR (●) vs. BR renatured in A8-35 in the presence (■) or absence (▲) of lipids. Data are expressed as a percentage of native BR at 20 °C (From Dahmane et al. 2013).

The higher stability of APol-trapped vs. detergent-solubilized BR is quite typical, similar effects having been reported with many other MPs. GPCRs, for instance, are also significantly stabilized once complexed by A8-35 as compared to their detergent-solubilized forms. As shown in Fig. 5.28, A8-35-trapped BLT1, one of the receptors of leukotriene LTB$_4$, loses no activity after 20 days at 4 °C, whether lipids are present or not, whereas, under the same conditions of buffer and temperature, BLT1 kept in fos-choline-16/asolectin solution loses about half of its activity over the same period (Fig. 5.28, *right*). In detergent/lipid mixed micelles, BLT1 denatures at ~27 °C, while it is stable up to ~35 °C when trapped in pure A8-35 and up to ~39 °C in the presence of A8-35 + lipids (Fig. 5.28, *left*; Dahmane et al. 2009).

Figure 5.29 illustrates that the insulin-responsive facilitative glucose transporter GLUT4, which is much more stable in the very mild detergent 2,2-didecylpropane-1,3-bis-$\beta$-D-maltopyranoside (LMNG) than in DDM (Fig. 5.29A, B), is further stabilized by transfer to A8-35 (Fig. 5.29C;

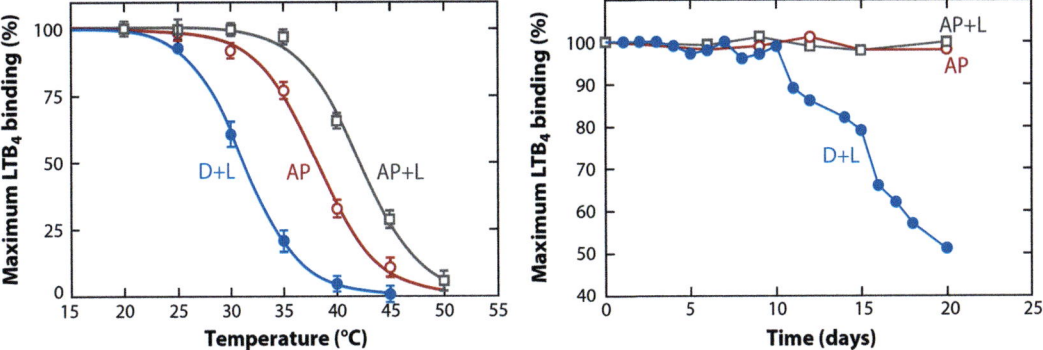

**Fig. 5.28** Stabilization of a G protein-coupled receptor by A8-35 in the presence or absence of lipids. The BLT1 receptor of leukotriene LTB$_4$ was either trapped by A8-35 or by an A8-35/asolectin mixture (5:1 w/w) or kept in a fos-choline-16/asolectin solution (2:1 w/w). *Left.* Stability upon incubation at increasing temperature. *Right.* Stability upon extended storage at 4 °C. D + L, fos-choline-16/asolectin mixed micelles; AP, A8-35; AP+L, A8-35 + asolectin (From Popot 2010, adapted with permission from Dahmane et al. 2009, © 2009 American Chemical Society).

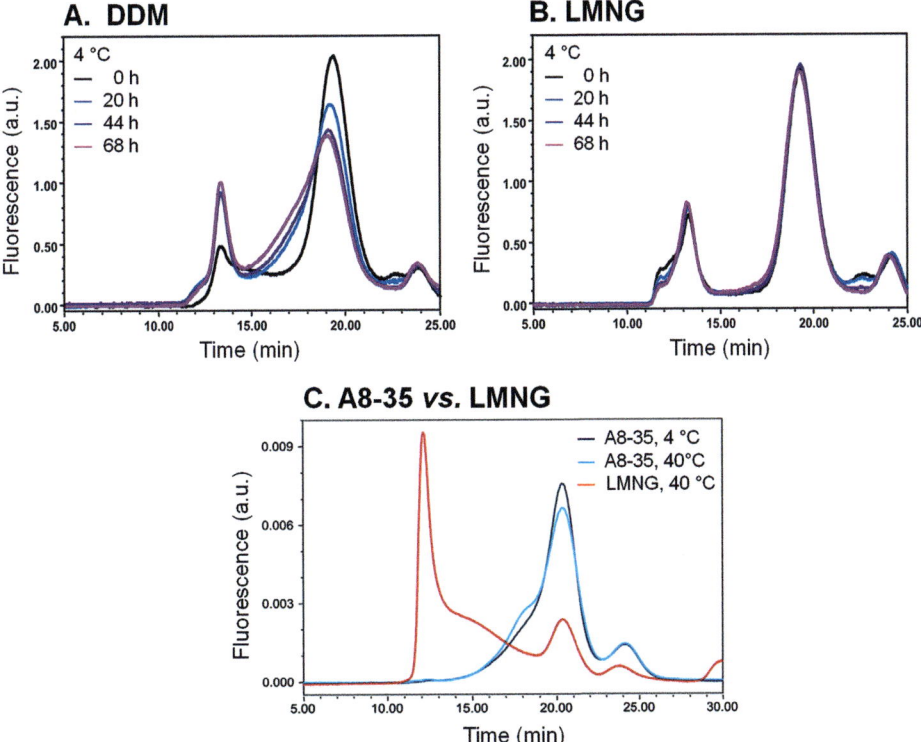

**Fig. 5.29** Stability of the insulin-responsive facilitative glucose transporter GLUT4 solubilized in dodecylmaltoside (DDM) or 2,2-didecylpropane-1,3-bis-$\beta$-D-maltopyranoside (LMNG) or trapped with A8-35, as determined by size exclusion chromatography (SEC). (**A, B**) Green fluorescent protein-tagged GLUT4 was solubilized in either DDM (**A**) or LMNG (**B**) and incubated at 4 °C for 68 h. Aliquots were analyzed by fluorescence SEC after 0, 20, 44, and 68 h. **C.** Stabilizing effect of A8-35 as compared to LMNG. A8-35-trapped GLUT4 incubated at 40 °C for 2 h and analyzed by fluorescence SEC before (*dark blue*) and after (*light blue*) incubation, as compared with LMNG-solubilized GLUT4 after incubation at 40 °C for 2 h (*red*) (From Kraft et al. 2015, © 2015 The Protein Society).

Kraft et al. 2015). Stabilization upon transfer from detergent to APols has been observed with several other MPs, including the sarcoplasmic $Ca^{2+}$ pump SERCA1a (Champeil et al. 2000; Picard et al. 2006; see § 5.6), the respiratory complex cytochrome $bc_1$ (Charvolin et al. 2014), and the prokaryotic voltage-gated sodium channel NavMs (Ireland et al. 2017). Less direct observations also suggest stabilization of yet other $\alpha$-helical MPs by APols as compared to detergents. A case in point is the ion channel TRPV1, which has been studied by electron microscopy after trapping in A8-35. Galleries of images of negatively stained particles indicate that their overall shape is much more reproducible in A8-35 than it is in DDM, suggesting stabilization (Cao et al. 2013; Liao et al. 2013; see Chap. 12, § 12.3.4.2, Fig. 12.26).

APol-induced stabilization has also been noted for several $\beta$-barrel MPs, such as the major outer membrane protein (MOMP) from *Chlamydia trachomatis*. At pH 7.4, the midpoint of the transition of heat-induced unfolding of the MOMP trimer in Zwittergent 3–14 is ~52 °C, whereas A8-35-trapped MOMP does not unfold at all up to 78 °C, indicating an exceptionally strong stabilization (Tifrea et al. 2011), which translates into a remarkable stability upon extended storage (Tifrea et al. 2011; Cocco et al. 2013; see Chap. 15, Figs. 15.2 and 15.3).

Note that, whereas APols are generally used to stabilize MPs, extended exposure (from 40 min to 24 h and more at 4 °C) to high concentrations (3.5%) of A8-35 has been exploited to gently and progressively disassemble the dimeric $F_1F_O$ ATP synthase from *Polytomella* sp. (Vázquez-Acevedo et al. 2016).

A detailed thermodynamic study of the stabilization of *E. coli* OmpA by A8-35 carried out by J.H. Kleinschmidt and colleagues has yielded interesting insights into stabilization mechanisms (Pocanschi et al. 2013). Equilibrium folding/unfolding by urea was compared following trapping with A8-35 vs. in LDAO solution, allowing the calculation and comparison of the free energies of unfolding. In line with an earlier study of the stability of OmpA in sonicated lipid vesicles (Hong and Tamm 2004), reversible folding/unfolding conditions were only achieved at pH 10. Unfolding and folding titration curves superimposed only after very long incubation times: 30–40 days for OmpA/A8-35 complexes and 18–25 days for OmpA/LDAO ones. It is the unfolding reaction that imposes these long equilibration times. In urea at pH 10, OmpA is thermodynamically more stable in LDAO than in A8-35. However, the activation free energy for unfolding OmpA is much higher in A8-35 than in LDAO, as indicated by the slower kinetics. At pH 10, OmpA (pI = 5.5) is strongly negatively charged, because of the deprotonation of some of its 17 lysine (pK $\approx$ 9.5–10.5) and 17 tyrosine (pK $\approx$ 9.5–10) residues. The resulting increased negative charge leads to intermolecular repulsion and stronger side-chain hydration, which prevents aggregation of the denatured proteins and ensures a better solubility and reversibility. However, the increased net negative charge of OmpA may also lead to less stable complexes with the negatively charged A8-35, because of charge-charge repulsion. This might be the reason for the reduced thermodynamic stability of folded OmpA observed in A8-35 as compared to LDAO. However, the activation energy of unfolding of OmpA being much higher in A8-35 than in LDAO, it takes longer for OmpA to unfold, despite its lower thermodynamic stability (Pocanschi et al. 2013). This is consistent with the view that part of the stabilizing properties of APols results from their damping the conformational excursions of MPs, which slows down denaturation (see § 5.6).

Trapping with SMA has been observed to strongly stabilize PagP (Knowles et al. 2009), bacterial photosynthetic reaction centers (Swainsbury et al. 2014; Fig. 5.30A), or, more modestly, the adenosine $A_{2A}$ receptor ($A_{2A}R$) (Jamshad et al. 2015a, b; Fig. 5.30B). On the basis of NMR data, the GPCR CRFR2$\beta$ has been reported to be stable after transfer to NVoy (Klammt et al. 2011), but no comparative studies of the stability of MPs complexed by this polymer vs. detergent-solubilized ones have been reported yet.

Comparative studies about the relative ability of various APols to stabilize MPs remain scarce, but they seem to point in at least one direction: the less charges APols bear, the more stabilizing they

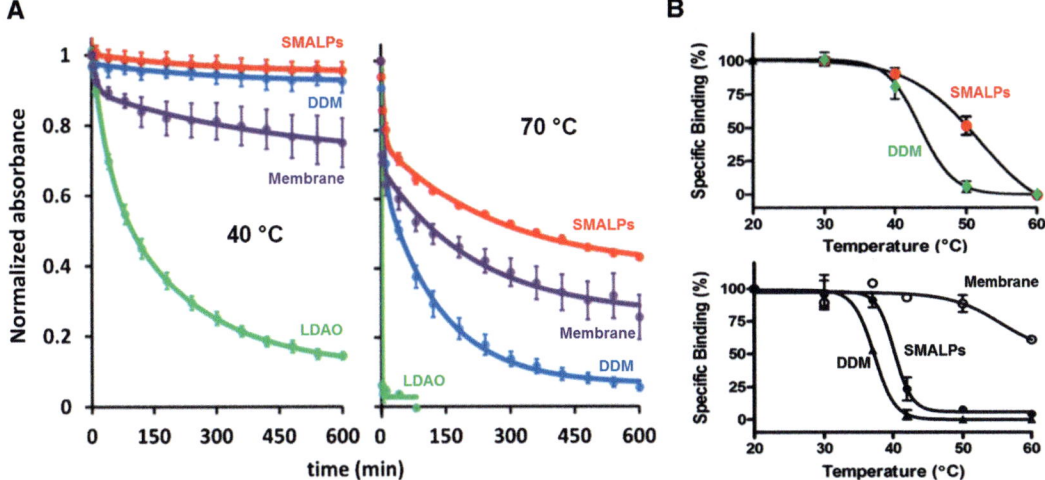

**Fig. 5.30** Stabilization of membrane proteins by trapping with SMA. (**A**) Thermal stability of the photosynthetic reaction center (RC) from *Rhodobacter sphaeroides* at 40 °C (*left*) or 70 °C (*right*). RCs were either kept in membrane fragments (*purple*), in lauryldimethylaminoxide (*LDAO*; *green*) or in dodecylmaltoside (*DDM*, *blue*) solution, or trapped in SMALPs (*red*). *Circles* show average data from three series of measurements, with standard errors. *Solid lines* show fits to a double exponential decay (From Swainsbury et al. 2014). (**B**) Thermostability of the adenosine $A_{2A}$ receptor ($A_{2A}R$) trapped in SMALPs or solubilized in DDM. The protein was expressed either in *Pichia pastoris* (*left*) or in HEK293T cells (*right*). *Left:* SMA-trapped (●) and DDM-solubilized (♦) $A_{2A}R$. *Right:* $A_{2A}R$ trapped in SMALPs (●), solubilized in DDM (▲), or kept in membrane fragments (○). Data are expressed as specific binding relative to the 20 °C data point (mean ± S.E.M. of three separate experiments, each performed in triplicate) (From Jamshad et al. (2015a, b), © The Biochemical Journal for Biosciences Report).

are. Thus, BR is more stable when trapped in A8-35 than in A8-75, a very similar poly(acrylic acid)-derived APol that carries ~75% free carboxylate groups rather than ~35% (Tribet et al. 1996, and C. Tribet, unpublished data). Similarly, SERCA1a is more stable in A8-35 than in SAPols, which also carry ~75% of charged groups (Picard et al. 2006; see Fig. 5.34). Cytochrome $b_6f$, a highly detergent-sensitive complex (Breyton et al. 1997), is not very stable in A8-35 or A8-75, particularly in the absence of lipids (Tribet et al. 1996, 1998; Bazzacco et al. 2012), but much more so in NAPols, whether glycosylated or not (Prata et al. 2001; Bazzacco et al. 2012). The stability of BR in SAPols is much greater in the presence of 100 mM NaCl than in its absence (Dahmane et al. 2009). A simple rationale for these effects is that, as may be the case for detergents, the presence of a net charge on the surfactant bound to a MP tends to favor, due to electrostatic repulsion, the formation of particles with a small radius of curvature. This would drive the opening of the protein's structure or, in the case of a multi-subunit assembly like cytochrome $b_6f$, fragmentation. Note, however, that the tetrameric ion channel TRAP1 has been reported to be *more* stable in SAPols than in A8-35 (Huynh et al. 2014). It is to be expected that, depending on the mechanism of denaturation of individual MPs, on the APol they are transferred to and on experimental conditions, such as the ionic strength and the presence or absence of lipids, different stabilization mechanisms will come into play to different extents, and it may be more or less relevant to favor one type of APol over another.

On the basis of currently available data, we can identify at least two, probably three, mechanisms as likely to contribute to the stabilizing effect of APols. The first two are classical:

(i)   APols are intrinsically less disruptive, to most proteins, than most detergents, as shown by the fact that, at equal mass ratios of protein and surfactant, MPs are, usually, much more stable in the presence of APols than in that of detergents (see e.g. Fig. 5.26). In other terms, for an equal volume of hydrophobic sink, APols are less destabilizing than detergents. This may have several causes, which may contribute differently to the stabilization of different proteins. For instance, the fact that lipids rebind (in the case of BR) or are likely to rebind (in the case of the nAChR) upon transfer of a MP from detergent solution to APols suggests that APols do not compete as efficiently as detergents do to displace lipids from TM protein surfaces. Because lipids play an important role in stabilizing MP TM regions (see Chap. 2), this, as observed, for example for BR (Fig. 5.27C) or for the leukotriene BLT1 receptor (Fig. 5.27), will by itself have a stabilizing effect. By the same token, APols may be less able to compete with the protein/protein interactions that keep together TM segments and TM subunits. In keeping with this hypothesis, A8-35 has indeed been found to be less efficient than detergents at dissociating, in the absence of lipids, the dimer formed by the TM $\alpha$-helical anchor of glycophorin A (Stangl et al. 2014). Similarly, BR exposed to A8-35 in the absence of lipids is much more stable than when exposed to OTG in their presence (compare Fig. 5.27A, C; Dahmane et al. 2013).

(ii) Because APols do not spontaneously desorb from MPs unless they are displaced by another surfactant (cf. Fig. 5.7), it is possible, circumstances permitting, to work in the presence of very little or no free APols. Such is the case, for instance, when working with dilute protein solutions. When using submicromolar concentrations of protein, for instance, as frequently happens for functional or spectroscopic measurements, the total concentration of APol (bound + free) can be kept in the range of tens of µg per mL without incurring protein aggregation. This is lower than can be achieved with most detergents. When aggregation is rendered impossible, as is the case for a MP immobilized on a chip or a column, or if a limited degree of aggregation is tolerable, as can be the case for EM single-particle imaging or for functional measurements, free APol can be dispensed with entirely. As seen in Fig. 5.26B, lowering the APol concentration can, by diminishing the volume of the hydrophobic sink, further slow down protein inactivation.

These two mechanisms are classical in the sense that they apply to any surfactant and are at the basis of strategies for improving MP stability when working with detergents (Chap. 2). Several observations suggest that a third mechanism may be involved in the case of APols, namely their damping of the dynamics of MPs, which would affect both their stability and, in some cases, their function. The evidence in favor of this (hypothetical) mechanism will be discussed in the next section.

## 5.6 Membrane Protein Dynamics and the Effects of Amphipols on Stability and Function

The idea that damping by APols of MP dynamics may be involved in their effects on stability and function (Popot et al. 2003) emerged as the outcome of studies on the sarcoplasmic calcium pump (SERCA1a) (Champeil et al. 2000; Picard et al. 2006), compared to those on BR (Gohon et al. 2008) and the nAChR (Martinez et al. 2002). It has received some support from MD studies of OmpX/A8-35 complexes (Perlmutter et al. 2014). We will start by discussing the experimental observations and then confront to the MD data the hypothesis they led to.

### 5.6.1    Functional Observations

As noted in § 5.4, most MPs are functional after trapping with APols. However, at variance with these observations, the ATPase activity of the sarcoplasmic calcium pump SERCA1a was found to be inhibited by APols as compared to what is observed in permeabilized membrane fragments or in detergent solutions (Fig. 5.31), a phenomenon that does not result from any interference with ATP nor calcium binding (Champeil et al. 2000; Picard et al. 2006). With the couple A8-35/$C_{12}E_8$, the ATPase activity decreases about exponentially as the APol/detergent mass ratio in the belt of protein-associated surfactant increases, the activity in pure APol extrapolating to ~10% of that in pure detergent (Fig. 5.31C). This would correspond to a relatively modest increase of ~1.5 kcal·mol$^{-1}$ in the free

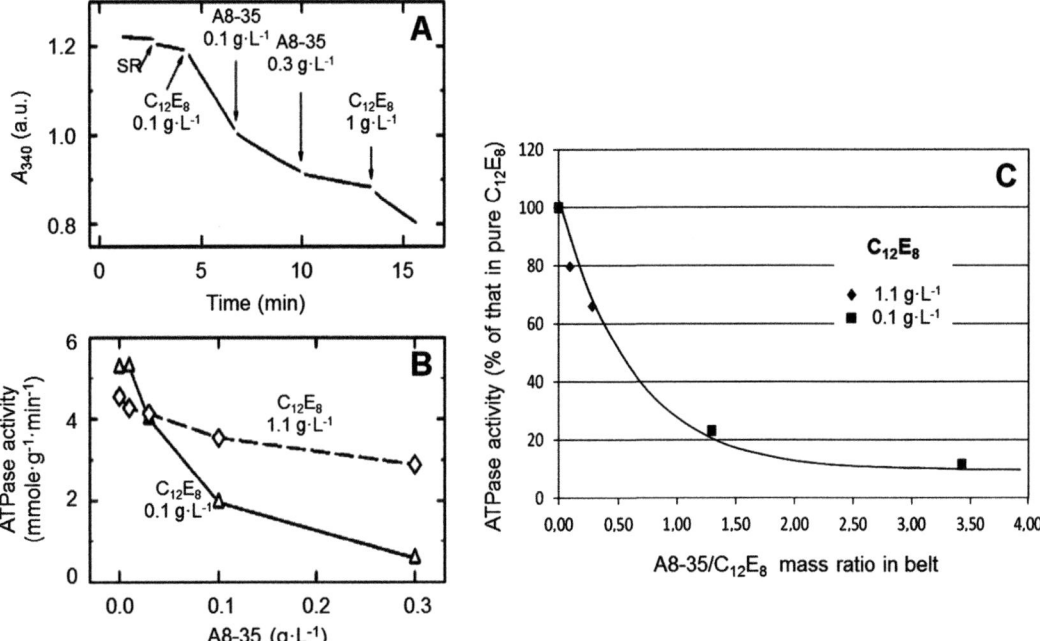

**Fig. 5.31** Reversible inhibition by A8-35 of the ATPase activity of the $C_{12}E_8$-solubilized sarcoplasmic reticulum (SR) calcium pump, SERCA1a. SR vesicles were diluted to 2 mg·L$^{-1}$ in the assay medium, and their hydrolytic activity monitored continuously with a coupled enzyme system. (**A**) A typical recording of NADH oxidation, showing the stimulatory effect of 0.1 g·L$^{-1}$ $C_{12}E_8$ (due to the loss of the back pressure exerted by Ca$^{2+}$ ions accumulating inside the vesicles), the immediate inhibition of ATPase activity upon addition of A8-35 in the presence of 0.1 g·L$^{-1}$ $C_{12}E_8$, and the rapid partial reversal of this inhibition upon addition of a higher concentration of $C_{12}E_8$. (**B**) ATPase activity measured after addition of either 0.1 g·L$^{-1}$ (Δ) or 1.1 g·L$^{-1}$ (◇) $C_{12}E_8$ (the latter concentration causing a slight inhibition due to delipidation), followed by increasing concentrations of A8-35 (This research was originally published in Champeil et al. 2000, © The American Society for Biochemistry and Molecular Biology). (**C**) The CMC of $C_{12}E_8$ being ~0.048 g·L$^{-1}$ (le Maire et al. 2000) and the lipid concentration negligible, ca. half of the detergent is micellar at 0.1 g·L$^{-1}$ $C_{12}E_8$ and ~95% of it at 1.1 g·L$^{-1}$. These values will increase in the presence of APol, as the micellar detergent gets diluted by the polymer and the concentration of monomeric detergent drops. Taking this effect into account, as described in Box 5.3, one can estimate that the APol/detergent mass ratio $r$ in the surfactant belt associated to the protein varies approximately as follows: for [$C_{12}E_8$] = [A8-35] = 0.1 g·L$^{-1}$, $r \approx 1.3{:}1$; for [$C_{12}E_8$] = 0.1 g·L$^{-1}$ and [A8-35] = 0.3 g·L$^{-1}$, $r \approx 3.4{:}1$; for [$C_{12}E_8$] = 1.1 g·L$^{-1}$ and [A8-35] = 0.1 g·L$^{-1}$, $r \approx 0.09{:}1$; for [$C_{12}E_8$] = 1.1 g·L$^{-1}$ and [A8-35] = 0.3 g·L$^{-1}$, $r \approx 0.28{:}1$. Using these values, the activity is seen to decrease exponentially as the APol/detergent mass ratio in the surfactant belt increases and to reach, in nearly pure APol, a plateau close to one-tenth of the activity in pure detergent.

energy of activation of the limiting step in the enzymatic cycle. The latter, which, under most circumstances, corresponds to the transition between the E1P and E2P states (Champeil et al. 1986) (step ③ in Fig. 1.32, Chap. 1), is accompanied by large transmembrane (and extramembrane) rearrangements (reviewed in Møller et al. 2010; for a recent description, see Norimatsu et al. 2017).

---

### Box 5.3   Estimating the Composition of Mixed Micelles and Protein-Bound Surfactant Belts in Amphipol/Detergent Mixtures

In order to understand the behavior of MPs exposed to amphipol/detergent mixtures, as is the case of several of the experiments described in this section, it is useful to estimate the composition of the surfactant belt they are exposed to. This is not simply given by the global mass ratio of APol to detergent in the sample, because some of the detergent will always be present as monomers. Some of the APol is as well, but the CAC is so low (~0.002 g·L-1 for A8-35; Giusti et al. 2012; see Chap. 4, § 4.3.1.1) that it can usually be neglected. Experiments carried out in the presence of APol/detergent mixtures, e.g. A8-35/DDM ones, show that (i) the two surfactants mix close to ideally and (ii) the composition of the protein-bound surfactant belt is related to that of protein-free mixed micelles by a partition coefficient $P$ with a relatively low value (Zoonens et al. 2007; Tribet et al. 2009). This is illustrated in Fig. 5.32, where the loss of FRET and the dequenching of the fluorescence of tOmpA's tryptophans upon displacing by DDM a fluorescent version of A8-35, FAPol$_{NBD}$, have been modeled according to these two hypotheses. The best fit was obtained by assuming that the mole fraction of APols in the surfactant belt is ~2.8× that in protein-free mixed particles.

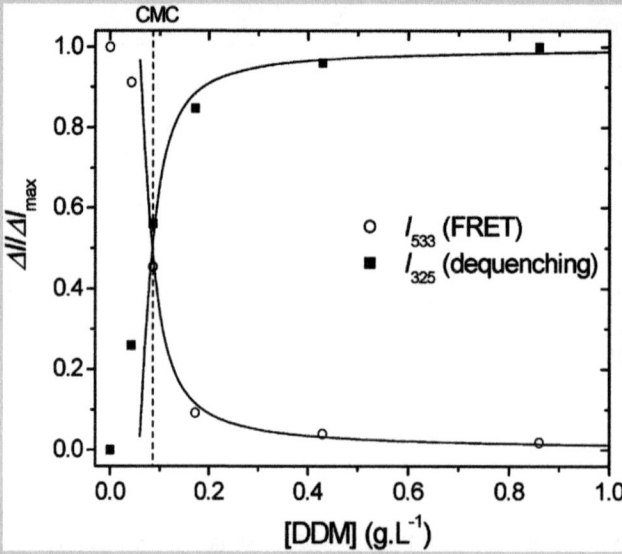

**Fig. 5.32** Steady-state fluorescence observed after addition of increasing concentrations of DDM to a tOmpA/FAPol$_{NBD}$ sample. Emission spectra were recorded a few minutes after adding DDM, that is, at equilibrium. The rise of dequenching at 325 nm and the drop of FRET at 533 nm are plotted as percentages of the extent of quenching and FRET observed in pure FAPol$_{NBD}$, respectively. Curves are fits calculated assuming a partition coefficient for the APol of 2.83 between the surfactant belt and protein-free mixed particles (Reprinted with permission from Zoonens et al. 2007, © 2007 American Chemical Society).

---

Fitting the equilibrium values for the FAPol/$C_{12}E_8$ exchange (experimental data are shown in Fig. 5.40) by use of the same equations required a value of $P \approx 1.15$ (Zoonens et al. 2007). In the following discussion, which deals with A8-35/$C_{12}E_8$ mixtures, we will neglect this very marginal

**Box 5.3  (continued)**

effect and consider that the associated surfactant mixture has the same composition whether adsorbed to the protein or free as protein-free particles. Finally, a last assumption is that the concentration of detergent monomers is given by the CMC multiplied by the volume fraction of the detergent in the belts and particles. This approximation rests on the double assumption that (i) the rate of collision of the monomers with the surface of the particles or belts, and the probability that the collision will lead to the monomer being incorporated, is about the same whatever the composition of the belt or particle, and (ii) that the residence time of a monomer in a particle or belt does not depend on the latter's composition. The first hypothesis is probably reasonable for an uncharged detergent like $C_{12}E_8$. The second one seems reasonable as well, given that it is the free energy cost of extracting the hydrophobic chain of the detergent and exposing it to water that determines the residence time (see e.g. Frindi et al. 1992a, b) and that this value is not going to change significantly with the composition of the mixture. It could change, however, if the mixture is not ideal and detergent polar heads, for instance, interact more or less favorably with the APol than they do between themselves.

The concentration of $C_{12}E_8$ monomers in an A8-35/$C_{12}E_8$ aqueous mixture, $[Det]_{mono}$, is equal to $[Det]_{tot} - [Det]_{mic}$, where $[Det]_{tot}$ is the total concentration of detergent in the sample and $[Det]_{mic}$ that of the aggregated detergent present in the mixed micelles and protein-bound belts. Given the assumptions discussed above, $[Det]_{mic}$ is given by the following quadratic equation:

$$a \cdot [Det]_{mic}^2 + b \cdot [Det]_{mic} + c = 0$$

with:

$a = \bar{v}_{det},$
$b = CMC \cdot \bar{v}_{det} - [Det]_{tot} \cdot \bar{v}_{det} + [A8\text{-}35]_{tot} \cdot \bar{v}_{A8\text{-}35},$
$c = - [Det]_{tot} \cdot [A8\text{-}35]_{tot} \cdot \bar{v}_{A8\text{-}35}.$

$\bar{v}_{A8\text{-}35} = 0.809 \ L \cdot g^{-1}$ (Gohon et al. 2004) and $\bar{v}_{det} = 0.973 \ L \cdot g^{-1}$ (le Maire et al. 2000) are respectively the specific volumes of A8-35 and $C_{12}E_8$ and $[A8\text{-}35]_{tot}$ the total concentration of APol in the sample. The composition of the protein-bound belt and mixed particles under the experimental conditions of Fig. 5.31 as calculated according to this equation are given in the legend to the figure and have been used to draw its Panel C.

The diminished activity of SERCA1a when its environment is enriched in APols is accompanied by biochemical stabilization. As shown in Fig. 5.33, the ATPase inactivates within minutes if it is deprived of $Ca^{2+}$ in the presence of detergent (*blue curve*). Upon transfer into an environment comprising, in mass, about 3 g of A8-35 per g of $C_{12}E_8$, the rate of inactivation is considerably slowed, with the time for half-inactivation reaching ~1 h (*green curve*). An intermediate situation is observed when the surfactant belt contains about equal masses of APol and detergent (*purple curve*) (Champeil et al. 2000). As shown in Fig. 5.31C, the first condition corresponds, in the presence of ATP and $Ca^{2+}$, to full activity, the second one to an activity reduced to ~10% of that in pure detergent, and the third one to ~20% activity.

An apparent anticorrelation between activity and stability was also observed when comparing the protective and inhibitory effects of various types of APols on SERCA1a (Picard et al. 2006). Four APols were tested, namely A8-35, PMAL-C12, PMALA-C12, and a SAPol. In addition to the inhibition of ATPase activity (Fig. 28A) and the stability over time upon calcium removal (Fig. 28C), the slowing down by APols of the rate of $Ca^{2+}$ release from TM binding site I was also examined (Fig. 5.34B). This process requires opening of the bundle of TM helices to create a way out for the ion (Toyoshima and Nomura 2002; Toyoshima and Inesi 2004; Obara et al. 2005; see Chap. 1, § 1.6.3). As a rule, little or no difference is observed, at a given APol concentration, between the effects of A8-35, PMAL-C12, and PMALA-C12. SAPol, on the contrary, was found to be both less inhibitory

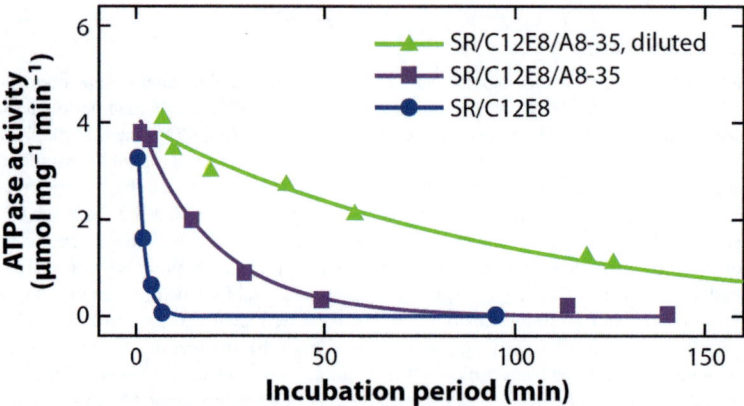

**Fig. 5.33** Stabilization of the sarcoplasmic reticulum calcium ATPase SERCA1a by A8-35. The destabilization of the ATPase was initiated by diluting solubilized sarcoplasmic reticulum (SR) into an EGTA-containing solution, thereby leaving the ATPase in a $Ca^{2+}$-deprived, detergent-solubilized state, which is known to lead to very rapid, irreversible inactivation: (*blue curve*) the dilution medium contained 5 g·L$^{-1}$ (9.3 mM) $C_{12}E_8$; (*purple curve*) same medium, but with the addition of 5 g·L$^{-1}$ A8-35, resulting in a mass ratio of A8-35/$C_{12}E_8$ in the surfactant belt very close to 1:1; (*green curve*) same medium as the latter sample, but incubation took place after a 250× dilution with surfactant-free buffer; under the latter conditions, the A8-35/$C_{12}E_8$ mass ratio in the surfactant belt, calculated as described in Box 5.3, is ~3.1:1 (From Popot 2010, adapted from research originally published in Champeil et al. 2000, © The American Society for Biochemistry and Molecular Biology).

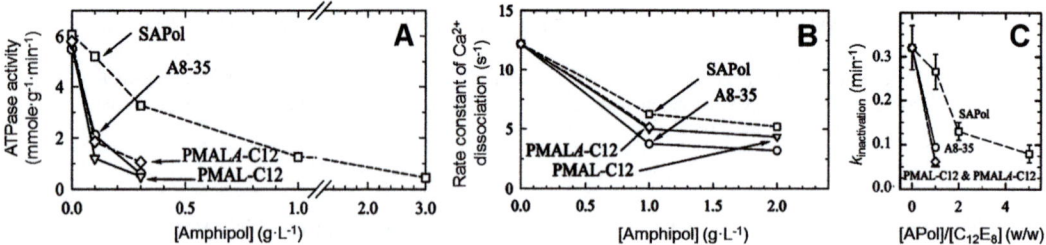

**Fig. 5.34** Comparative effects of various amphipols on the activity and stability of the sarcoplasmic reticulum (SR) calcium ATPase, SERCA1a. (**A**) Reversible inhibition by APols of the ATPase activity of detergent-solubilized SR membranes. SR vesicles were diluted to 0.004 g·L$^{-1}$ in a detergent-free assay medium, and ATPase activity was monitored continuously with a coupled enzyme system. Following a similar protocol to that in Fig. 5.31A, samples were successively supplemented with (i) 0.1 g·L$^{-1}$ (i.e. twice the CMC) $C_{12}E_8$, which solubilizes the SR membranes, thereby relieving the inhibition due to the accumulation of $Ca^{2+}$ within SR vesicles; (ii) various APols, namely either A8-35, PMAL-C12, PMAL*A*-C12, or SAPol, at final concentrations first of 0.1 g·L$^{-1}$ and then of 0.3 g·L$^{-1}$; finally (iii) a larger concentration of $C_{12}E_8$ added, partly relieving inhibition by APols (1 g·L$^{-1}$ was added to the 0.1 g·L$^{-1}$ already present). The ATPase activity measured during step ii is plotted as a function of APol concentration for each of the four APols tested. (**B**) Slowing down by APols of $Ca^{2+}$ dissociation from the detergent-solubilized ATPase. Summary of the rate constants found in parallel experiments with the four APols, plotted as a function of APol concentration. (**C**) Protection conferred by various APols against the irreversible inactivation of the ATPase upon calcium depletion in the presence of detergent. Experiments were carried out as described in Fig. 5.33. Rate constants for irreversible inactivation are plotted vs. the ratio of APol to $C_{12}E_8$ in the samples. □, SAPol; ○, A8-35; ◇, ▲, PMAL-C12 and PMAL*A*-C12 (Reprinted with permission from Picard et al. 2006, © 2006 American Chemical Society).

(Fig. 5.34A, B) and less protective (Fig. 5.34C) than the other APols tested. This raised again the intriguing question of a possible common mechanism underlying the two types of effects (Picard et al. 2006).

The P-type pumps to the family of which the SERCA1a belongs are remarkable by the extensive rearrangement of the transmembrane helix bundle (and extramembrane domains) that takes place during the enzymatic cycle (reviewed by Palmgren and Nissen 2011; see Norimatsu et al. 2017, Sweadner 2017, and Chap. 1, § 1.6.3). This has led to the suggestion that the inhibition of SERCA1a by APols could be due to an increase of the free energy of activation associated with SERCA1a transconformations, which would result from the rearrangement of the polymer around the protein. This hypothetical process has been dubbed the "Gulliver effect" (Popot et al. 2003, 2011; Picard et al. 2006) by reference to the movements of Swift's character being impeded by the tiny strings of the Lilliputians (Swift 1726; Fig. 5.35). It was further suggested that small (sub-nanometric) conformational changes, such as those that affect or may affect the TM surface of BR (see Hirai et al. 2009; Wickstrand et al. 2015; see Chap. 1, § 1.6.1) and that of the nAChR (see Hilf and Dutzler 2009; Corringer et al. 2010; Chap. 1, § 1.6.2), can probably be accommodated by displacements of the APol's alkyl chains. Those take place in the ns range (Perlmutter et al. 2011, 2014; see § 5.6.2) and would not necessarily entail an increased free energy of activation. This might explain why an inhibition of transitions between conformational states is observed neither with BR (Gohon et al. 2008; Dahmane et al. 2013) nor with the nAChR (Martinez et al. 2002; Fig. 5.30). Larger interfacial movements (nm), such as those undergone by the TM helix bundle of SERCA1a upon transiting between the E1 and E2 states (Chap. 1, § 1.6.3), on the contrary, may cause a reorganization of the polymer's backbone, which could entail a higher free energy penalty in APol than in detergent and, thereby, slow down the enzymatic cycle.

**Fig. 5.35** Lemuel Gulliver's movements being restrained by the tiny strings of the Lilliputians (Swift 1726; illustration by Frédéric Bouchot (1849)).

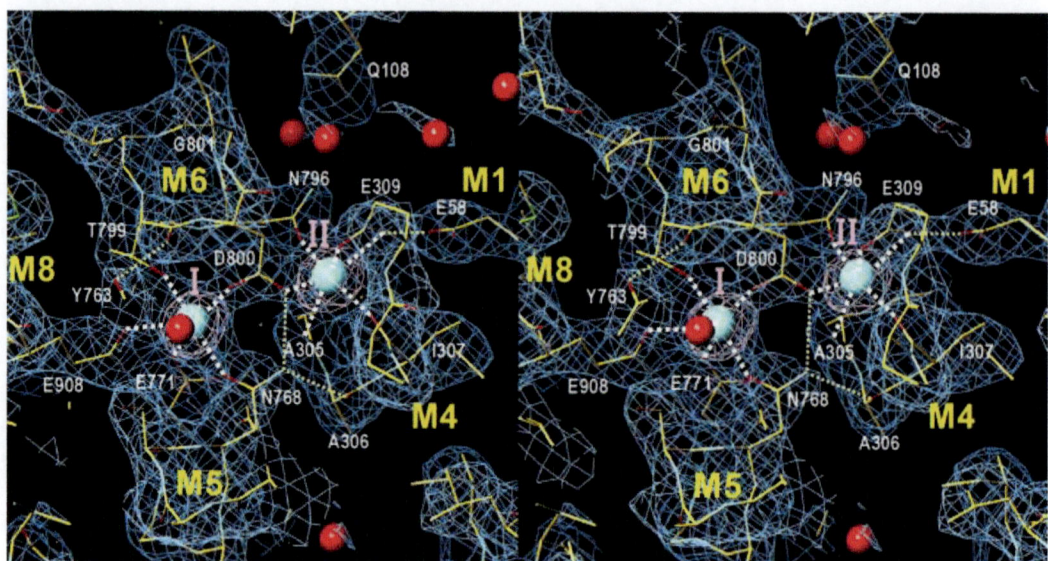

**Fig. 5.36** Details of the transmembrane $Ca^{2+}$ binding sites in the sarcoplasmic calcium pump SERCA1a. The refined model is superimposed onto a $2|F_o| - |F_c|$ composite-omit map (*blue* meshes, contoured at 1.5 σ). The meshes in *pink* show an omit $|F_o| - |F_c|$ map for the two $Ca^{2+}$ ions and a bound water molecule (cutoff at 3 σ). *Light blue spheres* represent $Ca^{2+}$, *red spheres* water molecules. Stereo view along an axis roughly normal to the membrane, from the cytoplasmic side. Coordination of oxygen atoms to $Ca^{2+}$ is indicated by *white dotted lines*, possible hydrogen bonds stabilizing the coordination geometry by *green dashed lines* (Reprinted from Toyoshima et al. 2000, permission from Nature © 2000, Macmillan publishers).

It is probably relevant to this issue that the detergent-induced denaturation of SERCA1a seems to start with the opening of the transmembrane helix bundle, with the $Ca^{2+}$-binding TM region being unfolded first (Merino et al. 1994). The fact that ATPase inactivation in the presence of detergent is faster in the absence of $Ca^{2+}$ than in its presence is probably related, at least in part, to the "cross-linking" of the TM helices by two $Ca^{2+}$ ions, which bind to negatively charged residues in TM helices M1, M4, M5, M6, and M8 (Toyoshima et al. 2000; Fig. 5.36; cf. Chap. 1, § 1.6.3): once these ions are gone, denaturation of the ATPase presumably starts with a limited, transient, reversible opening of the TM region, which would become irreversible upon further unfolding and/or formation of intermolecular contacts. By damping the dynamics of large-amplitude TM movements, APols may thus slow denaturation. This might explain why the "Gulliver effect" can both inhibit and stabilize SERCA1a, a result of the higher free energy of activation imposed, in both cases, on rearrangements of the protein's TM helix bundle (Popot et al. 2003, 2011; Picard et al. 2006). This proposal is consistent with the later observation, described in § 5.5, that, at least at pH 10, A8-35 stabilizes OmpA against urea-induced denaturation kinetically, by creating a higher barrier against unfolding, rather than thermodynamically, by stabilizing the native state as compared to the unfolded ones (Pocanschi et al. 2013). The proposal that APols may damp MP movements has received some further support from MD (Perlmutter et al. 2011) and neutron scattering (Tehei et al. 2014) estimates of the viscosity of A8-35 (see Chap. 4, § 4.3.1.2.4), as well as from a recent MD study of the dynamics of A8-35-trapped OmpX (Perlmutter et al. 2014), to which we will now turn.

## 5.6.2    Molecular Dynamics Simulations

The effects of APols on the dynamics of MPs have been examined by MD simulations, taking as a model a complex between OmpX and A8-35 (Perlmutter et al. 2014). The choice of this system was guided by the fact that a host of experimental data are available, particularly NMR data on the OmpX/A8-35 complex (Catoire et al. 2009, 2010b; Etzkorn et al. 2014; cf. § 5.3.3 and Chap. 10), as well as MD data on pure A8-35 (Perlmutter et al. 2011; Chap. 4, § 4.3.1.2.4). Experimental data can thus be confronted to the results of MD simulations, which helped validating the methodology adopted. Once a satisfying model had been built (§ 5.3.3), the dynamics of APol-trapped OmpX were studied and compared to those of the same protein either embedded in a DOPC bilayer or complexed by the detergent $diC_6PC$.

Figure 5.37 shows the protein root mean square fluctuations describing the conformational flexibility of OmpX's backbone at each amino acid position. The general profile shows lower values (restrained dynamics) for the $\beta$-strands and larger values (increased flexibility) for the loops (L) and turns (T). Comparison of the protein dynamics in the presence of either A8-35 (*green line*), $diC_6PC$ (*blue line*), or DOPC (*orange line*) shows reduced conformational flexibility in the presence of the APol. Whereas the dynamics of the barrel are very generally damped in A8-35, there is a differential effect on the loops, the dynamics of the relatively long extracellular loops being, as a rule, more strongly affected than those of the short periplasmic turns. In agreement with an earlier MD study (Choutko et al. 2011), there are also differences between OmpX solubilized in $diC_6PC$ and OmpX inserted into a lipid bilayer.

Principal component analysis was used to describe dynamics independently at longer and shorter length scales. In Fig 5.38A, the principal components of the protein dynamics are presented for OmpX in A8-35, $diC_6PC$, and DOPC. Each eigenvector describes a single degree of collective motion, with lower indices corresponding to larger length-scale motions. The corresponding eigenvalues describe the magnitude of fluctuations along that degree of motion – essentially a mean square fluctuation for each eigenvector. The results are consistent with those of Fig. 5.37: the protein dynamics are more restrained in the complex with A8-35 than they are in those with either $diC_6PC$ or DOPC (Fig. 5.38A).

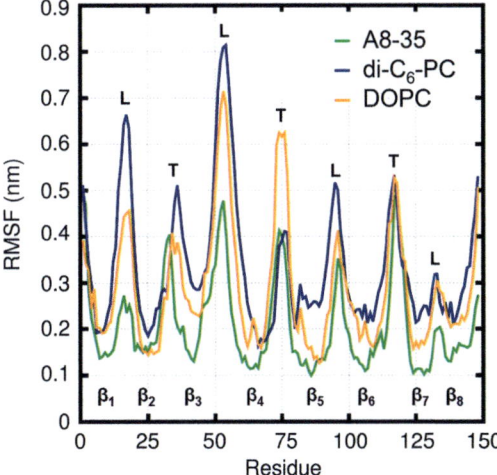

**Fig. 5.37** Root mean square fluctuations of the protein backbone, simulated without restraints, in complexes of OmpX with A8-35, with a dioleoylphosphatidylcholine (DOPC) bilayer, or with the detergent $diC_6PC$. The dips in the curves correspond to transmembrane $\beta$-strands, the peaks to extracellular loops (L), and periplasmic turns (T) (Adapted from Perlmutter et al. 2014).

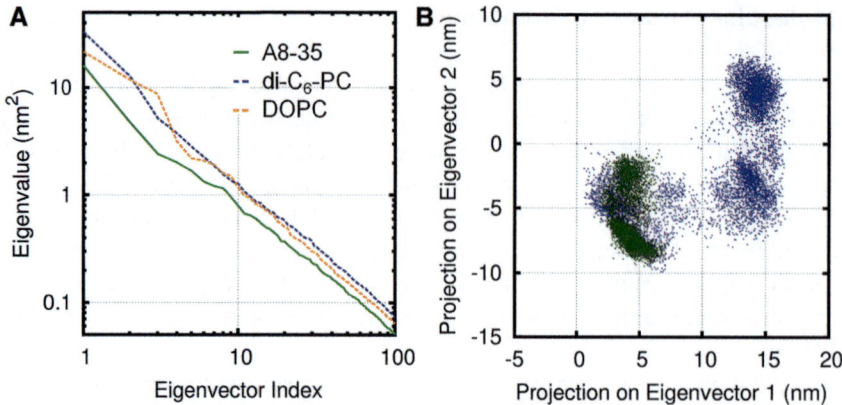

**Fig. 5.38** A comparison of the dynamics of OmpX in various environments using principal component analysis. (**A**) Eigenvalues for principal components of the protein dynamics in different environments. The dynamics of A8-35-trapped OmpX are restricted as compared to those in diC$_6$PC or DOPC whatever the amplitude of the movements. (**B**) Projections onto the first and second principal components for OmpX in complex with either A8-35 (*green dots*) or diC$_6$PC (*blue dots*). Note the larger amplitude of the fluctuations in diC$_6$PC (From Perlmutter et al. 2014).

This analysis indicates that modes of motion on all accessible length scales are restrained by APols: the eigenvalues for the protein in complex with A8-35 are lower than in diC$_6$PC or DOPC for all of the first 100 eigenvectors.

Figure 5.38B describes the restriction of the largest length-scale dynamics of OmpX in A8-35 vs. diC$_6$PC. Each *blue* point corresponds to one frame from the simulation of OmpX in diC$_6$PC projected onto a 2D grid, where the *x*-axis indicates the displacement along the lowest-index principal component and the *y*-axis that along the second lowest-index principal component. The first component appears as a twisting of the barrel and the second component as a radial widening and compressing type of motion. In *green*, the frames from the OmpX/A8-35 simulation are projected onto the same axes. Whereas there are substantial structural fluctuations for the protein in diC$_6$PC, these dynamics are largely restrained in APol.

To clarify the relationship between the dynamics of the protein and those of its environment, the duration for which the hydrophobic groups of A8-35 (octyl chains), diC$_6$PC (hexanoyl chains), and DOPC (oleoyl chains) are in contact with the hydrophobic domain of the protein was calculated. These residence times were fit to an exponential decay, revealing very similar decay constants for A8-35 and diC$_6$PC (0.87 and 1.06 ns, respectively), and a somewhat longer one for DOPC (2.16 ns). It does not seem, therefore, that the damped dynamics of OmpX in A8-35 vs. diC$_6$PC and DOPC be the result of a restricted motility of the hydrophobic chains directly in contact with the protein.

As summarized in § 5.6.1, an ensemble of experimental observations suggests that trapping with APols may damp the dynamics of MPs and that this may have the double effect of (i) stabilizing them, by creating a higher free energy barrier to unfolding, and (ii) slowing down functional cycles that require large rearrangements of the protein/polymer interface. The MD study of OmpX in complex with various surfactants brings some support to this hypothesis, inasmuch as it does show a damping effect of A8-35 as compared to diC$_6$PC or even a DOPC bilayer. It also provides some further insights into this phenomenon. In its original form, the "Gulliver effect" hypothesis proposed that only relatively large-scale (nanometric) transmembrane movements were affected, which would account for the fact that, whereas most MPs are stabilized by transfer from detergent solutions to APols, only two of those hitherto studied, SERCA1a and the F$_1$F$_O$ ATP synthase, see their enzymatic cycle inhibited (§ 5.4). It was therefore speculated that protein conformational changes that could be accommodated by small movements of the APol alkyl chains may not be strongly affected, whereas those requiring rearrangements of the polyacrylate backbone would (Popot et al. 2003, 2011; Picard et al. 2006). The MD data lead to a more nuanced view, inasmuch as (i) all movements appear to be damped, whatever their length scale, even though large-scale movements tend to be more strongly damped than small-scale ones (Fig. 5.38A); this may suggest that reasoning in terms of increased overall surface viscosity of the surfactant, as observed when comparing the polar surface of pure A8-35 particles with that of detergent micelles (Perlmutter et al. 2011; see Chap. 4, § 4.3.1.2.4), may be more appropriate than dissecting the molecular movements of the polymer; and (ii) the effect propagates to the extramembrane loops, even though those are not in contact with the polymer, or only minimally so (§ 5.3.3, Fig. 5.15B). This opens the possibility that functional effects may be indirect. Whatever the mechanism, it does not seem shocking that SERCA1a, the enzymatic cycle of which requires ample TM and extramembrane rearrangements (see Chap. 1, § 1.6.3), be detectably inhibited, whereas BR and the nAChR, whose transconformations are subtler (ibid., § 1.6.1 and 1.6.2), show no obvious evidence of perturbation.

It is worth noting that, on the basis of coarse-grained MD simulations, a "straightjacket effect" similar to the "Gulliver effect" has been postulated to account for the inhibition experienced by SERCA1a upon insertion into too thin lipid bilayers, such as those formed by $diC_{14:1}PC$ (Sonntag et al. 2011). These effects are reminiscent of the slowing down of dynamics and functional blockade of proteins by partial dehydration (Zaccai 2004). It is interesting to note that the stabilizing effect of some mutations on GPCRs seems to involve a stiffening of the TM region, thermostable mutants showing less relative TM helix movements than their respective wild-type receptors (Vaidehi et al. 2016). By totally different routes, trapping with APols and selecting thermostabilizing mutations may reach the same end, that of restricting conformational excursions that lead to denaturation.

Clearly, the degree of functional perturbation experienced by a number of MPs with different structures and functions will have to be studied in greater detail before a general picture can emerge. The OmpX/A8-35 simulations report on the structural fluctuations of a small compact protein around a single equilibrium structure and not on large conformational changes in a complex multi-domain protein such as SERCA1a. One should also note that, if the rate-limiting step in a functional cycle does not depend on transconformations occurring in the TM region, APols are unlikely to affect it. It should also be kept in mind that there is now good evidence that some of the functional differences observed between detergent-solubilized and APol-trapped MPs are due to lipid rebinding upon transfer to APols, as is clearly the case for the photocycle of BR (Dahmane et al. 2013), and is suspected of the allosteric equilibria of the nAChR (Martinez et al. 2002) (see § 5.5). Lipid rebinding is clearly a stabilizing factor, whether in the presence of detergents (Chap. 2, § 2.4.1) or in APols (this chapter, § 5.5). It may suffice, in some cases, to account for the stabilization observed upon transfer from detergent solution to APols, without invoking their effects on dynamics. Lipid rebinding, however, cannot explain all stabilizing effects of APols. In the case of SERCA1a, it cannot explain the observed correlation between stabilization and inhibition: given that delipidation is known to *diminish* the activity of the pump (de Foresta et al. 1989; Lund et al. 1989), lipid rebinding upon transfer from detergent to APols ought to be stimulatory, not inhibitory. In the case of BR, the protein is more stable in A8-35 in the total absence of lipids than it is in detergent solution with its whole quota of purple membrane lipids present (Dahmane et al. 2013; see Fig. 5.27). This is compatible with the APol-trapped protein experiencing a higher free energy barrier to unfolding even in the absence of lipids.

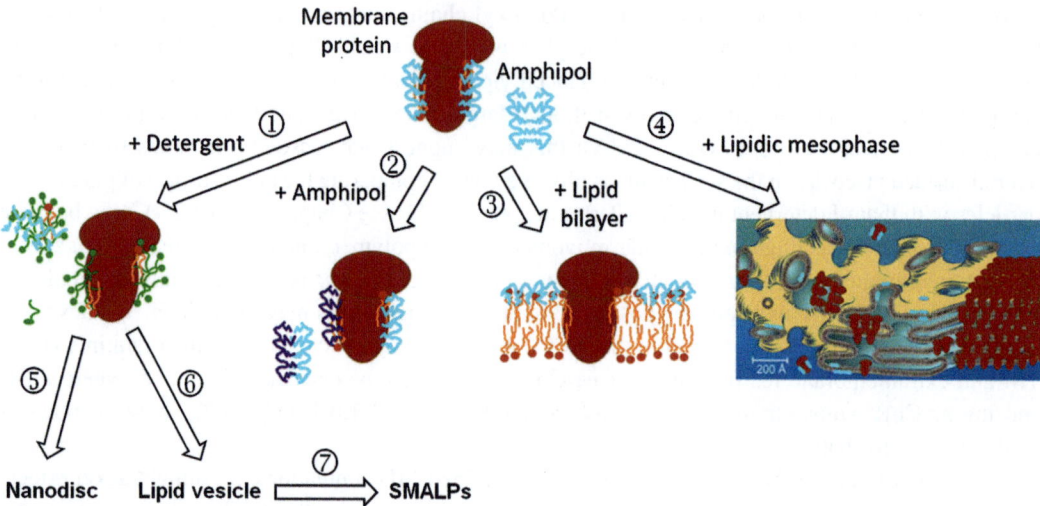

**Fig. 5.39** Displacement of membrane protein-bound amphipols by other surfactants, be they a detergent, another amphipol, preformed membranes (vesicles, black films, biological membranes, etc.), or a lipidic mesophase. For examples of each type of transfer, see e.g. ①②, Zoonens et al. (2007); ③, Pocanschi et al. (2006b) and Kyrychenko et al. (2012); ④, Polovinkin et al. (2014b); ⑤, Damian et al. (2012) and Casiraghi et al. (2016); and ⑥ ⑦, Logez et al. (2016). For further references, see Table 5.6 (Mesophase cartoon adapted from Cherezov et al. 2006).

## 5.7    Transferring Membrane Proteins from Amphipols to Other Environments

As will be described in Chaps. 8, 9, 10, 11, 12, 13, 14, and 15, MP/APol complexes lend themselves to a whole gamut of applications ranging from biophysical studies to biomedical applications. There are cases, however, where another environment is either mandatory or desirable. Such is the case, for instance, when transport or permeability measurements have to be carried out. The protein, in such cases, has to be transferred to a medium endowed with separate compartments, such as a vesicle, a black lipid film, or a cell. Similarly, studies of the functional effects of lipids can probably be carried out, to some extent, by trapping the target proteins in APols in the presence of various lipids, but it may be preferable to reconstitute a more bilayer-like environment, such as a lipid membrane or a nanodisc, possibly a lipid-rich SMALP or bicelle. Another case in point is that of crystallization: at this point, despite some fractional successes, crystallizing MP/APol complexes remains a difficult goal (Charvolin et al. 2014; see Chap. 11, § 11.3.1). On the contrary, excellent crystals have been grown after transferring an APol-trapped MP to a lipid cubic phase (Polovinkin et al. 2014b; ibid., § 11.3.2). In all of these cases, the APol has to be substituted, directly or indirectly, with another surfactant (Fig. 5.39). This does not raise any particularly difficult problems and has been achieved in many studies, some of which are listed in Table 5.6.

MP-adsorbed layers of A8-35 exchange with free A8-35 in solution (Zoonens et al. 2007), most likely, given the very low CAC and the near-absence of free individual APol molecules (Giusti et al. 2012), via a mechanism involving collisions between a MP/APol complex and a free APol particle, followed by fusion, mixing, and fission (for a discussion, see Chap. 4, § 4.3.1.2.4). As expected, the kinetics of exchange are highly dependent – from minutes to tens of hours – on the extent to which repulsive electrostatic interactions are screened (Zoonens et al. 2007). In contrast, A8-35 remains firmly associated to MPs upon exposure to large volumes of surfactant-free buffer, as occurs upon

**Table 5.6**  Some examples of transfer of membrane proteins from amphipols to other amphipols or other environments.

| Amphipol | Final environment | Comments | References |
|---|---|---|---|
| A8-35 | Another APol | An examination of the rate of exchange of MP-adsorbed APols when exposed to an excess of a competing APol. In the case of A8-35, the rate of exchange is strongly dependent on the ionic strength | Zoonens et al. (2007) |
| A8-35, A8-75, OAPA-20, PMAL-B | Detergent or mixed detergent/ lipid micelles | APols can be efficiently and very rapidly displaced from the TM surface of MPs by an excess of detergent (cf. Fig. 5.40) | Tribet et al. (1997, 2009), Nagy et al. (2001), and Zoonens et al. (2007) |
| OAPA-20, PMAL-B, A8-35, NAPols | Lipid vesicles | Transfer was initiated by directly exposing protein/APol complexes to preformed lipid vesicles. The first study deals with an $\alpha$-helical MP, diacylglycerol kinase (DAGK), the second one with the pore-forming domain of diphtheria toxin | Nagy et al. (2001) and Kyrychenko et al. (2012) |
| A8-35 | Black lipid film | The functionality of OmpA and FomA following folding in A8-35 was examined by applying the complexes to black lipid films. It was noted that recovery of native-like pore conductance requires that the APol be applied to both the *cis* and *trans* sides of the film (cf. Fig. 5.35) | Pocanschi et al. (2006b) |
| A8-35 | Live cells | A rhodamine-labeled peptide mimicking the transmembrane anchor of neuropilin-1 was trapped with NBD-labeled A8-35 and the distribution of the two species following delivery to COS-7 cells followed by confocal fluorescence microscopy. See Chap. 15, Fig. 15.23 | Popot et al. (2011) |
| A8-35 | Lipidic mesophase | Direct transfer of BR from A8-35 to a lipidic mesophase led to the formation of highly ordered BR crystals containing no APol. See Chap. 11, § 11.3.2 | Polovinkin et al. (2014b) |
| SMA | Lipidic mesophase | Same as above, using SMALP-trapped bacteriorhodopsin from *Haloquadratum walsbyi* | Broecker et al. (2017) |
| A8-35 | Nanodiscs | Two GPCRs were expressed in *E. coli* as inclusion bodies, solubilized in SDS, folded in A8-35, and transferred to DDM by immobilized-metal affinity chromatography and from DDM to nanodiscs | Damian et al. (2012) and Casiraghi et al. (2016) |
| A8-35 | SMALP | The ghrelin GHS-R1a receptor was expressed in *E. coli* as inclusion bodies, solubilized in SDS, folded in A8-35, transferred to DDM by immobilized-metal affinity chromatography, reconstituted into proteoliposomes, and the proteoliposomes solubilized with SMA | Logez et al. (2016) |
| A8-35, A8-75, A34-35, A34-75, SAPol, NAPol | Vacuum | APol-trapped MPs were transferred to vacuum for mass spectrometry (MS) either by electrospray ionization (ESI) or matrix-assisted laser desorption/ionization (MALDI). See Chap. 14 | Catoire et al. (2009), Bechara et al. (2012), Leney et al. (2012), Hopper et al. (2013), Calabrese et al. (2015), and Watkinson et al. (2015, 2017) |

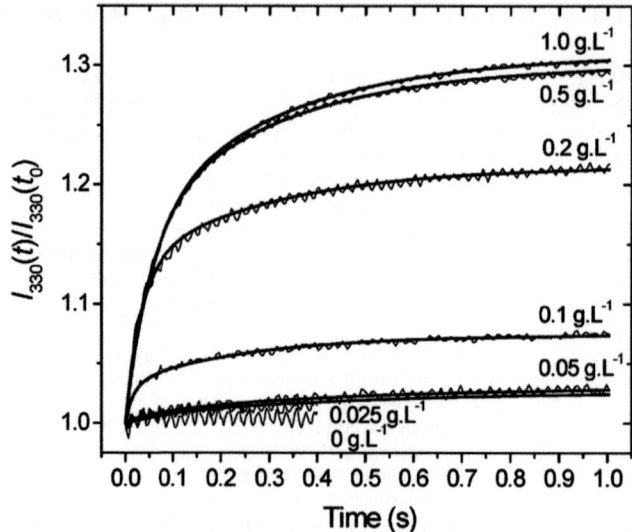

**Fig. 5.40** Displacement of a membrane protein-bound fluorescent amphipol by an excess of detergent. $C_{12}E_8$ was added to tOmpA/FAPol$_{NBD}$ complexes in a stopped-flow instrument, and the kinetics of dequenching of the protein's tryptophan residues followed as a function of time. The final concentrations of detergent extended from $2\times$ below the CMC ($0.025$ g·L$^{-1}$) to $20\times$ above it ($1$ g·L$^{-1}$). The final protein and APol concentrations were kept constant at $0.025$ and $0.1$ g·L$^{-1}$, respectively. Experimental data (*thin lines*) have been normalized to the intensity of tryptophan fluorescence at the first measurable point, taken as $I_{330}(t_0)$, and fitted with three exponentials (*thick lines*) (Reprinted with permission from Zoonens et al. 2007, © 2007 American Chemical Society).

extensive dilution (Zoonens et al. 2007; Tribet et al. 2009), or upon flushing of complexes attached to an SPR chip via a histidine tag carried by the protein (Popot et al. 2003; see Fig. 5.7). Consistent with these observations, MPs bound to a solid support via a biotinylated APol neither desorb nor become inactivated upon extensive washing of the chips with surfactant-free buffer (Charvolin et al. 2009; Basit et al. 2012; Della Pia et al. 2014; Ferrandez et al. 2014; see Chap. 13).

As mentioned in § 5.2.1, APols, whether present as free particles or as a MP-adsorbed layer, freely mix with detergents, in a nearly ideal manner (Zoonens et al. 2007; Tribet et al. 2009). This makes it very easy to exchange one type of surfactant for the other (see e.g. Tribet et al. 1997; Zoonens et al. 2007). The ease and speed (Fig. 5.40) with which detergents can wash APols away from the surface of MPs may seem contradictory with the strong retention of APols by MPs upon extensive dilution or flushing with surfactant-free buffers. This apparent paradox is due to the fact that there is little or no free energy cost to displacing APols from a MP hydrophobic transmembrane surface to a mixed detergent/APol particle while replacing it with detergent, whereas it is extremely costly to bare the same surface from any surfactant (Tribet et al. 2009; Giusti et al. 2012).

APols have been used by C.R. Sanders and colleagues to deliver diacylglycerol kinase (DAGK) (Nagy et al. 2001) and by A.S. Ladokhin to deliver the pore-forming domain of diphtheria toxin (Kyrychenko et al. 2012) to preformed lipid vesicles, as well as by J.H. Kleinschmidt and colleagues to deliver two outer membrane β-barrel proteins that had been refolded in A8-35, OmpA from *E. coli* and FomA from *Fusobacterium nucleatum*, to lipid black films (Pocanschi et al. 2006b; Fig. 5.41). In all cases, the native state of at least some of the proteins inserted was demonstrated by functional tests. APols have also been used to deliver a synthetic peptide mimicking the single transmembrane helix of a growth factor receptor to cells in culture. In the hours and days that followed, fluorescence imaging showed that the peptide (and the APol) was endocytosed (Popot et al. 2011; see Chap. 15, § 15.3).

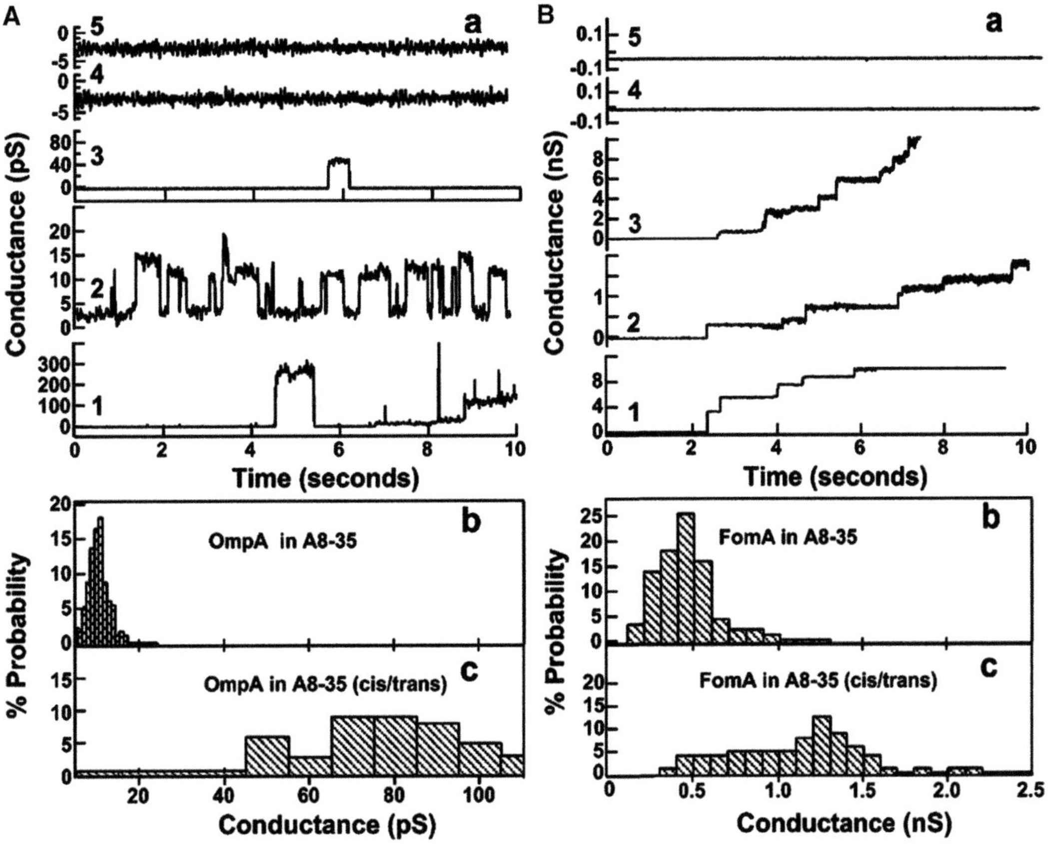

**Fig. 5.41** Delivery of two membrane proteins from A8-35 to a black lipid film. Single-channel recordings of refolded OmpA (from *E. coli*) and FomA (from *Fusobacterium nucleatum*) integrated into black lipid films. (**A**) OmpA. (**a**) Large-conductance (1) and small-conductance (2) states observed when A8-35-refolded OmpA was added to the *cis* side of a diphytanoylphosphatidylcholine film. The conductance of the small-conductance state increased when A8-35 was added to the *trans* side (3). No event was observed in the presence of A8-35 alone, whether added in *trans* (not shown), in *cis* (4), or on both sides of the film (not shown), or in that of unfolded OmpA (5). (**b, c**) Distribution of small-conductance states observed after addition of A8-35/OmpA complexes to the *cis* side in the absence (**b**) or presence (**c**) of A8-35 on the *trans* side. (**B**) FomA. (**a**) Channels formed by A8-35-refolded FomA added to the *cis* side with A8-35 present (1) or absent (2) in the *trans* compartment and control experiments with LDAO-refolded FomA (3), with A8-35 only (on the *cis* side) (4), and with denatured FomA (5). (**b, c**) Distribution of conductance states observed after addition of FomA/A8-35 complexes to the *cis* side without (**b**) and with (**c**) A8-35 added to the *trans* side (Reprinted with permission from Pocanschi et al. 2006a, b, © 2006 American Chemical Society).

A couple of caveats should be mentioned regarding the use of APols to deliver MPs to preformed membranes. First, the carrier APols will themselves become inserted into the target membrane (cf. Pocanschi et al. 2006b; Popot et al. 2011), even though, very likely, they will dissociate from the protein and migrate independently from it (cf. the separation of BR from A8-35 upon transfer of the complexes to a lipidic mesophase; see below). They may cause perturbations that have to be paid attention to. Indeed, the conductance of the pores formed by OmpA and FomA delivered to a black film was abnormally low until A8-35 was also added to the *trans* compartment, indicating that the asymmetry created by APol binding to one side only of the film perturbed the function of the channel (Fig. 5.41; Pocanschi et al. 2006a, b). Second, not all MPs can be expected to survive such a drastic procedure. For the protein to adopt a TM position, some of its hydrophilic regions have to somehow cross the bilayer, which can be a highly destabilizing process. It is reasonable to expect that the more robust (or the simpler) the protein is, the greater are the chances that it can be transferred without denaturation, or may be able to recover from it. Studying the transfer of a variety of MPs is clearly needed before a general view of the usefulness of this procedure can be formed.

As will be discussed in Chap. 11, A8-35 – and, presumably, all APols carrying a net charge – may not be suitable to forming well-diffracting MP crystals (Charvolin et al. 2014). However, A8-35 has been used by Valentin Gordeliy and colleagues to deliver BR to a lipidic mesophase, in which the protein organized into 3D crystals diffracting to better than 2-Å resolution (Polovinkin et al. 2014b). Crystallographic analysis showed that the APol had been totally displaced and was not present in the crystals, whereas the purple membrane lipids introduced into the mesophase along with the protein had remained associated to it. This experiment has been recently reproduced using SMA-trapped BR from *Haloquadratum walsbyi* (Broecker et al. 2017). Delivering MPs and, in particular, GPCRs to lipidic mesophases for crystallization is probably a very promising approach (see Chap. 11), particularly given the high rate of success of APol-induced folding of GPCRs expressed in inclusion bodies (Chap. 6).

Once an APol-trapped MP has been transferred to a detergent, any classical reconstitution procedure can of course be resorted to. This has been exploited to transfer GPCRs that had been folded in A8-35 to either nanodiscs (Damian et al. 2012; Casiraghi et al. 2016) or SMALPs (Logez et al. 2016). In the first case, the proteins were transferred from A8-35 to DDM, using IMAC to wash away the APol, after which the detergent solution was supplemented with lipids and scaffold proteins, and the nanodiscs formed by eliminating the detergent (Damian et al. 2012; Casiraghi et al. 2016; see § 5.9, Protocol 5.3). In the second case, the APol-folded receptor was similarly transferred to DDM by IMAC and then introduced into proteoliposomes, which were solubilized by SMA (Logez et al. 2016). These procedures aim at providing the protein with a more bilayer-like environment, particularly with the view of studying the regulatory effects of lipids.

Finally, a strange, totally foreign medium to which MPs have been transferred from an APol-trapped state is vacuum, that vacuum that reigns inside the cavity of mass spectrometers. Electron spray ionization MS (ESI-MS) coupled with ion mobility spectrometry (IMS), in particular, has shown that APols preserve better than detergents the folded state of MPs during the critical stage of transfer from an aqueous solution to the gas phase (Calabrese et al. 2015; Watkinson et al. 2015). The applications of APols to MS will be discussed in Chap. 14.

## 5.8    Conclusion

In this and the preceding chapter, I have tried to present an overview of the chemical structure and solution properties of APols, of the modes of formation of MP/APol complexes, and of their organization, dynamics, and solution properties. In the following chapters, we will examine, first, how APols may help to produce MPs amenable to in vitro studies, either by assisting their folding from a denatured state such as can be obtained by dissolving inclusion bodies in a denaturing medium (Chap. 6) or by acting as the folding medium for MPs expressed in vitro by cell-free expression (Chap. 7). Chapters 8, 9, 10, 11, 12, 13, 14, and 15 will then be devoted to presenting and discussing what has been and what can be done experimentally with MP/APol complexes, namely optical spectroscopic studies (Chap. 8); solution studies by such approaches as radiation scattering, analytical ultracentrifugation, etc. (Chap. 9); NMR spectroscopy (Chap. 10); radiocrystallography (Chap. 11); electron microscopy (Chap. 12); studies exploiting tagged APols, such as solid-state ligand-binding studies using APol-immobilized MPs (Chap. 13); proteomics, including MS (Chap. 14); and biomedical applications, including vaccination, using APols as a delivery vector for MPs or hydrophobic peptides, and so forth (Chap. 15).

## 5.9 Protocols

Three standard protocols are proposed below:

- Protocol 5.1. Transferring a membrane protein from a detergent to an amphipol environment.
- Protocol 5.2. Measuring the amount of protein-bound APol.
- Protocol 5.3. Transferring an APol-trapped MP to nanodiscs.

Comments are printed in italics and indicated by a pointing hand (☞), the protocols themselves being printed in roman.

### 5.9.1 Protocol 5.1. Transferring MPs from Detergents to APols

☞ *Even though direct extraction of MPs from biological membranes by APols has been observed (cf. Box 5.1), detergents are usually resorted to at the solubilization step, unless MPs are produced by cell-free expression system (Chap. 7) or folded directly in the presence of APols (Chap. 6). The transfer procedure consists in replacing detergents by APols in a sample of MPs, which, in general, but not necessarily, has already been purified. The protocol is simple, easy, fast, and requires no important biochemical optimization.*

#### 5.9.1.1 Preparation of a Stock Solution of APols

A8-35 is supplied as a white powder, which can be stored at room temperature. Note that most APols are very stable molecules, except for glucosylated NAPols and FAPols, which carry sugar groups and fluorescent probes, respectively. Storage of glucosylated NAPols at $-20\,°C$ is advisable whatever their conditioning, i.e. in powder or in solution, because sugars can be hydrolyzed. FAPols need to be protected from UV-visible light with aluminum foil. When needed, a stock solution of APols at $100\,g\cdot L^{-1}$, or 10% w/w, is prepared with Milli-Q water (water purified on an A10 Advantage Millipore system):

- Weigh some powder, for instance 20 mg, with an analytical balance in an Eppendorf tube or a small glass vial. (Note: the powder is sometimes very electrostatic and caution is required.)
- Add 180 μL of Milli-Q water in order to reach a final mass of 200 mg.
- Homogenize the solution with a vortex or by magnetic stirring. Incubate at least a couple of hours before use for a good rehydration and dispersion of the polymer. The solution is then kept at 4 °C or, if need be, stored frozen at $-20\,°C$.

#### 5.9.1.2 Determination of the Protein Concentration

The exchange of detergent for APols is carried out by supplying APols pre-solubilized in water to the sample of MPs. The amount of APol to add is calculated on the basis of the mass of MP present in the sample. The concentration of protein must therefore be known, at least approximately. It can be determined by its optical density from UV-visible spectra. If the epsilon coefficient of the protein is unknown, its concentration can be assessed by colorimetric measurements such as bicinchoninic acid (BCA) assay. Alternatively, amino acid analysis after HCl hydrolysis can also be used.

### 5.9.1.3    Determination of the Optimal MP/APol Mass Ratio

☞ *The sole optimization required is that of the mass ratio of APols needed to keep the MP soluble and well dispersed in aqueous solution after detergent removal. For that, the protein and detergent concentrations are kept unchanged, while increasing concentrations of APols are tested.*

- Determine the range of MP/APol mass ratios to be tested: when starting on a novel MP, the mass of APol each molecule will likely bind can be roughly estimated on the basis of the size of the TM region (the MW of the protein being irrelevant). No precise relationship has been worked out yet between the extent of the TM surface of the protein and the mass of APol that will bind to it, but an estimate can be derived from the data in Table 5.5 and Fig. 5.9. A range of $1\times$ to $10\times$ this estimate is a good starting point. If too little APol is used, aggregates will form, which can be detected, crudely, by precipitation upon ultracentrifugation, and, in a second stage and more precisely, by SEC. An excess of APol is useless and can be detrimental to fragile MPs (cf. § 5.5). The table below gives an example for relatively small MPs mainly composed by a bundle of TM $\alpha$-helices or a $\beta$-barrel, so that the total MW of the protein is used for the ratio calculations.

| Mass of MP (mg) | MP/APol mass ratio | Mass of APol (mg) | Volume of APol (μL) |
|---|---|---|---|
| 0.5 | 1:0 (control) | 0 | 0 |
| 0.5 | 1:0.5 | 0.25 | 2.5 |
| 0.5 | 1:1 | 0.5 | 5 |
| 0.5 | 1:2 | 1 | 10 |
| 0.5 | 1:5 | 2.5 | 25 |
| 0.5 | 1:10 | 5 | 50 |

Note that the mass of protein to trap can be smaller or larger than 0.5 mg and the interval between ratios can be narrower. The sample at ratio 1:0 will be used for both positive and negative controls and, thus, should be prepared twice.

- Pipet seven aliquots of equal volume of MP in detergent solution (if the concentration of MP is $1~g\cdot L^{-1}$, the volume of aliquots is 500 μL for each condition). Add the appropriate volume of APols indicated in the table above. Dilution effects can be neglected up to 10% of variation after adding APols. Keep aside the two control samples.
- Mix and incubate for 15–20 min at either room temperature or 4 °C depending on the stability of the protein of interest.

☞ *When APols are supplied to the samples, they mix freely with detergent molecules in micelles and at the transmembrane surface of the protein, as shown by fluorescence and isothermal calorimetry studies (see § 5.2.1). This leads to the formation of MP/detergent/APol ternary complexes (Box 5.2).*

### 5.9.1.4    Detergent Removal

This step can be carried out in various ways. Most often, detergent removal is achieved by adsorption onto polystyrene beads (Bio-Beads SM-2). Note that APols do not significantly adsorb onto Bio-Beads (cf. Fig. 5.5). The mass of beads to add is typically $20\times$ the mass of detergent present in the sample. For instance, if the concentration of detergent is $6~g\cdot L^{-1}$ in 500 μL, the amount of Bio-Beads to add is ~60 mg.

- Calculate the appropriate amount of beads according to the mass of detergent present in each sample. Weigh the beads and add them in the five samples containing APols and in one of the two APol-free control samples, which will become the negative control (detergent removal in the absence of APol). Note: Bio-Beads are usually washed out successively in ethanol and water prior to use and then stored in water. Before weighing, drop them off on a tissue paper for maximal water removal. The last sample, without beads, represents the positive control (MP kept in detergent solution).
- Incubate the samples for 2 h under gentle shaking at either room temperature or 4 °C.
- Remove the beads by pipetting the samples while excluding the beads. For that, apply the tip of the micropipette flush with the wall of the Eppendorf tube so that only the solution is sucked in. Put the samples in new Eppendorf tubes.

Alternatively, it is possible to eliminate the detergent micelles by dilution under the CMC of the detergent. Note that this method is more suitable to detergents with a high CMC rather than detergents with a low one, such as $n$-dodecyl-$\beta$-D-maltoside (DDM), because even under the CMC these detergents are still able to keep MPs soluble. If the dilution method is employed, dilute the five samples containing APols plus that of the negative control with a detergent-free buffer. Dilute the last sample, which becomes the positive control, with buffer containing detergent at the same concentration as initially present in the sample.

Whatever the protocol used, some detergent monomers can still be present in the samples. Usually, they are not problematic, as long as the negative control shows that the monomers cannot keep the MP in solution in the absence of APols, but if need be they can be eliminated by dialysis or by several cycles of dilution/concentration using ultrafiltration devices. Note that the presence of APols in the external dialysis buffer is not required as APols do not cross standard dialysis membranes of 12–14 kDa MW cutoff. Indeed, the MW of the particles of A8-35 is ~40 kDa (Gohon et al. 2006), and, because of its low CAC (~0.002 g·L$^{-1}$) (Giusti et al. 2012), there are very few free molecules in solution (see Chap. 4, § 4.3.1.1). Another procedure for detergent removal, albeit seldom used, is to adsorb it onto cyclodextrins (Althoff et al. 2011).

### 5.9.1.5   Identification of the Optimal MP/APol Ratio

- Measure the UV-visible spectrum of each sample.
- Centrifuge the samples at $100,000 \times g$ for 20 min. (Note: the speed and duration of the centrifugation step are given for a small protein of ~30 kDa. These parameters may have to be adjusted if the protein of interest is larger so as to make sure that close to 100% of the protein remains in the supernatant in the presence of detergent.)
- Take off the supernatants and measure again their UV-visible spectra.
- Calculate the percentage of protein kept in the supernatant for each condition. This experiment determines the minimal MP/APol mass ratio required to keep ≥90% of the MP soluble (Fig. 5.42). However, to establish the minimal MP/APol mass ratio required to obtain homogeneous complexes, which is somewhat higher, it is recommended to analyze the samples by SEC (cf. Chap. 9, § 9.5, Protocol 9.2).

The optimal MP/APol mass ratios for two model MPs of small MW like bacteriorhodopsin of *H. salinarum* (BR, 27 kDa) and the transmembrane domain of OmpA of *E. coli* (tOmpA, 19 kDa) are 1:5 and 1:4, respectively (Zoonens et al. 2007; Gohon et al. 2008). These ratios exceed by ≥2× the amount of A8-35 that binds to these MPs (see Protocol 5.3). This is because APols, which have a weak dissociating power, cannot prevent protein/protein interaction if they are not present in excess in the sample (see § 5.2.1). To keep MP/APol complexes homogeneously distributed, an excess of APols is

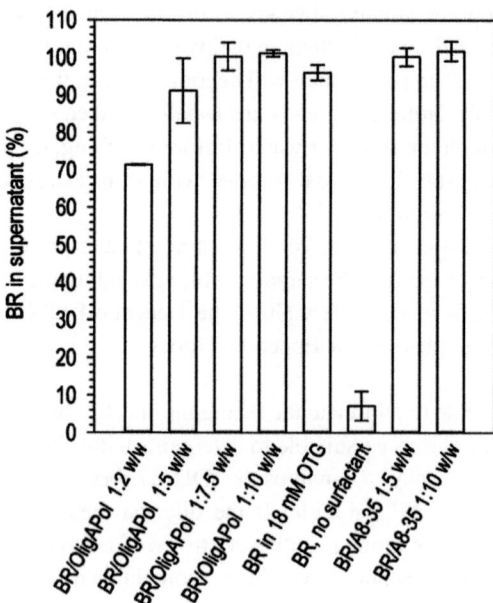

**Fig. 5.42** An example of determination of the minimal amount of a given amphipol needed to trap a given membrane protein. The APol tested was an oligonucleotide-grafted version of A8-35 ("OligAPol"), the protein chosen for the test bacteriorhodopsin (BR). The controls included keeping untreated a BR sample solubilized in 18 mM octylthioglucoside (OTG), in which case >95% of BR remained in the supernatant under the centrifugation conditions used, and depleting another of OTG in the absence of any APol, in which case ~95% of BR precipitated. Detergent removal in the presence of A8-35 at either a 1:5 or 1:10 BR/A8-35 mass ratio resulted in complete retention of BR in the supernatant. With the OligAPol, a 1:2 BR/OligAPol mass ratio was insufficient, a 1:5 ratio borderline, and the 1:7.5 and 1:10 mass ratios resulted in complete retention of BR in the supernatant (From Le Bon et al. 2014, © 2014 Oxford University Press).

therefore required. There is no need, however, to increase the concentration of APols beyond the minimal concentration yielding an acceptable monodispersity, because, due to the hydrophobic sink effect, this may compromise the stability of fragile MPs (§ 5.5).

*Protocol prepared by Manuela Zoonens on the basis of* Zoonens et al. (2007), Gohon et al. (2008), *and laboratory notes.*

### 5.9.2    Protocol 5.2. Determining the Amount of MP-Bound APol

☞ *APols specifically adsorb onto the transmembrane region of MPs, where they form a compact layer 1.5–2 nm thick (§ 5.3.2). The mass of APol constituting the protein-bound belt has been estimated in several studies (§ 5.3.1.1), the most detailed of which used BR, cytochrome $bc_1$, and tOmpA as model MPs. The first determination is mostly based on extensive physical measurements carried out on BR complexed with either plain or deuterated A8-35, using primarily SANS and AUC (Gohon et al. 2008). It is thought to give a relatively accurate measurement of the amount of A8-35 bound per BR monomer, but is extremely work-intensive. The second and third studies relied on the use of a radioactively labeled APol ($[^3H]A8-35$) and a fluorescently labeled one (FAPol), respectively. Free APol was separated from MP/APol complexes either by SEC or sucrose gradient centrifugation (Charvolin et al. 2014) or by separating polyhistidine-tagged MP/APol complexes from free APol by IMAC (Zoonens et al. 2007). Under the conditions used in Zoonens et al. (2007), it yielded, for reasons to be discussed below, what is thought to be a lower limit to the amount of A8-35 bound to the tOmpA* *monomer. We describe here* (i) *how best to express the amount of APol bound per MP,* (ii) *how to estimate* a priori *the amount of APol a given MP is likely to bind, and* (iii) *three protocols for measuring it using FAPols.*

### 5.9.2.1 Why Is It Preferable to Express the Amount of APols Bound per MP in Mass Rather Than as a Number of Molecules?

☞ *APols being highly polydisperse polymers, the size of individual molecule varies considerably, and their MW can be estimated only on average* (see Chap. 4, Box 4.1). *The number-average mass of A8-35 molecules is ~4.3 kDa, but they are widely polydisperse* (see Chap. 4, § 4.2.1). *Nevertheless, despite the variable mass of individual APol chains, the particles they form in solution migrate upon SEC with a size distribution almost as narrow as that of globular proteins. SANS and AUC analyses indicate that they feature a well-defined Stokes radius ($R_S \approx 3.15$ nm) and mass (~40 kDa)* (Gohon et al. 2006) (Chap. 4, Table 4.3). *The average mass of individual molecules being only a rough estimate, the amount of APol bound per MP is more meaningfully expressed in mass ratio rather than as a molar stoichiometry. Similarly, in the case of functionalized APols, to preserve accuracy and reliability, the number of fluorophores or tags is better expressed as their number per 40 kDa APol particle, which can be used as a mass reference, rather than as their number per APol chain, which has no great significance and is inaccurate.*

### 5.9.2.2 How to Estimate A Priori the Likely Amount of APols Bound per MP Based on Structural Data?

☞ *In the case of α-helical MPs, the most thoroughly studied MP/APol complexes are those of BR with A8-35. In the complexes, the protein/APol mass ratio is ~1:2, i.e. ~54 kDa of A8-35 per monomer of BR (27 kDa)* (Gohon et al. 2008). *Lipids (~9 kDa) are also present in the complexes. In the case of β-barrel MPs, the best characterized complexes, in terms of composition, are those of tOmpA with A8-35. The mass ratio that has been estimated is ~1:1.3, i.e. ~25 kDa of A8-35 per monomer of tOmpA (19 kDa)* (Zoonens et al. 2007). *This value should be considered as a minimal value, however, because the conditions under which the measurements were done (see below) led to some aggregation and, very likely, to the loss of some APol. MD calculations suggest an upper value of ~45 kDa A8-35 per tOmpA monomer* (Perlmutter et al. 2014) *(see § 5.3.3). The truth lies probably between these two estimates.*

*Based on these values, and assuming that the volume of the A8-35 belt surrounding a MP is roughly proportional to the perimeter of the TM domain to be covered, it is possible to estimate the amount of APols interacting with any other MP. The only information needed is the dimensions of the hydrophobic domain of the protein of interest, modeled as a cylinder filled up by the TM helices or delimited by the TM β-strands, whose perimeter increases roughly as the square root of the number of helices (α-helical bundles) or linearly with the number of β-strands. For instance, the TM domain of BR is a bundle of seven α-helices. If the MP of interest has a similar TM topology, like a GPCR, it can be expected to bind approximately the same amount of A8-35, i.e. ~54 kDa (BR/A8-35 complexes comprise ~9 kDa of bound lipids* (Gohon et al. 2008), *which increases slightly the TM perimeter, but the effect on APol binding is likely to be minor). On the other hand, if the MP contains twice more helices than BR and features a more or less homothetic shape, the volume of its TM domain doubles, while the TM surface increases by ~40%, and one can expect in the ballpark of ~75 kDa of bound A8-35. It is fair to say, however, that too few accurate measurements are available to date* (Table 5.5) *to gather how reliable such an approach is, the plot of whatever few binding data are available vs. the square root of the number of TM helices showing a considerable scatter* (Fig. 5.13). *The only other relatively precise estimate of bound A8-35 has been obtained with the cytochrome $bc_1$ dimer, which* *has 22 TM helices. On the basis of the above calculations, one would expect it to bind ~96 kDa A8-35. The experimental estimate is only 49–63 kDa (Popot et al. 2003; Charvolin et al. 2014). Note also that it is not unreasonable to expect that the ionic strength may affect the volume and mass of the belts of ionic APols, because it modulates the repulsion between charged polar groups. Despite these uncertainties, estimating* a priori *the probable mass ratio of MP to APol in complexes is useful to provide guidelines when planning trapping experiments, or when undertaking to measure experimentally the amount of bound APol.*

### 5.9.2.3   How to Experimentally Measure the Quantity of APols Bound per MP?

☞ *As previously mentioned, the mass of APol to add for trapping is in excess of that of APol that actually binds to the surface of the MP. After trapping, some APol remains present as free particles in the sample. Measuring the amount of bound APols can be carried out by several approaches. Initial experiments resorted to radioactively labeled APols (see e.g. Tribet et al. 1997; Gohon et al. 2008, and Chap. 9, Fig. 9.3). Later analyses were facilitated by using FAPols as tracers.*

- Prepare a stock solution of APol/FAPol mixture.

☞ *Several FAPols carrying various fluorescent probes are available (see Chap. 4, Table 4.5). The choice of FAPol depends on the absorbance spectrum of the protein of interest. For instance, if the protein absorbs only at 280 nm, FAPol$_{NBD}$, which shows a maximum absorbance at 490 nm, is suitable. On the other hand, if the protein also absorbs visible light, as BR does, another FAPol, like FAPol$_{AF657}$, may be chosen in order to avoid an overlap between the protein and FAPol absorbance bands.*

*Pure FAPols usually absorb too much at the peak of absorbance of the fluorophore, and possibly also at 280 nm, interfering with protein determination. They are better used diluted with nonfluorescent APol (A8-35). Because labeled and unlabeled APols mix freely and rapidly in salty aqueous solutions (Zoonens et al. 2007), they can be mixed from two stock solutions prepared at 100 g·L$^{-1}$. A convenient FAPol/A8-35 ratio is one at which the absorbance of the FAPol at its maximal absorbance wavelength is, in the complexes with the MP, ~25% of that of the protein at 280 nm. This ratio can be estimated* a priori *based on the extinction coefficients of the protein and FAPol, the estimate of the amount of APols bound per MP (cf. § 5.9.3.2), and the MP/APol mass ratio needed for trapping. If the protein possesses many tryptophan residues, its extinction coefficient may be high enough so that no dilution of the FAPol stock solution is necessary.*

- Measure the spectral absorbance of the FAPol/A8-35 mixture at 10 g·L$^{-1}$ if FAPol and A8-35 were mixed in a 1:9 ratio (or pure FAPol at 1 g·L$^{-1}$ if dilution with A8-35 is not necessary). Determine the relative contribution of APols at 280 nm and at the peak of absorbance. (Note: even if neither APol nor FAPol absorbs significantly at 280 nm, it is advisable to check on it.)

☞ *After the complexes have been formed, they must be separated from the excess APol used at the trapping step for the MP/APol ratio in the complexes to be estimated. As of today, three different separation methods have been resorted to.*

### Method 1. Size Exclusion Chromatography (SEC)

☞ *This approach is appropriate for MPs that are large enough – ≥40 kDa, say – for MP/APol complexes and free APol particles to be sufficiently resolved.*

- Wash the gel filtration column with three column volumes of running buffer. Note that, for such an analysis, APol is usually not required in the running buffer, in contrast to detergents, which must always be present above their CMC. This is also how samples for electron microscopy (EM) are prepared before being spread out on the EM grids. However, for this particular experiment, it cannot be excluded that a small amount of APol leaches from the protein as the complexes migrate into APol-free buffer. It might therefore be preferable to saturate the solution with which the column is equilibrated with "some" free APol, such as 5–10% of the concentration present in the sample, so as to prevent desorption, and to subtract the corresponding background. However, this modification to the procedure has not been carefully investigated yet. The composition of standard buffer is 20 mM Tris/HCl, pH 8.0, 150 mM NaCl, but it can be modified provided the pH is above 7.0 and divalent cations are absent.
- Inject an aliquot of FAPol/A8-35 mixture (or pure FAPol) at 10 g·L$^{-1}$. The elution profile is monitored at two wavelengths, 280 nm and the maximum absorbance wavelength of the fluorophore, for example 490 nm for FAPol$_{NBD}$. Determine the elution volume of APol particles.
- After trapping the protein in the FAPol/A8-35 mixture, inject an aliquot of the sample at an appropriate concentration in order to get a good signal-to-noise ratio of the elution peak. Follow the elution of MP/A8-35/FAPol$_{NBD}$ complexes at the two wavelengths, e.g. 280 nm and 490 nm. If the separation from free APol particles is good, calculate the amount of bound APols per MP as follows: integrate the peak area of MP/A8-35/FAPol$_{NBD}$ complexes at 280 nm and 490 nm in order to determine, respectively, the mass of MP and that of FAPol$_{NBD}$ which has comigrated with the protein. A subtraction of the APol contribution to the absorbance at 280 nm may have to be applied, based on the ratio of the surface of the peaks at 280 nm and 490 nm observed with the pure FAPol/A8-35 mixture or on the ratio of the optical densities at 280 nm and 490 nm measured from a UV-visible spectrum. The total mass of APol is then calculated taking into account the dilution of FAPol with A8-35. The ratio of APol and MP masses gives the amount of bound APols per MP. (Note: if the elution peaks of APol particles and MP/APol complexes overlap, use a more resolutive gel filtration column or try another separation procedure.)

**Method 2. Immobilized-Metal Affinity Chromatography (IMAC)**

☞ *The presence of a tag fused to the MP under study makes it possible to immobilize MP/APol complexes onto an affinity column and to eliminate the excess of APol particles. This procedure is particularly convenient when the protein is small and MP/APol complexes cannot be efficiently separated from free APol particles by SEC. Note that free APol particles are, however, required to keep homogeneous MP/APol complexes. Indeed, in the absence of free APol particles, small MP/APol oligomers tend to form, which is likely to be accompanied by some desorption of the MP-bound APol (Zoonens et al. 2007). Because of this effect, the MP/APol ratio determined by this method must be taken as an estimate by default unless buffers have been supplemented with some free APol.*

- After MP trapping in the FAPol/A8-35 mixture, inject the sample on an affinity resin. For instance, if the protein has a polyhistidine tag, load the sample on a Ni:NTA resin. The majority of the protein (~80%) will be retained on the resin (Zoonens et al. 2007; Giusti et al. 2015).
- Rinse the resin with equilibration buffer to wash out free FAPol/A8-35 particles. Elute the MP/FAPol/A8-35 complexes with a buffer containing imidazole. Note that, as noted above,

the presence of APol in equilibration and elution buffers at 5–10% that in the sample may be advisable, so as to prevent APol desorption.

- Desalt the sample to remove imidazole and measure the optical density of the sample at 280 nm and at the maximum absorbance wavelength of FAPol. The concentrations of MP and FAPol are calculated using their respective extinction coefficients. Subtract, if need be, the contribution of FAPol at 280 nm, and calculate the total mass of APol if FAPol was initially mixed with A8-35 before trapping. The ratio of APol and MP masses gives the amount of bound APol per MP.

**Method 3. Analytical Ultracentrifugation (AUC)**

☞ *The MP/APol mass ratio in complexes can be precisely determined by sedimentation velocity (SV) measurements using AUC. AUC is a priori applicable to any MP, because the density of A8-35 particles and that of MP/APol complexes are different enough for them to separate during the centrifugation run, even if their hydrodynamic radii are not very different (see Chap. 9, Protocol 9.1). For example, the sedimentation coefficients (s) of A8-35 particles ($R_S \approx 3.15$ nm; Gohon et al. 2006) and BR/A8-35 complexes ($R_S \approx 5.0$ nm; Gohon et al. 2008) are 1.6 S and 3.2 S, respectively, making them easily distinguishable. The specific volume of the sodium salt of A8-35, $\bar{v}_2$, is 0.809 $L \cdot g^{-1}$, its density, $\rho = 1/\bar{v}_2$, 1.236 $L \cdot g^{-1}$ (Gohon et al. 2004, 2006). The MP/APol mass ratio can be determined by sophisticated AUC measurements involving the comparison of sedimentation properties of complexes formed between the protein and unlabeled or deuterated A8-35 and/or simultaneous measurements of the absorbance and refractive index of the complexes (Gohon et al. 2008) (see Chap. 9). However, with the advent of FAPols, it is simpler to measure the respective absorbance of the protein and the APol in the complexes, as done above for the complexes separated by SEC or affinity chromatography.*

- After MP trapping in the FAPol/A8-35 mixture, adjust the sample concentration by dilution or concentration so that the protein absorbance at 280 nm, in the AUC cell, reaches ~0.5.
- Define the parameters of the SV run, namely time and speed, according to the sedimentation coefficient of the protein under study. For instance, in the case of small MPs, like BR or tOmpA, the SV experiment is carried out at 42,000 rpm during 4 h. The migration of the particles and complexes is followed at two wavelengths, 280 nm and the maximum absorbance wavelength of FAPol, and, if available, with interference optics, which give a measure of the refractive index.
- Measure the solvent density and viscosity.
- Analyze the SV profiles with Sedfit or an equivalent program (for details, see Gohon et al. 2008, and Chap. 9, Protocol 9.1). The distribution $c(s)$ of sedimentation coefficients ($s$) shows peaks reflecting the migration of MP/APol complexes and of free APol particles during the SV run. Integrate the peak areas at 280 nm and at the second wavelength. As noted above for SEC and IMAC experiments, the contribution of APols at 280 nm may have to be subtracted. The ratio of MP and APol in the complexes can be determined from their respective extinction coefficients.

*Protocol prepared by Manuela Zoonens on the basis of Zoonens et al. (2007), Gohon et al. (2008), Charvolin et al. (2014), and laboratory notes.*

### 5.9.3 Protocol 5.3. Transferring a MP from A8-35 to Nanodiscs

#### 5.9.3.1 Exchange of A8-35 for DDM

☞ *The interest of conducting MP studies in nanodiscs (NDs) rather than in APols lies in the ability of NDs to provide a bilayer-like environment of defined size and lipid composition (see Chap. 3, § 3.3). Even if the target MP is active and stable when trapped in APols, as is the case of the BLT2 GPCR used here (Dahmane et al. 2009; Catoire et al. 2010a), reintroducing it into a membrane environment is preferable for certain studies, especially those aimed at investigating the impact of the composition and biophysical properties of the lipid environment. When starting from a detergent-solubilized MP, the incorporation into NDs is straightforward, as NDs form spontaneously upon detergent removal from a mixture comprising the scaffold protein (MSP), lipids, and the target MP in a detergent solution. In the case of a MP trapped in APols, an additional step is required in order to remove the APol before proceeding to ND reconstitution. This is achieved by displacing the APol with an excess of detergent (Tribet et al. 1997; Zoonens et al. 2007). In our example, A8-35 is exchanged for DDM. Sodium cholate has been tested as well, as this detergent is frequently used to solubilize the ND reconstitution mixture. However, in our experience, sodium cholate is inefficient at displacing A8-35 from the transmembrane surface of BLT2. DDM, on the contrary, completely removes the APol (Zoonens et al. 2007), which has been checked by fluorescence spectroscopy, mass spectrometry, and NMR spectroscopy measurements (Casiraghi et al. 2016). The exchange is performed on a Ni:NTA-charged resin, to which BLT2 is bound via its polyhistidine tag. The addition of $Ca^{2+}$ facilitates APol removal, because it decreases A8-35 solubility (Picard et al. 2006; Diab et al. 2007) (Chap. 4, § 4.3.1.2.2). The exchange is conducted in the presence of 0.2 $g \cdot L^{-1}$ cholesteryl hemisuccinate (CHS), a soluble analog of cholesterol that has a stabilizing effect on GPCRs (Rosenbaum et al. 2007; Kuszak et al. 2009).*

*Briefly, the procedure is carried out as follows. Following A8-35-assisted folding of BLT2 (see Chap. 6), both the receptor/APol complex and the Ni:NTA resin are incubated in a buffer containing a high concentration of DDM (3 $g \cdot L^{-1}$). Following binding of BLT2 to the resin, two washing steps are applied, first at 2 $g \cdot L^{-1}$ DDM in the presence of $CaCl_2$, then with the same buffer but without $Ca^{2+}$. Once the exchange of surfactants is complete, the resin, carrying the bound BLT2/DDM complexes, is collected to proceed to ND reconstitution. Experiments conducted in the presence of $FAPol_{NBD}$, a fluorescent version of A8-35 (Zoonens et al. 2007) (see Chap. 4, § 4.4), whose absorbance was followed at 476 nm during FPLC and gel filtration analyses, attested to the complete removal of the APol.*

- Protocol

  1. Collect the APol-folded BLT2 (~1 mg for routine assays, ~6 mg for NMR samples) from the dialysis bag in which it was freed of the last traces of dodecyl sulfate (see Chap. 6, Protocol 6.1), and incubate at 4 °C for 2 h under gentle stirring with 3 $g \cdot L^{-1}$ DDM, 0.2 $g \cdot L^{-1}$ CHS in 50 mM Tris/HCl, 150 mM NaCl buffer, pH 8.
  2. It is advisable to use a FPLC chromatography system, if possible, so as to be able to monitor the $OD_{280}$ during the exchange of surfactants. Set up the chromatography instrument at 4 °C, pour in a XK 16/20 Column (GE Healthcare Life Sciences) the Ni:NTA resin (Ni-NTA Superflow, Qiagen, 2–3 mL of resin per mg of BLT2), wash with water purified on a Millipore Milli-Q Advantage A10 system (Milli-Q water), and equilibrate with 3 $g \cdot L^{-1}$ DDM, 0.2 $g \cdot L^{-1}$ CHS in 50 mM Tris/HCl, 150 mM NaCl buffer, pH 8.
  3. Pour the BLT2/A8-35/DDM solution incubated at stage 1 onto the chromatography column washed and equilibrated at stage 2. Wash the resin with 20 column volumes of

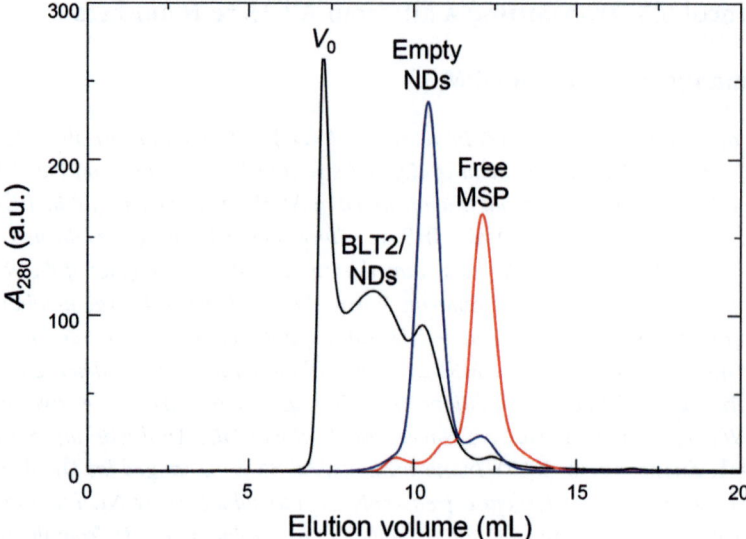

**Fig. 5.43** Gel filtration analysis of the product of a BLT2/ND reconstitution experiment. *Blue*, sample eluted from the Ni:NTA column with 10 mM imidazole; *black*, sample eluted with 400 mM imidazole; *red*, free MSP (control). In the void volume ($V_0$) elute liposomes, proteoliposomes, and aggregated material (From Casiraghi 2016).

  2 g·L$^{-1}$ DDM, 0.2 g·L$^{-1}$ CHS in 50 mM Tris/HCl, 150 mM NaCl, 2.5 mM CaCl$_2$ buffer, pH 8, followed by 30 volumes of the same buffer without CaCl$_2$.

4. Once the exchange has been completed, remove the column from the FLPC system and collect the resin, to proceed to ND reconstitution as described in the next section, steps 5–10.

### 5.9.3.2   Reconstitution into Nanodiscs

☞ *Incorporating a MP into NDs can be performed either in solution or with the protein attached to an Ni:NTA resin. In the first case, the protein is eluted from the resin after exchanging A8-35 for DDM; in the second, the resin is used directly. Whatever option is chosen, the protocol for ND formation is the same, except at the end of the procedure: if the reconstitution has been performed on the resin, MP-containing NDs can be separated from empty NDs by washing the resin with EDTA-free buffer, provided the target MP carries a polyhistidine tag and that of the MSP has been clipped off; if the reconstitution has been carried out in solution, empty NDs can be separated from those hosting a protein by SEC or IMAC.*

  *The lipids are solubilized with a detergent before being added to the reconstitution mixture. At the onset of the reconstitution process, the four partners are present as a solution containing detergent monomers and mixed MP/lipid/detergent, SMP/lipid/detergent, and lipid/detergent particles. Next, the detergent is removed, resulting in the formation of NDs, some of which incorporate the target MP. Whereas bile salts can be dialyzed away, DDM, due to its low CMC, cannot (see Chap. 2); it is eliminated by adsorption onto polystyrene beads. At the end of the procedure, the sample is analyzed by SEC (Fig. 5.43). If need be, the MSP/lipid ratio, which is a critical parameter (see Chap. 3, § 3.3.2), is adjusted to optimize disc formation.*

- Protocol

5. At the end of the APol/detergent exchange, the volume of the sample is estimated with a ruler, assuming the solution to represent ~30% of the volume of the resin (usually 5–7 mL). Purified MSP1D1 at ~15 $g \cdot L^{-1}$ is added at a 40:1 MSP1D1/BLT2 molar ratio. Aliquots of MSP1D1 (100 μL) are added to solubilized samples of DMPC (25 mM in 100 mM DDM) at ~20:1 DMPC/MSP1D1 molar ratio, final concentrations being ~4 mM DMPC and 200 μM MSP1D1. The CHS concentration is adjusted based on the total volume of the sample after the APol-to-detergent exchange. Different CHS/DMPC molar ratios were tested, so as to evaluate their effect on GPCR activity (Casiraghi et al. 2016), the CHS concentration being adjusted with 20-$g \cdot L^{-1}$ CHS solution in 50 mM Tris/HCl, pH 8, 100 mM DDM. Reconstitution can be conducted either directly in the column, immediately after the APol-to-detergent exchange, or the resin can be transferred to a glassware.

6. Once all the components are added, incubate the mixture for 1 h above the transition temperature of the lipids (23 °C in this specific case, as the transition temperature of the binary lipid mixture was estimated to be ~20.5 °C) under slight stirring in a thermostatically controlled device or a temperature-controlled room, so as to let the components mix (Ritchie et al. 2009).

7. NDs spontaneously form upon detergent removal, for which various methods can be used (cf. Chap. 3). If polystyrene beads are used, they must be washed several times in ethanol followed by several more washes in Milli-Q water and finally stored in water at 4 °C. Before use, dry away the excess of water by depositing the beads on filter paper. Typically, 60% w/v Bio-Beads SM2 (Bio-Rad) are directly added to the column or vial. Stir for 2 h at the gel-fluid phase temperature transition, then at 4 °C overnight.

8. At this point, BLT2/ND complexes are separated from empty discs by IMAC. If reconstitution was performed directly in the chromatographic column, this step can be performed by FPLC, by simply reconnecting the column to the system. For reconstitution conducted in a glassware, the resin is poured into a disposable plastic column. In both cases, empty discs are washed with 50 mM Tris/HCl, 150 mM NaCl, 10 mM imidazole, pH 8, after which BLT2/ND complexes are eluted with 50 mM Tris/HCl, 150 mM NaCl, 400 mM imidazole, pH 8.

9. Exchange the buffer of the eluate for 20 mM Tris/HCl, 100 mM NaCl, 5 mM EDTA, pH 8, using centrifugal filters (Amicon Ultra-4, 10 kDa cutoff, Merck), and concentrate by centrifugation at 3,500 × $g$, 4 °C (Allegra 64R centrifuge, Beckman Coulter), until reaching a final volume suitable for SEC.

10. Before SEC, the sample is centrifuged at 40,000 × $g$, 30 min, 4 °C (Beckman Coulter, Optima MAX-XP ultracentrifuge; rotor TLA 100), in order to remove most of the excess of lipids, proteoliposomes, and aggregates. SEC is performed on a Superose 12 10/300 GL column (GE Healthcare, Life Sciences) directly assembled on the Äkta purifier system (GE Healthcare) and equilibrated with 20 mM Tris/HCl, pH 8, 100 mM NaCl, 5 mM EDTA. The sample (~100 μL) is injected through a 100 μL loop. Typically, after reconstitution, a mixture of different particles is obtained (Fig. 5.43). The peak corresponding to BLT2/ND complexes is pooled and characterized (usually by SDS-PAGE, EM, mass spectrometry, etc.; see Fig. 5.44) and, if the sample is satisfying, used for NMR studies.

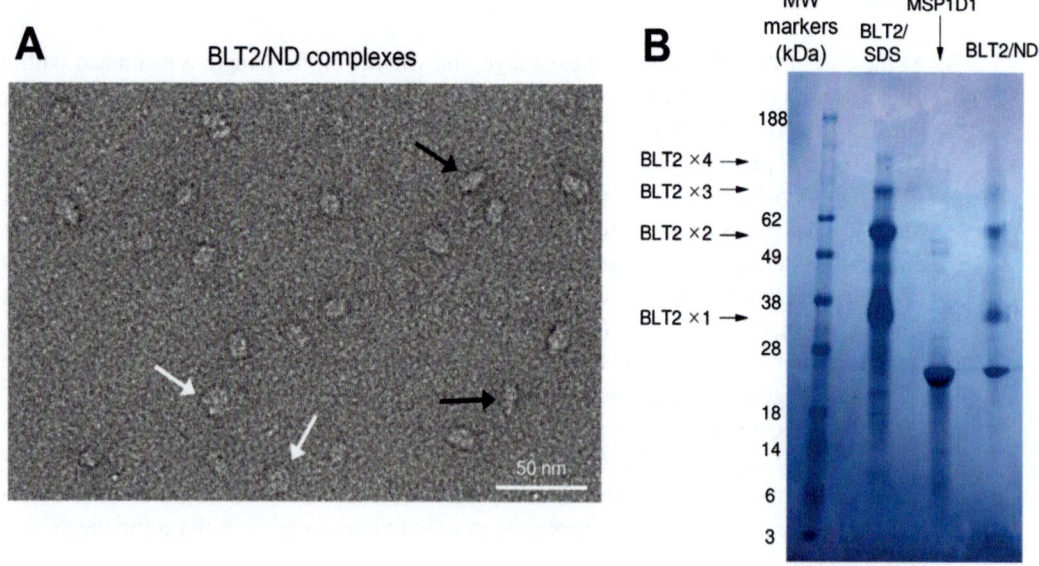

**Fig. 5.44** Characterization of MSP1D1 NDs hosting the leukotriene BLT2 receptor. (**A**) Electron microscopy image of negatively stained BLT2/ND complexes purified by SEC (cf. Fig. 5.43). The discs are 12–15 nm in diameter. *White arrows,* top views; *black arrows,* side views. (**B**) SDS-polyacrylamide gel electrophoresis analysis. BLT2/NDs, complexes purified by SEC; BLT2/SDS and MSP1D1, controls. Note that, as frequently occurs with polytopic MPs, BLT2 tends to form oligomers in SDS (Reprinted with permission from Casiraghi et al. 2016, © 2016 American Chemical Society).

*Protocol prepared by Marina Casiraghi and Laurent J. Catoire on the basis of* Damian et al. (2012), Casiraghi et al. (2016), *and laboratory notes.*

## References

Agosto, M.A., Zhang, Z., He, F., Anastassov, I.A., Wright, S.J., McGehee, J., Wensel, T.G. (2014) Oligomeric state of purified TRPM1, a protein essential for dim light vision. *J. Biol. Chem.* **289**:27019–27033.

Althoff, T., Mills, D.J., Popot, J.-L., Kühlbrandt, W. (2011) Assembly of electron transport chain components in bovine mitochondrial supercomplex $I_1III_2IV_1$. *EMBO J.* **30**:4652–4664.

Arunmanee, W., Harris, J.R., Lakey, J.H. (2014) Outer membrane protein F stabilised with minimal amphipol forms linear arrays and LPS-dependent 2D crystals. *J. Membr. Biol.* **247**:949–956.

Baenziger, J.E., Hénault, C.M., Therien, J.P.D., Sun, J. (2015) Nicotinic acetylcholine receptor-lipid interactions: Mechanistic insight and biological function. *Biochim. Biophys. Acta* **1848**:1806–1817.

Bai, X.-C., Yan, C., Yang, G., Lu, P., Ma, D., Sun, L., Zhou, R., Scheres, S.H.W., Shi, Y. (2015) An atomic structure of human γ-secretase. *Nature* **525**:212–218.

Baker, M.R., Fan, G., Serysheva, I.I. (2015) Single-particle cryo-EM of the ryanodine receptor channel in an aqueous environment. *Eur. J. Transl. Myol.* **25**:35–48.

Banères, J.-L., Popot, J.-L., Mouillac, B. (2011) New advances in production and functional folding of G protein-coupled receptors. *Trends Biotechnol.* **29**:314–322.

Basit, H., Sharma, S., Van der Heyden, A., Gondran, C., Breyton, C., Dumy, P., Winnik, F.M., Labbé, P. (2012) Amphipol-mediated surface immobilization of FhuA: a platform for label-free detection of the bacteriophage protein pb5. *Chem. Commun.* **48**:6037–6039.

Bazzacco, P. (2009) Non-ionic amphipols: new tools for in vitro studies of membrane proteins. Validation and development of biochemical and biophysical applications. Thèse de Doctorat, Université Paris-7, Paris, 176 p.

Bazzacco, P., Billon-Denis, E., Sharma, K.S., Catoire, L.J., Mary, S., Le Bon, C., Point, E., Banères, J.-L., Durand, G., Zito, F., Pucci, B., Popot, J.-L. (2012) Non-ionic homopolymeric amphipols: Application to membrane protein folding, cell-free synthesis, and solution NMR. *Biochemistry* **51**:1416–1430.

Bazzacco, P., Sharma, K.S., Durand, G., Giusti, F., Ebel, C., Popot, J.-L., Pucci, B. (2009) Trapping and stabilization of integral membrane proteins by hydrophobically grafted glucose-based telomers. *Biomacromolecules* **10**:3317–3326.

Bechara, C., Bolbach, G., Bazzacco, P., Sharma, S.K., Durand, G., Popot, J.-L., Zito, F., Sagan, S. (2012) MALDI mass spectrometry analysis of membrane protein/amphipol complexes. *Anal. Chem.* **84**:6128–6135.

Bersch, B., Dörr, J.M., Hessel, A., Killian, J.A., Schanda, P. (2017) Proton-detected solid-state NMR spectroscopy of a zinc diffusion facilitator protein in native nanodiscs. *Angew. Chem. Int. Ed.* **56**:2508–2512.

Booth, M., Peel, R., Partanen, R., Hondow, N., Vasilca, V., Jeuken, L.J.C., Critchley, K. (2013) Amphipol-encapsulated $CuInS_2/ZnS$ quantum dots with excellent colloidal stability. *RSC Adv.* **3**:20559–20566.

Botte, M., Zaccai, N.R., Lycklama A., Nijeholt, J., Martin, R., Knoops, K., Papai, G., Zou, J., Deniaud, A., Karuppasamy, M., Jiang, Q., Singha Roy, A., Schulten, K., Schultz, P., Rappsilber, J., Zaccai, G., Berger, I., Collinson, I., Schaffitzel, C. (2016) A central cavity within the holotranslocon suggests a mechanism for membrane protein insertion. *Sci. Rep.* **6**:38399.

Breyton, C., Tribet, C., Olive, J., Dubacq, J.-P., Popot, J.-L. (1997) Dimer to monomer conversion of the cytochrome $b_6 f$ complex: causes and consequences. *J. Biol. Chem.* **272**:21892–21900.

Broecker, J., Eger, B.T., Ernst, O.P. (2017) Crystallogenesis of membrane proteins mediated by polymer-bounded lipid nanodiscs. *Structure* **25**:384–392.

Brotherus, J.R., Jost, P.C., Griffith, O.H., Hokin, L.E. (1979) Detergent inactivation of sodium- and potassium-activated adenosinetriphosphatase of the electric eel. *Biochemistry* **18**:5043–5050.

Calabrese, A.N., Watkinson, T.G., Henderson, P.J.F., Radford, S.E., Ashcroft, A.E. (2015) Amphipols outperform dodecylmaltoside micelles in stabilizing membrane protein structure in the gas phase. *Anal. Chem.* **87**:1118–1126.

Cao, E., Liao, M., Cheng, Y., Julius, D. (2013) TRPV1 structures in distinct conformations reveal activation mechanisms. *Nature* **504**:113–118.

Casiraghi, M. (2016) Functional Modulation of a G Protein-Coupled Receptor Conformational Landscape in a Lipid Bilayer. Thèse de Doctorat, Paris-7 University, Paris, 249 p.

Casiraghi, M., Damian, M., Lescop, E., Point, E., Moncoq, K., Morellet, N., Levy, D., Marie, J., Guittet, E., Banères, J.-L., Catoire, L.J. (2016) Functional modulation of a GPCR conformational landscape in a lipid bilayer. *J. Am. Chem. Soc.* **138**:11170–11175

Catoire, L.J., Damian, M., Baaden, M., Guittet, E., Banères, J.-L. (2011) Electrostatically-driven fast association and perdeuteration allow detection of transferred cross-relaxation for G protein-coupled receptor ligands with equilibrium dissociation constants in the high-to-low nanomolar range. *J. Biomol. NMR* **50**:191–195.

Catoire, L.J., Damian, M., Giusti, F., Martin, A., van Heijenoort, C., Popot, J.-L., Guittet, E., Banères, J.-L. (2010a) Structure of a GPCR ligand in its receptor-bound state: leukotriene $B_4$ adopts a highly constrained conformation when associated to human BLT2. *J. Am. Chem. Soc.* **132**:9049–9057.

Catoire, L.J., Zoonens, M., van Heijenoort, C., Giusti, F., Guittet, E., Popot, J.-L. (2010b) Solution NMR mapping of water-accessible residues in the transmembrane $\beta$-barrel of OmpX. *Eur. Biophys. J.* **39**:623–630.

Catoire, L.J., Zoonens, M., van Heijenoort, C., Giusti, F., Popot, J.-L., Guittet, E. (2009) Inter- and intramolecular contacts in a membrane protein/surfactant complex observed by heteronuclear dipole-to-dipole cross-relaxation. *J. Magn. Res.* **197**:91–95.

Champeil, P., le Maire, M., Andersen, J.P., Guillain, F., Gingold, M., LundII, S., Møller, J.V. (1986) Kinetic characterization of the normal and detergent-perturbed reaction cycles of the sarcoplasmic reticulum calcium pump. Rate-limiting steps under different conditions. *J. Biol. Chem.* **261**:16372–16384.

Champeil, P., Menguy, T., Tribet, C., Popot, J.-L., le Maire, M. (2000) Interaction of amphipols with the sarcoplasmic reticulum $Ca^{2+}$-ATPase. *J. Biol. Chem.* **275**:18623–18637.

Changeux, J.-P., Giraudat, J., Heidmann, T., Popot, J.-L., Sobel, A. (1980) Functional properties of the acetylcholine receptor protein. *Neurochem. Int.* **2**:219–231.

Charvolin, D., Perez, J.-B., Rouvière, F., Giusti, F., Bazzacco, P., Abdine, A., Rappaport, F., Martinez, K.L., Popot, J.-L. (2009) The use of amphipols as universal molecular adapters to immobilize membrane proteins onto solid supports. *Proc. Natl. Acad. Sci. USA* **106**:405–410.

Charvolin, D., Picard, M., Huang, L.-S., Berry, E.A., Popot, J.-L. (2014) Solution behavior and crystallization of cytochrome $bc_1$ in the presence of amphipols. *J. Membr. Biol.* **247**:981–996.

Chen, Y., Clarke, O.B., Kim, J., Stowe, S., Kim, Y.-K., Assur, Z., Cavalier, C., Godoy-Ruiz, R., von Alpen, D.C., Manzini, C., Blaner, W.S., Frank, J., Quadro, L., Weber, D.J., Shapiro, L., Hendrickson, W.A., Mancia, F. (2016) Structure of the STRA6 receptor for retinol uptake. *Science* **353**:pii aad8266–8261.

Cherezov, V., J. C., Papiz, M.Z., Caffrey, M. (2006) Room to move: crystallizing membrane proteins in swollen lipidic mesophases. *J. Mol. Biol.* **357**:1605–1618.

Chiu, Y.H., Jin, X., Medina, C., Leonhardt, S.A., Kiessling, V., Bennett, B.C., Shu, S., Tamm, L.K., Yeager, M., Ravichandran, K.S., Bayliss, D.A. (2017) A quantized mechanism for activation of pannexin channels. *Nat. Commun.* **8**:14324.

Choutko, A., Glättli, A., Fernández, C., Hilty, C., Wüthrich, K., van Gunsteren, W.F. (2011) Membrane protein dynamics in different environments: simulation study of the outer membrane protein X in a lipid bilayer and in a micelle. *Eur. Biophys. J.* **40**:39–58.

Constantine, M., Liew, C.K., Lo, V., Macmillan, A., Cranfield, C.G., Sunde, M., Whan, R., Graham, R.M., Martinac, B. (2016) Heterologously-expressed and liposome-reconstituted human transient receptor potential melastatin 4 channel (TRPM4) is a functional tetramer. *Sci. Rep.* **6**:19352.

Corcelli, A., Colella, M., Mascolo, G., Fanizzi, F.P., Kates, M. (2000) A novel glycolipid and phospholipid in the purple membrane. *Biochemistry* **39**:3318–3326.

Corringer, P.-J., Baaden, M., Bocquet, N., Delarue, M., Dufresne, V., Nury, H., Prevost, M., Van Renterghem, C. (2010) Atomic structure and dynamics of pentameric ligand-gated ion channels: new insight from bacterial homologues. *J. Physiol.* **588**:565–572.

Cuevas Arenas, R., Danielczak, B., Martel, A., Porcar, L., Breyton, C., Ebel, C., Keller, S. (2017) Fast collisional lipid transfer among polymer-bounded nanodiscs. *Sci. Rep.* **7**:45875.

Cvetkov, T.L., Huynh, K.W., Cohen, M.R., Moiseenkova-Bell, V.Y. (2011) Molecular architecture and subunit organization of TRPA1 ion channel revealed by electron microscopy. *J. Biol. Chem.* **286**:38168–38176.

daCosta, C.J.B., Baenziger, J.E. (2009) A lipid-dependent uncoupled conformation of the acetylcholine receptor. *J. Biol. Chem.* **284**:17819–17825.

Dahmane, T. (2007) Protéines membranaires et amphipols : stabilisation, fonction, renaturation, et développement d'amphipols sulfonatés pour la RMN des solutions. Thèse de Docorat, Université Paris-7, Paris, 229 p.

Dahmane, T., Damian, M., Mary, S., Popot, J.-L., Banères, J.-L. (2009) Amphipol-assisted *in vitro* folding of G protein-coupled receptors. *Biochemistry* **48**:6516–6521.

Dahmane, T., Giusti, F., Catoire, L.J., Popot, J.-L. (2011) Sulfonated amphipols: Synthesis, properties and applications. *Biopolymers* **95**:811–823.

Dahmane, T., Rappaport, F., Popot, J.-L. (2013) Amphipol-assisted folding of bacteriorhodopsin in the presence and absence of lipids. Functional consequences. *Eur. Biophys. J.* **42**:85–101.

Damian, M., Marie, J., Leyris, J.-P., Fehrentz, J.-A., Verdié, P., Martinez, J., Banères, J.-L., Mary, S. (2012) High constitutive activity is an intrinsic feature of ghrelin receptor protein: a study with a functional monomeric GHS-R1a receptor reconstituted in lipid discs. *J. Biol. Chem.* **287**:3630–3641.

Davidson, A.L., Nikaido, H. (1991) Purification and characterization of the membrane-associated components of the maltose transport system from *Escherichia coli*. *J. Biol. Chem.* **266**:8946–8951.

de Foresta, B., le Maire, M., Orlowski, S., Champeil, P., Lund, S., Møller, J.V., Michelangeli, F., Lee, A.G. (1989) Membrane solubilization by detergent: use of brominated phospholipids to evaluate the detergent-induced changes in $Ca^{2+}$-ATPase/lipid interaction. *Biochemistry* **28**:2558–2567.

de Vitry, C., Diner, B.A., Popot, J.-L. (1991) Photosystem II particles from *Chlamydomonas reinhardtii*: purification, molecular weight, small subunit composition, protein phosphorylation. *J. Biol. Chem.* **266**:16614–16621.

Della Pia, E.A., Holm, J., Lloret, N., Le Bon, C., Popot, J.-L., Zoonens, M., Nygård, J., Martinez, K.L. (2014) A step closer to membrane protein multiplexed nano-arrays using biotin-doped polypyrrole. *ACS Nano* **8**:1844–1853.

Diab, C., Tribet, C., Gohon, Y., Popot, J.-L., Winnik, F.M. (2007) Complexation of integral membrane proteins by phosphorylcholine-based amphipols. *Biochim. Biophys. Acta* **1768**:2737–2747.

Dominguez Pardo, J.J., Dörr, J.M., Iyer, A., Cox, R.C., Scheidelaar, S., Koorengevel, M.C., Subramaniam, V., Killian, J.A. (2017) Solubilization of lipids and lipid phases by the styrene-maleic acid copolymer. *Eur. Biophys. J.* **46**:91–101.

Dörr, J.M., Koorengevel, M.C., Schäfer, M., Prokofyev, A.V., Scheidelaar, S., van der Cruijsenb, E.A.W., Dafforn, T.R., Baldus, M., Killian, J.A. (2014) Detergent-free isolation, characterization, and functional reconstitution of a tetrameric $K^+$ channel: The power of native nanodiscs. *Proc. Natl. Acad. Sci. USA* **111**:18607–18612.

Dörr, J.M., Scheidelaar, S., Koorengevel, M.C., Dominguez, J.J., Schäfer, M., van Walree, C.A., Killian, J.A. (2016) The styrene-maleic acid copolymer: a versatile tool in membrane research. *Eur. Biophys. J.* **45**:3–21.

Duarte, A.M.S., Wolfs, C.J.A.M., Koehorsta, R.B.M., Popot, J.-L., Hemminga, M.A. (2008) Solubilization of V-ATPase transmembrane peptides by amphipol A8-35. *J. Peptide Chem.* **14**:389–393.

Elter, S., Raschle, T., Arens, S., Viegas, A., Gelev, V., Etzkorn, M., Wagner, G. (2014) The use of amphipols for NMR structural characterization of 7-TM proteins. *J. Membr. Biol.* **247**:957–964.

Etzkorn, M., Raschle, T., Hagn, F., Gelev, V., Rice, A.J., Walz, T., Wagner, G. (2013) Cell-free expressed bacteriorhodopsin in different soluble membrane mimetics: biophysical properties and NMR accessibility. *Structure* **21**:394–401.

Etzkorn, M., Zoonens, M., Catoire, L.J., Popot, J.-L., Hiller, S. (2014) How amphipols embed membrane proteins: Global solvent accessibility and interaction with a flexible protein terminus. *J. Membr. Biol.* **247**:965–970.

Fan, G., Gonzalez, J., Popova, O.B., Wensel, T.G., Serysheva, I.I. (2014) A first look into the 3D structure of the TRPV2 channel by single-particle cryo-EM. *Biophys. J.* 106:600a-601a.

Feinstein, H.E., Tifrea, D., Sun, G., Popot, J.-L., de la Maza, L.M., Cocco, M.J. (2014) Long-term stability of a vaccine formulated with the amphipol-trapped major outer membrane protein from *Chlamydia trachomatis*. *J. Membr. Biol.* **247**:1053–1065.

Fernández, C., Adeishvili, K., Wüthrich, K. (2001) Transverse relaxation-optimized NMR spectroscopy with the outer membrane protein OmpX in dihexanoylphosphatidylcholine micelles. *Proc. Natl. Acad. Sci. USA* **98**:2358–2363.

Ferrandez, Y., Dezi, M., Bosco, M., Urvoas, A., Valério, M., Le Bon, C., Giusti, F., Broutin, I., Durand, G., Polidori, A., Popot, J.-L., Picard, M., Minard, P. (2014) Amphipol-mediated screening of molecular orthoses specific for membrane protein targets. *J. Membr. Biol.* **247**:925–940.

Flötenmeyer, M., Weiss, H., Tribet, C., Popot, J.-L., Leonard, K. (2007) The use of amphipathic polymers for cryo-electron microscopy of NADH:ubiquinone oxidoreductase (Complex I). *J. Microsc.* **227**:229–235.

Forman, S.A., Chiara, D.C., Miller, K.W. (2015) Anesthetics target interfacial transmembrane sites in nicotinic acetylcholine receptors. *Neuropharmacology* **96**:169–177.

Frindi, M., Michels, B., Zana, R. (1992a) Ultrasonic absorption studies of surfactant exchange between micelles and bulk phase in aqueous micellar solutions of nonionic surfactants with a short alkyl chain. 2. $C_6E_3$, $C_8E_4$ and $C_8E_8$. *J. Phys. Chem.* **96**:6095–6102.

Frindi, M., Michels, B., Zana, R. (1992b) Ultrasonic absorption studies of surfactant exchange between micelles and bulk phase in aqueous micellar solutions of nonionic surfactants with a short alkyl chain. 3. Surfactants with a sugar head group. *J. Phys. Chem.* **96**:8137–8141.

Gao, Y., Cao, E., Julius, D., Cheng, Y. (2016) TRPV1 structures in nanodiscs reveal mechanisms of ligand and lipid action. *Nature* **534**:347–351.

Ge, J., Li, W., Zhao, Q., Li, N., Chen, M., Zhi, P., Li, R., Gao, N., Xiao, B., Yang, M. (2015) Architecture of the mammalian mechanosensitive Piezo1 channel. *Nature* **527**:64–69.

Giusti, F., Kessler, P., Westh Hansen, R., Della Pia, E.A., Le Bon, C., Mourier, G., Popot, J.-L., Martinez, K.L., Zoonens, M. (2015) Synthesis of a polyhistidine-bearing amphipol and its use for immobilizing membrane proteins. *Biomacromolecules* **16**:3751–3761.

Giusti, F., Popot, J.-L., Tribet, C. (2012) Well-defined critical association concentration and rapid adsorption at the air/water interface of a short amphiphilic polymer, amphipol A8-35: A study by Förster resonance energy transfer and dynamic surface tension measurements. *Langmuir* **28**:10372–10380.

Giusti, F., Rieger, J., Catoire, L., Qian, S., Calabrese, A.N., Watkinson, T.G., Casiraghi, M., Radford, S.E., Ashcroft, A. E., Popot, J.-L. (2014) Synthesis, characterization and applications of a perdeuterated amphipol. *J. Membr. Biol.* **247**:909–924.

Goddard, A.D., Dijkman, P.M., Adamson, R.J., Inácio dos Reis, R., Watts, A. (2015) Reconstitution of membrane proteins: A GPCR as an example. *Meth. Enzymol.* **556**:405–424.

Gohon, Y. (1996) Etude des interactions entre un analogue du fragment transmembranaire de la glycophorine A et des polymères amphiphiles: les amphipols. Thèse de DEA, Université Paris VI, Paris, 28 p.

Gohon, Y. (2002) Etude structurale et fonctionnelle de deux protéines membranaires, la bactériorhodopsine et le récepteur nicotinique de l'acétylcholine, maintenues en solution aqueuse non détergente par des polymères amphiphiles. Thèse de Doctorat, Université Paris-VI, Paris, 467 p.

Gohon, Y., Dahmane, T., Ruigrok, R., Schuck, P., Charvolin, D., Rappaport, F., Timmins, P., Engelman, D.M., Tribet, C., Popot, J.-L., Ebel, C. (2008) Bacteriorhodopsin/amphipol complexes: structural and functional properties. *Biophys. J.* **94**:3523–3537.

Gohon, Y., Giusti, F., Prata, C., Charvolin, D., Timmins, P., Ebel, C., Tribet, C., Popot, J.-L. (2006) Well-defined nanoparticles formed by hydrophobic assembly of a short and polydisperse random terpolymer, amphipol A8-35. *Langmuir* **22**:1281–1290.

Gohon, Y., Pavlov, G., Timmins, P., Tribet, C., Popot, J.-L., Ebel, C. (2004) Partial specific volume and solvent interactions of amphipol A8-35. *Anal. Biochem.* **334**:318–334.

Gorzelle, B.M., Hoffman, A.K., Keyes, M.H., Gray, D.N., Ray, D.G., Sanders II, C.R. (2002) Amphipols can support the activity of a membrane enzyme. *J. Am. Chem. Soc.* **124**:11594–11595.

Goyal, P., Krasteva, P.V., Van Gerven, N., Gubellini, F., Van den Broeck, I., Troupiotis-Tsaïlaki, A., Jonckheere, W., Péhau-Arnaudet, G., Pinkner, J.S., Chapman, M.R., Hultgren, S.J., Howorka, S., Fronzes, R., Remaut, H. (2014) Structural and mechanistic insights into the bacterial amyloid secretion channel CsgG. *Nature* **516**:250–253.

Grethen, A., Glueck, D., Keller, S. (2018) Role of coulombic repulsion in collisional lipid transfer among SMA(2:1)-bounded nanodiscs. *J. Membr. Biol., in the press.*

Grethen, A., Oluwole, A.O., Danielczak, B., Vargas, C., Keller, S. (2017) Thermodynamics of nanodisc formation mediated by styrene/maleic acid (2:1) copolymer. *Sci. Rep.* **7**:11517.

Grigorieff, N., Ceska, T.A., Downing, K.H., Baldwin, J.M., Henderson, R. (1996) Electron-crystallographic refinement of the structure of bacteriorhodopsin. *J. Mol. Biol.* **259**:393–421.

Guild, K., Zhang, Y., Stacy, R., Mundt, E., Benbow, S., Green, A., Myler, P.J. (2011) Wheat germ cell-free expression system as a pathway to improve protein yield and solubility for the SSGCID pipeline. *Acta Crystallogr. Sect. F Struct. Biol. Cryst. Commun.* **67**:1027–1031.

Gulati, S., Jamshad, M., Knowles, T.J., Morrison, K.A., Downing, R., Cant, N., Collins, R., Koenderink, J.B., Ford, R. C., Overduin, M., Kerr, I.D., Dafforn, T.R., Rothnie, A.J. (2014) Detergent-free purification of ABC (ATP-binding-cassette) transporters. *Biochem. J.* **461**:269–278.

Han, S.G., Baek, S.I., Son, T.J., Lee, H., Kim, N.H., Yu, Y.G. (2017) Preparation of functional human lysophosphatidic acid receptor 2 using a P9* expression system and an amphipathic polymer and investigation of its in *vitro* binding preference to $G_\alpha$ proteins. *Biochem. Biophys. Res. Commun.* **487**:103–108.

Han, S.G., Na, J.H., Lee, W.K., Park, D., Oh, J., Yoon, S.H., Lee, C.K., Sung, M.H., Shin, Y.K., Yu, Y.G. (2014) An amphipathic polypeptide derived from poly-γ-glutamic acid for the stabilization of membrane proteins. *Prot. Sci.* **23**:1800–1807.

He, Y., Gao, X., Goswami, D., Hou, L., Pal, K., Yin, Y., Zhao, G., Ernst, O.P., Griffin, P., Melcher, K., Xu, H.E. (2017) Molecular assembly of rhodopsin with G protein-coupled receptor kinases. *Cell Res.* **2017**:1–20.

Hénault, C.M., Sun, J., Therien, J.P.D., daCosta, C.J.B., Carswell, C.L., Labriola, J.M., Juranka, P.F., Baenziger, J.E. (2015) The role of the M4 lipid-sensor in the folding, trafficking, and allosteric modulation of nicotinic acetylcholine receptors. *Neuropharmacology* **96**:157–168.

Hilf, R.J., Dutzler, R. (2009) Structure of a potentially open state of a proton-activated pentameric ligand-gated ion channel. *Nature* **457**:115–118.

Hirai, T., Subramaniam, S., Lanyi, J.K. (2009) Structural snapshots of conformational changes in a seven-helix membrane protein: lessons from bacteriorhodopsin. *Curr. Opin. Struct. Biol.* **19**:433–439.

Hong, H., Tamm, L.K. (2004) Elastic coupling of integral membrane protein stability to lipid bilayer forces. *Proc. Natl. Acad. Sci. USA* **101**:4065–4070.

Hopper, J.T.S., Yu, Y.T.-C., Li, D., Raymond, A., Bostock, M., Liko, I., Mikhailov, V., Laganowsky, A., Benesch, J.L. P., Caffrey, M., Nietlispach, D., Robinson, C.V. (2013) Detergent-free mass spectrometry of membrane protein complexes. *Nat. Meth.* **10**:1206–1208.

Huang, K.-S., Bayley, H., Liao, M.-J., London, E., Khorana, H.G. (1981) Refolding of an integral membrane protein. Denaturation, renaturation, and reconstitution of intact bacteriorhodopsin and two proteolytic fragments. *J. Biol. Chem.* **256**:3802–3809.

Huynh, K.W., Cohen, M.R., Moiseenkova-Bell, V.Y. (2014) Application of amphipols for structure-functional analysis of TRP channels. *J. Membr. Biol.* **247**:843–851.

Ireland, S.M., Sula, S., Wallace, B.A. (2017) Thermal melt circular dichroism spectroscopic studies for identifying stabilising amphipathic molecules for the voltage-gated sodium channel NavMs. *Biopolymers* **2017**:e23067.

Jamshad, M., Charlton, J., Lin, Y.-P., Routledge, S.J., Bawa, Z., Knowles, T.J., Overduin, M., Dekker, N., Dafforn, T.R., Bill, R.M., Poyner, D.R., Wheatley, M. (2015a) G protein-coupled receptor solubilization and purification for biophysical analysis and functional studies, in the total absence of detergent. *Biosc. Rep.* **35**:e00188.

Jamshad, M., Grimard, V., Idini, I., Knowles, T.J., Dowle, M.R., Schofield, N., Sridhar, P., Lin, Y., Finka, R., Wheatley, M., Thomas, O.R.T., Palmer, R.E., Overduin, M., Govaerts, C., Ruysschaert, J.-M., Edler, K.J., Dafforn, T.R. (2015b) Structural analysis of a nanoparticle containing a lipid bilayer used for detergent-free extraction of membrane proteins *Nano Res.* 8:774–789.

Jamshad, M., Lin, Y.P., Knowles, T.J., Parslow, R.A., Harris, C., Wheatley, M., Poyner, D.R., Bill, R.M., Thomas, O.R.T., Overduin, M., Dafforn, T.R. (2011) Surfactant-free purification of membrane proteins with intact native membrane environment. *Biochem. Soc. Trans.* **39**:813–818.

Jeong, H., Kim, J.-S., Song, S., Shigematsu, H., Yokoyama, T., Hyun, J., Ha, N.-C. (2016) Pseudoatomic structure of the tripartite multidrug efflux pump AcrAB-TolC reveals the intermeshing cogwheel-like interaction between AcrA and TolC. *Structure* **24**:272–276.

Jin, P., Bulkley, D., Guo, Y., Zhang, W., Guo, Z., Huynh, W., Wu, S., Meltzer, S., Cheng, T., Jan, L.Y., Jan, Y.-N., Cheng, Y. (2017) Electron cryo-microscopy structure of the mechanotransduction channel NOMPC. *Nature* **547**:118–122.

Joshi, M., Dracheva, S., Mukhopadhyay, A.K., Bose, S., Hendler, R.W. (1998) Importance of specific native lipids in controlling the photocycle of bacteriorhodopsin. *Biochemistry* **37**:14463–14470.

Kevany, B.M., Tsybovsky, Y., Campuzano, I.D.G., Schnier, P.D., Engel, A., Palczewski, K. (2013) Structural and functional analysis of the native peripherin-ROM1 complex isolated from photoreceptor cells. *J. Biol. Chem.* **288**:36272–36284.

Kievit, O., Brudvig, G.W. (2001) Direct electrochemistry of photosystem I. *J. Electroanal. Chem.* **497**:139–149.

Klammt, C., Perrin, M.-H., Maslennikov, I., Renault, L., Krupa, M., Kwiatkowski, W., Stahlberg, H., Vale, W., Choe, S. (2011) Polymer-based cell-free expression of ligand-binding family B G protein-coupled receptors without detergents. *Prot. Sci.* **20**:1030–1041.

Kleinschmidt, J.H., Popot, J.-L. (2014) Folding and stability of integral membrane proteins in amphipols. *Arch. Biochem. Biophys.* **564**:327–343.

Knowles, T.J., Finka, R., Smith, C., Lin, Y.-P., Dafforn, T.R., Overduin, M. (2009) Membrane proteins solubilized intact in lipid-containing nanoparticles bounded by styrene maleic acid copolymer. *J. Am. Chem. Soc.* **131**:7484–7485.

Kraft, T.E., Hresko, R.C., Hruz, P.W. (2015) Expression, purification, and functional characterization of the insulin-responsive facilitative glucose transporter GLUT4. *Protein Sci.* **24**:2008–2019.

Kumar, S., Bagchi, S., Prasad, S., Sharma, A., Kumar, R., Kaur, R., Singh, J., Bhondekar, A.P. (2016) Bacteriorhodopsin–ZnO hybrid as a potential sensing element for low-temperature detection of ethanol vapour. *Beilstein J. Nanotechnol.* **7**:501–510.

Kuszak, A.J., Pitchiaya, S., Anand, J.P., Mosberg, H.I., Walter, N.G., Sunahara, R.K. (2009) Purification and functional reconstitution of monomeric μ-opioid receptors: Allosteric modulation of agonist binding by $G_{i2}$. *J. Biol. Chem.* **284**:26732–26741.

Kyrychenko, A., Rodnin, M.V., Vargas, M.U., Sharma, S.K., Durand, G., Pucci, B., Popot, J.-L., Ladokhin, A.S. (2012) Folding of diphtheria toxin T-domain in the presence of amphipols and fluorinated surfactants: Toward thermodynamic measurements of membrane protein folding. *Biochim. Biophys. Acta* **1818**:1006–1012.

Ladavière, C., Toustou, M., Gulik-Krzywicki, T., Tribet, C. (2001) Slow reorganization of small phosphatidylcholine vesicles upon adsorption of amphiphilic polymers. *J. Colloid Interface Sci.* **241**:178–187.

Ladavière, C., Tribet, C., Cribier, S. (2002) Lateral organization of lipid membranes induced by amphiphilic polymer inclusions. *Langmuir* **18**:7320–7327.

Laursen, T., Borch, J., Knudsen, C., Bavishi, K., Torta, F., Martens, H.J., Silvestro, D., Hatzakis, N.S., Wenk, M.R., Dafforn, T.R., Olsen, C.E., Motawia, M.S., Hamberger, B., Møller, B.L., Bassard, J.-E. (2016) Characterization of a dynamic metabolon producing the defense compound dhurrin in sorghum. *Science* **354**:890–893.

Laursen, T., Naur, P., Møller, B.L. (2013) Amphipol trapping of a functional CYP system. *Biotechn. Applied Biochem.* **60**:119–127.

Le Bon, C., Della Pia, E.A., Giusti, F., Lloret, N., Zoonens, M., Martinez, K.L., Popot, J.-L. (2014) Synthesis of an oligonucleotide-derivatized amphipol and its use to trap and immobilize membrane proteins. *Nucleic Acids Res.* **42**:e83.

Le Bon, C., Marconnet, A., Masscheleyn, S., Popot, J.-L., Zoonens, M. (2018) Folding and stabilizing membrane proteins in amphipol A8-35. *Methods, in the press.*

le Maire, M., Champeil, P., Møller, J.V. (2000) Interaction of membrane proteins and lipids with solubilizing detergents. *Biochim. Biophys. Acta* **1508**:86–111.

Lee, S.C., Khalid, S., Pollock, N.L., Knowles, T.J., Edler, K., Rothnie, A.J., Thomas, O.R.T., Dafforn, T.R. (2016) Encapsulated membrane proteins: A simplified system for molecular simulation. *Biochim. Biophys. Acta* **1858**:2549–2557.

Lee, S.C., Pollock, N.L. (2016) Membrane proteins: is the future disc shaped? *Biochem. Soc. Trans.* **44**:1011–1018.

Leney, A.C., McMorran, L.M., Radford, S.E., Ashcroft, A.E. (2012) Amphipathic polymers enable the study of functional membrane proteins in the gas phase. *Anal. Chem.* **84**:9841–9847.

Leray, V., Hubert, P., Burgun, C., Staedel, C., Crémel, G. (1993) Reconstitution studies of lipid effects on insulin-receptor kinase activation. *Eur. J. Biochem.* **213**:277–284.

Leray, V., Hubert, P., Crémel, G., Staedel, C. (1992) Detergents affect insulin binding, tyrosine kinase activity and oligomeric structure of partially purified insulin receptors. *Arch. Biochem. Biophys.* **294**:22–29.

Li, D., Li, J., Zhuang, Y., Zhang, L., Xiong, Y., Shi, P., Tian, C. (2015a) Nano-size uni-lamellar lipodisq-improved in situ auto-phosphorylation analysis of *E. coli* tyrosine kinase using $^{19}$F nuclear magnetic resonance. *Protein Cell* **6**:229–233.

Li, J.P., Ouyang, Y.Y., Kong, X., Zhu, J.Y., Lu, D.N., Liu, Z. (2015b) A multi-scale molecular dynamics simulation of PMAL facilitated delivery of siRNA. *RSC Adv.* **5**:68227–68233.

Li, M., Zhou, X., Wang, S., Michailidis, I., Gong, Y., Su, D., Li, H., Li, X., Yang, J. (2017) Structure of a eukaryotic cyclic-nucleotide-gated channel. *Nature* **542**:60–65.

Liao, M., Cao, E., Julius, D., Cheng, Y. (2013) Structure of the TRPV1 ion channel determined by electron cryo-microscopy. *Nature* **504**:107–112.

Liao, M., Cao, E., Julius, D., Cheng, Y. (2014) Single particle electron cryo-microscopy of a mammalian ion channel. *Curr. Opin. Struct. Biol.* **27**:1–7.

Liguori, N., Roy, L.M., Opačić, M., Durand, G., Croce, R. (2013) Regulation of light-harvesting in the green alga *Chlamydomonas reinhardtii*: the *C*-terminus of LHCSR is the knob of a dimmer switch. *J. Am. Chem. Soc.* **135**:18339–18342.

Lim, S.J., Zahid, M.U., Le, P., Ma, L., Entenberg, D., Harney, A.S., Condeelis, J., Smith, A.M. (2015) Brightness-equalized quantum dots. *Nat. Commun.* **20**:161–169.

Lindhoud, S., Carvalho, V., Pronk, J.W., Aubin-Tam, M.E. (2016) SMA-SH: Modified styrene-maleic acid copolymer for functionalization of lipid nanodiscs. *Biomacromolecules* **17**:1516–1522.

Logez, C., Damian, M., Legros, C., Dupré, C., Guéry, M., Mary, S., Wagner, R., M'Kadmi, C., Nosjean, O., Fould, B., Marie, J., Fehrentz, J.A., Martinez, J., Ferry, G., Boutin, J.A., Banères, J.-L. (2016) Detergent-free isolation of functional G protein-coupled receptors into nanometric lipid particles. *Biochemistry* **55**:38–48.

Long, A.R., O'Brien, C.C., Malhotra, K., Schwall, C.T., Albert, A.D., Watts, A., Alder, N.N. (2013) A detergent-free strategy for the reconstitution of active enzyme complexes from native biological membranes into nanoscale discs. *BMC Biotechnol.* **13**:41.

Lu, P., Bai, X.-C., Ma, D., Xie, T., Yan, C., Sun, L., Yang, G., Zhao, Y., Zhou, R., Scheres, S.H.W., Shi, Y. (2014) Three-dimensional structure of human $\gamma$-secretase. *Nature* **512**:166–170.

Luccardini, C., Tribet, C., Vial, F., Marchi-Artzner, V., Dahan, M. (2006) Size, charge, and interactions with giant lipid vesicles of quantum dots coated with an amphiphilic macromolecule. *Langmuir* **22**:2304–2310.

Lund, S., Orlowski, S., de Foresta, B., Champeil, P., le Maire, M., Møller, J.V. (1989) Detergent structure and associated lipid as determinants in the stabilization of solubilized $Ca^{2+}$-ATPase from sarcoplasmic reticulum. *J. Biol. Chem.* **264**:4907–4915.

Ma, D., Martin, N., Tribet, C., Winnik, F.M. (2014) Quantitative characterization by asymmetrical flow field-flow fractionation of IgG thermal aggregation with and without polymer protective agents. *Anal. Bioanal. Chem.* **406**:7539–7547.

Malhotra, K., Alder, N.N. (2014) Advances in the use of nanoscale bilayers to study membrane protein structure and function. *Biotechnol. Genet. Eng. Rev.* **30**:79–93.

Marie, E., Sagan, S., Cribier, S., Tribet, C. (2014) Amphiphilic macromolecules on cell membranes: from protective layers to controlled permeabilization. *J. Membr. Biol.* **247**:861–881.

Martin, N., Ma, D., Herbet, A., Boquet, D., Winnik, F.M., Tribet, C. (2014) Prevention of thermally induced aggregation of IgG antibodies by noncovalent interaction with poly(acrylate) derivatives. *Biomacromolecules* **15**:2952–2962.

Martin, N., Ruchmann, J., Tribet, C. (2015) Prevention of aggregation during refolding of carbonic anhydrase via Coulomb and hydrophobic complexation with octadecyl-modified or azobenzene-modified poly(acrylate) derivatives. *Langmuir* **31**:338–349.

Martinez, K.L., Gohon, Y., Corringer, P.-J., Tribet, C., Mérola, F., Changeux, J.-P., Popot, J.-L. (2002) Allosteric transitions of *Torpedo* acetylcholine receptor in lipids, detergent and amphipols: molecular interactions *vs.* physical constraints. *FEBS Lett.* **528**:251–256.

Mary, S., Damian, M., Rahmeh, R., Marie, J., Mouillac, B., Banères, J.-L. (2014) Amphipols in G protein-coupled receptor pharmacology: What are they good for? *J. Membr. Biol.* **247**:853–860.

Merino, J.M., Møller, J.V., Gutiérrez-Merino, C. (1994) Thermal unfolding of monomeric $Ca^{2+},Mg^{2+}$-ATPase from sarcoplasmic reticulum of rabbit skeletal muscle. *FEBS Lett.* **343**:155–159.

Milder, S.J., Thorgeirsson, T.E., Miercke, L.J., Stroud, R.M., Kliger, D.S. (1991) Effects of detergent environments on the photocycle of purified monomeric bacteriorhodopsin. *Biochemistry* **30**:1751–1761.

Møller, J.V., le Maire, M. (1993) Detergent binding as a measure of hydrophobic surface area of integral membrane proteins. *J. Biol. Chem.* **268**:18659–18672.

Møller, J.V., Olesen, C., Winther, A.M.L., Nissen, P. (2010) The sarcoplasmic $Ca^{2+}$-ATPase: design of a perfect chemi-osmotic pump. *Q. Rev. Biophys.* **43**:501–566.

Murakami, S., Nakashima, R., Yamashita, E., Yamaguchi, A. (2002) Crystal structure of bacterial multidrug efflux transporter AcrB. *Nature* **419**:587–593.

Nagy, J.K., Kuhn Hoffmann, A., Keyes, M.H., Gray, D.N., Oxenoid, K., Sanders, C.R. (2001) Use of amphipathic polymers to deliver a membrane protein to lipid bilayers. *FEBS Lett.* **1**:115–120.

Ning, Z., Hawley, B., Seebun, D., Figeys, D. (2014) APols-aided protein precipitation: a rapid method for concentrating proteins for proteomic analysis. *J. Membr. Biol.* **247**:941–947.

Ning, Z., Seebun, D., Hawley, B., Chang, C.-K., Figeys, D. (2013) From cells to peptides: "One-stop" integrated proteomic processing using amphipols. *J. Proteome Res.* **12**:1512–1519.

Norimatsu, Y., Hasegawa, K., Shimizu, N., Toyoshima, C. (2017) Protein-phospholipid interplay revealed with crystals of a calcium pump. *Nature* **545**:193–198.

Nowaczyk, M., Oworah-Nkruma, R., Zoonens, M., Rögner, M., Popot, J.-L. (2004) Amphipols: strategies for an improved PS2 environment in aqueous solution, in: Miyake, J. (Ed.), *Biohydrogen III.* Elsevier, Dordrecht, The Netherlands, Kyoto, pp. 151–159.

Obara, K., Miyashita, N., Xu, C., Toyoshima, I., Sugita, Y., Inesi, G., Toyoshima, C. (2005) Structural role of countertransport revealed in $Ca^{2+}$ pump crystal structure in the absence of $Ca^{2+}$. *Proc. Natl. Acad. Sci. USA* **102**:14489–14496.

Oluwole, A.O., Danielczak, B., Meister, A., Babalola, J.O., Vargas, C., Keller, S. (2017) Solubilization of membrane proteins into functional lipid-bilayer nanodiscs using a diisobutylene/maleic acid copolymer. *Angew. Chem. Int. Ed. Engl.* **56**:1919–1924.

Opačić, M., Durand, G., Bosco, M., Polidori, A., Popot, J.-L. (2014a) Amphipols and photosynthetic light-harvesting pigment-protein complexes. *J. Membr. Biol.* **247**:1031–1041.

Opačić, M., Giusti, F., Broos, J., Popot, J.-L. (2014b) Isolation of *Escherichia coli* mannitol permease, EII^mtl, trapped in amphipol A8-35 and fluorescein-labeled A8-35. *J. Membr. Biol.* **247**:1019–1030.

Orwick-Rydmark, M., Lovett, J.E., Graziadei, A., Lindholm, L., Hicks, M.R., Watts, A. (2012) Detergent-free incorporation of a seven-transmembrane receptor protein into nanosized bilayer Lipodisq particles for functional and biophysical studies. *Nano Lett.* **12**:4687–4692.

Palmgren, M.G., Nissen, P. (2011) P-Type ATPases. *Annu. Rev. Biophys.* **40**:243–266.

Park, K.-H., Billon-Denis, E., Dahmane, T., Lebaupain, F., Pucci, B., Breyton, C., Zito, F. (2011) In the cauldron of cell-free synthesis of membrane proteins: Playing with new surfactants. *New Biotech.* **28**:255–261.

Paulin, S., Jamshad, M., Dafforn, T.R., Garcia-Lara, J., Foster, S.J., Galley, N.F., Roper, D.I., Rosado, H., Taylor, P.W. (2014) Surfactant-free purification of membrane protein complexes from bacteria: application to the staphylococcal penicillin-binding protein complex PBP2/PBP2a. *Nanotechnology* **25**:285101.

Paulsen, C.E., Armache, J.-P., Gao, Y., Cheng, Y., Julius, D. (2015) Structure of the TRPA1 ion channel suggests regulatory mechanisms. *Nature* **520**:511–517.

Periasamy, A., Shadiac, N., Amalraj, A., Garajová, S., Nagarajan, Y., Waters, S., Mertens, H.D.T., Hrmova, M. (2013) Cell-free protein synthesis of membrane (1,3)-$\beta$-D-glucan (curdlan) synthase: co-translational insertion in liposomes and reconstitution in nanodiscs. *Biochim. Biophys. Acta* **1828**:743–757.

Perlmutter, J.D., Drasler, W.J., Xie, W., Gao, J., Popot, J.-L., Sachs, J.N. (2011) All-atom and coarse-grained molecular dynamics simulations of a membrane protein-stabilizing polymer. *Langmuir* **27**:10523–10537.

Perlmutter, J.D., Popot, J.-L., Sachs, J.N. (2014) Molecular dynamics simulations of a membrane protein/amphipol complex. *J. Membr. Biol.* **247**:883–895.

Picard, M., Dahmane, T., Garrigos, M., Gauron, C., Giusti, F., le Maire, M., Popot, J.-L., Champeil, P. (2006) Protective and inhibitory effects of various types of amphipols on the Ca$^{2+}$-ATPase from sarcoplasmic reticulum: a comparative study. *Biochemistry* **45**:1861–1869.

Planchard, N., Point, E., Dahmane, T., Giusti, F., Renault, M., Le Bon, C., Durand, G., Milon, A., Guittet, E., Zoonens, M., Popot, J.-L., Catoire, L.J. (2014) The use of amphipols for solution NMR studies of membrane proteins: advantages and limitations as compared to other solubilizing media. *J. Membr. Biol.* **247**:827–842.

Pocanschi, C., Popot, J.-L., Kleinschmidt, J.H. (2013) Folding and stability of outer membrane protein A (OmpA) from *Escherichia coli* in an amphipathic polymer, amphipol A8-35. *Eur. Biophys. J.* **42**:103–118.

Pocanschi, C.L., Apell, H.-J., Puntervoll, P., Høgh, B.T., Jensen, H.B., Welte, W., Kleinschmidt, J.H. (2006a) Folding and membrane insertion of the major outer membrane protein of *Fusobacterium nucleatum* (FomA). *J. Mol. Biol.* **355**:548–561.

Pocanschi, C.L., Dahmane, T., Gohon, Y., Rappaport, F., Apell, H.-J., Kleinschmidt, J.H., Popot, J.-L. (2006b) Amphipathic polymers: tools to fold integral membrane proteins to their active form. *Biochemistry* **45**:13954–13961.

Polovinkin, V., Balandin, T., Volkov, O., Round, E., Borshchevskiy, V., Utrobin, P., von Stetten, D., Royant, A., Willbold, D., Arzumanyan, A., Popot, J.-L., Gordeliy, V. (2014a) Nanoparticle surface-enhanced Raman scattering of bacteriorhodopsin stabilized by amphipol A8-35. *J. Membr. Biol.* **247**:971–980.

Polovinkin, V., Gushchin, I., Balandin, T., Chervakov, P., Round, E., Shevchenko, V., Popov, A., Borshchevskiy, V., Popot, J.-L., Gordeliy, V. (2014b) High-resolution structure of a membrane protein transferred from amphipol to a lipidic mesophase. *J. Membr. Biol.* **247**:997–1004.

Popot, J.-L. (2010) Amphipols, nanodiscs, and fluorinated surfactants: Three non-conventional approaches to studying membrane proteins in aqueous solutions. *Annu. Rev. Biochem.* **79**:737–775.

Popot, J.-L. (2014) Folding membrane proteins *in vitro*: A table and some comments. *Arch. Biochem. Biophys.* **564**:314–326.

Popot, J.-L., Althoff, T., Bagnard, D., Banères, J.-L., Bazzacco, P., Billon-Denis, E., Catoire, L.J., Champeil, P., Charvolin, D., Cocco, M.J., Crémel, G., Dahmane, T., de la Maza, L.M., Ebel, C., Gabel, F., Giusti, F., Gohon, Y., Goormaghtigh, E., Guittet, E., Kleinschmidt, J.H., Kühlbrandt, W., Le Bon, C., Martinez, K.L., Picard, M., Pucci, B., Rappaport, F., Sachs, J.N., Tribet, C., van Heijenoort, C., Wien, F., Zito, F., Zoonens, M. (2011) Amphipols from A to Z. *Annu. Rev. Biophys.* **40**:379–408.

Popot, J.-L., Berry, E.A., Charvolin, D., Creuzenet, C., Ebel, C., Engelman, D.M., Flötenmeyer, M., Giusti, F., Gohon, Y., Hervé, P., Hong, Q., Lakey, J.H., Leonard, K., Shuman, H.A., Timmins, P., Warschawski, D.E., Zito, F., Zoonens, M., Pucci, B., Tribet, C. (2003) Amphipols: polymeric surfactants for membrane biology research. *Cell. Mol. Life Sci.* **60**:1559–1574.

Postis, V., Rawson, S., Mitchell, J.K., Lee, S.C., Parslow, R.A., Dafforn, T.R., Baldwin, S.A., Muench, S.P. (2015) The use of SMALPs as a novel membrane protein scaffold for structure study by negative stain electron microscopy. *Biochim. Biophys. Acta* **1848**:496–501.

Prabudiansyah, I., Kusters, I., Caforio, A., Driessen, A.J.M. (2015) Characterization of the annular lipid shell of the Sec translocon. *Biochim. Biophys. Acta* **1848**:2050–2056.

Prata, C., Giusti, F., Gohon, Y., Pucci, B., Popot, J.-L., Tribet, C. (2001) Non-ionic amphiphilic polymers derived from *tris*(hydroxymethyl)-acrylamidomethane keep membrane proteins soluble and native in the absence of detergent. *Biopolymers* **56**:77–84.

Qi, L., Gao, X. (2008) Quantum dot-amphipol nanocomplex for intracellular delivery and real-time imaging of siRNA. *ACS Nano* **2**:1403–1410.

Rahmeh, R., Damian, M., Cottet, M., Orcel, H., Mendre, C., Durroux, T., Sharma, K.S., Durand, G., Pucci, B., Trinquet, E., Zwier, J.M., Deupi, X., Bron, P., J.-L. B., Mouillac, B., Granier, S. (2012) Structural insights into biased G protein-coupled receptor signaling revealed by fluorescence spectroscopy. *Proc. Natl. Acad. Sci. USA* **109**:6733–6738.

Rajesh, S., Knowles, T.J., Overduin, M. (2011) Production of membrane proteins without cells or detergents. *New Biotechnol.* **28**:250–254.

Rehan, S., Paavilainen, V.O., Jaakola, V.P. (2017) Functional reconstitution of human equilibrative nucleoside transporter-1 into styrene maleic acid co-polymer lipid particles. *Biochim. Biophys. Acta* **1859**:1059–1065.

Reich-Slotky, R., Panagiotidis, C., Reyes, M., Shuman, H.A. (2000) The detergent-soluble maltose transporter is activated by maltose binding protein and verapamil. *J. Bact.* **182**:993–1000.

Renault, M. (2008) Etudes structurales et dynamiques de la protéine membranaire KpOmpA par RMN en phase liquide et solide. Thèse de Doctorat, Université Paul Sabatier, Toulouse, 180 p.

Renner, C., Kessler, B., Oesterhelt, D. (2005) Lipid composition of integral purple membrane by $^{1}$H and $^{31}$P NMR. *J. Lipid Res.* **46**:1755–1764.

Ritchie, T.K., Grinkova, Y.V., Bayburt, T.H., Denisov, I.G., Zolnerciks, J.K., Atkins, W.M., Sligar, S.G. (2009) Reconstitution of membrane proteins in phospholipid bilayer nanodiscs. *Meth. Enzymol.* **464**:211–231.

Rosenbaum, D.M., Cherezov, V., Hanson, M.A., Rasmussen, S.G., Thian, F.S., Kobilka, T.S., Choi, H.J., Yao, X.J., Weis, W.I., Stevens, R.C., Kobilka, B.K. (2007) GPCR engineering yields high-resolution structural insights into $\beta_{2}$-adrenergic receptor function. *Science* **318**:1266–1273.

Sahu, I.D., McCarrick, R.M., Troxel, K.R., Zhang, R., Smith, H.J., Dunagan, M.M., Swartz, M.S., Rajan, P.V., Kroncke, B.M., Sanders, C.R., Lorigan, G.A. (2013) DEER EPR measurements for membrane protein structures via bifunctional spin labels and lipodisq nanoparticles. *Biochemistry* **52**:6627–6632.

Sebai, S., Cribier, S., Karimi, A., Massotte, D., Tribet, T. (2010) Permeabilisation of lipid membranes and cells by a light-responsive copolymer. *Langmuir* **26**:14135–14141.

Sebai, S., Milioni, D., Walrant, A., Alves, I., Sagan, S., Huin, C., Auvray, L., Massotte, D., Cribier, S., Tribet, C. (2012) Photocontrol of the translocation of molecules, peptides, and quantum dots through cell and lipid membranes doped with azobenzene copolymers. *Angew. Chem. Int. Ed.* **51**:2132–2136.

Shao, J., Fu, Z., Ji, Y., Guan, X., Guo, S., Ding, Z., Yang, X., Cong, Y., Shen, Y. (2016) Leucine zipper-EF-hand containing transmembrane protein 1 (LETM1) forms a $Ca^{2+}/H^{+}$ antiporter. *Sci. Rep.* **6**:34174.

Sharma, K.S., Durand, G., Gabel, F., Bazzacco, P., Le Bon, C., Billon-Denis, E., Catoire, L.J., Popot, J.-L., Ebel, C., Pucci, B. (2012) Non-ionic amphiphilic homopolymers: Synthesis, solution properties, and biochemical validation. *Langmuir* **28**:4625–4639.

Sharma, K.S., Durand, G., Giusti, F., Olivier, B., Fabiano, A.-S., Bazzacco, P., Dahmane, T., Ebel, C., Popot, J.-L., Pucci, B. (2008) Glucose-based amphiphilic telomers designed to keep membrane proteins soluble in aqueous solutions: synthesis and physico-chemical characterization. *Langmuir* **24**:13581–13590.

Shen, P.S., Yang, X., DeCaen, P.G., Liu, X., Bulkley, D., Clapham, D.E., Cao, E. (2016) The structure of the Polycystic Kidney Disease channel PKD2 in lipid nanodiscs. *Cell* 167:763–773.e711.

Shinzawa-Itoh, K., Shimomura, H., Yanagisawa, S., Shimada, S., Takahashi, R., Oosaki, M., Ogura, T., Tsukihara, T. (2016) Purification of active respiratory supercomplex from bovine heart mitochondria enables functional studies. *J. Biol. Chem.* **291**:4178–4184.

Skaar, K., Korza, H.J., Tarry, M., Sekyrova, P., Högbom, M. (2015) Expression and subcellular distribution of GFP-tagged human tetraspanin proteins in *Saccharomyces cerevisiae*. *PLoS One* **10**:e0134041.

Smirnova, I.A., Sjöstrand, D., Li, F., Björck, M., Schäfer, J., Östbye, H., Högbom, M., von Ballmoos, C., Lander, G., Ädelroth, P., Brzezinski, P. (2016) Isolation of yeast Complex IV in native lipid nanodiscs. *Biochim. Biophys. Acta,* **1858**:2984–2992.

Sólyom, Z., Ma, P., Schwarten, M., Bosco, M.I., Polidori, A., Durand, G., Willbold, D., Brutscher, B. (2015) The disordered region of the HCV protein NS5A: conformational dynamics, SH3 binding, and phosphorylation. *Biophys. J.* **109**:1483–1496.

Sonntag, Y., Musgaard, M., Olesen, C., Schiøtt, B., Møller, J.V., Nissen, P., Thøgersen, L. (2011) Mutual adaptation of a membrane protein and its lipid bilayer during conformational changes. *Nat. Commun.* **2**:304.

Sousa, J.S., Mills, D.J., Vonck, J., Kühlbrandt, W. (2016) Functional asymmetry and electron flow in the bovine respirasome. *eLIFE* **5**:e21290.

Spear, J.M., Koborssy, D.A., Schwartz, A.B., Johnson, A.J., Audhya, A., Fadool, D.A., Stagg, S.M. (2015) Kv1.3 contains an alternative C-terminal ER exit motif and is recruited into COPII vesicles by Sec24a. BMC Biochem. 16:16.

Stangl, M., Unger, S., Keller, S., Schneider, D. (2014) Sequence-specific dimerization of a transmembrane helix in amphipol A8-35. *PLOS One* **9**:e110970.

Sverzhinsky, A., Qian, S., Yang, L., Allaire, M., Moraes, I., Ma, D., Chung, J.W., Zoonens, M., Popot, J.-L., Coulton, J.W. (2014) Amphipol-trapped ExbB-ExbD membrane protein complex from *Escherichia coli*: A biochemical and structural case study. *J. Membr. Biol.* **247**:1005–1018.

Swainsbury, D.J.K., Scheidelaar, S., van Grondelle, R., Killian, J.A., Jones, M.R. (2014) Bacterial reaction centers purified with styrene maleic acid copolymer retain native membrane functional properties and display enhanced stability. *Angew. Chemie Int. Ed.* **53**:11803–11807.

Sweadner, K.J. (2017) An ion-transport enzyme that rocks. *Nature* **545**:162–164.

Swift, J. (1726) *Travels into Several Remote Nations of the World. In Four Parts. By Lemuel Gulliver, First a Surgeon, and then a Captain of several Ships.* Benjamin Motte, London.

Tehei, M., Perlmutter, J.D., Giusti, F., Sachs, J.N., Zaccai, G., Popot, J.-L. (2014) Thermal fluctuations in amphipol A8-35 particles: A neutron scattering and molecular dynamics study. *J. Membr. Biol.* **247**:897–908.

Tifrea, D., Pal, S., Cocco, M.J., Popot, J.-L., de la Maza, L.M. (2014) Increased immunoaccessibility of MOMP epitopes in a vaccine formulated with amphipols may account for the very robust protection elicited against a vaginal challenge with *C. muridarum. J. Immunol.* **192**:5201–5213.

Tifrea, D.F., Pal, S., Le Bon, C., Giusti, F., Cocco, M.J., Zoonens, M., de la Maza, L.M. (2018a) Resiquimod conjugated with amphipols bound to the *Chlamydia muridarum* MOMP enhances protection against a mucosal challenge. *In preparation.*

Tifrea, D.F., Pal, S., Le Bon, C., Giusti, F., Popot, J.-L., Cocco, M.J., Zoonens, M., de la Maza, L.M. (2018b) Co-delivery of amphipol-conjugated adjuvant with antigen, and adjuvant combinations, enhance immune protection elicited by a membrane protein-based vaccine against a mucosal challenge with *Chlamydia. Submitted for publication.*

Tifrea, D.F., Sun, G., Pal, S., Zardeneta, G., Cocco, M.J., Popot, J.-L., de la Maza, L.M. (2011) Amphipols stabilize the *Chlamydia* major outer membrane protein and enhance its protective ability as a vaccine. *Vaccine* **29**:4623–4631.

Toyoshima, C., Inesi, G. (2004) Structural basis of ion pumping by $Ca^{2+}$-ATPase of the sarcoplasmic reticulum. *Annu. Rev. Biochem.* **73**:269–292.

Toyoshima, C., Nakasako, M., Nomura, H., Ogawa, H. (2000) Crystal structure of the calcium pump of sarcoplasmic reticulum at 2.6 Å resolution. *Nature* **405**:647–655.

Toyoshima, C., Nomura, H. (2002) Structural changes in the calcium pump accompanying the dissociation of calcium. *Nature* **418**:605–611.

Tribet, C., Audebert, R., Popot, J.-L. (1996) Amphipols: polymers that keep membrane proteins soluble in aqueous solutions. *Proc. Natl. Acad. Sci. USA* **93**:15047–15050.

Tribet, C., Audebert, R., Popot, J.-L. (1997) Stabilization of hydrophobic colloidal dispersions in water with amphiphilic polymers: Application to integral membrane proteins. *Langmuir* **13**:5570–5576.

Tribet, C., Diab, C., Dahmane, T., Zoonens, M., Popot, J.-L., Winnik, F.M. (2009) Thermodynamic characterization of the exchange of detergents and amphipols at the surfaces of integral membrane proteins. *Langmuir* **25**:12623–12634.

Tribet, C., Mills, D., Haider, M., Popot, J.-L. (1998) Scanning transmission electron microscopy study of the molecular mass of amphipol/cytochrome $b_6 f$ complexes. *Biochimie* **80**:475–482.

Tribet, C., Vial, F. (2008) Flexible macromolecules attached to lipid bilayers: impact on fluidity, curvature, permeability and stability of the membranes. *Soft Matter* **4**:68–81.

Tsybovsky, Y., Orban, T., Molday, R.S., Taylor, D., Palczewski, K. (2013) Molecular organization and ATP-induced conformational changes of ABCA4, the photoreceptor-specific ABC transporter. *Structure* **21**:854–860.

Vahedi-Faridi, A., Jastrzebska, B., Palczewski, K., Engel, A. (2013) 3D imaging and quantitative analysis of small solubilized membrane proteins and their complexes by transmission electron microscopy. *Microscopy (Oxf)* **62**:95–107.

Vaidehi, N., Grisshammer, R., Tate, C.G. (2016) How can mutations thermostabilize G protein-coupled receptors? *Trends Pharmacol. Sci.* **37**:37–46.

van Pee, K., Neuhaus, A., D'Imprima, E., Mills, D.J., Kühlbrandt, W., Yildiz, Ö. (2017) CryoEM structures of membrane pore and prepore complex reveal cytolytic mechanism of pneumolysin. *eLlife* **6**:e23644.

Vázquez-Acevedo, M., Vega-de Luna, F., Sánchez-Vásquez, L., Colina-Tenorio, L., Remacle, C., Cardol, P., Miranda-Astudillo, H., González-Halphen, D. (2016) Dissecting the peripheral stalk of the mitochondrial ATP synthase of chlorophycean algae. *Biochim. Biophys. Acta* **1857**:1183–1190.

Vial, F., Cousin, F., Bouteiller, L., Tribet, C. (2009) Rate of permeabilization of giant vesicles by amphiphilic polyacrylates compared to the adsorption of these polymers onto large vesicles and tethered lipid bilayers. *Langmuir* **25**:7506–7513.

Vial, F., Oukhaled, A.G., Auvray, L., Tribet, C. (2007) Long-living channels of well-defined radius opened in lipid bilayers by polydisperse, hydrophobically-modified polyacrylic acids. *Soft Matter* **3**:75–78.

Vial, F., Rabhi, S., Tribet, C. (2005) Association of octyl-modified poly(acrylic acid) onto unilamellar vesicles of lipids and kinetics of vesicle disruption. *Langmuir* **21**:853–862.

Wang, Z., Fan, G., Hryc, C.F., Blaza, J.N., Serysheva, I.I., Schmid, M.F., Chiu, W., Luisi, B.F., Du, D. (2017) An allosteric transport mechanism for the AcrAB-TolC multidrug efflux pump. *eLife* **6**:e24905.

Watkinson, T.G., Calabrese, A.N., Ault, J.R., Radford, S.E., Ashcroft, A.E. (2017) FPOP-LC-MS/MS suggests differences in interaction sites of amphipols and detergents with outer membrane proteins. *J. Am. Soc. Mass Spectrom.* **28**:50–55.

Watkinson, T.G., Calabrese, A.N., Giusti, F., Zoonens, M., Radford, S.E., Ashcroft, A.E. (2015) Systematic analysis of the use of amphipathic polymers for studies of outer membrane proteins using mass spectrometry. *Int. J. Mass Spectrom.* **391**:54–61.

Wei, R., Wang, X., Zhang, Y., Mukherjee, S., Zhang, L., Chen, Q., Huang, X., Jing, S., Liu, C., Li, S., Wang, G., Xu, Y., Zhu, S., Williams, A.J., Sun, F., Yin, C.C. (2016) Structural insights into $Ca^{2+}$-activated long-range allosteric channel gating of RyR1. *Cell Res.* **26**:977–994.

Wheatley, M., Charlton, J., Jamshad, M., Routledge, S.J., Bailey, S., La-Borde, P.J., Azam, M.T., Logan, R.T., Bill, R. M., Dafforn, T.R., Poyner, D.R. (2016) GPCR–styrene maleic acid lipid particles (GPCR–SMALPs): their nature and potential. *Biochem. Soc. Trans.* **44**:619–623.

Wickstrand, C., Dods, R., Royant, A., Neutze, R. (2015) Bacteriorhodopsin: Would the real structural intermediates please stand up? *Biochim. Biophys. Acta* **1850**:536–553.

Wilkens, S. (2000) $F_1F_O$-ATP synthase–stalking mind and imagination. *J. Bioenerg. Biomembr.* **32**:333–339.

Wilkens, S., Capaldi, R.A. (1998a) Electron microscopic evidence of two stalks linking the $F_1$ and $F_O$ parts of the *Escherichia coli* ATP synthase. *Biochim. Biophys. Acta* **1365**:93–97.

Wilkens, S., Capaldi, R.A. (1998b) ATP synthase's second stalk comes into focus. *Nature* **393**:29.

Wilkens, S., Zhou, J., Nakayama, R., Dunn, S.D., Capaldi, R.A. (2000) Localization of the δ subunit in the *Escherichia coli* $F_1F_O$-ATPsynthase by immuno-electron microscopy: The δ subunit binds on top of the $F_1$. *J. Mol. Biol.* **295**:387–391.

Wilkes, M., Madej, M.G., Kreuter, L., Rhinow, D., Heinz, V., De Sanctis, S., Ruppel, S., Richter, R.M., Joos, F., Grieben, M., Pike, A.C., Huiskonen, J.T., Carpenter, E.P., Kühlbrandt, W., Witzgall, R., Ziegler, C. (2017) Molecular insights into lipid-assisted $Ca^{2+}$ regulation of the TRP channel polycystin-2. *Nat. Struct. Mol. Biol.* **24**:123–130.

Wojtowicz, H., Prochnicka-Chalufour, A., de Amorim, G.C., Roudenko, O., Simenel, C., Malki, I., Pehau-Arnaudet, G., Gubellini, F., Koutsioubas, A., Pérez, J., Delepelaire, P., Delepierre, M., Fronzes, R., Izadi-Pruneyre, N. (2016) Structural basis of the signalling through a bacterial membrane receptor, HasR, deciphered by an integrative approach. *Biochem. J.* **473**:2239–2248.

Wu, Z.S., Cui, Z.C., Cheng, H., Fan, C., Melcher, K., Jiang, Y., Zhang, C.H., Jiang, H.L., Cong, Y., Liu, Q., Xu, H.E. (2015) High yield and efficient expression and purification of the human 5-$HT_{3A}$ receptor. *Acta Pharmacol. Sin.* **36**:1024–1032.

Zaccai, G. (2004) The effect of water on protein dynamics. *Phil. Trans. R. Soc. Lond. B* **359**:1269–1275.

Zhang, N., Tsybovsky, Y., Kolesnikov, A.V., Rozanowska, M., Swider, M., Schwartz, S.B., Stone, E.M., Palczewska, G., Maeda, A., Kefalov, V.J., Jacobson, S.G., Cideciyan, A.V., Palczewski, K. (2015a) Protein misfolding and the pathogenesis of ABCA4-associated retinal degenerations. *Hum. Mol. Genet.* **24**:3220–3237.

Zhang, R., Sahu, I.D., Liu, L., Osatuke, A., Comer, R.G., Dabney-Smith, C., Lorigan, G.A. (2015b) Characterizing the structure of lipodisq nanoparticles for membrane protein spectroscopic studies. *Biochim. Biophys. Acta* **1848**:329–333.

Zoonens, M. (2004) Caractérisation des complexes formés entre le domaine transmembranaire de la protéine OmpA et des polymères amphiphiles, les amphipols. Application à l'étude structurale des protéines membranaires par RMN à haute résolution. Thèse de Doctorat, Université Paris-6, 233 p.

Zoonens, M., Catoire, L.J., Giusti, F., Popot, J.-L. (2005) NMR study of a membrane protein in detergent-free aqueous solution. *Proc. Natl. Acad. Sci. USA* **102**:8893–8898.

Zoonens, M., Giusti, F., Zito, F., Popot, J.-L. (2007) Dynamics of membrane protein/amphipol association studied by Förster resonance energy transfer. Implications for in vitro studies of amphipol-stabilized membrane proteins. *Biochemistry* **46**:10392–10404.

Zoonens, M., Popot, J.-L. (2014) Amphipols for each season. *J. Membr. Biol.* **247**:759–796.

Zubcevic, L., Herzik, M.A., Jr., Chung, B.C., Liu, Z., Lander, G.C., Lee, S.-Y. (2016) Cryo-electron microscopy structure of the TRPV2 ion channel. *Nat. Struct. Mol. Biol.* **23**:180–186.

# Amphipol-Assisted Folding of Membrane Proteins

**6**

**Summary**

*Folding membrane proteins (MPs) in vitro is important from both practical and theoretical points of view. Amphipols (APols) have emerged as an interesting medium in which to fold MPs. Both α-helical and β-barrel MPs are amenable to folding in APols, using simple procedures and achieving high yields. All APols hitherto tested (A8-35, sulfonated APols, non-ionic APols) have proven to provide a medium favorable to MP folding. From a practical point of view, the approach is particularly interesting when applied to MPs that have been produced in large amounts in inclusion bodies, from which they can be recovered only under a denatured state. The resulting MP/APol complexes can be used as such or the protein transferred to other media more appropriate to a given type of experiment (e.g. to lipidic mesophases for crystallization). From a theoretical point of view, the observation that MPs can fold very efficiently in a medium so different from a lipid bilayer raises interesting questions about the forces that determine the final fold.*

## 6.1 Introduction

Folding a denatured or unfolded membrane protein (MP) to a functional form is one of the toughest challenges of membrane biochemistry (for reviews, see e.g. Booth 2003; Kiefer 2003; White 2003; Tamm et al. 2004; Bowie 2005; Bannwarth and Schulz 2003; Michaux et al. 2008; Stanley and Fleming 2008; Harris and Booth 2012; Otzen and Andersen 2013; Popot 2014; Kleinschmidt 2015). It is considered by many as akin to reaching an isolated spot of solid land in the middle of a treacherous swamp, spotted with potholes and quicksand, with no paths marked. Many a laboratory director hesitates before committing students or postdoctoral fellows to such a project, for fear of ruining their career. Yet, most MPs are too rare to be purified from natural sources in amounts sufficient for structural and extensive functional studies. Overexpressing them in large amounts under a functional form is tricky, costly, lengthy to set up and optimize, and, more often than not, frustrating. This is because targeting an overexpressed MP to a membrane, where it can insert and achieve its native structure, tends to be toxic, resulting in limited expression yields, and only a fraction of the proteins

© Springer International Publishing AG, part of Springer Nature 2018
J. -L. Popot, *Membrane Proteins in Aqueous Solutions*, Biological and Medical Physics, Biomedical Engineering, https://doi.org/10.1007/978-3-319-73148-3_6

produced end up correctly folded (for reviews, see e.g. Grisshammer and Tate 1995; Maeda and Schertler 2013; Goehring et al. 2014; Milić and Veprintsev 2015; Zorman et al. 2015).

Two alternatives are to express the target MP in vitro, in a cell-free expression (CFE) system, or to target it to the cytosol of the cell, where it precipitates as nontoxic inclusion bodies (IBs). Either approach can yield tens of mg of MPs, but coaxing these into adopting a functional structure remains a major difficulty. In CFE, MPs can be provided with a surfactant into which they will insert and, hopefully, fold during their synthesis, or they can be left to precipitate and be recovered later on by dissolution with a detergent. The recovered MP is often nonfunctional. This approach will be discussed in Chap. 7. In the second case, IBs are recovered by centrifugation, washed with a non-denaturing detergent such as Triton X-100, and dissolved either with a chaotropic agent or a denaturing detergent, typically urea for $\beta$-barrel MPs and sodium dodecyl sulfate (SDS) for $\alpha$-helical ones. Obtaining tens or hundreds of mg of MP in this way is relatively easy. Folding the recovered protein to a functional form is nothing short of nightmarish. Many procedures have been described that lead from an unfolded to a folded and functional MP, usually with limited yields (§ 6.2). Assisting the folding of MPs appears to be one of the most successful and promising applications of amphipols (APols).

## 6.2    Context: Existing Approaches to Folding Membrane Proteins In Vitro

The first successful attempts at folding MPs in vitro were reported in the late 1970s/early 1980s. They were largely a consequence of ongoing attempts at establishing the sequence of MPs by chemical methods. Because DNA sequencing did not exist at the time, this required cutting chemically or enzymatically the protein into short, overlapping peptides, sequencing them manually and putting together the whole puzzle to obtain the complete sequence. For $\alpha$-helical MPs, this raised particularly redoubtable problems, because most transmembrane (TM) helices correspond to long stretches of hydrophobic residues, which are highly insoluble in aqueous solutions and hard to handle. Organic solvents had to be identified in which to separate and purify them. The technology thus developed made it possible to obtain MPs in a fully unfolded state, as ascertained by spectroscopic methods. Beyond sequencing, this work thus sets the stage for tackling the question of whether such unfolded MPs could be brought back to a functional form.

The first MP refolding experiment on record appears to have been published by the group of Ulf Henning in 1978. It was carried out on outer membrane protein A (OmpA, then called OmpII*) from *E. coli* and took advantage of the fact that most $\beta$-barrel MPs migrate differently, upon SDS-polyacrylamide gel electrophoresis (SDS-PAGE), depending on whether they have been heated in SDS, and presumably unfolded, or kept at room temperature. Henning and coworkers showed that addition of the outer membrane lipid lipopolysaccharide to the boiled species after it had been cooled down caused it to migrate again at the same position as the unboiled species, strongly suggesting a denaturation/renaturation phenomenon (Schweizer et al. 1978).

In 1981, the group of H. Gobind Khorana at MIT published a memorable paper in which they demonstrated that bacteriorhodopsin (BR; see Chap. 1, § 1.6.1) could be refolded to a functional form starting from a denatured form in SDS (Huang et al. 1981). When overproduced, BR, a light-driven proton pump, accumulates in the plasma membrane of *Halobacterium salinarum* in the form of 2D protein/lipid crystals, the so-called purple membrane. Its chromophore, retinal, which is covalently but loosely bound to a lysine residue by a Schiff base, confers it a characteristic purple color (for an overview of BR structure and function, see Chap. 1, § 1.6.1). When purple membrane is solubilized in SDS, BR denatures, and the Schiff base hydrolyzes, releasing the retinal and bacterio-opsin (BO), the apoprotein. This causes an absorbance peak shift from ~555 nm (dark-adapted BR) to ~382 nm (free retinal). The protocols developed by Khorana and his coworkers are schematized in Fig. 6.1. In SDS, BO features a substantial amount of

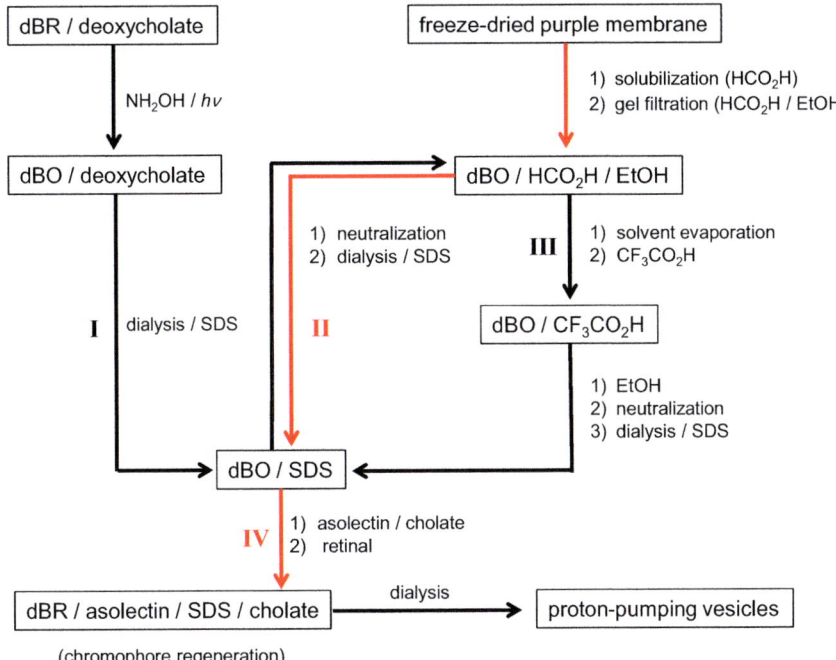

(chromophore regeneration)

**Fig. 6.1** The original protocols for refolding bacteriorhodopsin (BR) developed by H. Gobind Khorana and collaborators. *Abbreviations*: *dBR* delipidated BR, *dBO* delipidated bacterio-opsin (BO; the apoprotein), *DOC* deoxycholate, *SBL* soybean lipids. In pathway I, delipidated BO is denatured in SDS, where it retains about two-thirds of its helicity. In protocols II and III, organic solvents are used to completely unfold the protein, as ascertained by circular dichroism and NMR measurements. Following separation from lipids and retinal by gel filtration, the unfolded apoprotein is transferred to SDS, where it recovers a degree of helicity. In all three protocols, the preparation in SDS is then supplemented with retinal and with a large excess (relative to SDS) of cholate and lipids, which initiates refolding. Following dialysis to remove the detergents, vesicles form, in which light-driven proton pumping by the refolded BR can be demonstrated (The figure has been redrawn after Huang et al. 1981). The route shown in *red* has been used to fold delipidated BR in A8-35 or A8-35/lipid mixtures, used as the folding medium instead of the asolectin/cholate mixtures used by Khorana and coworkers (Pocanschi et al. 2006; Dahmane et al. 2013).

$\alpha$-helical structure (~50% vs. ~75% in the native protein; for a recent review on MP/SDS complexes, see Otzen 2015). To demonstrate refolding from a totally unfolded state, the MIT team solubilized BR in various organic solvents, in which spectroscopic data (circular dichroism and NMR) showed that unfolding was complete, transferred it to SDS, where it recovered ~50% helicity, and renatured it from the SDS solution. The renaturation procedure relied on diluting the SDS in a large excess of a mixture of bile salt and lipids, upon which BR renatured and rebound retinal, as evidenced by the recovery of the purple color characteristic of the native holoprotein. Upon removal of the two detergents by dialysis, vesicles formed, in which light-driven proton pumping by the refolded BR could be demonstrated.

These early experiments demonstrated that MPs, or at least some of them, obey Anfinsen's principle, which posits that the folded, functional state of a protein corresponds to the (or at least to a) free energy minimum of the system comprising the protein and its environment (see Box 6.1). In other terms, it results from the minimization of the free energy of interactions of the polypeptide chain with itself and with its environment and does not depend on information conferred onto it in vivo in the course of the synthesis.

**Box 6.1    Anfinsen's Principle**

Taking ribonuclease as a model, Christian Anfinsen and his colleagues, in the 1960s, showed that this soluble enzyme, after its eight disulfide bridges had been reduced and the polypeptide unfolded in urea, could recover full activity in vitro when the denaturant was removed and the bridges allowed to reoxidize (Anfinsen et al. 1961; Anfinsen 1973; Anfinsen and Scheraga 1975). This established that the native structure of the enzyme is not dictated by the biosynthetic apparatus, but by the interaction of the amino acid sequence with itself and with its environment, and that it corresponds to the (or a) free energy minimum of this ensemble. This has become known as Anfinsen's principle.

Until Khorana's work, however, it was totally uncertain whether the same principle could apply to MPs, whose synthesis and membrane insertion take place in a highly anisotropic medium and are catalyzed by a complex apparatus whose composition and role just started to be unraveled in the 1970s. That MPs could be kinetically blocked, for instance, because some of their regions cannot flip through the membrane, in a native conformation that does not correspond to the free energy minimum was a real possibility. In such a case, refolding in vitro would have been extremely complex, if not impossible. The work carried out at MIT in the early 1980s showed that it was not.

These two pioneering experiments were to have a rich posterity. Over the past 35 years, some 90 MPs have been successfully folded in vitro using a variety of unfolding and folding media and protocols (reviewed in Popot 2014). Some essential points are illustrated in Figs. 6.2, 6.3, and 6.4. They can be summarized in the following way:

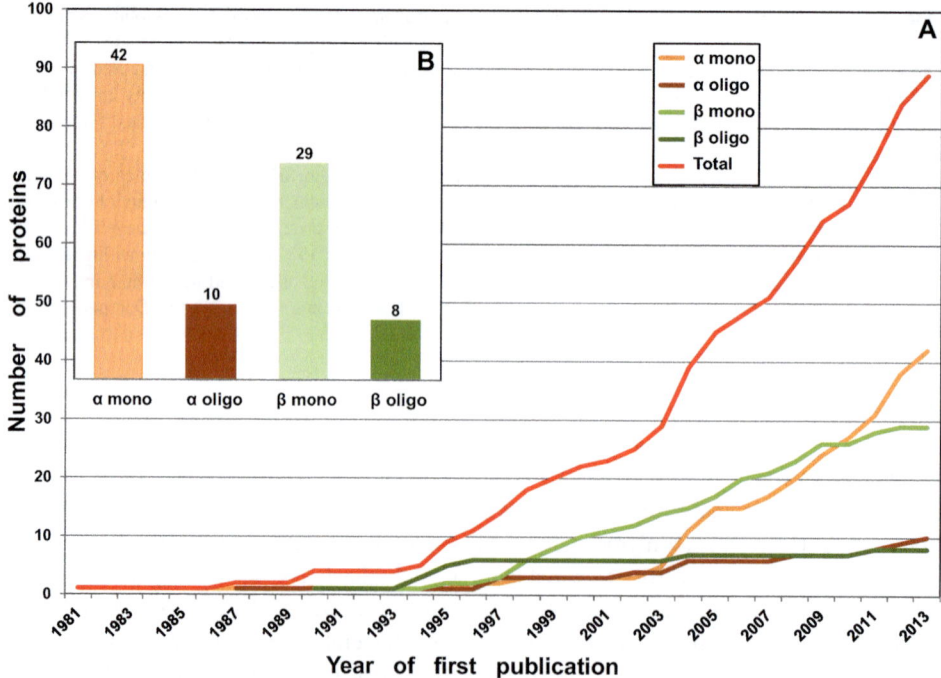

**Fig. 6.2**  Number and types of integral membrane proteins that have been either refolded or folded de novo in vitro. MPs are distributed according to the secondary and quaternary structure of their transmembrane domain: (i) monomeric $\alpha$-helical MPs ("$\alpha$ mono"), (ii) oligomeric $\alpha$-helical MPs ("$\alpha$ oligo"), (iii) monomeric $\beta$-barrel MPs ("$\beta$ mono"), and (iv) oligomeric $\beta$-barrel MPs ("$\beta$ oligo"). Each protein is counted only once, irrespective of the number of different ways it may have been (re)folded. In **A**, the cumulative number of MPs of each type that have been (re)folded is plotted as a function of time, each protein being entered only once, in the year of the first successful report, even if it has been (re)folded using various methods. In **B**, a histogram is shown of the total number of MPs of each type that had been (re)folded by at least one method by the end of 2013 (From Popot 2014, © 2014 Elsevier Inc. All rights reserved).

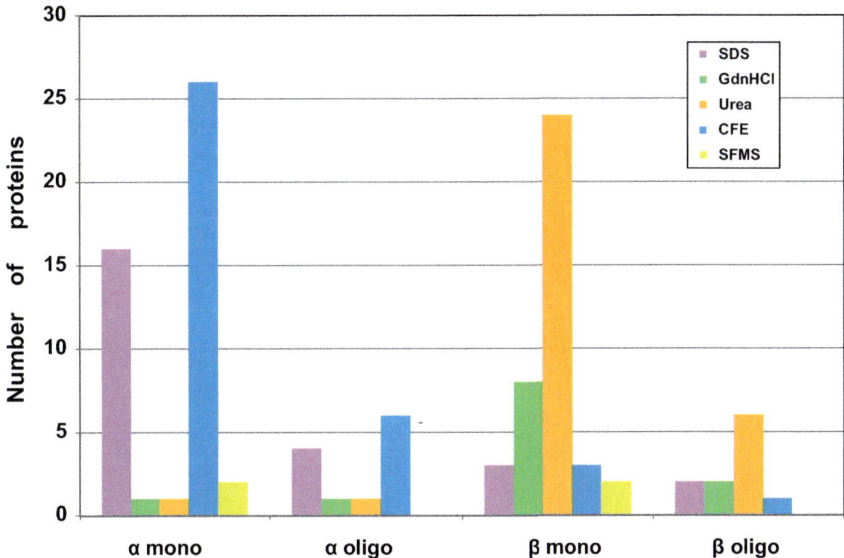

**Fig. 6.3** States from which membrane proteins have been folded de novo or refolded. Five cases have been distinguished: (i) denaturation in SDS or LDS, (ii) denaturation in guanidinium chloride (GdnHCl), (iii) denaturation in urea, (iv) cell-free synthesis (CFE), and (v) unfolding by mechanical traction in a single-molecule force spectroscopy (SMFS) experiment, followed by spontaneous refolding after the traction was released. For each type of unfolded state, MPs are distributed as a function of the secondary and quaternary structure of their transmembrane region, as defined in Fig. 6.2 (From Popot 2014, © 2014 Elsevier Inc. All rights reserved).

- All sorts of MPs have been shown to be amenable to folding in vitro: prokaryotic or eukaryotic, $\alpha$-helical bundles or $\beta$-barrels, monomeric or oligomeric, comprising or devoid of prosthetic groups, etc. (Fig. 6.2).
- A variety of unfolding methods have been used (Fig. 6.3). Note that not all denaturation methods yield preparations that are suitable for renaturation studies: heating in the absence of a chemical denaturant, for instance, will lead to irreversible aggregation.
- Some preferred procedures have emerged over the years. Thus, $\beta$-barrel MPs are nowadays most often denatured in urea, and renatured by diluting urea into a medium containing a surfactant, whereas $\alpha$-helical MPs tend to be denatured in SDS and the SDS either diluted or eliminated in the presence of a surfactant.
- The surfactant used as an acceptor medium can be lipid vesicles, lipid/detergent mixed micelles, or a mild detergent (Fig. 6.4).
- Using classic surfactants, the search for an efficient folding protocol tends to be a highly time-consuming endeavor, involving the exploration of a multidimensional matrix of compositions and conditions, plagued with dead ends and low yields. For G protein-coupled receptors (GPCRs), it is exceptional to exceed a 30% yield, and this using sophisticated procedures involving, for instance, immobilizing the refolding proteins on a column to prevent their aggregation. Developing such protocols often requires years of painstaking groping and optimization. This provided a strong incentive to examine whether APols could not provide a way toward simpler and more effective protocols.

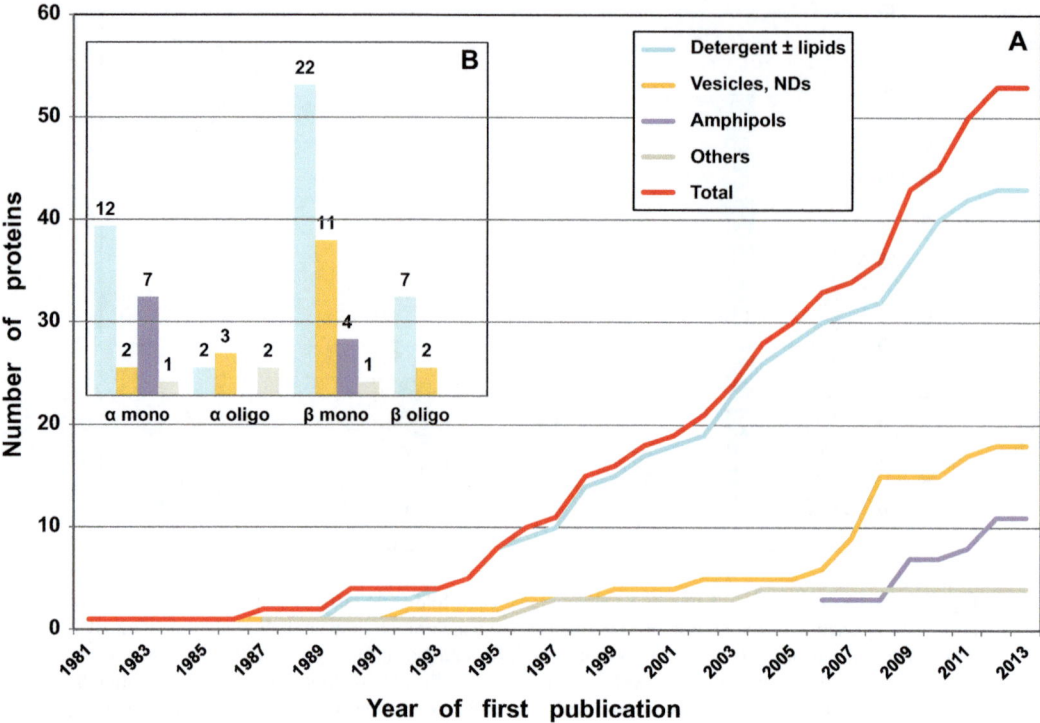

**Fig. 6.4** Environments that have proven favorable to (re)folding membrane proteins in vitro. Four types of environments have been considered: (i) detergent or mixed lipid/detergent micelles; (ii) lipid bilayers, in the form of either lipid vesicles or nanodiscs (NDs); (iii) amphipols and other amphipathic polymers; and (iv) other media. In **A**, the cumulative number of MPs that has been successfully (re)folded in each medium is plotted as a function of time, each protein being entered once for each curve, in the year of the first successful report. Because some MPs have been (re)folded in different media and they are counted only once in the curve summing all folding media, the latter shows fewer proteins than the sum of the individual curves. In **B**, histograms are shown of the types of media that have been used for the (re)folding of each structural type of MP (From Popot 2014, © 2014 Elsevier Inc. All rights reserved).

## 6.3     Amphipol-Assisted Folding of Membrane Proteins

The rationale for trying to use APols to assist the folding of MPs was that, being able to keep them in aqueous solution under a native form more efficiently than detergents, they could provide them with a less aggressive environment in which to seek and adopt their native structure. We shall first review what has been achieved to date for both $\alpha$-helical and $\beta$-barrel MPs (§ 6.3.1). Three short sections will then be devoted to (*i*) a discussion of the possible mechanisms that underlie the success of the approach (§ 6.3.2), (*ii*) challenges and prospects (§ 6.3.3), and (*iii*) some more general considerations about what conclusions can be tentatively derived regarding MP folding in general (Box 6.3).

### 6.3.1     Which Membrane Proteins Have Been Folded in Amphipols and How

APol-assisted folding of MPs was initially demonstrated for three model MPs, an $\alpha$-helical MP, BR, and two $\beta$-barrel MPs, OmpA and FomA (Pocanschi et al. 2006). It was then extended to two more $\beta$-barrel MPs (Leney et al. 2012) and six G protein-coupled receptors (GPCRs), each of which comprises seven TM $\alpha$-helices (Dahmane et al. 2009; Banères et al. 2011; Bazzacco et al. 2012) (Table 6.1). We provide here a summary of the most significant results. More detailed accounts can be found in reviews by Kleinschmidt and Popot (2014) and Le Bon et al. (2018), on which this section is based. Because $\alpha$-helical and $\beta$-barrel MPs fold according to different principles and different protocols, they are treated separately.

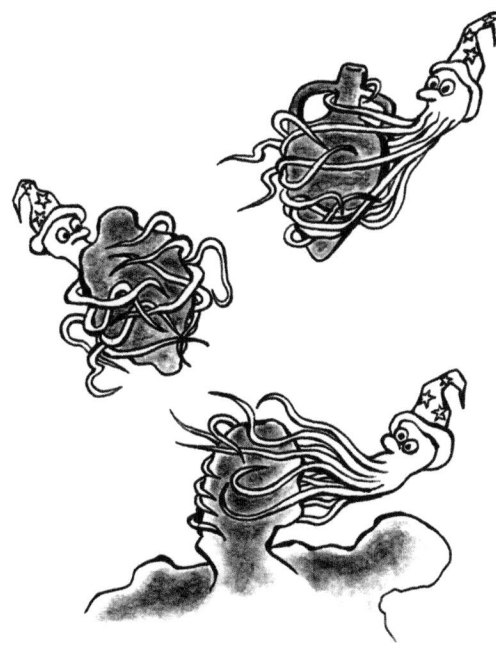

*Amphipol-assisted folding of membrane proteins*
*(© 2018 by Francis Haraux)*

## 6.3.1.1   Amphipol-Assisted Folding of $\alpha$-Helical Membrane Proteins

### 6.3.1.1.1   Folding of BR in A8-35

Because the recovery of its purple color makes it straightforward to observe and measure its renaturation, and it is easy to produce in large amounts, BR has served as a popular model protein in studies on MP folding ever since the seminal work of Khorana and coworkers (Huang et al. 1981) (for reviews, see e.g. Popot and Engelman 1990, 2000; Engelman et al. 2003; Popot 2014; Tastan et al. 2014). The protocol followed in the first experiments aiming at folding it in APols is schematized in Fig. 6.5A. As mentioned above (§ 6.2), when BR is solubilized in SDS, it denatures to BO, releasing its chromophore (because the Schiff base that associates the aldehyde function of the retinal with the amine function of a lysine residue hydrolyzes spontaneously once exposed to water), which causes an absorbance shift from ~555 nm (dark-adapted BR) to ~382 nm (free retinal). The removal of SDS from BO/retinal in the presence of APol A8-35 by precipitating dodecyl sulfate (DS) as potassium dodecyl sulfate (PDS) – a protocol initially developed to refold BR into minimal amounts of lipids (Popot et al. 1987) – results in

**Table 6.1** Folding of membrane proteins in amphipols.

| | MP | Structure | Denaturant | Amphipol | Additives | Method | Yield | References |
|---|---|---|---|---|---|---|---|---|
| 1 | BR[a] | 7-$\alpha$[b] | SDS 5% | A8-35 (2-25 g/g) | PM lipids, retinal | DS precipitation with KCl at pH 7, dialysis | Up to 87–92% | Pocanschi et al. (2006), Bazzacco et al. (2012), and Dahmane et al. (2013) |
| 2 | BR | 7-$\alpha$ | SDS 5% | A8-35 (5 or 10 g/g) | Retinal, no lipids | DS precipitation with KCl at pH 7, dialysis | 70–76% | Dahmane et al. (2013) |
| 3 | BR | 7-$\alpha$ | SDS 5% | A8-35 (5 or 10 g/g) | PM lipids, retinal | Dialysis at pH 7, no NaCl | 52–72% | Dahmane et al. (2013) |
| 4 | BR | 7-$\alpha$ | SDS 2.5% | A8-35 (5 or 10 g/g) | PM lipids, retinal | Dialysis at pH 7, no NaCl | 62–80% | Dahmane et al. (2013) |
| 5 | BR | 7-$\alpha$ | SDS 5% | A8-35 (5 or 10 g/g) | PM lipids, retinal | Dialysis at pH 7, 100 mM NaCl | ~76% | Dahmane et al. (2013) |
| 6 | BR | 7-$\alpha$ | SDS 2.5% | A8-35 (5 or 10 g/g) | PM lipids, retinal | Dialysis at pH 7, 100 mM NaCl | 74–80% | Dahmane et al. (2013) |
| 7 | BR | 7-$\alpha$ | SDS 0.25% | A8-35 (10 g/g) | PM lipids, retinal | 5× dilution at pH 7, no NaCl | 40–50% | Dahmane et al. (2013) |
| 8 | BR | 7-$\alpha$ | SDS 0.25% | A8-35 (10 g/g) | PM lipids, retinal | 5× dilution at pH 7, 100 mM NaCl | 70–80% | Dahmane et al. (2013) |
| 9 | BR | 7-$\alpha$ | TFE[c] | A8-35 (5 or 10 g/g) | Retinal, no lipids | Dialysis at pH 7 | ≤ 40% | Dahmane et al. (2013) |
| 10 | BLT1[d] | 7-$\alpha$ | SDS 0.8% | A8-35 (1–20 g/g) | None | DS precipitation with KCl at pH 8, dialysis | Up to ~50% | Dahmane et al. (2009) |
| 11 | BLT1 | 7-$\alpha$ | SDS 0.8% | A8-35 (5 g/g) | Asolectin/A8-35 1:5 w/w | DS precipitation with KCl at pH 8, dialysis | 65–70% | Dahmane et al. (2009) |
| 12 | BLT2[e] | 7-$\alpha$ | SDS 0.8% | A8-35 (5 g/g) | None | DS precipitation with KCl at pH 8, dialysis | ~50% | Dahmane et al. (2009) |
| 13 | BLT2 | 7-$\alpha$ | SDS 0.8% | A8-35 (5 g/g) | Asolectin/A8-35 1:5 w/w | DS precipitation with KCl at pH 8, dialysis | ~70% | Dahmane et al. (2009), Catoire et al. (2010), and Casiraghi et al. (2016) |
| 14 | CB1[e] | 7-$\alpha$ | SDS 0.8% | A8-35 (5 g/g) | None | DS precipitation with KCl at pH 8, dialysis | ~30% | Dahmane et al. (2009) |
| 15 | CB1 | 7-$\alpha$ | SDS 0.8% | A8-35 (5 g/g) | Asolectin/A8-35 1:5 w/w | DS precipitation with KCl at pH 8, dialysis | ~40% | Dahmane et al. (2009) |
| 16 | 5-HT$_{4(a)}$[e] | 7-$\alpha$ | SDS 0.8% | A8-35 (5 g/g) | None | DS precipitation with KCl at pH 8, dialysis | ~30% | Dahmane et al. (2009) |
| 17 | 5-HT$_{4(a)}$ | 7-$\alpha$ | SDS 0.8% | A8-35 (5 g/g) | Asolectin/A8-35 1:5 w/w | DS precipitation with KCl at pH 8, dialysis | ~60% | Dahmane et al. (2009) |
| 18 | GHSR-1a[d] | 7-$\alpha$ | SDS 0.8% | NAPol (10 g/g) | Asolectin/NAPol 1:5 w/w, 0.2% (w/v) cholesteryl hemisuccinate | DS precipitation with KCl at pH 8, dialysis | ~40% | Bazzacco et al. (2012) |
| 19 | FomA[f] | 14-$\beta$[g] | 10 M urea | A8-35 (8.5 g/g) | None | 10× urea dilution at pH 10 | ~90% | Pocanschi et al. (2006) |
| 20 | OmpA[h] | 8-$\beta$ | 8 M urea | A8-35 (8 g/g) | None | 20× urea dilution at pH 10 | ~100% | Pocanschi et al. (2006) |
| 21 | tOmpA[h,i] | 8-$\beta$ | 8 M urea | SAPol (4 g/g) | None | 9× urea dilution at pH7 | ~100% | Dahmane etal. (2009) |
| 22 | OmpT[h] | 10-$\beta$ | 8 M urea | A8-35 (5 g/g) | None | Dilution, then dialysis, pH 8, 4 °C | ~100% | Leney et al. (2012) |
| 23 | PagP[h] | 8-$\beta$ | 8 M urea | A8-35 (5 g/g) | None | Dilution, then dialysis, pH 8, 4 °C | ~40%[j] | Leney et al. (2012) |
| 24 | GLUT4[k] | 12-$\alpha$ | Exposure to pH10 in LMNG solution | A8-35 | None | n.s. | n.s. | Kraft et al. (2015) |

From Kleinschmidt and Popot (2014), with some updating

*n.s.* not specified

[a]From *Halobacterium salinarum*

[b]7-$\alpha$: bundle of seven transmembrane $\alpha$-helices

[c]Observation certain but could not be duplicated

[d]From *Homo sapiens*

[e]From *Mus musculus*

[f]From *Fusobacterium nucleatum*

[g]$n$-$\beta$: barrel of $n$ observed or predicted transmembrane $\beta$-strands

[h]From *Escherichia coli*

[i]Transmembrane domain of OmpA

[j]Misprinted 60% in the original paper

[k]From *Rattus norvegicus*

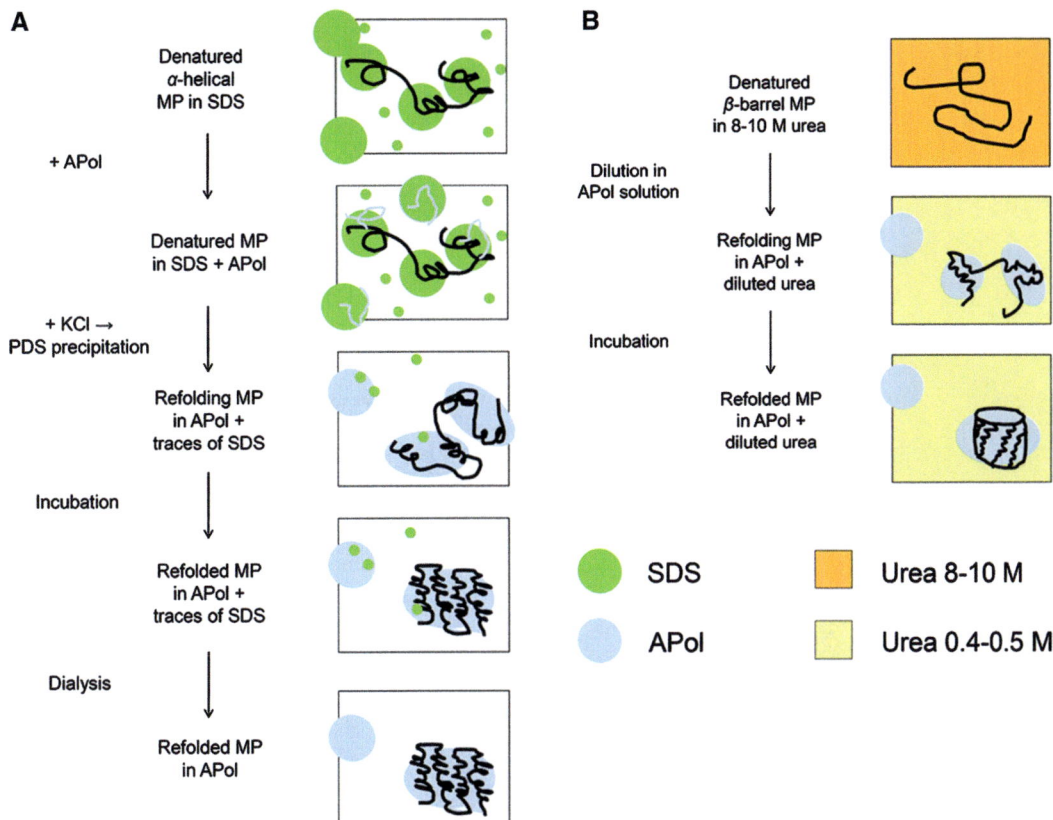

**Fig. 6.5** Schematic representation of typical protocols for amphipol-assisted folding of (**A**) α-helical and (**B**) β-barrel membrane proteins. *Green*, SDS; *light blue*, APol; *orange*, concentrated urea; *yellow*, dilute urea; PDS, potassium dodecyl sulfate. The cartoons on the right of each column show interpretations of what is going on in the samples, some aspects of which are solidly supported by experimental data, whereas others are more speculative.

the regeneration of the characteristic purple color of BR within minutes, indicating folding of BO and rebinding of the retinal (Pocanschi et al. 2006). After dialyzing to remove residual DS, almost quantitative refolding of BR ($\geq$90%) is observed for a mass ratio BR/A8-35 $\leq$1:5 (Pocanschi et al. 2006; Dahmane et al. 2013) (Fig. 6.6). When refolded at such mass ratios, BR migrates as an A8-35-trapped monomer, as determined by size exclusion chromatography (SEC). At a mass ratio of BR/A8-35 = 1:2, the yield is lower, and SEC reveals some aggregation of the renatured protein (Fig. 6.5). Control experiments carried out using the same protocol but with detergents instead of A8-35 show lower levels of BR refolding. Yields are in fact negligible with octylglucoside (OG), octylthioglucoside (OTG), or $C_8E_4$, but reach ~66% in dodecylmaltoside (DDM) (Dahmane et al. 2013). BR refolded in A8-35 displays a fully functional photocycle. The light-adapted protein is excited at 640 nm with a 5-ns laser flash. Transient absorbance changes, monitored from 10 ns to 100 ms in a spectral range from 370 to 500 nm, are found to be very similar to those observed with lipid-associated native BR trapped in A8-35 (Pocanschi et al. 2006; Dahmane et al. 2013).

More stringent experiments were performed in which BO was totally unfolded in formic acid and transferred to SDS after lipids and retinal had been removed by hydrophobic SEC, as described in path II of Fig. 6.1. They established that completely delipidated BO (dBO) refolds in pure A8-35 to 60–80% (Pocanschi et al. 2006; Dahmane et al. 2013).

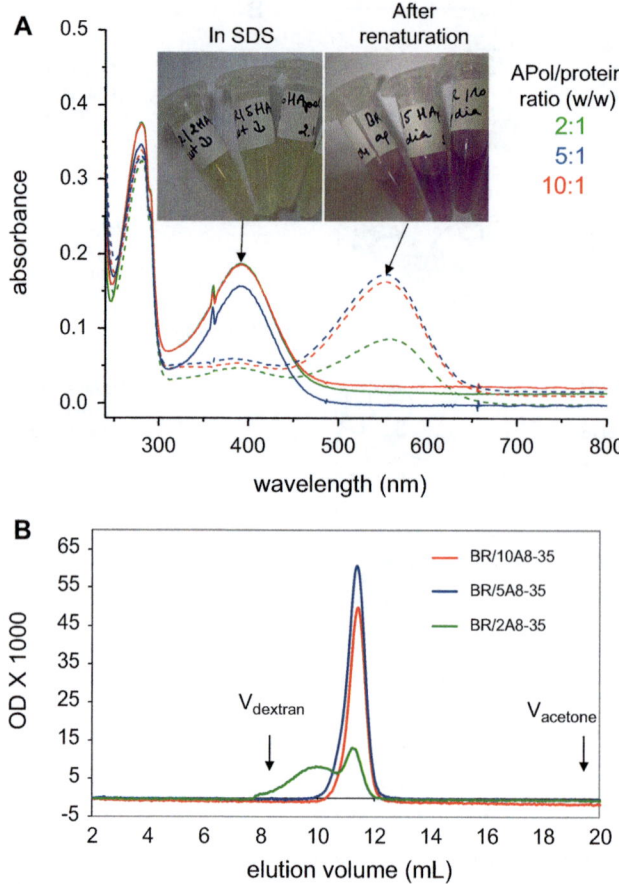

**Fig. 6.6** Renaturation of bacteriorhodopsin in amphipol A8-35. Bacteriorhodopsin (BR) was renatured from SDS-solubilized purple membrane according to the protocol shown in Fig. 6.5A, at three different amphipol/protein mass ratios. Samples were studied by absorbance spectroscopy before and after renaturation (**A**) and by size exclusion chromatography after renaturation (**B**) (Adapted with permission from Pocanschi et al. 2006, © 2006 American Chemical Society).

As summarized in the first two lines of Table 6.1, it has been systematically observed that lipids are not required for BR to fold in APols but that their presence increases the folding yield. The same observation was made again when folding G protein-coupled receptors (see below, § 6.3.1.1.2). This will be tentatively interpreted in § 6.3.2. It was also noted that larger amounts of A8-35 are necessary to reach optimal folding yields in the presence of lipids than in their absence. This suggests that lipids may favor BR aggregation at low A8-35 concentrations.

Interestingly, an attempt at direct refolding of dBO by transfer from trifluoroethanol (TFE) to A8-35, in aqueous buffer in the absence of SDS, resulted in a folding yield of ~40% (Dahmane et al. 2013) (Table 6.1, line 9). Despite many attempts, this observation could not be reproduced, but it is nevertheless worth mentioning, because chances that it is artefactual are about nil. It may be that a critical factor in the process of transfer from formic acid to TFE has not been identified. The approach may perhaps be worth exploring again for folding other MPs or MP-derived peptides when, for some reason, the use of SDS is to be avoided.

Various alternatives to PDS precipitation for transferring BO from SDS to A8-35 have been explored. Folding yields were somewhat lower when folding was initiated by dialysis rather than precipitation (Dahmane et al. 2013; Elter et al. 2014) (Table 6.1, lines 3–6 and 9). Upon SDS removal by dialysis, folding yields of BR increased with the A8-35/BO ratio, suggesting that aggregation is a limiting factor (Dahmane et al. 2013).

BR was also successfully refolded by diluting a BO/SDS/A8-35 mixture with SDS-free buffer (Table 6.1, lines 7–8), in which case larger mass ratios of A8-35/BO were required than observed using either PDS precipitation or SDS dialysis: when folding was initiated by dilution, yields were highest at a mass ratio of 25 (Dahmane et al. 2013). Two mechanisms may contribute to this effect. First, adding more A8-35 reduces the proportion of SDS in the environment of the refolding proteins. Second, the probability that the latter will establish intermolecular associations is reduced by diluting them with more APol, making aggregation less likely (cf. Zoonens et al. 2007). Folding kinetics were faster at higher concentrations of retinal, consistent with the view that refolded BO can pick up retinal very rapidly if it is present in its associated APol belt, whereas the process is much slower if retinal uptake depends on BO/APol complexes colliding with retinal-containing free particles of APol (cf. Zoonens et al. 2007).

Folding of dBO in A8-35 by PDS precipitation was also successful in the absence of retinal (Dahmane et al. 2013). However, retinal is known to stabilize BR (Kahn et al. 1992), and indeed refolded BO is less stable than refolded BR: when retinal was added shortly after the transfer from SDS to A8-35, the refolding yield was found to be as high as that observed when retinal is present during the transfer (~80%, which is in the upper range of that observed with lipid-free BO preparations; Table 6.1, line 2). If, on the other hand, retinal was first provided 3 days after the transfer, the percentage of holoprotein in the renatured preparation dropped to ~30%. The most straightforward interpretation of this observation is that, in the 3-day interval, ~2/3 of the BO that had initially refolded had denatured again. This observation confirms (*i*) that BO folds efficiently in the absence of retinal (renatured BO indeed features the same secondary structure as BR refolded from delipidated BO/SDS in the presence of retinal; Dahmane et al. 2013) and (*ii*) that early enough rebinding of retinal is critical in stabilizing the refolded state and, thereby, in determining the final yield of renaturation. This observation is consistent with the general view that folding yields of fragile MPs can be improved in the presence of their ligands.

Transfer from SDS to A8-35 can also be achieved by adsorbing SDS onto Bio-Beads, provided aggregation is limited by immobilizing polyhistidine-tagged BR onto a nickel-bearing column (Elter et al. 2014).

### 6.3.1.1.2  A8-35-Assisted Folding of GPCRs

The remarkable results achieved with BR prompted attempts at folding GPCRs, for which folding using classic detergent/lipid systems, when successful, is typically limited to yields of 30% or less (reviewed in Banères et al. 2011). The first GPCR whose folding in A8-35 was studied was BLT1, a leukotriene receptor involved in the control of inflammatory processes. The protocol used was the most successful of those tested on BR (Pocanschi et al. 2006; Dahmane et al. 2013) (Fig. 6.6A): BLT1, after being solubilized in SDS from inclusion bodies, was supplemented with A8-35, most of DS precipitated as PDS, and residual DS removed by dialysis. Ligand-binding assays indicated that the receptor thus folded was functional, with a dissociation constant, $K_D \approx 9$ nM, similar to that of native BLT1 expressed in membrane fractions (Damian et al. 2006). Based on the number of binding sites per mass of protein in the original SDS solution, the yield of folding was ~50% in pure A8-35 and 65–70% in the presence of soybean lipids (asolectin; BLT1/A8-35/lipid mass ratio 1:5:1) (Table 6.1, lines 10–11). A8-35-folded BLT1 showed a similar pharmacological profile to the membrane-bound receptor (Dahmane et al. 2009). Interestingly, in the absence of lipids, an optimal protein/APol ratio of ~1:5 w/w was observed for BLT1 (Fig. 6.8, *left*), the possible reasons for which are discussed in Box 6.2.

The study was then extended to three other GPCRs: another leukotriene receptor, BLT2, the serotonin receptor 5-HT$_{4(a)}$, and the cannabinoid receptor CB1 (Dahmane et al. 2009). In the presence of asolectin, A8-35 improved the folding yields of 5-HT$_{4(a)}$ previously observed by more than a factor of two, namely from 20–25% in detergent/asolectin micelles (Banères et al. 2005) to ~30% in pure A8-35 and ~60% in A8-35 + asolectin (Dahmane et al. 2009). For BLT2, which shares ~45% sequence identity with BLT1, the yield improved from 3–4% in detergent/asolectin micelles to ~50% in pure A8-35 and ~70% in A8-35 + asolectin (Dahmane et al. 2009). As regards CB1, which had not been folded to any significant extent in detergent/lipid mixtures at the time of these experiments (but has been folded to ~30% since; see Michalke et al. 2010), folding yields of ~30% were achieved in pure A8-35 and of ~40% in A8-35/asolectin mixtures. Altogether, these data, obtained without any extensive search for optimal folding conditions, suggest an interesting potential of APols as a generally useful new tool for the folding of GPCRs. Indeed, more recently, another GPCR, the ghrelin GHSR-1a receptor, has been successfully refolded in APols (Table 6.1, line 18, and § 6.3.1.1.3) (Banères et al. 2011; Bazzacco et al. 2012). Folding of a sixth GPCR, the type 2 arginine-vasopressin receptor, has been reported in Banères et al. (2011), but experimental details have yet to be published.

Whereas in no case was a 100% yield reached, separation of active from inactive 5-HT$_{4(a)}$ can be achieved using a GR113808 affinity column (Banères et al. 2005), yielding ~96% active receptor (Dahmane et al. 2009). In this context, it is worth noting that, whereas 3D crystallization of APol-trapped MPs remains a difficult challenge (see Chap. 11, § 11.3.1), A8-35-trapped BR has yielded highly organized crystals (diffracting to <2-Å resolution) following direct transfer to lipidic mesophases (Polovinkin et al. 2014) (Chap. 11, § 11.3.2). This observation has been recently reproduced using a BR variant trapped in styrene-maleic acid copolymer (SMA) (Broecker et al. 2017). Crystallization of GPCRs in mesophases has proven remarkably successful (see e.g. Cherezov et al. 2007; Rasmussen et al. 2011, and references therein), and it has been extended to a large variety of MPs (for reviews, see Cherezov et al. 2006; Caffrey 2011; Cherezov 2011; Ishchenko et al. 2014). Combining APol-assisted folding of GPCRs or other MPs with crystallization in mesophases, which does not require very pure MPs, might therefore open very interesting perspectives for MP crystallization in general (see Chap. 11).

---

**Box 6.2   Why Is There an Optimal Amphipol/Protein Ratio for Folding the BLT1 Receptor in Pure A8-35?**

As shown in Fig. 6.7 *(left)*, better yields are observed when folding the BLT1 receptor in pure A8-35 at ~1:5 BLT1/A8-35 mass ratio than either with less (1:1) or with more (1:10 or 1:20) APol present. How to account for this observation?

As regards the 1:1 w/w ratio, the most likely hypothesis is that the protein is not diluted enough in the surfactant "phase" to prevent unproductive intermolecular interactions from forming in the course of folding, preventing correct folding. Indeed, as shown in § 6.3.1.1.1, when BR is folded at too low an APol/protein ratio, the yield drops (Fig. 6.6A), and part of the folded protein is found in aggregated form (Fig. 6.6B).

As regards the drop of yield at high APol/BLT1 ratios, the explanation is less straightforward. If a stabilizing cofactor (prosthetic group, lipid, etc.) were diluted by the APol, one could indeed expect a drop in yield at high APol/protein ratios. However, the system here comprises only the protein, the APol, and, initially, dodecyl sulfate, the concentration of which in the aqueous solution and in the APol particles during the initial stages of folding is set by the concentration of free potassium ions. Dilution with further APol should not change the environment of the folding protein. We can consider two types of interpretation:

- BLT1 is known to dimerize (Mesnier and Banères 2004). Upon SEC analysis of the A8-35-folded BLT1, at least part of it does seem to migrate as a dimer rather than a

### 6.3.1.1.3  Folding of $\alpha$-Helical Membrane Proteins in Non-ionic Amphipols

Non-ionic amphipols (NAPols; see Chap. 4, Fig. 4.8) have been used successfully to fold both BR and the ghrelin GHSR-1a GPCR to their native state (Bazzacco et al. 2012). Because of their non-ionic character, NAPols may provide an even milder environment than polyanionic A8-35 (Bazzacco et al. 2012), and they present the advantage of being soluble over a broader pH range (Sharma et al. 2012), covering the mildly acidic regime favorable for NMR work (see Chap. 10). Folding of BR in homopolymeric NAPols was achieved using the same strategy as previously used for folding BR (Pocanschi et al. 2006) and GPCRs (Dahmane et al. 2009) in A8-35, i.e. by supplementing the SDS-solubilized PM with NAPols, followed by PDS precipitation and removal of residual DS by extensive dialysis (Fig. 6.5A). Quantitative analysis by UV-visible absorbance spectroscopy indicated a yield ≥90%. NAPol-refolded BR was homogeneous, as shown by SEC (Bazzacco et al. 2012).

Following the proof of principle achieved with BR as a model, NAPols were used for folding the GHSR-1a GPCR. The key advantage of NAPols over A8-35 for these experiments is their absence of charges. Indeed, ghrelin, the positively charged and amphipathic ligand of GHSR-1a, binds

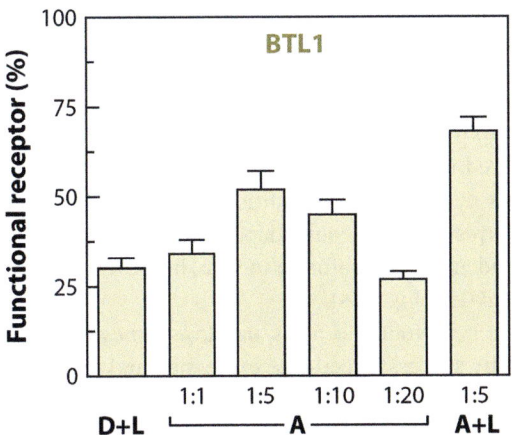
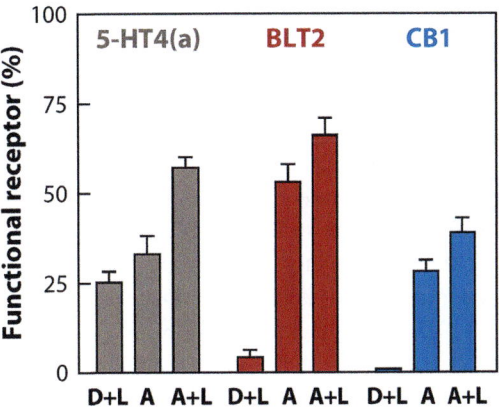

**Fig. 6.7** Yields of folding of four GPCRs under different conditions. The BLT1 leukotriene receptor (*left*) and the 5-HT$_{4(a)}$ serotonin receptor, the BLT2 leukotriene receptor, and the CB1 cannabinoid receptor (*right*), all of them class A (rhodopsin-like) GPCRs, were expressed in inclusion bodies and purified in an inactive form in sodium dodecyl sulfate (SDS) solution. They were folded by substituting SDS either with a lipid/detergent mixture (D+L), with pure A8-35 (A) at different protein/APol mass ratios (BLT1), with A8-35 (A) at a 1:5 mass ratio (BLT2, 5-HT$_{4(a)}$ and CB1), or with A8-35 supplemented with asolectin in a 1:5:1 protein/APol/asolectin mass ratio (A+L) (all receptors). The extent of correct folding is expressed as the percentage of total receptor (on the basis of the protein concentration in the SDS solution) that is able to bind a specific ligand (Figure from Popot 2010, data from Dahmane et al. 2009).

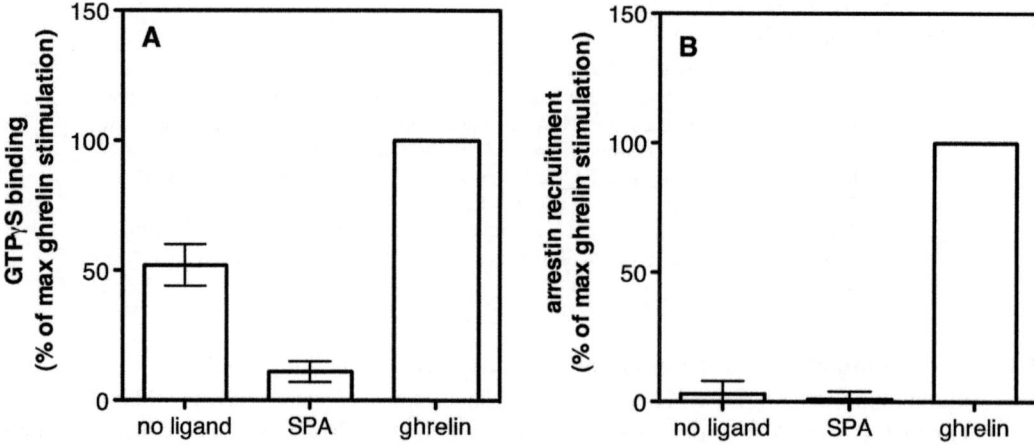

**Fig. 6.8** G protein activation and arrestin recruitment by the ghrelin receptor folded in NAPols. (**A**) BODIPY FL GTPγS binding to the $G_{\alpha q}$ protein induced by GHS-R1a in the absence of ligand, in the presence of 5 µM SPA ([D-Arg[1], D-Phe[5], D-Trp[7,9], Leu[11]] substance P, an inverse agonist), or in the presence of 5 µM ghrelin. Data are presented as the percentage of maximal BODIPY FL fluorescence change measured in the presence of ghrelin. (**B**) Changes in the emission intensity of bimane-labeled arrestin-2 induced by GHS-R1a in the absence of ligand or in the presence of either 5 µM SPA or 5 µM ghrelin. Data are presented as the percentage of maximal bimane fluorescence change measured in the presence of ghrelin. In panels A and B, the data represent the mean value ± the standard deviation from three independent experiments (Reprinted with permission from Bazzacco et al. 2012, © 2012 American Chemical Society).

non-specifically to A8-35 particles. This causes difficulties in assessing by ligand binding the extent of folding achieved in A8-35. Measurements are easier when the ghrelin receptor is folded in NAPols, where the background is lower. Ligand-binding measurements indicated a folding yield of ~40%, and a receptor ~97% active was obtained after affinity chromatography (Bazzacco et al. 2012). The binding properties of the folded GHSR-1a were then further examined to assess the quality of folding. Fluorescence energy transfer from GHSR-1a, labeled with Alexa Fluor 350, to a ghrelin peptide labeled with fluorescein isothiocyanate was recorded in competition experiments with synthetic antagonists. The competition profiles obtained by this method are within the same range as those previously inferred from radioactive and TagLite-based measurements on human embryonic kidney (HEK) cells transiently expressing GHSR-1a (Leyris et al. 2011). In addition, GHSR-1a folded in NAPols (*i*) is able to activate G proteins, (*ii*) recruits arrestin in an agonist-dependent manner, and (*iii*) adopts a very similar equilibrium between active and inactive conformations as in the membrane, confirming that it is fully functional (Bazzacco et al. 2012) (Fig. 6.8).

The polyanionic APols A8-35 and SAPols have been found to block in vitro synthesis of MPs (Park et al. 2011), but NAPols do not, as exemplified by successful cell-free expression and folding of BR (Bazzacco et al. 2012). This may open an interesting alternative route to producing hard-to-express MPs (see Chap. 7).

### 6.3.1.2  Amphipol-Assisted Folding of β-Barrel Membrane Proteins

In spite of all progress, finding the right conditions for folding unfolded β-barrel MPs obtained from inclusion bodies into detergents and/or lipids has remained a time-consuming and, more often than not, frustrating task (reviewed in Buchanan 1999; Buchanan et al. 2012; Otzen and Andersen 2013; Popot 2014). It is therefore of great interest to develop alternative methodologies.

### 6.3.1.2.1    Folding of OmpA and FomA in A8-35

Applying APols to (re)folding outer membrane proteins was first attempted using as models OmpA from *E. coli* and FomA from *Fusobacterium nucleatum*, two monomeric β-barrels with, respectively, 8 (observed) and 14 (predicted) TM β-strands. OmpA features, in addition, a periplasmic *C*-terminal domain (~17 kDa), which can be genetically removed, yielding the isolated TM domain (tOmpA). Folding was achieved according to the very simple scheme shown in Fig. 6.5B: The proteins, expressed as inclusion bodies, are dissolved in unfolded form in either 8 M (OmpA) or 10 M urea (FomA), and the preparations diluted ~20-fold in a solution of A8-35 (Pocanschi et al. 2006). Folding yields can be assessed by SDS-PAGE, taking advantage of the fact that the folded form resists unfolding by SDS at room temperature and migrates at a position different from unfolded ones (Fig. 6.9A, B).

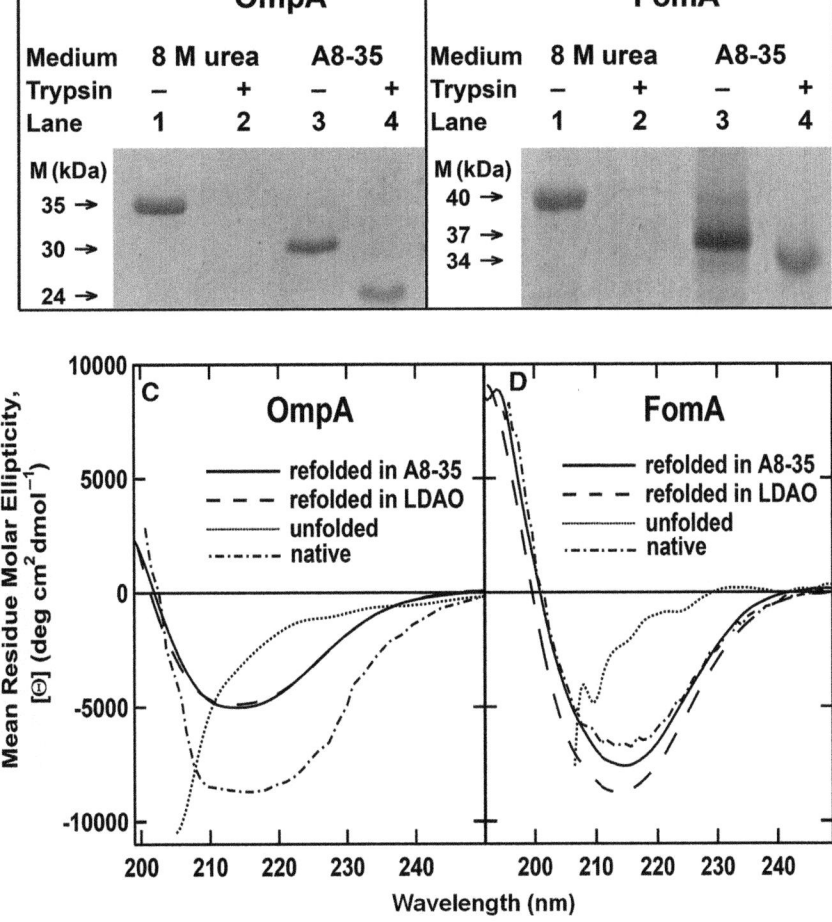

**Fig. 6.9**  Folding of OmpA and FomA in amphipol A8-35. (**A, B**) Migration of OmpA and FomA upon SDS-PAGE, indicating folding and protection of the folded proteins against trypsin digestion: lane *1*, denatured MP in 8 M urea; lane *2*, same incubated for 2 h with trypsin; lane *3*, refolded MP in A8-35; and lane *4*, same incubated for 2 h with trypsin. (**C, D**) Far-UV CD spectra of OmpA and FomA recorded before and after folding in A8-35. Spectra of OmpA and FomA folded in LDAO, which are similar to those of the native proteins, are shown for comparison (Adapted with permission from Pocanschi et al. 2006, © 2006 American Chemical Society).

Folding was observed to be very efficient, with yields of ~100% for OmpA and ~90% for FomA (Fig. 6.9A, B; Table 6.1, lines 19–20). It was completed within ~7 h for OmpA and ~24 h for FomA. Several criteria were used to assess that the two proteins had achieved their native state:

(i)  SDS-PAGE (Fig. 6.9A, B);
(ii)  Protection of the folded $\beta$-barrels against proteolysis (ibid.);
(iii)  Far-UV circular dichroism (CD) spectroscopy (Fig. 6.9C, D);
(iv)  Functional studies, that is single-channel conductance recordings after transfer of A8-35-refolded OmpA and FomA to black lipid films (Pocanschi et al. 2006).

In these experiments, the first three criteria did not reveal any differences between the detergent- and A8-35-refolded forms. However, in single-channel measurements of the conductance of the pores formed upon incorporation into black lipid bilayers, both OmpA and FomA initially displayed smaller conductance when inserted from complexes with A8-35 rather than from detergent-refolded forms. This difference was traced to an asymmetrical distribution of A8-35 in the black lipid films. When A8-35 was present at equal concentrations on both sides of the film, the same conductance was observed as with the native or detergent-folded proteins (Pocanschi et al. 2006; see Chap. 5, § 5.7, Fig. 5.34). The spontaneous transfer of refolded OmpA and FomA from A8-35 to black lipid bilayers is consistent with the transfer, observed earlier, of the $\alpha$-helical MP DAGK from another APol, OAPA-20 (similar to A8-75; see Chap. 4, Fig. 4.1), into lipid bilayers of 1-palmitoyl-2-oleoyl-phosphatidyl-choline (multilamellar vesicles) in a functionally active form (Nagy et al. 2001; see Chap. 5, § 5.7). The effect of the asymmetrical distribution of A8-35 on conductance levels is consistent with the adsorption of APols onto either artificial lipid vesicles (reviewed in Marie et al. 2014) or living cells (Popot et al. 2011; see Chap. 15). A direct interaction of the membrane-adsorbed APol with the proteins is possible, but seems rather unlikely, because both entropic considerations and observations made in lipid mesophases (Polovinkin et al. 2014; see Chap. 11) rather suggest that APol and protein dissociate one from another upon insertion into a lipid membrane (see Chap. 5, § 5.7). A physical effect seems more likely, such as could result from an APol-induced asymmetry in surface charge and/or lateral pressure. This would readily explain why the conductance levels go back to normal upon adding APol to the other side of the film.

### 6.3.1.2.2  Folding of tOmpA, OmpT, and PagP

Folding of a $\beta$-barrel MP has also been demonstrated with sulfonated APols (SAPols; see Chap. 4, Fig. 4.1), using a similar protocol (Dahmane et al. 2011). After 2 days of incubation, the genetically engineered TM domain of OmpA (tOmpA), isolated after expression into inclusion bodies, folded to yields approaching 100% (Table 6.1, line 21), as determined by the shift of its electrophoretic mobility from ~16 kDa for the unfolded form to ~19 kDa for the folded one.

OmpT, an integral outer membrane protease with ten TM $\beta$-strands, and PagP, an eight-stranded $\beta$-barrel MP that catalyzes the transfer of a palmitate chain from a phospholipid to lipid A, have been folded from their unfolded forms in 8 M urea by supplementing them with A8-35 at a 1:5 protein/APol mass ratio and dialyzing away the urea (Leney et al. 2012). Using electrospray ionization mass spectrometry coupled with ion mobility spectrometry (ESI-IM-MS; see Chap. 14), yields were shown to reach ~100% for OmpT and ~40% for PagP (Table 6.1, lines 22–23, and Fig. 6.11D, E). Folding was confirmed by electrophoretic mobility measurements (Fig. 6.10A, B), by far-UV CD spectroscopy, and by functional studies (Leney et al. 2012). SEC showed a single narrow peak for OmpT/A8-35 complexes, indicating the presence of a homogeneous single species of OmpT, whereas a broader peak was observed after folding of PagP, indicating the presence of a mixture of folded and unfolded species, consistent with the ~60% folding yield observed by SDS-PAGE and ESI-IMS-MS (Leney et al. 2012).

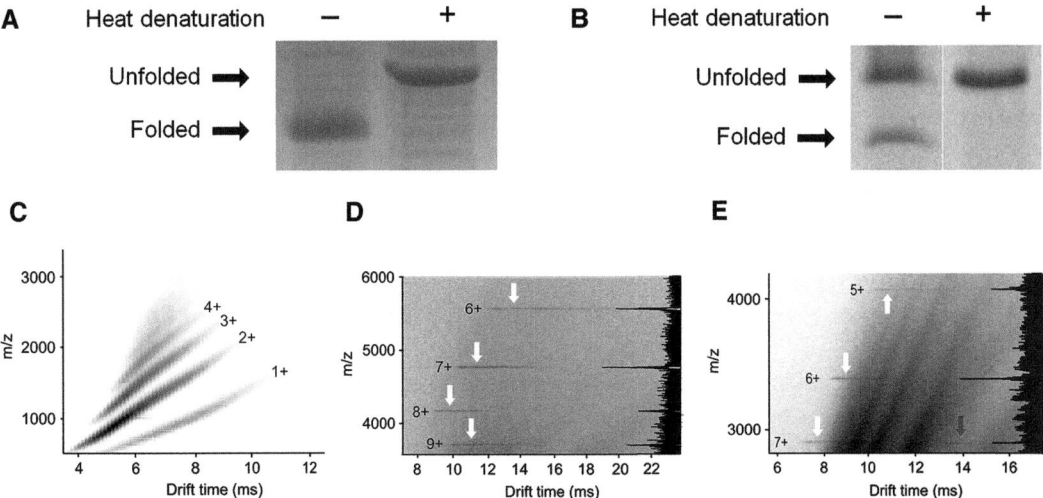

**Fig. 6.10** SDS-PAGE and mass spectrometry analysis of the folding in A8-35 of the $\beta$-barrel outer membrane proteins OmpT and PagP. Each of the two proteins, initially unfolded in 8 M urea, was supplemented with A8-35 in a 1:5 protein/APol mass ratio and dialyzed for 24 h. (**A, B**) The resulting preparations were analyzed by SDS-PAGE with and without heating in SDS. All of unheated OmpT migrates as the folded form (**A**), whereas ~60% of PagP has not reached its native conformation (**B**). (**C**) ESI-IM-MS driftscope plot of A8-35 alone, highlighting four different charge state ion series. The graph plots the $m/z$ value of the ions *vs.* their drift time (ms). Because an ion's drift time depends on both its shape and the number of charges it carries, and A8-35 is heterogeneous, each charge species yields a fuzzy slanted streak. (**D, E**) ESI-IM-MS analysis of folded preparations of OmpT and PagP, respectively. The proteins, whose mass is homogeneous, yield horizontal streaks, superimposed over the slanted streaks of the APol. OmpT (**D**), which is folded to ~100% (cf. **A**), yields a single series of streaks. PagP (**E**), which is folded only to ~40% (cf. **B**), yields two series, one corresponding to the compact folded form (*white arrows*), the other to the more extended unfolded one (*red arrow*) (Adapted with permission from Leney et al. 2012, © 2012 American Chemical Society (further permissions to reuse this material must be directed to http://pubs.acs.org/doi/abs/10.1021/ac302223s)). For more details about MS analyses of amphipol-trapped membrane proteins, see Chap. 14 .

### 6.3.1.2.3 Kinetics and Thermodynamics of Folding OmpA in A8-35

A detailed study of the kinetics and thermodynamics of folding and unfolding of OmpA in A8-35 has been carried out by Jörg H. Kleinschmidt and coworkers (Pocanschi et al. 2013). Folding was initiated by 18-fold dilution of urea in the presence of various mass ratios of A8-35, ranging from 0.5 to 16 g per g of OmpA and at various temperatures. Analyses of the time course of electrophoretic mobility shift from the unfolded form ($M_r \approx 35$ kDa) to the folded one ($M_r \approx 30$ kDa) indicated that folding takes ~6 to 8 h at pH 10 (a high pH used to ensure a good solubility of the unfolded protein). The minimum mass ratio of A8-35/OmpA required to achieve complete folding was 2:1. Kinetics of folding were obtained by fluorescence spectroscopy and by monitoring the formation of tertiary structure by SDS-PAGE. Whichever method was used, two parallel pathways of folding were observed, presumably due to the coexistence, at the time of urea dilution, of various forms of OmpA carrying different charges and, as a consequence, engaging in different folding paths. The rate constants of the two folding processes did not depend on the concentration of A8-35, indicating that intermolecular interactions between proteins are not involved. SDS-PAGE did not reveal any folding intermediates. (Such intermediates likely exist, but they do not survive exposure to SDS and revert to the unfolded form.)

The activation energy of folding was determined from the temperature dependence of the folding kinetics. The SDS-PAGE assay yielded an activation energy of ~5.9 $\pm$ 4.1 kJ·mol$^{-1}$ (~1.4 kcal·mol$^{-1}$)

for the fast process, ~36.5 $\pm$ 9.6 kJ·mol$^{-1}$ (~8.7 kcal·mol$^{-1}$) for the slow one. When monitored by fluorescence spectroscopy, these activation energies were ~8.8 $\pm$ 2.3 kJ·mol$^{-1}$ for the fast process and 28.9 $\pm$ 8.1 kJ·mol$^{-1}$ for the slow one. Within error margins, these values are comparable. They are smaller than the activation energy of 46 kJ·mol$^{-1}$ reported for folding OmpA in small unilamellar vesicles of dioleoylphosphatidylcholine (Kleinschmidt and Tamm 1996; Kleinschmidt 2015), which may reflect the fact that reorganization of the protein in APols faces lower free energy barriers than in lipids.

## 6.3.2    Why Are Amphipols a Good Medium for Membrane Protein Folding?

As described above, folding MPs in APols has been, to date, remarkably successful. This statement must be qualified by noting that only a limited range of structural types has been explored so far: 7-$\alpha$-helix bundles and single $\beta$-barrel outer membrane proteins (Table 6.1). In all cases that have been tested to date, folding yields were at least as good as and usually better than those obtained, using generally more complex procedures, in the presence of detergent or lipid/detergent mixed micelles. Furthermore, no lengthy search for optimization was required, as is usually needed when endeavoring to fold a new MP in a detergent environment. Indeed, two more or less universal protocols based on urea dilution ($\beta$-barrel MPs) or PDS precipitation ($\alpha$-helical ones) in the presence of APols give, as a rule, satisfying results without departing very much from the conditions initially established for model proteins, OmpA and FomA on the one hand, BR on the other (Pocanschi et al. 2006).

In a couple of cases where GPCRs did not fold efficiently in A8-35, the preparation in SDS appeared heterogeneous upon SDS-PAGE (unpublished data). It is unsurprising that preparations in which a large fraction of the material is either aggregated or misfolded do not fold in good yield. Indeed, in early experiments in which PDS precipitation was used to refold and reassociate two BR fragments in a lipid environment, satisfactory yields were not achieved before a way was devised to transfer the fragments from organic solution to SDS without inducing any aggregation (Popot et al. 1987). Every effort should therefore be spent to obtain starting material that yields a single band upon SDS-PAGE before investing too much time in refolding attempts. In difficult cases, one could consider resorting to organic solvents to achieve complete unfolding, followed by transfer to SDS (Huang et al. 1981; Popot et al. 1987; Pocanschi et al. 2006; Dahmane et al. 2013). Replacing SDS with tetradecyl sulfate, which is a stronger denaturant (Moosavi-Movahedi et al. 2003), is perhaps another option, which has not been tested yet.

It is worth noting that, in all experiments published to date, no special precaution was taken to control the reformation of disulfide bridges, if any. It is to be expected that control of the redox potential, e.g. by folding first in a reducing medium, followed by oxidation, or by folding in the presence of a mixture of reduced and oxidized glutathione, will turn out to be necessary in specific cases.

In all cases where it has been tested, the presence of lipids ($i$) was not necessary to obtaining good folding yields but ($ii$) improved the yield as compared to that observed in their absence. As lipids were used in small amounts as compared to APols (typically in a 1:5 mass ratio), an effect on the physical properties of the APol belt surrounding the refolding protein seems rather unlikely. More probably, molecular interactions are at work. Crystallographic structures of MPs have revealed well-defined binding sites for lipids, often at the protein/lipid interface, sometimes within protein TM domains (see Chap. 1, § 1.5.2). Lipids in biological membranes, should therefore not be considered as a mere two-dimensional solvent: they also play the role of cofactors (for discussions, see e.g. Popot and Engelman 2000; Lee 2003, 2011; Adamian et al. 2011; Aponte-Santamaríaa et al. 2012; Stansfeld et al. 2013). One way to understand their favorable effect on folding yields in APols is to assume that in the course of the conformational excursions experienced by a (re)folding protein, transient, partially folded states appear, some of which may exhibit lipid-binding sites. Binding of lipids to such an intermediate state will stabilize it, lengthening the period of time during which folding has a chance to proceed to completion. In other words, binding of lipids to partially folded states may steer folding toward a native-like conformation (Dahmane et al. 2013; Zoonens and Popot 2014). Once the native structure is achieved, the binding of lipids will stabilize it, diminishing the frequency of conformational excursions that could lead to denaturation, misfolding, and/or aggregation.

Why is it that folding of MPs in APols is, apparently, so efficient, even though this medium is so unlike a lipid bilayer, both in its chemistry and in its organization? Among the various mechanisms of stabilization that have been discussed in Chap. 5 (§ 5.6), the most relevant one may be their poorly dissociating character. When folding is carried out in detergent, or in lipid/detergent mixtures, the detergent – a surfactant that was initially selected for its dissociating properties – competes with reforming protein/protein and protein/lipid interactions. In the scheme of Fig. 6.11A, this means that folding to the native structure ① has to compete with partial folding ②, in which protein/surfactant interactions replace some protein/protein ones, misfolding ③, in which non-native-like interactions form intramolecularly, and aggregation ④, induced by intermolecular protein/protein interactions. APols, because of their low detergency, can be expected to favor the productive path ①, which, being more accessible, should more efficiently compete with the unproductive paths ②, ③, and ④ (Fig. 6.11B). Furthermore, most MPs are not particularly stable in detergent solutions, so that a protein that has managed to reach a native-like state is at risk of denaturing again at a later stage. In SDS or urea, it would go back to the "denatured" state, from which it can fold again. In a "non-denaturing" detergent, however, which is not so dissociating, it may well reach a misfolded ③ and/or aggregated ④ state, from which it may not recover (Fig. 6.11A). In an APol environment, chances that a protein that has reached a partially folded state ② will move to the native state ① can be expected to be higher, because of the lesser competition of protein/surfactant interactions with native-like protein/protein and protein/lipid ones. This, in turn, should diminish the risks of moving to the irreversible states ③ and ④ (Fig. 6.11B).

According to this view, APols could provide a good folding medium because, on the one hand, they adsorb onto hydrophobic surfaces, keeping unfolded MPs from aggregating (or slowing down their aggregation), while, on the other hand, they do not compete efficiently with the protein/protein and protein/lipid interactions that define the native structure. Thus, they would substitute for SDS or urea at the surface of the unfolded protein, keeping it soluble, but be progressively displaced from those protein surfaces that can form stronger interactions either with other proteic elements or with lipids. Although the term "molecular chaperones" has been overused and misused, it may be to some extent appropriate here, in the sense that APols may slow down the formation of non-specific, unproductive interactions between hydrophobic segments, which would lead to misfolding and/or aggregation, while moving out of the way when specific interactions establish themselves.

Three types of APols have led to successful (re)folding of MPs to date: A8-35, SAPols, and NAPols. Except for BR, for which comparable folding yields are achieved in A8-35 and in NAPols (Pocanschi et al. 2006; Bazzacco et al. 2012; Dahmane et al. 2013), no comparative studies of the folding yields achieved for one given MP using one or the other APol have been carried out yet. On the basis of early experiments, it was proposed that A8-35 formed around refolding MPs a sort of protective "bubble" that would allow folding to proceed while slowing down the formation of nonproductive intermolecular interactions (Pocanschi et al. 2006) (Fig. 6.11B). One possible mechanism providing relative isolation of refolding proteins from one another could be the electrostatic repulsion between complexes incorporating either A8-35 or SAPols: both of them are polyanions, whose interactions indeed strongly depend on ionic strength (Zoonens et al. 2007). The fact that the

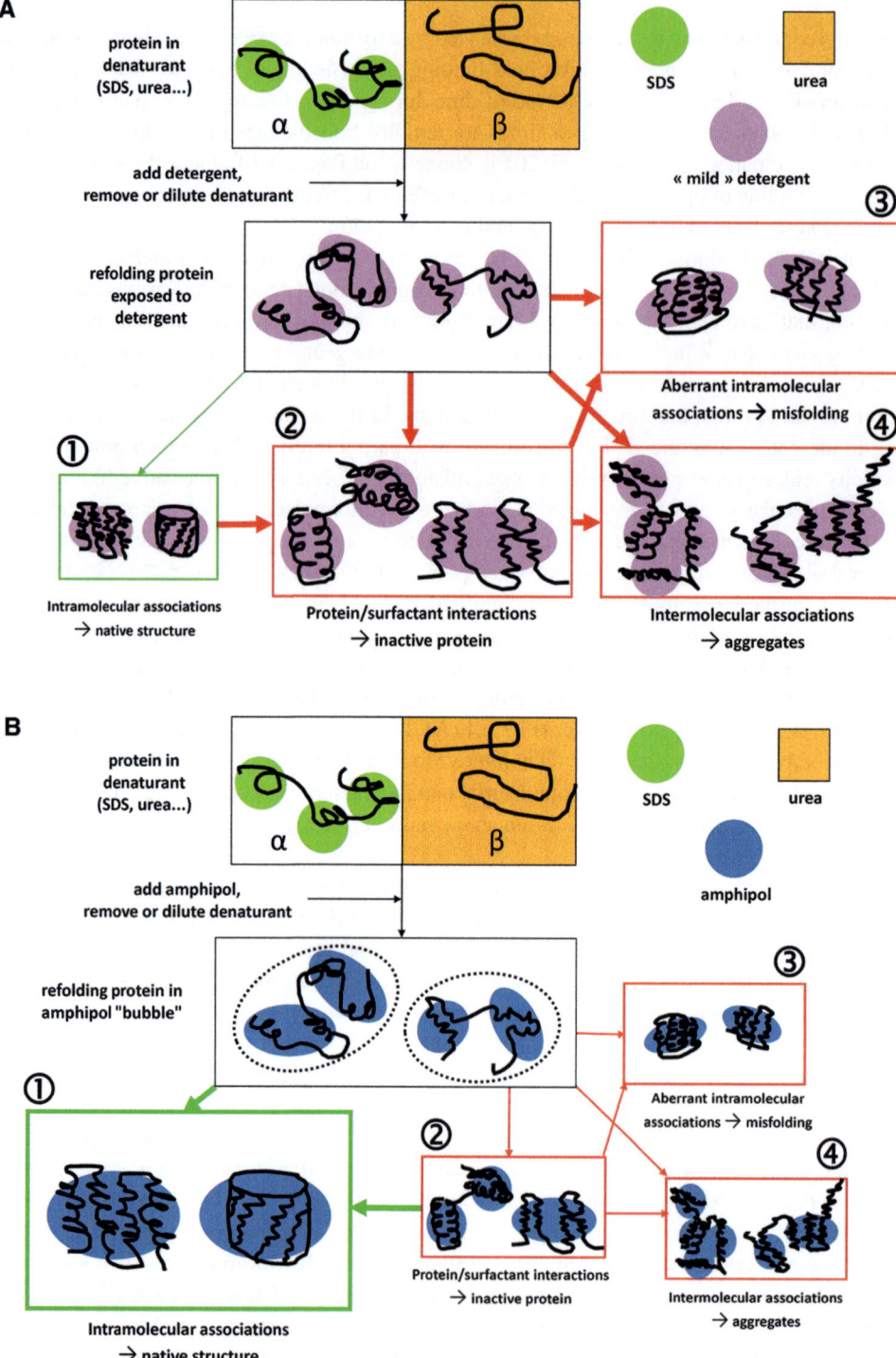

**Fig. 6.11** Hypothetical scheme of events leading either to (re)folding to the native state ① or to the formation of inactive forms due to incomplete (re)folding ②, misfolding ③, or aggregation ④, upon removal of the denaturant from a solution of denatured membrane protein in the presence of (**A**) detergent or mixed lipid/detergent micelles or (**B**) amphipols or amphipol/lipid mixtures. As compared to detergents, APols are postulated to favor pathway ① over, in particular, pathway ②, because of their lesser detergency. See text. (From Kleinschmidt and Popot (2014), © 2014 Elsevier Inc. All rights reserved).

totally uncharged NAPols also allow folding in good yield (Bazzacco et al. 2012) seems to indicate that such an electrostatic mechanism, if present, is not essential.

A related approach, cell-free synthesis of MPs using APols as the accepting medium, is described in Chap. 7.

A thought-provoking point is why APols seem to provide such a favorable folding medium for MPs, when neither their chemical structure nor the supramolecular structures they assemble into in aqueous solutions bear any similarity to lipid bilayers. This puzzling question is briefly discussed in Box 6.3.

---

**Box 6.3   Why Do Membrane Proteins Adopt Their Native Fold in the Absence of a Membrane?**

To the cell biologist, it may seem strange that such good MP folding yields can be obtained in APols. MPs are obviously adapted to a membrane environment, viz. the way they expose them to highly hydrophobic surfaces whose depth matches that of the membrane's hydrophobic core. (Re)folding experiments show that an unfolded MP can fold back to its native state if transferred either to lipid vesicles or to a "membrane-mimetic" environment (§ 6.2 and 6.3). Yet, can APols really be considered as "membrane-mimetic", when their chemical structure and the supramolecular aggregates they form in water (Chap. 4) are so different, chemically and physically, from a lipid bilayer? Does the high efficiency of MP folding in APols (Table 6.1) tell us something about the constraints MPs require to fold to their native state?

This question has been debated in a discussion paper to which the reader is referred (Popot and Engelman 2016). In brief, it is proposed that, because of the conditions in which MPs are inserted into membranes in vivo, relying, for folding, on the physical constraints provided by a lipid bilayer would be counterproductive and lead to low folding yields. This is because a folding MP in vivo is exposed to a highly complex environment that has little to do with a lipid bilayer, crowded as it is with proteins. Furthermore, physical constraints and chemical interactions differ from one membrane to the next, so that a given MP will usually be exposed during its synthesis and insertion to constraints different from those that prevail in the membrane compartment in which it will fulfill its function. These considerations lead to the proposal that MPs have evolved to be adapted to a membrane environment, but rely primarily, for their folding, on (*i*) their TM region being protected from the aqueous phase, without specific requirements about how this protection is afforded, and (*ii*) finding within their own sequence the information required to fold into a functional 3D structure, without guidance by physical constraints from their environment, such as bilayer thickness, lateral pressure gradient, etc. This is not to say, of course, that lipids have no influence on MP structure and function; much to the contrary (cf. Chap. 1). In some cases, the lipid composition can even modulate the TM topology adopted by a given MP (see e.g. Bogdanov et al. 2014; Vitrac et al. 2017; and references therein). For a full discussion of this intriguing question, the reader is referred to Popot and Engelman (2016).

---

### 6.3.3   Challenges and Prospects

One of the many open questions is which types of MPs will turn out to be amenable to APol-assisted folding and which not. MPs with extended extramembrane domains have not been tested yet (OmpA does feature such a domain, but without a function that could be easily tested to assess its folding). One may note, however, in this respect, that APols have been observed to protect denatured soluble proteins from aggregation (Ma et al. 2012; Martin et al. 2014, 2015), which should constitute a favorable factor.

A particularly challenging case is that of oligomeric MPs. There are, however, reasons to expect that, there as well, the use of APols might provide interesting perspectives:

(i) Several (currently nine) oligomeric MPs, both of the $\alpha$-helix bundle and of the $\beta$-barrel types, have been folded or expressed in vitro using detergents (Fig. 6.4; reviewed in Popot 2014) – a priori, because of their dissociating properties, much less favorable an environment than APols.

(ii) As illustrated by the delivery of retinal after BO has folded (Dahmane et al. 2013) and by the exchange of monomers between dimers of the TM $\alpha$-helix of glycophorin A (GpA) (Stangl et al. 2014), being trapped with APols does not prevent a MP from interacting and associating with a molecule delivered by another APol particle.

(iii) By increasing the MP/APol ratio, APol-trapped MPs can be induced to interact with one another (Zoonens et al. 2007; Gohon et al. 2008; Arunmanee et al. 2014).

(iv) There are indications that MPs folded in A8-35 can dimerize. Whereas no strong, direct demonstration has been provided yet, this is suggested, as regards GPCRs, by the fact that there is an optimum protein/A8-35 ratio for the BLT1 leukotriene receptor to fold (Dahmane et al. 2009). As discussed in Box 6.2, in the absence of lipids or any other cofactor that could become diluted, the most straightforward interpretation of this observation is that newly folded receptors become stabilized by dimerization, which occurs less efficiently in the presence of too large an excess of APol. SEC analysis of the solution behavior of the BLT1, BLT2, CB1, and 5-HT$_{4(a)}$ receptors folded in A8-35 does indeed suggest partial dimerization (Dahmane et al. 2009). Similarly, as noted above, dimers of the TM $\alpha$-helix of GpA can exchange monomers after being trapped in A8-35 (Stangl et al. 2014). This strongly suggests that monomers of GpA folded in APols should be able to dimerize.

One of the main challenges in folding oligomeric MPs is that, in most cases, unassembled monomers can be expected to be only marginally stable. Conditions must be found that favor their folding and assembly while discouraging the formation of improper intermolecular interactions. Given that it can be done in detergent or mixed lipid/detergent micelles (Fig. 6.4B), there is every reason to hope that APols will make it easier. It is likely that, in most cases, simultaneous (re)folding of all subunits will turn out to be the best strategy. In cases where unassembled monomers have very different stabilities, however, alternative routes could be experimented with, such as folding or expressing the most unstable subunit in the presence of its already folded partner(s).

## 6.4    Protocol 6.1. Amphipol-Assisted Folding of Membrane Proteins

APols have proven to be very helpful in folding MPs expressed as inclusion bodies in *E. coli*, like class A GPCRs or porins (Pocanschi et al. 2006, 2013; Dahmane et al. 2009, 2011, 2013; Banères et al. 2011; Bazzacco et al. 2012) (Fig. 6.3). The protocol used for $\alpha$-helical MPs is derived from one initially developed to refold BR in lipids (see Popot et al. 1987, in which many useful practical details can be found). It is quite simple but requires some optimization regarding the quantity of APol to add and the presence or not of lipids. For variants and the effect of various modifications to this protocol, see Dahmane et al. (2013). Comments are in italics and preceded with a pointing hand ($\sigma$).

### 6.4.1    Solubilization and Purification of MPs in Denaturing Conditions

$\sigma$ *Inclusion bodies are aggregates principally comprising misfolded forms of the protein of interest, but they can also contain some DNA and other bacterial proteins. They need to be solubilized and purified in denaturing conditions. For $\alpha$-helical MPs, the denaturing agent is usually SDS, whereas for $\beta$-barrel MPs it is urea. Purification is most often carried out by affinity chromatography. For instance, if the protein of interest is fused to a polyhistidine tag, purification can be carried out on* *an Ni:NTA resin. The concentration of SDS and urea tolerated by the resin is given by the supplier. Also, it is essential to work at room temperature, because both urea and SDS can crystallize in the cold room.*

1. Isolate the inclusion bodies by differential centrifugations and determine the concentration of protein by the BCA assay.
2. Prepare a solubilization buffer containing the appropriate denaturing agent. For example, for $\alpha$-helical MPs, the buffer contains 10 mM Tris/HCl pH 7.5, 100 mM $NaH_2PO_4$, 6 M urea, 0.8% SDS, 10% glycerol, and 4 mM $\beta$-mercaptoethanol (adapted from protocols described in Banères et al. 2005; Damian et al. 2006). For porins, the solubilization buffer contains 10 mM borate pH 10.0, 8 M urea, and 2 mM EDTA (Pocanschi et al. 2006, 2013). Notes: the presence of reducing agent is required only if cysteine residues are present. The solubility of urea can be increased to 10 M by heating.
3. Dissolve the inclusion bodies in the appropriate solubilization buffer at a final concentration of 10 $g·L^{-1}$, and incubate overnight at room temperature. Note: sonication pulses can be applied to speed up solubilization.
4. Centrifuge the sample for 20 min at 20,000 × *g* in order to remove insoluble material.
5. Proceed to the purification step. For purifying $\alpha$-helical MPs on Ni:NTA resin, the buffers are (*i*) equilibration buffer: 50 mM Tris/HCl pH 8.0, 300 mM NaCl, 0.8% SDS, and 4 mM $\beta$-mercaptoethanol; (*ii*) elution buffer: 50 mM Tris/HCl pH 8.0, 300 mM NaCl, 0.8% SDS, 400 mM imidazole, and 4 mM $\beta$-mercaptoethanol; (*iii*) desalting buffer: 50 mM Tris/HCl pH 8.0, 0.8% SDS, and 4 mM $\beta$-mercaptoethanol. For purifying $\beta$-barrel MPs, SDS in each buffer is replaced by 8 M urea.
6. Determine the concentration of protein by UV-absorbance or by the BCA assay.

## 6.4.2   Renaturation of $\alpha$-Helical MPs in APols

☞ *This step consists in exchanging SDS for APols (Fig. 6.6A). The optimal MP/APol mass ratio must be determined by carrying out folding tests with variable amounts of APols.*

1. Distribute 0.25 mg of the MP to be folded in three Eppendorf tubes. Add increasing volumes of APol – 5 µL, 12.5 µL, and 25 µL – from a stock solution at 100 $g·L^{-1}$ in order to obtain MP/APol mass ratios equal to 1:2, 1:5, and 1:10. Note that lipids generally help in the folding process – cf. Dahmane et al. (2009, 2013). Their usefulness can be tested by supplying them (e.g. soybean lipids) to the samples so that the APol/lipid mass ratio is 1:0.2. This ratio can be optimized, as well as the nature of the lipids.
2. Mix and incubate the samples for 30 min at room temperature.
3. SDS is eliminated by precipitating the dodecyl sulfate (DS) with KCl added from a 4 M stock solution so that the final concentration of KCl in the samples is equal to 150 mM plus the concentration of SDS. For example, if the volume of the sample is 1 mL and the concentration of SDS is 0.8% (28 mM), the final KCl concentration should be 178 mM. The volume of KCl to add is thus 44.5 µL.
4. Incubate for 30 min at room temperature under vigorous stirring.
5. Centrifuge the samples for 5 min at the maximum speed of a benchtop centrifuge at 20 °C.
6. Collect the supernatant and repeat the centrifugation step.
7. Measure the optical density of samples at 280 nm.

### 6.4.3    Renaturation of β-Barrel MPs in APols

This step relies on diluting urea in the presence of APols (Fig. 6.6B).

1. Set, for example, to 1 mg the mass of protein to fold. Dilute the sample by a 10× dilution factor into urea-free buffer containing 5 mg of APols, so that the final MP/APol mass ratio is 1:5. Test also ratios 1:2 and 1:10. Note that the dilution factor and speed of dilution can be optimized. If need be, incubate the samples at 40 °C for 24 h.
2. Concentrate the samples using an ultrafiltration device and measure the optical density of supernatants.

### 6.4.4    Completing the Renaturation

☞ *To increase the yield of folding, urea or DS traces can be further eliminated by a dialysis step.*

1. Dialyze the sample for 24 h at room temperature using a standard dialysis membrane of 12–14 kDa MW cutoff. Note that APols are not needed in the external bath, but the presence of 150 mM KCl is required to prevent redissolution of crystallites of PDS that may not have been totally removed by centrifugation. The volume of the external bath is ~500× larger than the volume of the samples.
2. Recover the sample and centrifuge it for 5 min at the maximum speed of a benchtop centrifuge.
3. Measure the optical density of the samples.
4. If the buffer needs to be exchanged, proceed to a second dialysis for 24 h at 4 °C.

☞ *The solubility of MP is not a criterion of folding. The simplest and most direct proof that the protein adopts its native conformation is to check its activity. If the activity assay is not easy to set up, the yield of folding can be assessed by other approaches such as ligand-binding experiments using equilibrium dialysis. In that case, ligand titration can be monitored by radioactivity measurements or by following changes in the intensity of fluorescence emission or light absorbance. If the protein is naturally colored in its native conformation due to the binding of a cofactor, such as retinal for BR, the native state can be quantified by spectral absorbance changes. It is also possible to check the homogeneity and size of the protein by SEC, its secondary structure by CD, the local environment of tryptophan residues by CD and fluorescence measurements, the melting temperature by differential scanning calorimetry, or by fluorescence thermal shift. In the case of porins, the folded state of the protein can usually be assessed by SDS-PAGE, upon which, as a rule, folded and unfolded forms exhibit different electrophoretic mobilities (cf. Figs. 6.9 and 6.10), by dot blots if an antibody recognizing the native state of the protein is available, by protease digestion, etc.*
    *Protocols prepared by Tassadite Dahmane and Manuela Zoonens, adapted from* Zoonens et al. (2014).

## References

Adamian, L., Naveed, H., Liang, J. (2011) Lipid-binding surfaces of membrane proteins: Evidence from evolutionary and structural analysis. *Biochim. Biophys. Acta* **1808**:1092–1102.

Anfinsen, C.B. (1973) Principles that govern the folding of protein chains. *Science* **181**:223–230.

Anfinsen, C.B., Harber, E., Sela, M. (1961) The kinetics of formation of native ribonuclease during oxidation of the reduced polypeptide chain. *Proc. Natl. Acad. Sci. USA* **47**:1309–1314.

Anfinsen, C.B., Scheraga, H.A. (1975) Experimental and theoretical aspects of protein folding. *Adv. Prot. Chem.* **29**:205–300.

Aponte-Santamaríaa, C., Brionesa, R., Schenk, A.D., Walz, T., de Groot, B.L. (2012) Molecular driving forces defining lipid positions around aquaporin-0. *Proc. Natl. Acad. Sci. USA* **109**:9887–9892.

Arunmanee, W., Harris, J.R., Lakey, J.H. (2014) Outer membrane protein F stabilised with minimal amphipol forms linear arrays and LPS-dependent 2D crystals. *J. Membr. Biol.* **247**:949–956.

Banères, J.-L., Mesnier, D., Martin, A., Joubert, L., Dumuis, A., Bockaert, J. (2005) Molecular characterization of a purified 5-HT$_4$ receptor. A structural basis for drug efficacy. *J. Biol. Chem.* **280**:20253–20260.

Banères, J.-L., Popot, J.-L., Mouillac, B. (2011) New advances in production and functional folding of G protein-coupled receptors. *Trends Biotechnol.* **29**:314–322.

Bannwarth, M., Schulz, G.E. (2003) The expression of outer membrane proteins for crystallization. *Biochim. Biophys. Acta* **1610**:37–45.

Bazzacco, P., Billon-Denis, E., Sharma, K.S., Catoire, L.J., Mary, S., Le Bon, C., Point, E., Banères, J.-L., Durand, G., Zito, F., Pucci, B., Popot, J.-L. (2012) Non-ionic homopolymeric amphipols: Application to membrane protein folding, cell-free synthesis, and solution NMR. *Biochemistry* **51**:1416–1430.

Bogdanov, M., Dowhan, W., Vitrac, H. (2014) Lipids and topological rules governing membrane protein assembly. *Biochim. Biophys. Acta* **1843**:1475–1488.

Booth, P.J. (2003) The trials and tribulations of membrane protein folding *in vitro*. *Biochim. Biophys. Acta* **1610**:51–56.

Bowie, J.U. (2005) Solving the membrane protein folding problem. *Nature* **438**:581–589.

Broecker, J., Eger, B.T., Ernst, O.P. (2017) Crystallogenesis of membrane proteins mediated by polymer-bounded lipid nanodiscs. *Structure* **25**:384–392.

Buchanan, S.K. (1999) β-barrel proteins from bacterial outer membranes: structure, function and refolding. *Curr. Opin. Struct. Biol.* **9**:455–461.

Buchanan, S.K., Yamashita, S., Fleming, K.G. (2012) Structure and folding of outer membrane proteins, in: Tamm, L.K. (Ed.), *Membranes*. Elsevier, Oxford:Academic Press, pp. 139–163.

Caffrey, M. (2011) Crystallizing membrane proteins for structure-function studies using lipidic mesophases. *Biochem. Soc. Trans.* **39**:725–732.

Casiraghi, M., Damian, M., Lescop, E., Point, E., Moncoq, K., Morellet, N., Levy, D., Marie, J., Guittet, E., Banères, J.-L., Catoire, L.J. (2016) Functional modulation of a GPCR conformational landscape in a lipid bilayer. *J. Am. Chem. Soc.* **138**:11170–11175

Catoire, L.J., Damian, M., Giusti, F., Martin, A., van Heijenoort, C., Popot, J.-L., Guittet, E., Banères, J.-L. (2010) Structure of a GPCR ligand in its receptor-bound state: leukotriene B$_4$ adopts a highly constrained conformation when associated to human BLT2. *J. Am. Chem. Soc.* **132**:9049–9057.

Cherezov, V. (2011) Lipidic cubic phase technologies for membrane protein structural studies. *Curr. Opin. Struct. Biol.* **21**:559–566.

Cherezov, V., J. C, Papiz, M.Z., Caffrey, M. (2006) Room to move: crystallizing membrane proteins in swollen lipidic mesophases. *J. Mol. Biol.* **357**:1605–1618.

Cherezov, V., Rosenbaum, D.M., Hanson, M.A., Rasmussen, S.G., Thian, F.S., Kobilka, T.S., Choi, H.J., Kuhn, P., Weis, W.I., Kobilka, B.K., Stevens, R.C. (2007) High-resolution crystal structure of an engineered human $\beta_2$-adrenergic G protein-coupled receptor. *Science* **318**:1258–1265.

Dahmane, T., Damian, M., Mary, S., Popot, J.-L., Banères, J.-L. (2009) Amphipol-assisted *in vitro* folding of G protein-coupled receptors. *Biochemistry* **48**:6516–6521.

Dahmane, T., Giusti, F., Catoire, L.J., Popot, J.-L. (2011) Sulfonated amphipols: Synthesis, properties and applications. *Biopolymers* **95**:811–823.

Dahmane, T., Rappaport, F., Popot, J.-L. (2013) Amphipol-assisted folding of bacteriorhodopsin in the presence and absence of lipids. Functional consequences. *Eur. Biophys. J.* **42**:85–101.

Damian, M., Martin, A., Mesnier, D., Pin, J.-P., Banères, J.-L. (2006) Asymmetric conformational changes in a GPCR dimer controlled by G-proteins. *EMBO J.* **13**:5693–5702.

Elter, S., Raschle, T., Arens, S., Viegas, A., Gelev, V., Etzkorn, M., Wagner, G. (2014) The use of amphipols for NMR structural characterization of 7-TM proteins. *J. Membr. Biol.* **247**:957–964.

Engelman, D.M., Chen, Y., Chin, C.-N., Curran, R., Dixon, A.M., Dupuy, A., Lee, A., Lehnert, U., Mathews, E., Reshetnyak, Y., Senes, A., Popot, J.-L. (2003) Membrane protein folding: beyond the two-stage model. *FEBS Lett.* **555**:122–125.

Goehring, A., Lee, C.-H., Wang, K.H., Michel, J.C., Claxton, D.P., Baconguis, I., Althoff, T., Fischer, S., Garcia, C., Gouaux, E. (2014) Screening and large-scale expression of membrane proteins in mammalian cells for structural studies. *Nat. Protoc.* **9**:2574–2585.

Gohon, Y., Dahmane, T., Ruigrok, R., Schuck, P., Charvolin, D., Rappaport, F., Timmins, P., Engelman, D.M., Tribet, C., Popot, J.-L., Ebel, C. (2008) Bacteriorhodopsin/amphipol complexes: structural and functional properties. *Biophys. J.* **94**:3523–3537.

Grisshammer, R., Tate, C.G. (1995) Overexpression of integral membrane proteins for structural studies. *Quart. Rev. Biophys.* **28**:315–422.

Harris, N.J., Booth, P.J. (2012) Folding and stability of membrane transport proteins *in vitro. Biochim. Biophys. Acta* **1818**:1055–1066.

Huang, K.-S., Bayley, H., Liao, M.-J., London, E., Khorana, H.G. (1981) Refolding of an integral membrane protein. Denaturation, renaturation, and reconstitution of intact bacteriorhodopsin and two proteolytic fragments. *J. Biol. Chem.* **256**:3802–3809.

Ishchenko, A., Abola, E., Cherezov, V. (2014) Lipidic cubic phase technologies for structural studies of membrane proteins, in: Mus-Veteau, I. (Ed.), *Membrane Proteins Production for Structural Analysis.* Springer, New York, pp. 289–314.

Kahn, T.W., Sturtevant, J.M., Engelman, D.M. (1992) Thermodynamic measurements of the contributions of helix-connecting loops and of retinal to the stability of bacteriorhodopsin. *Biochemistry* **31**:8829–8839.

Kiefer, H. (2003) *In vitro* folding of $\alpha$-helical membrane proteins. *Biochim. Biophys. Acta* **1610**:57–62.

Kleinschmidt, J.H. (2015) Folding of $\beta$-barrel membrane proteins in lipid bilayers – Unassisted and assisted folding and insertion. *Biochim. Biophys. Acta* **1848**:1927–1943.

Kleinschmidt, J.H., Popot, J.-L. (2014) Folding and stability of integral membrane proteins in amphipols. *Arch. Biochem. Biophys.* **564**:327–343.

Kleinschmidt, J.H., Tamm, L.K. (1996) Folding intermediates of a $\beta$-barrel membrane protein. Kinetic evidence for a multi-step membrane insertion mechanism. *Biochemistry* **35**:12993–13000.

Kraft, T.E., Hresko, R.C., Hruz, P.W. (2015) Expression, purification, and functional characterization of the insulin-responsive facilitative glucose transporter GLUT4. *Protein Sci.* **24**:2008–2019.

Le Bon, C., Marconnet, A., Masscheleyn, S., Popot, J.-L., Zoonens, M. (2018) Folding and stabilizing membrane proteins in amphipol A8-35. *Methods, in the press.*

Lee, A.G. (2003) Lipid-protein interactions in biological membranes: a structural perspective. *Biochim. Biophys. Acta* **1612**:1–40.

Lee, A.G. (2011) How to understand lipid-protein interactions in biological membranes, in: Yeagle, P. (Ed.), *Structure of Biological Membranes.* 3rd edition, Taylor and Francis, Boca Raton, Florida, USA, pp. 273–313.

Leney, A.C., McMorran, L.M., Radford, S.E., Ashcroft, A.E. (2012) Amphipathic polymers enable the study of functional membrane proteins in the gas phase. *Anal. Chem.* **84**:9841–9847.

Leyris, J.-P., Roux, T., Trinquet, E., Verdié, P., Fehrentz, J.A., Oueslati, N., Douzon, S., Bourrier, E., Lamarque, L., Gagne, D., Galleyrand, J.-C., M'kadmi, C., Martinez, J., Mary, S., Banères, J.-L., Marie, J. (2011) Homogeneous time-resolved fluorescence-based assay to screen for ligands targeting the growth hormone secretagogue receptor type 1a. *Anal. Biochem.* **408**:253–262.

Ma, D., Martin, N., Herbet, A., Boquet, D., Tribet, C., Winnik, F.M. (2012) The thermally induced aggregation of immunoglobulin G in solution is prevented by amphipols. *Chem. Lett.* **41**:1380–1382.

Maeda, S., Schertler, G.F.X. (2013) Production of GPCR and GPCR complexes for structure determination. *Curr. Opin. Struct. Biol.* **23**:381–392.

Marie, E., Sagan, S., Cribier, S., Tribet, C. (2014) Amphiphilic macromolecules on cell membranes: from protective layers to controlled permeabilization. *J. Membr. Biol.* **247**:861–881.

Martin, N., Ma, D., Herbet, A., Boquet, D., Winnik, F.M., Tribet, C. (2014) Prevention of thermally induced aggregation of IgG antibodies by noncovalent interaction with poly(acrylate) derivatives. *Biomacromolecules* **15**:2952–2962.

Martin, N., Ruchmann, J., Tribet, C. (2015) Prevention of aggregation during refolding of carbonic anhydrase via Coulomb and hydrophobic complexation with octadecyl-modified or azobenzene-modified poly(acrylate) derivatives. *Langmuir* **31**:338–349.

Mesnier, D., Banères, J.-L. (2004) Cooperative conformational changes in a G-protein coupled receptor dimer, the leukotriene $B_4$ receptor BLT1. *J. Biol. Chem.* **279**:49664–49670.

Michalke, K., Huyghe, C., Lichière, J., Gravière, M.E., Siponen, M., Sciara, G., Lepaul, I., Wagner, R., Magg, C., Rudolph, R., Cambillau, C., Desmyter, A. (2010) Mammalian G protein-coupled receptor expression in *Escherichia coli*: II. Refolding and biophysical characterization of mouse cannabinoid receptor 1 and human parathyroid hormone receptor 1. *Anal. Biochem.* **401**:74–80.

Michaux, C., Pomroy, N.C., Privé, G.G. (2008) Refolding SDS-denatured proteins by the addition of amphipathic cosolvents. *J. Mol. Biol.* **375**:1477–1488.

Milić, D., Veprintsev, D.B. (2015) Large-scale production and protein engineering of G protein-coupled receptors for structural studies. *Frontiers Pharmacol.* **6**: article 66.

Moosavi-Movahedi, A.A., Chamanil, J., Goto, Y., Hakimelahi, G.H. (2003) Formation of the molten globule-like state of cytochrome *c* induced by *n*-alkyl sulfates at low concentrations. *J. Biochem.* **133**:93–102.

Nagy, J.K., Kuhn Hoffmann, A., Keyes, M.H., Gray, D.N., Oxenoid, K., Sanders, C.R. (2001) Use of amphipathic polymers to deliver a membrane protein to lipid bilayers. *FEBS Lett.* **501**:115–120.

Otzen, D.E. (2015) Proteins in a brave new surfactant world. *Curr. Opin. Colloid Interface Sci.* **20**:161–169.

Otzen, D.E., Andersen, K.K. (2013) Folding of outer membrane proteins. *Arch. Biochem. Biophys.* **531**:34–43.

Park, K.-H., Billon-Denis, E., Dahmane, T., Lebaupain, F., Pucci, B., Breyton, C., Zito, F. (2011) In the cauldron of cell-free synthesis of membrane proteins: Playing with new surfactants. *New Biotech.* **28**:255–261.

Pocanschi, C., Popot, J.-L., Kleinschmidt, J.H. (2013) Folding and stability of outer membrane protein A (OmpA) from *Escherichia coli* in an amphipathic polymer, amphipol A8-35. *Eur. Biophys. J.* **42**:103–118.

Pocanschi, C.L., Dahmane, T., Gohon, Y., Rappaport, F., Apell, H.-J., Kleinschmidt, J.H., Popot, J.-L. (2006) Amphipathic polymers: tools to fold integral membrane proteins to their active form. *Biochemistry* **45**:13954–13961.

Polovinkin, V., Gushchin, I., Balandin, T., Chervakov, P., Round, E., Shevchenko, V., Popov, A., Borshchevskiy, V., Popot, J.-L., Gordeliy, V. (2014) High-resolution structure of a membrane protein transferred from amphipol to a lipidic mesophase. *J. Membr. Biol.* **247**:997–1004.

Popot, J.-L. (2010) Amphipols, nanodiscs, and fluorinated surfactants: Three non-conventional approaches to studying membrane proteins in aqueous solutions. *Annu. Rev. Biochem.* **79**:737–775.

Popot, J.-L. (2014) Folding membrane proteins *in vitro*: A table and some comments. *Arch. Biochem. Biophys.* **564**:314–326.

Popot, J.-L., Althoff, T., Bagnard, D., Banères, J.-L., Bazzacco, P., Billon-Denis, E., Catoire, L.J., Champeil, P., Charvolin, D., Cocco, M.J., Crémel, G., Dahmane, T., de la Maza, L.M., Ebel, C., Gabel, F., Giusti, F., Gohon, Y., Goormaghtigh, E., Guittet, E., Kleinschmidt, J.H., Kühlbrandt, W., Le Bon, C., Martinez, K.L., Picard, M., Pucci, B., Rappaport, F., Sachs, J.N., Tribet, C., van Heijenoort, C., Wien, F., Zito, F., Zoonens, M. (2011) Amphipols from A to Z. *Annu. Rev. Biophys.* **40**:379–408.

Popot, J.-L., Engelman, D.M. (2000) Helical membrane protein folding, stability and evolution. *Annu. Rev. Biochem.* **69**:881–923.

Popot, J.-L., Engelman, D.M. (2016) Membranes do not tell proteins how to fold. *Biochemistry* **55**:5–18.

Popot, J.-L., Gerchman, S.-E., Engelman, D.M. (1987) Refolding of bacteriorhodopsin in lipid bilayers: a thermodynamically controlled two-stage process. *J. Mol. Biol.* **198**:655–676.

Rasmussen, S.G.F., Choi, H.-J., Fung, J.J., Pardon, E., Casarosa, P., Chae, P.S., DeVree, B.T., Rosenbaum, D.M., Thian, F.S., Kobilka, T.S., Schnapp, A., Konetzki, I., Sunahara, R.K., Gellman, S.H., Pautsch, A., Steyaert, J., Weis, W.I., Kobilka, B.K. (2011) Structure of a nanobody-stabilized active state of the $\beta_2$ adrenoceptor. *Nature* **469**:175–180.

Schweizer, M., Hindennach, I., Garten, W., Henning, U. (1978) Major proteins of the *Escherichia coli* outer cell envelope membrane. Interaction of protein II with lipopolysaccharide. *Eur. J. Biochem.* **82**:211–217.

Sharma, K.S., Durand, G., Gabel, F., Bazzacco, P., Le Bon, C., Billon-Denis, E., Catoire, L.J., Popot, J.-L., Ebel, C., Pucci, B. (2012) Non-ionic amphiphilic homopolymers: Synthesis, solution properties, and biochemical validation. *Langmuir* **28**:4625–4639.

Stangl, M., Unger, S., Keller, S., Schneider, D. (2014) Sequence-specific dimerization of a transmembrane helix in amphipol A8-35. *PLOS One* **9**:e110970.

Stanley, A.M., Fleming, K.G. (2008) The process of folding proteins into membranes: Challenges and progress. *Arch. Biochem. Biophys.* **469**:46–66.

Stansfeld, P.J., Jeffreys, E.E., Sansom, M.S.P. (2013) Multiscale simulations reveal conserved patterns of lipid interactions with aquaporins. *Structure* **21**:810–819.

Tamm, L.K., Hong, H., Liang, B. (2004) Folding and assembly of β-barrel membrane proteins. *Biochim. Biophys. Acta* **1666**:250–263.

Tastan, O., Dutta, A., Booth, P., Klein-Seetharaman, J. (2014) Retinal proteins as model systems for membrane protein folding. *Biochim. Biophys. Acta* **1837**:656–663.

Vitrac, H., MacLean, D.M., Karlstaedt, A., Taegtmeyer, H., Jayaraman, V., Bogdanov, M., Dowhan, W. (2017) Dynamic lipid-dependent modulation of protein topology by post-translational phosphorylation. *J. Biol. Chem.* **292**:1613–1624.

White, S.H. (2003) Translocons, thermodynamics, and the folding of membrane proteins. *FEBS Lett.* **555**:116–121.

Zoonens, M., Giusti, F., Zito, F., Popot, J.-L. (2007) Dynamics of membrane protein/amphipol association studied by Förster resonance energy transfer. Implications for *in vitro* studies of amphipol-stabilized membrane proteins. *Biochemistry* **46**:10392–10404.

Zoonens, M., Popot, J.-L. (2014) Amphipols for each season. *J. Membr. Biol.* **247**:759–796.

Zoonens, M., Zito, F., Martinez, K.L., Popot, J.-L. (2014) Amphipols: a general introduction and some protocols, in: Mus-Veteau, I. (Ed.), *Membrane Proteins Production for Structural Analysis*. Springer, New York, Heidelberg, Dordrecht, London, pp. 173–203.

Zorman, S., Botte, M., Jiang, Q., Collinson, I., Schaffitzel, C. (2015) Advances and challenges of membrane-protein complex production. *Curr. Opin. Struct. Biol.* **32**:123–130.

# Amphipol-Assisted Cell-Free Expression of Membrane Proteins

**7**

**Summary**

*Cell-free expression of membrane proteins is a way to circumvent some of the problems encountered during their in vivo expression, among which is the small volume of membrane that is generally available for storing overexpressed membrane proteins and their toxicity when expressed in excessive amounts. Because of their mildness, amphipols appear as an attractive medium in which to solubilize and allow folding of membrane proteins expressed in vitro in cell lysates. Relatively few attempts at doing so have been described to date. They indicate (i) that all ionic polymers tested thus far interfere with the synthesis of α-helical membrane proteins and (ii) that two non-ionic polymers, NVoy and a glucosylated non-ionic amphipol, do allow their synthesis, folding, and solubilization in good yields.*

## 7.1 Introduction

Very few membrane proteins (MPs) are produced naturally in amounts sufficient for structural studies. As discussed in Chap. 1, this is mostly the consequence of two factors. First, the volume of the plasma membrane of a cell is small as compared to that of the cytosol. Even if both compartments are packed full with proteins (typically $\sim$200 g$\cdot$L$^{-1}$ in the cytosol, generally 1–3 g protein per g of lipid in the plasma membrane), the ratio in mass of MPs to proteins in general is necessarily small. Second, many MPs of great physiological importance are involved in signaling, which does not necessarily require them to be present in very many copies. In a muscle fiber, for instance, the nicotinic acetylcholine receptor (nAChR) is densely packed under the motor nerve terminals, but these cover only $\sim$0.1% of the total surface of the fiber. Similarly, most of the G protein-coupled receptors (GPCRs), which control so many regulatory mechanisms, need not be present in large amounts, because their response to a stimulus is enormously amplified by downstream events. This explains why the first MPs to be purified in sizable amounts were typically involved in energy production, such as bacteriorhodopsin (BR) or photosynthetic reaction centers, or in mass transport of solutes, such as the porins from bacterial outer membranes. The nicotinic acetylcholine receptor (nAChR) was first purified from the electric organs of either *Electrophorus* or *Torpedo*, where it is diverted from its usual signaling function and massively expressed so as to produce large electric discharges. In many cases, such as

© Springer International Publishing AG, part of Springer Nature 2018
J. -L. Popot, *Membrane Proteins in Aqueous Solutions*, Biological and Medical Physics,
Biomedical Engineering, https://doi.org/10.1007/978-3-319-73148-3_7

in photosynthesis or respiration, where the growth of the organism is limited by the production of energy and/or reducing equivalents by a certain set of MPs, the membrane surface available is greatly multiplied by the formation of tubes or sacculi that densely pack in the cytosol of the bacterium or the matrix of the chloroplast or mitochondrion. Beyond these favorable cases, however, gaining access to sizable (mg) amounts of MPs is generally impossible starting from natural sources. This has led, starting in the late 1980s (see Grisshammer and Tate 1995), to the development of overexpression methods.

As noted in the introduction to Chap. 6 (§ 6.1), overexpressing MPs in vivo in either homologous or heterologous systems is far from a trivial endeavor, because (i) in most cells, the volume of membrane that can be used to store overexpressed MPs is limited and (ii) overexpression of MPs tends to be toxic, either because it perturbs the membrane or it overloads the expression machinery, not to mention cases where the target proteins are intrinsically toxic. A fine line has to be walked between expressing too little protein for practical use and expressing too much of it and killing the cells, which also results in poor yields. In addition, the target proteins can undergo proteolysis, and they must be separated from the other proteins present in the cells. The various approaches that have been developed to circumvent these difficulties have been summarized in Chap. 1, § 1.7.2. The use of amphipols (APols) for folding MPs expressed in vivo in an inactive form accumulated in so-called inclusion bodies, so as to limit toxicity effects and increase yields, has been described in Chap. 6, § 6.3.

An alternative to in vivo overexpression is cell-free expression (CFE), in which the target protein is produced in vitro, in a lysate that contains all of the translation machinery (and, usually, the transcription machinery as well) and is supplied with amino acids, ATP, GTP, energy substrates, an energy-regenerating system, and, usually, T7 RNA polymerase, along with a plasmid encoding the target protein (see e.g. Nirenberg and Matthaei 1961; Zubay 1973; Spirin et al. 1988; Shirokov et al. 2007; Ge and Xu 2012; Kai et al. 2012; Shadiac et al. 2013; Zemella et al. 2015; and references therein). CFE can be carried out in batches, in which case all of the components are mixed in the same compartment, or by resorting to a regenerating system, in which diffusible components, including amino acids and energy substrates, on the one hand, and inhibitory by-products like phosphate, on the other, are exchanged between the reaction chamber and a feeding chamber through a dialysis membrane (Fig. 7.1). The single-batch approach is classically used to screen expression conditions and the regenerating system to improve yields and scale up production. Under good conditions, CFE can produce the target protein

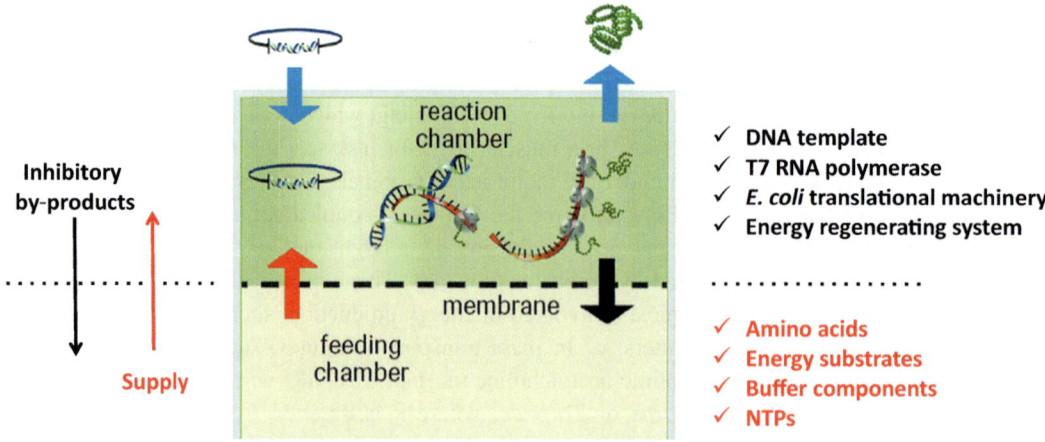

**Fig. 7.1** Scheme of a cell-free expression experiment with continuous feeding (Adapted from the Roche website. All rights reserved).

at a level of ≥1 mg per mL of lysate, an amount comparable to that which can be obtained in vivo in 1 L of *Escherichia coli* culture. Lysates are most often prepared from *E. coli*, but extracts from archaebacteria, protozoans, yeasts, wheat germs and other higher plant cells, insect cells, rabbit reticulocytes, or cultured human cell lines can also be employed with the view of facilitating either folding or posttranslational modifications of the target proteins. This comes, however, at the expense of simplicity, yield, and/or cost control (reviewed and critically compared in Zemella et al. 2015).

Cell-free expression offers many attractive features, among which are to do away with the toxicity issue and to allow labeling using limited amounts of isotopically labeled or non-natural amino acids for X-ray crystallography or for solution or solid-state NMR structural investigations, etc. (see e.g. Kigawa et al. 1999; Kigawa 2010; Maslennikov et al. 2010; Reckel et al. 2011; Abdine et al. 2012). In addition, it is possible to supply the lysate with lipids, detergents, or other surfactants, thereby providing the overexpressed proteins with an amphipathic environment destined to keeping it soluble and facilitating its folding. This makes CFE an extremely interesting approach for the production of MPs.

## 7.2    Context: Cell-Free Expression of Membrane Proteins

The putative advantages of CFE of MPs over in vivo expression are schematized in Fig. 7.2. On the left side of the figure are indicated the various steps required by in vivo approaches and the associated potential bottlenecks. Emphasized on the right side are the relative simplicity of in vitro expression and the many compounds that can be added to the lysate in order to stabilize the target protein, favor its folding, keep it soluble, or label it.

Application of CFE to MPs started in the mid-1990s (see e.g. Sonar et al. 1993; Van Gelder et al. 1994; Huppa and Ploegh 1997; Bogdanov and Dowhan 1998; for relatively recent reviews, the reader is referred, for example, to Rajesh et al. 2011; Ge and Xu 2012; Kai et al. 2012, 2015; Maeda and Schertler 2013; Hein et al. 2014; Rues et al. 2014, 2016; Sachse et al. 2014; Henrich et al. 2015; as well as to the list of MPs expressed by CFE compiled in Popot 2014). The most commonly used lysates are derived either from *E. coli* or from wheat germ, supplemented or not with surfactants (for a recent discussion of the pros and cons of prokaryotic vs. eukaryotic lysates, see Zemella et al. 2015). MP production by CFE has long remained marginal, but it picked up steam after, roughly, 2003 (Figs. 7.3 and 7.4A).

MPs obtained in vitro using CFE can be synthesized either in the presence of a surfactant – lipid vesicles, nanodiscs, bicelles, detergent or detergent-lipid mixed micelles, APols, etc. – in which case they probably insert and fold in the course of biosynthesis, or they can be left to precipitate from a surfactant-free lysate. The precipitates thus obtained are usually much easier to dissolve than inclusion bodies and can often be solubilized using a non-denaturing detergent (see e.g. Klammt et al. 2004, 2005, 2006; Keller et al. 2008; Junge et al. 2010; Hein et al. 2014; and references therein). In some cases, however, they are dissolved using either SDS or urea (see e.g. Focke et al. 2016), which makes it necessary to fold the target MP from the denatured state in which it is recovered (for an overview, see Chap. 6, § 6.2, or Popot 2014). *E. coli* lysates have yielded by far the largest amount of properly folded MPs (28 of them by the end of 2013), but since 2007, the use of wheat germ lysates has been progressing (Fig. 7.4A). Other lysates are being tested, e.g. the reticulocyte lysate supplemented with dog pancreas microsomes that has been used to express, fold, and assemble a T cell receptor-CD3 complex (Huppa and Ploegh 1997).

CFE has been much more often used to express $\alpha$-helical than $\beta$-barrel MPs (32 vs. 4 at the end of 2013; Fig. 7.4B). However, this situation may have a circumstantial rather than a real technical basis. Indeed, $\beta$-barrel MPs appear particularly easy to fold from inclusion bodies (see Buchanan et al.

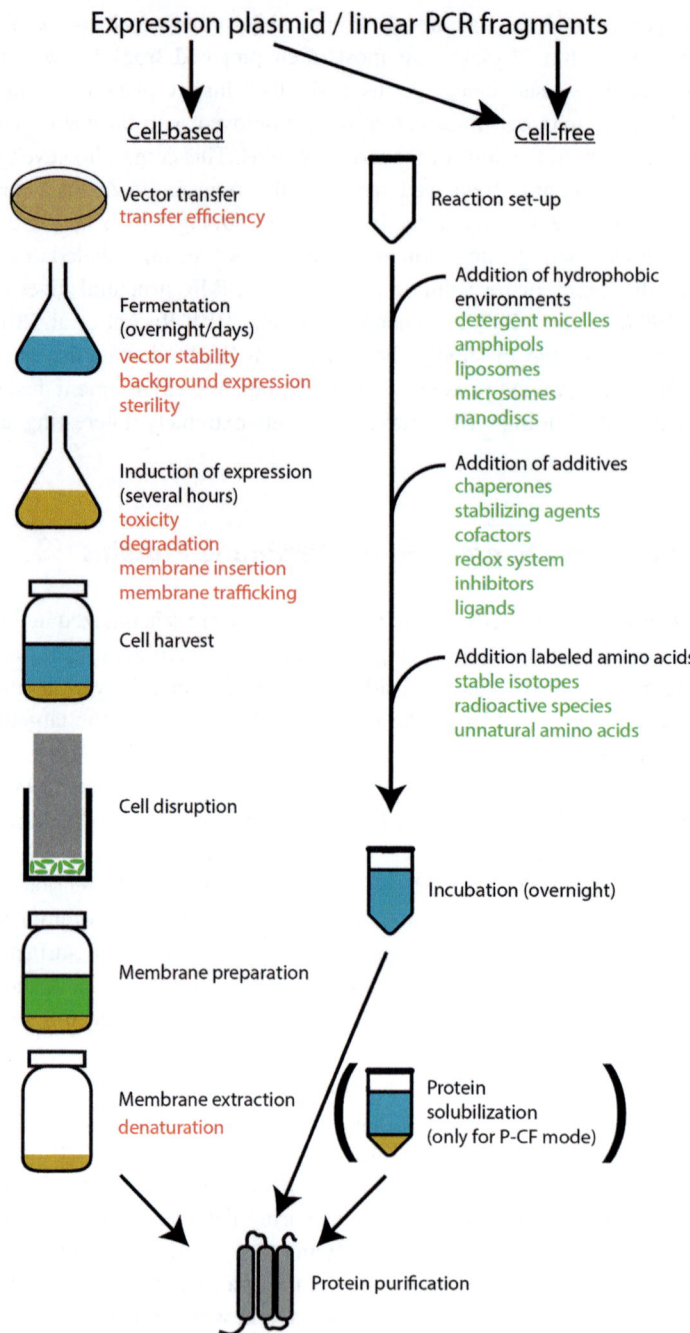

**Fig. 7.2** Flow chart comparison of in vivo vs. cell-free expression systems for membrane proteins. Typical individual steps starting from DNA templates are illustrated. Problematic bottlenecks of in vivo systems are indicated in *red*, examples of additives to the lysates that support in vitro synthesis in *green* (From Hein et al. 2014, © 2014 John Wiley & Sons, Inc. All Rights Reserved).

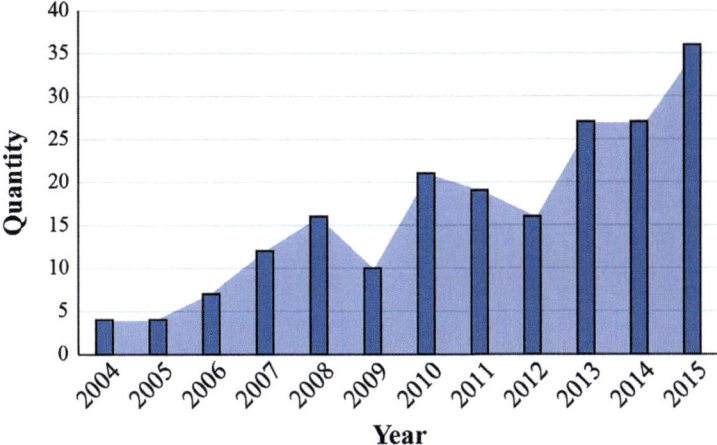

**Fig. 7.3** PubMed-referenced reports on cell-free expression of membrane proteins from 2004 to 2015 (From Henrich et al. 2015, © 2015 Federation of European Biomedical Societies).

2012; Otzen and Andersen 2013; Popot 2014; Kleinschmidt 2015), making it unnecessary, for most experiments, to turn to much more costly and technically more demanding CFE. In addition, the use of CFE has been strongly boosted by the demand of biological and pharmaceutical research for properly folded eukaryotic membrane receptors and channels, most of which feature α-helical transmembrane (TM) domains.

Experience has shown that, in most cases, only extremely mild surfactants – such as, as far as detergents are concerned, digitonin or detergents of the Brij family – allow MPs expressed by CFE to fold properly in good yield (see e.g. Hein et al. 2014 and references therein). Stronger surfactants may inhibit MP production by interfering either with the translation machinery or with protein folding. In most cases, the surfactant comprises detergent or mixed lipid/detergent micelles, lipid vesicles, bicelles, nanodiscs, native membrane fragments, etc. (for some examples of these various approaches, see Kaiser et al. 2008; Katzen et al. 2008; Wuu and Swartz 2008; Cappuccio et al. 2009; Kuruma et al. 2010; Reckel et al. 2011; Wada et al. 2011; Abdine et al. 2012; Isaksson et al. 2012; Lyukmanova et al. 2012; Uhlemann et al. 2012; Periasamy et al. 2013; Proverbio et al. 2013; Shenkarev et al. 2013; Matsubayashi et al. 2014; Kuruma and Ueda 2015; Niwa et al. 2015; Henrich et al. 2017; and references therein). Less classical surfactants, known or expected to be milder than detergents, have been occasionally resorted to, including fluorinated surfactants (Park et al. 2007, 2011; Breyton et al. 2009; Blesneac et al. 2012), amphipathic peptides (Corin et al. 2011; Wang et al. 2011), and, as will be described in the next section, amphipathic polymers (for an overview of folding media, see Popot (2014)).

## 7.3    Cell-Free Expression of Membrane Proteins Using Amphipols and Other Amphipathic Polymers

Amphipathic polymers that have been tested for CFE of MPs include A8-35, sulfonated APols (SAPols), glucosylated non-ionic APols (NAPols), NVoy, and styrene-maleic acid copolymers (SMA) (Table 7.1; for earlier reviews, see Popot et al. 2011; Popot 2014; Zoonens and Popot 2014).

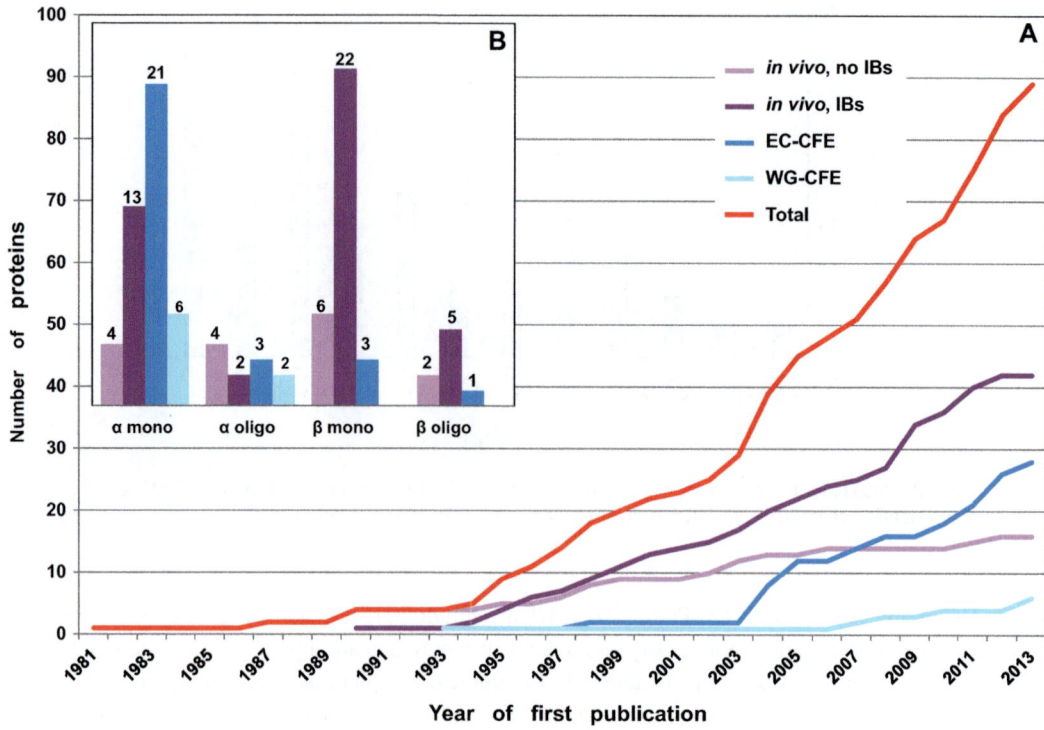

**Fig. 7.4** Origin of membrane proteins that had been successfully (re)folded in vitro by the end of 2013. Four types of origins are distinguished: (i) natural expression or overexpression in vivo, barring inclusion bodies (IBs); (ii) in vivo expression in IBs followed by in vitro folding; (iii) in vitro, cell-free expression in *E. coli* lysate (EC-CFE); and (iv) in vitro, cell-free expression in wheat germ lysate (WG-CFE). In **A**, the cumulative number of MPs obtained by each method that have been successfully (re)folded is plotted as a function of time, each protein being entered once for each curve, in the year of the first successful report. Because some MPs have been (re)folded by different methods and they are counted only once in the curve summing all production methods, the latter shows fewer proteins than the sum of the individual curves. Note the rapid increase in the number of folded MPs produced by CFE after 2003. In **B**, a histogram is shown of the total number of MPs of each type that have been (re)folded after being obtained by each method. α(β) *mono* and α(β) *oligo* refer, respectively, to α-helical (or β-barrel) MPs that are active either as monomers or as oligomers. Among MPs produced by CFE (*blue bars*), monomers with an α-helical TM region are by far the most numerous (27 out of a total of 36) (From Popot 2014, © 2004 Elsevier Inc. All rights reserved).

The first attempts, carried out with A8-35 and SAPols in an *E. coli* lysate, gave disappointing results (Study 7.1 in Table 7.1). Indeed, the two APols, used at 2 g·L$^{-1}$, were found to inhibit more or less completely the expression of three test MPs: BR, which comprises seven TM α-helices; the TM domain of the leader peptidase from *E. coli* (tLep), a two-TM helix preprotein; and the large-pore mechanosensitive channel MscL from *E. coli*, a pentameric protein whose protomers also comprise two TM α-helices each (Park et al. 2011). The synthesis of BR was found to be completely inhibited (Fig. 7.5A). The level of inhibition was variable for MscL (Fig. 7.5B) and tLep, but conditions producing acceptable yields in a reproducible manner could not be identified. A number of parameters were modified, using a homemade S30 lysate derived from the *E. coli* Rosetta (lDE3) strain, following

**Table 7.1** Publications relevant to polymer-assisted cell-free expression of membrane proteins.

| Study | Proteins | APol | Comments | References |
|---|---|---|---|---|
| 7.1 | BR, MscL, tLep | A8-35, SAPol | A8-35 and SAPol were found to inhibit more or less completely the synthesis of all three MPs, whereas that of GFP was unaffected. | Park et al. (2011) |
| 7.2 | Forty-four proteins or protein fragments | NVoy | Forty-four proteins or protein fragments from pathogenic organisms that express poorly or are poorly soluble upon overexpression in vivo in *E. coli* were expressed in wheat germ lysates in the presence or absence of NVoy. The polymer was observed not to block the synthesis and to favor solubility. However, whether any of these proteins contains a complete TM domain is not clear. | Guild et al. (2011) |
| 7.3 | GPCRs | NVoy, PMAL-B-100 | Eight GPCRs, belonging to families A, B, and C, were expressed in *E. coli* lysates in the presence or absence of NVoy or PMAL-B-100. NVoy did not inhibit CFE, whereas PMAL-B-100 did. A fraction (~10%) of CRFR1 and CRFR2$\beta$ expressed in the absence of surfactants, solubilized with a detergent, and transferred to NVoy was observed to bind specific agonists. | Klammt et al. (2011) |
| 7.4 | BR | NAPol, A8-35, SAPol | BR was expressed by CFE in an *E. coli* lysate in the presence of glucosylated non-ionic APols (NAPols). The proper folding of at least a fraction of the protein (estimated to ~2/3) was indicated by the purple color resulting from the binding of the cofactor retinal. The study confirmed the inhibition of BR synthesis by A8-35 and SAPols observed in Study 7.1. | Bazzacco (2009), Popot et al. (2011), Bazzacco et al. (2012), and Zoonens et al. (2014) |
| 7.5 | CrdS | A8-35, SMA | A8-35 and SMA were both observed to inhibit cell-free synthesis of CrdS in a wheat germ lysate. | Periasamy et al. (2013) |

*Abbreviations*: *BR* bacteriorhodopsin from *Halobacterium salinarum*, *CrdS* curdlan synthase from *Agrobacterium* sp., *CRFR* corticotrophin-releasing factor receptor, *GFP* green fluorescent protein, *GPCRs* G protein-coupled receptors, *MscL* large-pore mechanosensitive channel from *E. coli*, *NVoy* an amphipathic polysaccharide-based polymer (see Chap. 4, § 4.2.3.3), *SAPol* sulfonated amphipol, *SMA* styrene-maleic acid copolymer, *tLep* transmembrane domain of the leader peptidase from *E. coli*

the protocol of Zubay (1973), as modified by Kim et al. (2006): PEG-8000 was omitted during the S30 preparation, and CFE was carried out either in the presence or absence of 20 g·L$^{-1}$ PEG-8000. The effect of supplementing the lysate with 0.5 mM EDTA was also tested, so as to prevent the interaction of traces of Ca$^{2+}$ with A8-35 (Picard et al. 2006; Catoire et al. 2010; see Chap. 4, § 4.3.1.2.2). None of these modifications yielded any improvement in the production of the various test MPs.

***Amphipol-assisted cell-free expression of membrane proteins***
(© *2018 by Francis Haraux*)

In order to determine whether this inhibition was specific to MPs, expression of the soluble green fluorescent protein (GFP) was tested under the same conditions (Fig. 7.5C). No detectable inhibition was observed. This strongly suggests that the inhibition is not due to a general perturbation of the transcription/translation machineries but is specific to MPs. A possibility is that the inhibition is somehow related to the binding of APols to hydrophobic segments of MPs emerging from the ribosome tunnel: negative charges carried by A8-35 and SAPols bound to the emerging polypeptide may interact with basic side chains at the surface of the ribosome or within the mouth of the ribosome tunnel, thus sterically slowing down or blocking elongation, much as certain antibiotics seem to do by a different but equally steric mechanism (see e.g. Arévalo et al. 1988; Bulkley et al. 2010; and references therein). The incomplete inhibition observed with tLEP and MscL, at variance with the complete blockade of BR synthesis, may perhaps be related to their different number of TM helices. In $MLS_BK$ antibiotic-resistant *E. coli* strains that lack three critical basic residues in the tunnel-exposed $\beta$-hairpin loop of subunit L22, the mouth of the tunnel is more widely open, which appears to account for the resistance to erythromycin, given that the latter binds perfectly normally (Tu et al. 2005). The possibility that lysates derived from such mutants would allow CFE of MPs in the presence of polyanionic APols is a far-fetched hypothesis but might perhaps be worth testing (for further discussion, see § 7.4).

The hypothesis that the negative charges carried by A8-35 and SAPols are somehow responsible for the inhibition is generally although not fully consistent with later work described in Studies 7.2–5 (Table 7.1). In Study 7.5, a large number of surfactants were examined for their ability to support cell-free synthesis and solubility of CrdS, the curdlan synthase from *Agrobacterium* sp., an ~72 kDa MP predicted to comprise seven TM $\alpha$-helices and a single, large cytoplasmic region inserted between helices 3 and 4 (Hrmova et al. 2010). CFE in wheat germ lysates was found to be efficient in the presence of Brij-58 or of liposomes but was inhibited by 5–40 $g \cdot L^{-1}$ concentrations of either A8-35 or styrene-maleic acid (SMA) co-polymers, both of which are polyanionic polymers (Periasamy et al. 2013). It should be noted, however, that in Study 7.3, PMAL-B-100, an APol that is zwitterionic at and

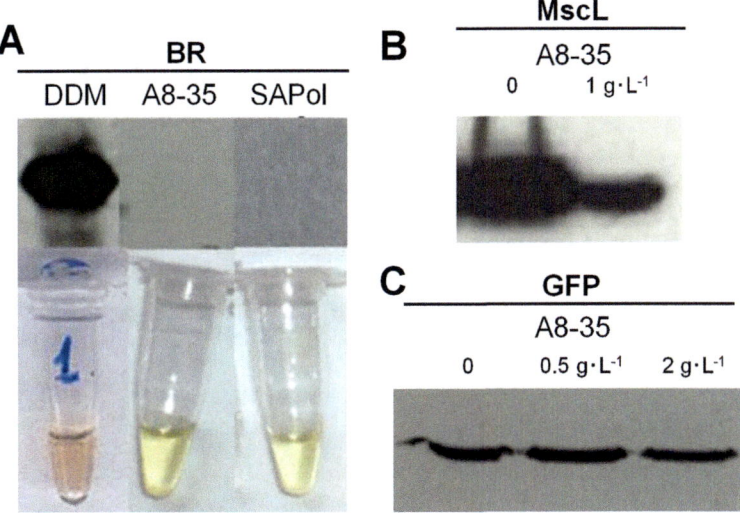

**Fig. 7.5** Cell-free expression of two membrane proteins, bacteriorhodopsin (BR) and the large-pore mechanosensitive channel (MscL), and a soluble protein, the green fluorescent protein (GFP), in either the presence or absence of polyanionic amphipols. All proteins wore polyhistidine tags. (**A**) Synthesis of BR in the presence of either detergent (2 mM DDM) A8-35 or SAPol (2 g·L$^{-1}$) and 50 µM all-*trans*-retinal. Note the red color of the DDM-containing sample, indicative of the formation of the holoprotein, whereas the two APol-containing samples exhibit only the yellow color of free retinal. (**B**, **C**) Synthesis of MscL (B) and GFP (C) in the presence of A8-35 at the indicated concentrations. Following CFE, the various lysates were submitted to electrophoresis on SDS-urea 12–18% polyacrylamide gels and the proteins detected by immunoblotting using an anti-(His)$_6$-tag antibody (From Park et al. 2011, © 2010 Elsevier B.V. All rights reserved).

above pH 7 (see Chap. 4, § 4.2.2.3), is reported to inhibit CFE of GFP at concentrations above 0.011 mM (~0.13 g·L$^{-1}$) and, at this concentration, to block GPCR synthesis (Klammt et al. 2011). This effect cannot be attributed to a nonexistent polyanionic character.

On the contrary, MP CFE was found to proceed efficiently in the presence of non-ionic amphipathic polymers, whether NVoy (Study 7.3 and, possibly, Study 7.2) or glucosylated non-ionic APols (NAPols; Study 7.4).

In Study 7.2, a large number of proteins or protein fragments from pathogenic organisms that are poorly expressed or poorly soluble upon overexpression in vivo in *E. coli*, a few of which only are MPs or fragments thereof, were expressed by CFE in a wheat germ lysate either in the presence or absence of NVoy, an uncharged polymer obtained by derivation with hydrophobic side chains of a carbohydrate (fructose) polymer (see Chap. 4, § 4.2.3.3). In most cases, NVoy was observed not to impede the synthesis and to favor solubility. However, it is unclear from the data provided whether any of the proteins or protein fragments tested comprised a TM domain. The bearing of these observations on CFE of MPs is therefore quite uncertain, beyond the fact that NVoy does not systematically block CFE and tends to improve the solubility of poorly soluble but not necessarily TM proteins.

In Study 7.3, an extensive investigation is reported on the production and characterization, using NVoy, of eight different polyhistidine-tagged membrane receptors. The MPs chosen are representative of the three GPCR families, each of which features seven TM $\alpha$-helices: the largest subgroup, family A (the rhodopsin-like family), comprises receptors for light, odorants, small molecules, and peptides and glycoprotein hormones; family B (secretin-like) is characterized by a relatively long amino terminus

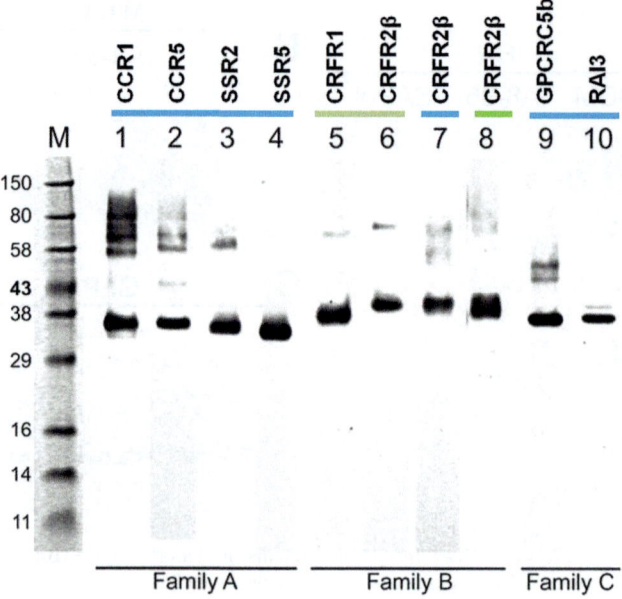

**Fig. 7.6** Cell-free expression of GPCRs of all three subfamilies in the presence or absence of NVoy. Western blot analysis of solubilized and purified GPCRs using an anti-His$_6$ (lanes 1, 2, 5, 6) or anti-T7 tag antibody (lanes 3, 4, 7–10). Three different modes of CFE were used: (i) CCR1, CCR5, SSR2, SSR5, GPCRC5b, and RAI3 were expressed in the presence of NVoy (*blue bars*); (ii) CRFR1 was expressed in the absence of surfactant, solubilized in the detergent 1-myristoyl-2-hydroxy-*sn*-glycero-3-[phosphor-RAC-(1-glycerol)] (LMPG), and transferred to NVoy (*celadon bar*); (iii) CRFR2$\beta$ was expressed using each of these two modes and, in addition, analyzed after expression in the absence of surfactant and solubilization in LMPG, without transfer to NVoy (*green bar*). *Abbreviations*: M molecular weight standards; CCR1 and CCR5, respectively, human CAC chemokine receptors types 1 and 5; SSR2 and SSR5, respectively, human somatostatin receptors types 2 and 5; CRFR1 and CRFR2$\beta$, respectively, mouse corticotropin-releasing factor receptors types 1 and 2$\beta$; GPCRC5b human GPCR family C group 5 member B; RAI3 human retinoic acid-induced protein 3. The GPCR family to which each protein belongs is indicated under the gel (Adapted from Klammt et al. 2011, © 2011 The Protein Society).

that contains several disulfide bridges; and family C (glutamate receptor-like) features a long amino terminus and a unique short and highly conserved third intracellular loop. Using *E. coli* lysates, three different CFE approaches were explored (Fig. 7.6): in approach (i), the proteins were expressed in the presence of 2.5 g·L$^{-1}$ NVoy (lanes marked with *blue bars* in Fig. 7.6); in approaches (ii) and (iii), they were expressed in the absence of any surfactant, leading to precipitation. The precipitates were insoluble in NVoy, even when the polymer was used at 25 g·L$^{-1}$, but they could be solubilized in a detergent, 1-myristoyl-2-hydroxy-*sn*-glycero-3-[phosphor-RAC-(1-glycerol)] (LMPG), and either (ii) transferred to NVoy (*celadon bar*) or (iii) purified and studied in LMPG (*green bar*). The yields typically varied between 0.5 and 1 mg per mL of lysate. Procedure *ii* resulted in >75% pure GPCRs, according to SDS-PAGE estimates. Following immobilized-metal affinity chromatography (IMAC) purification, ~0.5 mg of the CRFR2$\beta$ receptor could be obtained per mL of lysate, of which ~10% was active, according to ligand-binding measurements, and could be further purified on an affinity column. The homogeneity of NVoy-trapped GPCR samples was best using procedure *ii*, in which the precipitated protein is solubilized in LMPG before transfer to the polymer (see Fig. 9.9 in Chap. 9).

Purified CRFR2$\beta$/NVoy complexes were stable for weeks at 4 °C. They were characterized by their pharmacology, their migration upon size exclusion chromatography (see Chap. 9, Fig. 9.9), and their appearance in transmission electron microscopy following negative staining (Fig. 7.7A, B;

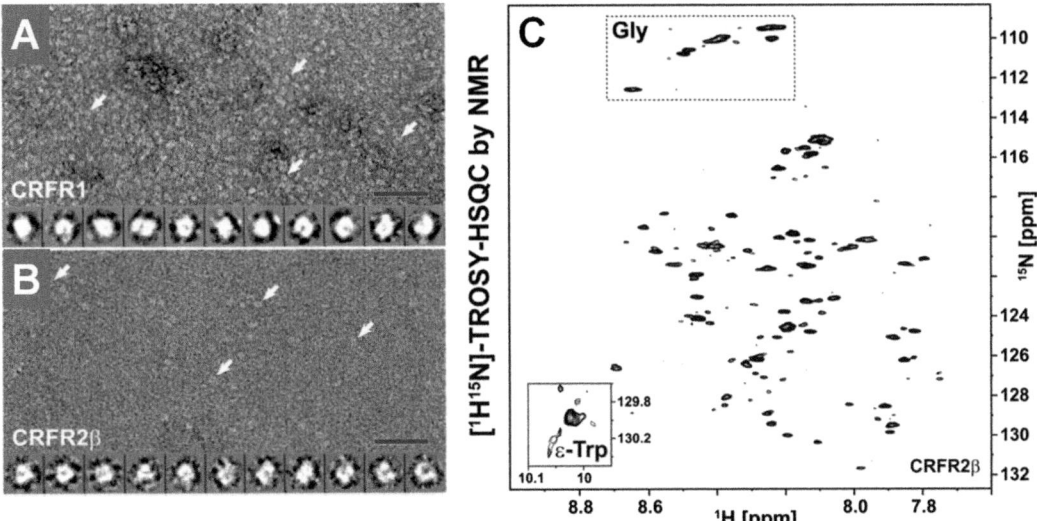

**Fig. 7.7** Electron microscopy and solution NMR characterization of two NVoy-trapped G protein-coupled receptors. (**A**, **B**) Electron micrographs of negatively stained CRFR1 (**A**) and CRFR2$\beta$ (**B**) expressed in vitro in the absence of surfactant, solubilized with LMPG, and transferred to NVoy show monodisperse, non-aggregated particles (*arrows*). Class averages are at the bottom. The scale bars (*red*) correspond to 60 nm; each average is in a $16 \times 16$ nm box. (**C**) [$^1$H,$^{15}$N]-TROSY-HSQC spectra of CRFR2$\beta$ expressed in vitro in the presence of NVoy. Approximately 20 $\mu$M U-$^{15}$N-CRFR2$\beta$ in ~1 mM NVoy, 10 mM NaCl, and 20 mM MES/Bis-Tris (pH 4.0), measured at 310 K on a 700 MHz spectrometer equipped with a cryogenic probe. Characteristic Trp N$^{\epsilon 1}$H and Gly H$^N$ cross peaks are indicated (From Klammt et al. 2011, © 2011 The Protein Society).

GPCRs expressed in the absence of surfactant, solubilized with LMPG, and transferred to NVoy). They were found to comprise monomers and dimers in variable proportions depending on the mode of synthesis and the method of observation. In solution NMR, the [$^1$H,$^{15}$N]-TROSY-HSQC spectra, measured at pH 4.0 and 310 K, of U-$^{15}$N-CRFR2$\beta$/NVoy complexes produced by CFE in the presence of NVoy, show a set of ~100 sharp peaks well dispersed between 7.7 and 8.7 ppm in the proton dimension (Fig. 7.7C). Given the size of the particles (~150 kDa), they most likely correspond to extramembrane, mobile peptide segments. The authors note that, when compared with CFE in the presence of NVoy, the samples obtained by precipitation in the absence of surfactant, solubilization in detergent, and transfer to NVoy "show similar (CRFR2$\beta$) or better (CRFR1) ligand binding, with the benefit of enhanced purity." Altogether, Study 7.3, the most comprehensive study yet published of polymer-trapped receptors obtained by CFE, appeared to bode well for the use of NVoy either during or after in vitro synthesis, but this approach does not seem to have been pursued.

In Study 7.4, it was examined whether recently developed glucosylated non-ionic APols (NAPols; see Chap. 4, § 4.2.3.2) could or could not support MP CFE more efficiently than the polyanionic APols tested with disappointing results in Study 7.1. Small-scale CFE of polyhistidine-tagged BR was tested at different concentrations of NAPols (0.5–4 g·L$^{-1}$), in the presence or absence of BR's cofactor, retinal. SDS-PAGE analysis followed by immunoblotting and revelation with an anti-His-tag antibody indicated that synthesis took place in all cases (Fig. 7.8). BR expressed in the presence of NAPols and retinal folded to a greater extent than in the presence of dodecylmaltoside (DDM) and retinal, as shown by the disappearance of the 380 nm absorbance peak of free retinal and the appearance of the visible peak at ~554 nm characteristic of native BR in its soluble state. The samples were centrifuged and the distribution of BR between pellet and supernatant determined by

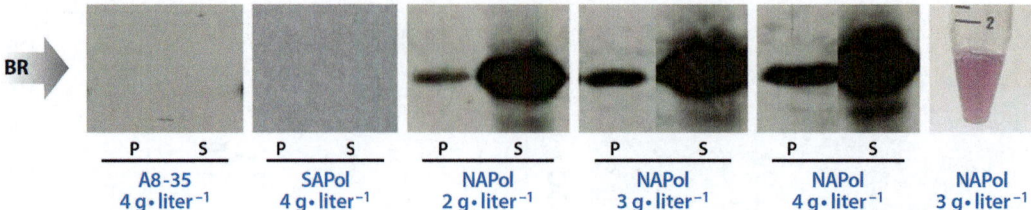

**Fig. 7.8** Amphipol-assisted cell-free expression of bacteriorhodopsin (BR). In vitro synthesis of polyhistidine-tagged BR in the presence of retinal and of A8-35, SAPols, or NAPols at the indicated concentrations. At the end of the synthesis, the samples were centrifuged at 16,000 × g for 20 min. The proteins present in the pellet (P) and supernatant (S) were analyzed by SDS-PAGE on a 12% polyacryl-amide gel and detected using an anti-His-tag antibody. *Far right panel*: BR after IMAC purification. The purple color of the solution indicates the presence of native BR (representing at least two-thirds of the protein present in the purified sample) (From Popot et al. 2011; Zoonens et al. 2014).

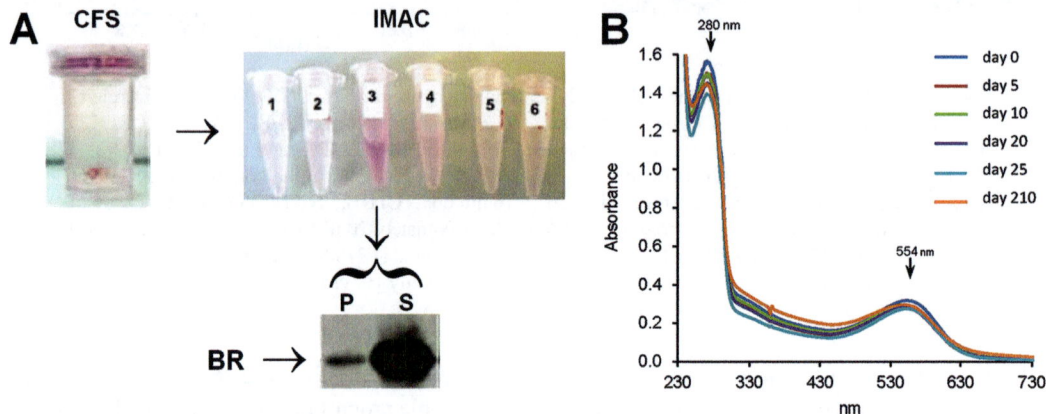

**Fig. 7.9** Cell-free synthesis of polyhistidine-tagged bacteriorhodopsin (BR) in the presence of glucosylated NAPols. The 1 mL *E. coli* lysate was supplemented with 70 μM retinal and 3 g·L$^{-1}$ NAPol. (**A**) After synthesis for 20 h and a one-step purification by IMAC (yielding fractions 1–6), one fraction (fraction 3) was found to contain most of the protein, the purple color of which indicated that at least part of it had folded correctly and bound its cofactor. Fraction 3 was centrifuged, and the pellet (P) and supernatant (S) were analyzed by SDS-PAGE. BR was detected by immunoblotting using an anti-His-tag antibody. (**B**) Stability of BR synthesized in the presence of NAPols was evaluated spectroscopically. Fraction 3 was stored for 7 months at 4 °C in the dark and its UV-visible absorbance spectrum monitored periodically. Note the persistence of the peak at ~554 nm due to the presence of the holoprotein (Reprinted with permission from Bazzacco et al. 2012, © American Chemical Society).

SDS-PAGE followed by immunolabeling (Fig. 7.8), so as to establish the optimal concentration of polymer to be used. Best results were obtained at 3 g·L$^{-1}$ NAPols in the presence of 70 μM retinal, under which conditions an estimated ~2/3 of BR was correctly folded and in a soluble form.

CFE was then scaled up to 1 mL batches (Fig. 7.9A), which yielded, after nickel-affinity purification, ~0.4 mg of pure BR, a majority of which (~90%) was present in the form of soluble BR/NAPol complexes (Fig. 7.9B). As observed for native BR trapped with NAPols in the presence of PM lipids (Chap. 5, § 5.5), BR synthesized in vitro in the presence of NAPols and in the absence of lipids was highly stable, inactivating only partially after over 7 months of storage at 4 °C (Fig. 7.9C). This stability is comparable to that observed after transferring native BR from detergent solution to A8-35 in the presence of archaebacterial lipids (Gohon et al. 2008; Dahmane et al. 2013). Whereas these observations bode well for the use of NAPols for CFE of MPs, difficulties in scaling their production up to a marketable level have hitherto prevented exploring the generality of the observations made with BR.

## 7.4   Conclusions and Perspectives

From the data presented in § 7.3, it will have been apparent that the use of APols for cell-free expression of MPs is only in its infancy: only a handful of studies have been published, some of which yielded negative results – inhibition of synthesis – and those in which synthesis proceeded normally have been limited, as of now, to two different polymers and a single structural type of MPs, featuring seven TM $\alpha$-helices.

Yet, extremely encouraging results have been obtained with the two non-ionic polymers tested, NVoy, shown to be compatible with the expression of seven GPCRs of various classes (Klammt et al. 2011), and a non-ionic, glucosylated NAPol, tested on the sole BR (Bazzacco et al. 2012). In both studies, reasonable yields were achieved (in the 0.4–1 g·L$^{-1}$ range), and most of the MP was recovered in the form of soluble MP/polymer complexes. For the one GPCR whose percentage of functional folding was determined, ~10% of the protein was found to bind its ligands and could be purified by affinity chromatography. For BR, the yield of functional folding was estimated to be at least two-thirds of the protein present in the IMAC-purified sample (Popot et al. 2011). It would be highly desirable, of course, to examine more in depth the factors that determine the folding yield, as well as to compare more systematically the yields obtained when expressing MPs in the presence of the polymer and when letting them precipitate and recovering them by detergent solubilization followed by transfer to the polymer (Klammt et al. 2011). It is also necessary to apply these two types of molecules to CFE of a wider variety of MPs, including oligomeric MPs and MPs that fold into $\beta$-barrels. For NVoy, carrying out these investigations ought to be relatively straightforward, as the molecule is simple to synthesize and commercially available. The synthesis of glucosylated NAPols is more cumbersome, and extensive studies of their use will have to wait for it to be scaled up and the product marketed, or for easier-to-synthesize NAPols to be developed.

It is worth noting that, with both non-ionic polymers, it has been shown that homogeneous or relatively homogeneous MP/polymer complexes can be obtained and studied by such structural methods as radiation scattering (BR/NAPol complexes), electron microscopy (GPCR/NVoy ones), and NMR (both types of complexes) (see Klammt et al. 2011; Bazzacco et al. 2012; Sharma et al. 2012). Furthermore, it can be expected that MPs complexed by non-ionic polymers might be easier to crystallize than charged ones (cf. Chap. 11), as well as more generally usable for ligand-binding studies, because of a lower background binding of cationic ligands (cf. Chaps. 5 and 13). The scattered results available on the stability of MPs trapped with either NVoy (Klammt et al. 2011) or NAPols (Bazzacco et al. 2012) suggest that the two polymers are particularly mild. MPs trapped in these polymers may therefore stand a good chance at being amenable to experiments that require them to remain properly folded for days or weeks, such as extensive NMR measurements or crystallization attempts. Furthermore, the possibility to express MPs directly in the presence of non-ionic polymers and, therefore, to avoid any contact with detergents opens up the prospect of gaining access to MPs that are too fragile to be either solubilized by detergents following precipitation or expressed in their presence.

As compared to charged APols and, in particular, A8-35, NVoy and NAPols suffer from the disadvantage that no or very few labeled or tagged derivatives exist yet (Chap. 4). Perdeuteration of any of these polymers would be difficult to carry out, and, if doable, would certainly be very expensive. No fluorescent derivatives of either polymer have been synthesized yet, and a single tagged one has been described (a biotinylated NAPol; see Ferrandez et al. 2014). There is, however, no particular reason that such derivatives could not be produced.

Charged polymers (A8-35, SAPols, PMAL-B-100, and SMA) have been shown to inhibit more or less completely MP CFE (Table 7.1). Yet inhibition is not always complete (Fig. 7.5). Given the very large number of A8-35 derivatives that are available to the experimenter (Chap. 4, § 4.4.1) and the extensive work invested in investigating the properties and developing the applications of A8-35 and MP/A8-35 complexes, it might perhaps be worth investing some more efforts in trying to understand what the basis of CFE inhibition by A8-35 is and whether it can be avoided. The fact that the synthesis of GFP proceeds normally in its presence strongly suggests that it is not the transcription/translation machinery that is affected. An obvious tentative conclusion is that the inhibition is somehow related to the presence of hydrophobic segments in the sequence of the MPs tested. One can imagine, for instance, that, upon binding to emerging hydrophobic segments, APols come in the close vicinity of the ribosome's surface and, by interacting with it, sterically or otherwise slow down or block elongation, much as erythromycin is thought to do (see Tu et al. 2005). It would be desirable to have a clearer idea of which polymers interfere with the synthesis of which proteins, why, and whether the blockade can be circumvented. Among the investigations that could be relatively easily set up would be to examine, using a given CFE lysate, a matrix comprising a selection of proteins and a selection of polymers. The polymers could include some APols that have not yet been tested in CFE, such as phosphorylcholine-based APols (Diab et al. 2007a, b; see Chap. 4, § 4.2.2.2). In addition to the proteins already tested, it would be desirable to include:

(i)  More than one soluble protein, in order to make sure that the unperturbed synthesis of GFP is not a special case.

(ii)  MPs with a variable number of $\alpha$-helices, including some with a single $\alpha$-helix; indeed, if the mechanism hypothesized above is correct, one may expect that the larger the number of helices and the closer they are spaced to one another, the greater the inhibition, which seems to be borne out by the comparison of the level of expression of BR, tLep, and MscL in the presence of A8-35 (Park et al. 2011).

(iii)  $\beta$-barrel MPs; indeed, the sequence of these proteins does not comprise strongly hydrophobic stretches, and, in most cases, the barrel is not expected to fold and expose a large hydrophobic surface before the synthesis is virtually completed; inhibition of CFE of such MPs might possibly not be impeded.

More farfetched hypotheses might be tested at an affordable cost. To take a single example, it is remarkable that the inhibition of elongation by erythromycin seems to be due not to a complete blockade of the extrusion of the growing chain through the ribosome tunnel but to a partial hindrance that presumably creates a higher free energy barrier to the progression of the polypeptide chain (Tu et al. 2005). Deletion of residues Met82, Lys83, and Arg84 from the conserved $C$-terminal $\beta$-hairpin of ribosomal protein L22 confers erythromycin resistance to $E.$ $coli$ ribosomes without reducing their affinity for the drug (Chittum and Champney 1994). The resistance therefore does not result from reduced erythromycin binding. It appears to be due to easing of the progression of the elongating chain by widening of the mouth of the tunnel downstream of the erythromycin binding site, due to the $\beta$-hairpin moving out of the way. Although admittedly a long shot, but one easy to test, it might perhaps be interesting to examine whether either a similar mechanism, or perhaps the loss of an interaction between APols and the basic region of the $\beta$-hairpin, would improve MP synthesis in lysates derived from erythromycin-resistant strains.

## 7.5    Protocol 7.1: Cell-Free Expression of Membrane Proteins in the Presence of Non-ionic Amphipols

### 7.5.1    Introduction

Obtaining sufficient quantities of functional MPs is often a major bottleneck that hinders structural and functional studies. A solution to this problem is to perform protein expression in an acellular system (CFE), which eliminates, at least in part, the limitations of in vivo expression systems resulting from toxicity or imperfect folding and membrane insertion (Zubay 1973; Rues et al. 2016). Among the many attractive features of CFE is the possibility it affords to label MPs using limited amounts of isotopically labeled or unnatural amino acids (Kigawa et al. 1999; LaGuerre et al. 2015). Whereas charged APols such as A8-35, SAPols, PMAL-B-100, or SMA inhibit the CFE of MPs (Park et al. 2011; Periasamy et al. 2013), glucosylated non-ionic APols (NAPols) are compatible with it (Bazzacco et al. 2012) (Figs. 7.8 and 7.9). BR expressed in vitro in the presence of NAPols is properly folded and stays stable over several months, whereas in DDM it tends to rapidly precipitate (Park et al. 2011; Bazzacco et al. 2012). Because APols tend to be much milder than detergents (see Chap. 5, § 5.5, and Kleinschmidt and Popot 2014), developing APol-assisted CFE of MPs appears as an attractive alternative to the use of classical detergents. CFE is carried out using commercial systems or a homemade lysate (Proverbio et al. 2014). The protocol described below was developed using *E.coli* lysate, but it should be extendable to eukaryotic ones.

Comments are in italics and preceded by a pointing hand (☞).

### 7.5.2    Protocol

- Plasmids for CFE

The RTS pIVEX *E. coli* vectors are designed for cell-free expression of proteins His$_6$-tagged either at the *N*- or *C*-terminus and contain all sequences needed for T7 RNA polymerase-mediated expression.

- CFE Small-Scale Reaction

Prior to performing a large-scale production of MP, it is essential to optimize the concentration of NAPol to be used. Small-scale syntheses are therefore carried out in the presence of a range of NAPol concentrations.

1. Prepare a stock solution of NAPols at $100 \text{ g} \cdot \text{L}^{-1}$ or 10% w/w. Weigh the powder with an analytical balance in an Eppendorf tube or, if possible, in a small glass vial, and add water purified on a Milli-Q Advantage A10 system in order to reach the final concentration.
2. Homogenize the solution with a vortex or by magnetic stirring for at least a couple of hours before use to reach full rehydration of the lyophilized powder. The solution is then kept at 4 °C or frozen at −20 °C.
3. Small-scale syntheses are carried out in the presence of 0.5 µg of plasmid and 3, 5, 8, or $10 \text{ g} \cdot \text{L}^{-1}$ NAPols in 25 or 50 µL of lysate. These quantities of NAPol are larger than is actually necessary because it is difficult to foresee the amount of expressed proteins. Incubate for 6 h in a ThermoMixer (Eppendorf) at 700 rpm and 25 °C.

☞ *Parameters such as temperature and Mg$^{2+}$ concentration can be optimized at this stage.*

4. In order to verify the solubility of the MP produced, 10 μL samples are diluted with the same volume of 10 mM Tris/HCl, pH 8.0 buffer, centrifuged for 10 min at 16,000 × g at room temperature, and the supernatant and pellet analyzed by Western blot after SDS-PAGE (cf. Figs. 7.8 and 7.9).

- CFE Large-Scale Reaction

Once the concentration of NAPols to be used has been optimized, the reaction can be scaled up to 1 mL of lysate:

1. The NAPol solution is prepared directly in the CFE reaction buffer used to resuspend the lyophilized components of the reaction (commercial system) or in the S30 buffer for the homemade lysate (Proverbio et al. 2014), so as to reach the desired final volume (1 mL) in the presence of the chosen concentration of NAPol. The solution is stirred overnight in the presence of 1% of sodium azide to guarantee optimal solubilization.
2. The 1 mL reaction is carried out in the presence of 15 μg of plasmid for 20–30 h in the ThermoMixer at the optimized temperature.

☞ *NAPols do not cross dialysis membranes and, therefore, their presence in the feeding chamber is not required.*

3. After CFE, the protein can be purified by IMAC, either in batch or using a prepacked column.

☞ *Again, it is not necessary to add NAPols in the purification buffers.*

*Protocol prepared by Francesca Zito on the basis of Bazzacco et al. (2012) and laboratory notes, with comments added.*

## References

Abdine, A., Park, K.H., Warschawski, D.E. (2012) Cell-free membrane protein expression for solid-state NMR. *Meth. Mol. Biol.* **831**:85–109.

Arévalo, M.A., Tejedor, F., Polo, F., Ballesta, J.P. (1988) Protein components of the erythromycin binding site in bacterial ribosomes. *J. Biol. Chem.* **263**:58–63.

Bazzacco, P. (2009) Non-ionic amphipols: new tools for *in vitro* studies of membrane proteins. Validation and development of biochemical and biophysical applications. Thèse de Doctorat, Université Paris-7, 176 p.

Bazzacco, P., Billon-Denis, E., Sharma, K.S., Catoire, L.J., Mary, S., Le Bon, C., Point, E., Banères, J.-L., Durand, G., Zito, F., Pucci, B., Popot, J.-L. (2012) Non-ionic homopolymeric amphipols: Application to membrane protein folding, cell-free synthesis, and solution NMR. *Biochemistry* **51**:1416–1430.

Blesneac, I., Ravaud, S., Juillan-Binard, C., Barret, L.A., Zoonens, M., Polidori, A., Miroux, B., Pucci, B., Pebay-Peyroula, E. (2012) Production of UCP1, a membrane protein from the inner mitochondrial membrane, using the cell-free expression system in the presence of a fluorinated surfactant. *Biochim. Biophys. Acta* **1818**:798–805.

Bogdanov, M., Dowhan, W. (1998) Phospholipid-assisted protein folding: phosphatidylethanolamine is required at a late step of the conformational maturation of the polytopic membrane protein lactose permease. *EMBO J.* **17**:5255–5264.

Breyton, C., Gabel, F., Abla, M., Pierre, Y., Lebaupain, F., Durand, G., Popot, J.-L., Ebel, C., Pucci, B. (2009) Micellar and biochemical properties of (hemi)fluorinated surfactants are controlled by the size of the polar head. *Biophys. J.* **97**:1077–1086.

Buchanan, S.K., Yamashita, S., Fleming, K.G. (2012) Structure and folding of outer membrane proteins, in: Tamm, L.K., ed., *Membranes*. Elsevier, Oxford:Academic Press, pp. 139–163.

Bulkley, D., Innis, A., Blaha, G., Steitz, T.A. (2010) Revisiting the structures of several antibiotics bound to the bacterial ribosome. *Proc. Natl. Acad. Sci. USA* **107**:17158–17163.

Cappuccio, J.A., Hinz, A.K., Kuhn, E.A., Fletcher, J.E., Arroyo, E.S., Henderson, P.T., Blanchette, C.D., Walsworth, V. L., Corzett, M.H., Law, R.J., Pesavento, J.B., Segelke, B.W., Sulchek, T.A., Chromy, B.A., Katzen, F., Peterson, T. C., Bench, G., Kudlicki, W., Hoeprich, P.D., Jr., Coleman, M.A. (2009) Cell-free expression for nanolipoprotein particles: building a high-throughput membrane protein solubility platform. *Methods Mol. Biol.* **498**:273–296.

Catoire, L.J., Zoonens, M., van Heijenoort, C., Giusti, F., Guittet, E., Popot, J.-L. (2010) Solution NMR mapping of water-accessible residues in the transmembrane $\beta$-barrel of OmpX. *Eur. Biophys. J.* **39**:623–630.

Chittum, H.S., Champney, W.S. (1994) Ribosomal protein gene sequence changes in erythromycin-resistant mutants of *Escherichia coli*. *J. Bact.* **176**:6192–6198.

Corin, K., Baaske, P., Ravel, D.B., Song, J., Brown, E., Wang, X., Wienken, C.J., Jerabek-Willemsen, M., Duhr, S., Luo, Y., Braun, D., Zhang, S. (2011) Designer lipid-like peptides: a class of detergents for studying functional olfactory receptors using commercial cell-free systems. *PLoS ONE* **6**:e25067.

Dahmane, T., Rappaport, F., Popot, J.-L. (2013) Amphipol-assisted folding of bacteriorhodopsin in the presence and absence of lipids. Functional consequences. *Eur. Biophys. J.* **42**:85–101.

Diab, C., Tribet, C., Gohon, Y., Popot, J.-L., Winnik, F.M. (2007a) Complexation of integral membrane proteins by phosphorylcholine-based amphipols. *Biochim. Biophys. Acta* **1768**:2737–2747.

Diab, C., Winnik, F.M., Tribet, C. (2007b) Enthalpy of interaction and binding isotherms of non-ionic surfactants onto micellar amphiphilic polymers (amphipols). *Langmuir* **23**:3025–3035.

Ferrandez, Y., Dezi, M., Bosco, M., Urvoas, A., Valério, M., Le Bon, C., Giusti, F., Broutin, I., Durand, G., Polidori, A., Popot, J.-L., Picard, M., Minard, P. (2014) Amphipol-mediated screening of molecular orthoses specific for membrane protein targets. *J. Membr. Biol.* **247**:925–940.

Focke, P.J., Hein, C., Hoffmann, B., Matulef, K., Bernhard, F., Dötsch, V., Valiyaveetil, F.I. (2016) Combining *in vitro* folding with cell-free protein synthesis for membrane protein expression. *Biochemistry* **55**:4212–4219.

Ge, X., Xu, J. (2012) Cell-free protein synthesis as a promising expression system for recombinant proteins. *Methods Mol. Biol.* **824**:565–578.

Gohon, Y., Dahmane, T., Ruigrok, R., Schuck, P., Charvolin, D., Rappaport, F., Timmins, P., Engelman, D.M., Tribet, C., Popot, J.-L., Ebel, C. (2008) Bacteriorhodopsin/amphipol complexes: structural and functional properties. *Biophys. J.* **94**:3523–3537.

Grisshammer, R., Tate, C.G. (1995) Overexpression of integral membrane proteins for structural studies. *Quart. Rev. Biophys.* **28**:315–422.

Guild, K., Zhang, Y., Stacy, R., Mundt, E., Benbow, S., Green, A., Myler, P.J. (2011) Wheat germ cell-free expression system as a pathway to improve protein yield and solubility for the SSGCID pipeline. *Acta Crystallogr. Sect. F Struct. Biol. Cryst. Commun.* **67**:1027–1031.

Hein, C., Henrich, E., Orbán, E., Dötsch, V., Bernhard, F. (2014) Hydrophobic supplements in cell-free systems: Designing artificial environments for membrane proteins. *Engin. Life Sci.* **14**:365–379.

Henrich, E., Hein, C., Dötsch, V., Bernhard, F. (2015) Membrane protein production in *Escherichia coli* cell-free lysates. *FEBS Lett.* **589**:1713–1722.

Henrich, E., Peetz, O., Hein, C., LaGuerre, A., Hoffmann, B., Hoffmann, J., Dötsch, V., Bernhard, F., Morgner, N. (2017) Analyzing native membrane protein assembly in nanodiscs by combined non-covalent mass spectrometry and synthetic biology. *eLife* **6**:e20954.

Hrmova, M., Stone, B.A., Fincher, G.B. (2010) High-yield production, refolding and molecular modelling of the catalytic module of (1,3)-$\beta$-D-glucan (curdlan) synthase from *Agrobacterium* sp. *Glycoconj. J.* **27**:461–476.

Huppa, J.B., Ploegh, H.L. (1997) *In vitro* translation and assembly of a complete T cell receptor-CD3 complex. *J. Exp. Med.* **186**:393–403.

Isaksson, L., Enberg, J., Neutze, R., Karlsson, B.G., Pedersen, A. (2012) Expression screening of membrane proteins with cell-free protein synthesis. *Protein Expr. Purif.* **82**:218–225.

Junge, F., Luh, L.M., Proverbio, D., Schäfer, B., Abele, R., Beyermann, M., Dötsch, V., Bernhard, F. (2010) Modulation of G-protein coupled receptor sample quality by modified cell-free expression protocols: A case study of the human endothelin A receptor. *J. Struct. Biol.* **172**:94–106.

Kai, L., Roos, C., Haberstock, S., Proverbio, D., Ma, Y., Junge, F., Karbyshev, M., Dötsch, V., Bernhard, F. (2012) Systems for the cell-free synthesis of proteins. *Methods Mol. Biol.* **800**:201–215.

Kai, L., Orbán, E., Henrich, E., Proverbio, D., Dötsch, V., Bernhard, F. (2015) Co-translational stabilization of insoluble proteins in cell-free expression systems. *Methods Mol. Biol.* **1258**:125–143.

Kaiser, L., Graveland-Bikker, J., Steuerwald, D., Vanberghem, M., Herlihy, K., Zhang, S. (2008) Efficient cell-free production of olfactory receptors: detergent optimization, structure, and ligand binding analyses. *Proc. Natl. Acad. Sci. USA* **105**:15726–15731.

Katzen, F., Fletcher, J.E., Yang, J.P., Kang, D., Peterson, T.C., Cappuccio, J.A., Blanchette, C.D., Sulchek, T., Chromy, B.A., Hoeprich, P.D., Coleman, M.A., Kudlicki, W. (2008) Insertion of membrane proteins into discoidal membranes using a cell-free protein expression approach. *J. Proteome Res.* **7**:3535–3542.

Keller, T., Schwarz, D., Bernhard, F., Dötsch, V., Hunte, C., Gorboulev, V., Koepsell, H. (2008) Cell-free expression and functional reconstitution of eukaryotic drug transporters. *Biochemistry* **47**:4552–4564.

Kigawa, T. (2010) Cell-free protein production system with the *E. coli* crude extract for determination of protein folds. *Methods Mol. Biol.* **607**:101–111.

Kigawa, T., Yabuki, T., Yoshida, Y., Tsutsui, M., Ito, Y., Shibata, T., Yokoyama, S. (1999) Cell-free production and stable-isotope labeling of milligram quantities of proteins. *FEBS Lett.* **442**:15–19.

Kim, T.W., Keum, J.W., Oh, I.S., Choi, C.Y., Park, C.G., Kim, D.M. (2006) Simple procedures for the construction of a robust and cost-effective cell-free protein synthesis system. *J. Biotechnol.* **126**:554–561.

Klammt, C., Lohr, F., Schäfer, B., Haase, W., Dötsch, V., Ruterjans, H., Glaubitz, C., Bernhard, F. (2004) High level cell-free expression and specific labeling of integral membrane proteins. *Eur. J. Biochem.* **271**:568–580.

Klammt, C., Schwarz, D., Fendler, K., Haase, W., Dötsch, V., Bernhard, F. (2005) Evaluation of detergents for the soluble expression of α-helical and β-barrel-type integral membrane proteins by a preparative scale individual cell-free expression system. *FEBS J.* **272**:6024–6038

Klammt, C., Schwarz, D., Löhr, F., Schneider, B., Dötsch, V., Bernhard, F. (2006) Cell-free expression as an emerging technique for the large scale production of integral membrane protein. *FEBS J.* **273**:4141–4153.

Klammt, C., Perrin, M.-H., Maslennikov, I., Renault, L., Krupa, M., Kwiatkowski, W., Stahlberg, H., Vale, W., Choe, S. (2011) Polymer-based cell-free expression of ligand-binding family B G protein-coupled receptors without detergents. *Prot. Sci.* **20**:1030–1041.

Kleinschmidt, J.H. (2015) Folding of β-barrel membrane proteins in lipid bilayers – Unassisted and assisted folding and insertion. *Biochim. Biophys. Acta* **1848**:1927–1943.

Kleinschmidt, J.H., Popot, J.-L. (2014) Folding and stability of integral membrane proteins in amphipols. *Arch. Biochem. Biophys.* **564**:327–343.

Kuruma, Y., Ueda, T. (2015) The PURE system for the cell-free synthesis of membrane proteins. *Nat. Protoc.* **10**:1328–1344.

Kuruma, Y., Suzuki, T., Ueda, T. (2010) Production of multi-subunit complexes on liposome through an *E. coli* cell-free expression system. *Methods Mol. Biol.* **607**:161–171.

LaGuerre, A., Löhr, F., Bernhard, F., Dötsch, V. (2015) Labeling of membrane proteins by cell-free expression. *Meth. Enzymol.* **563**:367–388.

Lyukmanova, E.N., Shenkarev, Z.O., Khabibullina, N.F., Kopeina, G.S., Shulepko, M.A., Paramonov, A.S., Mineev, K. S., Tikhonov, R.V., Shingarova, L.N., Petrovskaya, L.E., Dolgikh, D.A., Arseniev, A.S., Kirpichnikov, M.P. (2012) Lipid-protein nanodiscs for cell-free production of integral membrane proteins in a soluble and folded state: Comparison with detergent micelles, bicelles and liposomes. *Biochim. Biophys. Acta* **1818**:349–358.

Maeda, S., Schertler, G.F.X. (2013) Production of GPCR and GPCR complexes for structure determination. *Curr. Opin. Struct. Biol.* **23**:381–392.

Maslennikov, I., Klammt, C., Hwang, E., Kefala, G., Okamura, M., Esquivies, L., Mörs, K., Glaubitz, C., Kwiatkowski, W., Jeon, Y.H., Choe, S. (2010) Membrane domain structures of three classes of histidine kinase receptors by cell-free expression and rapid NMR analysis. *Proc. Natl. Acad. Sci. USA* **107**:10902–10907.

Matsubayashi, H., Kuruma, Y., Ueda, T. (2014) *In vitro* synthesis of the *E. coli* Sec translocon from DNA. *Angew. Chem. Int. Ed. Engl.* **53**:7535–7538.

Nirenberg, M.W., Matthaei, J.H. (1961) The dependence of cell-free protein synthesis in *E. coli* upon naturally occurring or synthetic polyribonucleotides. *Proc. Natl. Acad. Sci. USA* **47**:1588–1602.

Niwa, T., Sasaki, Y., Uemura, E., Nakamura, S., Akiyama, M., Ando, M., Sawada, S., Mukai, S.A., Ueda, T., Taguchi, H., Akiyoshi, K. (2015) Comprehensive study of liposome-assisted synthesis of membrane proteins using a reconstituted cell-free translation system. *Sci. Rep.* **5**:18025.

Otzen, D.E., Andersen, K.K. (2013) Folding of outer membrane proteins. *Arch. Biochem. Biophys.* **531**:34–43.

Park, K.-H., Berrier, C., Lebaupain, F., Pucci, B., Popot, J.-L., Ghazi, A., Zito, F. (2007) Fluorinated and hemifluorinated surfactants as alternatives to detergents for membrane protein cell-free synthesis. *Biochem. J.* **403**:183–187.

Park, K.-H., Billon-Denis, E., Dahmane, T., Lebaupain, F., Pucci, B., Breyton, C., Zito, F. (2011) In the cauldron of cell-free synthesis of membrane proteins: Playing with new surfactants. *New Biotech.* **28**:255–261.

Periasamy, A., Shadiac, N., Amalraj, A., Garajová, S., Nagarajan, Y., Waters, S., Mertens, H.D.T., Hrmova, M. (2013) Cell-free protein synthesis of membrane (1,3)-β-D-glucan (curdlan) synthase: co-translational insertion in liposomes and reconstitution in nanodiscs. *Biochim. Biophys. Acta* **1828**:743–757.

Picard, M., Dahmane, T., Garrigos, M., Gauron, C., Giusti, F., le Maire, M., Popot, J.-L., Champeil, P. (2006) Protective and inhibitory effects of various types of amphipols on the $Ca^{2+}$-ATPase from sarcoplasmic reticulum: a comparative study. *Biochemistry* **45**:1861–1869.

Popot, J.L. (2014) Folding membrane proteins *in vitro*: A table and some comments. *Arch. Biochem. Biophys.* **564**:314–326.

Popot, J.-L., Althoff, T., Bagnard, D., Banères, J.-L., Bazzacco, P., Billon-Denis, E., Catoire, L.J., Champeil, P., Charvolin, D., Cocco, M.J., Crémel, G., Dahmane, T., de la Maza, L.M., Ebel, C., Gabel, F., Giusti, F., Gohon, Y., Goormaghtigh, E., Guittet, E., Kleinschmidt, J.H., Kühlbrandt, W., Le Bon, C., Martinez, K.L., Picard, M., Pucci, B., Rappaport, F., Sachs, J.N., Tribet, C., van Heijenoort, C., Wien, F., Zito, F., Zoonens, M. (2011) Amphipols from A to Z. *Annu. Rev. Biophys.* **40**:379–408.

Proverbio, D., Roos, C., Beyermann, M., Orbán, E., Dötsch, V., Bernhard, F. (2013) Functional properties of cell-free expressed human endothelin A and endothelin B receptors in artificial membrane environments. *Biochim. Biophys. Acta* **1828**:2182–2192.

Proverbio, D., Henrich, E., Orbán, E., Dötsch, V., Bernhard, F. (2014) Membrane protein quality control in cell-free expression systems: Tools, strategies and case studies, in: Mus-Veteau, I., ed., *Membrane Proteins Production for Structural Analysis*. Springer, New York, pp. 45–70.

Rajesh, S., Knowles, T.J., Overduin, M. (2011) Production of membrane proteins without cells or detergents. *New Biotechnol.* **28**:250–254.

Reckel, S., Gottstein, G., Stehle, J., Löhr, F., Verhoefen, M.-K., Takeda, M., Silvers, R., Kainosho, M., Glaubitz, C., Wachtveitl, J., Bernhard, F., Schwalbe, H., Güntert, P.G., Dötsch, V. (2011) Solution NMR structure of proteorhodopsin. *Angew. Chem. Int. Ed.* **50**:11942–11946.

Rues, R.-B., Orbán, E., Dötsch, V., Bernhard, F. (2014) Cell-free expression of G protein-coupled receptors: new pipelines for challenging targets. *Biol. Chem.* **395**:1425–1434.

Rues, R.-B., Henrich, E., Boland, C., Caffrey, M., Bernhard, F. (2016) Cell-free production of membrane proteins in *Escherichia coli* lysates for functional and structural studies. *Methods Mol. Biol.* **1432**:1–21.

Sachse, R., Dondapati, S.K., Fenz, S.F., Schmidt, T., Kubick, S. (2014) Membrane protein synthesis in cell-free systems: From bio-mimetic systems to bio-membranes. *FEBS Lett.* **588**:2774–2781.

Shadiac, N., Nagarajan, Y., Waters, S., Hrmova, M. (2013) Close allies in membrane protein research: Cell-free synthesis and nanotechnology. *Mol. Membr. Biol.* **30**:229–245.

Sharma, K.S., Durand, G., Gabel, F., Bazzacco, P., Le Bon, C., Billon-Denis, E., Catoire, L.J., Popot, J.-L., Ebel, C., Pucci, B. (2012) Non-ionic amphiphilic homopolymers: Synthesis, solution properties, and biochemical validation. *Langmuir* **28**:4625–4639.

Shenkarev, Z.O., Lyukmanova, E.N., Butenko, I.O., Petrovskaya, L.E., Paramonov, A.S., Shulepko, M.A., Nekrasova, O.V., Kirpichnikov, M.P., Arseniev, A.S. (2013) Lipid-protein nanodiscs promote *in vitro* folding of transmembrane domains of multi-helical and multimeric membrane proteins. *Biochim. Biophys. Acta* **1828**:776–784.

Shirokov, V.A., Kommer, A., Kolb, V.A., Spirin, A.S. (2007) Continuous-exchange protein-synthesizing systems. *Methods Mol. Biol.* **375**:19–55.

Sonar, S., Patel, N., Fischer, W., Rothschild, K.J. (1993) Cell-free synthesis, functional refolding, and spectroscopic characterization of bacteriorhodopsin, an integral membrane protein. *Biochemistry* **32**:13777–13781.

Spirin, A.S., Baranov, V.I., Ryabova, L., Ovodov, S.Y., Alakhov, Y.B. (1988) A continuous cell-free translation system capable of producing polypeptides in high yield. *Science* **242**:1162–1164.

Tu, D., Blaha, G., Moore, P.B., Steitz, T.A. (2005) Structures of MLS$_B$K antibiotics bound to mutated large ribosomal subunits provide a structural explanation for resistance. *Cell* **121**:257–270.

Uhlemann, E.M., Pierson, H.E., Fillingame, R.H., Dmitriev, O.Y. (2012) Cell-free synthesis of membrane subunits of ATP synthase in phospholipid bicelles: NMR shows subunit fold similar to the protein in the cell membrane. *Prot. Sci.* **21**:279–288.

Van Gelder, P., De Cock, H., Tommassen, J. (1994) Detergent-induced folding of the outer-membrane protein PhoE, a pore protein induced by phosphate limitation. *Eur. J. Biochem.* **226**:783–787.

Wada, T., Shimono, K., Kikukawa, T., Hato, M., Shinya, N., Kim, S.Y., Kimura-Someya, T., Shirouzu, M., Tamogami, J., Miyauchi, S., Jung, K.H., Kamo, N., Yokoyama, S. (2011) Crystal structure of the eukaryotic light-driven proton-pumping rhodopsin, *Acetabularia* rhodopsin II, from marine alga. *J. Mol. Biol.* **411**:986–998.

Wang, X.Q., Corin, K., Baaske, P., Wienken, C.J., Jerabek-Willemsen, M., Duhr, S., Braun, D., Zhang, S.G. (2011) Peptide surfactants for cell-free production of functional G protein-coupled receptors. *Proc. Natl. Acad. Sci. USA* **108**:9049–9054.

Wuu, J.J., Swartz, J.R. (2008) High yield cell-free production of integral membrane proteins without refolding or detergents. *Biochim. Biophys. Acta* **1778**:1237–1250.

Zemella, A., Thoring, L., Hoffmeister, C., Kubick, S. (2015) Cell-free protein synthesis: Pros and cons of prokaryotic and eukaryotic systems. *ChemBioChem* **16**:2420–2431.

Zoonens, M., Popot, J.-L. (2014) Amphipols for each season. *J. Membr. Biol.* **247**:759–796.

Zoonens, M., Zito, F., Martinez, K.L., Popot, J.-L. (2014) Amphipols: a general introduction and some protocols, in: Mus-Veteau, I., ed., *Membrane Proteins Production for Structural Analysis*. Springer, New York, Heidelberg, Dordrecht, London, pp. 173–203.

Zubay, G. (1973) *In vitro* synthesis of protein in microbial systems. *Annu. Rev. Genet.* **7**:267–287.

# Optical Spectroscopy of Membrane Protein/Amphipol Complexes

**8**

**Summary**

*Most optical spectroscopy approaches can be applied to membrane protein/amphipol (MP/APol) complexes, namely UV-visible absorbance spectroscopy, light scattering, circular dichroism, static and time-resolved fluorescence measurements, fluorescence quenching and Förster resonance energy transfer studies, surface plasmon resonance measurements, etc. A notable exception is infrared absorbance studies in the peptide bond absorbance bands. Indeed, most APols comprise amide bonds, in which case absorbance spectroscopy studies of APol-trapped MPs at the peptide bond wavelengths have proven intractable. Resonance Raman studies however are possible, as well as surface-enhanced Raman spectroscopy.*

*A large number of APols carrying fluorescent labels have been developed, opening the way to a vast range of applications, from the study of the composition, organization, and dynamics of MP/APol complexes to topological and conformational studies of MPs and to imaging of the distribution and elimination of APols in cell cultures and live animals.*

## 8.1 Introduction

Optical spectroscopy has been used in most studies of amphipol (APol)-trapped membrane proteins (MPs), usually in the form of UV-visible absorbance spectroscopy, often in circular dichroism (CD) determination of the secondary structure of the trapped protein and/or dynamic light scattering (DLS) estimation of particle sizes. Fluorescence studies have also been numerous. As is the case for A8-35, most APols do not contain conjugated double bonds, making them transparent in the visible and near-UV regions of the spectrum. A8-35 absorbs only in the far UV (Fig. 8.1A), with a peak at ~219 nm, so that, at and above 280 nm, it interferes very little with absorbance measurements. Unlabeled A8-35 is readily detectable in the 220–230-nm region, with an extinction coefficient $\varepsilon_{219} \approx 1.3 \text{ g·L}^{-1}\text{·cm}^{-1}$. At 280 nm, $\varepsilon$ does not exceed $0.03 \text{ g·L}^{-1}\text{·cm}^{-1}$. Since APols are seldom used at APol/protein mass ratios exceeding 5:1 (see Chap. 5) and the extinction coefficient of proteins at 280 nm is typically $\geq 10 \text{ L·g}^{-1}\text{·cm}^{-1}$, the contribution of A8-35 to the absorbance at 280 nm of a preparation of MP/A8-35 complexes is generally <2%. The situation is probably similar for most other APols, except for the styrene-maleic acid (SMA)

© Springer International Publishing AG, part of Springer Nature 2018
J. -L. Popot, *Membrane Proteins in Aqueous Solutions*, Biological and Medical Physics, Biomedical Engineering, https://doi.org/10.1007/978-3-319-73148-3_8

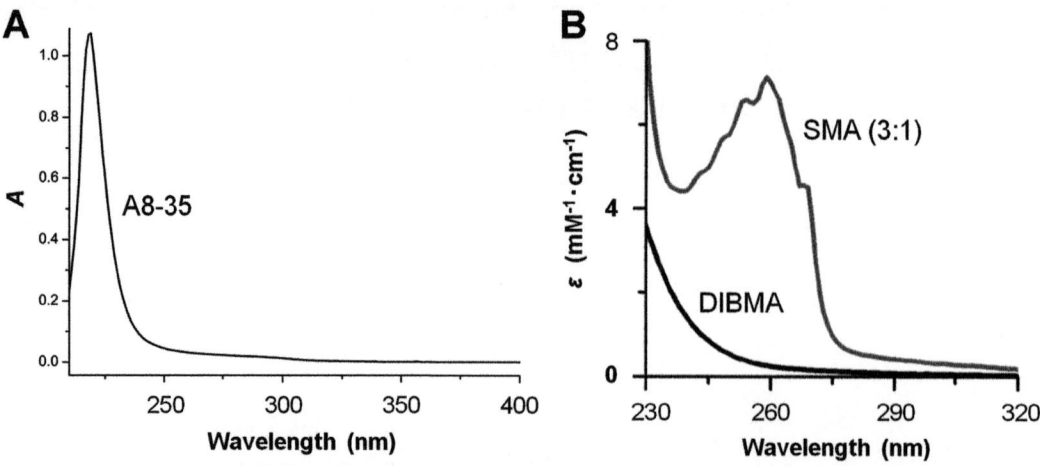

**Fig. 8.1** UV absorbance spectra of various amphipols. (**A**) UV absorbance spectrum of a 0.8-g·L$^{-1}$ solution of A8-35 (From Le Bon et al. 2014b). (**B**) Molar extinction coefficients $\varepsilon$ of SMA(3:1) and DIBMA as a function of wavelength (From Oluwole et al. 2017, © 2017 the authors. Published by Wiley –VCH).

copolymers used to form SMALPs, which, comprising styrene and maleic acid, are rich in aromatic groups and absorb strongly below 250 nm (Scheidelaar et al. 2015) (Table 8.1, Study 8.36; cf. Fig. 8.1B). This problem can be circumvented by replacing SMA with a diisobutylene-maleic acid copolymer (DIBMA) (Oluwole et al. 2017; Table 8.1, Study 8.44; Fig. 8.1B).

As regards CD, APols are generally not optically active, but for the non-ionic sugar-based APols (see Chap. 4, § 4.2.3), for which no CD spectra have been published yet. Peptide-based polymers (see Chap. 4, § 4.2.2.5) may be expected to interfere with CD studies of the secondary structure of the MPs they associate with.

Note that beyond studies exploiting the optical properties of MPs or tagged APols, interesting inroads have been made by Christophe Tribet and coworkers toward developing APols whose hydrophobicity can be modulated by light, thanks to the grafting of light-sensitive side chains. Light-controlled permeabilization of lipid vesicles such polymers had adsorbed to has been demonstrated (Sebai et al. 2010, 2012). The appealing idea of using such APols to "massage" refolding proteins so as to accelerate their renaturation, much like heat-shock proteins do using chemical energy (see e.g. Craig et al. 1993; Richter et al. 2010), has not been equally successful (Martin et al. 2015; Table 8.1, Study 8.34), but deserves to be further explored, e.g. by applying it to MPs.

## 8.2    Optical Spectroscopy Studies of Amphipols and Membrane Protein/Amphipol Complexes

Table 8.1 lists a selection of studies that present optical spectroscopy data on MP/APol complexes (or, when relevant, on pure APols). Excluded from the list are studies that present only static UV-visible absorbance data or light scattering or surface plasmon resonance (SPR) ones. Absorbance spectra of A8-35-trapped MPs appear in many publications. Above ~240 nm, the questions they raise, if any, are related to the physical interactions between the protein and its environment, not to the spectral properties of the APol. Examples of light scattering data have been presented in Chap. 4 (§ 4.3.1.2.2). SPR data, which are collected on MP/APol complexes adsorbed onto solid surfaces, will be treated in Chap. 13, § 13.2.1. We discuss below CD data (§ 8.2.1), IR and Raman data (§ 8.2.2), and fluorescence data (§ 8.2.3).

**Table 8.1**  Studies relevant to the optical spectroscopy of amphipols and amphipol-trapped membrane proteins. Studies in which only static absorbance data are shown are not listed, or studies relying only on light scattering (see Chap. 9) or on surface plasmon resonance measurements (see Chap. 13).

| Study | Protein | Amphipol | CD/ SRCD | Fluorescence | RRS/ SERS | TRAS | Comments | References |
|-------|---------|----------|----------|--------------|-----------|------|----------|------------|
| 8.1 | Nicotinic acetylcholine receptor (nAChR) from *Torpedo marmorata* | A8-35 | | | | | Comparative stopped-flow studies of the binding of a fluorescent agonist to the nAChR in native membranes, in detergent solution, or after trapping by A8-35 | Martinez et al. (2002) |
| 8.2 | Rabbit sarcoplasmic $Ca^{2+}$-ATPase (SERCA1a) | A8-35, SAPol, PMAL-C12, PMALA-C12 | | ● | | | Kinetics of dissociation of $Ca^{2+}$ from the detergent-solubilized or APol-trapped ATPase | Picard et al. (2006) |
| 8.3 | Bacteriorhodopsin (BR) from *Halobacterium salinarum* | A8-35, NAPols | | | | ● | Time-resolved absorbance spectroscopy (TRAS) studies of the photocycle of BR either in its native state (in purple membrane, detergent, or APols) or after refolding in either the presence or absence of lipids, and either free in solution or immobilized onto a solid support | Pocanschi et al. (2006), Gohon et al. (2008), Charvolin et al. (2009), Bazzacco et al. (2012), and Dahmane et al. (2013) |
| 8.4 | OmpA from *Escherichia coli* and FomA from *Fusobacterium nucleatum* | A8-35 | ● | | | | Assessment of refolding following transfer from urea to A8-35 | Pocanschi et al. (2006, 2013) |
| 8.5 | Transmembrane domain (tOmpA) of *E. coli* OmpA | A8-35, FAPol$_{NBD}$ | | ● | | | Förster resonance energy transfer (FRET) studies of the exchange of FAPol$_{NBD}$ for unlabeled A8-35 or detergent at the surface of tOmpA | Zoonens et al. (2007) and Tribet et al. (2009) |
| 8.6 | Transmembrane peptides from subunit *a* of the V-ATPase from *Saccharomyces cerevisiae* | A8-35 | ● | ● | | | CD and intrinsic fluorescence studies of the transfer of transmembrane peptides from trifluoroethanol to A8-35 | Duarte et al. (2008) |
| 8.7 | LTB1, LTB2, CB1, and 5-HT$_{4(a)}$ G protein-coupled receptors (GPCRs) | A8-35 | | ● | | | Fluorescence-based titration assays of ligand binding to GPCRs folded in A8-35 | Dahmane et al. (2009), Catoire et al. (2010), and Mary et al. (2014) |

(continued)

**Table 8.1** (continued)

| Study | Protein | Amphipol | CD/SRCD | Fluorescence | RRS/SERS | TRAS | Comments | References |
|---|---|---|---|---|---|---|---|---|
| 8.8 | PagP expressed in *E. coli* and BR from *Halobacterium salinarum* | SMA | ● | | | | CD studies of the secondary structure and thermal denaturation of PagP and BR in SMALPs, liposomes, or detergent | Knowles et al. (2009) |
| 8.9 | BR | A8-35 | ● | | | | Secondary structure investigations of BR in either detergent solution or trapped with A8-35, either native or refolded in the presence or absence of lipids | Popot et al. (2011) |
| 8.10 | Oleosins and caleosin (hydrophobic, non-membrane proteins) from *Arabidopsis thaliana* seed lipid bodies | A8-35 and other random or blocky PAA-derived APols | ● | | | | Secondary structure studies of oleosins and caleosin in various environments | Gohon et al. (2011) |
| 8.11 | Major outer membrane protein (MOMP) from *Chlamydia trachomatis* | A8-35 | ● | | | | CD studies of the thermostability of MOMP in a detergent vs. an APol environment | Tifrea et al. (2011) and Feinstein et al. (2014) |
| 8.12 | Cultures of COS-7 cells | FAPol$_{NBD}$ | | ● | | | Integration to the plasma membrane and internalization of a transmembrane peptide and its APol carrier | Popot et al. (2011) |
| 8.13 | Ghrelin receptor (GPCR) overexpressed in *E. coli* | NAPols | | ● | | ● | FRET measurements of the binding of ghrelin to the receptor folded in NAPols from SDS-solubilized inclusion bodies | Bazzacco et al. (2012) |
| 8.14 | No protein | A8-35, FAPol$_{NBD}$, FAPol$_{rhod}$ | | ● | | | A study of the critical aggregation concentration (CAC) of A8-35 using FRET between FAPol$_{NBD}$ and FAPol$_{rhod}$ | Giusti et al. (2012) |

Table 8.1 (continued)

| Study | Protein | Amphipol | CD/SRCD | Fluorescence | RRS/SERS | TRAS | Comments | References |
|---|---|---|---|---|---|---|---|---|
| 8.15 | Diphtheria toxin T-domain | A8-35, NAPol | ● | ● | | | CD and fluorescence studies of the conformation of diphtheria toxin T-domain and lipid vesicle permeabilization as a function of pH, temperature, and the environment (detergent, fluorinated surfactant, APols, or lipid vesicles) | Kyrychenko et al. (2012) |
| 8.16 | OmpT and PagP from *E. coli* | A8-35 | ● | | | | Amphipol-assisted folding and mass spectrometry of OmpT and PagP | Leney et al. (2012) |
| 8.17 | Immunoglobulin G (IgG; not a membrane protein) | A8-35 and other PAA-derived polymers | | ● | | | The thermal stability and solubility of IgG in water were monitored by differential scanning calorimetry (DSC), dynamic light scattering (DLS), fluorescence correlation spectroscopy (FCS), and asymmetrical flow field-flow fractionation (AF4) in the presence of various polymers | Ma et al. (2012, 2014) and Martin et al. (2014) |
| 8.18 | BR | SMA | ● | | | | Visible CD study of the association state of BR in the purple membrane, in SMALPs, and in lipid vesicles | Orwick-Rydmark et al. (2012) |
| 8.19 | Arginine-vasopressin Type 2 GPCR (V2R) | NAPol | | ● | | | Tryptophan intrinsic fluorescence spectroscopy and fluorescence quenching and lanthanide resonance energy transfer (LRET) studies of ligand and arrestin binding to V2R | Rahmeh et al. (2012) |

(continued)

**Table 8.1**  (continued)

| Study | Protein | Amphipol | CD/ SRCD | Fluorescence | RRS/ SERS | TRAS | Comments | References |
|---|---|---|---|---|---|---|---|---|
| 8.20 | Stress-related light-harvesting complex (LHCSR) from *Chlamydomonas reinhardtii* | NAPol | | ● | | | 77 K emission spectra and fluorescence decay kinetics of NAPol-trapped wild-type and mutant LHCSR were recorded under different pH conditions to elucidate the mechanism of pH sensing | Liguori et al. (2013) |
| 8.21 | In vivo studies | FAPol$_{rhod}$, FAPol$_{AF647}$ | | ● | | | The biodistribution and elimination of A8-35 were examined in mice, using FAPol$_{rhod}$ and FAPol$_{AF647}$, following intraperitoneal, intravenous, or subcutaneous injection | Fernandez et al. (2014) |
| 8.22 | Several eukaryotic ABC transporters: Pgp, MRP1, MRP4, ABCG2, and CFTR | SMA | ● | ● | | | Fluorescence studies of ligand binding and CD conformational studies | Gulati et al. (2014) |
| 8.23 | BR, tOmpA | A8-35, FAPol$_{NBD}$, FAPol$_{AF647}$, OligAPol | | ● | | | Fluorescence studies of OligAPol-mediated immobilization of BR and tOmpA onto gold beads and their recognition by specific antibodies | Le Bon et al. (2014a) |
| 8.24 | Reviews | Overview | | ● | | | Two general reviews of the properties and uses of functionalized APols, including fluorescent ones | Della Pia et al. (2014b) and Le Bon et al. (2014b) |
| 8.25 | LHCII from *Arabidopsis thaliana* | A8-35, NAPol | ● | ● | | | CD and time-resolved fluorescence studies of APol-trapped LHCII | Opačić et al. (2014a) |

**Table 8.1** (continued)

| Study | Protein | Amphipol | CD/ SRCD | Fluorescence | RRS/ SERS | TRAS | Comments | References |
|-------|---------|----------|----------|--------------|-----------|------|----------|-----------|
| 8.26 | Mannitol permease (EII$^{mtl}$) | A8-35, FAPol$_{fluo}$ | | ● | | | Studies of the quenching of the intrinsic fluorescence of single-tryptophane EII$^{mtl}$ mutants by FAPol$_{fluo}$ | Opačić et al. (2014b) |
| 8.27 | BR | A8-35 | | | ● | | Resonance Raman spectroscopy (RRS) and nanoparticle surface-enhanced Raman scattering (SERS) by A8-35-trapped BR in solution or immobilized onto silver nanoparticles | Polovinkin et al. (2014) |
| 8.28 | Glycophorin A (GpA) transmembrane anchor | p(HPMA)/LMA copolymers | ● | ● | | | CD, fluorescence emission, and FRET studies of p(HPMA)-co-p(LMA) copolymers and their complexes with GpA | Stangl et al. (2014a) |
| 8.29 | GpA transmembrane anchor | A8-35 | ● | ● | | | CD and FRET studies of the dimerization of GpA transmembrane α-helix in A8-35 and A8-35/SDS mixtures | Stangl et al. (2014b) |
| 8.30 | Cytochrome $bc_1$ complex | A8-35, [$^3$H]A8-35, FAPol$_{NBD}$ | | ● | | | Demonstrating the presence of A8-35 in 3D crystals of cytochrome $bc_1$/FAPol$_{NBD}$/ detergent complexes | Charvolin et al. (2014) |
| 8.31 | Human adenosine A$_{2A}$ GPCR (A$_{2A}$R) | SMA | ● | | | | Ligand binding and CD were used to examine the thermal stability of SMALP-integrated A$_{2A}$R | Jamshad et al. (2015a) |
| 8.32 | No protein | SMA | | | | | The structure of the SMALPs is studied by small-angle neutron scattering (SANS), electron microscopy (EM), attenuated total reflection Fourier transform infrared spectroscopy (ATR-FTIR), DSC, and NMR | Jamshad et al. (2015b) |

(continued)

**Table 8.1** (continued)

| Study | Protein | Amphipol | CD/ SRCD | Fluorescence | RRS/ SERS | TRAS | Comments | References |
|-------|---------|----------|----------|--------------|-----------|------|----------|-----------|
| 8.33 | Insulin-responsive facilitative glucose transporter (GLUT4) | A8-35 | | ● | | | Determination of the thermostability and refolding of GLUT4 in A8-35 vs. a detergent environment using fluorescence-quenching by an inhibitor | Kraft et al. (2015) |
| 8.34 | Carbonic anhdrase B (CAB; not a membrane protein) | PAA grafted with either $C_{18}$ or alkylamidoazobenzene groups | ● | | | | Prevention of aggregation during renaturation of urea-denatured CAB. CAB/polymer complexes were characterized by light scattering (LS) and FCS and the folding state of the protein examined by CD | Martin et al. (2015) |
| 8.35 | Multidrug transporter AcrB | SMA | | ● | | | Determination of ligand binding by fluorescence depolarization measurements | Postis et al. (2015) |
| 8.36 | No protein | SMA | | | | | DLS, EM, and size exclusion chromatography (SEC) were used to characterize the size of SMALPs | Scheidelaar et al. (2015) |
| 8.37 | No protein | SMA | | ● | | | DLS and EM were used to study the size of SMALPs, fluorescence techniques to show that the gel to liquid-crystalline phase transition temperature of DMPC is broadened relative to that in liposomes | Tanaka et al. (2015) |
| 8.38 | No protein | SMA | | | | | $^{19}F$ and $^{31}P$ NMR, static light scattering (SLS) and DLS, and isothermal titration calorimetry (ITC) were used to investigate the effects of exposing bilayer-forming phospholipids to SMA | Vargas et al. (2015) |

**Table 8.1** (continued)

| Study | Protein | Amphipol | CD/SRCD | Fluorescence | RRS/SERS | TRAS | Comments | References |
|---|---|---|---|---|---|---|---|---|
| 8.39 | No protein | SMA | | | | | The formation of SMALPs using different weight ratios of POPC/POPG lipids to SMA polymers was characterized via solid-state NMR and DLS | Zhang et al. (2015) |
| 8.40 | Melatonin MT1R and ghrelin GHS-R1a GPCRs | SMA | | ● | | | CHS-R1a was expressed in inclusion bodies, folded in A8-35, and transferred to lipid vesicles. MT1R was expressed in Chinese hamster ovary (CHO) cells, from which membrane fragments were purified. The two receptors were extracted with SMA and ligand binding studied by fluorescence spectroscopy | Logez et al. (2016) |
| 8.41 | Mitochondrial respirasome (I$_1$III$_2$IV$_1$ supercomplex) | A8-35 | | | ● | | Comparative absorbance and RRS spectra of mitochondrial complexes in isolation and integrated into a respiratory supercomplex | Shinzawa-Itoh et al. (2016) |
| 8.42 | Yeast cytochrome c oxidase (Complex IV) | SMA | | | | ● | The binding or unbinding of ligands (O$_2$, CO) to Complex IV was studied by TRAS | Smirnova et al. (2016) |
| 8.43 | No protein | SMA | | ● | | | Time-resolved FRET and SANS were used to determine the kinetics and mechanisms of phospholipid transfer among SMALPs | Cuevas Arenas et al. (2017) |

(continued)

**Table 8.1** (continued)

| Study | Protein | Amphipol | CD/SRCD | Fluorescence | RRS/SERS | TRAS | Comments | References |
|---|---|---|---|---|---|---|---|---|
| 8.44 | Outer membrane phospholipaseA (OmpLA) | DIBMA | | | ● | | Comparative UV absorbance spectra of DIBMA and SMA; comparative Raman spectra of lipids in vesicles, DIBMALPs, and SMALPs | Oluwole et al. (2017) |

*Abbreviations: ABCG2* ATP-binding cassette subfamily G member 2, *BLT1 and BLT2*, respectively, Type 1 and 2 leukotriene GPCRs, *BR* bacteriorhodopsin, *CAC* critical aggregation concentration, *CB1* Type 1 cannabinoid GPCR, *CD* circular dichroism, *CFTR* cystic fibrosis transmembrane conductance regulator, *CHO cells* Chinese hamster ovary cells, *DIBMA* diisobutylene-maleic acid copolymer, *DIBMALPs* IBMA/lipid particles, *DLS* dynamic light scattering, *DSC* differential scanning calorimetry, $EII^{mtl}$ mannitol permease, *EM* electron microscopy, *FAPol* a fluorescently labeled derivative of A8-35, $FAPol_{AF647}$ Alexa Fluor 647-labeled A8-35, $FAPol_{fluo}$ fluorescein-labeled A8-35, $FAPol_{NBD}$ NBD-labeled A8-35, $FAPol_{rhod}$ rhodamine-labeled A8-35, *FCS* fluorescence correlation spectroscopy, *FRET* Förster resonance energy transfer, *GLUT2* insulin-responsive facilitative glucose transporter, *GpA* glycophorin A, *HPMA* N-(2-hydroxypropyl)-methacrylamide, $5\text{-}HT_{4(a)}$ Type 4(a) serotonin GPCR, *IgG* immunoglobulin G, *LHC* light-harvesting complex, *LHCSR* stress-related light-harvesting complex, *LMA* lauryl methacrylate, *LS* light scattering, *MOMP* major outer membrane protein from *Chlamydia trachomatis*, *MRP1* and *MRP4* multidrug-resistance proteins 1 and 4, *nAChR* nicotinic acetylcholine receptor, *NBD* 7-nitrobenz-2-oxa-1,3-diazol-4-yl group, *OmpLA* outer membrane phospholipase A, *PAA* poly(acrylic acid), *PagP* lipid A palmitoyltransferase, *P-gp* P-glycoprotein 1, *RRS* resonance Raman spectroscopy, *SANS* small-angle neutron scattering, *SEC* size exclusion chromatography, *SERCA1a* sarcoplasmic $Ca^{2+}$-ATPase, *SERS* surface-enhanced Raman spectroscopy, *SLS* static light scattering, *SMA* styrene-maleic acid copolymer, *SMALPs* SMA/lipid particles, *SRCD* synchrotron radiation CD, *TRAS* time-resolved absorbance spectroscopy, *V2R* arginine-vasopressin Type 2 GPCR

### 8.2.1   Circular Dichroism and Synchrotron Radiation Circular Dichroism

Circular dichroism and synchrotron radiation circular dichroism (SRCD) have been resorted to in a score of studies (Table 8.1). A very frequent use is to assess the secondary structure of MPs that have been folded or refolded in APols. This is illustrated by the CD spectra of two $\beta$-barrel MPs shown in Fig. 8.2 (from Study 8.4) and the SRCD spectra of an $\alpha$-helical MP shown in Fig. 8.3 (from Study 8.9).

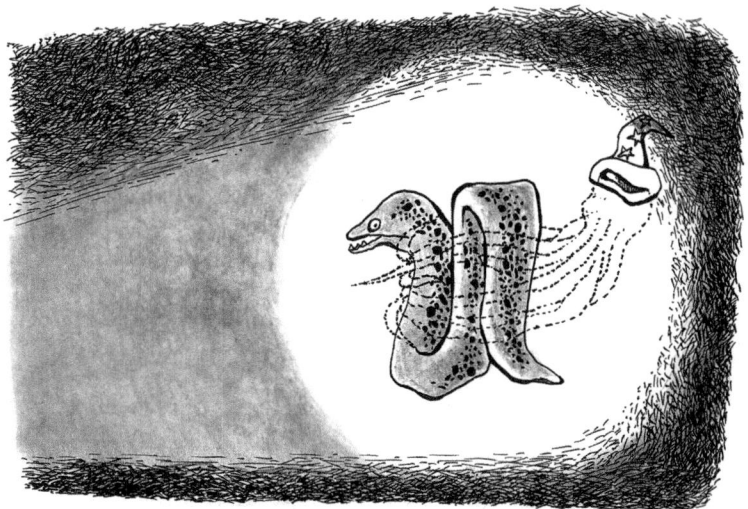

*Optical spectroscopy of membrane protein/amphipol complexes*
(© *2018 by Francis Haraux*)

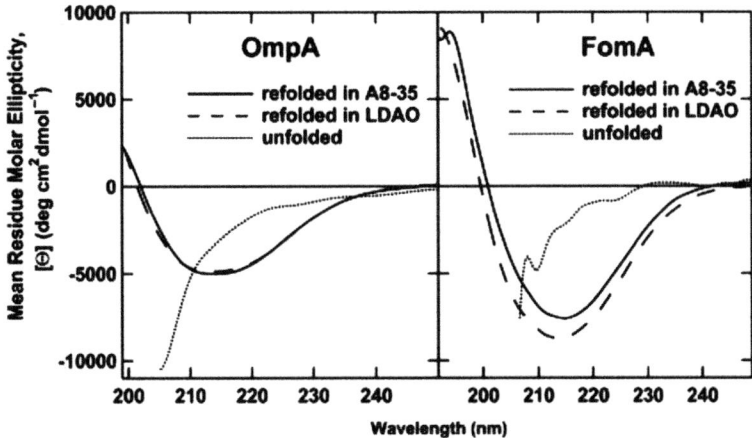

**Fig. 8.2** Circular dichroism spectra of two $\beta$-barrel MPs, OmpA (*left*) and FomA (*right*), either unfolded in urea or after refolding in A8-35. CD spectra of OmpA and FomA folded in lauryldimethylamine oxide (LDAO), which are similar to those of the native proteins, are shown for comparison (Reprinted with permission from Pocanschi et al. 2006, © 2006 American Chemical Society).

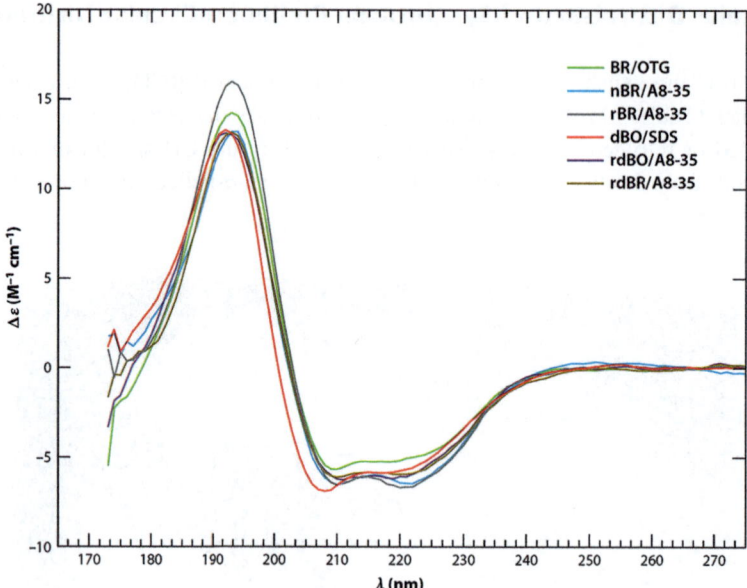

**Fig. 8.3** Synchrotron radiation circular dichroism (SRCD) spectra of SDS-denatured and A8-35-refolded bacteriorhodopsin. The spectra correspond to the following samples: BR/OTG, native BR (nBR) solubilized from purple membrane (PM) in octylthioglucoside (OTG), along with PM lipids; nBR/A8-35, nBR transferred from OTG to A8-35 (1:5 BR/APol mass ratio); rBR/A8-35, BR refolded in A8-35 from SDS-solubilized purple membrane; dBO/SDS, the apoprotein, bacterio-opsin (BO), delipidated and separated from retinal in organic solvents (dBO) and transferred to SDS; rdBO/A8-35, dBO refolded in A8-35 in the absence of retinal and lipids; rdBR/A8-35, dBO refolded in A8-35 in the presence of retinal and absence of lipids. SRCD spectra were recorded at 25 °C on the 3 m-nim Bessy and DISCO SOLEIL beamlines. They extend down to ~175 nm, clearly resolving the 190-nm π-π* transition of the exciton split originating from peptide bond electrons (Data from Dahmane 2007, Popot et al. 2011).

Another classical application of CD is to follow as a function of time and temperature the denaturation of an APol-trapped vs. a detergent-solubilized or a membrane-embedded MP. This is illustrated in Fig. 8.4 in the case of MOMP, the major outer membrane protein from the pathogenic bacterium *Chlamydia trachomatis* (Study 8.11).

Visible CD spectra have been exploited to explore interactions between chromophores in BR/SMALP complexes (Study 8.18; Fig. 8.5) and in APol-trapped light-harvesting complex II (LHCII; Study 8.25). The bilobed CD spectrum of BR in purple membranes is generally attributed to excitonic interactions between retinal molecules in the trimers that, associated with lipids, organize into two-dimensional BR crystals to form purple membrane patches. This feature disappears in detergent-solubilized BR (Heyn et al. 1975) and in SMALP-trapped BR, or BR reconstituted into lipid bilayers (Fig. 8.5), consistent with the monomerization of BR.

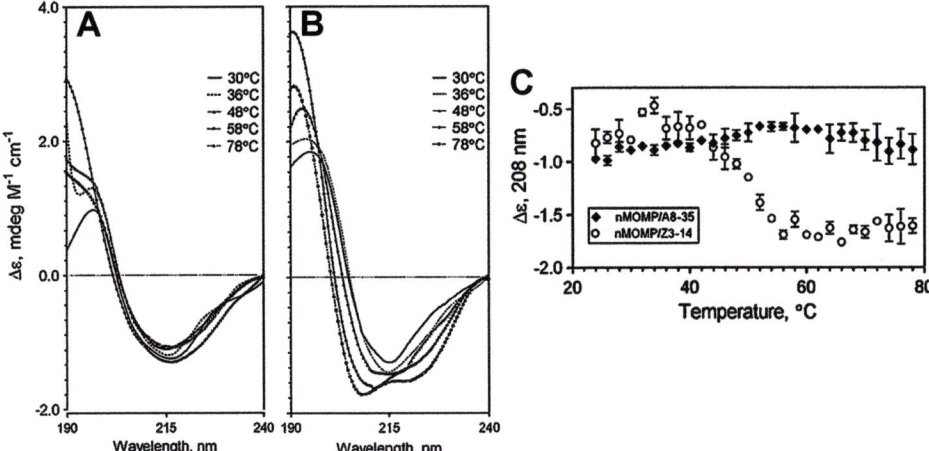

**Fig. 8.4** Circular dichroism spectra, as a function of temperature, of the major outer membrane protein (MOMP) from *Chlamydia trachomatis* either kept in Zwittergent 3–14 or transferred to A8-35. Preparations of MOMP (0.033 g·L$^{-1}$) in 20 mM sodium phosphate, pH 7.4, were incubated at temperatures ranging from 24 to 78 °C and analyzed by CD from 190 to 240 nm. Shown are spectra of (**A**) A8-35-trapped MOMP and (**B**) MOMP in 0.05% Zwittergent 3–14, recorded at 30, 36, 48, 58, and 78 °C. (**C**) A plot of $\Delta\varepsilon_{208}$ vs. temperature for each preparation, showing the remarkable stabilization afforded by the APol (From Tifrea et al. 2011, © 2011 Elsevier Ltd. All rights reserved).

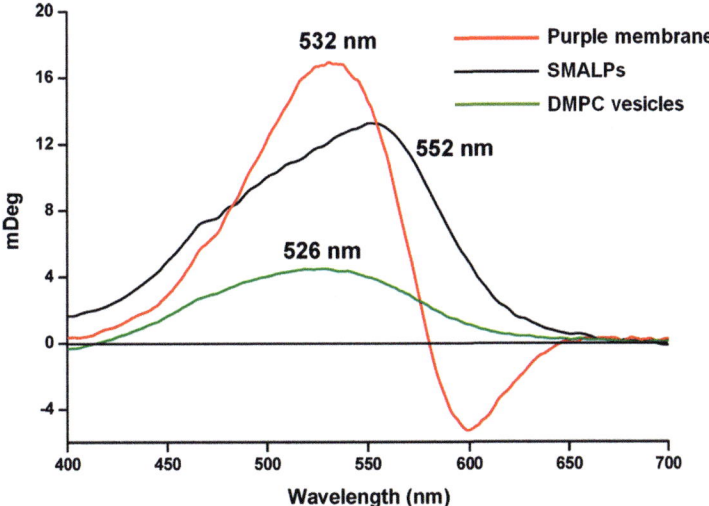

**Fig. 8.5** Visible CD spectra collected for bacteriorhodopsin in the native purple membrane (*red*), in SMALPs (*black*), and after reconstitution in dimyristoylphosphatidylcholine (DMPC) vesicles (*green*). The spectra represent the average of 60 scans and were recorded at 18 °C. The wavelength at the peak of each spectrum is indicated (Reprinted with permission from Orwick-Rydmark et al. 2012, © 2012 American Chemical Society).

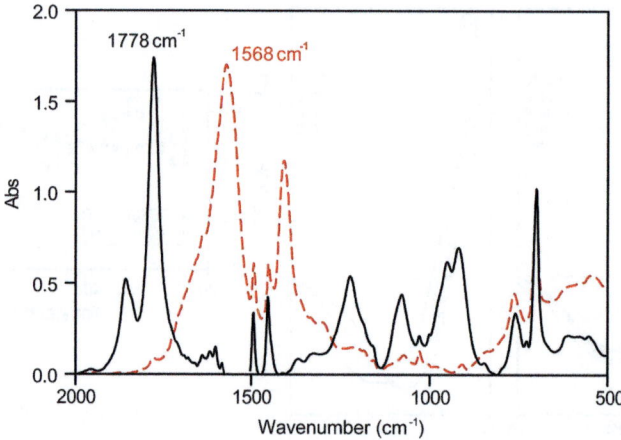

**Fig. 8.6** Fourier transform infrared (FTIR) spectrum of poly(styrene-*co*-maleic anhydride) (SMA) copolymer. The infrared spectra of both the non-hydrolyzed (*black*) and hydrolyzed sodium salt (*red*) forms of SMA are shown. The 1778-cm$^{-1}$ signal in non-hydrolyzed SMA is characteristic of the maleic anhydride carbonyl group. In the hydrolyzed form, which is the one used in membrane biochemistry, this signal is replaced by one at 1568 cm$^{-1}$ characteristic of the carboxylate carbonyl group (Reprinted with permission from Knowles et al. 2009, © 2009 American Chemical Society).

### 8.2.2    Infrared and Raman Spectroscopy

No infrared absorbance spectroscopy data on APol-trapped MPs has been reported yet. The reason is that the most widely used APols, including A8-35, all feature amide bonds (see Chap. 4). Their vibration frequencies superimpose those of the peptide bonds, whose relative frequencies and intensities are the main source of information that IR spectroscopy contributes to exploring the structure of proteins. In principle, it is possible to distinguish between protein and APol amide bonds by labeling one or the other partner with [15]N, which slightly changes some of the vibrational frequencies of the amide bonds. This approach has been attempted by Yann Gohon and Erik Goormaghtigh, using an [15]N-labeled MP trapped by unlabeled A8-35, but it has not succeeded in sufficiently resolving the contributions of the two partners. Indeed, in a BR/A8-35 complex, (i) there is ~2 g A8-35 per g BR (Chap. 5) and (ii) for each kDa of material, a protein contributes ~9.09 and A8-35 ~5.25 amide bonds. The ratio of the two contributions – assuming the complexes to have been totally freed of unbound A8-35 – would therefore be ~0.87:1. Isotopic labeling does not separate efficiently enough the bands contributed by the two partners to permit a clean resolution of the protein's bands. *N*-methylated A8-35 has recently been synthesized (F. Giusti, unpublished results), which could perhaps solve this problem.

The IR absorbance spectrum of poly(styrene-*co*-maleic acid) (SMA) copolymer (in *red* in Fig. 8.6) overlaps the peptide bond amide I bands (in the 1645–1675 cm$^{-1}$ region) but is weaker in the amide II band region (1230–1270 cm$^{-1}$). This suggests that IR data could be collected in the latter region more easily than with classical APols, but none have been reported yet.

**Box 8.1   Infrared Absorbance vs. Raman Scattering Spectroscopies**

Infrared absorbance spectroscopy and Raman scattering spectroscopy provide largely similar information on the vibrational modes present in the molecules studied. The wavelengths corresponding to the relatively small energy differences involved cover principally the 2.5–15-μm range or, expressed in wave numbers (the inverse of wavelengths), 4000–670 cm$^{-1}$. The physical mechanisms underlying the two types of measurements are different. In *IR absorbance spectroscopy*, photons are absorbed when their energy corresponds to the difference in energy between two levels of excitation of a given vibrational mode, for instance, the amide I band of the CONH group. Because the exact frequencies depend on the nature of the secondary structure elements the peptide bond is part of, the IR spectrum yields information about the organization in space of the polypeptide. In current FTIR instruments, illumination from a broadband IR source is modulated according to wavelength by an interferometer, from which the light is directed through the sample and into the IR detector. A computer converts the interferogram (as a function of wavelength) to an IR spectrum (as a function of frequency) via Fourier transformation, which considerably improves the sensitivity and signal to noise ratio compared to the early wavelength-dispersive instruments using wavelength scanning and diffraction gratings.

In *Raman scattering spectroscopy*, photons of higher energy are used, which are not absorbed, but scattered. Most of them are scattered elastically, meaning without any loss or gain of energy. They yield information about the size but not the internal organization of the particle. Some photons, however, exchange with the target molecule that amount of energy that will excite or de-excite a vibrational mode. This amount is the same as that ceded by an IR photon when exciting that mode, but now it manifests itself as a small energy difference, and therefore wavelength difference, between incident and inelastically scattered photons. The vibrational frequencies thus detected are identical to those observed in an IR absorbance spectrum, but, the underlying physical mechanisms being different, the relative intensities of the bands can vary widely between the two kinds of spectra. Indeed, IR absorbance is observed for vibrational modes which change the dipole moment of the molecule, whereas Raman scattering is associated with modes that produce a change in the polarizability of the molecule. Because of this, the Raman activity of a given vibrational mode may differ markedly from its IR activity. For example, a homonuclear molecule such as N$_2$ has no dipole moment either in its equilibrium position or when a stretching vibration causes a change in the distance between the two nuclei. Thus it shows no IR absorbance. On the other hand, the polarizability of the bond between the two atoms varies with the stretching vibrations, and a Raman shift is therefore observed at the frequency of the vibrational mode. Raman spectroscopy is not very sensitive, but it presents certain advantages, among which a relative insensitivity to the presence of water, whose absorbance of IR light is one of the principal difficulties in IR absorbance spectroscopy.

The sensitivity of Raman scattering spectroscopy is greatly increased in *resonance Raman spectroscopy* (RRS), which exploits the enhancement of the intensity of Raman bands due to the vibration of specific bonds when the incident light excites electronic transitions in their immediate vicinity. Thus, excitation at the absorbance wavelength of a chromophore will strongly enhance the signals due to the bonds of the chromophore and those connecting it to the protein. RRS is therefore extremely useful to gather detailed information about the conformation of a chromophore and the nature of the residues, atoms, and bonds involved in liganding it.

*Surface-enhanced Raman spectroscopy* (SERS) exploits the strong enhancement of Raman lines that is observed when the target molecules are adsorbed onto noble metal nanoparticles, particularly if they happen to be located in the cranks between two particles, where surface plasmon resonance effects are strongest. When combined with the enhancement resulting from RRS, this effect can be so strong (up to 10$^{14}$–10$^{15}$× as compared to conventional Raman measurements) that the detection of single molecules becomes possible.

For more detailed information about vibrational spectroscopy, the reader is referred to Cantor and Schimmel (1980) and Zaccai et al. (2017).

Raman spectroscopy is an alternative to IR absorbance spectroscopy. Based on a different physical mechanism (see Box 8.1), it also provides information about vibration frequencies in the target molecules. Raman spectroscopy by itself is a rather insensitive approach, and, as far as MP/APol complexes are concerned, it would face the same difficulty disentangling the contributions of the protein and the polymer. However, these two limitations can be circumvented using a combination of resonance Raman spectroscopy (RRS) and surface-enhanced Raman spectroscopy (SERS). SERS exploits the enhancement observed when the target molecule is in close proximity of noble metal nanoparticles, whereas RRS builds on that resulting from exciting the electronic transitions of a chromophore present in it. When combining the two approaches, signals from the vibrational modes of bonds within the chromophore or between the protein and the chromophore can be amplified up to $10^{14}$–$10^{15} \times$ as compared to conventional Raman measurements, making the detection of single molecules possible (Box 8.1). SERS of membrane-associated BR has been studied using purple membrane patches adsorbed onto silver electrodes (Nabiev et al. 1985, 1990).

The potential usefulness of APols for SERS studies of MPs has been explored in Study 8.27 (Table 8.1), taking BR/A8-35 complexes as a model system. Figure 8.7A, B shows that A8-35-trapped BR stands without denaturation being dried and resuspended, as indicated by its visible absorbance and RRS spectra. Figure 8.7D shows that resonance SERS spectra can be collected on dried mixtures of the complexes and silver nanoparticles. In particular, the band around 1530 cm$^{-1}$ due to C=C retinal stretches, characteristic of BR in its light-adapted and dark-adapted states (Smith et al. 1985), is

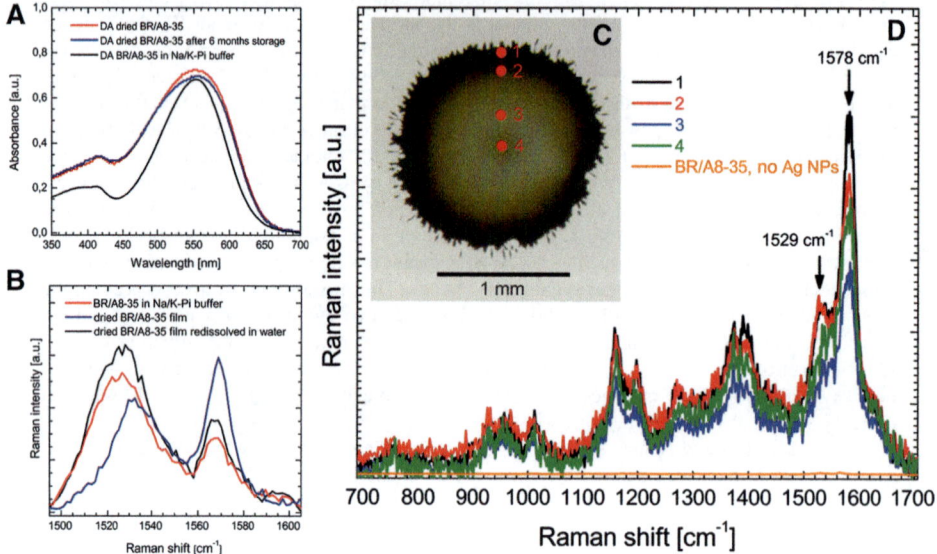

**Fig. 8.7** Vibrational spectroscopy of bacteriorhodopsin/A8-35 complexes using surface-enhanced resonance Raman spectroscopy. (**A**) UV-visible absorbance spectra of dark-adapted BR/A8-35 complexes in 20 mM Na/K phosphate buffer, pH 7.2 (*black*), of a dried BR/A8-35 film (*red*), and of the same film after 6-month storage in the dark at 22 °C at ~50% relative humidity (*blue*). (**B**) Scaled resonance Raman spectra of BR/A8-35 complexes in 20 mM Na/K phosphate buffer, pH 7.2 (*red*), of a dried BR/A8-35 film (*blue*), and of the same film after redissolution in water (*black*). (**C**) A bright-field optical image of a dried mixture of BR/A8-35 complexes and silver nanoparticles (5-μL drop at a BR concentration of 0.08 g·L$^{-1}$). (**D**) Four SERS spectra, marked as 1 (*black*), 2 (*red*), 3 (*blue*), and 4 (*green*), observed from the areas in **C** that are marked by red spots with corresponding numbers (collection time, 30 s; 514.5-nm laser, 10-μW power). The resonance Raman spectrum drawn as an *orange line* was collected from a dried 5-μL drop of A8-35-trapped BR (0.08 g·L$^{-1}$ BR) in the absence of nanoparticles (30 s, 514.5-nm laser, 1-mW power) and normalized to the acquisition parameters of the SERS spectra, neglecting the higher concentration of BR in the beam in the resonance Raman sample. The signal enhancement due to SERS is $>10^2 \times$ (From Polovinkin et al. 2014).

observed for both RRS and SERS spectra. The signal enhancement due to SERS is $>10^2\times$ over that due to RRS only. It probably reflects the contribution of only the very small fraction of complexes that find themselves in appropriate positions relative to the silver nanoparticles for SERS to occur and would be considerably greater if this proportion could be increased (see below, § 8.3).

### 8.2.3   Fluorescence Spectroscopy

With >20 studies published in >30 publications, fluorescence studies represent, along with UV-visible absorbance spectroscopy, the bulk of published experiments involving optical spectroscopy (Table 8.1). Many studies, e.g. Studies 8.1, 8.2, 8.7, and 8.22, rely on fluorescence intensity or depolarization measurements to determine the $K_D$ or kinetics of ligand binding to APol-trapped receptors or transporters (see Chap. 5, § 5.4). Others, like Studies 8.19, 8.20, or 8.29, have resorted to fluorescence lifetime or Förster resonance energy transfer (FRET) measurements to examine conformational changes, oligomerization, and/or aggregation of various APol-trapped MPs. All of these data have been collected in the presence of unlabeled A8-35, non-ionic APols (NAPols), or SMALPs (Table 8.1).

Other experiments have relied on the use of APols carrying fluorescent labels, six of which have been synthesized to date (Table 8.2 and Fig. 8.8).

Fluorescent APols (FAPols) have been used in a wide variety of experiments, among which:

- Examining the interactions of APols with lipid vesicles, using a naphthalene-labeled version of A8-75 (Vial et al. 2005, 2007, 2009).
- Studying the stability of MP/APol complexes and the exchange of NBD-labeled A8-35 (FAPol$_{NBD}$) for detergent or unlabeled APol at the transmembrane (TM) surface of a MP (Table 8.1, Study 8.5; see Chap. 5, § 5.7).
- Determining the critical aggregation concentration (CAC) of A8-35 as that concentration above which the formation of particles allows FRET to occur between FAPol$_{NBD}$ and a rhodamine-labeled version of A8-35 (FAPol$_{rhod}$; Study 8.14; see Chap. 4, § 4.3.1.1).
- Using FAPol$_{NBD}$ and a version of A8-35 labeled with Alexa Fluor 647 (FAPol$_{AF647}$) as tracers in studies of the immobilization of MPs via a version of A8-35 carrying an oligonu-cleotide tag (OligAPol). Indeed, various versions of A8-35 can be freely mixed, and, by choosing an appropriate ratio of OligAPol and a FAPol, MP/A8-35 complexes can be simultaneously tagged and labeled, the OligAPol mediating the immobilization while the FAPol reports on it (Study 8.23; see Chap. 13, § 13.2.2).
- Probing the topology of a polytopic MP. In Study 8.26, 5-fluorotryptophan (5-F-Trp) residues were introduced at unique positions in the TM domain of the mannitol permease, EII$^{mtl}$ (*black and red stars* in Fig. 8.9A), in which the four tryptophans present in the wild-type protein, all located in the transmembrane domain (*black circles*), had been replaced by phenylalanines. The transporter was purified and transferred either to unlabeled A8-35 or to fluorescein-labeled A8-35 (FAPol$_{fluo}$). As shown in Fig. 8.9B, C, the extent of quenching of 5-F-Trp fluorescence by FAPol$_{fluo}$ depended on the position of the residue in the sequence. The data suggests that position 188 is, on average, closer to the fluorescein groups carried by FAPol$_{fluo}$ than is position 167, resulting in stronger FRET and quenching. Whereas Study 8.26 was limited to a proof-of-principle demonstration, systematic investigations of this type could provide precious topological information about (i) the organization of the TM region and (ii) transconformations associated with the functional cycle of the protein.

**Table 8.2** Fluorescently tagged amphipols and the applications they have been put to. Abbreviations are defined in the legend to Table 8.1.

| Label | Carrier APol | Short name | Synthesis and properties | $\lambda_{exc}$ (nm) | $\lambda_{em}$ (nm) | Demonstrated applications | Studies |
|---|---|---|---|---|---|---|---|
| Naphthalene | A8-75 | – | Vial et al. (2005) | 290 | 310–370 | Studying A8-75 interactions with lipid vesicles. | Vial et al. (2005, 2007, 2009) |
| NBD | A8-35 | FAPol$_{NBD}$ | Zoonens et al. (2007) | 470 | 530 | FRET studies of the exchange of FAPol$_{NBD}$ for unlabeled A8-35 or detergent at the surface of a MP. Integration to the plasma membrane of COS-7 cells and internalization of a transmembrane peptide and its APol carrier. Determination of the CAC of A8-35 using FRET between FAPol$_{NBD}$ and FAPol$_{rhod}$. Demonstrating the presence of A8-35 in 3D crystals of cytochrome $bc_1$/FAPol$_{NBD}$/detergent complexes. Tracer in immobilization studies of OligAPol-trapped MPs. | Zoonens et al. (2007), Tribet et al. (2009), Popot et al. (2011), Giusti et al. (2012), Charvolin et al. (2014), and Le Bon et al. (2014a) |
| Fluorescein | A8-35 | FAPol$_{fluo}$ | Opačić et al. (2014) | 495 | 540 | Quenching of the intrinsic fluorescence of a MP. | Opačić et al. (2014) |
| Rhodamine | A8-35 | FAPol$_{rhod}$ | Giusti et al. (2012) | 555 | 575 | Study of the CAC of A8-35 using FRET between FAPol$_{NBD}$ and FAPol$_{rhod}$. Study of in vivo distribution and elimination of A8-35. Tracer in immobilization studies of MPs trapped with tagged APols. | Giusti et al. (2012) and Fernandez et al. (2014) |
| Atto 647 | A8-35 | FAPol$_{A647}$ | Unpublished results | 651 | 667 | Study of in vivo distribution and elimination of A8-35. | Unpublished results |
| Alexa Fluor 647 | A8-35 | FAPol$_{AF647}$ | Fernandez et al. (2014) | 651 | 668 | Study of in vivo distribution and elimination of A8-35. Tracer in immobilization studies of MPs trapped with tagged APols. | Della Pia et al. (2014a, b), Fernandez et al. (2014), and Le Bon et al. (2014a) |

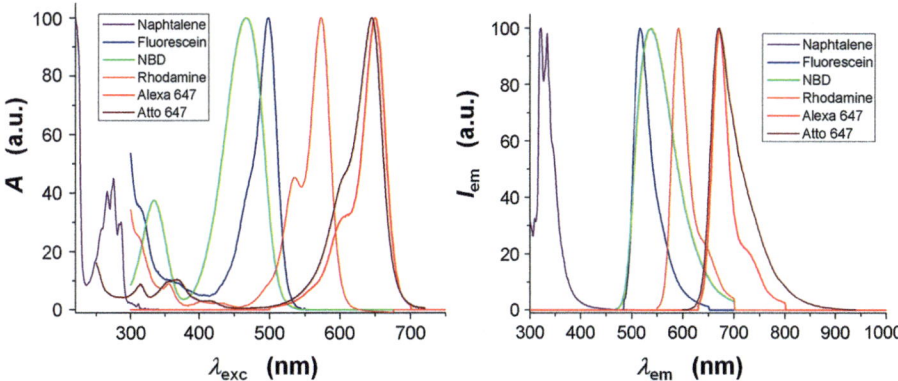

**Fig. 8.8** Normalized excitation (*left*) and emission (*right*) spectra of fluorescent dyes that have been covalently bound to either A8-75 (naphthalene) or A8-35 (all other dyes). The maxima of absorbance and emission are tabulated in Table 8.2, as well as references to papers describing the synthesis, characterization, and validated applications of each fluorescent APol (From Della Pia et al. 2014b).

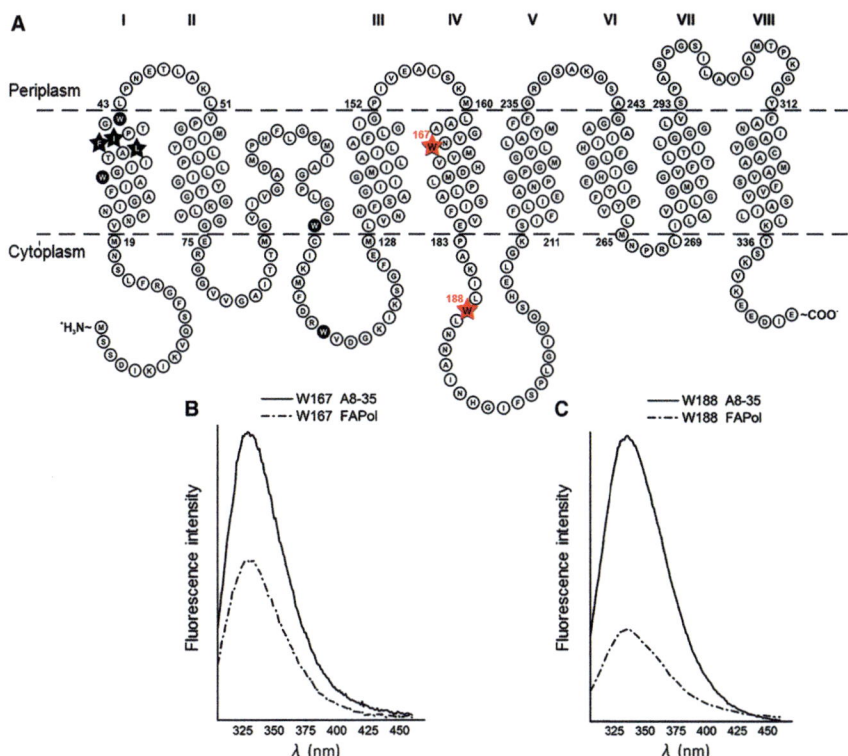

**Fig. 8.9** (**A**) Topology of the transmembrane region of the mannitol transporter EII$^{mtl}$ according to Vervoort et al. (2005). The four Trp positions in the wild-type protein are highlighted. Trp positions in the single-tryptophan EII$^{mtl}$ mutants used in Study 8.26 (W36, W37, W38, W167, and W188) are marked by a star. (**B, C**) Emission spectra of EII$^{mtl}$ mutants carrying a single 5-fluorotryptophan residue each at either position 167 (**B**) or W188 (**C**) (*red stars* in **A**) following trapping either with unlabeled A8-35 (*solid lines*) or with FAPol$_{fluo}$ (*dashed-dotted lines*) (From Opačić et al. 2014b).

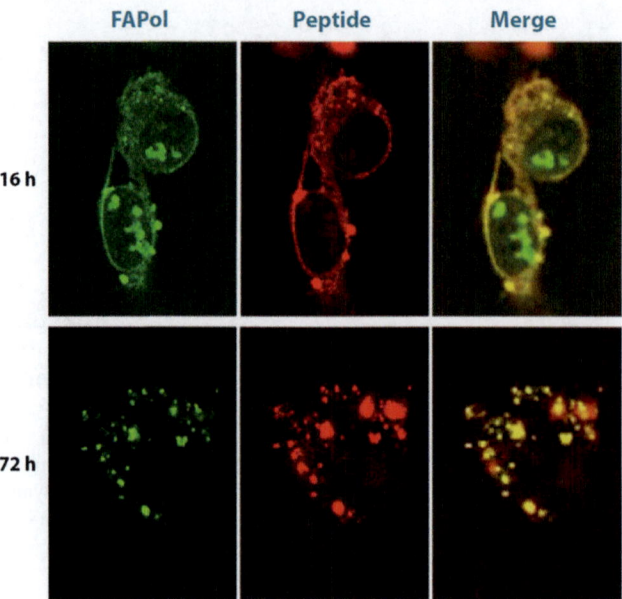

**Fig. 8.10** Using amphipols to deliver a transmembrane peptide to cells in culture. Confocal microscopic analysis of the distribution of NBD-labeled A8-35 (FAPol$_{NBD}$; *green*) and a rhodamine-labeled neuropilin-1 TM peptide (*red*) 16 and 72 h after application of peptide/FAPol$_{NBD}$ complexes to COS-7 cells. Note that the distribution of the two labels does not totally overlap, indicating at least partial segregation. After 72 h, the plasma membrane has been cleared of both labels (From Popot et al. 2011).

- Demonstrating the presence of the APol in three-dimensional crystals of cytochrome $bc_1$/A8-35/detergent ternary complexes (Study 8.30; see Chap. 11, Fig. 11.14).
- Examining in parallel the membrane integration and internalization of a TM peptide labeled with rhodamine and that of the FAPol$_{NBD}$ used to deliver it (Study 8.12; Fig. 8.10). Note the partial segregation of the two labels, with the internalization of some of the APol preceding that of the peptide. Because the preparation applied to the cells contained both MP/FAPol complexes and peptide-free FAPol particles, this segregation in itself constitutes no proof that the APol dissociates from the peptide upon interaction with the membrane, although this appears extremely likely. For a more extended discussion, see Chap. 15, § 15.3.

In Study 8.21, the long-wavelength absorbance and emission bands of FAPol$_{rhod}$ and FAPol$_{AF647}$ were exploited to examine in nude mice, as a function of the route of injection, the distribution and elimination of A8-35 in vivo (see Chap. 15, § 15.4).

## 8.3    Conclusions and Prospects

With the notable exception of infrared absorbance spectroscopy, the most common optical spectroscopy techniques have by now been validated for application to MP/APol complexes. Some of the recently published experiments suggest that further developments of the chemistry of APols could provide very interesting opportunities. I will take two examples.

Resonance SERS offers promising prospects for single-molecule spectroscopy (Study 8.27). The approach could probably benefit from the use of a recently developed sulfide-bearing version of A8-35

(F. Giusti, unpublished data), which ought to considerably increase the fraction of MP/APol complexes that adsorb onto silver or gold beads (cf. Henglein and Meisel 1998) and thereby find themselves in an appropriate position for SERS to occur.

Using labeled APols for topological or conformational studies of MPs is only in its infancy. Study 8.26 has opened the way. Trapping a MP that carries tryptophan – or, better, 5-fluorotryptophan (5-F-Trp), whose fluorescence decay is monoexponential – residues at single positions in its sequence using APols that can quench their fluorescence ought to permit mapping the relative distance of each position to the APol layer associated with the TM surface. The approach could be refined in many different ways. Trapping a given set of mutants with a series of APols that can quench the fluorescence from variable distances would be a way to improve distance estimates. This could be achieved by comparing the level of quenching achieved by fluorescent APols whose absorbance bands overlap more or less with the emission band of Trp or 5-F-Trp, thus modulating $R_0$, the distance at which 50% energy transfer takes place. A similar use could be made of a recently developed spin-labeled APol (F. Giusti, unpublished data) or, possibly, of APols carrying bromine atoms in either their hydrophobic or their polar head regions.

The same approach could conceivably be used to identify conformational transitions during the functional cycle of MPs and study their kinetics.

# References

Bazzacco, P., Billon-Denis, E., Sharma, K.S., Catoire, L.J., Mary, S., Le Bon, C., Point, E., Banères, J.-L., Durand, G., Zito, F., Pucci, B., Popot, J.-L. (2012) Non-ionic homopolymeric amphipols: Application to membrane protein folding, cell-free synthesis, and solution NMR. *Biochemistry* **51**:1416–1430.

Cantor, C.R., Schimmel, P.R. (1980) *Biophysical Chemistry. Part II: Techniques for the study of biological structure and function.* W.H. Freeman and company, San Francisco, 846 p.

Catoire, L.J., Damian, M., Giusti, F., Martin, A., van Heijenoort, C., Popot, J.-L., Guittet, E., Banères, J.-L. (2010) Structure of a GPCR ligand in its receptor-bound state: leukotriene B$_4$ adopts a highly constrained conformation when associated to human BLT2. *J. Am. Chem. Soc.* **132**:9049–9057.

Charvolin, D., Perez, J.-B., Rouvière, F., Giusti, F., Bazzacco, P., Abdine, A., Rappaport, F., Martinez, K.L., Popot, J.-L. (2009) The use of amphipols as universal molecular adapters to immobilize membrane proteins onto solid supports. *Proc. Natl. Acad. Sci. USA* **106**:405–410.

Charvolin, D., Picard, M., Huang, L.-S., Berry, E.A., Popot, J.-L. (2014) Solution behavior and crystallization of cytochrome $bc_1$ in the presence of amphipols. *J. Membr. Biol.* **247**:981–996.

Craig, E.A., Gambill, B.D., Nelson, R.J. (1993) Heat shock proteins: molecular chaperones of protein biogenesis. *Microbiol. Rev.* **57**:402–414.

Cuevas Arenas, R., Danielczak, B., Martel, A., Porcar, L., Breyton, C., Ebel, C., Keller, S. (2017) Fast collisional lipid transfer among polymer-bounded nanodiscs. *Sci. Rep.* **7**:45875.

Dahmane, T. (2007) Protéines membranaires et amphipols : stabilisation, fonction, renaturation, et développement d'amphipols sulfonatés pour la RMN des solutions. Thèse de Doctorat, Université Paris-7, Paris, 229 p.

Dahmane, T., Damian, M., Mary, S., Popot, J.-L., Banères, J.-L. (2009) Amphipol-assisted *in vitro* folding of G protein-coupled receptors. *Biochemistry* **48**:6516–6521.

Dahmane, T., Rappaport, F., Popot, J.-L. (2013) Amphipol-assisted folding of bacteriorhodopsin in the presence and absence of lipids. Functional consequences. *Eur. Biophys. J.* **42**:85–101.

Della Pia, E.A., Holm, J., Lloret, N., Le Bon, C., Popot, J.-L., Zoonens, M., Nygård, J., Martinez, K.L. (2014a) A step closer to membrane protein multiplexed nano-arrays using biotin-doped polypyrrole. *ACS Nano* **8**:1844–1853.

Della Pia, E.A., Westh Hansen, R., Zoonens, M., Martinez, K.L. (2014b) Functionalized amphipols: A versatile toolbox suitable for applications of membrane proteins in synthetic biology. *J. Membr. Biol.* **247**:815–826.

Duarte, A.M.S., Wolfs, C.J.A.M., Koehorsta, R.B.M., Popot, J.-L., Hemminga, M.A. (2008) Solubilization of V-ATPase transmembrane peptides by amphipol A8-35. *J. Peptide Chem.* **14**:389–393.

Feinstein, H.E., Tifrea, D., Sun, G., Popot, J.-L., de la Maza, L.M., Cocco, M.J. (2014) Long-term stability of a vaccine formulated with the amphipol-trapped major outer membrane protein from *Chlamydia trachomatis*. *J. Membr. Biol.* **247**:1053–1065.

Fernandez, A., Le Bon, C., Baumlin, N., Giusti, F., Crémel, G., Popot, J.-L., Bagnard, D. (2014) *In vivo* characterization of the biodistribution profile of amphipols *J. Membr. Biol.* **247**:1043–1051.

Giusti, F., Popot, J.-L., Tribet, C. (2012) Well-defined critical association concentration and rapid adsorption at the air/water interface of a short amphiphilic polymer, amphipol A8-35: A study by Förster resonance energy transfer and dynamic surface tension measurements. *Langmuir* **28**:10372–10380.

Gohon, Y., Dahmane, T., Ruigrok, R., Schuck, P., Charvolin, D., Rappaport, F., Timmins, P., Engelman, D.M., Tribet, C., Popot, J.-L., Ebel, C. (2008) Bacteriorhodopsin/amphipol complexes: structural and functional properties. *Biophys. J.* **94**:3523–3537.

Gohon, Y., Vindigni, J.-D., Pallier, A., Wien, F., Celia, H., Giuliani, A., Tribet, C., Chardot, T., Briozzo, P. (2011) High water solubility and fold in amphipols of proteins with large hydrophobic regions: Oleosins and caleosin from seed lipid bodies. *Biochim. Biophys. Acta* **1808**:706–716.

Gulati, S., Jamshad, M., Knowles, T.J., Morrison, K.A., Downing, R., Cant, N., Collins, R., Koenderink, J.B., Ford, R. C., Overduin, M., Kerr, I.D., Dafforn, T.R., Rothnie, A.J. (2014) Detergent-free purification of ABC (ATP-binding-cassette) transporters. *Biochem. J.* **461**:269–278.

Henglein, A., Meisel, D. (1998) Spectrophotometric observations of the adsorption of organosulfur compounds on colloidal silver nanoparticles. *J. Phys. Chem.* **102**:8364–8366.

Heyn, M.P., Bauer, P.-J., Dencher, N.A. (1975) A natural CD label to probe the structure of the purple membrane from *Halobacterium halobium* by means of exciton coupling effects. *Biochem. Biophys. Res. Commun.* **67**:897–903.

Jamshad, M., Charlton, J., Lin, Y.-P., Routledge, S.J., Bawa, Z., Knowles, T.J., Overduin, M., Dekker, N., Dafforn, T.R., Bill, R.M., Poyner, D.R., Wheatley, M. (2015a) G protein-coupled receptor solubilization and purification for biophysical analysis and functional studies, in the total absence of detergent. *Biosc. Rep.* **35**:e00188.

Jamshad, M., Grimard, V., Idini, I., Knowles, T.J., Dowle, M.R., Schofield, N., Sridhar, P., Lin, Y., Finka, R., Wheatley, M., Thomas, O.R.T., Palmer, R.E., Overduin, M., Govaerts, C., Ruysschaert, J.-M., Edler, K.J., Dafforn, T.R. (2015b) Structural analysis of a nanoparticle containing a lipid bilayer used for detergent-free extraction of membrane proteins. *Nano Res.* **8**:774–789.

Knowles, T.J., Finka, R., Smith, C., Lin, Y.-P., Dafforn, T., Overduin, M. (2009) Membrane proteins solubilized intact in lipid-containing nanoparticles bounded by styrene maleic acid copolymer. *J. Am. Chem. Soc.* **131**:7484–7485.

Kraft, T.E., Hresko, R.C., Hruz, P.W. (2015) Expression, purification, and functional characterization of the insulin-responsive facilitative glucose transporter GLUT4. *Protein Sci.* **24**:2008–2019.

Kyrychenko, A., Rodnin, M.V., Vargas, M.U., Sharma, S.K., Durand, G., Pucci, B., Popot, J.-L., Ladokhin, A.S. (2012) Folding of diphtheria toxin T-domain in the presence of amphipols and fluorinated surfactants: Toward thermodynamic measurements of membrane protein folding. *Biochim. Biophys. Acta* **1818**:1006–1012.

Le Bon, C., Della Pia, E.A., Giusti, F., Lloret, N., Zoonens, M., Martinez, K.L., Popot, J.-L. (2014a) Synthesis of an oligonucleotide-derivatized amphipol and its use to trap and immobilize membrane proteins. *Nucleic Acids Res.* **42**:e83.

Le Bon, C., Popot, J.-L., Giusti, F. (2014b) Labeling and functionalizing amphipols for biological applications. *J. Membr. Biol.* **247**:797–814.

Leney, A.C., McMorran, L.M., Radford, S.E., Ashcroft, A.E. (2012) Amphipathic polymers enable the study of functional membrane proteins in the gas phase. *Anal. Chem.* **84**:9841–9847.

Liguori, N., Roy, L.M., Opačić, M., Durand, G., Croce, R. (2013) Regulation of light-harvesting in the green alga *Chlamydomonas reinhardtii*: the *C*-terminus of LHCSR is the knob of a dimmer switch. *J. Am. Chem. Soc.* **135**:18339–18342.

Logez, C., Damian, M., Legros, C., Dupré, C., Guéry, M., Mary, S., Wagner, R., M'Kadmi, C., Nosjean, O., Fould, B., Marie, J., Fehrentz, J.A., Martinez, J., Ferry, G., Boutin, J.A., Banères, J.-L. (2016) Detergent-free isolation of functional G protein-coupled receptors into nanometric lipid particles. *Biochemistry* **55**:38–48.

Ma, D., Martin, N., Herbet, A., Boquet, D., Tribet, C., Winnik, F.M. (2012) The thermally induced aggregation of immunoglobulin G in solution is prevented by amphipols. *Chem. Lett.* **41**:1380–1382.

Ma, D., Martin, N., Tribet, C., Winnik, F.M. (2014) Quantitative characterization by asymmetrical flow field-flow fractionation of IgG thermal aggregation with and without polymer protective agents. *Anal. Bioanal. Chem.* **406**:7539–7547.

Martin, N., Ma, D., Herbet, A., Boquet, D., Winnik, F.M., Tribet, C. (2014) Prevention of thermally induced aggregation of IgG antibodies by noncovalent interaction with poly(acrylate) derivatives. *Biomacromolecules* **15**:2952–2962.

Martin, N., Ruchmann, J., Tribet, C. (2015) Prevention of aggregation during refolding of carbonic anhydrase via Coulomb and hydrophobic complexation with octadecyl-modified or azobenzene-modified poly(acrylate) derivatives. *Langmuir* **31**:338–349.

Martinez, K.L., Gohon, Y., Corringer, P.-J., Tribet, C., Mérola, F., Changeux, J.-P., Popot, J.-L. (2002) Allosteric transitions of *Torpedo* acetylcholine receptor in lipids, detergent and amphipols: molecular interactions *vs.* physical constraints. *FEBS Lett.* **528**:251–256.

Mary, S., Damian, M., Rahmeh, R., Marie, J., Mouillac, B., Banères, J.-L. (2014) Amphipols in G protein-coupled receptor pharmacology: What are they good for? *J. Membr. Biol.* **247**:853–860.

Nabiev, I.R., Efremóv, R.G., Chumanov, G.D. (1985) The chromophore-binding site of bacteriorhodopsin. Resonance Raman and surface-enhanced resonance Raman spectroscopy and quantum chemical study. *J. Biosci.* **8**:363–374.

Nabiev, I.R., Chumanov, G.D., Efremóv, R.G. (1990) Surface-enhanced Raman spectroscopy of biomolecules. Part II. Application of short- and long-range components of SERS to the study of the structure and function of membrane proteins. *J. Raman Spectrosc.* **21**:49–53

Oluwole, A.O., Danielczak, B., Meister, A., Babalola, J.O., Vargas, C., Keller, S. (2017) Solubilization of membrane proteins into functional lipid-bilayer nanodiscs using a diisobutylene/maleic acid copolymer. *Angew. Chem. Int. Ed. Engl.* **56**:1919–1924.

Opačić, M., Durand, G., Bosco, M., Polidori, A., Popot, J.-L. (2014a) Amphipols and photosynthetic light-harvesting pigment-protein complexes. *J. Membr. Biol.* **247**:1031–1041.

Opačić, M., Giusti, F., Broos, J., Popot, J.-L. (2014b) Isolation of *Escherichia coli* mannitol permease, EII^mtl, trapped in amphipol A8-35 and fluorescein-labeled A8-35. *J. Membr. Biol.* **247**:1019–1030.

Orwick-Rydmark, M., Lovett, J.E., Graziadei, A., Lindholm, L., Hicks, M.R., Watts, A. (2012) Detergent-free incorporation of a seven-transmembrane receptor protein into nanosized bilayer Lipodisq particles for functional and biophysical studies. *Nano Lett.* **12**:4687–4692.

Picard, M., Dahmane, T., Garrigos, M., Gauron, C., Giusti, F., le Maire, M., Popot, J.-L., Champeil, P. (2006) Protective and inhibitory effects of various types of amphipols on the Ca$^{2+}$-ATPase from sarcoplasmic reticulum: a comparative study. *Biochemistry* **45**:1861–1869.

Pocanschi, C.L., Dahmane, T., Gohon, Y., Rappaport, F., Apell, H.-J., Kleinschmidt, J.H., Popot, J.-L. (2006) Amphipathic polymers: tools to fold integral membrane proteins to their active form. *Biochemistry* **45**:13954–13961.

Pocanschi, C., Popot, J.-L., Kleinschmidt, J.H. (2013) Folding and stability of outer membrane protein A (OmpA) from *Escherichia coli* in an amphipathic polymer, amphipol A8-35. *Eur. Biophys. J.* **42**:103–118.

Polovinkin, V., Balandin, T., Volkov, O., Round, E., Borshchevskiy, V., Utrobin, P., von Stetten, D., Royant, A., Willbold, D., Arzumanyan, A., Popot, J.-L., Gordeliy, V. (2014) Nanoparticle surface-enhanced Raman scattering of bacteriorhodopsin stabilized by amphipol A8-35. *J. Membr. Biol.* **247**:971–980.

Popot, J.-L., Althoff, T., Bagnard, D., Banères, J.-L., Bazzacco, P., Billon-Denis, E., Catoire, L.J., Champeil, P., Charvolin, D., Cocco, M.J., Crémel, G., Dahmane, T., de la Maza, L.M., Ebel, C., Gabel, F., Giusti, F., Gohon, Y., Goormaghtigh, E., Guittet, E., Kleinschmidt, J.H., Kühlbrandt, W., Le Bon, C., Martinez, K.L., Picard, M., Pucci, B., Rappaport, F., Sachs, J.N., Tribet, C., van Heijenoort, C., Wien, F., Zito, F., Zoonens, M. (2011) Amphipols from A to Z. *Annu. Rev. Biophys.* **40**:379–408.

Postis, V., Rawson, S., Mitchell, J.K., Lee, S.C., Parslow, R.A., Dafforn, T.R., Baldwin, S.A., Muench, S.P. (2015) The use of SMALPs as a novel membrane protein scaffold for structure study by negative stain electron microscopy. *Biochim. Biophys. Acta* **1848**:496–501.

Rahmeh, R., Damian, M., Cottet, M., Orcel, H., Mendre, C., Durroux, T., Sharma, K.S., Durand, G., Pucci, B., Trinquet, E., Zwier, J.M., Deupi, X., Bron, P., Banères JL., Mouillac, B., Granier, S. (2012) Structural insights into biased G protein-coupled receptor signaling revealed by fluorescence spectroscopy. *Proc. Natl. Acad. Sci. USA* **109**:6733–6738.

Richter, K., Haslbeck, M., Buchner, J. (2010) The heat shock response: Life on the verge of death. *Molec. Cell* **40**:253–266.

Scheidelaar, S., Koorengevel, M.C., Pardo, J.D., Meeldijk, J.D., Breukink, E., Killian, J.A. (2015) Molecular model for the solubilization of membranes into nanodisks by styrene-maleic acid copolymers. *Biophys. J.* **108**:279–290.

Sebai, S., Cribier, S., Karimi, A., Massotte, D., Tribet, T. (2010) Permeabilisation of lipid membranes and cells by a light-responsive copolymer. *Langmuir* **26**:14135–14141.

Sebai, S., Milioni, D., Walrant, A., Alves, I., Sagan, S., Huin, C., Auvray, L., Massotte, D., Cribier, S., Tribet, C. (2012) Photocontrol of the translocation of molecules, peptides, and quantum dots through cell and lipid membranes doped with azobenzene copolymers. *Angew. Chem. Int. Ed.* **51**:2132–2136.

Shinzawa-Itoh, K., Shimomura, H., Yanagisawa, S., Shimada, S., Takahashi, R., Oosaki, M., Ogura, T., Tsukihara, T. (2016) Purification of active respiratory supercomplex from bovine heart mitochondria enables functional studies. *J. Biol. Chem.* **291**:4178–4184.

Smirnova, I.A., Sjöstrand, D., Li, F., Björck, M., Schäfer, J., Östbye, H., Högbom, M., von Ballmoos, C., Lander, G., Ädelroth, P., Brzezinski, P. (2016) Isolation of yeast Complex IV in native lipid nanodiscs. *Biochim. Biophys. Acta* **1858**:2984–2992.

Smith, S.O., Lugtenburg, J., Mathies, R.A. (1985) Determination of retinal chromophore structure in bacteriorhodopsin with resonance Raman spectroscopy. *J. Membr. Biol.* **85**:95–109.

Stangl, M., Hemmelmann, M., Allmeroth, M., Zentel, R., Schneider, D. (2014a) A minimal hydrophobicity is needed to employ amphiphilic *p*(HPMA)-*co*-*p*(LMA) random copolymers in membrane research. *Biochemistry* **53**:1410–1419.

Stangl, M., Unger, S., Keller, S., Schneider, D. (2014b) Sequence-specific dimerization of a transmembrane helix in amphipol A8-35. *PLOS One* **9**:e110970.

Tanaka, M., Hosotani, A., Tachibana, Y., Nakano, M., Iwasaki, K., Kawakami, T., Mukai, T. (2015) Preparation and characterization of reconstituted lipid-synthetic polymer discoidal particles. *Langmuir* **31**:12719–12626.

Tifrea, D.F., Sun, G., Pal, S., Zardeneta, G., Cocco, M.J., Popot, J.-L., de la Maza, L.M. (2011) Amphipols stabilize the *Chlamydia* major outer membrane protein and enhance its protective ability as a vaccine. *Vaccine* **29**:4623–4631.

Tribet, C., Diab, C., Dahmane, T., Zoonens, M., Popot, J.-L., Winnik, F.M. (2009) Thermodynamic characterization of the exchange of detergents and amphipols at the surfaces of integral membrane proteins. *Langmuir* **25**:12623–12634.

Vargas, C., Cuevas Arenas, R., Frotscher, E., Keller, S. (2015) Nanoparticle self-assembly in mixtures of phospholipids with styrene/maleic acid copolymers or fluorinated surfactants. *Nanoscale* **7**:20685–20696.

Vervoort, E.B., Bultema, J.B., Schuurman-Wolters, G.K., Geertsma, E.R., Broos, J., Poolman, B. (2005) The first cytoplasmic loop of the mannitol permease from *Escherichia coli* is accessible for sulfhydryl reagents from the periplasmic side of the membrane. *J. Mol. Biol.* **346**:733–743.

Vial, F., Rabhi, S., Tribet, C. (2005) Association of octyl-modified poly(acrylic acid) onto unilamellar vesicles of lipids and kinetics of vesicle disruption. *Langmuir* **21**:853–862.

Vial, F., Oukhaled, A.G., Auvray, L., Tribet, C. (2007) Long-living channels of well-defined radius opened in lipid bilayers by polydisperse, hydrophobically-modified polyacrylic acids. *Soft Matter* **3**:75–78.

Vial, F., Cousin, F., Bouteiller, L., Tribet, C. (2009) Rate of permeabilization of giant vesicles by amphiphilic polyacrylates compared to the adsorption of these polymers onto large vesicles and tethered lipid bilayers. *Langmuir* **25**:7506–7513.

Zaccai, N.R., Serdyuk, I.N., Zaccai, G. (2017) *Methods in Molecular Biophysics. Structure, dynamics, function,* 2nd ed. Cambridge University Press, Cambridge, 684 p.

Zhang, R., Sahu, I.D., Liu, L., Osatuke, A., Comer, R.G., Dabney-Smith, C., Lorigan, G.A. (2015) Characterizing the structure of lipodisq nanoparticles for membrane protein spectroscopic studies. *Biochim. Biophys. Acta* **1848**:329–333.

Zoonens, M., Giusti, F., Zito, F., Popot, J.-L. (2007) Dynamics of membrane protein/amphipol association studied by Förster resonance energy transfer. Implications for *in vitro* studies of amphipol-stabilized membrane proteins. *Biochemistry* **46**:10392–10404.

# Solution Studies of Membrane Protein/Amphipol Complexes

**9**

**Summary**

*Prior to using membrane protein/amphipol (MP/APol) complexes for structural or functional studies, their solution behavior must be characterized. Their mass, size, shape, composition, internal organization, and interactions can be examined by radiation scattering, ultracentrifugation, size exclusion chromatography, Blue-Native gel electrophoresis, and other biochemical and biophysical techniques. The implementation of most of these approaches is rather straightforward, as experimental protocols do not diverge from those used for soluble proteins. Attention should however be paid to at least two points:*

(i) *A minimum concentration of free APol is usually necessary to keeping preparations monodispersed, because total removal of free APol tends to induce a greater or lesser degree of aggregation.*

(ii) *With fragile MP complexes, a fine line may have to be walked between conditions, such as an excess of free APol, that may destabilize complexes and conditions, such as a deficit of it, that are conducive to aggregation.*

*Small-angle neutron scattering, a particularly powerful approach, is facilitated, as compared to studies in detergent solution, by the higher stability of APol-trapped MPs. It also benefits from the availability of deuterated APols, whose contribution to scattering can be abolished by matching their scattering length density to that of the solution while increasing the contrast of the protein with the solvent and reducing the background noise. However, it is mandatory, in order to collect exploitable data, to carry out a careful examination of the dispersity of the samples, a task to which analytical ultracentrifugation is particularly well suited.*

© Springer International Publishing AG, part of Springer Nature 2018          405
J. -L. Popot, *Membrane Proteins in Aqueous Solutions*, Biological and Medical Physics,
Biomedical Engineering, https://doi.org/10.1007/978-3-319-73148-3_9

## 9.1    Introduction

At the onset of any structural or functional study of an amphipol (APol)-trapped membrane protein (MP), it is advisable to check on the size and homogeneity of the complexes. It is, indeed, very useful and often essential to know whether one is working with monodisperse, homogeneous monomers or oligomers, with a mixture thereof, or with partially aggregated preparations. The interpretation of radiation scattering data, for instance, critically depends on the association state of the protein. Many types of solution studies can give access to the mass, shape, composition, homogeneity, and interparticle interactions of the complexes, and some of them can be used to gather information about their internal organization, such as the arrangement of subunits with respect to one another or to the transmembrane (TM) region, the disposition of the protein, lipid and APol components relative to each other, etc.

Solution studies on MP/APol complexes are generally straightforward. As shown in Fig. 9.1 and Table 9.1, some 60 studies have been published at the date of this writing, reported in close to 80 publications. Size exclusion chromatography (SEC) is by far the most widely used technique (~45 instances), followed by ultracentrifugation, either in the form of sucrose gradient sedimentation velocity (SGSV) measurements or analytical ultracentrifugation (AUC) (~16). Other approaches that have been implemented include small-angle neutron scattering (SANS), small-angle X-ray scattering (SAXS), and dynamic light scattering (DLS), Blue-Native polyacrylamide gel electrophoresis (BN-PAGE), chemical cross-linking, determination by solution NMR of the rotational correlation coefficient ($\tau_c$) of the protein, and asymmetric flow field-flow fractionation-multi-angle laser light scattering (AF4-MALLS). Not considered in the present chapter are solution NMR studies of the structure of the protein or of protein-bound ligands (see Chap. 10) nor studies by optical spectroscopy (Chap. 8). Neither are those approaches, like crystallography (Chap. 11), electron microscopy (Chap. 12), or mass spectrometry (Chap. 14), where the protein is no more in aqueous solution at the time when the data are collected.

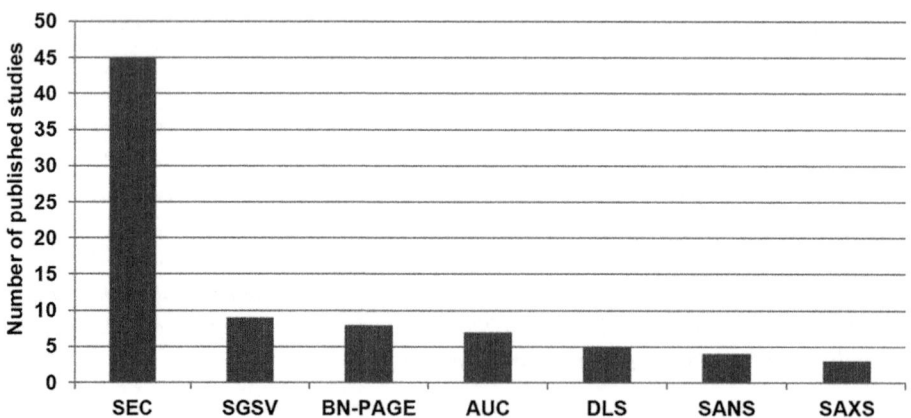

**Fig. 9.1** Principal techniques applied to studying the mass, size, shape, organization, and interactions of membrane protein/amphipol complexes in aqueous solutions. The graph indicates the number of studies listed in Table 9.1, each of which may comprise one or several publications. *Abbreviations*: *AUC* analytical ultracentrifugation, *BN-PAGE* Blue-Native polyacrylamide gel electrophoresis, *DLS* dynamic light scattering, *SANS* small-angle neutron scattering, *SAXS* small-angle X-ray scattering, *SEC* size exclusion chromatography, *SGSV* sucrose gradient sedimentation velocity measurements.

**Table 9.1** An overview of solution studies of the composition, mass, shape, and organization of membrane protein/amphipol complexes. Studies are listed in the chronological order of the first publication in a given study. The various approaches used to characterize the complexes are indicated. Methods are listed only if they bear directly on the solution state of the complexes. Solution NMR studies are discussed in Chap. 10 and are not listed here unless they also present other solution data. A few studies showing only SEC data have been omitted.

| Study | Membrane proteins (MW) | Amphipol | SANS | SAXS | SGSV | AUC | SEC | DLS | BN-PAGE | Comments | References |
|---|---|---|---|---|---|---|---|---|---|---|---|
| 9.1 | Bacteriorhodopsin (BR; ~27 kDa), cytochrome $b_6f$ superdimer (~211 kDa), OmpF trimer (~111 kDa), reaction center R-26 monomer (~100 kDa) | A8-35, A8-75, and [$^3$H]A8-75 | | | • | | | | | Formation, composition, and stability of MP/A8-35 and MP/A8-75 complexes in the presence or absence of free APol. APol binding and exchange. STEM mass determination. | Tribet et al. (1996, 1997, 1998) |
| 9.2 | Sarcoplasmic reticulum $Ca^{2+}$-ATPase (~110 kDa) | A8-35 | | | | • | • | | | Composition, size, stability, and activity of the $Ca^{2+}$-ATPase in the presence of detergent, APol, or detergent/APol mixtures. Effects of removing free APol and of long-term storage. | Champeil et al. (2000a, b) |
| 9.3 | Diacylglycerol kinase (DAGK) trimer (~39 kDa) | OAPA-20, PMAL-B | | | | | | | | Comparison of the oligomeric state of DAGK in lipids, detergents, and APols. Effect of removing free APol. | Nagy et al. (2001) |
| 9.4 | BR (~27 kDa) and cytochrome $b_6f$ superdimer (~211 kDa) | Non-ionic *Tris*-derived APols | | | • | | | | | Testing the ability of a variety of non-ionic *Tris*-derived APols to keep soluble and native BR and cytochrome $b_6f$. | Prata et al. (2001) |
| 9.5 | BR (~27 kDa) | A8-35, [$^3$H]A8-35, and partially deuterated A8-35 (DAPol) | • | | • | • | • | • | | Composition, size, shape, organization, and functionality of BR/A8-35 complexes. Effect of removing free APol and of long-term storage. APol binding. | Gohon (2001), Popot et al. (2003) and Gohon et al. (2008) |
| 9.6 | Monomeric and dimeric forms of the nicotinic acetylcholine receptor (nAChR) (~285 and ~570 kDa) | A8-35 and [$^{14}$C]A8-35 | | | | • | | | | Sedimentation coefficient and functionality of CHAPS-solubilized *vs.* A8-35-trapped nAChR. APol binding. | Martinez et al. (2002) |
| 9.7 | Cytochrome $bc_1$ superdimer (~490 kDa) | A8-35 and [$^3$H]A8-35 | • | • | • | | • | • | | Size and dispersity of $bc_1$/A8-35 complexes as a function of ionic strength, PEG concentration, presence or absence of free APol, and storage. APol binding. | Popot et al. (2003) and Charvolin et al. (2014) |
| 9.8 | Cyanobacterial photosynthetic photosystem II reaction center core complex (550 kDa) | A8-35 | | | | | • | | | Dispersity, stability, and activity of A8-35-trapped *vs.* DDM-solubilized Photosystem II. | Nowaczyk et al. (2004) |
| 9.9 | Transmembrane domain of OmpA (tOmpA) (~19 kDa) | A8-35, DAPol, and NBD-labeled A8-35 (FAPol$_{NBD}$) | | | | | • | | | Composition, size, stability, rotational correlation time, effect of free APol removal. APol binding and kinetics of exchange between protein-bound and free APol. | Zoonens (2004) and Zoonens et al. (2005, 2007) |
| 9.10 | Sarcoplasmic reticulum $Ca^{2+}$-ATPase (~110 kDa) | A8-35, sulfonated APol (SAPol), PMAL-C12, PMAL$A$-C12 | | | | | • | | | Comparative study of the dispersity, stability, and activity of the $Ca^{2+}$-ATPase in the presence of various APols. Effects of $Ca^{2+}$, $Mg^{2+}$ and EDTA. | Picard et al. (2006) |
| 9.11 | BR (~27 kDa) | A8-35 | | | | | • | | | Dispersity of BR refolded by transfer from SDS to A8-35. | Pocanschi et al. (2006) and Dahmane et al. (2013) |
| 9.12 | BR (~27 kDa), tOmpA (~19 kDa) | A8-35, A8-75, SAPol | | | | | • | | | Characterization of the particles formed between BR or tOmpA and a range of carboxylated and/or sulfonated APols. | Dahmane (2007) and Dahmane et al. (2011) |
| 9.13 | BR (~27 kDa), $b_6f$ (~211 kDa) | A8-35, phosphorylcholine-based APols (PC-APols) | | | • | | | | | Comparison of the solution behavior of MP/detergent, MP/PC-APol, and MP/A8-35 complexes. | Diab et al. (2007) |
| 9.14 | BR (~27 kDa), $b_6f$ (~211 kDa), tOmpA (~19 kDa), OmpX (~18.6 kDa), ghrelin receptor (~41 kDa) | A8-35, non-ionic APols (NAPols) | • | | | • | • | • | | Characterization of the mass, composition, size, and organization of MP/NAPol complexes. | Bazzacco (2009), Bazzacco et al. (2009, 2012), and Sharma et al. (2012) |
| 9.15 | OmpX (~18.6 kDa) | A8-35, DAPol | | | | | • | | | Determination and optimization of rotational correlation time of OmpX/APol complexes. | Catoire et al. (2009, 2010) |
| 9.16 | Leukotriene BLT1 and BLT2, serotonine 5-HT$_{4(a)}$, and cannabinoid CB1 GPCRs (respectively 37.5, 37.8, 43.9, and 52.8 kDa) | A8-35 | | | | | • | | | Characterization of GPCRs expressed in *E. coli* as inclusion bodies and folded by transfer from SDS to A8-35. | Dahmane et al. (2009) |
| 9.17 | BR (~27 kDa), lipid A palmitoyl transferase (PagP) (~22 kDa) | Poly(styrene-*co*-maleic acid) (SMA) | | | | | • | • | • | Characterization of the particles formed by direct extraction of BR or PagP from lipid vesicles with SMA. | Knowles et al. (2009) |
| 9.18 | Mitochondrial supercomplex B (~1.7 MDa) | A8-35 | | | • | | | | • | Identification and purification of supercomplex B (respirosome) prior to cryo-EM study. | Althoff (2011) and Althoff et al. (2011) |
| 9.19 | Transient receptor potential ankyrin 1 channel (TRPA1) (~525 kDa) | A8-35 | | | | | • | | | Characterization and purification of TRAP1/A8-35 complexes prior to EM study in negative stain. | Cvetkov et al. (2011) and Huynh et al. (2014) |
| 9.20 | Oleosins and caleosin (non-membrane hydrophobic proteins) | A8-35 and other random or blocky polyacrylate-derived polymers | | • | | | | • | • | Characterization of protein/APol complexes. | Gohon et al. (2011) |
| 9.21 | Corticotropin-releasing factor receptors 1 and 2β (CRFR1 and CRFR2β) (~47 and ~49 kDa, respectively) | NVoy | | | | | • | | | Characterization and mass determination of cell-free expressed, NVoy-trapped CRFR1 and CRFR2β prior to NMR and negative-stain EM studies. | Klammt et al. (2011) |
| 9.22 | NTT1 ATP/ADP transporter (~60 kDa) | A8-35 | | | | • | • | | | Determination of the oligomeric states of NTT1 overexpressed in *E. coli*. | Deniaud et al. (2012) |
| 9.23 | PagP (~22 kDa), OmpT (~33.5 kDa) | A8-35 | | | | | • | | | Characterization of MP/APol complexes following folding in A8-35 and prior to mass spectrometry studies. | Leney et al. (2012) |
| 9.24 | BR (~27 kDa) | SMA | | | | | • | • | | Characterization of BR/SMALP complexes. | Orwick-Rydmark et al. (2012) |
| 9.25 | Vasopressin 2 receptor (V2R) (~42 kDa) | NAPol | | | | | • | | • | Characterization of the oligomeric state of NAPol-trapped V2R. | Rahmeh et al. (2012) |
| 9.26 | Capsaicin (vanilloid) receptor transient receptor potential channel (TRPV1) (~300 kDa) | A8-35 | | | | | • | | | Characterization of the oligomeric state of A8-35-trapped TRPV1 prior to cryo-EM. | Liao et al. (2013) |
| 9.27 | BR (~27 kDa) | A8-35 | | | | | • | | | Characterization of cell-free expressed, A8-35-trapped BR prior to NMR measurements. | Etzkorn et al. (2013) |
| 9.28 | CYP79A1 and CYP71E1 P450 cytochromes (~59 and ~61 kDa) | A8-35 | | | | | • | | • | Examination of the trapping and function of two P450 cytochromes, CYP79A1 and CYP71E1, involved in the synthesis of dhurrin. | Laursen et al. (2013) |
| 9.29 | ATP-binding cassette transporter ABCA4 (~260 kDa + ~20 kDa sugars) | A8-35 | | | | | • | | | Characterization of ABCA4/A8-35 complexes prior to EM. | Tsybovsky et al. (2013) |

(continued)

**Table 9.1** (continued)

| Study | Membrane proteins (MW) | Amphipol | SANS | SAXS | SGSV | AUC | SEC | DLS | BN-PAGE | Comments | References |
|---|---|---|---|---|---|---|---|---|---|---|---|
| 9.30 | Transient receptor potential melastatin-1 (TRPM1) (~360 kDa) | A8-35 | | | | | ● | | | Determination of the oligomeric state of TRPM1/A8-35 complexes prior to EM. | Agosto et al (2014) |
| 9.31 | OmpF trimer (~111 kDa) | A8-35 | | | | | ● | | | Examination of the association state of OmpF in the presence or absence of free A8-35, lipopolysaccharide, and/or EDTA prior to negative-stain EM studies. | Arunmanee et al. (2014) |
| 9.32 | BR and two GPCRs, the melanocortin-2 and 4 receptors (~27, ~34 and ~36 kDa, respectively) | A8-35 | | | | | ● | | ● | Determination of the size and homogeneity and purification of MP/APol complexes following refolding in A8-35 and prior to NMR studies. | Elter et al. (2014) |
| 9.33 | OmpX (~18.6 kDa) | A8-35 | | | | | | | | Determination of rotational correlation time. | Etzkorn et al. (2014) |
| 9.34 | BLT2 (43.9 kDa) | A8-35 and perdeuterated A8-35 (perDAPol) | | | | | ● | | | Comparison of the size and dispersity of BLT2/APol complexes formed using either unlabeled or perdeuterated A8-35. | Giusti et al. (2014) |
| 9.35 | P-glycoprotein (PgP) (~170 kDa) | SMA | | | | ● | ● | | | Characterization of PgP/SMALP complexes prior to cryo-EM. | Gulati et al. (2014) |
| 9.36 | Lysophosphatidic acid receptor 2 (LPA2; fusion protein) | Poly-γ-glutamic acid grafted with octylamine and glucosamine (APG) | | | | | ● | | | In the frame of characterizing APG for use as an APol, LPA2/APG complexes were studied by SEC and AFM. | Han et al. (2014) |
| 9.37 | BR (~27 kDa) | A8-35 and oligonucleotide-tagged A8-35 (OligAPol) | | | | | ● | | | Comparison of A8-35- and OligAPol-trapped BR, binding of and immobilization via a complementary oligonucleotide. | Le Bon et al. (2014a) |
| 9.38 | Light-harvesting complexes II (LHCII) (~75 kDa) | A8-35, NAPol | | | ● | | ● | | | Comparison of DDM-solubilized and A8-35- or NAPol-trapped complexes. | Opačić et al. (2014a) |
| 9.39 | Mannitol permease (EII^mtl) (~68 kDa) | A8-35 and fluorescein-labeled A8-35 (FAPol_dns) | | | | | ● | | | Examination of the size, dispersity, and function of EII^mtl/APol complexes prior to fluorescence quenching experiments. | Opačić et al. (2014b) |
| 9.40 | BR (~27 kDa) | A8-35 | | | | | ● | | | Comparison of the dispersity of BR/A8-35 complexes in the presence or absence of free APol prior to crystallization by transfer to a lipid mesophase. | Polovinkin et al. (2014) |
| 9.41 | ExbB/ExbD complexes (monomers ~26 and ~15 kDa, respectively) | A8-35, DAPol, FAPol_SBD | ● | ● | | ● | ● | | ● | A broad array of biophysical and biochemical methods was harnessed to establish the mass, composition, shape, and organization of ExbB4/ExbD2 and ExbB6/ExbD4 complexes trapped in A8-35 or in deuterated or fluorescent derivatives thereof. | Sverzhinsky et al. (2014) |
| 9.42 | MexB and OprM from *Pseudomonas aeruginosa*, BR, cytochrome bc1 | Biotinylated A8-35 and NAPol | | | | | ● | | | Selection of αRep binders targeting MexB and other MPs. | Ferrandez et al. (2014) |
| 9.43 | tOmpA (~19 kDa) | A8-35 and polyhistidine-tagged A8-35 (HistAPol) | | | | | ● | | | Comparison of the solution behavior of tOmpA/A8-35 and tOmpA/HistAPol complexes prior to immobilization studies. | Giusti et al. (2015) |
| 9.44 | Human adenosine A2A receptor (A2AR) (44.7 kDa) | SMA | | | | ● | ● | | | Examination of the solution properties and stability of SMALP-trapped A2AR. | Jamshad et al. (2015) |
| 9.45 | Insulin-responsive facilitative glucose transporter GLUT4 (~43 kDa) | A8-35 | | | | | ● | | | GLUT4 overexpressed in HEK-293 cells was purified in LMNG and transferred to nanodiscs, A8-35, or proteoliposomes. GLUT4/A8-35 complexes were studied by SEC, denaturation and renaturation studies, and ligand binding. | Kraft et al. (2015) |
| 9.46 | Transient receptor potential ankyrin 1 channel (TRPA1) (~525 kDa) | PMAL-C8 | | | | | ● | | | Characterization of TRPA1/PMAL-C8 complexes prior to cryo-EM. | Paulsen et al. (2015) |
| 9.47 | Multidrug transporter AcrB trimer (~360 kDa) | SMA | | | | | | ● | | Examination as a function of ionic strength of the tendency of SMALP-trapped AcrB trimers to form SMALP-mediated "doublets". | Postis et al. (2015) |
| 9.48 | Human transient receptor potential melastatin 4 channel (TRPM4) (668 kDa; fusion protein) | A8-35 | | | | | ● | | | In the frame of a broad study of the oligomeric state of TRPM4, the dispersity of TRPM4/A8-35 complexes is examined by SEC. | Constantine et al. (2016) |
| 9.49 | BR (~27 kDa) | SMA, SMA-SH | | | | | ● | | | SEC is used to compare the size of the particles formed upon extracting BR from proteoliposomes either with SMA or with a thiolated version thereof (SMA-SH). | Lindhoud et al. (2016) |
| 9.50 | Melatonin MT1R (~39 kDa) and ghrelin GHS-R1a (~41 kDa) receptors | SMA | | | | | ● | | | CHS-R1a was expressed in inclusion bodies, folded in A8-35, and transferred to lipid vesicles. MT1R was expressed in Chinese hamster ovary (CHO) cells, from which membrane fragments were purified. The two GPCRs were extracted with SMA and the particles characterized by SEC. | Logez et al. (2016) |
| 9.51 | Polycystic Kidney Disease Channel-2 (PKD2) (~325 kDa) | A8-35 | | | | | ● | | | In the frame of a cryo-EM study, the size and distribution of PKD2/A8-35 and PKD2/nanodisc complexes are compared by SEC. | Shen et al. (2016) |
| 9.52 | Mitochondrial respirasome (I1III2IV1 supercomplex) (~1.7 MDa) | A8-35 | | | ● | | | | ● | A8-35-trapped mitochondrial complexes and supercomplexes are analyzed by BN-PAGE, SGSV, and spectroscopic approaches. | Shinzawa-Itoh et al. (2016) |
| 9.53 | Mitochondrial respirasome (I1III2IV1 supercomplex) (~1.7 MDa) | A8-35 | | | | | | | ● | In the context of a cryo-EM study, respirasome/A8-35 complexes are examined by BN-PAGE. | Sousa et al. (2016) |
| 9.54 | Dimeric mitochondrial ATP-synthase from *Polytomella* sp. (~1.6 MDa) | A8-35 | | | | | | | ● | High concentrations of A8-35 (up to 3.5%) are used to disrupt the F1Fo ATP-synthase into subcomplexes, which are analyzed by BN-PAGE and 2D gels. | Vázquez-Acevedo et al. (2016) |
| 9.55 | Bacteriorhodopsin from *Haloquadratum walsbyi* (HwBR) (~27 kDa) | SMA | | | | | ● | ● | | HwBR/SMA complexes were characterized by SEC, DLS, and UV-visible spectroscopy before being delivered to a lipidic mesophase for crystallization. | Broecker et al. (2017) |
| 9.56 | Lysophosphatidic acid receptor 2 (LPA2; fusion protein) | Poly-γ-glutamic acid grafted with octylamine, glucosamine, and diethyl aminopropylamine | | | | | ● | | | At variance with the SEC profile shown in Study 9.36, that shown here seems to indicate that the complexes formed by LPA2 and a variant of APG elute mostly in the void volume of the column, suggesting that they are very large or aggregated. | Han et al. (2017) |
| 9.57 | Transient receptor potential (TRP) channel NOMPC (~760 kDa) from *Drosophila melanogaster*. | A8-35 | | | | | ● | | | In the frame of a cryo-EM study, a comparative SEC analysis is presented of A8-35- and ND-trapped NOMPC. | Jin et al. (2017) |
| 9.58 | *E. coli* outer membrane phospholipase A (OmpLA) (~31 kDa) | Alternating diisobutylene/maleic acid copolymer (DIBMA) | | | | | ● | ● | | SEC and DLS profiles of OmpLA/DIBMA particles. | Oluwole et al. (2017) |

*Abbreviations*: *AFM* atomic force microscopy, *APG* poly-γ-glutamic acid grafted with octylamine and glucosamine, *AUC* analytical ultracentrifugation, *BN-PAGE* Blue-Native polyacrylamide gel electrophoresis, *cryo-EM* electron cryomicroscopy, *DIBMA* alternating diisobutylene/maleic acid copolymer,

*DLS* dynamic light scattering, *EM* electron microscopy, *FRET* Förster resonance energy transfer, *LMNG* 2,2-didecylpropane-1,3-bis-β-ᴅ-maltopyranoside, *MS* mass spectrometry, *ND* protein-stabilized nanodisc, *SANS* small-angle neutron scattering, *SAXS* small-angle X-ray scattering, *SEC* size exclusion chromatography, *SGSV* sucrose gradient sedimentation velocity, *SMA* poly(styrene-*co*-maleic acid), *SMALPs* SMA/lipid particles, *STEM* scanning transmission electron microscopy. Abbreviations for the proteins and for APols are defined in columns 2 and 3 of the table, respectively. The chemical structures of the various APols used are described in Chap. 4, § 4.2.

In the following, my purpose is not to critically examine everything that has been published in this particular field, but to provide the reader with (i) an overview and guide to the existing literature and (ii) some comments, suggestions, and caveats regarding the implementation of each of these techniques. Some stress is put on what is specific to the use of APols rather than other surfactants. In particular, at variance with small surfactant such as detergents (see Hiruma-Shimizu et al. 2015), APols can be relatively easily labeled either isotopically or with colored and/or fluorescent ligands. As a rule, because of the relatively large size of APol molecules, their colloidal properties are not significantly affected by the labeling (see Chap. 4, § 4.4). This makes it easy to use, for instance, radioactive or colored APols to follow the distribution of the polymer in a sucrose gradient or upon elution from a SEC column, or deuterated APols to contrast-match the surfactant and eliminate its contribution in SANS studies of MPs or MP complexes.

*Solution studies of membrane protein/amphipol complexes*
(© 2018 by Francis Haraux)

## 9.2 Overview of the Literature

A compact overview of the literature is presented in Table 9.1. Publications that present strongly complementary data have been lumped together into a single study, studies being listed in the order of appearance of the first publication. For each study, the following information is provided:

- The MP(s) that has or have been studied, with their approximate molecular mass. As regards the latter, no effort has been made to take close account of the contribution of prosthetic groups, posttranslational modifications, etc., the intent being simply to give the reader an idea of the size of the proteins that have been handled.
- The type(s) of APols used in the study. As discussed in Chap. 4, poly(styrene-*co*-maleic acid) (SMA), alternating diisobutylene/maleic acid copolymer (DIBMA), and NVoy (= NV-10) are listed here under the heading "Amphipols," as well as amphipathic derivatives of poly-$\gamma$-glutamic acid. Chemical formulae and solution properties of these various polymers are described in Chap. 4.
- The solution methods that have been used in the study, inasmuch as they fit into the subject matter of the present chapter. Many studies, for instance, include circular dichroism or electron microscopy data, which are not necessarily mentioned here (see Chaps. 8 and 12, respectively). In the "Comments" column, however, some indication is generally given of what the primary objective of the study was.

The aim of the table is to allow the reader to quickly identify those studies that bear on systems closest in size or complexity to the one she/he is studying and that have resorted to methods she/he is considering implementing. Note that the most informative studies tend to be those in which a variety of techniques have been harnessed to bear on the same system. Indeed, such multidisciplinary studies provide a good perspective on the limits of and complementarity between various approaches, and they give precious indications about how to use one to guard against the pitfalls of another. Such are the cases, for instance, of Study 9.5, with bacteriorhodopsin (BR) as the model system; Study 9.7, on the cytochrome $bc_1$ complex; or Study 9.41, on the complexes formed between ExbB and ExbD, two MPs involved in providing energy from the inner to the outer bacterial membrane to power nutrient import. BR is a particularly interesting model system, because essentially every polymer that has ever been developed for use in handling MPs has been tested on it, as well as novel detergents and other alternative surfactants, e.g. nanodiscs (see Chaps. 2 and 3). It has also been used to test optical spectroscopy (Chap. 8, § 8.2.2), NMR (Chap. 10, § 10.3.1), and crystallographic (Chap. 11, § 11.3.2) approaches to studying MPs complexed by APols. The information gathered by the techniques described in the present chapter has often been key to optimizing the samples to be studied by more resolving techniques.

## 9.3	A Walk Through Selected Solution Studies

In order to illustrate the application of the solution methods treated in this chapter to APol-trapped MPs, I will take as a main guide the first extensive study of this type to have been carried out, that of BR/A8-35 complexes (Studies 9.1 and 9.5). Key points and alternative approaches will be further illustrated by data taken from other studies, generally on other MPs. As described in more detail in Chap. 1, § 1.6.1, BR is a small MP (~27 kDa), folded into a bundle of seven TM $\alpha$-helices. Retinal, an aldehyde, is bound covalently but loosely, via a Schiff base, to the $\varepsilon$-amino group of a lysine residue located in the last TM helix. The Schiff base hydrolyzes spontaneously when the protein denatures, freeing the retinal, and it reforms, also spontaneously, when the protein refolds and the retinal rebinds. In the ground state of BR, the Schiff base is protonated, which shifts the absorbance peak of retinal from ~380 nm, which confers a yellow color to the free cofactor, to the vicinity of 560 nm, making the holoprotein purple. Upon illumination with visible light, the retinal isomerizes from all-*trans* to 13-*cis*, which initiates a complex cycle of conformational transitions during which the Schiff base proton is released in the extracellular medium and replaced by a proton taken from the cytosol (see Chap. 1,

Fig. 1.24). As a result, the photocycle of BR creates a TM proton gradient that the bacterium uses as a source of energy. In solution, the photocycle can be observed by time-resolved absorbance spectroscopy, which provides sensitive measurements of the rate at which each conformational state builds up and disappears (see e.g. Pocanschi et al. 2006; Dahmane et al. 2013). That the protein is functional can also be deduced from steady-state spectra, because photocycling results in a slight red shift of the ground-state absorbance as compared to that observed when the protein has been left to relax in the dark, due to the conversion of a mixture of states comprising the all-*trans* and 13-*cis* retinal isomers to pure all-*trans*. The detection of this light-induced bathochromic shift indicates that BR does undergo its photocycle.

When overproduced in its natural host, the archaebacterium *Halobacterium salinarum*, BR accumulates in the plasma membrane in the form of two-dimensional, hexagonal crystals, the so-called purple membrane (PM) patches, in which BR trimers associate one with another via well-organized, non-covalently bound lipids. PM patches, which are easily purified and in which BR is extraordinarily stable, are the most common starting material for biochemical and biophysical studies of BR. For studies in detergent solutions, BR is solubilized, along with *Halobacterium* lipids, with a detergent such as *n*-octyl-$\beta$-D-thioglucopyranoside (OTG).

In the following, examples are given for each of the various techniques that have been applied to solution studies of MP/APol complexes, going roughly from the most basic to the most information-rich approaches. The principle of each technique is briefly recalled either at the beginning of each section or in a box. For more in-depth descriptions, the reader is referred to biophysics textbooks, e.g. Cantor and Schimmel (1980) and Zaccai et al. (2017).

### 9.3.1   Sucrose Gradient Rate Zonal Ultracentrifugation

The principle of sucrose gradient rate zonal ultracentrifugation, or, more briefly, sucrose gradient sedimentation velocity (SGSV) measurements, is described in Box 9.1. In Study 9.1, BR in 10 mM OTG, whose critical micellar concentration (CMC) is ~9 mM, was diluted 2× with a detergent-free buffer containing amphipol A8-75, bringing the detergent under its CMC. The preparation was then layered on a surfactant-free sucrose gradient, containing neither detergent nor APol, and centrifuged for 10 h. Analysis of the distribution of the protein shows that it migrates as a single band, and none of it is found in the pellet (Fig. 9.2). The same observation was made with three other MPs solubilized in various detergents and transferred to A8-75 (Fig. 9.2). In the absence of APol, such experiments result in complete precipitation of the protein (see Chap. 5, § 5.9.1.5, Fig. 5.42).

---

**Box 9.1   Ultracentrifugation**

The principle of ultracentrifugation is to submit a solution of the particles or mixture of particles to be analyzed to a strong gravity field. There are four main approaches, depending on (i) whether the particles migrate in a density gradient or a homogeneous column of buffer and (ii) whether they are spun until an equilibrium distribution is reached or it is the rate at which they migrate that is analyzed:

- Sucrose gradient sedimentation velocity analysis (SGSV; § 9.3.1, Figs. 9.2, 9.3, 9.4, 9.5, and 9.6)

A gradient of a dense solute, typically 5–20% w/w sucrose, occasionally glycerol or iodinated solutes, is formed in centrifuge tubes and the sample deposited on the top. The tubes

**Box 9.1 (continued)**

are centrifuged and the centrifugation stopped before the particles, which are denser than the densest layer in the gradient, have reached the bottom of the tube (see Figs. 9.2, 9.3, 9.4, 9.5, and 9.6). As particles migrate, they slow down, because the difference between their density and that of the solution diminishes, while viscosity increases. This creates a stacking effect, opposing the broadening effect of translational diffusion. The position reached at any given time is related to the mass, the density, and the hydrodynamic radius ($R_H$) of the particles. Given proper abaci (McEwen 1967), it is possible to derive the sedimentation coefficient $s_{20,w}$ that the particle would exhibit in pure water at 20 °C, from which, knowing the specific volume $\bar{v}$ (inverse of the density) and $R_H$, the molecular mass can be deduced (see below). The fractions can be collected and their composition analyzed. SGSV is mostly used to examine the distribution and composition of particles in samples and as a preparative method, rather than for quantitative determination of solution properties.

- Equilibrium (isopycnic) sedimentation in sucrose gradients

Denser gradients, e.g. 20–40% w/w sucrose, are used to separate particles according to their density, the centrifugation being carried out until the particles have reached that position in the gradient where their density matches that of the solvent. The particles can be either sedimented or floated. This approach is used, for instance, to separate one from another membrane fragments whose density differs due to different protein/lipid ratios, such as bacterial inner and outer membrane fragments. It will not be considered in the present chapter.

- Sedimentation velocity analytical ultracentrifugation (SV-AUC) in the absence of a density gradient (§ 9.3.7; Fig. 9.12)

Alternatively, a homogeneous solution can be subjected to a gravity field. Those molecules closest to the bottom of the tube will rapidly pellet, whereas those close to the top will progressively sediment while diffusing. The distribution of the particles in the gradient is followed throughout the experiment by measuring the absorbance of the solution and its refractive index at each point in the buffer column as a function of time (see Fig. 9.12). Analysis of the distribution profiles yields the sedimentation coefficient $s$ and the translational diffusion coefficient $D_t$ of the particles, or of each of the particles in a mixture, a situation in which this approach is particularly powerful (cf. Figs. 9.12 and 9.14). Molecules with different absorbance spectra, such as protein and surfactant or protein and nucleic acid, can be followed at different wavelengths.

For an ideal solution, the sedimentation coefficient of a particle, $s$, in Svedberg ($1\ S = 10^{-13}\ s$), is related to its buoyant molar mass, $M_b$; its Stokes radius, $R_S$; and Avogadro's number, $N_A$, by the Svedberg equation:

$$s = M_b / (6\pi N_A \eta_0 R_S)$$

where $\eta_0$ is the viscosity of the solvent (in Pa·s$^{-1}$). (Note: the Stokes radius or Stokes-Einstein radius of a solute, $R_S$, is the radius of a hard sphere that diffuses at the same rate as the solute. In biophysics, $R_S$ and the hydrodynamic radius, $R_H$, tend to be used as synonyms of each other.)

$M_b$ is the mass of the particle minus the mass of the solvent it displaces. It is related to the particle molar mass, $M$, and its partial specific volume, $\bar{v}$ (in mL·g$^{-1}$), by the following equation:

$$M_b = M \left(1 - \rho_0 \bar{v}\right),$$

where $\rho_0$ is the density of the solvent.

- Equilibrium sedimentation analytical ultracentrifugation (Eq-AUC) in the absence of a density gradient (§ 9.3.7; Fig. 9.13)

**Box 9.1 (continued)**

For equilibrium sedimentation measurements, the hardware and procedure are similar to the previous case, except for the use of shorter buffer columns, lower gravity fields, and longer running times. The particles are left to reach their equilibrium distribution in the field, which depends only on their mass and $\bar{v}$ and not on their shape. Repeating the experiments at different solvent density, e.g. by replacing $H_2O$ with $D_2O$ or $D_2^{18}O$, yields a system of equations from which the buoyant molecular mass and $\bar{v}$ of the particles can be experimentally determined (cf. Fig. 9.13). This approach is particularly useful for homogeneous samples, or when studying equilibria, e.g. those due to oligomerization.

SGSV fractionation is a convenient way to separate MP/APol complexes from free APol, provided the protein is not too small. This is illustrated in Fig. 9.3, taken from Study 9.1, in which SGSV was used to estimate the amount of tritiated A8-75 bound per trimer of the porin OmpF (~111 kDa). Note that, in this experiment, the complexes between OmpF trimers and A8-75 appear essentially monodisperse, showing very little aggregation. A different distribution would probably have been observed if fractions 6–10, which contain most of the protein, had been collected, washed free of sucrose by dialysis or SEC, and centrifuged again under the same conditions: indeed, in the absence of free APol, OmpF would have aggregated (cf. Arunmanee et al. 2014; see Chap. 12, Fig. 12.13). Under the conditions of the experiment shown in Fig. 9.3, however, aggregation had no time to develop to a significant extent, even though some degree of it is suggested by the trailing of the peak toward the bottom of the gradient.

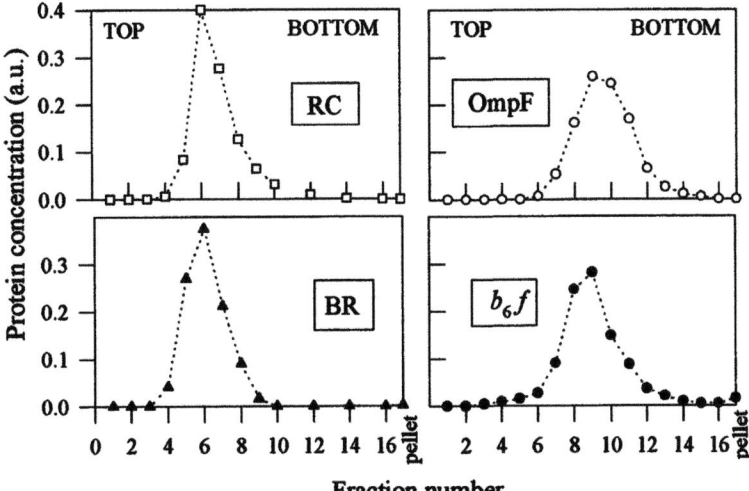

**Fig. 9.2** Sedimentation velocity analysis of four membrane protein/amphipol complexes in surfactant-free sucrose gradients. The four MPs are the photosynthetic reaction center from *Rhodobacter sphaeroides* R-26 (RC), OmpF from *Escherichia coli*, bacteriorhodopsin (BR) from *Halobacterium salinarum*, and the cytochrome $b_6f$ complex from *Chlamydomonas reinhardtii*. Aliquots (100 μL) of MPs in detergent solution were supplemented with 1 g·L$^{-1}$ A8-75 (plus, in the case of $b_6f$, 0.33 g·L$^{-1}$ egg phosphatidylcholine), diluted 2× with 100 mM ammonium phosphate buffer (pH 8.0), layered on top of 2 ml 5–20% (w/w) sucrose gradients in detergent- and APol-free buffer, and centrifuged at 54,000 rpm in the TLS-55 rotor of a Beckman Coulter TL100 ultracentrifuge. After 5.25 h (RC), 6.5 h (OmpF), 10 h (BR), or 5 h ($b_6f$), gradients were collected by 120 μL fractions. Protein concentrations in the 16 fractions and in the resuspended pellet were determined spectrophotometrically (From Tribet et al. 1996, © 1996 National Academy of Sciences).

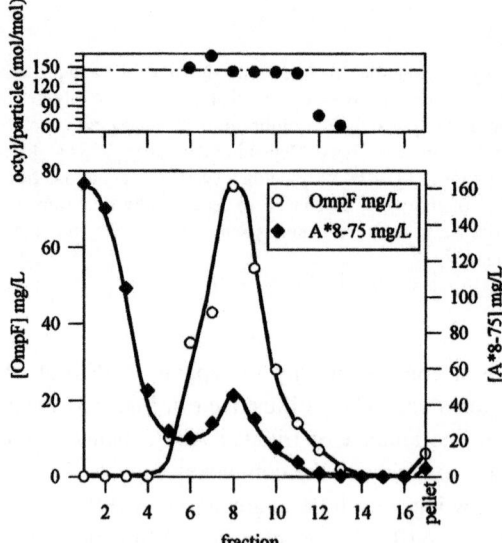

**Fig. 9.3** Sedimentation velocity analysis of a membrane protein/amphipol complex and determination of the protein/APol ratio in the complexes. OmpF trapped in [$^{14}$C]A8-75 (A*8-75) was deposited on a surfactant-free 5–20% w/w sucrose gradient and centrifuged at 4 °C for 6.5 h at 54,000 rpm (~250,000 × $g$) in the TLS-55 rotor of a Beckman Coulter TL100 ultracentrifuge. Fractions (120 µL) were collected from the top with a Hamilton syringe. The bottom of the tube was washed with 120 µL buffer ("pellet"). The concentration of OmpF was determined from its absorbance at 278 nm. *Bottom*: distribution of OmpF and [$^{14}$C]A8-75 in the gradient – (○) OmpF, (◆) [$^{14}$C]A8-75. Note that most of the free APol is found in fractions 1–5, at the top of the gradient. *Top*: number of A8-75 octyl groups per OmpF trimer, as deduced from the $^{14}$C/OmpF ratio (Reprinted with permission from Tribet et al. 1997, © 1997 American Chemical Society).

Separating free from MP-bound APol by SGSV is much more difficult when the protein is small, as is the case of BR. This is illustrated in Fig. 9.4, taken from Study 9.5, in which a sample of BR trapped in [$^{3}$H]A8-35 was subjected to two successive rounds of SGSV fractionation. In the first gradient (Panel **A**), a majority of the free [$^{3}$H]A8-35 (*short dashes*) is separated from BR/APol complexes, but enough of it comigrates with them to strongly affect the determination of the MP/APol mass ratio (**A′**). When the least contaminated BR fractions were pooled and applied to a second gradient (**B**), enough BR/APol fractions were reasonably uncontaminated (roughly from fraction 12 to 15) for an estimate of the BR/A8-35 mass ratio in the complexes to be obtained (**B′**). In the process, however, two new problems arose: (i) some of BR aggregated (compare the late fractions and the pellet in the first and second gradient); and (ii) AUC and SANS data (see §§ 9.3.7 and 9.3.8.2) showed that the mass of bound A8-35 measured by this procedure is underestimated by a factor close to two, very likely because aggregated BR had released part of the APol originally associated with it. Sucrose gradient fractionation obviously is not adapted to determining the mass ratio of APols associated to small MPs, unless the gradient contains enough free APols to prevent aggregation, which was not the case in these experiments.

As can be expected, the migration of a given protein in sucrose gradients is affected by the APol used to trap it: the hydrodynamic radius of the particle depends on the thickness of the APol belt, and its mass and density on the protein/APol mass ratio and the density of the APol. This is illustrated in Fig. 9.5 (taken from Study 9.37), where the migration of a light-harvesting complex (LHC) is compared in *n*-dodecyl-α-D-maltopyranoside (α-DDM) or after trapping with either A8-35 or a glycosylated, non-ionic APol, whose density is higher.

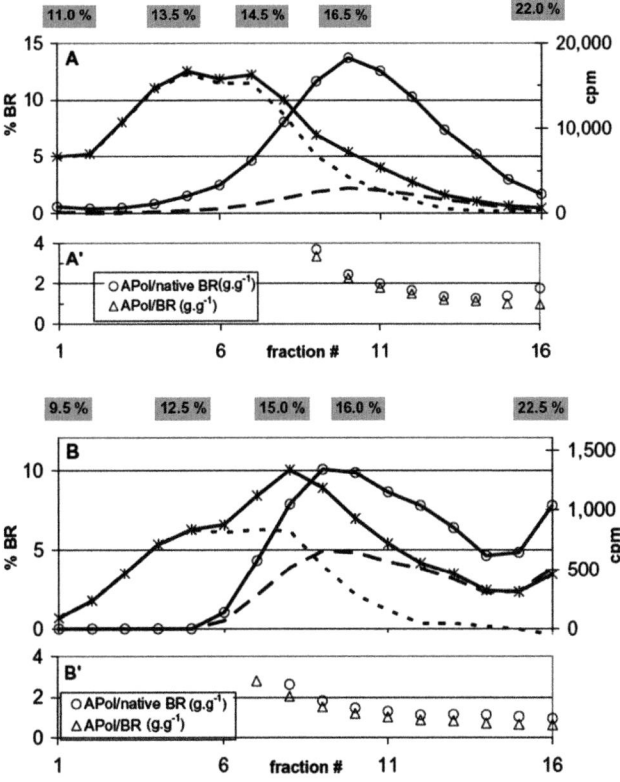

**Fig. 9.4** Distribution of BR and [$^3$H]A8-35 upon two successive 22-h centrifugations in 12–20% (w/w) sucrose gradients. Fractions 10–13 from a first gradient (panel **A**) were pooled, concentrated, and washed free of excess sucrose by microfiltration before re-loading and centrifuging onto a second, identical gradient (panel **B**). Experimental data: "% BR" (—○—), percentage of native BR in each fraction (on the basis of $A_{554}$), expressed with respect to the overall amount of BR in the whole gradient (from $A_{280}$); "cpm" (—□—), [$^3$H]A8-35 distribution. Breaking up the distribution of [$^3$H]A8-35 into free and bound fractions: inferred contributions of free (– – – –) and BR-bound (– – –) APol, assuming a constant ratio of [$^3$H]A8-35 to BR in the complexes. Panels **A′** and **B′**: mass ratio of (○) [$^3$H]A8-35/native BR (from $A_{554}$) and of (△) [$^3$H]A8-35/total protein (from $A_{280}$) (From Gohon et al. 2008, © 2008 The Biophysical Society. Published by Elsevier Inc. All rights reserved).

An interesting, unpublished observation, which would deserve closer investigation, was made by Yann Gohon during his Ph.D. work (Gohon 2002): as could be expected, the migration of cytochrome $b_6 f$/A3-70 complexes in sucrose gradients was found to depend on the nature of the counterion present in the buffer (in these experiments, either Li$^+$, Na$^+$, K$^+$, or Cs$^+$). A comparison between Li$^+$ and Cs$^+$ is shown in Fig. 9.6. Based on the mass ratio of bound A8-35 to BR determined by AUC and SANS (see § 9.3.8.2), it can be estimated that some 150 counterions comigrate with each BR/A8-35 complex, so that substituting Cs$^+$ for Li$^+$ in the buffer increases the dry mass of the complexes by >20%, assuming the binding of APol not to change. In Fig. 9.6, the complexes used are cytochrome $b_6 f$/A3-70 complexes, whose MP/APol mass ratio has not been determined but is certainly higher, leading to a lesser effect of the Li$^+$/Cs$^+$ substitution. Amphipol A3-70 was used because of its higher density of charges (twice more than A8-35), which is similar to that of A8-75 or SAPols. Qualitatively, this effect

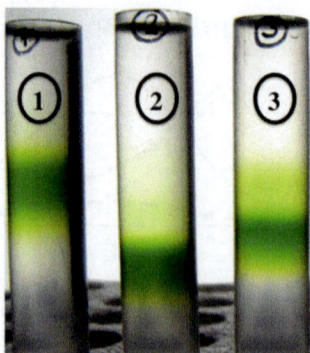

**Fig. 9.5** Sucrose gradient sedimentation velocity analysis of purified LHCII complexes from *Arabidopsis thaliana* in various surfactants: (1) A8-35-trapped; (2) trapped in a glycosylated, non-ionic APol; and (3) solubilized in 0.03% *n*-dodecyl-$\alpha$-D-maltopyranoside ($\alpha$-DDM). All gradients contained 0.1–1 M sucrose and were spun for 16 h at 280,000 × *g*. Gradients (1) and (2) contained no surfactant, gradient (3) 0.03% $\alpha$-DDM. Note the stronger release of chlorophyll (light green band on top of the main one) in the presence of detergent (3) (From Opačić et al. 2014a).

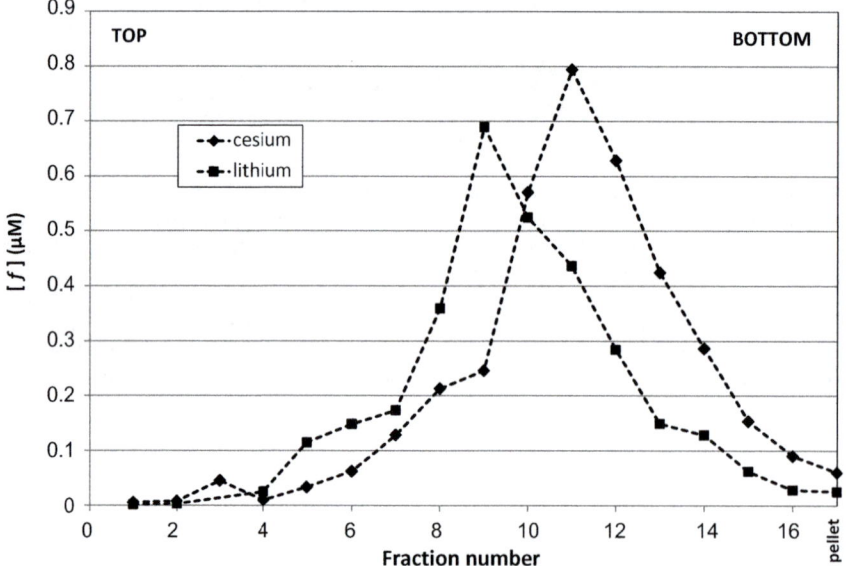

**Fig. 9.6** Sucrose gradient sedimentation velocity analysis of cytochrome $b_6 f$/A3-70 complexes in the presence of 300 mM LiCl (■) or CsCl (◆). Conditions were the same as in Fig. 9.2, *lower right*, but for the salt. The concentration of ascorbate-reduced cytochrome $f$ (in μM) was measured at 554 nm. Those are preliminary data, the fraction numbers having been recalculated to correct for the different size of the drops, presumably due to the different density of the two solutions. Note that, despite the higher density of the cesium-containing solution, $b_6 f$/APol complexes migrated faster in the presence of CsCl, due to the higher density conferred to the particles by the counterions. A3-70 was synthesized by C. Tribet starting from a 2 kDa sodium polyacrylate precursor. Only 30% of the carboxylates were grafted with an octyl chain, the remaining 70% being left ungrafted. This highly charged polymer was chosen so as to increase the density of particle-associated counterions (as compared to A8-35) in order to detect more easily the difference of density between the particles. A lower concentration of salt in the gradient, e.g. 20 mM, would have been preferable, as it would have had less influence on the density of the solution and led to a better separation (Y. Gohon, unpublished data).

of counterions could be exploited to distinguish MPs from soluble proteins in a rough detergent extract that has been trapped in polyanionic APols or to prevent comigration of proteins that bind or not APols. Quantitatively, running SGSV or AUC experiments in the presence of various alkaline cations could possibly provide a way to determine the mass of APol bound to a given MP and, therefore, the mass of the latter, without having to resort to labeled APols. This however would require assessing the extent to which this substitution may affect the amount of MP-bound APol and/or its hydration and, therefore, the mass and hydrodynamic radius of the particles. As discussed in Chap. 12 (§ 12.3.4.3), substituting $Li^+$ or $Na^+$ counterions with $Cs^+$ ones could also have interesting applications in electron cryomicroscopy.

## 9.3.2   Size Exclusion Chromatography

Size exclusion chromatography (SEC) relies on filtrating particles on a column of hollow beads, which offer a larger accessible volume of solution for small than for large particles to diffuse in. As a result, particles elute as a function of their hydrodynamic radius $R_H$, large ones eluting more rapidly than small ones. $R_H$ estimates are obtained by comparing the elution position of a particle to those of a set of soluble globular molecules of known $R_H$ (see e.g. Harlan et al. 1995). SEC is by far the most commonly used technique for characterizing the size and homogeneity of MP/APol complexes (Table 9.1).

Some typical examples are gathered in Fig. 9.6, each of which offers some specific interest. In Panel A, the elution pattern of BR/A8-35 complexes is compared to that of catalase, a soluble, globular protein with a molecular mass of 220 kDa. This panel illustrates several interesting points:

(i) When BR was trapped with the batch of A8-35 denoted "HAPol" (a plain, unlabeled batch), the complexes migrated with the same apparent size and dispersity as catalase, indicative of their homogeneity.

(ii) Their size is overestimated. Indeed, other experiments lead to conclude that the mass of the particles is actually close to 90 kDa, not to 220 kDa (see § 9.3.8.2). The origin of this effect is not clear, as it does not seem attributable to electrostatic repulsion between the column and the particles despite the relatively low ionic strength of the buffers used (typically 100–150 mM NaCl) (for a discussion, see § 9.3.8.2 and Zoonens et al. 2007).

(iii) When BR was trapped with the batch denoted "DAPol" (a partially deuterated batch), a degree of aggregation is perceptible, revealed by the shoulder that precedes the main peak. This was confirmed by AUC analyses (see Fig. 9.13). The formation of small oligomers is unrelated to the deuteration. It results from the fact that, due to problems that arose during the synthesis, this particular batch of A8-35 ended up being more hydrophobic than had been aimed for (see Chap. 4, § 4.3.1.2.2). This is a reminder that, whereas the synthesis of A8-35 may seem simple on paper, it has to be carefully controlled if optimal results are to be obtained (see Chap. 4, § 4.5, Protocol 4.1).

Panel B, from Study 9.7, provides one example of the effect of depleting a MP/APol preparation of free APol. Cytochrome $bc_1$/A8-35 complexes were purified on sucrose gradients containing either no or 0.1 g·L$^{-1}$ A8-35. The fractions containing the complexes were pooled, washed free of sucrose,

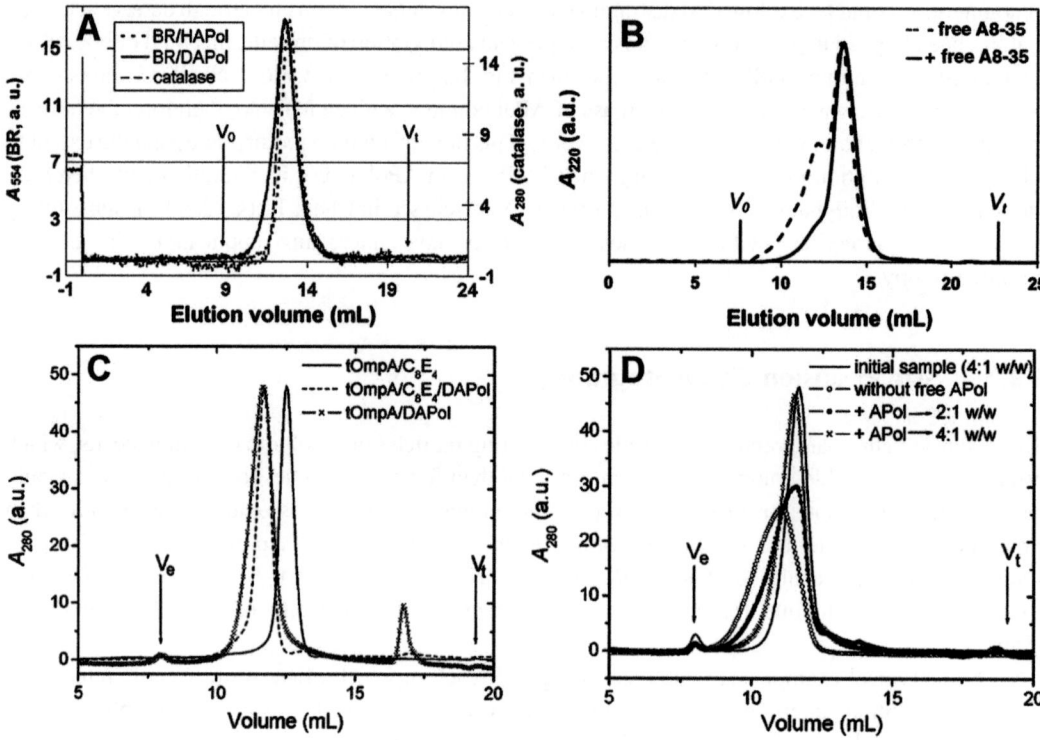

**Fig. 9.7** Examples of size exclusion chromatography analyses of membrane protein/amphipol complexes. (**A**) BR/A8-35 complexes. Arrows indicate the void ($V_0$) and total ($V_t$) volumes. The elution profiles of BR/HAPol complexes (3.3 $g \cdot L^{-1}$ BR), BR/DAPol complexes (1.2 $g \cdot L^{-1}$ BR), and catalase (220 kDa, 2 $g \cdot L^{-1}$) from a Superose 12 HR column were scaled to the same maximum (From Gohon et al. 2008, © 2008 The Biophysical Society. Published by Elsevier Inc. All rights reserved). (**B**) Cytochrome $bc_1$/A8-35 complexes purified on sucrose gradients containing either no (– – – –) or 0.1 $g \cdot L^{-1}$ (——) A8-35 were washed free of sucrose by dialysis and injected onto a Superose 6 column in surfactant-free buffer. Elution profiles were analyzed at 220 nm (From Charvolin et al. 2014). (**C**) SEC analysis of tOmpA/surfactant complexes. SEC elution profiles at various steps during the trapping procedure. An aliquot of a sample of tOmpA (0.22 $g \cdot L^{-1}$) in 6 $g \cdot L^{-1}$ (19.6 mM) $C_8E_4$ solution was injected onto a Superose 12 HR column equilibrated with a Tris-NaCl buffer containing the same concentration of detergent. Another aliquot of the tOmpA/$C_8E_4$ sample was supplemented with the deuterated form of A8-35 (DAPol) in a 4:1 APol/protein mass ratio and an aliquot of the ternary tOmpA/$C_8E_4$/DAPol complexes injected onto the column previously equilibrated with Tris-NaCl buffer containing a mixture of $C_8E_4$ and DAPol at 4 and 0.91 $g \cdot L^{-1}$, respectively. A second aliquot of the tOmpA/$C_8E_4$/DAPol complexes was incubated for 2 h with polystyrene beads, and the resulting tOmpA/DAPol complexes were injected onto the same column, equilibrated with surfactant-free Tris-NaCl buffer (the peak at ~17 mL in the latter profile is due to a contaminant released by the Bio-Beads). Profiles have been normalized to the same maximum (From Zoonens et al. 2007). (**D**) Effect of the presence of free APols on the dispersity of tOmpA/APol complexes. After polyhistidine-tagged tOmpA was trapped with DAPol at a 4:1 APol/protein mass ratio, an aliquot of this sample was injected onto a Superose 12 HR column equilibrated with Tris-NaCl buffer, pH 7.9. The rest of the sample was immobilized on a 1 mL nickel-chelating column and free APol particles washed away with surfactant-free buffer. The tOmpA/A8-35 complexes were eluted with 300 mM imidazole in the same buffer and desalted on a 5 mL desalting column. An aliquot of the fraction containing tOmpA/A8-35 was then injected onto the Superose 12 HR column. The full width at half-height (HHFW) of the peak of tOmpA/A8-35 complexes after IMAC (1.72 mL) is twice that of the initial sample (0.84 mL). Aliquots of the aggregated sample were then supplemented with A8-35 to reach a final overall APol/protein mass ratio of either 2:1 or 4:1. After one night of incubation, the aliquots were injected onto the Superose 12 HR column. The HHFW values decreased to 1.36 and 0.92 mL, respectively (Reprinted with permission from Zoonens et al. 2007, © 2007 American Chemical Society).

and injected onto a SEC column in surfactant-free buffer. The complexes purified in the presence of free APol migrated as reasonably monodisperse particles, whereas those depleted of it contained small aggregates, mainly dimers (Charvolin et al. 2014).

Panel C, from Study 9.9, compares the SEC profile of tOmpA solubilized in the detergent $C_8E_4$, of the same preparation supplemented with A8-35, and finally after depletion of $C_8E_4$ by adsorption onto polystyrene beads. One notes that tOmpA/A8-35 complexes elute both earlier and as a broader band than tOmpA/$C_8E_4$ ones, whereas the ternary tOmpA/A8-35/$C_8E_4$ complexes appear as large as the tOmpA/A8-35 ones but significantly more monodisperse. The origin of this phenomenon and its potential usefulness for crystallization attempts will be discussed in Chap. 11, § 11.3.1.

Finally, Panel D, also from Study 9.9, illustrates the reversibility of the aggregation induced by free APol depletion: an aliquot from a preparation of polyhistidine-tagged tOmpA trapped in A8-35 at a 1:4 protein/APol mass ratio was analyzed by SEC, and the rest of the sample separated from free A8-35 by immobilized metal affinity chromatography (IMAC). An aliquot of the resulting preparation, examined by SEC, exhibited a broad polydispersity, indicative of the formation of small aggregates. Two fractions from the depleted sample were supplemented with A8-35 so as to reach the final mass ratios of, respectively, 2 or 4 g of A8-35 per g tOmpA. SEC analysis revealed that, at 2 g per g tOmpA, aggregation was partially reversed, whereas at 4 g per g tOmpA, the reversal was essentially complete (Zoonens et al. 2007).

Other commented examples of SEC profiles are shown in § 9.5.2, Figs. 9.24, 9.25, and 9.26.

SEC has been exploited to examine the effect of the presence of divalent cations in buffers on the dispersity of the particles formed by various APols and, as a consequence, on that of the complexes formed between these APols and MPs (Study 9.10). A comparison is shown in Fig. 9.8 of the behavior of A8-35, SAPol, PMAL-C12, and PMALA-C12 particles in the presence or absence of low concentrations of $Ca^{2+}$ and/or $Mg^{2+}$. The distribution of the particles of A8-35 (from a rather poor batch, which formed heterogeneous particles) is slightly sensitive to the presence of 1 mM $Mg^{2+}$, which increases the heterogeneity, and much more strongly affected by that of 1 mM $Mg^{2+}$ + 0.5 mM $Ca^{2+}$, which entails an important degree of aggregation (Fig. 9.8A). The size and dispersity of particles of a sulfonated APol (SAPol; see Dahmane et al. 2011; formula shown in Chap. 4, Fig. 4.1) are essentially insensitive to either condition (Fig. 9.8B). The particles of PMAL-C12, a zwitterionic APol (Nagy et al. 2001; ibid.), are larger than those of A8-35 but behave essentially like them upon addition of either $Mg^{2+}$ or $Mg^{2+}$ + $Ca^{2+}$ (Fig. 9.8C). Finally, the particles of PMALA-C12, a sulfonated, zwitterionic APol with a net negative charge at the pH of the experiment, appear extremely large and polydisperse even in the absence of divalent cations, and their dispersity seems rather insensitive to their addition (Fig. 9.8D).

In addition to MP/A8-35 and MP/SAPol complexes (see above and Table 9.1), MP/polymer complexes that have been studied by SEC include those with non-ionic APols (Study 9.14, etc.), PMAL-C8 (Study 9.46), SMA (Study 9.17, etc.), DIBMA (Study 9.58), NVoy (Study 9.21; cf. Fig. 9.9), amphipathic derivatives of poly-$\gamma$-glutamic acid (Studies 9.36 and 9.55), as well as a variety of labeled or tagged versions of A8-35 (see further references in Table 9.1).

Analytical SEC can be combined with various optical techniques, as well as with the use of labeled APols, in order to determine the respective contributions of the protein and polymer to the mass of the particles (Fig. 9.9) or the Stokes radius of the complexes (Fig. 9.10).

### 9.3.3   Blue-Native Polyacrylamide Gel Electrophoresis

Blue-Native polyacrylamide gel electrophoresis (BN-PAGE) relies on supplementing MPs or MP complexes with the amphipathic dye Coomassie Brilliant Blue (CBB), which can substitute

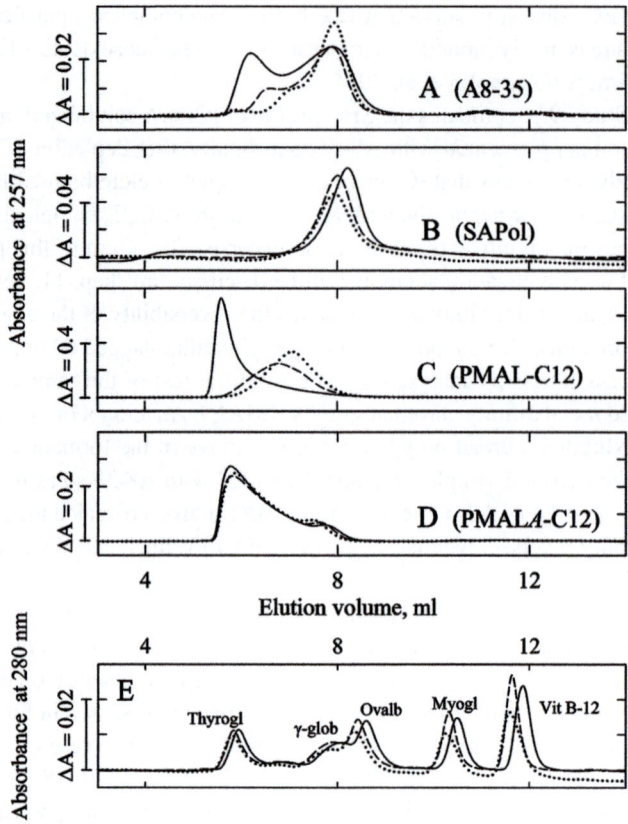

**Fig. 9.8** Effect of divalent cations on the aggregation of four different amphipols. Various APols (**A**, A8-35; **B**, SAPol; **C**, PMAL-C12; **D**, PMAL$A$-C12) and $R_H$ standards (**E**) were subjected to SEC using an Amersham TSK 3000 SW silica gel column. Experiments were run in 100 mM KCl and 20 mM $N$-[$tris$ (hydroxymethyl)methyl]-2-aminoethanesulfonic acid/NaOH buffer, pH 7, supplemented with either 0.5 mM EDTA (⋯⋯), 1 mM $Mg^{2+}$ (– – – – –), or 1 mM $Mg^{2+}$ + 0.5 mM $Ca^{2+}$ (——) (Reprinted with permission from Picard et al. 2006, © 2006 American Chemical Society).

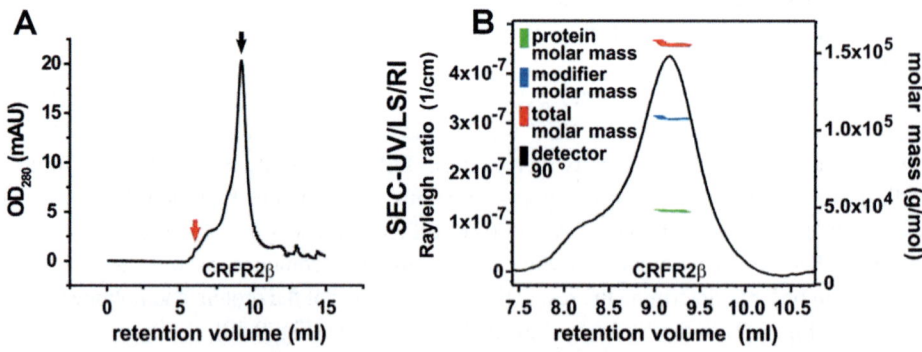

**Fig. 9.9** Size exclusion chromatography analysis of the corticotropin-releasing factor receptor CRFR2$\beta$ expressed in vitro in the absence of surfactant, solubilized in detergent, and transferred to NVoy. (**A**) Elution profile of CRFR2$\beta$ from a Shodex Protein KW-803 column. The aggregated and monomeric fractions are indicated by *red* and *black* arrows, respectively. (**B**) Analysis of a purified fraction using a combination of UV, light scattering at 90°, and refractive index measurements. The molar mass deduced for CRFR2$\beta$ (*green*), the polymer ("modifier," *blue*), and their complex (*red*) is indicated (From Klammt et al. 2011, © 2011 The protein Society).

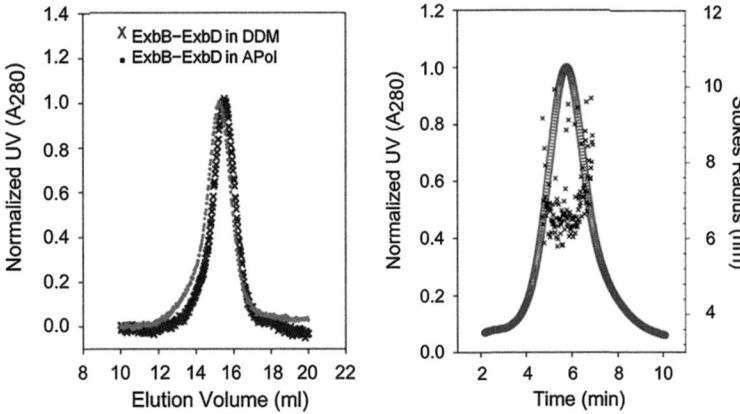

**Fig. 9.10** Size exclusion chromatography analysis of ExbB$_4$/ExbD$_2$ complexes. *Left.* Analytical SEC of the DDM-solubilized or A8-35-trapped complexes on a Superose 6 SEC column. *Right.* Independent measurements of the Stokes radius $R_S$ by asymmetric flow field-flow fractionation-multi-angle laser light scattering (AF4-MALLS) are consistent with the $R_H$ of protein/APol complexes determined by SEC (From Sverzhinsky et al. 2014).

detergents to keep MPs soluble and endow them with a charge proportional to the amount of bound dye. Upon electrophoresis on a polyacrylamide gel in non-denaturing conditions, MP/CBB complexes migrate as a function of their molecular mass, which can be roughly estimated, and of their charge, essentially carried by CBB. Following migration, their activity can be assayed in the gel (Schägger and von Jagow 1991).

BN-PAGE can be readily applied to APol-trapped MP complexes (Table 9.1). In Study 9.18, it was used to identify mitochondrial supercomplexes fractionated in sucrose gradients either in digitonin-solubilized form (Fig. 9.11, *left*) or after trapping with A8-35 (*right*). Other studies of either A8-35- or NAPol-trapped MPs or MP complexes have also resorted to BN-PAGE (Table 9.1). At the time of this writing, it has not been examined to which extent CBB mixes with APols or totally displaces them. However, the general similarity in the position of the bands in the gels on the left and right sides of Fig. 9.11, which correspond, respectively, to digitonin-solubilized and to A8-35-trapped supercomplexes, tends to suggest that displacement is more or less complete.

## 9.3.4   Chemical Cross-Linking

Trapping with APols does not prevent studying protein-protein interactions by chemical cross-linking either with glutaraldehyde (Studies 9.2 and 9.3) or with disuccinimidyl tartrate (Study 9.30).

## 9.3.5   Electron Resonance Spectroscopy

In Study 9.24, double electron-electron resonance (DEER) spectroscopy was applied to measure the distance between two nitroxide-labeled positions in SMA-trapped BR, so as to compare it with distances measured in BR crystals.

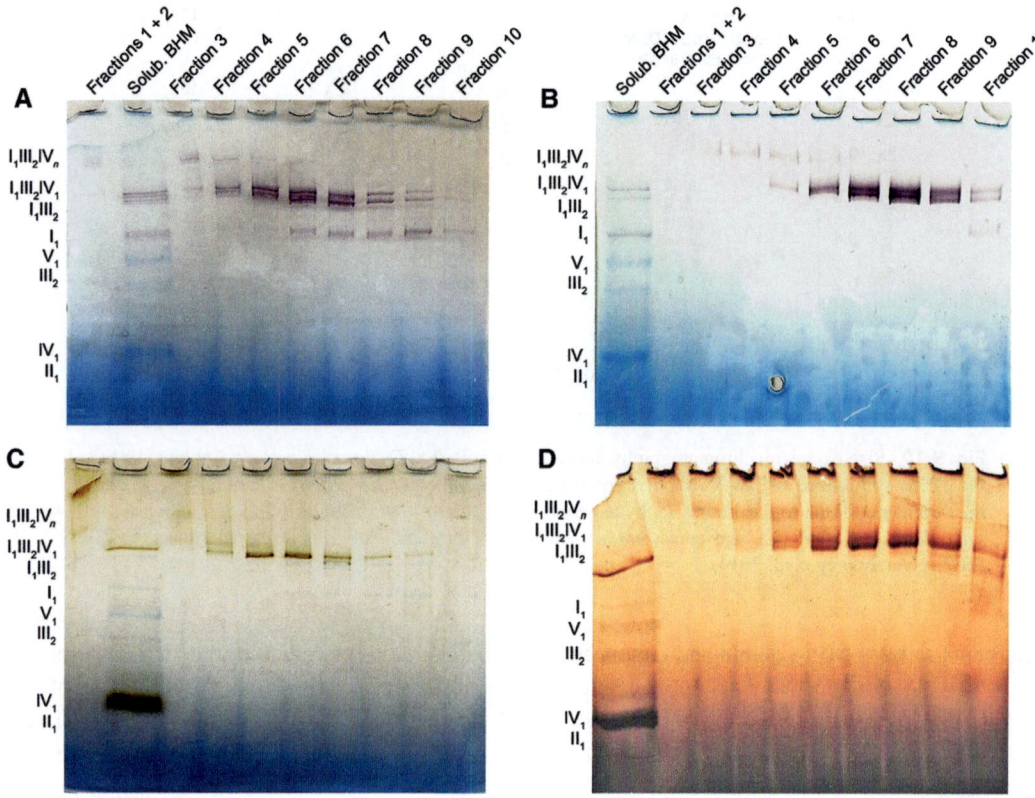

**Fig. 9.11** Identification by BN-PAGE of mitochondrial supercomplexes separated on sucrose gradients. Supercomplexes solubilized with digitonin (**A**, **C**) or trapped with A8-35 (**B**, **D**) were separated by SGSV. Fractions were collected and analyzed by 3–10% BN-PAGE. The gels were incubated either with *p*-nitrotetrazolium/*β*-NADH to reveal Complex I activity (**A**, **B**) or with 3,3′-diaminobenzidine tetrahydrochloride/cytochrome *c*/catalase to reveal Complex IV activity (**C**, **D**). The blue color is due to CBB stain from BN-PAGE (From Althoff et al. 2011, © 2011 European Molecular Biology Organization).

### 9.3.6    Determination of Rotational Correlation Times by Solution NMR

Solution NMR can be used to estimate the hydrodynamic radius of particles, thanks to the determination of their rotational correlation time, $\tau_c$, which can be deduced from relaxation measurements (see Chap. 10, Fig. 10.9). The two values are related by the following formula:

$$\tau_c = 4\pi\eta_0 R_H^{\,3}/3kT$$

where $\eta_0$ is the viscosity of the solvent.

This approach has been applied to tOmpA/A8-35 (Study 9.9) and OmpX/A8-35 complexes (Studies 9.14 and 9.33). It revealed that (i) the tumbling of OmpX/A8-35 complexes is significantly slower ($\tau_c = 31 \pm 4$ ns) (Catoire et al. 2010) than that of OmpX/diC$_6$PC ones ($\tau_c = 23 \pm 2$ ns) (Fernández et al. 2001), consistent with a larger $R_H$, and (ii) when EDTA is omitted from the buffer, the rate of tumbling of OmpX/A8-35 complexes slows down, their rotational correlation time increasing from $\tau_c = 31 \pm 4$ ns (Catoire et al. 2010) to $\tau_c = 39 \pm 5$ ns (Etzkorn et al. 2014). This observation suggests, in keeping with SEC data (§ 9.3.2; Fig. 9.8), that traces of $Ca^{2+}$ degrade the resolution of solution NMR spectra by bridging some of the particles via their A8-35 belt.

### 9.3.7    Analytical Ultracentrifugation

Analytical ultracentrifugation (AUC) is a particularly powerful method for determining the size and mass of solutes. It can be used at equilibrium, by subjecting the sample to a relatively low gravity field and examining its distribution throughout the buffer column once it has reached final equilibration, or as a sedimentation velocity technique, in which case the redistribution of the various solutes is followed throughout the experiment (see Box 9.1; for reviews, see e.g. Laue and Stafford 1999; Carruthers et al. 2000; Schuck 2000; Padrick et al. 2010; Ebel 2011; and references therein). The latter approach is particularly powerful when dealing with mixtures, as is usually the case of MP/surfactant preparations, which generally contain both MP/surfactant complexes and free surfactant particles. Detection can be achieved either by absorbance spectroscopy at multiple wavelengths or by measuring the refractive index of the solution, measurements being carried out at each point along the buffer column as a function of time. The data are then decomposed to yield the sedimentation coefficients and relative concentrations of the various species present in the solution, as well as, under favorable circumstances, the ratio of various components in a particle. AUC has been applied to eight APol-trapped MPs, ranging from small (BR; ~27 kDa) to moderately large (AcrB trimer; ~360 kDa), trapped in A8-35, in one of its fluorescent or deuterated derivatives, in NAPols, or in SMA (see Table 9.1).

As a first example of the kind of data obtained, Fig. 9.12 presents the results of AUC analyses of two BR/A8-35 samples (Study 9.5; Gohon et al. 2008). In one case, BR was trapped using a batch of unlabeled A8-35 (HAPol), which, upon SEC, was found to form monodisperse particles (cf. Fig. 4.14A). In keeping with SEC data (Fig. 9.7A), AUC showed that the preparation of HAPol-trapped BR comprised only two types of particles, pure HAPol particles, migrating with a sedimentation coefficient $s_{20,w} = 1.6$ S, as observed with pure A8-35 preparations (Gohon et al. 2006), and monomeric BR/APol particles, with $s_{20,w} = 3.2$ S, representing ~98% of the BR present in the sample (Fig. 9.12C; cf. the SEC analysis presented in Fig. 9.7A). The ratio $A_{280}/A_{550} = 1.6$ in the BR peak indicates that BR is in its native state. From a comparison of interference measurements (J), which detect all species with a refractive index different from that of the solution, with absorbance measurements at 280 nm (protein) and 550 nm (BR in its native state), one can deduce that BR/A8-35 complexes comprise ~2 g A8-35 per g of BR (taking into account the contribution of lipids, whose mass ratio in the complexes was determined by biochemical analysis; Gohon et al. 2008) (Table 9.2).

A second sample was prepared using a batch of deuterated A8-35 (DAPol), which, for reasons unrelated to the deuteration, turned out to be more hydrophobic than standard A8-35 and, upon SEC, formed slightly polydisperse particles. AUC of the preparation of BR trapped using this batch showed the presence of free DAPol particles, with $s_{20,w} \approx 2$ S, a majority of monomeric DAPol-trapped BR ($s_{20,w} \approx 4$), a small fraction (~11%) of dimers, and traces (~2%) of trimers (Fig. 9.12C′; see also SEC analysis in Fig. 9.7A). The ratio $A_{280}/A_{550} = 1.6$ in the BR peaks again indicates that the protein is in its native state. Some of the data deduced from these two series of experiments are compiled in Table 9.2.

Sedimentation velocity experiments were completed with equilibrium sedimentation ones (Fig. 9.13). Analyses carried out in buffers prepared in either $H_2O$, $D_2O$, or $D_2{}^{18}O$ yielded information about the composition of BR/A8-35 particles through the determination of their buoyant molar mass ($M_b$), which, in a given solvent, depends on the particle mass and density (see Box 9.1). Data were

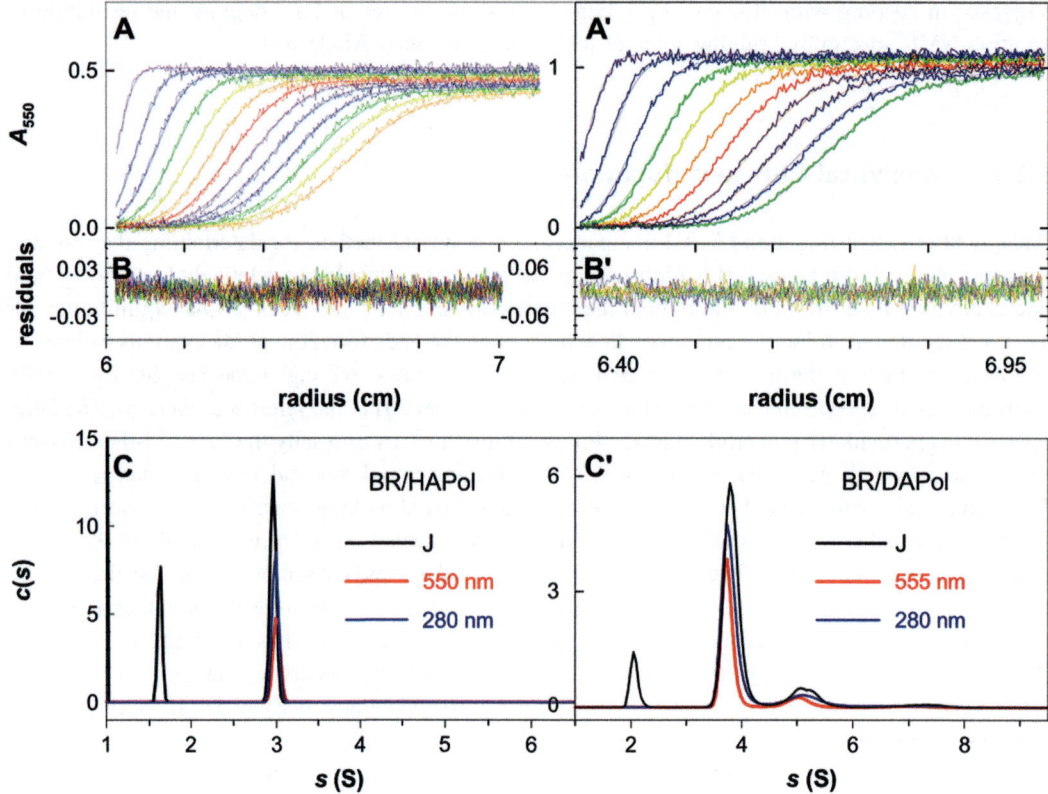

**Fig. 9.12** Sedimentation velocity analysis of bacteriorhodopsin (BR)/A8-35 complexes. (**A**) and (**A′**) superimposition of selected experimental and modeled profiles obtained at 20 °C in 3 mm optical path cells over 270 min at 42,000 rpm for a BR/HAPol preparation at 1 g·L$^{-1}$ (**A**) and over 90 min at 50,000 rpm for a BR/DAPol preparation at 2.3 g·L$^{-1}$ (**A′**). (**B**) and (**B′**) corresponding residuals. (**C**) and (**C′**) distribution $c(s)$ of the various components as a function of their sedimentation coefficient, $s$, determined using different optics. Data were collected (i) by absorbance spectroscopy at either 550 nm (BR in its native state, *red*) or 280 nm (any BR, whether native or not, *blue*) and (ii) by refractometry (J, *black*), which reveals both the protein and A8-35 particles, the latter having sedimentation coefficients of ~1.6 S (HAPol) or ~2 S (DAPol) (Gohon et al. 2006). HAPol is a batch of hydrogenated A8-35, which formed fairly monodispersed particles, as also found by SEC (see Chap. 4, § 4.3.1.2.2). DAPol is a batch of partially deuterated A8-35, which, for reasons related to the origin of the poly(acrylic acid) and not to the deuteration, was more hydrophobic than had been aimed for and whose particles showed a small degree of aggregation (ibid.). As a consequence, BR/HAPol complexes are monodisperse (**C**), whereas BR/DAPol ones show the presence of some small aggregates, mainly dimers (**C′**), as also observed by SEC (Fig. 9.7A) (From Gohon et al. 2008 Copyright © 2008 The Biophysical Society. Published by Elsevier Inc. All rights reserved).

acquired at 560 nm, so as to focus on the distribution of native BR. In D$_2$$^{18}$O buffer, the $M_b$ for BR/HAPol complexes is close to null, the particles being only slightly redistributed even at the highest angular velocity (Fig. 9.13C). Despite the slight heterogeneity of this sample (~9% dimer according to sedimentation velocity data), rather good fits were obtained assuming a single species of non-interacting particles (Fig. 9.13A–C). Two other samples of BR/HAPol and BR/DAPol were investigated in H$_2$O and D$_2$O buffers, using longer columns. Plots of $M_b$ values from all equilibrium sedimentation experiments as a function of solvent density indicate a good consistency of the results (Fig. 9.13D, E).

**Table 9.2** Sedimentation velocity analysis of two preparations of bacteriorhodopsin (BR)/A8-35 complexes. BR was trapped with either an unlabeled batch of A-35 (HAPol) or a batch in which the octyl and isopropyl side chains were perdeuterated (DAPol). The samples are those shown in Fig. 9.12. BR/HAPol complexes are 98% monomeric, whereas BR/DAPol complexes comprise a proportion of dimers (~11%) and trimers (~2%). $\varphi'$ and $\varphi'_R$ are two operational partial specific volumes, $\varphi'$ being calculated from the composition of the complexes and $\varphi'_R$ deduced from complementary sedimentation velocity measurements in $D_2O$ buffers (because the complex is a polyelectrolyte, $\varphi'_R$ is expected to be lower than $\varphi'$ by 0.01–0.02 mL·g$^{-1}$, in fair agreement with the observations). $s_1$, $s_2$, and $s_3$ are the sedimentation coefficients of the three forms of BR/DAPol complexes, interpreted as monomers, dimers, and trimers, respectively, and $A_1$, $A_2$, and $A_3$ their relative contribution to the total absorbance at 555 nm, $A_{tot}$. $M_w$ is the mass-averaged molar mass of all BR/DAPol species. Experimental errors are typically $\pm$ 0.1 S for $s$, $\pm$ 10% for APol/BR mass ratios, $\pm$ 0.01 mL·g$^{-1}$ for $\varphi'$ and $\varphi'_R$, and $\pm$ 0.05 for $M_w/M_1$ (From Gohon et al. 2008).

| | BR/HAPol | BR/DAPol |
|---|---|---|
| **Free APol** | | |
| $s_{20,w}$ (S) | 1.6 | 2.2 |
| Free APol/BR$_{(main\ species)}$ (g·g$^{-1}$) | 1.6 | 0.2 |
| **BR/APol complexes, major species** | | |
| $s_{20,w}$ (S) | 3.2 | 4 |
| $A_{280}/A_{555}$ | 1.6 | 1.6 |
| (Bound APol + lipids)/BR (g·g$^{-1}$) | 2.4 $\pm$ 0.2 | 2.1 $\pm$ 0.2 |
| Bound APol/BR (g·g$^{-1}$) | 2.0 $\pm$ 0.2 | 1.7 $\pm$ 0.2 |
| $\varphi'$ (mL·g$^{-1}$) | 0.868 | 0.824 |
| $\varphi'_R$ (mL·g$^{-1}$) | 0.84 | 0.80 |
| **BR/APol complexes, minor species** | | |
| $A_1/A_{tot}$ (at 555 nm) | 98% | 88% |
| $s_2/s_1$ ($A_2/A_{tot}$ at 555 nm) | | 1.37 (11%) |
| $s_3/s_1$ ($A_3/A_{tot}$ at 555 nm) | | 1.87 (2%) |
| $M_w/M_1$ | | 1.2 |

Figure 9.13, panels D and E, shows the expected individual contributions to $M_b$ of the different partners in a complex comprised of one monomer of BR and, per gram of protein, 0.38 g lipid (as inferred from chemical analysis) and 1.8 g APol (as deduced from sedimentation velocity experiments), as well as their sum (*solid red lines*). For BR/HAPol complexes, experimental values of $M_b$ are slightly larger than calculated ones. This small discrepancy may be related to the slight heterogeneity of the sample and/or to an underevaluation of the amount of bound HAPol – assuming 2.2 g HAPol/g BR (*dotted red line*) gives a better fit for purely monomeric BR – but the two phenomena of course could well contribute. For BR/DAPol complexes, a relatively good fit is obtained assuming the binding of 1.8 g DAPol/g BR and $M_w/M_1$ (defined in the legend to Table 9.2) to be 1.3 (*open symbols*), as found for this sample using sedimentation velocity analysis. Complementary analyses of BR/A8-35 complexes using SANS will be presented in § 9.3.8 and their consistency with AUC data discussed.

In Study 9.41, advantage was taken on the existence of a fluorescent, NBD-labeled form of A8-35, FAPol$_{NBD}$ (Zoonens et al. 2007; Le Bon et al. 2014b), used as a tracer, to follow the APol distribution at $\lambda = 490$ nm, whereas both the protein (ExbB/ExbD complexes) and the APol were detected at $\lambda = 280$ nm (Figs. 9.14 and 9.15).

In Study 9.47, sedimentation velocity AUC was applied to the analysis of complexes between SMA, lipids, and the trimeric AcrB multidrug transporter (Fig. 9.15) . The peaks are broad, which is thought to result from noise in the data more than from an intrinsic polydispersity of the samples, but they do reveal the existence of two population of particles, the smallest one a SMALP-trapped trimer ($s \approx 4.8$ S) and the other a doublet thereof ($s \approx 8.5$ S), whose relative proportion and exact

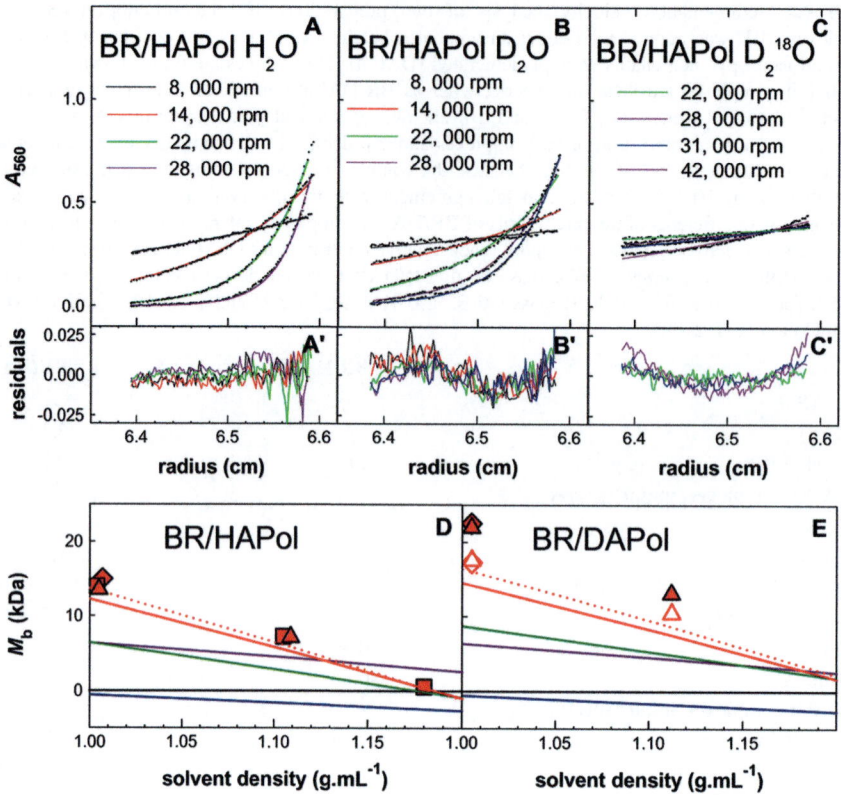

**Fig. 9.13** Sedimentation equilibrium analysis of bacteriorhodopsin/A8-35 complexes. (**A–C'**) Short-column sedimentation equilibrium centrifugation of BR/HAPol complexes. The complexes were transferred by SEC in 0.1 M NaCl, 0.2 M sodium phosphate, pH 7, before concentration (Centricon 30, Amicon) and 20× dilution in 0.1 M NaCl in $H_2O$, $D_2O$, or $D_2^{18}O$. For each buffer, data were collected at 1, 0.5, and 0.25 g·L$^{-1}$ and analyzed globally assuming the presence of a single type of particles. Data at 0.5 g·L$^{-1}$ (*black dots*) and fits (*colored lines*) (**A–C**) are shown above the corresponding residuals (**A'–C'**). (**D, E**) The buoyant mass, $M_b$, as a function of solvent density, $\rho°$, for BR/HAPol and BR/DAPol complexes. *Solid squares, triangles,* and *diamonds* refer to experimental sedimentation equilibrium data. *Open symbols* in E represent the buoyant mass $M_{b1}$ calculated for the monomeric BR/DAPol particle, derived from $M_b$ considering $M_w/M_1 = 1.3$, where $M_w$ is the mass-averaged average mass. Lines represent theoretical contributions from monomeric BR (*purple*), a full complement of lipids (*blue*), APol at 1.8 g/g BR (*green*), and the $M_b$ expected for a complex with either 1.8 (*solid red line*) or 2.2 g APol per g BR (*red dotted line*) (From Gohon et al. 2008, Copyright © 2008 The Biophysical Society. Published by Elsevier Inc. All rights reserved).

sedimentation coefficients depend on the ionic strength. Cryo-EM single-particle imaging revealed that the doublet is formed by two AcrB/SMALP complexes associated via their SMALP belts (see Chap. 12, Fig. 12.15).

## 9.3.8    Small-Angle X-ray and Neutron Solution Scattering

### 9.3.8.1    Introduction

Radiation scattering is a powerful approach to examining the mass, size, shape, and internal organization of solutes, as well as interactions between them. In addition to five DLS studies, which will be

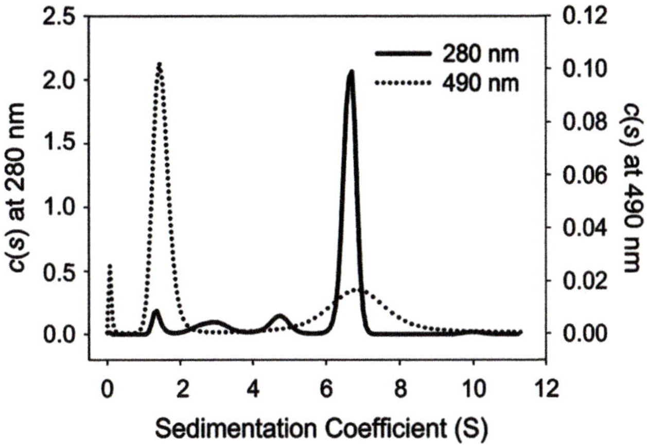

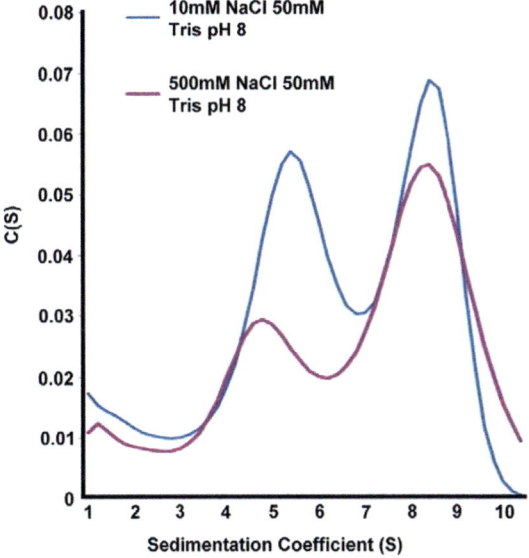

**Fig. 9.14**  $c(s)$ analyses of sedimentation velocity data following protein-trapping and supplementation with fluorescent A8-35 (FAPol$_{NBD}$) identify ExbB/ExbD/APol complex ($s \approx 6.9$ S) and the free APol ($s \approx$ 1.4–1.5 S). The spread of the $A_{490}$ peak around $s \approx 6.9$ S is probably due to experimental noise, resulting from the rather weak absorbance of the diluted FAPol$_{NBD}$, rather than to an actual broader distribution of the APol as compared to the protein to which it is bound (From Sverzhinsky et al. 2014).

**Fig. 9.15**  Sedimentation velocity analytical ultracentrifugation profiles of AcrB/SMA/lipid complexes at 10 mM and 500 mM NaCl in 50 mM Tris/HCl buffer, pH 8. Two distinct populations are seen, corresponding to singlets and doublets of SMALP-trapped trimers, with sedimentation coefficients of ~4.8 and ~8.5 S and molecular masses of ~405 kDa and ~810 kDa, respectively (From Postis et al. 2015).

presented in § 9.3.9, the solution properties of MP/APol complexes have been studied mostly by either small-angle neutron scattering (SANS; four studies) or small-angle X-ray scattering (SAXS; three studies) (Table 9.1). The principle and the principal characteristics and constraints of these two approaches are presented in Box 9.2.

**Box 9.2   X-ray and Neutron Scattering**

In broad terms, X-ray and neutron scattering, and which information can be gathered from them, are not conceptually difficult: in the same way as photons reflected by your environment, focused by your eye lens, detected by your retina, and analyzed by your brain will tell you whether you are facing a charging elephant or a pack of hamsters (and, hopefully, trigger the appropriate response), the photons or neutrons scattered by an ensemble of biological objects of nanometric dimensions contain information about their size, shape, composition, internal organization, interactions, and dynamics. The Devil, however, is in the details: the physics of radiation scattering is quite intricate, and its mathematics is discouraging to the normally constituted biochemist. The success and reliability of "small-angle" scattering experiments (in short, SAXS for X-rays and SANS for neutrons) therefore rely on the combined expertise involved in carefully preparing appropriate samples and rigorously collecting, correcting, and analyzing scattering data, which most often depends on setting up appropriate collaborations. It is essential, however, that the biochemist know what is doable and what to pay attention to and why, and the physicist or biophysicist have some notions of what dirty tricks biochemical samples may have up their sleeves. The present box has no other objective than to offer this basic background. For a more detailed and rigorous treatment, the reader is referred to such textbooks as Cantor and Schimmel (1980), Svergun et al. (2013), and Zaccai et al. (2017).

Here are a few key points:

- *Origin of the contrast and how to modify it.* Neutrons interact with nuclei and X-rays with electrons. This has various consequences that make the two approaches complementary. The contrast between particles and solvent depends, for X-rays, on their respective electron density and, for neutrons, on the properties of their nuclei, measured by a parameter, $\rho$, known as the "scattering length density." Elements with a high atomic number carry more electrons and scatter X-rays more strongly than lighter elements. Such is not the case with neutrons, the interaction with which depends on nuclear properties that do not vary monotonically. X-rays are insensitive to isotopic labeling, whereas in SANS, the contrast between various components of a particle and between them and the solvent can be modulated by it. In particular, $^1H$ nuclei scatter neutrons in opposition of phase with all other common nuclei, including $^2H$, in effect subtracting their contribution from that of most other atoms (which is why the scattering length density of $^1H_2O$ is negative; see Fig. 9.16A). This has two extremely useful consequences:

   - First, by preparing buffers in appropriately chosen $H_2O/D_2O$ mixtures, it is possible to "contrast-match" one or another component of a complex particle, e.g. to make "invisible," at least at low resolution, the detergent or the APol that keeps a MP soluble. Figure 9.16A illustrates how the contrast between the solvent and various molecules changes as a function of the percentage of $D_2O$ in the solvent (for an overview, see Jacrot 1976). Proteins, for instance, have a point of zero contrast ("contrast-match point," CMP) of ~40% $D_2O$ (*pink line*), whereas lipids, which are richer in $^1H$, need only ~10% $D_2O$ (*purple line*). A8-35 is contrast-matched at ~23.5% $D_2O$ (Gohon et al. 2004) (Fig. 9.16B). Note that the hydrophobic effect is stronger in $D_2O$, which tends to favor aggregation. It is therefore advisable, when transferring a sample to $D_2O$ or $H_2O/D_2O$ mixtures, to check on the absence of perturbations, e.g. by SAXS or AUC.
   - Second, by deuterating selectively one of the components of a particle, its contrast with $H_2O$ is enormously increased and its CMP displaced to much higher $D_2O$ concentrations (see deuterated protein and DNA in Fig. 9.16A). Two deuterated versions of A8-35 are available: one whose side chains only are deuterated, with a CMP of ~85% $D_2O$ (Gohon et al. 2004; Fig. 9.16B), and the other perdeuterated, with a CMP above 100% $D_2O$ (Giusti et al. 2014). Working in $D_2O$-rich buffers that contrast-match deuterated APols has two great advantages. First, it increases the contrast between the MP and the solvent; second, it diminishes considerably the background noise, a large fraction

**Box 9.2 (continued)**

of which originates from disordered ("incoherent") scattering of neutrons by the $^1$H nuclei of light water. As a result, much better signal-to-noise ratios are obtained when studying a hydrogenated MP trapped in a contrast-matched deuterated APol than if the two components are hydrogenated.

- Because X-rays interact with electrons, contrast variation implies chemical or physical-chemical changes to the sample, such as adding sugar or salts to the buffers in order to increase their electron density, and it is less widely used.

- *Interpretation of scattering curves.* In a SAXS or SANS experiment, the intensity scattered is measured as a function of the angle $2\theta$ between incident and scattered waves. This intensity depends on the size, shape, composition, etc. of the scattering particles, on their concentration, but also on the wavelength $\lambda$ of the radiation. It is useful to combine the wavelength and the scattering angle into a single parameter, which makes the analysis independent of $\lambda$. Inconveniently, but traditionally, different parameters are used in SANS and SAXS: SANS data are usually plotted as a function of $Q$ or $Q^2$, where $Q = (4\pi \sin\theta)/\lambda$, and SAXS data as a function of $s$ or $s^2$, with $s = (2 \sin\theta)/\lambda$, $Q$ and $s$ having dimensions of either $nm^{-1}$ or $Å^{-1}$. As to the intensity, $I(Q)$, which, in the case of independent particles, is proportional to their concentration, $C$, it is often normalized as $I(Q)/C$, where $C$ is expressed in $g \cdot L^{-1}$, so as to permit easy comparisons between data sets collected at different concentrations (see e.g. Fig. 9.19A or Fig. 4.14D in Chap. 4).

As in a diffraction pattern, to which scattering curves are closely related, data at high $Q$ or $s$ contain information about short distances between scattering centers in the sample, e.g. the distribution of individual atoms within a particle, and data at very small angles information about long distances, such as the distribution of the particles in the solution. The presentation of scattering data resorts principally to three types of graphs, each of which facilitates extracting certain types of information.

- Guinier plots (Guinier and Fournet 1955). A common representation is to plot $\ln I(Q)$ (or $\ln (I(Q)/C)$) vs. $Q^2$ (or $\ln I(s)$ vs. $s^2$) (Fig. 9.17). A single look at this plot tells the experimenter, at a glance, (i) whether the preparation is likely to be homogeneous or not and, if it is, (ii) what the molecular mass $M$ and radius of gyration $R_g$ of the particles are. $M$ is extracted, knowing the contrast of the particle with the solvent and its concentration in $g \cdot L^{-1}$, from the extrapolation of the plot at zero angle ($I(0)/C$). The slope of the curve at small angles yields $R_g$, the radius of gyration, which is determined by the spatial distribution of the scattering centers in the particle ("small angles," here, means that the Guinier approximation holds, typically, for $Q \cdot R_g \leq 1$, although for certain shapes it may extend further in $Q$).

Deviations from linearity at small angles mean that the preparation is not comprised of identical, independent particles. Repulsion between particles, for instance, will result in a downward deflection at low angles (Fig. 9.17A; see e.g. Fig. 9.19A and Fig. 11.13 in Chap. 11), whereas aggregation, or the presence of large contaminants, will create an upward deflection (Fig. 9.17B; see e.g. Fig. 9.19B and Fig. 4.14D in Chap. 4). These deviations are not too difficult to handle as long as they do not affect the higher-angle part of the Guinier region. Such is the case, for instance, for a limited contamination with much larger particles, because they do not scatter significantly beyond very small angles (for an example of this situation, see Fig. 4.14D in Chap. 4). On the contrary, contamination with relatively similar particles, such as a proportion of dimers in a population of monomers, can be difficult to detect, because the deviation from linearity may be lost in the experimental noise (cf. below; Fig. 9.19B). If this polydispersity is overlooked, it will bias the analysis toward an overestimation of both the mass and $R_g$ of the monomer. It is therefore essential to characterize carefully the dispersity of SANS and SAXS samples, e.g. using AUC.

- Shape information. Beyond the Guinier region – at wider angles, that is – the scattering curve contains information about the shape of the particles. This is illustrated in Fig. 9.18A, which compares the scattering curves of three particles of the same volume,

**Box 9.2 (continued)**

but different shapes, namely spheres (*black*), oblate (lenticular) ellipsoids (*green*), and prolate (cigar-like) ellipsoids (*blue*). The difference is due to interference between photons or neutrons scattered by different points within each particle. By Fourier transformation, one obtains the pair-distance distribution function, $p(r)$ (Fig. 9.18B), which describes the probability of observing a given distance $r$ between two points within a particle. Analysis of this function provides precious shape information. It is obvious, for instance, that if one distorts a sphere into a cigar, the function $p(r)$ will extend to longer distances, corresponding to those between scatterers located at the two ends of the cigar: compare the black curve (spheres) with the blue one (prolate ellipsoids) in Fig. 9.18B. Comparing experimental scattering curves and $p(r)$ functions with those of models is a powerful approach to studying the shape of particles and determining, for instance, the arrangement of two protomers into a dimer. Computer programs permit ab initio constructions of models by adjusting particle shapes created in silico to optimize the fit of predicted to experimental scattering curves (see e.g. Figs. 9.21 and 9.22).

- *Information on the dynamics of the particles.* The low-energy ("thermal") neutrons used in scattering experiments can exchange with the nuclei they interact with amounts of energy that are measurable as a change in their wavelength and speed. These contain information about the diffusion of the particles as a whole ("quasi-elastic" scattering), as well as about the vibrations of atoms or groups of atoms within them ("inelastic" scattering). This approach can be used to measure the translational diffusion coefficient $D_t$ of the particles, as well as to determine whether they are "hard" or "soft," such as below or above a phase transition from viscous to liquid-like (cf. § 4.3.1.2.4 in Chap. 4).

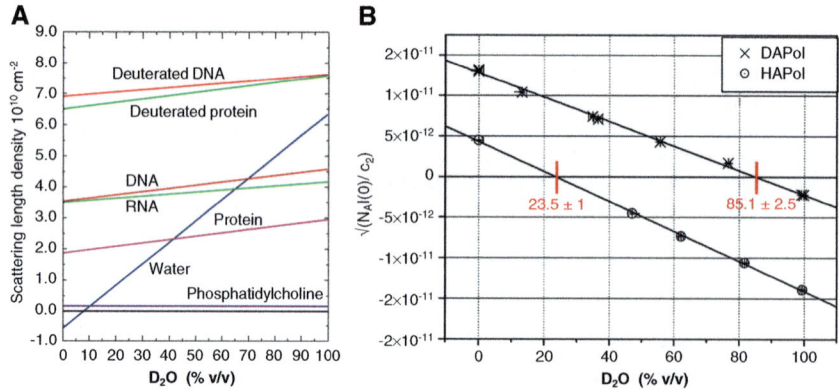

**Fig. 9.16** (**A**) Neutron scattering length densities of various natural-abundance and deuterated biological molecules as a function of the percentage of $D_2O$ in the solvent (From Zaccai et al. 2017). For any given percentage of $D_2O$ in the solvent, the contrast is given by the absolute difference between the scattering length density of the solvent and that of the particle. Note that the scattering length density of biological molecules increases more or less rapidly with the $D_2O$ content in the solution, depending on how rich they are in exchangeable hydrogens. Thus, the scattering length density of phosphatidylcholine, like that of most lipids and detergents, is essentially insensitive to the $D_2O$ content of the solvent, and due to its high content in non-exchangeable hydrogen atoms, it features a low contrast-match point (CMP) of ~10% $D_2O$. In contrast, the CMP of proteins lies around 40% $D_2O$, and their scattering length density increases with the $D_2O$ content of the solvent. (**B**) Unlabeled A8-35 (HAPol) and A8-35 whose main chain is hydrogenated and octyl and isopropyl side chains deuterated (DAPol) are contrast-matched at ~23.5% and ~85% $D_2O$, respectively (Adapted from Gohon et al. 2004). Perdeuterated A8-35 (perDAPol) is calculated to be contrast-matched at ~120% $D_2O$, meaning that, to be contrast-matched in 100% $D_2O$, it would have to be diluted with HAPol or DAPol (Giusti et al. 2014).

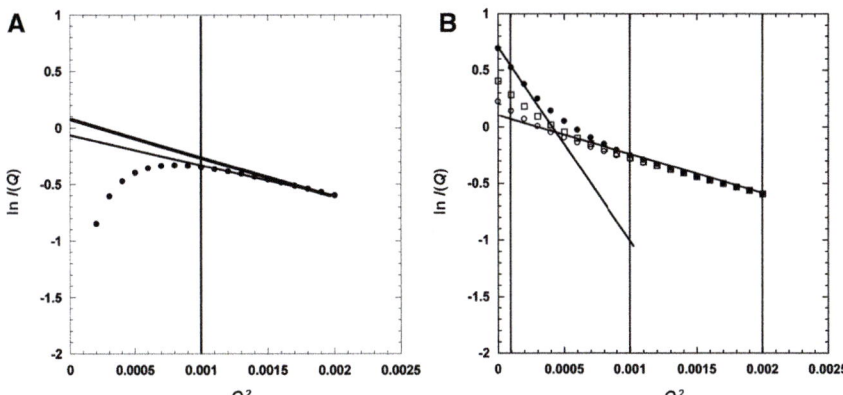

**Fig. 9.17** Guinier representation of scattering data. Within a range limited approximately by $Q \cdot R_g \leq 1$, the plot is linear if the particles in solution are all identical and randomly distributed in position and orientation. The particle radius of gyration is calculated from the slope of the plot, and its molecular mass from the extrapolation to $Q = 0$, provided particle composition and concentration are known. (**A**) The dip at low $Q$ values in the Guinier plot indicates that the assumption of random distribution is not justified: there is an *interparticle effect*. The vertical line is at $Q^2 \cdot R_g^2 = 1$, and the "correct" radius of gyration corresponds to the bold upper straight line. The thinner straight line below, which better fits the data, significantly underestimates the value of the radius of gyration. In such a case, all efforts should be undertaken to reduce the interparticle effect, for example by reducing the particle concentration or increasing the salt concentration in the case of charged particles. (**B**) The plot illustrates the effect of aggregation. Guinier representation data were calculated for a solution of particles with $R_g = 31.6$-Å contaminated with different proportions of 100-Å $R_g$ ones. The vertical lines at $Q^2 = 0.0001$ and $0.001$ are at $Q \cdot R_g = 1$ for the large and small particle, respectively. The slope of the straight line fitting the high-$Q$ points corresponds to $R_g = 31.6$ Å. Note how the data points rise progressively at low $Q$ values as the contribution of larger particles increases. The steep line is a fit to low $Q$ values for a mixture containing 2/3 small and 1/3 large particles. The corresponding apparent $R_g^2$ value is a weighted average of the $R_g^2$ of component particles. Similarly to the interparticle effect, if the data indicate aggregation, all efforts should be made to reduce it. The examples given in (**A**) and (**B**) illustrate large interparticle and aggregation effects, respectively. Where there are smaller effects, however, the Guinier plots may not show the downward or upward swings as clearly. The straight line fits would then provide underestimated (in the case of interparticle effects) or overestimated (in the case of aggregation) $R_g$ values. A careful experiment should include a concentration series to dispel the uncertainty (see e.g. Fig. 9.19A) (Figure and legend courtesy of G. Zaccai).

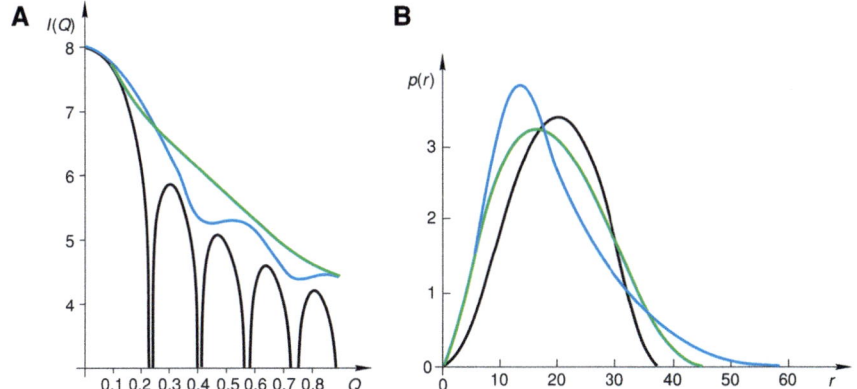

**Fig. 9.18** Scattering intensity $I(Q)$ (**A**) and pair-distance distribution function $p(r)$ (**B**) of a sphere (*black*), an oblate ellipsoid (*green*), or a prolate ellipsoid (*blue*) (From Zaccai et al. 2017).

In the following two sections, some of the results obtained by SANS and SAXS analyses of MP/APol complexes are presented, with particular attention being paid to the problems and opportunities specific to the use of APols.

### 9.3.8.2 Small-Angle Neutron Scattering

*SANS Studies of BR/A8-35 Complexes*

The first SANS study of an APol-trapped MP was that of BR complexed by either unlabeled A8-35 ("HAPol") or a partially deuterated version of it ("DAPol") obtained by derivation of unlabeled polyacrylic acid (PAA) with deuterated octyl- and isopropylamine (Gohon et al. 2004, 2006). This study (Study 9.5) is particularly instructive because, being the first of its kind, it had the privilege of discovering – by the virtue of falling into them – quite a few of the traps that should be paid attention to when carrying out such investigations. It is therefore worth presenting in some details, without smoothing away its wrinkles.

Figure 9.19A presents the variation of the neutron scattering intensity, $I(Q)$, by solutions of BR/DAPol complexes as a function of $Q^2$. In the particular case of Fig. 9.19A, $I(Q)$, which is directly proportional to the concentration of particles, has been divided by $c_{BR}$, the concentration of BR, so that the shape of the scattering curves collected at different BR concentrations can be compared. Three points are worth noting:

(i) The scattering curves appear linear over a large part of the Guinier region (marked by dashed vertical lines in Panel B), which is consistent with the preparations being monodisperse, but does not prove that they are strictly so. As a matter of fact, they are not, sedimentation velocity experiments similar to those of Fig. 9.12 showing that the preparation actually comprises ~75% monomers, ~20% dimers, and ~5% trimers. This polydispersity must be taken into account in the analysis, or conclusions about the mass and $R_g$ of the particles will be off the mark.

(ii) The slope of the linear region does not change with the concentration, which indicates that no association/dissociation process takes place as the concentration varies.

(iii) In the very small-angle region, a downward deflection is seen at high concentration (10 $g \cdot L^{-1}$), which progressively disappears as the preparation is diluted. This feature indicates that at high concentration, particles repulse each other. Electrostatic repulsion between the negatively charged MP/A8-35 complexes and A8-35 particles has indeed been observed to control the rate of exchange between free and protein-bound APols (Zoonens et al. 2007) and is also observed, using SAXS, between cytochrome $bc_1$/A8-35 complexes (Charvolin et al. 2014; see Chap. 11, Fig. 11.13). The data in Panel A indicates that 1.9 $g \cdot L^{-1}$, the lowest concentration tested, provides data of sufficient quality without any such perturbation in the Guinier region.

In the experiments presented in Fig. 9.19B, the concentration of BR has been kept at 1.9 $g \cdot L^{-1}$, and the percentage of $D_2O$ in the buffer has been varied between 0% and 100%, with intermediate concentrations of 40% (*diamonds*), at which BR is contrast-matched; 65% (*squares*), at which the whole complex is matched; and 84.5% (*inverted triangles*), at which DAPol is matched. The shape of the scattering curves is seen to vary with the $D_2O$ content, because the relative contributions of the complexes and of the free DAPol particles that are also present in the preparation depend on their respective contrast with the solvent. When BR (*diamonds*) or the BR/DAPol complexes (*squares*) are contrast-matched, increased scattering at very small angles ($Q^2 < 0.001$ Å$^{-2}$) reveals the presence of

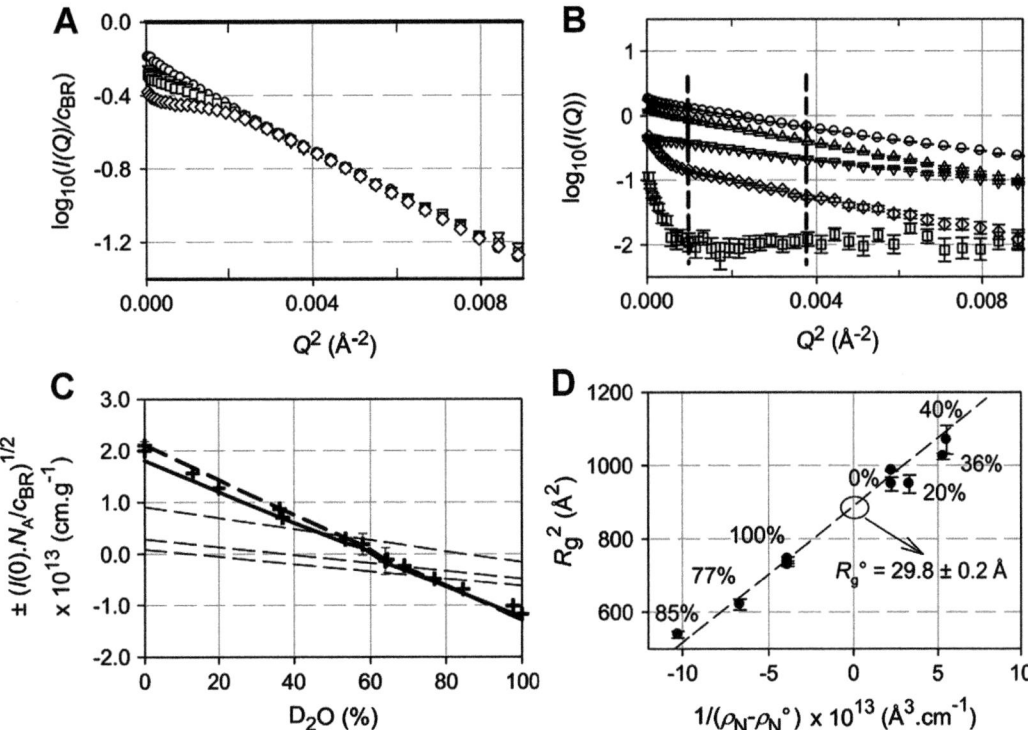

**Fig. 9.19** Small-angle neutron scattering study of bacteriorhodopsin trapped by a partially deuterated form of A8-35 (DAPol). (**A**) Guinier plots of a BR/DAPol preparation in $H_2O$ buffer, showing the effect of dilution (intensities are normalized by the concentration of BR). From bottom to top, the concentration of BR decreases from 10.0 g·L$^{-1}$ ($\lozenge$) to 5.0 ($\square$), 2.5 ($\triangledown$), and 1.9 g·L$^{-1}$ ($\circ$). The downward deflection from linearity in the small-angle region is due to the electrostatic repulsion between particles. It vanishes at the lowest concentration. (**B**) Guinier plots for the same sample at 1.9 g·L$^{-1}$ BR at various $D_2O$ contents in the buffer. Dashed vertical lines delimit the $Q^2$ domain (Guinier region) used to determine $R_g$ and the forward intensity, $I(0)$. From bottom to top, the $D_2O$ content varies from 65% ($\square$, close to the contrast-matching point of the BR/DAPol complex) to 40% ($\lozenge$, contrast-matching of BR), 84.5% ($\triangledown$, contrast-matching of DAPol), 0% ($\triangle$), and 100% ($\circ$). At 65% and 40% $D_2O$, the upward deflection of the curve at small angles is due to the presence of a small fraction of very large protein-free DAPol particles (Gohon et al. 2006; see Chap. 4, § 4.3.1.2, Fig. 4.14). (**C**) Forward intensities as a function of $D_2O$ content in the solvent. Crosses are experimental values for two sets of measurements at [BR] = 1.9 g·L$^{-1}$. Thick lines are calculations, assuming BR/DAPol complexes to comprise 75% monomers, 20% dimers, and 5% trimers, as indicated by sedimentation velocity analysis, with a full complement of purple membrane lipids and either 1.8 (model 1, ——) or 2.2 (model 2, – – – –) g bound DAPol per g BR, along with 0.38 g·L$^{-1}$ free DAPol particles. The thin dashed lines represent the individual contributions of the individual components of the monomeric BR/DAPol complex, namely from bottom to top, purple membrane lipids at 0.38 g per g BR, BR, and DAPol at 1.8 g per g BR. (**D**) Stuhrmann plot ($R_g^2$ vs. the inverse of the contrast; Stuhrmann 1970), where $\rho_N$ is the scattering length density of the buffer as a function of $D_2O$ content and $\rho_N^\circ$ that of BR/DAPol particles at their contrast-matching point (61.2% $D_2O$). The dashed line represents the linear regression on $R_g^2$ measurements ($\bullet$). $R_g^\circ$ is the radius of gyration of the BR/DAPol complex at "infinite contrast", that is not weighted by the different contrasts of BR, lipids, and DAPol with the solvent (From Gohon et al. 2008, © 2008 The Biophysical Society) (Published by Elsevier Inc. All rights reserved).

small amounts of very large aggregates of pure DAPol, as observed in pure DAPol preparations (Gohon et al. 2006; see Chap. 4, § 4.3.1.2, Fig. 4.14). The aggregates stand out because of their large size (> 1 MDa; Gohon et al. 2006), as compared to ~90 kDa (see below) for the BR/A8-35 complexes. Had their presence been known in advance, they could have been removed by either filtration or ultracentrifugation of the stock solution of DAPol. However, they barely affect the analysis, because of three favorable factors: (i) they are rare (typically ~0.1% of DAPol in mass; see Gohon et al. 2006 and Chap. 4, ibid.); (ii) they do not bind BR (note the absence of low-angle deflection when DAPol is contrast-matched; *inverted triangles*); and (iii) due to their large $R_g$, their contribution falls rapidly at high $Q$, so that they do not scatter significantly in nor beyond the Guinier region used for determining the mass and $R_g$ of the complexes, where the signal from BR/DAPol complexes and small DAPol particles is predominant.

At ~85% $D_2O$ (*inverted triangles*), where DAPol is contrast-matched (Gohon et al. 2006), Guinier plots are essentially linear, consistent with the limited polydispersity of BR/DAPol complexes already observed by SEC (§ 9.3.2) and AUC (§ 9.3.7). The curvature of the curve due to the presence of ~20% dimers and ~5% trimers (see below) is barely detectable.

Figure 9.19C, D shows the normalized forward intensities (Panel C) and radii of gyration (Panel D), respectively, derived from the extrapolated value at zero angle and from the slope of the Guinier plots. In Panel C, the value of $\pm (N_A I(0)/c_{BR})^{1/2}$ is plotted as a function of the percentage of $D_2O$ in the solvent. This term includes the contributions of both BR/DAPol complexes and small free DAPol particles. The contribution of the latter is small (~15% of $I(0)$ in $H_2O$), but, close to the contrast-matching point of the complexes, it prevents perfect matching, so that $I(0)$ never falls exactly to zero. The composition of the BR/lipid/DAPol complexes determines their contrast-matching point (experimental interpolated value, 61 $\pm$ 2% $D_2O$), whereas the presence of multimers affects slightly the slope of the Guinier plot, from which the $R_g$ of the complexes is derived, and the value of $I(0)$, used to calculate their mass (cf. Panel B). Predicted $I(0)$ values were calculated for various models and compared to experimental ones. The BR/DAPol sample was assumed to contain a major species, comprised of one molecule of BR, purple membrane lipids in the same ratio as in the membrane, and 1.8 or 2.2 g DAPol per g of BR (models 1 and 2, respectively), along with an adjustable number of multimers thereof. Scattering by free DAPol particles (~0.38 $g \cdot L^{-1}$, based on sedimentation velocity data) was taken to be identical to that of pure DAPol solutions (Gohon et al. 2006). The calculated contribution of the different components is shown (*thin dashed lines*). A good fit of the forward intensities is obtained, for this sample, for 75% monomers, 20% dimers, and 5% trimers (w/w), which is consistent with sedimentation velocity results.

SANS data also contain information about the relative arrangement of the protein and APol in the BR/DAPol complexes. In Panel D, a so-called Stuhrmann plot is shown, in which the square of the radius of gyration of BR/DAPol particles, $R_g^2$, is plotted vs. the inverse of the contrast, $\rho_N - \rho_0$, between the complex and the solvent. The plot is linear, which indicates that BR and the DAPol moiety of the complexes have the same center of mass (Stuhrmann 1970). The positive slope of the plot indicates that, as expected, the component of highest scattering density (DAPol) lies farther from the center of mass of the particle than BR. The point where the plot crosses $(\rho_N - \rho_0)^{-1} = 0$ (infinite contrast) yields $R_g^\circ$, the radius of gyration that the complex would exhibit were it chemically homogeneous, a very useful constraint for model building.

From the ensemble of chemical analysis, AUC and SANS data, a crude model of the complex was built (Fig. 9.20). It includes one BR monomer with the crystallographic structure of BR and nine protein-bound purple membrane lipids (Belrhali et al. 1999), to which 2 g DAPol per g BR has been added as a 1.7 nm thick belt surrounding the TM surface. The $R_g$ for monomeric BR calculated from the PDB file is 1.8 nm; the $R_g^\circ$ ($R_g$ at infinite contrast) calculated for the model is 2.6 nm. As expected, given the slight heterogeneity of the samples (see above), experimental values are somewhat larger, namely $R_g \approx 2.3$ nm at the match point of DAPol (signal due to BR and lipids) and $R_g^\circ \approx 3.0$ nm (Fig. 9.19D). The $R_g$ at the match point of BR (where scattering is due to DAPol and lipids) is 3.2 nm, which is, logically, larger than that measured for pure DAPol particles (2.4 nm) (Gohon et al. 2006).

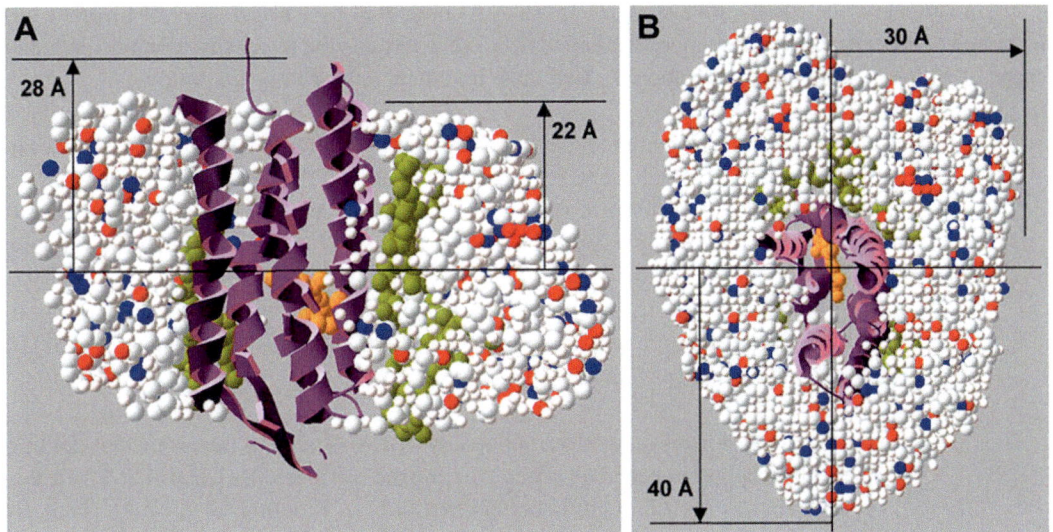

**Fig. 9.20** A model of BR/A8-35 complexes deduced from chemical analysis, AUC and SANS data. (**A**) Cross-section in a plane normal to the membrane plane. (**B**) View along an axis normal to the membrane plane. A monomer of BR (PDB code 1QHJ; Belrhali et al. 1999) is represented as a *purple* ribbon, with the retinal in *orange*, lipids in space-filling *olive green*, and DAPol in *white*, *red*, and *blue* beads (From Gohon et al. 2008, © 2008 The Biophysical Society. Published by Elsevier Inc. All rights reserved).

*SANS* vs. *AUC* vs. *SEC* vs. *EM Studies of BR/A8-35 Complexes: Consistency and Inconsistencies*

The various approaches used in Study 9.5 provide, in general, consistent estimates of the composition, size, and mass of monomeric BR/A8-35 complexes. Combining composition and $s$-values yields $R_S \approx 3.6$ nm for both BR/HAPol and BR/DAPol complexes, whereas combining sedimentation velocity and sedimentation equilibrium data gives $R_S \approx 3.8$ nm for BR/HAPol complexes. SANS yields an $R_g^\circ$ value for BR/DAPol complexes of ~3.0 nm, which, for spherical particles, would correspond to $R_S = R_g^\circ \times (5/3)^{1/2} \approx 3.8$ nm, in good agreement with AUC data. By EM, one observes half-widths and half-lengths of ~3.2 and ~4 nm for both BR/HAPol and BR/DAPol particles (Gohon et al. 2008).

The outlier is SEC, whose estimate of $R_S$ is significantly larger ($R_S \approx 5.0 \pm 0.15$ nm). Earlier observations had already suggested that, for reasons that are not clear yet, SEC tends to overestimate the hydrodynamic radius of MP/A8-35 complexes as compared to MP/detergent ones. Indeed, gel filtration (Zoonens et al. 2007) yields much larger apparent $R_S$ differences between tOmpA/A8-35 complexes ($R_S \approx 4.3$–$3.7$ nm) and tOmpA/di-$C_6$-phosphatidylcholine ones ($R_S \approx 2.6$ nm) than is indicated by an NMR study of their respective rotational correlation times, $\tau_c$ (Zoonens et al. 2005). The NMR data suggests that SEC overestimates the $R_S$ of tOmpA/A8-35 particles by as much as 30–50% (Zoonens et al. 2007), a difference close to that observed for BR in Study 9.5 (~30%). Electrostatic repulsion between the complexes and resin-bound A8-35 does not seem to account for this effect (for a discussion, see Zoonens et al. 2007). Note that, in keeping with the latter conclusion, SEC analysis of BR/non-ionic APol complexes also leads to an overestimation of $R_S$ as compared to AUC data (Sharma et al. 2012). This question is clearly in need of closer investigation.

In summary, the comparative study of BR/A8-35 complexes by a broad array of methods has identified some caveats in using biophysical approaches to determine the mass, size, composition, and shape of MP/APol complexes, of which the two most important appear to be the following:

- SEC is useful to examine the homogeneity of the preparations, but the $R_S$ values it provides in the conditions used in Studies 9.5 and 9.9 are overestimated. If the SEC-derived value of $R_S$ is used to derive the molecular mass of monomeric BR/A8-35 complexes from AUC sedimentation velocity data, it leads to strongly overestimating it (~130 kDa, whereas the actual mass is ~90 kDa).
- When using SANS or SAXS to study MP/APol complexes, great attention should be paid to the monodispersity of the samples. In Gohon et al. (2006, 2008) and Zoonens et al. (2007), two causes of polydispersity were identified:

  (i)   Unplanned-for excessive hydrophobicity of A8-35, due to the artifactual grafting of a hydrophobic secondary product during the synthesis (see Chap. 4, § 4.2.1); this problem has been brought under control by careful optimization of the synthesis protocol (see Gohon et al. 2006, and Protocol 4.1 in Chap. 4, § 4.5).
  (ii)  Oligomerization of the complexes upon complete removal of free APol (Zoonens et al. 2007; Gohon et al. 2008; Arunmanee et al. 2014; Charvolin et al. 2014; see Chap. 5, § 5.2.1). This problem can be avoided by working in the presence of a sufficient concentration of free APol (see e.g. Zoonens et al. 2007; Charvolin et al. 2014). For particularly fragile MP complexes, it may however be difficult to find conditions that will prevent oligomerization while avoiding any degree of disaggregation (see e.g. Study 9.41).

Whatever the cause of aggregation, if any, it must be carefully documented, so as to prevent any overestimation of the mass and dimensions of the monomeric complexes. AUC used in the sedimentation velocity mode is, by far, the most performing way to quantify the relative proportion of the various forms (cf. Fig. 9.12).

*Other SANS Studies of MP/APol Complexes*

Also relevant are the following SANS studies of MP/APol complexes:

- In Study 9.7, the solution behavior of cytochrome $bc_1$/A8-35 complexes has been studied by both SAXS and SANS (see § 9.3.8.3). SANS studies were conducted in buffers containing 17% $D_2O$, so as to eliminate most of the contribution to scattering of polyethylene glycol, detergent, and/or A8-35 (Charvolin et al. 2014).
- In Study 9.16, the mass, size, and shape of complexes between BR and a non-ionic APol have been studied by SANS. Scattering data at wide angles were exploited (cf. Box 9.2) to yield an envelope of the particle (Fig. 9.21).
- In Study 9.41, the composition and organization of ExbB/ExbD complexes have been studied by a broad array of techniques, including SANS and SAXS, and envelopes of the particles calculated from wide-angle data and compared to reconstructed EM envelopes (see § 9.3.8.3, Fig. 9.22).

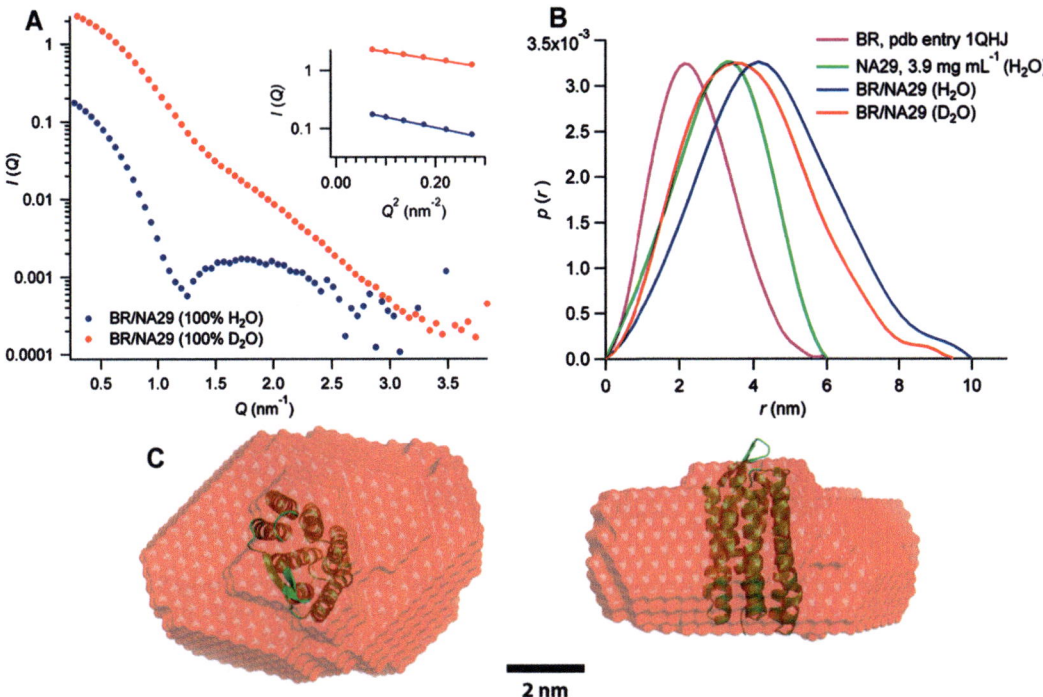

**Fig. 9.21** Small-angle neutron scattering (SANS) analysis of bacteriorhodopsin/non-ionic amphipol complexes and model of the particle. (**A**) SANS signal of BR/NAPol complexes in hydrogenated buffer (*blue*) or in deuterated buffer (*red*). The data sets were obtained by extrapolating *I/c* to infinite dilution and subtracting the signal from the free NAPol particles. The *inset* shows Guinier plots of the two curves. (**B**) Pair-distance distribution function $p(r)$ (normalized) for the BR monomer (calculated from PDB entry 1QHJ, *purple*), for free NAPol particles (batch NA29) in $H_2O$ (*green*), and for the BR/NAPol complex in either $H_2O$ (*blue*) or $D_2O$ (*red*) buffer. (**C**) Top and side views of the low-resolution envelope for BR/NAPol complexes in $D_2O$, derived from $p(r)$. Models giving the best fits to the extended scattering data were calculated using the program DAMMIF (Franke and Svergun 2009). A BR monomer (from Belrhali et al. 1999; PDB entry 1QHJ) was placed in the orientation providing the best fit between the height of its TM region and the thickness of the particle. Lipids have been omitted (Reprinted with permission from Sharma et al. 2012, © 2012 American Chemical Society).

### 9.3.8.3   Small-Angle X-ray Scattering

The principle of X-ray scattering is the same as that of neutron scattering, with the following main differences:

   (i) In SAXS, the contrast is due to differences in electron density between the solution and the particles, not to different interactions with the nuclei (see Box 9.2).

   (ii) Because the X-ray beams delivered by synchrotrons are much brighter than neutron beams from nuclear reactors or spallation sources, measuring times are shorter, making it possible to investigate many more conditions.

   (iii) On the downside, isotopic labeling cannot be used for contrast-matching, which is a serious disadvantage when studying MP/surfactant complexes.

SAXS has been used in two studies of MP/APol complexes (Studies 9.7 and 9.41) and one study of complexes formed between various APols and non-membrane, hydrophobic proteins extracted from seeds (Study 9.20).

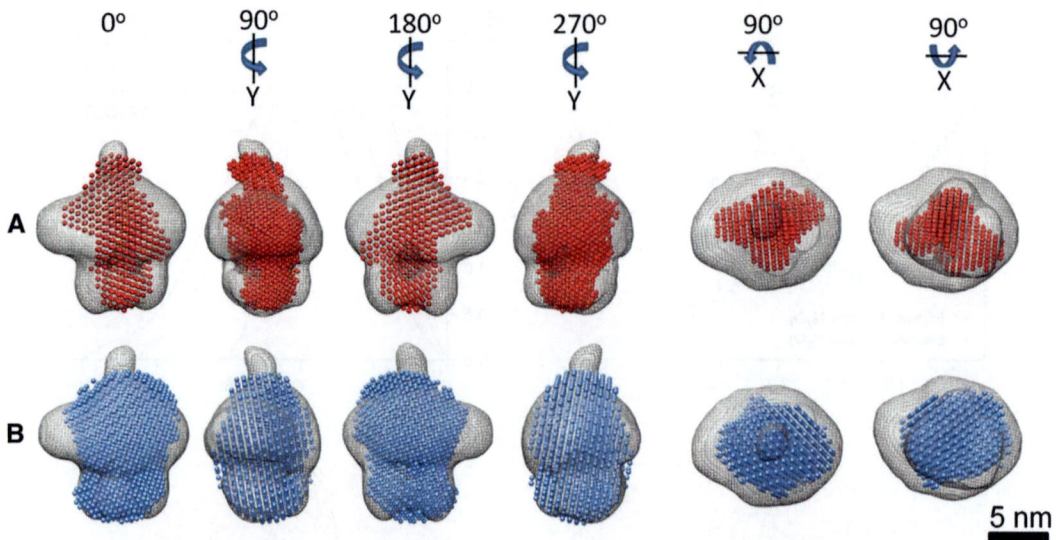

**Fig. 9.22** A comparison of the overall shape of the ExbB$_4$/ExbD$_2$/A8-35 complex as seen by SAXS (*blue dots*), SANS (*red dots*), and reconstruction of EM images of negatively stained single particles (*gray envelope*). After averaging multiple reconstructions (Svergun 1999) from SAXS and SANS data and selecting only high-occupancy positions, the resulting bead models were compared to a composite map of EM structural states. The SAXS and SANS models are consistent with each other: they have similar sizes along their long axis, and their overall dimensions are compatible with the EM composite map. The SAXS model (**B**) occupies a larger volume due to the contribution of the APol, which was contrast-matched and, therefore, invisible in the SANS study (**A**) (From Sverzhinsky et al. 2014).

- In Study 9.7, complementary use was made of SANS and SAXS to study the solution behavior of cytochrome $bc_1$/A8-35 complexes as a function of the concentration of salt and/or polyethylene glycol in the solution, with the view of providing guidelines for crystallization attempts (see Chap. 11, § 11.3.1, Fig. 11.13).
- In Study 9.41, ExbB/ExbD complexes were studied by a broad array of techniques, including SANS, SAXS, AF4-MALLS, and single-particle electron microscopy. Figure 9.22 presents a comparison of the data gathered by various methods about the size and shape of the complexes, with the negative-stain EM envelope in gray and the models selected for their fit with scattering data in *red* (SANS) and *blue* (SAXS). Note that the SANS model is thinner than the SAXS one, because the partially deuterated A8-35 bound to the TM region of the complex was contrast-matched in the first case and not in the second one.

### 9.3.9   Dynamic Light Scattering

Dynamic light scattering (DLS) has been abundantly used to examine the size and homogeneity of APol particles (see Chap. 4, § 4.3.1.2.2, and Table 4.2) but only rarely for the study of MP/APol complexes (Table 9.1). DLS provides a view of the distribution of particles as a function of their hydrodynamic radius $R_H$. As with any scattering technique, the scattered intensity increases, for a given composition, as the square of the molecular mass, so that large particles stand out even if their mass ratio is small (see e.g. Fig. 4.14B in Chap. 4). DLS is more useful to characterize mixtures of particles with widely different sizes than for those with relatively similar sizes, whose contributions

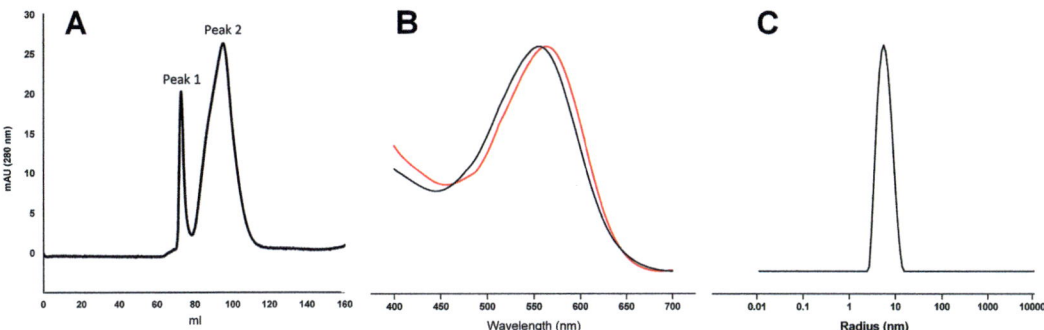

**Fig. 9.23** (**A**) Size exclusion chromatography purification of bacteriorhodopsin/SMALP nanoparticles, with the absorbance measured at 280 nm. (**B**) UV-visible spectra for Peak 1, with $A_{max}$ at 565 nm (*red*), and for Peak 2, with $A_{max}$ at 555 nm (*black*). (**C**) Dynamic light scattering data for Peak 2, yielding a particle diameter of $12 \pm 2$ nm (Reprinted with permission from Orwick-Rydmark et al. 2012, © 2012 American Chemical Society).

overlap. Examples of DLS studies of the complexes formed between various APols and non-membrane hydrophobic proteins are shown in Study 9.20.

In Studies 9.17 and 9.24, DLS has been used to characterize the complexes formed between BR or PagP and SMALPs. The SEC profile of the preparation obtained by transferring BR from lipid vesicles to SMALPs features two peaks absorbing at 280 nm (Fig. 9.23, Panel A). Peak 1 features a maximum of visible absorbance at 565 nm, suggesting that this population is mainly composed of purple membrane fragments, whereas Peak 2 absorbs maximally at 555 nm, consistent with BR being monomeric and incorporated into SMALP nanoparticles (Panel B). Analysis by DLS of fractions from Peak 2 gives a calculated diameter of ~12 nm, that is ~3 nm larger than the reported size of empty SMALPs (Panel C).

## 9.4   Conclusions and Prospects

From the data presented in the preceding section, one can draw the following conclusions:

- Sucrose gradient sedimentation velocity, SEC, BN-PAGE, DLS, AF4-MALLS, and AUC can all be used to examine the homogeneity of preparations of MP/APol complexes, with AUC providing the best resolution and DLS the worst one. SANS and SAXS also can be used to this end but may fail to clearly resolve species of similar sizes, such as a minor contamination of dimers in a large excess of monomers.
- SANS, SAXS, and AUC yield precise determinations of the hydrodynamic or gyration radius and of the molecular mass of the complexes. For reasons that need to be further investigated, SEC tends to overestimate $R_H$ and, therefore, the mass of MP/APol complexes. The acquisition of reliable data by SANS and SAXS critically depends on a careful analysis of the dispersity of the solutions, preferably by AUC.
- Thanks to the use of deuterated, radioactive, or fluorescently labeled APols, SANS, AUC, SEC, AF4-MALLS, and SGSV can provide data on the protein/APol mass ratio in the complexes. Attention should however be paid to the fact that separating MP/APol complexes from all free APol particles can result in protein aggregation and desorption of some of the

APols (Chap. 5, § 5.2.1, and Fig. 9.7D). Such measurements should therefore be carried out either in the presence of enough free APol to prevent these phenomena, or by focusing on the composition of the monomeric fraction of the preparations, as can be done, for instance, by multiwavelength analysis in AUC.

- Performed under carefully controlled conditions, SANS and SAXS have the potential to provide detailed data about the dimensions and shape of APol-trapped MPs. SANS is particularly powerful, because contrast-matching makes it possible to eliminate the contributions to scattering of both free and protein-bound APols. Comparing models built using data gathered with and without contrast-matching of the APol may help in their interpretation and that of electron microscopy images.

A detailed knowledge and control of the composition of MP/APol preparations is a strong asset when designing experiments aiming at collecting more detailed data by NMR, crystallography, or electron microscopy. In NMR, for instance (Chap. 10), the resolution of spectra can be improved by adding EDTA, a treatment that prevents traces of divalent ions present in the buffers from cross-linking MP/A8-35 complexes (Catoire et al. 2010). Homogeneity is a critical parameter for crystallization from solutions (Chap. 11, § 11.3.1). It may not be so important for crystallization in lipidic mesophases (ibid., § 11.3.2). However, the dispersity of the complexes before they are transferred to the mesophase will likely be a parameter worth controlling. In single-particle cryo-EM (Chap. 12), it is always possible to sort out the images of particles that are sufficiently different from one another, such as monomers and dimers, but this can become more difficult, for instance, with oligomers featuring different subunit stoichiometries. Previous knowledge of the homogeneity or heterogeneity of the preparations is definitely desirable.

The next three chapters will be devoted to a review and discussion of the studies of APol-trapped MPs by NMR, X-ray crystallography, and electron microscopy and the prospects they present.

## 9.5 Protocols

### 9.5.1 Protocol 9.1: Analytical Ultracentrifugation of Membrane Protein/ Amphipol Complexes

#### 9.5.1.1 Introduction

Analytical ultracentrifugation (AUC) examines the behavior of macromolecules in solution when subjected to a centrifugal field. Absorbance and interference fringe shifts measured as a function of the distance to the rotation axis reflect the evolution of the distribution of the particles. Sedimentation velocity (SV) experiments are performed at high speed, sedimentation profiles being measured as a function of time. The different types of macromolecules (in the case of MPs, MP/surfactant complexes, and free surfactant particles) sediment at different velocities depending on their sedimentation coefficient $s$. Analysis of AUC data yields the distribution $c(s)$ of the particles present in a sample as a function of $s$. SV-AUC is a very powerful tool to assess sample homogeneity, the amount of free and bound surfactant, and the occurrence of protein-protein association equilibria. If, as is the case for bacteriorhodopsin (BR), the UV-visible spectra of the native and denatured protein are different, the two states can be distinguished and the extent to which the protein has retained its native state can be precisely assessed.

AUC is routinely used for the characterization of MP/detergent complexes (for reviews, see e.g. Ebel 2011; Le Roy et al. 2014, 2015). It is a priori applicable to any MP, because surfactant particles and MP/surfactant complexes are easily discriminated, even for relatively small particles whose hydro-dynamic radii ($R_S$) are not necessarily very different from one another. For example, the sedimentation coefficients of A8-35 particles ($R_S \approx 3.15$ nm; Gohon et al. 2006) and BR/A8-35 complexes ($R_S \approx 5.0$ nm; Gohon et al. 2008) are 1.6 S and 3.2 S, respectively, making them easily distinguishable. The higher sensitivity of SV-AUC is the consequence of $s$ being proportional to the ratio of the buoyant molar mass over $R_S$, which is roughly proportional to $M^{2/3}$, with $M$ the molar mass, whereas size exclusion chromatography (SEC) separates particles according to $R_S$, which varies roughly as $M^{1/3}$. As regards determining the mass of APol bound per particle, AUC is technologically more demanding than SEC or IMAC, but it presents the advantage that MP/APol complexes are never separated from the free APol, which sediments more slowly, thus eliminating the risks of desorption and/or aggregation.

### 9.5.1.2    Instrumentation

AUC experiments require access to an analytical ultracentrifuge (XLI Beckman, USA), a rotor (AnTi 50 or 60), and cell assemblies with sapphire windows (e.g. from Nanolytik, Germany).

### 9.5.1.3    Designing the Samples and Experiments

There is some flexibility in terms of protein and APol concentrations. Typically, samples are examined at several different concentrations in order to look for the presence of eventual association/dissociation processes and to check on the results. A sample containing only APol can be included so as to probe its behavior.

*☞Available cell centerpieces have optical path lengths of 1.2, 0.3, or 0.15 mm, requiring volumes of 450, 150, and 80 μL, respectively. The sedimentation can be followed at a maximum of three wavelengths in the range 230–600 nm, preferably at maxima in the absorbance spectrum. Typically, sedimentation profiles are recorded by measuring the absorbance at 280 nm and at a second wavelength if relevant (e.g. if the protein contains a chromophore or binds a partner absorbing at a different wavelength, or if one resorts to a labeled APol), as well as by interference optics.*

*The absorbance of the stock sample has to be measured and an aliquot diluted to achieve a final absorbance in the centrifuge between 1.2 and 0.2 (it can be raised to 1.4 depending on AUC optics quality or lowered to 0.05 if the sample is highly homogeneous). Usually, we dilute the stock sample, which typically is in the $g \cdot L^{-1}$ range, by a factor of two to four. These concentrations are also convenient for interference measurements. In principle, for maintaining the chemical potential of APol constant in the dilution series, the buffer used for dilution should contain free APol at the same concentration as in the stock sample, but this is not necessarily an issue. Note that above $\sim 2$ $g \cdot L^{-1}$, samples behave non-ideally, which will have to be taken into account in the analysis of results obtained at different dilutions (Solovyova et al. 2001).*

*☞ The buffer should not absorb, or only slightly so, at the wavelengths of interest, and should contain typically 100 mM salt to avoid repulsive electrostatic interactions. AUC centerpieces have two channels, one of which is to be filled with the reference buffer. The reference buffer should have exactly the same composition as the sample buffer (it is particularly important for buffers containing large amount of, for instance, glycerol), but for the omission of APol.*

### 9.5.1.4   Compiling the Parameters That Will Be Required for the Analysis

The partial specific volumes $\bar{v}$, in $mL \cdot g^{-1}$; the extinction coefficients $\varepsilon$, in $L \cdot g^{-1} \cdot cm^{-1}$, at the wavelengths of measurement in the AUC; the refractive index increments $(\partial n / \partial c)$ of the components (protein and APol); the protein molar mass from its sequence; and the buffer density and viscosity should be known. Complementary information on the hydrodynamic radius of the complex, possible protein association states, amount of bound lipids, etc. should be compiled.

- Parameters for the protein can be calculated from its sequence, using, for example, the program Sedfit (http://www.analyticalultracentrifugation.com).

  ☞ *If the protein is glycosylated, glycosylation has to be taken into account* (see e.g. le Maire et al. 2008).

- Parameters for the APol. The specific volume $\bar{v}$ of the sodium salt of A8-35 is $0.809 \ mL \cdot g^{-1}$, its density, $\rho = 1/\bar{v} = 1.236 \ g \cdot mL^{-1}$. However, because of the polyelectrolyte nature of A8-35, the apparent specific volume to be considered in dilute salt solution is $0.866 \ mL \cdot g^{-1}$ (Gohon et al. 2004, 2006). $(\partial n / \partial c) = 0.15 \ g \cdot mL^{-1}$ was used as a likely value for A8-35 (Gohon et al. 2008). It can be experimentally measured, as well as the extinction coefficient, from SV experiments with APol samples of known concentrations in the $g \cdot L^{-1}$ range (cf. Salvay and Ebel 2006).
- Parameters for the buffer can be obtained from tabulated data, e.g. with the program SEDNTERP (http://www.jphilo.mailway.com) for usual or simple buffers, or measured with a viscometer and a densimeter.

### 9.5.1.5   Running SV Experiments

The parameters to be set are the temperature, the angular velocity, the interval between scans, and the total time of sedimentation.

- The temperature can be chosen between 4 and 30 °C.
- Angular velocities of 42,000 or 50,000 rpm (130,000 or 184,000 × $g$) are usually equally convenient. For very large complexes (above 500 kDa), the angular velocity should be decreased to, for example 35,000 rpm (90,000 × $g$).
- Interval between scans: fix 5-min intervals. The effective interval will be 5 min or larger, depending on the number of cells and number of wavelengths; on the XLI AUC, the interval between scans is 12–15 min for four cells measured at two wavelengths and with interference optics.
- Total duration: we typically request the acquisition of 999 scans (the maximum possible input) and run the AUC experiment overnight before stopping it manually after, typically, 100 scans. The relevant scans (before everything is pelleted) will be selected for the analysis: for instance, small MPs, like BR or tOmpA, will be pelleted in ~7 h.

### 9.5.1.6   *c(s)* Analysis

Analyze the SV profiles for each detection mode (different wavelengths and interference optics) with the $c(s)$ analysis model of the program Sedfit (http://www.analyticalultracentrifugation.com) (for details, see Gohon et al. 2008). The distribution $c(s)$ of sedimentation coefficients ($s$) shows peaks reflecting the migration of MP/APol complexes and of free APol particles. Save the curves in the program Gussi (http://biophysics.swmed.edu/MBR/software.html) for an easy superposition of the different $c(s)$ curves (for different optics and samples) and for integration of the peak areas (Le Roy et al. 2015).

   *☞The $c(s)$ analyses are generally robust, i.e. their results do not depend on the input parameters (partial specific volume, solvent density and viscosity, the frictional ratio being fitted).*
   *☞Comparing the $c(s)$ curves at different concentrations allows to determine whether the samples can be considered as containing non-interacting species (in which case the peaks are always at the same s-values, which can be interpreted in terms of molar mass), or if there are association/ dissociation equilibria (in which case s increases when increasing the concentration), or if non-ideality has to be considered (in which case s decreases at the higher concentrations).*

- The signals from the integration of the peak areas are interpreted in terms of bound APol, in g per g protein in the complexes, with the input of the extinction coefficients and refractive index increments.

   *☞The precision of the results depends on that of the extinction coefficients.*
   *☞If not taken into account, lipid binding can bias the results, leading to overestimating the mass of APol bound. Indeed, combining absorbance and interference optics data provides the amount of bound surfactants (APol + lipids), assuming lipids and APol, whose UV absorbance spectra are different, to contribute similarly to the interference signal.*

- The $s$-values from the integration of the peak areas are interpreted for non-interacting species using the Svedberg equation, taking into consideration the amount of bound APol and using either an estimate of $R_S$ (but beware that the latter is overestimated by SEC; see Gohon et al. 2008 and § 9.3.2) or assuming a shape. Typical values for the frictional ratio are 1.25 for a globular compact shape, 1.5 for a moderately anisotropic shape, 1.8 for a significantly anisotropic shape; glycosylated proteins are characterized by larger values, such as 1.5–1.8 for a moderately asymmetrical shape (Le Roy et al. 2014).

   *☞Alternatively, values of protein molar masses corresponding to different association states (e.g. monomer, dimer) can be used, along with the protein/APol mass ratio within the complex, to derive, from the values of the sedimentation coefficients, those of $R_S$ and/or of the frictional ratio. Only the putative association states corresponding to experimental values of $R_S$, or to reasonable – see above – values of frictional ratio, will be considered as plausible.*

Samples do not have to be homogeneous for such an analysis to be performed. If, from the $c(s)$ analysis, a sample is found to comprise several non-interacting species (at most three), such as APol particles and two different MP/APol complexes, their properties can be determined separately, the analysis providing, for each species, the values of its diffusion coefficient $D_t$ and sedimentation coefficient $s$. $D_t$ is directly related to $R_S$. Combining $D_t$ and $s$ yields the buoyant molar mass, $M_b = s \cdot RT \cdot D_t^{-1} = \sum M_i \left(1 - \rho^\circ \bar{v}_i\right)$, with $M_i$ and $\bar{v}_i$ the molar mass and partial specific volumes of the protein, APol, and, if present, lipids, and $\rho^\circ$ the solvent density. Thus, knowing the masses of bound APol and lipid in g per g protein, the buoyant molar mass provides the mass of the protein (see Gohon et al. 2008).

   *Protocol prepared by Christine Ebel on the basis of Gohon et al. (2008) and published reviews and protocols (Ebel 2011; Le Roy et al. 2014, 2015).*

## 9.5.2    Protocol 9.2: Size Exclusion Chromatography Analysis of Membrane Protein/Amphipol Complexes

### 9.5.2.1    Introduction

Size exclusion chromatography (SEC) provides a very convenient tool to evaluate the outcome of experiments in which membrane proteins (MPs) are trapped and/or folded in amphipols (APols). Analytical SEC will give information on (i) the homogeneity and the size of the MP/APol complexes (see e.g. Picard et al. 2006; Le Bon et al. 2014a), (ii) the fraction of free APol present in the preparation (see e.g. Zoonens et al. 2007; Charvolin et al. 2014), (iii) the presence of protein aggregates (see e.g. Pocanschi et al. 2006; Gohon et al. 2008), (iv) the mass ratio of MP and APol in the complexes (Zoonens et al. 2007; Charvolin et al. 2014; cf. in Chap. 5, § 5.9.2, Protocol 5.2), and (v) the stability of MP/APol complexes over time (Charvolin et al. 2014). In preparative SEC, peak fractionation provides a purification step in view of structural studies, e.g. for cryo-electron microscopy (see e.g. Chap. 12, § 12.4, Protocol 12.1, based on Paulsen et al. 2015, or, for a more recent piece of work, Fitzpatrick et al. 2017).

Bacteriorhodopsin (BR; see Chap. 1, § 1.6.1) was used as a model protein in several of the experiments to be described below. BR is a particularly convenient model as it provides a visual control of the protein's state: when it is correctly folded and linked to its cofactor retinal via a Schiff base, it presents a characteristic purple color; when it denatures and the Schiff base becomes hydrolyzed, retinal is released, and the color of the sample turns to yellow. The ratio of the absorbance at 554 and 280 nm is a convenient measure of the fraction of the protein that is in its native state. In the following, details of the experimental steps that are specific to BR are marked with a ■ symbol, those applying to other MPs with □. General comments and tips for optimization are in italics and marked with a pointing hand sign (☞).

### 9.5.2.2    Biochemistry

■ Purple membrane purification: see Gohon et al. (2008).

■ □ Preparation of MP/APol complexes. MP/APol complexes can be obtained: (i) by transferring to APols a detergent-solubilized MP (see Chap. 5, § 5.9.1, Protocol 5.1) or (ii) by folding in APol a denatured MP (see Chap. 6, § 6.4, Protocol 6.1). When first developing a trapping or folding protocol, a set of samples will usually be prepared at different MP/APol mass ratios. SEC is a convenient tool to determine the optimal ratio.

■ For BR/APol complexes, the usual mass ratios examined when first testing a new APol are typically 1:2, 1:5, and 1:10. With A8-35, 1:5 w/w is a good ratio; BR binds ~2 g A8-35 per g protein, the rest of the APol remaining free (Gohon et al. 2008).

□ For the transmembrane domain of outer membrane protein A from *Escherichia coli* (tOmpA), which, like BR, has most of its mass in a TM position, 1:4 w/w is a good MP/APol ratio (Zoonens et al. 2005). For MPs whose extramembrane domains are comparatively more extended, less APol per mass of protein is required, e.g. only 1:1.5 w/w MP/A8-35 in the case of cytochrome $bc_1$ (Charvolin et al. 2014). In principle, one would expect the mass of APol that binds to a given MP to be more or less proportional to its membrane-exposed hydrophobic TM surface, but existing data are quite scattered (see Chap. 5, Table 5.5 and Fig. 5.9).

### 9.5.2.3   Size Exclusion Chromatography System

■ For BR/amphipol complexes, SEC was performed at room temperature (RT) on an FPLC (fast protein liquid chromatography) system (Äkta Purifier 10, GE Healthcare) equipped with a Superose 12 10/300 GL column (total volume ~20 mL). UV-visible absorbance detection was performed at three wavelengths: 220, 280, and 554 nm. The volume of the sample loop was 100 μL. The flow rate was set at 0.5 mL·min$^{-1}$.

☞The choice of the column must be adapted to the expected mass of MP/APol complexes. In the case of BR/A8-35 complexes, the molar mass of the protein is 27 kDa, that of free APol particles ~40 kDa (Gohon et al. 2006; cf. Chap. 4, § 4.3.1.2), and that of BR/A8-35 complexes ~81 kDa (Gohon et al. 2008; see Chap. 5, § 5.3.1). The separation range of the Superose 12 10/300 GL column is 10–300 kDa.

☞For high-resolution separation, the volume of sample injected should be <2% of the total column volume.

☞When the protein is temperature-sensitive, experiments can be carried out in a cold room.

☞UV-visible absorbance detection. A8-35 does not significantly absorb at 280 nm (Le Bon et al. 2014b; see Chap. 8, Fig. 8.1). The signal at 280 nm therefore is mainly due to the MP, whereas that at 220 nm has contributions from both MP and the APol, allowing an evaluation of the concentration of free APol in the sample (if MP/APol complexes and free APol are well separated). Fluorescent APols (FAPols) carry covalently bound fluorescent dyes such as fluorescein, rhodamine, NBD, Alexa Fluor 647, etc. (see Chap. 4, § 4.4, Table 4.5, and Le Bon et al. 2014b), with absorbance peaks spread over the full visible range (see Fig. 8.9 in Chap. 8, § 8.2.3). FAPols can be very useful tools to quantify APols present either in MP/APol complexes or as free particles (Zoonens et al. 2014).

### 9.5.2.4   Buffers

■ ☐ Some of the experiments with BR described below were carried out in 20 mM sodium phosphate, 100 mM NaCl, pH 7. However, this is not the optimal buffer.

☞ When working with A8-35, do not work below pH 7, which results in polymer aggregation (Gohon et al. 2006); it is preferable to use pH 7.5 or pH 8 buffers.

☞ Multivalent cations, even present as traces, cause free A8-35 particles and MP/A8-35 complexes to aggregate (Picard et al. 2006; Catoire et al. 2010). SEC profiles are improved in the presence of ~1 mM EDTA.

☞ When working with FAPols, there can be some pH limitation to avoid dye degradation.

☞ There is, as a rule, no need to supplement the SEC buffer with free APol. However, complete separation of MP/APol complexes from free APol particles can result in a degree of aggregation (Zoonens et al. 2007; see Fig. 9.7D and Chap. 5, § 5.2.1). This can be avoided by working in the presence of some free APol (whose contribution to the absorbance profiles will have to be taken into account when determining MP/APol ratios in complexes). Because aggregation can be expected to depend on the protein, the buffer, etc., how much free APol to use must be determined experimentally. Note that, when using preparative SEC, for example for cryo-EM purposes, aggregates generated by the removal of the free APol, if any, can be dispersed by adding back free APol to the purified complexes (Zoonens et al. 2007).

### 9.5.2.5    Protocol

1. Wash the column with 4–5 column volumes of water purified on a Millipore Milli-Q Advantage A10 system ("Milli-Q water").
2. Equilibrate the column by washing it with 4 column volumes of SEC buffer.
3. Inject aliquots of the MP/APols complexes prepared at the different MP/APol ratios to be tested.
4. Inject control samples: an aliquot of an APol solution at a concentration equivalent to that present in the MP/APols samples and aliquots of dextran (determination of the excluded/void volume) and acetone (determination of the total volume).

☞ *The injection of an APol solution will establish the elution volume of free APols; unlabeled free APols present in the MP/APols samples will appear at the same position in the SEC profiles recorded at 220 nm.*

5. Examination of the chromatograms will provide the following information:

   • Elution volume ($V_{el}$) of MP/APols complexes: Stokes radius (see below);
   • Width at half-height (WHH) and symmetry or asymmetry of the MP/APol peak: size dispersity of the complexes and presence or not of small oligomers;
   • Area under the various peaks at each wavelength: relative amounts of MP/APol complexes and free APol particles; if using FAPols, MP/APol ratio in the complexes; in the case of a protein like BR, whose spectrum depends on its native state, proportion of native vs. denatured protein.

☞ *The elution volume* $V_{el}$ *of the MP/APols complexes can be correlated to their Stokes radius via a calibration curve, by calibrating the column with standard globular proteins (which can be obtained from GE Healthcare, Bio-Rad, etc.). See e.g. Harlan et al. (1995). Note however that interactions between the complexes and the column may bias this evaluation (see Gohon et al. 2008 and Chap. 5, § 5.3.2).*

### 9.5.2.6    Commented Examples

In the legends of the following three figures, a few examples of using SEC for analyzing MP/APol samples are commented upon (Figs. 9.24, 9.25, and 9.26).

   *Protocol prepared by Christel Le Bon based primarily on* Zoonens et al. (2007), Gohon et al. (2008), and Le Bon et al. (2014a), *and laboratory notes.*

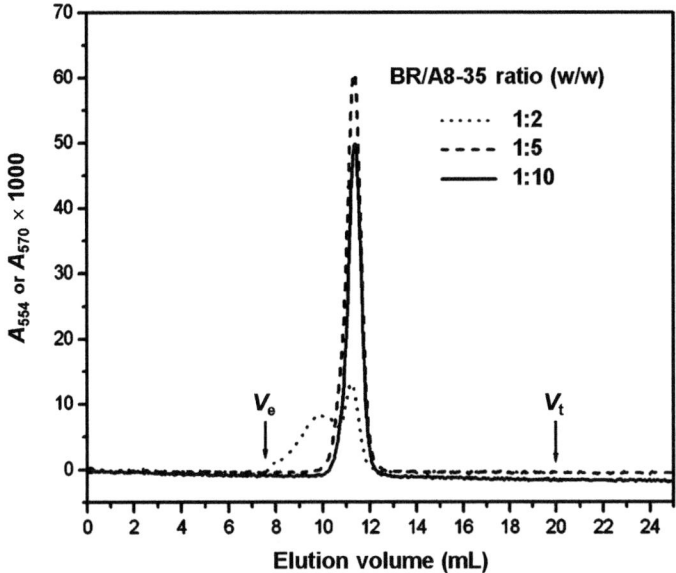

**Fig. 9.24** Determination of an optimal BR/A8-35 ratio for renaturing bacteriorhodopsin. Chromatograms of samples of BR refolded in A8-35 at 1:2, 1:5, and 1:10 BR/A8-35 mass ratios, recorded at 554 nm (1:2 and 1:10 w/w ratios) or 570 nm (1:5 w/w ratio), that is, at wavelengths where only the native protein absorbs. $V_e$ is the excluded volume and $V_t$ the total volume. For the two highest APol/PM ratios, the peak is well-defined and symmetrical, BR being mainly present as the A8-35-trapped monomer. At the lowest ratio, the renatured protein is partly aggregated (Reprinted with permission from Pocanschi et al. 2006, © 2006 American Chemical Society).

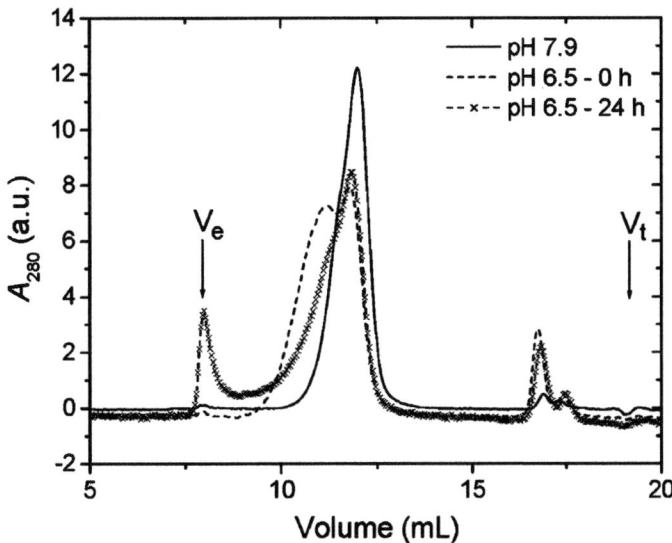

**Fig. 9.25** Effect of lowering the pH on the homogeneity of tOmpA/A8-35 complexes. The chromatogram of tOmpA/A8-35 complexes at pH 7.9 shows a homogeneous peak (albeit with a slight asymmetry, indicating the probable presence of a tiny fraction of small oligomers). When the pH is decreased below 7, some of the carboxylates that keep A8-35 soluble become protonated, which induces a decrease in solubility and the formation of aggregates (Gohon et al. 2008): the sample diluted at pH 6.9 and analyzed immediately afterward contains a fraction of small aggregates, which is reflected in the decrease in area of the major peak and the appearance of a second peak that elutes earlier; after 24 h of incubation, most of the small aggregates have converted into larger ones, which elute in the void volume ($V_e$) (Reprinted with permission from Zoonens et al. 2007, © 2007 American Chemical Society).

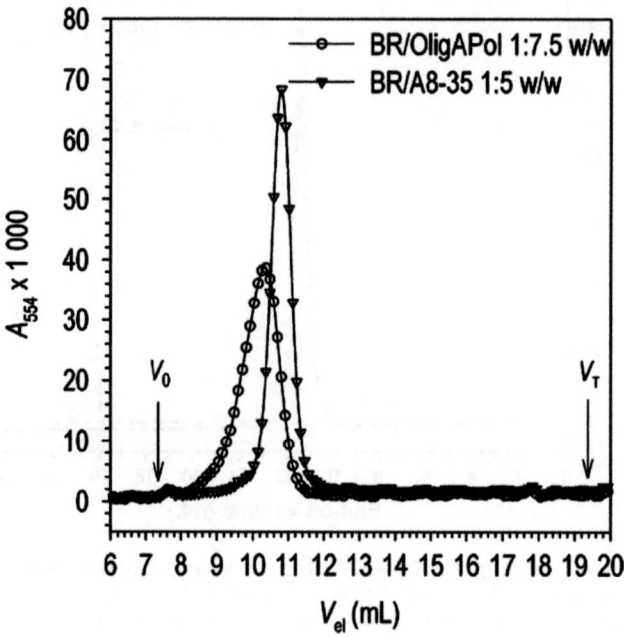

**Fig. 9.26** Trapping a membrane protein with a functionalized amphipol (From Le Bon et al. 2014a, © 2014 Oxford University Press). BR was trapped with an OligAPol, namely a preparation of A8-35 grafted with an oligodeoxynucleotide (ODN), to the level of ~0.5 ODN per APol particle. Particles that carry an ODN migrate more rapidly than ODN-free particles, most likely because the ODN, while it confers to the particles a negligible increase in mass (<20% for free A8-35 particles, <10% for BR/APol complexes), extends away from the surface of the APol particle or belt, increasing its Stokes radius. This effect is relatively complex, because a given ODN-labeled A8-35 molecule is transferred several times from one particle to another in the course of the SEC analysis, temporarily accelerating the migration of the particle it happens to be associated with (for a discussion, see Chap. 4, § 4.4.2).

# References

Agosto, M.A., Zhang, Z., He, F., Anastassov, I.A., Wright, S.J., McGehee, J., Wensel, T.G. (2014) Oligomeric state of purified TRPM1, a protein essential for dim light vision. *J. Biol. Chem.* **289**:27019–27033.

Althoff, T. (2011) Strukturelle Untersuchungen am Superkomplex $I_1III_2IV_1$ der Atmungskette mittels Kryoelektronen-mikroskopie, Fachbereich Biochemie, Chemie und Pharmazie. Ph. D. Dissertation, Johann Wolfgang Goethe-Universität, Frankfurt-am-Main, 248 p.

Althoff, T., Mills, D.J., Popot, J.-L., Kühlbrandt, W. (2011) Assembly of electron transport chain components in bovine mitochondrial supercomplex $I_1III_2IV_1$. *EMBO J.* **30**:4652–4664.

Arunmanee, W., Harris, J.R., Lakey, J.H. (2014) Outer membrane protein F stabilised with minimal amphipol forms linear arrays and LPS-dependent 2D crystals. *J. Membr. Biol.* **247**:949–956.

Bazzacco, P. (2009) Non-ionic amphipols: new tools for *in vitro* studies of membrane proteins. Validation and development of biochemical and biophysical applications. Thèse de Doctorat, Université Paris-7, Paris, 176 p.

Bazzacco, P., Sharma, K.S., Durand, G., Giusti, F., Ebel, C., Popot, J.-L., Pucci, B. (2009) Trapping and stabilization of integral membrane proteins by hydrophobically grafted glucose-based telomeres. *Biomacromolecules* **10**:3317–3326.

Bazzacco, P., Billon-Denis, E., Sharma, K.S., Catoire, L.J., Mary, S., Le Bon, C., Point, E., Banères, J.-L., Durand, G., Zito, F., Pucci, B., Popot, J.-L. (2012) Non-ionic homopolymeric amphipols: Application to membrane protein folding, cell-free synthesis, and solution NMR. *Biochemistry* **51**:1416–1430.

Belrhali, H., Nollert, P., Royant, A., Menzel, C., Rosenbusch, J.P., Landau, E.M., Pebay-Peyroula, E. (1999) Protein, lipid and water organization in bacteriorhodopsin crystals: a molecular view of the purple membrane at 1.9 Å resolution. *Structure* **7**:909–917.

Broecker, J., Eger, B.T., Ernst, O.P. (2017) Crystallogenesis of membrane proteins mediated by polymer-bounded lipid nanodiscs. *Structure* **25**:384–392.

Cantor, C.R., Schimmel, P.R. (1980) *Biophysical Chemistry. Part II: Techniques for the study of biological structure and function.* W.H. Freeman and company, San Francisco, 846 p.

Carruthers, L.M., Schirf, V.R., Demeller, B., Hansen, J.C. (2000) Sedimentation velocity analysis of macromolecular assemblies. *Meth. Enzymol.* **321**:67–80.

Catoire, L.J., Zoonens, M., van Heijenoort, C., Giusti, F., Popot, J.-L., Guittet, E. (2009) Inter- and intramolecular contacts in a membrane protein/surfactant complex observed by heteronuclear dipole-to-dipole cross-relaxation. *J. Magn. Res.* **197**:91–95.

Catoire, L.J., Zoonens, M., van Heijenoort, C., Giusti, F., Guittet, E., Popot, J.-L. (2010) Solution NMR mapping of water-accessible residues in the transmembrane $\beta$-barrel of OmpX. *Eur. Biophys. J.* **39**:623–630.

Champeil, P., Menguy, T., Tribet, C., Popot, J.-L., le Maire, M. (2000) Interaction of amphipols with the sarcoplasmic reticulum $Ca^{2+}$-ATPase. *J. Biol. Chem.* **275**:18623–18637.

Champeil, P., Menguy, T., Tribet, C., Popot, J.-L., le Maire, M. (2000b) Long-term protection of solubilized sarcoplasmic reticulum $Ca^{2+}$-ATPase by an amphipathic polymer, amphipol A8-35, in: Kaya, K.T.S, ed., *Na/K-ATPase and Related ATPases.* Elsevier Science B.V., Amsterdam, pp. 209–212.

Charvolin, D., Picard, M., Huang, L.-S., Berry, E.A., Popot, J.-L. (2014) Solution behavior and crystallization of cytochrome $bc_1$ in the presence of amphipols. *J. Membr. Biol.* **247**:981–996.

Constantine, M., Liew, C.K., Lo, V., Macmillan, A., Cranfield, C.G., Sunde, M., Whan, R., Graham, R.M., Martinac, B. (2016) Heterologously-expressed and liposome-reconstituted human transient receptor potential melastatin 4 channel (TRPM4) is a functional tetramer. *Sci. Rep.* **6**:19352.

Cvetkov, T.L., Huynh, K.W., Cohen, M.R., Moiseenkova-Bell, V.Y. (2011) Molecular architecture and subunit organization of TRPA1 ion channel revealed by electron microscopy. *J. Biol. Chem.* **286**:38168–38176.

Dahmane, T. (2007) Protéines membranaires et amphipols : stabilisation, fonction, renaturation, et développement d'amphipols sulfonatés pour la RMN des solutions. Thèse de Doctorat, Université Paris-7, Paris, 229 p.

Dahmane, T., Damian, M., Mary, S., Popot, J.-L., Banères, J.-L. (2009) Amphipol-assisted *in vitro* folding of G protein-coupled receptors. *Biochemistry* **48**:6516–6521.

Dahmane, T., Giusti, F., Catoire, L.J., Popot, J.-L. (2011) Sulfonated amphipols: Synthesis, properties and applications. *Biopolymers* **95**:811–823.

Dahmane, T., Rappaport, F., Popot, J.-L. (2013) Amphipol-assisted folding of bacteriorhodopsin in the presence and absence of lipids. Functional consequences. *Eur. Biophys. J.* **42**:85–101.

Deniaud, A., Panwar, P., Frelet-Barrand, A., Bernaudat, F., Juillan-Binard, C., Ebel, C., Rolland, N., Pebay-Peyroula, E. (2012) Oligomeric status and nucleotide binding properties of the plastid ATP/ADP transporter 1: Toward a molecular understanding of the transport mechanism. *PLoS ONE* **7**:e32325.

Diab, C., Tribet, C., Gohon, Y., Popot, J.-L., Winnik, F.M. (2007) Complexation of integral membrane proteins by phosphorylcholine-based amphipols. *Biochim. Biophys. Acta* **1768**:2737–2747.

Ebel, C. (2011) Sedimentation velocity to characterize surfactants and solubilized membrane proteins. *Methods* **54**:56–66.

Elter, S., Raschle, T., Arens, S., Viegas, A., Gelev, V., Etzkorn, M., Wagner, G. (2014) The use of amphipols for NMR structural characterization of 7-TM proteins. *J. Membr. Biol.* **247**:957–964.

Etzkorn, M., Raschle, T., Hagn, F., Gelev, V., Rice, A.J., Walz, T., Wagner, G. (2013) Cell-free expressed bacteriorhodopsin in different soluble membrane mimetics: biophysical properties and NMR accessibility. *Structure* **21**:394–401.

Etzkorn, M., Zoonens, M., Catoire, L.J., Popot, J.-L., Hiller, S. (2014) How amphipols embed membrane proteins: Global solvent accessibility and interaction with a flexible protein terminus. *J. Membr. Biol.* **247**:965–970.

Fernández, C., Adeishvili, K., Wüthrich, K. (2001) Transverse relaxation-optimized NMR spectroscopy with the outer membrane protein OmpX in dihexanoyl phosphatidylcholine micelles. *Proc. Natl. Acad. Sci. USA* **98**:2358–2363.

Ferrandez, Y., Dezi, M., Bosco, M., Urvoas, A., Valério, M., Le Bon, C., Giusti, F., Broutin, I., Durand, G., Polidori, A., Popot, J.-L., Picard, M., Minard, P. (2014) Amphipol-mediated screening of molecular orthoses specific for membrane protein targets. *J. Membr. Biol.* **247**:925–940.

Fitzpatrick, A.W.P., Llabrés, S., Neuberger, A., Blaza, J.N., Bai, X.-C., Okada, U., Murakami, S., van Veen, H.W., Zachariae, U., Scheres, S.H.W., Luisi, B.F., Du, D. (2017) Structure of the MacAB-TolC ABC-type tripartite multidrug efflux pump. *Nat. Microbiol.* **2**:17070.

Franke, D., Svergun, D.I. (2009) DAMMIF, a program for rapid *ab initio* shape determination in small-angle scattering. *J. Appl. Cryst.* **42**:342–346.

Giusti, F., Rieger, J., Catoire, L.J., Qian, S., Calabrese, A.N., Watkinson, T.G., Casiraghi, M., Radford, S.E., Ashcroft, A. E., Popot, J.-L. (2014) Synthesis, characterization and applications of a perdeuterated amphipol. *J. Membr. Biol.* **247**:909–924.

Giusti, F., Kessler, P., Westh Hansen, R., Della Pia, E.A., Le Bon, C., Mourier, G., Popot, J.-L., Martinez, K.L., Zoonens, M. (2015) Synthesis of a polyhistidine-bearing amphipol and its use for immobilizing membrane proteins. *Biomacromolecules* **16**:3751–3761.

Gohon, Y. (2002) Etude structurale et fonctionnelle de deux protéines membranaires, la bactériorhodopsine et le récepteur nicotinique de l'acétylcholine, maintenues en solution aqueuse non détergente par des polymères amphiphiles. Thèse de Doctorat, Université Paris-VI, Paris, 467 p.

Gohon, Y., Pavlov, G., Timmins, P., Tribet, C., Popot, J.-L., Ebel, C. (2004) Partial specific volume and solvent interactions of amphipol A8-35. *Anal. Biochem.* **334**:318–334.

Gohon, Y., Giusti, F., Prata, C., Charvolin, D., Timmins, P., Ebel, C., Tribet, C., Popot, J.-L. (2006) Well-defined nanoparticles formed by hydrophobic assembly of a short and polydisperse random terpolymer, amphipol A8-35. *Langmuir* **22**:1281–1290.

Gohon, Y., Dahmane, T., Ruigrok, R., Schuck, P., Charvolin, D., Rappaport, F., Timmins, P., Engelman, D.M., Tribet, C., Popot, J.-L., Ebel, C. (2008) Bacteriorhodopsin/amphipol complexes: structural and functional properties. *Biophys. J.* **94**:3523–3537.

Gohon, Y., Vindigni, J.-D., Pallier, A., Wien, F., Celia, H., Giuliani, A., Tribet, C., Chardot, T., Briozzo, P. (2011) High water solubility and fold in amphipols of proteins with large hydrophobic regions: Oleosins and caleosin from seed lipid bodies. *Biochim. Biophys. Acta* **1808**:706–716.

Guinier, A., Fournet, G. (1955) *Small-angle scattering of X-rays*. Wiley, New York, 268 p.

Gulati, S., Jamshad, M., Knowles, T.J., Morrison, K.A., Downing, R., Cant, N., Collins, R., Koenderink, J.B., Ford, R. C., Overduin, M., Kerr, I.D., Dafforn, T.R., Rothnie, A.J. (2014) Detergent-free purification of ABC (ATP-binding-cassette) transporters. *Biochem. J.* **461**:269–278.

Han, S.G., Na, J.H., Lee, W.K., Park, D., Oh, J., Yoon, S.H., Lee, C.K., Sung, M.H., Shin, Y.K., Yu, Y.G. (2014) An amphipathic polypeptide derived from poly-γ-glutamic acid for the stabilization of membrane proteins. *Prot. Sci.* **23**:1800–1807.

Han, S.G., Baek, S.I., Son, T.J., Lee, H., Kim, N.H., Yu, Y.G. (2017) Preparation of functional human lysophosphatidic acid receptor 2 using a P9* expression system and an amphipathic polymer and investigation of its *in vitro* binding preference to $G_\alpha$ proteins. *Biochem. Biophys. Res. Commun.* **487**:103–108.

Harlan, J.E., Picot, D., Loll, P.J., Garavito, R.M. (1995) Calibration of size-exclusion chromatography: use of a double Gaussian distribution to describe pore sizes. *Anal. Biochem.* **224**:557–563.

Hiruma-Shimizu, K., Shimizu, H., Thompson, G.S., Kalverda, A.P., Patching, S.G. (2015) Deuterated detergents for structural and functional studies of membrane proteins: Properties, chemical synthesis and applications. *Mol. Membr. Biol.* **32**:139–155.

Huynh, K.W., Cohen, M.R., Moiseenkova-Bell, V.Y. (2014) Application of amphipols for structure-functional analysis of TRP channels. *J. Membr. Biol.* **247**:843–851.

Jacrot, B. (1976) The study of biological structures by neutron scattering from solution. *Rep. Prog. Phys.* **39**:911–953.

Jamshad, M., Charlton, J., Lin, Y.-P., Routledge, S.J., Bawa, Z., Knowles, T.J., Overduin, M., Dekker, N., Dafforn, T.R., Bill, R.M., Poyner, D.R., Wheatley, M. (2015) G protein-coupled receptor solubilization and purification for biophysical analysis and functional studies, in the total absence of detergent. *Biosc. Rep.* **35**:e00188.

Jin, P., Bulkley, D., Guo, Y., Zhang, W., Guo, Z., Huynh, W., Wu, S., Meltzer, S., Cheng, T., Jan, L.Y., Jan, Y.-N., Cheng, Y. (2017) Electron cryo-microscopy structure of the mechanotransduction channel NOMPC. *Nature* **547**:118–122.

Klammt, C., Perrin, M.-H., Maslennikov, I., Renault, L., Krupa, M., Kwiatkowski, W., Stahlberg, H., Vale, W., Choe, S. (2011) Polymer-based cell-free expression of ligand-binding family B G protein-coupled receptors without detergents. *Prot. Sci.* **20**:1030–1041.

Knowles, T.J., Finka, R., Smith, C., Lin, Y.-P., Dafforn, T., Overduin, M. (2009) Membrane proteins solubilized intact in lipid containing nanoparticles bounded by styrene maleic acid copolymer. *J. Am. Chem. Soc.* **131**:7484–7485.

Kraft, T.E., Hresko, R.C., Hruz, P.W. (2015) Expression, purification, and functional characterization of the insulin-responsive facilitative glucose transporter GLUT4. *Protein Sci.* **24**:2008–2019.

Laue, T.M., Stafford, W.E., III (1999) Modern application of analytical ultracentrifugation. *Annu. Rev. Biophys. Biomol. Struct.* **28**:75–100.

Laursen, T., Naur, P., Møller, B.L. (2013) Amphipol trapping of a functional CYP system. *Biotechn. Applied Biochem.* **60**:119–127.

Le Bon, C., Della Pia, E.A., Giusti, F., Lloret, N., Zoonens, M., Martinez, K.L., Popot, J.-L. (2014a) Synthesis of an oligonucleotide-derivatized amphipol and its use to trap and immobilize membrane proteins. *Nucleic Acids Res.* **42**: e83.

Le Bon, C., Popot, J.-L., Giusti, F. (2014b) Labeling and functionalizing amphipols for biological applications. *J. Membr. Biol.* **247**:797–814.

Le Roy, A., Breyton, C., Ebel, C. (2014) Analytical ultracentrifugation and size-exclusion chromatography coupled with light scattering for the characterization of membrane proteins in solution, in: Mus-Veteau, I., ed., *Membrane Proteins Production for Structural Analysis*. Springer, New York, pp. 267–287.

Le Roy, A., Wang, K., Schaack, B., Schuck, P., Breyton, C., Ebel, C. (2015) AUC and small-angle scattering for membrane proteins. *Meth. Enzymol.* **562**:257–286.

Leney, A.C., McMorran, L.M., Radford, S.E., Ashcroft, A.E. (2012) Amphipathic polymers enable the study of functional membrane proteins in the gas phase. *Anal. Chem.* **84**:9841–9847.

Liao, M., Cao, E., Julius, D., Cheng, Y. (2013) Structure of the TRPV1 ion channel determined by electron cryo-microscopy. *Nature* **504**:107–112.

Lindhoud, S., Carvalho, V., Pronk, J.W., Aubin-Tam, M.E. (2016) SMA-SH: Modified styrene-maleic acid copolymer for functionalization of lipid nanodiscs. *Biomacromolecules* **17**:1516–1522.

Logez, C., Damian, M., Legros, C., Dupré, C., Guéry, M., Mary, S., Wagner, R., M'Kadmi, C., Nosjean, O., Fould, B., Marie, J., Fehrentz, J.A., Martinez, J., Ferry, G., Boutin, J.A., Banères, J.-L. (2016) Detergent-free isolation of functional G protein-coupled receptors into nanometric lipid particles. *Biochemistry* **55**:38–48.

le Maire, M., Arnou, B., Olesen, C., Georgin, D., Ebel, C., Møller, J.V. (2008) Gel chromatography and analytical ultracentrifugation to determine the extent of detergent binding and aggregation, and Stokes radius of membrane proteins using sarcoplasmic reticulum $Ca^{2+}$-ATPase as an example. *Nat. Protoc.* **3**:1782–1795.

Martinez, K.L., Gohon, Y., Corringer, P.-J., Tribet, C., Mérola, F., Changeux, J.-P., Popot, J.-L. (2002) Allosteric transitions of *Torpedo* acetylcholine receptor in lipids, detergent and amphipols: molecular interactions *vs.* physical constraints. *FEBS Lett.* **528**:251–256.

McEwen, C.R. (1967) Tables for estimating sedimentation through linear concentration gradients of sucrose solution. *Anal. Biochem.* **20**:114–119.

Nagy, J.K., Kuhn Hoffmann, A., Keyes, M.H., Gray, D.N., Oxenoid, K., Sanders, C.R. (2001) Use of amphipathic polymers to deliver a membrane protein to lipid bilayers. *FEBS Lett.* **501**:115–120.

Nowaczyk, M., Oworah-Nkruma, R., Zoonens, M., Rögner, M., Popot, J.-L. (2004) Amphipols: strategies for an improved PS2 environment in aqueous solution, in: Miyake, J., ed., Biohydrogen III. Elsevier, Dordrecht, The Netherlands, Kyoto, pp. 151–159.

Oluwole, A.O., Danielczak, B., Meister, A., Babalola, J.O., Vargas, C., Keller, S. (2017) Solubilization of membrane proteins into functional lipid-bilayer nanodiscs using a diisobutylene/maleic acid copolymer. *Angew. Chem. Int. Ed. Engl.* **56**:1919–1924.

Opačić, M., Durand, G., Bosco, M., Polidori, A., Popot, J.-L. (2014a) Amphipols and photosynthetic light-harvesting pigment-protein complexes. *J. Membr. Biol.* **247**:1031–1041.

Opačić, M., Giusti, F., Broos, J., Popot, J.-L. (2014b) Isolation of *Escherichia coli* mannitol permease, EII$^{mtl}$, trapped in amphipol A8-35 and fluorescein-labeled A8-35. *J. Membr. Biol.* **247**:1019–1030.

Orwick-Rydmark, M., Lovett, J.E., Graziadei, A., Lindholm, L., Hicks, M.R., Watts, A. (2012) Detergent-free incorporation of a seven-transmembrane receptor protein into nanosized bilayer Lipodisq particles for functional and biophysical studies. *Nano Lett.* **12**:4687–4692.

Padrick, S.B., Deka, R.K., Chuang, J.L., Wynn, R.M., Chuang, D.T., Norgard, M.V., Rosen, M.K., Brautigam, C.A. (2010) Determination of protein complex stoichiometry through multisignal sedimentation velocity experiments. *Anal. Biochem.* **407**:89–103.

Paulsen, C.E., Armache, J.-P., Gao, Y., Cheng, Y., Julius, D. (2015) Structure of the TRPA1 ion channel suggests regulatory mechanisms. *Nature* **520**:511–517.

Picard, M., Dahmane, T., Garrigos, M., Gauron, C., Giusti, F., le Maire, M., Popot, J.-L., Champeil, P. (2006) Protective and inhibitory effects of various types of amphipols on the $Ca^{2+}$-ATPase from sarcoplasmic reticulum: a comparative study. *Biochemistry* **45**:1861–1869.

Pocanschi, C.L., Dahmane, T., Gohon, Y., Rappaport, F., Apell, H.-J., Kleinschmidt, J.H., Popot, J.-L. (2006) Amphipathic polymers: tools to fold integral membrane proteins to their active form. *Biochemistry* **45**:13954–13961.

Polovinkin, V., Gushchin, I., Balandin, T., Chervakov, P., Round, E., Shevchenko, V., Popov, A., Borshchevskiy, V., Popot, J.-L., Gordeliy, V. (2014) High-resolution structure of a membrane protein transferred from amphipol to a lipidic mesophase. *J. Membr. Biol.* **247**:997–1004.

Popot, J.-L., Berry, E.A., Charvolin, D., Creuzenet, C., Ebel, C., Engelman, D.M., Flötenmeyer, M., Giusti, F., Gohon, Y., Hervé, P., Hong, Q., Lakey, J.H., Leonard, K., Shuman, H.A., Timmins, P., Warschawski, D.E., Zito, F., Zoonens, M., Pucci, B., Tribet, C. (2003) Amphipols: polymeric surfactants for membrane biology research. *Cell. Mol. Life Sci.* **60**:1559–1574.

Postis, V., Rawson, S., Mitchell, J.K., Lee, S.C., Parslow, R.A., Dafforn, T.R., Baldwin, S.A., Muench, S.P. (2015) The use of SMALPs as a novel membrane protein scaffold for structure study by negative stain electron microscopy. *Biochim. Biophys. Acta* **1848**:496–501.

Prata, C., Giusti, F., Gohon, Y., Pucci, B., Popot, J.-L., Tribet, C. (2001) Non-ionic amphiphilic polymers derived from *tris*(hydroxymethyl)-acrylamidomethane keep membrane proteins soluble and native in the absence of detergent. *Biopolymers* **56**:77–84.

Rahmeh, R., Damian, M., Cottet, M., Orcel, H., Mendre, C., Durroux, T., Sharma, K.S., Durand, G., Pucci, B., Trinquet, E., Zwier, J.M., Deupi, X., Bron, P., Banères J.-L., Mouillac, B., Granier, S. (2012) Structural insights into biased G protein-coupled receptor signaling revealed by fluorescence spectroscopy. *Proc. Natl. Acad. Sci. USA* **109**:6733–6738.

Salvay, A.G., Ebel, C. (2006) Analytical ultracentrifuge for the characterization of detergent in solution. *Progr. Colloid Polym. Sci.* **131**:74–82.

Schägger, H., von Jagow, G. (1991) Blue native electrophoresis for isolation of membrane protein complexes in enzymatically active form. *Anal. Biochem.* **1999**:223–231.

Schuck, P. (2000) Size-distribution analysis of macromolecules by sedimentation velocity ultracentrifugation and Lamm equation modeling. *Biophys. J.* **78**:1606–1619.

Sharma, K.S., Durand, G., Gabel, F., Bazzacco, P., Le Bon, C., Billon-Denis, E., Catoire, L.J., Popot, J.-L., Ebel, C., Pucci, B. (2012) Non-ionic amphiphilic homopolymers: Synthesis, solution properties, and biochemical validation. *Langmuir* **28**:4625–4639.

Shen, P.S., Yang, X., DeCaen, P.G., Liu, X., Bulkley, D., Clapham, D.E., Cao, E. (2016) The structure of the Polycystic Kidney Disease channel PKD2 in lipid nanodiscs. *Cell* **167**:763–773.e711.

Shinzawa-Itoh, K., Shimomura, H., Yanagisawa, S., Shimada, S., Takahashi, R., Oosaki, M., Ogura, T., Tsukihara, T. (2016) Purification of active respiratory supercomplex from bovine heart mitochondria enables functional studies. *J. Biol. Chem.* **291**:4178–4184.

Solovyova, A., Schuck, P., Costenaro, L., Ebel, C. (2001) Non-ideality by sedimentation velocity of halophilic malate dehydrogenase in complex solvents. *Biophys. J.* **81**:1868–1880.

Sousa, J.S., Mills, D.J., Vonck, J., Kühlbrandt, W. (2016) Functional asymmetry and electron flow in the bovine respirasome. *eLIFE* **5**:e21290.

Stuhrmann, H.B. (1970) Interpretation of small-angle scattering functions of dilute solutions and gases. A representation of the structures related to a one-particle scattering function. *Acta Cryst. Sect. A* **26**:297–306.

Svergun, D.I. (1999) Restoring low-resolution structure of biological macromolecules from solution scattering using simulated annealing. *Biophys. J.* **76**:2879–2886.

Svergun, D.I., Koch, M.H.J., Timmins, P.A., May, R.P. (2013) *Small Angle X-Ray and Neutron Scattering from Solutions of Biological Macromolecule.* Oxford University Press, Oxford, 368 p.

Sverzhinsky, A., Qian, S., Yang, L., Allaire, M., Moraes, I., Ma, D., Chung, J.W., Zoonens, M., Popot, J.-L., Coulton, J.W. (2014) Amphipol-trapped ExbB-ExbD membrane protein complex from *Escherichia coli*: A biochemical and structural case study. *J. Membr. Biol.* **247**:1005–1018.

Tribet, C., Audebert, R., Popot, J.-L. (1996) Amphipols: polymers that keep membrane proteins soluble in aqueous solutions. *Proc. Natl. Acad. Sci. USA* **93**:15047–15050.

Tribet, C., Audebert, R., Popot, J.-L. (1997) Stabilization of hydrophobic colloidal dispersions in water with amphiphilic polymers: Application to integral membrane proteins. *Langmuir* **13**:5570–5576.

Tribet, C., Mills, D., Haider, M., Popot, J.-L. (1998) Scanning transmission electron microscopy study of the molecular mass of amphipol/cytochrome $b_6f$ complexes. *Biochimie* **80**:475–482.

Tsybovsky, Y., Orban, T., Molday, R.S., Taylor, D., Palczewski, K. (2013) Molecular organization and ATP-induced conformational changes of ABCA4, the photoreceptor-specific ABC transporter. *Structure* **21**:854–860.

Vázquez-Acevedo, M., Vega-de Luna, F., Sánchez-Vásquez, L., Colina-Tenorio, L., Remacle, C., Cardol, P., Miranda-Astudillo, H., González-Halphen, D. (2016) Dissecting the peripheral stalk of the mitochondrial ATP synthase of chlorophycean algae. *Biochim. Biophys. Acta* **1857**:1183–1190.

Zaccai, N.R., Serdyuk, I.N., Zaccai, G. (2017) *Methods in Molecular Biophysics. Structure, dynamics, function,* 2nd ed. Cambridge University Press, Cambridge, 684 p.

Zoonens, M. (2004) Caractérisation des complexes formés entre le domaine transmembranaire de la protéine OmpA et des polymères amphiphiles, les amphipols. Application à l'étude structurale des protéines membranaires par RMN à haute résolution. Thèse de Doctorat, Université Paris-6, Paris, 233 p.

Zoonens, M., Catoire, L.J., Giusti, F., Popot, J.-L. (2005) NMR study of a membrane protein in detergent-free aqueous solution. *Proc. Natl. Acad. Sci. USA* **102**:8893–8898.

Zoonens, M., Giusti, F., Zito, F., Popot, J.-L. (2007) Dynamics of membrane protein/amphipol association studied by Förster resonance energy transfer. Implications for *in vitro* studies of amphipol-stabilized membrane proteins. *Biochemistry* **46**:10392–10404.

Zoonens, M., Zito, F., Martinez, K.L., Popot, J.-L. (2014) Amphipols: a general introduction and some protocols, in: Mus-Veteau, I., ed., *Membrane Proteins Production for Structural Analysis.* Springer, New York, Heidelberg, Dordrecht, London, pp. 173–203.

**Summary**

*Because they stabilize most membrane proteins as compared to detergent solutions and they trap them under the form of small particles, amphipols (APols) appear as an attractive potential medium for solution-state NMR (sNMR) studies. sNMR studies of APol-trapped membrane proteins have yielded information on the conformation of the proteins, their areas of contact with the polymer, their dynamics, their accessibility to water, and the structure of protein-bound ligands. They have benefited from the diversification of APol chemical structures and the availability of deuterated APols. The advantages and constraints of working with APols are discussed and compared to those associated with other nonconventional media.*

## 10.1 Introduction

Nuclear magnetic resonance (NMR) is, along with X-ray crystallography (Chap. 11) and electron microscopy (EM; Chap. 12), one of the three techniques that can provide an exhaustive description of the three-dimensional (3D) arrangement of atoms in biological macromolecules and, thereby, help understand their structure and function. The application of NMR to membrane proteins (MPs), while fraught with difficulties, has been progressing in a spectacular manner since the late 1990s to the early 2000s, when the first NMR structures of MPs were established (MacKenzie et al. 1997; Arora et al. 2001; Fernández et al. 2001a, b). Whereas much of the present chapter will be concerned with the challenges and opportunities of using APols for establishing complete structures of MPs by solution NMR, it would be misleading to limit to this particular application the contributions that NMR can make to understanding MPs. NMR signals are sensitive to the distance between nuclei and to the dynamics of their interactions, and they are usually collected at or slightly above room temperature, rather than on frozen samples as most X-ray and EM data are. They can therefore be exploited to derive information that neither X-ray crystallography nor EM can, or can easily, provide, such as conformational equilibria and the changes they undergo upon ligand binding, the 3D structure of a ligand bound to a non-crystallizable receptor, interactions with the membrane environment, etc. Such information can be more easily accessible than establishing de novo by NMR a whole MP structure, and it brings information that is complementary to that yielded by other approaches.

© Springer International Publishing AG, part of Springer Nature 2018 453
J. -L. Popot, *Membrane Proteins in Aqueous Solutions*, Biological and Medical Physics,
Biomedical Engineering, https://doi.org/10.1007/978-3-319-73148-3_10

There are two basic types of NMR studies, solution NMR (sNMR) and "solid-state" NMR (ssNMR) studies, the latter actually applying both to solid, e.g. frozen, samples and to viscous ones, such as a biological membrane at room temperature, as well as to large macromolecules that are beyond the reach of sNMR. A major difference between the two approaches lies in the way the influence of the magnetic field of the spectrometer is averaged over the various orientations the target macromolecule can adopt with respect to it. This will define the average field experienced by each nucleus, and, thereby, the frequency at which it resonates when subjected to an oscillating field. Solution NMR relies on the fast tumbling of the molecule, which must be rapid enough so that, on the time-scale of spin relaxation, each nucleus has experienced and averaged the influence of all possible orientations in the magnetic field. If this condition is not met, various equivalent nuclei in various molecules will experience a slightly different average field and resonate at slightly different frequencies, resulting in peak broadening. Because there are tens if not hundreds of similar nuclei, e.g. amide protons, in even a small protein, and many of them resonate at frequencies that are not spread over a very broad range, peak broadening prevents separating the signals from various nuclei from one another and identifying them, which is a prerequisite to any structure determination. In sNMR, averaging is left to the molecule itself, which, by rapidly tumbling, experiences all possible orientations. The tumbling rate depends strongly on the particle size: $\tau_c$, the rotational correlation time, is proportional to the third power of $R_S$, the Stokes radius of the particle, and to $\eta$, the viscosity of the solution:

$$\tau_c \approx 4\pi R_S^3 \eta / 3kT.$$

As $\tau_c$ increases, the transverse relaxation time $T_2$ decreases, and the width of the resonance peaks increases. In practice, using the most advanced instruments and approaches, it remains currently difficult to establish de novo the structure of a MP whose mass exceeds 30–40 kDa, and even that remains a challenge (see below).

In ssNMR, it falls to the experimenter to take care of the orientation problem, which can be done using one of two approaches. First, one can orient the molecules with respect to the magnetic field, e.g. by adsorbing stacks of membranes onto glass plates, or by letting MP-containing bicelles orient by themselves in the magnetic field (cf. Chap. 3, § 3.2.1). In this way, all equivalent nuclei experience (more or less) the same field. Second, one can rapidly rotate the sample at the so-called magic angle with respect to the instrument's magnetic field. The "magic angle" (54.7°) is the angle that the internal diagonal of a cube makes with any of the edges. The nuclei of a sample rotated at this angle thus experience all possible orientations with respect to the field of the spectrometer, which prevents peak broadening. ssNMR does away with the necessity for the sample to tumble rapidly. There is therefore, in ssNMR, no limit on the size of the particles, which renders accessible, for instance, the study of MPs imbedded into a lipid bilayer. However, because of resolution limits, the complexity of the samples must not be too high. ssNMR therefore is particularly adapted to the detailed study of small groups of atoms, such as those of a ligand, a cofactor, specific residues, or a very small subunit, integrated into or associated to a structure that can be as large as desirable. The potential of APols for ssNMR studies has not been explored yet. This chapter therefore will concentrate on sNMR, but *in fine* a word will be said about what APols could possibly contribute to ssNMR (see § 10.4).

Solution NMR is, along with EM, the structural biology application that has generated the most structural studies of APol-trapped MPs to date (see Table 10.1). The conditions under which well-resolved spectra can be collected have been defined and optimized, pitfalls have been recognized, APols specifically designed for NMR have been developed and validated, model MPs and their interactions with APols have been studied, and original biological results are starting to trickle in. It is fair to say, however, that although all of the tools appear to be at hand to solve de novo a MP structure in APols, the demonstration that it is actually doable has not yet been brought. Furthermore, the parallel development of nanodisc-based NMR studies of MPs (see Chap. 3, § 3.3.5) has considerably slowed down that of APol-based ones, so that their full potential remains to be realized. In § 10.2, we will briefly describe the state of the art of sNMR studies of MPs. In § 10.3, we will review existing NMR experiments using APols and discuss the advantages and constraints associated to their use. In § 10.4, we will compare them with those of other surfactants, and consider current challenges and prospects.

## 10.2   Context: Solution NMR Studies of Membrane Proteins

The technical difficulties associated with studying MPs by sNMR have been discussed in numerous reviews, to which the reader is referred (see e.g. Sanders and Oxenoid 2000; Opella and Marassi 2004, 2017; Page et al. 2006; Sanders and Sönnichsen 2006; Poget and Girvin 2007; Wang 2008; Kim et al. 2009; Raschle et al. 2010; Nietlispach and Gautier 2011; Patching 2011; Qureshi and Goto 2011; Tapaneeyakorn et al. 2011; Warschawski et al. 2011; Catoire et al. 2014; Kaplan et al. 2016; Mineev and Nadezhdin 2017; Sim et al. 2017). We will concentrate on the two main ones, seen from the biochemist's perspective: (i) the production of sufficient amounts (0.1–1 mg) of isotopically labeled, properly folded, functional MPs and (ii) their solubilization under a functional and stable form as small, rapidly tumbling particles.

### 10.2.1   Producing the Protein

 Most MPs are not abundant enough that it be practical to extract mg amounts of them from natural sources. Overproducing them is, as discussed in Chaps. 1 and 6, a major stumbling block in biochemical/biophysical studies. Furthermore, NMR studies require that appropriate isotopes be introduced during the synthesis, those most frequently used being $^2H$, $^{15}N$, and $^{13}C$. Various expression systems have been explored, the most prevalent ones being *Escherichia coli*, various yeasts, the combination of insect cells and baculovirus, animal or human cells, and in vitro synthesis. When dealing with eukaryotic MPs, there are rough, opposite gradients of convenience as regards, on the one hand, expression yield, cost-efficiency, and ease of isotopic labeling (these considerations favoring *E. coli*) and, on the other, folding efficiency and posttranslational modifications (favoring mammalian cells), with the other methods lying somewhere in between (for discussions, see e.g. Kim et al. 2009; Nietlispach and Gautier 2011; Tapaneeyakorn et al. 2011; for the special case of overproducing G protein-coupled receptors for NMR studies, see the recent review of Milić and Veprintsev 2015).

The impact of APols, as regards the production of MP samples for NMR studies, may lie in two principal directions, namely folding and stabilization (Fig. 10.1). As discussed in Chaps. 6 and 7, APols make it easier to fold to their native state MPs that have been produced in vivo in *E. coli* – or, conceivably, in yeasts – in an unfolded or misfolded form, such as in inclusion bodies. They can also be used in the frame of in vitro, cell-free synthesis, whether they serve as the recipient medium in which MPs fold to their native form, or they are used to fold a posteriori MPs that have been synthesized in the absence of surfactant and solubilized in a denaturant. It appears that, in many cases, posttranslational modifications, such as glycosylation, are not essential to the function of MPs (they may be more important for governing their quality control and traffic in eukaryotic cells) and that a functional MP can be obtained in *E. coli*, provided it can be convinced to adopt a native-like 3D structure, which APols seem good at (see Chap. 6 and, for general discussions, Popot 2014; Popot and Engelman 2016). Because good expression yields and the sophisticated isotopic labeling schemes that are needed to study large proteins are more easily achieved and less costly in *E. coli* and during cell-free expression, APols can probably be expected to be most helpful when using these two modes of synthesis, but they could also be used to stabilize MPs produced under a native form in eukaryotic cells.

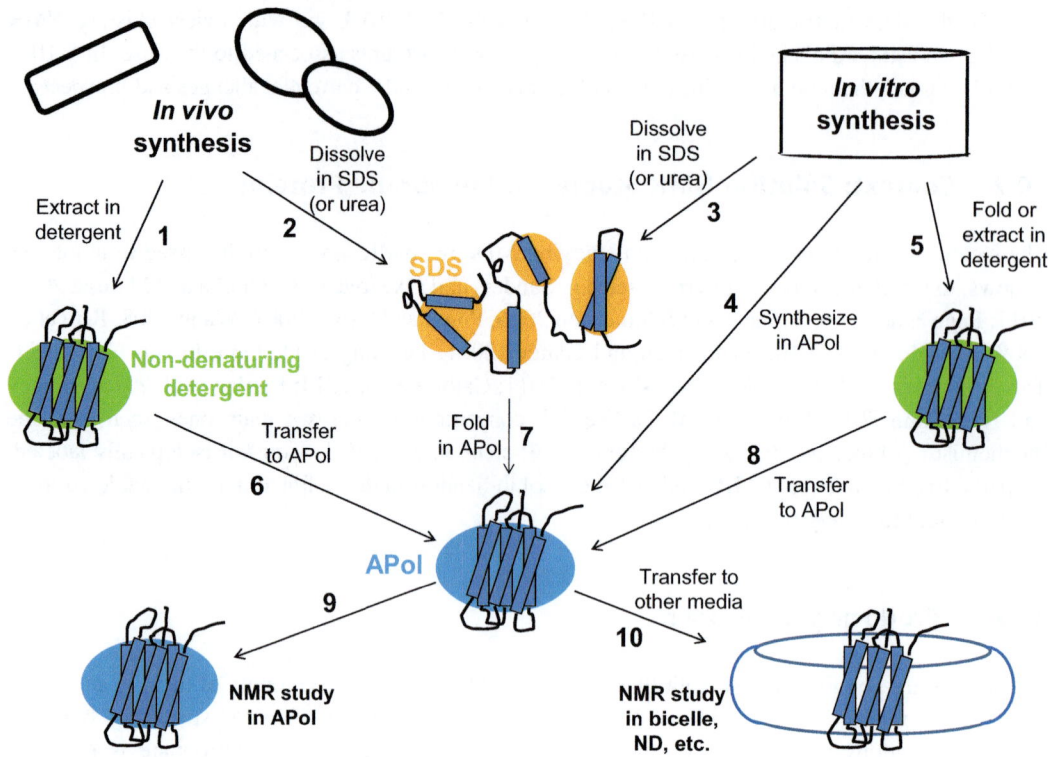

**Fig. 10.1** Using amphipols (APols) for preparing samples for solution NMR studies of membrane proteins (MPs). MPs can be produced in a properly folded form either in vivo or in vitro and extracted using non-denaturing detergents (**1**, **5**). Transfer to APols (**6**, **8**) yields samples that, as a rule, are more stable than those in detergent solution and can stand longer data collection times at higher temperature (**9**). MPs can also be directly expressed in vitro in the presence of APols (**4**) and either used as such (**9**) or transferred to a medium more appropriate to the experiences or protein at hand, such as nanodiscs (NDs), bicelles, a detergent solution, etc. (**10**). Alternatively, MPs can be recovered in a denatured form after solubilization of inclusion bodies (**2**) or of cell-free synthesis precipitates (**3**) into either urea (β-barrel MPs) or SDS (α-helical ones), and APols used to fold them to a functional state (**7**).

## 10.2.2   Membrane Protein/Surfactant Complexes Suitable for Solution NMR

The second major difficulty, once a conveniently labeled and properly folded MP is available in sufficient quantity, is to identify a surfactant system that will keep it in solution under an appropriate form. For sNMR studies, the protein must be (i) in a conformation as close as possible to that (or one of those) it adopts in its native membrane, (ii) stable for extended periods (several hours, if not days) at or above room temperature, and (iii) tumbling rapidly. These requirements are to some extent antagonistic, because as a rule the surfactants that yield particularly small particles are not those in which the MP is best folded and most stable. The experimenter therefore has to navigate an uncomfortable route between the Charybdis of structure perturbation and instability and the Scylla of peak broadening and unusable spectra (for discussions about arbitrating between these constraints, see e.g. Poget and Girvin 2007; Qureshi and Goto 2011).

Because of these difficulties, a great deal of imagination and exploration has been invested into trying to identify systems that offer the best compromise between these requirements, and many reviews have devoted important sections to discussing their relative advantages and constraints (see

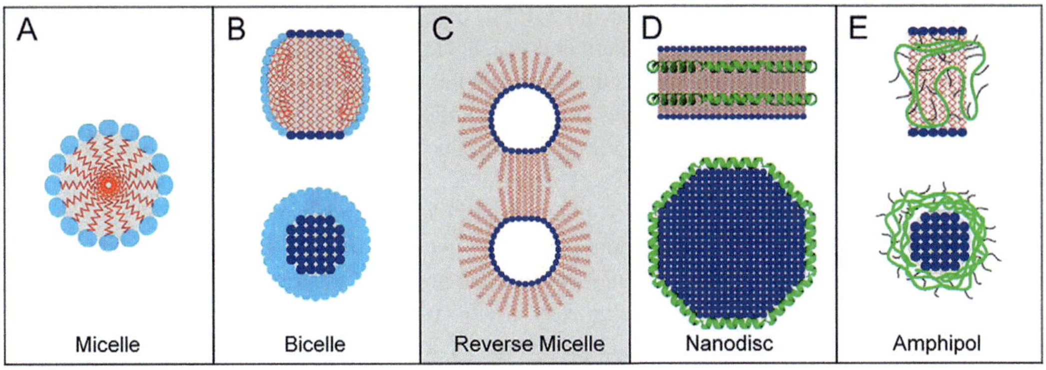

**Fig. 10.2** Schematic representation of various surfactant systems used for solution NMR spectroscopy studies of membrane proteins. (**A**) Detergent solution, (**B**) bicelle, (**C**) reverse micelle, (**D**) nanodisc, (**E**) amphipol. *Gray* shading designates nonpolar environments (From Kim et al. 2009, © 2009 Elsevier B.V. All rights reserved).

e.g. Page et al. 2006; Poget and Girvin 2007; Kim et al. 2009; Raschle et al. 2010; Kang and Li 2011; Nietlispach and Gautier 2011; Qureshi and Goto 2011; Tapaneeyakorn et al. 2011; Warschawski et al. 2011; Catoire et al. 2014; Mineev and Nadezhdin 2017; Opella and Marassi 2017). Only a brief overlook will be provided here. The main systems used are schematically illustrated in Fig. 10.2. They include non-denaturing detergents (Fig. 10.2A), by far the most frequently used medium, which can be supplemented with lipids for stabilization purposes. As will be briefly discussed below, the choice of the best detergent is essentially the result of an empiric and somewhat tedious search, which has to be started again, essentially from scratch, for each new MP to be studied. Some detergents however are used with particular frequency, among them dodecylphosphocholine (DPC) and dihexanoyl- or diheptanoylphosphatidylcholine (both of them frequently noted DHPC, which is confusing; hereafter we will use diC$_6$ and diC$_7$PC, respectively). These detergents frequently provide a good compromise between stabilizing a native-looking conformation and forming with the MP relatively small, fast-tumbling particles. The denaturing detergent SDS has been used in some studies. SDS denatures most TM $\alpha$-helix bundles at room temperature (and, as a rule, TM $\beta$-barrels at higher temperatures), but some of them do resist SDS (cf. Chill et al. 2006a, b), and, even when there is denaturation, some features may persist, such as TM $\alpha$-helices and, sometimes, contacts between them.

Two systems aim at providing MPs with an environment that is as similar as possible to that offered by lipid bilayers, namely bicelles (Fig. 10.2B) and nanodiscs (NDs; Fig. 10.2D). Bicelles are small discs of lipid bilayer whose rim is formed either by a detergent or a detergent-like short-chain lipid, which also partitions in the plane of the disc (see Chap. 3, § 3.2). The use of bicelles for sNMR (and ssNMR) has been discussed, for example in Marcotte and Auger (2005), Prosser et al. (2006), Sanders and Sönnichsen (2006), Poget et al. (2007), Kim et al. (2009), Nietlispach and Gautier (2011), Warschawski et al. (2011), Catoire et al. (2014), Beaugrand et al. (2016), and Mineev and Nadezhdin (2017) (cf. Chap. 3, § 3.2.1). Their main advantage is the bilayer nature of the MP environment, their main disadvantages the fact that they form larger particles than simple MP/detergent assemblies and that their formation and stability depends on the precise control of the ratio between the associated fraction of the two surfactants, which is affected by environmental factors such as concentration and temperature. This renders the handling of bicelles rather tricky. NDs represent a variation on the same theme, in which the rim of the disc is stabilized by "scaffold proteins" (see Chap. 3, § 3.3). Once formed, they are more stable and easier to handle than bicelles, but they are larger than MP/detergent complexes, and the formation of homogeneous MP/ND complexes is far from trivial (see e.g. Raschle et al. 2010; Warschawski et al. 2011; Etzkorn et al. 2013; Hagn et al. 2013; Catoire et al. 2014; Viegas et al. 2016). Variants of scaffold proteins that form smaller or more stable NDs have been developed (see Chap. 3, § 3.3.2). A succinct overview of NMR studies of MPs trapped in NDs is presented in Chap. 3 (§ 3.3.5). Several NMR laboratories have introduced the interesting suggestion that spectra collected on MPs trapped either in bicelles (Howell et al. 2005; Poget and Girvin 2007) or in NDs (Shenkarev et al. 2010), which are more complicated to implement and do not necessarily yield the best spectra, but which provide MPs with a more membrane-like environment, be used as a benchmark to evaluate, by comparing 2D $^1$H,$^{15}$N-TROSY spectra, the ability of detergents to maintain the native structure of target MPs for which there is no easy in vitro functional assay.

Organic solvents offer the potential advantage of a low viscosity, which can considerably accelerate tumbling, and the drawback of being poor substitutes for the membrane environment. They have been used occasionally, according to either of two configurations. In the first one, the MP is directly solubilized in a solvent like trifluoroethanol (TFE), at the risk of affecting its structure (see e.g. Shenkarev et al. 2010). In the second one, each MP binds two reverse micelles (Fig. 10.2C). The water phase inside the micelles interacts with the extramembrane regions of the MP, the organic solvent with the TM one. There is currently too little experience with this system to evaluate the risk of MP structure modification, but the latter should not be underestimated, as organic solvents can interfere with protein/protein and protein/lipid interactions (see e.g. Kang and Li 2011; Qureshi and Goto 2011, and refs. therein).

Within each of these systems, optimizing the choice of the surfactant and experimental conditions is an essential and difficult task (see e.g. Poget and Girvin 2007; Nietlispach and Gautier 2011; Qureshi and Goto 2011). This is illustrated in Figs. 10.3 and 10.4 by two examples taken from the work of Daniel Nietlispach and Antoine Gautier on sensory rhodopsin II (pSRII), a MP with seven TM $\alpha$-helices and a mass of 26.7 kDa (Gautier et al. 2010; Nietlispach and Gautier 2011). Figure 10.3 illustrates two effects, that of the choice of the detergent (compare the first four panels with the last two) and that of the temperature: the last two spectra were obtained in the same detergent, diC$_7$PC, but spectrum **E** was collected at 40 °C, spectrum **F** at 50 °C.

Figure 10.4 illustrates, on a given system – sensory rhodopsin in diC$_7$PC, the effect of varying the detergent concentration. As the concentration is increased from 16.5 mM (33× the CMC of diC$_7$PC) to 55 mM (110× the CMC), the resolution of the resonance peaks keeps increasing. At 82.5 mM (165× the CMC), it decreases again, because of the increasing viscosity of the detergent solution. These two series of experiments illustrate one of the difficulties associated with studying MPs by sNMR in detergent solutions: conditions that improve the resolution of the spectra, such as increasing the detergent concentration and raising the temperature, are also conditions which, as we have seen in Chap. 2, accelerate MP inactivation.

Despite these difficulties, rapid progress is being made, helped by the development of higher field NMR magnets, cryogenic probes, improved pulse sequences, and labeling schemes, so that solving de novo the structure of a 30–40 kDa MP is no longer out of reach (see e.g. Kim et al. 2009; Kang and Li 2011; Nietlispach and Gautier 2011; Qureshi and Goto 2011; Warschawski et al. 2011); for a recent review of MP structures studied by ssNMR, see Patching 2015; for up-to-date lists of NMR structures of MPs, see the sites maintained by Dror E. Warschawski (http://www.drolist.com/nmr/MPNMR.html) and by Stephen H. White (http://blanco.biomol.uci.edu/mpstruc/). A most interesting survey of relatively recent sNMR studies of $\alpha$-helical MPs is given in Nietlispach and Gautier 2011 (see also Kim et al. 2009; Qureshi and Goto 2011; Warschawski et al. 2011). Out of the seven MPs listed for which complete assignments were obtained, three are monomeric, the heaviest one being pSRII,

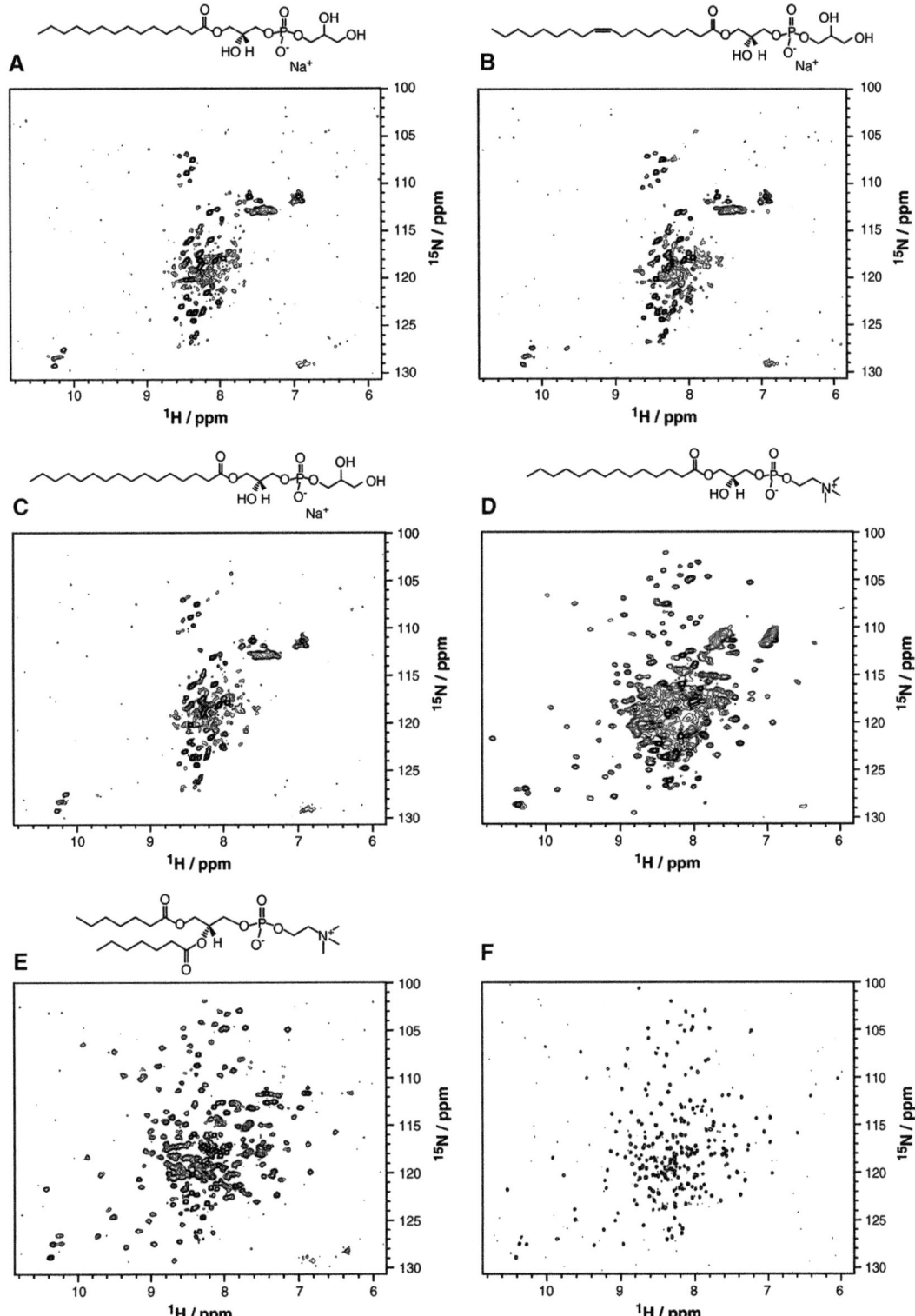

**Fig. 10.3**  Effect of detergent choice and temperature on the spectral appearance of sensory rhodopsin II (pSRII). Comparison of 2D $^1$H-$^{15}$N TROSY spectra of pSRII solubilized in a range of micelle-forming detergents using [u-$^{15}$N]pSRII: (**A**) 1-myristoyl-2-hydroxy-*sn*-glycero-3-[phosphor-*rac*-(1-glycerol)]

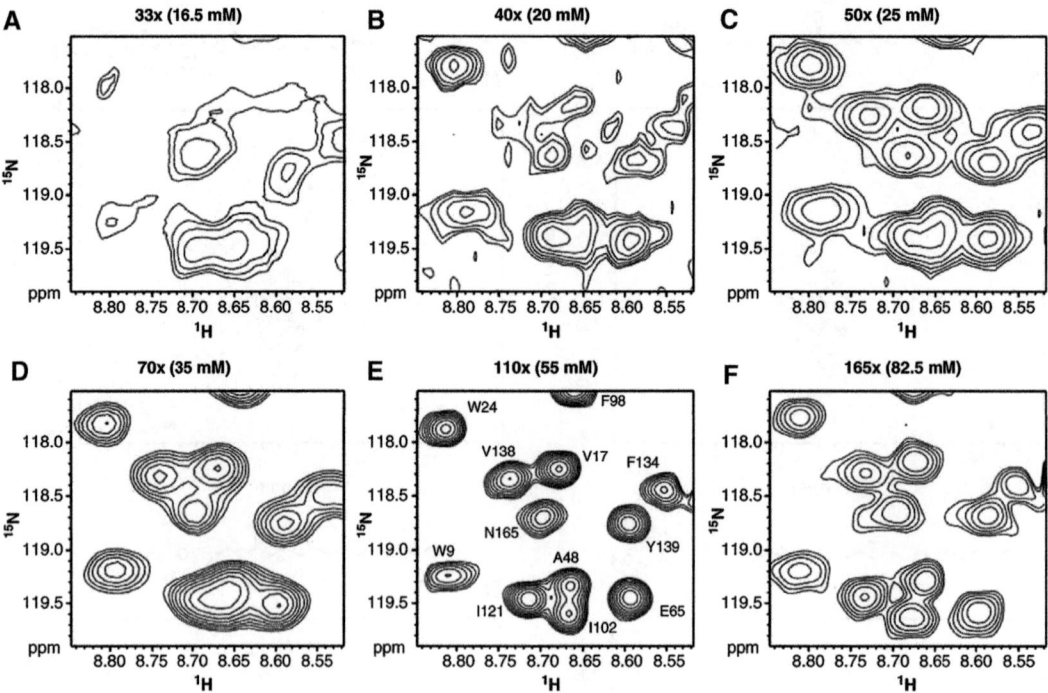

**Fig. 10.4** Preparation of highly homogeneous pSRII-detergent samples for NMR. 2D $^1$H-$^{15}$N- TROSY spectra of pSRII (0.5 mM, pH 6) were recorded using varying amounts of diC$_7$PC. All diC$_7$PC concentrations (16.5–82.5 mM) are at least 10× above the CMC of the detergent. Increasing the diC$_7$PC concentration initially improves sample homogeneity and spectral quality until an optimum (ca. 110× the CMC) is reached. At even higher concentrations, the NMR signals start to broaden as the increased viscosity leads to slower molecular reorientation (From Nietlispach and Gautier 2011, © 2011 Elsevier Ltd. All rights reserved).

at 26.7 kDa (Gautier et al. 2010). Oligomeric MPs, in which the number of distinct nuclei is divided by the number of protomers, reach 39.3 kDa for the diacylglycerol kinase (DAGK) trimer, i.e. 13.1 kDa per protomer (Van Horn et al. 2009), and 71.1 kDa for the KcsA tetramer, i.e. 18.8 kDa per protomer (Yu et al. 2005). All but one of these proteins have been expressed in *E. coli*, the last one by cell-free synthesis. All structures have been obtained in detergent solution, most often DPC, diC$_6$PC or diC$_7$PC. Experimental temperatures range from 30 to 50 °C. To this list, one may want to add VDAC-1, a 30.8-kDa, 19-strand eukaryotic $\beta$-barrel protein from the outer mitochondrial membrane. The protein was expressed in *E. coli*, folded in vitro, and spectra recorded in the detergent LDAO at 30 °C (Hiller et al. 2008). The use of NDs for sNMR studies, which is developing rapidly, is reviewed in Chap. 3 (§ 3.3.5, Table 3.1).

**Fig.  10.3** (continued)  (40  °C),  (**B**)  1-oleoyl-2-hydroxy-*sn*-glycero-3-[phosphor-*rac*-(1-glycerol)] (40 °C), (**C**) 1-palmitoyl-2-hydroxy-*sn*-glycero-3-[phosphor-*rac*-(1-glycerol)] (40 °C), (**D**) lysomyristoyl-phosphatidylcholine (50 °C), (**E**) diC$_7$PC (40 °C) and [u-$^2$H,$^{15}$N]pSRII, (**F**) same at 50 °C. All spectra were recorded at 800 MHz, and detergent concentrations were individually optimized for best protein signal intensity (pSRII 0.4 mM, experiment times 3 h) (From Nietlispach and Gautier 2011, © 2011 Elsevier Ltd. All rights reserved).

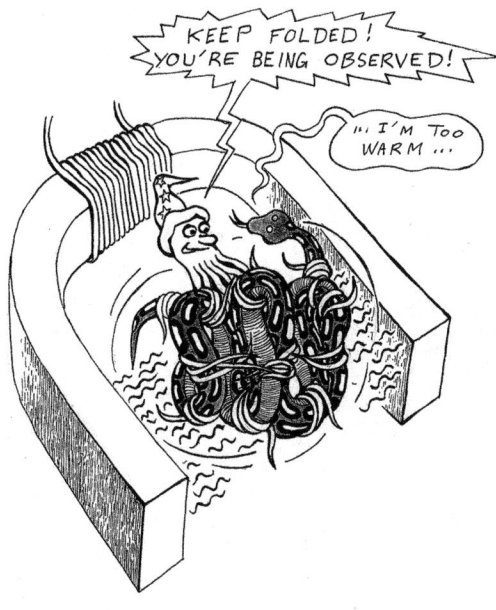

NMR   STUDIES   OF
AMPHIPOL-TRAPPED   MEMBRANE   PROTEINS

*Amphipols stabilize membrane proteins for solution NMR studies*
*(© 2018 by Francis Haraux)*

There are few unmodified MPs that can stand extended exposure to high concentrations of detergent at relatively high temperatures. This observation was one of the major incentives for examining whether trapping MPs with APols, which is known to stabilize most of them as compared to detergent solutions (Chap. 5), would not be compatible with sNMR. Where this endeavor stands now is described in the next section.

## 10.3   Solution NMR Studies of Amphipol-Trapped Membrane Proteins

An overview of all published sNMR studies of APol-trapped MPs to date is presented in Table 10.1. Many of them have focused on feasibility issues. Among the questions asked are the following ones:

- (i)   Can a resolution comparable to that obtained in detergent solutions be achieved?
- (ii)  Do APols have an edge over detergents in terms of the survival of the protein during lengthy data acquisition at relatively high temperature? How do APols compare with NDs in terms of ease of implementation, data quality, etc.?
- (iii) How to optimize experimental conditions?
- (iv)  Where does the APol lie in the complexes, how does it interact with the protein, and does it affect its structure and dynamics?
- (v)   Can specially designed APols facilitate NMR studies?

These various technological aspects, which constitute currently the bulk of NMR studies, will be summarized and discussed in § 10.3.1. Specialized APols – either pH-insensitive or deuterated – developed mainly in view of their use for sNMR studies will be presented in § 10.3.2. More biologically oriented studies have started to appear, involving such medically important MPs as the

**Table 10.1** An overview of articles presenting NMR spectra obtained using amphipol-trapped membrane proteins. Studies are listed in chronological order of publication in journals. Color coding of the cells: *tan*, observations on $\beta$-barrel MPs; *pink*, observations on $\alpha$-helical MPs.

| Study | MP | Structure | MW | APol | Main conclusions | References |
|---|---|---|---|---|---|---|
| 10.1 | tOmpA (transmembrane domain of *E. coli* OmpA) | 8-strand $\beta$-barrel | 19 kDa | A8-35 | [u-$^2$H,$^{15}$N]-labeled tOmpA trapped in partially deuterated A8-35 (DAPol) yields TROSY-HSQC spectra with a dispersity comparable to that observed in detergent solution, indicating that the protein is properly folded and accessible to an NMR study. Resonance peaks are somewhat broader than in diC$_6$PC, due to the slightly larger size of the particles. Comparison of $^1$H$^N$ peak width in DAPol and in unlabeled A8-35 (HAPol) shows that the APol interacts exclusively with the transmembrane (TM) surface of the protein. | Zoonens (2004) and Zoonens et al. (2005) |
| 10.2 | OmpX from *E. coli* | 8-strand $\beta$-barrel | 18.6 kDa | A8-35 | [u-$^2$H,$^{13}$C,$^{15}$N]- or [u-$^{13}$C,$^{15}$N]-labeled OmpX was trapped with HAPol or DAPol, respectively. 2D [$^1$H,$^{13}$C] heteronuclear NOE spectroscopy (HOESY) experiments reveal *(i)* intermolecular contacts between aromatic amino acids of OmpX and alkyl chains of A8-35, and *(ii)* intra-protein dipolar interactions, some of them involving OmpX carbonyl carbons and aliphatic $^1$H groups. | Catoire et al. (2009) |
| 10.3 | OmpX from *E. coli* | 8-strand $\beta$-barrel | 18.6 kDa | A8-35 | Hydrogen/deuterium exchange after prolonged equilibration was measured on [u-$^2$H,$^{13}$C,$^{15}$N]-labeled OmpX/diC$_6$PC and OmpX/A8-35 complexes, providing insights into the dynamics of the $\beta$-barrel, possibly relevant to its controversial function as a TM channel. The presence of EDTA was observed to substantially improve the resolution of 2D [$^{15}$N,$^1$H] spectra of OmpX/A8-35 complexes, presumably by preventing the formation of interparticle bridges by residual traces of Ca$^{2+}$ or other multivalent cations. At the long time scale used (eight weeks), the dynamics of OmpX in diC$_6$PC and A8-35, as reflected by H/D exchange, appear very similar, except that some $^1$H$^N$ nuclei at the edges of the barrel exchange more readily in detergent than in APol. | Catoire et al. (2010b) |
| 10.4 | Human BLT2 leukotriene receptor (GPCR) | 7-$\alpha$-helix bundle | 41.5 kDa | A8-35 | [u-$^2$H,$^{15}$N]-labeled human BLT2 leukotriene receptor, a G protein-coupled receptor (GPCR), was expressed in *E. coli* as inclusion bodies, solubilized in SDS, and folded by dodecylsulfate precipitation in the presence of DAPol. The structure of receptor-bound leukotriene B$_4$ (LTB$_4$) was solved by analyzing transferred NOE signals. Upon binding, LTB$_4$ adopts a highly constrained, seahorse conformation, at variance with the free state, where it explores a wide range of mainly extended ones. The approach appears applicable to ligands with affinities as low as the micromolar range, whose $k_{on}$, due to electrostatic interactions, can be higher than expected from pure diffusion limits. Comparative [$^{15}$N,$^1$H]CRINEPT spectra of DAPol-trapped [u-$^2$H,$^{15}$N]BLT2 before and after addition of LTB$_4$ show that binding of the ligand induces conformational changes in the receptor. | Catoire et al. (2010a, 2011) |
| 10.5 | Bacteriorhodopsin (BR) from *H. salinarum* | 7-$\alpha$-helix bundle + retinal | 27 kDa | A8-35 | A detailed comparison is presented of solution NMR studies of BR in dodecylmaltoside (DDM), APols (A8-35), and DMPC-based nanodiscs (NDs). In order to avoid the bias that lipids extracted from purple membrane along with BR could introduce and to facilitate specific labeling, the protein has been produced in a cell-free system, solubilized in SDS, and folded in each of the three environments. The thermostability of BR increases in the order DDM < A8-35 < NDs, whereas the resolution of 2D [$^{15}$N,$^1$H]-TROSY spectra decreases from DDM (216 resolved peaks) to A8-35 (178 peaks) to NDs (150 peaks). The nature of the environment has little effect on the chemical shift of residues in the inner core of BR, but affects residues in the loops and at the junction between loops and TM helices. | Raschle et al. (2010), Etzkorn et al. (2013), and Elter et al. (2014) |
| 10.6 | tOmpA (transmembrane domain of *E. coli* OmpA) | 8-strand $\beta$-barrel | 19 kDa | SAPol | The synthesis and solution properties of sulfonated APols (SAPols) are described. At variance with A8-35, SAPols make it possible to record NMR spectra at low pH. [$^{15}$N,$^1$H]-TROSY spectra obtained at pH 6.8 of tOmpA folded in SAPols show that the protein is correctly folded. The spectra have a resolution similar to that achieved with A8-35 at pH 8.0 and reveal water-exposed amide and indole protons whose resonance peaks are absent at pH 8.0. Because of the stability of SAPol-trapped tOmpA, spectra could be recorded up to 70 °C. | Dahmane (2007) and Dahmane et al. (2009) |
| 10.7 | OmpX from *E. coli* | 8-strand $\beta$-barrel | 18.6 kDa | NAPol | Two-dimensional [$^{15}$N,$^1$H]-HSQC spectra of OmpX trapped in a non-anionic APol were recorded at pH 6.8. They present a resolution similar to that of the spectra of OmpX/A8-35 complexes recorded at pH 8.0 and give access to signals from solvent-exposed, rapidly exchanging $^1$H$^N$ nuclei that are not observable at pH 8. | Bazzacco (2009) and Bazzacco et al. (2012) |

(continued)

**Table 10.1** (continued)

| Study | MP | Structure | MW | APol | Main conclusions | References |
|-------|-----|-----------|-----|------|------------------|-----------|
| 10.8 | Melanocortin-2 and -4 receptors (GPCRs) | 7-α-helix bundle + retinal | 27 kDa | A8-35 | The melanocortin-2 and -4 GPCRs were expressed in vitro, solubilized in SDS, and folded in A8-35 using various methods, the one giving the most satisfying results being dodecylsulfate precipitation by KCl after immobilization of the unfolded receptors onto Ni/NTA-carrying beads. Preliminary 2D [$^{15}$N,$^1$H]-TROSY spectra are shown, whose quality is currently limited by difficulties in scaling up the folding procedure. | Elter et al. (2014) |
| 10.9 | OmpX from *E. coli* | 8-strand β-barrel | 18.6 kDa | A8-35 | The accessibility to a water-soluble paramagnetic quenching agent of the surface of OmpX trapped in partially deuterated A8-35 (DAPol) has been examined. Of 71 assigned $^1$H$^N$ resonance peaks, 57 are protected and 14 are accessible, corresponding respectively to residues located in the TM β-barrel and in the turns and loops. The distribution thus obtained corresponds remarkably well with that of the detergent diC$_7$PC (Hilty et al. 2004), established using the same approach, and with the distribution of A8-35 deduced from 2D [$^1$H,$^{13}$C] heteronuclear NOE spectra (Catoire et al. 2010b) and from molecular dynamics calculations (Perlmutter et al. 2014). | Etzkorn et al. (2014) |
| 10.10 | Bacteriorhodopsin from *H. salinarum* | 7-α-helix bundle + retinal | 27 kDa | A8-35 | The interactions with the surfactant belt of a polyhistidine tag fused to the *N*-terminus of BR was examined in DDM, in DMPC-based nanodiscs (NDs), and in APol A8-35. The tag was observed to be partially protected from water in DDM and in NDs, whereas it appeared freely accessible in A8-35. The difference is tentatively attributed to the smaller volume of the polar head region of A8-35. | Etzkorn et al. (2014) |
| 10.11 | Major outer membrane protein (MOMP) from *Chlamydia trachomatis* | 16-strand (?) β-barrel trimer | ~120kDa | A8-35 | HSQC spectra of $^{15}$N-labeled MOMP, the major outer MP from the pathogenic bacterium *Chlamydia trachomatis*, were recorded *(i)* in the presence of DPC at 35 °C and *(ii)* in A8-35 at both 35 °C and 50°C. The spectra recorded in APol are of much better quality than those obtained in detergent solution. The spectra obtained in A8-35 at 35 *vs.* 50 °C provide complementary structural information, probably originating from the extramembrane loops of MOMP and the regions connecting the loops to the TM β-barrel. | Feinstein et al. (2014) |
| 10.12 | KpOmpA (transmembrane domain of OmpA from *Klebsiella pneumoniae*) | 8-strand β-barrel | 23.4 kDa | A8-35 | Intermolecular contacts between the TM domain of [u-$^2$H,$^{13}$C,$^{15}$N]-labeled OmpA from *K. pneumoniae* (KpOmpA) and A8-35 were deduced from 3D $^{15}$N-edited ($^1$H,$^1$H) HSQC-NOESY-TROSY spectra, allowing the identification of interactions between specific amide and indole $^1$H nuclei of KpOmpA and octyl and isopropyl chains of A8-35. | Planchard et al. (2014) |

low-affinity, BLT2 receptor of leukotriene – a GPCR involved in inflammatory diseases and allergic responses –, a couple of other GPCRs, as well as MOMP, the major outer membrane protein from *Chlamydia trachomatis*, a potential target in the development of vaccines against this pathogenic bacterium, as is also the case of KpOmpA, an outer MP from *K. pneumoniae*. These data will be presented in § 10.3.3. Perspectives and suggestions for future developments will be discussed in § 10.4.

## 10.3.1 Validating and Developing the Methodology

As is usual in such technological developments, the first MPs to be studied were workhorse proteins, sturdy, easy to produce, and already well studied by NMR in detergent solution (Table 10.1). This included three small, eight-stranded β-barrel MPs from bacterial outer membranes, namely OmpX (18.6 kDa) from *E. coli* and the transmembrane (TM) domains of OmpA from *E. coli* (tOmpA; 19 kDa) and from *Klebsiella pneumoniae* (Kp-tOmpA; 23.4 kDa), as well as the seven-α-helix bundle bacteriorhodopsin (BR; 27 kDa) from *Halobacterium salinarum*, which is frequently used to develop methods planned to be later applied to G protein-coupled receptors (GPCRs). Their 3D structures are shown in Fig. 10.5.

Most of these studies have been carried out with A8-35 and with a partially deuterated version of it, DAPol, which features an unlabeled main chain carrying deuterated octyl and isopropyl side chains (Gohon et al. 2004, 2006). Their chemical structures are shown in Fig. 10.6A, B and their 1D $^1$H and 2D $^1$H,$^1$H and $^1$H,$^{13}$C NMR spectra in Figs. 10.19 and 10.20. In addition, a perdeuterated version of

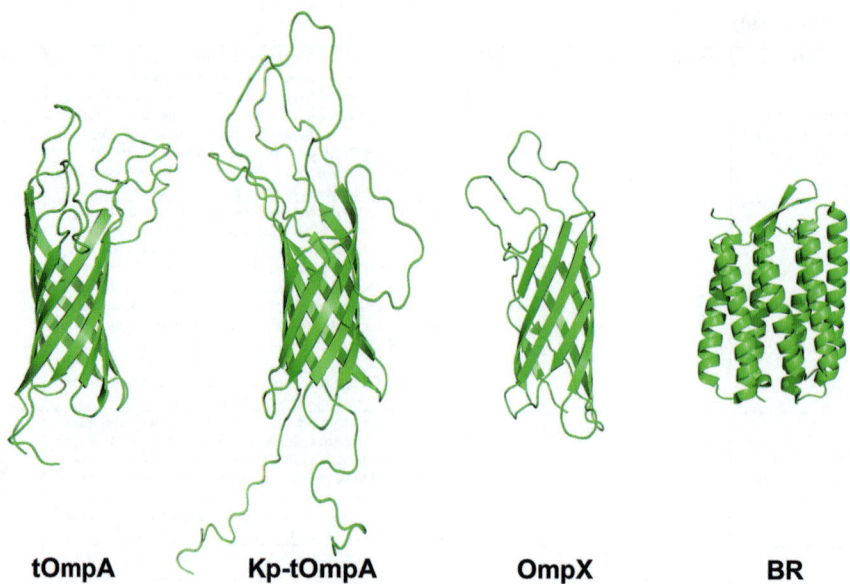

**Fig. 10.5** Three-dimensional structures of four model membrane proteins (MPs) used in sNMR studies of MP/amphipol complexes. *tOmpA* transmembrane (TM) domain of OmpA from *Escherichia coli* (1G90), *Kp-tOmpA* TM domain of OmpA from *Klebsiella pneumoniae* (2K0L), *OmpX* OmpX from *E. coli* (1Q9F), *BR* bacteriorhodopsin from *Halobacterium salinarum* (4HWL) (Figure courtesy of Laurent J. Catoire).

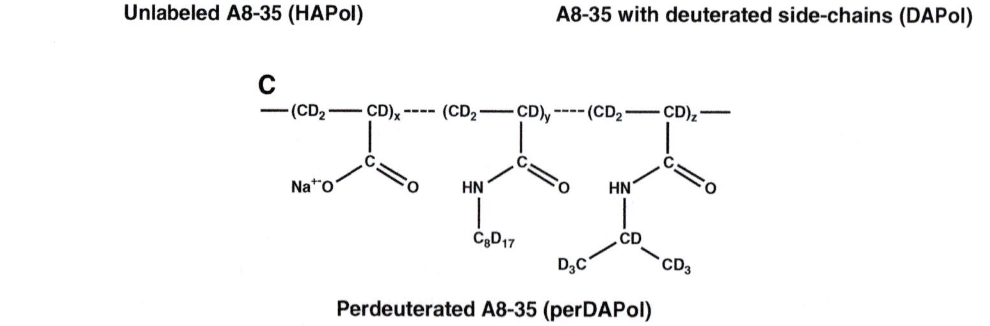

**A**

Unlabeled A8-35 (HAPol)

**B**

A8-35 with deuterated side-chains (DAPol)

**C**

Perdeuterated A8-35 (perDAPol)

**Fig. 10.6** Chemical structures of three forms of A8-35 used in solution NMR experiments. (**A**) Unlabeled A8-35 (HAPol), (**B**) a partially deuterated form thereof (DAPol), (**C**) perdeuterated A8-35 (perDAPol). Syntheses are described in the following publications: HAPol (Gohon et al. 2006; Tribet et al. 1996; see also the annotated protocol given in Chap. 4, § 4.5); DAPol (Gohon et al. 2004, 2006); perDAPol (Giusti et al. 2014). $x \approx 35\%$, $y \approx 25\%$, $z \approx 40\%$. 1D $^1$H and 2D $^1$H,$^1$H and $^1$H,$^{13}$C NMR spectra are shown in Figs. 10.19 and 10.20.

A8-35 and two types of APols with different chemical structures have also been validated for use in sNMR and will be discussed below (§ 10.3.2).

### 10.3.1.1  Validating the Use of Amphipols for sNMR and Resolution of the Spectra as Compared with Those Obtained in Detergent Solutions

The princeps demonstration that APols can be used for sNMR studies of MPs (Study 10.1 in Table 10.1) was achieved by Laurent J. Catoire and Manuela Zoonens using as a model the TM domain of *E. coli* OmpA (tOmpA) trapped either in unlabeled A8-35 (HAPol) or in DAPol (Zoonens 2004; Zoonens et al. 2005). It was followed by a couple of studies (Studies 10.2 and 10.3 in Table 10.1) using *E. coli* OmpX trapped in the same APols (Catoire et al. 2009, 2010b). Known to be highly stable in vitro, these two proteins, which express very well, were chosen because of the extensive sNMR data available from studies conducted in detergent solutions (Arora et al. 2001; Fernández et al. 2001a, b), which had led to 3D structures very similar to those obtained by X-ray diffraction (Vogt and Schulz 1999; Pautsch and Schulz 2000). As no activity tests are available for these two proteins in solution, the fact that they had maintained their conformation after transfer to APols was established by (i) examining their migration during SDS-PAGE, which indicates whether the $\beta$-barrel is present or not (Schweizer et al. 1978; Pautsch et al. 1999), and (ii) comparing their $^{1}$H and $^{15}$N NMR chemical shifts with those observed in detergent solutions (Fig. 10.7). In the case of full-length OmpA, it has also been shown that the protein, after refolding in A8-35, presents the same proteolysis pattern as native OmpA and that it induces the formation in lipid black films of ion channels

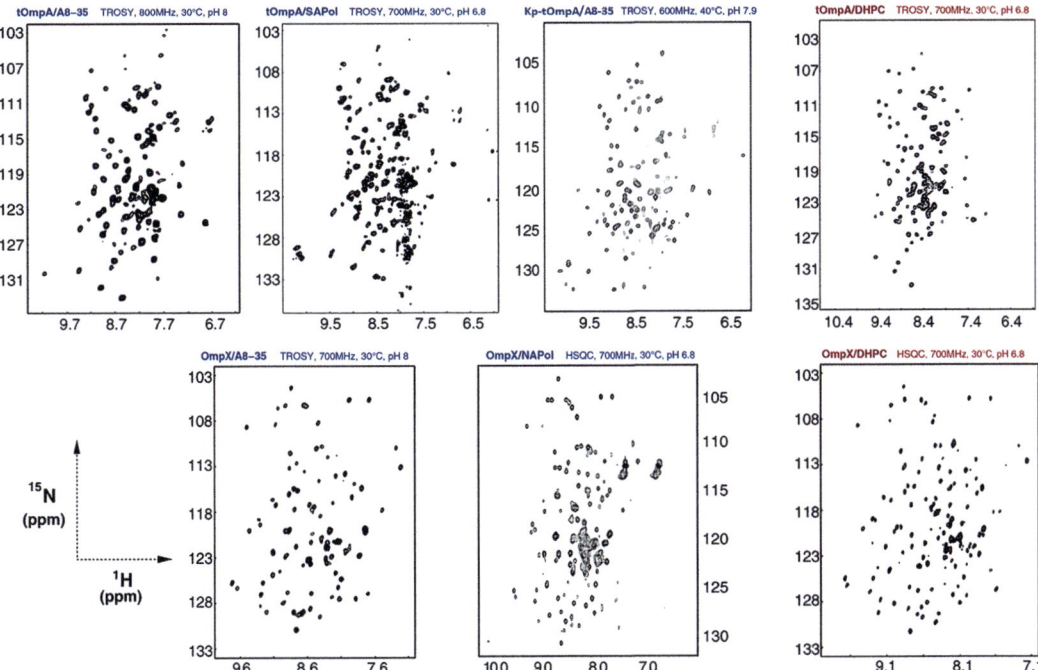

**Fig. 10.7**  A comparison of the 2D $^{1}$H-$^{15}$N NMR correlation spectra of three $\beta$-barrel membrane proteins kept water soluble by either amphipols or detergents (*SAPol* sulfonated APol, *NAPol* non-ionic APol, *DHPC* diC$_6$PC). *Top*, the transmembrane domain of OmpA from either *E. coli* (tOmpA) or *K. pneumoniae* (Kp-tOmpA); *bottom*, OmpX from *E. coli*. Original data are from the following publications: Zoonens et al. (2005), tOmpA/A8-35 and tOmpA/DHPC; Dahmane et al. (2009), tOmpA/SAPol; Planchard et al. (2014), KpOmpA/A8-35; Catoire et al. (2010b), OmpX/A8-35 and OmpX/DHPC; and Bazzacco et al. (2012), OmpX/NAPol. Experimental conditions are indicated above each spectrum. The composite figure is from Planchard et al. (2014).

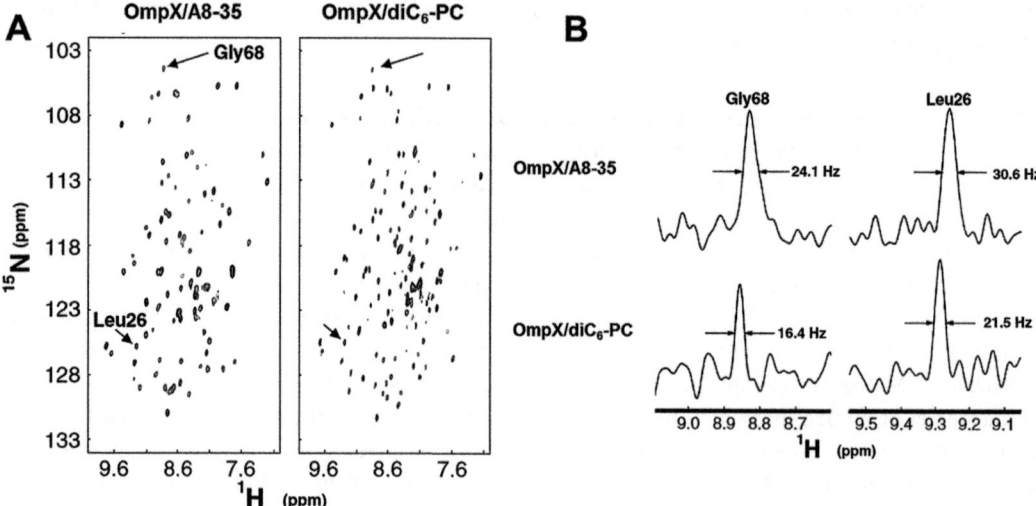

**Fig. 10.8** (**A**) 2D $^{15}$N-$^{1}$H TROSY spectra of [u-$^{2}$H,$^{13}$C,$^{15}$N]OmpX complexed either by A8–35 (*left*; [OmpX] = 1.3 mM, 8 transients/increment, pH 8.0) or by diC$_6$PC (*right*; [OmpX] = 0.7 mM, 16 transients/increment, pH 6.8). (**B**) Cross-sections of two selected resonance peaks. The *upper* and *lower* rows are taken from the *left* and *right* spectra of panel **A**, respectively. Experiments were carried out at 30 °C on a Bruker Avance II 700 spectrometer equipped with a 5-mm triple-resonance (TXI) gradient probe (From Catoire et al. 2010b).

with properties identical to those formed by the native protein (Pocanschi et al. 2006) (see Chap. 6, § 6.3.1.2). There is therefore no doubt that the structures of APol-trapped tOmpA and OmpX are extremely close if not identical to those of the proteins solubilized in detergent solutions that have been used in X-ray and earlier sNMR studies.

Early works aimed at investigating to which extent sNMR of MP/APol complexes was practical. They showed that, although MP/APol complexes are slightly bigger and, as a result, tumble slightly less rapidly than the best MP/detergent ones – typically formed with diC$_6$PC or DPC – leading to a somewhat degraded resolution (see Fig. 10.8), the latter is sufficient for structure determination (Zoonens et al. 2005; Catoire et al. 2009, 2010b; Planchard et al. 2014). A fairly rapid tumbling (for OmpX, $\tau_c \approx 31$ ns; Catoire et al. 2010b; see Fig. 10.9) is consistent with A8-35 forming a thin layer at the surface of the protein, rather than a diffuse corona, in keeping with conclusions from analytical ultracentrifugation (AUC) and small-angle neutron scattering (SANS) (Gohon et al. 2008; Chap. 9) and from EM (e.g. Althoff et al. 2011; Huynh et al. 2014; Liao et al. 2014; Chap. 12), as well as with MD calculations (Perlmutter et al. 2014; Chap. 5).

Compared to MPs in complex with detergents commonly used for solution-state NMR studies, such as diC$_6$PC or DPC, the overall correlation times ($\tau_c$) of small A8-35-trapped MPs, tOmpA and OmpX, are ~30–50% longer (Zoonens et al. 2005; Catoire et al. 2010b), consistent with the hydrodynamic radius of the complexes being somewhat larger (cf. Figs. 10.8 and 10.9). A limiting factor a decade ago, this range of $\tau_c$, a few tens of ns, does not, given the progress of NMR equipment and methodology and of isotopic labeling strategies (see e.g. Tugarinov et al. 2006; Poget and Girvin 2007), hamper the observation of well-resolved peaks nor the acquisition of 3D NMR data within a reasonable time, as long as there are not too many overlapping signals (for a discussion, see Planchard et al. 2014). Depending on the MP being studied, a slight loss of resolution might be more than offset by the improvement in stability, which allows longer collection times at higher temperatures. Striving to produce the smallest particles possible is still important, but it is not necessarily as compelling today as it formerly was, and should not be sought at the expense of distorting the protein's native state

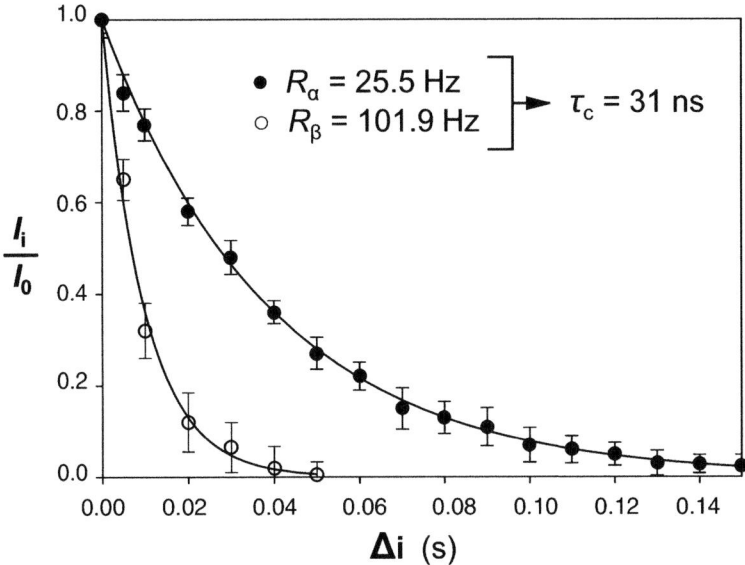

**Fig. 10.9** Effective correlation time, $\tau_c$, of [u-$^2$H,$^{13}$C,$^{15}$N]OmpX/A8–35 complexes estimated by a 1D $^{15}$N-$^1$H TROSY rotational correlation time (TRACT) experiment (Lee et al. 2006) carried out at 30 °C. $I_0$ and $I_i$ are integrations of the $^1$H$^N$ 1D spectrum region at $t = 0$ and $t = \Delta i$, respectively. The *upper* and *lower* curves correspond to the slow ($\alpha$-spin state of $^{15}$N) and fast ($\beta$-spin state of $^{15}$N) transverse relaxation decays. The exponential fits yield the $R_\alpha$ and $R_\beta$ values indicated, from which an estimate of $\tau_c$ can be derived as described in Lee et al. (2006) (From Catoire et al. 2010b).

and/or compromising its stability (see e.g. Poget and Girvin 2007; Zhou and Cross 2013; Zoonens et al. 2013).

APols being inherently polydisperse molecules (see Chap. 4), a recurrent concern has been that they could form with MPs populations of complexes with a wide distribution of sizes or NMR chemical shifts due to variable electronic environments. Indeed, complexes formed between a small MP, tOmpA, and A8-35 do migrate, upon size exclusion chromatography (SEC), as a broader band than tOmpA/octyltetraoxyethylene (tOmpA/C$_8$E$_4$) complexes (see Chap. 5). The origin of this apparent size dispersity deserves a discussion.

There are two sources of heterogeneity in most APol preparations, the length of the polymer and the random distribution of the hydrophilic and hydrophobic moieties (Chap. 4). Whether the size distribution is by itself a source of particle size polydispersity is uncertain, given that a version of A8-35 with restricted length polydispersity showed essentially the same behavior in SEC as the standard polydisperse mixtures (Fabrice Giusti and Christophe Tribet, unpublished results quoted in Giusti et al. 2014). It is possible, however, that the mere size of APol molecules favors a degree of heterogeneity. A8-35 molecules (on average ~4.3 kDa; Giusti et al. 2014; see Chap. 4, § 4.2.1) are ~10× bigger than detergent ones (respectively, 482, 351, and 511 Da for diC$_6$PC, DPC, and dodecylmaltoside (DDM)), and, therefore, a MP binds a lesser number of APol molecules than of detergent ones (Chap. 5). This results in a less efficient thermodynamic pressure toward optimization of the bulk of the APol vs. the detergent belt, which can generate a degree of size dispersity without necessarily affecting the chemical environment of the protein. The heterogeneity in alkyl chain distribution probably plays a role, in the sense that a given APol particle or MP/APol complex may recruit a smaller or larger mass of APol depending on the number and distribution of the chains carried by each molecule. This is suggested by the observation that homopolymeric non-ionic APols, which are also polydisperse in length but homogeneous as regards alkyl chain distribution, migrate upon size exclusion chromatography (SEC) as sharper bands than A8-35 (Sharma et al. 2012) (see Chap. 4, § 4.3.2, Fig. 4.28), which is not the case for their heteropolymeric counterparts (Sharma et al. 2008; Bazzacco et al. 2009). For further discussion of the source(s) of polydispersity of APol particles and MP/APol complexes and possible ways to improvements, see Chap. 4, § 4.3.1.2.2.

It is interesting to note that trapping tOmpA with an A8-35/$C_8E_4$ mixture results in the formation of particles that migrate upon SEC with the same apparent $R_S$ as pure tOmpA/A8-35 complexes, but are more homogeneous (Zoonens et al. 2007; see Chap. 9, § 9.3.2, Fig. 9.6). A tentative interpretation of this observation is that detergent molecules provide the "small change" that permits optimization of the volume of the surfactant belt, and relaxation toward a more uniform size. Because APols have a stabilizing effect on MPs even when mixed with detergents (Champeil et al. 2000; Chap. 5, § 5.6.1, Fig. 5.27), such mixtures could prove useful should particle homogeneity improve the quality of sNMR signals.

Close attention must be paid to the fact that many factors can contribute to increasing the polydispersity of MP/APol complexes, which can lead to signal degradation. Among those factors that have been identified to date, on may cite:

(i) Deviations from the nominal composition of A8-35: a slightly higher hydrophobicity leads to aggregation of APol particles and MP/APol complexes (see Gohon et al. 2004, 2006; Chap. 4, § 4.3.1.2.2).

(ii) Incubation for a long time (days) at a pH too close to pH 7 (ibid.).

(iii) The presence of traces of calcium or other multivalent cations, which, presumably, bridge MP/A8-35 complexes (Picard et al. 2006; Chap. 4, § 4.3.1.2.2, Fig. 4.16); indeed, adding EDTA improves the quality of solution NMR spectra (Catoire et al. 2009, 2010b).

(iv) Removal of the excess of free APol that is required for efficient MP trapping. This is a critical parameter. APols, as discussed in Chap. 5, are not very dissociating, and, in the absence of a slight excess of them, MPs tend to form small oligomers (see Arunmanee et al. 2014; Gohon et al. 2008; Zoonens et al. 2007, Chap. 5, § 5.2.1, and Chap. 12, § 12.3.1 and 12.3.2). Solutions of A8-35-trapped tOmpA, for instance, appear monodisperse in SEC for an overall MP/APol mass ratio of 1:4, whereas the protein is estimated to bind only about half of that (Zoonens et al. 2007; see Chap. 5, § 5.2.1). As discussed in more detail in Chap. 9, removing the excess APol by immobilized metal affinity chromatography leads to the formation of small oligomers, which redissolve if the free polymer is added back (Zoonens et al. 2007). The presence of oligomers does not preclude sNMR investigations, but it can degrade the intensity and quality of the signals.

As described in Chaps. 5 and 9, checking for polydispersity can be done by SEC, by small-angle X-ray or neutron scattering, or by AUC. AUC is a particularly sensitive and quantitative technique, as illustrated by a detailed analysis of BR/A8-35 complexes (Gohon et al. 2008; see Chap. 9, § 9.3.7).

Another factor leading to improved resolution is to substitute unlabeled with deuterated APols. Dipole-dipole coupling between amide protons ($^1H^N$) and nearby $^1H$ nuclei plays a major role in $^1H^N$ spin relaxation phenomena. Among those, transverse relaxation can be substantially slowed down by deuterium labeling (Anglister et al. 1993; Gardner and Kay 1998). Comparison of $^1H$-$^{15}N$ TROSY-HSQC spectra of tOmpA trapped in two forms of A8-35, HAPol, which is fully hydrogenated, and DAPol, whose isopropylamine and octylamine chains are perdeuterated (Fig. 10.6), shows substantial linewidth differences, whose analysis, as will be described in the next section, yielded the first experimental insights into the relative arrangements of the protein and the polymer in MP/APol complexes.

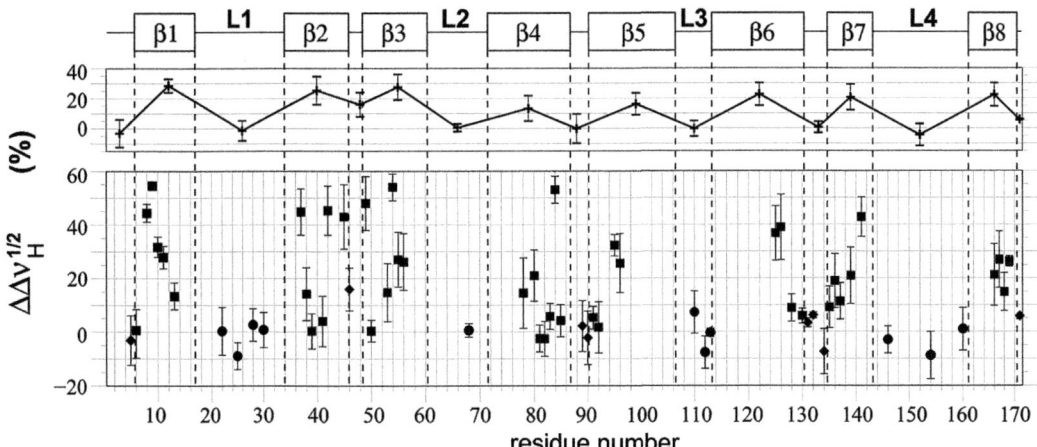

**Fig. 10.10** Dipolar broadening due to tOmpA/A8-35 interactions. Variations in $^1H^N$ linewidth, measured at mid-height ($\Delta\Delta\nu_H^{1/2}$), for tOmpA trapped with fully protonated A8-35 (HAPol) with respect to signals obtained after complexation with A8-35 with perdeuterated alkyl chains (DAPol) are plotted against residue number. Data from $^1H$-$^{15}N$ TROSY-HSQC spectra of [u-$^2H$,$^{15}N$]tOmpA/HAPol and /DAPol samples recorded at 30 °C, pH 7.9, and 800 MHz $^1H$ frequency. Experiments were performed at least three times. Average variations are shown $\pm$ SD. *Squares*, *diamonds*, and *circles* refer to amino acids belonging to $\beta$-strands, periplasmic turns, and external loops, respectively. Averages for each secondary structure element are shown at the *top*, under a schematic representation of the secondary structure. Sequence segments forming the extracellular loops are labeled L1-L4 (From Zoonens et al. 2005, © 2005 National Academy of Sciences).

### 10.3.1.2   Mapping Membrane Protein/Amphipol Interactions by NMR

As described in Chap. 5, a consistent picture of the composition, organization, and dynamics of MP/APol complexes has emerged from an ensemble of biochemical and biophysical studies including small-angle radiation scattering, NMR, EM, and mass spectrometry data, as well as MD calculations. NMR has been particularly instrumental in mapping MP/APol interactions and the accessibility to the solvent of the surface of APol-trapped proteins.

As noted above, Study 10.1 (Table 10.1) exploited differences between $^1H$-$^{15}N$ TROSY-HSQC spectra of tOmpA collected either in unlabeled A8-35 (HAPol) or in DAPol to map the distribution of the polymer at the surface of the protein. The broadening of the $^1H$ resonance lines of a subset of $^1H^N$ nuclei was indeed observed to correlate with their position in the 3D structure. This information is patchy, because less than half of the amide protons of tOmpA are assigned (colored residues in Fig. 10.11), but it nevertheless provided a first atomic-level description of the way an APol interacts with the MP it keeps soluble. In brief, two sets of $^1H$-$^{15}N$ TROSY-HSQC spectra were recorded at high magnetic field ($\nu_H = 800$ MHz), one with HAPol- and the other with DAPol-trapped [u-$^2H$,$^{15}N$]tOmpA. Half-height linewidths were compared in the $^1H$ dimension, because the coupling with remote protons is more pronounced with $^1H$ than with $^{15}N$ nuclei (Pervushin et al. 1997). The extent of relative broadening induced by substituting HAPol for DAPol is shown in Fig. 10.10, and their relationship with elements of secondary structure indicated. Color-coded representations are shown in Fig. 10.11, superimposed onto models of the secondary and tertiary structures. Strikingly, only residues belonging to $\beta$-strands displayed any dipolar broadening (on average, by $22 \pm 8\%$), whereas no statistically significant broadening was observed for amide protons located in the periplasmic turns or the extracellular loops (Fig. 10.10). Even though this experiment does not report directly on Van der Waals contacts between tOmpA amino acid side chains and the alkyl chains of A8-35, the latter are obviously close enough to the amide protons of the protein for the broadening effect of substituting HAPol for DAPol to be detected. Using the (simplifying) assumption that broadening is mainly due to interactions with the

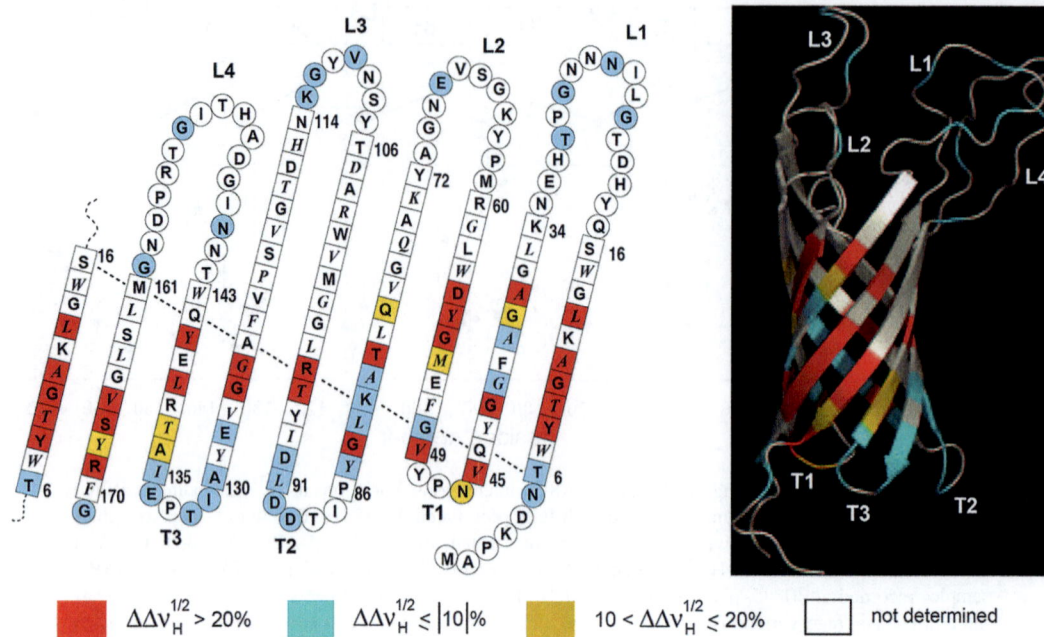

$$\blacksquare \quad \Delta\Delta v_H^{1/2} > 20\% \qquad \blacksquare \quad \Delta\Delta v_H^{1/2} \le |10|\% \qquad \blacksquare \quad 10 < \Delta\Delta v_H^{1/2} \le 20\% \qquad \square \quad \text{not determined}$$

**Fig. 10.11** Mapping of contacts with A8-35 alkyl chains onto the structure of tOmpA. Residues are color-coded depending on whether dipolar line broadening upon substituting HAPol for DAPol (Fig. 10.10) is strong (*red*), weak (*yellow*), or undetectable (*blue*). Residues giving rise to no assigned line are left *white*. The data are plotted on a topology sketch (Adapted from Pautsch and Schulz 1998; the side chains of residues shown in *italics* point toward the exterior of the barrel) and on a ribbon representation of the 3D structure (Protein Data Bank code 1G90; Arora et al. 2001). *Squares* indicate residues that are part of the β-barrel, *circles* residues in the extramembrane loops and turns (From Zoonens et al. 2005. © 2005 National Academy of Sciences).

terminal methyl group of a single alkyl chain of A8-35 (but see Study 10.2), numerical simulations indicate that broadening by $\ge 20\%$ should correspond to distances $\le 3.5$ Å.

Several subsequent studies confirmed that APols interact specifically, if not always exclusively, with the TM surface of the MPs they trap. Two studies used as model MP OmpX from *E. coli*, another eight-stranded β-barrel, while resorting to widely different approaches. In Study 10.2 (Table 10.1; Catoire et al. 2009), heteronuclear dipole-to-dipole cross-relaxation was applied to exploring inter-molecular interactions and intramolecular spatial proximities in the OmpX/A8-35 complexes. The experiments revealed the existence of intermolecular contacts between aromatic amino acid residues of the protein and specific groups of the polymer, in addition to intra-protein dipolar interactions.

In Study 10.9 (Table 10.1; Etzkorn et al. 2014), the surface accessibility of amide moieties in OmpX/A8-35 complexes was mapped using the paramagnetic water-soluble relaxation agent Gd (DOTA)⁻. In these experiments, the solvent accessibility is quantified by the intermolecular paramag-netic relaxation enhancement effect, as expressed by the relaxivity constant $\varepsilon$, an indicator of the minimal distance that the paramagnet can approach a given amide moiety (Caravan et al. 1999; Hilty et al. 2004). Large values of $\varepsilon$ indicate close minimal distances and thus a high solvent accessibility. For amide groups in MPs, differences in $\varepsilon$ reflect shielding by adsorbed surfactant molecules, either detergent or APol. Using the value of $\varepsilon = 2$ s⁻¹·mM⁻¹ as the threshold level for the classification of amide moieties into "protected" and "accessible" groups, it appears that, of the 71 $^1H^N$ nuclei for which resonance peaks can be assigned in spectra of OmpX/A8-35 complexes, 57 are protected and 14 are accessible. The accessible amide moieties are located in the turns and loops of OmpX, and the

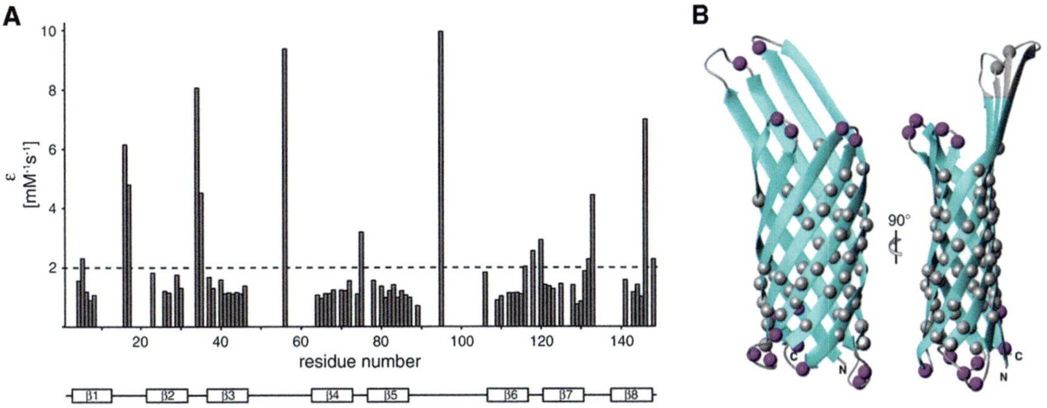

**Fig. 10.12** Surface accessibility of OmpX in OmpX/A8-35 complexes. (**A**) Paramagnetic relaxivity $\varepsilon$ of backbone amide moieties by Gd(DOTA)$^-$, plotted vs. the amino acid sequence of OmpX. The secondary structure elements of OmpX are indicated below. (**B**) Structure of OmpX (PDB 1QJ8; Vogt and Schulz 1999), where all amide moieties of OmpX in A8-35 that could be unambiguously assigned are shown as spheres. *Gray* and *magenta* colors indicate protected and accessible amide moieties, respectively, as classified by the threshold level $\varepsilon = 2\ \text{s}^{-1} \cdot \text{mM}^{-1}$ (*dashed line* in panel A) (From Etzkorn et al. 2014).

protected ones in the $\beta$-barrel region (Fig. 10.12). The relaxivities observed for OmpX in A8-35 correlate remarkably well with those previously measured in diC$_6$PC (Hilty et al. 2004; Fig. 10.13). Overall, the solvent accessibility mapping shows that both surfactants adsorb specifically onto the hydrophobic surface of OmpX, with a very similar distribution.

In Study 10.10 (Table 10.1), the accessibility to water of a polyhistidine tag fused at the *N*-terminus of BR was compared in three environments, A8-35, DDM, and DMPC-based NDs. Solvent-accessible imidazole protons of the histidine ring experience proton exchange rates with the surrounding water molecules in the fast chemical exchange regime and thus are not detectable in the NMR spectrum (Plesniak et al. 2011). 2D TROSY NMR spectra of His-tagged BR in DDM solution and in NDs feature intense correlation cross-peaks in the imidazole spectral region (Etzkorn et al. 2014). These resonance peaks could be unambiguously assigned to the tag, because BR does not contain any histidine residue in its natural amino acid sequence and because these peaks disappeared in equivalent preparations of the protein after selective cleavage of the tag. The presence of histidine resonance peaks indicates that the tag is at least partially protected from fast exchange with water in DDM and NDs. The resonances are not detected in BR/A8-35 complexes, suggesting that the poly-His tag is directly accessible to the solvent and not embedded in the APol belt (Etzkorn et al. 2014). This difference between APols, detergent, and NDs may be due to the relatively small volume of polar moieties in A8-35 (carboxylate groups), as compared with glycerophosphatidylcholine in NDs and maltoside in DDM. It is interesting, in this context, to recall that His-tagged MPs trapped in A8-35 do interact with beads or chips carrying Ni/NTA groups (Zoonens et al. 2007; Giusti et al. 2015; see Chap. 13, § 13.3.1), and so do His-tags carried by the APol itself (Giusti et al. 2015; Perry et al. 2018; ibid., and Chap. 12, § 12.3.4.3), indicating that in neither case is the tag buried into the APol belt.

The general picture of a belt of APol limited to the TM surface of MPs is consistent with MD and EM data (see Chaps. 5 and 12, respectively). Further studies have provided more details about the nature of the contacts. In Study 10.2 (Table 10.3; Catoire et al. 2009), 2D [$^1$H,$^{13}$C] heteronuclear NOE spectroscopy (HOESY) experiments (Rinaldi 1983; Yu and Levy 1983) were carried out on [u-$^2$H,$^{13}$C,$^{15}$N]OmpX/HAPol or [$^{13}$C,$^{15}$N]OmpX/DAPol complexes. Analysis of the various sets of data revealed $^1$H-$^{13}$C interactions within the protein, within the APol, and between the APol and the protein (the high content of methylene and methyl groups of HAPol makes such interactions detectable in the

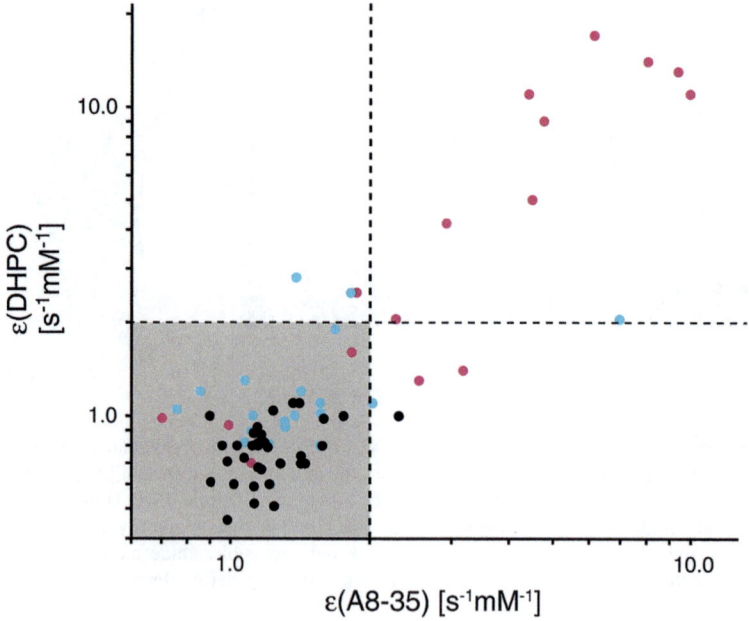

**Fig. 10.13** Comparison of the surface accessibility of OmpX amide groups to Gd(DOTA)$^-$ in OmpX/A8-35 and OmpX/diC$_6$PC complexes. Relaxivities ($\varepsilon$) of amide $^1$H$^N$ nuclei in OmpX/A8-35 complexes determined in Etzkorn et al. (2014), for 71 residues are compared with the corresponding values in OmpX/diC$_6$PC complexes (Hilty et al. 2004). *Dashed lines* denote the threshold level of $\varepsilon = 2$ s$^{-1}$·mM$^{-1}$ used for classifying the residues as exposed or protected. The *gray* area comprises the residues classified as protected under both conditions. Residues are colored according to their location in the 3D structure of OmpX (PDB 1QJ8; Vogt and Schulz 1999): *black*, in $\beta$-strand, within >2 positions from end of the strand; *blue*, in $\beta$-strand, within ≤2 positions from end; *magenta*, in loops and turns (From Etzkorn et al. 2014).

HOESY spectrum even with natural abundance $^{13}$C level). In the aromatic $^{13}$C chemical shift range (116–135 ppm), numerous cross-peaks are observed, resulting from contacts between $^{13}$C nuclei of aromatic rings in the protein and $^1$H nuclei in the side chains of the APol. This is consistent with the structure of OmpX, which features aromatic rings exposed at the TM surface, potentially accessible to APol alkyl chains. Interestingly, interactions of the protein with A8-35 octyl chains seem to involve more extensively the last five CH$_2$ groups than the terminal CH$_3$ one, nor the first two CH$_2$ groups (Catoire et al. 2009), suggesting that the alkyl chain may interact sideways rather than end on with the TM surface of the protein.

In Study 10.12 (Table 10.3; Planchard et al. 2014), 3D $^{15}$N-edited $^1$H-$^1$H HSQC-NOESY-TROSY spectra of [u-$^2$H,$^{13}$C,$^{15}$N]KpOmpA/HAPol complexes were exploited to map the contacts of the various CH$_3$ and CH$_2$ groups in octylamine and isopropylamide side chains of the APol with the surface of the protein (the TM domain of OmpA from *K. pneumoniae*). The distribution of the two types of side chains at the surface of KpOmpA is found to largely overlap. Once again, contacts are observed between the APol and some of the aromatic residues that mark the limits of the TM region.

Study 10.5 (Table 10.1; Etzkorn et al. 2013) presents an interesting comparative study of BR in complex with either A8-35, DDM, or DMPC-based NDs. TROSY-HSQC NMR spectra recorded at 40 °C were found to be marginally better in A8-35 than in DDM, and significantly better than in NDs (Fig. 10.14), despite a suboptimal dispersity of the BR/A8-35 complexes (presence of small aggregates; Etzkorn et al. 2013). It was noted that, whereas the TM region of BR adopts the same structure in all three environments, some loop residues show chemical shift differences, indicating a different conformation or environment. In particular, in some loop regions, the data collected on

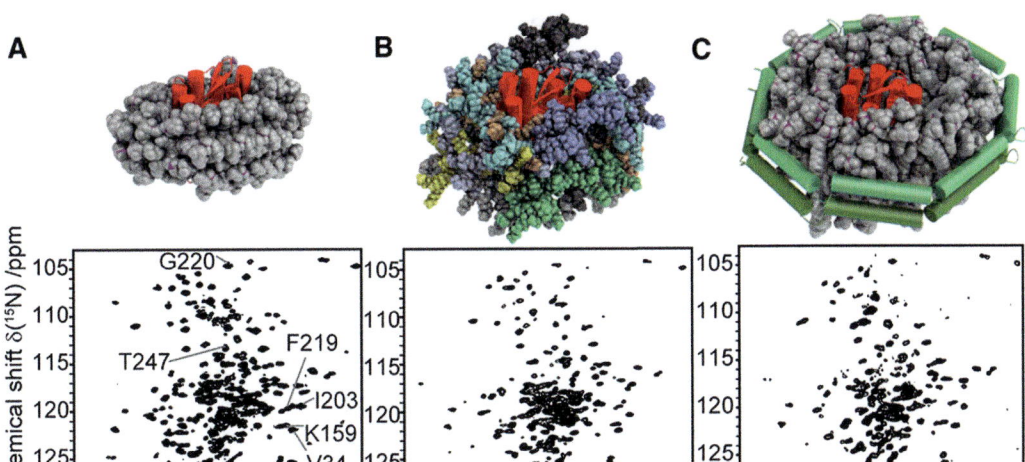

**Fig. 10.14** $^{1}$H-$^{15}$N TROSY-HSQC NMR spectra of bacteriorhodopsin in different environments. All spectra were recorded at 40 °C on cell-free expressed and refolded BR. In the top row, the illustration in (**A**) shows BR (*red*) surrounded by 126 molecules of DDM. In (**B**), BR is represented trapped by eight A8-35 molecules shown in different hues. In (**C**), it is incorporated into a ND comprising 100 molecules of DMPC (*gray*) and two copies of the scaffold protein MSP1D1 (*green*). The particles are drawn to scale and are not energy-minimized (From Etzkorn et al. 2013, © 2013 Elsevier Ltd. All rights reserved).

A8-35-trapped BR are similar to data obtained by ssNMR on BR in its native purple membrane, as opposed to BR in DDM or in NDs (Etzkorn et al. 2013). Along with other sNMR data collected on OmpX (Catoire et al. 2010b; § 10.3.3.1) and on *Chlamydia trachomatis* MOMP (Feinstein et al. 2014; § 10.3.3.3), this observation suggests that the nature of the surfactant may affect the conformation and/or dynamics of segments connecting the TM region and extramembrane loops of MPs.

## 10.3.2    Developing pH-Insensitive and Fully Deuterated Amphipols

All APol-based studies described above resorted to A8-35 (HAPol) or a partially deuterated variant thereof (DAPol) (Fig. 10.6). With the evidence solidly established that APols can be used for sNMR, it became timely to try to overcome two of the limitations of HAPol and DAPol: (i) they cannot be used below pH 7, which is a serious limitation in NMR because it prevents studying water-exposed amide protons, whose rate of exchange with water is too fast at this pH (Wüthrich 1986), and (ii) the residual $^{1}$H nuclei in DAPol can be a nuisance for certain measurements. These two problems were addressed by developing (i) pH-insensitive APols and (ii) a fully deuterated version of A8-35.

### 10.3.2.1    pH-Insensitive Amphipols and Their Use in sNMR

As described in Chaps. 4 and 5, three types of pH-insensitive APols have been developed to date, namely a phosphorylcholine-based APol (PC-APol) (Diab et al. 2007a, b), a sulfonated APol (SAPol) (Picard et al. 2006; Dahmane et al. 2011), and a series of non-ionic, glucose-based APols (NAPols) (Sharma et al. 2008, 2012; Bazzacco et al. 2009, 2012). Their synthesis and properties have been described in Chaps. 4 and 5. We are concerned here with their use for NMR studies, which, as of now,

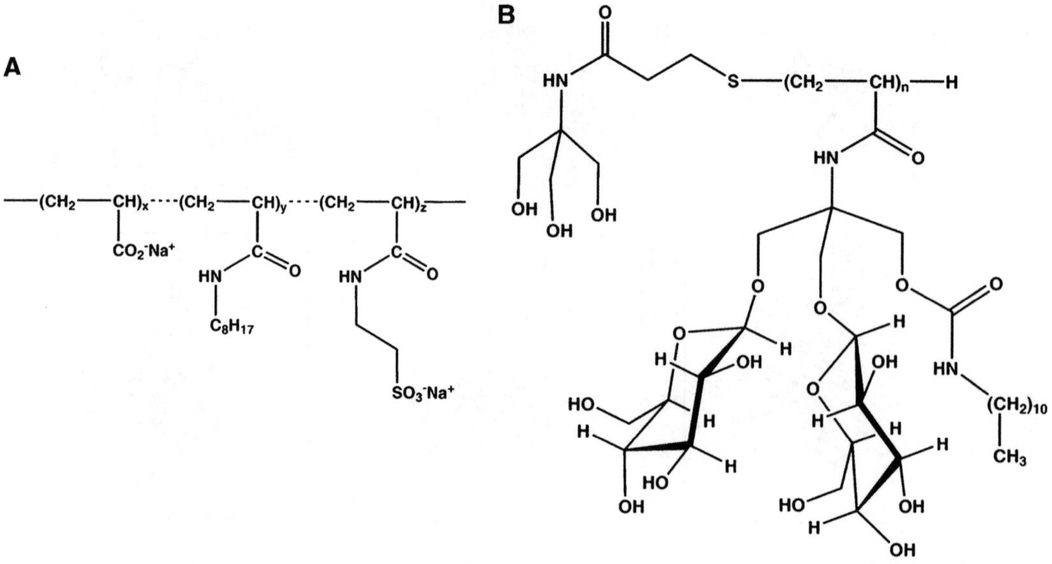

**Fig. 10.15** Chemical structures of two pH-insensitive amphipols (APols). (**A**) A sulfonated APol (SAPol), obtained by random derivatization of a polyacrylate precursor with octylamine and with taurine (Dahmane et al. 2009). $x \approx 35\%$, $y \approx 25\%$, $w \approx 40\%$. (**B**) A non-anionic APol (NAPol) obtained by homotelomerization of a precursor carrying one undecyl chain and two glucose moieties (Bazzacco et al. 2012; Sharma et al. 2012).

has been validated only for SAPols and NAPols. For convenience, the chemical structures of SAPols and of the latest version of NAPols are recalled in Fig. 10.15.

Sulfonated APols (Fig. 10.15A) are obtained by random derivatization of a polyacrylate precursor with octylamine and with taurine. Their structure is very similar to that of A8-35, except that the isopropyl groups in A8-35 are replaced by sulfonate ones in SAPols. Their charge density is similar to that of A8-75 (Tribet et al. 1996), the precursor of A8-35 (see Chap. 4). Due to the presence of the sulfonates, however, SAPols and MP/SAPol complexes remain soluble at any pH and, at variance with A8-35, their solubility is not affected by the presence of calcium ions (Picard et al. 2006; Dahmane et al. 2009).

The applicability of SAPols to sNMR studies has been explored using tOmpA as the model MP (Study 10.6 in Table 10.1; Dahmane et al. 2011). tOmpA was directly folded in SAPol (see Chap. 6) and its structure examined by SDS-PAGE, which evidenced the formation of the $\beta$-barrel, and by sNMR. As described above (§ 10.3.1.1), the 2D [$^1$H,$^{15}$N]-TROSY spectrum of A8-35-trapped tOmpA features very similar chemical shifts to those observed in detergent solution, indicating that the protein is in its native state (Zoonens et al. 2005). A very similar spectrum was observed for tOmpA folded in SAPols, diagnostic of correct folding (Fig. 10.16).

Because MP/SAPol complexes remain soluble and monodisperse at low pH, it is possible to observe solvent-accessible amide protons, whose exchange with the solution is too rapid at the slightly basic pH (8.0) used in NMR studies of tOmpA/A8-35 complexes. At pH 6.8, many correlation peaks belonging to loop residues (upfield in the $^1$H dimension, i.e. between 7.5 and 8.5 ppm) indeed become observable. Downfield in the $^1$H dimension (~10.1 ppm), five indole protons are now detected, compared with only one in A8-35 at pH 8.0 (Zoonens et al. 2005). Also, correlation peaks in the center of the spectrum that are enlarged at pH 8.0 due to unfavorable fast chemical exchange become much sharper at pH 6.8 (see Fig. 10.16). tOmpA/A8-35 and tOmpA/SAPol TROSY spectra feature similar linewidths, in keeping with the similar particle sizes observed by SEC (Dahmane et al. 2009).

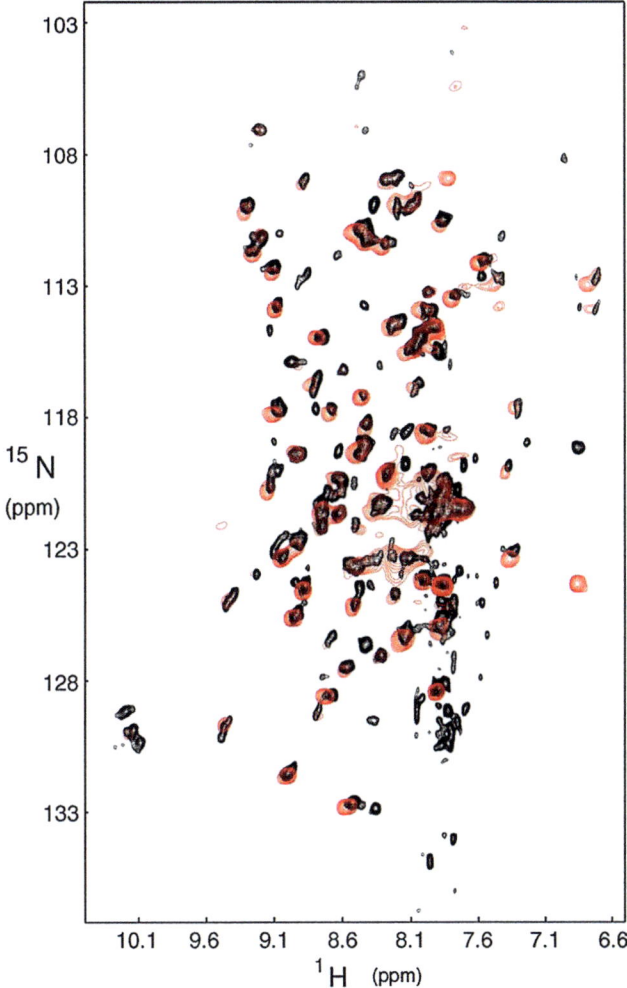

**Fig. 10.16** Superimposed 2D $^{15}$N-$^{1}$H TROSY spectra of [u-$^{2}$H,$^{15}$N]tOmpA trapped either in a sulfonated amphipol (SAPol; Fig. 10.15A; in *black*; 20 mM NaPi buffer, pH 6.8, and 100 mM NaCl) or in A8-35 (in *red*; 20 mM Tris buffer, pH 8.0, and 100 mM NaCl). The experiments were carried out at 30 °C. Note the extra peaks that can be observed at pH 6.8, for example indole protons at ~10 ppm in the $^{1}$H dimension and amide protons belonging to exposed parts of extracellular loops, mostly between 7.5 and 8.0 ppm (From Dahmane et al. 2011, © 2011 John Wiley & Sons, Inc. All rights reserved).

The stability of tOmpA/SAPol complexes was tested by recording 2D $^{15}$N-$^{1}$H TROSY spectra at 30, 50, and 70 °C (Fig. 10.17). The very similar and well-dispersed correlation peaks indicate that SAPol-trapped tOmpA retains its native 3D structure for days even at 70 °C. As expected, the line shape and intensity of signals originating from loop residues change more with temperature than those from $\beta$-barrel residues, due to faster conformational exchange in the loops. Similarly, it has been observed that transferring the insulin-responsive facilitative glucose transporter GLUT4 from the highly stabilizing detergent lauryl maltose neopentyl glycol (LMNG) to APol A8-35 increased the accessible NMR acquisition time at 35 °C from 2 to 36 h without inducing any detectable protein aggregation (Kraft et al. 2015). Having access to a broad range of temperatures is of interest not only for speeding up the tumbling rate, thereby increasing the resolution, but also for modulating the rate of

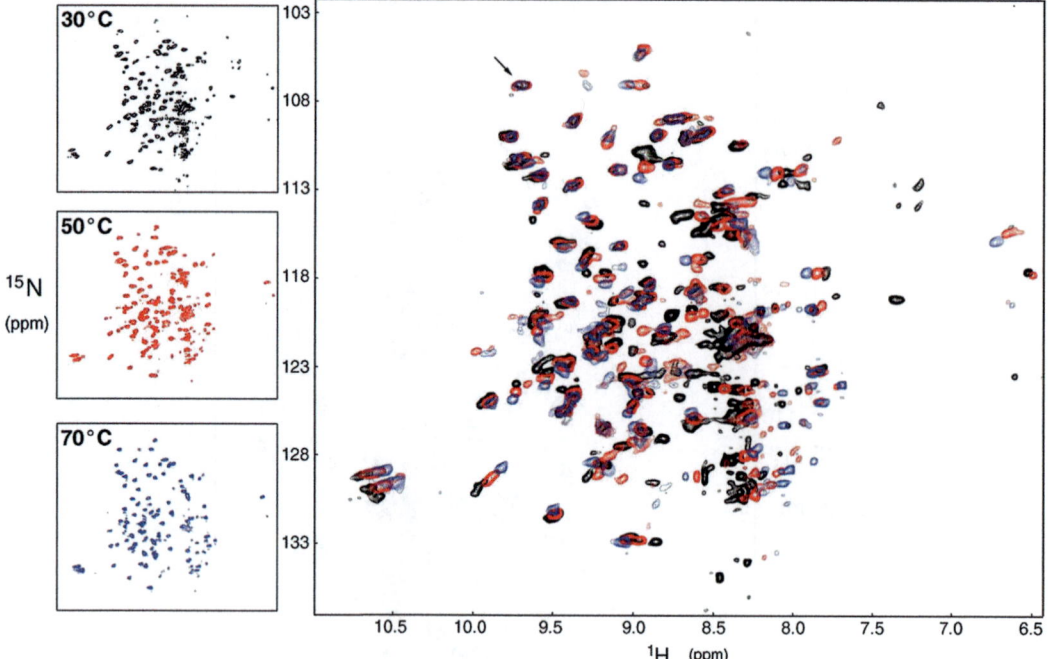

**Fig. 10.17** Comparison of 2D $^{15}$N-$^{1}$H TROSY spectra of SAPol-trapped [u-$^{2}$H,$^{15}$N]tOmpA acquired at 30 °C (*black*), 50 °C (*red*), and 70 °C (*blue*). In the panel to the *right*, the spectra recorded at 50 and 70 °C have been shifted so as to superimpose the G41 peak (*arrow*) to that observed at 30 °C in order to facilitate the comparison of the chemical shifts of residues participating in the β-barrel structure (downfield in the $^{1}$H dimension) (From Dahmane et al. 2011, © 2011 John Wiley & Sons, Inc. All rights reserved).

exchange between interconverting conformations, which can create line broadening and make detection impossible in multidimensional experiments (Sanders and Sönnichsen 2006).

tOmpA is a very stable MP, whose NMR structure has been studied in detergent solution at 50 °C (Arora et al. 2001). In comparative studies of MP stability in SAPols vs. A8-35, SAPols appeared to be less stabilizing than A8-35 (Picard et al. 2006; Dahmane et al. 2011), presumably due to their higher charge density (see Chap. 5). This drawback could probably be alleviated by bringing the charge density of SAPols in line with that of A8-35. The synthesis of an APol that would comprise 35% of sulfonate groups, 25% of octyl chains, and 40% of isopropylamine ones would achieve this goal, but it would require the development of a different synthesis protocol, because it is impossible to derivatize the totality of the carboxylate groups carried by a polyacrylate.

The deuteration or perdeuteration of SAPols, if needed, could be carried out following the same routes as used for obtaining partially and totally deuterated forms of A8-35.

As described in Chap. 4, developing a totally non-ionic APol (NAPol) has been a protracted endeavor that went through a succession of chemical structures and synthesis protocols. The major difficulty is to endow the molecules with enough non-ionic but polar groups to confer them a high solubility in water despite the many alkyl chains they have to carry. The most satisfying structure obtained to date, which was designed and developed in collaboration with the organic chemistry laboratory of Bernard Pucci, is shown in Fig. 10.15B. Its solubility is due to the presence of two glucose residues per undecyl chain, a sugar residue to alkyl chain ratio identical to that in undecylmaltoside. This homotelomeric NAPol has been extensively validated for trapping, folding, and stabilizing MPs (Bazzacco et al. 2012; Sharma et al. 2012; see Chap. 4, § 4.2.3.1, Chap. 5, § 5.5, and Chaps. 6 and 7).

In Study 10.7 (Table 10.1; Bazzacco et al. 2012), OmpX was used as a model MP to examine the usefulness of NAPols for sNMR studies. 2D $^1$H-$^{15}$N HSQC spectra of OmpX/NAPol complexes feature a high resolution and a wide spectral dispersion in both dimensions, equivalent to those observed with OmpX/A8-35 (Catoire et al. 2009, 2010b; Fig. 10.17A) and tOmpA/SAPol complexes (Dahmane et al. 2011; Figs. 10.16 and 10.17). The small size and monodispersity of MP/NAPol complexes, observed in SEC, are comparable to MP/A8-35 ones (Sharma et al. 2012), which gives rise to a good sensitivity and resolution of the 2D HSQC spectrum at a $^1$H Larmor frequency of 700 MHz, without the need for transverse relaxation-optimized spectroscopy. Both amide $^1$H and $^{15}$N chemical shifts are very similar in the spectra of the NAPol- and A8-35-trapped samples recorded at pH 6.8 and 8.0, respectively (Fig. 10.18A), consistent with the view that, in both cases, it is mainly the alkyl chains that interact with the protein (see Chap. 5, § 5.3.3). On the basis of the assumption that the closest peaks observed in NAPols and A8-35 vs. those in diC$_6$PC correspond to the same residue, and if one excepts residue Ala-10 (located toward the end of the first TM $\beta$-strand), the $^1$H and $^{15}$N weight-average chemical shift differences between NAPol- and A8-35-complexed OmpX lie between 0.001 and 0.20 ppm, with an average of 0.04 ppm (Bazzacco et al. 2012). Keeping in mind that (i) the chemical shifts of OmpX/A8-35 samples are themselves very close to those observed in diC$_6$PC (Catoire et al. 2010b), in which the NMR structure was obtained (Fernández et al. 2001a, b), and (ii) the NMR structure is similar to the X-ray one, obtained in C$_8$E$_4$ (Vogt and Schulz 1999), this indicates that NAPol-trapped OmpX also is properly folded.

For residues that are not exposed to the solvent, the linewidth of the peaks in the 2D spectrum of OmpX/NAPol complexes at pH 6.8 is very similar to that for OmpX/A8-35 complexes at pH 8.0 (see e.g. residue G81 in Fig. 10.18B, *left*). On the contrary, peak intensities are substantially higher at pH 6.8 for solvent-exposed residues, such as those located in the loops or turns, whose $^1$H chemical shifts lie roughly between 7 and 8.5 ppm (see e.g. residue G22 in Fig. 10.18B, *right*). As a result, the spectrum of the OmpX/NAPol sample at pH 6.8 shows additional cross-peaks in this region (Fig. 10.18A, B, *right*). NAPols have also been used for a sNMR study on a non-transmembrane, intrinsically disordered viral protein (Sólyom et al. 2015).

The high quality of NMR signals observed with OmpX/NAPol complexes opens perspectives of complete structural determination of MP/NAPol complexes, provided the size of the MP of interest is compatible with sNMR studies. APols thus join bicelles (Vold and Prosser 1996; Sanders and Prosser 1998; see Chap. 3, § 3.2) and NDs (Bayburt et al. 2002; Denisov et al. 2004; Etzkorn et al. 2013; Hagn et al. 2013; Chap. 3, § 3.3) as useful alternatives to detergents for sNMR studies of MPs.

Media milder than detergents, whether they provide a bilayer-like environment, like bicelles and NDs, or favor the retention of MP-bound lipids, as APols seem to do (Martinez et al. 2002; Gohon et al. 2008; Dahmane et al. 2013; see Chap. 5, § 5.3.1.2), are likely to better preserve native-like structural features. This is suggested by differences in protein signal chemical shifts between MP/detergent and either MP/bicelle (Chou et al. 2002; Lee et al. 2008) or MP/ND (Glück et al. 2009; Raschle et al. 2009, 2010; Shenkarev et al. 2010; Etzkorn et al. 2013) complexes. It will be interesting to examine how well APols fare in this respect. A notable advantage of APols is the simplicity of preparation and handling of MP/APol complexes. MP/bicelle complexes indeed require a strict control of the ratio of aggregated long-chain vs. short-chain lipids or detergents all along the preparation and data collection to ensure the presence of small and monodisperse bicellar particles (Triba et al. 2005; Sanders and Sönnichsen 2006; cf. Chap. 3, § 3.2). This control is far from trivial when buffer exchange or protein concentration steps are required. MP/ND complexes are simpler to handle once they have been formed; however, their preparation involves the production and purification of substantial amounts of lipoproteins,

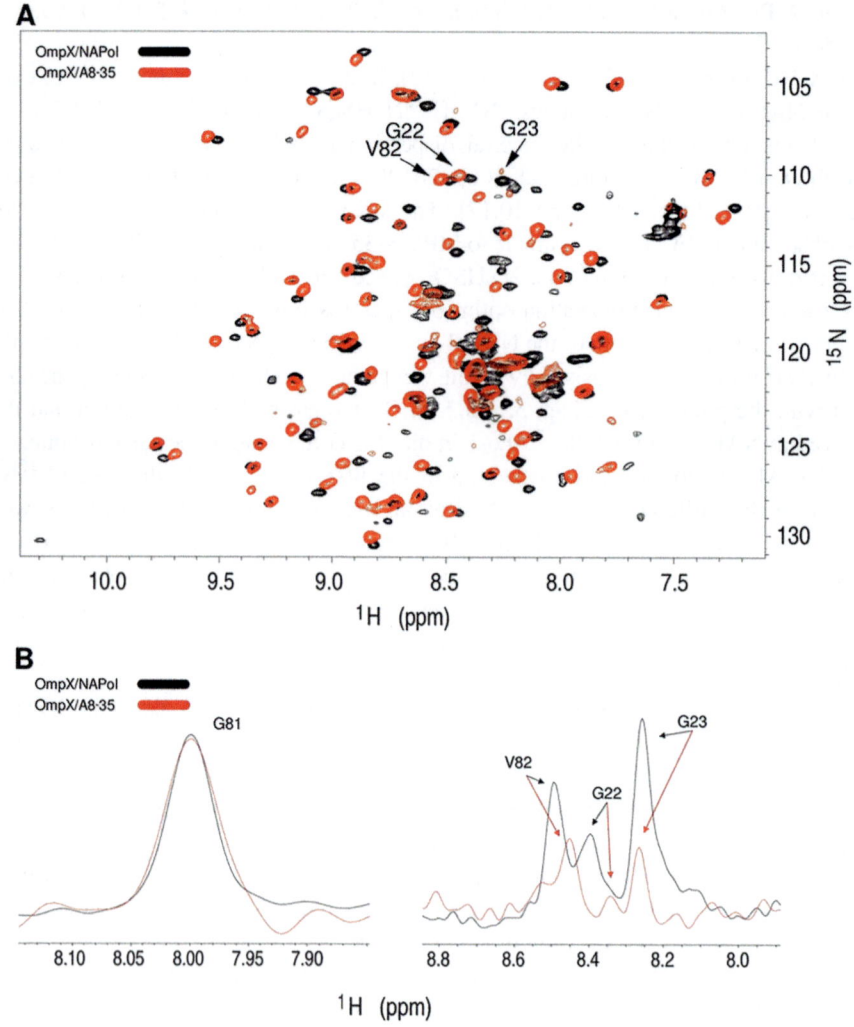

**Fig. 10.18** Solution NMR spectra of OmpX stabilized by either A8-35 or a non-ionic amphipol (homotelomeric NAPol; Fig. 10.15B). (**A**) Superimposed NMR 2D $^{15}$N-$^{1}$H HSQC spectra of [u-$^{2}$H,$^{13}$C, $^{15}$N]OmpX complexed by either A8-35 (pH 8.0, *red*) or NAPol (pH 6.8, *black*). (**B**) Comparative extracted rows at the same signal/noise ratio at the $^{15}$N Larmor frequencies for G81 (*left*) and G23 (*right*) (from spectra in panel **A**). G81, located in the middle of the transmembrane β-barrel, is one of the least solvent-exposed residues in A8-35-trapped OmpX, whereas G23 is very accessible (Catoire et al. 2010b). The peaks for G22 and V82, which have $^{15}$N chemical shifts similar to those of G23, are also visible. In the *left* panel, to allow for a better comparison of the intensity and linewidth of the G81 resonance, the two peaks have been superimposed at the proton chemical shift observed for OmpX/NAPol complexes (Adapted with permission from Bazzacco et al. 2012, © 2012 American Chemical Society).

especially if one aims to capture a single MP per ND, and some purification work is required to obtain monodisperse preparations (see e.g. Ritchie et al. 2009, and Chap. 3, § 3.3).

From a practical point of view, a marked advantage of A8-35 or SAPols over NAPols is that they are relatively easy to deuterate, which is a great asset for certain NMR studies (see § 10.3.3). They would have to be preferred to NAPols should it be necessary to work with unprotonated APols, because perdeuteration of glucosylated NAPols would be difficult and costly. NAPols are likely to be

**Table 10.2** Amphipols that have been validated for sNMR, with their advantages and drawbacks. NMR studies in which each type of APol has been used are listed in Table 10.1 (Adapted from Planchard et al. 2014).

| Short name | Polar moieties | Apolar moieties | Advantages | Drawbacks |
|---|---|---|---|---|
| **A8-35** | Carboxylate groups | Octyl and isopropyl groups | Best characterized APol | Aggregates at acidic pH and in the presence of multivalent cations |
| | | | Exists in partially deuterated (apolar groups) and perdeuterated forms | |
| | | | Hydrogenated form commercially available | |
| **SAPols** | Carboxylate and sulfonate groups | Octyl groups | Insensitive to acidic pH and to multivalent cations | Time-consuming purification |
| | | | | Probably harsher than A8-35 |
| | | | Would be easy to deuterate | Not commercially available yet |
| **NAPols** | Glucose moieties | Undecyl groups | Insensitive to acidic pH and to multivalent cations | Difficult synthesis |
| | | | | Perdeuteration would be very costly |
| | | | Probably milder than A8-35 | The current synthesis protocol makes large-scale production difficult |

milder than SAPols (see Chap. 5, § 5.5) and may permit to collect data for longer periods and/or at higher temperature. However, their current synthesis protocol makes them difficult to market in large amounts, even in their unlabeled form. The relative advantages and drawbacks of A8-35, SAPols, and NAPols for sNMR studies of MPs are summarized in Table 10.2.

After this chapter was written, a review appeared in which the application of NDs and styrene-maleic acid copolymer (SMA)/lipid/MP complexes (SMALPs; see Chap. 4, § 4.2.2.4) to MP sNMR is discussed (Puthenveetil et al. 2017; cf. Chap. 3). SMALPs would suffer from the same limitations as A8-35 and any APol whose solubility depends on the presence of free carboxylate moieties as regards their sensitivity to acidic pH and divalent cations (Postis et al. 2015). To my knowledge, no sNMR spectra of SMA-associated MPs have yet been published, but a publication describing the use of SMALPs for solid-state NMR studies of a MP has recently appeared (Bersch et al. 2017). A [$^1$H,$^{15}$N]-TROSY-HSQC spectrum of the GPCR CRFR2$\beta$ expressed and trapped in NVoy, a polyfructose-based amphipathic polymer (see Chap. 4, § 4.2.3.3), has been presented in Klammt et al. 2011 (see Chap. 7, Fig. 7.7C), but this work does not seem to have been pursued.

### 10.3.2.2   Deuterated Amphipols

Deuterated surfactants have many uses in biophysical MP studies (as regards detergents, see Hiruma-Shimizu et al. 2015). In SANS and AUC experiments, their contrast with the solvent can be matched by setting appropriately the scattering length density of the latter, by playing on its H$_2$O/D$_2$O ratio, or its hydrodynamic density. Thereby, it becomes easier to separate the contributions of the protein and the surfactant in the complexes and, for instance, to study the mass and shape of the protein more or less independently of those of the surfactant associated to it (see examples in Chaps. 5 and 9). In sNMR, deuteration permits to eliminate more or less completely the contribution of the APol to $^1$H-$^1$H NOE data (Catoire et al. 2010a) (§ 10.3.3.2), as well as resonance broadening due to spin diffusion between $^1$H nuclei of the protein and those of the surfactant (see e.g. Zoonens et al. 2005; § 10.3.1.2). As already mentioned, two deuterated versions of A8-35 have been synthesized and tested to date, one of them deuterated on its side chains (DAPol; Fig. 10.6B), the other perdeuterated (perDAPol; Fig. 10.6C). In the present section, their properties and uses will be further discussed.

DAPol was obtained by grafting perdeuterated octyl and isopropyl chains onto hydrogenated polyacrylate, using the standard protocol used for synthesizing unlabeled A8-35 (HAPol; Gohon et al. 2004, 2006; Chap. 4). As the starting polymer is the same, the chain length distribution is also the same and the only chemical difference between HAPol and DAPol comes from the slight batch-to-batch variations of grafting ratios that are inherent to the synthesis protocol. As described in Chap. 4, the physical and biochemical properties of HAPol and DAPol are indeed indistinguishable as long as they are not affected by the isotopic labeling. The size, mass, solution behavior, SANS contrast match point, etc. of free DAPol particles and individual molecules have been thoroughly studied by SEC, DLS, SANS, AUC, and MS (see Chap. 4).

1D $^1$H spectra of HAPol and DAPol are shown in Fig. 10.19A. They show, as expected, the nearly complete disappearance of contributions from the $^1$H nuclei of the octyl and isopropyl chains, whereas contributions from the methylene and methine $^1$H nuclei in the main chain persist. Outside the field of NMR, DAPol has been used, in particular, to study by SANS and AUC the composition, mass, and spatial organization of MP/APol complexes (Gohon et al. 2008; Sverzhinsky et al. 2014; Chaps. 5 and 9), as well as to distinguish, in INS experiments, the dynamics of the backbone of A8-35 from that

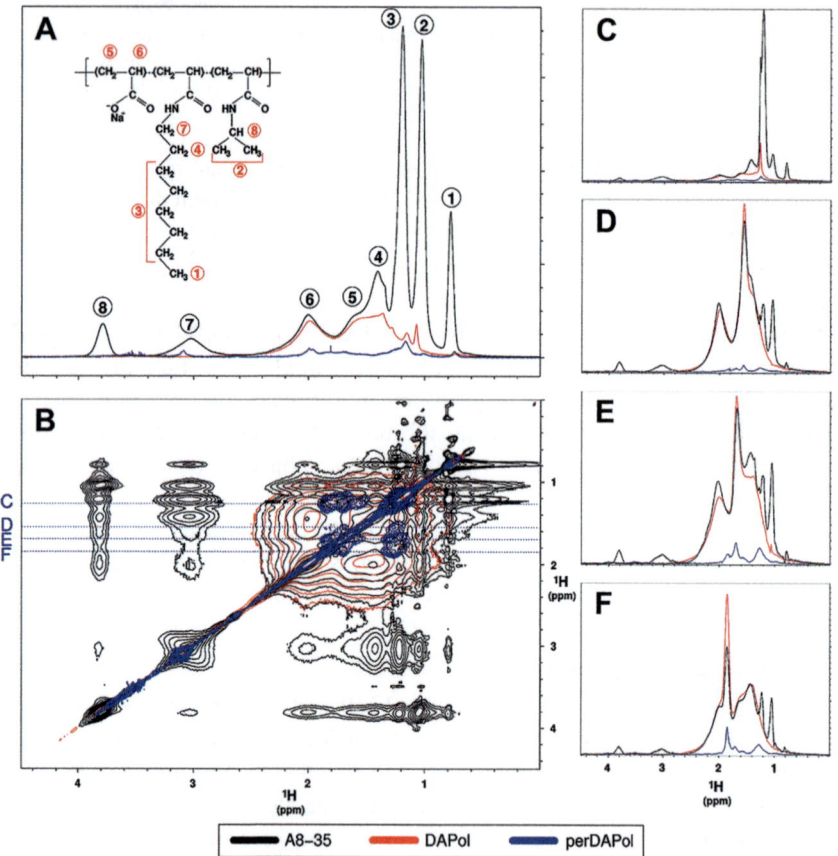

**Fig. 10.19** $^1$H 1D (**A**) and 2D NOESY (**B**) homonuclear NMR spectra of HAPol, DAPol, and perDAPol. In **A**, the inset describes the chemical structure of APol A8-35. *Circled* numbers refer to chemically equivalent $^1$H whose resonances are shown in the 1D spectra displayed in **A**. One-dimensional spectra corresponding to the rows labeled C–F in **B** are shown to the *right* (panels **C–F**, respectively). All data were acquired at 400 MHz $^1$H Larmor frequency, 25 °C, at the same concentration of polymers (30 g·L$^{-1}$) solubilized in pure D$_2$O (From Giusti et al. 2014).

of its side chains (Tehei et al. 2014; Chap. 4). In sNMR, it has been used, as described above, to improve spectra and to map MP/APol contacts (Zoonens et al. 2005; Catoire et al. 2010b; Planchard et al. 2014). Its use has been critical in experiments involving the measurement of heteronuclear (Catoire et al. 2009) and homonuclear (Catoire et al. 2010a) NOE signals.

Because contributions from the $^1H$ nuclei in the main chain remain a hindrance both in INS studies of the dynamics of APol-trapped MPs and in NMR experiments involving the use of $^1H$-$^1H$ or $^1H$-$^{13}C$ NOE signals, a perdeuterated version of A8-35 (perDAPol) has been developed (Fig. 10.6C). From the point of view of chemical synthesis, this was more of a challenge, because it necessitated the synthesis of a perdeuterated polyacrylate with the same average length and dispersity as the commercial hydrogenated product used for the syntheses of HAPol and DAPol (see Chap. 4). Once obtained, the perdeuterated polyacrylate was grafted with perdeuterated octylamine and isopropylamine according to the regular procedure for synthesizing HAPol and DAPol (Giusti et al. 2014).

Figures 10.19 and 10.20 illustrate to which extent perdeuteration diminishes the contribution of the APol to both homonuclear $^1H$-$^1H$ (Fig. 10.19) and heteronuclear $^1H$-$^{13}C$ 2D spectra (Fig. 10.20). As will be shown below (§ 10.3.3.2), determination of the structure of the ligand LTB$_4$ bound to the BLT2 G protein-coupled receptor was possible using DAPol because a sufficient number of protons in the ligand have chemical shifts that differ from those of the backbone protons of A8-35 (Catoire et al. 2010a). Nevertheless, the conformation of some regions of the molecule remained less precisely defined, because NOE signals between aliphatic protons of the ligand were masked by those from the APol. This problem would have been crippling for other ligands, such as the heptadecanoid 12-HHT (12S-hydroxyheptadeca-5Z, 8E, 10E-trienoic acid; Catoire et al. 2011). For such studies, the use of perDAPol ought to be a decisive asset, given the very low intensity of the signals originating from the few remaining $^1H$ nuclei. In contrast, the protonated backbone of DAPol gives rise to intense

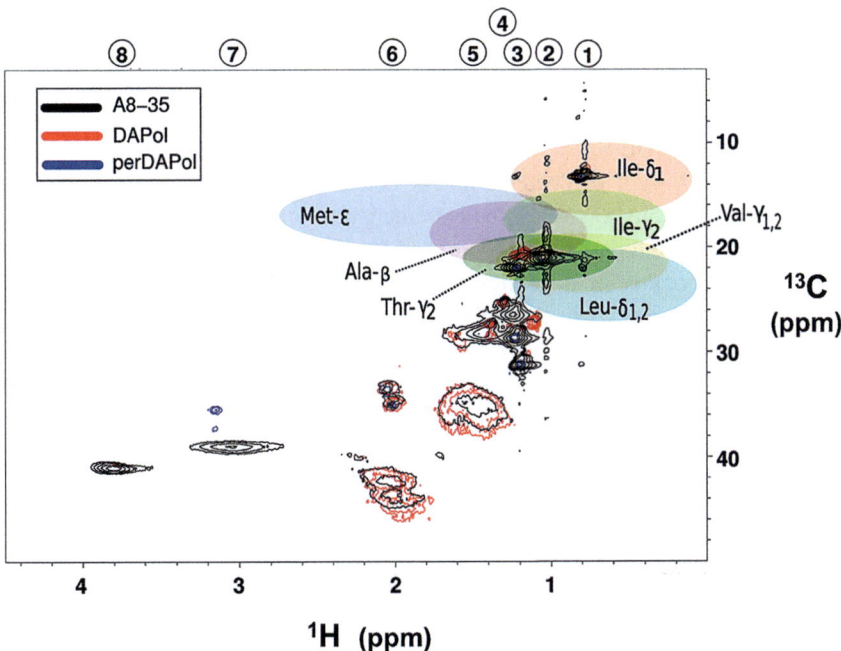

**Fig. 10.20** Two-dimensional $^1H$,$^{13}C$ HSQC spectra of HAPol, DAPol, and perDAPol in D$_2$O. Circled numbers have the same meaning as in Fig. 10.19A. The areas where most $^{13}CH_3$ correlation peaks lie in methyl-TROSY-based experiments are indicated by colored ellipses (Adapted from Plevin and Boisbouvier 2012) (From Giusti et al. 2014).

signals between 0.8 and 2.6 ppm (Fig. 10.19B), creating important distortions of the baseline of the NOESY spectrum. Baseline corrections carried out after the Fourier transform can affect the volumes of weak ligand cross-peaks that are located close to DAPol ones, which affects the accuracy of peak volume measurements. With perDAPol, no such baseline treatment will be needed.

As regards MP structural studies, APol signals can interfere with those from $\alpha$-carbon-linked protons and $CH_n$ moieties of amino acid side chains, as illustrated in Fig. 10.20. The use of perDAPol should be very advantageous, especially when protein $^{13}C$ signals need to be observed above ~25 ppm (Fig. 10.20). Among various approaches that emerged during the past 10 years to study large proteins or protein complexes, one of the best strategies is the use of $^{13}CH_3$ methyl groups immersed in a perdeuterated environment (Sprangers et al. 2007; Plevin and Boisbouvier 2012). Unfortunately, most of the $^{13}CH_3$ methyl groups in amino acids resonate between 0 and 2.5 ppm in the $^1H$ dimension and between 10 and 30 ppm in the $^{13}C$ one (Fig. 10.20), i.e. in a region where many signals from the $CH_n$ groups of the alkyl chains of surfactants are localized. In the case of methyl-TROSY-based experiments, the use of perDAPol rather than DAPol would not make much of a difference, because in this region, i.e. between 0.6 and 1.4 ppm in the $^1H$ dimension, the remaining $^1H$ signals are more or less the same for the two polymers.

Previous work has shown that the resolution of 2D $^1H$-$^{15}N$ TROSY spectra collected on APol-trapped MPs is sufficient to obtain insights into membrane protein structure and dynamics at least up to the size of BR (27 kDa) (see § 10.3.1). A8-35 permits the identification of most amide protons, but not, however, of those whose exchange with the solvent is too rapid at the slightly alkaline pH (typically, pH 8) that is optimal for use of this APol. Hence, sequence-specific assignment of backbone $^1H^N$, $^{15}N$, and $^{13}C^\alpha$ chemical shifts using TROSY-based HNCA (Salzmann et al. 1999) and HNCOCA (Kay et al. 1990) experiments should be realized with MPs trapped in a pH-insensitive APol, such as SAPol or NAPol, at least for amide protons located in solvent-exposed loops. Concerning the identification of intramolecular $^1H$ dipolar interactions through 3D $^{15}N$-edited and $^{13}C$-edited NOESY experiments, the choice of the APol to be used would again depend on how exchangeable the protons investigated are. DAPol and perDAPol would be highly recommended to observe methyl-methyl interactions, for instance, whereas NAPol or SAPol would be needed to observe NOEs involving loop $^1H^N$ backbone protons, provided that the NOE cross-peaks are not hidden by NAPol or SAPol signals. This begs the question of SAPol and NAPol perdeuteration. As noted in § 10.3.2.1, because their solubility is ensured by sugar residues, NAPols would be extremely costly to perdeuterate. SAPols, on the other hand, could be perdeuterated following the route used for obtaining perDAPol. A remaining concern is that SAPols, probably because of their high charge density, appear less stabilizing than A8-35 or NAPol (see Chap. 5, § 5.5). They will probably be adequate for many MPs, but, as mentioned above, a higher stabilizing power could probably be achieved by lowering their charge density to the level of that of A8-35. This, however, would require the development of a different synthesis protocol.

## 10.3.3    Biologically Oriented Studies

Most of the sNMR studies involving APols that have been published to date – 8 out of the 12 listed in Table 10.1 – have been methodological in their intent, aiming to understand the structure and dynamics of MP/APol complexes, define optimal protocols for their preparation and study, develop APols optimized for NMR applications, clarify the advantages, pitfalls, and limitations of the approach, etc. Studies in which a biological problem is the main issue are however starting to trickle in. Four of them, of unequal breadth, are listed in Table 10.1. The first of them (Study 10.3) investigates the dynamics of OmpX, whose role as an outer membrane ion channel is disputed. The second and most extensive study (Study 10.4) exploits APols to fold a GPCR and to establish, using transferred NOE signals, the structure of a hydrogenated ligand bound to the deuterated receptor trapped in DAPol. The other two studies are more preliminary: Study 10.8 examines the feasibility of using A8-35 to study two more GPCRs, the melanocortin-2 and -4 receptors, Study 10.11 that of using sNMR to gather information on the extramembrane loops of *C. trachomatis* MOMP, an outer MP used as an immunogen in the formulation of vaccines (see Chap. 15).

### 10.3.3.1   Dynamics of the Transmembrane $\beta$-Barrel of OmpX

Time-averaged atomic structures of proteins often leave it uncertain how they achieve some of their functions. OmpX from *E. coli* is a case in point. While electrophysiological studies suggest that OmpX can translocate ions (Dupont et al. 2004; Arnold et al. 2007), structural studies describe this MP as a rigid TM $\beta$-barrel, whose lumen is too densely packed with amino acid side chains to admit any solute (Vogt and Schulz 1999; Fernández et al. 2001a, b, 2004). However, low-energy, ground-state conformations do not preclude the existence of time-dependent conformational fluctuations that may allow the transient formation of a channel.

In Study 10.3 (Table 10.1; Catoire et al. 2010b), the solvent accessibility of amide protons ($^1H^N$) of OmpX kept water-soluble either by A8-35 or by diC$_6$PC was investigated by measuring the extent of hydrogen/deuterium (H/D) exchange after extensive equilibration, providing insights into the dynamics of the $\beta$-barrel. The resolution was improved by adding EDTA to the buffers, presumably preventing traces of $Ca^{2+}$ ions from bridging some MP/A8-35 complexes. The accessibility of $^1H^N$ protons involved in the $\beta$-barrel hydrogen bond network was assessed by comparing $^{15}N$-$^1H$ TROSY spectra collected after extensive equilibration (8 weeks at 4 °C) of OmpX/A8-35 complexes in either 5 or 100% D$_2$O (Fig. 10.21). Unsurprisingly, the most protected $^1H^N$ protons (in red in Fig. 10.22) are found in the TM region of the barrel, with a few residues at the ends of each strand showing higher accessibility. The number of protected $^1H^N$ protons differs from one $\beta$-strand to the other, the first two strands, in particular, displaying only a few partially shielded $^1H^N$ protons in the central part of the TM region.

H/D exchange measurements carried out in diC$_6$PC at pH 8.0 yielded essentially the same results. Most of those $^1H^N$ protons that do exchange do it to comparable degrees in the two environments, suggesting that the overall dynamics of the barrel is similar. However, in diC$_6$PC, residues belonging to the end of the barrel that faces the periplasmic space tend to exchange somewhat more than in A8-35. This is reminiscent of the observations reported by Gerhard Wagner and colleagues on BR (Etzkorn et al. 2013; see § 10.3.1) and by Melanie J. Cocco and colleagues on *C. trachomatis* MOMP (Feinstein et al. 2014; see § 10.3.3.3).

Overall, the observations on OmpX show that the barrel does not behave as a solid block, some of its strands appearing more mobile (and/or accessible) than others. While it cannot be excluded that the water molecules involved in the exchange originate from within the surfactant layer, they may also come from the bulk solution via the lumen of the barrel. H/D exchange, in this hypothesis, would provide insights into conformational changes that affect the packing of side chains in the lumen. One interesting working hypothesis could be that ion flux is associated with small relative movements of strands 1 and 2 that would disrupt the intraluminal network of hydrogen bonds (Böckmann and Caflisch 2005), letting a transient channel form. A similar channel-gating mechanism has been proposed for OmpA (Hong et al. 2006). Alternatively, the dynamics of the $\beta$-strands suggested by H/D exchange measurements could reflect a transport mechanism distinct from ion flux, of which the latter would only be a side effect (for a discussion, see Catoire et al. 2010b).

### 10.3.3.2   Conformation of LTB$_4$ Bound to the BLT2 Leukotriene Receptor

G protein-coupled receptors (GPCRs) mediate a wide range of physiological responses. Producing them under a functional form and determining their 3D structure, their transconformations, and the

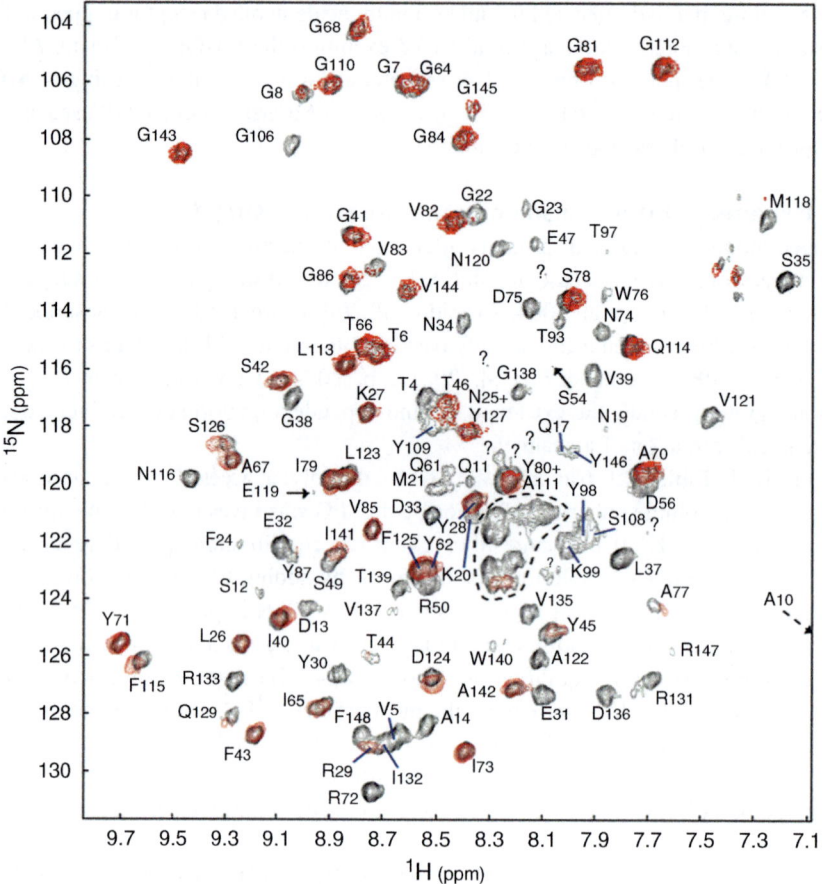

**Fig. 10.21** Superimposed 2D $^{15}$N-$^1$H TROSY spectra of [u-$^2$H,$^{13}$C,$^{15}$N]OmpX/A8-35 complexes following or not hydrogen/deuterium exchange. The sample was prepared in D$_2$O and stored at 4 °C either in 5% (in *black*, eight transients per increment) or in 100% D$_2$O (in *red*, 128 transients per increment) for 8 weeks before NMR experiments. The *dashed line* delineates a crowded area including correlation peaks of residues Y9, Q15, Y57, N58, K59, N60, Y63, Q91, E94, D101, and D104 (From Catoire et al. 2010b).

structure adopted by ligands while bound to them is a major goal of pharmacological research. APols have proven to be an efficient medium into which fold GPCRs from a denatured state, some of which had never been folded to any appreciable extent using more classical approaches (Dahmane et al. 2009; Banères et al. 2011; Bazzacco et al. 2012; see Chap. 6). The low-affinity leukotriene receptor (BLT2) is involved in inflammatory diseases and allergic responses. BLT2 is activated by leukotriene B$_4$ (5$S$,12$R$-dihydroxy-6$Z$,8$E$,10$E$,14$Z$-eicosatetraenoic acid; LTB$_4$), a member of the family of eicosanoids. Before its APol-assisted folding was developed (Dahmane et al. 2009; see Chap. 6, § 6.3.1.1.2), BLT2 had never been obtained in amounts suitable for structural studies, and the structure of LTB$_4$ bound to BLT2 was unknown. LTB$_4$ is characterized by a triene motif and the presence of two hydroxyl groups and an acidic function (Fig. 10.23). The presence of conjugated and unconjugated double bonds and that of hydroxyl and carboxyl groups creates a wide variety of chemical environment, resulting in a broad range of $^1$H chemical shifts, which is a great asset in establishing an NMR structure.

In Study 10.4 (Table 10.1; Catoire et al. 2010a), BLT2 was expressed in perdeuterated, $^{15}$N-labeled form in *E. coli* inclusion bodies, solubilized in SDS, folded in DAPol, and the structure of BLT2-bound LTB$_4$ deduced from intramolecular transferred NOE signals. In order to separate the

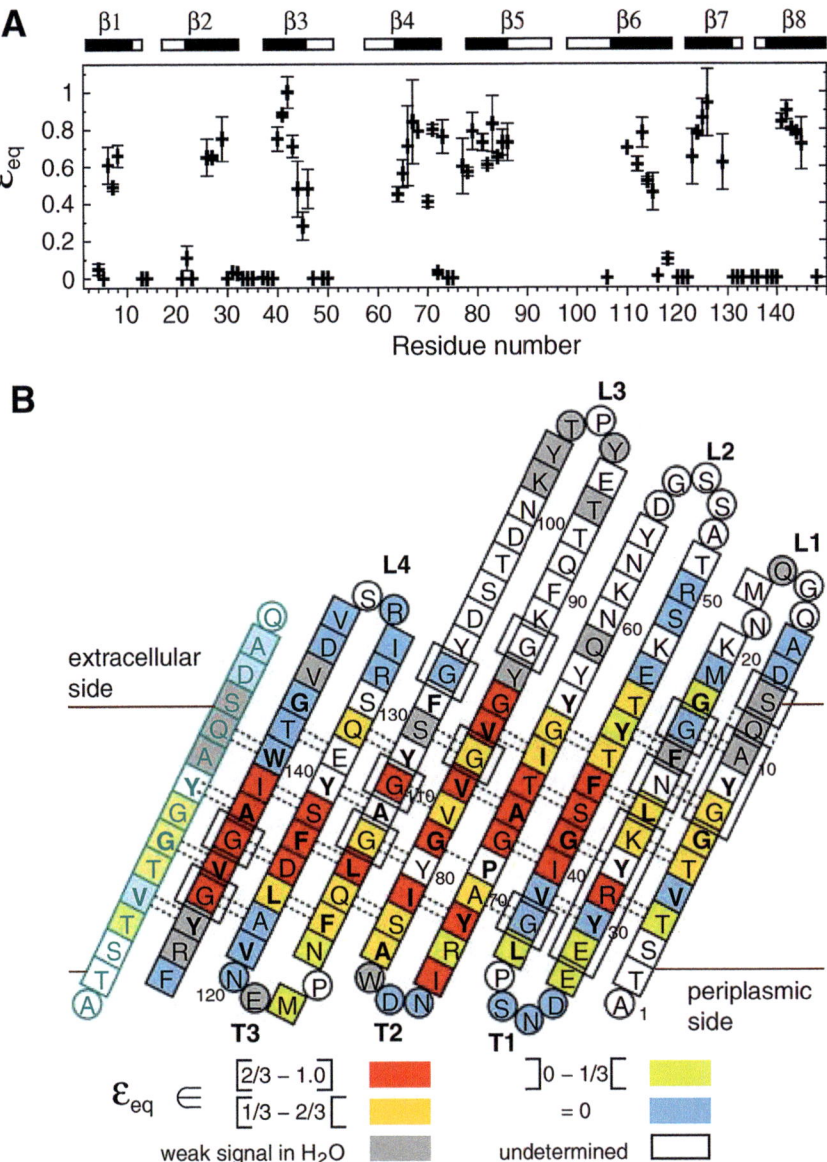

**Fig. 10.22** Solvent accessibility of amide protons in OmpX/A8-35 complexes revealed by hydrogen/deuterium exchange. (**A**) $^1H^N$ protection factor $\varepsilon_{eq}$ plotted against residue number. $\varepsilon_{eq}$ represents the ratio of peak volumes for OmpX/APol complexes incubated in 100% vs. 5% $D_2O$ (data from the TROSY spectra shown in Fig. 10.21). $\varepsilon_{eq}$ is normalized relative to the ratio observed for the most protected residue, S42. The position of $\beta$-strands is indicated by bars above the graph, with *black* and *white* sections noting the TM and extramembrane regions, respectively. (**B**) Topology diagram of the protein. Residues belonging to the $\beta$-barrel are shown in *squares*, with *bold* letters indicating that the side chain points toward the surfactant. Hydrogen bonds in the TM region are indicated by *dotted lines*. High-homology sequences and conserved residues in the eight-stranded $\beta$-barrel family are framed by a *solid line*. Colors refer to the ranges of $\varepsilon_{eq}$ indicated in the figure. *Colorless* residues are prolines or residues for which $^1H^N$ and $^{15}N$ chemical shifts are not known; residues that were not unambiguously assignable in the OmpX/APol spectrum due to chemical shift differences with the spectrum in diC$_6$PC, overcrowding of the lines, and/or lines that were too broadened by the high pH (residues belonging to the loops or to the protruding part of $\beta$-strands), or residues with indistinguishable $^1H^N/^{15}N$ chemical shifts. Residues whose correlation peaks are too weak in H$_2$O to permit a reliable estimation of $\varepsilon_{eq}$ are shown in *gray* (From Catoire et al. 2010b).

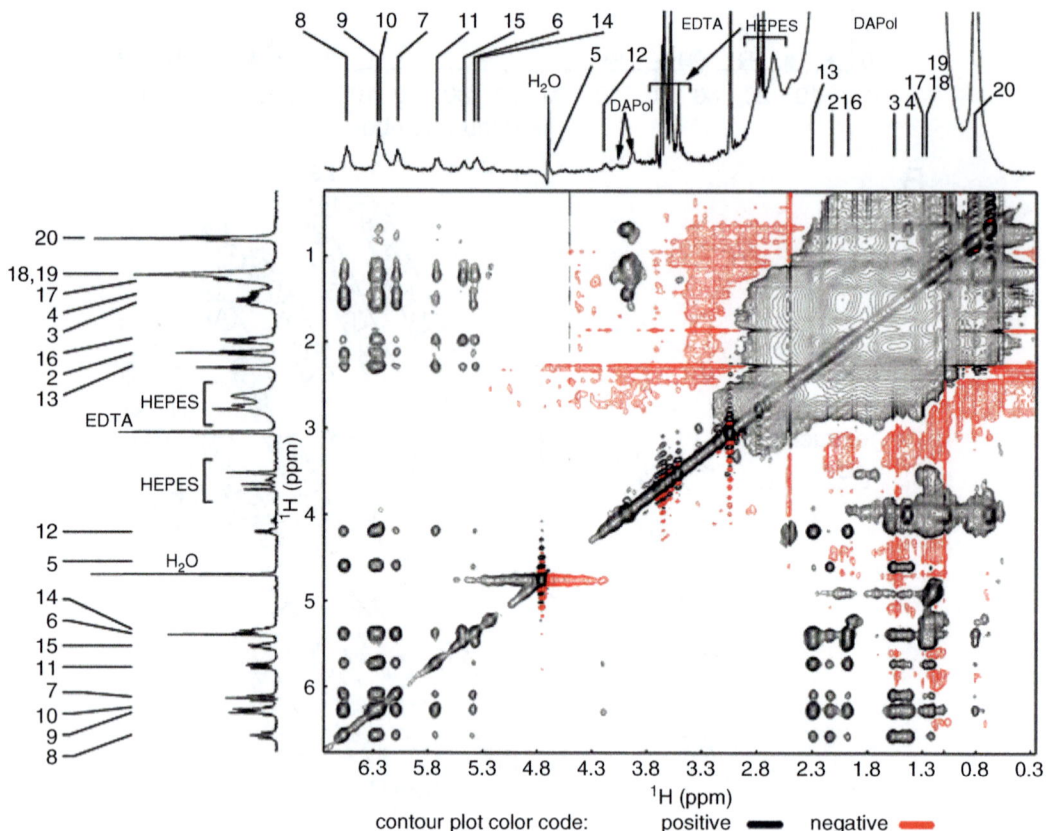

**Fig. 10.23**   Chemical structure of leukotriene $B_4$.

**Fig. 10.24**   Dipolar interactions in the $LTB_4/u$-$^2H$-BLT2/DAPol complex observed in a 2D NOESY spectrum ($\tau_m = 0.5$ s, $\nu_H = 600$ MHz, 25 °C). The corresponding 1D $^1H$ spectrum is shown above the 2D spectrum. In order to help in the identification of cross-peaks, the 1D spectrum of free $LTB_4$ in solution is displayed on the *left* side. Numbers refer to the protons annotated on the $LTB_4$ chemical structure in Fig. 10.23 (Reprinted with permission from Catoire et al. 2010a, © 2010 American Chemical Society).

contributions of $LTB_4$ specifically bound to its physiological site from those of ligand molecules free in the aqueous solution, dissolved in BLT2-bound and free DAPol, or non-specifically bound to the protein, the study required numerous controls, including the use of a BLT2 mutant that does not specifically bind $LTB_4$.

Some of the primary data are shown in Fig. 10.24. In the lower left-hand corner of the spectrum are found cross-peaks due to magnetization transfer between $LBT_4$ protons, whose intensity can be used to derive distance constraints between them (Fig. 10.25A). Note that a large portion of the 2D NOESY spectrum is rendered hard or impossible to interpret because of contributions originating from the backbone of DAPol, which is hydrogenated. In the case of $LTB_4$, many protons, due to their environment, have chemical shifts that are different from each other and different from those of DAPol

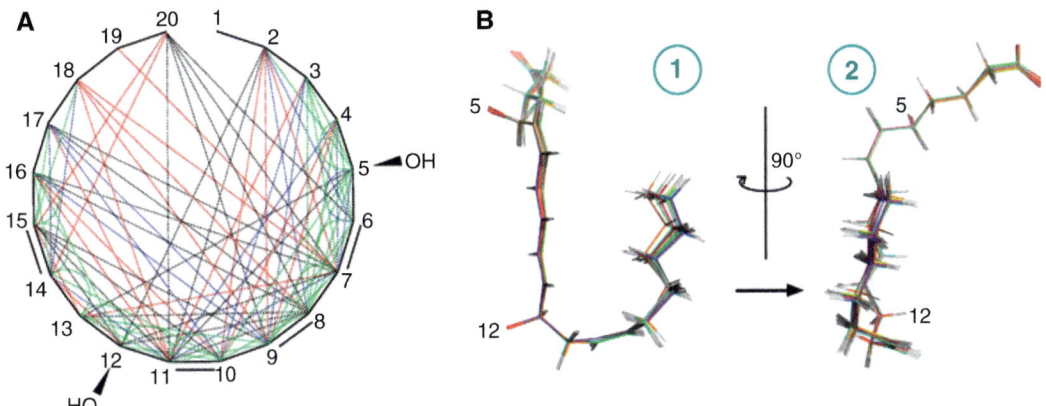

**Fig. 10.25** Three-dimensional structure of LTB$_4$ associated to A8-35-trapped wild-type BLT2. (**A**) Experimental NOE-based distance restraints used in the structure calculation, represented by dotted lines on a regular icosagon symbolizing the eicosanoid (in *green, blue, red,* and *black* distant restraints obtained with $\tau_m = 50, 100, 200,$ and 500 ms, respectively). (**B**) Two different views of an ensemble of ten energy-minimized conformers (in *light gray,* hydrogen atoms; in *red,* oxygen atoms; carbon atoms are assigned a different color for each conformer). Numbers indicate the two hydroxyl groups on positions 5 and 12 (Reprinted with permission from Catoire et al. 2010a, © 2010 American Chemical Society).

protons, so that enough information is available to reconstitute the 3D structure of the bound ligand (Fig. 10.25B). In less favorable cases, the recourse to perDAPol – which was not available at the time of Study 10.4 – would have been mandatory (Catoire et al. 2011; cf. Fig. 10.19).

The 2D NOESY experiments yielded 89 NOE-based distance restraints, from which the structure of BLT2-bound LTB$_4$ could be calculated. On the whole, the ensemble of converged structures depicts a highly constrained LTB$_4$ molecule, adopting a seahorse conformation (Fig. 10.25, *bottom*). Because most putative interaliphatic $^1$H-$^1$H NOEs are masked by signals from the main chain of DAPol, the structure is more loosely defined at both extremities, especially between carbons 1 and 4. This precludes any conclusion about the mobility of this part of the molecule as compared to that of the rest of the skeleton. Nonetheless, several important features can be deduced from the data. First, the ensemble of converged structures suggests that, at variance with that of LTB$_4$ free in solution (Sugiura et al. 1984), the triene motif is not planar. Second, in the same region of the molecule, associated to the deformation of the triene, there is a steric interaction between H5 and H8, characterized experimentally by a strong NOE. Another striking characteristic of the structure of LTB$_4$ in its BLT2-bound state is the tail folding back along the triene motif, as revealed by the many NOEs observed between protons H7-H11 and protons H16-H20 (Figs. 10.24 and 10.25, *top*), yielding the seahorse conformation. This conformation, at variance with extended conformations, is not significantly populated in solution (Sugiura et al. 1984). In other words, this fold results from strong constraints applied by amino acids residues lining the binding pocket. Conversely, such strains are likely to modulate the equilibrium between conformational states of the receptor, as qualitatively observed by comparing $^{15}$N-$^1$H CRINEPT spectra of wild-type [$^2$H,$^{15}$N] BLT2 trapped in DAPol before and after addition of LTB$_4$ (Catoire et al. 2010a; for a more detailed discussion, see Catoire et al. 2011).

In subsequent experiments, in order to study the allosteric effects of cholesterol present in the environment of BLT2 on its conformational equilibria, the A8-35-folded receptor was transferred from APol to NDs (Casiraghi et al. 2016). The transfer protocol is described in Chap. 5, § 5.9.

### 10.3.3.3   Preliminary Studies of *Chlamydia trachomatis* MOMP and Two Melanocortin Receptors

Attempts have been reported in Study 10.8 (Table 10.1; Elter et al. 2014) to fold in A8-35 two other GPCRs, melanocortin receptors 2 and 4. The best folding yields were obtained by precipitating SDS in the presence of A8-35, after the denatured His-tagged receptors had been immobilized onto Ni/NTA-carrying beads (cf. Chap. 6). NMR studies, however, were hampered by difficulties upscaling the procedure (Elter et al. 2014).

*C. trachomatis* is a major bacterial pathogen throughout the world, responsible for many sequelae including infertility and trachoma (see Chap. 15, § 15.2.1). Although antibiotherapy is effective when the infection is detected early, a majority of the cases are asymptomatic, which requires the development of preventive measures. Efforts have focused on the production of a vaccine using *C. trachomatis* major outer membrane protein (MOMP) as an immunogen. MOMP is purified in its native trimeric form (nMOMP) using the zwitterionic detergent Z3-14, but its stability in detergent solutions is limited. Preservation of protein structure and optimization of exposure of the most effective antigenic regions can avoid vaccination with misfolded, poorly protective protein. It has been shown that A8-35 preserves nMOMP secondary structure, that it stabilizes it, and that nMOMP/APol vaccine formulations elicit better protection against *C. trachomatis* than formulations using either recombinant MOMP or nMOMP solubilized in Z3-14 (Tifrea et al. 2011, 2014; see Chap. 15). In Study 10.11 (Table 10.1; Feinstein et al. 2014), $^1$H-$^{15}$N HSQC NMR spectra of $^{15}$N-labeled nMOMP were recorded in either DPC or A8-35. The extramembrane loops of MOMP, whose conformation is probably critical in eliciting a protective vaccinal response, are detectable in APols, but not in detergent, and the protein can be studied at higher temperature (50 °C; see Chap. 15, Fig. 15.5).

It is worth noting that the study of Kp-tOmpA described in § 10.3.1.2 (Study 10.12 in Table 10.1; Planchard et al. 2014) can lead to the same application, inasmuch as Kp-tOmpA is a potential immunogen for vaccination against *K. pneumoniae* (Chap. 15).

## 10.4   Challenges and Prospects

At the time of writing this chapter, it is somewhat difficult to anticipate which of the many media in which MPs can be made water-soluble will turn out to be best adapted to which type of NMR studies. Table 10.3 attempts to grade the advantages and drawbacks of the four systems that have been validated for sNMR studies, namely detergent solutions (Chap. 2), bicelles (Chap. 3, § 3.2), amphipols (Chaps. 4 and 5), and nanodiscs (Chap. 3, § 3.3), according to the following criteria:

- *Particle size.* As regards aiming for the smallest possible particle size, small detergents like diC$_6$PC or DPC are hard to beat. MP/APol complexes are somewhat larger, and MP/ND

**Table 10.3** A rough assessment of the advantages and disadvantages of various environments for sNMR studies of membrane protein. Color code: *green*, excellent; *yellow*, good; *orange*, suboptimal; *red*, problematical; *gray*, not well documented.

| Environment → | Detergent | Bicelles | Amphipols | Nanodiscs |
|---|---|---|---|---|
| Particle size | | | | |
| Ease of preparation | | | | |
| Membrane-like environment | | | | |
| Protein stability | | | | |
| Deuteration | | | | |

complexes even more (see § 10.3.1). Smaller NDs have been engineered, but this may come at the cost of moving MP/ND complexes away from a bilayer-like environment and imposing new constraints on the protein (see Chap. 3, § 3.3.2). As regards the size of bicelles, it can be modulated by playing on the proportion of plane and ring surfactant (Chap. 3, § 3.2).

- *Sample preparation.* MP/detergent complexes are the form under which most MPs are initially obtained, and their transfer to APols is straightforward (Chap. 5). Alternatively, the protein can be directly folded (Chap. 6) or expressed (Chap. 7) in the presence of APols. Handling bicelles is more difficult, and the preparation of homogeneous preparations of MP/ND complexes is work-intensive (Ritchie et al. 2009).

- *Stability.* Because of their dissociating properties, detergents provide the most unfriendly, least membrane-like environment, and, as a result, the stability of most MPs is often poor, limiting the duration of the experiments and the temperature at which they can be carried out. The stability is almost always increased, most often dramatically so, upon transfer to APols. This is in part due to the rebinding of lipids (see Chap. 5), which reconstitutes, to some extent, interactions more akin to those the MP experiences in its native membrane environment than it does in detergent solution. Bicelles offer a bilayer-like environment, but contaminated by the ring surfactant, either a detergent or short-chain, detergent-like lipids, which partitions in the bilayer. There are data suggesting that they provide a stabilizing environment (see e.g. Poget et al. 2007), but those are scarce. NDs offer the environment closest to a lipid bilayer, and the few comparative data available indicate that ND-trapped MPs can be even more stable than APol-trapped ones (see e.g. Popot 2010; Etzkorn et al. 2013).

- *Isotopic labeling.* Finally, isotopic labeling is straightforward in detergent solution and bicelles – that is, provided the protein is stable enough in the presence of those few surfactants that exist in deuterated form (Hiruma-Shimizu et al. 2015). It is also relatively easy in the case of APols, within the constraints discussed in § 10.3.2. There is nothing to prevent the preparation of NDs formed of deuterated DMPC and deuterated scaffold proteins, but for the cost and work involved.

One of the assets of APols is their great chemical versatility. Deuteration has already been implemented as regards A8-35 and, as noted, could easily be extended to SAPols. SAPols themselves, as discussed in § 10.3.2.1, could probably be improved by lowering the density of charges along the chain, e.g. by replacing the carboxylate groups currently present along with the sulfonate ones (Fig. 10.15) by isopropylamine. A SAPol whose charge density would be comparable to that of A8-35 would likely be as mild as it toward fragile MPs, while remaining soluble at all pH, and it could easily be obtained in perdeuterated form. It ought also to be possible, should it be desirable, to synthesize $^{13}$C-free APols.

Chemical modifications of APols cover a very wide range (Le Bon et al. 2014b; see Chap. 4, § 4.4, Table 4.5). Some of them could be exploited for NMR studies. Polyhistidine-tagged A8-35 (HistAPol), whose synthesis has been described and which has been validated for trapping and immobilizing MPs (Giusti et al. 2015), could conceivably be used to bind traces of paramagnetic ions in the vicinity of the TM region of HistAPol-trapped MPs. A spin-labeled APol, which has been synthesized (F. Giusti and M. Zoonens, unpublished data), provides an alternative. Spin-labeled APols could be of use to simplify NMR spectra and facilitate their interpretation by selectively suppressing, in a controlled and tunable manner, the resonance peaks from TM residues. They could possibly also be used to improve the sensitivity of NMR measurements by resorting to dynamic nuclear polarization (magnetization transfer from the spin label to nuclei; see Griesinger et al. 2012). APols that carry tags such as polyhistidine (Giusti et al. 2015), biotin (Charvolin et al. 2009), or an oligonucleotide (Le Bon et al. 2014a), all of which have been validated for attaching MPs onto solid beads or chips, could conceivably be used to associate MPs to fibers for partial orientation in the magnetic field in view of residual dipolar coupling measurements (see e.g. Prestegard et al. 2000; Bax et al. 2001; Kay 2001; Torchia 2015, and refs. therein), or to immobilize them onto resins for the purpose of screening banks of ligands (cf. Früh et al. 2011).

As mentioned at the beginning of this chapter, no report has yet appeared of applying ssNMR to APol-trapped MPs. This is to some extent logical: ssNMR is particularly well adapted to studying MPs in their native membrane environment, and little interest may be found in applying it to MPs that have been extracted from it. Yet, as has been discussed in Chap. 5, there is some evidence that MPs trapped in APol in the presence of lipids rebind them, forming ternary complexes in which proteins may exhibit functional properties closer to those observed in native membranes than in detergent solution (Martinez et al. 2002; Dahmane et al. 2013). Whereas APols themselves are definitely no membrane mimics, the interactions experienced by MPs trapped in APols in the presence of lipids may resemble those experienced in the membrane more than one might think when looking at the chemical structure of APols. Sensitivity is a major issue in ssNMR. APol-trapped BR can be precipitated without denaturation either by extensive ultracentrifugation (Gohon et al. 2008) or by deliberately aggregating the complexes, e.g. by adding $Ca^{2+}$ ions to A8-35-trapped BR, or multivalent avidin to BR trapped with a biotinylated APol (Bazzacco 2009). The concentration of BR in the paste thus obtained is much higher than can be achieved in membrane samples, a highly favorable factor for ssNMR studies. $^{19}F$ ssNMR has been applied to studying the phosphorylation of an *E. coli* tyrosine kinase incorporated into SMALPs (Li et al. 2015).

More generally, one may note that there is a dearth of comparative NMR studies of the same MP in different environments, on the model of those reported in Etzkorn et al. (2013) (§ 10.3.1.2). A very interesting suggestion is to use data collected in environments closest to the natural one, such as ssNMR data collected on membrane-integrated MPs or sNMR data collected on ND-trapped ones, to assess to which extent other surfactants do or do not affect the structure and dynamics of the MP under study (Howell et al. 2005; Poget and Girvin 2007; Shenkarev et al. 2010; Etzkorn et al. 2013). Such comparisons would help delineate which perturbations may or may not be expected from one or the other surfactant, and help to choose, for each experiment, the type of samples that offers the best compromise between experimental convenience and reliable information.

# References

Althoff, T., Mills, D.J., Popot, J.-L., Kühlbrandt, W. (2011) Assembly of electron transport chain components in bovine mitochondrial supercomplex $I_1III_2IV_1$. *EMBO J.* **30**:4652–4664.

Anglister, J., Grzesiek, S., Ren, H., Klee, C.B., Bax, A. (1993) Isotope-edited multidimensional NMR of calcineurin B in the presence of the non-deuterated detergent CHAPS. *J. Biomol. NMR* **3**:121–126.

Arnold, T., Poynor, M., Nussberger, S., Lupas, A.N., Linke, D. (2007) Gene duplication of the eight-stranded β-barrel OmpX produces a functional pore: a scenario for the evolution of transmembrane β-barrels. *J. Mol. Biol.* **366**:1174–1184.

Arora, A., Abildgaard, F., Bushweller, J.H., Tamm, L.K. (2001) Structure of outer membrane protein A transmembrane domain by NMR spectroscopy. *Nat. Struct. Biol.* **8**:334–338.

Arunmanee, W., Harris, J.R., Lakey, J.H. (2014) Outer membrane protein F stabilised with minimal amphipol forms linear arrays and LPS-dependent 2D crystals. *J. Membr. Biol.* **247**:949–956.

Banères, J.-L., Popot, J.-L., Mouillac, B. (2011) New advances in production and functional folding of G protein-coupled receptors. *Trends Biotechnol.* **29**:314–322.

Bax, A., Kontaxis, G., Tjandra, N. (2001) Dipolar couplings in macromolecular structure determination. *Meth. Enzymol.* **339**:127–174.

Bayburt, T.H., Grinkova, Y.V., Sligar, S.G. (2002) Self-assembly of discoidal phospholipid bilayer nanoparticles with membrane scaffold proteins. *Nano Lett.* **2**:853–856.

Bazzacco, P. (2009) Non-ionic amphipols: new tools for *in vitro* studies of membrane proteins. Validation and development of biochemical and biophysical applications. Thèse de Doctorat, Université Paris-7, 176 p.

Bazzacco, P., Billon-Denis, E., Sharma, K.S., Catoire, L.J., Mary, S., Le Bon, C., Point, E., Banères, J.-L., Durand, G., Zito, F., Pucci, B., Popot, J.-L. (2012) Non-ionic homopolymeric amphipols: Application to membrane protein folding, cell-free synthesis, and solution NMR. *Biochemistry* **51**:1416–1430.

Bazzacco, P., Sharma, K.S., Durand, G., Giusti, F., Ebel, C., Popot, J.-L., Pucci, B. (2009) Trapping and stabilization of integral membrane proteins by hydrophobically grafted glucose-based telomers. *Biomacromolecules* **10**:3317–3326.

Beaugrand, M., Arnold, A.A., Juneau, A., Balieiro Gambaro, A., Warschawski, D.E., Williamson, P.T.F., Marcotte, I. (2016) Magnetically oriented bicelles with monoalkylphosphocholines: versatile membrane mimetics for nuclear magnetic resonance applications. *Langmuir* **32**:13244–13251.

Bersch, B., Dörr, J.M., Hessel, A., Killian, J.A., Schanda, P. (2017) Proton-detected solid-state NMR spectroscopy of a zinc diffusion facilitator protein in native nanodiscs. *Angew. Chem. Int. Ed.* **56**:2508–2512.

Böckmann, R.A., Caflisch, A. (2005) Spontaneous formation of detergent micelles around the outer membrane protein OmpX. *Biophys. J.* **88**:3191–3204.

Caravan, P., Ellison, J.J., McMurry, T.J., Lauffer, R.B. (1999) Gadolinium(III) chelates as MRI contrast agents: Structure, dynamics, and applications. *Chem. Rev.* **99**:2293–2352.

Casiraghi, M., Damian, M., Lescop, E., Point, E., Moncoq, K., Morellet, N., Levy, D., Marie, J., Guittet, E., Banères, J.-L., Catoire, L.J. (2016) Functional modulation of a GPCR conformational landscape in a lipid bilayer. *J. Am. Chem. Soc.* **138**:11170–11175

Catoire, L.J., Damian, M., Baaden, M., Guittet, E., Banères, J.-L. (2011) Electrostatically-driven fast association and perdeuteration allow detection of transferred cross-relaxation for G protein-coupled receptor ligands with equilibrium dissociation constants in the high-to-low nanomolar range. *J. Biomol. NMR* **50**:191–195.

Catoire, L.J., Damian, M., Giusti, F., Martin, A., van Heijenoort, C., Popot, J.-L., Guittet, E., Banères, J.-L. (2010a) Structure of a GPCR ligand in its receptor-bound state: leukotriene B$_4$ adopts a highly constrained conformation when associated to human BLT2. *J. Am. Chem. Soc.* **132**:9049–9057.

Catoire, L.J., Warnet, X.L., Warschawski, D.E. (2014) Micelles, bicelles, amphipols, nanodiscs, liposomes or intact cells: The hitch-hiker guide to the study of membrane proteins by NMR, in: Mus-Veteau, I., ed., Membrane Protein Production for Structural Analysis. Springer, pp. 315–346.

Catoire, L.J., Zoonens, M., van Heijenoort, C., Giusti, F., Guittet, E., Popot, J.-L. (2010b) Solution NMR mapping of water-accessible residues in the transmembrane $\beta$-barrel of OmpX. *Eur. Biophys. J.* **39**:623–630.

Catoire, L.J., Zoonens, M., van Heijenoort, C., Giusti, F., Popot, J.-L., Guittet, E. (2009) Inter- and intramolecular contacts in a membrane protein/surfactant complex observed by heteronuclear dipole-to-dipole cross-relaxation. *J. Magn. Res.* **197**:91–95.

Champeil, P., Menguy, T., Tribet, C., Popot, J.-L., le Maire, M. (2000) Interaction of amphipols with the sarcoplasmic reticulum Ca$^{2+}$-ATPase. *J. Biol. Chem.* **275**:18623–18637.

Charvolin, D., Perez, J.-B., Rouvière, F., Giusti, F., Bazzacco, P., Abdine, A., Rappaport, F., Martinez, K.L., Popot, J.-L. (2009) The use of amphipols as universal molecular adapters to immobilize membrane proteins onto solid supports. *Proc. Natl. Acad. Sci. USA* **106**:405–410.

Chill, J.H., Louis, J.M., Baber, J.L., Bax, A. (2006a) Measurement of $^{15}$N relaxation in the detergent-solubilized tetrameric KcsA potassium channel. *J. Biomol. NMR* **36**:123–136.

Chill, J.H., Louis, J.M., Miller, M., Bax, A. (2006b) NMR study of the tetrameric KcsA potassium channel in detergent micelles. *Protein Sci.* **15**:684–698.

Chou, J.J., Kaufman, J.D., Stahl, S.J., Wingfield, P.T., Bax, A. (2002) Micelle-induced curvature in a water-insoluble HIV-1 Env peptide revealed by NMR dipolar coupling measurement in stretched polyacrylamide gel. *J. Am. Chem. Soc.* **124**:2450–2451.

Dahmane, T. (2007) Protéines membranaires et amphipols : stabilisation, fonction, renaturation, et développement d'amphipols sulfonatés pour la RMN des solutions. Thèse de Doctorat, Université Paris-7, 229 p.

Dahmane, T., Damian, M., Mary, S., Popot, J.-L., Banères, J.-L. (2009) Amphipol-assisted *in vitro* folding of G protein-coupled receptors. *Biochemistry* **48**:6516–6521.

Dahmane, T., Giusti, F., Catoire, L.J., Popot, J.-L. (2011) Sulfonated amphipols: Synthesis, properties and applications. *Biopolymers* **95**:811–823.

Dahmane, T., Rappaport, F., Popot, J.-L. (2013) Amphipol-assisted folding of bacteriorhodopsin in the presence and absence of lipids. Functional consequences. *Eur. Biophys. J.* **42**:85–101.

Denisov, I.G., Grinkova, Y.V., Lazarides, A.A., Sligar, S.G. (2004) Directed self-assembly of monodisperse phospholipid bilayer nanodiscs with controlled size. *J. Am. Chem. Soc.* **126**:3477–3487.

Diab, C., Tribet, C., Gohon, Y., Popot, J.-L., Winnik, F.M. (2007a) Complexation of integral membrane proteins by phosphorylcholine-based amphipols. *Biochim. Biophys. Acta* **1768**:2737–2747.

Diab, C., Winnik, F.M., Tribet, C. (2007b) Enthalpy of interaction and binding isotherms of non-ionic surfactants onto micellar amphiphilic polymers (amphipols). *Langmuir* **23**:3025–3035.

Dupont, M., Dé, E., Chollet, R., Chevalier, J., Pagès, J.-M. (2004) *Enterobacter aerogenes* OmpX, a cation-selective channel mar- and osmo-regulated. *FEBS Lett.* **569**:27–30.

Elter, S., Raschle, T., Arens, S., Viegas, A., Gelev, V., Etzkorn, M., Wagner, G. (2014) The use of amphipols for NMR structural characterization of 7-TM proteins. *J. Membr. Biol.* **247**:957–964.

Etzkorn, M., Raschle, T., Hagn, F., Gelev, V., Rice, A.J., Walz, T., Wagner, G. (2013) Cell-free expressed bacteriorhodopsin in different soluble membrane mimetics: biophysical properties and NMR accessibility. *Structure* **21**:394–401.

Etzkorn, M., Zoonens, M., Catoire, L.J., Popot, J.-L., Hiller, S. (2014) How amphipols embed membrane proteins: Global solvent accessibility and interaction with a flexible protein terminus. *J. Membr. Biol.* **247**:965–970.

Feinstein, H.E., Tifrea, D., Sun, G., Popot, J.-L., de la Maza, L.M., Cocco, M.J. (2014) Long-term stability of a vaccine formulated with the amphipol-trapped major outer membrane protein from *Chlamydia trachomatis*. *J. Membr. Biol.* **247**:1053–1065.

Fernández, C., Adeishvili, K., Wüthrich, K. (2001a) Transverse relaxation-optimized NMR spectroscopy with the outer membrane protein OmpX in dihexanoyl phosphatidylcholine micelles. *Proc. Natl. Acad. Sci. USA* **98**:2358–2363.

Fernández, C., Hilty, C., Bonjour, S., Adeishvili, K., Pervushin, K., Wüthrich, K. (2001b) Solution NMR studies of the integral membrane proteins OmpX and OmpA from *Escherichia coli*. *FEBS Lett.* **504**:173–178.

Fernández, C., Hilty, C., Wider, G., Guntert, P., Wüthrich, K. (2004) NMR structure of the integral membrane protein OmpX. *J. Mol. Biol.* **336**:1211–1221.

Früh, V., IJzerman, A.P., Siegal, G. (2011) How to catch a membrane protein in action: a review of functional membrane protein immobilization strategies and their applications. *Chem. Rev.* **111**:640–656.

Gardner, K.H., Kay, L.E. (1998) The use of $^2$H,$^{13}$C,$^{15}$N multidimensional NMR to study the structure and dynamics of proteins. *Annu. Rev. Biophys. Biomol. Struct.* **27**:357–406.

Gautier, A., Mott, H.R., Bostock, M.J., Kirkpatrick, J.P., Nietlispach, D. (2010) Structure determination of the seven-helix transmembrane receptor sensory rhodopsin II by solution NMR spectroscopy. *Nat. Struct. Mol. Biol.* **17**:768–774.

Giusti, F., Kessler, P., Westh Hansen, R., Della Pia, E.A., Le Bon, C., Mourier, G., Popot, J.-L., Martinez, K.L., Zoonens, M. (2015) Synthesis of a polyhistidine-bearing amphipol and its use for immobilizing membrane proteins. *Biomacromolecules* **16**:3751–3761.

Giusti, F., Rieger, J., Catoire, L., Qian, S., Calabrese, A.N., Watkinson, T.G., Casiraghi, M., Radford, S.E., Ashcroft, A. E., Popot, J.-L. (2014) Synthesis, characterization and applications of a perdeuterated amphipol. *J. Membr. Biol.* **247**:909–924.

Glück, J.M., Wittlich, M., Feuerstein, S., Hoffmann, S., Willbold, D., Koenig, B.W. (2009) Integral membrane proteins in nanodiscs can be studied by solution NMR spectroscopy. *J. Am. Chem. Soc.* **131**:12060–12061.

Gohon, Y., Dahmane, T., Ruigrok, R., Schuck, P., Charvolin, D., Rappaport, F., Timmins, P., Engelman, D.M., Tribet, C., Popot, J.-L., Ebel, C. (2008) Bacteriorhodopsin/amphipol complexes: structural and functional properties. *Biophys. J.* **94**:3523–3537.

Gohon, Y., Giusti, F., Prata, C., Charvolin, D., Timmins, P., Ebel, C., Tribet, C., Popot, J.-L. (2006) Well-defined nanoparticles formed by hydrophobic assembly of a short and polydisperse random terpolymer, amphipol A8-35. *Langmuir* **22**:1281–1290.

Gohon, Y., Pavlov, G., Timmins, P., Tribet, C., Popot, J.-L., Ebel, C. (2004) Partial specific volume and solvent interactions of amphipol A8-35. *Anal. Biochem.* **334**:318–334.

Griesinger, C., Bennati, M., Vieth, H.M., Luchinat, C., Parigi, G., Höfer, P., Engelke, F., Glaser, S.J., Denysenkov, V., Prisner, T.F. (2012) Dynamic nuclear polarization at high magnetic fields in liquids. *Prog. Nucl. Magn. Reson. Spectrosc.* **64**:4–28.

Hagn, F., Etzkorn, M., Raschle, T., Wagner, G. (2013) Optimized phospholipid bilayer nanodiscs facilitate high-resolution structure determination of membrane proteins. *J. Am. Chem. Soc.* **135**:1919–1925.

Hiller, S., Garces, R.G., Malia, T.J., Orekhov, V.Y., Colombini, M., Wagner, G. (2008) Solution structure of the integral human membrane protein VDAC-1 in detergent micelles. *Science* **321**:1206–1210.

Hilty, C., Wider, G., Fernández, C., Wüthrich, K. (2004) Membrane protein-lipid interactions in mixed micelles studied by NMR spectroscopy with the use of paramagnetic reagents. *ChemBioChem* **5**:467–473.

Hiruma-Shimizu, K., Shimizu, H., Thompson, G.S., Kalverda, A.P., Patching, S.G. (2015) Deuterated detergents for structural and functional studies of membrane proteins: Properties, chemical synthesis and applications. *Mol. Membr. Biol.* **32**:139–155.

Hong, H., Szabo, G., Tamm, L.K. (2006) Electrostatic couplings in OmpA ion-channel gating suggest a mechanism for pore opening. *Nat. Chem. Biol.* **2**:627–635.

Howell, S.C., Mesleh, M.F., Opella, S.J. (2005) NMR structure determination of a membrane protein with two transmembrane helices in micelles: MerF of the bacterial mercury detoxification system. *Biochemistry* **44**:5196–5206.

Huynh, K.W., Cohen, M.R., Moiseenkova-Bell, V.Y. (2014) Application of amphipols for structure-functional analysis of TRP channels. *J. Membr. Biol.* **247**:843–851.

Kang, C.B., Li, Q. (2011) Solution NMR study of integral membrane proteins. *Curr. Opin. Struct. Biol.* **15**:560–569.

Kaplan, M., Pinto, C., Houben, K., Baldus, M. (2016) Nuclear magnetic resonance (NMR) applied to membrane-protein complexes. *Quart. Rev. Biophys.* **49**:1–25.

Kay, L.E. (2001) Nuclear magnetic resonance methods for high molecular weight proteins: a study involving a complex of maltose binding protein and betacyclodextrin. *Meth. Enzymol.* **339B**:174–203.

Kay, L.E., Ikura, M., Tschudin, R., Bax, A. (1990) Three-dimensional triple resonance NMR spectroscopy of isotopically enriched proteins. *J. Magn. Reson.* **89**:496–514.

Kim, H.M., Howell, S.C., Van Horn, W.D., Jeon, Y.H., Sanders, C.R. (2009) Recent advances in the application of solution NMR spectroscopy to multi-span integral membrane proteins. *Progr. Nucl. Magn. Reson. Spectrosc.* **55**:335–360.

Klammt, C., Perrin, M.-H., Maslennikov, I., Renault, L., Krupa, M., Kwiatkowski, W., Stahlberg, H., Vale, W., Choe, S. (2011) Polymer-based cell-free expression of ligand-binding family B G-protein coupled receptors without detergents. *Prot. Sci.* **20**:1030–1041.

Kraft, T.E., Hresko, R.C., Hruz, P.W. (2015) Expression, purification, and functional characterization of the insulin-responsive facilitative glucose transporter GLUT4. *Protein Sci.* **24**:2008–2019.

Le Bon, C., Della Pia, E.A., Giusti, F., Lloret, N., Zoonens, M., Martinez, K.L., Popot, J.-L. (2014a) Synthesis of an oligonucleotide-derivatized amphipol and its use to trap and immobilize membrane proteins. *Nucleic Acids Res.* **42**:e83.

Le Bon, C., Popot, J.-L., Giusti, F. (2014b) Labeling and functionalizing amphipols for biological applications. *J. Membr. Biol.* **247**:797–814.

Lee, D., Hilty, C., Wider, G., Wüthrich, K. (2006) Effective rotational correlation times of proteins from NMR relaxation interference. *J. Magn. Reson.* **178**:72–76.

Lee, D., Walter, K.F., Brückner, A.K., Hilty, C., Becker, S., Griesinger, C. (2008) Bilayer in small bicelles revealed by lipid-protein interactions using NMR spectroscopy. *J. Am. Chem. Soc.* **130**:13822–13823.

Li, D., Li, J., Zhuang, Y., Zhang, L., Xiong, Y., Shi, P., Tian, C. (2015) Nano-size uni-lamellar lipodisq-improved *in situ* auto-phosphorylation analysis of *E. coli* tyrosine kinase using $^{19}$F nuclear magnetic resonance. *Protein Cell* **6**:229–233.

Liao, M., Cao, E., Julius, D., Cheng, Y. (2014) Single particle electron cryo-microscopy of a mammalian ion channel. *Curr. Opin. Struct. Biol.* **27**:1–7.

MacKenzie, K.R., Prestegard, J.H., Engelman, D.M. (1997) A transmembrane helix dimer: structure and implications. *Science* **276**:131–133.

Marcotte, I., Auger, M. (2005) Bicelles as model membranes for solid- and solution-state NMR studies of membrane peptides and proteins. *Concepts Magn. Reson.* **24A**:17–37.

Martinez, K.L., Gohon, Y., Corringer, P.-J., Tribet, C., Mérola, F., Changeux, J.-P., Popot, J.-L. (2002) Allosteric transitions of *Torpedo* acetylcholine receptor in lipids, detergent and amphipols: molecular interactions *vs.* physical constraints. *FEBS Lett.* **528**:251–256.

Milić, D., Veprintsev, D.B. (2015) Large-scale production and protein engineering of G protein-coupled receptors for structural studies. *Frontiers Pharmacol.* **6**:66.

Mineev, K.S., Nadezhdin, K.D. (2017) Membrane mimetics for solution NMR studies of membrane proteins. *Nanotech. Rev.* **6**:15–32.

Nietlispach, D., Gautier, A. (2011) Solution NMR studies of polytopic alpha-helical membrane proteins. *Curr. Opin. Struct. Biol.* **21**:497–508.

Opella, S.J., Marassi, F.M. (2004) Structure determination of membrane proteins by NMR spectroscopy. *Chem. Rev.* **104**:3587–3606.

Opella, S.J., Marassi, F.M. (2017) Applications of NMR to membrane proteins. *Arch. Biochem. Biophys.* **628**:92–101.

Page, R.C., Moore, J.D., Nguyen, H.B., Sharma, M., Chase, R., Gao, F.P., Mobley, C.K., Sanders, C.R., Ma, L., Sönnichsen, F.D., Lee, S., Howell, S.C., Opella, S.J., Cross, T.A. (2006) Comprehensive evaluation of solution nuclear magnetic resonance spectroscopy sample preparation for helical integral membrane proteins. *J. Struct. Func. Genom.* **7**:51–64.

Patching, S.G. (2011) NMR structures of polytopic integral membrane proteins. *Mol. Membr. Biol.* **28**:370–397.

Patching, S.G. (2015) Solid-state NMR structures of integral membrane proteins. *Mol. Membr. Biol.* **32**:156–178.

Pautsch, A., Schulz, G.E. (1998) Structure of the outer membrane protein A transmembrane domain. *Nat. Struct. Biol.* **5**:1013–1017.

Pautsch, A., Schulz, G.E. (2000) High-resolution structure of the OmpA membrane domain. *J. Mol. Biol.* **298**:273–282.

Pautsch, A., Vogt, J., Model, K., Siebold, C., Schulz, G.E. (1999) Strategy for membrane protein crystallization exemplified with OmpA and OmpX. *Proteins* **34**:167–172.

Perlmutter, J.D., Popot, J.-L., Sachs, J.N. (2014) Molecular dynamics simulations of a membrane protein/amphipol complex. *J. Membr. Biol.* **247**:883–895.

Perry, T., Souabni, H., Rapisarda, C., Fronzes, R., Giusti, F., Popot, J.-L., Zoonens, M., Gubellini, F. (2018) Visualizing transmembrane regions of protein complexes by electron microscopy using biotinylated amphipols, *submitted for publication*.

Pervushin, K.V., Riek, R., Wider, G., Wüthrich, K. (1997) Attenuated $T_2$ relaxation by mutual cancellation of dipole-dipole coupling and chemical shift anisotropy indicates an avenue to NMR structures of very large biological macromolecules in solution. *Proc. Natl. Acad. Sci. USA* **94**:12366–12371.

Picard, M., Dahmane, T., Garrigos, M., Gauron, C., Giusti, F., le Maire, M., Popot, J.-L., Champeil, P. (2006) Protective and inhibitory effects of various types of amphipols on the $Ca^{2+}$-ATPase from sarcoplasmic reticulum: a comparative study. *Biochemistry* **45**:1861–1869.

Planchard, N., Point, E., Dahmane, T., Giusti, F., Renault, M., Le Bon, C., Durand, G., Milon, A., Guittet, E., Zoonens, M., Popot, J.-L., Catoire, L.J. (2014) The use of amphipols for solution NMR studies of membrane proteins: advantages and limitations as compared to other solubilizing media. *J. Membr. Biol.* **247**:827–842.

Plesniak, L.A., Mahalakshmi, R., Rypien, C., Yang, Y., Racic, J., Marassi, F.M. (2011) Expression, refolding, and initial structural characterization of the *Y. pestis* Ail outer membrane protein in lipids. *Biochim. Biophys. Acta* **1808**:482–489.

Plevin, M.J., Boisbouvier, J. (2012) Isotope-labelling of methyl groups for NMR studies of large proteins, in: Clore, M., Potts, J., eds., *Recent Developments in Biomolecular NMR*, Royal Society of Chemistry, pp. 1–24.

Pocanschi, C.L., Dahmane, T., Gohon, Y., Rappaport, F., Apell, H.-J., Kleinschmidt, J.H., Popot, J.-L. (2006) Amphipathic polymers: tools to fold integral membrane proteins to their active form. *Biochemistry* **45**:13954–13961.

Poget, S.F., Cahill, S.M., Girvin, M.E. (2007) Isotropic bicelles stabilize the functional form of a small multidrug-resistance pump for NMR structural studies. *J. Am. Chem. Soc.* **129**:2432–2433.

Poget, S.F., Girvin, M.E. (2007) Solution NMR of membrane proteins in bilayer mimics: small is beautiful, but sometimes bigger is better. *Biochim. Biophys. Acta* **1768**:3098–3106.

Popot, J.-L. (2010) Amphipols, nanodiscs, and fluorinated surfactants: Three non-conventional approaches to studying membrane proteins in aqueous solutions. *Annu. Rev. Biochem.* **79**:737–775.

Popot, J.-L. (2014) Folding membrane proteins *in vitro*: A table and some comments. *Arch. Biochem. Biophys.* **564**:314–326.

Popot, J.-L., Engelman, D.M. (2016) Membranes do not tell proteins how to fold. *Biochemistry* **55**:5–18.

Postis, V., Rawson, S., Mitchell, J.K., Lee, S.C., Parslowc, R.A., Dafforn, T.R., Baldwin, S.A., Muench, S.P. (2015) The use of SMALPs as a novel membrane protein scaffold for structure study by negative stain electron microscopy. *Biochim. Biophys. Acta* **1848**:496–501.

Prestegard, J.H., Al-Hashimi, H.M., Tolman, J.R. (2000) NMR structures of biomolecules using field-oriented media and residual dipolar couplings. *Q. Rev. Biophys.* **33**:371–424.

Prosser, R.S., Evanics, F., Kitevski, J.L., Al-Abdul-Wahid, M.S. (2006) Current applications of bicelles in NMR studies of membrane-associated amphiphiles and proteins. *Biochemistry* **45**:8453–8465.

Puthenveetil, R., Nguyen, K., Vinogradova, O. (2017) Nanodiscs and solution NMR: preparation, application and challenges. *Nanotech. Rev.* **6**:111–126.

Qureshi, T., Goto, N.K. (2011) Contemporary methods in structure determination of membrane proteins by solution NMR. *Top. Curr. Chem.*

Raschle, T., Hiller, S., Etzkorn, M., Wagner, G. (2010) Nonmicellar systems for solution NMR spectroscopy of membrane proteins. *Curr. Opin. Struct. Biol.* **20**:471–479.

Raschle, T., Hiller, S., Yu, T.Y., Rice, A.J., Walz, T., Wagner, G. (2009) Structural and functional characterization of the integral membrane protein VDAC-1 in lipid bilayer nanodiscs. *J. Am. Chem. Soc.* **131**:17777–17779.

Rinaldi, P.L. (1983) Heteronuclear 2D-NOE spectroscopy. *J. Am. Chem. Soc.* **105**:5167–5168.

Ritchie, T.K., Grinkova, Y.V., Bayburt, T.H., Denisov, I.G., Zolnerciks, J.K., Atkins, W.M., Sligar, S.G. (2009) Reconstitution of membrane proteins in phospholipid bilayer nanodiscs. *Meth. Enzymol.* **464**:211–231.

Salzmann, M., Wider, G., Pervushin, K., Wüthrich, K. (1999) Improved sensitivity and coherence selection for [$^{15}$N,$^1$H]-TROSY elements in triple resonance experiments. *J. Biomol. NMR* **15**:181–184.

Sanders, C.R., Oxenoid, K. (2000) Customizing model membranes and samples for NMR spectroscopic studies of complex membrane proteins. *Biochim. Biophys. Acta* **1508**:129–145.

Sanders, C.R., Prosser, R.S. (1998) Bicelles: a model membrane system for all seasons? *Structure* **6**:1227–1234.

Sanders, C.R., Sönnichsen, F. (2006) Solution NMR of membrane proteins: practice and challenges. *Magn. Reson. Chem.* **44**:S24-S40.

Schweizer, M., Hindennach, I., Garten, W., Henning, U. (1978) Major proteins of the *Escherichia coli* outer cell envelope membrane. Interaction of protein II with lipopolysaccharide. *Eur. J. Biochem.* **82**:211–217.

Sharma, K.S., Durand, G., Gabel, F., Bazzacco, P., Le Bon, C., Billon-Denis, E., Catoire, L.J., Popot, J.-L., Ebel, C., Pucci, B. (2012) Non-ionic amphiphilic homopolymers: Synthesis, solution properties, and biochemical validation. *Langmuir* **28**:4625–4639.

Sharma, K.S., Durand, G., Giusti, F., Olivier, B., Fabiano, A.-S., Bazzacco, P., Dahmane, T., Ebel, C., Popot, J.-L., Pucci, B. (2008) Glucose-based amphiphilic telomers designed to keep membrane proteins soluble in aqueous solutions: synthesis and physico-chemical characterization. *Langmuir* **24**:13581–13590.

Shenkarev, Z.O., Lyukmanova, E.N., Paramonov, A.S., Shingarova, L.N., Chupin, V.V., Kirpichnikov, M.P., Blommers, M.J., Arseniev, A.S. (2010) Lipid-protein nanodiscs as reference medium in detergent screening for high-resolution NMR studies of integral membrane proteins. *J. Am. Chem. Soc.* **132**:5628–5629.

Sim, D.-W., Lu, Z., Won, H.-S., Lee, S.-N., Seo, M.D., Lee, B.-J., Kim, J.-H. (2017) Application of solution NMR to structural studies on α-helical integral membrane proteins. *Molecules* **22**:1347.

Sólyom, Z., Ma, P., Schwarten, M., Bosco, M.I., Polidori, A., Durand, G., Willbold, D., Brutscher, B. (2015) The disordered region of the HCV protein NS5A: conformational dynamics, SH3 binding, and phosphorylation. *Biophys. J.* **109**:1483–1496.

Sprangers, R., Velyvis, A., Kay, L.E. (2007) Solution NMR of supramolecular complexes: providing new insights into function. *Nat. Meth.* **4**:697–703.

Sugiura, M., Beierbeck, H., Bélanger, P.C., Kotovych, G. (1984) Conformational analysis of leukotriene B$_4$ in solution based on high-field nuclear magnetic resonance measurements. *J. Am. Chem. Soc.* **106**:4021–4025.

Sverzhinsky, A., Qian, S., Yang, L., Allaire, M., Moraes, I., Ma, D., Chung, J.W., Zoonens, M., Popot, J.-L., Coulton, J.W. (2014) Amphipol-trapped ExbB-ExbD membrane protein complex from *Escherichia coli*: A biochemical and structural case study. *J. Membr. Biol.* **247**:1005–1018.

Tapaneeyakorn, S., Goddard, A.D., Oates, J., Willis, C.L., Watts, A. (2011) Solution- and solid-state NMR studies of GPCRs and their ligands. *Biochim. Biophys. Acta* **1808**:1462–1475.

Tehei, M., Perlmutter, J.D., Giusti, F., Sachs, J.N., Zaccai, G., Popot, J.-L. (2014) Thermal fluctuations in amphipol A8-35 particles: A neutron scattering and molecular dynamics study. *J. Membr. Biol.* **247**:897–908.

Tifrea, D.F., Pal, S., Cocco, M.J., Popot, J.-L., de la Maza, L.M. (2014) Increased immunoaccessibility of MOMP epitopes in a vaccine formulated with amphipols may account for the very robust protection elicited against a vaginal challenge with *C. muridarum*. *J. Immunol.* **192**:5201–5213.

Tifrea, D.F., Sun, G., Pal, S., Zardeneta, G., Cocco, M.J., Popot, J.-L., de la Maza, L.M. (2011) Amphipols stabilize the *Chlamydia* major outer membrane protein and enhance its protective ability as a vaccine. *Vaccine* **29**:4623–4631.

Torchia, D.A. (2015) NMR studies of dynamic biomolecular conformational ensembles. *Prog. Nucl. Magn. Reson. Spectrosc.* **84-85**:14–32.

Triba, M.N., Warschawski, D.E., Devaux, P.F. (2005) Reinvestigation by phosphorus NMR of lipid distribution in bicelles. *Biophys. J.* **88**:1887–1901.

Tribet, C., Audebert, R., Popot, J.-L. (1996) Amphipols: polymers that keep membrane proteins soluble in aqueous solutions. *Proc. Natl. Acad. Sci. USA* **93**:15047–15050.

Tugarinov, V., Kanelis, V., Kay, L.E. (2006) Isotope labeling strategies for the study of high-molecular-weight proteins by solution NMR spectroscopy. *Nat. Protoc.* **1**:749–754.

Van Horn, W.D., Kim, H.J., Ellis, C.D., Hadziselimovic, A., Sulistijo, E.S., Karra, M.D., Tian, C., Sönnichsen, F.D., Sanders, C.R. (2009) Solution nuclear magnetic resonance structure of membrane-integral diacylglycerol kinase. *Science* **324**:1726–1729.

Viegas, A., Viennet, T., Etzkorn, M. (2016) The power, pitfalls and potential of the nanodisc system for NMR-based studies. *Biol. Chem.* **397**:1335–1354.

Vogt, J., Schulz, G.E. (1999) The structure of the outer membrane protein OmpX from *Escherichia coli* reveals possible mechanisms of virulence. *Structure* **7**:1301–1309.

Vold, R.R., Prosser, R.S. (1996) Magnetically oriented phospholipid bilayered micelles for structural studies of polypeptides. Does the ideal bicelle exist? *J. Magn. Reson.* **B113**:267–271.

Wang, G. (2008) NMR of membrane-associated peptides and proteins. *Curr. Protein Pept. Sci.* **9**:50–69.

Warschawski, D.E., Arnold, A.A., Beaugrand, M., Gravel, A., Chartrand, E., Marcotte, I. (2011) Choosing membrane mimetics for NMR structural studies of transmembrane proteins. *Biochim. Biophys. Acta* **1808**:1957–1974.

Wüthrich, K. (1986) *NMR of Proteins and Nucleic Acids*. Wiley, New York, NY, USA, 320 p.

Yu, C., Levy, G.C. (1983) Solvent and intramolecular proton dipolar relaxation of the three phosphates of ATP: a heteronuclear 2D study. *J. Am. Chem. Soc.* **105**:6994–6996.

Yu, L., Sun, C., Song, D., Shen, J., Xu, N., Gunasekera, A., Hajduk, P.J., Olejniczak, E.T. (2005) Nuclear magnetic resonance structural studies of a potassium channel-charybdotoxin complex. *Biochemistry* **44**:15834–15841.

Zhou, H.X., Cross, T.A. (2013) Influences of membrane mimetic environments on membrane protein structures. *Annu. Rev. Biophys.* **42**:361–392.

Zoonens, M. (2004) Caractérisation des complexes formés entre le domaine transmembranaire de la protéine OmpA et des polymères amphiphiles, les amphipols. Application à l'étude structurale des protéines membranaires par RMN à haute résolution. Thèse de Doctorat, Université Paris-6, 233 p.

Zoonens, M., Catoire, L.J., Giusti, F., Popot, J.-L. (2005) NMR study of a membrane protein in detergent-free aqueous solution. *Proc. Natl. Acad. Sci. USA* **102**:8893–8898.

Zoonens, M., Comer, J., Masscheleyn, S., Pebay-Peyroula, E., Chipot, C.J., Miroux, B., Dehez, F. (2013) Dangerous liaisons between detergents and membrane proteins. The case of mitochondrial uncoupling protein 2. *J. Am. Chem. Soc.* **135**:15174–15182.

Zoonens, M., Giusti, F., Zito, F., Popot, J.-L. (2007) Dynamics of membrane protein/amphipol association studied by Förster resonance energy transfer. Implications for *in vitro* studies of amphipol-stabilized membrane proteins. *Biochemistry* **46**:10392–10404.

# Amphipols and Membrane Protein Crystallization

# 11

**Summary**

*X-ray crystallography is the field of structural biology to which, to date, amphipols have contributed the least. Complexes formed from a membrane protein (MP) and the best characterized amphipol, A8-35, have stubbornly refused to crystallize, whereas ternary MP/A8-35/detergent complexes yielded crystals diffracting to low resolution. Plausible causes of these difficulties and possible ways to alleviate them will be discussed in relation to solution measurements and other data. New tools are being developed, including more adequate amphipols. An alternative approach relying on the use of amphipols to deliver membrane proteins to a lipidic mesophase, where they do crystallize, has been explored with very promising results.*

## 11.1 Introduction

At the onset of the development of amphipols (APols), their use for membrane protein (MP) crystallization was deemed a particularly desirable and perhaps realistic objective. The main rationales for hope were the asset provided by the stabilization that APols afford to most MPs, the good behavior of MP/APol complexes in solution, and the fact that – it was thought – MP/APol complexes could be handled in surfactant-free solutions as though they were soluble proteins, in the total or near-total absence of free surfactant. This would avoid having to expose MPs to high concentrations of detergent, which is, as discussed in Chap. 2, a major cause of inactivation, as well as having to cope with the phase transitions detergent solutions experience with changes of ionic strength, concentration, or temperature.

It was of course realized that the charges carried by the first APols to be developed, A8-35 and its congeners (Tribet et al. 1996; see Chap. 4), were far from being a favorable factor. Indeed, detergents that carry a net charge are known not to be conducive to MP crystallization (see e.g. Privé 2007; Parker and Newstead 2012). Work was soon initiated in two directions: (*i*) examining, by radiation scattering, which salt conditions could be used to screen the electrostatic repulsion arising from these charges (Popot et al. 2003) and (*ii*) developing non-ionic APols (Prata et al. 2001). Nevertheless, despite years of efforts, not a single MP has yet been crystallized under the form of pure MP/APol complexes.

In the process, however, much has been learned about the difficulties to be overcome and insights obtained into possible routes to do so. In particular:

(i) Three-dimensional (3D) crystals of ternary MP/APol/detergent complexes have been produced, which do diffract X-rays, even though, at this point, to limited resolution (Charvolin et al. 2014).

(ii) 3D crystals diffracting to very high resolution have been obtained by using APols as a shuttle to introduce MPs into a lipidic mesophase (Polovinkin et al. 2014; Broecker et al. 2017).

(iii) Satisfying non-ionic APols have been designed, synthesized, and validated (Bazzacco et al. 2012; Sharma et al. 2012; see Chap. 4), although not yet tested for MP crystallization.

These progresses will be reviewed in § 11.3, and challenges and prospects discussed. In § 11.2, a brief overview will be first provided of the current status of MP crystallization and crystallography, with the accent placed on developing technologies, particularly inasmuch as they could potentially benefit from being combined with the use of APols.

## 11.2  Context: Classical and Less Classical Approaches to Crystallizing Membrane Proteins

Until the 1980s, MP crystallization, much like MP folding (see Chap. 6), was considered by most membrane biochemists as an almost hopeless endeavor, and few people were daring enough to seriously engage in such risky if not doomed attempts. The first well-diffracting crystals were reported in the early 1980s, by two different groups, for two bacterial MPs, *E. coli*'s outer membrane porin OmpF (Garavito and Rosenbusch 1980) and the photosynthetic reaction center from the purple bacterium *Rhodopseudomonas viridis* (Michel 1982). Phasing problems delayed the resolution of the structure of OmpF until the 1990s (Cowan et al. 1992), whereas a 3-Å resolution electron density map of the reaction center was established in 1985, allowing the construction of the first atomic molecular model ever to be built of an integral MP in its entirety, i.e. including its transmembrane (TM) region (Deisenhofer et al. 1984, 1985; Michel et al. 1986; Deisenhofer and Michel 1989).

### 11.2.1  Factors Contributing to the Steady Growth of the Number of MP Structures Solved

As every general review on MP crystallization will recall (see e.g. Ostermeier and Michel 1997; Wiener 2001, 2004; Caffrey 2003, 2009, 2015; Nollert 2004, 2005; Cherezov et al. 2006; Tate 2010; Cherezov 2011; Ujwal and Bowie 2011; Parker and Newstead 2012; Kang et al. 2013; Ishchenko et al. 2014, 2017; Loll 2014; Moraes et al. 2014), the initial growth in new MP structures was excruciatingly slow, before progressively picking up steam and embarking onto an exponential course (Fig. 11.1). The main reason for the initial slow growth was the dearth of MPs that could be obtained in detergent solution in sufficient amount and under a stable enough form for crystallization trials to be launched. Most MPs, indeed, are not produced *en masse* in cells and tissues (Chap. 1), and most of them are unstable in detergent solutions (Chap. 2). The progressive acceleration in the rate of 3D structures solved is due to a large number of factors among which are psychological ones (the realization that the task after all was not as desperate as it had looked earlier, which dramatically increased the number of people willing to tackle it and the bulk of resources allotted to it), biochemical ones, which we will

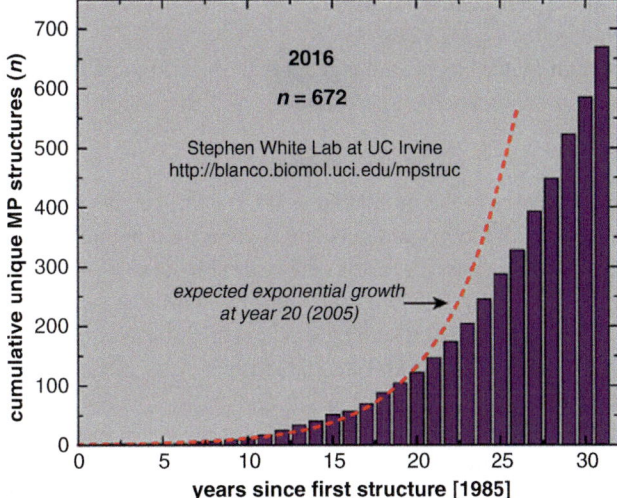

**Fig. 11.1** Cumulative number of unique membrane protein structures deposited in the Protein Data Bank as a function of the number of years elapsed since the publication of the first structure (Deisenhofer et al. 1985) (From http://blanco.biomol.uci.edu/mpstruc/).

briefly discuss below, and factors linked to crystallography itself, such as the development of crystallization robots, of ever more powerful synchrotrons with ever brighter and finer beams, and, along with it, the ability to prepare, recover, identify, select, and analyze smaller and smaller crystals in a more and more automated manner. Technological progress in these directions is constant, with the use of micrometric beams and crystals becoming almost routine.

The most recent major technical progress, currently at its onset and in full development, is the use of free-electron laser X-ray (X-FEL) sources, which deliver very short (~50 to 100 fs) pulses whose brightness – many orders of magnitude above that of synchrotron sources – makes it possible to record fragmentary diffraction patterns from thousands if not millions of micrometric to submicrometric crystals and, by merging them, to reconstruct the full pattern and calculate the electron density map of the target protein. Data are collected in the first tens of femtoseconds of exposure, i.e. before radiation damage sets in (see Neutze et al. 2000, 2015; Chapman et al. 2011; Boutet et al. 2012; Johansson et al. 2012; Kang et al. 2013; Martin-Garcia et al. 2016). Despite the limited number of X-FEL sources, their high running costs, and the considerable technical difficulties involved, this approach, which has been proven to be applicable to MPs (Chapman et al. 2011; Johansson et al. 2012; Kupitz et al. 2014; Suga et al. 2017), has the potential of yielding the structures of important MP targets that are reluctant to form crystals large enough for conventional analysis using synchrotron X-ray sources.

Among biochemical progresses, a decisive one was the ability to produce, by overexpression, MPs that are too rare in natural organisms to be purified in amounts commensurate to the needs of a crystallographic study (see Chap. 1), combined with the miniaturization and robotization of the crystallization assays, which multiplies the number of trials that can be conducted with a given amount of protein. Another useful progress has been the diversification of surfactants (reviewed in Zhang et al. 2011; Sadaf et al. 2015; cf. Chap. 2, § 2.5.1), even though, compared to the tens of novel molecules that have been designed and validated biochemically, their contribution to the determination of novel MP structures remains, to date, relatively modest (see Parker and Newstead 2012; Loll 2014; Moraes et al. 2014). An example of success story is the use of "MNG" (maltose-neopentyl glycol) detergents (see Chap. 2, Fig. 2.15) for handling and crystallizing G protein-coupled receptors (GPCRs) (Rasmussen et al. 2007, 2011; Chae et al. 2010). A highly significant progress has been the development of nonconventional approaches to crystallization. Because those bear more directly on the question of whether APols can contribute usefully to MP crystallography, they will be presented somewhat more in detail below (§ 11.2.2).

A critically important factor has been the genetic engineering and biochemical manipulation of MPs, so as to make them more amenable to crystallization. Schematically, this engineering has taken two main paths:

- Modifying the sequence of the protein so as (*i*) to eliminate as much flexibility as possible – e.g. by removing mobile loops and tails and separating domains that tend to move one with respect to the other; an approach that is classical with soluble proteins – and (*ii*) to make the target MPs more stable by selecting detergent-resistant mutants (see e.g. Zhou and Bowie 2000; Bowie 2001; Magnani et al. 2008; Sarkar et al. 2008; Serrano-Vega et al. 2008; Shibata et al. 2009; Tate 2010; Miller and Tate 2011; Lluis et al. 2013; Egloff et al. 2014; Vaidehi et al. 2016, and discussion in Chap. 2, § 2.5.2).
- Decorating the target protein with water-soluble domains in order to (*i*) increase the hydrophilic surface available for establishing crystal contacts, (*ii*) move the proteins farther one from another in the crystal, which can limit interference by the detergent belts, (*iii*) stabilize the protein, and (*iv*) possibly select among conformational states. There are two ways to introduce these domains:

  - One is to co-crystallize the target protein along with soluble binders, such as natural or engineered antibody $F_{ab}$ or $F_v$ fragments or camelid single-chain antibody fragments (nanobodies) (see e.g. Ostermeier et al. 1995, 1997; Hunte et al. 2000; Zhou et al. 2001; Uysal et al. 2009; Hibbs and Gouaux 2011; Rasmussen et al. 2011; see, below, Fig. 11.4). Alternatively, one can resort to designed artificial proteins such as DARPins (Binz et al. 2003; Sennhauser and Grütter 2008; Boersma and Plückthun 2011; Jost and Plückthun 2014; Plückthun 2015) or $\alpha$Reps (Urvoas et al. 2012; Ferrandez et al. 2014).
  - Another is to insert genetically a soluble domain, such as T4 lysozyme, into the sequence of the target protein itself, e.g. by substituting it for a flexible loop (see e.g. Cherezov et al. 2007; Rosenbaum et al. 2007 and, below, Fig. 11.3).
- A third approach, which requires no tinkering with the target protein but can be combined with it, is to screen orthologs from many species in order to identify those that express well and are particularly stable (see e.g. Love et al. 2010).

As for any protein, it is of course good politics to crystallize a fragile and/or flexible MP along with a high-affinity ligand, whose binding will stabilize it and shift the conformational equilibrium toward a single state.

## 11.2.2  The Formation of Membrane Protein Crystals

In order to form a clear view of the advantages and disadvantages APols can present for MP crystallization, it is necessary to examine the structure of MP crystals and the various media in which they can form. The first MP crystals were obtained in aqueous solutions, by inducing MP/detergent or MP/lipid/detergent complexes to associate into 3D crystals, at first glance much like soluble proteins do (§ 11.2.2.1). A second, more recent approach is to induce the formation of the crystals within 3D lipidic phases, in which MPs diffuse not as soluble complexes but as membrane-embedded entities (§ 11.2.2.2).

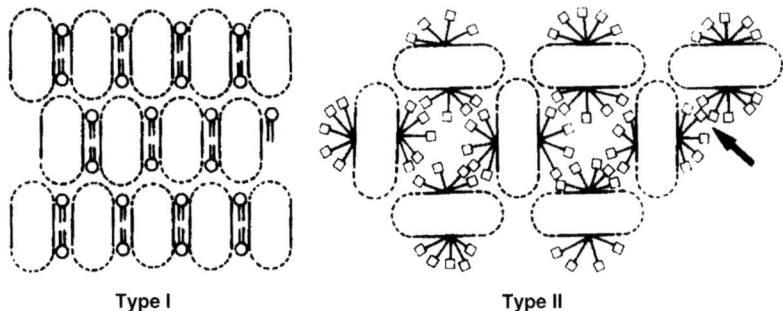

**Fig. 11.2** The two types of MP crystals discussed by H. Michel in his 1983 mini-review. The surface of the MP is represented *dashed* where it is hydrophilic (extramembrane surfaces) and *solid* where it is hydrophobic (transmembrane one). Lipids are represented as having one polar head and two hydrophobic tails (*left*), detergent molecules as having one head and one tail (*right*). The *arrow* points to contacts between detergent belts (From Michel 1983, © 1983 Published by Elsevier Ltd).

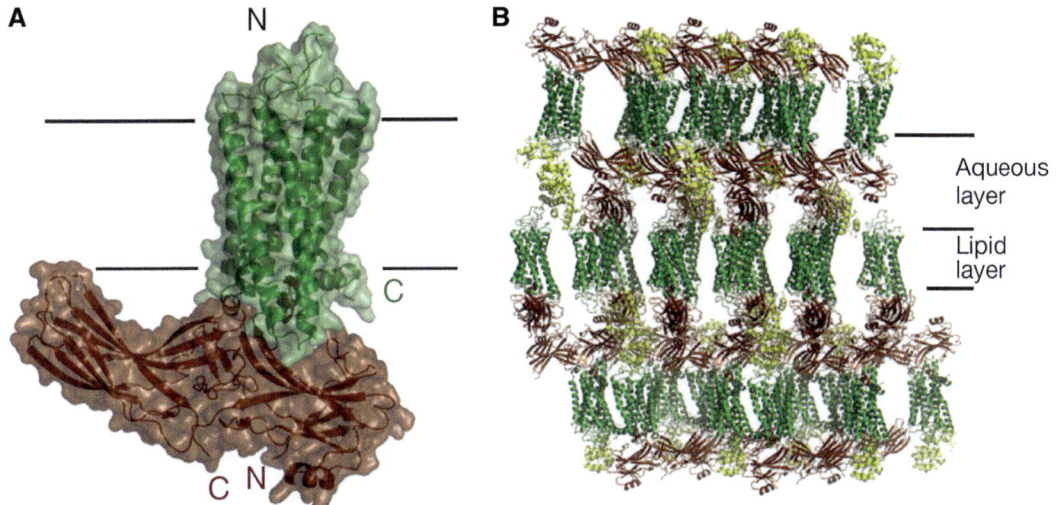

**Fig. 11.3** An example of Type I crystal. Packing interactions in the crystals of rhodopsin (*green*)/arrestin (*brown*) complexes. (**A**) An overall view of the rhodopsin/arrestin complex shown with transparent solid surface. T4 lysozyme (T4L), which is fused in one of the loops of rhodopsin, is omitted from this view. (**B**) Crystal packing diagram of the rhodopsin/arrestin complex with T4L as a *yellow* ribbon model (From Kang et al. 2015, Copyright © 2015, Rights Managed by Nature Publishing Group).

Before moving on, it is useful to say a word of which types of MP crystals could be a priori expected. Those were distinguished early on by Hartmut Michel in an insightful mini-review (Michel 1983; for a more recent, extensive analysis of the organization of the protein and detergent in MP crystals, see Schulz 2011). H. Michel described two types of 3D crystals according to the kinds of contacts the proteins would make between themselves and with their environment. Type I crystals (Fig. 11.2, *left*) consist of stacks of two-dimensional (2D) crystals, such as had been observed to form in some electron microscopy (EM) samples. Two different types of contacts would stabilize the structure of such crystals. Within each plane, contacts would be primarily hydrophobic, possibly mediated by ordered lipids, as is the case between bacteriorhodopsin (BR) trimers in the purple

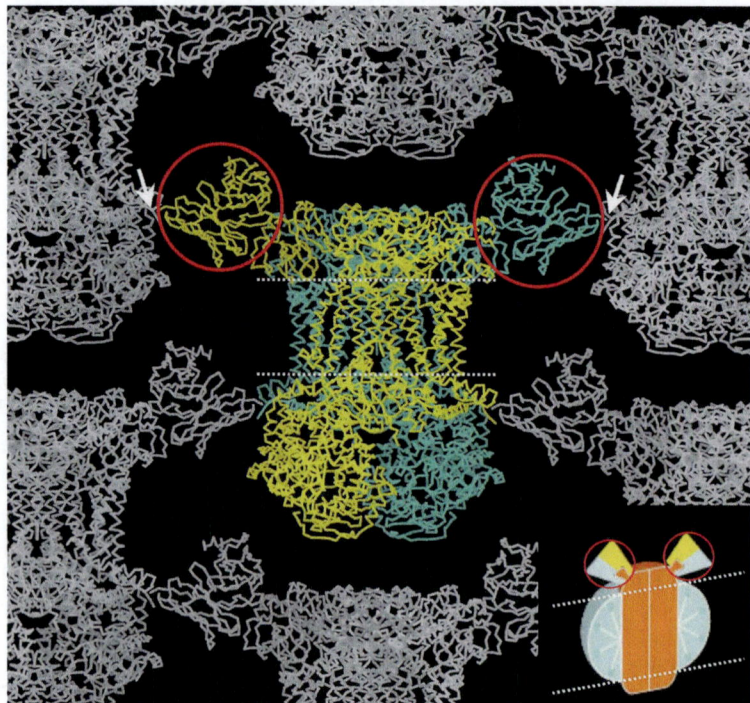

**Fig. 11.4** Organization in a Type II crystal of yeast cytochrome $bc_1$ dimer (*green* and *yellow*, in the center) decorated with two $F_v$ fragments. Essential crystal contacts (*white arrows*) are mediated by the $F_v$ fragment. Each fragment (encircled in *red*) binds to the extramembrane domain of one of the $bc_1$ subunits, the Rieske protein, and, while interacting with a nearby complex (*light gray*), keeps it at a distance, providing space for the (invisible) detergent belt that surrounds the transmembrane portion of the complex (marked by *dotted lines*), as schematically depicted in the inset (From Hunte 2001, © 2001 Federation of European Biomedical Societies).

membrane 2D crystals (Grigorieff et al. 1996). This is schematically shown in the cartoon of Fig. 11.2 (*left*). Between planes, contacts would be established between the hydrophilic, extramembrane surfaces of the proteins, as in soluble protein crystals. H. Michel deemed the prospect of growing such crystals to a size useful for X-ray crystallography as dim, because the crystallization procedure would have to simultaneously optimize both hydrophobic and hydrophilic contacts. Indeed, they are seldom observed when MP crystals are grown from detergent solutions.

Type II crystals, which were the ones obtained for the photosynthetic reaction center (see Roth et al. 1989), resemble crystals of soluble proteins, and their ordering relies mainly on protein/protein contacts at the level of the hydrophilic, extramembrane protein surfaces, the role of the detergent belt adsorbed onto their transmembrane (TM) surface being essentially to keep the protein water-soluble and prevent precipitation (Fig. 11.2, *right*). It is immediately clear, however, that, except for MPs with very large extramembrane regions and a small TM one, interactions between the detergent belts will occur. If repulsive, they may prevent MPs from coming close enough one to another to interact via their extramembrane surfaces; if attractive, they may induce them to do so, provided the belt is not too bulky and the solution is close enough to conditions where detergent micelles start attracting each other (the consolution limit; see Chap. 2). This led H. Michel to propose that minimizing the thickness of the detergent belt, either by using small detergents or by supplementing them with small surfactants that would modulate their surface curvature and could substitute for bulkier detergent molecules at points of contact, could be a decisive factor in permitting the growth of well-ordered crystals. Indeed, contacts between detergent belts surrounding neighboring MPs are revealed by neutron diffraction

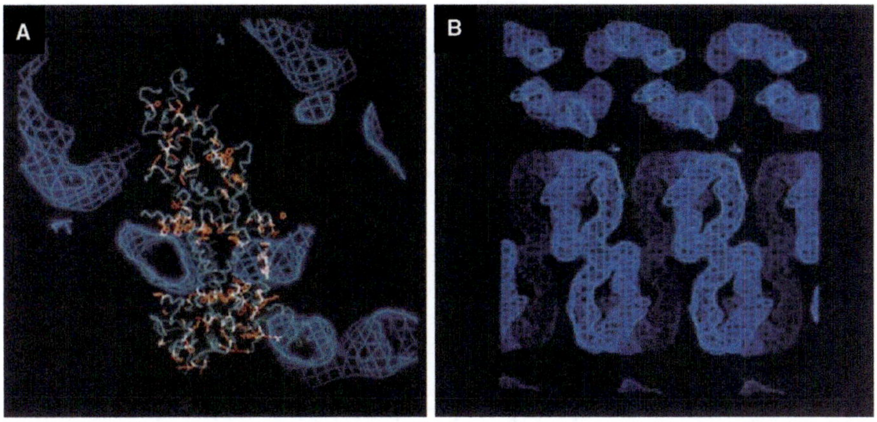

**Fig. 11.5** Arrangement of the detergent belts in 3D crystals of the photosynthetic reaction center from *Rhodopseudomonas viridis*. (**A**) Composite image showing the X-ray structure of the backbone and aromatic side chains of the reaction center (respectively, in *green* and *orange*) and the distribution of the detergent (*N,N*-dimethyl-dodecylamine-*N*-oxide; LDAO) as determined by neutron diffraction (*blue* mesh). A ring of LDAO, seen in cross-section, surrounds the whole transmembrane region of the complex. (**B**) Contacts between detergent rings surrounding neighboring complexes (From Roth et al. 1989).

studies of 3D crystals of photosynthetic reaction centers (see Fig. 11.5). Keeping in mind this type of considerations is critical when trying to crystallize APol-trapped MPs.

There are two main routes to crystallizing MPs: (*i*) from a detergent solution (§ 11.2.2.1) and (*ii*) from a lipidic mesophase (§ 11.2.2.2). Other approaches have succeeded as well, such as crystallizing MPs from bicelles (Faham and Bowie 2002; Luecke et al. 2008; Ujwal et al. 2008; Vinothkumar 2011; reviewed in Johansson et al. 2009; Ujwal and Bowie 2011; Agah and Faham 2012; Loll 2014; Poulos et al. 2015; see Chap. 3, § 3.2.2), from proteoliposomes (Liu et al. 2004), and even from a dodecylsulfate/cosolvent mixture (Cuesta-Seijo et al. 2010), but they are of less relevance to the use of APols and will not be discussed here.

### 11.2.2.1 Crystals Formed in Detergent Solutions

Forming 3D crystals of MPs in solution relies on the same principle as for soluble proteins, namely to bring the solution to supersaturation. This is achieved by starting from a soluble preparation and concentrating it and/or diffusing into it precipitants, typically polyethylene glycol (PEG) or salts, which will decrease the solubility of the species to be crystallized. This can be achieved by a variety of methods, among which vapor diffusion (hanging or sitting drop), microdialysis, batch crystallization, and free interface diffusion. Whereas all of them have been used in early days, and all have yielded diffraction-quality crystals (for reviews, see e.g. Michel 1991; Garavito et al. 1996), most attempts nowadays rely on vapor diffusion, in which a droplet of a solution of MP in detergent plus precipitant is left to equilibrate with a more concentrated reservoir of precipitant. At variance with globular proteins, however, MPs, to remain water-soluble, must be associated with a surfactant, typically a detergent, which will provide the interface between their hydrophobic TM surface and the aqueous solution (see Chaps. 2 and 3). This is where the trouble begins.

For a detailed discussion of the problems created by the presence of the detergent, the reader is referred to specialized reviews, e.g. those by Kühlbrandt (1988), Michel (1991), Garavito et al. (1996), Ostermeier and Michel (1997), Wiener (2001, 2004), Berger et al. (2005), Tate (2010), Gourdon et al. (2011), Schulz (2011), Kang et al. (2013), and Loll (2014). The following points should be noted:

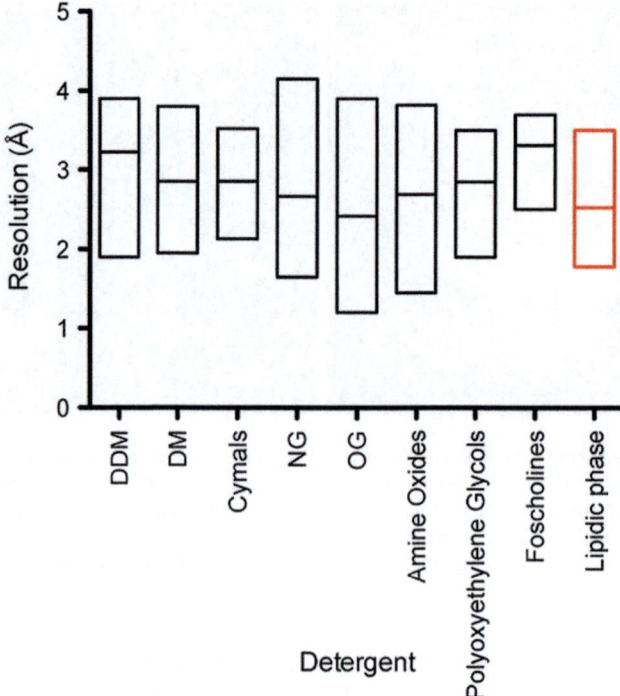

**Fig. 11.6** Analysis of reported resolution for $\alpha$-helical membrane protein X-ray structures. *Box* plots display the range of resolution achieved for eight of the most successful crystallization detergent classes; the middle *line* in each *box* represents the mean value for each class. Only structures crystallized using a single detergent are included. *DDM* dodecylmaltoside, *DM* decylmaltoside, *NG* nonylglucoside, *OG* octylglucoside. Detergents with the smallest polar heads (NG, OG, aminoxides) tend to yield the best diffracting crystals. In *red*, the same analysis is used for structures determined using *in meso* crystallization (From Parker and Newstead 2012, © 2012 The protein Society).

- Detergents, as discussed in Chap. 2, tend to destabilize MPs and to do so the more rapidly when they are more concentrated. In typical crystallization experiments by vapor diffusion, the initial solution is already quite concentrated in MP and detergent, and the evaporation of water will increase both concentrations.
- Detergents have their own phase diagrams: when concentrating a micellar solution of detergent – or warming it or cooling it – separation can occur between a detergent-poor and a detergent-rich phase, in which micelles interact with one another (Chap. 2). Phase separation is generally to be avoided (see however Garavito et al. 1983, 1996), but coming close to it can help, because slightly attractive interactions between the detergent belts surrounding the proteins can favor the formation of the crystal (cf. Fig. 11.5). Interactions between detergent micelles indeed are a good predictor of crystallization conditions (Hitscherich et al. 2000).
- The bulk of the detergent belt is also a source of problems, as it may interfere with the formation of the crystal and make the formation of contacts between the hydrophilic surfaces of the protein more difficult. Indeed, when crystals do form, they tend to be more water-rich and less well ordered in the presence of detergents that form large micelles, and there is indeed a tendency for detergents with small polar heads and short hydrophobic tails to yield crystals that diffract to slightly higher resolution (Parker and Newstead 2012). Compare, for instance, the best resolutions obtained in either *n*-dodecyl-$\beta$-D-maltopyranoside (DDM) or *n*-octyl-$\beta$-D-glucopyranoside (OG) in Fig. 11.6. The flip side is that MPs, for reasons that have

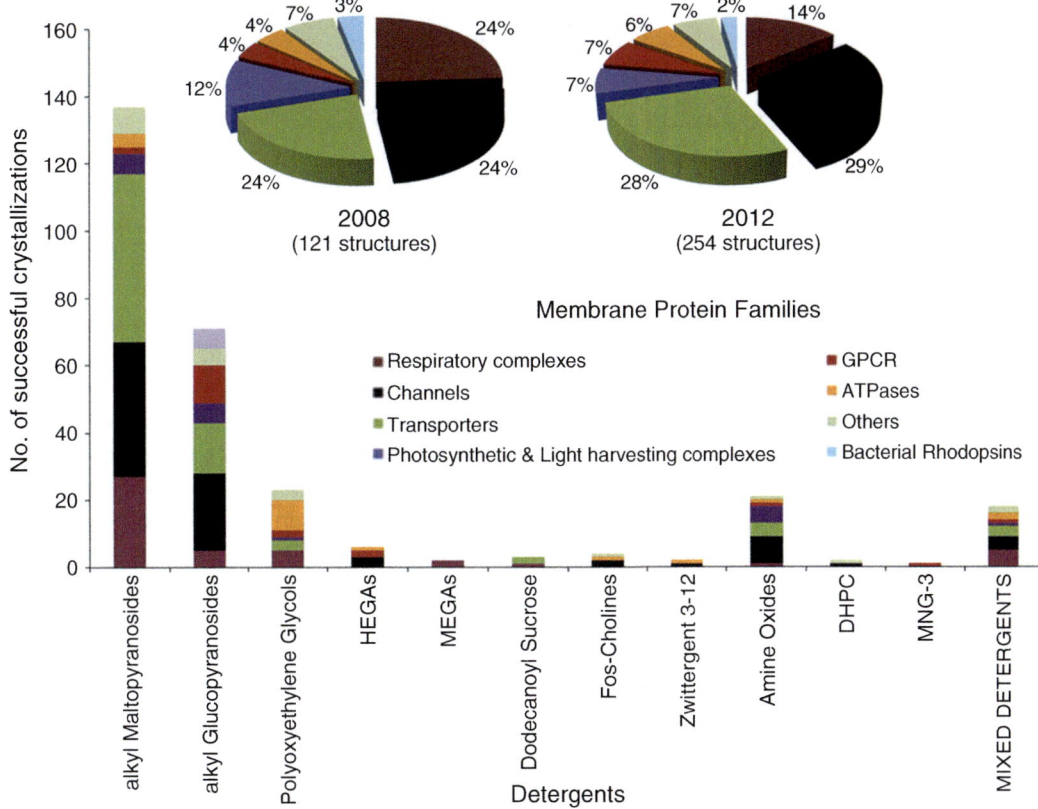

**Fig. 11.7** Detergents successfully used for α-helical membrane protein crystallization. Numbers of deposited PDB structures for each detergent class are shown, subdivided into the eight MP families used in the analysis, namely respiratory complexes (*brown*), channels (*black*), transporters (*green*), photosynthetic and light-harvesting complexes (*purple*), G protein-coupled receptors (GPCRs; *red*), ATPases (*orange*), bacterial rhodopsins (*blue*), and the "other" categories (*olive*), those not fitting into the seven main groupings. *Inset*: Pie charts showing the change in the proportion of structures belonging to the eight MP families between 2008 and 2012. The percentage contribution made by each family is indicated (From Parker and Newstead 2012, © 2012 The protein Society).

been discussed in Chap. 2 (§ 2.4), tend to denature more rapidly in detergents with small hydrophobic chains, which have a high CMC, than in those with long chains and low CMC. The same search for minimal encumbrance by the surfactant belt has led to the use of small amphiphilic additives like pentanetriol, which, by increasing the spontaneous curvature of the surfactant/water interface, favor the formation of small micelles and can substitute for bulky detergent molecules at points of close approach between surfactant belts (see Michel 1983; Parker and Newstead 2012).

Figure 11.7 presents an interesting survey, broken down by MP families, of which detergents have led to crystals good enough for a molecular model to be built into the electron density map and deposited in the Protein Data Bank (Parker and Newstead 2012). Note that, as mentioned above, whereas zwitterionic detergents have been used successfully (zwittergents, foscholines, aminoxides, diC$_7$PC), detergents carrying a net charge have still to yield a single structure. In considering this diagram, one should also keep in mind that detergents that can boast of a long record of successful crystallization attempts, like alkyl malto- and alkyl glucopyranosides, are among the first to be used in crystallization screens and will keep for a long while yielding more structures even if they are no better than more recent detergents with a more limited track record.

### 11.2.2.2  Forming Crystals in Lipid Three-Dimensional Phases

Twenty years ago, Ehud M. Landau and Jürg P. Rosenbusch introduced what they rightly called "a novel concept for the crystallization of membrane proteins" and demonstrated its validity using BR as a model (Landau and Rosenbusch 1996). Their proposal was to use as a medium for MP crystallization, rather than a detergent solution, some of the organized 3D phases that specific lipids can form under specific conditions and, in particular, the so-called cubic phases. As described in Box 11.1, these "mesophases" are comprised of a 3D network of interconnected tubes. The wall of the tubes has a bilayer structure, while two distinct networks of aqueous channels pervade the whole sample. A macroscopic sample of lipidic cubic phase appears as a transparent, highly viscous blob, which can nevertheless be handled using syringes, as well as overlaid with solutions. The walls of the tubes allow the diffusion of membrane-inserted molecules throughout the sample, and so do the two networks of aqueous channels for water-soluble molecules. Cubic phases are resistant to low concentrations of detergent, as well as to most precipitants, and they tolerate the insertion, up to a certain concentration, of biological lipids (see Caffrey 2015, and refs. therein). Below ~17 °C, the cubic phase formed by mono-olein (MO; Fig. 11.8), which is the most commonly used lipid, is metastable and occasionally converts to the lamellar crystalline or solid phase, which does not allow 3D crystallization. This has prompted the development of alternative lipids, such as 7.9 MAG or phytane derivatives, whose phase diagrams make it possible to work at lower temperatures (reviewed in Caffrey 2015).

---

**Box 11.1  Lipidic Mesophases**

Lipids and, more generally, surfactants organize in space so as to satisfy the hydrophobicity, hydrophilicity, bulk, charge, etc. of their various moieties. Hydration, salinity, and temperature play a critical role in determining which arrangement (phase) is the most stable for a given molecule or mixture of molecules under a given set of conditions (Chap. 1, § 1.2; for general discussions, see e.g. Tanford 1980; Israelachvili 2011). A *mesophase* is "a normal phase in a thermodynamic sense, but one that is structurally more complex than a simple liquid or solid phase. It can contain many small molecular aggregates that can be monodisperse or polydisperse, or it can have convoluted lamellar or tubular structures that link up with each other to form a repeat three-dimensional network that extends indefinitely throughout the phase" (Israelachvili 2011).

Certain lipids such as monoacylglycerols and, in particular, mono-olein (Fig. 11.8) form, under appropriate conditions, 3D networks of tubes, the walls of which have locally the molecular arrangement of a bilayer, among them the "cubic" Pn3m and Ia3d phases (Fig. 11.9) and the "sponge" phase (not shown) . The latter contains more water, features larger water-filled channels, and is more flexible and disordered (Caffrey 2015; Wöhri et al. 2008). MPs inserted in the wall of the tubes forming the cubic or sponge phases experience a lipid bilayer-like environment and, at variance with the 2D confinement that prevails in lamellar phases, can diffuse in three dimensions throughout the macroscopic volume of the phase. Because these phases are isotropic, they are non-birefringent and optically clear, making it possible, under good conditions, to identify even micrometric crystals.

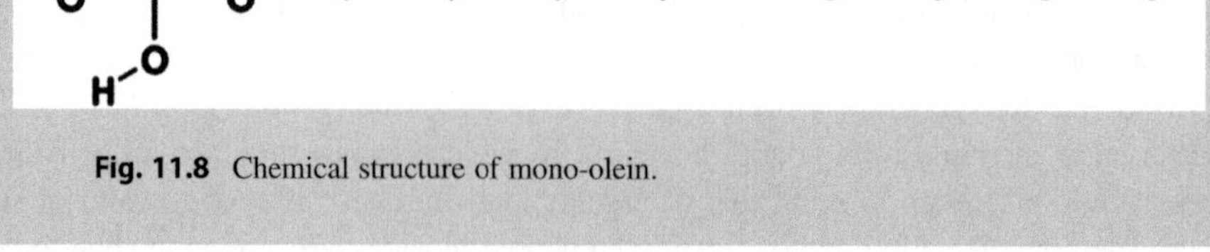

**Fig. 11.8**  Chemical structure of mono-olein.

---

**Box 11.1 (continued)**

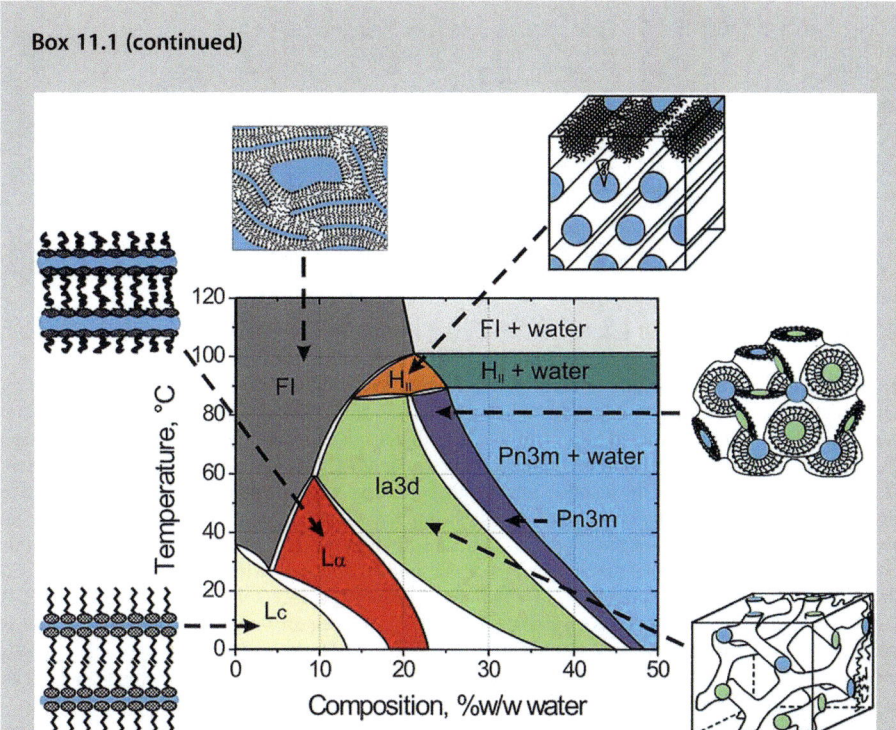

**Fig. 11.9** Temperature-composition phase diagram of the mono-olein/water system, determined under conditions of use by starting from 20 °C and heating or cooling the samples. A schematic representation of the various phase states is included in which colored zones represent water channels or layers. The cubic phases are metastable below ~17 °C (Qiu and Caffrey 2000). *Abbreviations*: *Fl* fluid isotropic phase, *H_{II}* inverted hexagonal phase, *L_α* lamellar liquid crystalline phase, *L_c* lamellar crystal phase (From Cherezov et al. (2006), © 2006 Elsevier Ltd. All rights reserved. Redrawn from Qiu and Caffrey (2000), © 1999 Elsevier Science Ltd. All rights reserved).

When a MP in detergent solution is mixed with a lipidic cubic phase, it inserts into the walls of the tubes, where it experiences a membrane-like environment. It can diffuse freely throughout the whole volume of the phase. Addition of precipitants causes, under appropriate conditions, the formation of nucleation sites and then of microcrystals, which are fed by the 3D network, grow to crystallographically useful sizes, and can diffract to high resolution (Fig. 11.10). The water channels in the MO cubic phase are only ~50 Å in diameter, which is a limitation to the diffusion of MPs with large extramembrane regions. Some compounds, like PEG or butanediol, cause the cubic phase to pick up more water, moving in the direction of a "sponge" phase. The water-filled channels swell, and the whole phase becomes more flexible and less ordered, but it retains its 3D continuity and can be used, in particular, to crystallize MPs with extended extramembrane domains (see Wadsten et al. 2006; Wöhri et al. 2008; Caffrey 2015). The term *in meso*, which is frequently used to refer to MP crystallization in mesophases, is therefore more generally applicable than the more restrictive alternative *in cubo*.

It is noteworthy that all 3D crystals grown *in meso* to date are of Type I (Fig. 11.2), which is formed from stacks of 2D crystals. A proposed mechanism for their growth from the 3D cubic or sponge phases is shown in Fig. 11.10 and discussed in its legend (Li et al. 2013).

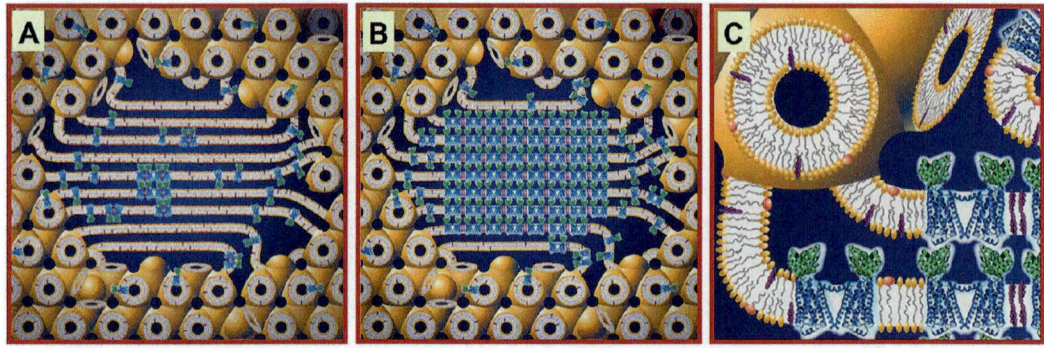

**Fig. 11.10** Cartoon representation of the events proposed to take place during the crystallization of an integral membrane protein from the lipidic cubic mesophase. The process begins with the protein (*blue/green*) reconstituted into the curved bilayer of the bicontinuous cubic phase (*yellow*). Added precipitants shift the equilibrium away from stability in the cubic membrane. This leads to phase separation, wherein protein molecules diffuse from the bicontinuous bilayered reservoir of the cubic phase into a sheet-like or lamellar domain (**A**) and locally concentrate therein in a process that progresses to nucleation and crystal growth (**B**). Co-crystallization of the protein with native lipid (in this case, cholesterol) is shown in this illustration. As much as possible, the dimensions of the lipid (*yellow oval* with tail), detergent (*pink oval* with tail), cholesterol (*purple*), protein (*blue* and *green*; $\beta_2$-adrenergic receptor-T4 lysozyme fusion; PDB code 2RH1), and bilayer and aqueous channels (*dark blue*) have been drawn to scale. The lipid bilayer is ~40 Å thick. Panels **A** and **B** are reprinted from Li et al. (2011). An expanded view of the various components in the system is shown in (**C**) (The composite figure is reprinted with permission from Li et al. 2013, © 2013 American Chemical Society).

After relatively slow beginnings – due, in part, to practical difficulties in handling the samples, to the initial absence of robots able to prepare them, and to the rather cumbersome recovery of the crystals once they have formed – the approach has become strongly established as a major tool in MP crystallogenesis (for reviews, see e.g. Rummel et al. 1998; Nollert 2004; Wiener 2004; Cherezov et al. 2006; Hunte and Richers 2008; Caffrey 2009, 2011, 2015; Johansson et al. 2009; Cherezov 2011; Ishchenko et al. 2014, 2017; Moraes et al. 2014; Bogorodskiy et al. 2015). As of September 2014, close to 200 MPs, or about 10% of all published crystallographic MP structures, had been crystallized *in meso* (Caffrey 2015). Their distribution by function is shown in Fig. 11.11. Success has been spectacular in the field of GPCRs, whose crystallization from detergent solutions is notoriously difficult.

## 11.3   Using Amphipols for Crystallizing Membrane Proteins

As has been discussed in the case of NMR (cf. Chap. 10, Fig. 10.1), APols can be used upstream of crystallization itself, so as to produce the target MPs, e.g. by folding them from inclusion bodies or by expressing them in vitro. APol-trapped MPs thus obtained can be (*i*) used as such for crystallization attempts; (*ii*) supplemented with a controlled amount of detergent so as to form ternary MP/APol/detergent complexes; (*iii*) transferred to a detergent solution, from which to crystallize it using conventional approaches; or (*iv*) transferred to a lipid mesophase. Approach (*i*) has been extensively tried using the complexes formed between mitochondrial cytochrome $bc_1$ and A8-35, without ever yielding any crystals, the possible reasons for which failure will be discussed below (§ 11.3.1). Approach (*ii*) did yield crystals, although those have not been optimized yet (§ 11.3.1). Approach (*iii*) has not yet been used, to our knowledge, and is essentially beyond the frame of this chapter, since APols would be removed before starting crystallization attempts. Approach (*iv*) has been experimented with, using BR from *Halobacterium salinarum* or from *Haloquadratum walsbyi* as the test protein, and

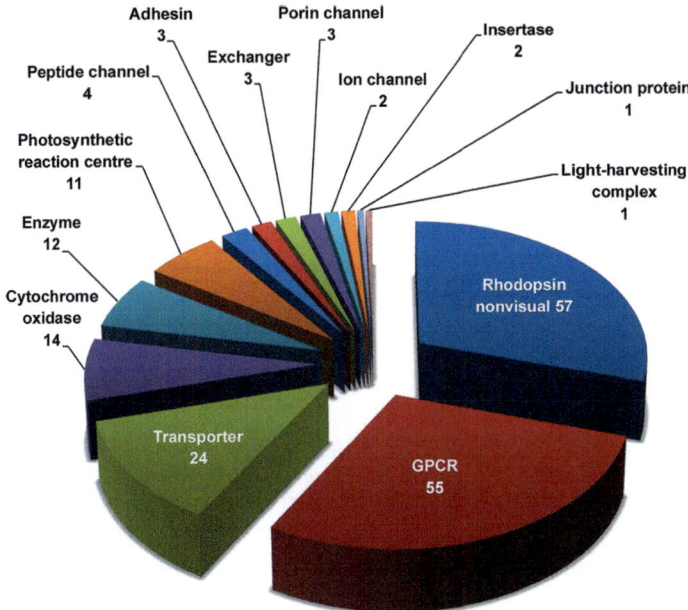

**Fig. 11.11** Distribution by biological function of integral membrane proteins and peptides crystallized *in meso* that, as of September 2014, had yielded crystal structures deposited in the Protein Data Bank. *GPCR* G protein-coupled receptors (From Caffrey 2015, reproduced with permission of the International Union of Crystallography).

has yielded excellent crystals, from which the structure could be solved at high resolution (§ 11.3.2). We will discuss successively approaches (*i*), (*ii*), and (*iv*), on the basis of data presented in the studies listed in Table 11.1. The results of the crystallization trials of Study 11.3 will be presented first, and their outcome discussed with the help of solution studies described in Studies 11.1–11.3. The most recent and very encouraging results of Studies 11.4 and 11.5 will then be presented, before moving on to § 11.3.3, which discusses challenges and perspectives.

## 11.3.1   Crystallization of Cytochrome $bc_1$/A8-35 Complexes in Aqueous Solutions: Failures and Partial Successes and the Possible Reasons Underlying Them

Study 11.3 reports the results of the first attempts at crystallizing MP/APol complexes (Charvolin et al. 2014). The model MP chosen is the cytochrome $bc_1$ complex extracted from beef heart mitochondria. Cytochrome $bc_1$ presents numerous advantages for such a study: the complex is abundant, colored, rugged, and relatively easy to crystallize from detergent solutions under a variety of conditions, and it features extended extramembrane domains, which represent ~4/5 of its mass (Berry et al. 2000) (Fig. 11.12A). As a consequence, and despite the large number of $\alpha$-helices and cofactors that compose its TM region, Type II crystals can form in which protein/protein contacts are solely established between extramembrane surfaces, and there is no or very little interaction between surfactant layers (Fig. 11.12B). In Study 11.3, the formation, composition, dispersity, and stability of $bc_1$/A8-35 complexes have been examined, as well as their solution behavior as a function of such variables as the presence of excess APol or the salt concentration. Crystallization attempts were carried out both of binary $bc_1$/A8-35 complexes and of ternary $bc_1$/A8-35/detergent ones.

   Complexation of cytochrome $bc_1$ by A8-35 took place efficiently under conditions similar to those described for other MPs, e.g. the TM domain of OmpA (tOmpA) (Zoonens et al. 2005) or BR

**Table 11.1** Five studies relevant to the use of amphipols for MP crystallization. Color code: *blue*, background studies; *green*, studies where diffraction results are reported.

| Study | Membrane protein | Amphipol | Methods | Main observations | References |
|---|---|---|---|---|---|
| 11.1 | Cytochrome $bc_1$ | A8-35 | SAXS, SANS, SEC | Small-angle X-ray and neutron scattering studies show that cytochrome $bc_1$/A8-35 complexes repulse each other at low salt concentration. Between 300 and 500 mM NaCl, or upon the addition of PEG, the complexes start to associate. | Popot et al. (2003) and Charvolin et al. (2014) |
| 11.2 | tOmpA | A8-35 | SEC, FRET | In SEC analyses, tOmpA/A8-35 complexes migrate as larger and less monodisperse particles than tOmpA/$C_8E_4$ ones. Ternary tOmpA/A8-35/$C_8E_4$ complexes appear as large as tOmpA/A8-35 complexes, but as monodisperse as tOmpA/$C_8E_4$ ones. Removal of free A8-35 causes the formation of small aggregates, which redisperse upon adding back free APol. | Zoonens et al. (2007) |
| 11.3 | Cytochrome $bc_1$ | A8-35 | SANS, crystallization in aqueous solution, X-ray diffraction | The dispersity and stability of cytochrome $bc_1$/A8-35 complexes have been examined. Binary $bc_1$/A8-35 complexes did not yield any crystals under any of the experimental conditions tested. Ternary $bc_1$/A8-35/detergent ones yielded crystals containing both the protein and the APol, which diffracted to ~20-Å resolution. | Charvolin et al. (2014) |
| 11.4 | Bacteriorhodopsin (BR) | A8-35 | SAXS, crystallization *in meso*, X-ray diffraction | BR/A8-35 complexes were mixed with a 3D mesophase of mono-olein. Crystals formed, which diffracted to better than 2-Å resolution. The structure was solved to this resolution and shown to be identical to that of BR crystallized *in meso* from a detergent solution. The crystals do not contain any APol. | Polovinkin et al. (2014) |
| 11.5 | Bacteriorhodopsin from *Haloquadratum walsbyi* (HwBR) | SMA | Dynamic light scattering, crystallization *in meso*, X-ray diffraction | HwBR was expressed in *E. coli*, solubilized with SMA, purified by immobilized metal affinity chromatography followed by size-exclusion chromatography, and transferred to a lipidic mesophase, where it formed crystals diffracting to better than 2-Å resolution. The 2.0-Å structure thus obtained is virtually identical to a 2.2-Å structure obtained after transfer from a detergent solution. | Broecker et al. (2017) |

(Tribet et al. 1996; Gohon et al. 2008; Zoonens et al. 2014) (see Chap. 5). As observed for these proteins, efficient retention into solution following dilution of the detergent under its CMC required that the $bc_1$ be trapped with A8-35 at APol/protein mass ratios slightly higher than that eventually measured in the final complexes. The mass ratio of bound APol, however, is much lower for cytochrome $bc_1$ (~0.11 g bound A8-35 per g protein) than for BR (~2 g per g) (Gohon et al. 2008). This is consistent with the different distribution of masses between TM and extramembrane regions: in BR (27 kDa), the TM region represents ~75% of the mass of the protein, i.e. ~20 kDa, against only

~22% (~107 kDa) for the $bc_1$ complex (486 kDa in its native dimeric form). It is surprising, however, that the $bc_1$ dimer, with 26 TM $\alpha$-helices, should bind about the same amount of A8-35 (~53 to 54 kDa) as monomeric BR, which contains only seven helices + bound lipids (cf. Chap. 5, § 5.3.1.1). Given that these measurements are difficult to carry out with good accuracy (if only because MPs tend to oligomerize when separated from free APols, probably releasing APols; see Chap. 5), they clearly have to be taken as provisional.

Functional measurements show that replacing the detergent belt with A8-35 does not affect the electron transfer activity of the $bc_1$ complex (Charvolin et al. 2014).

Solution studies of $bc_1$/A8-35 complexes showed them to be essentially monodisperse. Upon SEC, they migrated as though they were slightly larger than $bc_1$/detergent (DDM) ones. This behavior is consistent with that observed for the calcium pump SERCA1a (A8-35 vs. DDM; Champeil et al. 2000), tOmpA (A8-35 vs. $C_8E_4$; Zoonens et al. 2005, 2007), and OmpX (A8-35 vs. diC$_6$PC; Catoire et al. 2010) (see Chaps. 5 and 9). In the case of cytochrome $bc_1$, a large MP with a relatively narrow TM "waist" (Fig. 11.12A), it seems likely that the earlier elution has more to do with interactions with the column than with an actual increase of hydrodynamic radius (for a discussion of these effects, see Zoonens et al. 2007; Gohon et al. 2008, and Chap. 5, Box 5.2).

Detailed studies have shown that, upon SEC, tOmpA/A8-35 complexes appear more broadly distributed than tOmpA/$C_8E_4$ ones, suggesting that the amount of bound surfactant varies more from one tOmpA/APol complex to another than it does in detergent solution (Zoonens et al. 2007) (Study 11.2; see Chaps. 5 and 9). In the case of cytochrome $bc_1$, the dispersity of MP/detergent and MP/A8-35 complexes appears to be similar (Fig. 11.13). This discrepancy is not too surprising: because a large

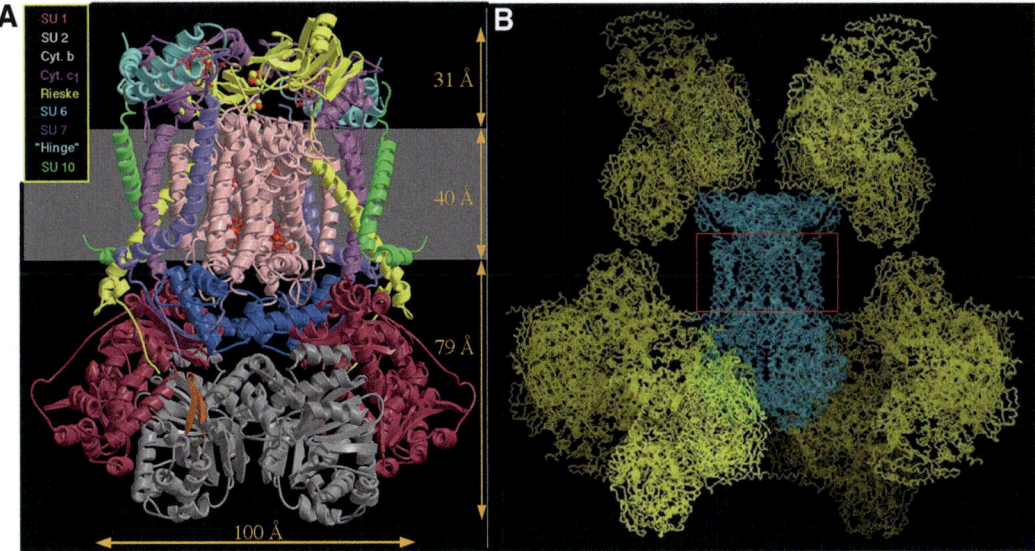

**Fig. 11.12** Structure and crystal packing of cytochrome $bc_1$. (**A**) Ribbon diagram of the chicken $bc_1$ complex dimer. The molecular twofold axis runs vertically between the two monomers. The key for the color coding of each subunit is given in the inset. Quinones, phospholipids, and detergent molecules are not shown for the sake of clarity. The presumed position of the membrane bilayer is represented by a gray band (From Zhang et al. 1998, Reprint by permission from Macmillan Publishers Ltd: Nature, © 1998). (**B**) Packing of beef heart $bc_1$ dimers in the $P6_522$ crystal form. The central dimer is in *green*, the surrounding ones in *yellow*. Note the protein-free space around the transmembrane region (in the *red rectangle*), onto the surface of which the detergent (dodecylmaltoside) is adsorbed. PDB structure 1BE3 from Iwata et al. (1998), figure prepared using the molecular graphics program "O" (Jones et al. 1991) (Courtesy of E.A. Berry).

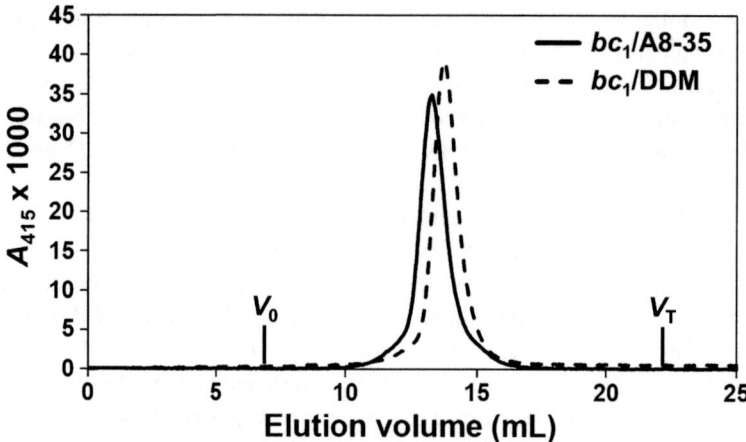

**Fig. 11.13** Comparison of the SEC elution profiles of $bc_1$/A8-35 (——) and $bc_1$/dodecylmaltoside (DDM) (– – –) complexes, analyzed at 415 nm, a wavelength at which only the protein absorbs. The equilibration and running buffer for all experiments was 20 mM Tris/HCl buffer, pH 8.5, 100 mM NaCl, 5 mM EDTA, supplemented, in the case of $bc_1$/DDM complexes, with 0.2 mM DDM. $V_0$ excluded volume, $V_T$ total volume (From Charvolin et al. 2014).

fraction of the surface of tOmpA is covered with surfactant and the protein has only small extramembrane loops, differences in the amount of bound APol from particle to particle can be expected to significantly affect the $R_S$ distribution. With cytochrome $bc_1$, on the contrary, most of the bulk of which is extramembrane, slight variations of the amount of APol surrounding the TM region of the protein are unlikely to strongly affect its apparent $R_S$, which is more sensitive to its overall length.

The homogeneity of $bc_1$/A8-35 particles, an essential factor in crystallization attempts, can be affected by two further factors: on the one hand, the biochemical stability of the complexes, and, in particular, the absence of monomerization; on the other, the propensity of $bc_1$ dimers to form higher-order oligomers. As regards the first point, substituting DDM with A8-35 drastically improved the biochemical stability of cytochrome $bc_1$ (Charvolin et al. 2014). Over the course of several weeks at 4 °C, however, a small degree of monomerization did develop. If previous experience with the homologous $b_6f$ complex (Breyton et al. 1997, and unpublished observations) and with GPCRs (Dahmane et al. 2009) is any guide, trapping some lipids along with the protein might be expected to improve stability. As regards oligomerization, observations with the $bc_1$ complex are reminiscent of those made with tOmpA (Zoonens et al. 2007), BR (Gohon et al. 2008), and OmpF (Arunmanee et al. 2014) (see Chaps. 5, § 5.2.1, and 12, § 12.3.1): namely the presence of a small excess of APol is needed to prevent $bc_1$ dimers from forming higher-order oligomers (Fig. 11.14). This effect has been attributed to the poorly dissociating character of APols (Chap. 5, § 5.2.1): upon removal of free APol, APol-complexed MPs tend to form small aggregates, which, in the case of tOmpA, redisperse upon adding back some free APol (Zoonens et al. 2007). The observations with the $bc_1$ complex, added to those with OmpF (Arunmanee et al. 2014), further support the view that this may be a general phenomenon. Experiments that benefit from monodispersity or require it, such as radiation scattering or crystallization in aqueous solutions, should therefore be carried out in the presence of a slight excess of APols. In the case of cytochrome $bc_1$, the required amount of free APol is much lower (~0.1 g free A8-35 per g of bound one) than in those of tOmpA (Zoonens et al. 2007) and BR (Gohon et al. 2008) (~1:1). Since the propensity to aggregate is determined by the competition between protein/protein and protein/APol interactions, the optimal ratio of free to bound APol that ensures monodispersity can be

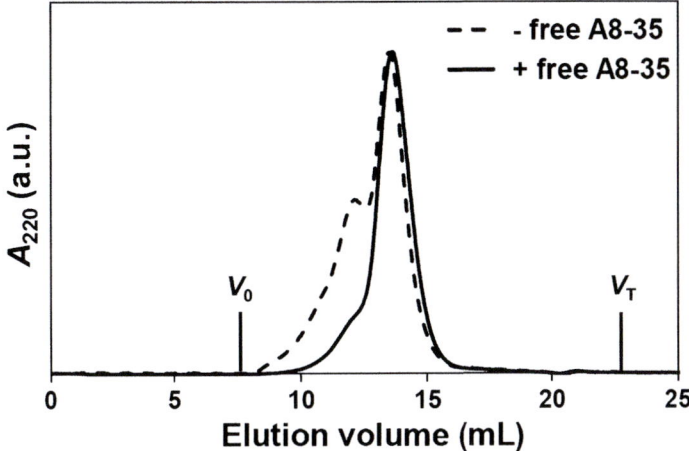

**Fig. 11.14** Size exclusion chromatography analysis of a preparation of cytochrome $bc_1$/A8-35 complexes purified on a sucrose gradient containing (——) or not (– – –) 0.2 g·L$^{-1}$ free A8-35. Note the shoulder present in the absence of free APol, due to the formation of small aggregates. Data have been normalized to the same maximal absorbance at 220 nm. $V_0$ excluded volume, $V_T$ total volume (From Charvolin et al. 2014).

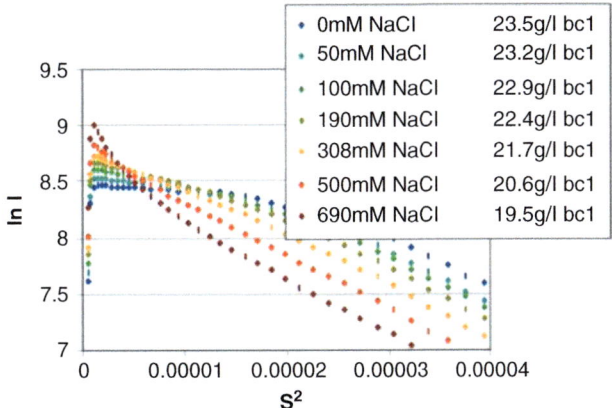

**Fig. 11.15** Guinier plots of SAXS data collected on a concentrated sample of cytochrome $bc_1$/A8-35 complexes as a function of salt concentration. Deviations from linearity in the small-angle region of the plot show the transition from repulsive interactions at low ionic strength (downward deflection) to aggregation at high ionic strength (upward deflection). The logarithm of the intensity $I$ measured as a function of the scattering angle $2\theta$ is plotted against $s^2 = (2\lambda^{-1} \cdot \sin\theta)^2$, where $\lambda$ is the wavelength of the X-rays (From Popot et al. 2003).

expected to vary depending on the tendency of the protein to self-associate, as well as on the nature of the APol and buffer conditions.

SANS (Charvolin et al. 2014) and SAXS (Popot et al. 2003) studies show that, as expected, interactions between the highly charged $bc_1$/A8-35 particles are repulsive at low salt concentration. They become attractive upon increasing the ionic strength and/or adding PEG. In the case of cytochrome $bc_1$, repulsive interactions disappear and attractive interactions develop above 300 mM salt in the absence of PEG (Fig. 11.15). Under these conditions, Guinier plots (see Chap. 9, Box 9.2) become linear, which is compatible with perfect monodispersity and neutral interactions. It cannot be excluded, however, that the onset of association masks lingering repulsive interactions. These observations suggest, nevertheless, that conditions can be found where the charged character of A8-35 ceases to prevent $bc_1$ complexes from interacting one with another. The shape of the $bc_1$ complex facilitates the formation of crystals in which the dimers interact exclusively via their extramembrane domains, without bringing TM regions into close proximity. Such is indeed the case of the crystals ($P6_522$) that are usually obtained under the conditions used in Study 11.3 (Zhang et al. 1998) (Fig. 11.12B). It seemed therefore reasonable to hope that crystallization of $bc_1$/A8-35 complexes might occur.

*Particles carrying a net charge do not crystallize easily*
(© 2018 by Francis Haraux)

Extensive attempts at crystallizing pure $bc_1$/A8-35 complexes nevertheless consistently yielded negative results. They were conducted either blindly, by using crystallization kits designed either for soluble or for membrane proteins, or by searching the crystallization space around those conditions that give good crystals in detergent solution (Zhang et al. 1998), while varying such factors as the ionic strength, temperature, pH, PEG nature and concentration, or the presence of traces of divalent cations. None of these approaches led to the formation of even "hopeful precipitates."

Better success was achieved using a stepwise approach. Starting from conditions that consistently yield crystals in the presence of detergent, the detergent phase was progressively enriched (some would say poisoned) by supplementing it with increasing ratios of APols, while adjusting the crystallization conditions so as to keep obtaining crystals. This approach did lead to the formation of ternary crystals containing the protein (based on their color and unit cell dimensions), A8-35 (based on the fluorescence observed when crystals were formed in the presence of FAPol$_{NBD}$, an NBD-labeled derivative of A8-35), and detergent (Fig. 11.16). The presence of detergent in the crystals was not directly established, but can be inferred from previous studies that show that, in mixtures containing a MP, detergent, and APols, the composition of the surfactant layer adsorbed onto the protein reflects that of APol/detergent mixed particles (Zoonens et al. 2007; Tribet et al. 2009; see Chap. 5, Box 5.3). On the basis of the mass ratio of $bc_1$/DDM in the starting material, the amounts of A8-35 and OG supplemented, and the assumption of a free distribution of the detergent between APol/detergent

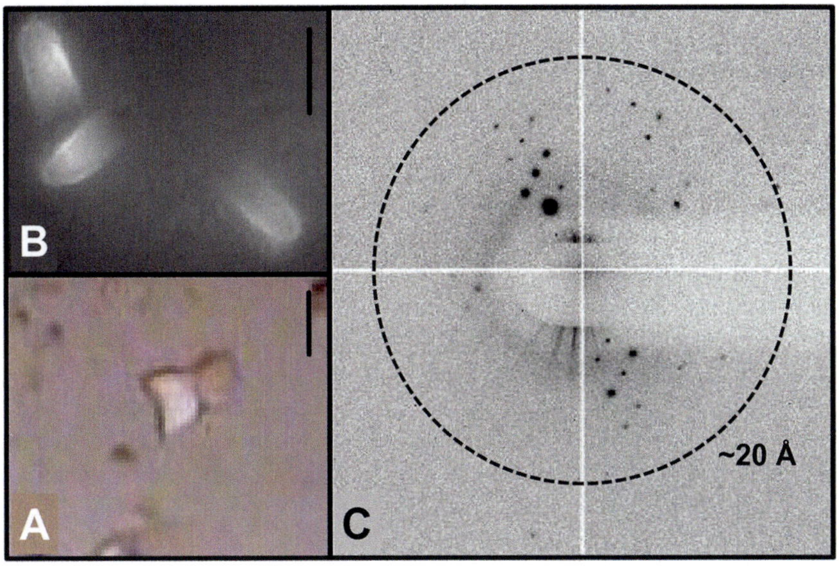

**Fig. 11.16** Crystallization of cytochrome $bc_1$/A8-35/detergent ternary complexes. (**A**) Crystals were obtained under the following conditions: sample, 17 g·L$^{-1}$ $bc_1$, 0.3 g A8-35 per g protein; reservoir, 12% PEG4k, 100 mM NaCl, 10 mM $n$-octyl-$\beta$-D-glucopyranoside. Samples were incubated for 10 days at 20 °C. Bar = 100 μm. (**B**) Crystals obtained under the same condition as in A, but after replacing plain A8-35 by FAPol$_{NBD}$, a fluorescent, NBD-labeled version thereof (Zoonens et al. 2007; Le Bon et al. 2014), observed by fluorescence microscopy. Bar = 100 μm. (**C**) Diffraction pattern recorded on a synchrotron light source (ESRF, beamline ID14-1) (From Charvolin et al. 2014).

mixed particles, soluble $bc_1$/A8-35/detergent complexes, and crystals, it can be estimated that the mass ratio of APol to detergent in the crystals most enriched in APols reached about 1:1 w/w (Charvolin et al. 2014). A protocol for the crystallization of $bc_i$/A8-35/detergent ternary complexes is described in § 11.4 (Protocol 11.1).

Why ternary complexes crystallized whereas pure $bc_1$/APol complexes did not can have several causes:

- First, diluting A8-35 with a neutral detergent effectively lowers the charge density of the surface of the surfactant layer, diminishing global electrostatic repulsion.
- Second, it offers a way for the system to locally avoid close approach between carboxylate groups of APol molecules adsorbed onto neighboring $bc_1$ dimers, since charged groups can move away from each other and be replaced by neutral detergent polar heads, much in the same way as small surfactants are thought to facilitate the formation of Type II crystals by substituting for bulkier detergent molecules to avoid steric clashes (§ 11.2.2.1).
- Third, ternary tOmpA/A8-35/detergent complexes, as already mentioned, appear more homogeneous, upon gel filtration, than binary tOmpA/A8-35 ones (Zoonens et al. 2007; see Chap. 9, § 9.3.2). This effect is probably due to the fact that MPs bind only a limited number of APol molecules and that the latter are relatively large and heterogeneous. This may prevent the mixture from relaxing to a state where each MP is surrounded by exactly the same amount of surfactant. Detergents may provide the "small change" that helps adjusting the surfactant layer to about the same size from one particle to another, obviously a favorable factor for crystallization.
- Fourth, it is worth noting that some EM (Althoff et al. 2011) and MD (Perlmutter et al. 2014) studies suggest that MP-bound A8-35 layers are not smooth, but bumpy (see Chaps. 5, § 5.3.3

and 12, § 12.3.3). The origin of this phenomenon is uncertain, and it has not been observed in all EM studies (see Liao et al. 2013, 2014; for discussions, see Huynh et al. 2014; Zoonens and Popot 2014, and Chap. 12, § 12.3.4.1). If real, it may conceivably interfere with crystallization. It would be of interest to examine whether mixed APol/detergent belts exhibit or not the same feature.

Whereas crystals of ternary complexes were obtained, the few of them that could be tested diffracted only to ~20-Å resolution. It is currently not known whether this represents an intrinsic limit or is simply due to the small number of crystals that could be examined and/or to difficulties in handling and/or freezing them. Nevertheless, the mere fact that 3D crystals of APol-trapped MPs could be obtained is an encouragement to pursue such attempts. Those could be carried out using either A8-35 or more recently developed APols with different chemical structures, e.g. those described in Diab et al. (2007a, b), Bazzacco et al. (2012), and Sharma et al. (2012), which can be used under a globally neutral or totally uncharged form (see Chap. 4). Of particular interest is the coming of age of non-ionic APols, whose latest versions have shown an excellent ability to trap and stabilize MPs (Bazzacco et al. 2012; Sharma et al. 2012).

Finally, one may note two other ways in which APols may help to obtain the crystallographic structure of MPs. First, it is now established that APol/detergent mixtures can (*i*) stabilize MPs as compared to pure detergent (Champeil et al. 2000) (see Chap. 5, § 5.6.1) and (*ii*) yield 3D crystals (Charvolin et al. 2014). APols therefore could possibly be used as additives, to stabilize MPs whose instability in pure detergent solution prevents crystals from growing to a sufficient size. Second, APols have proven an excellent medium in which to fold MPs to their native state (Chap. 6), as well as to synthesize them in vitro (Chap. 7). They could therefore be used as mere tools to produce the folded protein, whereas crystallization would be attempted after transferring the latter either to a detergent, a bicelle, or a lipidic mesophase (Fig. 11.17). The latter approach, which is described in the next section, has permitted to obtain crystals diffracting to <2 Å of BR transferred from A8-35 or SMA to a lipid mesophase (Polovinkin et al. 2014; Broecker et al. 2017).

### 11.3.2  Using Amphipols as a Vehicle to Crystallize Membrane Proteins in Meso

One of the best investigated examples of MP trapping and stabilization by APols is that of BR. Of relevance to the present section are the following observations, which have been described in the preceding chapters:

- BR can be transferred in its native state from BR/lipid/detergent complexes to various APols, an environment where it is functional (Tribet et al. 1996, 1997; Gohon et al. 2008; Bazzacco et al. 2012; Sharma et al. 2012; Dahmane et al. 2013) (Chap. 5).
- It can be renatured by transfer from sodium dodecyl sulfate to either A8-35 or non-ionic APols either in the presence or absence of lipids (Pocanschi et al. 2006; Bazzacco et al. 2012; Dahmane et al. 2013) (Chap. 6) or synthesized in vitro in the presence of non-ionic APols (Bazzacco et al. 2012) (Chap. 7).
- In the course of these works, it was noted that archaebacterial lipids displaced by detergents rebind to BR upon its transfer to APols, with functional consequences on the photocycle (Gohon et al. 2008; Dahmane et al. 2013).

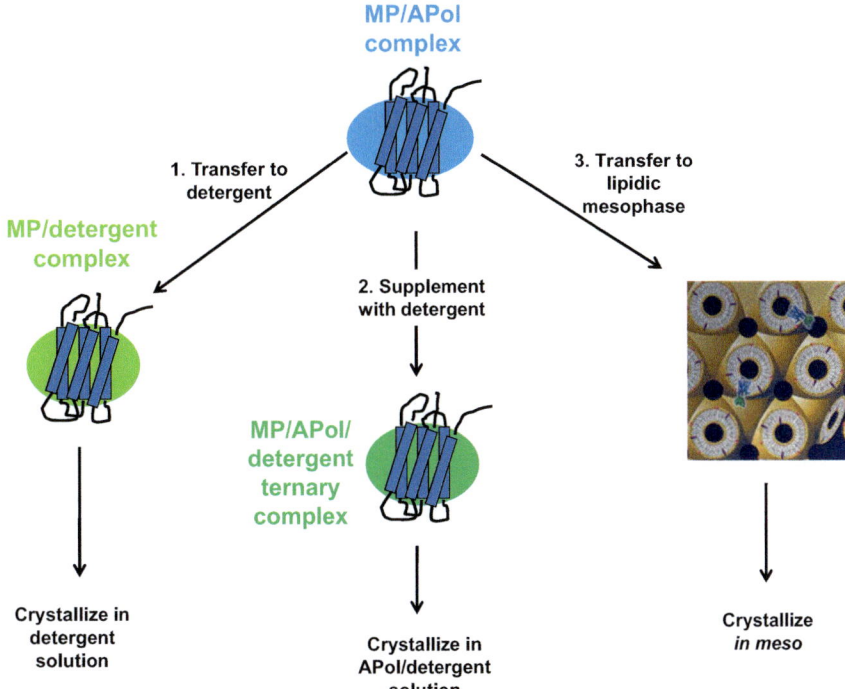

**Fig. 11.17** Three routes for crystallizing MPs starting from MP/APol complexes. In Route 1, APol-trapped MPs, obtained by any of the routes shown in Fig. 10.1, are transferred to detergent solution and crystals obtained in any of the traditional ways. In Route 2, the complexes are supplemented with enough detergent to make crystallization of the ternary complexes possible. In Route 3, MP/APol complexes are directly inserted into a lipidic mesophase. Route 1 has not yet been demonstrated to work, but the only unknown is the extent to which APols must be eliminated so that they do not interfere with crystallization. The (relative) success of Route 2 using ~1:1 APol/detergent mass ratios suggests that extensive removal will not be needed. Route 2 has been shown to permit the formation of crystals of ternary complexes, but how well they can be made to diffract remains to be established (§ 11.3.1). Route 3 has been shown to lead to the formation of crystals that diffract to high resolution (§ 11.3.2) (The mesophase cartoon is reprinted with permission from Li et al. 2013, © 2013 American Chemical Society).

- As described above (Chap. 6; for reviews, see Kleinschmidt and Popot 2014; Popot 2014; Zoonens and Popot 2014), APol-assisted folding has been extended to many other MPs, among which the pharmacologically important class of GPCRs.
- BR is a classical model for MP studies and has been instrumental in the development of MP crystallization in detergent solution (Michel and Oesterhelt 1980; Michel 1982), in lipidic mesophases (Landau and Rosenbusch 1996; Takeda et al. 1998), or from bicelles (Faham and Bowie 2002; see Chap. 3, § 3.2.2).

In Study 11.4, the laboratory of Valentin Gordeliy used BR/A8-35 complexes to develop an approach for crystallizing APol-trapped MPs by direct transfer to a lipidic mesophase.

Following solubilization of *H. salinarum* purple membrane with OG, BR was trapped in A8-35, along with archaebacterial lipids, using a ~2.5× excess of APol over the amount that actually binds to the protein (Gohon et al. 2008). Crystallization attempts were carried out at 20 °C either in the presence or after removal of the excess APol. Experiments in which the crude sample, containing free APol, was mixed directly with mono-olein (MO) did not yield any high-quality crystals. On the contrary, after

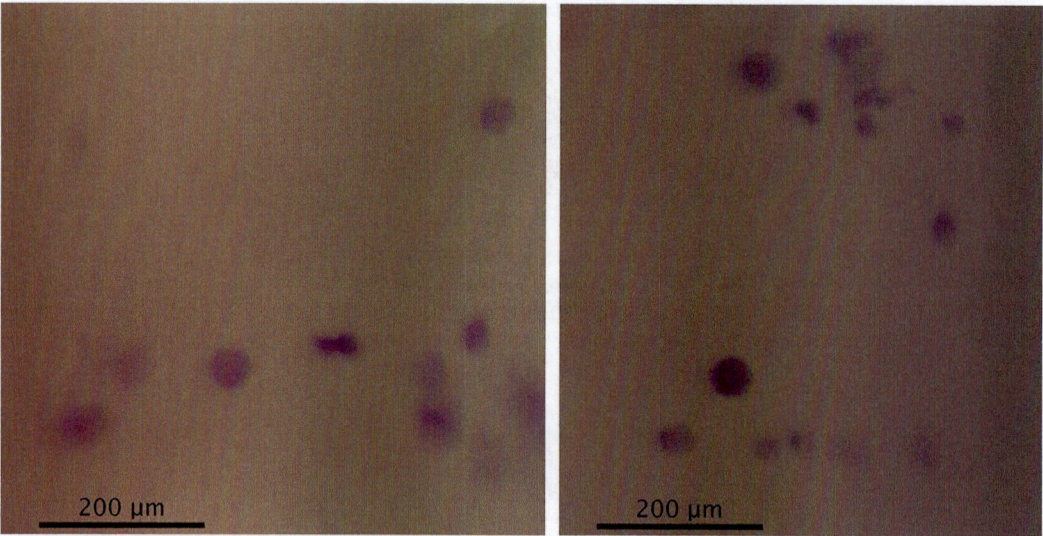

**Fig. 11.18** Crystals of A8-35-trapped BR grown *in meso* following transfer of BR/A8-35 complexes to mono-olein in the absence of free APol (From Polovinkin et al. 2014).

most of the free APol had been removed by several cycles of dilution/ultracentrifugation (Gohon et al. 2008), well-shaped crystals of $P6_3$ symmetry were obtained (Fig. 11.18), which diffracted beyond 2.0 Å (Table 11.2). The crystals were of Type I (see § 11.2), being formed of stacked planar layers, each layer resembling the native purple membrane, where BR trimers are organized in a hexagonal 2D crystalline lattice (Henderson 1975).

*Crystallizing membrane proteins in a lipidic cubic phase*
(© *2018 by Francis Haraux*)

The crystallization conditions for APol-trapped BR are nearly the same as those used for crystallizing *in meso* BR solubilized in octylglucoside. This suggests that, at least up to a certain concentration, the presence of A8-35 in the mesophase matrix does not interfere with MP crystallization. This view is supported by the fact that, according to SAXS data, MO/APol/buffer (20 mM $Na_2HPO_4$/$KH_2PO_4$, pH 5.6) systems form a cubic phase with Pn3m symmetry (Lindblom and Rilfors 1989), and no significant lattice parameter changes are observed upon increasing the APol/MO mass ratio up to 7% with respect to MO (Fig. 11.19). The best crystals were obtained at an APol/MO mass

**Table 11.2** Data collection and refinement statistics for BR crystals grown by transfer from A8-35 to mono-olein mesophase (From Polovinkin et al. 2014).

| Data collection | |
|---|---|
| Space group | $P6_3$ |
| Cell dimensions | |
| $a, b, c$ (Å) | 60.999, 60.999, 108.174 |
| $\alpha, \beta, \gamma$ (°) | 90, 90, 120 |
| Resolution (Å) | 52.83–2.00 (2.11–2.00)[a] |
| $R_{merge}$ (%) | 12.1 (81.4) |
| $I/\sigma_I$ | 9.7 (2.2) |
| Completeness (%) | 99.9 (100.0) |
| Redundancy | 6.4 (6.5) |
| **Refinement** | |
| Resolution (Å) | 52.83–2.00 |
| Number of reflections | 15,418 |
| $R_{work}/R_{free}$ | 17.9/20.9% |
| Number of atoms | |
| Protein | 1661 |
| Retinal | 20 |
| Water | 14 |
| $B$-factors (Å$^2$) | |
| Protein | 20.7 |
| Ligand/ion | 13.7 |
| Water | 19.1 |
| RMS deviations | |
| Bond lengths (Å) | 0.008 |
| Bond angles (°) | 1.5 |

[a]Highest resolution shell is shown in parentheses

ratio of ~5%. Increasing the APol/MO mass ratio above 7% induces a transition to a cubic phase with Ia3d symmetry, and then the phase transforms to a lamellar phase at mass ratios higher than 20% (Sintsov et al. unpublished results). This is likely the reason why no high-quality BR crystals were observed in the presence of an excess of APol. Indeed, under the conditions used, the APol/MO mass ratio was ~12%, which is above the ratio compatible with a Pn3m cubic phase.

As already mentioned, removal of free APol from preparations of MP/APol complexes, including BR/A8-35 ones, leads to a variable degree of aggregation (Zoonens et al. 2007; Gohon et al. 2008; Arunmanee et al. 2014; Charvolin et al. 2014) (cf. Chaps. 5, § 5.2.1 and 12, § 12.3). Indeed, after removal of free APol, the main peak observed in the size exclusion chromatography elution profile of the BR/A8-35 complexes became slightly broader and eluted slightly earlier than before removal (Polovinkin et al. 2014). This behavior is similar to that observed with tOmpA/A8-35 complexes (Zoonens et al. 2007) (see Chap. 5, Fig. 5.6). It reflects the formation of small oligomers, as well as a very small amount of larger aggregates, which elute in the void volume. At the time of mixing with MO, BR/A8-35 complexes depleted of free APol therefore are not homogeneous. This does not prevent BR from reorganizing into well-ordered 3D crystals once inserted into the lipidic mesophase. Similarly, in the presence of lipopolysaccharide, OmpF/A8-35 filaments can evolve toward 2D crystals (Arunmanee et al. 2014).

The structure of BR crystallized from APol-trapped complexes was solved to 2.0-Å resolution (Fig. 11.20 and Table 11.2). Besides the atoms of the protein, the structure reveals the covalently bound all-*trans*-retinal (Fig. 11.20A). The cluster of hydrogen-bonded water molecules (W401, W402,

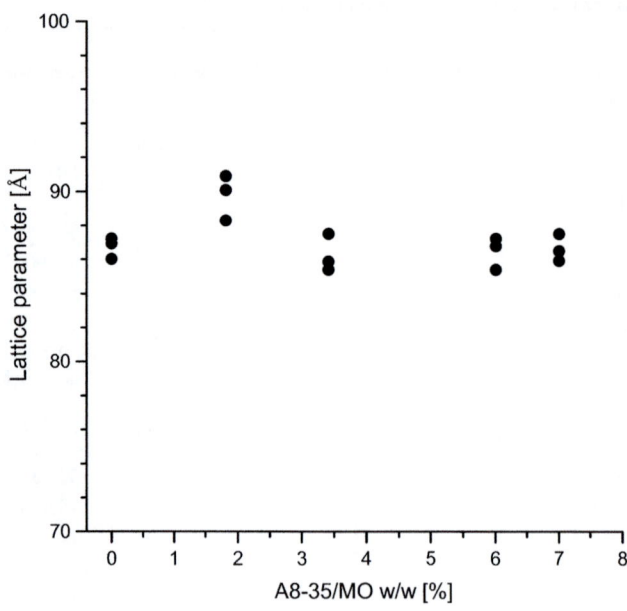

**Fig. 11.19** Variation of the cubic Pn3m phase lattice parameter with A8-35 concentration in mono-olein (MO)/APol/buffer (1.5 M $Na_2HPO_4/KH_2PO_4$, pH 5.6). Data were collected on frozen samples at $-100\,°C$. Three independent SAXS experiments were performed at each APol/MO ratio (From Polovinkin et al. 2014).

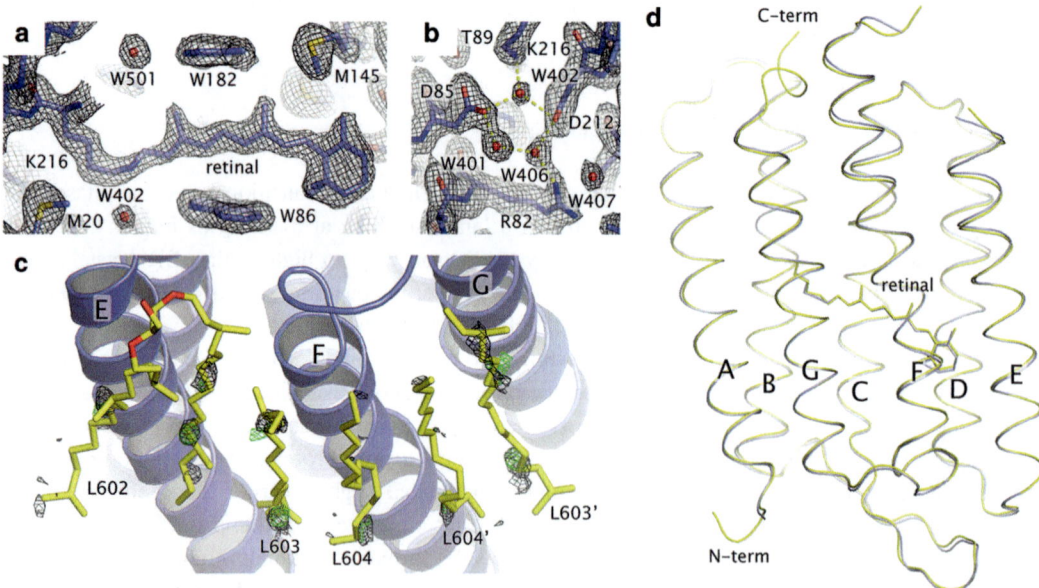

**Fig. 11.20** Structure of bacteriorhodopsin (BR) in 3D crystals grown from BR/A8-35 complexes. (**A**) $2F_o$-$F_c$ electron density map around the retinal drawn at the level of 1.5 $\sigma$. (**B**) Position of the water molecules close to the retinal's Schiff base. The configuration is identical to that observed in crystals grown by traditional *in meso* technique. (**C**) Residual $2F_o$-$F_c$ (*black*) and $F_o$-$F_c$ (*green*) electron densities in the lipid tail region at the levels of 1 and 3 $\sigma$, respectively. The $F_c$ coefficients and phases were calculated in the absence of any lipid molecules in the model. The densities are overlaid with the lipid molecule positions from a 1.55-Å resolution structure (*yellow*, PDB ID 1C3W; Luecke et al. 1999). (**D**) Comparison of the overall BR crystal structure obtained from BR/A8-35 complexes (*gray*) with the structure observed in crystals grown by traditional *in meso* technique (*yellow*, PDB ID 1C3W; Luecke et al. 1999). The structures are essentially identical. The root mean square deviation (RMSD) of atomic positions is 0.86 Å for all atoms and 0.38 Å for backbone atoms (From Polovinkin et al. 2014).

and W406) close to the retinal's Schiff base is easily discernible (Fig. 11.20B). These features are absolutely identical to those observed in crystals grown by the traditional *in meso* technique, starting from BR in detergent solution (Luecke et al. 1999; Lanyi 2004). Apart from the electron densities corresponding to BR or ordered water molecules, some residual densities are seen in the lipid tail region. The position of these densities corresponds closely to that of the lipid molecules identified at 1.55-Å resolution in crystals obtained by classical *in meso* crystallization (Luecke et al. 1999). Typical densities along the extracellular side of TM helices E, F, and G are shown in Fig. 11.20C. It is highly likely that these densities also correspond to lipids and not to A8-35 molecules. This is consistent with the observation that purple membrane lipids initially solubilized along with BR are present in BR/A8-35 complexes (Gohon et al. 2008) and modulate the photocycle of BR (Dahmane et al. 2013). Their presence is probably essential to the formation of the crystals.

Apart from those that can be attributed to lipids, no unaccounted-for densities are observed. Thus, it seems that molecules of A8-35 are not present in the crystals grown from APol-trapped BR, or, if they are, that they are significantly disordered. Taking into account that the average molecular weight of A8-35 is ~4.3 kDa (Giusti et al. 2014; Chap. 4, § 4.3.1.2.1) and that the packing of BR molecules in Type I crystals is extremely tight (Pebay-Peyroula et al. 1997; Luecke et al. 1999), the presence of APol molecules in the crystals seems highly unlikely. Indeed, the lattice constants are essentially the same for the crystals obtained from BR/APol complexes (61.0, 61.0, and 108.2 Å; Polovinkin et al. 2014) and from detergent-solubilized BR (61.0, 61.0, and 110.0 Å; Borshchevskiy et al. 2011). It has been shown before that MPs can be transferred to lipid bilayers upon exposure of MP/APol complexes either to lipid vesicles (Nagy et al. 2001) or to black films (Pocanschi et al. 2006) (Chap. 5, § 5.7). It seems safe to assume that upon mixing BR/A8-35 complexes with mono-olein, the APol dissociates from BR and dissolves in the mesophase, probably adsorbing at the interface between water channels and lipid bilayer. MP molecules and lipids likely diffuse on their own in the lipid bilayer and assemble into crystals without interference by the APol, as schematized in Fig. 11.21. The high electrostatic charge density of A8-35 therefore cannot oppose the formation of protein crystal contacts, as may be the case during conventional vapor diffusion crystallization of MP/A8-35 complexes (§ 11.3.1).

In conclusion, using the *in meso* crystallization approach, BR can be crystallized from an APol-trapped state. The crystals obtained are of high X-ray diffracting quality, suitable for high-resolution protein structure determination. The lipid bilayer of the crystallization matrix likely displaces A8-35 molecules from the TM surface of the protein and dissolves them. Thereby, they do not interfere with crystallization. The crystals obtained from A8-35-trapped BR diffract to high resolution, and the structure was solved to 2 Å. There is essentially no difference between BR structures obtained from these crystals and from crystals grown from the protein solubilized in detergents and transferred to the mesophase, and the crystals do not appear to contain any APol. In these experiments, APols therefore act as a vehicle to deliver the MP to the mesophase and, probably, play no role in the crystallization itself.

The approach to MP crystallization explored in Study 11.4 could well be very general. Indeed, similar results have been recently obtained following transfer to a mesophase of SMA-trapped BR from *Haloquadratum walsbyi* (Broecker et al. 2017; Study 11.5) and of *H. salinarum* BR trapped in nanodiscs (Nikolaev et al. 2017). Given that (*i*) crystallization *in meso* has proven applicable to a large variety of MPs (Caffrey 2009; Cherezov 2011; Ishchenko et al. 2014, 2017; Caffrey 2015) (Fig. 11.11), and (*ii*) APols have proven efficient both to stabilize MPs in aqueous solution (Chap. 5) and to produce them, either by assisting their folding from a denatured state, such as that obtained by solubilization of inclusion bodies in dodecyl sulfate or urea (Chap. 6), or by acting as the host medium during cell-free synthesis (Chap. 7), combining APol-assisted production and/or stabilization of MPs with *in meso* crystallization seems to offer a promising new route to MP crystallization.

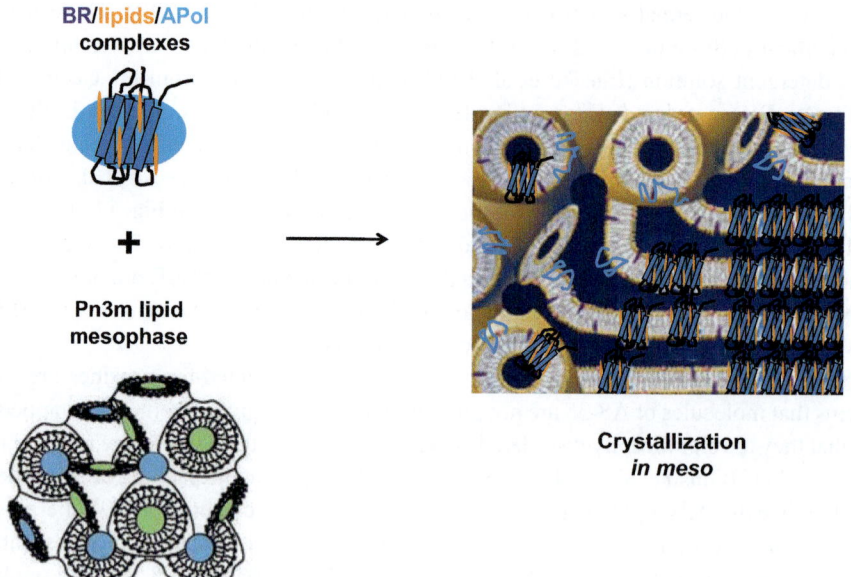

**BR/lipids/APol complexes**

**+**

**Pn3m lipid mesophase**

⟶

**Crystallization in meso**

**Fig. 11.21** Hypothetical scheme of the behavior of A8-35 (*light blue*), BR (*purplish*), and lipids (*orange*) upon transfer to a lipid cubic mesophase (*yellow*). In this scheme, it has been assumed that the archaebacterial lipids that are known to be associated to BR in BR/APol/lipid preparations (Dahmane et al. 2013; Gohon et al. 2008) do not dissociate upon transfer to mono-olein and diffuse in the mesophase along with the protein. However, there are currently no experimental data to exclude that they may dissociate, diffuse on their own, and rebind only during the formation of the crystals. A8-35 molecules (*light blue scribbles*) are presumed to dissociate from BR/lipid complexes and to diffuse independently, presumably adsorbed at the lipid/water interface. Crystallographic data indicate that they are excluded from the crystals (Polovinkin et al. 2014). Mesophase cartoons are borrowed from Cherezov et al. (2006) and Li et al. (2013).

### 11.3.3  Challenges and Prospects

The general impression one derives from the data analyzed in this chapter is that, whereas APols to date have not permitted to solve any new crystallographic MP structure, they may hold an interesting potential that is only starting to be investigated:

1. Experiments with cytochrome $bc_1$/A8-35 complexes have led to the conclusion that A8-35 itself is unlikely to lead to the formation of crystals of binary MP/APol complexes, but that it can be used in ternary MP/APol/detergent complexes, in which the APol may represent as much as half the mass of the protein-bound surfactant, and that these complexes can crystallize. It is much to be regretted that circumstances made it impossible to establish up to which resolution such crystals can be made to diffract. Current results, however, suggest that APols, because they can be stabilizing even when used in mixtures with detergents, could be tentatively used as additives to stabilize MPs that are too unstable in solutions of pure detergent to yield good-quality crystals.

2. The same experiments strongly suggest that traces of APols, even of charged APols such as A8-35, can be tolerated in solutions used for crystallization of MP/detergent complexes. This implies that preparations of APol-trapped MPs can be transferred to detergent solutions for the purpose of crystallization without the need for an extensive removal of traces of APols.

3. It is remarkable that crystals can be formed from ternary $bc_1$/A8-35/detergent complexes, in which the surfactant belt carries a high density of surface charges, whereas no crystallographic MP structure has ever been obtained using pure detergents carrying a net charge. On the one hand, this may suggest that such detergents could possibly be used, if need be, in combination with neutral or zwitterionic ones. On the other, it bodes well for the use of non-ionic APols for MP crystallization. Nothing is preventing such studies to be carried out but for the current absence of large-scale production of these molecules ... and the will of funding agencies.

4. Particularly promising are the results achieved upon transferring BR from its complexes with A8-35 to a mono-olein cubic phase. As tens of MPs have been shown to survive the transfer from MP/detergent complexes to lipidic mesophases, in which they crystallize, it is reasonable to expect that many MPs purified, folded, or produced in APols will do the same. Given the high yields observed upon folding MPs in general and GPCRs in particular using APols (Chap. 6), APol-assisted folding from solubilized inclusion bodies followed by transfer to lipidic mesophases could well provide a very useful novel route for MP crystallization, particularly of those MPs that do not stand well being exposed to "non-denaturing" detergents.

## 11.4   Protocol: Crystallizing $bc_1$/A8-35/Detergent Complexes from Aqueous Solutions

Given the stabilizing properties of APols, crystallizing MP/APol complexes could give access to the structure of MPs that cannot be handled in detergents. It has proven itself, however, to be a very frustrating endeavor. Still, some positive results have been obtained. Whereas attempts at crystallizing MPs trapped in pure A8-35 have failed, partial success has been achieved with ternary complexes (§ 11.3.1). We describe here a protocol for the crystallization of cytochrome $bc_1$/A8-35/detergent complexes. The purified $bc_1$ is first pre-crystallized from detergent solution. Microcrystals are collected and stored. After redissolution, they provide the starting material for crystallization attempts. The latter are conducted in a mixed APol/detergent environment. In the following, the protocol itself is printed in roman, whereas comments are in italics and marked with a pointing hand (☞).

### 11.4.1   Pre-crystallization

*☞At low ionic strength, the $bc_1$ complex from beef heart mitochondria purified according to* Berry *et al.* (1991) *is only sparingly soluble in detergent solution, especially if the pH is slightly acidic. Precipitated microcrystals can be collected at high concentration at low ionic strength and buffer concentration, so that, following redissolution, crystallization conditions will be primarily determined by the nature and concentration of precipitants, surfactants, and additives rather than by those of the solution the protein was in before precipitation.*

**Procedure**
1. Beef heart mitochondria are purified according to Smith (1967). At the end of $bc_1$ purification, conducted according to Berry et al. (1991), run a sizing column to remove aggregates and

exchange the buffer to 20 mM KMOPS, pH 7.2, 100 mM NaCl, 0.5 mM EDTA, $0.1 \text{ g} \cdot \text{L}^{-1}$ DDM. Pool the peak fractions and concentrate to $\sim100 \text{ µM}$.

2. Titrate with the following precipitant buffer: 100 mM KMES, pH 6.4, $0.1 \text{ g} \cdot \text{L}^{-1}$ DDM, 10% w/w PEG 4 k, 0.5 mM EDTA. Turbidity appears after 0.3–0.5 volumes have been added. Spin down, transfer the supernatant to a clean tube, and discard the first small pellet. This step eliminates aggregated complexes. Under stirring, add more precipitant ($\sim0.2$ original volumes) and incubate several hours at 4 °C to let precipitate. Microscopic examination reveals that the turbidity results from the presence of minute hexagonal crystals. Centrifuge again so as to obtain a large pellet, while leaving some color in the supernatant (if the supernatant is very red, add more precipitant and spin down again). Decant the pellet and discard the supernatant.

3. Weigh the pellet, and resuspend in $\sim10$ pellet volumes of 10 mM KMES, pH 6.4, 1% w/w PEG 4 k, 0.5 mM EDTA, $0.1 \text{ g} \cdot \text{L}^{-1}$ DDM. Centrifuge and resuspend the pellet in about ten volumes of cold distilled water. This preparation is then divided into aliquots and can be stored for several weeks at 4 °C or for longer periods at or below $-20$ °C.

☞*We suppose that resuspending in water before freezing minimizes damage due to peroxide and/or free radicals from PEG, but have not documented this point.*

## 11.4.2    Crystallization of Ternary Protein/Amphipol/Detergent Complexes

**Procedure**

1. The pre-crystallized $bc_1$ is resuspended at $10 \text{ g} \cdot \text{L}^{-1}$ in a solution of 100 mM Tris/HCl, pH 7.5, containing increasing concentrations of A8-35 (the best crystals are obtained at an A8-35/$bc_1$ ratio of 0.3 g per g), and incubated on ice overnight.

2. On the following day, the sample is ultracentrifuged at 72,000 rpm for 12 min in the TLA 100.2 rotor of a Beckman TL100 centrifuge so as to pellet the aggregated material. The supernatant is used directly to set up crystallization drops. The drops are set at room temperature by combining the $bc_1$ sample, at a 1:1 v/v ratio, with increasing concentrations of PEG, at low ionic strength. As a first screen, the concentrations of NaCl and PEG 4 k were varied in the absence of added detergent: sucrose, 30 mM; KMES, pH 6.4, 100 mM; NaCl, 60–120 mM; PEG 4 k, 4–12% w/w.

   ☞*Although hits were readily obtained in the absence of added detergent (but presence of residual DDM), the size of the crystals increased upon including 5–15 mM n-octyl-β-D-glucoside (OG) in the precipitant solution (which is under the CMC of ~25 mM). Under these conditions, it has been estimated that the ratio of detergent (OG + residual DDM) to A8-35 in the crystals most enriched in APol is ~1:1 w/w* (Charvolin et al. 2014).

3. Crystals were obtained after 10 days. They had a wedge shape and often grew head-to-tail. They measured about 120 µm in length.

   Protocol prepared by Martin Picard based on Charvolin et al. (2014), unpublished data, and laboratory notes, with comments added.

# References

Agah, S., Faham, S. (2012) Crystallization of membrane proteins in bicelles. *Methods Mol. Biol.* **914**:3–16.

Althoff, T., Mills, D.J., Popot, J.-L., Kühlbrandt, W. (2011) Assembly of electron transport chain components in bovine mitochondrial supercomplex $I_1III_2IV_1$. *EMBO J.* **30**:4652–4664.

Arunmanee, W., Harris, J.R., Lakey, J.H. (2014) Outer membrane protein F stabilised with minimal amphipol forms linear arrays and LPS-dependent 2D crystals. *J. Membr. Biol.* **247**:949–956.

Bazzacco, P., Billon-Denis, E., Sharma, K.S., Catoire, L.J., Mary, S., Le Bon, C., Point, E., Banères, J.-L., Durand, G., Zito, F., Pucci, B., Popot, J.-L. (2012) Non-ionic homopolymeric amphipols: Application to membrane protein folding, cell-free synthesis, and solution NMR. *Biochemistry* **51**:1416–1430.

Berger, B.W., Gendron, C.M., Robinson, C.R., Kaler, E.W., Lenhoff, A.M. (2005) The role of protein and surfactant interactions in membrane-protein crystallization. *Acta Crystallogr. D* **61**:724–730.

Berry, E.A., Guergova-Kuras, M., Huang, L.-S., Crofts, A.R. (2000) Structure and function of cytochrome $bc_1$ complexes. *Annu. Rev. Biochem.* **69**:1005–1075.

Berry, E.A., Huang, L.-S., DeRose, V. (1991) Ubiquinol-cytochrome $c$ oxidoreductase from higher plants. Isolation and characterization of the $bc_1$ complex from potato tuber mitochondria. *J. Biol. Chem.* **266**:9064–9077.

Binz, H.K., Stumpp, M.T., Forrer, P., Amstutz, P., Plückthun, A. (2003) Designing repeat proteins: well-expressed, soluble and stable proteins from combinatorial libraries of consensus ankyrin repeat proteins. *J. Mol. Biol.* **332**:489–503.

Boersma, Y.L., Plückthun, A. (2011) DARPins and other repeat protein scaffolds: advances in engineering and applications. *Curr. Opin. Biotechnol.* **22**:849–857.

Bogorodskiy, A., Frolov, F., Mishin, A., Round, E., Polovinkin, V., Cherezov, V., Gordeliy, V., Büldt, G., Gensch, T., Borshchevskiy', V. (2015) Nucleation and growth of membrane protein crystals *in meso*–A fluorescence microscopy study. *Cryst. Growth Des.* **15**:5656–5660.

Borshchevskiy, V.I., Round, E.S., Popov, A.N., Büldt, G., Gordeliy, V.I. (2011) X-ray-radiation-induced changes in bacteriorhodopsin structure. *J. Mol. Biol.* **409**:813–825.

Boutet, S., Lomb, L., Williams, G.J., Barends, T.R.M., Aquila, A., Doak, R.B., Weierstall, U., DePonte, D.P., Steinbrener, J., Shoeman, R.L., Messerschmidt, M., Barty, A., White, T.A., Kassemeyer, S., Kirian, R.A., Seibert, M.M., Montanez, P.A., Kenney, C., Herbst, R., Hart, P., Pines, J., Haller, G., Gruner, S.M., Philipp, H.T., Tate, M. W., Hromalik, M., Koerner, L.J., Bakel, N., Morse, J., Ghonsalves, W., Arnlund, D., Bogan, M.J., Caleman, C., Fromme, R., Hampton, C.Y., Hunter, M.S., Johansson, L.C., Katona, G., Kupitz, C., Liang, M., Martin, A.V., Nass, K., Redecke, L., Stellato, F., Timneanu, N., Wang, D., Zatsepin, N.A., Schafer, D., Defever, J., Neutze, R., Fromme, P., Spence, J.C.H., Chapman, H.N., Schlichting, I. (2012) High-resolution protein structure determination by serial femtosecond crystallography. *Science* **337**:362–364.

Bowie, J.U. (2001) Stabilizing membrane proteins. *Curr. Opin. Struct. Biol.* **11**:397–402.

Breyton, C., Tribet, C., Olive, J., Dubacq, J.-P., Popot, J.-L. (1997) Dimer to monomer conversion of the cytochrome $b_6f$ complex: causes and consequences. *J. Biol. Chem.* **272**:21892–21900.

Broecker, J., Eger, B.T., Ernst, O.P. (2017) Crystallogenesis of membrane proteins mediated by polymer-bounded lipid nanodiscs. *Structure* **25**:384–392.

Caffrey, M. (2003) Membrane protein crystallization. *J. Struct. Biol.* **142**:108–132.

Caffrey, M. (2009) Crystallizing membrane proteins for structure determination: use of lipidic mesophases. *Annu. Rev. Biophys.* **38**:29–51.

Caffrey, M. (2011) Crystallizing membrane proteins for structure-function studies using lipidic mesophases. *Biochem. Soc. Trans.* **39**:725–732.

Caffrey, M. (2015) A comprehensive review of the lipid cubic phase or *in meso* method for crystallizing membrane and soluble proteins and complexes. *Acta Crystallogr. F* **71**:3–18.

Catoire, L.J., Zoonens, M., van Heijenoort, C., Giusti, F., Guittet, E., Popot, J.-L. (2010) Solution NMR mapping of water-accessible residues in the transmembrane β-barrel of OmpX. *Eur. Biophys. J.* **39**:623–630.

Chae, P.S., Rasmussen, S.G.F., Rana, R., Gotfryd, K., Chandra, R., Goren, M.A., Kruse, A.C., Nurva, S., Loland, C.J., Pierre, Y., Drew, D., Popot, J.-L., Picot, D., Fox, B.G., Guan, L., Gether, U., Byrne, B., Kobilka, B.K., Gellman, S.H. (2010) Maltose-neopentyl glycol (MNG) amphiphiles for solubilization, stabilization and crystallization of membrane proteins. *Nat. Methods* **7**:1003–1008.

Champeil, P., Menguy, T., Tribet, C., Popot, J.-L., le Maire, M. (2000) Interaction of amphipols with the sarcoplasmic reticulum $Ca^{2+}$-ATPase. *J. Biol. Chem.* **275**:18623–18637.

Chapman, H.N., Fromme, P., Barty, A., White, T.A., Kirian, R.A., Aquila, A., Hunter, M.S., Schulz, J., DePonte, D.P., Weierstall, U., Doak, R.B., Maia, F.R.N.C., Martin, A.V., Schlichting, I., Lomb, L., Coppola, N., Shoeman, R.L., Epp, S.W., Hartmann, R., Rolles, D., Rudenko, A., Foucar, L., Kimmel, N., Weidenspointner, G., Holl, P., Liang, M.,

Barthelmess, M., Caleman, C., Boutet, S., Bogan, M.J., Krzywinski, J., Bostedt, C., Bajt, S., Gumprecht, L., Rudek, B., Erk, B., Schmidt, C., Hömke, A., Reich, C., Pietschner, D., Strüder, L., Hauser, G., Gorke, H., Ullrich, J., Herrmann, S., Schaller, G., Schopper, F., Soltau, H., Kühnel, K.-U., Messerschmidt, M., Bozek, J.D., Hau-Riege, S. P., Frank, M., Hampton, C.Y., Sierra, R.G., Starodub, D., Williams, G.J., Hajdu, J., Timneanu, N., Seibert, M.M., Andreasson, J., Rocker, A., Jönsson, O., Svenda, M., Stern, S., Nass, K., Andritschke, R., Schröter, C.-D., Krasniqi, F., Bott, M., Schmidt, K.E., Wang, X.-Y., Grotjohann, I., Holton, J.M., Barends, T.R.M., Neutze, R., Marchesini, S., Fromme, R., Schorb, S., Rupp, D., Adolph, M., Gorkhover, T., Andersson, I., Hirsemann, H., Potdevin, G., Graafsma, H., Nilsson, B., Spence, J.C.H. (2011) Femtosecond X-ray protein nanocrystallography. *Nature* **470**:73–77.

Charvolin, D., Picard, M., Huang, L.-S., Berry, E.A., Popot, J.-L. (2014) Solution behavior and crystallization of cytochrome $bc_1$ in the presence of amphipols. *J. Membr. Biol.* **247**:981–996.

Cherezov, V. (2011) Lipidic cubic phase technologies for membrane protein structural studies. *Curr. Opin. Struct. Biol.* **21**:559–566.

Cherezov, V., J. C, Papiz, M.Z., Caffrey, M. (2006) Room to move: crystallizing membrane proteins in swollen lipidic mesophases. *J. Mol. Biol.* **357**:1605–1618.

Cherezov, V., Rosenbaum, D.M., Hanson, M.A., Rasmussen, S.G., Thian, F.S., Kobilka, T.S., Choi, H.J., Kuhn, P., Weis, W.I., Kobilka, B.K., Stevens, R.C. (2007) High-resolution crystal structure of an engineered human $\beta_2$-adrenergic G protein-coupled receptor. *Science* **318**:1258–1265.

Cowan, S.W., Schirmer, T., Rummel, G., Steiert, M., Ghosh, R., Pauptit, R.A., Jansonius, J.N., Rosenbusch, J.P. (1992) Crystal structures explain functional properties of two *E. coli* porins. *Nature* **358**:727–733.

Cuesta-Seijo, J.A., Neale, C., Khan, M.A., Moktar, J., Tran, C.D., Bishop, R.E., Pomès, R., Privé, G.G. (2010) PagP crystallized from SDS/cosolvent reveals the route for phospholipid access to the hydrocarbon ruler. *Structure* **18**:1210–1219.

Dahmane, T., Damian, M., Mary, S., Popot, J.-L., Banères, J.-L. (2009) Amphipol-assisted *in vitro* folding of G protein-coupled receptors. *Biochemistry* **48**:6516–6521.

Dahmane, T., Rappaport, F., Popot, J.-L. (2013) Amphipol-assisted folding of bacteriorhodopsin in the presence and absence of lipids. Functional consequences. *Eur. Biophys. J.* **42**:85–101.

Deisenhofer, J., Epp, O., Miki, K., Huber, R., Michel, H. (1984) X-ray structure analysis of a membrane protein complex. Electron density map at 3 Å resolution and a model of the chromophores of the photosynthetic reaction center from *Rhodopseudomonas viridis*. *J. Mol. Biol.* **180**:385–398.

Deisenhofer, J., Epp, O., Miki, K., Huber, R., Michel, H. (1985) Structure of the protein subunits in the photosynthetic reaction center of *Rhodopseudomonas viridis* at 3 Å resolution. *Nature* **318**:618–624.

Deisenhofer, J., Michel, H. (1989) The photosynthetic reaction center from the purple bacterium *Rhodopseudomonas viridis*. *EMBO J.* **8**:2149–2170.

Diab, C., Tribet, C., Gohon, Y., Popot, J.-L., Winnik, F.M. (2007a) Complexation of integral membrane proteins by phosphorylcholine-based amphipols. *Biochim. Biophys. Acta* **1768**:2737–2747.

Diab, C., Winnik, F.M., Tribet, C. (2007b) Enthalpy of interaction and binding isotherms of non-ionic surfactants onto micellar amphiphilic polymers (amphipols). *Langmuir* **23**:3025–3035.

Egloff, P., Hillenbrand, M., Klenk, C., Batyuk, A., Heine, P., Balada, S., Schlinkmann, K.M., Scott, D.J., Schütz, M., Plückthun, A. (2014) Structure of signaling-competent neurotensin receptor 1 obtained by directed evolution in *Escherichia coli*. *Proc. Natl. Acad. Sci. USA* **111**:E655–E662.

Faham, S., Bowie, J.U. (2002) Bicelle crystallization: a new method for crystallizing membrane proteins yields a monomeric bacteriorhodopsin structure. *J. Mol. Biol.* **316**:1–6.

Ferrandez, Y., Dezi, M., Bosco, M., Urvoas, A., Valério, M., Le Bon, C., Giusti, F., Broutin, I., Durand, G., Polidori, A., Popot, J.-L., Picard, M., Minard, P. (2014) Amphipol-mediated screening of molecular orthoses specific for membrane protein targets. *J. Membr. Biol.* **247**:925–940.

Garavito, R.M., Jenkins, J., Jansonius, J.N., Karlsson, R., Rosenbusch, J.P. (1983) X-ray diffraction analysis of matrix porin, an integral membrane protein from *Escherichia coli* outer membranes. *J. Mol. Biol.* **164**:313–327.

Garavito, R.M., Picot, D., Loll, P.J. (1996) Strategies for crystallizing membrane proteins. *J. Bioenerg. Biomembr.* **28**:13–27.

Garavito, R.M., Rosenbusch, J.P. (1980) Three-dimensional crystals of an integral membrane protein: an initial X-ray analysis. *J. Cell Biol.* **86**:327–329.

Giusti, F., Rieger, J., Catoire, L., Qian, S., Calabrese, A.N., Watkinson, T.G., Casiraghi, M., Radford, S.E., Ashcroft, A. E., Popot, J.-L. (2014) Synthesis, characterization and applications of a perdeuterated amphipol. *J. Membr. Biol.* **247**:909–924.

Gohon, Y., Dahmane, T., Ruigrok, R., Schuck, P., Charvolin, D., Rappaport, F., Timmins, P., Engelman, D.M., Tribet, C., Popot, J.-L., Ebel, C. (2008) Bacteriorhodopsin/amphipol complexes: structural and functional properties. *Biophys. J.* **94**:3523–3537.

Gourdon, P., Andersen, J.L., Langmach, K., Hein, K.L., Bublitz, M., Pedersen, B.P., Liu, X.-Y., Yatime, L., Nyblom, M., Nielsen, T.T., Olesen, C., Møller, J.V., Nissen, P., Morth, J.P. (2011) HiLiDe–systematic approach to membrane protein crystallization in lipid and detergent. *Cryst. Growth Des.* **11**:2098–2106.

Grigorieff, N., Ceska, T.A., Downing, K.H., Baldwin, J.M., Henderson, R. (1996) Electron-crystallographic refinement of the structure of bacteriorhodopsin. *J. Mol. Biol.* **259**:393–421.

Henderson, R. (1975) The structure of purple membrane from *Halobacterium halobium*: analysis of the X-ray diffraction pattern. *J. Mol. Biol.* **93**:123–138.

Hibbs, R.E., Gouaux, E. (2011) Principles of activation and permeation in an anion-selective Cys-loop receptor. *Nature* **474**:54–60.

Hitscherich, C.J., Kaplan, J., Allaman, M., Wiencek, J., Loll, P.J. (2000) Static light scattering studies of OmpF porin: implications for integral membrane protein crystallization. *Protein Sci.* **9**:1559–1566.

Hunte, C. (2001) Insights from the structure of the yeast cytochrome $bc_1$ complex: crystallization of membrane proteins with antibody fragments. *FEBS Lett.* **504**:126–132.

Hunte, C., Koepke, J., Lange, C., Roßmanith, T., Michel, H. (2000) Structure at 2.3 Å resolution of the cytochrome $bc_1$ complex from the yeast *Saccharomyces cerevisiae* co-crystallized with an antibody $F_v$ fragment. *Structure* **8**:669–684.

Hunte, C., Richers, S. (2008) Lipids and membrane protein structures. *Curr. Opin. Struct. Biol.* **18**:406–411.

Huynh, K.W., Cohen, M.R., Moiseenkova-Bell, V.Y. (2014) Application of amphipols for structure-functional analysis of TRP channels. *J. Membr. Biol.* **247**:843–851.

Ishchenko, A., Abola, E., Cherezov, V. (2014) Lipidic cubic phase technologies for structural studies of membrane proteins, in: Mus-Veteau, I., ed., *Membrane Proteins Production for Structural Analysis*. Springer, New York, pp. 289–314.

Ishchenko, A., Abola, E.E., Cherezov, V. (2017) Crystallization of membrane proteins: An overview. *Methods Mol. Biol.* **1607**:117–141.

Israelachvili, J.N. (2011) *Intermolecular and Surface Forces*. 3rd ed., Elsevier/Academic Press, London, 450 p.

Iwata, S., Lee, J.W., Okada, K., Lee, J.K., Iwata, M., Rasmussen, B., Link, T., Ramaswamy, S., Jap, B.K. (1998) Complete structure of the 11-subunit bovine mitochondrial cytochrome $bc_1$ complex. *Science* **281**:64–71.

Johansson, L.C., Arnlund, D., White, T.A., Katona, G., DePonte, D.P., Weierstall, U., Doak, R.B., Shoeman, R.L., Lomb, L., Malmerberg, E., Davidsson, J., Nass, K., Liang, M., Andreasson, J., Aquila, A., Bajt, S., Barthelmess, M., Barty, A., Bogan, M.J., Bostedt, C., Bozek, J.D., Caleman, C., Coffee, R., Coppola, N., Ekeberg, T., Epp, S.W., Erk, B., Fleckenstein, H., Foucar, L., Graafsma, H., Gumprecht, L., Hajdu, J., Hampton, C.Y., Hartmann, R., Hartmann, A., Hauser, G., Hirsemann, H., Holl, P., Hunter, M.S., Kassemeyer, S., Kimmel, N., Kirian, R.A., Maia, F.R.N.C., Marchesini, S., Martin, A.V., Reich, C., Rolles, D., Rudek, B., Rudenko, A., Schlichting, I., Schulz, J., Seibert, M.M., Sierra, R.G., Soltau, H., Starodub, D., Stellato, F., Stern, S., Strüder, L., Timneanu, N., Ullrich, J., Wahlgren, W.Y., Wang, X., Weidenspointner, G., Wunderer, C., Fromme, P., Chapman, H.N., Spence, J.C.H., Neutze, R. (2012) Lipidic phase membrane protein serial femtosecond crystallography. *Nat. Meth.* **9**:263–265.

Johansson, L.C., Wöhri, A.B., Katona, G., Engström, S., Neutze, R. (2009) Membrane protein crystallization from lipidic phases. *Curr. Opin. Struct. Biol.* **19**:372–378.

Jones, T.A., Zhou, J.-Y., Cowan, S.J., Kjeldgaard, M. (1991) Improved methods for building protein models in electron density maps and the location of errors on these models. *Acta Crystallogr. A* **47**:110–119.

Jost, C., Plückthun, A. (2014) Engineered proteins with desired specificity: DARPins, other alternative scaffolds and bispecific IgGs. *Curr. Opin. Struct. Biol.* **27**:102–112.

Kang, H.J., Lee, C., Drew, D. (2013) Breaking the barriers in membrane protein crystallography. *Int. J. Biochem. Cell Biol.* **45**:636–644.

Kang, Y., Zhou, X.E., Gao, X., He, Y., Liu, W., Ishchenko, A., Barty, A., White, T.A., Yefanov, O., Han, G.W., Xu, Q., deWaal, P.W., Ke, J., Tan, M.H.E., Zhang, C., Moeller, A., West, G.M., Pascal, B.D., Van Eps, N., Caro, L.N., Vishnivetskiy, S.A., Lee, R.J., Suino-Powell, K.M., Gu, X., Pal, K., Ma, J., Zhi, X., Boutet, S., Williams, G.J., Messerschmidt, M., Gati, C., Zatsepin, N.A., Wang, D., James, D., Basu, S., Roy-Chowdhury, S., Conrad, C.E., Coe, J., Liu, H., Lisova, S., Kupitz, C., Grotjohann, I., Fromme, R., Jiang, Y., Tan, M., Yang, H., Li, J., Wang, M., Zheng, Z., Li, D., Howe, N., Zhao, Y., Standfuss, J., Diederichs, K., Dong, Y., Potter, C.S., Carragher, B., Caffrey, M., Jiang, H., Chapman, H.N., Spence, J.C.H., Fromme, P., Weierstall, U., Ernst, O.P., Katritch, V., Gurevich, V.V., Griffin, P. R., Hubbell, W.L., Stevens, R.C., Cherezov, V., Melcher, K., Xu, E. (2015) Crystal structure of rhodopsin bound to arrestin by femtosecond X-ray laser. *Nature* **523**:561–567.

Kleinschmidt, J.H., Popot, J.-L. (2014) Folding and stability of integral membrane proteins in amphipols. *Arch. Biochem. Biophys.* **564**:327–343.

Kühlbrandt, W. (1988) Three-dimensional crystallization of membrane proteins. *Quat. Rev. Biophys.* **21**:429–477.

Kupitz, C., Basu, S., Grotjohann, I., Fromme, R., Zatsepin, N.A., Rendek, K.N., Hunter, M.S., Shoeman, R.L., White, T. A., Wang, D., James, D., Yang, J.H., Cobb, D.E., Reeder, B., Sierra, R.G., Liu, H., Barty, A., Aquila, A.L., Deponte, D., Kirian, R.A., Bari, S., Bergkamp, J.J., Beyerlein, K.R., Bogan, M.J., Caleman, C., Chao, T.C., Conrad, C.E.,

Davis, K.M., Fleckenstein, H., Galli, L., Hau-Riege, S.P., Kassemeyer, S., Laksmono, H., Liang, M., Lomb, L., Marchesini, S., Martin, A.V., Messerschmidt, M., Milathianaki, D., Nass, K., Ros, A., Roy-Chowdhury, S., Schmidt, K., Seibert, M., Steinbrener, J., Stellato, F., Yan, L., Yoon, C., Moore, T.A., Moore, A.L., Pushkar, Y., Williams, G. J., Boutet, S., Doak, R.B., Weierstall, U., Frank, M., Chapman, H.N., Spence, J.C., Fromme, P. (2014) Serial time-resolved crystallography of photosystem II using a femtosecond X-ray laser. *Nature* **513**:261–265.

Landau, E.M., Rosenbusch, J.P. (1996) Lipid cubic phases: A novel concept for the crystallization of membrane proteins. *Proc. Natl. Acad. Sci. USA* **93**:14532–14535.

Lanyi, J.K. (2004) X-ray diffraction of bacteriorhodopsin photocycle intermediates. *Mol. Membr. Biol.* **21**:143–150.

Le Bon, C., Popot, J.-L., Giusti, F. (2014) Labeling and functionalizing amphipols for biological applications. *J. Membr. Biol.* **247**:797–814.

Li, D., Lee, J., Caffrey, M. (2011) Crystallizing membrane proteins in lipidic mesophases. A host lipid screen. *Cryst. Growth Des.* **11**:530–537.

Li, D., Shah, S.T.A., Caffrey, M. (2013) Host lipid and temperature as important screening variables for crystallizing integral membrane proteins in lipidic mesophases. Trials with diacylglycerol kinase. *Cryst. Growth Des.* **13**:2846–2857.

Liao, M., Cao, E., Julius, D., Cheng, Y. (2013) Structure of the TRPV1 ion channel determined by electron cryo-microscopy. *Nature* **504**:107–112.

Liao, M., Cao, E., Julius, D., Cheng, Y. (2014) Single particle electron cryo-microscopy of a mammalian ion channel. *Curr. Opin. Struct. Biol.* **27**:1–7.

Lindblom, G., Rilfors, L. (1989) Cubic phases and isotropic structures formed by membrane lipids–possible biological relevance. *Biochim. Biophys. Acta* **988**:221–256.

Liu, Z., Yan, H., Wang, K., Kuang, T., Zhang, J., Gui, L., An, X., Chang, W. (2004) Crystal structure of spinach major light-harvesting complex at 2.72 Å resolution. *Nature* **428**:287–291.

Lluis, M.W., Godfroy, J.I., III, Yin, H. (2013) Protein engineering methods applied to membrane protein targets. *Prot. Eng. Des. Sel.* **26**:91–100.

Loll, P.J. (2014) Membrane proteins, detergents and crystals: what is the state of the art? *Acta Crystallogr. F* **70**:1576–1583.

Love, J., Mancia, F., Shapiro, L., Punta, M., Rost, B., Girvin, M., Wang, D.N., Zhou, M., Hunt, J.F., Szyperski, T., Gouaux, E., Mackinnon, R., McDermott, A., Honig, B., Inouye, M., Montelione, G., Hendrickson, W.A. (2010) The New York Consortium on Membrane Protein Structure (NYCOMPS): a high-throughput platform for structural genomics of integral membrane proteins. *J. Struct. Funct. Genomics* **11**:191–199.

Luecke, H., Schobert, B., Richter, H.-T., Cartailler, J.-P., Lanyi, J.K. (1999) Structure of bacteriorhodopsin at 1.55 Å resolution. *J. Mol. Biol.* **291**: 899–911.

Luecke, H., Schobert, B., Stagno, J., Imasheva, E.S., Wang, J.M., Balashov, S.P., Lanyi, J.K. (2008) Crystallographic structure of xanthorhodopsin, the light-driven proton pump with a dual chromophore. *Proc. Natl. Acad. Sci. USA* **105**:16561–16565.

Magnani, F., Shibata, Y., Serrano-Vega, M.J., Tate, C.G. (2008) Co-evolving stability and conformational homogeneity of the human adenosine $A_{2a}$ receptor. *Proc. Natl. Acad. Sci. USA* **105**:10744–10749.

Martin-Garcia, J.M., Conrad, C.E., Coe, J., Roy-Chowdhury, S., Fromme, P. (2016) Serial femtosecond crystallography: A revolution in structural biology. *Arch. Biochem. Biophys.* **602**:32–47.

Michel, H. (1982) Three-dimensional crystals of a membrane protein complex. The photosynthetic reaction centre from *Rhodopseudomonas viridis*. *J. Mol. Biol.* **158**:567–572.

Michel, H. (1983) Crystallization of membrane proteins. *Trends Biochem. Sci.* **8**:56–59.

Michel, H. (1991) Crystallization of Membrane Proteins. First ed. CRC Press, Boca Raton, 224 p.

Michel, H., Epp, O., Deisenhofer, J. (1986) Pigment-protein interactions in the photosynthetic reaction centre from *Rhodopseudomonas viridis*. *EMBO J.* **5**:2445–2451.

Michel, H., Oesterhelt, D. (1980) Three-dimensional crystals of membrane proteins: bacteriorhodopsin. *Proc. Natl. Acad. Sci. USA* **77**:1283–1285.

Miller, J.L., Tate, C.G. (2011) Engineering an ultra-thermostable $\beta_1$-adrenoceptor. *J. Mol. Biol.* **413**:628–638.

Moraes, I., Gwyndaf Evans, G., Sanchez-Weatherby, J., Newstead, S., Stewart, P.D.S. (2014) Membrane protein structure determination–The next generation. *Biochim. Biophys. Acta* **1838**:78–87.

Nagy, J.K., Kuhn Hoffmann, A., Keyes, M.H., Gray, D.N., Oxenoid, K., Sanders, C.R. (2001) Use of amphipathic polymers to deliver a membrane protein to lipid bilayers. *FEBS Lett.* **501**:115–120.

Neutze, R., Brändén, G., Schertler, G.F. (2015) Membrane protein structural biology using X-ray free electron lasers. *Curr. Opin. Struct. Biol.* **33**:115–125.

Neutze, R., Remco, W., van der Spoel, D., Weckert, E., Hajdu, J. (2000) Potential for biomolecular imaging with femtosecond X-ray pulses. *Nature* **406**:752–757.

Nikolaev, M., Round, E., Gushchin, I., Polovinkin, V., Balandin, T., Kuzmichev, P., Shevchenko, V., Borshchevskiy, V., Kuklin, A., Round, A., Bernhard, F., Willbold, D., Büldt, G., Gordeliy, V. (2017) Integral membrane proteins can be crystallized directly from nanodiscs. *Cryst. Growth Des.* **17**:945–948.

Nollert, P. (2004) Lipidic cubic phases as matrices for membrane protein crystallization. *Methods* **34**:348–353.

Nollert, P. (2005) Membrane protein crystallization in amphiphile phases: practical and theoretical considerations. *Prog. Biophys. Mol. Biol.* **88**:339–357.

Ostermeier, C., Harrenga, A., Ermler, U., Michel, H. (1997) Structure at 2.7 Å resolution of the *Paracoccus denitrificans* two-subunit cytochrome *c* oxidase complexed with an antibody $F_v$ fragment. *Proc. Natl. Acad. Sci. USA* **94**:10547–10553.

Ostermeier, C., Iwata, S., Ludwig, B., Michel, H. (1995) $F_v$ fragment-mediated crystallization of the membrane protein bacterial cytochrome *c* oxidase. *Nature Struct. Biol.* **2**:842–846.

Ostermeier, C., Michel, H. (1997) Crystallization of membrane proteins. *Curr. Opin. Struct. Biol.* **7**:697–701.

Parker, J.L., Newstead, S. (2012) Current trends in α-helical membrane protein crystallization: An update. *Prot. Sci.* **21**:1358–1365.

Pebay-Peyroula, E., Rummel, G., Rosenbusch, J.P., Landau, E. (1997) X-ray structure of bacteriorhodopsin at 2.5 Å from microcrystals grown in lipidic cubic phases. *Science* **277**:1676–1881.

Perlmutter, J.D., Popot, J.-L., Sachs, J.N. (2014) Molecular dynamics simulations of a membrane protein/amphipol complex. *J. Membr. Biol.* **247**:883–895.

Plückthun, A. (2015) Designed ankyrin repeat proteins (DARPins): binding proteins for research, diagnostics, and therapy. *Annu. Rev. Pharmacol. Toxicol.* **55**:489–511.

Pocanschi, C.L., Dahmane, T., Gohon, Y., Rappaport, F., Apell, H.-J., Kleinschmidt, J.H., Popot, J.-L. (2006) Amphipathic polymers: tools to fold integral membrane proteins to their active form. *Biochemistry* **45**:13954–13961.

Polovinkin, V., Gushchin, I., Balandin, T., Chervakov, P., Round, E., Shevchenko, V., Popov, A., Borshchevskiy, V., Popot, J.-L., Gordeliy, V. (2014) High-resolution structure of a membrane protein transferred from amphipol to a lipidic mesophase. *J. Membr. Biol.* **247**:997–1004.

Popot, J.-L. (2014) Folding membrane proteins *in vitro*: A table and some comments. *Arch. Biochem. Biophys.* **564**:314–326.

Popot, J.-L., Berry, E.A., Charvolin, D., Creuzenet, C., Ebel, C., Engelman, D.M., Flötenmeyer, M., Giusti, F., Gohon, Y., Hervé, P., Hong, Q., Lakey, J.H., Leonard, K., Shuman, H.A., Timmins, P., Warschawski, D.E., Zito, F., Zoonens, M., Pucci, B., Tribet, C. (2003) Amphipols: polymeric surfactants for membrane biology research. *Cell. Mol. Life Sci.* **60**:1559–1574.

Poulos, S., Morgan, J.L., Zimmer, J., Faham, S. (2015) Bicelles coming of age: an empirical approach to bicelle crystallization. *Meth. Enzymol.* **557**:393–416.

Prata, C., Giusti, F., Gohon, Y., Pucci, B., Popot, J.-L., Tribet, C. (2001) Non-ionic amphiphilic polymers derived from *tris*(hydroxymethyl)-acrylamidomethane keep membrane proteins soluble and native in the absence of detergent. *Biopolymers* **56**:77–84.

Privé, G.G. (2007) Detergents for the stabilization and crystallization of membrane proteins. *Methods* **41**:388–397.

Qiu, H., Caffrey, M. (2000) Phase diagram of the monoolein/water system: metastability and equilibrium aspects. *Biomaterials* **21**:223–234.

Rasmussen, S.G., Choi, H.J., Rosenbaum, D.M., Kobilka, T.S., Thian, F.S., Edwards, P.C., Burghammer, M., Ratnala, V.R., Sanishvili, R., Fischetti, R.F., Schertler, G.F., Weis, W.I., Kobilka, B.K. (2007) Crystal structure of the human $β_2$ adrenergic G protein-coupled receptor. *Nature* **450**:383–387.

Rasmussen, S.G.F., Choi, H.-J., Fung, J.J., Pardon, E., Casarosa, P., Chae, P.S., DeVree, B.T., Rosenbaum, D.M., Thian, F.S., Kobilka, T.S., Schnapp, A., Konetzki, I., Sunahara, R.K., Gellman, S.H., Pautsch, A., Steyaert, J., Weis, W.I., Kobilka, B.K. (2011) Structure of a nanobody-stabilized active state of the $β_2$ adrenoceptor. *Nature* **469**:175–180.

Rosenbaum, D.M., Cherezov, V., Hanson, M.A., Rasmussen, S.G., Thian, F.S., Kobilka, T.S., Choi, H.J., Yao, X.J., Weis, W.I., Stevens, R.C., Kobilka, B.K. (2007) GPCR engineering yields high-resolution structural insights into $β_2$-adrenergic receptor function. *Science* **318**:1266–1273.

Roth, M., Lewitt-Bentley, A., Michel, H., Deisenhofer, J., Huber, R., Oesterhelt, D. (1989) Detergent structure in crystals of a bacterial photosynthetic reaction center. *Nature* **340**:659–662.

Rummel, G., Hardmeyer, A., Widmer, C., Chiu, M.L., Nollert, P., Locher, K.P., Pedruzzi, I.I., Landau, E.M., Rosenbusch, J.P. (1998) Lipidic cubic phases: new matrices for the three-dimensional crystallization of membrane proteins. *J. Struct. Biol.* **121**:82–91.

Sadaf, A., Cho, K.H., Byrne, B., Chae, P.S. (2015) Amphipathic agents for membrane protein study. *Meth. Enzymol.* **557**:57–94.

Sarkar, C.A., Dodevski, I., Kenig, M., Dudli, S., Mohr, A., Hermans, E., Plückthun, A. (2008) Directed evolution of a G protein-coupled receptor for expression, stability, and binding selectivity. *Proc. Natl. Acad. Sci. USA* **105**:14808–14813.

Schulz, G.E. (2011) A new classification of membrane protein crystals. *J. Mol. Biol.* **407**:640–646.

Sennhauser, G., Grütter, M.G. (2008) Chaperone-assisted crystallography with DARPins. *Structure* **16**:1443–1453.

Serrano-Vega, M.J., Magnani, F., Shibata, Y., Tate, C.G. (2008) Conformational thermostabilization of the $\beta_1$-adrenergic receptor in a detergent-resistant form. *Proc. Natl. Acad. Sci. USA* **105**:877–882.

Sharma, K.S., Durand, G., Gabel, F., Bazzacco, P., Le Bon, C., Billon-Denis, E., Catoire, L.J., Popot, J.-L., Ebel, C., Pucci, B. (2012) Non-ionic amphiphilic homopolymers: Synthesis, solution properties, and biochemical validation. *Langmuir* **28**:4625–4639.

Shibata, Y., White, J.F., Serrano-Vega, M.J., Magnani, F., Aloia, A.L., Grisshammer, R., Tate, C.G. (2009) Thermostabilization of the neurotensin receptor NTS1. *J. Mol. Biol.* **390**:262–277.

Smith, A.L. (1967) Preparation, properties, and conditions for assay of mitochondria: slaughterhouse material, small scale. *Methods Enzymol.* **10**:81–86.

Suga, M., Akita, F., Sugahara, M., Kubo, M., Nakajima, Y., Nakane, T., Yamashita, K., Umena, Y., Nakabayashi, M., Yamane, T., Nakano, T., Suzuki, M., Masuda, T., Inoue, S., Kimura, T., Nomura, T., Yonekura, S., Yu, L.J., Sakamoto, T., Motomura, T., Chen, J.H., Kato, Y., Noguchi, T., Tono, K., Joti, Y., Kameshima, T., Hatsui, T., Nango, E., Tanaka, R., Naitow, H., Matsuura, Y., Yamashita, A., Yamamoto, M., Nureki, O., Yabashi, M., Ishikawa, T., Iwata, S., Shen, J.R. (2017) Light-induced structural changes and the site of O=O bond formation in PSII caught by XFEL. *Nature* **543**:131–135.

Takeda, K., Sato, H., Hino, T., Kono, M., Fukuda, K., Sakurai, I., Okada, T., Kouyama, T. (1998) A novel three-dimensional crystal of bacteriorhodopsin obtained by successive fusion of the vesicular assemblies. *J. Mol. Biol.* **283**:463–474.

Tanford, C. (1980) *The Hydrophobic Effect: Formation of Micelles and Biological Membranes.* 2nd ed., John Wiley & Sons, New York, 233 p.

Tate, C.G. (2010) Practical considerations of membrane protein instability for purification and crystallisation, in: Mus-Veteau, I., ed., *Membrane Protein Expression.* The Humana Press, Totowa, New Jersey, USA, pp. 187–203.

Tribet, C., Audebert, R., Popot, J.-L. (1996) Amphipols: polymers that keep membrane proteins soluble in aqueous solutions. *Proc. Natl. Acad. Sci. USA* **93**:15047–15050.

Tribet, C., Audebert, R., Popot, J.-L. (1997) Stabilization of hydrophobic colloidal dispersions in water with amphiphilic polymers: Application to integral membrane proteins. *Langmuir* **13**:5570–5576.

Tribet, C., Diab, C., Dahmane, T., Zoonens, M., Popot, J.-L., Winnik, F.M. (2009) Thermodynamic characterization of the exchange of detergents and amphipols at the surfaces of integral membrane proteins. *Langmuir* **25**:12623–12634.

Ujwal, R., Bowie, J.U. (2011) Crystallizing membrane proteins using lipidic bicelles. *Methods* **55**:337–341.

Ujwal, R., Cascio, D., Colletier, J.-P., Faham, S., Zhang, J., Toro, L., Ping, P., Abramson, J. (2008) The crystal structure of mouse VDAC1 at 2.3 Å resolution reveals mechanistic insights into metabolite gating. *Proc. Natl. Acad. Sci. USA* **105**:17742–17747.

Urvoas, A., Valerio-Lepiniec, M., Minard, P. (2012) Artificial proteins from combinatorial approaches. *Trends Biotechnol.* **30**:512–520.

Uysal, S., Vásquez, V., Tereshko, V., Esaki, K., Fellouse, F.A., Sidhu, S.S., Koide, S., Perozo, E., Kossiakoff, A. (2009) Crystal structure of full-length KcsA in its closed conformation. *Proc. Natl. Acad. Sci. USA* **106**:6644–6649.

Vaidehi, N., Grisshammer, R., Tate, C.G. (2016) How can mutations thermostabilize G protein-coupled receptors? *Trends Pharmacol. Sci.* **37**:37–46.

Vinothkumar, K.R. (2011) Structure of rhomboid protease in a lipid environment. *J. Mol. Biol.* **407**:232–247.

Wadsten, P., Wöhri, A.B., Snijder, A., Katona, G., Gardiner, A.T., Cogdell, R.J., Neutze, R., Engström, S. (2006) Lipidic sponge phase crystallization of membrane proteins. *J. Mol. Biol.* **364**:44–53.

Wiener, M.C. (2001) Existing and emergent roles for surfactants in the three-dimensional crystallization of integral membrane proteins. *Curr. Opin. Colloid Interface Sci.* **6**:412–419.

Wiener, M.C. (2004) A pedestrian guide to membrane protein crystallization. *Methods* **34**:364–372.

Wöhri, A.B., Johansson, L.C., Wadsten-Hindrichsen, P., Wahlgren, W.Y., Fischer, G., Horsefield, R., Katona, G., Nyblom, M., Oberg, F., Young, G., Cogdell, R.J., Fraser, N.J., Engström, S., Neutze, R. (2008) A lipidic-sponge phase screen for membrane protein crystallization. *Structure* **16**:1003–1009.

Zhang, Q., Tao, H., Hong, W.-X. (2011) New amphiphiles for membrane protein structural biology. *Methods* **55**:318–323.

Zhang, Z., Huang, L.-S., Shulmeister, V.M., Chi, Y.-I., Kim, K.K., Hung, L.-W., Crofts, A.R., Berry, E.A., Kim, S.-H. (1998) Electron transfer by domain movement in cytochrome $bc_1$. *Nature* **392**:677–684.

Zhou, Y., Bowie, J.U. (2000) Building a thermostable membrane protein. *J. Biol. Chem.* **275**:6975–6979.

Zhou, Y., Morais-Cabral, J.H., Kaufman, A., MacKinnon, R. (2001) Chemistry of ion coordination and hydration revealed by a K$^+$ channel-F$_{ab}$ complex at 2.0 Å resolution. *Nature* **414**:43–48.

Zoonens, M., Catoire, L.J., Giusti, F., Popot, J.-L. (2005) NMR study of a membrane protein in detergent-free aqueous solution. *Proc. Natl. Acad. Sci. USA* **102**:8893–8898.

Zoonens, M., Giusti, F., Zito, F., Popot, J.-L. (2007) Dynamics of membrane protein/amphipol association studied by Förster resonance energy transfer. Implications for *in vitro* studies of amphipol-stabilized membrane proteins. *Biochemistry* **46**:10392–10404.

Zoonens, M., Popot, J.-L. (2014) Amphipols for each season. *J. Membr. Biol.* **247**:759–796.

Zoonens, M., Zito, F., Martinez, K.L., Popot, J.-L. (2014) Amphipols: a general introduction and some protocols, in: Mus-Veteau, I., ed., *Membrane Proteins Production for Structural Analysis*. Springer, New York, Heidelberg, Dordrecht, London, pp. 173–203.

# The Use of Amphipols for Electron Microscopy

**12**

**Summary**

*Electron microscopy is, to date, the field of structural biology to which amphipols have contributed most, to the point that testing amphipols has become a common practice at the onset of any single-particle high-resolution study of a membrane protein (MP) by electron cryomicroscopy (cryo-EM). Yet, few methodological studies have been published. In many cases, it has been shown that the use of amphipols facilitates cryo-EM studies as compared to detergent solutions and results in better data, but the origins of this improvement seem to be multiple and have not all been sorted out. Mechanisms that have some degree of credibility include (i) biochemical stabilization and, at least in some cases, reduction of the variability of the images of the target MP, presumably due to limitation of its dynamics; (ii) improved contrast, due to the absence or near-absence of free surfactant in the solution; and (iii) better spread of the particles in the water film stretched over the holes in the supporting carbon film, probably related to surface tension issues.*

## 12.1 Introduction

The application of amphipols (APols) to electron microscopy (EM) developed relatively late. The first structural biology study of an APol-trapped membrane protein (MP) was actually an EM one (Tribet et al. 1998; Study 12.1 in Table 12.1 below). However, it did not resort to standard transmission EM (TEM), but to scanning TEM (STEM). STEM of freeze-dried, unstained specimens does not aim at obtaining high-resolution images, but at mapping the distribution of mass over the supporting carbon film. By integration, the mass of each particle can be determined (for a recent review, see Vahedi-Faridi et al. 2013). The aim of this study was to determine, using preparations of dimers and monomers of the cytochrome $b_6f$ complex, whether the mass distribution of APol-trapped particles reflected that characterized in detergent solution by analytical ultracentrifugation (AUC) (it did; see § 12.3.1). A series of TEM studies of negatively stained, APol-trapped $F_1F_O$ ATP synthase were published in 1998–2000 (Wilkens and Capaldi 1998a, b; Wilkens 2000; Wilkens et al. 2000; Study 12.2). They long remained unnoticed by the APol community because APols were mentioned neither in the title nor in the keywords or abstracts of the papers, the samples being simply described as "detergent-free."

The first study by TEM of vitrified specimens observed at cryogenic temperatures (electron cryomicroscopy, cryo-EM) of an APol-trapped MP, that of mitochondrial Complex I, was started in 1997 but appeared in print only in 2007 (Flötenmeyer et al. 2007; Study 12.3). It established the feasibility of the approach, but, for reasons that had nothing to do with the use of APols (see below, § 12.3.1), the three-dimensional (3D) reconstruction was limited to a very low resolution (58 Å), which may have, mistakenly, dissuaded potential users from giving APols a try. A study of negatively stained bacteriorhodopsin (BR) appeared in 2008 as part of a larger-scale characterization of BR/APol complexes (Gohon et al. 2008; Study 12.4). EM studies however really picked up steam only after the publication of the respirasome structure by cryo-EM in 2011 (Althoff et al. 2011; Study 12.5; § 12.3.2). They have been making up for lost time ever since, with, by mid-2017, >40 studies published, representing as many proteins or protein complexes and >50 publications (see below, Table 12.1, and additional publications listed in §§ 12.3.2 and 12.3.3).

In the following, we will first examine the context, namely which EM approaches are applied to MPs and, in particular, how the samples are prepared (§ 12.2). By a fortunate coincidence, at the time when electron microscopists were finally discovering the potential interest of using APols for studying MPs, great technological progress was being accomplished as regards the collection, correction, and analysis of cryo-EM images, which have revolutionized the field (see below, § 12.2.3). APols bring their little stone to this progress by improving specimen preparation and visualization (§ 12.3).

## 12.2 Context: Electron Microscopy Studies of Membrane Proteins

### 12.2.1 The Various Approaches to Studying Membrane Proteins by Electron Microscopy

Electron microscopy can be applied to small MPs (e.g. BR, 27 kDa), to huge complexes (e.g. the respirasome, 1.7 MDa), to subcellular structures, such as mitochondria or synaptic terminals, or even to whole cells. The nature of the samples and techniques varies widely depending on the target. For small objects (typically $\leq 150$ kDa), one will principally resort to two-dimensional (2D) crystals, in the form either of sheets or tubes, which will be studied by electron crystallography and by Fourier transform analysis of EM images (for recent reviews, see Raunser and Walz 2009; Ubarretxena-Belandia and Stokes 2010; Fujiyoshi 2011; Wisedchaisri et al. 2011; Kühlbrandt 2013; Stahlberg et al. 2015). Above ~150 kDa, one can either resort to 2D crystals, if the protein or complex accepts to crystallize, or reconstruct 3D images by combining tens or hundreds of thousands of projection images of single particles (see below, § 12.2.2). For whole cells or large subcellular structures, one will use electron tomography, in which a 3D reconstruction of a single object is calculated from images obtained at various tilts with respect to the electron beam (for reviews, see e.g. Leis et al. 2009; Lučič et al. 2013; Villa et al. 2013).

One of the major difficulties of EM of biological samples is to control radiation-induced damage (for a detailed discussion and comparison with X-ray crystallography, see Henderson 1995). This limits the doses that can be applied and, therefore, the amount of information afforded by each image. There are two main ways around this problem. One is to freeze the sample at cryogenic temperatures, typically those of liquid nitrogen or helium. As this is not sufficient to permit collecting complete data on a single molecule, data must be averaged from many thousands of individual molecules, each of which has interacted with a limited number of electrons. This is achieved either by crystallographic analysis of 2D crystals, in which the molecules are oriented with respect to each other and significant data can be separated from noise by Fourier analysis, or by aligning images of single particles and averaging them – the aligning procedure being a complex one. As pointed out early on by Richard

Henderson in a landmark review, the latter procedure cannot be applied to biological molecules smaller than 50–100 kDa (Henderson 1995, 2015; Henderson and McMullan 2013). Alternatively, the contrast can be increased, and the biological sample made electron-resistant, by embedding it into an electron-dense metallic stain, typically a uranium or tungsten salt. This approach can be applied to both crystalline and single-particle samples. It is most often the first step when embarking on an EM study and developing protocols for the preparation of samples, because it is quick and easy and yields within minutes a first view of their nature and quality. What is imaged, in that case, is not the biological sample itself, but the crust of metal salt that has deposited around it. Due to the grain of the salt, this approach, known as "negative staining" (NS) because molecules appear as white objects that have excluded the salt, is limited to resolutions of, at best, ~15 to 18 Å (Ohi et al. 2004; for a recent overview of "direct imaging methods" relying on the use of stains or metal shadowing and the way they relate to X-ray and cryo-EM methods, see Miyaguchi 2014).

Except under very special circumstances, which have not been exploited yet and will be briefly discussed below (§ 12.3.4.3), there is little probability that APols can contribute much to EM studies of 2D crystals. This is because the order in these crystals comes from lateral contacts that, as a rule, involve mostly the transmembrane (TM) regions of MPs. As discussed in Chap. 11 about Type 1 3D crystals (see § 11.2.2), APol-coated TM surfaces are unlikely to allow the formation of regular contacts leading to the formation of well-ordered crystals, unless APols move out of the way and allow direct protein/protein contacts (§ 12.3.4.3). We will therefore focus on the practice and constraints of single-particle EM studies.

## 12.2.2   Single-Particle Electron Microscopy Studies of Biological Molecules: The Basics

Single-particle EM studies of biological molecules (for general reviews, see e.g. Frank 2006; Rubinstein 2007; Cheng et al. 2015) require a more moderate purification of the sample than other structural techniques such as, say, radiation scattering, because averaging of the data over a population of particles will be done only after the particles to be studied have been selected, based on the appearance of their projections. Thus, it is perfectly possible, provided that the different forms can be efficiently distinguished one from another, to work on a mixture of monomeric and dimeric complexes, whose 3D structure will be reconstructed separately (see e.g. Postis et al. 2015; Study 12.22 below), or on preparations containing a given molecule in several different conformations (see e.g. Cheng 2015). The particles are spread on an EM grid covered with a carbon film which can be either continuous or, as is most often the case in cryo-EM, holey, and has been rendered hydrophilic, usually with a glow discharger. In the case of a continuous carbon film, the particles adhere to it. For NS, the grid, after rinsing, is exposed to a solution of heavy metal salt, blotted, and air-dried before being introduced in the electron microscope (for a recent review, see Vahedi-Faridi et al. 2013).

In cryo-EM (for recent reviews, see e.g. Frank 2006; Orlova and Saibil 2011; Lau and Rubinstein 2013; Milne et al. 2013; Bai et al. 2015a; Cheng 2015; Cheng et al. 2015; Nogales and Scheres 2015; De Zorzi et al. 2016; Vinothkumar and Henderson 2016), the particles are most often suspended in the aqueous films that form over holes in a carbon film when the grid is blotted (Fig. 12.1A). The grids are then rapidly plunged (within seconds) into a freezing medium, generally a slurry of liquid ethane cooled to liquid nitrogen temperature, in which liquid water turns to vitreous ice (Dubochet et al. 1988; Dubochet 2015). They are dipped in liquid nitrogen and introduced in the microscope in a grid holder that is kept at either liquid nitrogen or liquid helium temperature. Important features of vitrified specimens include (*i*) an amorphous ice layer sufficiently thick to accommodate the particles, but not much thicker so as not to lower the contrast; (*ii*) particles that form a dense population, but no or not too many aggregates, and are well distributed across the field of view (crowding toward the edges of the film is a frequent problem); and (*iii*) in general, particles that adopt random orientations, so that each projection angle can be thoroughly sampled.

In all cases, the grid holder can be tilted, so that the same specimen can be imaged at various angles with respect to the electron beam. This feature can be used, in particular, to reconstruct 3D models from pairs of projection images collected at a known angle to each other (random conical tilt reconstruction method; see Radermacher et al. 1987; Frank 2006).

Images of NS specimens have enough contrast so that they can be easily identified and studied individually. They provide very useful information about the purity of the sample and its homogeneity: presence of aggregates or, on the contrary, fragmentation of the complexes, overall shape, internal symmetry, conformational variability, etc. (reviewed in Ohi et al. 2004; Vahedi-Faridi et al. 2013). Studies in NS can stop here or proceed to a limited analysis such as the search for rotational symmetry. Most of the time, however, the noise is averaged out by combining numerous images of projections at a similar angle, and tomography is applied to reconstructing a 3D structure: computer programs class images according to the direction of the projection, orient them in-plane with respect to one another, average them, and reconstruct a 3D volume by combining all orientations (see e.g. Cheng et al. 2015). The resolution of the resulting 3D structures is limited to the 15–20 Å range, and no information is obtained on the internal variations of electron density of the object, its shape only being preserved. In addition, NS introduces artifacts, such as specimen flattening. 3D reconstructions of negatively stained specimens are, these limitations notwithstanding, an extremely useful first step in the EM analysis of a macromolecule.

In cryo-EM, the contrast is very limited (the electron density of the ice film is typically ~80% that of the particle; Henderson and McMullan 2013), and individual images are not interpretable in details (Fig. 12.3B). Numerous factors contribute to degrading their quality, among which artifacts arising from aberrations in the electron optics, the need to defocus to generate contrast, the high levels of noise arising from the finite number of electrons used (typically $\leq 20$ electrons per $\text{Å}^2$, most of which go through unperturbed), movement of the particles during exposure, and the limited efficiency of the photographic film or charge-coupled device (CCD) detectors used until very recently. Nevertheless, image correction, computerized identification, and analysis of hundreds of thousands if not millions of images (Fig. 12.4) permit, provided the objects are large enough to be aligned – theoretically $\geq 50$–100 kDa (Henderson 1995; Henderson and McMullan 2013), in practice, $\geq 150$ kDa – to reconstruct 3D structures to a resolution that, until recently, could reach, in the best cases, 6–7 Å (for recent reviews, see e.g. Bai et al. 2015a; Carazo et al. 2015; Cheng 2015; Cheng et al. 2015; Nogales and Scheres 2015; De Zorzi et al. 2016; Vinothkumar and Henderson 2016). Recent technical progress (including the use of a Volta phase plate) shows that this size limit can be lowered (see Khoshouei et al. 2017), whereas the resolution has now reached the critical level at which large amino acid side-chains can be identified (see below, § 12.2.3).

Critical assessment of the resolution achieved in cryo-EM reconstructions is a complicated issue. It seems to be best addressed by comparing quantitatively the Fourier transforms of two models extracted independently from two sets of images (the so-called "gold standard" procedure), taking as a criterion the resolution at which Fourier shell correlation (FSC) falls below 0.143 (Rosenthal and Henderson 2003; Scheres and Chen 2012; see Box 12.1), a procedure that has been agreed upon by microscopists (Henderson et al. 2012) and is now in common use. Note that it is not rare that, either as a consequence of the alignment procedure or because they are more or less disordered, some regions of the maps are better resolved than others.

### Box 12.1    Fourier Transforms, Thon Rings, and Fourier Shell Correlation

The Fourier transform is a mathematical operation that decomposes a function of time or space (a signal) into the frequencies that make it up. Fourier transforms are used widely in biophysics, in domains as varied as crystallography, small-angle radiation scattering, NMR, and infrared spectroscopy. Very often, the signal that is recorded by the experimenter is the Fourier transform of some property of the object under study, such as the distribution of electron density. In X-ray or neutron crystallography, the diffraction patterns are the transform of the electron or neutron scattering length density distribution in the crystals, in electron crystallography that of the potential map. They are formed of discrete peaks or rings because of the crystalline nature of the sample. EM images of 2D crystals are Fourier-transformed mathematically in order to exploit the crystalline nature of the sample to extract signals from the noise. In small-angle X-ray or neutron scattering, the sample is disordered and the signal measured is a continuous Fourier transform, which, at small angles, contains information about long-distance correlations, such as the distribution of molecules in the solution (attraction, repulsion, random distribution...), at larger angles, about their size and mass, and, at still larger angles, about their shape and internal structure (see Chap. 9, Box 9.2).

In the present chapter, we are mostly concerned with the use of Fourier transforms to estimate resolution, that is the smallest distance between structural details that can be distinguished in an EM image or a reconstruction. In the first case, one can calculate a Fourier transform of the image of a thin amorphous film. The extension away from the center of the transform of the so-called Thon rings gives an indication of the instrumental resolution (compare the data collected using a direct-electron detection camera and a scintillator-based camera, Fig. 12.1; see also Fig. 12.21 in the main text).

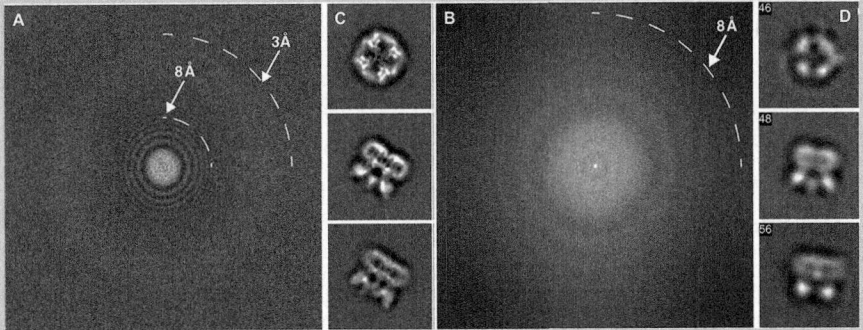

**Fig. 12.1** Thon rings give an indication of the resolution of a micrograph. Cryo-EM data are significantly improved with direct-detection cameras and motion correction, as illustrated here. (**A**) Fourier transform of a typical cryo-EM image of TRPV1 ion channel embedded in thin layer of vitreous ice recorded with a direct-electron detection camera, K2 Summit, after motion correction. Thon rings are visible up to ~3-Å resolution. The position corresponding to 8 Å is also marked. (**B**) Fourier transform of a similar image recorded using a phosphor scintillator-based CMOS camera, TemF816. Thon rings are visible up to ~8-Å resolution. (**C**) Three typical 2D class averages of TRPV1 particle images recorded with the K2 camera. (**D**) Similar 2D class averages of images recorded with the TemF816 one (From Liao et al. 2014, © 2014 Elsevier Ltd. All rights reserved).

Fourier transforms are also used to assess the resolution of 3D reconstructions. According to the "gold standard" approach (see main text, § 12.2.2), two 3D models of the distribution of density within the particles are reconstructed totally independently, and the Fourier transforms of their density distribution are compared. The farther a significant correlation extends – a Fourier shell correlation (FSC) of 0.143 being generally taken as the limit – the higher the resolution (Fig. 12.2).

**Box 12.1 (continued)**

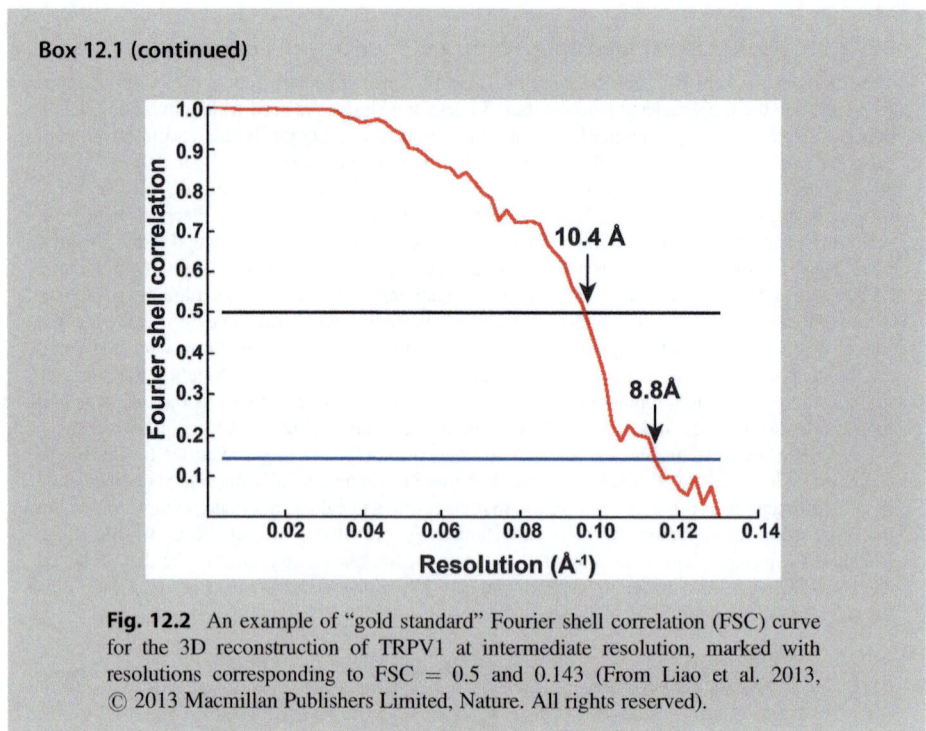

**Fig. 12.2** An example of "gold standard" Fourier shell correlation (FSC) curve for the 3D reconstruction of TRPV1 at intermediate resolution, marked with resolutions corresponding to FSC = 0.5 and 0.143 (From Liao et al. 2013, © 2013 Macmillan Publishers Limited, Nature. All rights reserved).

### 12.2.3   Single-Particle Cryo-EM Studies of Biological Molecules: The Resolution Revolution

It is no exaggeration to say that, since 2012, the field of single-particle cryo-EM has undergone a revolution (for analyses and comments, see e.g. Henderson 2013, 2015; Agard et al. 2014; Kühlbrandt 2014a, b; Liao et al. 2014; Bai et al. 2015a; Callaway 2015; Cheng 2015; Schröder 2015; Vinothkumar 2015; De Zorzi et al. 2016; Vinothkumar and Henderson 2016). This is principally due to two main factors, which combine to produce much better images. First, direct-electron detector device (DDD) cameras have been developed, which, not having to convert electrons into photons to be able to detect them, have a higher detective quantum efficiency and better signal-to-noise ratio than the CCD cameras used previously. Second, the underlying complementary metal-oxide semiconductor (CMOS) technology makes it possible to collect series of views of a given field during illumination, rather than a single averaged one (Fig. 12.5Bi). This permits to implement much more efficient corrections of the movements that the specimen undergoes under the electron beam (due, in particular, to annealing of the stresses in the vitrified water film) (Fig. 12.5Bii). The higher quality of the images, in turn (cf. Fig. 12.1C), improves the performances of the alignment and 3D reconstruction programs, which, themselves, have undergone considerable progress made implementable by the constant increase in computing power. Also of significant help are advances in computer control of EM multi-specimen handling and stage movement, which allow "driverless" running of the microscope

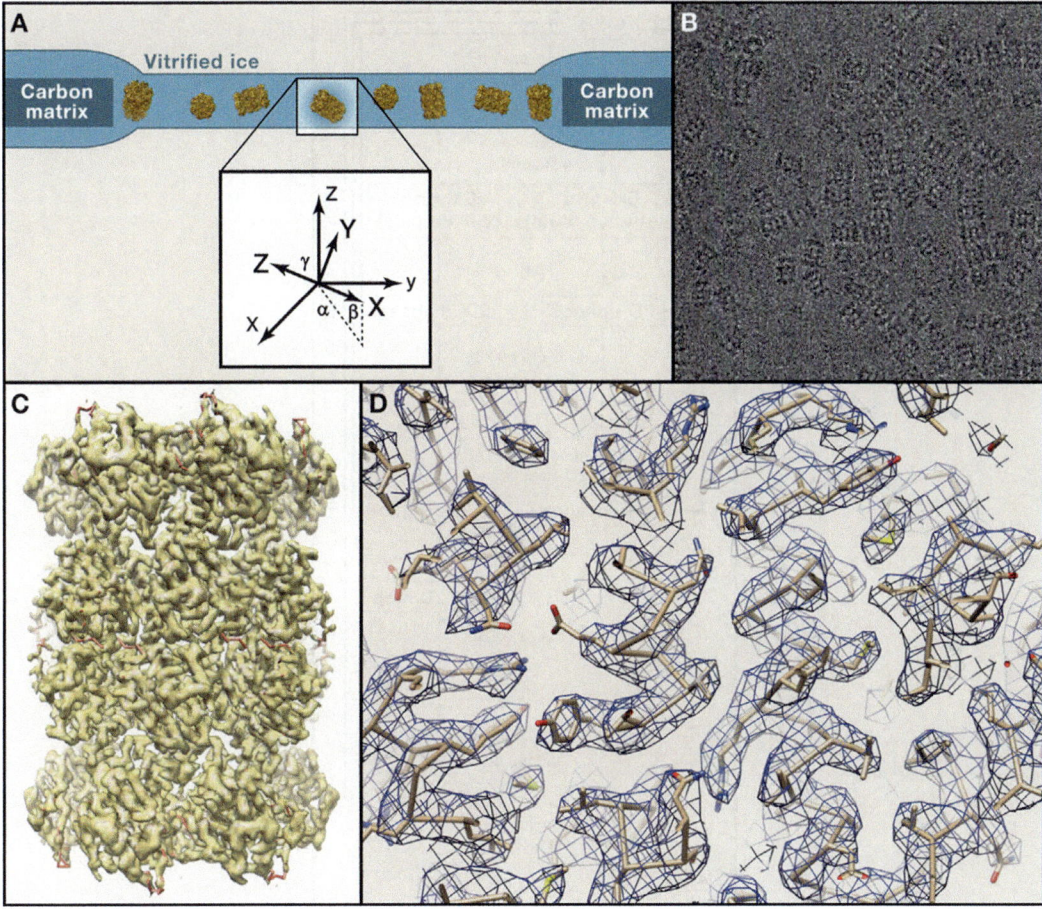

**Fig. 12.3** Principle of single-particle cryo-EM. (**A**) Purified biological molecules are embedded in a thin layer of vitreous ice, in which they ideally adopt random orientations. The orientations are specified by the in-plane position parameters, $x$ and $y$; depth with respect to the focal plane, $z$; and three Euler angles $\alpha$, $\beta$, and $\gamma$, which are refined iteratively to high accuracies (cf. Fig. 12.4). (**B**) Typical image of frozen-hydrated archaeal 20 S proteasomes. (**C**) 3D reconstruction of the 20 S proteasome at 3.3-Å resolution. (**D**) Side-chain densities of the map shown in C are comparable to those seen in maps determined by X-ray crystallography at a similar resolution (From Cheng 2015, © 2015 Elsevier Inc. All rights reserved).

for hours at a time in order to collect the massive amounts of data required. This progress, which makes it possible to collect and selectively average those images that are least affected by radiation damage and drift, has had the dramatic consequence that 3D structures derived from single-particle cryo-EM images have now reached or exceeded a resolution of ~4 Å. At this critical resolution, bulky amino acid side chains can be identified, and it becomes possible to directly fit atomic models into density maps, without having to rely on extraneous data such as X-ray models of individual subunits or domains (see Figs. 12.1, 12.6, and 12.7).

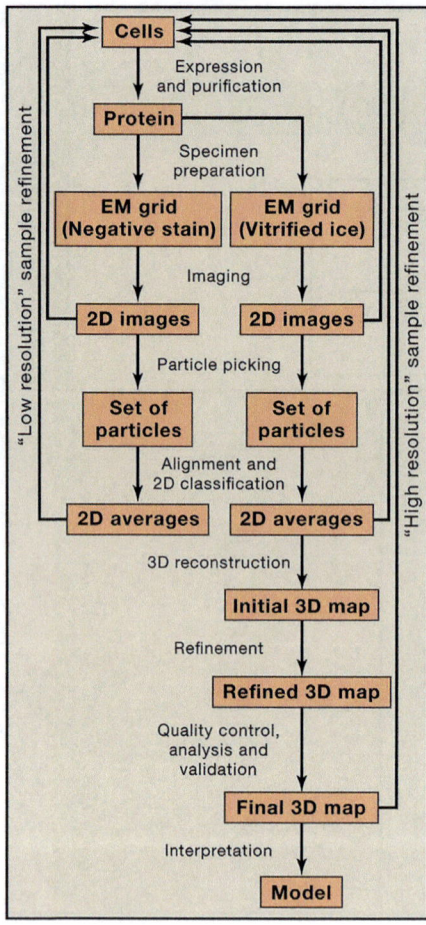

**Fig. 12.4** The steps involved in structure determination by single-particle cryo-EM. A single-particle project should start with a characterization of the specimen in negative stain (NS) (*left arm of the workflow*). Averaged NS images yield low-resolution (typically ~18-Å) two-dimensional (2D) projections or three-dimensional (3D) reconstructions of the envelope of the particles. Only once the EM images, or potentially 2D class averages, are satisfactory, i.e. the particles are monodisperse and show little aggregation and a manageable degree of heterogeneity ("low-resolution" sample refinement), is the sample ready for analysis by cryo-EM (*right arm of the workflow*). The images, 2D class averages, and 3D maps obtained with vitrified specimens may indicate that the sample requires further improvement to reach near-atomic resolution ("high-resolution" sample refinement) (From Cheng et al. 2015, © 2015 Elsevier Inc. All rights reserved).

## 12.2.4 Single-Particle Electron Microscopy Studies of Membrane Proteins

Cryo-EM studies of MPs present special challenges, essentially due to the presence of detergent. Among difficulties attributed to the detergent are the following:

(i) Biochemical instability of the target protein (see Chap. 2).
(ii) Loss of contrast (cf. Mazhab-Jafari and Rubinstein 2016; Stark and Chari 2016); this is probably not a serious problem with detergents with a low critical micellar concentration (CMC), like maltose neopentyl glycol (MNG) detergents or digitonin, which do not need to be present at a high concentration for MPs to remain soluble, but it can become one with detergents like CHAPS or CHAPSO, whose concentration must be kept above their high CMC of ~8 mM (~5 g·L$^{-1}$).

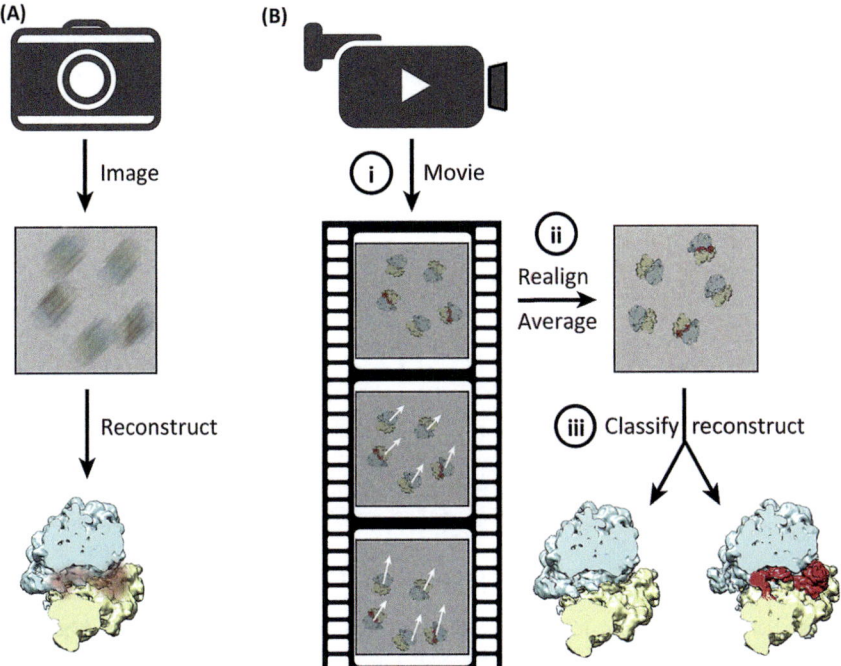

**Fig. 12.5** A schematic rendering of the impact of recent technological advances in cryo-EM on the quality of 2D images and 3D reconstructions. (**A**) Previously, noisier images were recorded on photographic film or CCD detectors, beam-induced sample motion led to image blurring, and images of structurally different particles could happen to be mixed in a single reconstruction, as, in this example, images of a complex with or without a ligand (*red*) bound. (**B**) Three recent advances yield better reconstructions: (*i*) digital direct-electron detectors yield data of unprecedented quality and allow recording movies during exposure; (*ii*) computer programs to realign the movie frames may correct for sample movements that are induced by the electron beam; it is also possible to eliminate frames that show excessive drift (early ones) or too much radiation damage (last ones); and (*iii*) powerful classification methods permit to sort out and separately reconstruct multiple structures from a heterogeneous sample, such as, here, the particle with and without bound ligand (From Bai et al. 2015a, © 2015 Elsevier Ltd. All rights reserved).

(iii) Uneven dispersion of the particles, which tend to be pushed away from the thin central region of the aqueous film and to crowd in the meniscus at the rim of the holes, where they lie in a thicker region of the film, lowering the contrast, and tend to interact with one another (see e.g. Flötenmeyer et al. 2007; Baker et al. 2015; Jeong et al. 2016; cf. below Figs. 12.9 and 12.26).

It could be hoped that APols would help with the first two problems, because (*i*) they tend to stabilize most MPs and MP complexes as compared to detergent solutions, and (*ii*) free APols do not need to be present in high concentration, and can even be totally removed if the risk of a degree of aggregation is assumed (see Chap. 5). It turned out that, unexpectedly, APols also seem to help with problem (*iii*).

## 12.3   The Use of Amphipols for Membrane Protein Electron Microscopy Studies

The literature in the field is somewhat hard to follow and analyze, for two reasons. First, at variance with other structural approaches such as NMR and crystallography, there have been few systematic

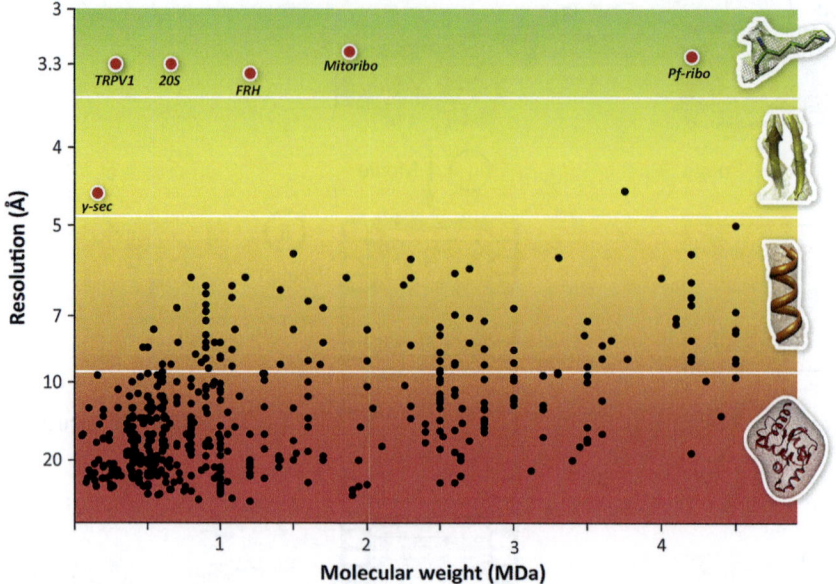

**Fig. 12.6** Recent technical progress has revolutionized cryo-EM single-particle analysis. The *black dots* represent single-particle cryo-EM structures that were released from the Electron Microscopy Data Bank (EMDB) between 2000 and 2012. The *red dots* are examples of post-2012 recent progress in the field: γ-secretase (γ-sec; Lu et al. 2014), the transient receptor potential cation channel subfamily V member 1 (TRPV1; Cao et al. 2013b; Liao et al. 2013), the 20 S proteasome (20 S; Li et al. 2013), the $F_{420}$-reducing [NiFe] hydrogenase (FRH; Allegretti et al. 2014), the large subunit of the yeast mitochondrial ribosome (mitoribo; Amunts et al. 2014), and the cytoplasmic ribosome of *Plasmodium falciparum* in complex with emetine (Pf-ribo; Wong et al. 2014). Whereas previously many structures only resolved protein domains (*red area*) or α-helices (*orange area*), recent structures are detailed enough to distinguish β-strands (*yellow area*) or even amino acid side chains (*green area*) (From Bai et al. 2015a, © 2015 Elsevier Ltd. All rights reserved).

studies of the factors that come into play and can make APols useful or useless auxiliaries for EM studies. Second, APols are often considered by electron microscopists as just an alternative "detergent," and no specific mention is made of their use in abstracts or keywords, nor, in many cases, is the APol literature referred to. A bibliographical search based on titles, keywords, or literature cited can therefore miss interesting studies – hence the absence of references to the Wilkens/Capaldi work on the $F_1F_O$ ATP synthase (Study 12.2 in Table 12.1) in our early general reviews on APols and their applications. In the following, I have done my best to offer an exhaustive survey of relevant publications, but I cannot guarantee that some potentially instructive studies have not been overlooked.

EM studies of MPs stabilized by APols have yielded much more novel biological information than have, to date, NMR or crystallographic studies. As a result, authors have tended to focus on the biological implications of their work and to spend little time on experimental details, or on the discussion of what APols contributed to their work and why. The description of experimental methods is often skimpy, with the source and exact nature of the APol used sometimes remaining unspecified and the protocol of transfer not described in detail. One can be left uncertain, for instance, whether free APol was separated from the complexes before EM grids were prepared, or about how much time elapsed between blotting the EM grid and freezing it, even though these can be critical parameters influencing the thickness of the film and particle dispersity (see below, § 12.3.4.2). This dearth of data is much to be regretted, because it complicates the interpretation of the results and slows down the identification of pitfalls and good practices in a field to which APols appear to have so much to

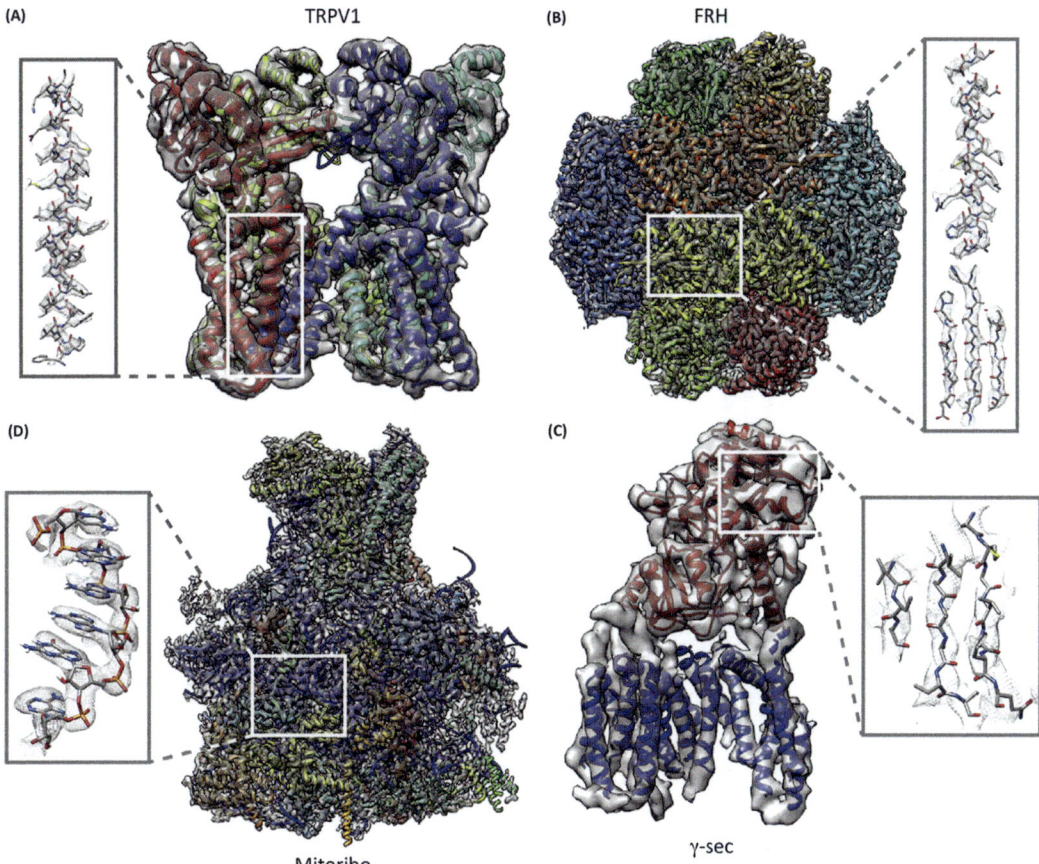

**Fig. 12.7** High-resolution cryo-EM maps with de novo-built atomic models. Atomic models and cryo-EM density maps are shown for the transient receptor potential cation channel subfamily V member 1 (TRPV1) ion channel (**A**) (Cao et al. 2013b; Liao et al. 2013), the $F_{420}$-reducing [NiFe] hydrogenase (FRH) (**B**) (Allegretti et al. 2014), the large subunit of the yeast mitochondrial ribosome (mitoribo) (**C**) (Amunts et al. 2014), and $\gamma$-secretase ($\gamma$-sec) (**D**) (Sun et al. 2015). The resolution of the first three structures is close to 3 Å (Fig. 12.6), and density for many side chains and individual RNA bases is visible in these maps; the resolution of the $\gamma$-secretase structure is 4.5 Å, at which bulky side chains are visible and $\beta$-strands are well resolved (*insets*) (The composite figure is from Bai et al. 2015a, © 2015 Elsevier Ltd. All rights reserved).

contribute. In the following, I will leave aside the biological implications of the works presented, which can be fascinating but are beyond the scope of the present text, and try to extract from published papers technical information that can be relevant to the use of APols for EM studies of other biological systems. Table 12.1 lists the 27 MP studies that had come to my attention by the time the writing of this chapter was completed, plus one study (Study 12.20) on a non-membrane system that may shed some light on what is going on with MPs. The speed at which SP-EM studies of APol-trapped MPs are published has become very rapid. A dozen more studies that were published or came to my attention too late to be included in the table and discussions are mentioned at the beginning of § 12.3.2 (NS studies) and the end of 12.3.3 (cryo-EM studies). In the following, biological background and experimental details are kept to a minimum except when they appear particularly relevant to a rational use of APols. A quick summary of each study and some complementary information, such as the molecular masses of the complexes studied, can be found in the table.

**Table 12.1** Electron microscopy single-particle studies of membrane proteins using amphipols. Color code: *buff*, scanning transmission EM (STEM); *blue*, cryo-EM (most cryo-EM studies also present NS data); *green*, transmission electron microscopy (TEM) after negative staining (NS); *white*: a non-membrane protein. The resolution indicated is that of averaged three-dimensional maps and corresponds, when it is available, to the "gold standard" resolution at the 0.143 threshold of the Fourier shell correlation curve. When several figures are given, they refer to different structures established in the same study.

| Study | Membrane protein (MW) | Amphipol | EM methods | Resolution (Å) | Main observations relevant to the use of amphipols for EM | References |
|---|---|---|---|---|---|---|
| 12.1 | Cytochrome $b_6f$ complex from *Chlamydomonas reinhardtii* (211 and 83–86 kDa) | A8-35, A8-75 | Cryo-STEM | | Cytochrome $b_6f$ purified in the detergent Hecameg was trapped by A8-35 either under its native state, a superdimer comprising 14 subunits, or as a delipidated, lighter, monomeric form lacking some subunits. A good consistency was observed between the masses of the two forms calculated from their biochemical composition and those determined by cryo-STEM, showing that trapping in A8-35 preserved the original aggregation state of the protein in detergent solution. | Tribet et al. (1998) |
| 12.2 | $F_1F_0$ ATP synthase from *Escherichia coli* (~560 kDa) | A8-75 | Negative staining + cryo-EM | | The $F_1F_0$ ATP synthase from *Escherichia coli* was purified in the detergent taurodeoxycholate, transferred to A8-75 either by affinity chromatography or by ultracentrifugation, and imaged after negative staining with uranyl acetate. A 3D model was reconstructed at an unspecified resolution. Preliminary cryo-EM data are presented. | Wilkens and Capaldi (1998a,b), Wilkens (2000), and Wilkens et al. (2000) |
| 12.3 | Mitochondrial NADH:ubiquinone oxidoreductase (Complex I) from *Neurospora crassa* (~1.12 MDa) | A8-35 | Cryo-EM | ~58 | Complex I from *Neurospora crassa* was solubilized and purified in Triton X-100 and transferred to A8-35 by Bio-Beads adsorption of the detergent. Cryo-EM images of quick-frozen Complex I/A8-35 samples were used for computer-based single-particle averaging and 3D reconstruction to ~58-Å resolution. The reconstruction thus obtained is very similar to that previously obtained in *n*-dodecyl-$\beta$-D-maltoside (DDM) after negative staining, filtered at the same resolution. The potential of APols for cryo-EM is discussed. | Popot et al. (2003) and Flötenmeyer et al. (2007) |
| 12.4 | Bacteriorhodopsin from *Halobacterium salinarum* (27 kDa) | A8-35 | Negative staining | | Bacteriorhodopsin (BR) was solubilized in octylthioglucoside, along with the archaebacterial lipids present in the purple membrane, and transferred to A8-35 using various protocols. The functionality of BR was established by measuring its photocycle. The composition, size, and organization of the BR/lipid/A8-35 complexes were studied by a combination of biochemical analyses, spectroscopy, SANS, SEC, AUC, and TEM after negative staining. In the presence of excess A8-35, BR is monomeric and surrounded by a belt of A8-35. After removal of free APol, BR is observed to aggregate and, after several months of storage, to self-associate side-by-side into long filaments. | Gohon et al. (2008) |
| 12.5 | Mitochondrial respirasome from bovine mitochondrion (~1.7 MDa) | A8-35 | Cryo-EM | 19 | The respiratory supercomplex $I_1III_2IV_1$ was solubilized in digitonin from mitochondrial inner membranes, transferred to A8-35 in the presence of $\gamma$-cyclodextrin, and purified on a surfactant-free sucrose gradient. Cryo-EM images were collected and a 3D reconstruction built to 19-Å resolution. After fitting the three subcomplexes (Complex I, the $bc_1$ superdimer, and cytochrome $c$ oxidase) into the overall envelope, a band of unaccounted for electron density ~2 nm thick and ~4 nm wide, corresponding to the APol belt, is seen to follow the TM surface of the super-complex. In a subsequent study (Study 12.24), the resolution was pushed to 9Å. | Althoff (2011) and Althoff et al. (2011) |
| 12.6 | Transient receptor potential ankyrin 1 channel (TRPA1) (~525 kDa) | A8-35 | Negative staining | ~16 | The mouse TRPA1 channel, a member of the transient receptor potential (TRP) channel super-family, was expressed in *Saccharomyces cerevisiae*, solubilized and purified using the detergent Fos-Choline-12, and transferred to A8-35 by dilution and dialysis in the presence of Bio-Beads. TEM images of uranyl acetate-stained TRAP1/A8-35 particles were averaged to yield a 16-Å resolution 3D structure into which were fitted molecular models of the *N*- and *C*-termini of the channel. | Cvetkov et al. (2011, 2014) |
| 12.7 | Human CRFR1 and mouse CRFR2β (~47 and ~49 kDa) | NVoy | Negative staining | | In the frame of a more general study of the use of NVoy (NV10) as a recipient medium for membrane protein (MP) cell-free expression, images are presented of two NVoy-complexed, negatively-stained GPCRs (corticotropin-releasing factor receptors CRFR1 and CRFR2β) averaged to an unspecified resolution. | Klammt et al. (2011) |

(continued)

**Table 12.1** (continued)

| Study | Membrane protein (MW) | Amphipol | EM methods | Resolution (Å) | Main observations relevant to the use of amphipols for EM | References |
|---|---|---|---|---|---|---|
| 12.8 | Human arginine-vasopressin type 2 receptor (V2R) (~40 kDa) | Non-ionic amphipol | Negative staining | | In the frame of a fluorescence study of ligand-induced GPCR conformation transitions, images are shown of a negatively-stained arginine-vasopressin type 2 receptor (V2R) complexed by a non-ionic amphipol, averaged to an unspecified resolution. | Rahmeh et al. (2012) |
| 12.9 | Capsaicin (vanilloid) receptor transient receptor potential channel (TRPV1) (~300 kDa) | A8-35 | Cryo-EM | 3.4-4.2 | TRPV1, the capsaicin (vanilloid) receptor TRP channel, is a heat-activated cation channel that is modulated by inflammatory agents and contributes to acute and persistent pain. The rat receptor was expressed in HEK cells, solubilized and purified in DDM, and transferred to A8-35 in the presence of Bio-Beads. Cryo-EM images were collected using direct electron detection and new image processing algorithms to correct motion-induced image blurring and improve signal and contrast. 3D reconstructions of TRPV1 in the presence and absence of ligands were built to 3.4-4.2-Å resolution, allowing direct fitting into the electron density maps of molecular models. | Cao et al. (2013) and Liao et al. (2013, 2014) |
| 12.10 | Peripherin-ROM1 tetramer (~153 kDa) | A8-35 | Negative staining | | Tetramers of the peripherin-ROM1 complex were purified from bovine retinas using DDM, supplemented with A8-35, and the detergent and excess APol removed by SEC. Uranyl acetate-stained samples were imaged by TEM and a 3D reconstruction calculated to ~18-Å resolution. | Kevany et al. (2013) |
| 12.11 | ATP-binding cassette transporter ABCA4 (~260 kDa + ~20 kDa sugars) | A8-35 | Negative staining | | ABCA4, a member of the ATP-binding cassette (ABC) transporter family, is involved in the transport of N-retinylidene-phosphatidylethanolamine to the outer side of retina disc membranes. It was purified from bovine retinas using DDM and transferred to A8-35 by SEC. Images of uranyl acetate-stained samples were averaged to yield a 3D reconstruction at 18-Å resolution. | Tsybovsky et al. (2013) |
| 12.12 | Transient receptor potential melastatin-1 (TRPM1) (~360 kDa) | A8-35 | Cryo-EM | 22 | Study 12.10 presents a medium-resolution cryo-EM study of yet another TRP, transient receptor potential melastatin-1 (TRPM1), as part of a broader study of the oligomeric structure of this protein involved in dim-light vision. TRPM1 was expressed in sf9 insect cells, solubilized and purified in Fos-Choline-14, and transferred to A8-35 using Bio-Beads. No SEC step is mentioned, so the samples used for EM probably still contained excess APol. They were applied to holey grids with a thin continuous carbon film coating, blotted, frozen in liquid ethane, and imaged using an electron microscope equipped with a conventional CCD camera. Models generated independently from even and odd halves of the data were used to generate an FSC curve, which indicated a resolution of 22 Å at the 0.143 threshold. The reconstructed structure exhibited C2 symmetry, consistent with the biochemical indications that the complex is a homodimer. This is, however, somewhat puzzling given the homology of the putative pore-forming region with that of tetrameric TRPs such as TRPA1 or TRPV1. | Agosto et al. (2014) |
| 12.13 | OmpF trimer (~111 kDa) | A8-35 | Negative staining | | Native, trimeric OmpF samples, containing lipopolysaccharide (LPS), were produced by solubilization of E. coli outer membranes and purified in octyl-polyoxyethylene ($C_8$-POE). LPS-free samples were produced as inclusion bodies and folded in $C_8$-POE. Both types of samples were transferred to A8-35 using Bio-Beads and separated from free APol by SEC. EM samples were stained with uranyl acetate. LPS-free samples depleted of unbound APol were found to be organized into long filaments, in which OmpF trimers are associated side by side, adsorbed onto the carbon film by their transmembrane side. The filaments dissociated upon adding back free APol. LPS-containing samples tended to organize into 2D sheets under the influence of LPS-LPS interactions mediated by multivalent cations, which were disrupted upon addition of EDTA. | Arunmanee et al. (2014) |

| Study | Membrane protein (MW) | Amphipol | EM methods | Resolution (Å) | Main observations relevant to the use of amphipols for EM | References |
|---|---|---|---|---|---|---|
| 12.14 | CsgG/CsgE complexes (~410 kDa, + covalently bound lipids) | A8-35 | Cryo-EM | ~24 | Curli are functional amyloid fibers that constitute the major protein component of the extracellular matrix in the biofilms formed by various bacteria. Their formation requires a dedicated protein secretion machinery comprising the outer membrane lipoprotein CsgG and two soluble accessory proteins, CsgE and CsgF. CsgG was transferred from LDAO to A8-35 by SEC, supplemented with CsgE, and uncomplexed CsgG removed by IMAC. A8-35-trapped CsgG was imaged after negative staining with uranyl acetate and the CsgG/CsgE complexes studied by cryo-EM. The resolution of the cryo-EM reconstruction was estimated to be ~24 Å at FSC = 0.5. | Goyal et al. (2014) |
| 12.15 | γ-secretase (~170 kDa, + oligosaccharides) | A8-35 | Cryo-EM | 3.4–4.5 | The γ-secretase complex, comprising presenilin 1 (PS1), PEN-2, APH-1, and nicastrin, is a membrane-embedded protease that controls a number of important cellular functions through substrate cleavage. In Lu et al. (2014) the four components of the human γ-secretase complex were co-expressed in mammalian HEK293F cells, extracted and purified in CHAPSO, and transferred to digitonin (yielding poorly exploitable cryo-EM images), and then to A8-35, using Bio-Beads. Free APol was removed by SEC. The cryo-EM structure was solved to 4.5-Å resolution. In a subsequent paper (Bai et al. 2015c), improved data collection and the use of a larger sample pushed the resolution to 3.4 Å. In Bai et al. (2015b), image classification procedures are used to characterize molecular plasticity at the secondary structure level, revealing the presence of three distinct conformations and that of an additional transmembrane helix. The structure of γ-secretase in complex with a dipeptidic inhibitor is established. In ref. (Sun et al. 2015), T4 lysozyme was fused to the N-terminus of PS1 and imaging carried out in digitonin. The resolution of the complex as a whole was only marginally better (4.32 Å) than in (Lu et al. 2014), but that of the TM region improved sufficiently for all TM α-helices to be convincingly identified. The structures obtained in A8-35 and in digitonin are nearly identical. | Lu et al. (2014), Bai et al. (2015a,b) and Sun et al. (2015) |
| 12.16 | ExbB–ExbD complex (139 kDa) | A8-35 | Negative staining | | Nutrient import across the outer membrane of Gram-negative bacteria is powered by the proton-motive force, delivered by the plasma membrane MP complex ExbB-ExbD-TonB. The ExbB₅–ExbD₂ complex from E. coli was extracted and purified with DDM and either studied in DDM or transferred to A8-35 using Bio-Beads. 3D structures were reconstructed to ~21–27-Å resolution from images of samples stained with uranyl acetate. | Sverzhinsky et al. (2014) |
| 12.17 | P-glycoprotein (PgP) (~170 kDa) | SMALPs | Cryo-EM | ~35 | The ABCB1 P-glycoprotein (PgP), an ABC drug transporter, was expressed in High Five insect cells and directly extracted from membrane fragments into SMALPs. The particles were imaged in vitreous ice and a 3D model reconstructed to ~35-Å resolution. | Gulati et al. (2014) |
| 12.18 | TRPV2 channel (~385 kDa) | A8-35 | Cryo-EM | ~15 | The TRPV2 channel is a high-threshold thermosensor involved in calcium signaling during inflammation and nerve injury. It was expressed in Saccharomyces cerevisiae, purified in detergent, and transferred to A8-35. The TRPV2/APol complexes were imaged in vitreous ice and a 3D model reconstructed at ~15-Å resolution. To-date seems to have been reported only in abstract form, but see Study 12.25. | Fan et al. (2014) |
| 12.19 | Type 1 ryanodine receptor (RyR1) (~2.3 MDa) | A8-35 | Cryo-EM | | Ryanodine receptors (RyRs) are tetrameric ligand-gated channels. They are responsible for the release of Ca²⁺ from the sarcoplasmic reticulum and the resulting increase of cytosolic Ca²⁺ concentration, leading to muscle contraction. The enormous size and peculiar shape of RyRs make its structure more amenable to cryo-EM than to X-ray crystallography or NMR. This review of cryo-EM studies of RyRs presents original data collected on Type 1 RyR trapped in A8-35. Substituting A8-35 to CHAPS results in i) more evenly distributed particles throughout the water film formed on holey-carbon grids and ii) higher contrast. However, the particles tend to adopt a uniform orientation, with their long axis normal to the plane of the film. Supplementing the preparation with a submicellar concentration of OG (~10% of the cmc) results in a random distribution of orientations. This is one of the few studies where hypotheses about the reasons for improved cryo-EM data of APol-trapped vs. detergent-solubilized MPs are critically discussed. | Baker et al. (2015) |

(continued)

**Table 12.1**  (continued)

| Study | Membrane protein (MW) | Amphipol | EM methods | Resolution (Å) | Main observations relevant to the use of amphipols for EM | References |
|---|---|---|---|---|---|---|
| 12.20 | Dynactin (not a MP) | A8-35 | Cryo-EM | | Dynactin is not a MP. Nevertheless, mixing 0.025% (w/v) A8-35 with the sample immediately before applying it on the grid was observed to aid in dispersing dynactin particles across the aqueous films stretched over the holes in the carbon film, which may bear on similar observations made with MPs. | Chowdhury et al. (2015) |
| 12.21 | TRPA1 (~525 kDa) | PMAL-C8 | Cryo-EM | 3.9–4.7 | The TRPA1 ion channel (also known as the wasabi receptor) is a detector of noxious chemical agents encountered in the environment or produced endogenously during tissue injury or drug metabolism. A TRPA1-maltose binding protein fusion protein was transferred from MNG-3 to PMAL-C8 using Bio-Beads and imaged by cryo-EM. The resolutions of the reconstructed models (with one agonist or one or two antagonists bound) were estimated to be 4.24 Å, 3.9 Å and 4.7 Å, respectively, at FSC = 0.143. | Paulsen et al. (2015) |
| 12.22 | AcrB trimer (~360 kDa) | SMALPs | Negative staining | 23 | A polyhistidine-tagged version of the multidrug transporter AcrB was expressed in *E. coli*, directly extracted from membrane fragments into SMALPs, purified by IMAC, and imaged after staining with uranyl acetate. The particles appeared as either "singlets" or "doublets," the images of which were separately analyzed. A 3D model of "singlets," reconstructed to a final resolution of 23 Å, is consistent with the known X-ray structure of AcrB trimer, with SMAL forming a belt around the transmembrane region. Class analyses of the "doublets" show a variety of association modes, all of them seeming to involve contacts between the SMAL belts of two AcrB trimers. Control experiments suggest that the formation of doublets occurs in solution, and not on the EM grid. The lack of any preferred doublet interface suggests that particle association occurs non-specifically via the SMAL belts and does not involve any defined protein–protein interface. | Postis et al. (2015) |
| 12.23 | Holotranslocon (SecYEG, SecDF-YajC; YidC; *E. coli*) (~250 kDa) | A8-35 | Cryo-EM | 14–15 | An integrative approach combining small-angle neutron scattering (SANS), low-resolution electron microscopy, and biophysical analyses was used to determine the arrangement of the proteins and lipids within the holotranslocon super-complex. Cryo-EM images were collected on both cross-linked and native holotranslocon complexes transferred from DDM to A8-35 using Bio-Beads. The data were analyzed to 14–15-Å resolution (FSC = 0.143), revealing a hollow, lipid-filled interior, consistent with SANS data, whose functional role in MP insertion is discussed. | Botte et al. (2016) |
| 12.24 | Zebrafish retinol-binding protein receptor (STRA6) (~180 kDa) | A8-35 | Cryo-EM | 3.9 | The STRA6 receptor mediates cellular uptake of vitamin A by recognizing retinol-binding protein/retinol and triggering release and internalization of retinol. The structure of zebrafish STRA6 has been determined to 3.9-Å resolution by single-particle cryo-EM. STRA6 features one intramembrane and nine transmembrane helices in an intricate dimeric assembly. The presence of a deep lipophilic cleft that is open to the membrane suggests a possible mode for internalization of retinol through direct diffusion into the lipid bilayer. | Chen et al. (2016) |
| 12.25 | Human transient receptor potential channel type melastatin 4 channel (TRPM4) (~550 kDa) | A8-35 | Negative staining | | Mutations in the transient receptor potential channel type melastatin 4 (TRPM4) have been linked to various cardiovascular diseases. In the frame of a broader study, NS images of TRPM4 illustrate its tetrameric organization. | Constantine et al. (2016) |
| 12.26 | Engineered AcrAB-TolC multidrug efflux pump (~760 kDa) | A8-35 | Cryo-EM | 8.2 | The structure of an A8-35-trapped complex consisting of an AcrAB fusion protein and a chimeric TolC was solved to 8.2-Å resolution by single-particle cryo-EM. The authors note that, in detergent (DDM) solution, the particles were heavily concentrated near the edge of the holes of carbon EM grids, yielding images unfit for processing. The particle distribution was "tremendously improved" when the detergent was replaced by A8-35. | Jeong et al. (2016) |

| Study | Membrane protein (MW) | Amphipol | EM methods | Resolution (Å) | Main observations relevant to the use of amphipols for EM | References |
|---|---|---|---|---|---|---|
| 12.27 | Mitochondrial respirasome from bovine mitochondrion (~1.7 MDa) | A8-35 | Cryo-EM | 9.1–10.4 | A sequel to Study 12.5, where the resolution was pushed to ~9-Å resolution. Sample preparation was improved by performing the initial purification in the detergent PCC-a-M, rather than digitonin, before transfer to A8-35, yielding biochemically more homogeneous samples, the images of which could be distributed into three classes, whose structures were solved respectively to 9.1, 10.4, and 9.9-Å resolution. The improved structure shows that most protein-protein contacts between complexes I, III, and IV in the membrane are mediated by supernumerary subunits. Of the two Rieske iron-sulfur cluster domains in the complex III dimer, one is resolved, indicating that this domain is immobile and therefore unable to transfer electrons, The functional asymmetry of complex III provides strong evidence for directed electron flow from complex I to complex IV through the active complex III monomer. | Sousa et al. (2016) |
| 12.28 | HasR heme transporter and HasR/HasA complex from *Serratia marcescens*. (~95 and ~115 kDa) | A8-35 | Negative staining | 20 | HasR, a β-barrel MP from the outer membrane of *Serratia marcescens*, imports heme delivered by HasA, an extracellular heme carrier protein. Following trapping with A8-35 and negative staining, the structures of HasR and the HasR/HasA complex were solved to 20-Å resolution, revealing that the HasR periplasmic domain responsible for signal transfer *i*) is highly flexible in two stages of signalling; *ii*) extends into the periplasm approximately 70–90 Å away from the TM β-barrel; and *iii*) exhibits local conformational changes in response to the binding of heme-loaded HasA. | Wojtowicz et al. (2016) |
| 12.29 | Rabbit transient receptor potential vanilloid cation channel 2 (TRPV2) (~280 kDa) | A8-35 | Cryo-EM | ~4 | TRPV2, a member of the TRPV family, is regulated by temperature, by ligands, such as probenecid and cannabinoids, and by lipids. It has been implicated in many biological functions, including somatosensation, osmosensation, and innate immunity. TRPV2 was transferred from a DDM/lipid mixture to A8-35 using Bio-Beads. Cryo-EM single-particle images were analyzed to a resolution of ~4 Å, with regions resolved to 3.3 Å. Comparison with the TRPV1 structures suggests mechanisms for activation and desensitization. Putative bound lipids are identified. | Zubcevic et al. (2016) |
| 12.30 | *Caenorhabditis elegans* cyclic-nucleotide-gated channel (TAX-4) | A8-35 | Cryo-EM | 3.5 | Cyclic-nucleotide-gated channels are essential for vision and olfaction. They belong to the voltage-gated ion channel superfamily but their activities are controlled by intracellular cyclic nucleotides instead of transmembrane voltage. The structure of a cyclic-nucleotide-gated channel from *Caenorhabditis elegans* in the cyclic guanosine monophosphate (cGMP)-bound open state trapped in A8-35 has been solved to 3.5-Å-resolution by single-particle cryo-EM. | Li et al. (2017) |
| 12.31 | Pneumolysin pore (*Streptococcus pneumoniae*) (~2.2 MDa) | A8-35 | Cryo-EM | 4.5 | Pneumolysin, the main virulence factor from *Streptococcus pneumoniae*, forms large pores in the plasma membrane of target cells, resulting in cell lysis. The toxin was expressed under its soluble, monomeric form, incubated with liposomes, and the resulting proteoliposomes solubilized in Cymal-6. Following addition of A8-35, the detergent was removed by dialysis. Cryo-EM images were analyzed to 4.5-Å resolution. The pores comprise 42 monomers assembled in a ring, each monomer featuring four TM β-strands, resulting in a 168-strand, 260-Å-wide β-barrel. A model for the step-wise transition from the soluble monomer to the final pore is proposed. | van Pee et al. (2017) |
| 12.32 | AcrABZ-TolC multidrug efflux pump (*E. coli*) (~770 kDa) | A8-35 | Cryo-EM | 3.6–6.5 | Bacterial efflux pumps confer multidrug resistance by expelling diverse antibiotics from the cell. The components of the AcrABZ-TolC pump were co-expressed and the complex solubilized and purified in DDM solution. The purified complex was supplemented with A8-35 and the detergent removed using Bio-Beads. Cryo-EM images of AcrABZ-TolC in the presence or absence of an inhibitor were analyzed respectively to 3.6-Å and 6.5-Å resolution (FSC = 0.143). Comparison of the two models reveals a quaternary structural switch that allosterically couples and synchronizes initial ligand binding with channel opening. | Wang et al. (2017) |
| 12.33 | Human transient receptor potential channel polycystin-2 (PC2) (~440 kDa) | A8-35 | Cryo-EM | 4.2–4.3 | Polycystin-2 (PC2), a calcium-activated cation TRP channel, is involved in diverse $Ca^{2+}$ signaling pathways. PC2 was expressed in HEK cells and solubilized and purified in LMNG/cholesterol hemisuccinate solutions. Purified PC2 was supplemented with A8-35 and the detergent removed using Bio-Beads. Analysis of cryo-EM images yielded two conformations, with resolutions of 4.2 and 4.3 Å, corresponding to a more or less open pore binding more or less $Ca^{2+}$ ions. | Wilkes et al. (2017) |

### 12.3.1 The Early Years: 1998–2008

As recalled in the introduction, the first EM study of MP/APol complexes (Study 12.1 in Table 12.1) was a STEM study, aimed at establishing whether trapping with APols respected the dispersity of the protein in detergent solution. The test MP, cytochrome $b_6 f$, was prepared in solutions of the detergent Hecameg under two forms: an intact superdimer, comprising bound lipids, and a delipidated, lighter form, which is a super-monomer that has lost some of the subunits present in the dimeric form (Breyton et al. 1997). The two forms were transferred to APols A8-35 or A8-75 by mixing, dilution under the CMC of Hecameg, and ultracentrifugation in surfactant-free sucrose gradients. Expected masses for each APol-trapped form were calculated on the basis of the known subunit composition and stoichiometry and the measured binding of lipids and A8-75 to the heavy form, the delipidated light form being assumed to bind half the amount of APol measured for the heavy form (probably an underestimate) (Tribet et al. 1997). STEM mass measurements yielded a mass of $312 \pm 60$ kDa for the heavy form trapped in A8-75 (expected, $332 \pm 13$ kDa) vs. $133 \pm 21$ kDa for the light one (expected, $108 \pm 5$ kDa). Measurements on $b_6 f$/A8-35 complexes yielded a similar agreement (Tribet et al. 1998). This study established (*i*) that APols appear to trap MPs under the oligomeric state that pre-existed in detergent solution and (*ii*) that MP/APol complexes could be dispersed on glow-discharged carbon-coated grids for EM observation.

In Study 12.2 (Wilkens 2000; Wilkens and Capaldi 1998a, b; Wilkens et al. 2000), polyhistidine-tagged $F_1 F_O$ ATP synthase from *Escherichia coli* was solubilized in the detergent taurodeoxycholate, bound onto a column of immobilized nickel, and eluted in the presence of A8-75. The particles were applied to carbon-coated EM grids and stained with uranyl acetate. The stain revealed a second stalk, connecting the stator to the rotor, which had been expected on theoretical grounds but had not been observed by cryo-EM imaging of detergent-solubilized $F_1 F_O$ or in the X-ray structure of $F_1$ (Fig. 12.8). Preliminary cryo-EM data are also shown in Wilkens (2000). This series of studies

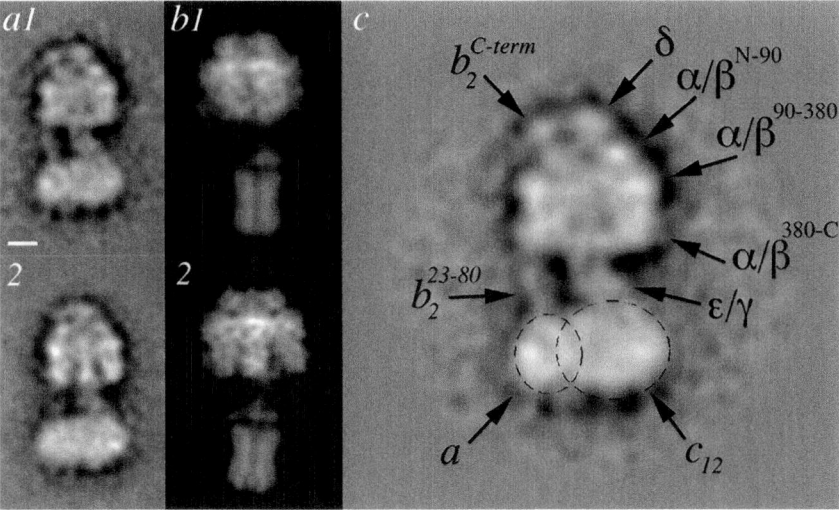

**Fig. 12.8** Image analysis of A8-75-trapped ATP synthase from *Escherichia coli*. A data set of ~3000 $F_1 F_O$ molecules, negatively stained with uranyl acetate, was analyzed by alignment and classification procedures. **a.** The bi- and trilobed projections of the enzyme. Bar: 5 nm. **b.** Projection images of the 4.5-Å X-ray structure (From Stock et al. 1999; PDB code 1QO1) showing the bi- and trilobed views of the $F_1$ domain. **c.** Interpretation of the structural features seen in the EM projections. Letters refer to ATP-synthase subunits (From Wilkens 2000).

established, for the first time, that APol-trapped MPs could be studied by both NS and cryo-EM approaches. These observations, however, seem to have remained largely unnoticed outside of the ATP-synthase community.

A cryo-EM study with reconstruction of a 3D model of a MP complex was first presented in Study 12.3, taking the mitochondrial Complex I as a test MP (Flötenmeyer et al. 2007). Complex I (NADH:ubiquinone oxidoreductase) was purified from the fungus *Neurospora crassa* in the presence of the detergent Triton X-100 and transferred to A8-35 using Bio-Beads SM-2 and ultrafiltration. It is likely that the latter procedure removed most if not all of the free APol, a significant point in view of some aspects of the cryo-EM observations (see below). A comparison of cryo-EM grids prepared either with the Triton X-100-solubilized or with the A8-35-trapped preparation is shown in Fig. 12.9. Panels a, c, and e show samples in Triton X-100, which were blotted on perforated carbon films and fast-frozen in liquid ethane. The resulting ice films show a pronounced thickness gradient across the holes (Fig. 12.9E), and the protein tends to clump and aggregate at the edges (Fig. 12.9C). With samples of A8-35-trapped Complex I (Panels b, d, f), an ice film much more uniform in thickness is obtained (Panel f), in which individual Complex I particles are better dispersed and can be more easily identified (Panel d) and picked for image processing. This phenomenon considerably facilitated the collection of a sufficient number of images of isolated particles. Its cause and possible ways to control it will be discussed below (§ 12.3.4.2).

*Amphipols and electron cryomicroscopy of membrane proteins*
(© 2018 Francis Haraux)

A total of 1200 particles were selected, normalized, and subjected to multivariate analysis (van Heel and Frank 1981). The top 16 class averages are shown in Fig. 12.10A. The class average projections were then cross-correlated with a set of angular projections of a 3D model taken from the reconstruction of negatively stained Complex I (Guénebaut et al. 1998). The matching projections of the model are shown in Fig. 12.10B. There is good agreement between the class average projections of the cryo-EM images and the corresponding projections of the negative-stain model, making it possible to use the latter as a reference for the angular reconstitution method (Serysheva et al. 1995).

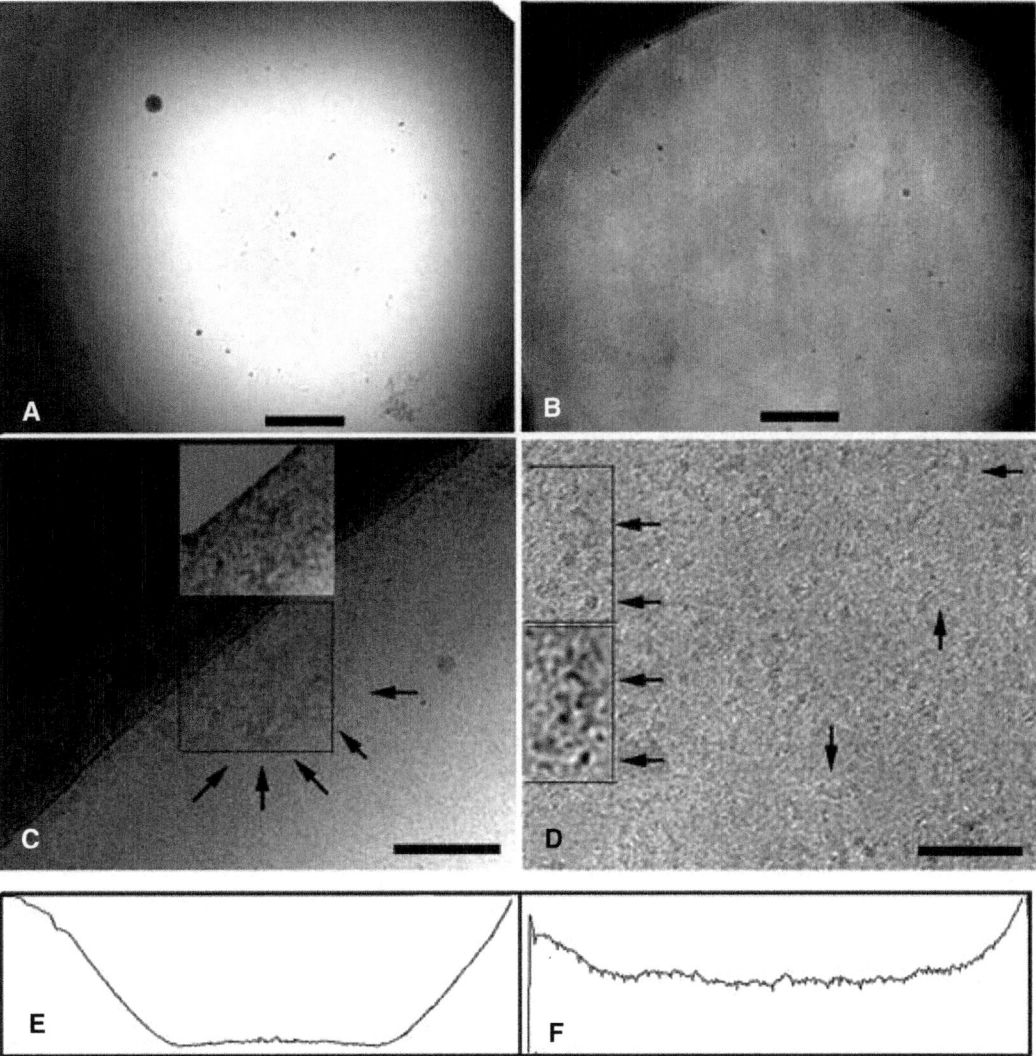

**Fig. 12.9** Comparison of ice thickness and particle distribution in samples of detergent-solubilized vs. A8-35-trapped Complex I. (**A**) Low-magnification image of an ice film covering a fenestrated grid hole in the presence of Triton X-100; scale bar: 0.7 μm. (**B**) A similar image of the ice film for an A8-35-trapped sample; scale bar: 0.7 μm. (**C**) Higher-magnification image of a sample in detergent showing clumping of Complex I particles around the edge of the hole, where the ice is thicker (*arrows*); scale bar: 100 nm. (**D**) Image of frozen A8-35-trapped Complex I, showing a more even distribution of particles across the hole (*arrows* mark some of the particles); scale bar: 100 nm. The *insets* in (**C, D**) are low-pass filtered images of the adjacent boxed regions, in which the Complex I particles can be seen more clearly. (**E, F**) Density scans for equal areas across the center of the images shown in (**A, B**). In the case of the detergent solution **E**, there is a marked thinning of the ice toward the center of the hole, which is much less pronounced for the APol-trapped preparation (**F**). The horizontal scale in (**E, F**) is the same as in (**A, B**). The vertical scale is the same in (**E, F**) and is in arbitrary ice density units. If the thickness of the carbon film at the edge of the hole is taken to be 30 nm (Karlsson 2001), the thickness of the ice at the center of the hole can be estimated to be ~2 nm for the detergent-solubilized preparation and ~15 nm for the APol-trapped one. The apparently higher noise in (**F**) is density variation caused by the presence of randomly distributed Complex I particles in (**B**) (From Flötenmeyer et al. 2007, © 2007 John Wiley & Sons, Inc. All Rights Reserved).

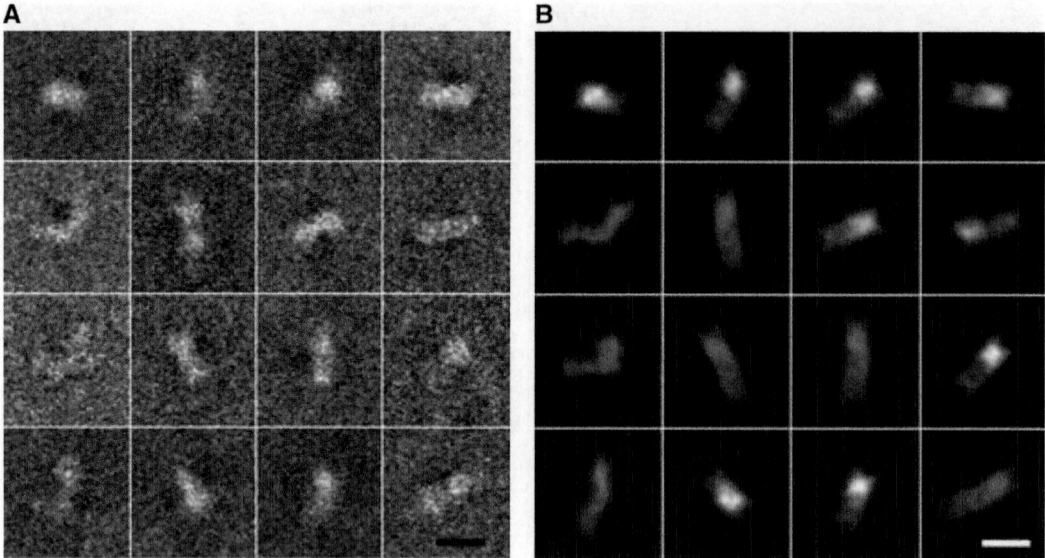

**Fig. 12.10**  Averaged cryo-EM images of A8-35-trapped Complex I. (**A**) The top 16 class averages of images for the full set of Complex I/A8-35 particles. (**B**) Corresponding 2D projections of the 3D model from negative staining (From Guénebaut et al. 1998); scale bars: 20 nm (From Flötenmeyer et al. 2007, © 2007 John Wiley & Sons, Inc. All Rights Reserved).

A 3D reconstruction by this method was carried out using the full data set of 1200 particles, with, as a reference to determine the tilt angles for the individual images, >3000 projections of the 2.8 nm resolution 3D reconstruction made from negatively stained Complex I (Guénebaut et al. 1998). These were cross-correlated with the single-particle cryo-EM images to obtain a set of projections at known Euler angles, which were then combined by weighted back-projection to give a 3D model (Fig. 12.11A). The resolution estimated by Fourier shell correlation (Saxton and Baumeister 1982) for reconstructions from two half data sets was 5.8 nm. A low-pass filtered model of the negatively stained reconstruction (Guénebaut et al. 1998) at 6 nm was calculated for comparison (Fig. 12.11B).

The low resolution of the 3D reconstruction obtained can be explained partly by the small number of particles sampled. Another factor may have been the inherent flexibility of the structure of Complex I. A hinge movement of the two arms relative to each other, which was not taken into account when carrying out the reconstruction from the cryo-EM images, may have resulted in some blurring. The membrane arm of Complex I, which has the greater contrast, tends to dominate the alignment procedure. The result is that the TM arm is similar in shape and size to that observed in negative stain, whereas the cytoplasmic arm is reduced (Fig. 12.11). However, why this problem should have arisen in Study 12.3 and not in more recent ones (Zickermann et al. 2009; Vinothkumar et al. 2014) is not clear.

The fact that the use of APols resulted in ice films with a more uniform thickness and therefore a more even distribution of particles compared to that observed in the presence of detergents is a particularly interesting observation, whose origin will be discussed below (§ 12.3.4.2).

The period under consideration closes with Study 12.4, in which BR/A8-35 complexes were examined by negative staining in the frame of a broader study of their composition and properties (Gohon et al. 2008). This study identified some of the factors that could lead to heterogeneous preparations, such as (*i*) problems arising during the synthesis of the polymers, making them more hydrophobic than had been aimed for (Gohon et al. 2006, 2008; see Chap. 4), and (*ii*) removal of free APol (Zoonens et al. 2007; see Chap. 5). EM views of preparations that had been identified as homogeneous or heterogeneous by small-angle neutron scattering (SANS) and AUC are shown in

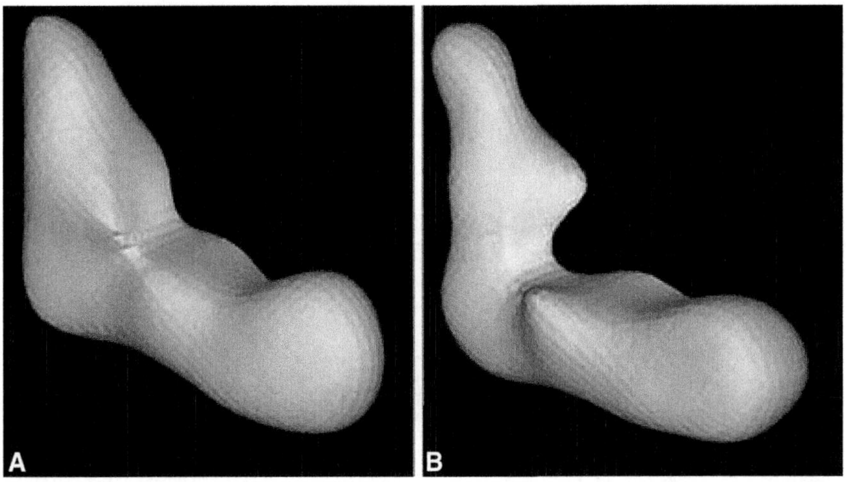

**Fig. 12.11** Reconstructed images of Complex I. (**A**) 3D reconstruction of unstained A8-35-trapped Complex I imaged in ice. (**B**) 3D reconstruction of the negatively stained complex in detergent (From Guénebaut et al. 1998) with a similar resolution cutoff (6 nm). Scale bar: 10 nm (From Flötenmeyer et al. 2007, © 2007 John Wiley & Sons, Inc. All Rights Reserved).

Fig. 12.12A, B, respectively. The diameter of individual particles was ~6.3 ± 0.8 nm, with a rather narrow distribution. The distribution of lengths of the particles presented a first maximum at ~8 nm, corresponding to the major, globular species (monomeric BR/A8-35 complexes), consistent with SANS and AUC data, followed by maxima at 13, 19, 25, and 31 nm for the more elongated particles present in heterogeneous preparations. These views suggested that BR/A8-35 monomers have a propensity to auto-assemble in a linear way.

This was spectacularly confirmed by observations on a sample of BR/A8-35 complexes depleted from free APol that had been kept for 2 years at 4 °C (A8-35-trapped BR is pretty stable; see Chap. 5). This preparation was found to comprise a mixture of monomers and small oligomers, similar to those observed in fresh preparations (Fig. 12.12C, E), but also much longer filaments (Fig. 12.12D). The filaments showed a fairly regular longitudinal segmentation (Fig. 12.12D). The length of the segments corresponds roughly to tetra- or pentamers of the globular species. Because the stain formed rather thick puddles in filament-rich regions (Fig. 12.12C), it is difficult to decide from these micrographs whether the segmented appearance results from the lengthwise association of smaller oligomers or from the filaments adopting a helical structure that rises periodically above the carbon film. Related observations were to be done later on preparations of OmpF/A8-35 complexes (see below, Study 12.13), as well as on PgP/SMAL ones (Study 12.17), and will be commented upon below (§ 12.3.4.2).

Study 12.4 closes what could be termed the prehistory of APol-based EM studies. From then on, APols and their relatives will be used more and more with biological objectives in mind, with highly rewarding results, and technological considerations will fade into the background. This second period opens with the cryo-EM study of the A8-35-trapped mitochondrial respirasome (Althoff et al. 2011), moving toward higher and higher resolutions thanks to the technological progress of cryo-EM summarized in § 12.2.3, with, in parallel, the development of negative-stain works. From 2011 on, APols have been considered more and more as just another surfactant to be tested when launching an EM study. Whereas the interest in the biological conclusions derived from studying MP/APol complexes by EM grew, scant attention was paid to the peculiarities of APol physical-chemical properties and the technical aspects of their use. Our description of the two dozen studies published in this second period will therefore remain rather sketchy, with the focus placed on those observations

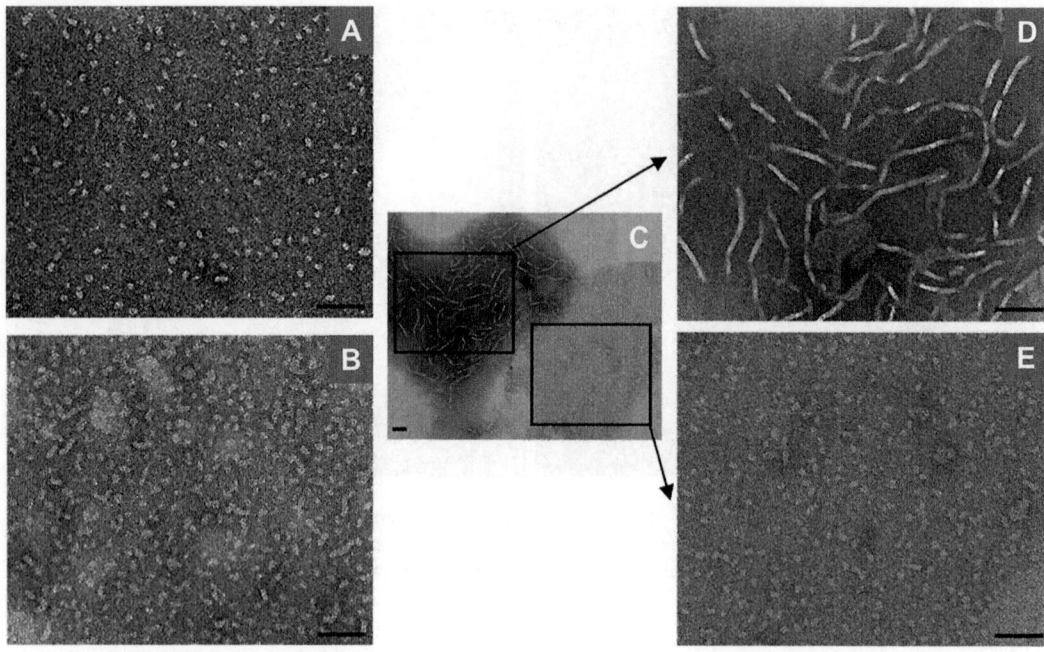

**Fig. 12.12** EM views of negatively stained preparations of bacteriorhodopsin (BR)/A8-35 complexes. (**A**) A fairly homogeneous BR/A8-35 sample. (**B**) A slightly heterogeneous BR/A8-35 sample. (**C–E**) A sample depleted of free A8-35 and stored for 2 years at 4 °C. Scale bars: 50 nm (From Gohon et al. 2008, © 2008 The Biophysical Society. Published by Elsevier Inc. All rights reserved).

that may be of general interest to microscopists working on different systems rather than on the biological relevance of each specific study.

We will consider first recent studies carried on negatively stained samples of APol-trapped MPs.

## 12.3.2   Recent Studies of Negatively Stained MP/APol Complexes (2011–Present)

The regulated channels known under the rather awkward name of "transient receptor potential" (TRP) channels form a large family of tetrameric, $Ca^{2+}$-permeable, nonselective cation channels, which are implicated in a broad range of cellular processes including pain and temperature sensing, neuronal development, angiogenesis, cardiac and pulmonary function, and cancer (reviewed in Venkatachalam and Montell 2007). They play key roles in such cellular processes as vision, hearing, touch, pain, and thermo- and mechanosensory transduction and, as such, are important pharmacological targets. Mammalian TRP channels share ~20% sequence similarity and are distributed into six subfamilies: TRPC (canonical), TRPV (vanilloid), TRPA (ankyrin), TRPM (melastatin), TRPP (polycystin), and TRPML (mucolipin). They have resisted 3D crystallization but appear to be particularly amenable to single-particle EM studies, particularly as APol-trapped complexes. Seven such studies have been published to date, one in negative stain, that of TRPA1 (Cvetkov et al. 2011; Huynh et al. 2014b; Study 12.6 in Table 12.1), and six by cryo-EM, those of TRPV1 (Cao et al. 2013b; Liao et al. 2013, 2014; Study 12.9), TRPM1 (Agosto et al. 2014; Study 12.12), TRPV2 (Fan et al. 2014; Study 12.18), TRPA1 (Paulsen et al. 2015; Study 12.21), TRPV2 again (Zubcevic et al. 2016; Study 12.29), and polycystin-2 (Wilkes et al. 2017; Study 12.33). To the studies listed in Table 12.1 and discussed below should be added seven further relevant studies, which were published or located too late to be integrated in the table and the discussion, namely those by Spear et al. (2015), Wu et al. (2015), Zhang et al. (2015),

Shao et al. (2016), Abeyrathne and Grigorieff (2017), Bausewein et al. (2017), Chiu et al. (2017), He et al. (2017), and Zhou et al. (2017). For a recent review on TRP cryo-EM studies, see Madej and Ziegler 2018.

The negative-stain study of TRPA1 by Vera Y. Moiseenkova-Bell and colleagues (Cvetkov et al. 2011; Huynh et al. 2014a, b; Study 12.6 in Table 12.1) was the first to introduce the use of APols for EM studies of the TRP channel family and was to have a rich posterity. Mouse TRPA1 was expressed in *Saccharomyces cerevisiae*, solubilized and purified in Fos-Choline-12 (FC12), and transferred to A8-35 by dilution under the CMC of FC12 and dialysis in the presence of Bio-Beads. Under these conditions, most of the free A8-35 must have remained present in the samples. The TRPA1/A8-35 complexes were adsorbed onto carbon film-coated copper grids, washed with pure water, and stained with uranyl acetate (Fig. 12.13A). A 3D model was reconstructed to ~16-Å resolution from ~8500 projection images, and computer-generated models of the TM and two cytosolic domains of the TRPA1 tetramer fitted into it (Fig. 12.13B). An uneven band of stain exclusion was observed around the TM region, which likely is due to the APol belt (Fig. 12.13B), as also reported in subsequent negative-stain or cryo-EM studies of APol-trapped MPs (see below).

An unexpected observation was made in the course of this work, namely that TRPA1 appeared to be more stable in sulfonated APols (SAPols; see Chap. 4) than in A8-35 (Huynh et al. 2014b). This is unusual, since hitherto all comparisons had pointed to SAPols, presumably because of their higher density of charges, tending to be less stabilizing than A8-35 (Picard et al. 2006; Dahmane et al. 2011; see Chap. 5, § 5.5). Being able to handle a MP complex in SAPols rather than in A8-35 is an asset, because it makes it possible to work below pH 7 and/or in the presence of multivalent cations or at high ionic strength, which the physical-chemical properties of A8-35 do not allow (see Chap. 4, § 4.3.2).

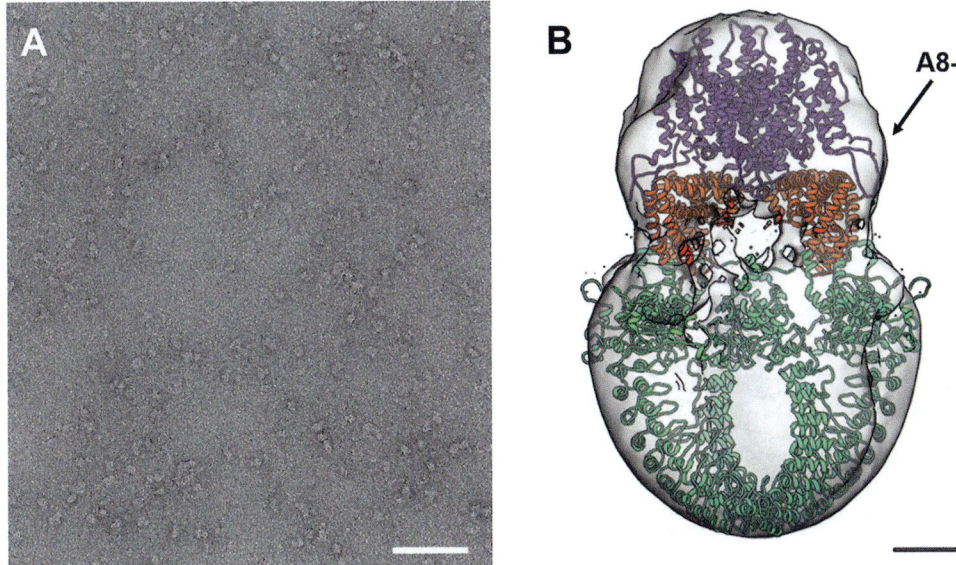

**Fig. 12.13** Negative-stain single-particle electron microscopy study of the A8-35-trapped mouse TRPA1 tetramer. (**A**) Representative micrograph of negatively stained TRPA1 trapped in A8-35. Scale bar: 70 nm (From Cvetkov et al. 2011). (**B**) Three-dimensional reconstruction of A8-35-trapped TRPA1. The predicted structure of the TRPV1 transmembrane (TM) domain (*purple*) (Fernández-Ballester and Ferrer-Montiel 2008) was docked into the putative TM domain of the TRPA1 EM structure. Models of the TRPA1 *N*-terminus (*green*) and *C*-terminus (*orange*) were fitted into the putative cytosolic density (Cvetkov et al. 2011). The likely position of the APol belt is indicated. Scale bar: 50 Å (The composite figure is from Huynh et al. 2014b)

In the same review (Huynh et al. 2014b), it is noted that another TRP channel, TRPV2, is stabilized in its tetrameric state upon being transferred from DDM to A8-35, whereas exchange to SAPols promotes the aggregation of TRPV2 into higher oligomeric structures. This initiates a theme that will run throughout the rest of this chapter: it is currently impossible to predict which APols, or other surfactants, will turn out to provide the best preparations for an EM study. It is therefore advisable, at the onset of such a study, to screen several of them to identify the molecule and conditions that yield the best stability, dispersity, and homogeneity of the particles. In the case of TRPV2, it turned out that it is one of the MNG detergents that yielded the most homogeneous preparations as determined by SEC, Blue-Native gel, and NS EM, and it is in this surfactant that the cryo-EM study was performed, reaching a resolution of 13.6 Å (Huynh et al. 2014a). Such comparisons, however, should be taken with caution, as there is a degree of chance involved: once satisfactory images have been obtained with one surfactant, investigators will not necessarily invest months of work examining whether an apparently inferior one would not yield even better data under slightly different conditions, so the order in which surfactants and experimental conditions are tested plays a role in the final choice. TRPV2 is a good example: in the same year one group recommended decyl MNG over A8-35, another reported having reached essentially the same resolution (~15 Å) in A8-35 (Fan et al. 2014; Study 12.18), and high-resolution structures were finally obtained both in A8-35 (to ~4 Å) (Zubcevic et al. 2016; Study 12.29) and in decyl MNG (to ~5 Å) (Huynh et al. 2016). In the case of the mechanosensitive channel Piezo1, NS images were recorded in $C_{12}E_{10}$, $C_{12}E_8$, and A8-35. $C_{12}E_{10}$ was chosen for cryo-EM analysis because it produced slightly better micrographs (Ge et al. 2015). The N-type rotary ATPase from *Burkholderia pseudomallei* yielded much better images in lauryldimethy-laminoxide than in A8-35 (Schulz et al. 2017), whereas better images of the mechanotransduction channel NOMPC were obtained in nanodiscs than in A8-35 (Jin et al. 2017).

*In the world of membrane protein electron microscopy, the "one-fits-all" concept does not apply*
(© *2018 by Francis Haraux*)

Anticipating somewhat on the next section, we note that the structure of the γ-secretase was solved both in digitonin (to 4.3 Å) (Sun et al. 2015) and in A8-35 (to 3.4 Å) (Lu et al. 2014; Bai et al. 2015b, c). The same complex, studied by NS EM in LMNG, is seen to adopt a variety of conformations (Elad et al. 2015). A recent review concludes that, for this particular complex, among the many surfactants tested, only digitonin and A8-35 work well (Yang et al. 2017). The ryanodine receptor RyR1 has been studied by cryo-EM both in A8-35 (Baker et al. 2015; Study 12.19) and in nanodiscs (NDs), where its structure was solved to 6.1-Å resolution (Efremov et al. 2015). The structure of TRPV1 has been solved both in A8-35 (to 3.4 Å) (Cao et al. 2013b; Liao et al. 2013; Study 12.9) and in NDs (to 2.9–3.4 Å) (Gao et al. 2016). The structure of the respirasome (from different sources) has been established to 9 Å in A8-35 (Sousa et al. 2016; Study 12.27) and to 5.8–6.7 Å (Letts et al. 2016) and 5.4 Å (Gu et al. 2016) in digitonin. Similarly, the cryo-EM structures of the Polycystic Kidney Disease 2 (Pkd2) channel obtained in NDs (to 3-Å resolution) or in A8-35 (to 4 Å) are superimposable (Shen et al. 2016). Whenever a comparison is possible, the structures solved in A8-35 vs. digitonin or A8-35 vs. NDs appear extremely similar, suggesting that they all reflect the conformation (or one of the conformations) the MP or MP complex adopts in the membrane. Which system to choose depends both on the protein's preferences and on experimental priorities and convenience.

Subsequent works on APol-trapped TRP channels were carried out by cryo-EM and will be presented in § 12.3.3.

In 2013, two further studies resorted to trapping with A8-35 followed by NS and yielded 3D reconstructions consistent with a belt of APol surrounding the TM region. Study 12.10 (Kevany et al. 2013) bore on a heterotetrameric complex of peripherin and ROM1, two retina-specific tetraspanins that are required for the morphogenesis and maintenance of photoreceptor outer segments. The complexes were purified from bovine retinas in DDM, supplemented with A8-35, and the detergent and excess APol removed by SEC. Grids were stained with uranyl acetate and images collected in both tilted and untilted modes. A 3D reconstruction was calculated to ~18-Å resolution (Fig. 12.14A). Consideration of the predicted volumes occupied by the TM and extramembrane domains led to their assignment to the thicker and thinner regions of the model, respectively, the TM region being surrounded by a 1.5–2-nm-thick belt of APol (*dark blue* band in Fig. 12.14A; Kevany et al. 2013).

Study 12.11 (Tsybovsky et al. 2013) bore on ABCA4, a member of the ATP-binding cassette (ABC) family of transporters. ABCA4 is predominantly expressed in the outer segments of rod and cone photoreceptors and is involved in the clearance from photoreceptors of all-*trans*-retinal, the product of light-induced isomerization of 11-*cis*-retinal. ABCA4 was purified from bovine retinas in DDM and transferred to A8-35 by SEC in the absence of surfactant, a process that must have removed both the detergent and free APol. EM grids were negatively stained with uranyl acetate and imaged at both 0° and 45° angles to the electron beam. An initial 3D model at 38-Å resolution was calculated from pairs of tilted and untilted images (random conical tilt method) and used as a starting point for classifying ~25,000 untilted projections, from which a 18-Å-resolution 3D model was calculated (Fig. 12.14C, D). The intradiscal domain, which is glycosylated, was identified by its binding of concanavalin. The TM domain was estimated to have a ~2-nm-thick belt of A8-35 bound to its periphery, in good agreement with the expected dimensions of the TM region (Fig. 12.14E).

Study 12.15 examined the behavior of the trimeric *E. coli* outer membrane protein OmpF following transfer from the detergent $C_8$-POE to A8-35 using Bio-Beads and elimination of free APol by SEC, in the presence or absence of lipopolysaccharide (LPS), the main lipidic component of the outer leaflet of *E. coli*'s outer membrane (Arunmanee et al. 2014). Within minutes, OmpF trimers depleted of free APol in the absence of LPS started to organize into long filaments (Fig. 12.15C), which elongated with time (Fig. 12.15D–F). The thickness of the filaments, ~6 nm, corresponds to the height of OmpF trimers, normal to the membrane, indicating that they associate side by side (Fig. 12.15B), as proposed for BR/A8-35 filaments (Gohon et al. 2008; cf. Fig. 12.12 in § 12.3.1). This suggests that, upon APol depletion, MPs tend to interact hydrophobically via their TM surfaces, probably releasing some APol in the process. The SEC data show that association occurs in solution, prior to adsorption onto the EM grids. The filaments interact mostly sideways with the carbon film, that is, again, via the APol-coated TM surface of OmpF. As observed with the small oligomers formed by

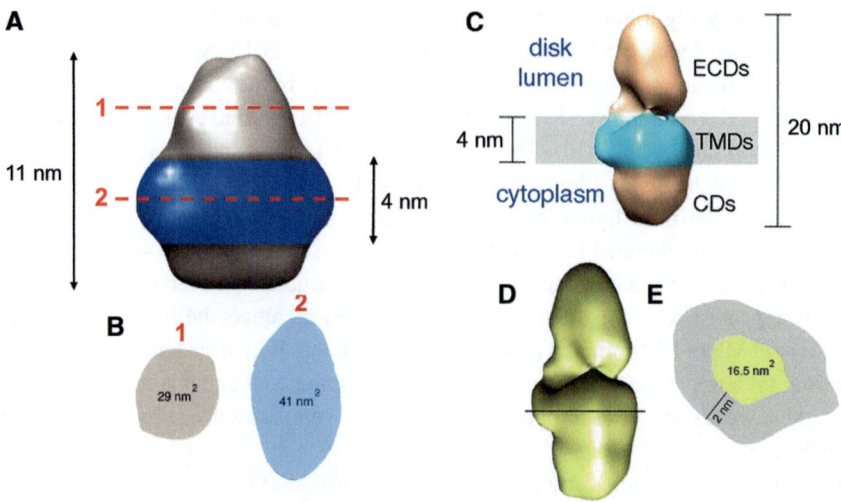

**Fig. 12.14** 3D reconstruction of negatively stained A8-35-trapped peripherin-ROM1 complex. (**A**) 3D reconstruction, with the proposed location of the A8-35 belt near the bottom of the model shown in *dark blue*. *Arrowheads* indicate positions where horizontal cross sections were taken to calculate the cross-sectional areas shown in (**B**). Adapted from a figure kindly provided by (**A**) (Engel and K. Palczewski; for details see Kevany et al. (2013)). (**C**) 3D reconstruction of negatively stained A8-35-trapped ABCA4 transporter with assigned luminal (extracellular; *ECDs*), transmembrane (*TMDs*), and cytoplasmic (*CDs*) domains indicated. The proposed A8-35 belt around the TM region is colored in *cyan*. The *gray rectangle* indicates the probable position of the lipid bilayer in situ. (**D**), (**E**) A cross section (**E**) was taken in the model shown in (**C**) at the position marked in (**D**) (*black line*), perpendicularly to the vertical axis, in the center of the proposed TM region. (**E**) A 2-nm-thick A8-35 belt (*gray*) was subtracted from the cross section. The remaining area (*yellow*) was assumed to be occupied by the TM region of the protein and matches its expected size (From Tsybovsky et al. 2013, © 2013 Elsevier Ltd. All rights reserved).

APol-depleted tOmpA (Zoonens et al. 2007; see Chap. 5), the filaments disaggregate upon back addition of free APol (Arunmanee et al. 2014).

The experiments shown in Fig. 12.15 were carried out with LPS-free OmpF folded in vitro from urea-denatured inclusion bodies. Since LPS molecules are known to associate strongly with OmpF (see Baboolal et al. 2008, and references therein) and LPS molecules can cross-link to adjacent LPS via divalent cations such as $Ca^{2+}$ and $Mg^{2+}$ (Schneck et al. 2010), the OmpF/LPS interaction could affect lateral OmpF association. LPS-free OmpF in detergent solution was therefore supplemented with LPS and transferred to A8-35 in the presence or absence of 2,2′,2″,2‴-(ethane-1,2-diyldinitrilo)tetraacetic acid (EDTA). LPS-free samples and LPS-supplemented samples ± EDTA were examined by SEC and TEM after NS (Fig. 12.16). TEM observations confirmed the formation of OmpF/APol filaments following removal of free APol (Fig. 12.16C). The filaments showed a beaded structure which probably corresponds to the repetitive pattern of OmpF trimers along the chain. OmpF/APol complexes in the presence of LPS, however, tended to assemble into sheetlike structures rather than filaments (Fig. 12.16D), leading to small areas of apparent 2D crystallization. Once divalent cations were chelated by EDTA, a combination of small sheetlike and filamentous structures was observed (Fig. 12.16E). These results indicate that LPS disrupts the linear (1D) filaments formed by APol-depleted OmpF/A8-35 complexes and provokes a 2D behavior reminiscent of that in the natural membrane, as well as of the 2D crystals formed in vitro by OmpF/lipid mixtures (Hoenger et al. 1993; Schabert and Engel 1994; Baboolal et al. 2008). OmpF/APol complexes appear to preferably form a type of interaction where a side-by-side, linear association of trimers takes precedence over all other forms. However, this arrangement is easily disrupted, by LPS to form 2D sheets or by excess APol to

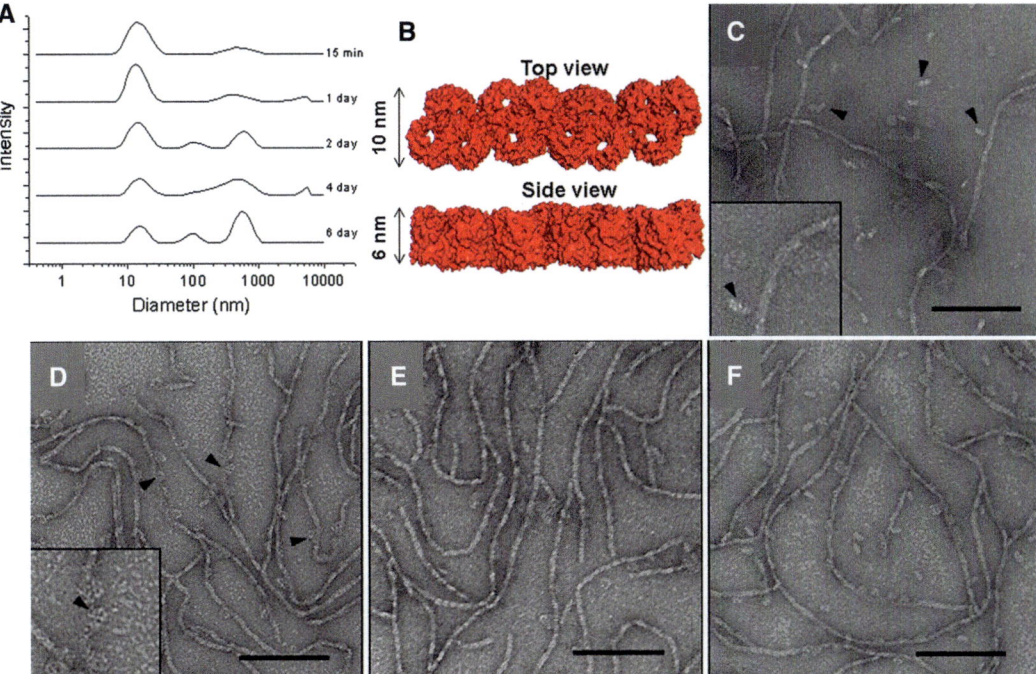

**Fig. 12.15** Organization into filaments of OmpF/A8-35 complexes following removal of free APol. (**A**) Size distribution of OmpF/A8-35 complexes measured at 30 °C by dynamic light scattering from 15 min to 6 days after removal of free APol by SEC. (**B**) Structural model of OmpF filaments showing "top" and "side" views and their expected dimensions. (**C–F**) EM views of negatively stained preparations of OmpF/A8-35 complexes studied at different times after SEC: (**C**) After 10 min. Single particles of OmpF trimer are present (*arrowheads*). (**D**) After 10 min. OmpF pores (*arrowheads*) can be seen on the filaments. (**E**) After 1 day. (**F**) After 1 week. The filaments are ~6-nm thick. Scale bars: 100 nm. The *insets* in C and D are at 2× higher magnification (From Arunmanee et al. 2014).

form isolated complexes. In the presence of a carbon film, one hydrophobic face of the linear filament, which, in solution, is stabilized by APol, adsorbs onto the film's surface. The APol/OmpF filaments are thus marginally stable, such that they can be influenced by small changes in their environment. They were proposed as possible starting points for 2D crystallization studies of OmpF complexes (Arunmanee et al. 2014).

Nutrient import across the outer membrane of Gram-negative bacteria is powered by the proton-motive force delivered by the cytoplasmic membrane protein complex ExbB-ExbD-TonB. In Study 12.16 (Sverzhinsky et al. 2014), the $ExbB_4$-$ExbD_2$ and $ExbB_6$-$ExbD_4$ complexes were purified in DDM, transferred to A8-35 using Bio-Beads, and studied by a variety of techniques including size exclusion chromatography (SEC), SANS, AUC, light scattering, asymmetric flow field-flow fractionation, and TEM after NS with uranyl acetate. 3D models were reconstructed to 21–27-Å resolution and molecular models fitted into them (Sverzhinsky et al. 2014; see Chap. 9, § 9.3.8.3, Fig. 9.20).

Study 12.22, one of the latest published negative-stain TEM studies at the time of this writing, is interesting in that it resorts to a styrene maleic acid copolymer (SMA) rather than classical APols. As described in Chap. 5, § 5.2.2.1, SMA permits to directly extract MPs from membranes. The resulting particles (SMALPs) comprise SMA, the protein, and part of their surrounding lipids (see Chap. 5, § 5.3.1.2). In Study 12.22, a polyhistidine-tagged version of the multidrug transporter AcrB was expressed in *E. coli*, directly extracted from membrane fragments into SMALPs, purified by

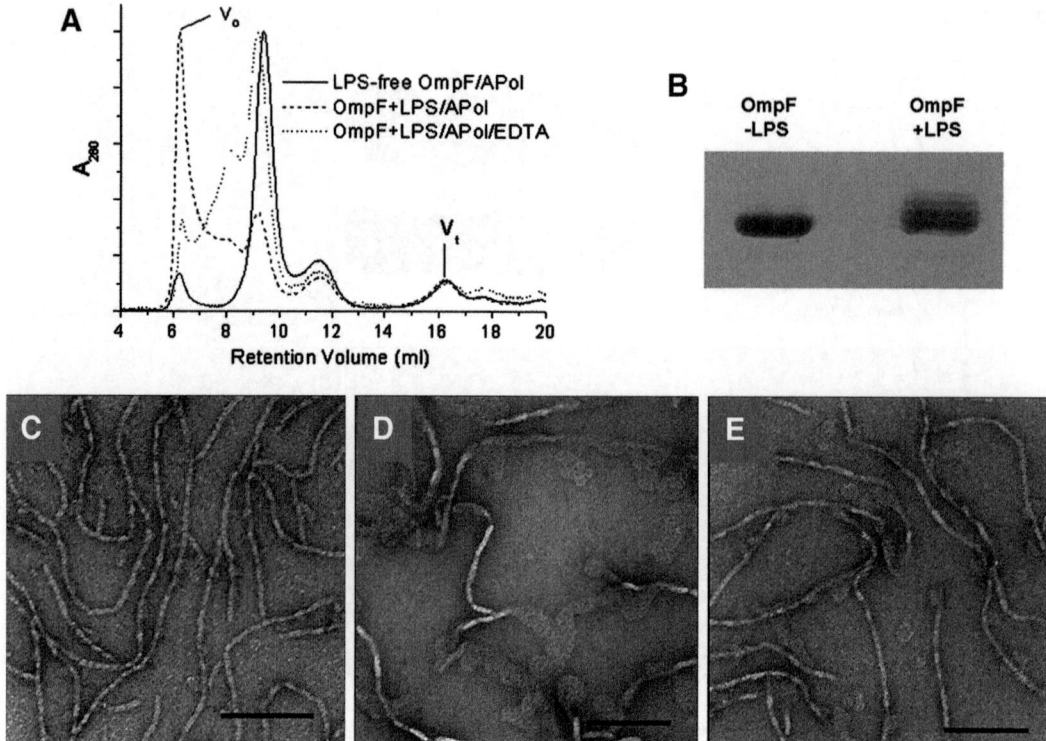

**Fig. 12.16** Effect of lipopolysaccharides (LPS) on the structure of OmpF/APol complexes. (**A**) SEC analysis of the size and homogeneity of OmpF/APol complexes in various conditions: LPS-free OmpF (——), OmpF/LPS (- - - -), and OmpF/LPS in 1 mM EDTA (·······). In each case, the SEC column was pre-equilibrated with detergent-free buffer. Profiles have been normalized to the same maximum. (**B**) SDS-PAGE analysis of LPS-free OmpF vs. OmpF with added LPS, showing the characteristic ladder of OmpF bands resulting from LPS binding. (**C, D**) EM study of OmpF/APol complexes depleted of free APol in the presence and absence of LPS. (**C**) LPS-free OmpF/APol complexes. (**D**) OmpF/LPS/APol complexes. (**E**) OmpF/LPS/APol complexes in the presence of EDTA. Scale bars: 100 nm (From Arunmanee et al. 2014).

immobilized-metal affinity chromatography (IMAC), and imaged after staining with uranyl acetate (Postis et al. 2015). The particles appeared as either "singlets" or "doublets," the images of which were separately analyzed. A 3D model of "singlets," reconstructed to a final resolution of 23 Å, is consistent with the known X-ray structure of the AcrB trimer (Murakami et al. 2002), with SMA + lipids (SMAL) forming a belt around the TM region. Class analyses of the doublets show a variety of association modes, all of them seeming to involve contacts between the SMAL belts of two AcrB trimers. Control experiments suggest that the formation of doublets occurs in solution and not on the EM grid. The lack of any preferred doublet interface suggests that particle association occurs non-specifically via the SMAL belts and does not involve any defined protein/protein interface. These observations suggest that any SMAL-trapped MP is likely to behave in the same way. Note that, as also observed with OmpF/A8-35 complexes (Study 12.13; Arunmanee et al. 2014), AcrB/SMAL ones tend to adsorb onto the carbon film sideways, with their TM surface in contact with the carbon, even though the latter has been made hydrophilic by glow-discharge.

In Table 12.1 are given references to four further studies presenting NS EM data on MP/NVoy (Study 12.7; cf. Chap. 7, Fig. 7.7A, B), MP/non-ionic APol (Study 12.8), and MP/A8-35 (Studies 12.25 and 12.28) complexes. A recent atomic-force microscopy study of MotPS/A8-35 complexes should also be mentioned (Terahara et al. 2017).

### 12.3.2.1  Conclusions from Negative-Stain Studies

From this survey of the most informative studies in which TEM was used to examine negatively stained MP/APol complexes, one can tentatively draw the following conclusions:

- MP/APol (and MP/SMA) complexes are amenable to such studies, and 3D models can be routinely reconstructed to ~16 to 18-Å resolution (Cvetkov et al. 2011; Kevany et al. 2013; Tsybovsky et al. 2013; Huynh et al. 2014b) or close to it (Sverzhinsky et al. 2014; Postis et al. 2015), that is to the best resolution that can be achieved using NS.
- Attention should be paid to the fact that removal of free APol may induce the formation of aggregates, which, in the two studies where they have been examined by TEM/NS (Gohon et al. 2008; Arunmanee et al. 2014), take the form of filaments, occasionally of sheets (Arunmanee et al. 2014). Whether these structures form, and at which rate, is likely to be strongly dependent on the MP, on the nature of the APol, and on conditions such as the residual concentration of free APol and/or detergent, the ionic strength, and the presence of divalent cations. Note that the presence of traces of divalent cations degrades the quality of NMR spectra of MP/A8-35 complexes (Catoire et al. 2009, 2010), certainly due to calcium-mediated association of some of the complexes via their APol belts (see § 10.3.1.1 in Chap. 10). It also affects, as shown in Study 12.13, the nature of the aggregates, filaments, or sheets, formed by APol-depleted OmpF/A8-35 complexes in the presence of lipopoly-saccharide (Arunmanee et al. 2014). Adding EDTA to solutions of MP/A8-35, MP/A8-75, or MP/SMA complexes, all of which are sensitive to the presence of $Ca^{2+}$ (Champeil et al. 2000; Postis et al. 2015), is certainly a good precaution if aggregation is to be avoided.

The formation of superstructures could perhaps be harnessed for the reconstruction of 3D models by helical reconstruction of filaments or by inducing the formation of 2D crystals. Analysis of the internal periodicity of the filaments formed by BR (Gohon et al. 2008) or OmpF (Arunmanee et al. 2014) ought to indicate whether protein/protein contacts are involved, which could generate some order, or only polymer-polymer contacts, as seems to be the case in AcrB/SMA "doublets" (Postis et al. 2015; see Fig. 12.17), in which case disorder is probable. This point will be further discussed in § 12.3.4.3.

- There may be a tendency for the complexes to adhere to the carbon film via the TM region of the protein (Arunmanee et al. 2014; Postis et al. 2015), which may be a nuisance if a random distribution of the particles is desired. Ways could be devised to control this phenomenon, as found in a cryo-EM study of the ryanodine receptor (see below, Study 12.19 in § 12.3.3).

Whereas negatively stained samples are easy to prepare and comparatively easy to study, only unstained samples trapped in vitreous ice have the potential to yield density maps that, thanks to recent progress in the collection, correction, and analysis of cryo-EM images, can be interpreted in terms of atomic structures (see § 12.2.3). As described in the next section, APols are bringing their modest but useful contribution to such studies by facilitating the preparation and observation of EM samples.

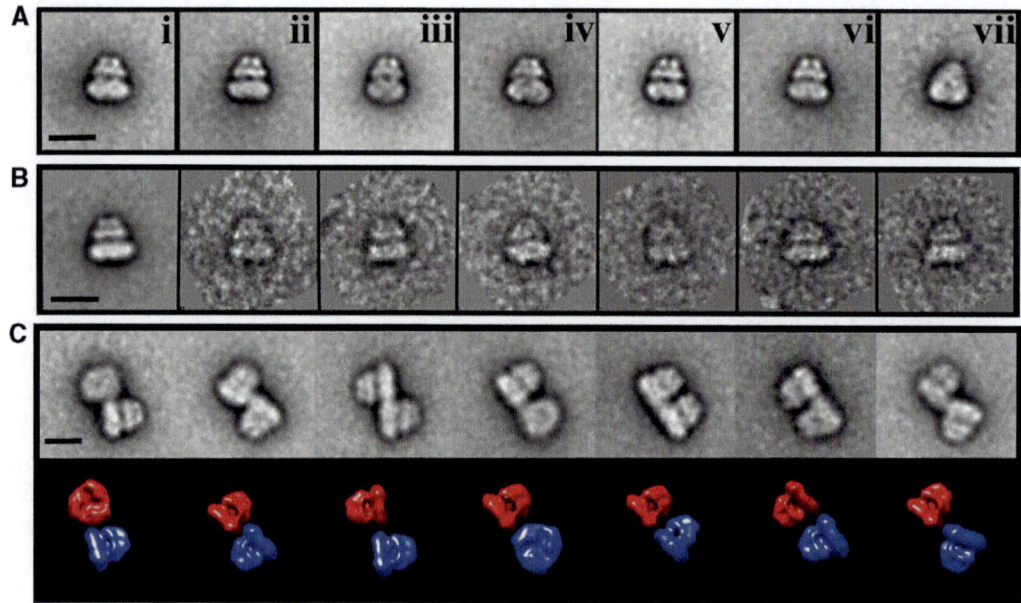

**Fig. 12.17** Transmission negative-stain electron microscopy study of trimeric AcrB directly extracted from *Escherichia coli* inner membrane fragments by a styrene maleic acid copolymer (SMA). (**A**) Representative classes of images of negatively stained AcrB/lipid/SMA complexes. The classes are predominantly of a "side" view, equivalent to that viewed from the plane of the bilayer, with an example of a "base" view from the cytoplasmic surface of the protein shown in the far right panel (*vii*). (**B**) Representative AcrB "singlet" averaged image of a class on the *left*, with some of the raw particles that make up the class shown to its right. (**C**) Classes of the AcrB trimer "doublets" (*top panel*), with the orientations shown below (*bottom panel*) using the AcrB 3D model reconstruction. Scale bars: 15 nm (From Postis et al. 2015).

### 12.3.3   Cryo-EM Studies of MP/APol Complexes (2011–Present)

A decisive impetus was given to the use of APols for cryo-EM studies of MPs and MP complexes by Study 12.5, carried out by the laboratory of Werner Kühlbrandt, in which supercomplex B from the inner membrane of mitochondria was solubilized using digitonin, transferred to A8-35, purified by sucrose gradient centrifugation, and imaged after freezing in vitreous ice (Althoff 2011; Althoff et al. 2011). Supercomplex B, the "respirasome," comprises one copy of Complex I (NADH:ubiquinone oxidoreductase), a superdimer of Complex III (cytochrome $bc_1$), and one copy of Complex IV (cytochrome *c* oxidase), for a total mass of ~1.7 MDa. Complex I transfers electrons from NADH to ubiquinone, Complex III from ubiquinol to cytochrome *c*, and Complex IV from cytochrome *c* to dioxygen, the whole process generating water and a proton gradient, which the $F_1F_O$ ATP synthase uses for synthesizing ATP (for a review, see Kühlbrandt 2015).

Mitochondria were isolated from bovine heart tissue and supercomplexes solubilized with digitonin. However, digitonin-solubilized supercomplexes did not partition well into the holes of holey carbon films. Following solubilization, digitonin was therefore replaced by A8-35 using $\gamma$-cyclodextrin, and the various APol-trapped supercomplexes separated from one another and from free APol by sucrose gradient ultracentrifugation in the absence of surfactant. The gradient purification worked better than in digitonin, due to larger differences in apparent molecular mass between the supercomplexes, which resulted in improved separation. Fractions containing APol-trapped supercomplex B were identified on the basis of their migration in Blue-Native gels and enzymatic activity and chosen for EM studies. Samples negatively stained with uranyl acetate showed a high density of homogeneous, well-preserved particles, of which ~5600 were selected for initial single-particle processing. Class averages showed the characteristic triangular and F-shaped views of supercomplex B that had been previously observed in negative stain (Schäfer et al. 2006). Other class averages, which had not been reported before, were round or oval. As a control, ~4700 digitonin-solubilized supercomplexes were examined after NS. They were less uniform but also revealed a number of round or oval class averages, indicating that the rounded shapes were not due to the APol. As no significant differences were observed between supercomplexes trapped by A8-35 or solubilized in digitonin, the A8-35-trapped material was used for the 3D reconstruction, considering the generally higher stability of membrane proteins in APols, the better separation on density gradients, and the absence of detergent, which facilitates cryo-EM grid preparation.

For cryo-EM, samples were vitrified in a thin layer of buffer on continuous carbon support films. This is not often done in single-particle cryo-EM, because the ice thickness is more difficult to control than on holey carbon films, and the support film adds background noise. However, the preferential orientations of the complexes on the support film help in particle selection and classification, and it becomes easier to use the random conical tilt method of 3D reconstruction (Radermacher et al. 1987). This approach has the advantage that angular relationships between the particle images are known, providing important constraints in the initial rounds of reconstruction, which facilitated sorting the particles into subpopulations. Two images of each area were recorded at tilt angles of $0°$ and $-45°$ (Fig. 12.18A, B). Five characteristic views out of the final set of 150 class averages from untilted images are shown in Fig. 12.18C. A 3D map was generated and refined by projection matching to a resolution of 19 Å as determined at FSC $= 0.143$. Projections of the final 3D map are compared with the corresponding class averages in Fig. 12.18D.

The X-ray structures of the three individual respiratory chain complexes were docked manually into the 19-Å map (Fig. 12.19). The fit of the three complexes left unoccupied an irregular, belt-shaped region of density around the perimeter of the complex. This band (*brownish* in Fig. 12.19A) correlates closely with the hydrophobic TM surface of Complexes I, III, and IV and accordingly was assigned to A8-35, providing the first direct view of a MP-associated APol belt. The belt is on average ~2 nm thick, and ~4 nm wide, comparable in width to the hydrophobic region of a lipid bilayer (Lewis and Engelman 1983). Its thickness is consistent with the estimates of 1.5–2 nm that had been obtained from SANS and other indirect techniques (Zoonens et al. 2005; Gohon et al. 2008; see Chap. 5, § 5.3.2) and with that observed in subsequent EM studies (see e.g. Figs. 12.14, 12.23, and 12.25). The belt showed a number of ~3 nm protrusions, spaced about 6–10 nm apart, suggesting an irregular distribution along the edge of the membrane region. These bumps could correspond either to deformations of the belt, or, perhaps, to the attachment of A8-35 particles (which are, however, somewhat larger: ~6.3 nm in diameter; Gohon et al. 2006; Chap. 4, § 4.3.1.2.2). Whatever their origin, comparison of supercomplex 3D reconstructions generated from two equal halves of the data set indicates that the bumps are a distinct map feature, implying that they attach to, or form at, the same positions in each supercomplex.

The cryo-EM structure of supercomplex B made it possible to map the arrangement of the three complexes relative to each other and to delineate possible paths for the transfer of ubiquinol/ubiquinone between Complex I and the $bc_1$ complex, as well as that of cytochrome $c$ between the $bc_1$ and cytochrome $c$ oxidase, and to determine the distances between enzymatic sites. The resolution was later pushed to 9 Å, providing further functional insights into the transfer of electrons from Complex III to Complex IV (Study 12.27; Sousa et al. 2016).

The publication of the respirasome structure seems to have finally attracted the attention of microscopists to the possible interest of using APols for cryo-EM studies; it was followed by a flurry of other works, 22 at the date of this writing (mid-2017), plus one study, also listed in Table 12.1,

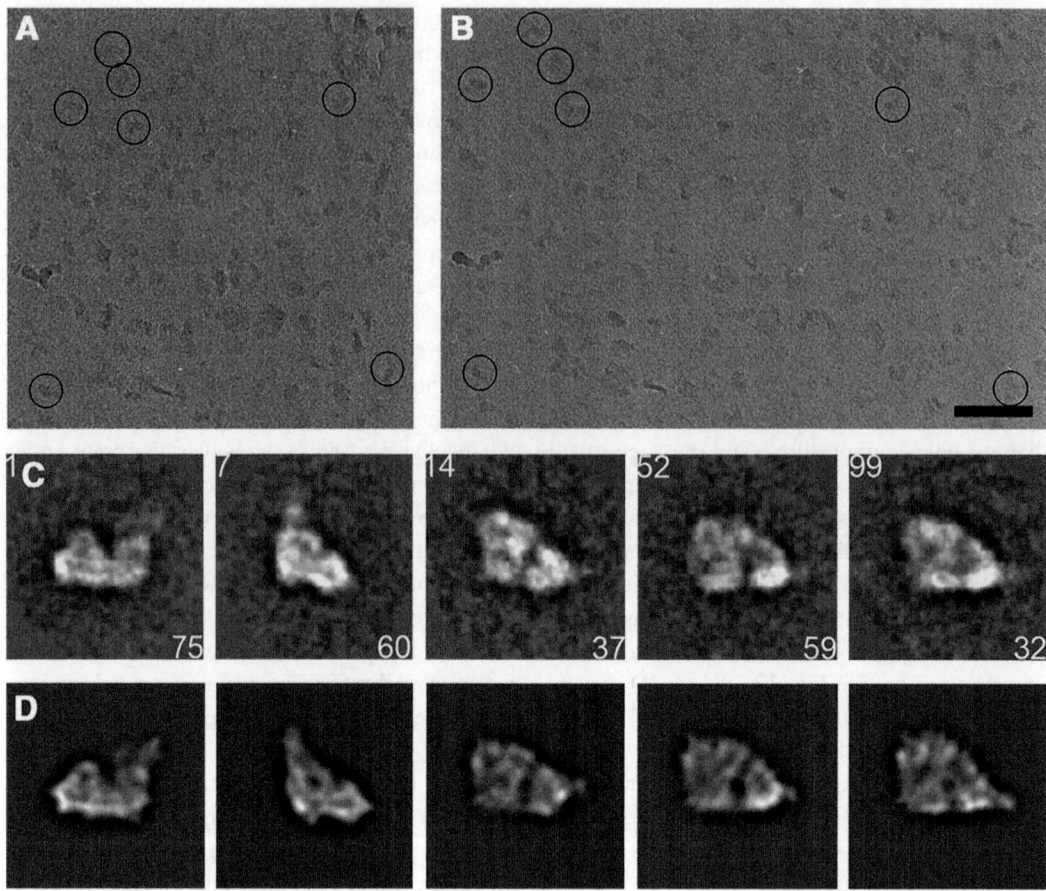

**Fig. 12.18** (A, B) Cryo-EM imaging and single-particle analysis of A8-35-trapped supercomplex B in vitrified buffer on continuous carbon support film recorded at a tilt angle of −45° **A** or 0° **B**. Scale bar: 100 nm. (C) Five out of 150 selected class averages of untilted images filtered to 50 Å after the final round of multi-reference alignment and multivariate statistical analysis classification (class number in *upper left corner*, number of particles in *lower right corner*). (**D**) Reprojections of the final 19-Å reconstruction corresponding to the class averages shown in **C** (From Althoff et al. 2011, © 2011 European Molecular Biology Organization).

applying APols to a non-membrane system (Study 12.20). Seven of these MP studies reached resolutions close to or better than 4 Å, at which molecular models can be directly fitted into density maps (Fig. 12.20). A rapid survey will be presented, focusing, as before, on observations that can be of technological interest for cryo-EM studies of MP/APol complexes rather than on the biological aspects.

The first single-particle cryo-EM study ever to breach the 4-Å barrier bore on another channel of the TRP family, TRPV1, studied by David Julius, Yifan Cheng, and colleagues (Study 12.9 in Table 12.1; Cao et al. 2013b; Liao et al. 2013). Solving the structure of TRPV1 to high resolution was achieved thanks to the technological progress described in § 12.2.3. It was facilitated by stabilizing TRPV1 by transferring it from DDM to A8-35 (for reviews, see Huynh et al. 2014b; Liao et al. 2014; Madej and Ziegler 2018 and, for a detailed protocol, § 12.4 below). TRPV1, the Type 1 vanilloid receptor TRP channel, is the target of capsaicin, the agent from chili peppers that elicits burning pain. It is the prototype of a subfamily of thermosensitive TRP channels that enable somatosensory neurons, or

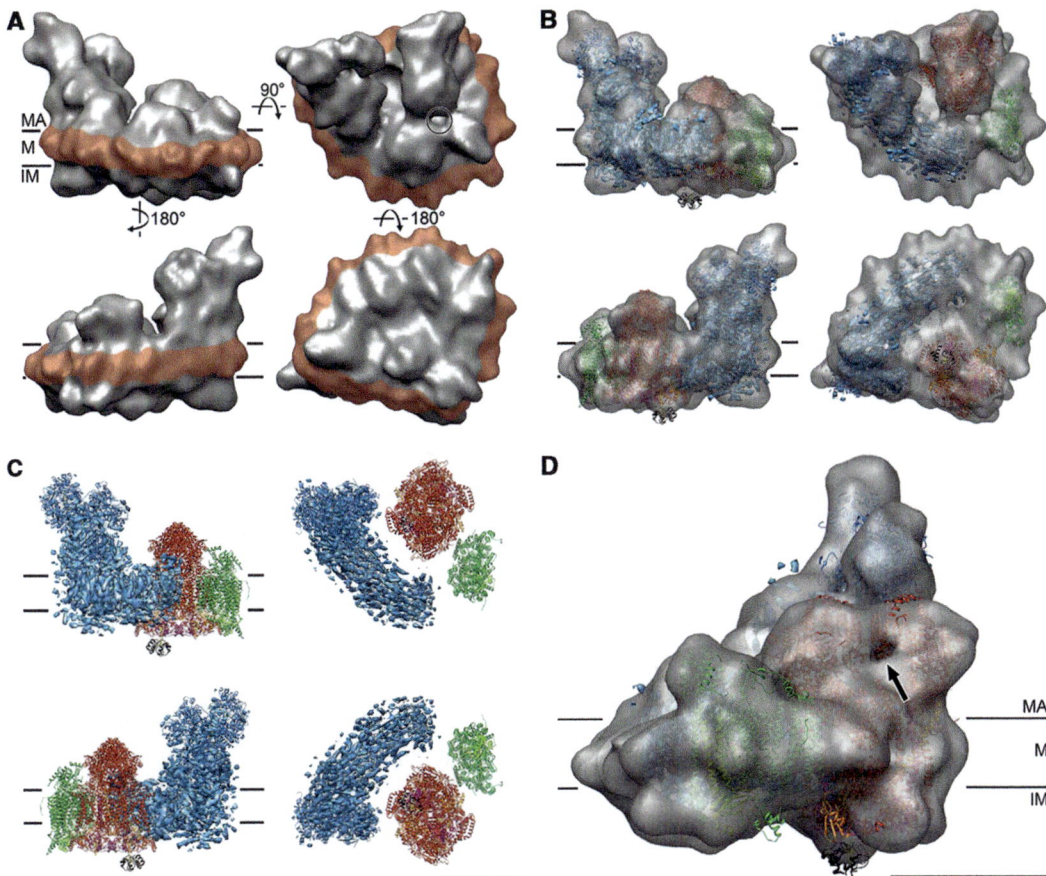

**Fig. 12.19** 3D map of the respirasome (mitochondrial supercomplex B) and fitted X-ray structures. (**A**) Cryo-EM 3D map as seen from two opposite sides (*left*), from the matrix (*top right*), and from the intermembrane space (*lower right*). The amphipol belt is shown in *brown*. The *circle* marks the gap between Complex I and Complex III. (**B**) X-ray structures of component complexes (*blue*, Complex I; *red*, Complex III; *green*, Complex IV) and cytochrome *c* (*black*) fitted to the 3D cryo-EM map. (**C**) Docked X-ray structures without map. (**D**) Enlarged view with docked X-ray structures. The mobile, electron-shuffling domains of the two Rieske proteins are shown in *orange*. The *arrow* points to the gap between the two matrix domains of Complex III, which was used for positioning the X-ray structure. *MA* matrix, *M* membrane, *IM* intermembrane space. Scale bars: 10 nm (From Althoff et al. 2011, © 2011 European Molecular Biology Organization).

other cell types, to detect changes in ambient temperature. TRPV1 is activated by noxious heat and modulated by inflammatory agents, such as extracellular protons and bioactive lipids, which contribute to pain hypersensitivity. TRPV1 and other somatosensory TRP channels are considered important targets for analgesic drugs. A truncated but functional version of the rat receptor was expressed in HEK cells, solubilized and purified in DDM, and transferred to A8-35 using Bio-Beads, followed by SEC, which must have eliminated most of the free APol. Freezing in vitreous ice on holey carbon films was carried out 6 s after blotting. Cryo-EM images were collected using a direct-electron detector and image processing algorithms to correct motion-induced image blurring and improve signal and contrast (§ 12.2.3). A 3D reconstruction of TRPV1 in the absence of ligands was built to 3.4-Å resolution (FSC = 0.143) (Fig. 12.21). This high resolution allowed direct fitting into density maps of molecular models, particularly in the TM region (Fig. 12.22) (Liao et al. 2013). In a companion paper,

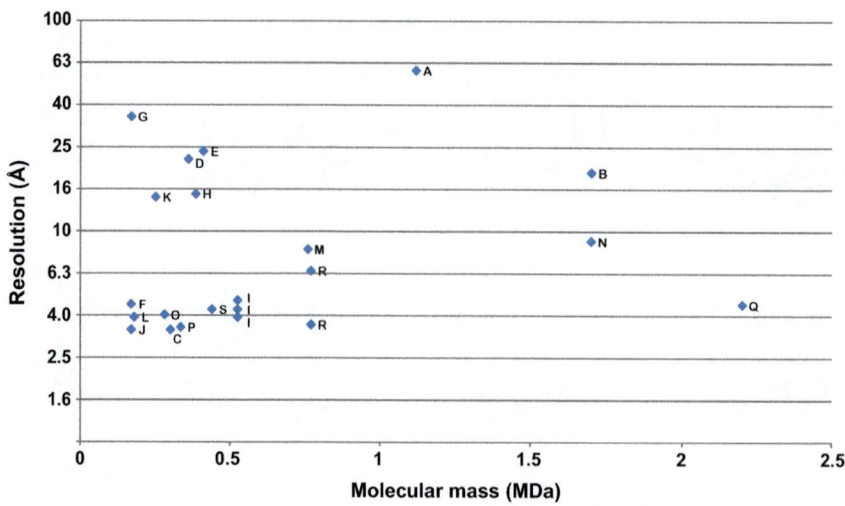

**Fig. 12.20** Single-particle cryo-EM studies of membrane protein/amphipol complexes whose resolution of the final three-dimensional reconstruction is stated, published as of April 2017. The data are plotted as a function of the molecular mass of the proteins, in abscissa, and the resolution achieved (at the standard limit FSC = 0.143, when stated), on a logarithmic scale, in ordinate. (**A**) Mitochondrial Complex I (Study 12.3 in Table 12.1; Flötenmeyer et al. 2007). (**B**) Respirasome (mitochondrial supercomplex $I_1III_2IV_1$; Study 12.5; Althoff et al. 2011). (**C**) TRPV1 (Study 12.9; Cao et al. 2013b; Liao et al. 2013, 2014). (**D**) TRPM1 (Study 12.12; Agosto et al. 2014). (**E**) CsgG/CsgE (curli secretion machinery; Study 12.14; Goyal et al. 2014). (**F**) γ-secretase (Study 12.15; Lu et al. 2014; Sun et al. 2015). (**G**) P-glycoprotein (Study 12.17; Gulati et al. 2014). (**H**) TRPV2 (Study 12.18; Fan et al. 2014). (**I**) TRPA1 (Study 12.21; Paulsen et al. 2015). (**J**) γ-secretase (Study 12.15; Bai et al. 2015c). (**K**) Holotranslocon (Study 12.23; Botte et al. 2016). (**L**) STRA6 receptor (Study 12.24; Chen et al. 2016). (**M**) Engineered AcrAB-TolC complex (Study 12.26; Jeong et al. 2016). (**N**) Respirasome (Study 12.27; Sousa et al. 2016). (**O**) TRPV2 (Study 12.29; Zubcevic et al. 2016). (**P**) Cyclic nucleotide-gated TAX-4 channel (Study 12.30; Li et al. 2017b). (**Q**) Pneumolysin (Study 12.31; van Pee et al. 2017). (**R**) AcrABZ-TolC (Study 12.32; Wang et al. 2017). (**S**) Polycystin-2 (Study 12.33; Wilkes et al. 2017). Note that the resolutions achieved are not all comparable to one another, the criteria used varying from one study to the next (see Table 12.1). The APol used is A8-35 in all cases, but for Studies 12.17 (SMA; **G**) and 12.21 (PMAL-C8; **I**). For further details about each study, see Table 12.1 and text. A list of additional, more recent structures that were published or located too late to be incorporated into the figure is given at the end of § 12.3.3.

two further structures of A8-35-trapped TRPV1 were solved, with either of two ligands bound, a peptide toxin and capsaicin, respectively, at 3.8-Å and 4.2-Å resolution, visualizing the conformational changes that lead to channel opening (Cao et al. 2013b).

It is interesting to note that TRPV1 was subsequently studied in nanodiscs (NDs) to a somewhat higher resolution (unliganded, agonist-bound, and antagonist-bound states at resolutions of 3.2, 2.9, and 3.4 Å, respectively), in order to examine protein/lipid interactions (Gao et al. 2016). Improvements in resolution were not limited to TM regions, but also extended to cytoplasmic domains. These improved density features may reflect enhanced stability of the channel in NDs, but other technical advances also contribute. The ND- and APol-stabilized structures of a given conformational state are essentially identical, albeit with some specific differences that may relate to lipid and/or ligand binding (Gao et al. 2016).

Detailed views of the APol belt surrounding the TM region of TRPV1 are presented in Liao et al. (2014) (Fig. 12.23). They feature two interesting differences with the APol belt visualized in the respirasome study (Althoff et al. 2011) (Fig. 12.19A). First, the belt does not present the characteristic bumps observed in the earlier study. This point will be commented upon in § 12.3.4.1. Second, there

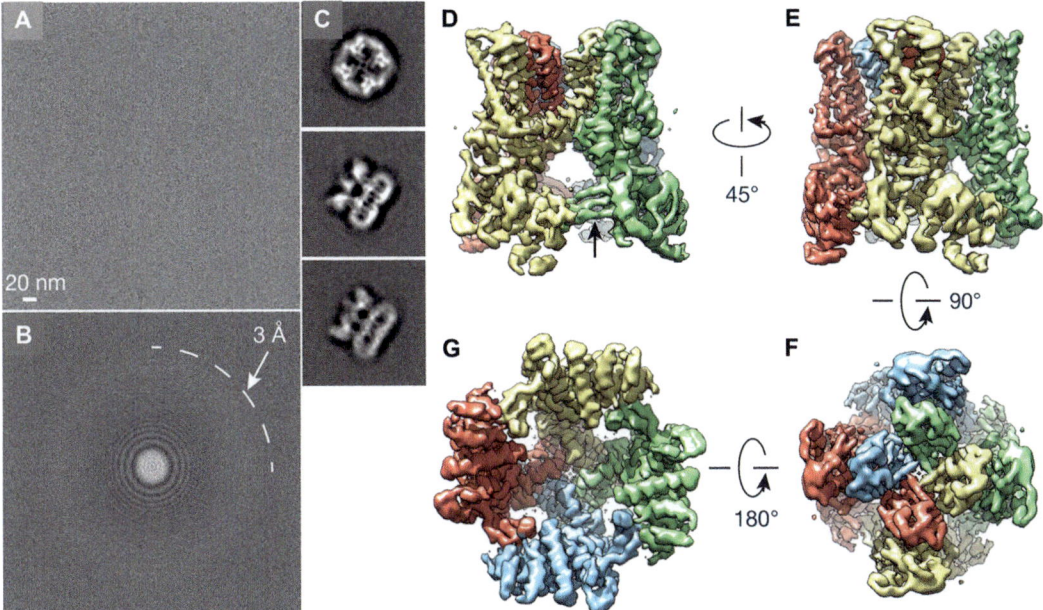

**Fig. 12.21** Three-dimensional (3D) reconstruction of A8-35-stabilized TRPV1 determined by single-particle cryo-EM. (**A**) Representative electron micrograph of TRPV1/A8-35 complexes embedded in a thin layer of vitreous ice recorded at a defocus of 1.7 μm. (**B**) Fourier transform of the micrograph shown in (**A**), with Thon rings extending to nearly 3 Å. (**C**) Enlarged views of three representative two-dimensional class averages showing fine features of the homotetrameric channel. (**D–G**) 3D density map of the TRPV1 channel filtered to a resolution of 3.4 Å, with each subunit color-coded. Four different views of the channel are shown, from *side* **D**, **E**, *top* **F**, and *bottom* **G**. The *arrow* in panel **D** indicates β-sheet structure in the cytosolic domain (From Liao et al. 2013, © 2013 Macmillan Publishers Limited, Nature. All rights reserved).

seems, as noted by Huynh et al. 2014b, to be a gap of low density at the immediate contact of the protein surface, giving the impression that the APol is hovering some distance above the surface of the protein. This of course makes no physical-chemical sense, and it does fly in the face of molecular dynamics (MD) models of MP/A8-35 complexes (Perlmutter et al. 2014; see Chap. 5, § 5.3.3). A likely explanation seems to be that, at the high resolution achieved, the map distinguishes between the relatively electron-dense layer comprising the carboxylates, counterions, and amide moieties of A8-35, which form the interface with the surrounding water, and its low-density octyl chains, which are expected to lie in contact with the TM surface of the protein, as is indeed observed in MD calculations (Perlmutter et al. 2014). This effect could be reinforced by the presence of low-density lipid acyl chains in contact with the protein.

Following the study of TRPV1, a number of other cryo-EM studies of APol-trapped MP complexes appeared in 2014–2017 (Fig. 12.20). Most of them, apart from their biological importance, do not bring much new information about the implementation of APols for cryo-EM studies, except for the fact that APols now tend to appear as routine tools to be tested, among others, whenever embarking upon such a project. The following 13 low- to high-resolution studies are summarized in Table 12.1 and will not be further discussed here:

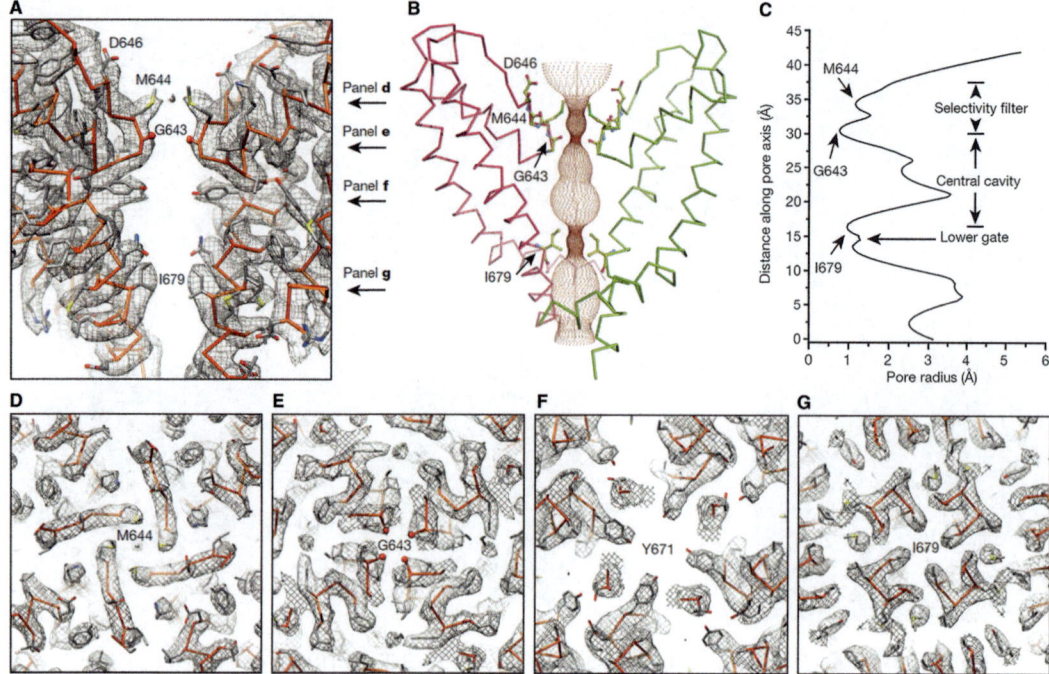

**Fig. 12.22** The ion permeation pathway of TRPV1. (**A**) Cryo-EM densities of the pore in longitudinal cross section are superimposed on an atomic model. Only two diagonally opposed subunits are shown for clarity. Several residues along the pore are labeled for orientation. *Arrows* denote positions of density maps for horizontal cross sections shown in panels (**D–G**), as indicated. (**B**) Solvent-accessible pathway along the pore mapped using the HOLE program. Residues located at the selectivity filter and lower gate are rendered as sticks. (**C**) Radius of the pore calculated with HOLE. (**D–G**) Cryo-EM densities of several residues along the pore are superimposed on the atomic model; all panels represent views along the fourfold axis, showing residues from each subunit of the homotetrameric channel (From Liao et al. 2013, © 2013 Macmillan Publishers Limited, Nature. All rights reserved).

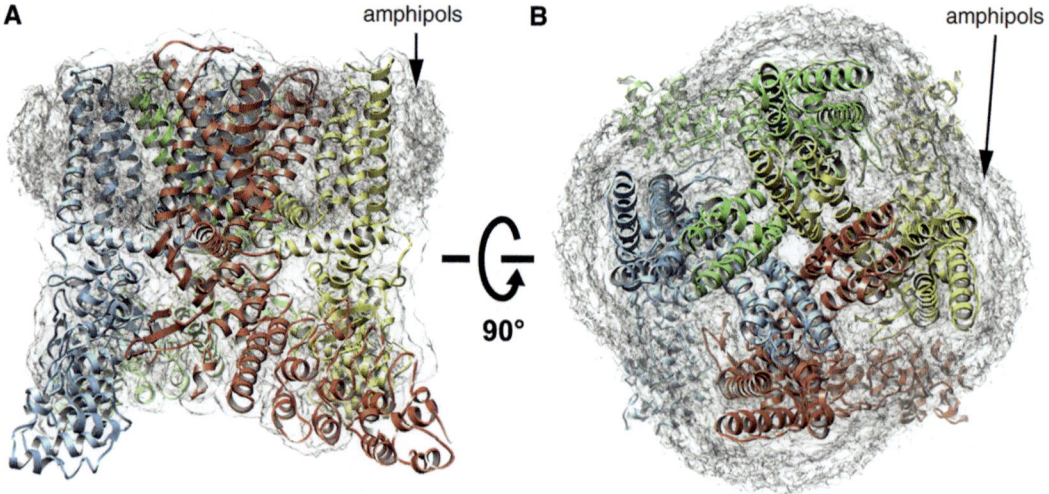

**Fig. 12.23** Arrangement of A8-35 around the transmembrane region of rat TRPV1. (**A**) Side and (**B**) top views of TRPV1 in ribbon diagram showing the belt of APol surrounding the TM region (From Liao et al. 2014, © 2014 Elsevier Ltd. All rights reserved).

- Study 12.12: TRPM1, the transient receptor potential melastatin-1, solved in A8-35 to a resolution of 22 Å at FSC = 0.143 (Agosto et al. 2014).
- Study 12.14: the CsgG/CsgE complex involved in the formation of bacterial curli, solved in A8-35 to a resolution of 24 Å at FSC = 0.5 (Goyal et al. 2014).
- Study 12.17: the ABCB1 P-glycoprotein (PgP), an ABC drug transporter, solved in SMALPs at a resolution of ~35 Å (Gulati et al. 2014).
- Study 12.18: the TRPV2 thermosensor channel, solved in A8-35 at ~15-Å resolution (Fan et al. 2014).
- Study 12.23: the structure of *E. coli* holotranslocon was deduced from an ensemble of biophysical data, including 14–15-Å resolution cryo-EM images of the translocon stabilized with A8-35 (Botte et al. 2016).
- Study 12.24: the structure of the zebrafish STRA6 receptor, which mediates the internalization of retinol, solved to 3.9-Å resolution following trapping with A8-35 (Chen et al. 2016).
- Study 12.26: the structure of an engineered AcrAB-TolC complex, trapped in A8-35, solved to 8.2-Å resolution (Jeong et al. 2016); the authors show a comparison of particle distribution illustrating the crowding at the edge of the holes in detergent solution as compared to a much more even distribution in A8-35.
- Study 12.27: the resolution of A8-35-stabilized respirasome was pushed from 19 Å (Study 12.5; Althoff et al. 2011) to ~9 Å (Sousa et al. 2016).
- Study 12.29: TRPV2 is regulated by temperature; by ligands, such as probenecid and cannabinoids; and by lipids. Following transfer from a DDM/lipid mixture to A8-35, cryo-EM single-particle images were analyzed to a resolution of ~4 Å (Zubcevic et al. 2016). Putative bound lipids are identified.
- Study 12.30: cyclic-nucleotide-gated channels belong to the voltage-gated ion channel superfamily, but their activities are controlled by intracellular cyclic nucleotides. The structure of a cyclic-nucleotide-gated channel trapped in A8-35 has been solved to 3.5-Å resolution (Li et al. 2017b).
- Study 12.31: the structure of the TM pore formed by the *Streptococcus* toxin pneumolysin was solved to 4.5-Å resolution following extraction from liposomes with Cymal-6 and transfer to A8-35 (van Pee et al. 2017).
- Study 12.32: the bacterial efflux pump AcrABZ-TolC was transferred from DDM to A8-35 and its structure in the presence or absence of an inhibitor solved, respectively, to 3.6-Å and 6.5-Å resolution (Wang et al. 2017).
- Study 12.33: the structure of yet another member of the TRP channel family, polycystin-2, was solved to 4.2–4.3-Å resolution after transfer from LMNG/cholesterol hemisuccinate solution to A8-35 (Wilkes et al. 2017).

Two further high-resolution studies of APol-trapped MPs have appeared, that of γ-secretase, to 3.4–4.5 Å (Study 12.15; Lu et al. 2014; Bai et al. 2015b,c), and that of the TRPA1 channel (Study 12.21; three structures at 4.24, 3.9, and 4.7 Å, depending on the ligand bound; Paulsen et al. 2015). They are briefly described in Table 12.1 but each of them deserves a special comment here.

The four components of the human γ-secretase were co-expressed in mammalian HEK293F cells (for a recent review about co-expression of multisubunit complexes, see Zorman et al. 2015). The complex was extracted and purified in CHAPSO and transferred to digitonin. However, the complexes in digitonin yielded poorly exploitable cryo-EM images (Fig. 12.24A). The complexes were therefore transferred to A8-35 using Bio-Beads and free APol removed by SEC, resulting in significant improvement (Fig. 12.24B). Combined with the use of a direct-electron detector in single-electron

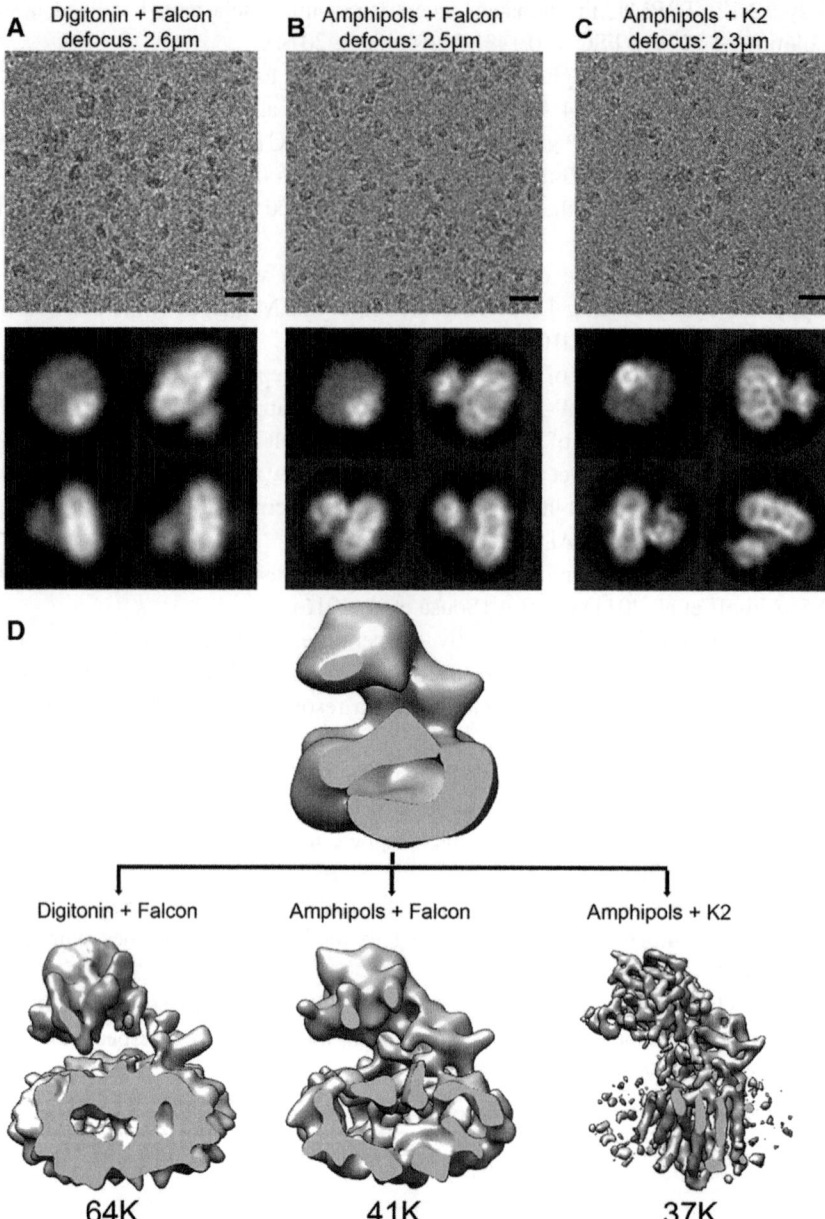

**Fig. 12.24** Cryo-EM analysis of the human γ-secretase complex. (**A**) Analysis of γ-secretase in digitonin imaged using a back-thinned Falcon II detector. A representative electron micrograph (scale bar: 20 nm) and reference-free two-dimensional class averages are shown in the *top* and *bottom* panels, respectively. (**B**) Analysis of A8-35-trapped γ-secretase imaged on the same detector. (**C**) Analysis of A8-35-trapped γ-secretase imaged on a direct-detection K2 Summit detector. (**D**) Comparison of reconstructed 3D models for the three methods of imaging described above. *Numbers* indicate the number of particles used, in thousands. The lowest number of particles (37,310) was used for the generation of higher-resolution images for A8-35-trapped samples imaged using the K2 Summit detector (From Lu et al. 2014, © 2014 Macmillan Publishers Limited, Nature. All rights reserved).

counting mode so as to achieve higher signal-to-noise ratios at the lower spatial frequencies, which are crucial for particle alignment, and with statistical image classification and movie processing, this approach produced a markedly improved map with an overall resolution of 4.5 Å (Fig. 12.24C, D) (Lu et al. 2014; Bai et al. 2015b, c). Nineteen $\alpha$-helices were identified in the TM region. However, in contrast to the density for the TM helices, that for the connecting sequences between neighboring helices was often weak or absent. Only seven TM helices appeared to be connected by strong density, suggesting their order of linkage.

In a subsequent work (Sun et al. 2015), T4 lysozyme was fused to the *N*-terminus of one of the subunits, presenilin-1, and imaging carried out in digitonin. The resolution of the complex as a whole was only marginally better than that achieved in the first study (4.32 Å instead of 4.5), but that of the TM region improved sufficiently for all connections between TM $\alpha$-helices to be convincingly identified. The structures obtained in A8-35 and in digitonin are nearly identical. It would be interesting to sort out exactly what, in the differences between the two structures, is due to the use of a different surfactant and what to the fusion with T4 lysozyme. These two studies offer, in any case, a reminder that a surfactant that is inadequate to study a given target can prove to be effective for another, very similar one. The study in A8-35 was pushed to 3.4-Å resolution (Bai et al. 2015c), permitting atomic modeling of the side chains, and extended to an inhibitor-bound form (Bai et al. 2015b).

Cryo-EM Study 12.21 (Paulsen et al. 2015) bore on APol-trapped TRPA1, the same receptor that had been examined in negative stain in Study 12.4 in complex with A8-35 (Cvetkov et al. 2011; Huynh et al. 2014b). Maltose-binding protein-tagged human TRPA1 was expressed in HEK cells and solubilized and purified using MNG-3 as the detergent. For cryo-EM, MNG-3 was exchanged for PMAL-C8, an APol that is zwitterionic at the pH (8.0) used in these experiments (Gorzelle et al. 2002; Nagy et al. 2001; see Chap. 4), using Bio-Beads followed by SEC. They were vitrified on holey carbon grids in the presence of either an agonist or one of two antagonists and imaged using a microscope equipped with a K2 Summit direct-electron detector camera operated in counting mode, each image being fractionated into 30 subframes. The final resolutions of the three reconstructions were estimated, according to the FSC = 0.143 criterion, to be 4.24, 3.9, and 4.7 Å, respectively (Paulsen et al. 2015). Contouring the density map at a low level visualizes the APol belt surrounding the TM region of the complex (Fig. 12.25).

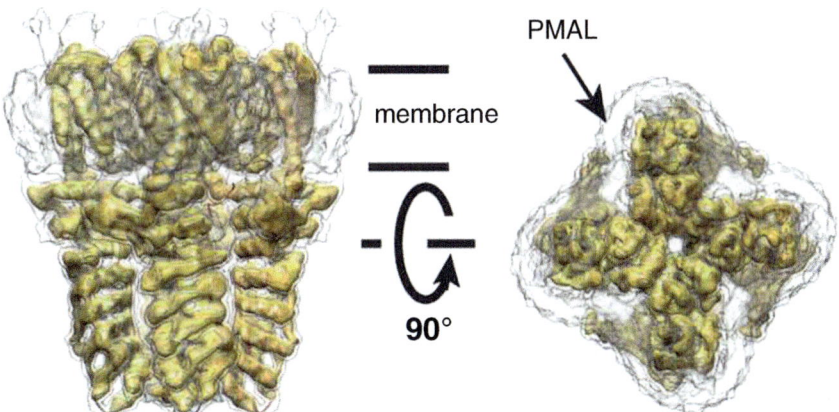

**Fig. 12.25**  A single-particle cryo-EM 3D reconstruction of PMAL-C8-trapped TRPA1. Two views of the density map filtered to 6-Å resolution and displayed at two different isosurface levels (high in *yellow* and low in *gray*). At low-density level, the belt of PMAL-C8 surrounding the transmembrane region of the complex is visible (*gray*) (From Paulsen et al. 2015, © 2015 Macmillan Publishers Limited, Nature. All rights reserved).

It is notable that, for this particular study, PMAL-C8 provided better data than A8-35 (Y. Cheng, personal communication), perhaps because of its lower net charge density, illustrating once more that the concept of "one-fits-all" does not work with MPs.

A brief mention should be made of Study 12.20 (Chowdhury et al. 2015). It does not bear on a MP or MP complex, but on dynactin, a cytosolic protein involved in the transport of cellular cargoes by microtubules. The cryo-EM protocol used mentions that purified dynactin was applied to freshly glow-discharged holey grids. Immediately before application of the protein sample on the grid, a dilute (0.25 g·L$^{-1}$) solution of A8-35 was mixed with the sample to aid in dispersing dynactin particles across the holes in the carbon film. Excess sample was manually blotted with filter paper for ~5 to 7 s, and the sample was immediately vitrified by plunge-freezing in liquid-ethane slurry at −179 °C. These observations are interesting, because it seems unlikely that A8-35 bind to dynactin. Therefore, the better dispersion of the particles, which reminds one of the observations made with Complex I (Flötenmeyer et al. 2007; see Fig. 12.9), the ryanodine receptor (Baker et al. 2015; see below, Fig. 12.27), and an engineered AcrAB-TolC complex (Jeong et al. 2016), is probably related to either the thickness of the film or its surface properties (see below, § 12.3.4.2). Similarly, A8-35 was used in the cryo-EM study of a (non-membrane) nuclease as a substitute to detergent, in order to make the holey carbon film hydrophilic (Wilkinson et al. 2016a, b). It is indeed not unusual for microscopists working on non-membrane objects to add some detergent at submicellar concentrations to their samples in order to modify the distribution or orientation of the particles (for a recent discussion of the formation and effects of air/water surfactant monolayers in cryo-EM, see Glaeser et al. 2016).

Finally, it is worth mentioning observations reported in Study 12.19, a review of cryo-EM studies of the ryanodine receptor (RyR) (Baker et al. 2015) written shortly before the structure of this huge, homotetrameric complex (~2.3 MDa) was solved to 3.8 Å (Yan et al. 2015) and 4.8 Å (Zalk et al. 2015) by cryo-EM studies in detergent solutions. The review presents original data collected on Type 1 RyR trapped in A8-35. Substituting A8-35 to CHAPS results in (*i*) more evenly distributed particles throughout the water film formed on holey carbon grids (Fig. 12.26A vs. B) and (*ii*) higher contrast

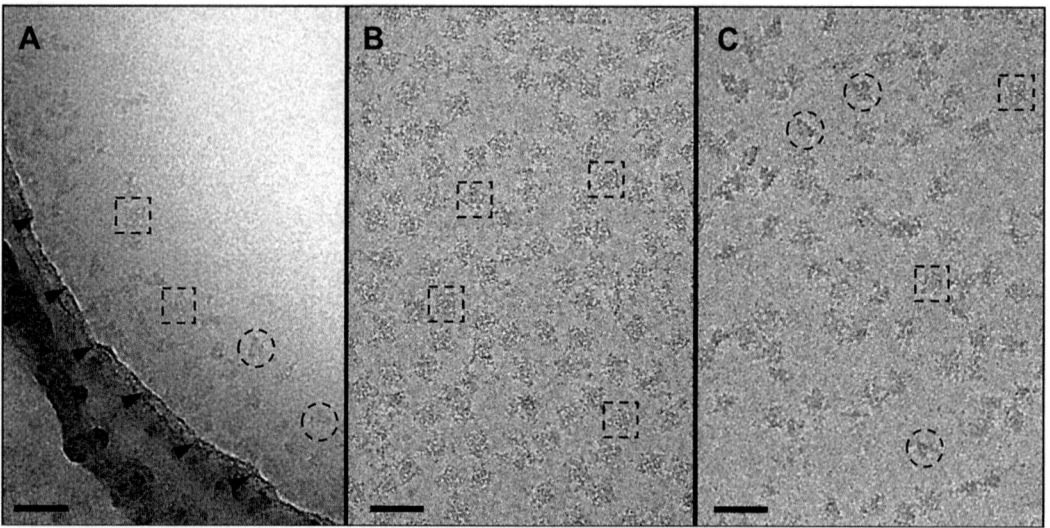

**Fig. 12.26** Cryo-EM images of ice-embedded purified rabbit ryanodine receptor-1. (**A**) In the presence of 0.4% CHAPS. (**B**) After trapping with A8-35. (**C**) Same as B + *n*-octyl-*β*-D-glucopyranoside at one-tenth the CMC. Note the crowding of the RyR1 particles in the meniscus of the film in (**A**) and the good distribution but preferred orientation they adopt in (**B**), with their long axis perpendicular to the plane of the film (*dashed squares*), whereas in (**C**) they are more randomly oriented within the vitreous ice (*dashed squares* and *circles*). Scale bars: 500 Å (From Baker et al. 2015).

(images of RyR1/A8-35 complexes are reported to have an ~2 to 6% contrast at 5–10-Å resolution as compared to 0–2% in the presence of CHAPS, using a traditional CCD detector). However, the RyR1/A8-35 particles tend to adopt a uniform orientation, with their long axis normal to the plane of the film (Fig. 12.26B). Supplementing the preparation with a submicellar concentration of octylglucoside (OG) (~10% of the CMC) results in a random distribution of orientations (Fig. 12.26C). This is one of the few studies where hypotheses about the reasons for improved cryo-EM data collected with APol-trapped vs. detergent-solubilized MPs are critically discussed.

As already mentioned, cryo-EM studies of APol-stabilized MPs are being published at a rapid rate. Studies published or located after this chapter was written and that could not be included in Table 12.1 or discussed here include Wei et al. 2016; Xu et al. 2016; Arnaud et al. 2017; Bausewein et al. 2017; Beckham et al. 2017; Chiu et al. 2017; Ekiert et al. 2017; Fitzpatrick et al. 2017; Li et al. 2017a; Punjani et al. 2017; Schoebel et al. 2017; Terahara et al. 2017; and Zhou et al. 2017.

## 12.3.4 Conclusions and Perspectives

From the overview of published works presented above, it appears that APols have integrated the field of EM and look likely to continue to be helpful in the preparation of improved samples and in their imaging. A discussion of their plusses and minuses and of useful steps that could be taken can be split into two main issues: (*i*) the solution properties of APol-trapped vs. detergent-solubilized samples and (*ii*) issues related to imaging.

### 12.3.4.1 Solution Properties of the Samples

The effects of APols on the MPs they trap have been discussed in Chap. 5. They relate essentially to three distinct mechanisms, whose impact on EM studies, however, is difficult to fathom from the published literature:

(i) APol-trapped MPs are generally more stable than their detergent-solubilized counterparts (Chap. 5, § 5.5). For MP oligomers, this means, for instance, that the risk of monomerization or the loss of subunits is reduced. By the same token, a monomeric MP stands more chance of being observed in its native conformation. This effect will reduce the heterogeneity of the preparations and the risk that the painfully reconstructed 3D model does not correspond to a native state.

   Not all MPs are stabilized by APols. As discussed in more detail in Chap. 5, when stability is an issue, steps that can be taken include comparing the effects of different APols (there are suggestions that the least charged APols are more stabilizing – see e.g. Picard et al. 2006; Dahmane et al. 2009; Bazzacco et al. 2012 – but this may not be universal, cf. Huynh et al. 2014b), supplementing the preparation with a small amount of lipids (see e.g. Dahmane et al. 2013), and/or, perhaps, playing on the ionic strength: whereas this point has not yet been experimentally examined, it seems likely that electrostatic repulsion between MP-adsorbed APols plays a role in destabilization and that raising the ionic strength could mitigate this effect. Needless to say, working in the presence of high-affinity ligands is a stabilizing factor.

(ii) There are indications that APols may affect the dynamics of MPs, through the "Gulliver effect" discussed in Chap. 5, namely a (hypothetical) increase in the free energy barrier between various conformational states, which would slow down or block transitions. There are some indications, discussed in more detail in Chap. 5 (§ 5.4), that this effect may not apply to MPs whose function does not involve large conformational changes at the interface between the protein and the membrane. Thus, (*i*) the nicotinic acetylcholine receptor undergoes, following trapping with A8-35, ligand-induced conformational changes with kinetics similar to that in its native membrane environment (Martinez et al. 2002); (*ii*) BR recovers, following transfer from detergent to A8-35, a photocycle very similar to that observed in its native purple membrane (Gohon et al. 2008; Dahmane et al. 2013); and (*iii*) TRPV1 was imaged in different conformational states after adding agonists and antagonists to the APol-trapped channel (Cao et al. 2013b). Yet, some MPs, such as the sarcoplasmic calcium pump (Champeil et al. 2000; Picard et al. 2006) or the $F_1F_O$ ATP synthase (Wilkens 2000), are reversibly inhibited by APols, which could be related to APols slowing down or preventing transconformations required by the enzymatic cycle (see Chap. 5, § 5.6). In such cases, it could be desirable either to leave the system enough time to relax to the state induced by one or the other ligand before freezing it or to trap it after the ligand has been added.

As with detergents, the question arises whether the APol-trapped state is identical to that – or to one of those – adopted by the protein in its native membrane environment. An interesting observation concerns BR, whose TM region looks, according to solution-state NMR, very similar in DDM, A8-35, or nanodiscs, whereas the conformation of the loops resembles more that in the membrane following APol trapping than in DDM solution (Etzkorn et al. 2013). This cannot be due to the thickness of the polar head region, which is closest to that in the membrane in DDM than in A8-35. It suggests, rather, a change in dynamics (see Chap. 5, § 5.6). This is a very interesting issue that deserves to be studied in more depth.

(iii) A point that remains to be clarified is that of the nature and origin of the "bumps" of the APol belt observed at medium resolution (19 Å) with the respirasome/A8-35 complex (Fig. 12.19A; Althoff et al. 2011), but not at higher resolution (~4 Å) in subsequent cryo-EM studies of TRP channels (Figs. 12.23 and 12.25) (Liao et al. 2014; Paulsen et al. 2015). One possibility that may perhaps deserve to be kept in mind is that the formation of bumps may result from the frustration created by the adsorption of the APol onto the nearly flat surface of the respirasome, which forces the APol/water interface away from its natural curvature. This situation is not reproduced, or not to the same extent, with the much smaller TRP channels.

### 12.3.4.2   Imaging

The reported effects of APol-trapping vs. detergent solubilization on imaging fall in three categories: the contrast, the dispersity of the particles, and their orientation.

(i) *Contrast.* In the case of the ryanodine receptor, it was noted that the contrast was significantly better with the A8-35-trapped complex than with that kept soluble by CHAPS (Baker et al. 2015). Although this is not discussed by other authors, one should expect that this effect can be significant with high-CMC detergents such as CHAPS or CHAPSO, which need to be present at relatively high concentration (>8 mM) to keep the protein from aggregating, probably much less so with low-CMC detergents such as Triton X-100, DDM, digitonin, or the MNG detergents.

(ii) *Dispersity.* It has been noted in several studies (e.g. Flötenmeyer et al. 2007; Baker et al. 2015; Jeong et al. 2016) that particles that, in detergent samples, tend to crowd at the rim of the holes in the carbon film are more evenly dispersed in APol-trapped samples, so that collecting a large number of well-separated images becomes easier. In Study 12.3 (Flötenmeyer et al. 2007), this effect was tentatively attributed to the surface tension of the air/water interface being possibly higher for the APol-trapped sample than for the detergent-solubilized one. A high surface tension, such as that between pure water and air, makes it energetically costly for the film to form a meniscus with a small radius of curvature, whereas the adsorption of surfactants at the interface lowers this energetic cost. As a result, thinning of the film is easier in detergent solution, as actually observed (cf. Fig. 12.9E, F), and the large particles of mitochondrial Complex I that were the target of Study 12.3 are more likely to be drained toward the meniscus that forms at the contact between the liquid film and the supporting carbon in detergent solution than after trapping with APols.

Later studies showed that, at equilibrium, the surface tension of A8-35 solutions is in fact not much different from that of detergent solutions (Giusti et al. 2012; see Chap. 4, § 4.3.1.2.4). However, kinetic effects may come into play. In a typical cryo-EM experiment, (*i*) the free APol is separated from the MP/APol complexes by SEC, and (*ii*) the excess solution deposited on the grid is blotted a few seconds before freezing (Dobro et al. 2010). Blotting must remove most of the surfactant layer adsorbed at the air/water interface, bringing the surface tension close to that of pure water. In detergent solutions, which are above the CMC, the interfacial monolayers can reform quickly due to the rapid diffusion of monomeric detergent molecules, and the surface tension may drop again before the film is frozen. In solutions of MP/APol complexes, where the concentration of free APol is very low, the monolayers will reform due to the migration to the surface of occasional free APol particles and of MP/APol complexes. This process, however, is necessarily slower than the adsorption of small detergent molecules, and it seems possible that, at the time of freezing, the surface tension is still higher, opposing the formation of a sharply bent meniscus and the thinning of the film. Note, however, that such a mechanism falls short of explaining why dynactin is more evenly spread in the presence of A8-35 than in the absence of any surfactant (Study 12.20).

These processes deserve a systematic examination, because understanding them would likely be of importance in achieving a better control of the distribution of the particles across the aqueous film, particularly for large complexes.

(iii)  *Orientation.* A preferential orientation of the particles with respect to the plane of the film can be an advantage or a disadvantage depending on the reconstruction procedure used. In the conical tilt reconstruction method, which relies on imaging the same field of particles at two different angles, it is an advantage. When the samples are not tilted and reconstruction is based on reorienting with respect to one another averaged 2D projections covering all possible angles, it becomes a handicap. Little information is available about the way MP/APol complexes interact with carbon films. In Study 12.13, it was noted that the OmpF filaments that form when OmpF/A8-35 complexes are depleted of free APol tend to orient with the TM region of BR facing the carbon film (Arunmanee et al. 2014). In Study 12.9, on the contrary, it seems that replacing DDM with A8-35 results in a preferential, although not systematic, orientation of the particle with their fourfold symmetry axis normal to the film, that is with the APol layer away from the carbon surface (Liao et al. 2013; compare, above, images and averages in DDM, Fig. 12.27A, B, with those of APol-trapped complexes, Fig. 12.27C, D).

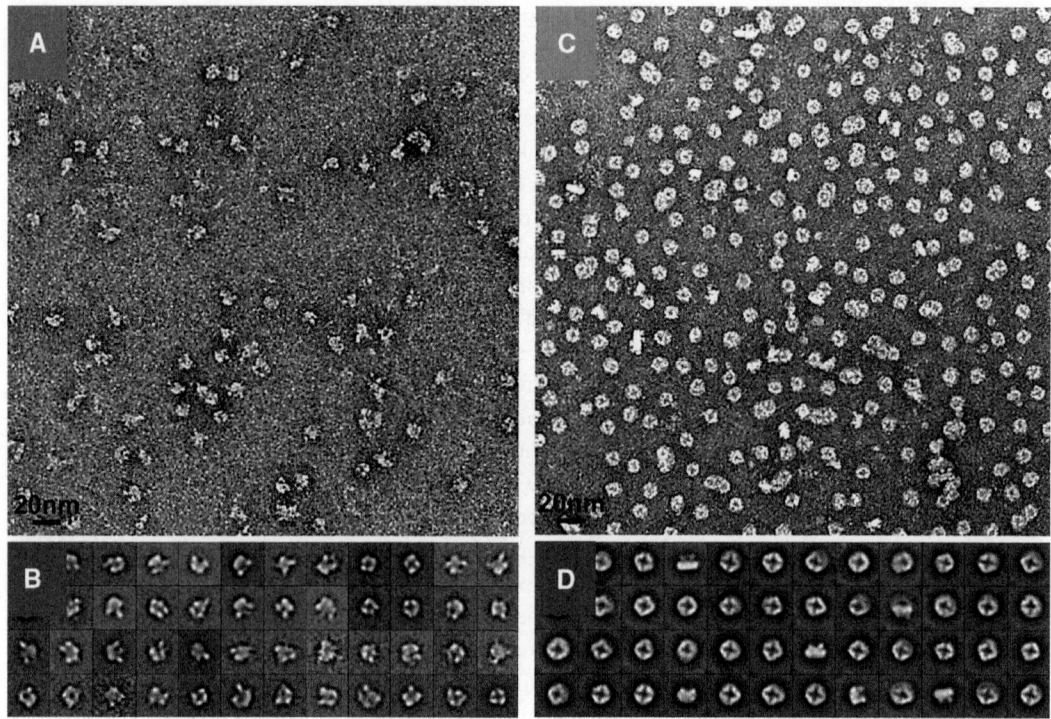

**Fig. 12.27** Negative-stain EM images of TRPV1. (**A**) Representative negative-stain image of purified truncated TRPV1 in DDM. (**B**) 2D class averages of negatively stained DDM-solubilized TRPV1. (**C**) Representative negative-stain images of truncated TRPV1 trapped in A8-35. (**D**) 2D class averages of negatively stained TRPV1/A8-35 particles (From Liao et al. 2013, reprint by permission from Macmillan Publishers Ltd.: Nature © 2013).

There are relatively few statements in published papers regarding the orientation behavior of MP/APol complexes in liquid films. In Study 12.19, it was noted that the ryanodine receptor-1 tended, when complexed with pure A8-35, to orient with its long axis normal to the plane of the film, that is with the APol belt away from the interface (see above, Fig 12.26B). Supplementing the sample with a submicellar concentration of OG brought about a more random distribution of orientations (Fig. 12.26C; Baker et al. 2015). The origin of this phenomenon is not clear. At one-tenth its CMC, one can expect that OG will represent, in mass, ca. 10% of the surfactant surrounding the TM region of RyR1 (see Tribet et al. 2009; Zoonens et al. 2007, and discussion in Chap. 5, Box 5.3). This will lower by about as much the surface charge density of the belt, and make it possible for it to adjust the distribution of surface charges in response to the local electric field, but it does not seem enough of an effect to have such major consequences. Another possible effect is that the presence of OG will permit the rapid reformation of a surface film following blotting, lowering the surface tension and, perhaps, bringing about some thinning, which could have a mechanical effect on the orientation of the particles. Whatever the mechanism, this is a very interesting observation, which could be of practical use in the EM study of other complexes and deserves further study.

### 12.3.4.3   Perspectives, Suggestions, and Speculations

In concluding this chapter, I would like to point to some areas where there seems to be room for progress.

(i)  A couple of practical suggestions

- Documentation of experimental conditions and methodological studies. It would be desirable that the exact conditions under which the samples are prepared be more precisely documented. As discussed in the preceding section, experimental details such as the extent of removal of free APol or the time that elapses between blotting the cryo-EM grid and freezing it may well play important roles in the distribution and orientation of the particles in the aqueous film. Now that the usefulness of APols for EM studies is well-established, it would be helpful to carry out more systematic studies of the importance of the various parameters that may affect this behavior, including ionic strength, the nature of the APol, and the presence of traces of detergent. As APols with various chemical structures, charge densities, etc. have been developed and validated for handling MPs (see Chaps. 4 and 5), it would be desirable that their usefulness for EM studies be investigated. Note that APols are miscible with detergents (Zoonens et al. 2007; Tribet et al. 2009; see Chap. 4) and, very likely, at least for some of them, with one another, so that many parameters, such as charge density, could be varied in a continuous manner and their effect investigated without having to multiply syntheses.
- A related question is that of the commercial availability of a range of APols. We have repeatedly noted, in our survey of the literature, that a surfactant that gives excellent results with a MP complex may be unsatisfactory with another. It would be highly desirable that, at the onset of an EM study, a range of different APols be tested. This is more easily said than done, however, because it is not straightforward to convince a private company to market a variety of molecules, each of which calls for upscaling and optimizing a synthesis protocol, when the volumes of sales will, according to all probability, remain small. Perhaps a consortium of potential users should be set up with the aim of contracting for the synthesis of medium- to large-scale batches of a selection of molecules, which could then be made available to the community as a whole (see the general "Conclusions and Perspectives" (§ 12.3.4) at the end of this chapter).

(ii)  Some more or less wild speculations
   Many opportunities offered by APols have not been exploited for EM yet. Let me list a few of them:

- Visualizing the APol belt. This could be useful, particularly in low- or medium-resolution studies, when the position of the TM region in the averaged images can be ambiguous. One can think of several approaches. Several tagged APols are by now available, carrying a biotin (Charvolin et al. 2009; Ferrandez et al. 2014), an oligonucleotide (Le Bon et al. 2014a), or a polyhistidine tag (Giusti et al. 2015) (reviewed in Le Bon et al. 2014b, and in Chap. 4, § 4.4). They can be used to tie gold nanobeads or monomeric avidin to the TM region of the protein under study. It is not impossible that the oligonucleotides, which are 6–7 kDa in mass, could be directly detected or hybridized to a larger marker. The position of the tags will necessarily vary from one particle to the next, as there is no reason that they should adopt specific positions with respect to the protein, but they would nevertheless indicate where the APol belt lies. This approach, which has been validated for the couple biotinylated A8-35/monomeric streptavidin, using the mitochondrial cytochrome $bc_1$ complex and a bacterial secretion protein, CsgG, as test MPs, will be described in an upcoming article (Perry et al. 2018). The thiomorpholine-tagged version of A8-35 (see Chap. 4) could be used in the same manner, in combination with gold nanoparticles (cf. Suárez-Suárez et al. 2013).

In cryo-EM, one could conceivably play on the nature of the counterions. A belt of A8-35 can be expected to carry about 0.88 carboxylates per $nm^2$ of APol/water interface, or, in projection, for a belt ~4 nm high and 2 nm thick, ~4.4 carboxylates per $nm^2$. AUC experiments have shown that MP/A8-35 complexes can be handled in the presence of any alkaline cation from $Li^+$ to $Cs^+$ (Gohon 2002). Rough calculations indicate that, were two otherwise identical samples of MP/A8-35 complexes to be imaged in the presence of 50–100 mM of either LiCl or CsCl, the increase in contrast of the APol belt over the surrounding solution would be of ~2.3× when substituting $Cs^+$ to $Na^+$, ~3× when comparing $Li^+$ and $Cs^+$, which ought to be easily detected.

- Inducing the complexes to adopt preferred orientations. Tagged APols could possibly be used to bias the orientation of MP/APol complexes. For instance, a carbon film onto which would have been adsorbed a layer of Ni/NTA-carrying lipids (Kelly et al. 2008) could possibly be used to induce MPs trapped with a poly-histidine-tagged APol to lie with their TM region in contact with the carbon film. We have speculated elsewhere (Le Bon et al. 2014a) on the possibility to use DNA scaffolds to organize in space MPs trapped with an oligonucleotide-tagged APol (cf. Selmi et al. 2011).

- In the same order of ideas, it would be interesting to push somewhat further the study of the filaments and, if applicable, the 2D sheets that some MP/APol complexes tend to form upon depletion of free APol (see above, §§ 12.3.1 and 12.3.2). It would be important, in particular, to establish whether the formation of filaments is mediated by APol/APol or by protein/protein contacts, which could be done by studying the periodicity of the protein along the filaments. In the first case, not much order can be expected from the proteins. In the second case, on the contrary, a regular helix could form, which could be amenable to helical reconstruction. This would be especially valuable for MPs whose small size renders them difficult to orient for single-particle image reconstruction and that refuse to yield 2D or 3D crystals.

It is not clear why the side-by-side association into filaments does not proceed further to form 2D sheets, as one might have expected. One possibility is that electrostatic repulsion between APol belts opposes it, favoring the addition of new molecules at the extremities rather than at the side of filaments. It would be of great interest to examine the effect of factors such as the charge density of the APol (SAPols = A8-75 > A8-35 > NAPols; see Chap. 4), the ionic strength, or the addition of small amounts of neutral detergent. In the case of *E. coli* OmpF, one factor has already been identified: addition of LPS in the presence of divalent cations favors the transition from a 1D organization, filaments, to a 2D one, sheets or crystals (Arunmanee et al. 2014; see above, § 12.3.2). As suggested in the latter work, it would be intriguing to carry such studies further and examine whether APols could not be used as modulators in controlling the formation of MP 2D crystals.

- Finally, another intriguing possibility, which might perhaps deserve examination, would be to trap MPs with an APol carrying free-radical scavenger groups, whose synthesis has been developed (F. Giusti and M. Zoonens, unpublished data), in the hope of slowing down radiation damage during illumination.

## 12.4   Protocol 12.1: Preparation of Amphipol-Trapped Membrane Proteins for Cryo-EM Studies

### 12.4.1   Introduction

Recent advancements in single-particle cryo-EM, collectively termed the "resolution revolution," have made it possible to determine near-atomic resolution structures of a number of MPs, including ion channels (see above, § 12.2.3, and Table 12.1). To characterize MPs using structural methods, including cryo-EM, they must first be extracted from the membrane and stabilized in aqueous solution by a surfactant, typically a detergent (Chap. 2), nanodiscs (NDs; Chap. 3, § 3.3), or APols (Chaps. 4 and 5). As a rule, MPs are initially solubilized and purified in detergents, e.g. DDM or neopentyl glycol maltosides (see Chap. 2), after which they are generally transferred to a more stabilizing surfactant, such as digitonin, NDs, or APols (Fig. 12.28).

As there are many different types of detergents, NDs, and APols, many options can be screened for a given target MP. Screening is necessary because not all MPs yield the best resolution with the same surfactant (see above, § 12.2.3). In recent years, the laboratory of David Julius has had success determining the near-atomic resolution structures of two mammalian ion channels involved in pain sensation: the capsaicin receptor, TRPV1, and the wasabi receptor, TRPA1 (Liao et al. 2013; Paulsen et al. 2015; Studies 12.7 and 12.19 in Table 12.1). In both cases, the ion channels were stabilized in APols: A8-35 for TRPV1 and PMAL-C8 for TRPA1. For TRPV1, the Julius lab was also able to resolve agonist-induced, gating-associated conformational changes in A8-35. It later resolved nearly identical conformational changes for TRPV1 in NDs, demonstrating that both media could yield physiologically relevant structures (Cao et al. 2013b; Gao et al. 2016). To date, close to 20 near-atomic resolution structures of MPs stabilized in APols have been determined by cryo-EM (cf. Table 12.1,

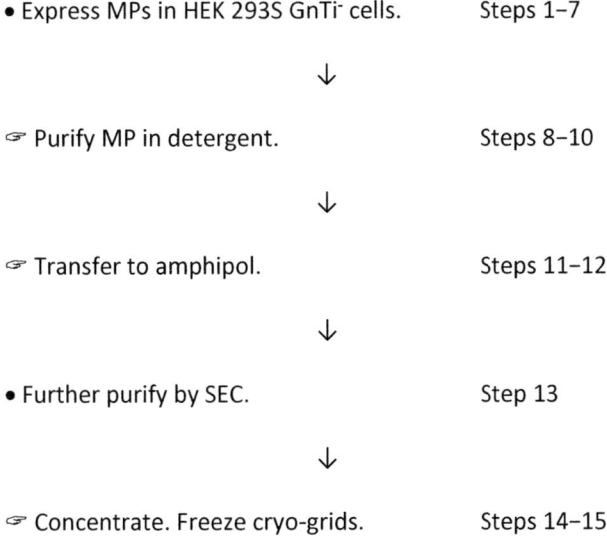

**Fig. 12.28** Flow chart of the preparation of membrane protein/amphipol complexes for cryo-EM studies. Pointing hands denote the main protocol optimization points.

Fig. 12.20, and additional references given at the end of § 12.3.3). The following is an annotated protocol for purifying TRPV1 and TRPA1 in detergent solution and transferring them to APols for structural studies. Experimental steps that are common to both MPs are marked with a black bullet (•) and details that are specific to TRPV1 and TRPA1 with ■ and □ symbols, respectively; general comments and tips for optimization are in italics and marked with a pointing hand sign (☞).

## 12.4.2   Vectors and Cell Lines

- pFastBac1 plasmid containing the gene of interest in frame with an *N*-terminal maltose binding protein (MBP) affinity tag and a TEV protease site cloned downstream of an engineered human cytomegalovirus (CMV) promoter for baculovirus transduction-based expression in HEK293S GnTi⁻ cells

*☞MPs can be expressed in a number of expression systems including suspension of HEK cells, Sf 9 insect cells, yeast, and bacteria. The most appropriate expression system for each protein of interest needs to be determined empirically. For TRPV1, well-behaved, homogenous protein can be purified from both suspensions of HEK and Sf 9 cells (Cao et al. 2013a). For TRPA1, well-behaved, homogeneous protein could only be purified from suspensions of HEK cells.*

- HEK293S GnTi⁻ cells (ATCC, catalogue number CRL-3022)
- Sf 9 cells (Life Technologies, catalogue number 12659017)

*☞Detailed protocols for the growth and maintenance of Sf 9 cells for baculovirus generation and HEK293S GnTi⁻ cells for the expression of MPs are given in Goehring et al. (2014).*

## 12.4.3   Buffers

- Lysis buffer:
  ■ **TRPV1**: 50 mM Tris/HCl buffer, 36.5 mM sucrose, 2 mM *tris*(2-carboxyethyl)phosphine (TCEP), pH 8.0, supplemented with 1 mM phenylmethanesulphonylfluoride (PMSF), 3 mg·L$^{-1}$ leupeptin, 1 mg·L$^{-1}$ pepstatin.
  □ **TRPA1**: 50 mM Tris/HCl buffer, 36.5 mM sucrose, 1 mM β-mercaptoethanol (β-ME), 5 mM EDTA, 1 mM inositol hexakisphosphate (IP$_6$), pH 8.0, supplemented with 1× protease inhibitor cocktail tablet (Roche). 1× protease inhibitor cocktail is produced by dissolving one tablet in 50 mL of buffer.
- Membrane resuspension buffer:
  ■ **TRPV1**: 50 mM HEPES, 200 mM NaCl, 2 mM TCEP, 10% glycerol, pH 8.0, supplemented with 1 mM PMSF, 3 mg·L$^{-1}$ leupeptin, 1 mg·L$^{-1}$ pepstatin.
  □ **TRPA1**: 50 mM HEPES, 150 mM NaCl, 1 mM dithiothreitol (DTT), 1 mM IP$_6$, 10% glycerol, pH 8.0, supplemented with 1× protease inhibitor cocktail tablet (Roche). 1× protease inhibitor cocktail is produced by dissolving one tablet in 50 mL of buffer.
- Wash buffer:

- **TRPV1**: 20 mM HEPES, 150 mM NaCl, 2 mM TCEP, 10% glycerol, $0.2 \text{ g} \cdot \text{L}^{-1}$ soybean lipids (asolectin), 0.5 mM DDM, pH 8.0.
- ☐ **TRPA1**: 20 mM HEPES, 150 mM NaCl, 1 mM DTT, 1 mM $IP_6$, 0.5 mM Lauryl Maltose Neopentyl Glycol (LMNG, MNG-3), pH 8.0.
- Elution buffer:
  - **TRPV1**: 20 mM HEPES, 150 mM NaCl, 2 mM TCEP, 10% glycerol, $0.2 \text{ g} \cdot \text{L}^{-1}$ asolectin, 0.5 mM DDM, 20 mM maltose, pH 8.0.
  - ☐ **TRPA1**: 20 mM HEPES, 150 mM NaCl, 1 mM DTT, 1 mM $IP_6$, 0.5 mM MNG-3, 40 mM maltose, pH 8.0.
- Size exclusion chromatography buffer:
  - **TRPV1**: 20 mM HEPES, 150 mM NaCl, 2 mM TCEP, pH 7.4.
  - ☐ **TRPA1**: 20 mM HEPES, 150 mM NaCl, 1 mM DTT, 1 mM $IP_6$, pH 8.0.

### 12.4.4   Protocol

1. Transduce HEK293S GnTi$^-$ cells grown in suspension at 37 °C in an orbital shaker at a density of $1.5\text{--}2 \times 10^6$ cells per mL with 1:10 (v/v)–1:50 (v/v) baculovirus encoding the protein of interest.

   ☞*The volume of baculovirus to use to transduce cells can be readily determined by a Western blot-based dilution series expression test. Briefly, transduce HEK293T or HEK293S GnTi$^-$ cells in a 6-well dish with 0, 1:10, 1:20, 1:30, 1:40, or 1:50 (v/v) baculovirus. Add sodium butyrate at a final concentration of 10 mM. After 16–24 h, collect cells, generate cell lysates, and use for anti-MBP Western blot. Choose the lowest volume of virus that yields the maximum protein production.*

2. Twenty-four hours after transduction, add sodium butyrate to the culture at a final concentration of 5–10 mM to boost protein expression.

   ☞*An alternative expression protocol that is commonly used and can serve as a protocol optimization point is to transduce HEK293S GnTi$^-$ cells with baculovirus and add sodium butyrate at the same time.*

3. Harvest cells 48 h after transduction by centrifugation (3000 rpm, 15 min at 4 °C).

   ☞*The length of transduction time is another protocol optimization point. Typically, cells are transduced for 30–48 h. When cells are transduced for 30 h, sodium butyrate is added at the same time as baculovirus. Typically, the cells are transduced in the presence of sodium butyrate for at least 24 h. Though not routinely done in our lab, growing HEK293 cells at 30 °C after adding sodium butyrate can often enhance protein expression (Goehring et al. 2014).*

4. Resuspend cell pellets in lysis buffer (50 mL lysis buffer for each liter of cells) and break cells by passing through an emulsifier twice.

☞*An alternative and quicker purification protocol that serves as a protocol optimization point and which works equally well for TRPV1 and TRPA1 is to directly flash-freeze cell pellets (before resuspension) and store at −80 °C before purification. The cell pellets are thawed in a water bath at room temperature, resuspended in membrane resuspension buffer without glycerol (25 mL per liter of cells), further resuspended with a Dounce homogenizer, and directly solubilized by adding the appropriate detergent as in step 8. The subsequent purification protocol is then continued forward.*

5.  Clear cell debris by centrifugation (8000× g, 20 min, 4 °C).
6.  Transfer supernatant to ultracentrifuge tubes and collect membranes by ultracentrifugation in a Beckman 45 Ti rotor (200,000× g, 60 min, 4 °C).
7.  Discard supernatant and resuspend membrane pellets in membrane resuspension buffer (25 mL buffer for each liter of cells) on ice. Use a Dounce homogenizer to aid membrane resuspension.

☞*Membranes can be prepared in batch, flash-frozen in liquid nitrogen, and stored at −80 °C at this stage for future purifications. Of note, we have found that resuspended membranes are not as stable over time (longer than 1 year) as cell pellets.*

8.  Thaw resuspended membranes in a room-temperature water bath. Solubilize membranes with 20 mM DDM or 10 mM MNG-3 on an end-to-end rotator for 2 h at 4 °C.

☞*The optimal detergent for purification needs to be empirically determined for each MP of interest. The fluorescence size exclusion chromatography (FSEC) procedure developed by the laboratory of Eric Gouaux represents an excellent screening strategy (Kawate and Gouaux 2006). Detergent extraction time and extraction temperature (4 °C vs. room temperature) are additional points of optimization. Often times, complete extraction can be accomplished in 1–1.5 h at 4 °C.*

9.  Remove detergent-insoluble material by ultracentrifugation in a Beckman 45 Ti rotor (30,000× g, 30 min, 4 °C) and mix the supernatant with amylose resin (1 mL of 50% slurry per liter of cells) on an end-to-end rotator for 2 h at 4 °C.

☞*The amylose resin is supplied in an ethanol solution. A working stock of resin is prepared as a 50% slurry in water by washing the resin 3× with water. Pellet resin at 1000 rpm between washes. Store working resin stock at 4 °C.*

10. Collect resin in a disposable Poly-Prep column at room temperature. Wash resin with ten column volumes of wash buffer and elute bound protein with elution buffer (200 μL fractions for each 1 mL of 50% resin slurry). Measure $A_{280}$ and pool fractions containing protein.

☞*Measuring A$_{280}$ can be achieved in a number of ways. We use a NanoDrop.*

11. Mix pooled protein with APols (Anatrace, A8-35 or PMAL-C8, 100 g·L$^{-1}$ or 50 g·L$^{-1}$ stock solutions freshly prepared in water, respectively) at a 1:3 (w/w) dilution (MP:APol) on an end-to-end rotator overnight at 4 °C. Remove detergent with Bio-Beads SM-2 (Bio-Rad)

added 4 h after APol addition or the following day by mixing on an end-to-end rotator at 4 °C.

☞*The time at which to add Bio-Beads is a protocol optimization point that needs to be empirically determined. For TRPV1, Bio-Beads are added 4 h after APols. For TRPA1, Bio-Beads are added the following morning because earlier Bio-Bead addition results in detectable loss of protein features as assessed by negative stain electron microscopy. If it is necessary to remove the MBP affinity tag, add an empirically determined amount of TEV protease (~1:25 w/w ratio to the MP) along with the APol and mix overnight on an end-to-end rotator.*

12. Remove Bio-Beads over a disposable Poly-Prep column.
13. Clear eluent by centrifugation (14,000 rpm, 5 min, 4 °C) before further purification on a Superose 6 or Superdex 200 column pre-equilibrated in gel filtration buffer.

☞*For TEV-digested TRPV1 (300 kDa tetramer) and MBP-tagged TRPA1 (690 kDa tetramer), biochemically well-behaved, homogeneous protein elutes around 11 mL on a Superdex 200 and around 13 mL on a Superose 6 column, respectively. Both columns have a 25 mL bed volume. When PMAL-C8 is used as the APol, excess PMAL-C8 elutes as a separate peak around 17 mL on a Superose 6 column as detected with the in-line $A_{280}$ detector. A corresponding peak is not seen with excess A8-35, which does not absorb at this wavelength (see Chaps. 4 and 8).*

14. Combine protein-containing fractions and concentrate to 0.5–3 $g \cdot L^{-1}$ in a 100 kDa cutoff Amicon filter by centrifugation (3000 rpm, 7 min, 4 °C).
15. Use sample to prepare grids for cryo-EM analysis. Many aspects of freezing cryo-grids must be empirically determined including the recourse to additives (e.g. glycerol, agonists, antagonists), the optimal protein concentration applied to the grids (could be anywhere from 0.5 to 3 $g \cdot L^{-1}$), the volume of sample applied to the grids (e.g. 2.5 μL), blotting time, type of cryo-grids used (e.g. 200-mesh or 400-mesh Quantifoil grids), glow discharge time, and relative humidity in the blotting chamber.

Wishing you all the very best of luck on your membrane protein adventures,

Candice E. Paulsen, Department of Physiology, University of California, San Francisco. Protocol based on Cao et al. (2013a), Liao et al. (2013), and Paulsen et al. (2015) (Studies 12.9 and 12.21), with added comments.

# References

Abeyrathne, P.D., Grigorieff, N. (2017) Expression, purification, and contaminant detection for structural studies of *Ralstonia metallidurance* ClC protein rm1. *PLoS One* **12**:e0180163.

Agard, D.A., Cheng, Y., Glaeser, R.M., Subramaniam, S. (2014) Single-particle cryo-electron microscopy (cryo-EM): progress, challenges, and perspectives for future improvement. *Adv. Imaging Electron Phys.* **185**:113–137.

Agosto, M.A., Zhang, Z., He, F., Anastassov, I.A., Wright, S.J., McGehee, J., Wensel, T.G. (2014) Oligomeric state of purified TRPM1, a protein essential for dim light vision. *J. Biol. Chem.* **289**:27019–27033.

Allegretti, M., Mills, D.J., McMullan, G., Kühlbrandt, W., Vonck, J. (2014) Atomic model of the $F_{420}$-reducing [NiFe] hydrogenase by electron cryo-microscopy using a direct electron detector. *eLife* **3**:e01963.

Althoff, T. (2011) Strukturelle Untersuchungen am Superkomplex $I_1III_2IV_1$ der Atmungskette mittels Kryoelektronen-mikroskopie, Fachbereich Biochemie, Chemie und Pharmazie. Johann Wolfgang Goethe-Universität, Frankfurt-am-Main, 248 p.

Althoff, T., Mills, D.J., Popot, J.-L., Kühlbrandt, W. (2011) Assembly of electron transport chain components in bovine mitochondrial supercomplex $I_1III_2IV_1$. *EMBO J.* **30**:4652–4664.

Amunts, A., Brown, A., Bai, X.-C., Llácer, J.L., Hussain, T., Emsley, P., Long, F., Murshudov, G., Scheres, S.H.W., Ramakrishnan, V. (2014) Structure of the yeast mitochondrial large ribosomal subunit. *Science* **343**:1485–1489.

Arnaud, C.-A., Effantin, G., Vivès, C., Engilberge, S., Bacia, M., Boulanger, P., Girard, E., Schoehn, G., Breyton, C. (2017) Bacteriophage T5 tail tube structure suggests a trigger mechanism for *Siphoviridae* DNA ejection. *Nat. Comm.* **8**:1953.

Arunmanee, W., Harris, J.R., Lakey, J.H. (2014) Outer membrane protein F stabilised with minimal amphipol forms linear arrays and LPS-dependent 2D crystals. *J. Membr. Biol.* **247**:949–956.

Baboolal, T.G., Conroy, M.J., Gill, K., Ridley, H., Visudtiphole, V., Bullough, P.A., Lakey, J.H. (2008) Colicin N binds to the periphery of its receptor and translocator, outer membrane protein F. *Structure* **16**:371–379.

Bai, X.-C., McMullan, G., Scheres, S.H.W. (2015a) How cryo-EM is revolutionizing structural biology. *Trends Biochem. Sci.* **40**:49–57.

Bai, X.-C., Rajendra, E., Yang, G., Shi, Y., Scheres, S.H.W. (2015b) Sampling the conformational space of the catalytic subunit of human $\gamma$-secretase. *eLIFE* **4**:e11182.

Bai, X.-C., Yan, C., Yang, G., Lu, P., Ma, D., Sun, L., Zhou, R., Scheres, S.H.W., Shi, Y. (2015c) An atomic structure of human $\gamma$-secretase. *Nature* **525**:212–218.

Baker, M.R., Fan, G., Serysheva, I.I. (2015) Single-particle cryo-EM of the ryanodine receptor channel in an aqueous environment. *Eur. J. Transl. Myol.* **25**:35–48.

Bausewein, T., Mills, D.J., Langer, J.D., Nitschke, B., Nussberger, S., Kühlbrandt, W. (2017) Cryo-EM structure of the TOM core complex from *Neurospora crassa*. *Cell* **170**:693–700.e697.

Bazzacco, P., Billon-Denis, E., Sharma, K.S., Catoire, L.J., Mary, S., Le Bon, C., Point, E., Banères, J.-L., Durand, G., Zito, F., Pucci, B., Popot, J.-L. (2012) Non-ionic homopolymeric amphipols: Application to membrane protein folding, cell-free synthesis, and solution NMR. *Biochemistry* **51**:1416–1430.

Beckham, K.S., Ciccarelli, L., Bunduc, C.M., Mertens, H.D., Ummels, R., Lugmayr, W., Mayr, J., Rettel, M., Savitski, M.M., Svergun, D.I., Bitter, W., Wilmanns, M., Marlovits, T.C., Parret, A.H., Houben, E.N. (2017) Structure of the mycobacterial ESX-5 type VII secretion system membrane complex by single-particle analysis. *Nat. Microbiol.* **2**:17047.

Botte, M., Zaccai, N.R., Lycklama à Nijeholt, J., Martin, R., Knoops, K., Papai, G., Zou, J., Deniaud, A., Karuppasamy, M., Jiang, Q., Singha Roy, A., Schulten, K., Schultz, P., Rappsilber, J., Zaccai, G., Berger, I., Collinson, I., Schaffitzel, C. (2016) A central cavity within the holotranslocon suggests a mechanism for membrane protein insertion. *Sci. Rep.* **6**:38399.

Breyton, C., Tribet, C., Olive, J., Dubacq, J.-P., Popot, J.-L. (1997) Dimer to monomer conversion of the cytochrome $b_6 f$ complex: causes and consequences. *J. Biol. Chem.* **272**:21892–21900.

Callaway, E. (2015) The revolution will not be crystallized. *Nature* **525**:172–174.

Cao, E., Cordero-Morales, J.F., Liu, B., Qin, F., Julius, D. (2013a) TRPV1 channels are intrinsically heat sensitive and negatively regulated by phosphoinositide lipids. *Neuron* **77**:667–679.

Cao, E., Liao, M., Cheng, Y., Julius, D. (2013b) TRPV1 structures in distinct conformations reveal activation mechanisms. *Nature* **504**:113–118.

Carazo, J.M., Sorzano, C.O.S., Otón, J., Marabini, R., Vargas, J. (2015) Three-dimensional reconstruction methods in single-particle analysis from transmission electron microscopy data. *Arch. Biochem. Biophys.* **581**:39–48.

Catoire, L.J., Zoonens, M., van Heijenoort, C., Giusti, F., Popot, J.-L., Guittet, E. (2009) Inter- and intramolecular contacts in a membrane protein/surfactant complex observed by heteronuclear dipole-to-dipole cross-relaxation. *J. Magn. Res.* **197**:91–95.

Catoire, L.J., Zoonens, M., van Heijenoort, C., Giusti, F., Guittet, E., Popot, J.-L. (2010) Solution NMR mapping of water-accessible residues in the transmembrane β-barrel of OmpX. *Eur. Biophys. J.* **39**:623–630.

Champeil, P., Menguy, T., Tribet, C., Popot, J.-L., le Maire, M. (2000) Interaction of amphipols with the sarcoplasmic reticulum $Ca^{2+}$-ATPase. *J. Biol. Chem.* **275**:18623–18637.

Charvolin, D., Perez, J.-B., Rouvière, F., Giusti, F., Bazzacco, P., Abdine, A., Rappaport, F., Martinez, K.L., Popot, J.-L. (2009) The use of amphipols as universal molecular adapters to immobilize membrane proteins onto solid supports. *Proc. Natl. Acad. Sci. USA* **106**:405–410.

Chen, Y., Clarke, O.B., Kim, J., Stowe, S., Kim, Y.-K., Assur, Z., Cavalier, C., Godoy-Ruiz, R., von Alpen, D.C., Manzini, C., Blaner, W.S., Frank, J., Quadro, L., Weber, D.J., Shapiro, L., Hendrickson, W.A., Mancia, F. (2016) Structure of the STRA6 receptor for retinol uptake. *Science* **353**:pii aad 8266–8261.

Cheng, Y. (2015) Single-particle cryo-EM at crystallographic resolution. *Cell* **161**:450–457.

Cheng, Y., Grigorieff, G., Penczek, P.A., Walz, T. (2015) A primer to single-particle cryo-electron microscopy. *Cell* **161**:438–449.

Chiu, Y.H., Jin, X., Medina, C., Leonhardt, S.A., Kiessling, V., Bennett, B.C., Shu, S., Tamm, L.K., Yeager, M., Ravichandran, K.S., Bayliss, D.A. (2017) A quantized mechanism for activation of pannexin channels. *Nat. Commun.* **8**:14324.

Chowdhury, S., Ketcham, S.A., Schroer, T.A., Lander, G.C. (2015) Structural organization of the dynein-dynactin complex bound to microtubules. *Nat. Struct. Mol. Biol.* **22**:345–347.

Constantine, M., Liew, C.K., Lo, V., Macmillan, A., Cranfield, C.G., Sunde, M., Whan, R., Graham, R.M., Martinac, B. (2016) Heterologously-expressed and liposome-reconstituted human transient receptor potential melastatin 4 channel (TRPM4) is a functional tetramer. *Sci. Rep.* **6**:19352.

Cvetkov, T.L., Huynh, K.W., Cohen, M.R., Moiseenkova-Bell, V.Y. (2011) Molecular architecture and subunit organization of TRPA1 ion channel revealed by electron microscopy. *J. Biol. Chem.* **286**:38168–38176.

Dahmane, T., Giusti, F., Catoire, L.J., Popot, J.-L. (2011) Sulfonated amphipols: Synthesis, properties and applications. *Biopolymers* **95**:811–823.

Dahmane, T., Rappaport, F., Popot, J.-L. (2013) Amphipol-assisted folding of bacteriorhodopsin in the presence and absence of lipids. Functional consequences. *Eur. Biophys. J.* **42**:85–101.

De Zorzi, R., Liao, M., Walz, T. (2016) Single-particle electron microscopy in the study of membrane protein structure. *Microscopy (Oxf)* **65**:81–96.

Dobro, M.J., Melanson, L.A., Jensen, G.J., McDowall, A.W. (2010) Plunge freezing for electron cryomicroscopy. *Meth. Enzymol.* **481**:63–82.

Dubochet, J., Adrian, M., Chang, J.J., Homo, J.C., Lepault, J., McDowall, A.W., Schultz, P. (1988) Cryo-electron microscopy of vitrified specimens. *Q. Rev. Biophys.* **21**:129–228.

Dubochet, J. (2015) A reminiscence about early times of vitreous water in electron cryomicroscopy. *Biophys. J.* **109**:812–817.

Efremov, R.G., Leitner, A., Aebersold, R., Raunser, S. (2015) Architecture and conformational switch mechanism of the ryanodine receptor. *Nature* **517**:39–43.

Ekiert, D.C., Bhabha, G., Isom, G.L., Greenan, G., Ovchinnikov, S., Henderson, I.R., Cox, J.S., Vale, R.D. (2017) Architectures of lipid transport systems for the bacterial outer membrane. *Cell* **169**:273–285.

Elad, N., De Strooper, B., Lismont, S., Hagen, W., Veugelen, S., Arimon, M., Horré, K., Berezovska, O., Sachse, C., Chávez-Gutiérrez, L. (2015) The dynamic conformational landscape of γ-secretase. *J. Cell Sci.* **128**:589–598.

Etzkorn, M., Raschle, T., Hagn, F., Gelev, V., Rice, A.J., Walz, T., Wagner, G. (2013) Cell-free expressed bacteriorhodopsin in different soluble membrane mimetics: biophysical properties and NMR accessibility. *Structure* **21**:394–401.

Fan, G., Gonzalez, J., Popova, O.B., Wensel, T.G., Serysheva, I.I. (2014) A first look into the 3D structure of the TRPV2 channel by single-particle cryo-EM. *Biophys. J.* **106**:600a–601a.

Fernández-Ballester, G., Ferrer-Montiel, A. (2008) Molecular modeling of the full-length human TRPV1 channel in closed and desensitized states. *J. Membr. Biol.* **223**:161–172.

Ferrandez, Y., Dezi, M., Bosco, M., Urvoas, A., Valério, M., Le Bon, C., Giusti, F., Broutin, I., Durand, G., Polidori, A., Popot, J.-L., Picard, M., Minard, P. (2014) Amphipol-mediated screening of molecular orthoses specific for membrane protein targets. *J. Membr. Biol.* **247**:925–940.

Fitzpatrick, A.W.P., Llabrés, S., Neuberger, A., Blaza, J.N., Bai, X.-C., Okada, U., Murakami, S., van Veen, H.W., Zachariae, U., Scheres, S.H.W., Luisi, B.F., Du, D. (2017) Structure of the MacAB–TolC ABC-type tripartite multidrug efflux pump. *Nat. Microbiol.* **2**:17070.

Flötenmeyer, M., Weiss, H., Tribet, C., Popot, J.-L., Leonard, K. (2007) The use of amphipathic polymers for cryo-electron microscopy of NADH:ubiquinone oxidoreductase (Complex I). *J. Microsc.* **227**:229–235.

Frank, J. (2006) *Three-dimensional Electron Microscopy of Macromolecular Assemblies: Visualization of Biological Molecules in their Native State.* Oxford Univerity Press, New York, 342 p.

Fujiyoshi, Y. (2011) Electron crystallography for structural and functional studies of membrane proteins. *J Electron Microsc (Tokyo)* **60 Suppl 1**:S149-S159.

Gao, Y., Cao, E., Julius, D., Cheng, Y. (2016) TRPV1 structures in nanodiscs reveal mechanisms of ligand and lipid action. *Nature* **534**:347–351.

Ge, J., Li, W., Zhao, Q., Li, N., Chen, M., Zhi, P., Li, R., Gao, N., Xiao, B., Yang, M. (2015) Architecture of the mammalian mechanosensitive Piezo1 channel. *Nature* **527**:64–69.

Giusti, F., Kessler, P., Westh Hansen, R., Della Pia, E.A., Le Bon, C., Mourier, G., Popot, J.-L., Martinez, K.L., Zoonens, M. (2015) Synthesis of a polyhistidine-bearing amphipol and its use for immobilizing membrane proteins. *Biomacro-molecules* **16**:3751–3761.

Giusti, F., Popot, J.-L., Tribet, C. (2012) Well-defined critical association concentration and rapid adsorption at the air/water interface of a short amphiphilic polymer, amphipol A8-35: A study by Förster resonance energy transfer and dynamic surface tension measurements. *Langmuir* **28**:10372–10380.

Glaeser, R.M., Han, B.G., Csencsits, R., Killilea, A., Pulk, A., Cate, J.H.D. (2016) Factors that influence the formation and stability of thin, cryo-EM specimens. *Biophys. J.* **110**:749–755.

Goehring, A., Lee, C.-H., Wang, K.H., Michel, J.C., Claxton, D.P., Baconguis, I., Althoff, T., Fischer, S., Garcia, C., Gouaux, E. (2014) Screening and large-scale expression of membrane proteins in mammalian cells for structural studies. *Nat. Protoc.* **9**:2574–2585.

Gohon, Y. (2002) Etude structurale et fonctionnelle de deux protéines membranaires, la bactériorhodopsine et le récepteur nicotinique de l'acétylcholine, maintenues en solution aqueuse non détergente par des polymères amphiphiles. Université Paris-VI, Paris, 467 p.

Gohon, Y., Dahmane, T., Ruigrok, R., Schuck, P., Charvolin, D., Rappaport, F., Timmins, P., Engelman, D.M., Tribet, C., Popot, J.-L., Ebel, C. (2008) Bacteriorhodopsin/amphipol complexes: structural and functional properties. *Biophys. J.* **94**:3523–3537.

Gohon, Y., Giusti, F., Prata, C., Charvolin, D., Timmins, P., Ebel, C., Tribet, C., Popot, J.-L. (2006) Well-defined nanoparticles formed by hydrophobic assembly of a short and polydisperse random terpolymer, amphipol A8-35. *Langmuir* **22**:1281–1290.

Gorzelle, B.M., Hoffman, A.K., Keyes, M.H., Gray, D.N., Ray, D.G., Sanders II, C.R. (2002) Amphipols can support the activity of a membrane enzyme. *J. Am. Chem. Soc.* **124**:11594–11595.

Goyal, P., Krasteva, P.V., Van Gerven, N., Gubellini, F., Van den Broeck, I., Troupiotis-Tsaïlaki, A., Jonckheere, W., Péhau-Arnaudet, G., Pinkner, J.S., Chapman, M.R., Hultgren, S.J., Howorka, S., Fronzes, R., Remaut, H. (2014) Structural and mechanistic insights into the bacterial amyloid secretion channel CsgG. *Nature* **516**:250–253.

Gu, J., Wu, M., Guo, R., Yan, K., Lei, J., Gao, N., Yang, M. (2016) The architecture of the mammalian respirasome. *Nature* **537**:639–643.

Guénebaut, V., Schlitt, A., Weiss, H., Leonard, K., Friedrich, T. (1998) Consistent structure between bacterial and mitochondrial NADH:ubiquinone oxidoreductase (complex I). *J. Mol. Biol.* **276**:105–112.

Gulati, S., Jamshad, M., Knowles, T.J., Morrison, K.A., Downing, R., Cant, N., Collins, R., Koenderink, J.B., Ford, R. C., Overduin, M., Kerr, I.D., Dafforn, T.R., Rothnie, A.J. (2014) Detergent-free purification of ABC (ATP-binding-cassette) transporters. *Biochem. J.* **461**:269–278.

He, Y., Gao, X., Goswami, D., Hou, L., Pal, K., Yin, Y., Zhao, G., Ernst, O.P., Griffin, P., Melcher, K., Xu, H.E. (2017) Molecular assembly of rhodopsin with G protein-coupled receptor kinases. *Cell Res.* **2017**:1–20.

Henderson, R. (1995) The potential and limitations of neutrons, electrons and X-rays for atomic resolution microscopy of unstained biological molecules. *Quart. Rev. Biophys.* **28**:171–193.

Henderson, R. (2013) Ion channel seen by electron microscopy. *Nature* **504**:93–94.

Henderson, R. (2015) Overview and future of single particle electron cryomicroscopy. *Arch. Biochem. Biophys.* **581**:19–24.

Henderson, R., McMullan, G. (2013) Problems in obtaining perfect images by single-particle electron cryomicroscopy of biological structures in amorphous ice. *Microscopy (Oxf)* **62**:43–50.

Henderson, R., Sali, A., Baker, M.L., Carragher, B., Devkota, B., Downing, K.H., Egelman, E.H., Feng, Z., Frank, J., Grigorieff, N., Jiang, W., Ludtke, S.J., Medalia, O., Penczek, P.A., Rosenthal, P.B., Rossmann, M.G., Schmid, M.F., Schröder, G.F., Steven, A.C., Stokes, D.L., Westbrook, J.D., Wriggers, W., Yang, H., Young, J., Berman, H.M., Chiu, W., Kleywegt, G.J., Lawson, C.L. (2012) Outcome of the first electron microscopy validation task force meeting. *Structure* **20**:205–214.

Hoenger, A., Pagès, J.-M., Fourel, D., Engel, A. (1993) The orientation of porin OmpF in the outer membrane of *Escherichia coli. J. Mol. Biol.* **233**:400–413.

Huynh, K.W., Cohen, M.R., Chakrapani, S., Holdaway, H.A., Stewart, P.L., Moiseenkova-Bell, V.Y. (2014a) Structural insight into the assembly of TRPV channels. *Structure* **22**:260–268.

Huynh, K.W., Cohen, M.R., Jiang, J., Samanta, A., Lodowski, D.T., Zhou, H., Moiseenkova-Bell, V. (2016) Structure of the full-length TRPV2 channel by cryo-EM. *Nat. Commun.* **7**:1130.

Huynh, K.W., Cohen, M.R., Moiseenkova-Bell, V.Y. (2014b) Application of amphipols for structure-functional analysis of TRP channels. *J. Membr. Biol.* **247**:843–851.

Jeong, H., Kim, J.-S., Song, S., Shigematsu, H., Yokoyama, T., Hyun, J., Ha, N.-C. (2016) Pseudoatomic structure of the tripartite multidrug efflux pump AcrAB-TolC reveals the intermeshing cogwheel-like interaction between AcrA and TolC. *Structure* **24**:272–276.

Jin, P., Bulkley, D., Guo, Y., Zhang, W., Guo, Z., Huynh, W., Wu, S., Meltzer, S., Cheng, T., Jan, L.Y., Jan, Y.-N., Cheng, Y. (2017) Electron cryo-microscopy structure of the mechanotransduction channel NOMPC. *Nature* **547**:118–122.

Karlsson, G. (2001) Thickness measurements of lacey carbon films. *J. Microsc.* **203**:326–328.

Kawate, T., Gouaux, E. (2006) Fluorescence-detection size-exclusion chromatography for precrystallization screening of integral membrane proteins. *Structure* **14**:673–681.

Kelly, D.F., Abeyrathne, P.D., Dukovski, D., Walz, T. (2008) The Affinity Grid: a pre-fabricated EM grid for monolayer purification. *J. Mol. Biol.* **382**:423–433.

Kevany, B.M., Tsybovsky, Y., Campuzano, I.D.G., Schnier, P.D., Engel, A., Palczewski, K. (2013) Structural and functional analysis of the native peripherin-ROM1 complex isolated from photoreceptor cells. *J. Biol. Chem.* **288**:36272–36284.

Khoshouei, M., Radjainia, M., Baumeister, W., Danev, R. (2017) Cryo-EM structure of haemoglobin at 3.2 Å determined with the Volta phase plate. *Nat. Commun.* **8**:16099.

Klammt, C., Perrin, M.-H., Maslennikov, I., Renault, L., Krupa, M., Kwiatkowski, W., Stahlberg, H., Vale, W., Choe, S. (2011) Polymer-based cell-free expression of ligand-binding family B G-protein coupled receptors without detergents. *Prot. Sci.* **20**:1030–1041.

Kühlbrandt, W. (2013) Introduction to electron crystallography. *Methods Mol. Biol.* **955**:1–16.

Kühlbrandt, W. (2014a) Cryo-EM enters a new era. *eLife*:e03678.

Kühlbrandt, W. (2014b) The resolution revolution. *Science* **343**:1443–1444.

Kühlbrandt, W. (2015) Structure and function of mitochondrial membrane protein complexes. *BMC Biol.* **13**:89.

Lau, W.C., Rubinstein, J.L. (2013) Single-particle electron microscopy. *Methods Mol. Biol.* **955**:401–426.

Le Bon, C., Della Pia, E.A., Giusti, F., Lloret, N., Zoonens, M., Martinez, K.L., Popot, J.-L. (2014a) Synthesis of an oligonucleotide-derivatized amphipol and its use to trap and immobilize membrane proteins. *Nucleic Acids Res.* **42**:e83.

Le Bon, C., Popot, J.-L., Giusti, F. (2014b) Labeling and functionalizing amphipols for biological applications. *J. Membr. Biol.* **247**:797–814.

Leis, A., Rockel, B., Andrees, L., Baumeister, W. (2009) Visualizing cells at the nanoscale. *Trends Biochem. Sci.* **34**:60–70.

Letts, J.A., Fiedorczuk, K., Sazanov, L.A. (2016) The architecture of respiratory supercomplexes. *Nature* **537**:644–648.

Lewis, B.A., Engelman, D.M. (1983) Lipid bilayer thickness varies linearly with acyl chain length in fluid phosphatidylcholine vesicles. *J. Mol. Biol.* **166**:211–217.

Li, M., Zhang, W.K., Benvin, N.M., Zhou, X., Su, D., Li, H., Wang, S., Michailidis, I.E., Tong, L., Li, X., Yang, J. (2017a) Structural basis of dual $Ca^{2+}$/pH regulation of the endolysosomal TRPML1 channel. *Nat. Struct. Mol. Biol.* **24**:205–213.

Li, M., Zhou, X., Wang, S., Michailidis, I., Gong, Y., Su, D., Li, H., Li X., Yang, J. (2017b) Structure of a eukaryotic cyclic-nucleotide-gated channel. *Nature* **542**:60–65.

Li, X., Mooney, P., Zheng, S., Booth, C.R., Braunfeld, M.B., Gubbens, S., Agard, D.A., Cheng, Y. (2013) Electron counting and beam-induced motion correction enable near-atomic-resolution single-particle cryo-EM. *Nat. Meth.* **10**:584–590.

Liao, M., Cao, E., Julius, D., Cheng, Y. (2013) Structure of the TRPV1 ion channel determined by electron cryo-microscopy. *Nature* **504**:107–112.

Liao, M., Cao, E., Julius, D., Cheng, Y. (2014) Single-particle electron cryo-microscopy of a mammalian ion channel. *Curr. Opin. Struct. Biol.* **27**:1–7.

Lu, P., Bai, X.-C., Ma, D., Xie, T., Yan, C., Sun, L., Yang, G., Zhao, Y., Zhou, R., Scheres, S.H.W., Shi, Y. (2014) Three-dimensional structure of human γ-secretase. *Nature* **512**:166–170.

Lučič, V., Rigort, A., Baumeister, W. (2013) Cryo-electron tomography: the challenge of doing structural biology *in situ*. *J. Cell Biol.* **202**:407–419.

Madej, M.G., Ziegler, C.M. (2018) Dawning of a new era in TRP channel structural biology by cryo-electron microscopy. *Pflügers Archiv - Eur. J. Physiol.* **470**:213–225.

Martinez, K.L., Gohon, Y., Corringer, P.-J., Tribet, C., Mérola, F., Changeux, J.-P., Popot, J.-L. (2002) Allosteric transitions of *Torpedo* acetylcholine receptor in lipids, detergent and amphipols: molecular interactions *vs.* physical constraints. *FEBS Lett.* **528**:251–256.

Mazhab-Jafari, M.T., Rubinstein, J.L. (2016) Cryo-EM studies of the structure and dynamics of vacuolar-type ATPases. *Sci. Adv.* **2**:e1600725.

Milne, J.L., Borgnia, M.J., Bartesaghi, A., Tran, E.E., Earl, L.A., Schauder, D.M., Lengyel, J., Pierson, J., Patwardhan, A., Subramaniam, S. (2013) Cryo-electron microscopy—a primer for the non-microscopist. *FEBS J.* **280**:28–45.

Miyaguchi, K. (2014) Direct imaging electron microscopy (EM) methods in modern structural biology: Overview and comparison with X-ray crystallography and single-particle cryo-EM reconstruction in the studies of large macromolecules. *Biol. Cell* **106**:323–345.

Murakami, S., Nakashima, R., Yamashita, E., Yamaguchi, A. (2002) Crystal structure of bacterial multidrug efflux transporter AcrB. *Nature* **419**:587–593.

Nagy, J.K., Kuhn Hoffmann, A., Keyes, M.H., Gray, D.N., Oxenoid, K., Sanders, C.R. (2001) Use of amphipathic polymers to deliver a membrane protein to lipid bilayers. *FEBS Lett.* **501**:115–120.

Nogales, E., Scheres, S.H.W. (2015) Cryo-EM: A unique tool for the visualization of macromolecular complexity. *Mol. Cell* **58**:677–689.

Ohi, M., Li, Y., Cheng, Y., Walz, T. (2004) Negative staining and image classification – powerful tools in modern electron microscopy. *Biol. Proced. Online* **6**:23–34.

Orlova, E.V., Saibil, H.R. (2011) Structural analysis of macromolecular assemblies by electron microscopy. *Chem. Rev.* **111**:7710–7748.

Paulsen, C.E., Armache, J.-P., Gao, Y., Cheng, Y., Julius, D. (2015) Structure of the TRPA1 ion channel suggests regulatory mechanisms. *Nature* **520**:511–517.

Perlmutter, J.D., Popot, J.-L., Sachs, J.N. (2014) Molecular dynamics simulations of a membrane protein/amphipol complex. *J. Membr. Biol.* **247**:883–895.

Perry, T., Souabni, H., Rapisarda, C., Fronzes, R., Giusti, F., Popot, J.-L., Zoonens, M., Gubellini, F. (2018) Visualizing transmembrane regions of protein complexes by electron microscopy using biotinylated amphipols, *submitted for publication*.

Picard, M., Dahmane, T., Garrigos, M., Gauron, C., Giusti, F., le Maire, M., Popot, J.-L., Champeil, P. (2006) Protective and inhibitory effects of various types of amphipols on the Ca$^{2+}$-ATPase from sarcoplasmic reticulum: a comparative study. *Biochemistry* **45**:1861–1869.

Popot, J.-L., Berry, E.A., Charvolin, D., Creuzenet, C., Ebel, C., Engelman, D.M., Flötenmeyer, M., Giusti, F., Gohon, Y., Hervé, P., Hong, Q., Lakey, J.H., Leonard, K., Shuman, H.A., Timmins, P., Warschawski, D.E., Zito, F., Zoonens, M., Pucci, B., Tribet, C. (2003) Amphipols: polymeric surfactants for membrane biology research. *Cell. Mol. Life Sci.* **60**:1559–1574.

Postis, V., Rawson, S., Mitchell, J.K., Lee, S.C., Parslowc, R.A., Dafforn, T.R., Baldwin, S.A., Muench, S.P. (2015) The use of SMALPs as a novel membrane protein scaffold for structure study by negative stain electron microscopy. *Biochim. Biophys. Acta* **1848**:496–501.

Punjani, A., Rubinstein, J.L., Fleet, D.J., Brubaker, M.A. (2017) cryoSPARC: algorithms for rapid unsupervised cryo-EM structure determination. *Nat. Meth.* **14**:290–296.

Radermacher, M., Wagenknecht, T., Verschoor, A., Frank, J. (1987) Three-dimensional reconstruction from a single-exposure, random conical tilt series applied to the 50S ribosomal subunit of *Escherichia coli*. *J. Microsc.* **146**:113–136.

Rahmeh, R., Damian, M., Cottet, M., Orcel, H., Mendre, C., Durroux, T., Sharma, K.S., Durand, G., Pucci, B., Trinquet, E., Zwier, J.M., Deupi, X., Bron, P., J.-L B, Mouillac, B., Granier, S. (2012) Structural insights into biased G protein-coupled receptor signaling revealed by fluorescence spectroscopy. *Proc. Natl. Acad. Sci. USA* **109**:6733–6738.

Raunser, S., Walz, T. (2009) Electron crystallography as a technique to study the structure on membrane proteins in a lipidic environment. *Annu. Rev. Biophys.* **38**:89–105.

Rosenthal, P.B., Henderson, R. (2003) Optimal determination of particle orientation, absolute hand, and contrast loss in single-particle electron cryomicroscopy. *J. Mol. Biol.* **333**:721–745.

Rubinstein, J.L. (2007) Structural analysis of membrane protein complexes by single-particle electron microscopy. *Methods* **41**:409–416.

Saxton, W.O., Baumeister, W. (1982) The correlation averaging of a regularly arranged bacterial cell envelope protein. *J. Microsc.* **127**:127–138.

Schabert, F.A., Engel, A. (1994) Reproducible acquisition of *Escherichia coli* porin surface topographs by atomic-force microscopy. *Biophys. J.* **67**:2394–2403.

Schäfer, E., Seelert, H., Reifschneider, N.H., Krause, F., Dencher, N.A., Vonck, J. (2006) Architecture of active mammalian respiratory chain supercomplexes. *J. Biol. Chem.* **281**:15370–15375.

Scheres, S.H.W., Chen, S. (2012) Prevention of overfitting in cryo-EM structure determination. *Nat. Meth.* **9**:853–854.

Schneck, E., Schubert, T., Konovalov, O.V., Quinn, B.E., Gutsmann, T., Brandenburg, K., Oliveira, R.G., Pink, D.A., Tanaka, M. (2010) Quantitative determination of ion distributions in bacterial lipopolysaccharide membranes by grazing-incidence X-ray fluorescence. *Proc. Natl. Acad. Sci. USA* **107**:9147–9151.

Schoebel, S., Mi, W., Stein, A., Ovchinnikov, S., Pavlovicz, R., DiMaio, F., Baker, D., Chambers, M.G., Su, H., Li, D., Rapoport, T.A., Liao, M. (2017) Cryo-EM structure of the protein-conducting ERAD channel Hrd1 in complex with Hrd3. *Nature* **548**:352–355.

Schröder, R.S. (2015) Advances in electron microscopy: A qualitative view of instrumentation development for macromolecular imaging and tomography. *Arch. Biochem. Biophys.* **581**:25–38.

Schulz, S., Wilkes, M., Mills, D.J., Kühlbrandt, W., Meier, T. (2017) Molecular architecture of the N-type ATPase rotor ring from *Burkholderia pseudomallei*. *EMBO Rep.* **18**:526–535.

Selmi, D.N., Adamson, R.J., Attrill, H., Goddard, A.D., Gilbert, R.J.C., Watts, A., Turberfield, A.J. (2011) DNA-templated protein arrays for single-molecule imaging. *Nano Lett.* **11**:657–660.

Serysheva, I.I., Orlova, E.V., Chiu, W., Sherman, M.B., Hamilton, S.L., van Heel, M. (1995) Electron cryomicroscopy and angular reconstitution used to visualize the skeletal muscle calcium release channel. *Nat. Struct. Biol.* **2**:18–24.

Shao, J., Fu, Z., Ji, Y., Guan, X., Guo, S., Ding, Z., Yang, X., Cong, Y., Shen, Y. (2016) Leucine zipper-EF-hand containing transmembrane protein 1 (LETM1) forms a Ca$^{2+}$/H$^+$ antiporter. *Sci. Rep.* **6**:34174.

Shen, P.S., Yang, X., DeCaen, P.G., Liu, X., Bulkley, D., Clapham, D.E., Cao, E. (2016) The structure of the Polycystic Kidney Disease channel PKD2 in lipid nanodiscs. *Cell* **167**:763–773.e711.

Sousa, J.S., Mills, D.J., Vonck, J., Kühlbrandt, W. (2016) Functional asymmetry and electron flow in the bovine respirasome. *eLIFE* **5**:e21290.

Spear, J.M., Koborssy, D.A., Schwartz, A.B., Johnson, A.J., Audhya, A., Fadool, D.A., Stagg, S.M. (2015) Kv1.3 contains an alternative *C*-terminal ER exit motif and is recruited into COPII vesicles by Sec24a. *BMC Biochem.* **16**:16.

Stahlberg, H., Biyani, N., Engel, A. (2015) 3D reconstruction of two-dimensional crystals. *Arch. Biochem. Biophys.* **581**:68–77.

Stark, H., Chari, A. (2016) Sample preparation of biological macromolecular assemblies for the determination of high-resolution structures by cryo-electron microscopy. *Microscopy (Oxf)* **65**:23–34.

Stock, D., Leslie, A.G.W., Walker, J.E. (1999) Molecular architecture of the rotary motor in ATP synthase. *Science* **286**:1700–1705.

Suárez-Suárez, S., Carriedo, G.A., Presa Soto, A. (2013) Gold-decorated chiral macroporous films by the self-assembly of functionalised block copolymers. *Chem. Eur. J.* **19**:15933–15940.

Sun, L., Zhao, L., Yang, G., Yan, C., Zhou, R., Zhou, X., Xie, T., Zhao, Y., Wu, S., Li, X., Shi, Y. (2015) Structural basis of human $\gamma$-secretase assembly. *Proc. Natl. Acad. Sci. USA* **112**:6003–6008.

Sverzhinsky, A., Qian, S., Yang, L., Allaire, M., Moraes, I., Ma, D., Chung, J.W., Zoonens, M., Popot, J.-L., Coulton, J.W. (2014) Amphipol-trapped ExbB-ExbD membrane protein complex from *Escherichia coli*: A biochemical and structural case study. *J. Membr. Biol.* **247**:1005–1018.

Terahara, T., Kodera, N., Uchihashi, T., Ando, T., Namba, K., Minamino, T. (2017) $Na^+$-induced structural transition of MotPS for stator assembly of the Bacillus flagellar motor. *Sci. Adv.* **3**:eaao4119.

Tribet, C., Audebert, R., Popot, J.-L. (1997) Stabilization of hydrophobic colloidal dispersions in water with amphiphilic polymers: Application to integral membrane proteins. *Langmuir* **13**:5570–5576.

Tribet, C., Diab, C., Dahmane, T., Zoonens, M., Popot, J.-L., Winnik, F.M. (2009) Thermodynamic characterization of the exchange of detergents and amphipols at the surfaces of integral membrane proteins. *Langmuir* **25**:12623–12634.

Tribet, C., Mills, D., Haider, M., Popot, J.-L. (1998) Scanning transmission electron microscopy study of the molecular mass of amphipol/cytochrome $b_6 f$ complexes. *Biochimie* **80**:475–482.

Tsybovsky, Y., Orban, T., Molday, R.S., Taylor, D., Palczewski, K. (2013) Molecular organization and ATP-induced conformational changes of ABCA4, the photoreceptor-specific ABC transporter. *Structure* **21**:854–860.

Ubarretxena-Belandia, I., Stokes, D.L. (2010) Present and future of membrane protein structure determination by electron crystallography. *Adv. Protein Chem. Struct. Biol.* **81**:33–60.

Vahedi-Faridi, A., Jastrzebska, B., Palczewski, K., Engel, A. (2013) 3D imaging and quantitative analysis of small solubilized membrane proteins and their complexes by transmission electron microscopy. *Microscopy (Oxf)* **62**:95–107.

van Heel, M., Frank, J. (1981) Use of multivariate statistics in analysing the images of biological macromolecules. *Ultramicroscopy* **6**:187–194.

van Pee, K., Neuhaus, A., D'Imprima, E., Mills, D.J., Kühlbrandt, W., Yildiz, Ö. (2017) CryoEM structures of membrane pore and prepore complex reveal cytolytic mechanism of pneumolysin. *eLlife* **6**:e23644.

Venkatachalam, K., Montell, C. (2007) TRP channels. *Annu. Rev. Biochem.* **76**:387–417.

Villa, E., Schaffer, M., Plitzko, J.M., Baumeister, W. (2013) Opening windows into the cell: focused-ion-beam milling for cryo-electron tomography. *Curr. Opin. Struct. Biol.* **23**:771–777.

Vinothkumar, K.R., Zhu, J., Hirst, J. (2014) Architecture of mammalian respiratory complex I. *Nature* **515**:80–84.

Vinothkumar, K.R. (2015) Membrane protein structures without crystals, by single-particle electron cryomicroscopy. *Curr. Opin. Struct. Biol.* **33**:103–114.

Vinothkumar, K.R., Henderson, R. (2016) Single-particle electron cryomicroscopy: trends, issues and future perspective. *Q. Rev. Biophys.* **49**:e13.

Wang, Z., Fan, G., Hryc, C.F., Blaza, J.N., Serysheva, I.I., Schmid, M.F., Chiu, W., Luisi, B.F., Du, D. (2017) An allosteric transport mechanism for the AcrAB-TolC multidrug efflux pump. *eLife* **6**:e24905.

Wei, R., Wang, X., Zhang, Y., Mukherjee, S., Zhang, L., Chen, Q., Huang, X., Jing, S., Liu, C., Li, S., Wang, G., Xu, Y., Zhu, S., Williams, A.J., Sun, F., C.C. Y. (2016) Structural insights into $Ca^{2+}$-activated long-range allosteric channel gating of RyR1. *Cell Res.* **26**:977–994.

Wilkens, S. (2000) $F_1F_0$-ATP synthase–stalking mind and imagination. *J. Bioenerg. Biomembr.* **32**:333–339.

Wilkens, S., Capaldi, R.A. (1998a) Electron microscopic evidence of two stalks linking the $F_1$ and $F_O$ parts of the *Escherichia coli* ATP synthase. *Biochim. Biophys. Acta* **1365**:93–97.

Wilkens, S., Capaldi, R.A. (1998b) ATP synthase's second stalk comes into focus. *Nature* **393**:29.

Wilkens, S., Zhou, J., Nakayama, R., Dunn, S.D., Capaldi, R.A. (2000) Localization of the δ subunit in the *Escherichia coli* $F_1F_0$-ATPsynthase by immuno-electron microscopy: The δ subunit binds on top of the $F_1$. *J. Mol. Biol.* **295**:387–391.

Wilkes, M., Madej, M.G., Kreuter, L., Rhinow, D., Heinz, V., De Sanctis, S., Ruppel, S., Richter, R.M., Joos, F., Grieben, M., Pike, A.C., Huiskonen, J.T., Carpenter, E.P., Kühlbrandt, W., Witzgall, R., Ziegler, C. (2017) Molecular insights into lipid-assisted $Ca^{2+}$ regulation of the TRP channel polycystin-2. *Nat. Struct. Mol. Biol.* **24**:123–130.

Wilkinson, M., Chaban, Y., Wigley, D.B. (2016a) Mechanism for nuclease regulation in RecBCD. *eLIFE* **5**:e18227.

Wilkinson, M., Troman, L., Wan Nur Ismah, W.A.K., Chaban, Y., Avison, M.B., Dillingham, M.S., Wigley, D.B. (2016b) Structural basis for the inhibition of RecBCD by Gam and its synergistic antibacterial effect with quinolones. *eLIFE* **5**:e22963.

Wisedchaisri, G., Reichow, S.L., Gonen, T. (2011) Advances in structural and functional analysis of membrane proteins by electron crystallography. *Structure* **19**:1381–1393.

Wojtowicz, H., Prochnicka-Chalufour, A., de Amorim, G.C., Roudenko, O., Simenel, C., Malki, I., Pehau-Arnaudet, G., Gubellini, F., Koutsioubas, A., Pérez, J., Delepelaire, P., Delepierre, M., Fronzes, R., Izadi-Pruneyre, N. (2016) Structural basis of the signalling through a bacterial membrane receptor HasR deciphered by an integrative approach. *Biochem. J.* **473**:2239–2248.

Wong, W., Bai, X.-C., Brown, A., Fernandez, I.S., Hanssen, E., Condron, M., Tan, Y.H., Baum, J., Scheres, S.H.W. (2014) Cryo-EM structure of the *Plasmodium falciparum* 80 S ribosome bound to the anti-protozoan drug emetine. *eLife* **3**:e03080.

Wu, Z.S., Cui, Z.C., Cheng, H., Fan, C., Melcher, K., Jiang, Y., Zhang, C.H., Jiang, H.L., Cong, Y., Liu, Q., Xu, H.E. (2015) High yield and efficient expression and purification of the human 5-$HT_{3A}$ receptor. *Acta Pharmacol. Sin.* **36**:1024–1032.

Xu, J., Gui, M., Wang, D., Xiang, Y. (2016) The bacteriophage Φ29 tail possesses a pore-forming loop for cell membrane penetration. *Nature* **534**:544–547.

Yan, Z., Bai, X.-C., Yan, C., Wu, J., Li, Z., Wei, T.X., Peng, W., Yin, C.-C., Li, X., Scheres, S.H.W., Shi, Y., Yan, N. (2015) Structure of the rabbit ryanodine receptor RyR1 at near-atomic resolution. *Nature* **517**:50–55.

Yang, G., Zhou, R., Shi, Y. (2017) Cryo-EM structures of human γ-secretase. *Curr. Opin. Struct. Biol.* **46**:55–64.

Zalk, R., Clarke, O.B., Georges, A., Grassucci, R.A., Reiken, S., Mancia, F., Hendrickson, W.A., Frank, J., Marks, A.R. (2015) Structure of a mammalian ryanodine receptor. *Nature* **517**:44–49.

Zhang, N., Tsybovsky, Y., Kolesnikov, A.V., Rozanowska, M., Swider, M., Schwartz, S.B., Stone, E.M., Palczewska, G., Maeda, A., Kefalov, V.J., Jacobson, S.G., Cideciyan, A.V., Palczewski, K. (2015) Protein misfolding and the pathogenesis of ABCA4-associated retinal degenerations. *Hum. Mol. Genet.* **24**:3220–3237.

Zhou, X., Li, M., Su, D., Jia, Q., Li, H., Li, X., Yang, J. (2017) Cryo-EM structures of the human endolysosomal TRPML3 channel in three distinct states. *Nat. Struct. Mol. Biol.* **24**:1146–1154.

Zickermann, V., Kerscher, S., Zwicker, K., Tocilescua, M.A., Radermacher, M., Brand, U. (2009) Architecture of complex I and its implications for electron transfer and proton pumping. *Biochim. Biophys. Acta.* **1787**:574–583.

Zoonens, M., Catoire, L.J., Giusti, F., Popot, J.-L. (2005) NMR study of a membrane protein in detergent-free aqueous solution. *Proc. Natl. Acad. Sci. USA* **102**:8893–8898.

Zoonens, M., Giusti, F., Zito, F., Popot, J.-L. (2007) Dynamics of membrane protein/amphipol association studied by Förster resonance energy transfer. Implications for *in vitro* studies of amphipol-stabilized membrane proteins. *Biochemistry* **46**:10392–10404.

Zorman, S., Botte, M., Jiang, Q., Collinson, I., Schaffitzel, C. (2015) Advances and challenges of membrane-protein complex production. *Curr. Opin. Struct. Biol.* **32**:123–130.

Zubcevic, L., Herzik, M.A., Jr., Chung, B.C., Liu, Z., Lander, G.C., Lee, S.-Y. (2016) Cryo-electron microscopy structure of the TRPV2 ion channel. *Nat. Struct. Mol. Biol.* **23**:180–186.

# Amphipol-Mediated Immobilization of Membrane Proteins and Its Applications

# 13

**Summary**

*Tagged amphipols (APols) are obtained by chemically grafting APols with moieties that can interact with specific partners. They combine in a single molecule three highly useful properties: keeping membrane proteins (MPs) water-soluble, stabilizing them, and functionalizing them. MPs trapped with tagged APols can be attached to solid supports and other scaffolds in a gentle, non-covalent but permanent manner, without interacting themselves with the support nor having to be genetically nor chemically modified. Furthermore, a single tagged APol can be used for any number of MPs. Three types of tags have been thoroughly validated to date – biotin, polyhistidine, and an oligonucleotide – and three more are under development, randomly distributed imidazoles, sulfides, and sulfhydrides. Between them, they cover a very broad range of properties, from low to extremely high affinity and from easy reversibility to near-irreversibility. They have been applied to immobilizing membrane proteins onto a variety of supports, including avidin-coated or nucleotide-bearing beads, chips, and culture plates. MPs immobilized via tagged APols remain native and functional and can be recognized by ligands big and small, such as antibodies, toxins, or neurotransmitter analogs, which can be directly or indirectly detected by such methods as surface plasmon resonance (SPR) and fluorescence microscopy. They have been used to select protein binders and to study the interaction of a bacterial outer membrane protein with a bacteriophage tail protein. They lend themselves well to multiplexing approaches permitting to immobilize multiple membrane proteins at defined positions onto chips or electrodes. Tagged APols have other possible applications, e.g. in structural biology and for medical uses.*

## 13.1 Introduction

One of the most useful characteristics of amphipols (APols) is that, at variance with detergents, they do not spontaneously desorb from the transmembrane (TM) surface of the membrane proteins (MPs) they keep water-soluble: APols can be readily removed, but they have to be exchanged for another surfactant, such as a detergent or another APol (Chap. 5, § 5.7). In solution, extensive washing with surfactant-free buffer results in MP aggregation, which is certainly accompanied by some APol desorption, part of the original MP/APol contacts being replaced by MP/MP ones (§ 5.2.1). If, however, a biotinylated MP trapped in APol is immobilized onto an avidin-coated surface plasmon resonance (SPR) chip and the chip extensively washed with surfactant-free buffer, no desorption is observed (ibid., Fig. 5.7). Under such conditions, the association between the protein and the polymer is permanent, even though it is not covalent. Because APols can be easily functionalized (Chap. 4, § 4.4), they provide a way to indirectly functionalize any MP they bind to, without the protein itself having to be chemically or genetically modified. This observation is at the basis of a great many applications. In the present chapter, we will describe those tagged APols that have been thoroughly characterized or are under development and consider those applications that have already been validated or represent likely prospects.

In the past few decades, protein microarrays, protein sensor chips, and microbeads have emerged as tools of choice for obtaining information about protein functions and interactions (see e.g. Weinrich et al. 2009). Protein arrays combined with sensitive optical detection systems such as SPR or fluorescence microscopy give access to multiplexing and high-throughput analyses, which have a high potential for screening studies. These techniques can also provide important information regarding protein affinities and kinetics of interaction using minimal amounts of reagents. All surface-based assays however require that the target proteins be immobilized under such conditions as to preserve their native structure and biological activity (see e.g. Jonkheijm et al. 2008). Functional immobilization of MPs can be achieved by covering bare or modified surfaces with native (Yang and Lundahl 1994; Gottschalk et al. 2000; Martinez et al. 2003; Perez et al. 2006) or reconstituted (Bieri et al. 1999; Früh et al. 2011) membrane fragments. These methods have the advantage of allowing the investigation of MPs in their native lipid environment, but they suffer from high levels of non-specific binding and from a relatively unstable immobilization, which limit their applications (Früh et al. 2011). More stable immobilization can be achieved by fusing genetically or chemically small peptide or protein tags to the target MPs for mediating reversible or irreversible immobilization of MP/detergent complexes (Schmid et al. 1998; Friedrich et al. 2004; Harding et al. 2006). These strategies require a covalent modification of the MPs under study, and they suffer from technical hurdles due to the use of detergents. APols offer a highly versatile alternative for MP immobilization. Their advantages include the following:

(i) No modification of the target MP is required.
(ii) APol-trapped MPs are, very generally, stabilized as compared to handling them in detergent solutions (Chap. 5, § 5.5).
(iii) The same APol can be used for any number of MPs, or mixtures of MPs, greatly simplifying the chemistry work.
(iv) Using mixtures of APols, MPs can be simultaneously tagged and labeled, e.g. with fluorescent APols, so as to visualize the immobilization process.
(v) Analyses can be conducted in detergent-free solutions.

At the time of this writing, more than 20 labeled or functionalized APols have been synthesized and characterized, including eight tagged ones (see Chap. 4, § 4.4, Table 4.5). Biotin has been attached to A8-35 (Charvolin et al. 2009), to phosphorylcholine-based APols (Basit et al. 2012), to non-ionic APols (Ferrandez et al. 2014), and to a styrene-maleic acid copolymer (SMA) (Lindhoud et al. 2016). A8-35 has also been derivatized with an oligonucleotide (ODN; Le Bon et al. 2014a), with hexahistidine tags (Giusti et al. 2015), with randomly distributed imidazole moieties (M. Zoonens, F. Giusti, and C. Le Bon, unpublished results), and with thiomorpholine (F. Giusti, unpublished results) (see below Table 13.1; for reviews, see Della Pia et al. (2014b), Le Bon et al. (2014b), and

Zoonens and Popot (2014)). SMA has been functionalized with cysteamine, whose thiol group has been used as a reactive moiety for further derivatization, but could also be exploited for immobilization (Lindhoud et al. 2016). Tagged APols make it possible to immobilize MPs onto beads or chips carrying appropriate groups, namely (i) avidin, streptavidin (SA), or neutravidin for biotin-tagged APols; (ii) a complementary nucleic acid for ODN-tagged ones; (iii) immobilized $Ni^{2+}$ or $Co^{2+}$ ions for polyhistidine and distributed imidazole tags; and (iv) gold or silver surfaces for thiomorpholine and sulfhydryl ones.

*Amphipol-Mediated Immobilization of Membrane Proteins and Its Applications*
*(© 2018 by Francis Haraux)*

A marked advantage of APols is that, because each protein binds several molecules of it, multiple functions can be associated to the same protein by the simple device of mixing APols. For instance, a tagged APol can be mixed with a fluorescent one, so as to easily visualize the immobilization process. Seven fluorescent APols (FAPols) derived from either A8-35 or A8-75 have been synthesized and characterized to date (see Chap. 4, § 4.4, Table 4.5), whose excitation bands range from 290 to 651 nm and emission bands from 310 to 668 nm (see Chap. 8, Fig. 8.9). Thiolated SMA has been used to bind Alexa Fluor 488 (Lindhoud et al. 2016). This broad panel of spectral characteristics greatly facilitates the choice of adequate experimental conditions, as well as multiplexing.

Figure 13.1 illustrates schematically the basic principle of using tagged APols for immobilizing MPs onto solid supports. In § 13.2, we will review data pertaining to the use of biotinylated APols for MP immobilization, in § 13.3 the current state of development of other tagged APols. In § 13.4, a few other validated or potential applications of tagged APols will be evoked. Table 13.1 presents an overview of publications that have resorted to the use of tagged APols. Biotin-mediated MP immobilization has been used in five studies using three different APols (Studies 13.1–13.3, 13.5, and 13.6). Biotinylated A8-35 (BA8-35) has also been used in an EM study (Study 13.4). The other tags, all of them grafted onto A8-35, have been the object of a single validation study or are still under examination (Studies 13.7–13.11).

**Table 13.1**  An overview of studies involving the use of tagged amphipols.

| Study | Tag | Amphipol (short name of derivative) | Synthesis | Comments | References |
|---|---|---|---|---|---|
| 13.1 | Biotin | A8-35 (BAPol or BA8-35) | | Describes the synthesis and characterization of a biotinylated version of A8-35 (BA8-35) and its use to immobilize MPs onto chips or beads and follow the binding of ligands (antibodies, a toxin, and a small ligand) by SPR and fluorescence microscopy (§ 13.2.1) | Charvolin et al. (2009) |
| 13.2 | | | ● | Gold electrodes are selectively functionalized by electropolymerization of biotin-doped polypyrrole and treated with tetravalent (strept)avidin. BA8-35 is used to anchor MPs onto selected electrodes and the selective binding of anti-MP antibodies to immobilized MPs demonstrated by fluorescence microscopy (§ 13.2.2) | Della Pia et al. (2014a) |
| 13.3 | | | | BA8-35 is used to immobilize target MPs to streptavidin-coated plates in order to select bacteriophages exposing soluble protein binders (§ 13.2.3) | Ferrandez et al. (2014) |
| 13.4 | | | | Visualization of the distribution of APol bound to a MP complex by binding monovalent streptavidin to BA8-35 (§ 13.4) | Perry et al. (2018) |
| 13.5 | Biotin | PC-APol (B-PCAPol) | ● | A biotinylated phosphorylcholine-based APol is used to immobilize a bacterial outer membrane protein onto SPR chips and analyze its interactions with a bacteriophage tail protein (§ 13.2.1) | Basit et al. (2012) |
| 13.6 | Biotin | Glucosylated NAPol (BNAPol) | ● | BNAPol is used to immobilize target MPs to streptavidin-coated plates in order to select bacteriophages exposing soluble protein binders (§ 13.2.3) | Ferrandez et al. (2014) |
| 13.7 | Polyhistidine | A8-35 (HistAPol) | ● | Synthesis of an APol bearing hexahistidine tags and validation of its use for reversibly immobilizing MPs onto $Ni^{2+}$:NTA-bearing chips or beads (§ 13.3.1) | Giusti et al. (2015) |
| 13.8 | Imidazole moieties | A8-35 (ImidAPol) | ● | Synthesis of an APol-bearing distributed imidazole moieties and demonstration of its use for reversibly immobilize MPs onto $Ni^{2+}$:NTA-bearing supports (§ 13.3.2) | Unpublished data cited in Le Bon et al. (2014b) |
| 13.9 | Oligodeoxynucleotide | A8-35 (OligAPol) | ● | Synthesis of an oligodeoxynucleotide (ODN)-bearing APol and its use to target MPs to specific spots on chips. (§ 13.3.3). The ODN chosen can also be used as an adjuvant in vaccine formulations using APol-stabilized MPs (see Chap. 15, § 15.2.3) | Le Bon et al. (2014a) |
| 13.10 | Cysteamine | SMA (SMA-SH) | ● | Cysteamine was grafted onto SMA carboxylates via its amine function. The thiol group was used for derivatization with Alexa Fluor 488 or biotin. No immobilization studies have been described yet | Lindhoud et al. (2016) |
| 13.11 | Thiomorpholine | A8-35 (SulfidAPol) | ● | Synthesis of a thiomorpholine-bearing APol and its use for immobilizing MPs onto gold surfaces, e.g. for surface plasmon resonance or surface-enhanced resonance spectroscopy (cf. Chap. 8) (§ 13.3.4). SulfidAPol can potentially be used to attach gold beads to APol belts for EM applications (§ 13.4) | Unpublished data |

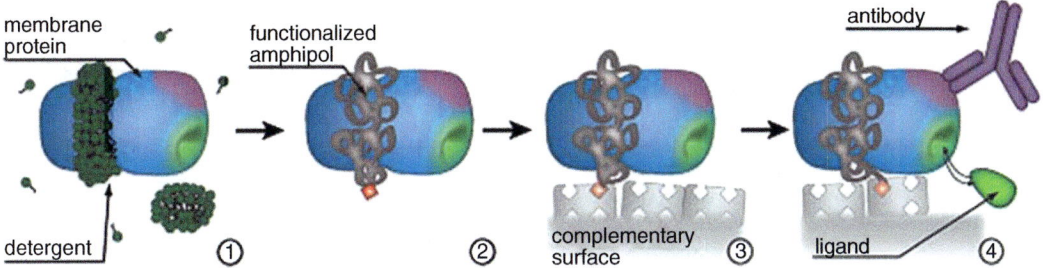

**Fig. 13.1** Experimental principle of amphipol-mediated immobilization of membrane proteins. A MP solubilized in detergent ① is transferred to a tagged APol ②. The complex thus formed is immobilized onto a support that exposes a functional group complementary to the tag ③. Putative ligands of the protein are then flushed over the support in surfactant-free buffer ④ and their interaction with the immobilized protein detected by any convenient method (From Charvolin et al. 2009. © 2009 National Academy of Sciences, USA).

## 13.2   Immobilization of Membrane Proteins Using Biotinylated Amphipols and Its Applications

### 13.2.1   Validation of the Approach Using Biotinylated A8-35

Biotinylated A8-35 (hereafter, BAPol or BA8-35) was synthesized as described in Chap. 4, § 4.4.1 (Charvolin et al. 2009). The feasibility of using it for immobilizing MPs onto solid supports was established in Study 13.1 (Table 13.1), using five model MPs, namely bacteriorhodopsin (BR) from *Halobacterium salinarum* (described in Chap. 1, § 1.6.1), the nicotinic acetylcholine receptor (nAChR) from *Torpedo marmorata* (Chap. 1, § 1.6.2), the cytochrome $b_6f$ complex from *Chlamydomonas reinhardtii* (Pierre et al. 1995; Stroebel et al. 2003), beef heart cytochrome $bc_1$ complex (Berry et al. 2000), and the transmembrane (TM) domain of *Escherichia coli* outer membrane protein A (tOmpA) (Pautsch and Schulz 2000; Pervushin et al. 2000; Arora et al. 2001). Each of these proteins is known to remain functional (or, in the case of tOmpA, to retain its native conformation) following trapping with A8-35 (see Chap. 5, § 5.4, and Chap. 10, § 10.3.1).

Each protein was trapped with an excess of BA8-35. The resulting preparations therefore comprised a mixture of MP/BA8-35 complexes and free BA8-35 particles. Upon flushing SA-coated chips either with BA8-35 alone (Fig. 13.2A) or with any of the MP/BA8-35 preparations (Fig. 13.2B, C), an SPR signal developed, a large fraction of which resisted washing with buffer. As observed with pure APols (Fig. 13.2A), SPR signals were much higher for MP/BA8-35 complexes than observed for the same MP trapped with underivatized A8-35, indicative of biotin-mediated immobilization (Fig. 13.2B, C). For comparative purposes, injections of the four proteins were carried out at constant overall APol concentration. For the two small MPs (tOmpA and BR), where the overall protein/polymer mass ratio in the preparation is relatively low (0.2:1 and 0.25:1, respectively), the SPR signal was comparable (Fig. 13.2C) and close to that observed with pure BA8-35; for larger complexes ($b_6f$ and $bc_1$, PM/BA8-35 ≈ 1:3 and 1:1.5 w/w, respectively), it was significantly higher (Fig. 13.2C), strongly suggesting that the protein contributed to the signal.

The presence of the four MPs on the chips and their ability to recognize specific ligands were demonstrated using protein-specific antibodies. Antisera were raised in rabbits against each target protein and the IgG fraction purified by affinity chromatography. After saturating non-specific binding sites with pre-immune antibodies, each flow cell was flushed with either pre-immune or protein-specific antibodies at several dilutions. Strong, specific responses were obtained at 1/100 dilution

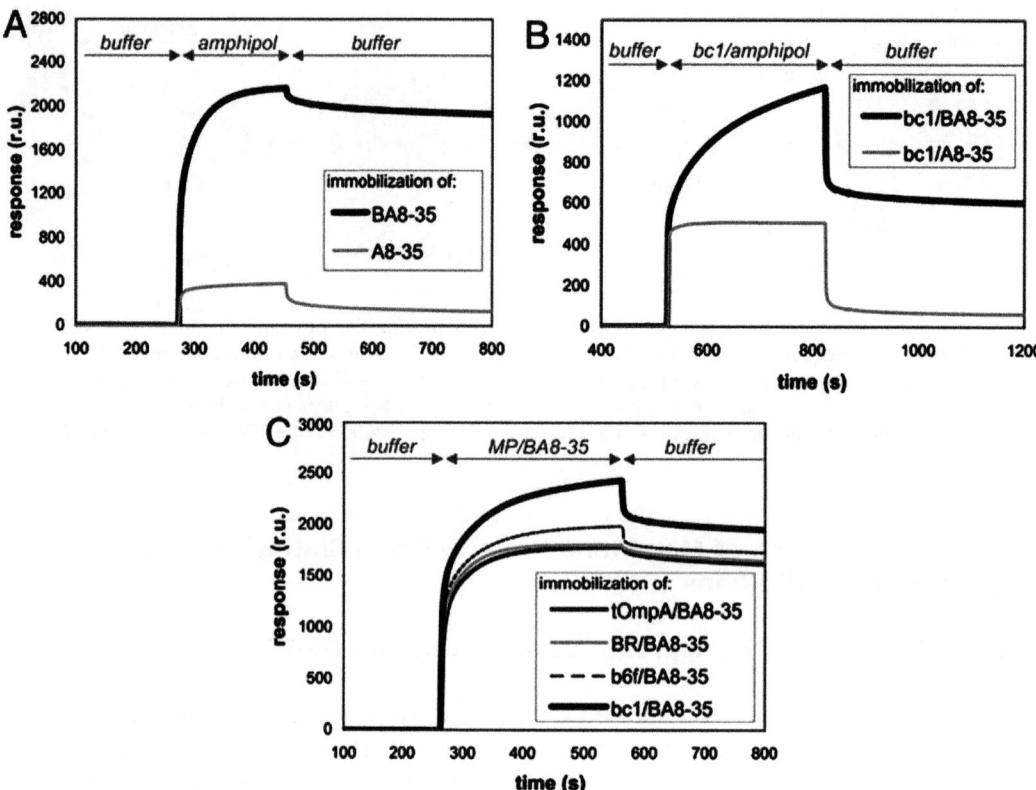

**Fig. 13.2** Immobilization of biotinylated A8-35 (BA8-35) and of membrane protein/BA8-35 complexes onto streptavidin (SA)-coated chips followed by surface plasmon resonance (SPR). (**A**) Thirty microliters of a solution of BA8-35 (*black line*) or of non-biotinylated A8-35 (*gray line*) were flushed over an SA-coated sensor chip (SA chip) at a concentration of 0.3 $g \cdot L^{-1}$ in 100 mM NaCl, 5 mM sodium phosphate buffer (pH 7.4), 1 mM EDTA (NaPh buffer). SPR signals were recorded at 25 °C in a Biacore 2000 instrument. (**B**) Fifty microliters of a solution of cytochrome $bc_1$ complex trapped either in BA8-35 (*black line*) or in non-biotinylated A8-35 (*gray line*) were flushed over an SA chip at a protein concentration of 2 μM in NaPh buffer. (**C**) Fifty microliters of solutions of tOmpA (*thin black line*), bacteriorhodopsin (BR; *thin gray line*), cytochrome $b_6 f$ (*dashed thin black line*), or cytochrome $bc_1$ (*thick black line*), each of them trapped in BA8-35, were flushed over various flow cells of two SA chips at a concentration of 0.3 $g \cdot L^{-1}$ BA8-35. The cytochrome $b_6 f$ sensorgram, which was recorded in a different session, has been normalized by using as a reference the response to cytochrome $bc_1$, which was measured each time. Experimental conditions differed in A, B, and C (in particular, two distinct BA8-35 batches were used, resulting in different binding capacities), so that sensorgrams shown in different panels cannot be directly compared (From Charvolin et al. 2009. © 2009 National Academy of Sciences, USA).

(Fig.13.3A). Responses to protein-specific antibodies were typically $10 \times$ higher or more than those to non-specific ones (Fig. 13.3B). Comparative antibody-binding experiments were carried out with proteins trapped with either BA8-35 or plain A8-35. Higher signals were observed with BA8-35-trapped proteins, with desorption kinetics compatible with the washing away of antibodies, whereas the signal observed when using ungrafted A8-35 was much weaker. This confirms that most, if not all, of the response is due to the binding of antibodies to MPs immobilized via the biotin/SA interaction rather than via non-specific interactions with the support. Taken together, these experiments establish (i) that each of the four MPs tested can be immobilized onto SA-coated surfaces via the biotinylated APol and (ii) that, once bound, each of them can be specifically recognized by antibodies directed against it.

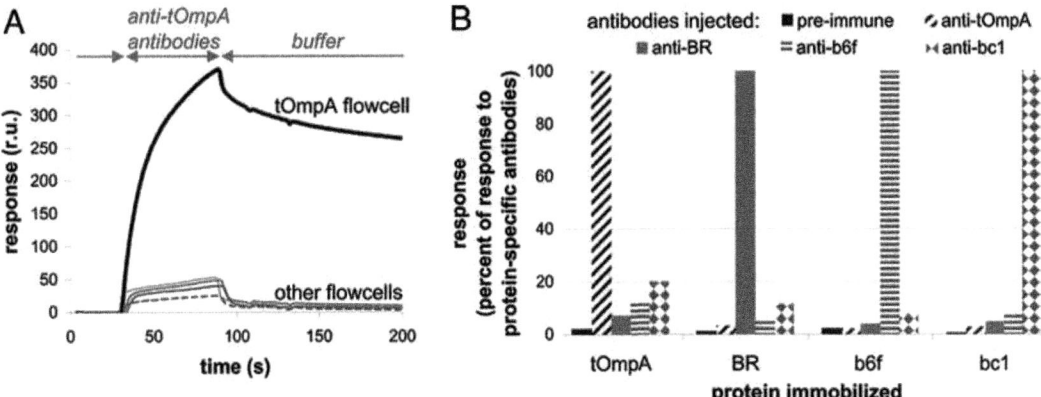

**Fig. 13.3** Specific recognition by antibodies of immobilized membrane protein/BA8-35 complexes. Ten microliters of polyclonal antibodies purified either from a pre-immune serum or from post-immune sera raised against tOmpA, BR, cytochrome $b_6 f$, or cytochrome $bc_1$ were flushed over SA chips onto which either BA8-35 or each of the four MP/ BA8-35 complexes had been immobilized. Purified antibodies were used at a 1/100 dilution in NaPh buffer. (**A**) An example of the SPR responses obtained, in this case by using anti-tOmpA antibodies. (**B**) Overview of the results. The intensity of the SPR response 10 s after the end of each injection is shown as the percentage of the specific response (that to antibodies raised against the protein immobilized on the flow cell under consideration) (From Charvolin et al. 2009. © 2009 National Academy of Sciences, USA).

Some interesting additional observations were made in the course of Study 13.1. In particular:

(i) Removing free BA8-35 from the trapped preparations, which suppresses the competition between MP/BA8-35 complexes and free BA8-35 particles for binding to SA, results, as expected, in increasing the protein fraction in the material adsorbed and, thereby, the signal-to-noise ratio observed upon challenging the chip with antibodies.

(ii) Despite a lower signal-to-noise ratio, specific recognition of antibodies can also be observed with crude sera.

(iii) The fact that A8-35 is insoluble in acidic solutions (Chap. 4, § 4.3) does not prevent regenerating chips carrying BA8-35-immobilized PMs by washing away PM-bound antibodies at pH 2.2 (Charvolin et al. 2009).

The functionality of immobilized MPs was tested on BR and the nAChR. The photocycle of A8-35-trapped BR was examined using BR trapped with BA8-35 and adsorbed onto SA-coated beads. Its kinetics were found to be undistinguishable from those of BR trapped in untagged A8-35 and kept free in solution. Comparison of the amplitude of the photo-induced signals measured before and after separation of free from immobilized BR showed the immobilization yield to be 70–75% (Charvolin et al. 2009).

The nAChR was solubilized with CHAPS from postsynaptic membranes of *T. marmorata* electric organs, purified, and trapped with a mixture of BA8-35 and A8-35 as described for non-functionalized A8-35 (Martinez et al. 2002), conditions under which the nAChR is known to remain functional (see Chap. 5, § 5.4). Free APols were removed by size exclusion chromatography (SEC), and the nAChR/APol complexes immobilized onto SA-coated beads. The binding kinetics of $\alpha$-bungarotoxin ($\alpha$-Bgt) fluorescently labeled with Alexa Fluor 647 ($\alpha$-Bgt-647) was monitored by confocal fluorescence microscopy. In the absence of nAChR, or in the presence of an excess of nonfluorescent $\alpha$-Bgt, no fluorescence signal was detected at the bead surface (Fig.13.4A, *lower panels*), whereas an intense signal appeared when $\alpha$-Bgt-647 was added to beads preincubated with

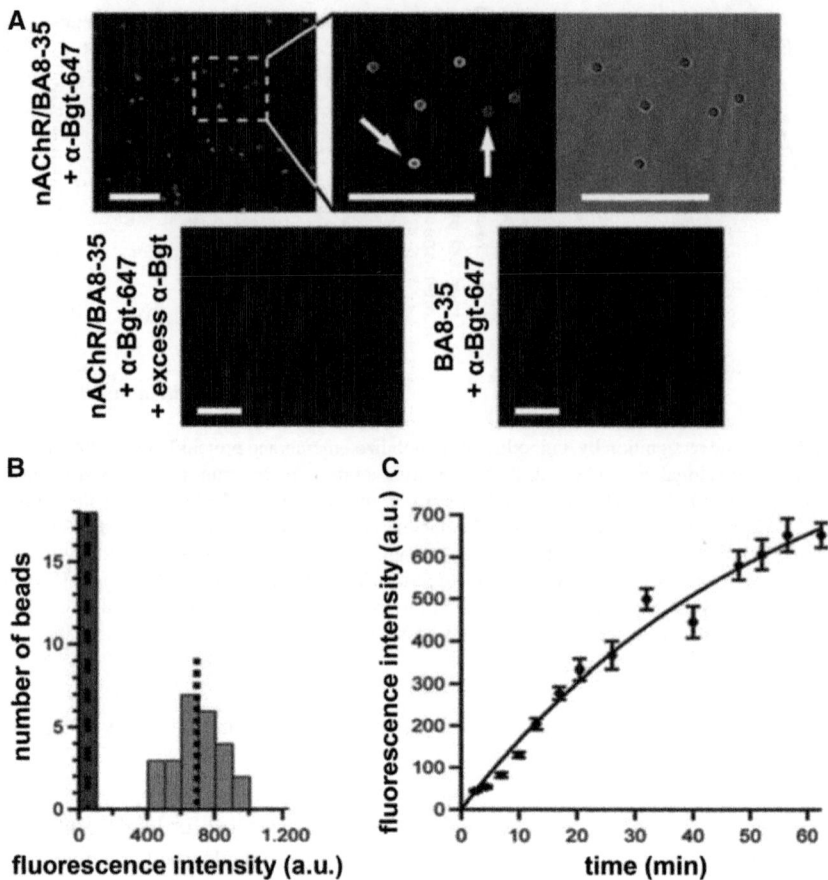

**Fig. 13.4** Immobilization of the nicotinic acetylcholine receptor (nAChR) via BA8-35 onto streptavidin-coated beads and application to monitoring the kinetics of ligand binding by fluorescence microscopy. (**A**) SA-coated polystyrene beads (1.8 μm diameter) preincubated with nAChR/BA8-35/A8-35 complexes and washed with buffer were imaged after 1-h incubation with 10 nM α-bungarotoxin (α-Bgt) fluorescently labeled with Alexa Fluor 647 (α-Bgt-647) (*upper left*). A homogeneous fluorescence signal was observed at their surface (enlarged view, *upper center*; the *upper right* panel shows the same field seen by light microscopy), whose intensity varied from bead to bead (arrows point to extreme cases). Under similar conditions, only very weak signals could be detected on nAChR-carrying beads preincubated with an excess of nonfluorescent α-Bgt (*lower left*) or on beads carrying only BA8-35 (*lower right*). Scale bars: 25 μm. (**B**) Distribution of the average fluorescence intensity per bead for nAChR/BA8-35- (*light gray*) or BA8-35-carrying beads (*dark gray*) after 1-h incubation with 10 nM α-Bgt-647 (data extracted from the experiments shown in A, *upper left* and *lower right*. *Dotted* and *dashed* lines indicate the average value for each population (respectively, 695 ± 29 and 49 ± 2 arbitrary units, SEM). (**C**) Binding kinetics of α-Bgt-647 to immobilized nAChR. The average intensity of the bead population (defined as shown in B) was determined at different times after addition of 10 nM α-Bgt-647 and corrected for non-specific binding (error bars, SEM; $n = 8$–25). The *solid line* is a fit with a first-order reaction model with $k_{obs} = 0.018 ± 0.004$ min$^{-1}$ (SD). This value yields $k_{on} = (3.0 ± 0.7) \times 10^4$ M$^{-1}$·s$^{-1}$ (SD) for the bimolecular reaction (From Charvolin et al. 2009. © 2009 National Academy of Sciences, USA).

BA8-35-trapped nAChR (Fig. 13.4A, *upper panels*), reflecting the interaction of the toxin with the agonist-binding sites of the receptor (Nirthanan and Gwee 2004). For a given concentration of α-Bgt-647, each bead exhibited a homogeneous signal, indicating the absence of protein clusters at the surface and confirming the quality of the immobilization strategy. However, the analysis of populations of beads indicated that the intensity of fluorescence varied from bead to bead, most probably because of a variable degree of coating with SA (Fig. 13.4B). Binding kinetics of $\alpha$-Bgt-647 were recorded online and data collected at the single-bead level. Their analysis yielded an observed association constant $k_{obs} = 0.018$ min$^{-1}$ at 10 nM $\alpha$-Bgt-647, assuming a pseudo-first-order reaction (Fig.13.4C), which corresponds to $k_{on} = (3.0 \pm 0.7) \times 10^4$ M$^{-1}$·s$^{-1}$ (SD) for the bimolecular reaction. This rate is within the range of those reported in the literature for the binding of radiolabeled $\alpha$-Bgt to purified Torpedo membranes or detergent-solubilized Torpedo nAChR (($0.8$–$33) \times 10^4$ M$^{-1}$ ·s$^{-1}$; see e.g. Franklin and Potter 1972; Blanchard et al. 1979; Lukas et al. 1981). The binding of $\alpha$-Bgt-647 could be inhibited by carbamylcholine, a small analog of acetylcholine (unpublished data), opening the way to determining the affinity constant of small, unlabeled ligands.

Study 13.1 thus established the feasibility of using tagged APols to immobilize MPs onto chips or beads and measure the binding to the immobilized proteins of large (antibodies, ~150 kDa), medium-sized ($\alpha$-Bgt, 8 kDa), or small (ACh analogs, a few hundred Da) ligands. In addition to the features noted above (absence of detergent, biochemical stabilization, absence of MP modification, use of a single tagged APol for any MP, possibility of mixing tagged and labeled APols), the approach presents other advantages: (i) unlike in protein-mediated immobilization methods, all extramembrane domains of the protein are expected to remain free from interactions with the support and to be comparably exposed to the solution; (ii) at variance with experimental schemes in which the target MP is embedded in membrane vesicles, all of its extramembrane regions are simultaneously accessible from the same aqueous compartment.

In most cases, trapping with APols is unlikely to interfere with ligand binding. As discussed in Chap. 5, § 5.4, it cannot be excluded, however, that, even though APols mostly confine themselves to the TM surface of trapped MPs, some of their tails or loops may explore MP extramembrane surfaces and bind to hydrophobic ligand-binding sites. This is not a purely theoretical possibility, because similar polymers have been shown to interact with serum albumin, a soluble protein, presumably via the hydrophobic groove that binds fatty acids (Tribet et al. 1998). We have noted in Chap. 5, however, that no interference of A8-35 is observed with the binding of LTB$_4$ to the BLT1 or BLT2 receptors (Dahmane et al. 2009; Catoire et al. 2010), even though, given the hydrophobicity of LTB$_4$, its binding site must be itself quite hydrophobic and could in principle attract APol octyl chains (§ 5.4). In the frontier region between the TM and extramembrane surfaces of the protein, APols could sterically or otherwise hinder the binding of macromolecular ligands. Furthermore, polyacrylate-based APols such as A8-35, being polyanionic, could also bind to strongly cationic pockets or patches: A8-35 has been shown to aggregate lysozyme (Champeil et al. 2000), and molecular dynamics simulations indicate that it does interact with extramembrane basic residues of OmpX (Perlmutter et al. 2014; see Chap. 5, § 5.3.3). Altogether, however, as noted in Chap. 5, § 5.4, no interference has been noted to date between APols and ligand binding, with the exception of that between anionic APols and G protein and arrestin binding to G protein-coupled receptors (see Chap. 5, § 5.4). The latter problem can be avoided by resorting to non-ionic APols. All things considered, this ensemble of data suggests that the possibility that APols interfere with ligand binding to APol-immobilized MPs cannot be ignored, but that, in most cases, it is unlikely to be a serious hindrance.

It is worth noting that, because APols bind to membranes (see Chap. 5, § 5.2.2.2), native membrane fragments or lipid vesicles could presumably be anchored to solid supports by the same strategy. This possibility has not yet been exploited.

Biotinylated APols are the only tagged APols that have been used to date in several publications. BA8-35 has been used to immobilize MPs in Studies 13.2 (Della Pia et al. 2014a; see § 13.2.2) and 13.3 (Ferrandez et al. 2014; see § 13.2.3). A biotinylated phosphorylcholine-based APol has been used in Study 13.5 to study the interaction between a bacterial outer membrane protein, FhuA, and a protein from the tail of a bacteriophage (Basit et al. 2012).

BA8-35 suffers from the disadvantage that high levels of non-specific binding can be observed when the ligands to be tested are cationic or expose cationic patches. This has led to the development of a biotinylated non-ionic APol (Ferrandez et al. 2014), whose use will be described in § 13.2.3. In § 13.3, we will consider the use of alternative tags. Right now, we will turn to Study 13.2 and the possibility of using tagged APols in multiplexing experiments, taking BA8-35 as a prototype.

## 13.2.2  Multiplexing

The development of protein arrays with micrometer features is of great interest because of their potential applications in life science and biomedical diagnostics (Christman et al. 2006a; Wingren and Borrebaeck 2007). Indeed, this technology can provide high-throughput screening of protein functions and biomolecular interactions while requiring very small amounts of analytes and short processing times (Bano et al. 2009). Further miniaturization of protein arrays to the nanometer scale could further improve detection sensitivity and give access to specific biomolecular reactions that are not accessible to micrometer-scale structures (Coyer et al. 2007). Arrays of proteins with micrometer and/or submicrometer features have been fabricated using patterning techniques such as microcontact printing (Coyer et al. 2007; Hoff et al. 2004); scanning near-field optical (Leggett 2007), dip-pen (Lee et al. 2002), and photo- and ion beam lithography (Kannan et al. 2004; Christman et al. 2006a, b; Zhang et al. 2007); atomic force microscopy (AFM) (Gu et al. 2004), and electrochemical-based approaches (Hoover et al. 2007). However, only a handful of examples of anchoring simultaneously multiple soluble proteins onto nanoarrays has been reported (Christman et al. 2006b), and, to date, none of the techniques mentioned above has been applied to nanopatterning multiple purified MPs (cf. Valiokas et al. 2006). Whereas there is a strong interest in the production of MP micro- and nanoarrays for both basic and applied research, the insolubility of MPs in surfactant-free aqueous solutions and their instability in the presence of detergents strongly complicate the fabrication of MP biochips (Bieri et al. 1999; Fang et al. 2002; Hong et al. 2005; Majd and Mayer 2008). Techniques involving drying steps and/or harsh conditions such as high temperatures and mechanical shear may denature MPs (Fang et al. 2002). Furthermore, as noted in Chap. 2, MP stability often depends on the retention of associated lipid molecules, which is difficult to ensure in the presence of detergents. This requirement sets further constraints on the use of the solvents that are normally used to prevent evaporation and keep the protein hydrated in scanning probe lithography techniques (Zheng et al. 2009). All of these problems can be circumvented or, at least, mitigated by the use of APols.

Study 13.2 describes a novel strategy to fabricate micro- and nanoarrays carrying multiple MPs, based on the combined use of BA8-35 and of the selective functionalization of gold electrodes by electropolymerization of a biotin-doped polypyrrole (PPy-biotin) film. The electrosynthesis of conducting polymers allows the controlled and reproducible functionalization of electrodes at the micro- and nanoscale (Cosnier et al. 1998). Pyrrole functionalized with different biotin groups has been synthesized (Cosnier et al. 1998, 1999; Torres-Rodriguez et al. 1998; Yang et al. 1998; Darmanin et al. 2012), and polymeric films have been prepared and used as a support for the immobilization of biotin groups to electrode surfaces (Cosnier et al. 1998, 1999; Bidan et al. 2000). After incubation with avidin, biotin-tagged biomolecules can be anchored to the free sites of avidin to construct biosensors (Cosnier et al. 1998). As biotin is negatively charged under the conditions used for polymerization of PPy, it can also be incorporated in the polymeric film when used as a doping anion. This method is particularly appealing, as it does not require any chemical synthesis. In Study 13.2, it has been used to establish a straightforward approach to MP multiplexing on gold micro- and nanoelectrodes. As schematically illustrated in Fig. 13.5, the key to the approach is the selective electropolymerization of conducting surfaces with a PPy film containing biotin, followed by the immobilization of avidin, neutravidin, or SA and the binding of BA8-35-trapped MPs.

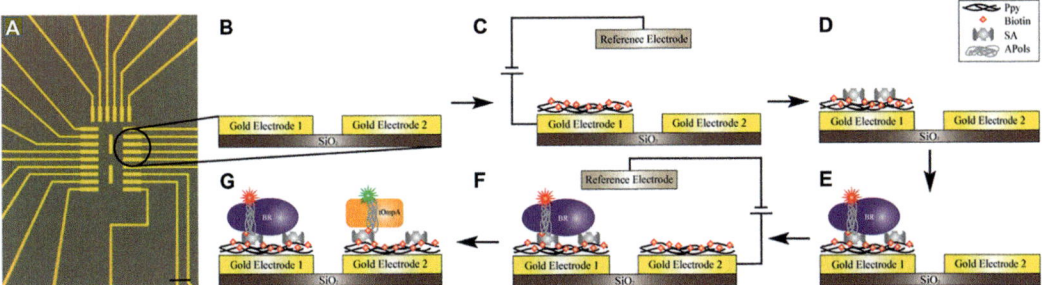

**Fig. 13.5** Schematic representation of a multiplexing immobilization procedure based on polypyrrole electropolymerization. (**A**) Differential interference contrast image of the array of electrodes used in the experiments. Scale bar is 50 μm. The array consists of 24 gold electrodes deposited on top of an insulating SiO$_2$ layer on a silicon substrate (only two electrodes are drawn in Panels B-G). (**C**) The array is immersed into a solution of 0.1 M pyrrole, 1 mM biotin, and 10 mM NaCl, and an oxidizing potential is applied to one of the electrodes. (**D**) After functionalization of the surface of a single electrode with a polypyrrole (PPy) biotin-doped film, the entire array is incubated into a solution of streptavidin (SA). (**E**) The array is extensively washed with buffer before being incubated with a solution containing the first target MP, here bacteriorhodopsin (BR, shown in *purple*) trapped in a mixture of BA8-35 and A8-35 labeled with a red fluorophore. (**F**) A second electrode is activated using the same electropolymerization procedure as in C. (**G**) After incubation of the array with SA, a second MP (tOmpA, shown in *orange*), trapped in a mixture of BA8-35 and A8-35 carrying a green fluorescent fluorophore, can be immobilized onto the electrode. Biotin-carrying molecules are colored in *gray* and biotin in *red*; a mesh of *black curved lines* represents the PPy film. The two MPs are shown trapped in APols (*gray*), with fluorophores represented as *red* and *green* stars (From Della Pia et al. 2014).

An example of the results obtained by the group of Karen L. Martinez using this approach is shown in Fig. 13.6. BR was trapped in a 1:1 mixture of BA8-35 and A8-35 labeled with Alexa Fluor 647 (FAPol$_{AF647}$; see Chap. 4, Table 4.5), partially separated from free APols by SEC, and applied to an electrode functionalized with a PPy-biotin film and SA. The high and homogeneous fluorescence signal measured on the functionalized electrode (Fig. 13.6A) confirms that SA is still active after immobilization and exposes free sites that can bind BAPols and MP/BAPol complexes. Next, a PPy-biotin film was formed over a second electrode, and the chip incubated with BR trapped in a mixture of untagged A8-35 and FAPol$_{AF647}$ (Fig. 13.6B). The fluorescence signal observed indicated that there is some interaction between the complexes and the film, but at a level at least 3× lower than with BA8-35. The biotin/avidin interaction is therefore essential to obtaining a high level of MP immobilization. These two experiments also show (i) that there is very little non-specific interaction with bare gold nor SiO$_2$ surfaces and (ii) that the immobilization achieved in the first step is stable: there is no significant decrease in fluorescence signal for the BR-functionalized electrode upon incubation with the pyrrole-biotin mixture or the SA solution. Given the prior demonstration that BA8-35-trapped BR immobilized onto avidin-carrying beads is perfectly functional (Charvolin et al. 2009), there is no reason to think that using the same procedure to attach it to a gold nanoelectrode coated with avidin will affect its structure nor function. In keeping with this expectation, immobilized BR was recognized by an anti-BR antibody labeled with Alexa Fluor 488 (Fig. 13.6C).

In Fig. 13.7, the sequential functionalization of electrodes with a PPy-biotin layer and SA has been exploited to attach two distinct MPs to two adjacent electrodes. As previously, BR was trapped in a mixture of BA8-35 and FAPol$_{AF647}$, whereas a second MP, tOmpA, was trapped in a mixture of BA8-35 and A8-35 fluorescently labeled with 7-nitrobenz-2-oxa-1,3-diazol-4-yl (FAPol$_{NBD}$; see Chap. 4, Table 4.5). SA was immobilized on a PPy-biotin layer electrogenerated on electrode 2. The chip was incubated in a solution of tOmpA/BA8-35/ FAPol$_{NBD}$ complexes, resulting in binding of tOmpA onto electrode 2 only (Fig. 13.7A). After rinsing with buffer, a PPy-biotin film was generated

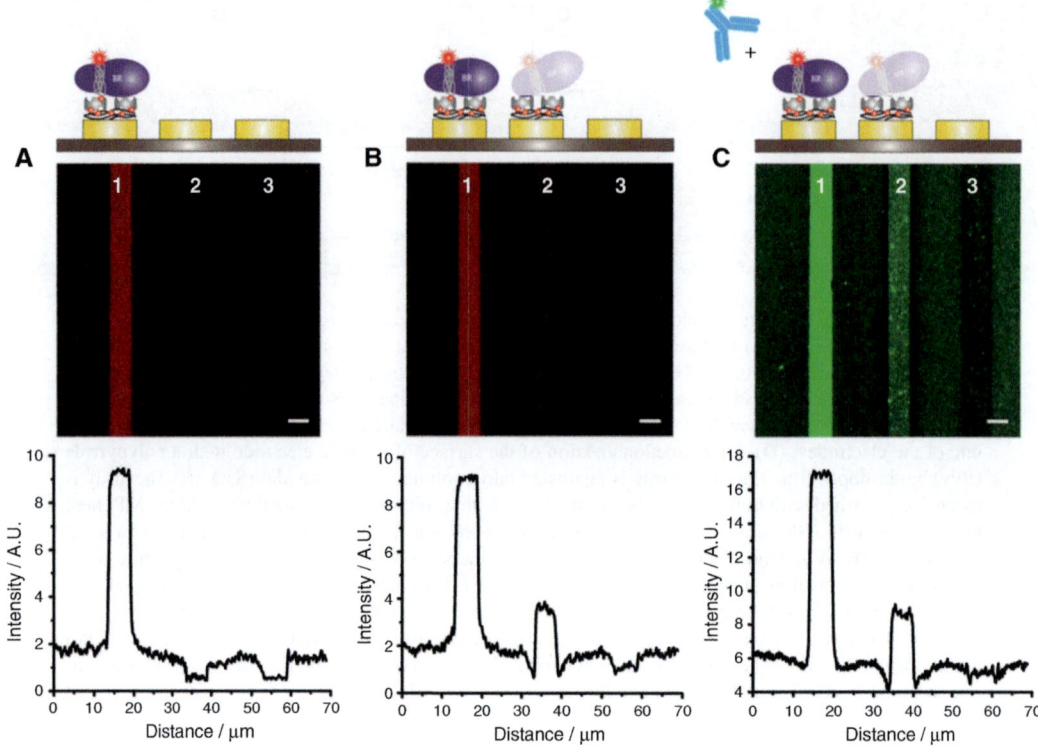

**Fig. 13.6** Membrane protein anchoring to a polypyrrole (PPy)-biotin surface functionalized with streptavidin (SA) via the avidin/biotinylated amphipol interaction, and specific recognition by an antibody. (**A**) Electrode 1 was modified with PPy-biotin. The chip was incubated with 1 µM SA, washed with buffer, and incubated with 1 µM bacteriorhodopsin (BR) trapped in a 1:1 mixture of BA8-35 and A8-35 labeled with Alexa Fluor 647 (FAPol$_{AF647}$) and partially separated from free APols by SEC. After washing with buffer, the electrodes were imaged by fluorescence microscopy ($\lambda_{exc} = 620 \pm 30$ nm, detection at $700 \pm 35$ nm). (**B**) Electrode 2 was modified with PPy-biotin, after which the chip was incubated with 1 µM SA, washed with buffer, and incubated with 1 µM BR trapped in a mixture of biotin-free A8-35 and FAPol$_{AF647}$. Non-specific interactions between the PPy-biotin film and A8-35 generated a degree of background on electrode 2. Electrode 3 was not modified at any time. (**C**) The chip was incubated for 1 h with 5 nM of an anti-BR antibody labeled with Alexa Fluor 488, washed with buffer for 30 min, and imaged ($\lambda_{exc} = 470 \pm 20$ nm, detection at $525 \pm 25$ nm). Scale bar is 5 µm (Adapted from Della Pia et al. 2014a).

on electrode 3 (in which process the remaining free avidin sites on the chip were blocked by biotin) and the chip exposed successively first to SA and then to BR/BA8-35/FAPol$_{A647}$ complexes. As shown in Fig. 13.7B, the procedure resulted in MP adsorption onto electrode 3 only, with minimal contamination of the electrode that had not been functionalized (electrode 1), or of that where a MP had already been immobilized and the biotin-binding sites blocked (electrode 2). These results further demonstrate that BA8-35-trapped MPs remain attached after incubation with the monomeric solution of pyrrole and are not detached by the application of the electric field needed for the activation of the adjacent electrode, nor of the second solution of MP/APol complexes. The procedure ought to be scalable to any number of different MPs and/or other biotin-tagged compounds.

Because of the highly localized nature of electric fields, this approach can be implemented on submicrometer electrodes. This is shown in Fig. 13.7C, D, where two electrodes with features as small as ∼50 nm were functionalized with BR and tOmpA.

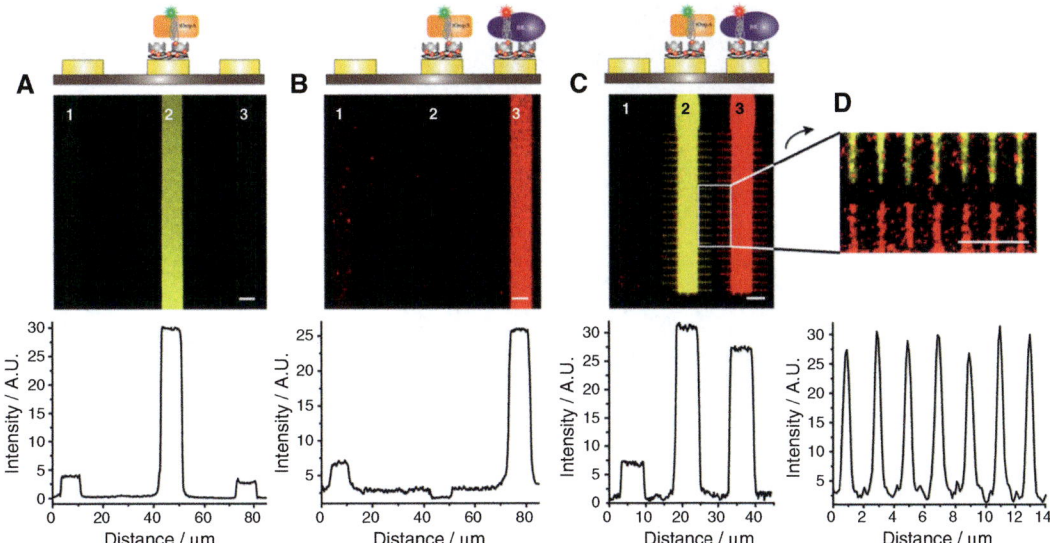

**Fig. 13.7** Sequential and addressable functionalization of micro- and nanometer surfaces with amphipol-trapped membrane proteins. (**A**) Electrode 2 was functionalized with a polypyrrole (PPy)-biotin film, and the chip incubated for 30 min with 1 μM streptavidin (SA) and for 30 min with 1 μM tOmpA trapped in a 1:1 mixture of BA8-35 and NBD-labeled A8-35 (FAPol$_{NBD}$), washed with buffer, and imaged ($\lambda_{exc} = 531 \pm 20$ nm, detection at $593 \pm 20$ nm). (**B**) Electrode 3 was functionalized with a PPy-biotin film (free avidin sites on electrode 2 being blocked by biotin in the process). The chip was incubated for 30 min with 1 μM SA and for 30 min with 2 μM bacteriorhodopsin (BR) trapped in a 1:1 mixture of BA8-35 and an Alexa Fluor 647-labeled A8-35 (FAPol$_{AF647}$), washed with buffer, and imaged ($\lambda_{exc} = 620 \pm 30$ nm, detection at $700 \pm 37$ nm). Electrode 1 was not functionalized. (**C**) Two electrodes with 50 nm wide extensions were selectively functionalized with (i) tOmpA trapped in BA8-35/FAPol$_{NBD}$ (electrode 2) and (ii) BR trapped in BA8-35/FAPol$_{AF647}$ (electrode 3) using the same procedure as in B. (**D**) Zoom on the nanosurfaces of Panel C. The line profiles show the fluorescence intensity measured across the nanometer-wide surfaces of both electrodes 2 and 3. Scale bar is 5 μm (From Della Pia et al. 2014).

To demonstrate the scalability of APol-mediated adsorption of MPs to the nanometer scale, 200-nm wide, 300-nm distant gold nanoelectrodes were functionalized with tOmpA trapped in a BA8-35/FAPol$_{NBD}$ mixture (Fig. 13.8). Due to the resolution limit of the fluorescence microscope, the fine structure of the nanosurfaces cannot be resolved, but the images show a strong fluorescence signal only on the electrodes and not on the SiO$_2$ insulating surfaces. This experiment shows that it is possible to biofunctionalize submicrometer surfaces, the only limit of this technique being the availability of an electrically contacted surface. The use of electrochemical polymerization to functionalize surfaces with MPs provides an important advantage over lithography-based techniques, which require challenging alignment procedures. As an example, nanowire and carbon nanotube arrays can be fabricated with diameters and pitches as small as 10 nm (Melosh et al. 2003). Selective functionalization of these nanostructures cannot be accomplished by stamping techniques (Wong and Melosh 2009), but it could be easily achieved by the electrochemical approach described in Study 13.2.

Comparative experiments revealed a marked difference in the respective ability of avidin, SA, and neutravidin to bind BA8-35-trapped MPs. When a chip bearing electrodes coated with each of these three proteins was exposed to a solution of BR/BA8-35/FAPol$_{AF647}$ complexes, the electrode carrying avidin showed a fluorescence intensity almost double that recorded on the electrodes functionalized with SA and neutravidin, suggesting a higher binding capacity. However, the fluorescence image of a second protein array immersed in a solution of BR trapped in a 1:1 mixture of

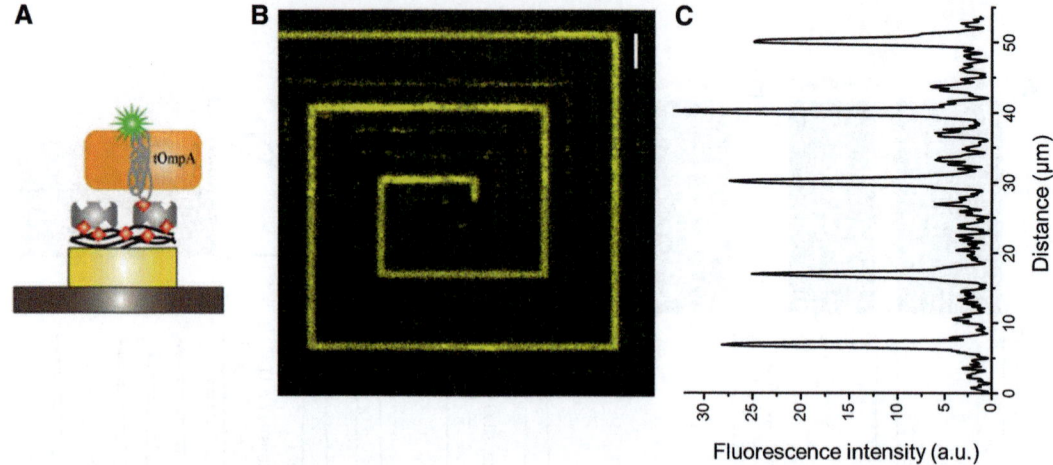

**Fig. 13.8** Functionalization of a nanosurface with a membrane protein trapped with a biotinylated amphipol. (**A**) Experimental scheme: a 200-nm thick electrode was functionalized with a PPy-biotin film and incubated first for 30 min with 1 μM streptavidin and then for another 30 min with 1 μM tOmpA trapped in a BA8-35/FAPol$_{NBD}$ mixture. (**B**) Fluorescence image taken at $\lambda_{exc} = 531 \pm 20$ nm, detection at $593 \pm 20$ nm. Scale bar is 5 μm. (**C**) Fluorescence intensity evaluated along a vertical line crossing the center of the figure (From Della Pia et al. 2014).

unfunctionalized A8-35 and FAPol$_{AF647}$ also showed a signal on the avidin-decorated surface, whereas no fluorescence signal was detected on the electrodes bearing SA or neutravidin (Della Pia et al. 2014a). The higher non-specific binding to the avidin surface can likely be attributed to electrostatic interactions taking place in the low ionic strength buffer solution. A8-35-trapped BR bears a high negative charge due to the many carboxylates carried by the APol belt, all of which are ionized at pH 7.2 (Gohon et al. 2004; see Chap. 4, § 4.3.1.2.2). As a result, it adsorbs non-specifically onto the positively charged avidin substrate (isoelectric point ∼10.5). In contrast, SA- and neutravidin-modified surfaces have a slightly negative charge at pH 7.2 (isoelectric points 6.1 and 6.3, respectively; Petrlova et al. 2007) and exhibit much less non-specific binding of BR/A8-35 complexes than avidin. Non-specific binding of cationic ligands by BA8-35 is a recurrent concern (Ferrandez et al. 2014; Mary et al. 2014, and unpublished data by various groups), which has led to the development of biotinylated non-ionic APols, as will be described in § 13.2.3.

### 13.2.3   Using Ionic vs. Non-ionic Biotinylated Amphipols for Phage Display Selection of Membrane Protein Binders

Resorting to specific and tight-binding protein partners can considerably facilitate the challenging process of MP crystallization (cf. Chap. 11, § 11.2.2). The binders are sometimes referred to as "crystallization chaperones" (Koide 2009; Lieberman et al. 2011; not necessarily a perfectly appropriate term given that they are part of the final crystals, and not temporary facilitators as chaperones are supposed to be), or "orthoses" (Ferrandez et al. 2014). Natural binders include monoclonal antibody F$_v$ fragments, an approach pioneered by Hartmut Michel (Ostermeier et al. 1995), heterologously produced versions thereof, and camelid nanobodies. Among the many types of artificial binders that have been proposed (for a general review, see Skerra 2007), a very promising class based on repeat proteins has been introduced by Andreas Plückthun and colleagues under the name of "designed

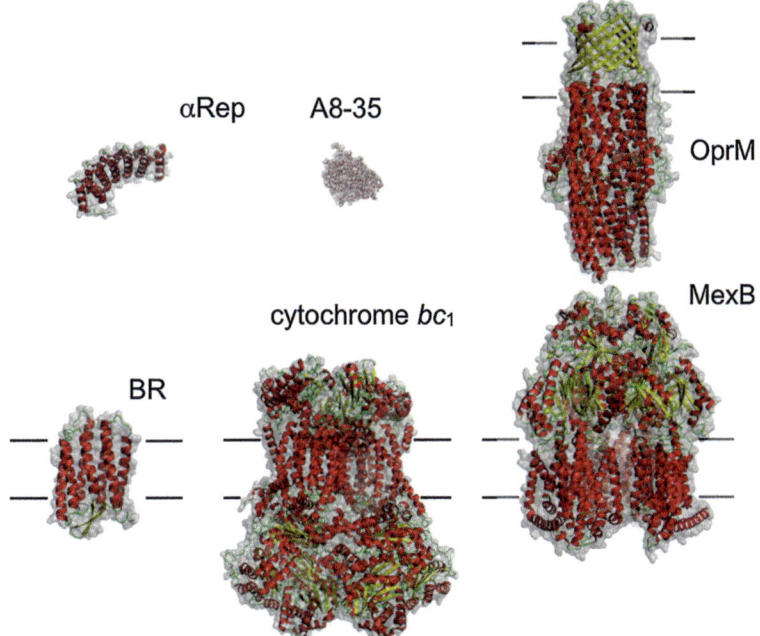

**Fig. 13.9** Structure and relative size of the various partners involved in amphipol-assisted selection of αReps (Study 13.3). PDB accession numbers of atomic coordinates: BR, 1AT9; cytochrome $bc_1$, 2A06; MexB, 2V50; OprM, 3D5K; αRep, 3LTJ (From Ferrandez et al. 2014).

ankyrin repeat proteins" (DARPins) (Binz et al. 2003; Stumpp et al. 2003; Sennhauser and Grütter 2008). Other types of repeats based on Armadillo, leucine-rich repeats, or tetratricopeptide repeats have also produced specific binders (Boersma and Plückthun 2011).

Philippe Minard and coworkers have introduced a new family of artificial repeat proteins named "αReps." αReps are derived from a subgroup of HEAT-like repeats found in several cytoplasmic proteins, the names of four of which are at the origin of the acronym (Huntingtin, elongation factor 3 (EF3), protein phosphatase 2A (PP2A), and the yeast kinase TOR1). αReps are efficiently expressed and folded, and they are very stable. Large libraries have been generated (Guellouz et al. 2013), from which tight and specific binders have been selected against several soluble proteins, with dissociation constants, $K_D$, in the nanomolar to micromolar range. Crystallographic structures of αRep/target protein complexes show conformation-specific recognition (Guellouz et al. 2013).

For the selection of binding proteins as putative crystallization helpers, it is critical that, at all stages of the selection process, the purified MPs used as targets be immobilized on a solid support under their native state. In Studies 13.3/13.6, biotinylated APols were used to this end. The approach was tested with four model MPs of known structures, MexB and OprM from the Gram$^-$ bacterium *Pseudomonas aeruginosa* efflux pump system, the cytochrome $bc_1$ complex from beef heart mitochondria, and BR from *H. salinarum*. MexB and OprM had not been previously trapped with APols, whereas BR and the $bc_1$ complex had already been used in APol-mediated immobilization experiments (§ 13.2.1 and 13.2.2). The structures and relative sizes of the partners involved in Studies 13.3/13.6 are illustrated in Fig. 13.9.

In the phage display selection procedure (Fig. 13.10), the target MP is immobilized onto the avidin-coated bottom of culture plate wells via a biotinylated APol and incubated with the phages produced from the αRep library, each phage exposing at its surface a different αRep. Several washing

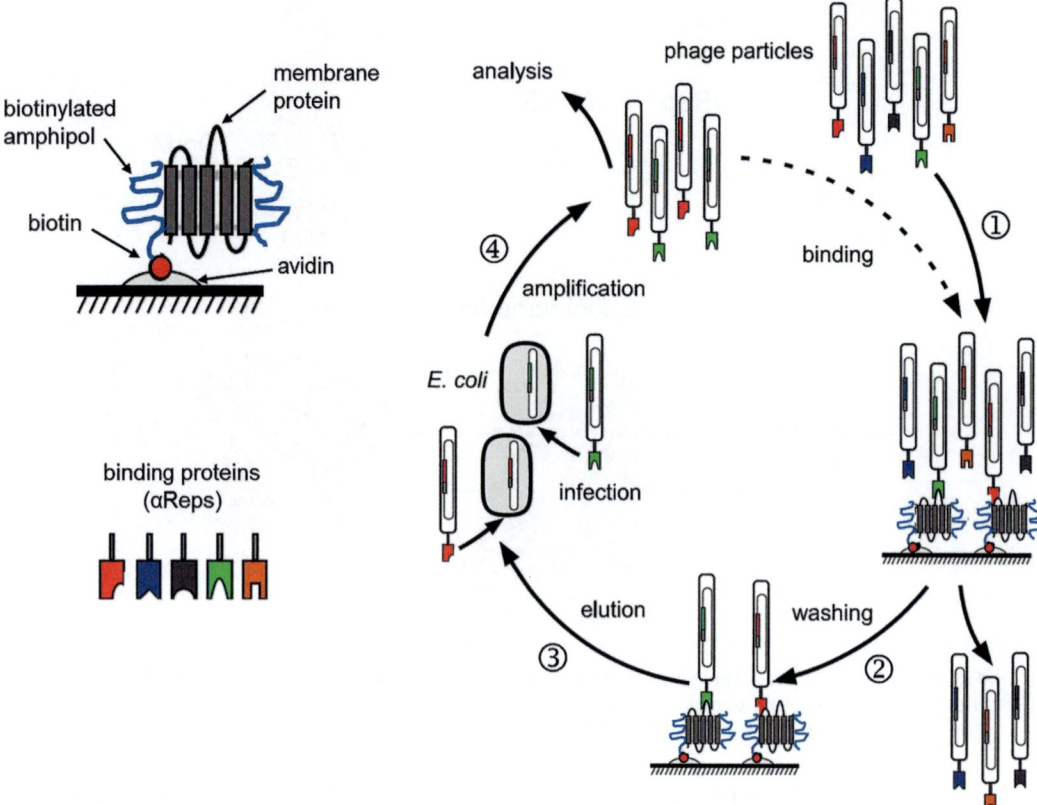

**Fig. 13.10** Schematic representation of a phage display selection cycle using amphipol-immobilized membrane proteins (MPs) as targets. MPs are immobilized via the interaction between avidin bound to a plastic well and biotin moieties carried by the APol. A phage library displaying variants of αReps on their surface while having the respective gene encapsulated inside the phage particle is used for the selection ①. After formation of the complexes between αReps and the target MP, unbound phage particles are washed off ②. Strongly bound phages are eluted from the immobilized target by a pH shift ③ and used for infecting *E. coli*. Phage particles are amplified ④ and can then be either analyzed or used as input for the next selection round (Adapted from Huber et al. 2007, © 2007 Elsevier Inc. All rights reserved).

steps are then carried out to wash off the phages that bind non-specifically, and the strongly bound phages are eluted using an acidic glycine solution. For the immobilization, two forms of biotinylated APols were used, BA8-35 (Charvolin et al. 2009) and a biotinylated version of a glucosylated non-ionic homotelomeric APol (Sharma et al. 2012; see Chap. 4, Fig. 4.2.3) developed for this study ("BNAPol") (Ferrandez et al. 2014).

In brief, these experiments showed that (i) the clones selected always displayed affinity for the APol used in the selection, even if they had been, in principle, depleted of APol binders on APol-coated wells, and (ii) only in the case of MP/BNAPol complexes did some of the clones also exhibit binding to the detergent-solubilized target protein. It had been anticipated that phages expressing αReps that interact with APols would be eliminated by the prescreening procedure. The results however show that this negative selection step is insufficient to prevent the selection of composite binders, which recognize both the protein and the APol. The failure of this preliminary counterselection may have various origins, perhaps the most likely of which is that all phages containing a plasmid copy coding for a given αRep in a library will not necessarily express it at their surface in a given round of selection, whereas they may do so during a subsequent screening step. Hence, a significant proportion of phages coding for APol binders might escape the negative selection step if it is not repeated. Once a specific library pruned of APol binders has been established by several rounds of depletion, selection of protein-specific binders using MP/APol complexes ought to become easier. Another option perhaps worth investigating could be, as has been done for DARPins (Seeger et al. 2013), to engineer a library of $\alpha$Reps poorer in hydrophobic residues. In the meanwhile, the use of tagged APols in the phage display process should probably be limited to MPs that have too limited a stability in detergent solutions to survive a detergent-based procedure or that cannot be easily tagged.

## 13.3  Alternative Tags

The biotin/avidin couple presents the advantage of a very high affinity, leading to the formation of non-covalent but essentially irreversible complexes. When biotin is chemically bound to the protein, the latter cannot be eluted. In the case of MPs trapped with a biotinylated APol, on the contrary, one has the option of washing it away with an excess of either detergent or a non-biotinylated APol: the chip or beads will retain the biotinylated APol and will not be regenerated, but the protein will elute. Similarly, an APol grafted with desthiobiotin (nonextant yet) could be displaced from avidin-coated surfaces by biotin. It is nevertheless worth examining the feasibility and applications of using other tag/binder combinations. In Studies 13.7–13.10 (see Table 13.1), the synthesis of four alternative tagged variants of A8-35 has been developed and their validation undertaken. The tags explored are (i) polyhistidine, yielding "HistAPol" (Study 13.7); (ii) randomly distributed imidazoles ("ImidAPol," Study 13.8); (iii) an oligodeoxynucleotide ("OligAPol," Study 13.9); and (iv) a sulfide-bearing moiety, thiomorpholine ("SulfidAPol," Study 13.11). HistAPol and ImidAPol can be used to immobilize MPs onto surfaces carrying immobilized $Ni^{2+}$ or $Co^{2+}$ ions. OligAPols interact with complementary oligonucleotides, SulfidAPols with gold or silver surfaces. Figure 13.11 and Table 13.2 compare the affinities of the various tags for their binders and the reversibility of the attachment.

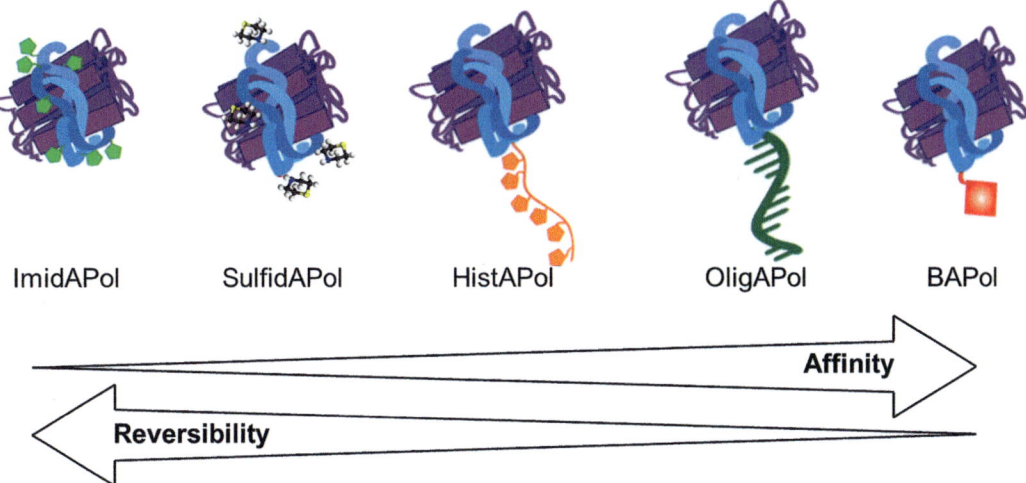

**Fig. 13.11** A schematic representation of the relative affinity and reversibility of the couples formed by biotin/avidin, oligonucleotide/complementary sequence, polyhistidine/$Ni^{2+}$, randomly distributed imidazoles/$Ni^{2+}$, and SulfidAPol/gold (Adapted from Della Pia et al. 2014b).

**Table 13.2** Properties of various immobilization tags grafted onto amphipols.

| Tag | $K_D$ (M) | Specificity | Reversibility | Studies[a] |
|---|---|---|---|---|
| Biotin | $10^{-14}$–$10^{-16b}$ | High | Low | 13.1–13.6 |
| Polyhistidine | $10^{-7}$–$10^{-8c,}$ | Medium | High | 13.7 |
| Imidazole | [d] | Medium | High | 13.8 |
| Oligonucleotide | $\sim 10^{-9e}$ | High | High | 13.9 |
| Sulfhydryl | $10^{-6}$–$10^{-7f}$ | High | High | 13.10 |
| Sulfide | $>10^{-6f}$ | High | High | 13.11 |

Adapted from Della Pia et al. (2014b)
[a]See Table 13.1
[b]Jonkheijm et al. (2008)
[c]Laitinen et al. (2007); depends on the sequence and length see Knecht et al. (2009)
[d]Depends on the density and distribution
[e]Niemeyer et al. (1998); depends on sequence and length
[f]Estimated based on data from Bain et al. (1989a, b), Lavrich et al. (1998), Lim et al. (2006), Ansar et al. (2011), and Ravi et al. (2013)

In the following sections, we review rapidly the most interesting characteristics of these new tagged APols.

## 13.3.1 Polyhistidine-Tagged A8-35

One of the most widely used tags is a peptide comprised of six or more histidine residues (His$_6$-tag, His$_8$-tag, etc.), which can specifically interact with $Ni^{2+}$ or $Co^{2+}$ ions bound to nitrilotriacetic acid (NTA)-functionalized surfaces. The selectivity and reversibility of this mode of immobilization make it very useful in biochemistry and biophysics for protein purification (Hochuli et al. 1987; Block et al. 2009) and protein immobilization onto chips (Nieba et al. 1997), nanoparticles (NPs) (Xu et al. 2004; Patel et al. 2007), or other surfaces (Kang et al. 2007; Liu et al. 2010). Study 13.7 describes the synthesis and validation of a derivative of A8-35, dubbed HistAPol, onto which His$_6$-tags have been bound via a short polyethylene glycol linker (Giusti et al. 2015; see Chap. 4, § 4.4). This modification necessitated first the chemical synthesis of hexahistidine peptides, which is far from a trivial endeavor. Chap. 4, § 4.4.2, also reports on the solution properties of HistAPol. The pH-dependence of the solubility, in particular, is markedly different from that of plain A8-35, because at low pH the histidine residues protonate, making the polymer zwitterionic and water-soluble, whereas, under the same conditions, A8-35 and MP/A8-35 complexes become insoluble and precipitate (Chap. 4, § 4.3.1.2.5).

At pH 8, the SEC behavior of HistAPol particles, tOmpA/HistAPol and BR/HistAPol complexes was comparable to that observed with plain A8-35, with a marginally broader distribution of hydrodynamic radii (Giusti et al. 2015). The average number of His$_6$-tags per MP/HistAPol complex can be evaluated based on estimates of the mass of APol bound by each protein. In the case of tOmpA, it is likely to be $\sim 4.5$, or slightly less, in that of BR $\sim 5.4$. As for HistAPol particles, the number of tags per MP/HistAPol complex is expected to follow a Poisson distribution and to fluctuate over time, as MP/HistAPol complexes collide with each other and with free HistAPol particles and exchange APol molecules, leading to a permanent redistribution of the tags. Under the ionic conditions used in Study 13.7 (100–150 mM NaCl), redistribution occurs on the time scale of several minutes (Zoonens et al. 2007; see Chap. 5, § 5.7), whereas SEC analysis takes $\sim 45$ min (see Chap. 9, § 9.5.2, Protocol 5.2).

Immobilized metal affinity chromatography (IMAC) was used to explore whether HistAPol can be used for immobilizing MPs onto nickel-activated surfaces. The elution profile of tOmpA/HistAPol

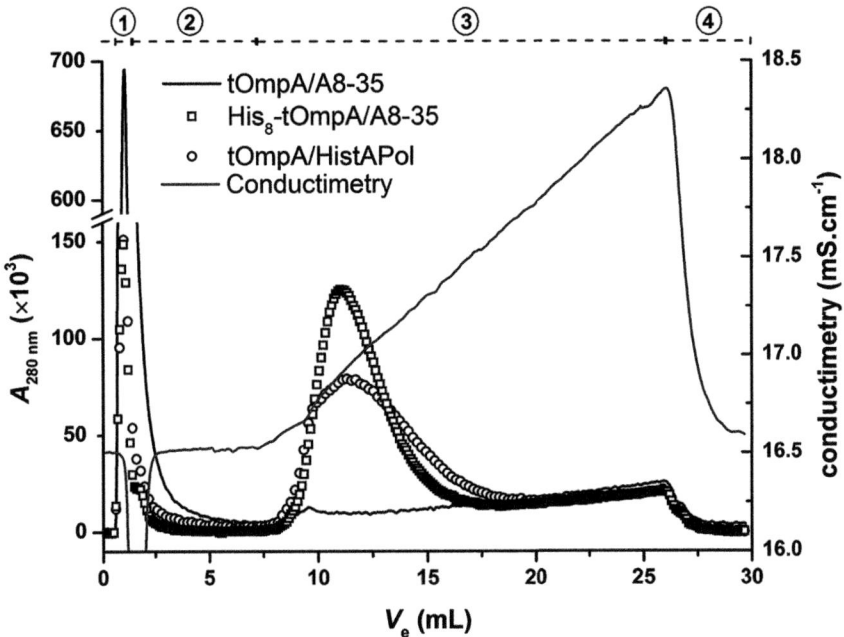

**Fig. 13.12** Immobilization of tOmpA/HistAPol and $His_8$-tOmpA/A8-35 onto and elution from an immobilized metal affinity chromatography column. Two versions of tOmpA, with and without a $His_8$-tag fused at the *N*-terminus, were trapped in A8-35 or HistAPol, respectively. A control sample, untagged tOmpA trapped with plain A8-35, shows the absence of non-specific interaction of the untagged complex with the Ni:NTA resin. Each sample (0.25 mg of protein) was injected onto a 1-mL HisTrap FF column (GE Healthcare) pre-equilibrated with 0.02 M Tris/HCl, pH 8.0, 0.15 M NaCl. After loading the sample ① and washing out the unbound material ②, elution was carried out with a linear gradient of imidazole (from 0 to 0.3 M; total volume 20 mL) ③ before washing the column ④. Protein elution was monitored at 280 nm (Reprinted with permission from Giusti et al. 2015, © 2015 American Chemical Society).

complexes injected onto a Ni:NTA column was compared to those obtained with two versions of tOmpA, one of them carrying a $His_8$-tag at its *N*-terminus, the other one devoid of tag, both of them trapped in non-functionalized A8-35. A8-35-trapped untagged tOmpA was not retained by the IMAC column, whereas the two tagged complexes were (Fig. 13.12). Upon application of a gradient of imidazole, the elution profile of tOmpA/HistAPol complexes appeared asymmetrical (Fig. 13.12), most of the complexes eluting in the same range of imidazole concentration as $His_8$-tOmpA/A8-35 complexes, whereas a minority of them eluted later. An asymmetric elution profile was also obtained for BR/HistAPol complexes (Giusti et al. 2015). Geometrical considerations suggest that delayed elution is likely to result from the rebinding during elution of particles carrying several $His_6$-tags, rather than from their simultaneous binding to several nickel ions (see Giusti et al. 2015). The percentage of tOmpA/HistAPol complexes retained on the column is very similar to that of $His_8$-tOmpA/A8-35 (81% and 84%, respectively). The recovery of tOmpA after elution with free imidazole shows the ability of HistAPol to mediate the reversible immobilization of MPs.

SPR was used to characterize the binding kinetics of MPs trapped in HistAPol to a sensor surface functionalized with Ni:NTA motifs and to investigate the potential of HistAPol-trapped MPs for surface-based detection methods. After activating the NTA sensor surface by injection of $NiCl_2$, tOmpA/HistAPol (Fig. 13.13) and BR/HistAPol (Giusti et al. 2015) complexes were immobilized onto the surface. The specificity of the binding of HistAPol-trapped tOmpA was confirmed by the absence of signal after injection of either (i) tOmpA/HistAPol complexes on an NTA-dextran surface

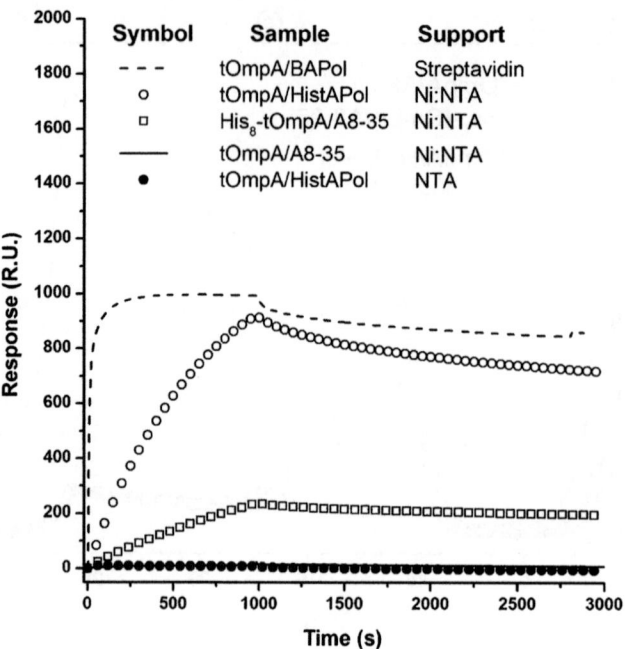

**Fig. 13.13** Sensorgrams of the binding of tOmpA trapped in various APols and injected onto a Ni:NTA-carrying surface plasmon resonance (SPR) chip. Binding was observed upon injection of 1 μM tOmpA trapped in HistAPol (O), but not upon injection of 1 μM tOmpA trapped in A8-35 (—), nor upon injection of 1 μM tOmpA/HistAPol onto a nickel-free NTA surface (●). Binding was observed when 1 μM His$_8$-tOmpA/A8-35 was applied to a Ni:NTA surface (□). For comparison, 1 μM tOmpA trapped in BA8-35 was applied to a streptavidin-coated sensor surface (- - -) (Reprinted with permission from Giusti et al. 2015, © 2015 American Chemical Society).

in the absence of nickel ions or (ii) untagged tOmpA/A8-35 complexes on a Ni:NTA-dextran surface (Fig. 13.13). These control experiments show that the non-specific binding of tOmpA/HistAPol to the NTA surface is low.

Immobilization of MPs on Ni$^{2+}$-activated surfaces using HistAPol was compared to that mediated by BA8-35 interacting with an SA-coated surface (Fig. 13.13). The values of the maximum responses were comparable, even though the two modes of interaction are very different. The shapes of the binding curves differed from each other, due to different association and dissociation kinetic constants. Even at prolonged injection times (>15 min), the binding of tOmpA/HistAPol complexes did not reach equilibrium, indicating that saturation of the surface is achieved much more slowly than for tOmpA/BAPol complexes and that, at saturation, a larger number of complexes would be bound.

After all injections of proteins trapped in HistAPol, the complexes could be washed from the Ni:NTA sensor surface by injecting EDTA. This led to full regeneration of the surface, as shown by the response returning to the baseline (Giusti et al. 2015).

To demonstrate that immobilization of MPs via HistAPol can be used for the analysis of protein-protein interactions, anti-tOmpA or anti-BR antibodies were applied in surfactant-free buffer to sensor surfaces onto which either tOmpA/HistAPol or BR/HistAPol complexes had been immobilized (Fig. 13.14). A specific response was detected for each MP/antibody pair. No signal was observed when, for example, anti-BR antibody was injected onto a tOmpA/HistAPol-derivatized surface (Fig. 13.14A), nor when anti-BR was injected onto a bare Ni:NTA surface (Fig. 13.14B). This result establishes the compatibility of MP immobilization using HistAPol with label-free detection of ligand interactions with MPs.

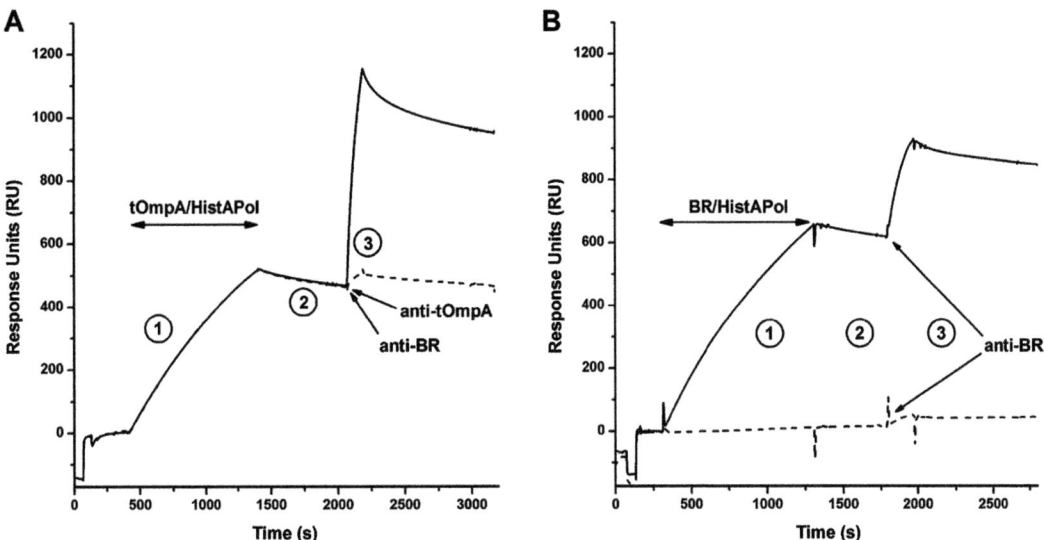

**Fig. 13.14** Detection of specific antibody binding to membrane proteins immobilized via HistAPol. **A.** The Ni:NTA surface was reacted with tOmpA/HistAPol complexes ①, washed with surfactant-free buffer ②, after which 30 nM of antibody raised against either tOmpA (—) or BR (- - -) were injected ③. **B.** The Ni:NTA surface was exposed (—) or not (- - -) to BR/HistAPol complexes ①, washed with surfactant-free buffer ②, after which 30 nM of anti-BR antibody were injected ③ (Reprinted with permission from Giusti et al. 2015, © 2015 American Chemical Society).

In summary, Study 13.7 validates the use of APols derivatized with polyhistidine tags for immobilizing specifically and reversibly untagged MPs onto solid supports carrying metal ions. Compared to MPs carrying a single His-tag, the presence of at least four $His_6$-tags per MP/HistAPol complex, distributed around the transmembrane domain of the protein, increases the affinity of the complexes for Ni:NTA-coated SPR chips, which results in a higher surface density of immobilized MPs and a more stable interaction. HistAPol-mediated immobilization provides an attachment of MPs to Ni:NTA surfaces almost as stable as that mediated by the biotin/avidin couple, with the major advantage of the reusability of the SPR biosensing chips. Thereby, HistAPol widens the spectrum of applications that require both a high stability of the target MPs and the reversibility of their immobilization.

## 13.3.2 Imidazole-Tagged Amphipols

Synthesizing HistAPol is costly and work-intensive, particularly because it is not straightforward to prepare the hexahistidine peptide in good yield (Giusti et al. 2015). HistAPol therefore is better adapted to applications requiring small amounts of polymer, such as immobilizing MPs onto chips for the purpose of ligand screening, than for uses that would require mass production. This led to investigating the synthesis and properties of a version of A8-35 onto which single imidazole moieties would be grafted at random (Study 13.8). The expectation was that, statistically, some pairs of imidazole groups would come close enough to interact simultaneously with immobilized $Ni^{2+}$ or $Co^{2+}$ cations. Following a procedure described in Chap. 4 (§ 4.4.1, Fig. 4.30A), 1-(3-aminopropyl)imidazole moieties were grafted onto A8-75 concomitantly with isopropylamine ones (Fig. 13.15). The level of grafting of the resulting "ImidAPol" reached ~6 imidazole moieties per 100 acrylic acid units, or ~20 per 40 kDa particle.

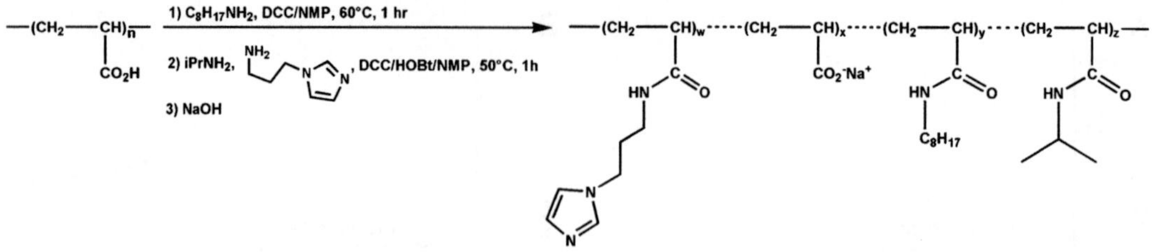

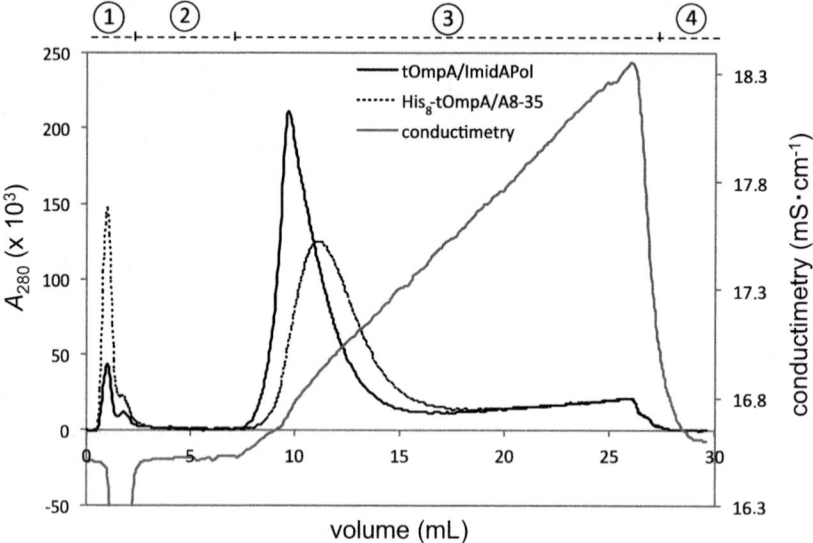

**Fig. 13.15** Synthesis of a derivative of A8-35 carrying randomly distributed imidazole moieties (ImidAPol). The synthesis was carried out according to the classical poly(acrylic acid) (PAA) hydrophobization procedure used to synthesize A8-35 (Gohon et al. 2004, 2006; cf. Chap. 4, Fig. 4.2). Briefly, amines were grafted in two steps onto a PAA precursor in the presence of the dicyclohexylcarbodiimide (DCC) coupling reagent with $N$-methyl-2-pyrrolidone (NMP) as solvent. Octylamine was coupled in Step 1, yielding A8-75; in Step 2, 1(3-aminopropyl)imidazole and isopropylamine were incorporated concomitantly. Cycles of solubilization-precipitation performed in basic and acidic aqueous media yielded ImidAPol after dialysis and freeze-drying (F. Giusti, unpublished data).

**Fig. 13.16** Immobilization of tOmpA/ImidAPol and $His_8$-tOmpA/A8-35 onto and elution from an Ni: NTA affinity chromatography column. Two versions of tOmpA, with and without a $His_8$-tag fused at the $N$-terminus, were trapped in plain A8-35 or in ImidAPol, respectively. Each sample (0.25 mg of protein) was injected onto a 1-mL HisTrap FF column (GE Healthcare) pre-equilibrated with 0.02 M Tris/HCl, pH 8.0, 0.15 M NaCl. After loading the sample ① and washing out the unbound material ②, elution was carried out with a linear gradient of imidazole (from 0 to 0.3 M; total volume 20 mL) ③ before washing the column ④. Protein elution was monitored at 280 nm (M. Zoonens, unpublished data).

The solution properties of ImidAPols were characterized. It self-associates into particles similar to those formed by A8-35. However, as already observed for HistAPol (Giusti et al. 2015; Chap. 4, § 4.4.2), ImidAPol does not precipitate efficiently at any low pH, because the protonation of both carboxylate and imidazole moieties turns it from a polyanionic to, depending on the pH, a zwitterionic or a cationic polymer. ImidAPol was as efficient as A8-35 at trapping MPs and keeping them soluble (C. Le Bon, F. Giusti & M. Zoonens, unpublished data). As shown in Fig. 13.16, MP/ImidAPol complexes can be immobilized onto a nickel-loaded NTA column and released by flushing it with an imidazole solution.

The applications of ImidAPol remain to be explored. One could consider, for instance, using it for selectively extracting MPs from a crude supernatant that also contains soluble proteins or for immobilizing a purified MP onto Ni:NTA beads with the view of screening drugs or identifying or purifying biological molecules that bind to the immobilized MP. As mentioned above, the presence of many imidazole moieties implies that, at acidic pH, such as exists in the lumen of endosomes, ImidAPol particles will become zwitterionic. As discussed in Chap. 15 (§ 15.2.3), this could be of interest for the cytosolic release of drugs or in the formulation of vaccines.

### 13.3.3 Oligonucleotide-Tagged Amphipol

In Study 13.9, Christel Le Bon and colleagues examined the possibility of functionalizing A8-35 with oligodeoxynucleotides (ODNs) (Le Bon et al. 2014a). The reversible immobilization of MPs via an ODN tag presents major assets for the regeneration and reconfiguration of biosensors and biochip surfaces. In addition, ODNs offer rich possibilities of modulating the stability and specificity of the attachment, by playing on their length and sequence, and they lend themselves particularly well to multiplexing. Indeed, they provide a chemically mild method for the site-selective immobilization of multiple MPs onto solid supports: traditional, well-established deoxyribonucleic acid (DNA) microarrays could be easily adapted as immobilization platforms. Previous studies have demonstrated sequence-specific and site-selective immobilization of semisynthetic protein-ODN conjugates onto surfaces modified with complementary nucleotide sequences (Zdyrko et al. 2003; Bano et al. 2009). However, these methods require protein engineering or chemical modification and cannot be easily adapted to MPs, which tend to be fragile and hard to handle. An alternative strategy is to first reconstitute the target MP into liposomes containing a cholesterol-modified ODN and then bind the proteoliposomes onto surfaces modified with the complementary ODN (cODN) (Bailey et al. 2009). Reconstitution of MPs into artificial vesicles, however, is a delicate, protein-specific process, and its optimization is time-consuming, whereas trapping MPs with APols is straightforward.

In Study 13.9 is described a route for synthesizing and purifying ODN-grafted A8-35, as well as the solution properties of the well-behaved particles into which the resulting "OligAPol" assembles (see Chap. 4, § 4.4.2, Fig. 4.31). The ability of OligAPol to trap and anchor MPs onto cODN-carrying surfaces was demonstrated using two different types of solid supports, magnetic beads and gold NPs. In both cases, adsorption was shown to be specifically mediated by the formation of the DNA duplex. Magnetic beads gave access to measuring adsorption yields, gold NPs to the homogeneity and reversibility of the adsorption. Anti-BR antibodies were shown to bind specifically to MP/OligAPol complexes immobilized onto gold NPs via the cODN (Le Bon et al. 2014a).

Figure 13.17 illustrates some of these experiments. Gold NPs immobilized onto poly-L-lysine-modified glass slides were used to (i) explore the possibility of using OligAPol for anchoring MPs onto gold surfaces and (ii) directly observe, by fluorescence microscopy, the hybridization of MP/OligAPol complexes and cODN onto solid supports and the binding of antibodies to immobilized MPs. The strong electrostatic interaction between the positively charged polylysine and the negatively charged surface of cODN-coated gold NPs ensured a stable immobilization. Differential interference contrast images showed well-dispersed particles (Fig. 13.17A). BR was trapped in a 1:1 mixture of OligAPol and Alexa Fluor 647-labeled A8-35 (FAPol$_{AF647}$; Fernandez et al. 2014; see Chap. 4, Table 4.5). After incubation with these complexes, cODN-functionalized gold NPs exhibited a strong and homogeneous fluorescence signal (Fig. 13.17B). Single-particle analysis of the fluorescence images showed a ratio between the average signal recorded on gold NPs and the background of $\sim$10 (Fig. 13.17G), indicative of a low level of non-specific binding of the BR/OligAPol/FAPol$_{AF647}$ complexes and free OligAPol/FAPol$_{AF647}$ particles to the polylysine-coated glass. Similar results were obtained upon incubation of

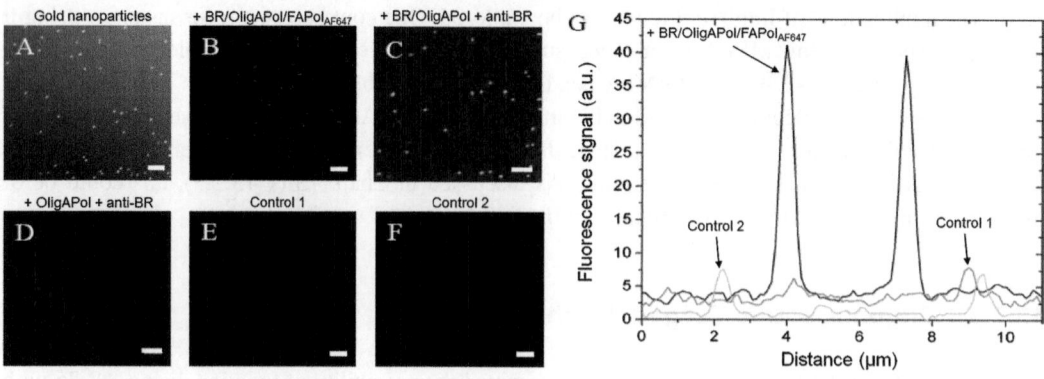

**Fig. 13.17** Immobilization of bacteriorhodopsin (BR) via OligAPol hybridization with a thiol-tagged cODN adsorbed onto gold nanoparticles (NPs) and specific recognition by an anti-BR antibody. (**A**) Differential interference contrast image of gold NPs (100 nm diameter) deposited on a polylysine-coated glass slide and washed with 10 mM potassium phosphate buffer, pH 7.2 (PPB). (**B**) Fluorescence image ($\lambda_{\text{exc}} = 620 \pm 30$ nm, detection at $700 \pm 37$ nm) of gold NPs preincubated with cODN, washed with PPB, and incubated for 2 h with BR trapped in a 1:1 OligAPol/ FAPol$_{\text{AF647}}$ mixture. (**C**) and (**D**) Fluorescence images ($\lambda_{\text{exc}} = 470 \pm 20$ nm, detection at $525 \pm 25$ nm) of gold NPs preincubated with cODN, washed with PPB, and incubated for 2 h either with BR trapped in a 1:1 OligAPol/ FAPol$_{\text{AF647}}$ mixture (C) or, as a control, with the OligAPol/FAPol$_{\text{AF647}}$ mixture alone (D), after which both samples were washed with PPB and incubated for 1 h with a BR-specific antibody labeled with Alexa Fluor 488. The fluorescence image showed a much smaller signal when BR/OligAPol/FAPol$_{\text{AF647}}$ complexes were incubated with particles coated with noncomplementary poly-C DNA strands ((**E**); Control 1) or with gold NPs functionalized with the cODN, but hybridized with an excess of Cy3-labeled ODN prior to the addition of the BR/OligAPol/FAPol$_{\text{AF647}}$ complexes ((**F**); Control 2). Brightness and contrast settings were the same for each set of images taken at the same wavelength (**B, E, F**, on the one hand, and **C, D**, on the other). Scale bar is 5 μm. (**G**) Line scans extracted from Panels **B, E**, and **F**. Each scan went through two representative NPs (From Le Bon et al. 2014, © 2014 Oxford University Press).

cODN-functionalized beads with another MP, tOmpA, trapped with the same 1:1 OligAPol/ FAPol$_{\text{AF647}}$ mixture (Le Bon et al. 2014a).

The presence of the MPs at the surface of gold NPs was confirmed by incubation with Alexa Fluor 488- or Alexa Fluor 647-labeled antibodies directed against BR (Fig. 13.17C) or against tOmpA (Le Bon et al. 2014a). Several control experiments were designed to confirm the specificity of the interaction between cODN-coated gold NPs and OligAPol-trapped MPs. No binding was observed, for instance, when gold NPs were coated with thiolated polycytosine, which is not complementary to the ODN (Control 1, Fig. 13.17E, G), nor onto cODN-coated gold NPs that had been preincubated with an excess of free, Cy3-labeled ODN (Control 2, Fig. 13.17F, G), hybridization of the free ODN being confirmed by the Cy3 fluorescence signal. The FAPol$_{\text{AF647}}$ signals measured in these control experiments were much lower ($\sim$10%) and barely emerged from the background noise (Fig. 13.17G).

Immobilization of MPs via OligAPols offers the advantage of being reversible. Indeed, most complexes were released from gold NPs that had been functionalized with cODN and BR trapped in OligAPol/FAPol$_{\text{AF647}}$ when the beads were exposed for 10 min to 8 M urea or when they were brought for 10 min under shaking at 72 °C, that is above the estimated melting temperature of the duplex (Le Bon et al. 2014a).

OligAPols are highly complementary to biotinylated or polyhistidine-tagged APols: their binding is strong, but it is reversible, the melting temperature can be modulated at will by playing on the sequence of the ODN and/or the cODN, and it features an exceptional flexibility in choosing complementary partners: by hybridizing OligAPol-trapped MPs with di-block oligonucleotides, one part of which recognizes the OligAPol and the other has any desirable sequence, each individual MP

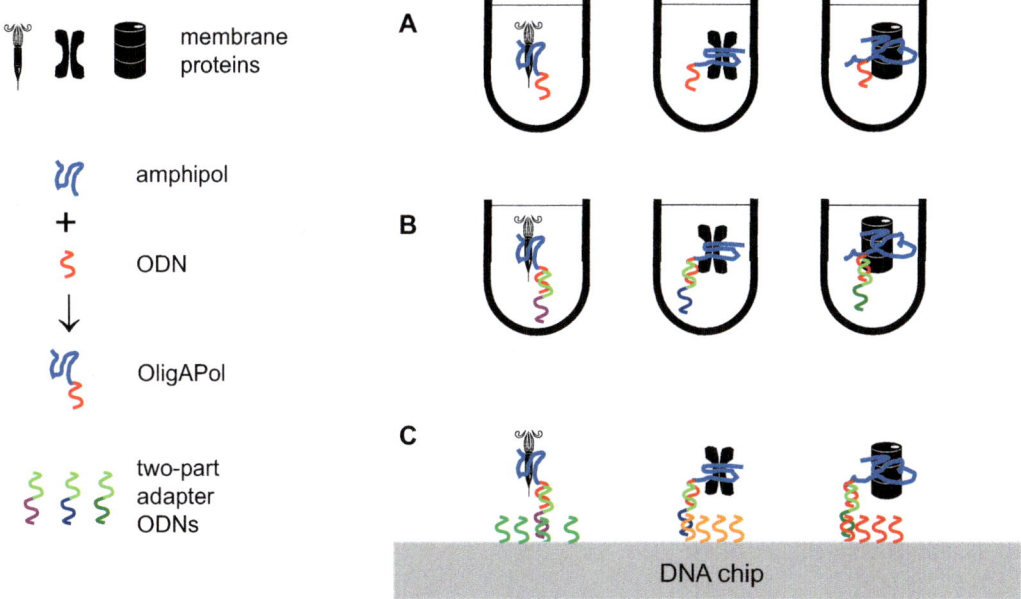

**Fig. 13.18** A strategy for targeting multiple membrane proteins (MPs) to specific spots on a DNA chip using a single oligonucleotide (ODN)-tagged amphipol (OligAPol). In Step (**A**), each individual MP is trapped with the same OligAPol. In Step (**B**), each complex is combined with a specific two-part adapter ODN, one part of which recognizes the ODN carried by the OligAPol, whereas the other is different for each protein. In Step (**C**), the complexes are applied, simultaneously or successively, to a DNA chip, the free part of the adapters directing each protein to a specific spot. The application must be rapid and the conditions optimized to prevent the exchange of OligAPols between MPs during the time they can collide with each other, which can be achieved by appropriately modulating the ionic strength (Zoonens et al. 2007; see Chap. 5, § 5.7). Exchange becomes impossible once the complexes are immobilized and the chip has been washed with surfactant-free buffer.

can be directed to a specific DNA sequence, which is particularly suitable for multiplexing (Fig. 13.18). The fact that MP/OligAPol complexes readily adsorb onto appropriately functionalized gold surfaces, and that the surfaces can be regenerated, offers rich perspectives for the construction of nanobiosensors. OligAPols could also conceivably be used to organize one or more MPs in space by attaching them at specific points of DNA filaments or 2D or 3D lattices (see e.g. LaBean and Li 2007; Selmi et al. 2011). This approach could be of use in structural biology, for example for electron microscopy (EM), or for measuring residual dipolar coupling in NMR experiments (cf. Bellot et al. 2013), as well as for building bioreactors (cf. Nowaczyk et al. 2004). Finally, it is worth noting that the ODN used for the present study, CpG-1826, was chosen because it is an efficient adjuvant in vaccines (Moyle and Toth 2013). Formulating vaccines with immunogenic MP/OligAPol complexes so as to improve the level of protection they confer is an intriguing potential development of ongoing attempts at using APol-stabilized MPs for vaccination (see Tifrea et al. 2018a, b and Chap. 15, § 15.2.3).

### 13.3.4   Sulfide- and Sulfhydryl-Tagged Amphipols

OligAPols are extremely polyvalent, but their synthesis is somewhat demanding. For some applications, such as nontargeted adsorption onto gold surfaces, simpler structures could suffice and even present certain advantages. A noted in Chap. 4, § 4.4.1, synthesizing a thiol-bearing derivative of

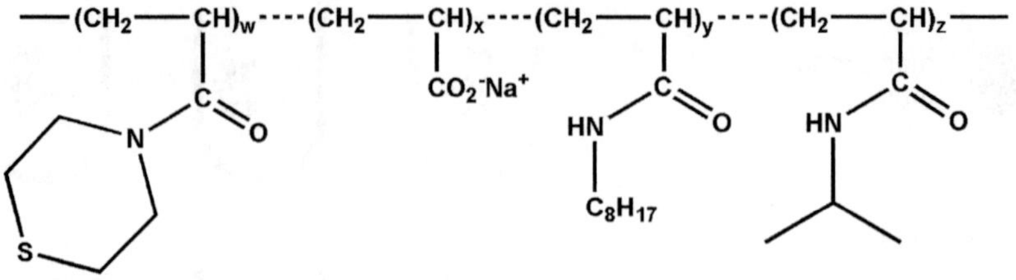

**Fig. 13.19** Chemical structure of SulfidAPol (F. Giusti, unpublished results).

A8-35 in good yield turned out to be a difficult and frustrating endeavor. Sulfides, which are chemically much easier to handle and also adsorb onto gold or silver surfaces, albeit with a lower affinity (see e.g. Troughton et al. 1988; Lenk et al. 1993; Beulen et al. 1996; Jung et al. 1998; Lim et al. 2006), offer an interesting alternative.

This has led to the development of a variant of A8-35 tagged with thiomorpholine (Fig. 13.19; F. Giusti, unpublished results). This novel tagged APol, dubbed "SulfidAPol", is still in the course of validation; it would be premature to draw at length on its putative applications. However, one may note that it could conceivably be of use:

   (i) For adsorbing MPs onto gold chips or beads, e.g. for SPR or fluorescence microscopy applications, as described above for OligAPols hybridizing with a thiol-tagged cODN adsorbed onto gold NPs, but at a much lower cost;
   (ii) For adsorption onto gold NPs in the formulation of vaccines (see Chap. 15, § 15.2.3);
   (iii) For surface-enhanced Raman spectroscopy (SERS), where it is essential that the target protein be as close as possible to the gold or silver surface (see Polovinkin et al. 2014, and Chap. 8, § 8.2.2);
   (iv) For electrochemistry, where it would ensure immediate proximity between MPs and supporting electrodes;
   (v) For labeling the APol belt with gold NPs in EM studies (cf. § 13.4).

A version of SMA carrying thiol moieties has been recently obtained by grafting cysteamine onto the polymer via its amine function (Lindhoud et al. 2016). Its application to immobilizing SMALP-trapped MPs has not yet been reported.

## 13.4    Other Applications of Tagged Amphipols

Up till now, tagged APols have been used essentially for immobilizing MPs onto solid surfaces for the sake, mostly, of ligand detection. As mentioned at the end of § 13.3.3, OligAPol-mediated immobilization onto DNA scaffolds could also be considered, e.g. for EM or NMR applications. ImidAPol, HistAPol, and SulfidAPol could prove useful in the formulation of vaccines (see Chap. 15, § 15.2.3). ImidAPol could be used to separate MPs from soluble proteins (§ 13.3.2). Other applications are conceivable. One of them is to locate the APol belt surrounding MPs in low- and medium-resolution EM studies. This could possibly be achieved by labeling a SulfidAPol or SMA-SH belt with gold nanoparticles (cf. Suárez-Suárez et al. 2013, where this approach has been used to visualize pores in films formed of block copolymers functionalized with thiomorpholine). The use of tagged APols for EM has been validated with a biotinylated APol, BA8-35, whose localization in single-particle images of MP/APol complexes was visualized, thanks to the binding of monovalent streptavidin (Perry et al.

2018) (see Chap. 12, § 12.3.4.3). Incorporation of MPs into bioelectronic devices is a very active field (see e.g. Mahyad et al. 2015; Park et al. 2015; Kumar et al. 2016, and references therein), in which APols, in particular, tagged APols, have probably a part to play (cf. Pérez-Mitta et al. 2016).

## 13.5 Protocol 13.1. Amphipol-Mediated Immobilization of Membrane Proteins for Surface Plasmon Resonance Experiments

Surface plasmon resonance (SPR) (see Rich and Myszka 2005) is a label-free technique based on the propagation of an evanescent wave along a gold-coated surface. It permits the detection of changes of refractive index in the vicinity of the surface that result from the binding of partners. This approach requires the immobilization of one of the interacting partners onto the surface. Various affinity tags can be used for protein immobilization, including polyhistidine tags, which bind to Ni:NTA motifs, affinity tags recognized by antibodies, and biotin or streptags binding to streptavidin. Functionalized APols fulfill several criteria for a specific immobilization of functional MPs onto surfaces: (i) they stabilize MPs, providing a longer lifetime to the isolated protein, (ii) they form with MPs permanent complexes, which do not dissociate in the absence of free APols; biosensors can thus be used in surfactant-free buffers, which simplifies their implementation and considerably limits the risk of missing weak or moderate interactions with ligands, and, finally, (iii) the complexes formed between MPs and functionalized APols will bind onto the solid surface of the biosensor without the need for any genetic or chemical modification of the protein (Charvolin et al. 2009).

The protocol described here is for MPs trapped in biotinylated A8-35 (BAPol), but it can be adapted to MPs trapped in other functionalized APols, such as HistAPol, ImidAPol, or OligAPol. If the MP possesses a tag, its trapping with unfunctionalized APol, e.g. plain A8-35, will improve its stability, and the immobilization of the complexes can be carried out following the same approach. This protocol is adapted for Biacore instruments (GE Healthcare), which are the most widely distributed commercial instruments. Specific conditions are indicated in the case of a Biacore X100 instrument. The signal can be optimized by adjusting experimental conditions (Karlsson and Fält 1997). For more details about data analysis, see Rich and Myszka (2005). In the following, comments and tips are in italics and marked with a pointing hand sign (☞).

### 13.5.1 Experimental Setup

MP/BAPols complexes (called ligands in Biacore terminology) are immobilized onto the Biacore sensor chip before addition of the interacting molecules, for instance, antibodies (called analyte in Biacore terminology), to reveal specific interactions.

☞*Choose the adequate sensor chips. The binding capacity of most Biacore chips is improved by pre-coupling of a dextran matrix. CM5 chips, which are the most widely used, exist already premodified for coupling of various proteins. If the immobilization is carried out via BAPols, the SA sensor chip (CM5 chips premodified with streptavidin) is used. Note that alternatives to SA sensor chips are the use of a standard CM5 chip, followed by aminocoupling of streptavidin, or the SA Capture chip, which is regenerable. In the case of charge repulsion between the sample to be immobilized and the surface, it is possible to improve the detection signal by using commercial chips with shorter dextran matrices. Note that two flow cells are required for the measurements: the sample flow cell, where the total signal will result from specific and non-specific binding onto surfaces, and the reference flow cell, used to evaluate the non-specific binding of proteins onto surfaces.*

## 13.5.2   Preparation of Samples

☞*Note that it is important to exclude compounds that affect the refractive index of the solutions (e.g. glycerol). All samples and buffers should be filtered on 0.22-μm filters or centrifuged 10 min at 10,000 × g.*

- Trap the protein in BAPol as described in Chap. 5, Protocol 5.1. The final protein concentration to use for a good signal depends on the protein under study. For example, in the case of BR, the protein concentration should be ~0.03 g·L$^{-1}$. Typical quantities required for the experiments are 0.06–0.1 μg of total protein.
- Prepare the solutions of antibodies (a non-specific antibody and a specific one for negative and positive controls, respectively), both solutions at two different concentrations (0.055 and 0.01 g·L$^{-1}$). Typical quantities required for the experiments are 6–10 μg of total protein. Note that these quantities are indicative and should be optimized depending on sample purity.
- Prepare the running buffer, HBS-N, which contains 10 mM HEPES pH 7.4, 150 mM NaCl. Note that the dextran matrix and the APols are both negatively charged, and so the ionic strength of buffer can be optimized for an optimal signal with minimal repulsion.
- Prepare the surface regeneration solution (1 M NaCl, 50 mM NaOH).

## 13.5.3   Measurements

The signal observed during an SPR experiment is proportional to the refractive index, which itself is proportional to the mass of molecules with the same refractive index. All proteins have approximately the same refractive index, and the empirical conversion factor (1 RU ≈ 1 pg·mm$^{-2}$) can be used for estimating the mass bound to the surface (Stenberg et al. 1991).

- Follow the instructions of the manufacturer for conditioning the sensor surface.
- After obtaining a stable baseline, immobilize the proteins onto the sensor surface, which becomes the sample flow cell. For this step, start a new cycle using only the second flow cell and inject the sample. In the case of qualitative analysis, it is recommended to try to saturate the surface with the MP/BAPol complexes so as to obtain the highest possible response from antibody binding in the next step. This can be achieved by repeating consecutive injections until stabilization of the signal (Note: for a flow rate of 10 μL·min$^{-1}$, use consecutive 60-s injections of MP/BAPol complexes at 0.03 g·L$^{-1}$ to saturate the surface). In the case of kinetic studies, it is usually preferable to limit the density of immobilized ligand to avoid diffusion-limited data, analyte depletion, and analyte rebinding.
- Prepare the reference flow cell (the first flow cell), which can either be left unmodified or be modified with either BAPol alone or a different MP trapped in BAPol.
- Stop the manual run after obtaining a stable signal in both cells.
- Measure the binding of the analyte, for instance, that of an antibody raised against the MP under study. It is recommended to use two different concentrations of antibodies (0.05 g·L$^{-1}$ for the first injection and 0.01 g·L$^{-1}$ for the second one) and one regeneration solution in each cycle. The solutions will be added according to the following series of the cycles:

  – Cycle 1: blank cycle with running buffer
  – Cycle 2: cycle with a non-specific antibody to be used for negative control
  – Cycle 3: cycle with a specific antibody raised against the MP under study

The sample parameters are: 30 μL·min$^{-1}$ flow rate, 180 s contact time, and 120 s dissociation time. The regeneration parameters are: 30 μL·min$^{-1}$ flow rate and 60 s contact time, wash in running buffer.

*Protocol adapted by Manuela Zoonens from* Zoonens et al. (2014)

# References

Ansar, S.M., Haputhanthri, R., Edmonds, B., Liu, D., Yu, L., Sygula, A., Zhang, D. (2011) Determination of the binding affinity, packing, and conformation of thiolate and thione ligands on gold nanoparticles. *J. Phys. Chem. C* **115**:653–660.

Arora, A., Abildgaard, F., Bushweller, J.H., Tamm, L.K. (2001) Structure of outer membrane protein A transmembrane domain by NMR spectroscopy. *Nat. Struct. Biol.* **8**:334–338.

Bailey, K., Bally, M., Leifert, W., Vörös, J., McMurchie, T. (2009) G protein-coupled receptor array technologies: Site directed immobilisation of liposomes containing the $H_1$-histamine or $M_2$-muscarinic receptors. *Proteomics* **9**:2052–2063.

Bain, C.D., Biebuyck, H.A., Whitesides, G.M. (1989a) Comparison of self-assembled monolayers on gold: Coadsorption of thiols and disulfides. *Langmuir* **5**:723–727.

Bain, C.D., Evall, J., Whitesides, G.M. (1989b) Formation of monolayers by the coadsorption of thiols on gold: Variation in the head group, tail group, and solvent. *J. Am. Chem. Soc.* **111**:7155–7164.

Bano, F., Fruk, L., Sanavio, B., Glettenberg, M., Casalis, L., Niemeyer, C.M., Scoles, G. (2009) Toward multiprotein nanoarrays using nanografting and DNA-directed immobilization of proteins. *Nano Lett.* **9**:2614–2618.

Basit, H., Sharma, S., Van der Heyden, A., Gondran, C., Breyton, C., Dumy, P., Winnik, F.M., Labbé, P. (2012) Amphipol-mediated surface immobilization of FhuA: a platform for label-free detection of the bacteriophage protein pb5. *Chem. Commun.* **48**:6037–6039.

Bellot, G., McClintock, M.A., Chou, J.J., Shih, W.M. (2013) DNA nanotubes for NMR structure determination of membrane proteins. *Nat. Protoc.* **8**:755–770.

Berry, E.A., Guergova-Kuras, M., Huang, L.-S., Crofts, A.R. (2000) Structure and function of cytochrome $bc_1$ complexes. *Annu. Rev. Biochem.* **69**:1005–1075.

Beulen, M.W.J., Huisman, B.-H., van der Heijden, P.A., van Veggel, F.C.J.M., Simons, M.G., Biemond, E.M.E.F., de Lange, P.J., Reinhoudt, D.N. (1996) Evidence for nondestructive adsorption of dialkyl sulfides on gold. *Langmuir* **12**:6170–6172.

Bidan, G., Billon, M., Galasso, K., Livache, T., Mathis, C., Roget, A., Torres-Rodriguez, L.M., Vieil, E. (2000) Electropolymerization as a versatile route for immobilizing biological species onto surfaces – Application to DNA biochips. *Appl. Biochem. Biotechnol.* **89**:183–193.

Bieri, C., Ernst, O.P., Heyse, S., Hofmann, K.P., Vogel, H. (1999) Micropatterned immobilization of a G protein-coupled receptor and direct detection of G protein activation. *Nat. Biotech.* **17**:1105–1108.

Binz, H.K., Stumpp, M.T., Forrer, P., Amstutz, P., Plückthun, A. (2003) Designing repeat proteins: well-expressed, soluble and stable proteins from combinatorial libraries of consensus ankyrin repeat proteins. *J. Mol. Biol.* **332**:489–503.

Blanchard, S.G., Quast, U., Reed, K., Lee, T., Schimerlik, M.I., Vandlen, R., Claudio, T., Strader, C.D., Moore, H.P., Raftery, M.A. (1979) Interaction of $^{125}$I-α-bungarotoxin with acetylcholine receptor from *Torpedo californica*. *Biochemistry* **18**:1875–1883.

Block, H., Maertens, B., Spriestersbach, A., Brinker, N., Kubicek, J., Fabis, R., Labahn, J., Schäfer, F. (2009) Immobilized-metal affinity chromatography (IMAC): a review. *Methods Enzymol.* **463**:439–473.

Boersma, Y.L., Plückthun, A. (2011) DARPins and other repeat protein scaffolds: advances in engineering and applications. *Curr. Opin. Biotechnol.* **22**:849–857.

Catoire, L.J., Damian, M., Giusti, F., Martin, A., van Heijenoort, C., Popot, J.-L., Guittet, E., Banères, J.-L. (2010) Structure of a GPCR ligand in its receptor-bound state: leukotriene $B_4$ adopts a highly constrained conformation when associated to human BLT2. *J. Am. Chem. Soc.* **132**:9049–9057.

Champeil, P., Menguy, T., Tribet, C., Popot, J.-L., le Maire, M. (2000) Interaction of amphipols with the sarcoplasmic reticulum $Ca^{2+}$-ATPase. *J. Biol. Chem.* **275**:18623–18637.

Charvolin, D., Perez, J.-B., Rouvière, F., Giusti, F., Bazzacco, P., Abdine, A., Rappaport, F., Martinez, K.L., Popot, J.-L. (2009) The use of amphipols as universal molecular adapters to immobilize membrane proteins onto solid supports. *Proc. Natl. Acad. Sci. USA* **106**:405–410.

Christman, K.L., Enriquez-Rios, V.D., Maynard, H.D. (2006a) Nanopatterning proteins and peptides. *Soft Matter* **2**:928–939.

Christman, K.L., Requa, M.V., Enriquez-Rios, V.D., Ward, S.C., Bradley, K.A., Turner, K.L., Maynard, H.D. (2006b) Submicron streptavidin patterns for protein assembly. *Langmuir* **22**:7444–7450.

Cosnier, S., Galland, B., Gondran, C., Le Pellec, A. (1998) Electrogeneration of biotinylated functionalized polypyrroles for the simple immobilization of enzymes. *Electroanalysis* **10**:808–813.

Cosnier, S., Stoytcheva, M., Senillou, A., Perrot, H., Furriel, R.P.M., Leone, F.A. (1999) A biotinylated conducting polypyrrole for the spatially controlled construction of an amperometric biosensor. *Anal. Chem.* **71**:3692–3697.

Coyer, S.R., García, A.J., Delamarche, E. (2007) Facile preparation of complex protein architectures with sub-100-nm resolution on surfaces. *Angew. Chem. Int. Ed.* **46**:6837–6840.

Dahmane, T., Damian, M., Mary, S., Popot, J.-L., Banères, J.-L. (2009) Amphipol-assisted *in vitro* folding of G protein-coupled receptors. *Biochemistry* **48**:6516–6521.

Darmanin, T., Bellanger, H., Guittard, F., Lisboa, P., Zurn, M., Colpo, P., Gilliland, D., Rossi, F. (2012) Structured biotinylated poly(3,4-ethylenedioxypyrrole) electrodes for biochemical applications. *RSC Adv.* **2**:1033–1039.

Della Pia, E.A., Holm, J., Lloret, N., Le Bon, C., Popot, J.-L., Zoonens, M., Nygård, J., Martinez, K.L. (2014a) A step closer to membrane protein multiplexed nano-arrays using biotin-doped polypyrrole. *ACS Nano* **8**:1844–1853.

Della Pia, E.A., Westh Hansen, R., Zoonens, M., Martinez, K.L. (2014b) Functionalized amphipols: A versatile toolbox suitable for applications of membrane proteins in synthetic biology. *J. Membr. Biol.* **247**:815–826.

Fang, Y., Frutos, A.G., Lahiri, J. (2002) Membrane protein microarrays. *J. Am. Chem. Soc.* **124**:2394–2395.

Fernandez, A., Le Bon, C., Baumlin, N., Giusti, F., Crémel, G., Popot, J.-L., Bagnard, D. (2014) *In vivo* characterization of the biodistribution profile of amphipols *J. Membr. Biol.* **247**:1043–1051.

Ferrandez, Y., Dezi, M., Bosco, M., Urvoas, A., Valério, M., Le Bon, C., Giusti, F., Broutin, I., Durand, G., Polidori, A., Popot, J.-L., Picard, M., Minard, P. (2014) Amphipol-mediated screening of molecular orthoses specific for membrane protein targets. *J. Membr. Biol.* **247**:925–940.

Franklin, G.I., Potter, L.T. (1972) Studies of the binding of α-bungarotoxin to membrane-bound and detergent-dispersed acetylcholine receptors from *Torpedo* electric tissue. *FEBS Lett.* **28**:101–106.

Friedrich, M.G., Giess, F., Naumann, R., Knoll, W., Ataka, K., Heberle, J., Hrabakova, J., Murgida, D.H., Hildebrandt, P. (2004) Active site structure and redox processes of cytochrome *c* oxidase immobilised in a novel biomimetic lipid membrane on an electrode. *Chem. Commun.* **21**:2376–2377.

Früh, V., IJzerman, A.P., Siegal, G. (2011) How to catch a membrane protein in action: a review of functional membrane protein immobilization strategies and their applications. *Chem. Rev.* **111**:640–656.

Giusti, F., Kessler, P., Westh Hansen, R., Della Pia, E.A., Le Bon, C., Mourier, G., Popot, J.-L., Martinez, K.L., Zoonens, M. (2015) Synthesis of a polyhistidine-bearing amphipol and its use for immobilizing membrane proteins. *Biomacromolecules* **16**:3751–3761.

Giusti, F., Rieger, J., Catoire, L., Qian, S., Calabrese, A.N., Watkinson, T.G., Casiraghi, M., Radford, S.E., Ashcroft, A. E., Popot, J.-L. (2014) Synthesis, characterization and applications of a perdeuterated amphipol. *J. Membr. Biol.* **247**:909–924.

Gohon, Y., Giusti, F., Prata, C., Charvolin, D., Timmins, P., Ebel, C., Tribet, C., Popot, J.-L. (2006) Well-defined nanoparticles formed by hydrophobic assembly of a short and polydisperse random terpolymer, amphipol A8-35. *Langmuir* **22**:1281–1290.

Gohon, Y., Pavlov, G., Timmins, P., Tribet, C., Popot, J.-L., Ebel, C. (2004) Partial specific volume and solvent interactions of amphipol A8-35. *Anal. Biochem.* **334**:318–334.

Gottschalk, I., Li, Y.M., Lundahl, P. (2000) Chromatography on cells: analyses of solute interactions with the glucose transporter Glut1 in human red cells adsorbed on lectin-gel beads. *J. Chromatogr. B* **739**:55–62.

Gu, J.H., Yam, C.M., Li, S., Cai, C.Z. (2004) Nanometric protein arrays on protein-resistant monolayers on silicon surfaces. *J. Am. Chem. Soc.* **126**:8098–8099.

Guellouz, A., Valerio-Lepiniec, M., Urvoas, A., Chevrel, A., Graille, M., Fourati-Kammoun, Z., Desmadril, M., van Tilbeurgh, H., Minard, P. (2013) Selection of specific protein binders for pre-defined targets from an optimized library of artificial helicoidal repeat proteins (alphaRep). *PLoS One* **8**:e71512.

Harding, P.J., Hadingham, T.C., McDonnell, J.M., Watts, A. (2006) Direct analysis of a GPCR-agonist interaction by surface plasmon resonance. *Eur. Biophys. J. Biophys. Lett.* **35**:709–712.

Hochuli, E., Döbeli, H., Schacher, A. (1987) New metal chelate adsorbent selective for proteins and peptides containing neighbouring histidine residues. *J. Chromatogr.* **411**:177–184.

Hoff, J.D., Cheng, L.J., Meyhofer, E., Guo, L.J., Hunt, A.J. (2004) Nanoscale protein patterning by imprint lithography. *Nano Lett.* **4**:853–857.

Hong, Y.L., Webb, B.L., Su, H., Mozdy, E.J., Fang, Y., Wu, Q., Liu, L., Beck, J., Ferrie, A.M., Raghavan, S., Mauro, J., Carre, A., Mueller, D., Lai, F., Rasnow, B., Johnson, M., Min, H.S., Salon, J., Lahiri, J. (2005) Functional GPCR microarrays. *J. Am. Chem. Soc.* **127**:15350–15351.

Hoover, D.K., Lee, E.J., Chan, E.W.L., Yousaf, M.N. (2007) Electroactive nanoarrays for biospecific ligand mediated studies of cell adhesion. *ChemBioChem* **8**:1920–1923.

Huber, T., Steiner, D., Röthlisberger, D., Plückthun, A. (2007) *In vitro* selection and characterization of DARPins and Fab fragments for the co-crystallization of membrane proteins: The Na$^+$-citrate symporter CitS as an example. *J. Struct. Biol.* **159**:206–221.

Jonkheijm, P., Weinrich, D., Schröder, H., Niemeyer, C.M., Waldmann, H. (2008) Chemical strategies for generating protein biochips. *Angew. Chem. Int. Ed.* **47**:9618–9647.

Jung, C., Dannenberger, O., Xu, Y., Buck, M., Grunze, M. (1998) Self-assembled monolayers from organosulfur compounds: A comparison between sulfides, disulfides, and thiols. *Langmuir* **14**:1103–1107.

Kang, E., Park, J.-W., McClellan, S.J., Kim, J.-M., Holland, D.P., Lee, G.U., Franses, E.I., Park, K., Thompson, D.H. (2007) Specific adsorption of histidine-tagged proteins on silica surfaces modified with Ni$^{2+}$/NTA-derivatized poly(ethylene glycol). *Langmuir* **23**:6281–6288.

Kannan, B., Kulkarni, R.P., Majumdar, A. (2004) DNA-based programmed assembly of gold nanoparticles on lithographic patterns with extraordinary specificity. *Nano Lett.* **4**:1521–1524.

Karlsson, R., Fält, A. (1997) Experimental design for kinetic analysis of protein-protein interactions with surface plasmon resonance biosensors. *J. Immunol. Methods* **200**:121–133.

Knecht, S., Ricklin, D., Eberle, A.N., Ernst, B. (2009) Oligo-his-tags: mechanisms of binding to Ni$^{2+}$-NTA surfaces. *J. Mol. Recognit.* **22**:270–279.

Koide, S. (2009) Engineering of recombinant crystallization chaperones. *Curr. Opin. Struct. Biol.* **19**:449–457.

Kumar, S., Bagchi, S., Prasad, S., Sharma, A., Kumar, R., Kaur, R., Singh, J., Bhondekar, A.P. (2016) Bacteriorhodopsin–ZnO hybrid as a potential sensing element for low-temperature detection of ethanol vapour. *Beilstein J. Nanotechnol.* **7**:501–510.

LaBean, T.M., Li, H. (2007) Constructing novel materials with DNA. *Nano Today* **2**:26–35.

Laitinen, O.H., Nordlund, H.R., Hytonen, V.P., Kulomaa, M.S. (2007) Brave new (strept)avidins in biotechnology. *Trends Biotechnol.* **25**:269–277.

Lavrich, D.J., Wetterer, S.M., Bernasek, S.L., Scoles, G. (1998) Physisorption and chemisorption of alkanethiols and alkyl sulfides on Au(111). *J. Phys. Chem. B* **102**:3456–3465.

Le Bon, C., Della Pia, E.A., Giusti, F., Lloret, N., Zoonens, M., Martinez, K.L., Popot, J.-L. (2014a) Synthesis of an oligonucleotide-derivatized amphipol and its use to trap and immobilize membrane proteins. *Nucleic Acids Res.* **42**: e83.

Le Bon, C., Popot, J.-L., Giusti, F. (2014b) Labeling and functionalizing amphipols for biological applications. *J. Membr. Biol.* **247**:797–814.

Lee, K.B., Park, S.J., Mirkin, C.A., Smith, J.C., Mrksich, M. (2002) Protein nanoarrays generated by dip-pen nanolithography. *Science* **295**:1702–1705.

Leggett, G.J. (2007) Bionanofabrication by near-field optical methods. *NanoBiotechnology* **3**:223–240.

Lenk, T.J., Hallmark, V.M., Rabolt, J.F., Häussling, L., Ringsdorf, H. (1993) Formation and characterization of self-assembled films of sulfur-derivatized poly(methyl methacrylates) on gold. *Macromolecules* **26**:1230–1237.

Lieberman, R.L., Culver, J.A., Entzminger, K.C., Pai, J.C., Maynard, J.A. (2011) Crystallization chaperone strategies for membrane proteins. *Methods* **55**:293–302.

Lim, J.K., Kim, I.-H., Kim, K.-H., Shin, K.S., Kang, W., Choo, J., Joo, S.W. (2006) Adsorption of dimethyl sulfide and methanethiolate on Ag and Au surfaces: Surface-enhanced Raman scattering and density functional theory calculation study. *Chem. Phys.* **330**:245–252.

Lindhoud, S., Carvalho, V., Pronk, J.W., Aubin-Tam, M.E. (2016) SMA-SH: Modified styrene-maleic acid copolymer for functionalization of lipid nanodiscs. *Biomacromolecules* **17**:1516–1522.

Liu, Y.C., Rieben, N., Iversen, L., Sørensen, B.S., Park, J., Nygård, J., Martinez, K.L. (2010) Specific and reversible immobilization of histidine-tagged proteins on functionalized silicon nanowires. *Nanotechnology* **21**:245105.

Lukas, R.J., Morimoto, H., Hanley, M.R., Bennett, E.L. (1981) Radiolabeled α-bungarotoxin derivatives: kinetic interaction with nicotinic acetylcholine receptors. *Biochemistry* **20**:7373–7378.

Mahyad, B., Janfaza, S., Hosseini, E.S. (2015) Bio-nano hybrid materials based on bacteriorhodopsin: Potential applications and future strategies. *Adv. Coll. Interf. Sci.* **225**:194–202.

Majd, S., Mayer, M. (2008) Generating arrays with high content and minimal consumption of functional membrane proteins. *J. Am. Chem. Soc.* **130**:16060–16064.

Martinez, K.L., Gohon, Y., Corringer, P.-J., Tribet, C., Mérola, F., Changeux, J.-P., Popot, J.-L. (2002) Allosteric transitions of *Torpedo* acetylcholine receptor in lipids, detergent and amphipols: molecular interactions *vs.* physical constraints. *FEBS Lett.* **528**:251–256.

Martinez, K.L., Meyer, B.H., Hovius, R., Lundstrom, K., Vogel, H. (2003) Ligand binding to G protein-coupled receptors in tethered cell membranes. *Langmuir* **19**:10925–10929.

Mary, S., Damian, M., Rahmeh, R., Marie, J., Mouillac, B., Banères, J.-L. (2014) Amphipols in G protein-coupled receptor pharmacology: What are they good for? *J. Membr. Biol.* **247**:853–860.

Melosh, N.A., Boukai, A., Diana, F., Gerardot, B., Badolato, A., Petroff, P.M., Heath, J.R. (2003) Ultrahigh-density nanowire lattices and circuits. *Science* **300**:112–115.

Moyle, P.M., Toth, I. (2013) Modern subunit vaccines: development, components, and research opportunities. *ChemMedChem* **8**:360–376.

Nieba, L., Nieba-Axmann, S.E., Persson, A., Hämäläinen, M., Edebratt, F., Hansson, A., Lidholm, J., Magnusson, K., Frostell Karlsson, Å., Plückthun, A. (1997) BIACORE analysis of histidine-tagged proteins using a chelating NTA sensor chip. *Anal. Biochem.* **252**:217–228.

Niemeyer, C.M., Burger, W., Hoedemakers, R.M.J. (1998) Hybridization characteristics of biomolecular adaptors, covalent DNA-streptavidin conjugates. *Bioconjugate Chem.* **9**:168–175.

Nirthanan, S., Gwee, M.C.E. (2004) Three-finger α-neurotoxins and the nicotinic acetylcholine receptor, forty years on. *J. Pharmacol. Sci.* **94**:1–17.

Nowaczyk, M., Oworah-Nkruma, R., Zoonens, M., Rögner, M., Popot, J.-L. (2004) Amphipols: strategies for an improved PS2 environment in aqueous solution, in: Miyake, J., ed., Biohydrogen III. Elsevier, Dordrecht, The Netherlands, Kyoto, Japan, pp. 151–159.

Ostermeier, C., Iwata, S., Ludwig, B., Michel, H. (1995) $F_v$ fragment-mediated crystallization of the membrane protein bacterial cytochrome *c* oxidase. *Nature Struct. Biol.* **2**:842–846.

Park, S., Kang, Y.J., Majd, S. (2015) A review of patterned organic bioelectronic materials and their biomedical applications. *Adv. Mater.* **27**:7583–7619.

Patel, J.D., O'Carra, R., Jones, J., Woodward, J.G., Mumper, R.J. (2007) Preparation and characterization of nickel nanoparticles for binding to his-tag proteins and antigens. *Pharm. Res.* **24**:343–352.

Pautsch, A., Schulz, G.E. (2000) High-resolution structure of the OmpA membrane domain. *J. Mol. Biol.* **298**:273–282.

Pérez-Mitta, G., Burr, L., Tuninetti, J.S., Trautmann, C., Toimil-Molares, M.E., Azzaroni, O. (2016) Noncovalent functionalization of solid-state nanopores via self-assembly of amphipols. *Nanoscale* **8**:1470–1478.

Perez, J.B., Martinez, K.L., Segura, J.M., Vogel, H. (2006) Supported cell-membrane sheets for functional fluorescence imaging of membrane proteins. *Adv. Funct. Mater.* **16**:306–312.

Perlmutter, J.D., Popot, J.-L., Sachs, J.N. (2014) Molecular dynamics simulations of a membrane protein/amphipol complex. *J. Membr. Biol.* **247**:883–895.

Perry, T., Souabni, H., Rapisarda, C., Fronzes, R., Giusti, F., Popot, J.-L., Zoonens, M., Gubellini, F. (2018) Visualizing transmembrane regions of protein complexes by electron microscopy using biotinylated amphipols, *submitted for publication*.

Pervushin, K., Braun, D., Fernàndez, C., Wüthrich, K. (2000) [$^{15}$N,$^1$H]/[$^{13}$C,$^1$H]-TROSY for simultaneous detection of backbone $^{15}$N-$^1$H, aromatic $^{13}$C-$^1$H and side-chain $^{15}$N-$^1$H$_2$ correlations in large proteins. *J. Biomol. NMR* **17**:195–202.

Petrlova, J., Masarik, M., Potesil, D., Adam, V., Trnkova, L., Kizek, R. (2007) Zeptomole detection of streptavidin using carbon paste electrode and square-wave voltammetry. *Electroanalysis* **19**:1177–1182.

Pierre, Y., Breyton, C., Kramer, D., Popot, J.-L. (1995) Purification and characterization of the cytochrome $b_6f$ complex from *Chlamydomonas reinhardtii*. *J. Biol. Chem.* **270**:29342–29349.

Polovinkin, V., Balandin, T., Volkov, O., Round, E., Borshchevskiy, V., Utrobin, P., von Stetten, D., Royant, A., Willbold, D., Arzumanyan, A., Popot, J.-L., Gordeliy, V. (2014) Nanoparticle surface-enhanced Raman scattering of bacteriorhodopsin stabilized by amphipol A8-35. *J. Membr. Biol.* **247**:971–980.

Ravi, V., Binz, J.M., Rioux, R.M. (2013) Thermodynamic profiles at the solvated inorganic−organic interface: The case of gold-thiolate monolayers. *Nano Lett.* **13**:4442–4448.

Rich, R.L., Myszka, D.G. (2005) Survey of the year 2004 commercial optical biosensor literature. *J. Mol. Recognit.* **18**:431–478.

Schmid, E.L., Tairi, A.P., Hovius, R., Vogel, H. (1998) Screening ligands for membrane protein receptors by total internal reflection fluorescence: The 5-HT$_3$ serotonin receptor. *Anal. Chem.* **70**:1331–1338.

Seeger, M.A., Zbinden, R., Flutsch, A., Gutte, P., Engeler, S., Roschitzki-Voser, H., Grütter, M.G. (2013) Design, construction, and characterization of a second-generation DARPin library with reduced hydrophobicity. *Prot. Sci.* **22**:1239–1257.

Selmi, D.N., Adamson, R.J., Attrill, H., Goddard, A.D., Gilbert, R.J.C., Watts, A., Turberfield, A.J. (2011) DNA-templated protein arrays for single-molecule imaging. *Nano Lett.* **11**:657–660.

Sennhauser, G., Grütter, M.G. (2008) Chaperone-assisted crystallography with DARPins. *Structure* **16**:1443–1453.

Sharma, K.S., Durand, G., Gabel, F., Bazzacco, P., Le Bon, C., Billon-Denis, E., Catoire, L.J., Popot, J.-L., Ebel, C., Pucci, B. (2012) Non-ionic amphiphilic homopolymers: Synthesis, solution properties, and biochemical validation. *Langmuir* **28**:4625–4639.

Skerra, A. (2007) Alternative non-antibody scaffolds for molecular recognition. *Curr. Opin. Biotechnol.* **18**:295–304.

Stenberg, E., Persson, B., Roos, H., Urbaniczky, C. (1991) Quantitative determination of surface concentration of protein with surface plasmon resonance using radio-labeled proteins. *J. Colloid Interface Sci.* **143**:513–526.

Stroebel, D., Choquet, Y., Popot, J.-L., Picot, D. (2003) An atypical haem in the cytochrome $b_6f$ complex. *Nature* **426**:413–418.

Stumpp, M.T., Forrer, P., Binz, H.K., Plückthun, A. (2003) Designing repeat proteins: modular leucine-rich repeat protein libraries based on the mammalian ribonuclease inhibitor family. *J. Mol. Biol.* **332**:471–487.

Suárez-Suárez, S., Carriedo, G.A., Presa Soto, A. (2013) Gold-decorated chiral macroporous films by the self-assembly of functionalised block copolymers. *Chem. Eur. J.* **19**:15933–15940.

Tifrea, D.F., Pal, S., Le Bon, C., Giusti, F., Cocco, M.J., Zoonens, M., de la Maza, L.M. (2018a) Resiquimod conjugated with amphipols bound to the *Chlamydia muridarum* MOMP enhances protection against a mucosal challenge. *In preparation.*

Tifrea, D.F., Pal, S., Le Bon, C., Giusti, F., Popot, J.-L., Cocco, M.J., Zoonens, M., de la Maza, L.M. (2018b) Co-delivery of amphipol-conjugated adjuvant with antigen, and adjuvant combinations, enhance immune protection elicited by a membrane protein-based vaccine against a mucosal challenge with Chlamydia. *Submitted for publication.*

Torres-Rodriguez, L.M., Roget, A., Billon, M., Bidan, G. (1998) Synthesis of a biotin functionalized pyrrole and its electropolymerization: Toward a versatile avidin biosensor. *Chem. Commun.* **1998**:1993–1994.

Tribet, C., Porcar, I., Bonnefont, P.A., Audebert, R. (1998) Association between hydrophobically modified polyanions and negatively charged bovine serum albumin. *J. Phys. Chem. B* **102**:1327–1333.

Troughton, E.B., Bain, C.D., Whitesides, G.M., Nuzzo, R.G., Allara, D.L., Porter, M.D. (1988) Monolayer films prepared by the spontaneous self-assembly of symmetrical and unsymmetrical dialkyl sulfides from solution onto gold substrates: Structure, properties, and reactivity of constituent functional groups. *Langmuir* **4**:365–385.

Valiokas, R., Vaitekonis, Š., Klenkar, G., Trinkunas, G., Liedberg, B. (2006) Selective recruitment of membrane protein complexes onto gold substrates patterned by dip-pen nanolithography. *Langmuir* **22**:3456–3460.

Weinrich, D., Jonkheijm, P., Niemeyer, C.M., Waldmann, H. (2009) Applications of protein biochips in biomedical and biotechnological research. *Angew. Chem. Int. Ed. Engl.* **48**:7744–7751.

Wingren, C., Borrebaeck, C.A. (2007) Progress in miniaturization of protein arrays-a step closer to high-density nanoarrays. *Drug Discov. Today* **12**:813–818.

Wong, I.Y., Melosh, N.A. (2009) Directed hybridization and melting of DNA linkers using counterion-screened electric fields. *Nano Lett.* **9**:3521–3526.

Xu, C., Xu, K., Gu, H., Zhong, X., Guo, Z., Zheng, R., Zhang, X., Xu, B. (2004) Nitrilotriacetic acid-modified magnetic nanoparticles as a general agent to bind histidine-tagged proteins. *J. Am. Chem. Soc.* **126**:3392–3393.

Yang, Q., Lundahl, P. (1994) Steric immobilization of liposomes in chromatographic gel beads and incorporation of integral membrane proteins into their lipid bilayers. *Anal. Biochem.* **218**:210–221.

Yang, S.T., Witkowski, A., Hutchins, R.S., Scott, D.L., Bachas, L.G. (1998) Biotin-modified surfaces by electrochemical polymerization of biotinyl-tyramide. *Electroanalysis* **10**:58–60.

Zdyrko, B., Klep, V., Luzinov, I. (2003) Synthesis and surface morphology of high-density poly(ethylene glycol) grafted layers. *Langmuir* **19**:10179–10187.

Zhang, G.J., Tanii, T., Kanari, Y., Ohdomari, I. (2007) Production of nanopatterns by a combination of electron beam lithography and a self-assembled monolayer for an antibody nanoarray. *J. Nanosci. Nanotechnol.* **7**:410–417.

Zheng, Z.J., Daniel, W.L., Giam, L.R., Huo, F.W., Senesi, A.J., Zheng, G.F., Mirkin, C.A. (2009) Multiplexed protein arrays enabled by polymer pen lithography: Addressing the inking challenge. *Angew. Chem., Int. Ed.* **48**:7626–7629.

Zoonens, M., Giusti, F., Zito, F., Popot, J.-L. (2007) Dynamics of membrane protein/amphipol association studied by Förster resonance energy transfer. Implications for *in vitro* studies of amphipol-stabilized membrane proteins. *Biochemistry* **46**:10392–10404.

Zoonens, M., Popot, J.-L. (2014) Amphipols for each season. *J. Membr. Biol.* **247**:759–796.

Zoonens, M., Zito, F., Martinez, K.L., Popot, J.-L. (2014) Amphipols: a general introduction and some protocols, in: Mus-Veteau, I., ed., *Membrane Proteins Production for Structural Analysis*. Springer, New York, Heidelberg, Dordrecht, London, pp. 173–203.

# The Use of Amphipols in Mass Spectrometry

**14**

### Summary

*Mass spectrometry (MS) analyses of membrane proteins (MPs) have progressed spectacularly over the past decade. Yet, the presence of detergents remains a source of problems. In matrix-assisted laser desorption ionization (MALDI), only certain detergents are acceptable, which are not necessarily those best tolerated by the proteins under study. In electrospray ionization-mass spectrometry (ESI-MS) studies, where MPs can be brought to the gas phase under their native oligomeric state and conformation, a fine line has to be navigated between the Charybdis of retaining too many detergent molecules, yielding unexploitable spectra, and the Scylla of using too much collisional energy to strip away the detergent and denaturing the protein. The use of amphipols (APols) as alternatives to detergents has been validated in these two experimental configurations. MALDI-time-of-flight (TOF) experiments have been carried out using either A8-35 or non-ionic APols and have yielded mass determinations, identification of posttranslational modifications and associated lipids, and sequence data. In ESI-MS experiments, often combined with ion mobility (IM) measurements of the collisional cross section of the protein, which yield information about its conformation, A8-35 has proven particularly convenient, protecting MPs against unfolding better than the much used detergent dodecylmaltoside. Fast photochemical oxidation has revealed details of the way the two surfactants interact with MPs that tally well with the conclusions from molecular dynamics and NMR investigations. Whether APols can be used to preserve the oligomeric state of MPs in the gas phase, however, is uncertain. Finally, A8-35 has been used as a powerful tool to simplify whole proteome investigations.*

## 14.1 Introduction

The application of mass spectrometry (MS) to membrane proteins (MPs) is a relatively recent development. As with most technologies, it has lagged behind application to soluble proteins, but it is now progressing and diversifying at an amazing pace (for recent reviews, see e.g. Barrera and

Robinson 2011; Whitelegge 2013; Konijnenberg et al. 2015; Mehmood et al. 2015; Rajabi et al. 2015; Marty et al. 2016b). MS practitioners distinguish "top-down" experiments, in which the protein or complex under study is first transferred intact (or as intact as possible) to vacuum, and can subsequently be more or less drastically fragmented, and "bottom-up" experiments, in which the target, which can be a whole proteome, is first proteolytically or chemically cut into peptides, which can be further fragmented inside the spectrometer in tandem (MS/MS) experiments and whose identification leads to that of the proteins they derive from, as well as to an examination of their posttranslational modifications (see e.g. Mann and Jensen 2003). The use of MP/amphipol (APol) complexes has been validated in both configurations.

MS relies on measuring the mass of positively or negatively charged ions derived from the molecules or fragments under study. A critical step is to transfer the targets from their original environment, typically an aqueous solution supplemented or not with surfactants, to a gas phase, under the form of molecular ions carrying one or several charges. The ions are then accelerated in a vacuum chamber, and the ratio of their mass $m$ to their charge $z$ is measured.

For biological macromolecules, there are two main approaches to producing the ions. In matrix-assisted laser desorption/ionization (MALDI) (Fig. 14.1) (see Zenobi and Knochenmuss 1998), the target molecule is mixed with a matrix, such as $\alpha$-cyano-4-hydroxycinnamic acid, and the mixture dried on a metal support. The function of the matrix is to absorb the energy delivered by a laser pulse and, upon vaporizing, to project the target molecule in the gas phase while conferring it one or more charges. MALDI is generally associated with time-of-flight (TOF) determination of the mass-to-charge ($m/z$) value for the ions. In tandem MS (MS/MS), selected fragments are further broken down by collisions with a gas, in order to facilitate their identification and/or permit sequencing. This approach is well adapted to identifying and measuring the mass of molecules or fragments thereof, but it does not preserve most non-covalent associations.

In electrospray ionization (ESI) (Fig. 14.2), no matrix is used. A solution of the target molecule (Fig. 14.2A) is slowly extruded from a fine metal-coated capillary and transferred to the gas phase as small droplets under the effect of an electric field (Fig. 14.2B). In the nano-ESI version of the approach, which relies on very slow (a few tens of $nL \cdot min^{-1}$) extrusion of the solution, very small droplets are formed (<200 nm in diameter) and aqueous buffers can be used (Wilm and Mann 1994; Juraschek et al. 1999). The droplets evaporate, while transferring to the target molecules the charge they acquired during extrusion, and the molecular ions are progressively stripped of adsorbed molecules, such as water or detergent molecules, due either to heating with a laser beam (Morgner et al. 2007) or, more commonly, to collisions with the gas present in a "collision cell" (Fig. 14.2C). Gentle stripping will

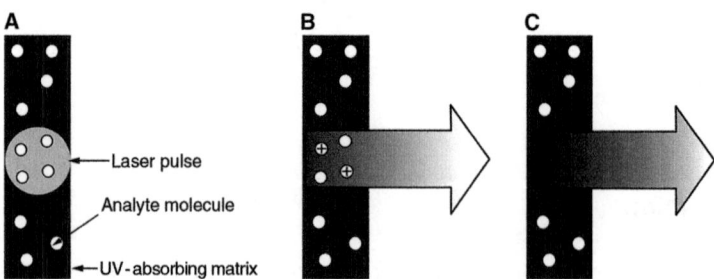

**Fig. 14.1** Matrix-assisted laser desorption/ionization (MALDI). (**A**) Absorbance of radiation by the matrix. (**B**) Dissociation of the matrix, phase change to super-compressed gas, and transfer of charges to sample molecules. (**C**) Expansion of the matrix at supersonic velocity, entrainment of target molecules or fragments in an expanding matrix plume, and transfer of the targets to the vacuum of the spectrometer (From Zaccai et al. 2017).

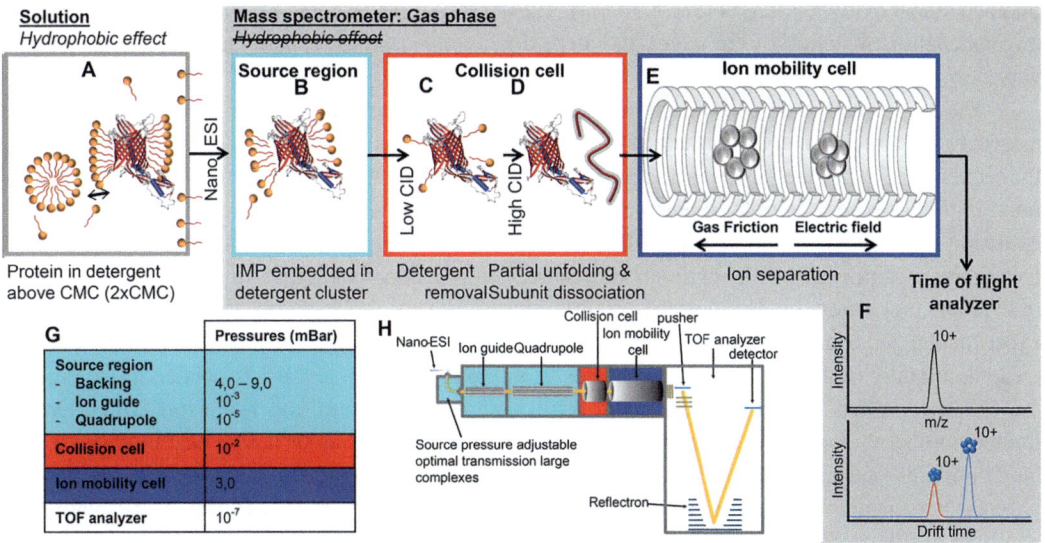

**Fig. 14.2** Schematic view of the fate of a detergent-solubilized membrane protein (MP) under non-denaturing electrospray ionization (ESI) mass spectrometry conditions. After release from solution by nano-ESI (**A**), the analyte enters the vacuum (**B**) in the source region where ion beam focusing and possible mass/charge selection in the quadrupole occur. In the collision cell (**C**), collisional activation causes stripping of detergent molecules (collision-induced dissociation, CID). The amount of energy needed in the gas phase, in the absence of the otherwise stabilizing hydrophobic effect, depends on the detergent used, Triton X-100 being released at low energy (10–50 V acceleration), whereas dodecylmaltoside (DDM) needs up to 100–200 V. Using excess energy (**D**) may initiate unfolding. The ion mobility (IM) cell (**E**) separates the protein complexes according to their mass, charge, and global structure (rotationally averaged size and shape, also called "collisional cross section"): due to the friction with gas, more extended ions will travel more slowly (drift plot, **F** – lower panel) even at the same charge and molecular mass (**F** – *upper panel*). Typical pressures in the various regions of the instrument are shown in Table **G**; colors match those in scheme **H** of an ion mobility quadrupole TOF mass spectrometer (Figure courtesy of J.F. van Dyck and F. Sobott. For more details, see Konijnenberg et al. 2015; van Dyck et al. 2017).

leave some molecules associated, such as ligands, water, or, for MPs, detergent or lipid molecules. As stripping gets more vigorous, protein oligomers fall apart; individual proteins unfold and, finally, fragment (Fig. 14.2D). Further information regarding the shape and physical size of the ions can be gathered by driving them through an "ion mobility (IM) cell" in the interim region between the ESI capillary and the vacuum regions of the mass spectrometer (Fig. 14.2E), in which they drift through a cell filled with a neutral gas under the influence of an electric field. For a given mass and $m/z$ ratio, ions that have larger rotationally averaged dimensions will experience more collisions and move slower, so that conformational changes, such as those induced by ligands or by denaturation, can also be explored. The ion mobility cell is situated inside the mass spectrometer, so that both mass and shape data can be obtained in a single experiment. Conformational differences that change the rotationally averaged "collisional cross section" (CCS) by as little as 2–5% can be detected (see Konijnenberg et al. 2014, 2015, and references therein).

Prior to vaporization, the target proteins can be subjected to various treatments. They can be exposed to ligands, which, if they are retained, will change the mass of the subunits they bind to, and which can modulate the conformation of the targets and the strength of the interaction between their components. They can also be exposed to treatments that will affect the mass of accessible regions, such as hydrogen/deuterium exchange, which reflects the exposure to water and the dynamics of regions containing exchangeable hydrogens, or various strategies leading to covalent labeling (e.g. fast photochemical oxidation; see below, § 14.3.3). These procedures provide information about the three-dimensional organization of the target, its interactions with surfactants, its dynamics, and how it can vary during the functional cycle.

MS can therefore be used for a vast variety of analyses, from ligand binding to exploring the assembly and stoichiometry of protein complexes, detecting conformational changes, and determining the sequence of individual proteins and posttranslational modifications. However, the target molecules must initially be kept soluble, either before mixing with the MALDI matrix and drying or before extrusion during ESI. In the case of MPs, this implies resorting to surfactants, typically detergents. Detergents complicate MS experiments because, on the one hand, they migrate themselves, as individual molecules or in clusters, contributing a heavy background that can compromise experiments, and, on the other hand, a variable number of them tend to stick to the target protein, creating a family of ions with different masses. If too many of them remain associated to the protein, the MS spectra become unpredictable (for a discussion, see Barrera and Robinson 2011). Only a limited number of detergents are compatible with MALDI, which may or not be well tolerated by the target proteins (see e.g. Cadene and Chait 2000 and references therein). In the case of ESI, a decisive progress was made when it was realized that, for a given mass, detergent-associated MPs withstand exposure to higher collision energies than soluble proteins do (Barrera et al. 2008) (Fig. 14.3). As a consequence, it is possible to vaporize MP/detergent complexes starting from a solution above the CMC of the detergent, where the protein does not aggregate, and to progressively strip it from the detergent until well-defined ions can be identified (Fig. 14.4). A fine line has to be walked, however, because at higher collisional energies, as the last detergent molecules leave the protein, MP complexes start to disassemble and individual proteins to unfold. The reason why MPs resist collisions better than soluble ones is not entirely clear, but an attractive hypothesis is that the evaporation of detergent molecules dissipates energy and cools the complex, retarding unfolding (Ruotolo et al. 2008; Borysik and Robinson 2012; Borysik et al. 2013). One merit of this proposal is to provide a plausible explanation for the fact that the onset of MP denaturation tends to coincide with the loss of the last detergent molecules (see e.g. Reading et al. 2015).

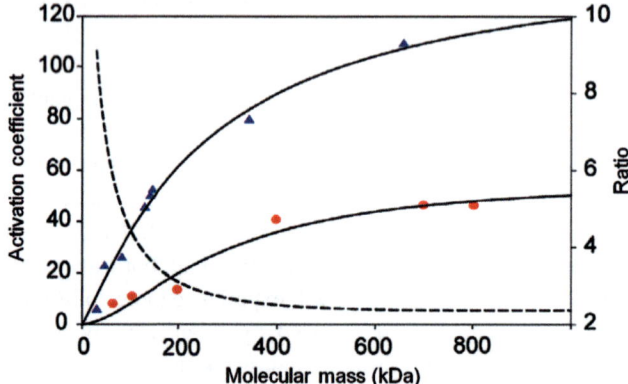

**Fig. 14.3** Plot of the activation coefficient required to maintain subunit interactions for membrane and soluble complexes as a function of molecular mass. Activation coefficients were computed from the values of the cone and collision cell voltages and multiplied by the average charge state of the complex. The ratio of activation coefficient between membrane and soluble complexes is plotted as a dashed line. Soluble (●) and membrane (▲) complexes in order of ascending molecular mass: avidin, concanavalin A, glycogen phosphorylase B, exosome, RNA pol III, GroEL and EmrE, LmrP, MscL, BtuCD, LmrCD, MacB, MexB, and ATP synthase, respectively. The activation coefficient limit was calculated for protein complexes from the experimental conditions at which at least 90% of the total ion current is assigned to the intact protein complex (From Barrera et al. 2009, © 2009 Macmillan Publishers Limited, Nature Methods. All rights reserved).

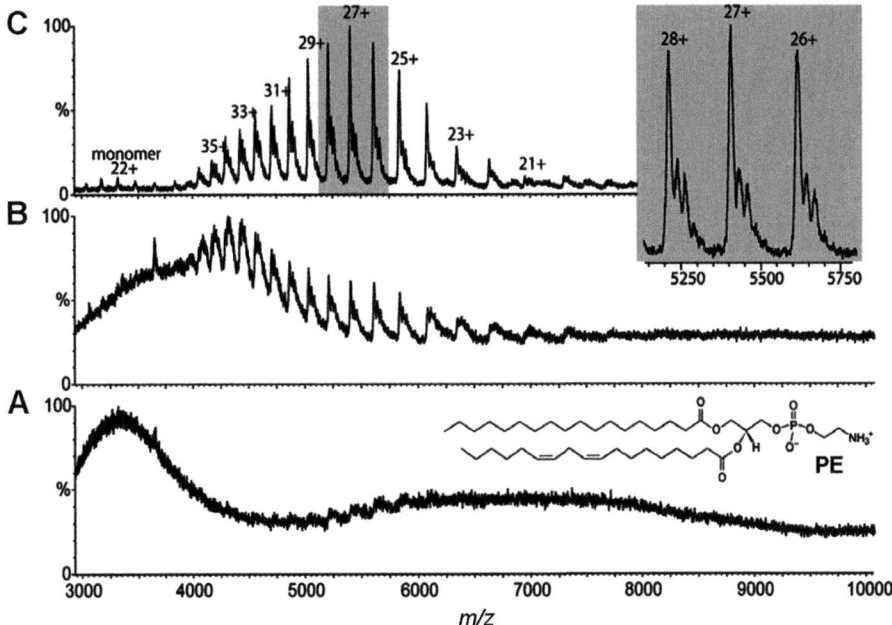

**Fig. 14.4** Stripping a membrane protein (MacB, an ATP-binding cassette transporter) from detergent (dodecylmaltoside) by increasing the activation energy. As the activation energy is increased from a coefficient of 42 (cf. Fig. 14.3) in the lower spectrum (**A**) to 48 (**B**) and 52 (**C**) in the upper spectra, resolved charge states are observed for the dimer of MacB. Under the highest activation conditions, a proportion of monomeric protein becomes apparent. *Inset*: Expansion across three charge states reveals binding to the dimer of phosphatidylethanolamine (PE) (From Barrera et al. 2009, © 2009 Macmillan Publishers Limited, Nature Methods. All rights reserved).

*How will membrane protein/amphipol complexes behave in vacuum?*
(© *2018 by Francis Haraux*).

It may seem strange that MP/detergent complexes do not immediately dissociate once exposed to vacuum, a medium in which the hydrophobic effect is absent and where there is, therefore, little free energy gain to be expected from adsorbing the detergent's alkyl tails onto the TM surface of the protein, whereas entropy favors dissociation (note however that Van der Waals forces do persist and, at variance with what would occur in a hydrocarbon solvent, can make significant contributions; see Liu et al. 2009). In a low dielectric environment and in the absence of competition with water, polar interactions between detergent polar heads are reinforced. It is certainly significant that, during an ESI-MS experiment, Triton X-100 is stripped from MPs at much lower energies than dodecylmaltoside (DDM) (Konijnenberg et al. 2014) (cf. legend to Fig. 14.2): the polar moieties of Triton, which consist of polyethylene glycol, cannot establish hydrogen bonds between themselves, whereas the maltosyl headgroups of DDM can. A similar effect is seen when comparing $C_8E_4$ and octylglucoside (Reading et al. 2015). Molecular dynamics (MD) studies (Rouse et al. 2013) do show that, upon simulated transfer of a MP/DDM complex from water to vacuum (and at variance with what is observed with ionic detergents), very little reorganization of the complex is observed (Fig. 14.5A, B). In the course of the simulation, most (but not all) water molecules evaporate (Fig. 14.5C, D), and most of the hydrogen bonds that the detergent formed with water disappear (Fig. 14.5E), to be replaced by DDM-DDM ones (Fig. 14.5F).

MS studies of MP/APol complexes are only at their beginning, but they already point to several interesting applications. Published studies are compiled in Table 14.1 (for earlier overviews, see Popot et al. 2011; Zoonens and Popot 2014). They fall into three categories:

(i) MALDI-TOF studies of the mass of MPs and associated cofactors, such as lipids and hemes (Studies 14.1 and 14.2);

(ii) ESI and ESI-IM studies of MPs and MP complexes, aiming, in particular, to examine whether using APols for stabilizing fragile MPs in aqueous solutions (Chap. 5, § 5.6) is compatible with MS studies (Studies 14.3 and 14.4, and 14.6–14.8);

(iii) Proteomic studies, where APols are exploited to simplify the production and recovery of MPs and proteolytic peptides (Studies 14.5 and 14.9).

***Amphipols and mass spectrometry of membrane proteins***
*(© 2018 by Francis Haraux).*

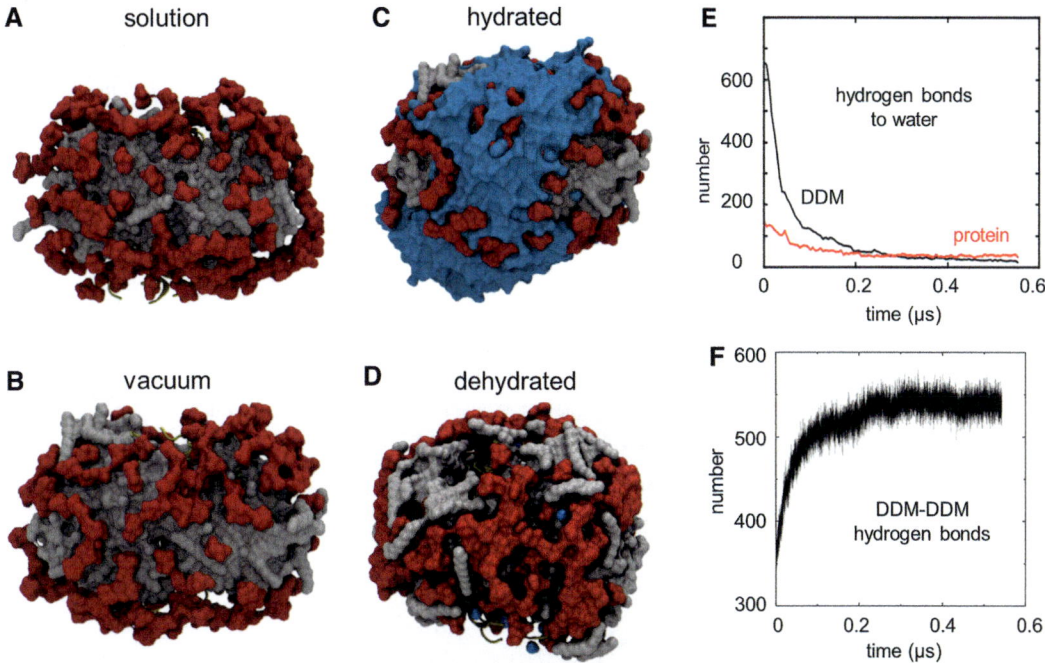

**Fig. 14.5** Transferring a membrane protein/detergent complex from aqueous solution to vacuum, as simulated by molecular dynamics (MD). The complex consists of influenza virus BM2 (a homotetramer of transmembrane $\alpha$-helices) and 150 molecules of $n$-dodecyl-$\beta$-D-maltoside (DDM). (**A–D**) The protein-detergent complexes are shown at the beginning ("solution," "hydrated") and end ("vacuum") of steered molecular dynamics trajectories and following extended simulation under dehydrating conditions ("dehydrated"). Bound water (*cyan*) is omitted in the left two panels for clarity. The headgroups of DDM are shown in *red* and its alkyl chains in *grey*. The protein (*yellow*) is entirely hidden by the detergent from the beginning to the end of the simulation. (**E**) Evaporation of water molecules from the complex during extended simulations under dehydrating conditions in vacuo. The number of hydrogen-bonding interactions of water molecules with DDM (*black lines*) and with the protein (*red lines*) is plotted as a function of time. (**F**) Evolution of the number of hydrogen bonds between DDM molecules during extended simulation under dehydrating conditions. The criteria for the presence of a hydrogen bond were an acceptor-donor distance $\leq 3.5$ Å and an acceptor-donor-hydrogen angle $\leq 30°$. Hydrogen bonds between detergent sugar-based headgroups increase on the same time scale as the loss of water molecules. (The figure has been assembled from data in Rouse et al. 2013, DOI: https://doi.org/10.1016/j.bpj.2013.06. 025, article available under the terms of the Creative Commons Attribution License (CC BY)).

**Table 14.1** An overview of studies in which mass spectrometry has been applied to membrane protein/amphipol complexes.

| Study | MP(s) | APol(s) | Method | Comments | References |
|---|---|---|---|---|---|
| **14.1** | *E. coli* OmpX | A8-35 | MALDI-TOF | The first demonstration that MP/APol complexes are amenable to MS, discreetly hidden in the supplementary information of an NMR study where MS was used to assess the extent of isotopic labeling of OmpX. | Catoire et al. (2009) |
| **14.2** | *E. coli* tOmpA, *H. salinarum* BR, *C. reinhardtii* cytochrome $b_6 f$, beef heart cytochrome $bc_1$ | A8-35, NAPol | MALDI-TOF | Four MPs of increasing complexity were examined by MALDI-TOF starting from complexes with either DDM, A8-35 or a glucosylated, homotelomeric non-ionic APol. Most but not all MPs or subunits could be observed, and bound lipids and cofactors and post-translational modifications identified. | Bechara et al. (2012) |
| **14.3** | *E. coli* OmpT and PagP | A8-35 | ESI-IM | The first demonstration that APols can replace detergents for ESI-MS of MPs. | Leney et al. (2012) |
| **14.4** | *E. coli* DAGK | A8-35 | ESI | Following ESI-MS, the trimeric A8-35-trapped diacylglycerol kinase was recovered as a mixture of monomers, trimers, and some dimers. | Hopper et al. (2013) |
| **14.5** | Total and membrane or secreted proteome from human cell cultures | A8-35 | HPLC-ESI-MS/MS | In bottom-up investigations of whole proteomes, A8-35 presents the advantages of not interfering with trypsinolysis and of being easily separated from tryptic peptides by acid precipitation. It can also be used to co-precipitate a whole proteome. | Ning et al. (2013, 2014) |
| **14.6** | PagP, OmpT, Mhp1 and GalP | A8-35 | ESI-IM | A8-35 is observed to better preserve than DDM the native state of MPs during ESI-MS. | Calabrese et al. (2015) |
| **14.7** | *E. coli* PagP, OmpT and tOmpA | A8-35, A8-75, SAPol, A34-35, A34-75, NAPol | ESI-IM | Comparative studies of six APols and three MPs show A8-35 to be the most generally useful APol and provide hints as to which APol properties favor observation of the native state of MPs. | Watkinson et al. (2015) |
| **14.8** | *E. coli* OmpT | A8-35 | FPOP-LC-MS/MS | Fast photochemical oxidation reveals subtle differences in the accessibility of some residues depending on whether OmpT is complexed by DDM or by A8-35. | Watkinson et al. (2017) |
| **14.9** | HEK cells | A8-35 | Trypsinolysis, LC-MS/MS | A construct comprising a tagged HIV Gag protein and the target protein is expressed in HEK cells. The construct and the partners of the target protein are recovered in virus-like particles, affinity-purified, and A8-35 used for solubilization and purification by co-precipitation at low pH. Following resolubilization and trypsinolysis, A8-35, non-proteolyzed material and trypsin are eliminated by co-precipitation at low pH and the peptides identified by LC-MS/MS. | Titeca et al. (2017) |

Studies bearing on individual MPs examined by MALDI-TOF or by ESI-MS are on a *blue* or a *green* background, respectively, proteomic studies on a *buff* background. *Abbreviations: BR* bacteriorhodopsin, *DAGK* diacylglycerol kinase, *ESI* electrospray ionization, *FPOP* fast photochemical oxidation of proteins, *IM* ion mobility spectrometry, *HPLC* high-pressure liquid chromatography, *MALDI* matrix-assisted laser desorption/ionization, *MS/MS* tandem mass spectrometry, *NAPol* non-ionic, glucosylated, homotelomeric APol, *TOF* time of flight, *tOmpA* transmembrane domain of outer membrane protein A

## 14.2   MALDI-TOF Studies

The first demonstration that APol-trapped MPs are amenable to MS studies was almost accidental: in the course of an NMR study (see Chap. 10, § 10.3.1.2, Study 10.2), Laurent J. Catoire and colleagues subjected to MALDI-TOF-MS a sample of *Escherichia coli* OmpX that had been universally labeled with $^2$H, $^{13}$C, and $^{15}$N, with the view of checking on the extent of isotopic labeling and the intactness of the protein (Catoire et al. 2009). The sample yielded an excellent spectrum (Fig. 14.6).

This led to a more thorough examination of the amenability to MALDI-MS studies of MPs of various complexities trapped with various APols (Study 14.2) (Bechara et al. 2012). The proteins tested were (i) the TM domain of *E. coli* outer membrane protein A, tOmpA, a small, monomeric, eight-strand $\beta$-barrel; (ii) bacteriorhodopsin (BR) from *Halobacterium salinarum*, a small, monomeric, seven-TM $\alpha$-helix MP with a covalently but loosely bound cofactor, retinal; and two large complexes: (iii) cytochrome $b_6 f$ from the thylakoids of *Chlamydomonas reinhardtii*, and (iv) beef heart mitochondrial cytochrome $bc_1$, each of which comprises many subunits of various topologies and several covalently or non-covalently bound cofactors. tOmpA was folded from a denatured preparation obtained by solubilizing inclusion bodies in urea and was not expected to contain any lipids, whereas the other three proteins were of natural origin and are known to comprise bound lipids. The two APols tested were a polyanionic, poly (acrylic acid)-derived APol, A8-35 (Tribet et al. 1996), and a glucosylated, homotelomeric non-ionic APol (NAPol) (Bazzacco et al. 2012; Sharma et al. 2012) (see Chap. 4, Figs. 4.1 and 4.8).

An overview of the results obtained without tryptic treatment is shown in Table 14.2. tOmpA was detected whatever the APol used. A curious case is that of BR, whose full-length ions could be observed after trapping in NAPols, but not in A8-35, whatever the matrix used, even though tryptic degradation of the same samples yielded peptides that could be readily observed. This behavior may result from some large-scale effect on the sample/matrix mixture. Indeed, tOmpA trapped in A8-35, which by itself is perfectly detectable, is not detected anymore if the same sample is mixed with BR/A8-35 complexes. A possibility is that the phenomenon is related to demixing of the samples on the MALDI target, perhaps facilitated by the large amount of lipids retained by BR (see Gohon et al. 2008), but the matter remains undecided. BR/NAPol samples gave rise to ions corresponding to the

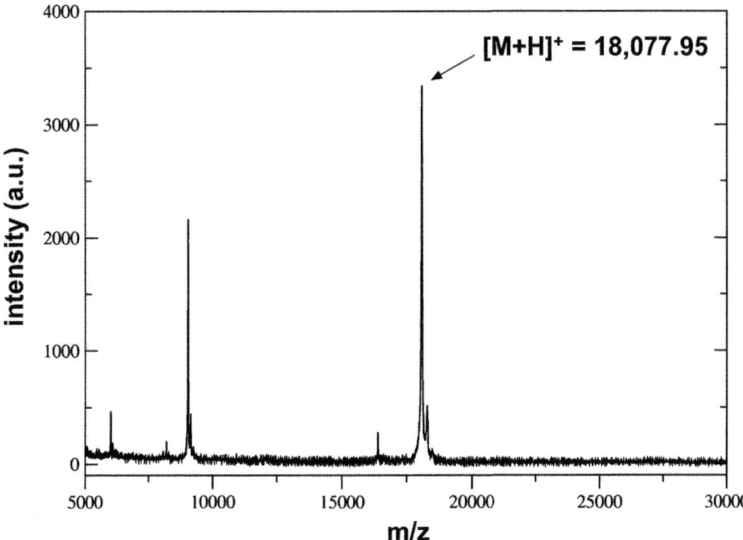

**Fig. 14.6**  MALDI-TOF spectrum of A8-35-trapped [u-$^2$H,$^{13}$C,$^{15}$N]OmpX from *Escherichia coli* (From Catoire et al. 2009, Copyright © 2009 Elsevier Inc. All rights reserved).

**Table 14.2**  MALDI-TOF analysis of four membrane proteins trapped in either A8-35 or a non-ionic amphipol (NAPol).

| Protein (subunits, cofactors, lipids) | Mass (Da) | TM segments | Detected in A8-35 | Detected in NAPol | Comments |
|---|---|---|---|---|---|
| **tOmpA** (his-tagged) | 20,189 | 8 $\beta$ | + | + | Mixing BR/A8-35 and tOmpA/A8-35 complexes resulted in the loss of the tOmpA signal (see text) |
| **Bacteriorhodopsin (BR)** | 26,783 | 7 $\alpha$ | − | + | |
| Retinal | | | | + (bound) | Both the apo- and the holoprotein are detected |
| Lipids | | | | +? (bound) | Deduced from the presence of a BR peak heavier by ~4.2 kDa |
| **Cytochrome $b_6 f$** | 220-kDa superdimer | | | | |
| PetA (cytochrome $f$) (his-tagged) | 32,624 (including $c$-type heme) | 1 $\alpha$ | + | + | Covalently bound $c$-type heme detected |
| PetB (cytochrome $b_6$) | 24,780 | 4 $\alpha$ | + | + | Covalently bound $c$-type heme detected in both APols; non-covalently bound $b$-type heme probably detected in A8-35. Weaker signal than for the one-TMH proteins of the same complex |
| PetC (Rieske protein) | 18,333 | 1 $\alpha$ | + | + | |
| PetD (subunit IV) | 17,442 | 3 $\alpha$ | + | − | Weaker signal in A8-35 than for the one-TMH proteins of the same complex. Not detected in NAPol, detected in DDM |
| PetG | 3981 | 1 $\alpha$ | + | − | Not detected in NAPol |
| PetL | 4875 | 1 $\alpha$ | − | − | The only subunit to remain undetected in either DDM or APols, even though its presence is certain |
| PetM | 4034 | 1 $\alpha$ | + | + | Sequence checked by MS/MS |
| PetN | 3282 | 1 $\alpha$ | + | + | |
| **Cytochrome $bc_1$** | 487-kDa superdimer | | | | Not examined in NAPol nor DDM |
| Cytochrome $b$ | 42,734 | 8 $\alpha$ | − | | Not detected |
| Cytochrome $c_1$ | 27,288 (including $c$-type heme) | 1 $\alpha$ | + | | With bound $c$-type heme |
| Rieske protein | 21,610 | 1 $\alpha$ | + | | |
| Core I | 49,212 | 0 | + | | Weak signal |
| Core II | 46,525 | 0 | + | | Weak signal |
| Hinge | 9176 | 0 | + | | |
| SU 6 | 13,345 | 0 | + | | |
| QP-C | 9532 | 1 $\alpha$ | + | | |
| SU 9 | 7998 | 0 | + | | |
| SU 10 | 7327 | 1 $\alpha$ | + | | |
| SU 11 | 6520 | 1 $\alpha$ | + | | |
| Lipids | | | + (free) | | Cf. Fig. 14.7 |

Adapted from Bechara et al. (2012). *Abbreviations: TM* transmembrane, *TMH* TM $\alpha$-helix

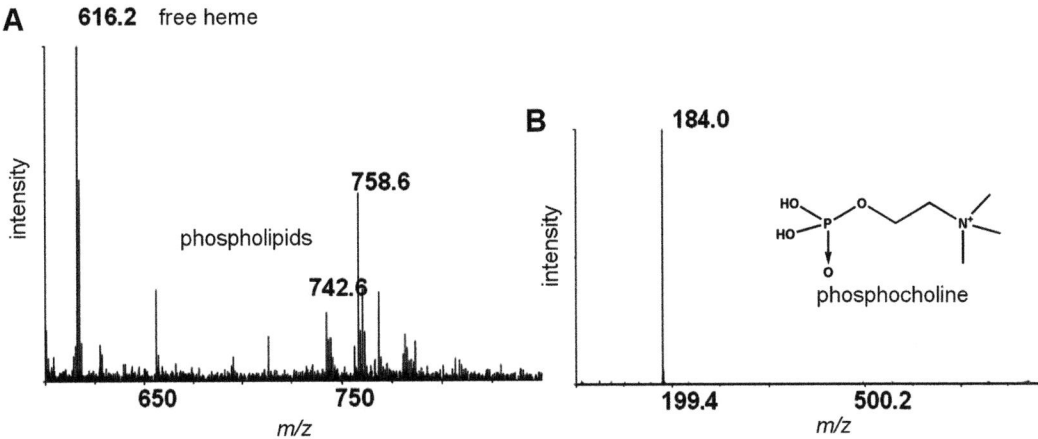

**Fig. 14.7** MS analysis of lipids and heme in A8-35-trapped cytochrome $bc_1$. (**A**) Low-mass region of the MALDI-MS spectrum. The ion at $m/z = 616.2$ corresponds to free heme released from cytochrome $b$, to which it is non-covalently bound, those at 742.6 and 758.6 to phospholipids. (**B**) MS/MS analysis of the ion at $m/z = 758$, showing the presence of phosphocholine, a phosphatidylcholine-derived fragment, at $m/z = 184$ (Reprinted with permission from Bechara et al. 2012, © 2012 American Chemical Society).

naked protein, but also to a collection of heavier ions that are too light to contain the APol and most likely correspond to BR/lipid complexes. The rate of proteolysis decreased in the order DDM $\geq$ A8-35 > NAPols, which correlates with the stabilizing potential of the three types of surfactants (see Chap. 5, § 5.5). These observations suggest that resorting to APols does not improve the sequence coverage of MPs, but may possibly help in distinguishing between cleavage sites that are always exposed and sites that become so only upon conformational excursions of the protein.

Most of the subunits present in cytochrome $b_6f$ were detected whatever the APol used. However, subunits PetD (three TM $\alpha$-helices) and PetG (one TM helix) were not detected in the NAPol-trapped preparation. Subunit PetL, although known to be present (Stroebel et al. 2003), was detected neither in A8-35, NAPol, nor DDM. Limited tryptic digestion did not show any significant surfactant-dependent differences between samples in terms of the fragments obtained. In terms of kinetics, however, peptide digests appeared more quickly when the $b_6f$ complex was solubilized in DDM as compared to being trapped in A8-35. As observed with BR, the $b_6f$ was most resistant to tryptic digestion when trapped in NAPols, only a few peptides being obtained from such samples. All tryptic peptides that were identified by MS/MS belonged to the three largest subunits, namely cytochromes $f$ and $b_6$ and, to a lesser extent, the Rieske protein. This is not surprising, given that those are the subunits most exposed to the aqueous solution. Subunit IV is mostly buried, while the small subunits PetG, PetL, PetM, and PetN are short, mainly transmembrane peptides, with no or few arginyl or lysyl residues (see Stroebel et al. 2003). Intact small subunits were occasionally detected in digested samples that had not been micropurified.

Cytochrome $bc_1$ was analyzed only in A8-35. Ten out of the 11 subunits were detected. The undetected subunit is cytochrome $b$, which comprises eight TM helices (see Berry et al. 2000). This is reminiscent of the fact that, in the case of the $b_6f$ complex, the two subunits with multiple TM helices, cytochrome $b_6$ and subunit IV, yield weaker signals than those with a single TM helix. As for the $b_6f$ complex, the presence of a $c$-type heme (bound to cytochrome $c_1$) was detected, as well as that of free hemes derived from cytochrome $b$ (Fig. 14.7A). Phospholipids were observed at $m/z = 742$, 758, 760, 768, and 780 (Fig. 14.7A). Consistent with this interpretation, MS/MS analyses showed a very intense peak at $m/z = 184$, the mass of phosphocholine, which is a clear signature of the presence of phosphatidylcholine (Fig. 14.7B).

In conclusion, Study 14.2 shows that MALDI-MS is applicable to APol-trapped MPs, with or without trypsinolysis. It does not reveal more proteins nor more peptides than detergent-solubilized preparations do, but it presents, nevertheless, a number of conceivable advantages. For one thing, the mildness of APols should facilitate MS studies of fragile complexes and supercomplexes. In particular, it may help identify subunits, cofactors, or lipids that are easily lost during MP purification in detergent solution. More hypothetically, comparison of proteolytic fragments produced in a detergent vs. an APol environment may provide hints as to which cleavage sites are naturally accessible and which become so under the destabilizing effect of the detergent.

## 14.3 Electrospray Ionization-Mass Spectrometry Studies

ESI-MS studies were undertaken to answer the following questions: (i) Are MP/APol complexes amenable to such studies, and what advantages and drawbacks do they present as compared to MP/detergent ones? (ii) Which APols should preferably be used? (iii) Can such studies reveal differences in the way detergents and APols interact with MPs?

### 14.3.1 Membrane Protein/Amphipol Complexes Are Amenable to Electrospray Mass Spectrometry

The groups of Alison E. Ashcroft, Sheena E. Radford, and Peter J. F. Henderson have explored the application of APols to ESI-MS, taking as models four monomeric bacterial MPs, two of them featuring TM $\beta$-barrels, the other two TM $\alpha$-helix bundles, whose structures are shown in Fig. 14.8.

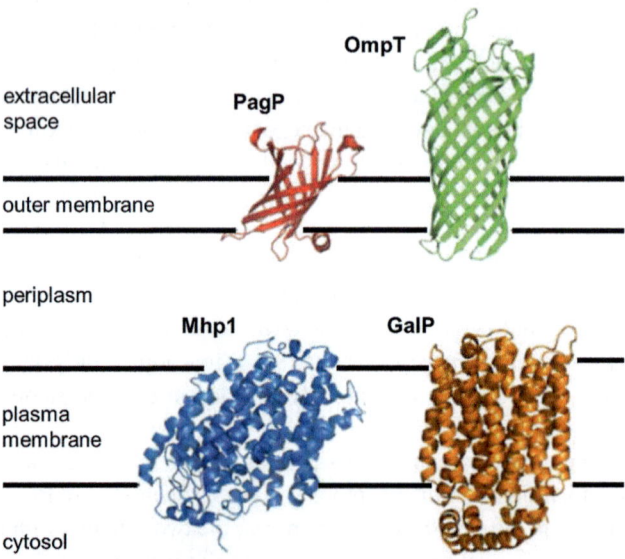

**Fig. 14.8** Three-dimensional structures and cellular localization of four bacterial membrane proteins used to explore the amenability of membrane protein/amphipol complexes to electrospray mass spectrometry. Crystal structures of PagP (PDB file 1THQ), OmpT (PDB file 1I78), and Mhp1 (PDB file 2X79) and a homology model of GalP based on the crystal structure of XylE, which shares 30% sequence identity with GalP (Adapted from Calabrese et al. 2015).

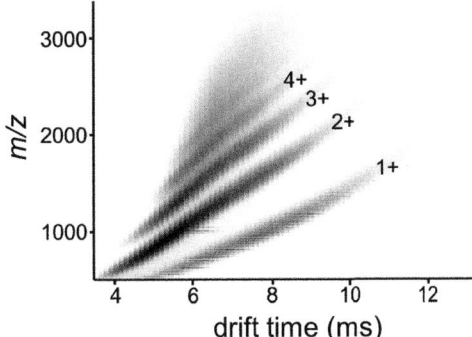

**Fig. 14.9** ESI-IM-MS driftscope plot of pure A8-35, showing the different charge state ion series that result from the wide distribution of molecular mass and size. The value of *m/z* of the ions is plotted vs. their drift time in the IM cell (Reprinted with permission from Leney et al. 2012, © 2012 American Chemical Society. For reproduction rights, contact http://pubs.acs.org/doi/abs/10.1021/ac302223s).

In a first study (Study 14.3), the acyl transferase PagP and the proteinase OmpT were trapped in A8-35, and their functionality demonstrated by enzymatic measurements (Leney et al. 2012). Prior to examining the MP/APol complexes themselves, pure A8-35 was subjected to ESI-IM-MS measurements (see § 14.1). Because measurement of the drift time adds one more dimension to the analysis, it is possible to separate the contributions of the variously charged ions, which would overlap in a classical ESI-MS experiment (cf. Kanu et al. 2008; Weidner and Trimpin 2010). Using "soft" ionization conditions, multiple species are observed from which 1+, 2+, 3+, and 4+ charge state ions can be separated clearly and identified (Fig. 14.9). A8-35 ions cover a wide range of masses and collisional cross sections, as expected from the polydisperse nature of the poly(acrylic acid) used for the synthesis (see Chap. 4, § 4.2.1 and Box 4.1). As a consequence, APols contribute, in these two-dimensional measurements, a relatively diffuse background, which is compatible with the detection of protein ions. ESI-IM-MS measurements have been instrumental in characterizing a perdeuterated version of A8-35 developed for NMR and SANS purposes (Giusti et al. 2014) (see Chap. 4, § 4.3.1.2.1, Fig. 4.13).

Under soft ionization conditions, MP/APol complexes were undetectable by ESI-MS, presumably because the proteins remained in complex with the polymer. However, when the settings for the trap and transfer regions of the mass spectrometer were increased to 150 and 100 V, respectively, MPs were released from the APol, and ions corresponding to the multiply charged MPs were detected (Fig. 14.10A, B). The ESI-MS spectrum of OmpT (Fig. 14.10A) shows a narrow charge state distribution corresponding to the 6+, 7+, and 8+ ions, together with slightly more expanded 9+ ions, giving an experimentally determined mass of 33,462 ± 5 Da, which is within 0.01% of the calculated mass based on the amino acid sequence (33,460 Da). ESI-IM-MS was able not only to effect the release of PagP from A8-35 but also to separate the folded and unfolded PagP conformers that SDS-PAGE had indicated to be present (Leney et al. 2012). The ESI-MS spectrum of PagP released from its complex with A8-35 shows a narrow charge distribution (5+, 6+, and 7+ charge state ions) (Fig. 14.10B, *white arrows*), corresponding to a compact structure with the expected molecular mass (20,175 ± 1 Da compared with 20,175 Da predicted from the amino acid sequence). However, significantly more expanded 7+ charge state ions indicative of a second conformer could also be detected in the ESI-IM-MS driftscope plot (Fig. 14.10B, *red arrow*). Higher charge states (8+, 9+, and 10+) were also observed for this more expanded PagP conformation (Leney et al. 2012). A correlation between higher ion charge state and protein unfolding is a classical, albeit not systematic observation (Hall and Robinson 2012).

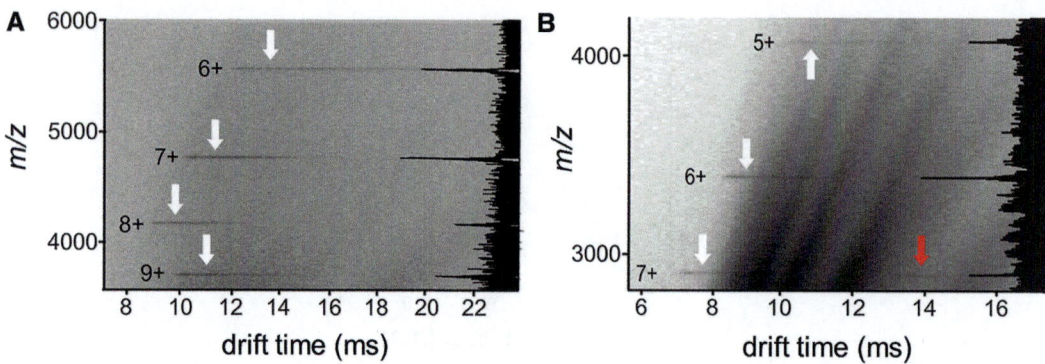

**Fig. 14.10** ESI-IM-MS driftscope plots of (**A**) OmpT/A8-35 complexes and (**B**) PagP/A8-35 complexes. The charge state of the ions is indicated, and the summed *m/z* spectrum for each sample is displayed on the right-hand side. *White arrows* highlight the most compact protein conformers. A second, more expanded conformer is observed for the 7+ ions in the PagP/A8-35 driftscope plot (*red arrow*) (Reprinted with permission from Leney et al. 2012, © 2012 American Chemical Society. For reproduction rights, contact http://pubs.acs.org/doi/abs/10.1021/ac302223s).

ESI-IM-MS separates ions according to their drift time through the mobility cell. By calibrating the arrival time distributions of protein ions of known structure, the collisional cross sections (CCS) of unknown proteins can be estimated. These experimental CCS values can then be compared with modeled values calculated from the PDB structures of the proteins of interest. If the experimentally estimated and theoretically determined CCS values are in agreement, it can be inferred that the protein retains a native-like structure in the gas phase (Ruotolo et al. 2008). The experimentally estimated CCS for the lowest charge state ions of PagP and OmpT (1790 and 2601 $\text{Å}^2$, respectively) are consistent within experimental error with the values predicted from their PDB crystal structure coordinates (1732 and 2718 $\text{Å}^2$, respectively). These data suggest that both PagP and OmpT released from A8-35 remain in a native-like conformation in the gas phase. The observation of an additional conformer of PgP, about 70% more expanded than the native form (3131 $\text{Å}^2$), is consistent with SDS-PAGE and SEC data indicating that PagP was not 100% folded (Leney et al. 2012), confirming the ability of ESI-IM-MS to transfer and separate folded and partially folded solution structures.

In two further studies (Studies 14.6 and 14.7), the same groups examined (i) whether the approach could be extended to $\alpha$-helical MPs, which are notoriously more fragile than $\beta$-barrel ones, and how do the results compare with those obtained with DDM, a classical detergent for such experiments (Calabrese et al. 2015), and (ii) which of the available APols are best suited to the task (Watkinson et al. 2015).

Study 14.6 (Calabrese et al. 2015) aimed at establishing whether substituting detergent with APols leads to differences in the conformational states of MPs in the gas phase. ESI-IM-MS was applied to four different MPs kept soluble either in DDM solution or by trapping with A8-35, namely the two $\beta$-barrel proteins already used in Study 14.3, PagP and OmpT, and two $\alpha$-helical MPs, Mhp1, which belongs to the 5-helical inverted repeat transporter superfamily, and GalP, a member of the major facilitator superfamily (Fig. 14.8). Overall, the data indicate that A8-35 leads to less charging of MPs as compared with DDM, which may be one reason why more native-like conformations are observed, as shown by CCS measurements. It has been proposed that upon exposure of MP/detergent complexes to vacuum, detergent removal must take place rapidly enough, so that detergent-free protein ions can be detected, but not so fast as to expose the native protein structure to the harsh conditions of the collision cell for an extended period, as this may lead to unfolding or structural collapse (Borysik and Robinson 2012; Reading et al. 2015). APols may likely desorb more slowly than detergents, which

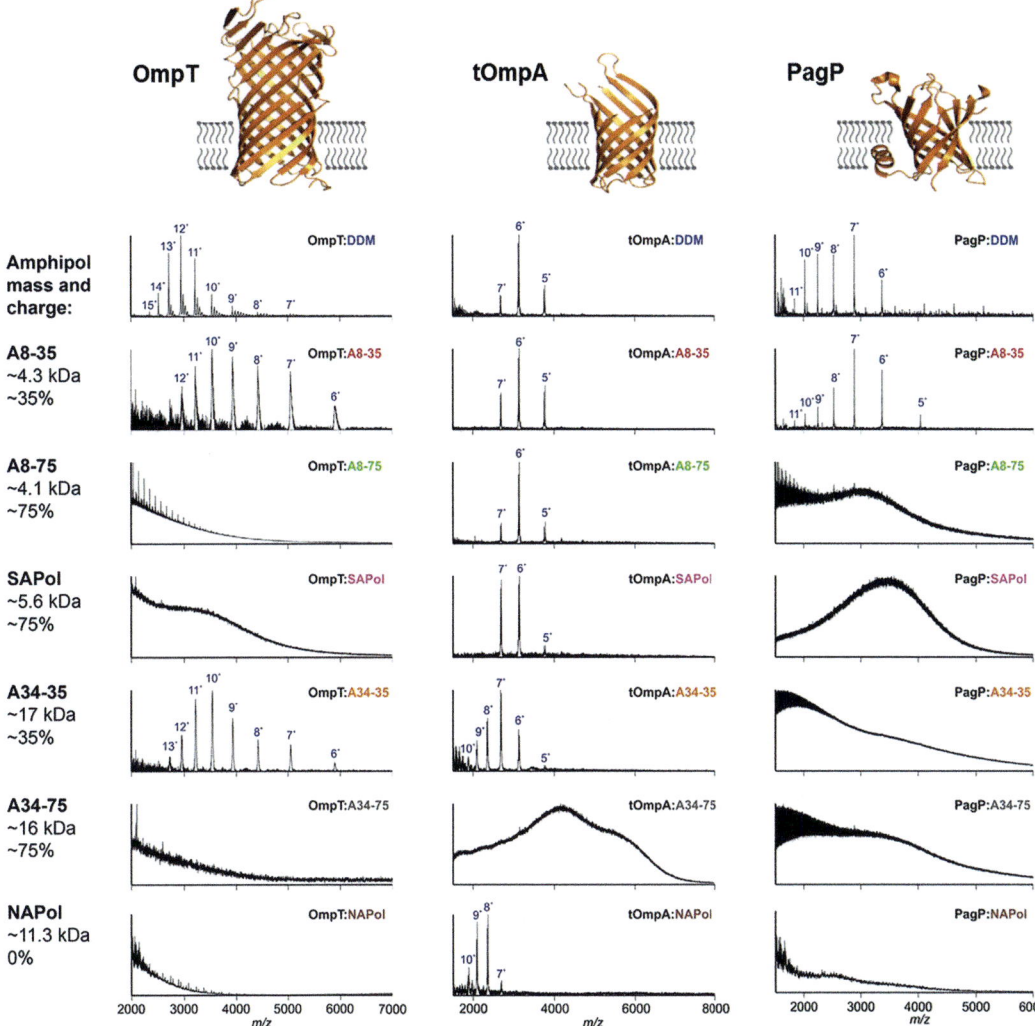

**Fig. 14.11** ESI-MS spectra of OmpT (*left*), tOmpA (*center*), and PagP (*right*) kept in DDM solution or transferred to A8-35, A8-75, SAPol, A34-35, A34-75, or NAPol, as indicated. The tOmpA/A34-35 spectrum was acquired following purification by SEC. All other spectra shown were obtained without prior SEC. Collision voltages in the trap (PagP and OmpT = 100–150 V; tOmpA = 50–100 V) and transfer (50–100 V) regions prior to and immediately following the drift cell, respectively, were varied to optimize liberation of each OMP with minimal impact on its structure. Origin of the crystal structures: OmpT, PDB file 1I78; tOmpA, PDB file 1QJP; PagP, PDB file 1THQ (Adapted from Watkinson et al. 2015. https://doi.org/10.1016/j.ijms.2015.06.017, article available under the terms of the Creative Commons Attribution License (CC BY)).

could play a role in protecting the proteins from unfolding in the gas phase. Study 14.6 also shows that A8-35-solubilized MPs are transferred to the gas phase along with bound lipids, thus permitting MP/lipid interactions to be explored by ESI-IM-MS. In a broad study of the holotranslocon from *E. coli*, lipids were extracted by chloroform/methanol from the A8-35-trapped translocon before being analyzed by liquid chromatography-ESI-MS (Botte et al. 2016).

In Study 14.7 (Watkinson et al. 2015), a systematic study was carried out of the transfer from solution to vacuum of three β-barrel MPs, OmpT, tOmpA, and PagP (Fig. 14.11, *top*), using as the surfactant either DDM or any of six different APols. The average mass ($\langle M_n \rangle$) and percentage of charged units in each APol are given in the left column of Fig. 14.11. As described in Chap. 4, § 4.2.1, A34-35 has the same chemical structure and charge density as A8-35, but its molecules are, on average, ~4× longer. A8-75 has the same length distribution as A8-35, but it has not been grafted with isopropylamine, and, therefore, it retains ~75% rather than ~35% of free carboxylates. A34-75 is a ~4× longer version of A8-75. These four APols were initially described in Tribet et al. 1996. The sulfonated APol (SAPol) is a version of A8-75 that has been grafted with ~40% taurine (Dahmane et al. 2011) (Chap. 4, § 4.2.2.1, Fig. 4.1). Its charge density and average length are the same as A8-75, but, because it carries sulfonates, its solubility is insensitive to the pH and to the presence of divalent cations. The last APol tested is a glucosylated, homotelomeric, non-ionic APol ("NAPol"; see Chap. 4, § 4.2.3.2, Fig. 4.5) (Bazzacco et al. 2012; Sharma et al. 2012). Taken together, these six APols provide an opportunity to test the effect of APol charge density and length on the transfer of MPs to the gas phase, and the use of three MPs gives a chance to assess the generality of the conclusions (Fig. 14.11).

The first conclusion from this analysis is the idiosyncrasy of OMPs. Despite their structural similarities (Fig. 14.11, *top*), they do not necessarily behave in the same way: tOmpA is relatively indifferent to the surfactant it is ejected from, whereas OmpT and PagP are more picky about it, meaning that they separate less easily without denaturation. The second point is that the chemical structure of the APol does influence the efficiency of the release of protein ions carrying a minimum amount of charges, which correlates with minimal denaturation (Watkinson et al. 2015): a longer and more densely charged APol is less favorable than a shorter and less charged one. Nevertheless, a short, totally uncharged APol (NAPol) does not give excellent results, the best results overall (on this restricted set of MPs) being obtained with A8-35. These observations will be further discussed below (§ 14.3.4). They are patchy, but they seem to be consistent with the view that the less polar interactions there are between MP-associated APols, the higher the probability that they will desorb without the protein denaturing, which may point the way to designing APols specifically optimized for MS.

A detailed protocol for ESI-MS-IM examination of MP/APol complexes is given in § 14.5.

### 14.3.2   Amphipols, ESI-MS, and Oligomeric Membrane Proteins

Very little data are available about using APols to examine oligomeric MPs by ESI-MS. In Study 14.4, a comparison was made of the release of MPs from detergents (DDM and *n*-decyl-$\beta$-D-malto-pyranoside), APols (A8-35), nanodiscs, and bicelles (Hopper et al. 2013). The MPs studied were *E. coli* lactose permease, a monomeric MP, *E. coli* diacylglycerol kinase (DAGK), a trimeric one, and sensory rhodopsin II (pSRII) from *Natronomonas pharaonis*, a monomeric 7-TM-helix photoreceptor. Only DAGK was studied as a MP/APol complex. As regards the identification of the correct oligomeric state, only bicelles performed reasonably well, releasing a mixture of DAGK trimers, monomers, and some dimers (Fig. 14.12C). DAGK released from A8-35 was mostly monomeric, with some trimers and occasional dimers (Fig. 14.12B), and DAGK released from DDM a mixture of monomers and dimers (Fig. 14.12A). Nanodiscs released a mixture of monomers, dimers, and trimers (Hopper et al. 2013). Whether the inability of A8-35 to release DAGK specifically as the native, physiological oligomer is specific to this protein, to A8-35, and/or to the conditions explored remains to be examined. As discussed in § 14.3.4, the use of other APols, in particular, might be worth considering.

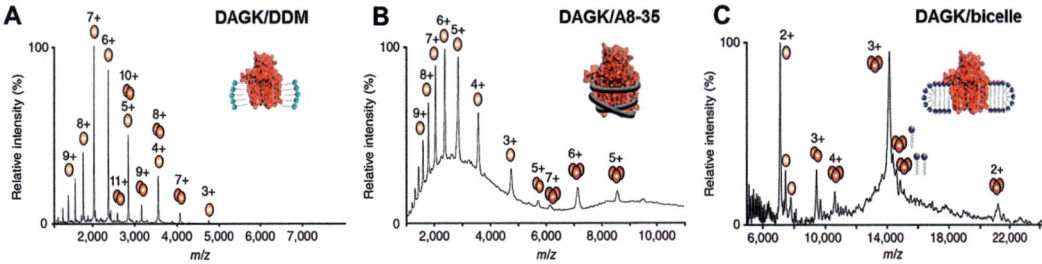

**Fig. 14.12** Comparison of the mass spectra of DAGK released from detergent, amphipols, or bicelles. Mass spectra of DAGK released from (**A**) a 0.02% DDM solution, (**B**) A8-35, and (**C**) a lipid bicelle. Charge states and oligomerization state are indicated above the peaks (From Hopper et al. 2013, © 2013 Macmillan Publishers Limited, Nature Methods. All rights reserved).

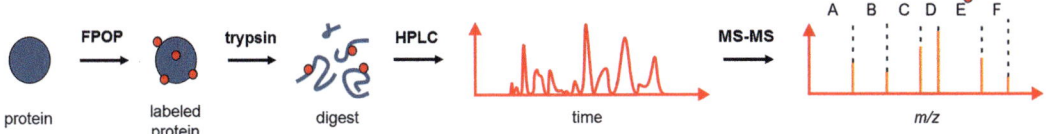

**Fig. 14.13** Principle of fast photochemical oxidation of proteins (FPOP) followed by mass spectrometry analysis. Following a laser flash, the hydroxyl radical generated by decomposition of $H_2O_2$ irreversibly label solvent-accessible sites on the side chains of the target protein. The covalently modified protein is digested to generate modified and unmodified tryptic peptides. LC-MS separation of the peptides is followed by MS/MS sequencing, from which each modified residue can be identified. Modification sites are symbolized by *red circles* (Adapted from Watkinson et al. 2017).

### 14.3.3 Interactions of Detergents vs. Amphipols with Membrane Proteins as Probed by Mass Spectrometry

In Study 14.8, the interaction of detergent (DDM) and APol (A8-35) with a model MP, OmpT, was examined using fast photochemical oxidation of proteins (FPOP) followed by tryptic digestion and MS-MS sequencing (Watkinson et al. 2017). The flow chart of the experiments is summarized in Fig. 14.13. In brief, FPOP uses a KrF excimer laser to generate hydroxyl radicals from $H_2O_2$, which is added to the medium in which the protein is incubated. FPOP labeling can result in a number of covalent chemical modifications, the most common of which being the addition of an oxygen atom, accompanied by a mass increase of 16 Da. Due to the presence of a quencher, the hydroxyl radicals have a short lifetime of ~1 μs. They therefore irreversibly oxidize the solvent-accessible side chains of the protein residues on a faster time scale than most protein folding/unfolding events. The labeling pattern therefore provides a snapshot of the distribution of conformations accessible to the protein under a given set of conditions. As compared to hydrogen/deuterium exchange, for instance, the advantages of FPOP include the short labeling times and the irreversible nature of the chemical modifications. The latter permits a comprehensive downstream analysis using liquid chromatography (LC)-MS/MS methods: following labeling, the protein is subjected to proteolysis and the resulting peptides separated and sequenced using reverse-phase LC-MS/MS, leading to the identification of the labeled residues (Fig. 14.13).

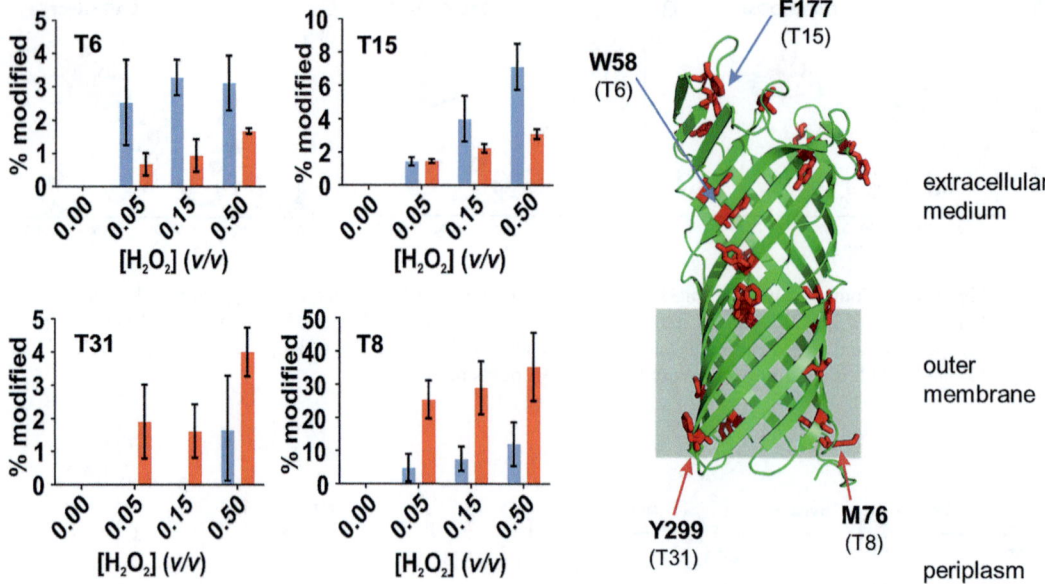

**Fig. 14.14** Some residues are oxidized to different degrees depending on whether OmpT is kept in DDM solution or transferred to A8-35. *Right:* bar graphs showing the percentage of four tryptic peptides of particular interest that is modified either in DDM (*blue bars*) or in A8-35 (*red bars*) following a single laser pulse, as a function of the concentration of $H_2O_2$ in the medium. *Right: red* and *blue arrows* indicate the position in the three-dimensional structure (PDB 1I78) of the residues that are modified in each peptide. Aromatic amino acid residues are shown in *red*. A couple of residues located at the periplasmic boundary of the TM region are less readily labeled in DDM (*lower bar graphs*), whereas two residues in the extracellular region appear less accessible in A8-35 (*upper bar graphs*) (Adapted from Watkinson et al. 2017).

ESI-LC-MS/MS analysis of tryptic digests following FPOP of both DDM-solubilized and A8-35-trapped OmpT yielded sequence coverages of 90–95%. Of the 31 predicted tryptic peptides, 13 were found to be modified, with a total of 20 modification sites being identified in both surfactants. The modified residues were either sulfur-containing or aromatic ones, as expected from their known propensity to undergo oxidative labeling (Xu and Chance 2007). The oxidation sites detected for OmpT are the same regardless of the surfactant employed, but the degree of modification of certain residues varied depending on the protein's solvent-accessible surface area in each surfactant: some tryptic peptides with modification sites in the extra-membrane region (T6 and T15) underwent more oxidation in DDM than in A8-35, whereas some sites located at the periplasmic boundary of the TM region (T8 and T31) were oxidized more readily in A8-35 than in DDM (Fig. 14.14).

The FPOP observations suggest differential interactions of A8-35 and DDM with OmpT. On the one hand, DDM, perhaps because of its much larger polar moieties, appears to better protect the periplasmic boundary region of the TM domain, limiting the oxidation of residues M76 (peptide T8) and Y299 (peptide T31). This is somewhat reminiscent of NMR observations showing one such "borderline" residue in OmpX to be much more accessible to a water-soluble quencher in OmpX/A8-35 complexes than in diC$_6$-PC solution (Etzkorn et al. 2014) (see Chap. 10, Fig. 10.13). On the other hand, A8-35 seems to limit the accessibility of two residues located in the rather large extramembrane region of OmpT, W58 (in peptide T6), and F177 (in peptide T15). This partial protection could result from local contacts between this region and the APol bound to the TM region (Watkinson et al. 2017), as observed in molecular dynamics (MD) simulations carried out on

OmpX/A-35 complexes (Perlmutter et al. 2014) (see Chap. 5, Fig. 5.11B). One may note, however, that F177 is located in the most distal part of an extracellular loop, whereas the side chain of W58 faces the lumen of the $\beta$-barrel (Fig. 14.14), two positions where an interaction with the loops or tails of A8-35 molecules adsorbed onto the TM surface is not particularly likely (at least in the crystallographic conformation). One can probably not exclude an interaction with other APol particles (Watkinson et al. 2017). Another possibility is that the protective effect of A8-35 is indirect. In the case of OmpX, MD data indicate that trapping with A8-35 severely restricts the mobility of the extramembrane loops (Perlmutter et al. 2014) (see Chap. 5, Fig. 5.31), so that certain conformations become much less accessible in APol than in detergent solution (Fig. 5.32B). One cannot exclude that such an APol-induced stiffening (for a discussion, see Chap. 5, § 5.7) shifts the distribution of OmpF conformers away from those in which W58 and F177 are most accessible to peroxide-derived free radicals.

## 14.3.4   Can Amphipols Better Suited to ESI-MS Be Developed?

Vacuum is a strange medium for a membrane biochemist to maneuver in, when the basis of our thinking is the way chemical groups organize with respect to water. I therefore feel highly incompetent to put forward suggestions about possible directions for the future of APols in the field of ESI-MS. However, my perceived incompetence has seldom prevented me from indulging in speculations, and I hope the reader will bear with me for a few more. If there turns out to be a grain of truth in them, perfect. If not, at least we will have had our half-page of fun.

It may seem bizarre that monomeric MPs resist better transfer from aqueous solution to vacuum when associated to APols than to detergent (Calabrese et al. 2015; Watkinson et al. 2015), whereas the trimeric state of DAGK is not particularly well preserved (Hopper et al. 2013). A single case is not a demonstration, and it does not prove that A8-35 cannot be used for studying MP oligomers. One may wonder, however, what could be done to improve the usefulness of APols in preserving the native state of MPs during their transfer from the solution to the gas state. One can probably speculate that the problem, if any, may lie in (i) the polymeric nature of APols and (ii) the ionic nature of A8-35's polar moieties. The polymeric nature of APols implies that, as compared to DDM, less entropy will be generated when they are shed by the protein. The "cooling" effect of surfactant evaporation (Ruotolo et al. 2008; Borysik and Robinson 2012; Reading et al. 2015) therefore ought to be less efficient with APols than with detergents. The ionic nature of the polar moieties of A8-35 implies that they will establish strong interactions between themselves and with the protein as the complexes are dehydrated. The larger the molecules, the more polar contacts they will establish with their neighbors and with the MP surface. As is the case of DDM vs. Triton X-100 (Konijnenberg et al. 2014) or octylglucoside vs. $C_8E_4$ (Reading et al. 2015), more collisional energy will probably be needed to disassemble the APol belt than if the hydrophilic moieties were less polar. The same situation probably holds for glucosylated NAPols, whose polar moieties can establish multiple hydrogen bonds between them upon dehydration, much as those of DDM do (Rouse et al. 2013).

*Optimizing amphipols for ESI-mass spectrometry of membrane proteins*
(© *2018 by Francis Haraux*).

Can we deduce from the above some suggestions for the future? First, it may be reasonable to expect, and this seems to be supported by experimental data (Watkinson et al. 2015) (§ 14.3.1), that shorter APols will require less collisional energy to desorb and should have more of a cooling effect than longer ones. A5-75, a shorter version of A8-75, has been validated as an efficient APol (Gohon 1996; Popot et al. 2003) (see Chap. 5, Table 5.1). It could be worth comparing, for chemically equivalent structures, the protecting effects of A5-35 vs. A8-35 vs. A34-35. If the above speculation is correct and the trend observed in Study 14.7 when comparing the latter two APols (Watkinson et al. 2015) is significant, A5-35 might yield better results than A8-35. Second, it seems likely that APols whose polar moieties would not strongly interact one with another upon dehydration would desorb more easily, which could help better preserve the native state of the MPs they associate with. By analogy with the differences observed between Triton X-100 and DDM (Konijnenberg et al. 2014), or between $C_8E_4$ and octylglucoside (Reading et al. 2015), one may wonder, for instance, whether APols whose polar moieties would be comprised of short polyethylene glycol (PEG) moieties – which remain to be developed – would not be better suited to ESI-MS than the current ionic or glycosylated APols. Shorter APol molecules and PEG moieties could presumably be combined.

## 14.4   Proteomics

APols can potentially be used for some proteomics tasks, such as separating MPs and MP complexes using two-dimensional gels (P. Bazzacco, unpublished observations). This application, however, has yet to be developed. One application that has been efficiently implemented (Studies 14.5 and 14.9) is the use of A8-35 to facilitate the production, recovery, and analysis of peptides for proteomics investigations (Ning et al. 2013, 2014). Somewhat ironically, one of the drawbacks of A8-35, its insolubility at low pH, can be turned into an asset for these types of investigations.

The nature of the problems encountered when attempting to identify all proteins from a tissue or cell culture, or a membrane fraction thereof, is summarized in Ning et al. (2013). In brief, the study of the proteome by mass spectrometry requires the lysis and recovery of proteins from cells and tissues. This requires the disruption of plasma and intracellular membranes, in order to release intracellular and intraorganelle proteins. In addition, most proteins associated with membranes, be it through TM domains, non-covalent interactions with TM proteins or lipids, or anchoring side groups (see Chap. 1), are insoluble in water. Detergents are generally used to dissolve membranes and keep hydrophobic proteins water-soluble. Detergents, however, at the concentration used to disrupt membranes, tend to interfere with the proteolytic enzymes, such as trypsin, used to produce peptides for MS analysis. They must therefore be removed prior to proteolysis. Furthermore, most of them interfere with the chromatography and MS of peptides by high-pressure LC/ESI tandem mass spectrometry (HPLC-ESI-MS). Removing the detergent causes most MPs to precipitate. Chaotrophic agents are therefore required to keep them soluble for the reduction, alkylation, and proteolysis steps. The activity of proteolytic enzymes is typically tolerant of low concentration of chaotropes. However, most chaotropic agents are not suitable for MS and, therefore, have to be removed following proteolytic digestion. As a consequence of these various constraints, preparing peptides for identification and sequencing by MS classically involves a large number of separation and transfer steps, which are time-consuming and lead to loss of material (Fig. 14.15A, *left*; for a recent review, see e.g. Zhang 2015).

Typically, lysates are obtained by exposing the cells to a mixture of detergents, such as that present in radioimmunoprecipitation assay (RIPA) buffers (see legend to Fig. 14.15B). In Study 14.5,

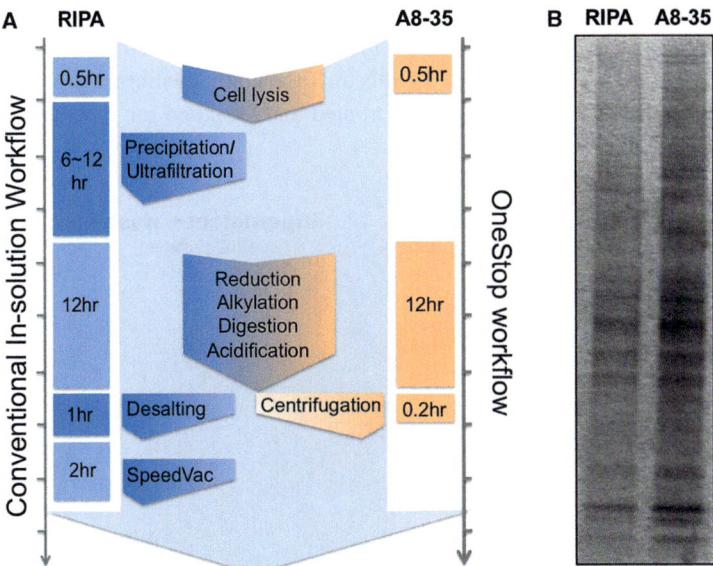

**Fig. 14.15** (**A**) A comparison of conventional (*left*) and A8-35-assisted (*right*) digestion strategies. The length of the color bars beside the axes is proportional to the duration of each step. Common steps are listed in the middle. The two steps colored in *dark blue* are absent in the APol-aided workflow. (**B**) Extraction efficiency test using either A8-35 (1 g·L$^{-1}$ in 50 mM ammonium bicarbonate buffer) or radioimmuno-precipitation assay (RIPA) buffer (50 mM Tris/HCl, pH 8, 150 mM NaCl, 1% NP-40, 0.5% sodium deoxycholate, 0.1% SDS) for extracting the total proteome from equal aliquots of HuH7 hepatocarcinoma cells. Both samples were vortexed and sonicated to speed up solubilization. SDS-PAGE comparison of A8-35 and RIPA extracts after precipitation (Reprinted with permission from Ning et al. 2013, © 2013 American Chemical Society).

A8-35 was used instead of detergents and chaotropes to extract and keep proteins soluble up to and during the proteolysis step. It was then separated from the tryptic peptides before ESI-MS/MS analysis by the simple device of acidifying the solution and removing the APol precipitate by table-top centrifugation. Compared to classical protocols, the use of APols eliminates the need for protein precipitation or ultrafiltration and sample desalting prior to and after proteolysis (Fig. 14.15A, *left*), greatly simplifying and speeding up the procedure (Fig. 14.15A, *right*). Furthermore, this strategy reduces sample loss compared with conventional approaches (Fig. 14.15B).

The efficiency of A8-35 at solubilizing MPs from whole cells observed in Study 14.5 contrasts with the poor detergency reported in early works (see e.g. Champeil et al. 2000). This issue deserves further investigation. However, two likely factors can be noted. For one thing, early solubilization experiments, using principally A8-35, happened to be carried out on membranes densely packed with proteins, such as the sarcoplasmic reticulum, thylakoids, or the purple membrane from *H. salinarum*. In other cases (bacterial and animal plasma membrane), direct extraction of MPs by A8-35 has been observed (see Chap. 5, § 5.2.2.2, Box 5.1). In the case of styrene-maleic acid copolymers (SMAs), it has been noted that rigid and protein-rich membranes, such as bacterial outer membranes, resist solubilization (Dörr et al. 2016) (Chap. 5, § 5.2.2.1). The choice of test membranes may thus have led to underestimating the detergency of APols. For another thing, an important difference between Study 14.5 and earlier solubilization attempts is the use of sonication. Sonication not only breaks up the target membranes into smaller fragments, but it also gives APol molecules access to the two faces of the membrane, which, in the case of detergent, is considered a critical step that determines the rate of solubilization (for a discussion, see le Maire et al. 2000). It may well be that sonication considerably accelerates the process of solubilization by A8-35 and that the higher detergency observed in Study 14.5 as compared to earlier studies is at least partially due to faster kinetics.

In a further publication (Ning et al. 2014), the same group observed that acid-induced precipitation of A8-35 carried out *before* trypsinolysis pulls down not only membrane proteins but also soluble ones, so that the whole proteome can be concentrated and prepared for proteolysis in a single step (Fig. 14.16).

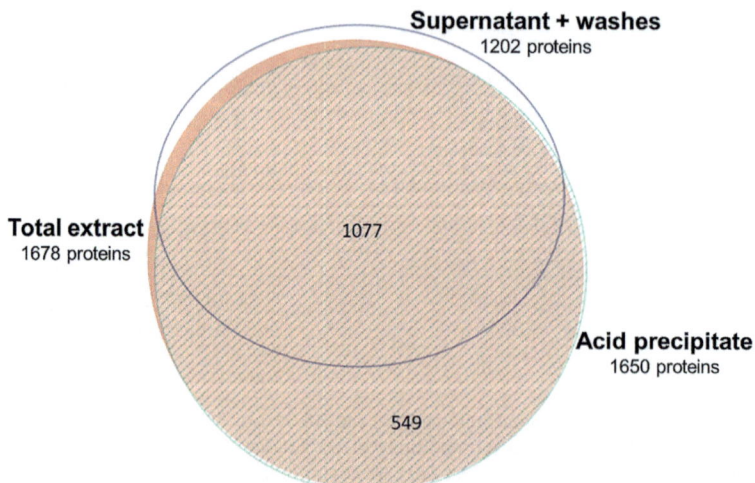

**Fig. 14.16**  Overlaps between the proteins identified in extracts from human embryonic kidney (HEK) 293T cells. Proteins were identified by proteolysis and LC-ESI-MS/MS in three types of samples: total extract obtained using 1 g·L$^{-1}$ A8-35 in 50 mM ammonium bicarbonate buffer, washed pH-3 precipitate, and supernatant + washes. The area of each circle is proportional to the number of proteins identified (Adapted from Ning et al. 2014).

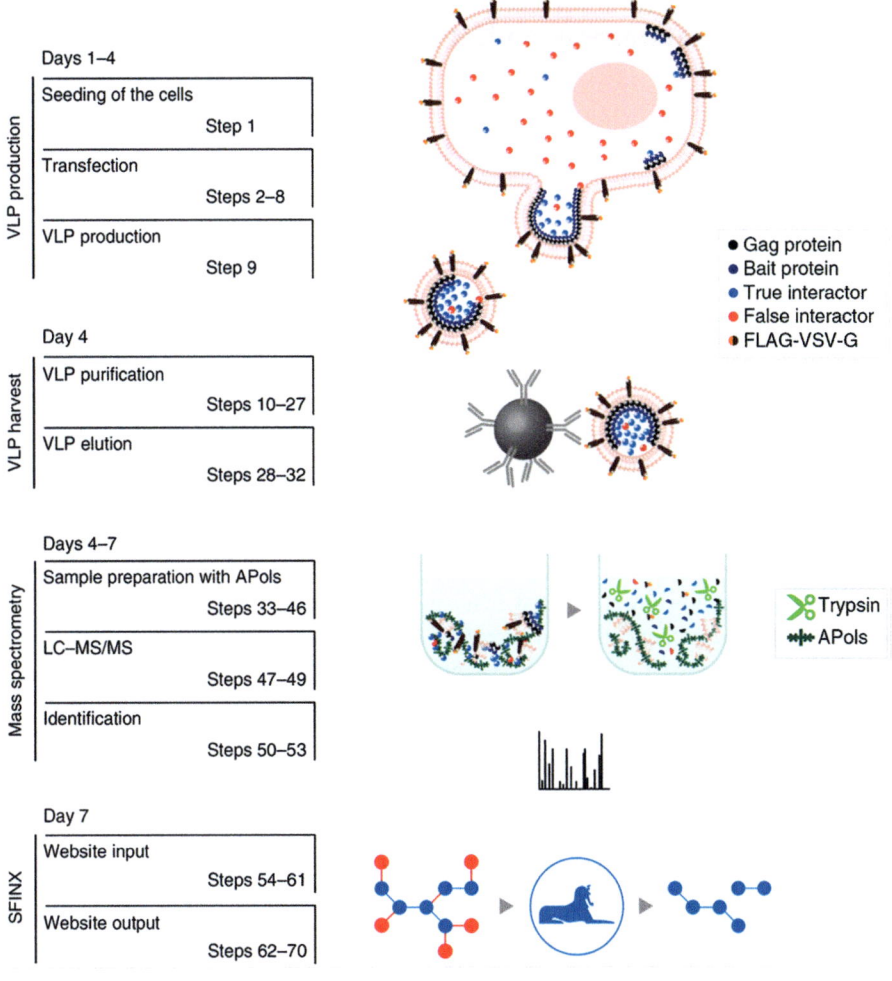

**Fig. 14.17** Overview of the Virotrap/SFINX protocol described in Study 14.9. The protocol consists of virus-like particle (VLP) production, VLP harvest, sample preparation using A8-35 (APols), liquid chromatography-tandem mass spectrometry (LC-MS/MS), peptide and protein identification, and elimination of false-positive interactions with the SFINX website interface (http://sfinx.ugent.be) (From Titeca et al. 2017, © 2013 Macmillan Publishers Limited, Nature Protocols. All rights reserved).

This approach has been extended to recovering and analyzing whole interactomes (Study 14.9; Titeca et al. 2017). The procedure relies in expressing in human embryonic kidney 293T (HEK293T) cell fusion proteins comprising a tagged version of the Gag protein from HIV, the *C*-terminus of which is fused to the *N*-terminus of the target protein. The expression of Gag suffices to induce the budding of vesicles (virus-like particles, VLPs) that trap the fusion protein along with any partner interacting with it (Fig. 14.17). VLPs are purified from the medium by affinity chromatography and solubilized with A8-35. As in Study 14.5, precipitation of A8-35 at low pH is used both to clean the preparation before trypsinolysis and to eliminate the APol after it, along with undigested proteins, including trypsin. The robust performance of APols on small sample volumes with dilute protein concentrations (Ning et al. 2014), combined with reduced sample transfer steps (Fig. 14.16), allowed downscaling of the protocol well below the minimal volumes required by detergent-based methods (Titeca et al. 2017).

## 14.5 Protocol 14.1: Characterization of Amphipol-Trapped Membrane Proteins by Electrospray Ionization-Ion Mobility Spectrometry-Mass Spectrometry

### 14.5.1 Introduction

Non-covalent (or "native") mass spectrometry (MS) is emerging as an invaluable method to study membrane proteins (MPs) and MP complexes, as well as their interactions with lipids and small molecules (Laganowsky et al. 2014; Bechara et al. 2015; Bechara and Robinson 2015; Konijnenberg et al. 2015; Rajabi et al. 2015; Gault et al. 2016). The approach requires samples to be maintained in their native conformations in solution before ionization, typically by the use of electrospray ionization (ESI). Implementation of non-covalent ESI-MS with ion mobility (IM) spectrometry, i.e. ESI-IM-MS, allows measurement of the ions' rotationally averaged collisional cross sections (CCSs), together with their mass-to-charge ($m/z$) ratios, in a single experiment. Ion CCSs are related to shape, size, and charge, and can yield structural information about biomolecules (Ruotolo et al. 2008).

Typically, MP analysis using ESI-IM-MS has been performed by solubilizing the protein of interest with detergents and subsequently dissociating the ionized MP/detergent complex in vacuo by collisional activation, leading to the ejection of the MP from its solubilizing media for mass analysis (Barrera et al. 2008; Laganowsky et al. 2013). This is key to successful MS analyses, as MP/surfactant complexes are highly heterogeneous, and therefore measurement of a complex's intact mass is not possible. We, and others, have developed detergent-free methods for the study of MPs using ESI-MS, by introducing MPs solubilized with amphipols (APols), bicelles, or nanodiscs into the gas phase for interrogation (Hopper et al. 2013; Calabrese et al. 2015; Watkinson et al. 2015; Marty et al. 2016a). Here, a general approach is described for the study of APol-solubilized MPs using ESI-IM-MS, highlighting key stages of the sample preparation and MS analyses that must be optimized for each system.

In the following, the protocol itself is printed in roman, background and comments in italics, preceded by a pointing hand (☞).

### 14.5.2 Materials

- Purified, detergent-solubilized MPs
- APol, usually A8-35, but we have also investigated the utility of A8-75, A34-35, A34-75, SAPol, and NAPol (see Chap. 4) for ESI-IM-MS, with mixed results (Watkinson et al. 2015)
- Caesium iodide solution (2 g·L$^{-1}$) in 50% (v/v) water/propan-2-ol
- IM calibrants (cytochrome $c$, bovine serum albumin, concanavalin A, alcohol dehydrogenase)
- ESI-MS compatible buffers: ammonium acetate and ammonium bicarbonate
- Buffer-exchange devices (e.g. desalting columns, dialysis apparatus)
- Gold/palladium-coated nano-ESI needles
- GELoader tips (Eppendorf)
- Synapt HDMS mass spectrometer, equipped with static nanoflow-ESI (nano-ESI) source (Waters Corporation)

### 14.5.3 Protocol

1. Preparation of samples for ESI-IM-MS

   ☞ *Conventional buffers used for biochemical analyses, such as >1 mM Tris/HCl or phosphate, are unsuitable for ESI-MS as they are nonvolatile and result in ion suppression and/or adduct formation. Therefore, MPs (and all biomolecules undergoing native ESI-MS analysis) must be transferred to a MS-compatible buffer before analysis. ESI-MS-compatible, volatile buffers include ammonium acetate or ammonium bicarbonate, which are routinely used at concentrations up to 1 M, depending on the desired pH and ionic strength (Hernandez and Robinson 2007). Buffer exchange can be achieved by a number of means, including dialysis, desalting columns, size exclusion chromatography, or ultrafiltration; however, it should be ensured that stringent buffer exchange is employed, as nonvolatile buffer contaminants can result in poor spectral quality. Nano-ESI is often used for non-covalent MS analyses, where sample flow rates of a few $nL \cdot min^{-1}$ are used (compared with $\mu L \cdot min^{-1}$ for conventional ESI), because the method is more sensitive than conventional ESI, is more tolerant to buffer contaminants, and requires lower amounts of sample (~1 $\mu L$ of sample can be used to obtain a mass spectrum). Alternatively, matrix-assisted laser desorption ionization (MALDI) can be used to ionize proteins for MS. APols have been shown to be compatible with MALDI-MS (Bechara et al. 2012); however, ionization by this method does not permit native structure and interactions to be maintained in vacuo.*

   *Typically, we have used detergent-solubilized MPs for subsequent trapping in APols (Calabrese et al. 2015). However, β-barrel outer membrane proteins refolded directly into APols have also been used for ESI-MS analysis (Leney et al. 2012). For this work, a range of APols has been screened. Among them, A8-35 showed the most promise as a generic solubilizing agent, allowing ESI-IM-MS spectra to be recorded for the greatest range of MPs tested (Watkinson et al. 2015).*

   In our study of both outer membrane and inner membrane proteins, the following sample preparation procedure was performed:

   - Detergent-solubilized MP samples were prepared in 10 mM Tris/HCl, pH 8.0, 0.02% (w/v) DDM. For ESI-MS, protein concentrations in the low μM range are required. We routinely prepare 50–500-μL samples (depending on protein availability) at protein concentrations ca. 10 μM.
   - APol was added to the detergent-solubilized membrane protein at a MP/APol ratio of 1:2 up to 1:5 (w/w), and the samples were incubated on ice for 30 min.

     ☞ *In our experience, lower quantities of free APol in solution result in increased spectral quality, so it is recommended that the amount of added APol be minimized for each sample, without compromising sample integrity. This should be optimized empirically for each protein by determining the lowest amount of APol to add without protein aggregation occurring.*

   - Bio-Beads were prepared by washing with methanol (3×), followed by water (3×) and the sample buffer (10 mM Tris/HCl, pH 8.0, with no added detergent). The pre-wet, washed, Bio-Beads were added (20 g wet beads per g of detergent) to remove the detergent and the sample mixture incubated for 1 h at 4 °C with gentle agitation.

- APol-trapped MPs were then dialyzed against 100 mM ammonium bicarbonate, pH 8.0, at 4 °C for 24 h.

  ☞*Ammonium bicarbonate was chosen as the buffer, because below pH 7 MP/APol complexes aggregate (see Chap. 4, § 4.3.1.2.2). The optimal buffering range for ammonium bicarbonate is pH 8.2–11.3. Ammonium acetate, the other commonly used MS-compatible buffer, is most efficient as a buffer at pH 3.8–5.8, and hence not suitable for analyses of MP/ A8-35 complexes.*

  ☞*It may be necessary to perform size exclusion chromatography (SEC) of APol-trapped MPs prior to ESI-IM-MS analysis. SEC enables removal of free APol, which, in our experience, enhances the detectability of MPs. For example, when the transmembrane β-barrel domain of OmpA (tOmpA) was trapped in A34-35, a mass spectrum could only be acquired after SEC (Watkinson et al. 2015).*

- Preparation of IM calibrants: cytochrome *c*, bovine serum albumin, concanavalin A, and alcohol dehydrogenase (Bush et al. 2010) were solubilized in 200 mM ammonium acetate solution (concentration ~10 μM) and desalted using spin desalting columns (e.g. Micro BioSpin 6, Bio-Rad) according to the manufacturer's instructions.

2. ESI-IM-MS analysis

  ☞*Several IM-enabled mass spectrometers are commercially available. Our analyses were performed on a Waters Synapt HDMS instrument, a hybrid quadrupole-travelling wave ion mobility-time-of-flight mass spectrometer (Fig. 14.17) (Pringle et al. 2007). The travelling wave (T-wave) sector of the instrument comprises an IM cell, sandwiched between two collision cells (termed the "trap" and "transfer" cells). Our instrument is fitted with a static nano-ESI source, and a quadrupole capable of mass selection up to m/z = 32,000 (Sobott et al. 2002). For nano-ESI, samples are ionized by loading into an electrically conductive capillary, which is placed near the instrument's sampling cone. An electric potential is applied to the capillary, resulting in a Taylor cone, and the formation of ions which are drawn into the mass spectrometer as a result of the potential difference between the capillary and sampling cone.*

- Capillaries are commercially available (e.g. from Waters Corporation and Proxeon Biosystems) or can be prepared in-house. We use borosilicate glass capillaries that are pulled to a tapered tip using a micropipette puller (Model P-97, Sutter Instruments) and coated using a sputter coater fitted with a gold/palladium target (Emitech Sc7620). Capillary tips are trimmed (with tweezers, the capillary being visualized with a light microscope) before use to obtain a stable spray. The cut capillary has often an internal diameter of 1–10 μm (this is a trial and error process) (Hernandez and Robinson 2007).

  ☞*For evidence that ionization by nano-ESI leads to native protein conformations and interactions being retained in vacuo see e.g. Ruotolo et al. (2005), Ruotolo and Robinson (2006), and Hilton and Benesch (2012).*

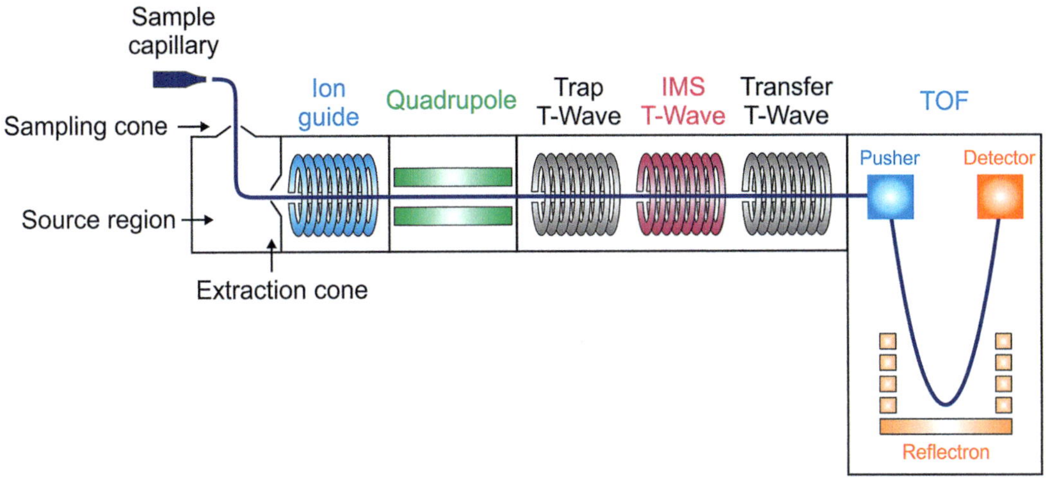

**Fig. 14.18** Schematic of a Waters Synapt HDMS hybrid quadrupole-travelling wave ion mobility-time-of-flight mass spectrometer equipped with nano-ESI. Important components are labelled, along with several key regions where parameters must be optimized for native MS of APol-trapped MPs. The ion beam is shown in *purple*.

- The *m/z* scale can be calibrated by spraying a solution of caesium iodide (see *Materials*) and processing the data according to the instrument manufacturer's instructions.

☞ *For non-covalent MS of soluble proteins, instrument voltages must be kept gentle to ensure that protein unfolding is minimized* (Hernandez and Robinson 2007; Sobott et al. 2002). *However, for APol-trapped MPs, instrument voltages must be increased and carefully tuned to ensure that APol molecules are dissociated from the complex and the protein is released for mass analysis* (Fig. 14.18, *inset*). *Optimization of a number of instrument parameters is essential to ensuring transmission and detection of MP complexes. In our experience, not all MPs could be observed when trapped with any APol. For example, we have shown that the outer membrane proteins OmpT, PagP, and tOmpA could all be released from A8-35, but that release from other APols (A8-75, A34-35, A34-75, SAPol, and NAPol) was variable* (Watkinson et al. 2015). *This is likely because the internal energies imparted on the MP/APol assemblies are insufficient to cause dissociation. In an attempt to overcome this, instruments have been modified to access higher voltages. Several key parameters to optimize on the Synapt HDMS are discussed below.*
☞ *Capillary voltage. The sample capillary* (Fig. 14.18) *is mounted on the stage, placed ~10 mm from the sampling cone, and a voltage is applied to the capillary to initiate a spray. Typical capillary voltages range from 1 to 1.8 kV for nano-ESI. It may be necessary to apply a gas pressure from behind the sample (0–2 bar), termed nanoflow gas, to initiate a spray. However, optimal spectra are typically obtained without nanoflow gas. Once a spray is initiated, the capillary voltage is gradually decreased to the lowest possible value that gives a stable spray (often in the region of 1.2–1.4 kV).*

☞*Source pressure. The transmission of ions into the mass spectrometer, and the maintenance of tertiary and quaternary protein structure, is heavily influenced by the so-called "backing" pressure in the source region of the instrument (Fig. 14.18). An increased source pressure results in collisional cooling and collisional focusing of ions, and can aid in the transmission of ions with high m/z. The optimal source pressure should be determined empirically for each sample, and can be adjusted using the isolation valve (Speedivalve) of the roughing pump, which changes the conductance of the vacuum line between the source and the pump. In our experience, source pressures in the region of 3–5 mbar were required to observe well-resolved spectra of APol-solubilized MPs. Pressures up to ~8 mbar can be achieved on the instrument, but care should be taken when adjusting the Speedivalve at such pressures to ensure the instrument does not automatically vent.*

☞*Sample cone and extraction cone. Increased cone voltages (sample and extraction cone, Fig. 14.18) can improve desolvation (leading to increases in signal-to-noise and mass accuracy) and the transmission of high m/z ions. Additionally, increasing the cone voltages imparts higher internal energies on analyte ions. This can be considered a double-edged sword for MP analysis, as increased energy can aid the dissociation of APols and ejection of "naked" MPs, but also potentially lead to gas phase unfolding and charge stripping of the protein. Unfolding can be discerned by analysis of the charge state distribution observed (more highly charged ions are more unfolded) or the CCS values of the ion species observed (increased CCSs indicate protein unfolding) (Benesch 2009). Cone voltages should be optimized to ensure optimal ion transmission and APol dissociation, without resulting in protein unfolding. In our experience, sampling cone voltages >100 V are required to observe APol-trapped MPs. The extraction cone should be set between 0 and 10 V.*

☞*Trap DC bias. The voltages in the IM sector of the instrument can also have a significant influence on ion heating. Again this can be considered a double-edged sword, as described above for the cone voltages. The trap DC bias voltage, the voltage which accelerates ions into the higher-pressure IM sector of the instrument from the lower-pressure trap cell (Fig. 14.18), is particularly important to optimize. For soluble proteins, the trap DC bias is typically set to 20 V for optimal transmission without inducing unfolding. In our experience, voltages of 80–100 V are required to observe APol-trapped MPs.*

☞*Trap/transfer collision voltages. The trap and transfer sectors of the Synapt HDMS (Fig. 14.18) are filled with an inert buffer gas, typically argon. Analyte ions are accelerated into the trap/transfer cells and undergo collisions with the buffer gas. This can be used to dissociate the MP from the APol. The trap and transfer collision voltages should be optimized empirically for each sample, but typically the trap collision voltage was set to 50–100 V, while the transfer collision voltage was set to 10–150 V.*

☞*IM parameters. IM separates ions based on their mass, shape, and charge. More extended ions experience more collisions with the buffer gas (in our case nitrogen, but helium can also be used; see Bush et al. 2010), relative to more compact ions, and therefore exit the ion mobility cell more slowly. The IM sector of the Synapt HDMS comprises the trap collision cell, ion mobility cell, and transfer collision cell, each of which is a stacked-ring ion guide (Fig. 14.18). Ion packets are pulsed periodically into the central ion mobility cell from the trap for IM separation. The mobility (drift time) of each ion is recorded, and the value of m/z is determined by time-of-flight (TOF) analysis. For each ion species, an arrival time distribution (ATD) can be extracted from the three-dimensional m/z, drift time, and intensity data set (Fig. 14.19). The width of ATDs of a protein or protein assembly can be used to study conformational heterogeneity and conformational changes (Konijnenberg et al. 2014; Zhou et al. 2014; Martens et al. 2016; Schiffrin et al. 2016). Because each conformation presents a Gaussian distribution, ATDs*

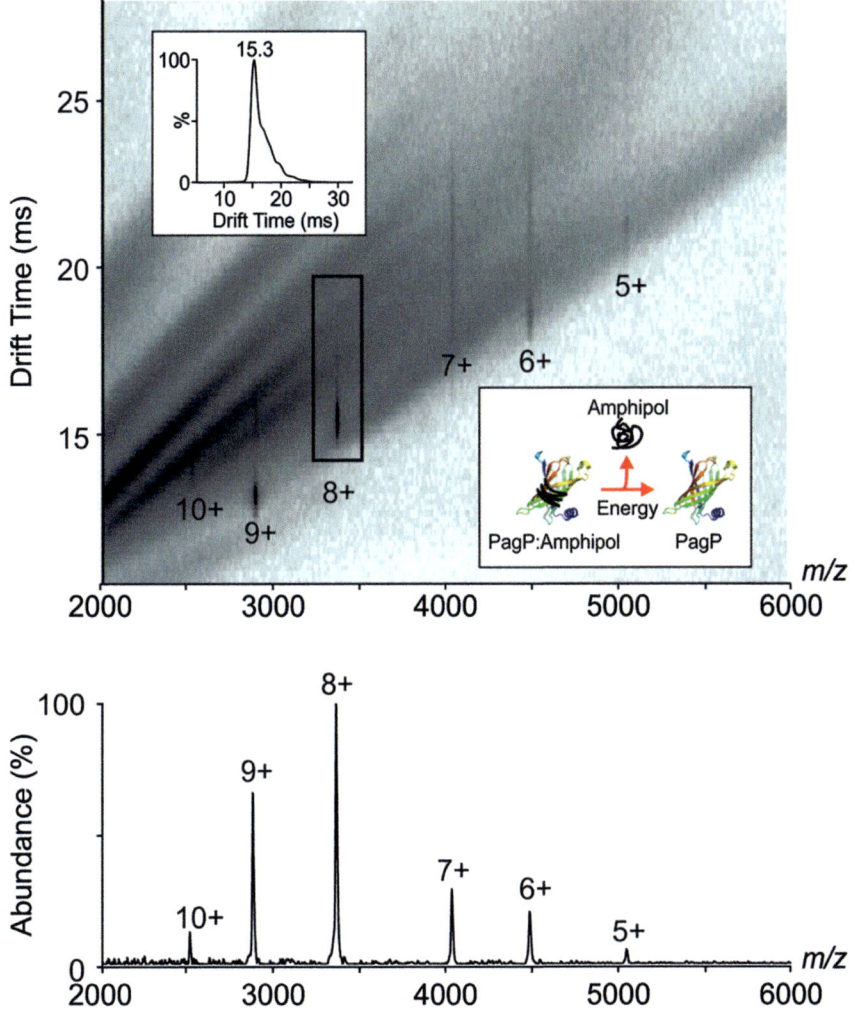

**Fig. 14.19** Nano-ESI-IM-MS analysis of a PagP/A8-35 sample (1:5 w/w) using a Waters Synapt G1 HDMS. Elevated trap and transfer collision energies and trap DC bias were used to observe free PagP ions that were not visible when in complex with heterogeneous A8-35. A typical mass spectrum is shown below an IM-MS heatmap. Ions of PagP are labelled with their charge state. The lower inset on the IM-MS heatmap displays a schematic of APol removal from PagP by collisional activation of the PagP/APol complex. The upper inset displays an example arrival time distribution for the 8+ charge state (which is boxed) (Figure courtesy of A. Calabrese. For more details, see Calabrese et al. 2015).

*must be deconvoluted to determine the number of conformations present* (Smith et al. 2009). *Additionally, collisional activation of the ion species can be performed and the ATDs scrutinized as a function of input internal energy to monitor unfolding. This can be used to discern if ligand binding has a stabilizing effect on structure* (Allison et al. 2015; Laganowsky et al. 2014; Eschweiler et al. 2015). *IM parameters must be optimized to ensure good ion separation and to minimize artifacts caused, for example by rollover (where ions from the previous IM pulse have not exited the mobility cell and reached the TOF before the next ion pulse). The travelling wave height and velocity parameters in the IM cell should be optimized for each analyte to ensure rollover does not occur and that good ion separation is achieved over the ms time scale of the IM experiment* (Ruotolo et al. 2008). *The gas pressure in the IM cell is typically set to 0.5 mbar;*

*wave heights in the region of 5–25 V and wave velocities between 200–400 m·s⁻¹ are typically* *used to achieve optimal separation. Further details of these parameters are beyond the scope* *of this protocol. The reader is referred to other published methods for more information* (Ruotolo et al. 2008).

3. Collisional cross section calibration and comparison with structural data

- ESI-IM-MS spectra of calibrant proteins must be acquired under the same instrument conditions used for the MP/APol complexes for all parameters downstream of the trap ion guide. When using native protein standards, the trap collision energy and cone voltages should be decreased, relative to those needed to observe MP/APol complexes, to prevent protein unfolding.

☞ *The calibration of CCSs from travelling wave IM ATD data is well established, and the CCSs* *of the calibrant proteins are known. The reader is referred to several recent publications for* *more information on the calibration procedure, which is now relatively routine* (Ruotolo et al. 2008; Bush et al. 2010).

☞ *The power of IM lies in the ability to compare experimentally determined CCS data with those* *of models or high-resolution structures. Several software packages, which use different* *methodologies, have been developed to determine in silico CCS values for comparison with* *IM data. The most reliable CCS calculation method is the trajectory method, which is* *implemented in the software MOBCAL and other packages (see* Mesleh et al. 1996; Shvartsburg *and Jarrold 1996, and MOBCAL–A Program to Calculate Mobilities. http://www.indiana.edu/* *~nano/software/). Recent work has demonstrated the use of alternative, less computationally* *intensive approaches that can provide data of similar quality, e.g. the corrected projection* *approximation implemented in IMPACT* (Marklund et al. 2015) *or the projection superposition* *approximation* (Bleiholder et al. 2011).

### 14.5.4   Outlook

MS-based methods, including native MS, but also fast photochemical oxidation (FPOP)-MS, hydrogen deuterium exchange (HDX)-MS, chemical crosslinking-MS, and other footprinting approaches, are becoming increasingly used to study membrane protein structure. We and others have demonstrated the utility of nonconventional surfactants to solubilize MPs of both $\beta$-barrel and $\alpha$-helical structures for native ESI-IM-MS (Hebling et al. 2010; Leney et al. 2012; Hopper et al. 2013; Calabrese et al. 2015; Lu et al. 2016; Marty et al. 2016; Watkinson et al. 2017). We have also performed a systematic analysis into the utility of a range of amphipols (A8-35, A8-75, A34-35, A34-75, SAPol, and NAPol) for ESI-IM-MS analysis (Watkinson et al. 2015). Continued development, application, and refinement of novel solubilization methods, including the use of APols, for native MS and other structural MS approaches will open new avenues to gain insights into the functional mechanisms of a range of MPs.

Protocol prepared by Antonio N. Calabrese, Thomas G. Watkinson, Sheena E. Radford, and Alison E. Ashcroft, Astbury Centre for Structural Molecular Biology, School of Molecular and Cellular Biology University of Leeds, based on Calabrese et al. (2015), and Watkinson et al. (2015), with comments added.

# References

Allison, T.M., Reading, E., Liko, I., Baldwin, A.J., Laganowsky, A., Robinson, C.V. (2015) Quantifying the stabilizing effects of protein-ligand interactions in the gas phase. *Nat. Comm.* **6**:8551.

Barrera, N.P., Di Bartolo, N., Booth, P.J., Robinson, C.V. (2008) Micelles protect membrane complexes from solution to vacuum. *Science* **321**:243–246.

Barrera, N.P., Isaacson, S.C., Zhou, M., Bavro, V.N., Welch, A., Schaedler, T.A., Seeger, M.A., Miguel, R.N., Korkhov, V.M., van Veen, H.W., Venter, H., Walmsley, A.R., Tate, C.G., Robinson, C.V. (2009) Mass spectrometry of membrane transporters reveals subunit stoichiometry and interactions. *Nat. Meth.* **6**:585–587.

Barrera, N.P., Robinson, C.V. (2011) Advances in the mass spectrometry of membrane proteins: from individual proteins to intact complexes. *Annu. Rev. Biochem.* **80**:247–271.

Bazzacco, P., Billon-Denis, E., Sharma, K.S., Catoire, L.J., Mary, S., Le Bon, C., Point, E., Banères, J.-L., Durand, G., Zito, F., Pucci, B., Popot, J.-L. (2012) Non-ionic homopolymeric amphipols: Application to membrane protein folding, cell-free synthesis, and solution NMR. *Biochemistry* **51**:1416–1430.

Bechara, C., Bolbach, G., Bazzacco, P., Sharma, S.K., Durand, G., Popot, J.-L., Zito, F., Sagan, S. (2012) MALDI mass spectrometry analysis of membrane protein/amphipol complexes. *Anal. Chem.* **84**:6128–6135.

Bechara, C., Nöll, A., Morgner, N., Degiacomi, M.T., Tampé, R., Robinson, C.V. (2015) A subset of annular lipids is linked to the flippase activity of an ABC transporter. *Nat. Chem.* **7**:255–262.

Bechara, C., Robinson, C.V. (2015) Different modes of lipid binding to membrane proteins probed by mass spectrometry. *J. Am. Chem. Soc.* **137**:5240–5247.

Benesch, J.L. (2009) Collisional activation of protein complexes: picking up the pieces. *J. Am. Soc. Mass Spectrom.* **20**:341–348.

Berry, E.A., Guergova-Kuras, M., Huang, L.-S., Crofts, A.R. (2000) Structure and function of cytochrome $bc_1$ complexes. *Annu. Rev. Biochem.* **69**:1005–1075.

Bleiholder, C., Wyttenbach, T., Bowers, M.T. (2011) A novel projection approximation algorithm for the fast and accurate computation of molecular collision cross sections. (I) Method. *Int. J. Mass Spectrom.* **308**:1–10.

Borysik, A.J., Hewitt, D.J., Robinson, C.V. (2013) Detergent release prolongs the lifetime of native-like membrane protein conformations in the gas-phase. *J. Am. Chem. Soc.* **135**:6078–6083.

Borysik, A.J., Robinson, C.V. (2012) The 'sticky business' of cleaning gas-phase membrane proteins: a detergent oriented perspective. *Phys. Chem. Chem. Phys.* **14**:14439–14449.

Botte, M., Zaccai, N.R., Lycklama à Nijeholt, J., Martin, R., Knoops, K., Papai, G., Zou, J., Deniaud, A., Karuppasamy, M., Jiang, Q., Singha Roy, A., Schulten, K., Schultz, P., Rappsilber, J., Zaccai, G., Berger, I., Collinson, I., Schaffitzel, C. (2016) A central cavity within the holotranslocon suggests a mechanism for membrane protein insertion. *Sci. Rep.* **6**:38399.

Bush, M.F., Hall, Z., Giles, K., Hoyes, J., Robinson, C.V., Ruotolo, B.T. (2010) Collision cross-sections of proteins and their complexes: A calibration framework and database for gas-phase structural biology. *Anal. Chem.* **82**:9557–9565.

Cadene, M., Chait, B.T. (2000) A robust, detergent-friendly method for mass spectrometric analysis of integral membrane proteins. *Anal. Chem.* **72**:5655–5658.

Calabrese, A.N., Watkinson, T.G., Henderson, P.J.F., Radford, S.E., Ashcroft, A.E. (2015) Amphipols outperform dodecylmaltoside micelles in stabilizing membrane protein structure in the gas phase. *Anal. Chem.* **87**:1118–1126.

Catoire, L.J., Zoonens, M., van Heijenoort, C., Giusti, F., Popot, J.-L., Guittet, E. (2009) Inter- and intramolecular contacts in a membrane protein/surfactant complex observed by heteronuclear dipole-to-dipole cross-relaxation. *J. Magn. Res.* **197**:91–95.

Champeil, P., Menguy, T., Tribet, C., Popot, J.-L., le Maire, M. (2000) Interaction of amphipols with the sarcoplasmic reticulum $Ca^{2+}$-ATPase. *J. Biol. Chem.* **275**:18623–18637.

Dahmane, T., Giusti, F., Catoire, L.J., Popot, J.-L. (2011) Sulfonated amphipols: Synthesis, properties and applications. *Biopolymers* **95**:811–823.

Dörr, J.M., Scheidelaar, S., Koorengevel, M.C., Dominguez, J.J., Schäfer, M., van Walree, C.A., Killian, J.A. (2016) The styrene-maleic acid copolymer: a versatile tool in membrane research. *Eur. Biophys. J.* **45**:3–21.

Eschweiler, J.D., Rabuck-Gibbons, J.N., Tian, Y., Ruotolo, B.T. (2015) CIUSuite: A quantitative analysis package for collision induced unfolding measurements of gas-phase protein ions. *Anal. Chem.* **87**:11516–11522.

Etzkorn, M., Zoonens, M., Catoire, L.J., Popot, J.-L., Hiller, S. (2014) How amphipols embed membrane proteins: Global solvent accessibility and interaction with a flexible protein terminus. *J. Membr. Biol.* **247**:965–970.

Gault, J., Donlan, J.A.C., Liko, I., Hopper, J.T.S., Gupta, K., Housden, N.G., Struwe, W.B., Marty, M.T., Mize, T., Bechara, C., Zhu, Y., Wu, B., Kleanthous, C., Belov, M., Damoc, E., Makarov, A., Robinson, C.V. (2016) High-resolution mass spectrometry of small molecules bound to membrane proteins. *Nat. Methods* **13**:333–336.

Giusti, F., Rieger, J., Catoire, L., Qian, S., Calabrese, A.N., Watkinson, T.G., Casiraghi, M., Radford, S.E., Ashcroft, A. E., Popot, J.-L. (2014) Synthesis, characterization and applications of a perdeuterated amphipol. *J. Membr. Biol.* **247**:909–924.

Gohon, Y. (1996) Etude des interactions entre un analogue du fragment transmembranaire de la glycophorine A et des polymères amphiphiles: les amphipols. Rapport de DEA, Université Paris VI, Paris, 28 p.

Gohon, Y., Dahmane, T., Ruigrok, R., Schuck, P., Charvolin, D., Rappaport, F., Timmins, P., Engelman, D.M., Tribet, C., Popot, J.-L., Ebel, C. (2008) Bacteriorhodopsin/amphipol complexes: structural and functional properties. *Biophys. J.* **94**:3523–3537.

Hall, Z., Robinson, C.V. (2012) Do charge state signatures guarantee protein conformations? *J. Am. Soc. Mass Spectrom.* **23**:1161–1168.

Hebling, C.M., Morgan, C.R., Stafford, D.W., Jorgenson, J.W., Rand, K.D., Engen, J.R. (2010) Conformational analysis of membrane proteins in phospholipid bilayer nanodiscs by hydrogen exchange mass spectrometry. *Anal. Chem.* **82**:5415–5419.

Hernandez, H., Robinson, C.V. (2007) Determining the stoichiometry and interactions of macromolecular assemblies from mass spectrometry. *Nat. Protocols* **2**:715–726.

Hilton, G.R., Benesch, J.L. (2012) Two decades of studying non-covalent biomolecular assemblies by means of electrospray ionization mass spectrometry. *J. R. Soc. Interface* **9**:801–816.

Hopper, J.T.S., Yu, Y.T.-C., Li, D., Raymond, A., Bostock, M., Liko, I., Mikhailov, V., Laganowsky, A., Benesch, J.L. P., Caffrey, M., Nietlispach, D., Robinson, C.V. (2013) Detergent-free mass spectrometry of membrane protein complexes. *Nat. Meth.* **10**:1206–1208.

Juraschek, R., Dülcks, T., Karas, M. (1999) Nanoelectrospray–more than just a minimized-flow electrospray ionization source *J. Am. Soc. Mass Spectrom.* **10**:300–308.

Kanu, A.B., Dwivedi, P., Tam, M., Matz, L., Hill Jr., H.H. (2008) Ion mobility-mass spectrometry. *J. Mass Spectrom.* **43**:1–22.

Konijnenberg, A., van Dyck, J.F., Kailing, L.L., Sobott, F. (2015) Extending native mass spectrometry approaches to integral membrane proteins. *Biol. Chem.* **396**:991–1002.

Konijnenberg, A., Yilmaz, D., Ingólfsson, H.I., Dimitrova, A., Marrink, S.J., Li, Z., Vénien-Bryan, C., Sobott, F., Koçer, A. (2014) Global structural changes of an ion channel during its gating are followed by ion mobility mass spectrometry. *Proc. Natl. Acad. Sci. USA* **111**:17170–17175.

Laganowsky, A., Reading, E., Allison, T.M., Ulmschneider, M.B., Degiacomi, M.T., Baldwin, A.J., Robinson, C.V. (2014) Membrane proteins bind lipids selectively to modulate their structure and function. *Nature* **510**:172–175.

Laganowsky, A., Reading, E., Hopper, T.S., Robinson, C.V. (2013) Mass spectrometry of intact membrane protein complexes. *Nat. Protoc.* **8**:639–651.

le Maire, M., Champeil, P., Møller, J.V. (2000) Interaction of membrane proteins and lipids with solubilizing detergents. *Biochim. Biophys. Acta* **1508**:86–111.

Leney, A.C., McMorran, L.M., Radford, S.E., Ashcroft, A.E. (2012) Amphipathic polymers enable the study of functional membrane proteins in the gas phase. *Anal. Chem.* **84**:9841–9847.

Liu, L., Bagal, D., Kitova, E.N., Schnier, P.D., Klassen, J.S. (2009) Hydrophobic protein–ligand interactions preserved in the gas phase. *J. Am. Chem. Soc.* **131**:15980–15981.

Lu, Y., Zhang, H., Niedzwiedzki, D.M., Jiang, J., Blankenship, R.E., Gross, M.L. (2016) Fast photochemical oxidation of proteins maps the topology of intrinsic membrane proteins: Light-Harvesting Complex 2 in a nanodisc. *Anal. Chem.* **88**:8827–8834.

Mann, M., Jensen, O.N. (2003) Proteomic analysis of post-translational modifications. *Nat. Biotechnol.* **21**:255–261.

Marklund, E.G., Degiacomi, M.T., Robinson, C.V., Baldwin, A.J., Benesch, J.L. (2015) Collision cross-sections for structural proteomics. *Structure* **23**:791–799.

Martens, C., Stein, R.A., Masureel, M., Roth, A., Mishra, S., Dawaliby, R., Konijnenberg, A., Sobott, F., Govaerts, C., McHaourab, H.S. (2016) Lipids modulate the conformational dynamics of a secondary multidrug transporter. *Nat. Struct. Mol. Biol.* **23**:744–751.

Marty, M.T., Hoi, K.K., Gault, J., Robinson, C.V. (2016a) Probing the lipid annular belt by gas-phase dissociation of membrane proteins in nanodiscs. *Angew. Chem. Int. Ed.* **55**:550–554.

Marty, M.T., Hoi, K.K., Robinson, C.V. (2016b) Interfacing membrane mimetics with mass spectrometry. *Acc. Chem. Res.* **49**:2459–2467.

Mehmood, S., Allison, T.M., Robinson, C.V. (2015) Mass spectrometry of protein complexes: from origins to applications. *Annu. Rev. Phys. Chem.* **66**:453–474.

Mesleh, M.F., Hunter, J.M., Shvartsburg, A.A., Schatz, G.C., Jarrold, M.F. (1996) Structural information from ion mobility measurements: Effects of the long-range potential. *J. Phys. Chem.* **100**:16082–16086.

Morgner, N., Kleinschroth, T., Barth, H.D., Ludwig, B., Brutschy, B. (2007) A novel approach to analyze membrane proteins by laser mass spectrometry: from protein subunits to the integral complex. *J. Am. Soc. Mass Spectrom.* **18**:1429–1438.

Ning, Z., Hawley, B., Seebun, D., Figeys, D. (2014) APols-aided protein precipitation: a rapid method for concentrating proteins for proteomic analysis. *J. Membr. Biol.* **247**:941–947.

Ning, Z., Seebun, D., Hawley, B., Chang, C.-K., Figeys, D. (2013) From cells to peptides: "One-stop" integrated proteomic processing using amphipols. *J. Proteome Res.* **12**:1512–1519.

Perlmutter, J.D., Popot, J.-L., Sachs, J.N. (2014) Molecular dynamics simulations of a membrane protein/amphipol complex. *J. Membr. Biol.* **247**:883–895.

Popot, J.-L., Berry, E.A., Charvolin, D., Creuzenet, C., Ebel, C., Engelman, D.M., Flötenmeyer, M., Giusti, F., Gohon, Y., Hervé, P., Hong, Q., Lakey, J.H., Leonard, K., Shuman, H.A., Timmins, P., Warschawski, D.E., Zito, F., Zoonens, M., Pucci, B., Tribet, C. (2003) Amphipols: polymeric surfactants for membrane biology research. *Cell. Mol. Life Sci.* **60**:1559–1574.

Popot, J.-L., Althoff, T., Bagnard, D., Baneres, J.L., Bazzacco, P., Billon-Denis, E., Catoire, L.J., Champeil, P., Charvolin, D., Cocco, M.J., Cremel, G., Dahmane, T., de la Maza, L.M., Ebel, C., Gabel, F., Giusti, F., Gohon, Y., Goormaghtigh, E., Guittet, E., Kleinschmidt, J.H., Kuhlbrandt, W., Le Bon, C., Martinez, K.L., Picard, M., Pucci, B., Sachs, J.N., Tribet, C., van Heijenoort, C., Wien, F., Zito, F., Zoonens, M. (2011) Amphipols from A to Z. *Annu. Rev. Biophys.* **40**:379–408.

Pringle, S.D., Giles, K., Wildgoose, J.L., Williams, J.P., Slade, S.E., Thalassinos, K., Bateman, R.H., Bowers, M.T., Scrivens, J.H. (2007) An investigation of the mobility separation of some peptide and protein ions using a new hybrid quadrupole/travelling wave IMS/oa-ToF instrument. *Int. J. Mass Spectrom.* **261**:1–12.

Rajabi, K., Ashcroft, A.E., Radford, S.E. (2015) Mass spectrometric methods to analyze the structural organization of macromolecular complexes. *Methods* **89**:13–21.

Reading, E., Liko, I., Allison, T.M., Benesch, J.L.P., Laganowsky, A., Robinson, C.V. (2015) The role of the detergent micelle in preserving the structure of membrane proteins in the gas phase. *Angew. Chem. Int. Ed.* **54**:4577–4581

Rouse, S.L., Marcoux, J., Robinson, C.V., Sansom, M.S.P. (2013) Dodecylmaltoside protects membrane proteins *in vacuo*. *Biophys. J.* **105**:648–656.

Ruotolo, B.T., Benesch, J.L.P., Sandercock, A.M., Hyung, S.J., Robinson, C.V. (2008) Ion mobility-mass spectrometry analysis of large protein complexes. *Nat. Protoc.* **3**:1139–1152.

Ruotolo, B.T., Giles, K., Campuzano, I., Sandercock, A.M., Bateman, R.H., Robinson, C.V. (2005) Evidence for macromolecular protein rings in the absence of bulk water. *Science* **310**:1658–1661.

Ruotolo, B.T., Robinson, C.V. (2006) Aspects of native proteins are retained in vacuum. *Curr. Opin. Chem. Biol.* **10**:402–408.

Schiffrin, B., Calabrese, A.N., Devine, P.W., Harris, S.A., Ashcroft, A.E., Brockwell, D.J., Radford, S.E. (2016) Skp is a multivalent chaperone of outer-membrane proteins. *Nat. Struct. Mol. Biol.* **23**:786–793.

Sharma, K.S., Durand, G., Gabel, F., Bazzacco, P., Le Bon, C., Billon-Denis, E., Catoire, L.J., Popot, J.-L., Ebel, C., Pucci, B. (2012) Non-ionic amphiphilic homopolymers: Synthesis, solution properties, and biochemical validation. *Langmuir* **28**:4625–4639.

Shvartsburg, A.A., Jarrold, M.F. (1996) An exact hard-spheres scattering model for the mobilities of polyatomic ions. *Chem. Phys. Lett.* **261**:86–91.

Smith, D.P., Knapman, T.W., Campuzano, I., Malham, R.W., Berryman, J.T., Radford, S.E., Ashcroft, A.E. (2009) Deciphering drift time measurements from travelling wave ion mobility spectrometry-mass spectrometry studies. *Eur. J. Mass Spectrom.* **15**:113–130.

Sobott, F., Hernandez, H., McCammon, M.G., Tito, M.A., Robinson, C.V. (2002) A tandem mass spectrometer for improved transmission and analysis of large macromolecular assemblies. *Anal. Chem.* **74**:1402–1407.

Stroebel, D., Choquet, Y., Popot, J.-L., Picot, D. (2003) An atypical haem in the cytochrome $b_6 f$ complex. *Nature* **426**:413–418.

Titeca, K., Van Quickelberghe, E., Samyn, N., De Sutter, D., Verhee, A., Gevaert, K., Tavernier, J., Eyckerman S. (2017) Analyzing trapped protein complexes by Virotrap and SFINX. *Nat. Protoc.* **12**:881–898.

Tribet, C., Audebert, R., Popot, J.-L. (1996) Amphipols: polymers that keep membrane proteins soluble in aqueous solutions. *Proc. Natl. Acad. Sci. USA* **93**:15047–15050.

van Dyck, J.F., Konijnenberg, A., Sobott, F. (2017) Native mass spectrometry for the characterization of structure and interactions of membrane proteins. *Meth. Mol. Biol.* **1635**:205–232.

Watkinson, T.G., Calabrese, A.N., Ault, J.R., Radford, S.E., Ashcroft, A.E. (2017) FPOP-LC-MS/MS suggests differences in interaction sites of amphipols and detergents with outer membrane proteins. *J. Am. Soc. Mass Spectrom.* **28**:50–55.

Watkinson, T.G., Calabrese, A.N., Giusti, F., Zoonens, M., Radford, S.E., Ashcroft, A.E. (2015) Systematic analysis of the use of amphipathic polymers for studies of outer membrane proteins using mass spectrometry. *Int. J. Mass Spectrom.* **391**:54–61.

Weidner, S.M., Trimpin, S. (2010) Mass spectrometry of synthetic polymers. *Anal. Chem.* **82**:4811–4829.

Whitelegge, J.P. (2013) Integral membrane proteins and bilayer proteomics. *Anal. Chem.* **85**:2558–2568.

Wilm, M.S., Mann, M. (1994) Electrospray and Taylor-Cone theory, Dole's beam of macromolecules at last? *Int. J. Mass Spectrom. Ion Processes* **136**:167–180.

Xu, G., Chance, M.R. (2007) Hydroxyl radical-mediated modification of proteins as probes for structural proteomics *Chem. Rev.* **107**:3514–3543.

Zaccai, N.R., Serdyuk, I.N., Zaccai, G. (2017) *Methods in Molecular Biophysics. Structure, dynamics, function,* 2nd ed., Cambridge University Press, Cambridge, 684 p.

Zenobi, R., Knochenmuss, R. (1998) Ion formation in MALDI mass spectrometry. *Mass Spectrom. Rev.* **17**:337–366

Zhang, X. (2015) Less is more: membrane protein digestion beyond urea-trypsin solution for next-level proteomics. *Mol. Cell. Proteom.* **14**:2441–2453.

Zhou, M., Politis, A., Davies, R.B., Liko, I., Wu, K.J., Stewart, A.G., Stock, D., Robinson, C.V. (2014) Ion mobility-mass spectrometry of a rotary ATPase reveals ATP-induced reduction in conformational flexibility. *Nat. Chem.* **6**:208–215.

Zoonens, M., Popot, J.-L. (2014) Amphipols for each season. *J. Membr. Biol.* **247**:759–796.

# Biomedical Applications

# 15

**Summary**

*Biomedical applications of amphipols (APols) are still in their infancy. However, some projects have been initiated that appear promising and ought to stimulate further developments. First, it has been demonstrated that a membrane protein from a pathogenic bacterium responsible for a widely distributed sexually transmitted infection, the major outer membrane protein (MOMP) from* Chlamydia trachomatis, *can be used as immunogen in the formulation of vaccines after being trapped with A8-35. A8-35-trapped MOMP is considerably stabilized as compared to its detergent-solubilized form, extending the thermal resistance and shelf life of the vaccine and increasing the protection it confers to mice as compared to detergent-based formulations. Versions of A8-35 carrying covalently linked adjuvants for co-delivery have been developed and, according to the first results, appear to stimulate the immune response better than free adjuvants. Second, internalization of A8-35 and its payload have been observed in cell cultures. Third, the absence of toxicity of A8-35 and its distribution and elimination when injected in mice by various routes have been characterized using fluorescent variants of A8-35. These data provide a basis for the use of APols for delivering drugs, e.g. hydrophobic anticancer peptides mimicking the transmembrane anchor of growth factor receptors. The rich chemistry of APols and the fact that several APol molecules bind to a given membrane protein provide exceptional opportunities for associating a drug or antigen with adjuvants and/or targeting moieties.*

## 15.1 Introduction

In Chap. 13, we have evoked one particular biomedical application of amphipols (APols), namely using tagged APols to immobilize putative drug targets onto solid supports for the purpose of in vitro drug screening or diagnostics. There are also prospects of in vivo uses for human therapeutics. The absence of apparent toxicity of A8-35 when injected in mice or applied to cell cultures was established quite early (Popot et al. 2003), raising the possibility that APols could be used for delivering drugs or immunogens. Exploration of this avenue has been quite slow and is only at its beginning, but the first observations are promising.

© Springer International Publishing AG, part of Springer Nature 2018

J. -L. Popot, *Membrane Proteins in Aqueous Solutions*, Biological and Medical Physics, Biomedical Engineering, https://doi.org/10.1007/978-3-319-73148-3_15

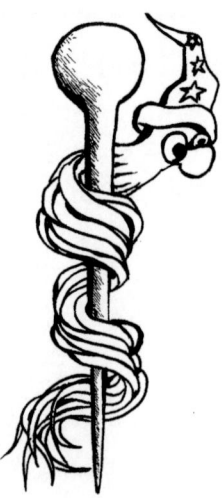

*Looking ahead to medical applications*
(© *2018 by Francis Haraux*).

The results available to date fall into two categories: (i) demonstrating that APols can be advantageously used to formulate subunit vaccines that present a better efficacy, thermal resistance, and shelf life than their detergent-based analogs, and (ii) examining the fate of APols and their cargoes when applied to cells in culture and their distribution and elimination when injected in mice. In addition, APols carrying various adjuvants have been synthesized. Relevant studies are listed in Table 15.1.

**Table 15.1**  Some studies relevant to biomedical uses of amphipols.[1]

| Study | Field | Comments | References |
|---|---|---|---|
| **15.1** | Cell biology | Visualization of the internalization of A8-35 and a passenger transmembrane peptide by cells in culture and of the access of a fluorescent APol to tumors in live mice | Popot et al. (2011) |
| **15.2** | Vaccinology | Compared protections afforded by detergent- and A8-35-based vaccines against *Chlamydia trachomatis* | Tifrea et al. (2011) |
| **15.3** | Biochemistry/ biophysics | Stability, solution properties, and shelf life of the major outer membrane protein from *C. trachomatis* in detergent solution vs. an A8-35-trapped form | Feinstein et al. (2014) |
| **15.4** | Vaccinology | Investigations into the origins of the better protection afforded by an A8-35- vs. a detergent-based vaccine against *C. muridarum* | Tifrea et al. (2014) |
| **15.5** | Formulation | Distribution and elimination of A8-35 following injection in mice by various routes | Fernandez et al. (2014) |
| **15.6** | Chemistry/ physical chemistry | Synthesis and characterization of several adjuvant-carrying versions of A8-35 | Le Bon et al. (2014a, b), and Tifrea et al. (2018a, b) |
| **15.7** | Vaccinology | Synthesis of an APol carrying a covalently bound peptidic adjuvant and demonstration that using the resulting "vaccipol" to trap and deliver a MP used as an immunogen can improve the immune response as compared to the use of the free adjuvant | Tifrea et al. (2018b) |

[1]Published too late to be discussed here is a study in which a membrane-associated amyloid-$\beta$ oligomer potentially relevant in Alzheimer's disease  was stabilized by non-ionic APols (Serra-Batiste et al. 2018)

## 15.2 Using Amphipols to Formulate a Subunit Vaccine Against *Chlamydia trachomatis*

In the following, we will briefly consider the results of attempts at using APols to improve the efficiency of a vaccine against the intracellular pathogenic bacterium *Chlamydia trachomatis*, before listing a few suggestions about how APols could contribute, more generally, to improving the formulation of vaccines. As will be discussed in the conclusion, the exploitation of APols in this area of medical research has been much neglected given its prospects, for reasons that, in my view, are more cultural than factual.

### 15.2.1 Background

Historically, vaccines have been prepared using whole pathogenic microorganisms that have been rendered harmless by either "attenuating" or killing them. Attenuation is achieved by repeatedly growing the organism in a foreign host, until it loses its virulence to humans. Killing is achieved by treating the organisms with chemicals, heat, or radiation. For a number of reasons, these vaccines tend to be the most efficient, but they cannot always be produced, viz. the cases of HIV, malaria, tuberculosis in adults, *Staphylococcus aureus*, or the dengue virus (for recent reviews, see e.g. Levine et al. 2010; Delany et al. 2014; Doolan et al. 2015; Plotkin et al. 2017). Furthermore, they do have their problems, such as safety and storage issues with attenuated pathogens. Vaccines based on whole organisms present a risk of batch-to-batch variations, and subjects are injected with a host of ill-defined molecules besides the desirable immunogens, some of which can cause side effects, such as hyper-sensitivity reactions. This has led to the development of alternative approaches, among which are recombinant DNA-based vaccines and "subunit" vaccines, in which purified immunogens can be injected either free or attached to a carrier (for recent reviews, see e.g. Moyle and Toth 2013; Bobbala and Hook 2016; Vartak and Sucheck 2016).

In the context of vaccine formulation, the term "subunit" does not refer to a protein that is part of an oligomer, as it does in biochemistry, but to an immunogen comprising subcellular elements that have been extracted from the pathogenic organism or synthesized, rather than whole cells or viral particles. These elements can be proteins, nucleic acids, lipids, polysaccharides, subcellular fragments, inactivated toxins (toxoids), etc. The use of highly purified molecules strongly improves, if not eliminates, the safety risk associated with biologics production. Subunit vaccines are generally easier to standardize and to store than those based on attenuated or killed pathogens, which improves product consistency, but they are also much less immunogenic (Moyle and Toth 2013; Bobbala and Hook 2016). Approaches to increasing their immunogenicity include binding them to a carrier, with the risk of potential batch-to-batch variations, and co-injecting them along with agents that will stimulate the immune response (adjuvants). "Adjuvant" is a broad term that can refer either to a chemical that will stimulate the immune response, e.g. a bacterial lipid or an oligodeoxynucleotide (ODN) mimicking bacterial DNA, or a physical support that improves the presentation of the immunogen to the organism, such as aluminum salts or an oil/water emulsion.

Purified membrane proteins (MPs) are not typically used in subunit vaccines, because of their well-deserved reputation of being difficult to produce and unstable in detergent solutions (cf. Chap. 2). APols could be game changers, for several reasons: (i) APols can facilitate large-scale production of MPs, e.g. by helping to fold them from inclusion bodies (Chap. 6); (ii) they stabilize MPs (Chap. 5, § 5.5); (iii) the physical-chemical properties of APols can be modulated, which, as will be discussed in § 15.2.3, could play an important role in optimizing the presentation of APol-trapped MPs to the immune system; and (iv) they can be derivatized (Chap. 4, § 4.4). APols can therefore carry adjuvants along with the proteins used as immunogen and/or tags that will direct them toward specific cells or associate them with particles used as carrier, so as to improve the uptake of target MPs. APol particles resemble, to an extent, the polymeric micelles already used in some vaccine formulations (Trimaille and Verrier 2015), some of which are based on hydrophobized poly($\gamma$-glutamic acid), a category of polymers which, as we have seen in Chap. 5 (Table 5.1), can be used as APols (Han et al. 2014). This intriguing set of potential advantages prompted attempts at using APols in the formulation of vaccines directed against a widely distributed pathogenic bacterium, *C. trachomatis*.

Chlamydiae are Gram-negative (Gram$^-$) eubacteria, obligate intracellular pathogens, and symbionts of diverse organisms, ranging from humans to amoebae (reviewed in Bachmann et al. 2014; Elwell et al. 2016). The best-studied group in the *Chlamydiae* phylum is the *Chlamydiaceae* family, which comprises 11 species that are pathogenic to humans or animals. *C. trachomatis* and *C. pneumoniae*, the major species that infect humans, are responsible for a wide range of diseases (Schachter and Dawson 1978; Centers for Disease Control and Prevention 1999; Schachter 1999). *C. trachomatis* is one of the most common human pathogens and is found in all regions of the world, where it is the leading cause of sexually transmitted bacterial infections. It can produce ocular, gastrointestinal, and respiratory infections, which can affect persons of all ages. In young individuals, *C. trachomatis* is the most common sexually transmitted bacterial pathogen. Genital infections can remain asymptomatic, but others can produce acute symptomatology. In women, long-term sequelae such as infertility and ectopic pregnancy can develop. At birth, newborns can become infected in the eyes and lungs if the mother has a genital tract infection at the time of delivery. In countries with poor hygienic conditions, young children can have multiple ocular infections that result in the development of trachoma later in life. *C. trachomatis* has also been isolated from the lungs of adults, in particular from immunocompromised patients. Infection with *C. trachomatis* facilitates the transmission of HIV and is associated with cervical cancer. *C. pneumoniae* causes respiratory infections, accounting for ~10% of community-acquired pneumonia, and is linked to a number of chronic diseases, including asthma, atherosclerosis, and arthritis (for reviews, see e.g. Schachter 1999; Elwell et al. 2016). The mouse pathogen *C. muridarum* – previously called *C. trachomatis* mouse pneumonitis (MoPn) – is a useful model of genital tract infections in humans.

Although chlamydial infections are treatable with antibiotics, many individuals go untreated, no drug is sufficiently cost-effective for the elimination of the bacterium in developing nations, even patients that are treated may develop chronic sequelae, and, somewhat paradoxically, countries that have established screening programs, followed by antibiotherapy, have observed an increase in the prevalence of the infection (Götz et al. 2002). This increase is thought to be due to a block in the development of natural immunity as a result of antibiotherapy (Götz et al. 2002; Brunham et al. 2005). The implementation of a vaccine is probably the best approach to control and eradicate these diseases (de la Maza and Peterson 2002; Brunham and Rey-Ladino 2005; Rockey et al. 2009; Farris and Morrison 2011), but producing an effective vaccine has thus far remained elusive (for a recent review, see de la Maza et al. 2017).

All chlamydiae share a developmental cycle in which they alternate between extracellular, infectious elementary bodies (EBs) and intracellular, noninfectious reticulate bodies (Fig. 15.1). EBs enter mucosal cells and differentiate into reticulate bodies in a membrane-enclosed compartment – the inclusion. After several rounds of replication, reticulate bodies redifferentiate into EBs and are released from the host cell, ready to infect neighboring cells (Elwell et al. 2016).

Attempts to produce a vaccine against *C. trachomatis* were initiated in the 1960s (see Grayston and Wang 1978, Igietseme et al. 2011, and references therein). Vaccines formulated with whole inactivated and viable organisms were tested in humans and in nonhuman primates for protection against trachoma. Several conclusions were reached from these studies. Some vaccine protocols

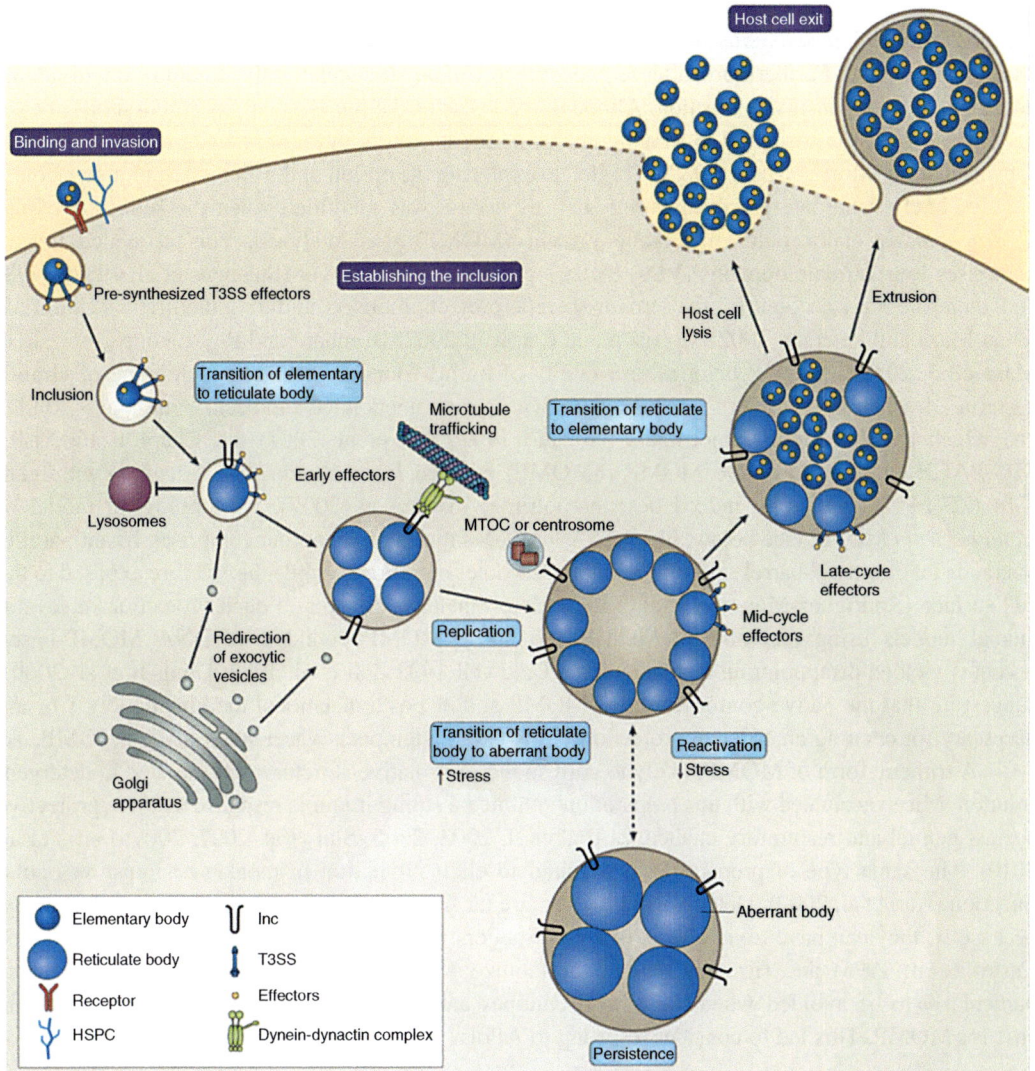

**Fig. 15.1**   The life cycle of *Chlamydia trachomatis*. The binding of elementary bodies (EBs) to host cells is initiated by the formation of a trimolecular bridge between bacterial adhesins, host receptors, and host heparan sulfate proteoglycans (HSPGs). Next, pre-synthesized type III secretion system (T3SS) effectors are injected into the host cell, some of which initiate cytoskeletal rearrangements to facilitate internalization and/or initiate mitogenic signaling to establish an antiapoptotic state. The EB is endocytosed into a membrane-enclosed compartment, known as the inclusion, which escapes the canonical endolysosomal pathway. Bacterial protein synthesis begins, EBs convert to reticulate bodies, and newly secreted inclusion membrane proteins (Incs) promote nutrient acquisition by hijacking exocytic vesicles that are in transit from the Golgi apparatus to the plasma membrane. The nascent inclusion is transported, probably by an Inc, along microtubules to the microtubule-organizing center (MTOC) or centrosome. During mid-cycle, the reticulate bodies replicate exponentially and secrete additional effectors that modulate processes in the host cell. Under conditions of stress, the reticulate bodies enter a persistent state as enlarged aberrant bodies. The bacteria can be reactivated upon removal of the stress. During the late stages of infection, reticulate bodies secrete late-cycle effectors and synthesize EB-specific effectors before differentiating back to EBs. EBs exit the host through lysis or extrusion (From Elwell et al. 2016, © 2016 Macmillan Publishers Limited, Nature Reviews Microbiology. All rights reserved).

induced protection, but the latter lasted only 1–2 years. In addition, it appeared to be serovar-specific, i.e., of the four *C. trachomatis* ocular isolates, the protection was effective only against the serovar used in the vaccine. Furthermore, after re-exposure to *Chlamydia*, some of the immunized individuals developed a hypersensitivity reaction. Although the cause of the hypersensitivity reaction has not yet been elucidated, it is attributed to an antigenic component present in the whole *Chlamydia* cells used for vaccination (Morrison et al. 1989). This led to exploring the option of developing a subunit vaccine.

A likely candidate for formulating such a vaccine was identified when the sequence of the *C. trachomatis* major outer membrane protein (MOMP) was analyzed. This sequence indeed comprises four variable domains (VDs) that are unique to each serovar (Stephens et al. 1987, 1998) and, therefore, may account for the serovar-specific protection observed during the first vaccine trials (de la Maza and Peterson 2002; Morrison and Caldwell 2002; Brunham and Rey-Ladino 2005; de la Maza et al. 2017). MOMP belongs to a family of porins found in the outer membrane of Gram$^-$ bacteria, of which *E. coli* OmpF is a prototype, whose monomers have a molecular mass of $\sim$40 kDa and which assemble into homotrimers (Nikaido 1992; Sun et al. 2007) (cf. Chap. 1, Fig. 1.9). SDS-PAGE analyses of native MOMP (nMOMP) purified from *C. muridarum* using Zwittergent 3-14 (Z3-14) showed it to indeed be a homotrimer (Sun et al. 2007). A topological model of *C. muridarum* MOMP (see below, Fig. 15.5A) proposes that each monomer comprises 16 antiparallel $\beta$-strands that form a $\beta$-barrel structure spanning the outer membrane, while the VDs are exposed to the cell surface (Rodriguez-Maranon et al. 2002). Unfortunately, attempts to elicit protection in several animal models using recombinant MOMP (rMOMP), MOMP peptides, or DNA MOMP-based vaccines yielded disappointing results (Su and Caldwell 1992; Pal et al. 1999; Dong-Ji et al. 2000), suggesting that the native conformation of MOMP and/or posttranslational modification(s) is or are necessary for eliciting an efficient protection. This led to attempts at vaccination using nMOMP.

A trimeric form of MOMP, likely to correspond to its native structure, was isolated in detergent solution. Mice vaccinated with this preparation mounted a strong immune response that was protective against genital and respiratory challenges (Pal et al. 2002, 2005; Sun et al. 2007, 2009; Farris et al. 2010). The same type of preparation was found to elicit protection in monkeys against an ocular infection (Kari et al. 2009). Detergents, however, are far from ideal surfactants to formulate a vaccine, because (i) they can have toxic effects (see e.g. Speijers et al. 1988, 1989; Blake and Wetzler 1995; Castro et al. 1995) and (ii) they tend to destabilize MPs (Chap. 2). The latter phenomenon is particularly to be avoided when the critical epitopes are conformational ones, as seems to be the case for MOMP. This led to consider resorting to APols.

### 15.2.2   Developing an Amphipol-Based Vaccine

In Studies 15.2–15.4 (Table 15.1), Luis M. de la Maza and his colleagues examined (i) whether transfer of nMOMP from Z3-14 to A8-35 would stabilize it and (ii) whether the efficiency of the vaccine would improve if it was formulated with MOMP/A8-35 rather than MOMP/Z3-14 complexes. These two expectations were vindicated (Tifrea et al. 2011, 2014; Feinstein et al. 2014).

The stability of the secondary structure of nMOMP in the presence of either Z3-14 or A8-35 was examined by Melanie J. Cocco and her colleagues by monitoring changes in the far-UV CD spectrum as a function of temperature (Fig. 15.2A). At 24 °C, the CD spectra of the two types of samples were indistinguishable. In keeping with earlier data (Cai et al. 2009), the CD spectrum of Z3-14-solubilized nMOMP changed with temperature, indicating a loss of $\beta$-strands and the formation of $\alpha$-helices, with a transition around 52 °C (Fig. 15.2B). In contrast, the secondary structure of A8-35-trapped nMOMP was not visibly affected up to 78 °C, the highest temperature tested (Fig. 15.2B). MP stabilization upon

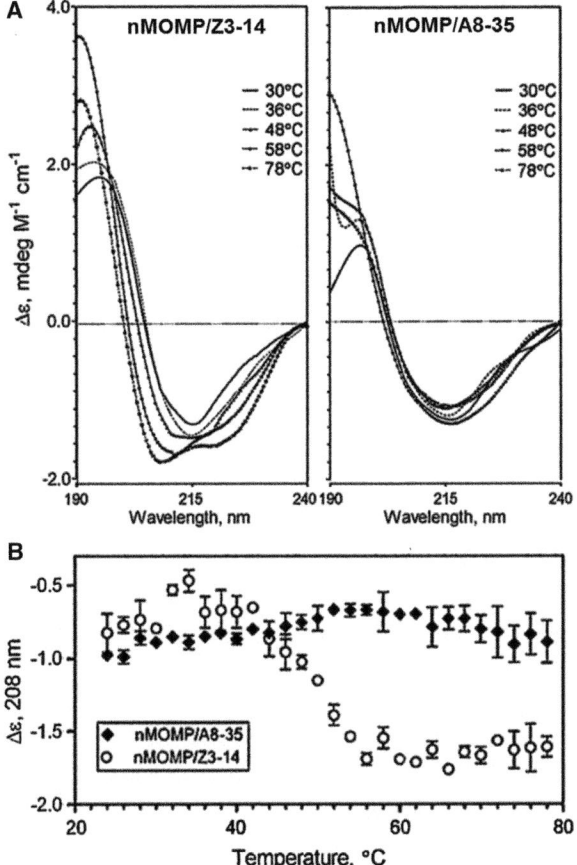

**Fig. 15.2** Circular dichroism (CD) spectra and thermostability of native MOMP (nMOMP) from *Chlamydia trachomatis* either kept in Zwittergent 3-14 (Z3-14) or transferred to A8-35. Preparations of nMOMP (0.033 g·L$^{-1}$ in 20 mM sodium phosphate buffer, pH 7.4) were incubated at temperatures ranging from 24 to 78 °C and analyzed by CD from 190 to 240 nm. (**A**) CD spectra of nMOMP in 0.05% Z3-14 and of nMOMP/A8-35 complexes at 30, 36, 48, 58, and 78 °C. (**B**) Plot of $\Delta\varepsilon$ at 208 nm for detergent-solubilized and A8-35-trapped nMOMP preparations as a function of temperature (From Tifrea et al. 2011, Copyright © 2011 Elsevier Ltd. All rights reserved).

transfer from detergent to APols is more the rule than the exception (see Chap. 5, § 5.5), but a transition temperature shift by >30 °C is exceptionally high.

Thermostabilization usually translates into a longer shelf life (see e.g. Chap. 5, Figs. 5.21 and 5.22). A higher stability of the native structure of nMOMP upon extended storage was indeed observed in Study 15.3, and this whatever criterion was used, be it the preservation of secondary structure (Fig. 15.3A), the absence of aggregation (Fig. 15.3B), the preservation of the trimeric state (Fig. 15.3C), or the recognition by a conformation-specific monoclonal antibody (Fig. 15.3D) (Feinstein et al. 2014). In addition, as noted in Chap. 5, § 5.5, for bacteriorhodopsin/A8-35 complexes, nMOMP was resistant to several cycles of freeze/thawing when trapped in A8-35, whereas the same treatment broke up the trimer kept in detergent solution (Fig. 15.3E) (Feinstein et al. 2014). Thermo- and cryostability, as well as resistance to freeze/thawing, are important practical features of vaccines for efficient distribution and storage, particularly in low-income tropical countries, where *C. trachomatis* is endemic and infrastructure is often limited.

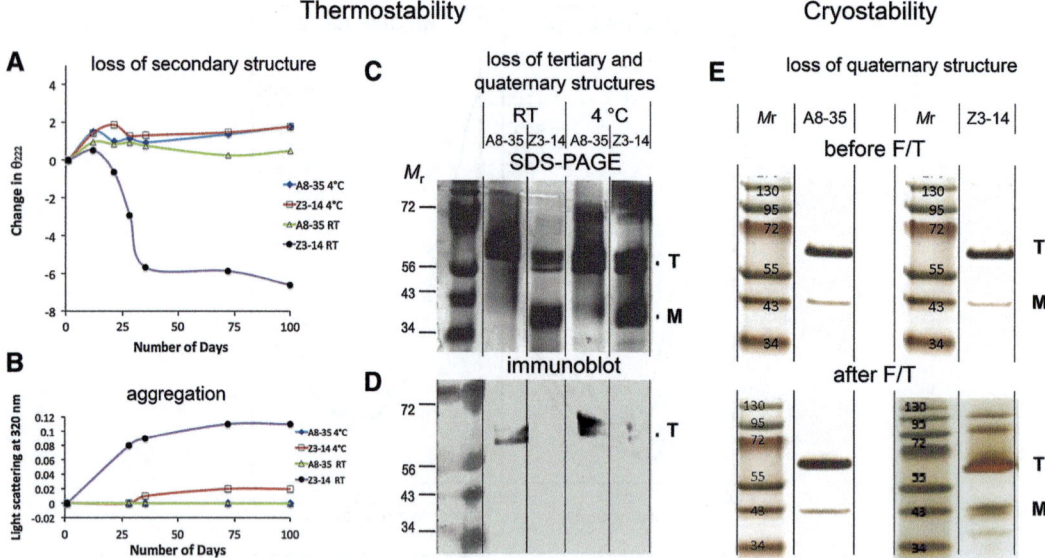

**Fig. 15.3** Thermo- and cryostability of native MOMP (nMOMP) either kept in detergent solution (Zwittergent 3-14 0.5%; Z3-14) or transferred to A8-35. (**A**) Preservation of the secondary structure. Evolution of the ellipticity at 222 nm as a function of time over a period of 100 days during which samples were kept either at 4 °C or a room temperature (RT). *Brown* and *purple lines*, nMOMP/Z3-14 complexes kept at 4 °C and RT, respectively; *blue* and *green lines*, nMOMP/A8-35 complexes kept at 4 °C and RT, respectively. (**B**) Protein aggregation as assessed from light scattering at 320 nm. Same color code. (**C, D**) Integrity of the tertiary and quaternary structures of nMOMP samples in either Z3-14 or A8-35, stored either at RT or at 4 °C for 100 days, as monitored by SDS-PAGE (**C**) and Western blotting using a conformation-sensitive monoclonal antibody (**D**). Native trimers and denatured monomers are indicated by *T* and *M*, respectively. By 100 days, loss of the trimer and formation of the monomer are detected in the detergent-solubilized sample, but not in the APol-trapped one. (**E**) Cryostability. Silver-stained SDS-PAGE analysis of nMOMP/Z3-14 and nMOMP/A8-35 samples before (*top*) and after (*bottom*) ten freeze/thaw (F/T) cycles. A band of monomer and two bands of aggregates appear in the detergent-solubilized sample, but not in the APol-trapped one (Adapted from Feinstein et al. 2014).

Mice were vaccinated with nMOMP or rMOMP by the intramuscular (i.m.; 7 μg protein/mouse/immunization) and the subcutaneous (s.c.; 3 μg protein/mouse/immunization) routes. CpG-1826 (10 μg/mouse/dose; 5′-TCCATGACGTTCCTGACGTT-3′) and Montanide ISA 720 at a 30:70 volume ratio of MOMP + CpG to Montanide were used as adjuvants (Sun et al. 2009). The mice were boosted two times at 2-week intervals with the same vaccine formulation. A negative control group of mice was inoculated intranasally (i.n.) with buffer (MEM-0). Positive control mice were immunized i.n. once with $1 \times 10^4$ inclusion forming units (IFU) of *C. muridarum* (Sun et al. 2009). Following an intranasal challenge of mice with live bacteria, the protection against infection afforded by nMOMP/A8-35 was more robust than that achieved with nMOMP/Z3-14, as assessed by various criteria including the loss of body weight (Fig. 15.4A) or the number of IFUs of *C. trachomatis* recovered in the lungs (Fig. 15.4B). The protection observed was almost as high as that of mice that had recovered from a previous infection with *C. muridarum* (noted CT MoPn in Fig. 15.4). It was higher with nMOMP than with a misfolded, monomeric recombinant preparation of MOMP (rMOMP) purified from inclusion bodies, consistent with the view that the native state of the protein is an important factor in determining the level of protection achieved. Interestingly, no difference in protection was observed between rMOMP preparations formulated with either detergent or APol.

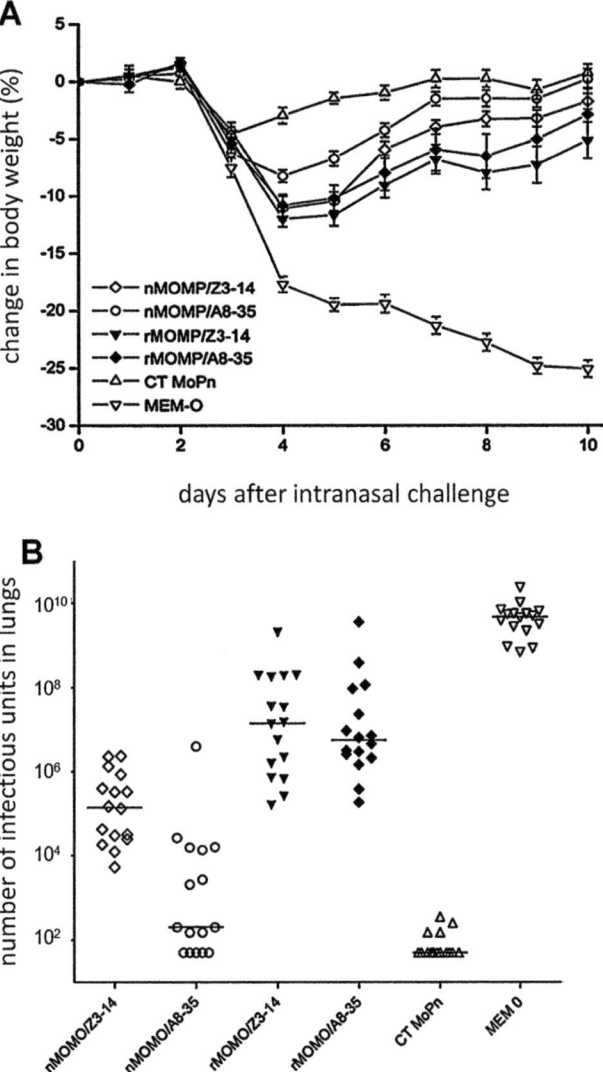

**Fig. 15.4** Protection afforded by preparations of native MOMP (nMOMP) or recombinant MOMP (rMOMP), formulated either with a detergent (Z3-14) or with A8-35, against an intranasal (i.n.) challenge with live *C. muridarum*. CT MoPn, mice protected as the result of a previous infection with *C. muridarum*; Mem-0, mice having received a mock injection of buffer. (**A**) Percentage change in mean body weight of mice following the i.n. challenge. (**B**) Number of *C. muridarum* infectious units recovered from the lungs at day 10 following the i.n. challenge. *Dots* represent individual animals and the horizontal bars the medians for the different groups. Mice vaccinated with the nMOMP/A8-35 preparation (○) are almost as well protected against lung infection as those that have recovered from a previous infection with *C. muridarum* (△). Note the log scale (From Tifrea et al. 2011, © 2011 Elsevier Ltd. All rights reserved).

The enhanced protection elicited by the nMOMP/A8-35-formulated vaccine as compared to the nMOMP/Z3-14 one therefore results from a better preservation of the native structure of nMOMP and/or from a more efficient presentation of the immunogen to the immune system, not from an unspecific adjuvant effect of A8-35.

In Study 15.4, the feasibility of eliciting protection against a vaginal challenge with *Chlamydia* was examined using the mouse/*C. muridarum* system, and the immune mechanisms involved in protection were characterized, using in parallel nMOMP/A8-35 and nMOMP/Z3-14 preparations. The vaccine formulated with nMOMP/A8-35 complexes was found to induce a very robust protection against a vaginal infection and to preserve fertility. The protection is dependent on CD4$^+$ T cells and may result from the increased accessibility to the immune system of epitopes in the nMOMP/A8-35 formulation as compared to the nMOMP/Z3-14 one. The latter is suggested by two observations: first, more MOMP peptides are recognized by circulating antibodies in the first case than in the second; second, solution NMR spectra show that MOMP tryptophan residues, most of which are thought to be located at the surface of the TM $\beta$-barrel that faces lipid polar heads (Fig. 15.5A), are more exposed to the solvent in nMOMP/A8-35 than in nMOMP/detergent (dodecylphosphocholine; DPC) complexes (Tifrea et al. 2014). Figure 15.5B, C shows regions where Trp indole protons typically resonate. In the nMOMP/detergent spectrum, a single weak signal is detected (Fig. 15.5B), which is consistent with the mobility of these side chains being restricted by the detergent. In contrast, there are five strong Trp signals in the spectrum of nMOMP/A8-35 (Fig. 15.5C). Because nMOMP/A8-35 complexes are relatively large, the only signals detected correspond to flexible, solvent-accessible regions. The presence of Trp signals in the nMOMP/A8-35 NMR spectrum indicates that these groups have increased exposure compared with the nMOMP/detergent sample. This observation is reminiscent of that made by fast photochemical oxidation-liquid chromatography-MS/MS studies of APol-trapped vs. detergent-solubilized OmpT (discussed in Chap. 14, § 14.3.3), according to which some residues located at the limit between the TM $\beta$-barrel of OmpT and the periplasm are more exposed to the

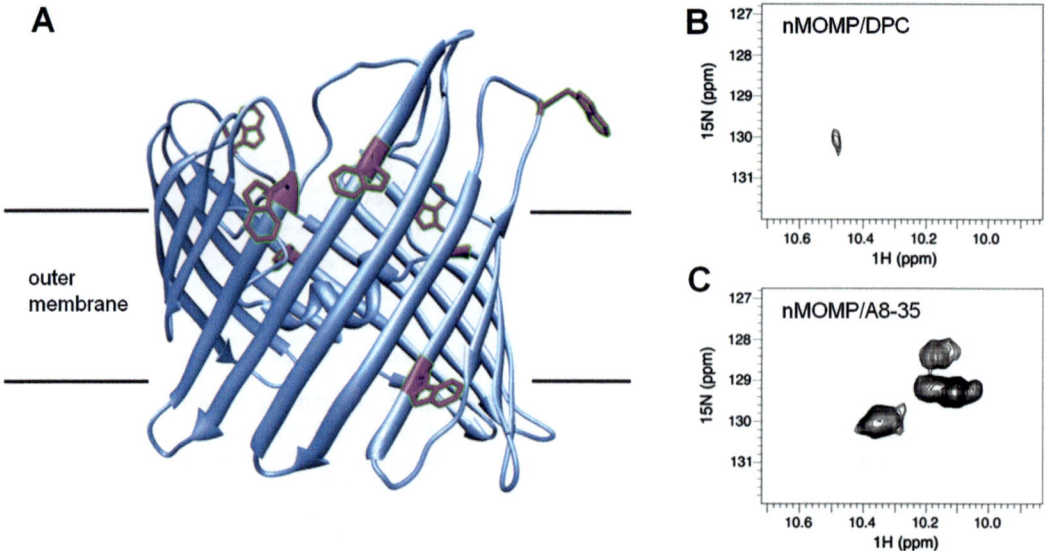

**Fig. 15.5** Tryptophan side chains are more mobile when nMOMP is trapped with A8-35 rather than kept soluble by a detergent. (**A**) Model of a MOMP monomer (Feher et al. 2013) showing the positions of seven of the eight tryptophan (Trp) side chains, close to the membrane surface. One Trp occurs in variable domain 4, which is not shown in this structural model. (**B, C**) 800-MHz NMR $^{15}$N HSQC spectra of 0.3 mM nMOMP solubilized in dodecylphosphocholine (DPC; B) or trapped with A8-35 (C) in 90% H$_2$O/ 10% D$_2$O, 100 mM sodium phosphate, pH 7.4. Only Trp indole signals resonate in the [$^1$H] region from 10 to 10.6 ppm. NMR data of $^{15}$N-labeled MOMP in DPC shows only one very weak Trp signal. In contrast, at least five strong Trp signals are seen when the protein is trapped in A8-35 (From Tifrea et al. 2014, © The American Association of Immunologists, Inc).

solvent in OmpT/A8-35 than in OmpT/dodecylmaltoside complexes (Watkinson et al. 2017). To which extent such differences are relevant to the presentation of MPs to the immune system remains uncertain, however given that the detergent is expected to desorb once diluted under its critical micellar concentration (Chap. 2), as happens following injection, and the APol to remain associated only as long as it is not displaced by other amphiphiles, as is expected to occur upon integration of the protein into the plasma membrane of a cell (see Chap. 5, § 5.7, and below, § 15.3). A further complicating element is that most of the immunization trials included Montanide as an adjuvant. Montanide is an oily emulsion, and whatever happens to MP/APol complexes when mixed with it remains to be examined.

### 15.2.3   How Could Amphipols Help in Formulating More Efficient Vaccines?

The results summarized in the previous section indicate that an A8-35-trapped MP used as immunogen in a vaccine preparation can elicit a stronger protection than the same protein kept in detergent solution. It remains to be determined whether this effect results from a higher stability; from the fact that the APol-trapped protein, at variance with the detergent-solubilized one, is expected to remain soluble following injection and dilution of the preparation in body fluids; from a different presentation to the immune system; or from an ensemble of these factors. A second important point is that transfer to A8-35 considerably stabilizes the protein used in these studies, MOMP, against inactivation by either heat treatment or freeze/thawing. This is of considerable practical importance for storing vaccines and using them in the field. Relatively few vaccines to date resort to MPs, but a few such examples have been reported, e.g. against hepatitis B (Ionescu-Matiu et al. 1983) or papillomavirus (Suzich et al. 1995). It would be of great interest to examine whether these vaccines could be improved by the use of APols.[1]

The two major obstacles to a broader use of MPs in formulating vaccines are (i) the difficulty of producing them in bulk at an affordable cost and (ii) their instability in detergent solution. As described in Chap. 6, APols can be used to fold to their native state many MPs after they have been expressed, simply and economically, as inclusion bodies. Identifying efficient folding conditions appears relatively straightforward for monomeric MPs, like monomeric porins, bacteriorhodopsin, or G protein-coupled receptors (reviewed in Popot 2014). For oligomeric MPs like MOMP, case-by-case development of efficient protocols will be required. As regards the stability issue, most APol-trapped MPs are stabilized, and often strongly stabilized, against thermal denaturation (see Chap. 5, § 5.5), and the two of them that have been submitted to freeze/thaw cycles, BR and MOMP, were found to be resistant to this treatment (respectively C. Le Bon and M. Zoonens, unpublished data cited in Chap. 5, § 5.5, and Feinstein et al. 2014). There therefore seems to be a case for taking advantage of these developments to explore more vigorously the use of MPs in the formulation of subunit vaccines.

The potential advantages of APols do not stop there. Indeed, they present two further properties that could be usefully harnessed for improving the efficiency of vaccines based on MP/APol complexes: first, they can be chemically modified so as to carry almost any desirable moieties (Chap. 4, § 4.4); second, because a given protein binds several APol molecules (Chap. 5, § 5.3.1.1), two or more APols carrying different groups can be mixed so as to provide MP/APol complexes with multiple functionalities. This has been illustrated in Chap. 13, § 13.2, by the use of complexes that carried both a tag, allowing their immobilization onto solid supports, and a fluorescent group, allowing their visualization (see e.g. Della Pia et al. 2014). Among the various APols and various functionalities that have been developed, three features appear to present a particular interest in the context of vaccine

---

[1] Published too late to be discussed here are vaccine trials in which A8-35-trapped BamA was used to boost the immune response against *Escherichia coli* (Vij et al. 2018)

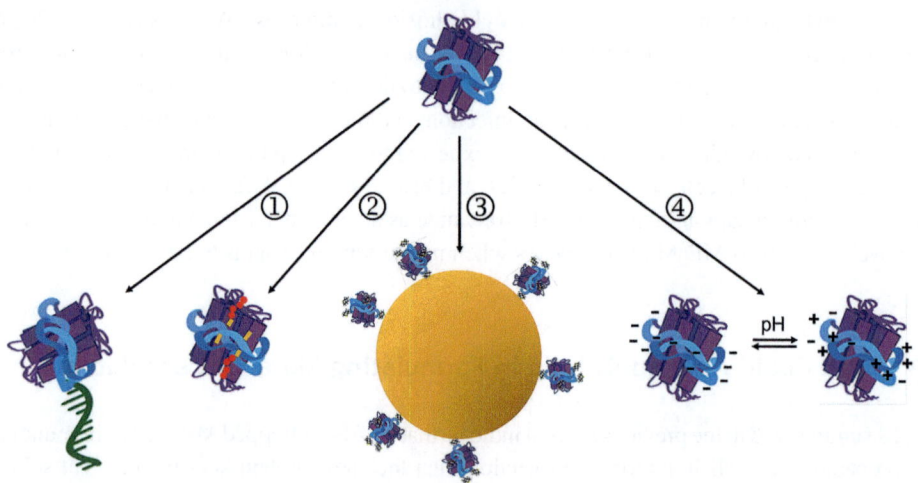

**Fig. 15.6** Various ways in which amphipols (APols) could be used to improve the efficiency of vaccines: ① covalently derivatizing APols with adjuvants; ② co-trapping a lipophilic adjuvant along with an immunogenic membrane protein (MP); ③ attaching MP/APol complexes to a particulate support via a tagged APol; ④ modulating the charge of the APol as a function of pH. All of these approaches could be combined. See text.

formulation: (i) adjuvant-carrying APols, (ii) tagged APols, and (iii) APols presenting different sensitivities to pH. Examples of the approaches that could be developed are presented in Fig. 15.6.

As noted in § 15.2.1, subunit-based vaccines tend to be weaker than whole-cell-based vaccines (Bachmann and Jennings 2010; Pulendran and Ahmed 2011; Bobbala and Hook 2016). This is because they contain a limited number of components capable of stimulating the immune system or against which immune responses are elicited, and, at variance with attenuated pathogens, they do not replicate, which limits exposure of the organism. Subunit-based vaccines therefore have to incorporate adjuvants (Moyle and Toth 2013). Trapping a MP protein with an adjuvant-carrying APol (Fig. 15.6, ①) will ensure that a cell that takes up a protein will also take up the adjuvant, which is known to lead to more vigorous immune responses (see e.g. Tighe et al. 2000; Maurer et al. 2002; Mutwiri et al. 2011; Zom et al. 2012; Ilyinskii et al. 2014). To date, three different adjuvant-carrying versions of A8-35 have been synthesized and their solution properties characterized (see Chap. 4, Table 4.5). They carry either a CpG ODN ("OligAPol"; Le Bon et al. 2014), which activates Toll-like receptors (TLRs) 9; peptide EP67 (Sanderson et al. 2012), which targets the C5a G protein-coupled receptor (Tifrea et al. 2018b); or resiquimod (Ilyinskii et al. 2014; Sachan et al. 2015; Wua et al. 2004), which targets TLRs 7/8 (Tifrea et al. 2018a). The first results of vaccination attempts using MOMP trapped with an EP67-A8-35 conjugate (Study 15.7) are encouraging, inasmuch as more robust a protection was observed in mice immunized with nMOMP/EP67-A8-35 complexes than with nMOMP/A8-35 complexes supplemented with free EP67 (Tifrea et al. 2018b). A similar improvement has been observed using the Resiquimod-tagged APol (Tifrea et al., 2018a).

APols carrying different adjuvants could of course be mixed if deemed desirable. Lipopolysaccharide, or lipopeptide adjuvants (see Moyle and Toth 2008, Beck et al. 2015, and references therein), or lipophilic drugs (see e.g. Liu et al. 2015), could be trapped along with the protein, forming ternary complexes (Fig. 15.6, ②).

Tagged APols are of interest in the context of vaccination because they make it possible to modulate the size of the particles that carry the immunogen, an important factor in controlling their diffusion throughout the organism and the efficiency with which they are internalized by cells of the immune system, in particular by dendritic cells (DCs). Whereas large particles (500–2000 nm) are picked up by DCs at the injection site, small (20–200 nm) nanoparticles and virus-like particles (~30 nm) are also found in lymph node (LN)-resident DCs and macrophages, suggesting free drainage of these particles to the LNs, where most DCs are concentrated. DCs appear to be strictly required for the transport of large particles from the injection site to LNs (Manolova et al. 2008), with smaller (25 nm) particles being more efficiently accumulated by DCs than larger ones (100 nm) (Reddy et al. 2007). Particles that are too small, however (less than ~10 nm, which is the case of most MP/APol complexes), are taken up by antigen-presenting cells less efficiently than those that are close in size to a virus (typically a few tens of nm) or a bacterium (typically in the μm range) (see e.g. Bachmann and Jennings 2010, Kushnir et al. 2012, Yue and Ma 2015, and references therein; note, however, that the amphiphilicity of MP/APol complexes may perhaps compensate for their small size; see Seong and Matzinger 2004). This dependence has led to the development of vaccines in which the immunogen is either aggregated into or carried by particles of the appropriate size, for instance, gold nanoparticles (reviewed in Vartak and Sucheck 2016). Some of the tagged APols would be well adapted to this use. For instance, the thiomorpholine-carrying version of A8-35 (Chap 4., Table 4.5; F. Giusti and M. Zoonens, unpublished data) could be used to adsorb MP/APol complexes onto gold beads of optimal size (Fig. 15.6, ③; cf. Suárez-Suárez et al. 2013).

Finally, the charge carried by endocytosed particles is known to play an important role in determining their fate. A positive charge, such as that conferred by cationic lipids to DNA-delivering liposomes, improves the cytosolic release of antigens by affecting the integrity of the endosomal membrane (see e.g. Li and Szoka 2007; Conwell and Huang 2005; Vartak and Sucheck 2016). The pH-sensitive character of the amphipathic polymer PP-50 – poly (L-lysine iso-phthalamide) grafted with phenylalanine – appears to favor its internalization (Mercado et al. 2016). The currently available panel of APols offers a wide diversity of behaviors, whose impact on vaccine efficiency deserves to be explored:

(i) A8-35, A8-75, and styrene-maleic acid co-polymers (SMA) are negatively charged at the neutral pH of the serum and will protonate and become insoluble at the lower pH (5.0–5.5) of the lumen of endosomes, which, in the case of methacrylate-based particles, has been shown to enhance CD8$^+$ T cell responses (see Keller et al. 2014, Trimaille and Verrier 2015, and references therein).

(ii) Sulfonated APols (SAPols; Dahmane et al. 2011) will remain negatively charged following internalization.

(iii) Non-ionic APols (NAPols; Bazzacco et al. 2012; Sharma et al. 2012) will remain uncharged.

(iv) Phosphorylcholine-based APols (PC-APols; Diab et al. 2007a, b), due to the presence of a secondary amine, carry a positive charge at neutral and acidic pH.

(v) Perhaps most interestingly, imidazole-carrying derivatives of A8-35 (ImidAPol; F. Giusti and M. Zoonens, unpublished data discussed in Chap. 4, § 4.4.2) will turn from anionic to zwitterionic and then cationic as the pH drops, a process that can be modulated by the density of imidazole moieties (Fig. 15.6, ④).

This range of physical-chemical properties provides the vaccinologist with a broad array of new tools to modulate the interactions of MP/APol particles with cells and the release of the immunogens into the cytosol (it is worth noting, in this respect, that delivery to anionic liposomes of siRNA associated with the zwitterionic APol PMAL-B-100 has been modeled by molecular dynamics and experimentally tested; see Li et al. 2015). It would perhaps also be conceivable to trap, along with the MP used as an immunogen, a pH-sensitive fusogenic lipid, whose integration into liposomes has been shown to strengthen the immune response (Sato et al. 2012; Miyabe et al. 2014).

Note that all of the approaches shown in Fig. 15.6 could easily be combined.

Needless to say, more sophisticated approaches could be devised, which would exploit the presence on DCs of specific receptors, such as Fcγ or C-type lectin receptors (see Vartak and Sucheck 2016 and references therein), or resort to DC-specific peptides or antibodies (see Moyle and Toth 2013 and references therein), to target to DCs the APol and, thereby, the APol-trapped MP used at an immunogen.

As shown below (§ 15.4), choosing the route of injection of MP/APol complexes can be used to favor the formation of long-lasting depots providing a sustained release of the immunogen, an important feature of long-term protection (Bobbala and Hook 2016).

***Biomedical applications of amphipols are still in their infancy***
(© *2018 by Francis Haraux*).

## 15.3    Uptake of A8-35 and a Passenger Peptide by Cells in Culture

Upon injection of MP/A8-35 complexes in mice, no toxicity is observed. Antibodies are produced against MPs, but not against APols (Popot et al. 2003). In cell cultures, A8-35 is not cytolytic, at least at the concentrations needed to deliver MPs to the plasma membrane (Popot et al. 2003). This has led to the examination of APols as vectors for immunization, as discussed in § 15.2, or for the delivery of MPs or other hydrophobic or amphipathic molecules for therapeutic purposes, as described here.

The transmembrane (TM) domain of bitopic MPs, which is made up of a single hydrophobic α-helix, often plays a key role in their oligomerization (see e.g. MacKenzie 2006, Matthews et al. 2006, Hubert et al. 2010, Stangl et al. 2014, and references therein). Peptides mimicking the TM domain of neuropilin-1 (NRP1), a receptor controlling various biological effects ranging from cell migration to cell proliferation and cell death (Roth et al. 2009), have been shown to be able to block its biological functions (Roth et al. 2008). In addition to helping to dissect signaling mechanisms, these peptides may also offer novel opportunities in cancer therapy. Preclinical studies have demonstrated that interfering with the function of NRP1 in tumor cells or in the endothelial cells establishing the neovascular network supporting tumor growth by hybridizing it with TM peptides largely reduces brain and breast tumor expansion and dissemination both in vitro and in vivo (Nasarre et al. 2010; Arpel et al. 2016). Similar results have been achieved against glioblastoma using a TM peptide derived from Plexin-A1 (Jacob et al. 2016). TM peptides, however, are insoluble and prone to aggregation, which makes it difficult to ensure optimal distribution and efficacy in whole organisms. As described in Chap. 5 (Table 5.1), APols have been shown to keep various TM peptides soluble (Gohon 1996; Duarte et al. 2008; Stangl et al. 2014), suggesting that they could be used as solubilizing and vectorizing agents. This has led to examining the ability of APols to distribute throughout the organism of mice and to deliver TM peptides to cells in culture.

In Study 15.1, a preliminary analysis has been reported of the biodistribution of APol-trapped TM peptides following their application to COS cells (fibroblast-like cells derived from monkey kidney tissue). The NRP1 peptide was labeled with rhodamine and trapped with 7-nitrobenz-2-oxa-1,3-diazol-4-yl-labeled A8-35 (FAPol$_{NBD}$; see Chap. 4, § 4.4), making it possible to follow separately the fate of the two partners. Confocal microscopy showed that the complexes reach the plasma membrane and associate with it within minutes. Both peptides and APols are endocytosed after a few hours (Fig. 15.7), leading to a total clearance of the plasma membrane in 3 days (Popot et al. 2011).

These results call for some comments. As visible in the top row of images in Fig. 15.7, taken after 16 h, the APol appears to be endocytosed more rapidly than the peptide, leading to a distinct distribution of the two markers. Given that the peptide was trapped with an excess of APol, a simplistic explanation would be that the free APol is internalized more rapidly than the peptide-bound one. This is unlikely to be correct. It is hard to see, indeed, why the APol would remain associated with the

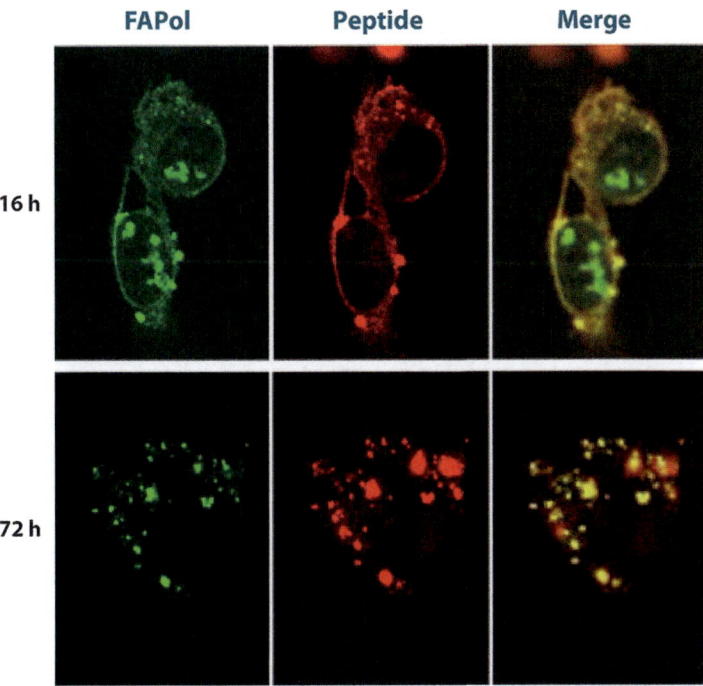

**Fig. 15.7**  Using amphipols to deliver a transmembrane (TM) peptide to cells. Confocal microscopic analysis of the distribution of NBD-labeled A8-35 (FAPol$_{NBD}$; *green*) and a rhodamine-labeled neuropilin-1 TM peptide (*red*) 16 and 72 h after application of peptide/FAPol$_{NBD}$ complexes to COS-7 cells. After 72 h, the plasma membrane has been cleared of both labels (From Popot et al. 2011).

peptide once each partner has the option of integrating into a membrane, where lipids compete with APols for the hydrophobic surface of the peptide and entropy works to separate the two partners. Förster resonance energy transfer experiments would permit direct probing of this issue. They have not been carried out in Study 15.1. However, as discussed in more detail in Chap. 5, § 5.7, and 11, § 11.3.2, indirect evidence in favor of such a dissociation is provided by the experiments of Polovinkin et al. (2014), in which bacteriorhodopsin/A8-35 complexes were mixed with a lipidic mesophase. The protein diffused and crystallized in the mesophase, without being accompanied by any APol (see Chap. 11, § 11.3.2). It is most likely that, in the experiment shown in Fig. 15.7, the two partners separated upon binding to the plasma membrane and diffused separately at the surface of the cell before being endocytosed. Whether internalized separately or not, they ended up in the same endocytic compartments, as shown by the images collected after 72 h.

## 15.4    Biodistribution and Elimination of Amphipols Following Injection in Mice

To examine whether APols could potentially be used to deliver anticancer peptides in vivo, mice bearing subcutaneous tumors were injected with FAPol$_{NBD}$. Histological examination of tumor slices revealed strong staining of intratumor blood vessels (Fig. 15.8).

A thorough analysis of the distribution of fluorescent derivatives of A8-35 was carried out in Study 15.5 (Fernandez et al. 2014). Two fluorescent versions of A8-35 were used, carrying, respectively, the far-red fluorophore Alexa Fluor 647 (FAPol$_{AF647}$) or the red fluorophore rhodamine (FAPol$_{rhod}$; see Chap. 4, Table 4.5, and spectra in Chap. 8, Fig. 8.9). This made it possible to follow over time, by bioimaging, their distribution and elimination following injection in nude mice. In addition, some animals were sacrificed at 4 h, 24 h, 72 h, 10 days, and 20 days after injection and dissected in order to collect organs (liver, kidney, spleen, fat pads, heart, lungs, and brain) and determine their fluorescence content. Injections were carried out through three different routes: intravenous (IV), intraperitoneal (IP), and subcutaneous (SC).

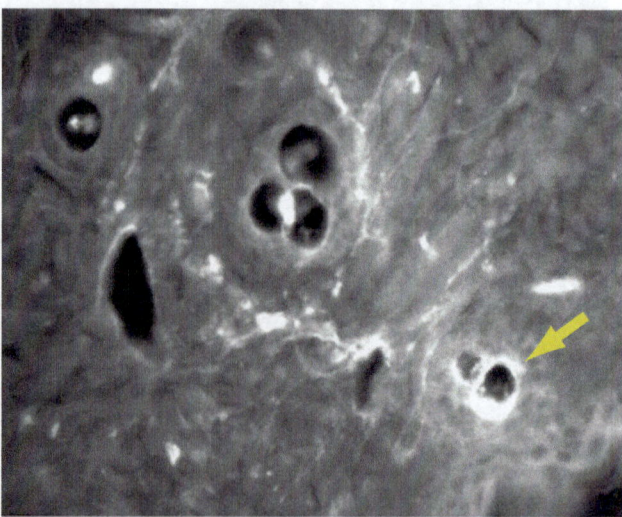

**Fig. 15.8** Distribution of A8-35 in tumor slices. Mice bearing subcutaneous tumors (C6 glioma) were injected intraperitoneally with FAPol$_{NBD}$ 24 h before tissue collection. Vascular and perivascular fluorescence demonstrates tissue penetration (*yellow arrow*) (From Popot et al. 2011).

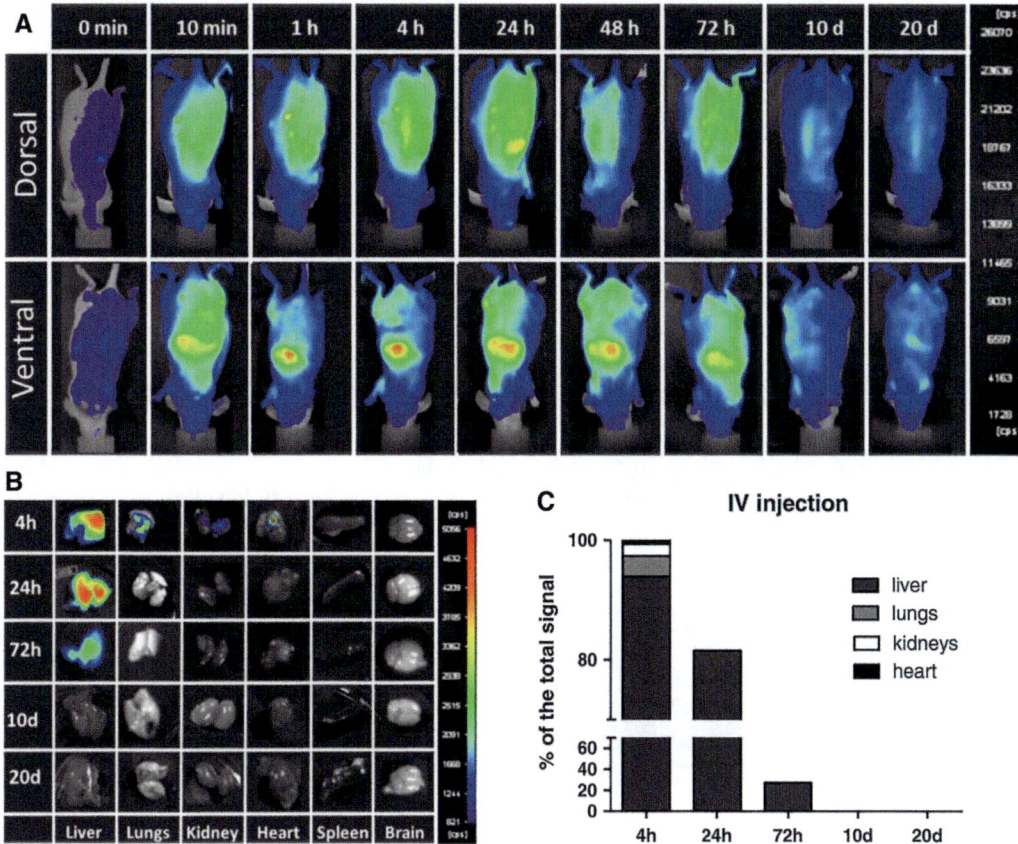

**Fig. 15.9** Biodistribution of FAPol$_{AF647}$ following intravenous injection. (**A**) Time-series images of a representative animal before (0 min) and after intravenous injection of 10 μg FAPol$_{AF647}$. *Upper row* dorsal views, *lower row* ventral views. (**B**) Time-series images of organs collected at different time points. (**C**) Quantitation of fluorescent signals measured in organs over time (From Fernandez et al. 2014).

IV injection is one of the most common routes of administration and is considered to give the best bioavailability. Indeed, following injection of FAPol$_{AF647}$ in the retro-orbital sinus, the fluorescent signal propagated rapidly throughout the whole body of the animals (Fig 15.9A). The signal was extremely stable, profiles obtained 10 min postinjection being similar to those obtained 72 h later. A significant decrease was first seen at 10 days postinjection. Ventral views of the animals revealed, as expected from IV injection, a rapid concentration in the liver. This hepatic signal was clearly detectable up to 72 h, with a maximal intensity seen at 4 h. The concentration of the signal in the liver was confirmed when collecting organs at different time points in some of the animals (Fig. 15.9B). As shown in Fig. 15.9C, 94% of the signal was measured in the liver, the rest being distributed among the lungs, the heart, and the kidneys, the latter signal being consistent with an excretion process. There was no detectable signal in the spleen or in the brain. Since all organs (except liver) were negative at 24 h when removed from the body, the observed fluorescent signal in the whole body at this time point is mainly due to the persistence of APols in the circulation.

IP injections are widely used in animal studies because of the ease of the procedure. Following injection of FAPol$_{AF647}$, the fluorescent signal remained concentrated in the peritoneal cavity for ~1 h (Fig. 15.10). It then reached the circulation to produce a uniform signal in the body, concomitant with a persistent and intense staining of the liver for up to 72 h. As for the IV route, the signal progressively

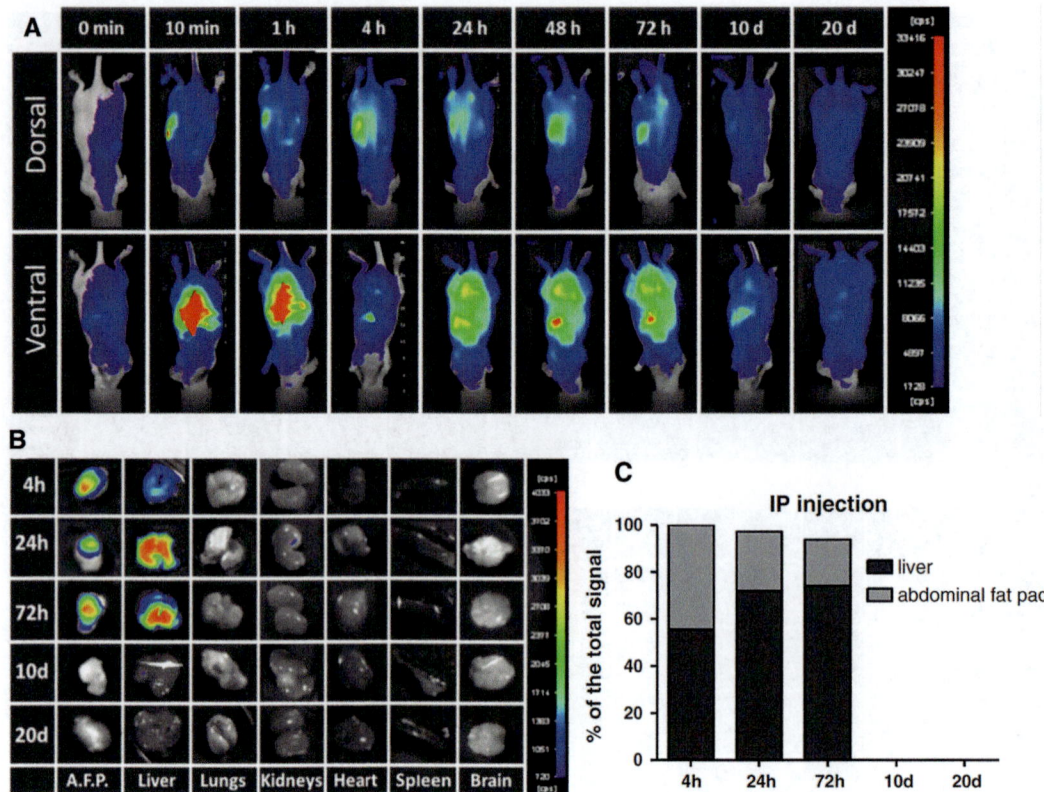

**Fig. 15.10** Biodistribution of FAPolAF647 following intraperitoneal injection. Same procedure as in Fig. 15.9 but for the injection route. *AFP* abdominal fat pad (From Fernandez et al. 2014).

disappeared at 10 and 20 days postinjection, consistent with a slow release of APols from the body. The collection of organs at different time points showed a large amount of signal in abdominal fat pads (45%) and in the liver (55%) up to 72 h. With the exception of a faint signal in the liver, FAPol$_{AF647}$ was undetectable at 10 and 20 days, suggesting complete elimination. In all cases, no signal was observed in the lung, heart, spleen, and brain, implying no accumulation of APols in these tissues. A very modest signal in the kidneys presumably reflected some renal elimination. It was difficult to monitor due to the slow release from fat and liver, generating very low doses of APol to eliminate at any given time point. Hence, as for the IV route, a majority of APols remains circulating once diffused from fat and is progressively eliminated.

The SC route is known to lead to slow but complete absorbance of drugs. Indeed, following injection of the FAPol$_{AF647}$ bolus, a compact and restricted signal was observed at the injection site for the entire period of analysis (Fig. 15.11). The intensity of the spot decreased very slowly and was still detectable after 20 days. This high site-specific concentration effect impeded detection of APols in the rest of the body, including in ventral views. Consistently, fluorescence was found associated with the dorsal fat pads, which appear to be a very efficient natural reservoir for APols. With the exception of the liver, which showed significant staining, suggesting hepatic elimination of APols, APols could not be detected in any of the collected organs, due to their low circulating concentration.

To validate the biodistribution profiles, experiments were repeated using the rhodamine-labeled version of A8-35 (FAPol$_{rhod}$). As shown in Fig 15.12, distribution profiles fully matched those obtained for FAPol$_{AF647}$. The profiles are presented for each administration route with the

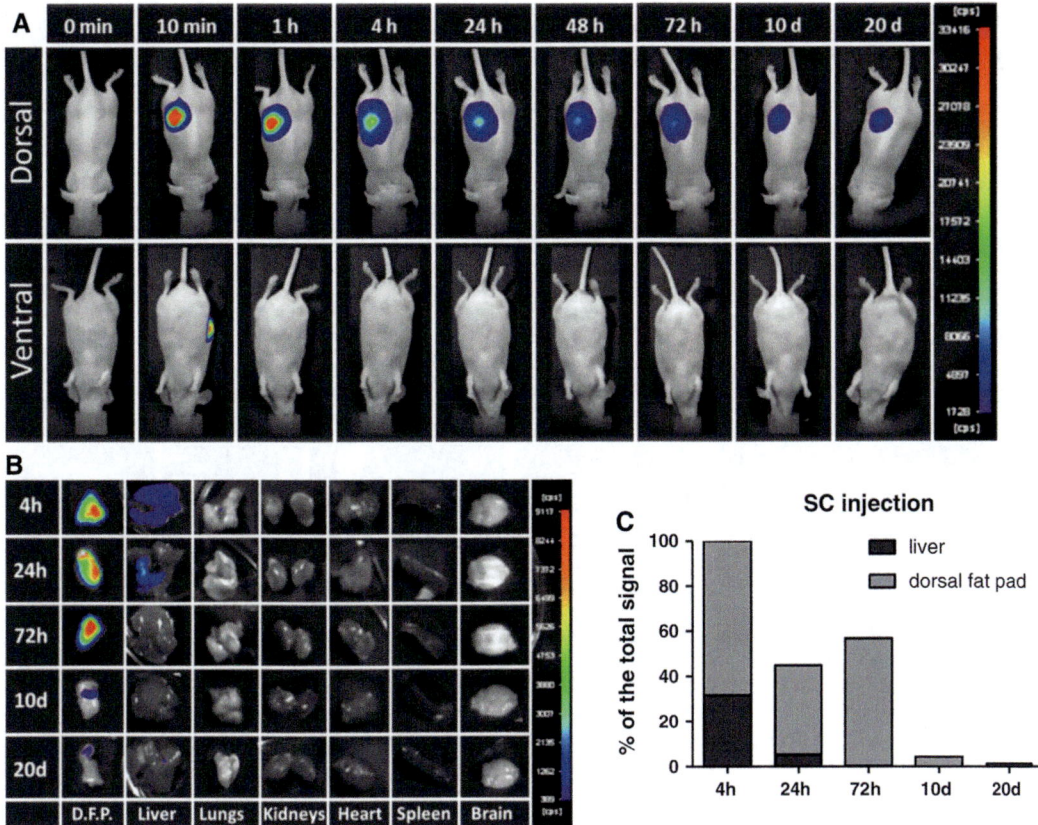

**Fig. 15.11** Biodistribution of FAPol$_{AF647}$ following subcutaneous injection. Same procedure as in Fig. 15.9 but for the injection route. DFP dorsal fat pad (From Fernandez et al. 2014).

corresponding distribution of free rhodamine, which served as a control. This confirmed that the observed persistent signals do correspond to FAPol$_{rhod}$, given that the signal of the free dye (even though it was ~100× more concentrated than the label carried by FAPol$_{rhod}$) reaches a peak around 1 h and then rapidly vanishes.

In summary, the systematic evaluation of three of the major administration routes (IV, IP, SC) carried out in Study 15.5 revealed that following IV or IP injection, APol A8-35 exhibits a long-lasting distribution in the whole body circulation up to 10–20 days. Such would also presumably be the case for MP/A8-35 complexes. It also showed that fat pads can trap APols, which are subsequently slowly released over time. The hepatic signal seen 10 min postinjection is consistent with the detoxification function of this organ. The fluorescent signals in the liver parenchyma were confirmed when imaging dissected organs, whereas most of the other organs were not fluorescent. This suggests that the majority of the detectable FAPols remains in blood circulation. When using IP or SC routes, long-term body retention of FAPols was also observed. Fat pads, in particular, were found to trap FAPols released from the injection sites. This intriguing property may reflect the amphipathic nature of APols, which may find a natural sink in lipophilic areas. This unforeseen result opens interesting opportunities to use APols for the delivery of anti-obesity drugs such as CLA (conjugated linoleic acid; see Blankson et al. 2000). Indeed, recent work has shown that nano-emulsions enhanced the bioavailability of CLA and its anti-obesity effect (Kim et al. 2013). The stable localization under the skin observed following IC injection suggests the interesting possibility of using APols for long-term and slow release of drugs

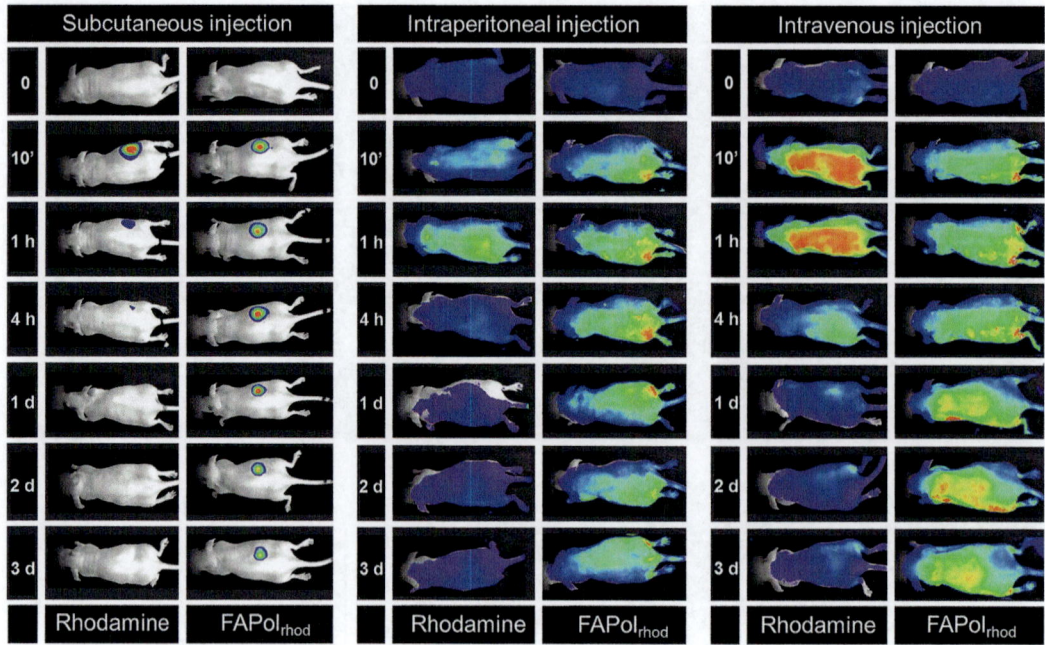

**Fig. 15.12** Biodistribution of rhodamine-labeled A8-35 (FAPol$_{rhod}$) *vs.* free rhodamine. Time-series dorsal views of representative animals following a subcutaneous injection of 10 μg free rhodamine (*first column*) or 10 μg FAPol$_{rhod}$ (*second column*), an intraperitoneal injection of 10 μg free rhodamine (*third column*) or 10 μg FAPol$_{rhod}$ (*fourth column*), or an intravenous injection of 10 μg free rhodamine (*fifth column*) or 10 μg FAPol$_{rhod}$ (*sixth column*). Bioimaging was performed before injection (0) and 10 min, 1 h, 4 h, 24 h, 48 h, and 72 h after injection (From Fernandez et al. 2014).

and immunogens. This is of great interest for vaccination, and it also opens the possibility to use APols for the delivery of hydrophobic drugs in the context of chronic diseases requiring long-term delivery of drugs, such as chronic pain, psychiatric disorders, hypertensive diseases, AIDS, or cancer.

In conclusion, the biodistribution profile of APols is compatible with many in vivo applications. Depending on the administration route, APol-associated MPs or drugs can be delivered rapidly or slowly, eventually benefiting from natural reservoirs such as fat. One should not overlook the fact that, because it is possible to functionalize APols with virtually any desirable molecule (Chap. 4, § 4.4), targeting MPs, peptides, or drugs using APols tagged with organ- or cell-specific ligands, toxins, or antibodies is a realistic proposition.

# References

Arpel, A., Gamper, C., Spenlé, C., Fernandez, A., Jacob, L., Baumlin, N., Laquerriere, P., Orend, G., Crémel, G., Bagnard, D. (2016) Inhibition of primary breast tumor growth and metastasis using a neuropilin-1 transmembrane domain interfering peptide. *Oncotarget* **7**:54723–54732.

Bachmann, M.F., Jennings, G.T. (2010) Vaccine delivery: a matter of size, geometry, kinetics and molecular pattern. *Nat. Rev. Immunol.* **10**:787–796.

Bachmann, N.L., Polkinghorne, A., Timms, P. (2014) Chlamydia genomics: providing novel insights into chlamydial biology. *Trends Microbiol.* **22**:464–472.

Bazzacco, P., Billon-Denis, E., Sharma, K.S., Catoire, L.J., Mary, S., Le Bon, C., Point, E., Banères, J.-L., Durand, G., Zito, F., Pucci, B., Popot, J.-L. (2012) Non-ionic homopolymeric amphipols: Application to membrane protein folding, cell-free synthesis, and solution NMR. *Biochemistry* **51**:1416–1430.

Beck, Z., Matyas, G.R., Jalah, R., Rao, M., Polonis, V.R., Alving, C.R. (2015) Differential immune responses to HIV-1 envelope protein induced by liposomal adjuvant formulations containing monophosphoryl lipid A with or without QS21. *Vaccine* **33**:5578–5587.

Blake, M.S., Wetzler, L.M. (1995) Vaccines for gonorrhea: where are we on the curve? *Trends Microbiol.* **3**:469–474.

Blankson, H., Stakkestad, J.A., Fagertun, H., Thom, E., Wadstein, J., Gudmundsen, O. (2000) Conjugated linoleic acid reduces body fat mass in overweight and obese humans. *J. Nutr.* **130**:2943–2948.

Bobbala, S., Hook, S. (2016) Is there an optimal formulation and delivery strategy for subunit vaccines? *Pharm. Res.* **33**:2078–2097.

Brunham, R.C., Rey-Ladino, J. (2005) Immunology of Chlamydia infection: implications for a *Chlamydia trachomatis* vaccine. *Nature Rev. Immunol.* **5**:149–161.

Cai, S., He, F., Samra, H.S., de la Maza, L.M., Bottazzi, M., Joshi, S.B., Middaugh, C.R. (2009) Biophysical and stabilization studies of the *Chlamydia trachomatis* mouse pneumonitis major outer membrane protein. *Mol. Pharm.* **6**:1553–1561.

Brunham, R.C., Pourbohloul, B., Mak, S., White, R., Rekart, M.L. (2005) The unexpected impact of a *Chlamydia trachomatis* infection control program on susceptibility to reinfection. *J. Infect. Dis.* **192**:1836–1844.

Castro, C.A., Hogan, J.B., Benson, K.A., Shehata, C.W., Landauer, M.R. (1995) Behavioral effects of vehicles: DMSO, ethanol, Tween-20, Tween-80, and Emulphor-620. *Pharmacol. Biochem. Behav.* **50**:521–526.

Conwell, C.C., Huang, L. (2005) Recent advances in non-viral gene delivery. *Adv. Genet.* **53**:1–18.

Centers for Disease Control and Prevention (1999) Summary of notifiable diseases. United States. 1998. *Morb. Mortal Wkly Rep.* **47**:11-23.

Dahmane, T., Giusti, F., Catoire, L.J., Popot, J.-L. (2011) Sulfonated amphipols: Synthesis, properties and applications. *Biopolymers* **95**:811–823.

de la Maza, L.M., Peterson, E.M. (2002) Vaccines for *Chlamydia trachomatis* infections. *Curr. Opin. Invest. Drugs* **3**:980–986.

Diab, C., Tribet, C., Gohon, Y., Popot, J.-L., Winnik, F.M. (2007a) Complexation of integral membrane proteins by phosphorylcholine-based amphipols. *Biochim. Biophys. Acta* **1768**:2737–2747.

Diab, C., Winnik, F.M., Tribet, C. (2007b) Enthalpy of interaction and binding isotherms of non-ionic surfactants onto micellar amphiphilic polymers (amphipols). *Langmuir* **23**:3025–3035.

Dong-Ji, Z., Yang, X., Shen, C., Lu, H., Murdin, A., Brunham, R.C. (2000) Priming with *Chlamydia trachomatis* major outer membrane protein (MOMP) DNA followed by MOMP ISCOM boosting enhances protection and is associated with increased immunoglobulin A and Th1 cellular immune responses. *Infect. Immun.* **68**:3074–3078.

Doolan, D.L., Apte, S.H., Proietti, C. (2015) Genome-based vaccine design: the promise for malaria and other infectious diseases. *Int. J. Parasitol.* **44**:901–913.

Duarte, A.M.S., Wolfs, C.J.A.M., Koehorsta, R.B.M., Popot, J.-L., Hemminga, M.A. (2008) Solubilization of V-ATPase transmembrane peptides by amphipol A8-35. *J. Peptide Chem.* **14**:389–393.

Elwell, C., Mirrashidi, K., Engel, J. (2016) Chlamydia cell biology and pathogenesis. *Nat. Rev. Microbiol.* **14**:385–400.

Farris, C.M., Morrison, R.P. (2011) Vaccination against Chlamydia genital infection utilizing the murine *C. muridarum* model. *Infect. Immun.* **79**:986–996.

Farris, C.M., Morrison, S.G., Morrison, R.P. (2010) CD4[+] T cells and antibody are required for optimal major outer membrane protein vaccine-induced immunity to *Chlamydia muridarum* genital infection. *Infect. Immun.* **78**:4374–4383

Feher, V.A., Randall, A., Baldi, P., Bush, R.M., de la Maza, L.M., Amaro, R.R. (2013) A three-dimensional trimeric β-barrel model for Chlamydia MOMP contains conserved and novel elements of Gram-negative bacterial porins. *PLoS One* **8**:e68934.

Feinstein, H.E., Tifrea, D., Sun, G., Popot, J.-L., de la Maza, L.M., Cocco, M.J. (2014) Long-term stability of a vaccine formulated with the amphipol-trapped major outer membrane protein from *Chlamydia trachomatis*. *J. Membr. Biol.* **247**:1053–1065.

Fernandez, A., Le Bon, C., Baumlin, N., Giusti, F., Crémel, G., Popot, J.-L., Bagnard, D. (2014) *In vivo* characterization of the biodistribution profile of amphipols *J. Membr. Biol.* **247**:1043–1051.

Gohon, Y. (1996) *Etude des interactions entre un analogue du fragment transmembranaire de la glycophorine A et des polymères amphiphiles: les amphipols*. Rapport de DEA, Université Paris VI, Paris, 28 p.

Götz, H., Lindback, J., Ripa, T., Arneborn, M., Ramsted, K., Ekdahl, K. (2002) Is the increase in notifications of *Chlamydia trachomatis* infections in Sweden the result of changes in prevalence, sampling frequency or diagnostic methods? *Scand. J. Infect. Dis.* **34**:28–34.

Grayston, J.T., Wang, S.P. (1978) The potential for vaccine against infection of the genital tract with *Chlamydia trachomatis*. *Sex. Transm. Dis.* **5**:73–77.

Han, S.G., Na, J.H., Lee, W.K., Park, D., Oh, J., Yoon, S.H., Lee, C.K., Sung, M.H., Shin, Y.K., Yu, Y.G. (2014) An amphipathic polypeptide derived from poly-γ-glutamic acid for the stabilization of membrane proteins. *Prot. Sci.* **23**:1800–1807.

Hubert, P., Sawma, P., Duneau, J.-P., Khao, J., Henin, J., Bagnard, D., Sturgis, J. (2010) Single-spanning transmembrane domains in cell growth and cell-cell interactions: More than meets the eye? *Cell Adh. Migr.* **4**:313–324.

Igietseme, J.U., Eko, F.O., Black, C.M. (2011) Chlamydia vaccines: recent developments and the role of adjuvants in future formulations. *Expert Rev. Vaccines* **10**:1585–1596.

Ilyinskii, P.O., Roy, C.J., O'Neil, C.P., Browning, E.A., Pittet, L.A., Altreutera, D.H., Alexis, F., Tonti, E., Shi, J., Basto, P.A., Iannaconec, M., Radovic-Moreno, A.F., Langer, R.S., Farokhzad, O.C., von Andrian, U.H., Johnston, L.P.M., Kishimoto, T.K. (2014) Adjuvant-carrying synthetic vaccine particles augment the immune response to encapsulated antigen and exhibit strong local immune activation without inducing systemic cytokine release. *Vaccine* **32**:2882–2895.

Ionescu-Matiu, I., Kennedy, R.C., Sparrow, J.T., Culwell, A.R., Sanchez, Y., Melnick, J.L., Dreesman, G.R. (1983) Epitopes associated with a synthetic hepatitis B surface antigen peptide. *J. Immunol.* **130**:1947–1952.

Jacob, L., Sawma, P., Garnier, N., Meyer, L.A.T., Fritz, J., Hussenet, T., Spenlé, C., Goetz, J., Vermot, J., Fernandez, A., Baumlin, N., Aci-Sèche, S., Orend, G., Roussel, G., Crémel, G., Genest, M., Hubert, P., Bagnard, D. (2016) Inhibition of PlexA1-mediated brain tumor growth and tumor-associated angiogenesis using a transmembrane domain targeting peptide. *Oncotarget* **7**:57851–57865.

Kari, L., Whitmire, W.M., Crane, D.D., Reveneau, N., Carlson, J.H., Goheen, M.M., Peterson, E.M., Pal, S., de la Maza, L.M., Caldwell, H.D. (2009) *Chlamydia trachomatis* native major outer membrane protein induces partial protection in nonhuman primates: implication for a trachoma transmission-blocking vaccine. *J. Immunol.* **182**:8063–8070.

Keller, S., Wilson, J.T., Patilea, G.I., Kern, H.B., Convertine, A.J., Stayton, P.S. (2014) Neutral polymer micelle carriers with pH-responsive, endosome-releasing activity modulate antigen trafficking to enhance CD8$^+$ T cell responses. *J. Control. Release* **191**:24–33.

Kim, D., Park, J.-H., Kweon, D.-J., Han, G.D. (2013) Bioavailability of nanoemulsified conjugated linoleic acid for an antiobesity effect. *Int. J. Nanomedicine* **8**:451–459.

Kushnir, N., Streatfield, S.J., Yusibov, V. (2012) Virus-like particles as a highly efficient vaccine platform: diversity of targets and production systems and advances in clinical development. *Vaccine* **31**:58–83.

Le Bon, C., Della Pia, E.A., Giusti, F., Lloret, N., Zoonens, M., Martinez, K.L., Popot, J.-L. (2014a) Synthesis of an oligonucleotide-derivatized amphipol and its use to trap and immobilize membrane proteins. *Nucleic Acids Res.* **42**: e83.

Le Bon, C., Popot, J.-L., Giusti, F. (2014b) Labeling and functionalizing amphipols for biological applications. *J. Membr. Biol.* **247**:797–814.

Levine, M.M., Dougan, G., Good, M.F., Liu, M.A., Nabel, G.J., Nataro, J.P., Rappuoli, R. (2010) *New Generation Vaccines,* 4th ed., Informa Healthcare, New York, N.Y., 1040 p.

Li, W., Szoka, F., Jr. (2007) Lipid-based nanoparticles for nucleic acid delivery. *Pharm. Res.* **24**:438–449.

Li, J.P., Ouyang, Y.Y., Kong, X., Zhu, J.Y., Lu, D.N., Liu, Z. (2015) A multi-scale molecular dynamics simulation of PMAL facilitated delivery of siRNA. *RSC Adv.* **5**:68227–68233.

Liu, G., Luo, Q., Gao, H., Chen, Y., Wei, X., Dai, H., Zhang, Z., Ji, J. (2015) Cell membrane-inspired polymeric micelles as carriers for drug delivery. *Biomater Sci.* **3**:490–499.

MacKenzie, K.R. (2006) Folding and stability of alpha-helical integral membrane proteins. *Chem. Rev.* **106**:1931–1977.

Manolova, V., Flace, A., Bauer, M., Schwarz, K., Saudan, P., Bachmann, M.F. (2008) Nanoparticles target distinct dendritic cell populations according to their size. *Eur. J. Immunol.* **38**:1404–1413.

Matthews, E.E., Zoonens, M., Engelman, D.M. (2006) Dynamic helix interactions in transmembrane signaling. *Cell* **127**:447–450.

Maurer, T., Heit, A., Hochrein, H., Ampenberger, F., O'Keeffe, M., Bauer, S., Lipford, G.B., Vabulas, R.M., Wagner, H. (2002) CpG-DNA aided cross-presentation of soluble antigens by dendritic cells. *Eur. J. Immunol.* **32**:2356–2364.

de la Maza, L.M., Zhong, G., Brunham, R.C. (2017) Update on *Chlamydia trachomatis* vaccinology. *Clinic. Vacc. Immunol.* **24**:e00543–16.

Delany, I., Rappuoli, R., De Gregorio, E. (2014) Vaccines for the 21st century. *EMBO Mol. Med.* **6**:708–720.

Della Pia, E.A., Holm, J., Lloret, N., Le Bon, C., Popot, J.-L., Zoonens, M., Nygård, J., Martinez, K.L. (2014) A step closer to membrane protein multiplexed nano-arrays using biotin-doped polypyrrole. *ACS Nano* **8**:1844–1853.

Mercado, S.A., Orellana-Tavra, C., Chen, A., Slater, N.K. (2016) The intracellular fate of an amphipathic pH-responsive polymer: Key characteristics towards drug delivery. *Mater. Sci. Eng. C. Mater. Biol. Appl.* **69**:1051–1057.

Miyabe, H., Hyodo, M., T.N.. Sato Y., Hayakawa, Y., Harashima, H. (2014) A new adjuvant delivery system 'cyclic di-GMP/YSK05 liposome' for cancer immunotherapy. *J. Control. Release* **184**:20–27.

Morrison, R.P., Caldwell, H.D. (2002) Immunity to murine chlamydial genital infection. *Infect. Immun.* **70**:2741–2751.

Morrison, R.P., Belland, R.J., Lyng, K., Caldwell, H.D. (1989) Chlamydial disease pathogenesis. The 57-kD chlamydial hypersensitivity antigen is a stress response protein. *J. Exp. Med.* **170**:1271–1283.

Moyle, P.M., Toth, I. (2008) Self-adjuvanting lipopeptide vaccines. *Curr. Med. Chem* **15**:506–516.

Moyle, P.M., Toth, I. (2013) Modern subunit vaccines: development, components, and research opportunities. *ChemMedChem* **8**:360–376.

Mutwiri, G., Gerdts, V., Van Drunen Littel-van den Hurk, S., Auray, G., Eng, N., Garlapati, S., Babiuk, L.A., Potter, A. (2011) Combination adjuvants: the next generation of adjuvants? *Expert Rev. Vaccines* **10**:95–107.

Nasarre, C., Roth, M., Jacob, L., Roth, L., Koncina, E., Thien, A., Labourdette, G., Poulet, P., Hubert, P., Crémel, G., Roussel, G., Aunis, D., Bagnard, D. (2010) Peptide-based interference of the transmembrane domain of neuropilin-1 inhibits glioma growth *in vivo*. *Oncogene* **29**:2381–2392.

Nikaido, H. (1992) Porins and specific channels of bacterial outer membranes. *Molec. Microbiol.* **6**:435–442.

Pal, S., Barnhart, K.M., Wei, Q., Abai, A.M., Peterson, E.M., de la Maza, L.M. (1999) Vaccination of mice with DNA plasmids coding for the *Chlamydia trachomatis* major outer membrane protein elicits an immune response but fails to protect against a genital challenge. *Vaccine* **17**:459–465.

Pal, S., Davis, H.L., Peterson, E.M., de la Maza, L.M. (2002) Immunization with the *Chlamydia trachomatis* mouse pneumonitis major outer membrane protein by use of CpG oligodeoxynucleotides as an adjuvant induces a protective immune response against an intranasal chlamydial challenge. *Infect. Immun.* **70**:4812–4817.

Pal, S., Peterson, E.M., de la Maza, L.M. (2005) Vaccination with the *Chlamydia trachomatis* major outer membrane protein can elicit an immune response as protective as that resulting from inoculation with live bacteria. *Infect. Immun.* **73**:8153–8160.

Plotkin, S.A., Orenstein, W.A., Offit, P.A., Edwards, K.M. (2017) *Vaccines,* 7th ed., Elsevier, New York, N.Y., 1720 p.

Polovinkin, V., Gushchin, I., Balandin, T., Chervakov, P., Round, E., Shevchenko, V., Popov, A., Borshchevskiy, V., Popot, J.-L., Gordeliy, V. (2014) High-resolution structure of a membrane protein transferred from amphipol to a lipidic mesophase. *J. Membr. Biol.* **247**:997–1004.

Popot, J.-L. (2014) Folding membrane proteins *in vitro*: A table and some comments. *Arch. Biochem. Biophys.* **564**:314–326.

Popot, J.-L., Berry, E.A., Charvolin, D., Creuzenet, C., Ebel, C., Engelman, D.M., Flötenmeyer, M., Giusti, F., Gohon, Y., Hervé, P., Hong, Q., Lakey, J.H., Leonard, K., Shuman, H.A., Timmins, P., Warschawski, D.E., Zito, F., Zoonens, M., Pucci, B., Tribet, C. (2003) Amphipols: polymeric surfactants for membrane biology research. *Cell. Mol. Life Sci.* **60**:1559–1574.

Popot, J.-L., Althoff, T., Bagnard, D., Banères, J.-L., Bazzacco, P., Billon-Denis, E., Catoire, L.J., Champeil, P., Charvolin, D., Cocco, M.J., Crémel, G., Dahmane, T., de la Maza, L.M., Ebel, C., Gabel, F., Giusti, F., Gohon, Y., Goormaghtigh, E., Guittet, E., Kleinschmidt, J.H., Kühlbrandt, W., Le Bon, C., Martinez, K.L., Picard, M., Pucci, B., Rappaport, F., Sachs, J.N., Tribet, C., van Heijenoort, C., Wien, F., Zito, F., Zoonens, M. (2011) Amphipols from A to Z. *Annu. Rev. Biophys.* **40**:379–408.

Pulendran, B., Ahmed, R. (2011) Immunological mechanisms of vaccination. *Nat. Immunol.* **12**:509–517.

Reddy, S.T., van der Vlies, A.J., Simeoni, E., Angeli, V., Randolph, G.J., O'Neil, C.P., Lee, L.K., Swartz, M.A., Hubbell, J.A. (2007) Exploiting lymphatic transport and complement activation in nanoparticle vaccines. *Nat. Biotechnol.* **25**:1159–1164.

Rockey, D.D., Wang, J., Lei, L., Zhong, G. (2009) Chlamydia vaccine candidates and tools for chlamydial antigen discovery. *Expert Rev. Vaccines* **8**:1365–1377.

Rodriguez-Maranon, M.J., Bush, R.M., Peterson, E.M., Schirmer, T., de la Maza, L.M. (2002) Prediction of the membrane-spanning β-strands of the major outer membrane protein of Chlamydia. *Protein Sci.* **11**:1854–1861.

Roth, L., Nasarre, C., Dirrig-Grosch, S., Aunis, D., Crémel, G., Hubert, P., Bagnard, D. (2008) Transmembrane domain interactions control biological functions of neuropilin-1. *Mol. Biol. Cell* **19**:646–654.

Roth, L., Koncina, E., Satkauskas, S., Crémel, G., Aunis, D., Bagnard, D. (2009) The many faces of semaphorins: From development to pathology. *Cell. Mol. Life Sci.* **66**:649–666.

Sachan, S., Ramakrishnan, S., Annamalai, A., Sharma, B.K., Malik, H., Saravanan, B.C., Jain, L., Saxena, M., Kumar, A., Krishnaswamy, N. (2015) Adjuvant potential of resiquimod with inactivated Newcastle disease vaccine and its mechanism of action in chicken. *Vaccine* **33**:4526–4532.

Sanderson, S.D., Thoman, M.L., Kis, K., Virts, E.L., Herrera, E.B., Widmann, S., Sepulveda, H., Phillips, J.A. (2012) Innate immune induction and influenza protection elicited by a response-selective agonist of human C5a. *PLoS One* **7**:e40303.

Sato, Y., Hatakeyama, H., Sakurai, Y., Hyodo, M., Akita, H., Harashima, H. (2012) A pH-sensitive cationic lipid facilitates the delivery of liposomal siRNA and gene silencing activity *in vitro* and *in vivo*. *J. Control. Release* **163**:267–276.

Schachter, J. (1999) Infection and disease epidemiology, in: Stephens, R.S., ed., *Chlamydia: Intracellular Biology, Pathogenesis and Immunity,* ASM, Washington, pp. 139–170.

Schachter, J., Dawson, C.R. (1978) *Human Chlamydial Infections*. PSG Pub. Co., Littleton, Mass., 273 p.

Seong, S.Y., Matzinger, P. (2004) Hydrophobicity: an ancient damage-associated molecular pattern that initiates innate immune responses. *Nat. Rev. Immunol.* **4**:469–478.

Serra-Batiste, M., Tolchard, J., Giusti, F., Zoonens, M., Carrula, N. (2018) Stabilization of a membrane-associated amyloid-β oligomer for its validation in Alzheimer's disease. *Front. Mol. Biosci.* **5**:38.

Sharma, K.S., Durand, G., Gabel, F., Bazzacco, P., Le Bon, C., Billon-Denis, E., Catoire, L.J., Popot, J.-L., Ebel, C., Pucci, B. (2012) Non-ionic amphiphilic homopolymers: Synthesis, solution properties, and biochemical validation. *Langmuir* **28**:4625–4639.

Speijers, G.J., Danse, L.H., Beuvery, E.C., Derks, H.J., Vos, J.G. (1988) Local reactions of Zwittergent-containing meningococcal vaccine after intramuscular injection in rats: comparison with the effect of diphtheria-pertussis-tetanus-polio vaccine. *Vaccine* **6**:419–422.

Speijers, G.J., Danse, L.H., Krajnc-Franken, M.A., van Leeuwen, F.X., Helleman, P.W., Beuvery, E.C., Vos, J.G., vd Heijden, C.A. (1989) Subacute toxicity of Zwittergent administered intramuscularly. *Vaccine* **7**:364–368.

Stangl, M., Unger, S., Keller, S., Schneider, D. (2014) Sequence-specific dimerization of a transmembrane helix in amphipol A8-35. *PLOS One* **9**:e110970.

Stephens, R.S., Sanchez-Pescador, R., Wagar, E.A., Inouye, C., Urdea, M.S. (1987) Diversity of *Chlamydia trachomatis* major outer membrane protein genes. *J. Bact.* 169:3879–3885.

Stephens, R.S., Kalman, S., Lammel, C., Fan, J., Marathe, R., Aravind, L., Mitchell, W., Olinger, L., Tatusov, R.L., Zhao, Q., Koonin, E.V., Davis, R.W. (1998) Genome sequence of an obligate intracellular pathogen of humans: *Chlamydia trachomatis*. *Science* **282**:754–759.

Su, H., Caldwell, H.D. (1992) Immunogenicity of a chimeric peptide corresponding to T helper and B cell epitopes of the *Chlamydia trachomatis* major outer membrane protein. *J. Exp. Med.* **175**:227–235.

Suárez-Suárez, S., Carriedo, G.A., Presa Soto, A. (2013) Gold-decorated chiral macroporous films by the self-assembly of functionalised block copolymers. *Chem. Eur. J.* **19**:15933–15940.

Sun, G., Pal, S., Sarcon, A.K., Kim, S., Sugawara, E., Nikaido, H., Cocco, M.J., Peterson, E.M., de la Maza, L.M. (2007) Structural and functional analyses of the major outer membrane protein of *Chlamydia trachomatis*. *J. Bacteriol.* **189**:6222–6235.

Sun, G., Pal, S., Weiland, J., Peterson, E.M., de la Maza, L.M. (2009) Protection against an intranasal challenge by vaccines formulated with native and recombinant preparations of the *Chlamydia trachomatis* major outer membrane protein. *Vaccine* **27**:5020–5025.

Suzich, J.A., Ghim, S.J., Palmer-Hill, F.J., White, W.I., Tamura, J.K., Bell, J.A., Newsome, J.A., Jenson, A.B., Schlegel, R. (1995) Systemic immunization with papillomavirus L1 protein completely prevents the development of viral mucosal papillomas. *Proc. Natl. Acad. Sci. USA* **92**:11553–11557.

Tifrea, D.F., Sun, G., Pal, S., Zardeneta, G., Cocco, M.J., Popot, J.-L., de la Maza, L.M. (2011) Amphipols stabilize the *Chlamydia* major outer membrane protein and enhance its protective ability as a vaccine. *Vaccine* **29**:4623–4631.

Tifrea, D., Pal, S., Cocco, M.J., Popot, J.-L., de la Maza, L.M. (2014) Increased immunoaccessibility of MOMP epitopes in a vaccine formulated with amphipols may account for the very robust protection elicited against a vaginal challenge with *C. muridarum*. *J. Immunol.* **192**:5201–5213.

Tifrea, D.F., Pal, S., Le Bon, C., Giusti, F., Cocco, M.J., Zoonens, M., de la Maza, L.M. (2018a) Resiquimod conjugated with Amphipols bound to the *Chlamydia muridarum* MOMP enhances protection against a mucosal challenge. *In preparation*.

Tifrea, D.F., Pal, S., Le Bon, C., Giusti, F., Popot, J.-L., Cocco, M.J., Zoonens, M., de la Maza, L.M. (2018b) Co-delivery of amphipol-conjugated adjuvant with antigen, and adjuvant combinations, enhance immune protection elicited by a membrane protein-based vaccine against a mucosal challenge with *Chlamydia*. *Submitted for publication*.

Tighe, H., Takabayashi, K., Schwartz, D., Marsden, R., Beck, L., Corbeil, J., Richman, D.D., Eiden, J.J., Jr., Spiegelberg, H.L., Raz, E. (2000) Conjugation of protein to immunostimulatory DNA results in a rapid, long-lasting and potent induction of cell-mediated and humoral immunity. *Eur. J. Immunol.* **30**:1939–1947.

Trimaille, T., Verrier, B. (2015) Micelle-based adjuvants for subunit vaccine delivery. *Vaccines (Basel)* **3**:803–813.

Vartak, A., Sucheck, S.J. (2016) Recent advances in subunit vaccine carriers. *Vaccines (Basel)* **4**:pii: E12.

Vij, R., Lin, Z., Chiang, N., Vernes, J.M., Storek, K.M., Park, S., Chan, J., Meng, G., Comps-Agrar, L., Luan, P., Lee, S., Schneider, K., Bevers III, J., Zilberleyb, I., Tam, C., Koth, C.M., Xu, M., Gill, A., Auerbach, M.R., Smith, P.A., Rutherford, S.T., Nakamura, G., Seshasayee, D., Payandeh, J., Koerber, J.T. (2018) A targeted boost-and-sort immunization strategy using Escherichia coli BamA identifies rare growth inhibitory antibodies. *Sci. Rep.* **8**:7136.

Watkinson, T.G., Calabrese, A.N., Ault, J.R., Radford, S.E., Ashcroft, A.E. (2017) FPOP-LC-MS/MS suggests differences in interaction sites of amphipols and detergents with outer membrane proteins. *J. Am. Soc. Mass Spectrom.* **28**:50–55.

Wua, J.J., Huang, D.B., Tyring, S.K. (2004) Resiquimod: a new immune response modifier with potential as a vaccine adjuvant for Th1 immune responses. *Antiviral Res.* **64**:79–83.

Yue, H., Ma, G. (2015) Polymeric micro/nanoparticles: Particle design and potential vaccine delivery applications. *Vaccine* **33**:5927–5936.

Zom, G.G., Khan, S., Filippov, D.V., Ossendorp, F. (2012) TLR ligand-peptide conjugate vaccines: Toward clinical application. *Adv. Immunol.* **114**:177–201.

# Final Comments

In the course of this book, we have examined and discussed the various methods by which a membrane protein can be kept water-soluble after being extracted from its native environment. Particular emphasis has been put on amphipols, with the design and development of which my colleagues and I have been most closely associated. At the time of concluding, one must ask what, after more than 20 years of development, is the current situation of amphipols in the general landscape of membrane protein biochemistry and biophysics. It is fair to say that it is both rewarding and frustrating.

It is rewarding on two main counts. First, the use of amphipols has been validated for a vast variety of membrane proteins and hosts of applications. Their general properties and those of membrane protein/amphipol complexes have been studied in detail with a variety of amphipols and proteins, providing the novice experimenter with a vast body of data to peer upon in order to decide which amphipol to use, how best to proceed, what to expect, and what to be wary of. At the date of this writing, close to 20 distinct amphipols have been synthesized, tested, and validated for biochemical use, to a more or less detailed extent, and close to 100 distinct membrane proteins have been trapped in amphipols and, when assayed, shown to be in their native state and, if testable, functional. A score of labeled or functionalized derivatives adapted to an amazing variety of uses have been synthesized and validated. The toolbox has thus become increasingly diversified. In several fields, when launching a new experimental study, testing amphipols along with other surfactants is becoming increasingly routine.

Second, amphipols have started to yield biological dividends. To cite only two examples, they have been used to fold hard-to-express membrane proteins, such as G protein-coupled receptors, and make those available for biochemical and biophysical studies, either in amphipols or following transfer to other environments. Amphipols have become routine tools in electron microscopy, where some half of the high-resolution single-particle cryo-EM structures of MPs that are currently published in a steady stream resort to them. More scattered contributions have also appeared in the fields of NMR, radiation scattering, mass spectrometry, membrane protein folding, synthesis or immobilization, biomedical applications, etc. Perspectives for the development of amphipols and their applications are extremely rich. I have attempted to discuss them at the end of each technical chapter. These suggestions and speculations will not be repeated here, but it should be stressed that almost any kind of labeling or tagging is accessible to chemical synthesis, and that whenever the physical-chemical or biochemical properties of the reference amphipol, A8-35, make it unsuitable for a given application, alternative amphipols can be resorted to. Tagged or labeled versions of these alternative amphipols have also started to appear.

© Springer International Publishing AG, part of Springer Nature 2018    683
J. -L. Popot, *Membrane Proteins in Aqueous Solutions*, Biological and Medical Physics,
Biomedical Engineering, https://doi.org/10.1007/978-3-319-73148-3

*Telling the Future of Amphipols*
(© 2008 by Francis Haraux)

In other respects, however, the current situation is quite frustrating, because some of the inroads that have been made and some of the tools that have been developed at great pains have not been used to the level one could have expected. Here also, I will take two examples. One of them is NMR. The application of amphipols to solution-state NMR studies of membrane proteins has been validated more than 10 years ago, and its advantages and limits carefully delimited. The usefulness of combining solution NMR studies of amphipol-trapped membrane proteins with upstream amphipol-assisted protein folding has been demonstrated by establishing the structure of a ligand bound to a previously inaccessible G protein-coupled receptor. Deuterated and pH-insensitive amphipols have been developed in order to facilitate these studies. The tools were there to seize, and yet they have not entered routine use. The same could be said of using amphipols to fold membrane proteins. Their efficiency has been demonstrated on a dozen different proteins, including half-a-dozen G protein-coupled receptors. If one considers that, downstream, the use of amphipol-trapped proteins has been validated in such important fields as electron microscopy, NMR, and crystallization following transfer to lipidic mesophases, one can be genuinely surprised that no more investigators have taken the jump and exploited amphipol-assisted folding to produce and study their target proteins. What are the reasons for the reluctance or slowness of some communities to embrace these novel tools? We can probably discern three of them.

A first obstacle to a more rapid dissemination is the classical hesitation at the prospect of adding, to the difficulties of any biological investigation, those of mastering a novel methodology. Handling membrane proteins in detergent solutions has the well-deserved reputation of being a tricky business, but it is also a well-trodden path, followed by two generations of membrane biochemists, and its dangers and tricks can be felt as mapped and somewhat foreseeable. Having to understand the behavior and master the use of bizarre, heterogeneous polymers before even collecting one's first biological data can be felt as dissuasive. Amphipol workshops are regularly organized, in which novice users can make their first hands-on contact with amphipols under the guidance of experienced users. They have certainly helped to lower this activation barrier, but, obviously, not efficiently enough.

A second obstacle is more cultural, if not societal, in nature. Biologists aim at understanding how a live organism functions. For this reason, they strive to provide the proteins they work on with as natural-like an environment as possible. Synthetic polymers do not look very natural, and the fact that they nevertheless manage to preserve to a large degree one of the key features of the membrane environment – lipid/protein interactions – is underestimated, as is, in my view, overestimated the importance of the physical forces exerted by lipid bilayers. There is an illusion in believing that any in vitro system, even the most seductive of them, that of protein-bound nanodiscs, provides a solubilized membrane protein with a perfectly natural environment. One will always deal with a more or less satisfying approximation thereof. This fear of straying too far away from nature is compounded by the legitimate but paralyzing concern that, even if one has carefully weighed the pros and cons of the various experimental systems and chosen that providing the best compromise between biological relevance and experimental expediency, some systems will make publication or grant approval easier than others. Journal referees and granting agencies tend to look more favorably to works where one's protein is expressed under a functional form in eukaryotic cells rather than folded from inclusion bodies produced in bacteria, and where it is studied in a supposedly native lipid bilayer rather than an "artificial" environment. All this is understandable, but it tends to push experimenters to use very costly and complex procedures where cheaper, simpler, and more resolutive ones would have led them more rapidly to the same conclusions, at least at a first level of description. Yet, detailed information – e.g. an atomic model – gathered thanks to the simple and convenient but more "artificial" system provided by amphipols can be checked against data, possibly less accurate or more fragmentary, obtained in more sophisticated and work-intensive in vitro systems, such as nanodiscs, and/or spot measurements carried out in vivo. Fortunately, there is a growing realization that each experimental system has its advantages and limitations, that none of them is systematically the best, and that the risk of falling victim to distortions and perturbations induced by one particular medium can be reduced by confronting results obtained in different media. The fact that identical results have been obtained, e.g. in electron cryomicroscopy, using amphipols and nanodiscs ought to make the former more widely accepted as a medium in which to gather biologically relevant data.

A third obstacle, which is of a very practical nature, is due to the difficulties in providing the community of biologists with a supply of diverse and well-controlled molecules. Polymer chemistry is not always easy, and the fact that the final product is necessarily a mixture does not simplify the standardization of the batches. Commercially speaking, amphipols are a "niche" market in the sense that, as long as no large-scale investigations rely on their use, the amounts sold will always remain relatively minute. This does not encourage private companies to invest in scaling up synthesis routes and developing quality control procedures. It explains why, whereas a large number of amphipol structures have been synthesized by organic chemists and validated by biochemists and biophysicists, only a handful of them are commercially available. Having to rely on the generosity of academic colleagues can be uncomfortable. For one thing, it entails the risk that a change in scientific priorities, the loss of a key coworker, or the drying up of funding cause production to be discontinued before a given research project is completed. For another, it can be compromised by unreasonable demands in terms of publication signatures. The simplest route out of this predicament is probably that networks of laboratories interested in exploiting a given type of amphipol or amphipol derivative combine their resources to have sufficiently large batches synthesized and distributed, making it profitable for a company to develop the production. Once the activation barrier of setting up the procedures has been overcome and the initial investment has paid off, maintaining a steady production is relatively straightforward. Among the many important roles that organic chemists still have to play in this field, one will be to develop molecules whose synthesis is increasingly simpler, cheaper, and more robust, so as to facilitate their dissemination.

Except for nanodiscs, whose acceptance is, deservedly, very broad, the above remarks could apply to any of the nonconventional systems for handling membrane proteins in vitro that have been described in the opening chapters. If this book encourages membrane biologists to further exploit the rich potential offered by amphipols and other out-of-the-way surfactants, basic and applied research ought to reap more benefits from the hefty amount of work the development of these methodologies has required, and the not-inconsiderable time invested into writing the present work will not have been wasted.

Suzon, France, October 2017

# Index

© Springer International Publishing AG 2018
J. -L. Popot, *Membrane Proteins in Aqueous Solutions*, Biological and Medical Physics,
Biomedical Engineering, https://doi.org/10.1007/978-3-319-73148-3

Printed by Printforce, the Netherlands